W9-DCM-516

Human Anatomy and Physiology second edition

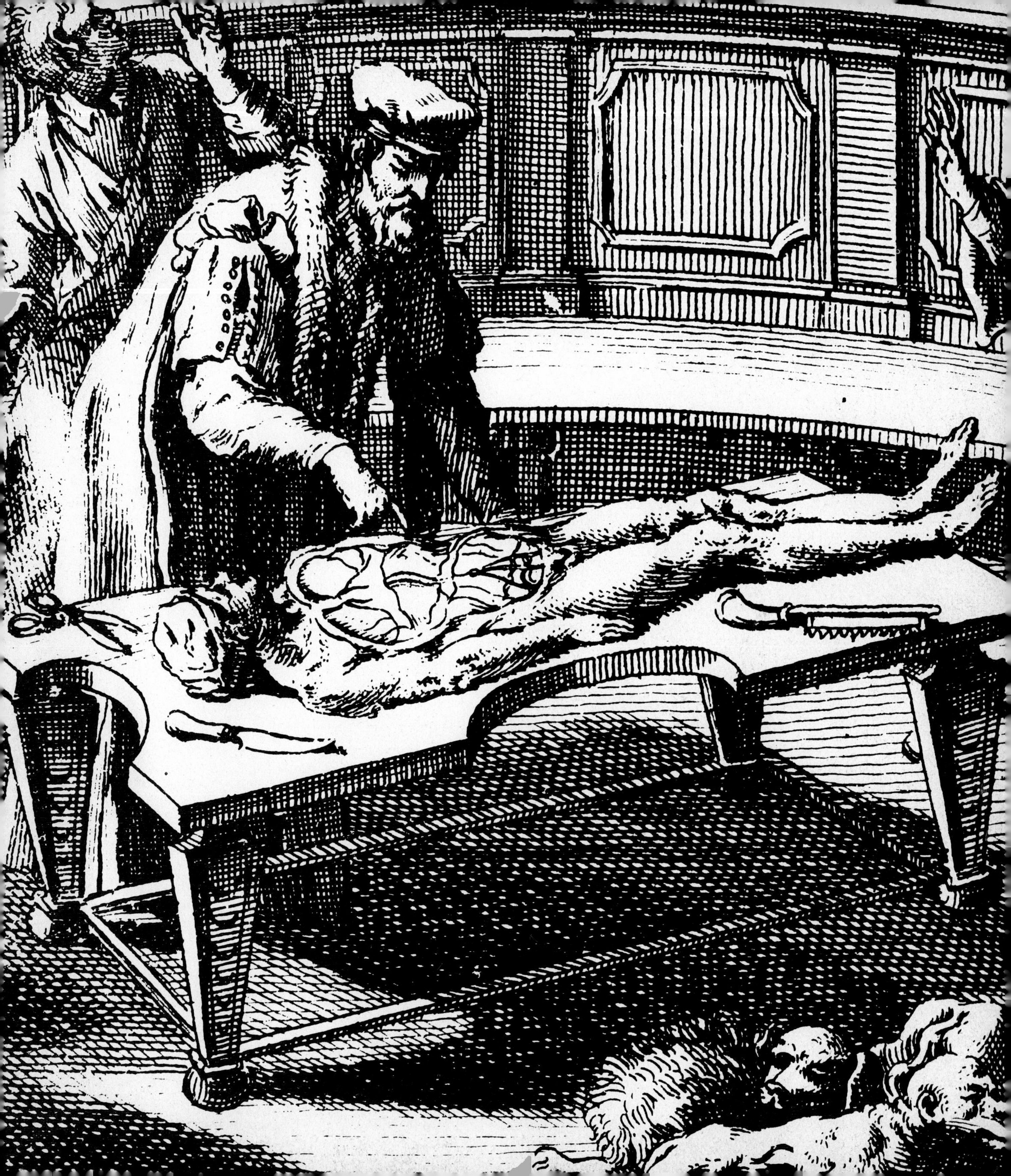

second edition

Human Anatomy and Physiology

James E. Crouch, Ph.D.
California State University–San Diego, Emeritus

J. Robert McClintic, Ph.D.
California State University–Fresno

JOHN WILEY & SONS, INC. New York London Sydney Toronto

Front cover: Image Enhancement by Aeronutronic Ford

This book was printed and bound by Kingsport Press.

The typesetter was Graphic Arts Composition.

The illustrations were designed and executed
by Ursula Wolf-Rottkay, and John Balbalis
with the assistance of the Wiley Illustration Dept.

The copyeditor was Susan Giniger.

Ben Kann was the text designer.

The production manager was Marion Palen.

The cover was designed by E. A. Burke

Library of Congress Cataloging in Publication Data

Crouch, James Ensign, 1908–
Human Anatomy and Physiology, second edition.

Includes bibliographies and index.
1. Human physiology. 2. Anatomy, Human.
I. McClintic, J. Robert, 1928– joint author.
II. Title. [DNLM: 1. Anatomy. 2. Physiology.
QS4 C952h]
QP34.5.C76 1976 612 76-868
ISBN 0-471-18918-9

Printed in the United States of America

10 9 8 7 6 5 4 3 2 1

To my students
for their inspiration
and understanding

J.E.C.

To my wife
Peggy
for her counsel and patience
during the preparation of this text
and to our daughters
Cathleen, Colleen, Marlene

J.R.McC.

Preface

The basic aims of the second edition of Human Anatomy and Physiology remain the same as those stated for the first edition, that is, to fill a part of the knowledge gap in our understanding of our body and its behavior, and to give a variety of students a sound and interesting background in human development, anatomy, physiology, and elementary pathology.

The theme that unifies the chapters is that of homeostasis, and the mechanisms that serve to maintain it, or to return it to within normal limits if it has been disturbed. Many new "flow sheets," tables, and diagrams emphasize this aspect of body activity.

To further the achievement of the aims stated above, the first several chapters of the second edition have been nearly completely rewritten and modernized. They include topics on: cellular activity and the energy sources for and genetic and hormonal control of activity; membrane potentials; and cellular reproduction and body development. A new chapter on muscle structure and physiology precedes the description of the muscular system. The nervous system has been divided into several smaller chapters that lead the reader from the more basic to more complicated concepts. These chapters include concepts of interest to behavioral scientists, and the entire nervous system has been moved to a position early in the text. Such a placement enables a greater understanding of nervous mechanisms involved in the homeostatic control of body functions as individual body systems are discussed.

To aid in the correlation of the various aspects of the anatomy, physiology, and pathology of the human organism, embryological discussions have been removed from a separate chapter, rewritten, and placed at the beginning of each appropriate chapter. Thus, the reader achieves an appreciation of how congenital anomalies may develop, and how to recognize deviation from normal, be they structural or functional.

Entirely new materials include the chapters on body defenses and immunity, circadian rhythms, and the concluding chapter on growth, development, and aging. Sections on prostaglandins, the pineal gland, hypertension, shock, venereal diseases, and a drug chart provide additional new and relevant text material. New art has been provided in all chapters to aid in elucidating the text material.

Other new features of this edition include: expansion of physiology in all chapters; chapter summaries that provide a review of the essential concepts within the chapter; and a comprehensive glossary that includes basic pronunciation and derivation of words.

Every attempt has been made to assure that the text is a working, interesting, and well illustrated guide to learning about our bodies.

We thank the staff of Wiley, and especially Robert L. Rogers, biology editor, for encouragement and support provided during preparation of manuscript. We wish to acknowledge Mrs. Helen Morris for her expert typing. Special thanks go to Peggy McClintic, MSN, for her constructive criticisms of manuscript content, typing, and proofreading, which have contributed to the relevancy of the text.

Responsibility for errors and omissions is ours.

J.E.C.
J.R.McC.

Contents

Chapter 5 Cell Reproduction and Body Development

Chapter 6 Epithelia, Connective Tissues, and the Skin

Chapter 7 The Skeletal System

Chapter 8 Articulations

Chapter 9 The Structure and Properties of Skeletal and Smooth Muscle

Chapter 10 The Skeletal Muscles

Chapter 11 The Basic Organization and Properties of the Nervous System

Chapter 12 The Spinal Cord and Spinal Nerves

Chapter 13 The Brain and Cranial Nerves

Chapter 14 Blood Supply of the Central Nervous System; Ventricles and Cerebrospinal Fluid

Chapter 15 The Autonomic Nervous System

Chapter 16 Sensation

Chapter 17 The Eye and Vision

Chapter 18 Hearing, Equilibrium, Taste and Smell

Chapter 19 Electrical Activity in the Brain; Wakefulness and Sleep; Association Areas

Chapter 20 Emotions, Learning, and Memory

Chapter 21 Body Fluids and Acid-Base Balance

Chapter 22 The Blood and Lymph

Chapter 23 Tissue Response to Injury; Immunity; Blood Groups

Chapter 24 The Heart

Chapter 25 The Blood and Lymphatic Vessels; Regulation of Blood Pressure

Chapter 26 The Respiratory System

Chapter 27 The Digestive System

Chapter 28 Intermediary Metabolism and Nutrition

Chapter 29 The Urinary System

Chapter 30 The Reproductive Systems

Chapter 31 The Endocrine System

Chapter 32 Circadian Rhythms

Chapter 33 Human Growth and Development; an Overview

Human Anatomy and Physiology second edition

1

An Introduction to the Human Organism

The place of the human organism in nature

What is a human? To most biologists, a human is an animal organism, the result of a long evolution and therefore related to all other animals and indeed to all life. Much has been learned or better understood about human anatomy, physiology, and behavior through comparative studies. It is a fortunate circumstance that humans have been able to use other animals for experimentation and study and that the knowledge gained proves applicable to humans in most instances. The human is used for experimentation in the disciplines where there is likely to be no damage to the individual. Knowledge is also gained by the study of sick and abnormal individuals or through surgery. The growing volume of research in the science of ethology *(behavior)* has contributed abundantly to our understanding of the behavior of animals, including, of course, humans.

Studies made on cells and tissues reveal also the common characteristics of all organisms in structure and function. Much emphasis in modern biology is on the molecular and atomic structure of the cell, particularly on the nucleus and its chromosomes and genes. The DNA and RNA complex, and its significance for the understanding of genetics and heredity and of all life, opens new possibilities for the further understanding of biology.

The question "What is a human?" is not, however, fully answered in the above statements. Humans, like all species of organisms, have their own UNIQUE CHARACTERISTICS, not the least of which is that they are concerned with such matters as the nature of life, their relationship to it, and their relationship to the physical environment. As G. W. Corner has said, "After all, if he is an ape, he is the only ape that is debating what kind of ape he is."

The place of humans in nature is objectively described by placing them, along with other living organisms, in the accepted scheme of classification as formulated by biologists. On this basis humans are members of the ANIMAL KINGDOM which, like the plant kingdom, is divided into a number of major categories, the PHYLA. The phyla are each critically defined and organisms are assigned to them on the basis of certain characteristics. The study of the phyla is called PHYLOGENY. The animal phyla range from the relatively simple Protozoa, mostly single-celled forms, to the Chordata, which encompasses backboned animals, including humans. This scheme of classification is based on the concept of the EVOLUTION of life, and the placement of organisms indicates the closeness or the distance of their relationship.

The PHYLUM CHORDATA includes all of the animals that have a notochord, a dorsal hollow nerve cord, and pharyngeal pouches (Fig. 1.1). These structures are apparent at some point in the life cycle *(ontogeny)* of these animals although they may be lost at other stages. Thus, it is important in defining an animal to consider not just the adult stage, but the embryo as well. The notochord, for example, an obvious structure in the human embryo, is not apparent, as such, in the adult.

The NOTOCHORD is a flexible rod along the dorsal side of the animal ventral to the dorsal hollow nerve cord. It is composed of vacuolated cells enclosed in a fibrous sheath and serves as a central axis of support for the animal. It persists throughout the life cycle of some of the lower chordates such as the lamprey eel, but is replaced struc-

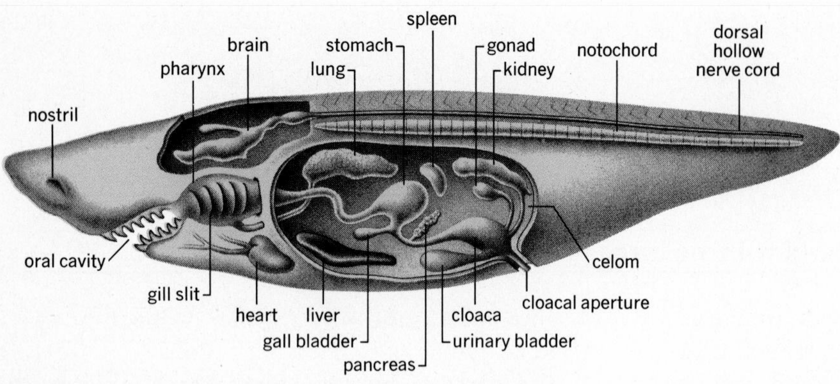

FIGURE 1.1 Schematic sketch of chordate anatomy.

turally and functionally in most vertebrates by the vertebral column. In human adults and other mammals the notochord is often considered to be represented by a small mass of tissue in the center of the intervertebral disks known as the NUCLEUS PULPOSUS (Fig. 1.2).

The DORSAL HOLLOW NERVE CORD lies along the back or posterior side of the animal and comes about as an infolding of the mid-dorsal part of the ectoderm germ layer of the embryo. It becomes the spinal cord and brain, which remain throughout life as hollow organs, there being a central canal in the spinal cord and ventricles in the brain.

The segmentally arranged PHARYNGEAL POUCHES are outpocketings or evaginations of the lateral walls of the pharynx of chordates. Over each evag-

FIGURE 1.2 Median section of vertebrae and intervertebral disks.

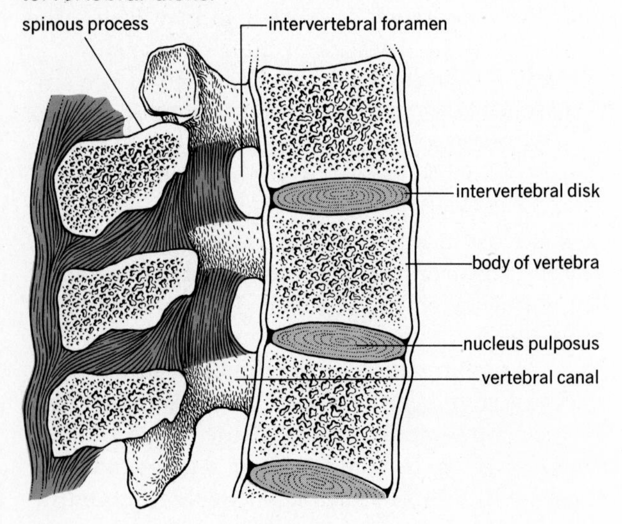

FIGURE 1.3 Schematic sketch showing origin of gill slits.

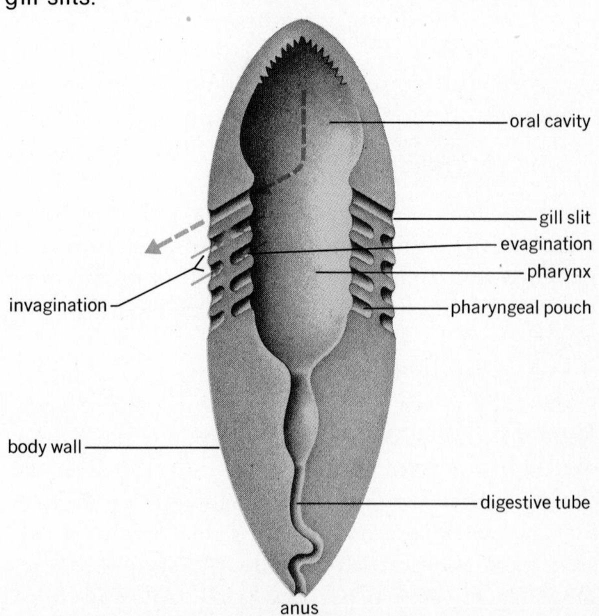

ination of the pharyngeal wall the body wall invaginates, the two make contact, and then break through to form a GILL SLIT or PHARYNGEAL CLEFT (Fig. 1.3). In lower vertebrates, such as fishes and some amphibians, gill slits are formed as described and gills for gas exchange between environment and bloodstream are produced in this area. In the lung-breathing vertebrates (reptiles, birds, and mammals), gills are not formed and gill slits are transitory structures in embryological development. In humans and other higher vertebrates the pharyngeal pouches give rise to other structures such as the AUDITORY *(Eustachian)* TUBE and MIDDLE EAR CAVITY. In this case the corresponding invagination of the body wall becomes the external auditory meatus or canal; the membrane separating the two is the eardrum. Breaking an eardrum is analogous to forming a gill slit. Infants are sometimes born with openings or fistulae in the sides of the neck that are in a sense gill slits. Such reminders of one's vertebrate ancestry can be closed surgically.

Although the above characteristics of Chordates are said to be diagnostic for the phylum, others of importance may be found also in members of other phyla. Body cavities are among these.

The BODY CAVITIES are closed spaces within the body; there are DORSAL and VENTRAL CAVITIES (Fig. 1.4). The DORSAL CAVITY is divided into CRANIAL and VERTEBRAL PORTIONS, the former housing the BRAIN, the latter the SPINAL CORD.

The VENTRAL CAVITY, also called the celom, is divided into a THORACIC portion superior to the DIAPHRAGM and an ABDOMINOPELVIC portion inferior to the diaphragm. The THORACIC CAVITY is divided into two PLEURAL CAVITIES, one around each lung, and the PERICARDIAL CAVITY around the heart. Between the pleural cavities there is also a potential cavity, the MEDIASTINUM. It is not considered a celomic cavity. The abdominopelvic cavity is divided arbitrarily into ABDOMINAL and PELVIC portions although no wall lies between them. Stomach, spleen, liver, pancreas, small intestine, and most of the large intestine lie within the abdominal cavity; the urinary bladder, sigmoid colon, rectum, and the ovaries, uterine tubes, and uterus of the female, and the prostate, seminal vesicles, and part of the ductus deferens of the male lie in the pelvic cavity.

The Phylum Chordata is divided into a number of SUBPHYLA; the most important for our purposes is the VERTEBRATA, characterized by the presence of a backbone or vertebral column. To this subphylum belong the classes of fishes, amphibians, reptiles, birds, and mammals, the latter including humans. The CLASS MAMMALIA is characterized by the presence of hair and mammary glands. The class is in turn divided into orders, humans being placed in the ORDER OF PRIMATES. Primates are characterized by the development of a large brain relative to the size of the body and particularly the growth of the cerebral cortex. In cerebral cortex

FIGURE 1.4 Body cavities, median section.

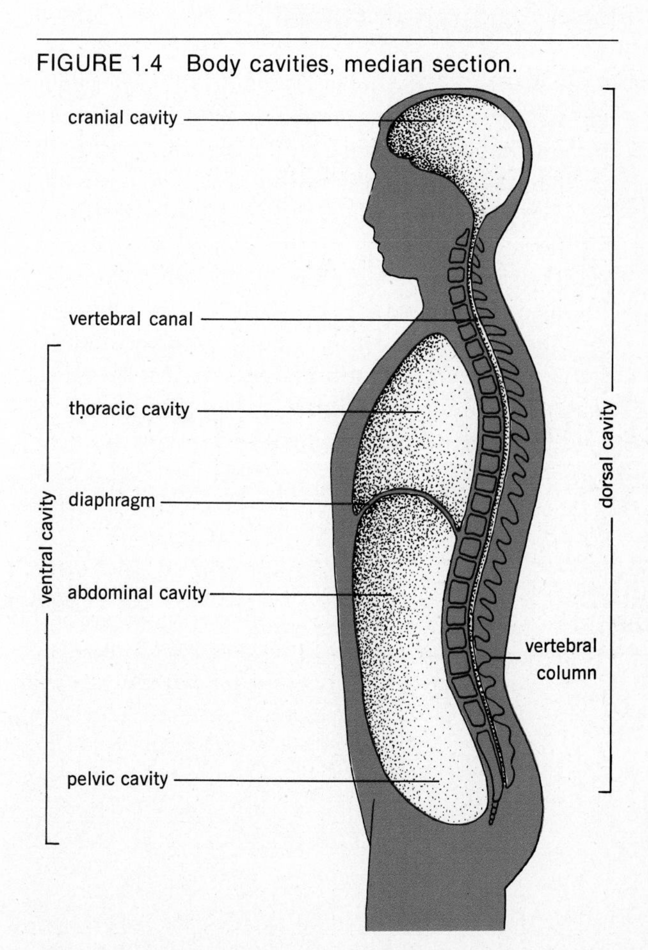

growth and development humans surpass all other primates. The human brain weighs about 1350 grams while those of apes weigh about 400 grams or less.

Finally, humans belong to the SUBORDER ANTHROPOIDEA, the FAMILY HOMINIDAE, the GENUS HOMO and the SPECIES SAPIENS. For further characterization of human beings, see R. J. Harrison's *Man the Peculiar Animal.*

> "Man hath all that Nature hath, but more,
> And in that more lie all his hopes of good."
> *Arnold*

Human relationships to other animals are summarized in the classification presented below.

Humans, unique animals

Humans are related most closely to the other vertebrates. It should be emphasized, however, that humans have their own unique qualities and their phylogenetic position is the highest of all animals. This high position, the result of a long evolution, is due not only to the presence of a LARGE CEREBRAL CORTEX surpassing that of all other animals, but also to BIPEDAL LOCOMOTION and UPRIGHT POSTURE, which free the upper limbs and hands better to serve the cerebrum. Humans are the most adaptable of all the animals, capable of living in a wide range of environments. Humans are also capable, more than any other animal, of changing the environment to satisfy their needs. Human reproductive efficiency is the highest among animals, particularly in the area of postnatal care. Humans are reproductively active throughout the year. The female after menopause, at about the age of 50, is no longer capable of producing children, but may remain sexually active. The male shows no definite point of cessation of sexual activity, and is capable of producing viable sperm throughout most of his life.

George Gaylord Simpson in the *Meaning of Evolution* says, "In the basic diagnosis of Homo sapiens the most important features are probably interrelated factors of intelligence, flexibility, individualization, and socialization. . . . In man all four are carried to a degree incomparably greater than in any other sort of animal." Of these four features of humans, INTELLIGENCE is the most significant and the one on which the other three hinge. Humans can learn; they are educable and thus are capable of the most varied responses to external and internal stimuli. These responses may be based on the experiences of others, even those who lived in the remote past and who left signs or manuscripts of their findings. The responses may be based on speculations into the future. Simpson refers to this as a "new sort of evolution" which has its basis in "the inheritance of learning." Unlike organic evolution, which rejects the inheritance of acquired characteristics,

Kingdom—Animalia.
- Phylum—Chordata—notochord; dorsal hollow nerve cord; pharyngeal pouches.
  - Subphylum—Vertebrata—backbone.
    - Class—Mammalia—hair, mammary glands.
      - Order—Primates—generalized anatomy; highly developed cerebrum; in some, five digits with nails; opposable thumb in some.
        - Suborder—Anthropoidea—apes, monkeys, and humans.
          - Family—Hominidae—terrestrial biped, highly developed cerebrum and eyesight.
            - Genus—Homo—steep facial angle, forehead, nose and chin prominent.
              - Species—sapiens—largest cerebrum; articulate speech.
                - Scientific name—Homo sapiens.

THE NEW EVOLUTION OPERATES BY THE INHERITANCE OF ACQUIRED CHARACTERISTICS—the knowledge and activities of those living in the past. Although in organic evolution new factors arise as MUTATIONS and occur purely by chance, in the new evolution the new factors arise CONSCIOUSLY; they are influenced by the needs of individuals, and individuals in turn are influenced by their relationship to a social group. THE NEW EVOLUTION IS CULTURAL.

While the "old" and the "new" evolutions are so different in kind—the one occurring without purpose or design and the other by conscious effort on the part of individuals and society—it does help to keep the comparison in mind as we attempt to understand human uniqueness in the living world.

Humans are said to be ETHICAL ANIMALS, indeed the only ethical animals. As such they are concerned with values, with right and wrong, with the "good life," with individual worth and responsibility, with the relationship of humans to each other and to the environment. Humans have sought for an adequate ethic over the centuries and are still searching. Most of their attempts have resulted in systems of ethics built on revelation. The ethics of humans have adhered to an ideal of absolutism. They have been considered as having eternal validity, for all forms of life. Ethics, as such, have not satisfied the needs of humankind. Perhaps this is true because such an ethic fails to incorporate the concept that change is the essence of life and that humans, while a product of the old evolution, are subject to the "new" evolution and it is only at this point that the need for an ethic appears. Perhaps **Albert Schweitzer s** definition of ethics is a good one. He said, "Let me give you a definition of ethics: it is good to maintain life and further life; it is bad to damage and destroy life. And this ethic, profound, universal, has the significance of religion. It *is* religion." **J. Bronowski,** in an interview by George Draper in *The American Scholar,* in response to the question "How does our search for knowledge shape our conducts?" said ". . . the very activity of trying to refine and enhance knowledge—of discovering 'what is'—imposes on us certain norms of conduct. The prime condition for its success is a scrupulous rectitude of behavior based on a set of values like truth, trust, dignity, dissent, and so on. . . . In societies where these values did not exist, science has had to create them, to make the practice of science possible."

Humans' most serious problems center in three areas: OVERPOPULATION, DESTRUCTION OF THE ENVIRONMENT, and WAR. Humans can control population so that human life can continue on this earth and with some chance for a good life for all people. By using the knowledge and skills of ecologists along with those of economists, engineers, and others, we can use more wisely the natural resources of our world and avoid the destruction of the environment which goes on so rapidly today in the United States and around the world.

The study of living organisms

Levels of organization—cells, tissues, organs, systems

We have looked at the human in the light of characteristics such as a notochord, pharyngeal pouches, and a dorsal hollow nerve cord in an effort to relate this unique animal to the rest of the living world.

Let us now use another approach to the humans' nature; one even more basic to their identity as organisms and also more important in our efforts to understand their anatomy and physiology. Like other living things, the basic units of human structure and function are CELLS. Cells have been known for over 300 years, and their significance as components of most living organisms has been appreciated for about 135 years. What this means is that advances in our knowledge of humans, particularly their physiology, must to a degree wait upon a more complete knowledge of these building blocks—the cells. Beyond the organization and functioning of cells as units, there is the fact that they become specialized in various ways, and similar cells join together along with intercellular materials to form TISSUES. Four tissue groups are recognized:

Epithelial tissues. Epithelia cover the FREE SURFACES of the body, and their cells lie close together since intercellular materials are at a minimum. They serve a wide variety of functions depending on their location, such as protection, absorption, excretion, secretion, and reception of stimuli.

Connective tissues. These tissues are characterized by having cells scattered in an extensive INTERCELLULAR MATERIAL, which ranges from hard in bone to liquid in blood. Other connective tissues are cartilage, ligaments, and tendons. They serve the body in support, connecting one part to another, and for protection.

Muscular tissues. Muscle is specialized in the function of CONTRACTILITY. It is responsible for the movement of the body as a whole and for the movement of materials through the body as in digestion and circulation.

Nervous tissue. This tissue is specialized in the properties of EXCITABILITY and CONDUCTIVITY. It forms a communication system within the body and is activated by stimuli from the environment.

Two or more tissues that are joined and have specific functions constitute an ORGAN. Organs are numerous in the body, and their names will become common to the student during the study of the various systems of the body. The stomach, kidney, lung, and heart are examples.

Organs are joined together to form the SYSTEMS of the body that have some broader function to perform.

Integumentary system. The skin and its accessory organs such as hair, glands, and nails.

Skeletal system. The bones, cartilages, and articulations of the body.

Muscular system. The skeletal or voluntary muscles of the body.

Circulatory system. The heart, blood, and blood vessels.

Lymphatic system. The lymph, lymph vessels, and the lymphoid organs such as thymus, spleen, tonsils, and lymph nodes. It may be considered a part of the circulatory system.

Respiratory system. The lungs, the air passages leading to them, and accessory structures such as the larynx.

Digestive system. The alimentary tract, the associated glands such as the salivary glands, liver, and pancreas, and other associated organs such as the tongue.

Urinary system. The kidneys and the accessory organs such as ureters, urinary bladder, and urethra.

Reproductive system. The essential organs such as the testes, ovaries, and associated organs such as uterus, ductus deferens, and vagina.

Nervous system. The brain, spinal cord, and peripheral nerves.

Endocrine system. The glands of internal secretion (ductless glands) that produce hormones.

Reticuloendothelial system. The phagocytic and plasma cells that serve a protective function.

It is interesting to note that in this approach to the study of the human we are using steps to be seen as one watches the embryological development of humans; that is, from one cell to multicellular organism, or—as in the evolution of life as most biologists conceive it—from a formless protoplasm to celled animals (Protozoans), to tissue animals (Coelenterates, such as hydra), to increasingly complex creatures with well-developed organs and systems. This phenomenon has been referred to as the RECAPITULATION THEORY or the BIOGENETIC LAW, which, as often stated, says, "ONTOGENY RECAPITULATES PHYLOGENY"—ONTOGENY referring to the development of the individual *(embryology)*, PHYLOGENY to the evolution of species. Whatever the true significance of this generalization may be, it has caused a lot of thought and debate among biologists since the latter part of the nineteenth century.

One final thought should be considered here: the whole is greater than the sum of the parts. Humans are not merely a result of the summated potentialities of cells, tissues, organs, and systems. The organization of these parts in some way not fully understood gives an organism its potential to make the adjustments necessary for survival.

Humans, creatures of dynamic balance—homeostasis

regulation of internal enviornment

It is a well-known physiological principle that animals at almost all evolutionary levels maintain a near constancy of the internal environment and that this is essential to the life of the cells and hence the life of the total organism. The INTERNAL ENVIRONMENT is that portion of the body lying inside a living membrane, as distinguished from those parts of the body having open passage to the external environment. The blood, lymph, and tissue fluid are all a part of the internal environment, while the lumina of the bladder, the alimentary tract, the uterus, the female peritoneal cavity, and other organs all maintain connections to the body surface, and materials leaving them do not cross living membranes. They are a part of the EXTERNAL ENVIRONMENT.

Knowing that organisms are constantly taking in food, oxygen, water, and other materials, and further realizing that the cells are always producing carbon dioxide and other wastes, the question arises as to how the internal fluids of the body—the internal environment—can maintain nearly constant levels of sugars, salts, and water, and an osmotic balance and acid-base (pH) relationship that show only small variations. Add to this the stress from the environment to which all organisms are subject and which must be met if the organism is to survive, and one begins to realize the complexity of the adaptive and control processes that must be involved. This near constancy or steady state of the internal environment of organisms was designated by **Walter B. Cannon** of Harvard University as HOMEOSTASIS. In consideration of the constant adjustments and the dynamic character of such adjustments required to maintain a steady state, the terms HOMEOKINESIS and HOMEODYNAMICS are coming into favor among physiologists.

Absolute values cannot be given for this "constancy" of the internal environment. They are often, instead, average figures, such as 98.6°F for normal body temperature. They are subject to variation within a species and even within one individual from one time to another. The values are more constant among organisms higher in the evolutionary scale. In general, higher animals can adjust more readily to a wider range of external environments than can the lower forms. At the same time, higher forms are more dependent on homeostasis for survival. Stress, such as disease which disrupts control systems, can be very dangerous. Humans, who are among these higher animals are universal in distribution in the world. Humans can adapt to almost any external environment but always are vulnerable to even the slightest deviations in the various physiological constants of the internal environment.

PHYSIOLOGICAL REGULATION. HOMEOSTASIS involves the action of regulatory systems, and a large segment of physiology is devoted to the discovery and study of these systems. The study of control systems is called CYBERNETICS, a term coined by **Norbert Weiner** in 1948. The basic concepts of this technology were developed and applied by Weiner to both artificial and biological systems, although mostly to the former. Modern automation was the outcome of this early work. A start has been made toward the application of this technology to the integrative systems of the living body, but much remains to be accomplished.

To achieve homeostasis, an organism must have the capacity to sense change—that is, to receive stimuli and to take appropriate action as a result. Appropriate action, in this case, is to perpetuate the physiological constant whether it is the hydrogen ion concentration of the blood or the glu-

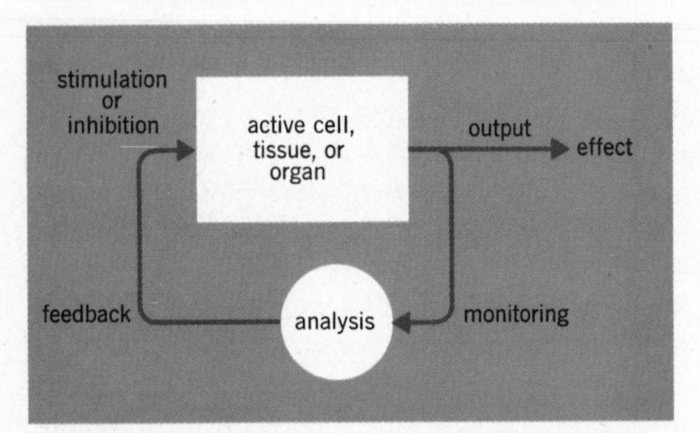

FIGURE 1.5 Cybernetic control by feedback.

cose level. This might be diagrammed as follows to show the direction of flow of information:

change (stimulus)→	organism	→ response
(input)	(system)	(output)

The principle is not different from the use of a thermostat to control the temperature in a house. The thermostat is set at a given temperature. When the temperature in the house falls below this level, the thermostat reports it to the furnace, which responds by turning on. Conversely, if the temperature goes above the level for which the thermostat is set, it will cause the furnace to be shut off and may start an air conditioner if one is built into the system.

Feedback. In such systems a FEEDBACK CONCEPT is used. A feedback system is one in which the output of the system is fed back to the system (Fig. 1.5). In the above example, the heat caused by the furnace being turned on (output) is fed back to the thermostat (system). It then becomes part of the input for a new stimulus-response cycle; that is, when the room gets too hot, the furnace will be turned off. Where the feedback change is in the reverse direction of the original condition, as in this particular example, it is called NEGATIVE FEEDBACK. Negative feedback makes for stability in the system or organism (in this case, the maintenance of a constant temperature). This kind of feedback is found in most physiological regulatory systems. POSITIVE FEEDBACK systems are those in which the feedback signal reinforces the input signal and the change in output is increased. Positive feedback results in instability and an increase of output rather than regulation; this occurs infrequently in physiological systems. It is involved, for example, in the control of the pupil of the eye (pupillary reflex).

In HOMEOSTASIS we are therefore concerned primarily with negative feedback systems. It is not our purpose at this point to look in depth at regulation but only to introduce the principles. Regulation will be treated in greater detail in the chapters on the nervous and endocrine systems, the regulating-integrating systems of the body. Abundant examples are also found in the chapters on circulation, respiration, and metabolism. One example, the regulation of blood sugar (glucose), may help to clarify our discussion of regulation up to this point. Glucose is the chief energy source of our cells, and its concentration in the blood at any given time is critical. Excesses of glucose are converted into glycogen and stored in the liver, only to be reconverted to glucose and returned to the circulation when needed. Two hormones from the islets of Langerhans of the pancreas are the primary agents controlling these metabolic processes. A fall in blood-sugar level initiates secretion of the hormone GLUCAGON from the alpha cells of the islets. This hormone influences the enzymes which stimulate conversion of GLYCOGEN to GLUCOSE *in the liver,* and the concentration of blood sugar is raised. INSULIN, a hormone from the beta cells of the islets, is secreted in response to an elevation in blood sugar (for example, after a meal) and stimulates the uptake of glucose by the cells as well as the conversion of glucose to glycogen for storage in the liver. These two hormones working together normally maintain the blood sugar at levels compatible with the health of the organism. Other factors, to be sure, are involved in this regulation. It is also a well-known fact that the metabolism of proteins and fats is related to and dependent on the proper utilization and control of sugar metabolism. A more elaborate description of these relationships is given in Chapter 31.

Levels of integration and control. The above section showed the importance of maintaining optimal operating conditions within the internal environment of the organism and suggested certain means of control. In addition, it should be pointed out that all changes, both external and internal, require control mechanisms for adaptation and that most parts of the organisms are involved in the process. An organism that cannot respond effectively to the ever-changing external environment has little chance of survival and a steady state internally will be of no help if the animal fails in this other, though not separate, area of activity. Life, in a very real sense, is the capacity to respond to change or stress in a controlled and effective way.

Control of organisms goes back basically to the function of the GENE. Life began with the appearance of the first nucleic acids, of which genes are the modern descendants. Not only are nucleic acids the ultimate basis for control in all organisms, but they also provide the blueprint for the perpetuation of life from generation to generation and for the creation of new life. Genes are variously defined: in the physical sense as simply a unit or portion of a chromosome; chemically as the DNA fraction of a nucleoprotein; or operationally as a unit of a chromosome that controls one reaction in metabolism. Genes may be considered as units that *(by mutation)* alter just one characteristic of an organism, or that in the reproductive process may cross from one chromosome to another. Genes account for the conservatism resulting in the continuity of species from one generation to another, for the variation causing individuality within species, and for the changes that, given time, are the basis for the evolution of life (Fig. 1.6).

Genes are not the only controlling agents of the cell. There are numerous such agents, and they

FIGURE 1.6 The role of the gene in controlling the organism's function.

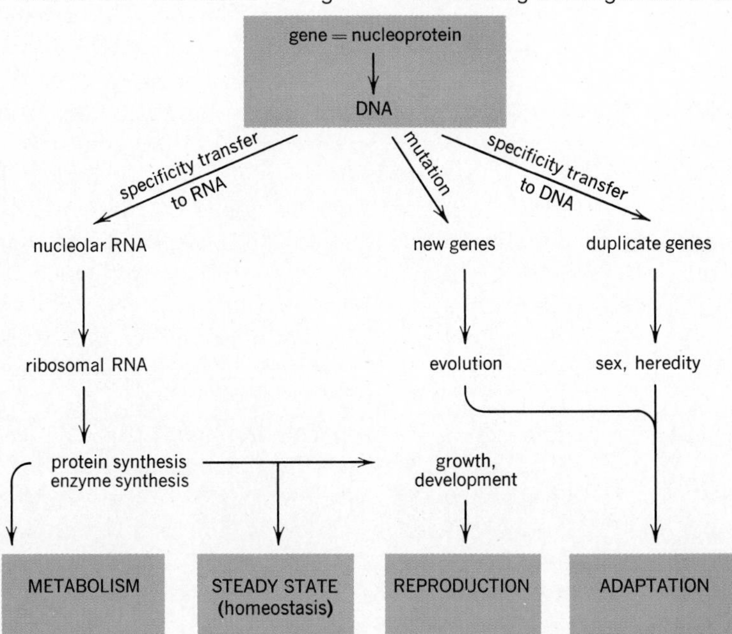

play many roles in the integrative processes of the cell. These agents act sometimes as receptors, other times as effectors, and sometimes as selective agents making decisions—they are often called modulators in cybernetic systems. The cell membrane is a good example. It is a receptor when stimulated by a molecule, a modulator when it "decides" what to do about it, an effector when it responds to the stimulation. If the molecule is glucose, the membrane may allow it to pass through into the cell. Many cell molecules free in the cytoplasm or complex cellular entities such as ribosomes play important control functions within the cell. The cell also extends its influence outward to affect its external environment or other cells with which it associated in a tissue or in a more complex organism. The control and integration of many relatively simple organisms are managed on this cell-to-cell basis without the benefit of special integrative systems.

As multicellular organisms evolved, communication among cells was at first cell-to-cell contact, and their stimulus-modulator-response mechanism was essentially that within the cells. Some animals later evolved transportation systems *(such as the circulatory system)* that enabled cells to extend their influence more readily to distant parts of the organism. But these were not controlling and integrating systems in the usual sense. These appeared later in the endocrine and nervous systems, which serve these functions primarily, although not exclusively. Locomotion places additional stresses on most animals to which sedentary organisms are not subjected. These animals not only must adjust to a changing external environment as do other organisms, but they also move quickly from one part of the external environment to another, placing extra stress on integrative systems that must maintain the essential constancy of the internal environment.

The structural unit of the nervous system is the NEURON, a specialized cell; the functional unit is the REFLEX ARC. Reflex arcs have their receptors either on the body surface or internally; sensory or afferent neurons transmit impulses (messages) to the brain or spinal cord that act as integrative centers or modulators; efferent neurons carry impulses to effectors or responding organs such as muscle tissue or glands.

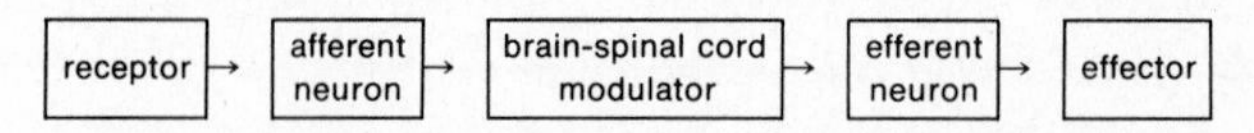

This system is presented in detail in Chapters 11–14. The endocrine system, which produces hormones and releases them in the blood to reach their target organs, is elaborated in Chapter 31.

Even in the higher animals such as humans, although special integrative systems are present, almost all organs of the body and their products exert some integrative or controlling influence. Carbon dioxide, looked on primarily as a waste product eliminated from the lungs, is an important regulator of respiration. The kidneys are not just excretory organs; they are necessary in regulating acid-base balance and osmotic pressure, and in other ways they help to maintain the constancy of the internal environment. The liver is far more than an organ contributing to the digestion and absorption of food. It is involved in sugar, protein, and fat metabolism. One should remember, too, that there is a complex cooperative action among organs and tissues in multicellular organisms. What happens in one influences others. The control of body temperatures in birds and mammals, which we take so much for granted, is a result of the coordinated actions of many organs of the body. One should never forget, either, that in all of this the basic controlling influence of cells is always going on.

One final level of integration and control must be emphasized—the brain, particularly the human brain. This brings us full circle to the questions discussed in the introductory part of this chapter. What is a human being? What is its nature? These questions bring us to the reason for the chapters that are to follow, which will help us to understand better the human organism and its anatomy and physiology in the broadest sense.

Terms of direction

In order to describe structures and their relationships in organisms, it is essential that there be common agreement as to the position of the animal when the descriptions are written. Such an agreed position is called the ANATOMICAL POSITION. The anatomical position for humans is one in which the body is standing erect with eyes level and directed forward, the arms at the sides with the palms forward, and the feet parallel with the heels approximated (Fig. 1.7).

Assuming a human to be in the anatomical position, the following terms have universal meaning.

Anterior or ventral. The front or belly side of the body.

Posterior or dorsal. The back of the body.

Superior. Above, or something on a higher portion

FIGURE 1.7 · Anatomical position and terms of direction.

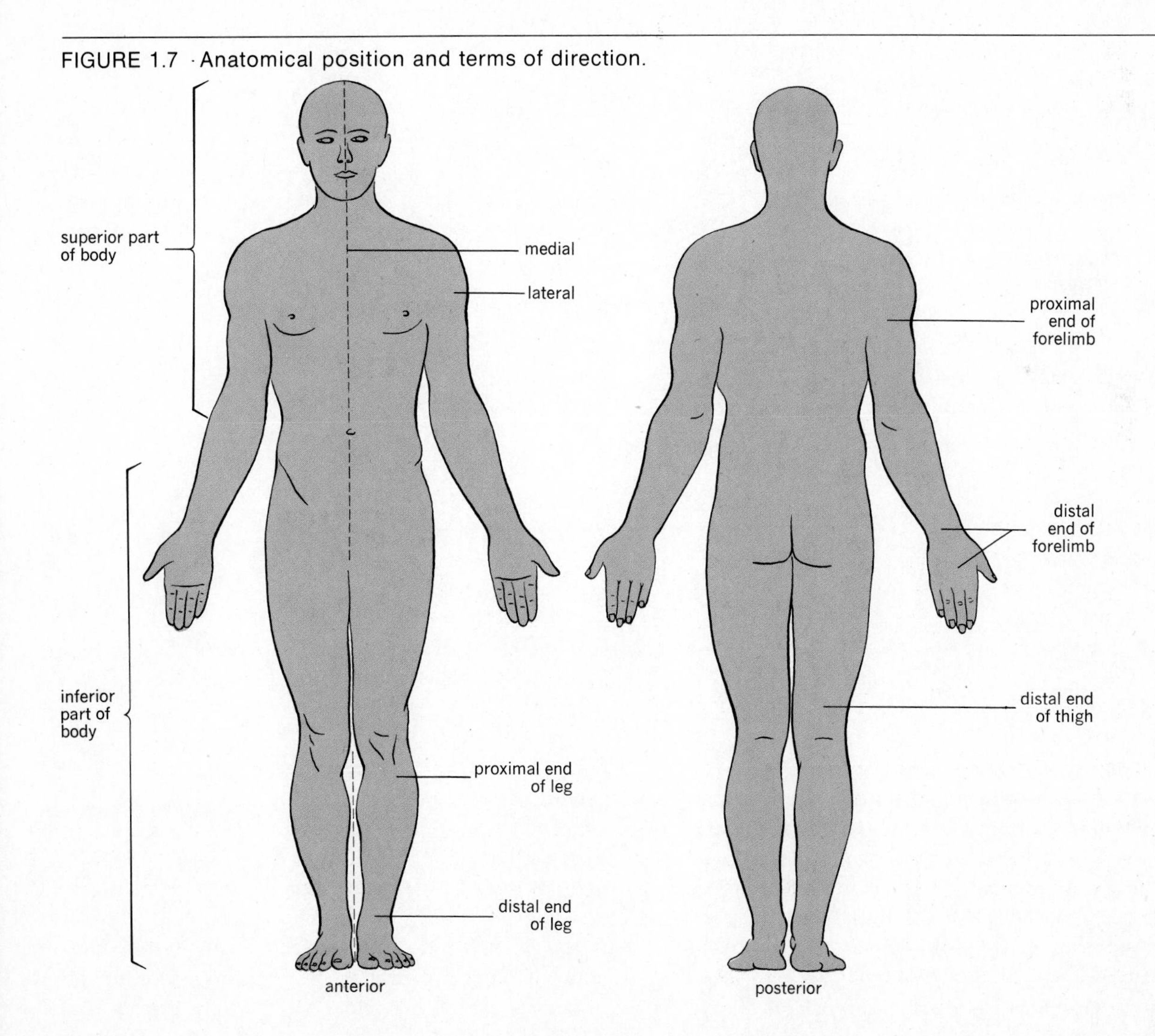

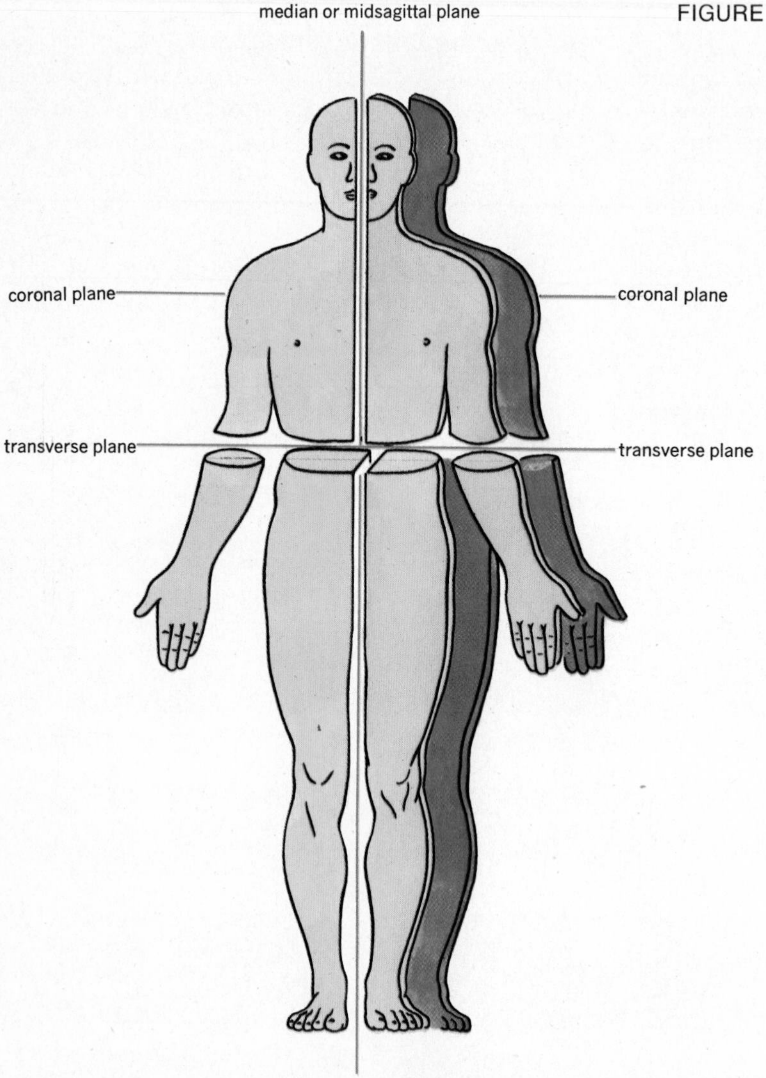

FIGURE 1.8 Planes of section.

of the body than the original point of reference.

Inferior. Below, or something on a lower portion of the body than the original point of reference.

Medial. A line running from the center of the forehead to between the feet defines the centerline of the body. Medial implies a position closer to the centerline. The term MEDIAD is used to denote movement toward the midline.

Lateral. A structure toward the side of the body. LATERAD denotes movement away from the midline.

External. The general meaning of this term is "outside of." It is most commonly employed in referring to the outer surface of the body as a whole or to a position further removed from the interior of a hollow organ.

Internal. The general meaning of this term is "inside." It is used to refer to structures lying within the body or closer to the interior of a hollow organ.

Superficial. This term refers to something closer to the external body surface. It carries a similar meaning to external, but is a preferable term when referring to a structure that approaches, but does not reach, the surface.

Deep. This term refers to structures removed inward from a surface. It suggests that a structure lies under other structures.

Proximal. Strictly defined, proximal means closest to the point of attachment of a part to the body or to the midline. For example, the shoulder is proximal to the elbow; the wrist is proximal to the fingers.

Distal. Strictly defined, distal means further from the point of attachment or the midline. The wrist would be distal to the elbow. Proximal and distal are most commonly employed to designate opposite ends of bones or appendages. The terms lose their significance if applied to the trunk, head, or neck.

Planes of section (Fig. 1.8)

Additional information about the placement and relationships of organs can be gained by cutting

FIGURE 1.9 General descriptive areas of the body.

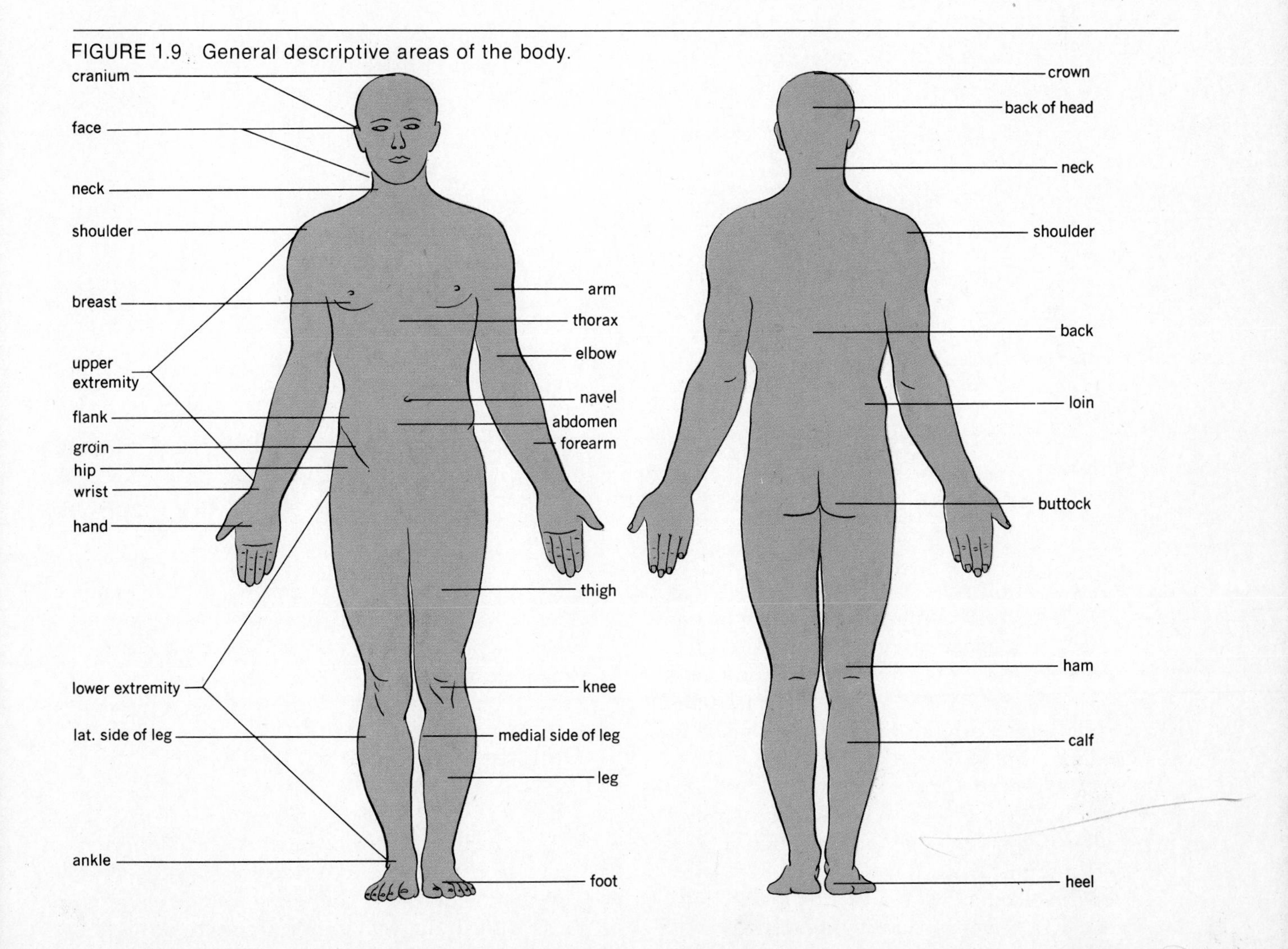

FIGURE 1.10 Male and female adults.

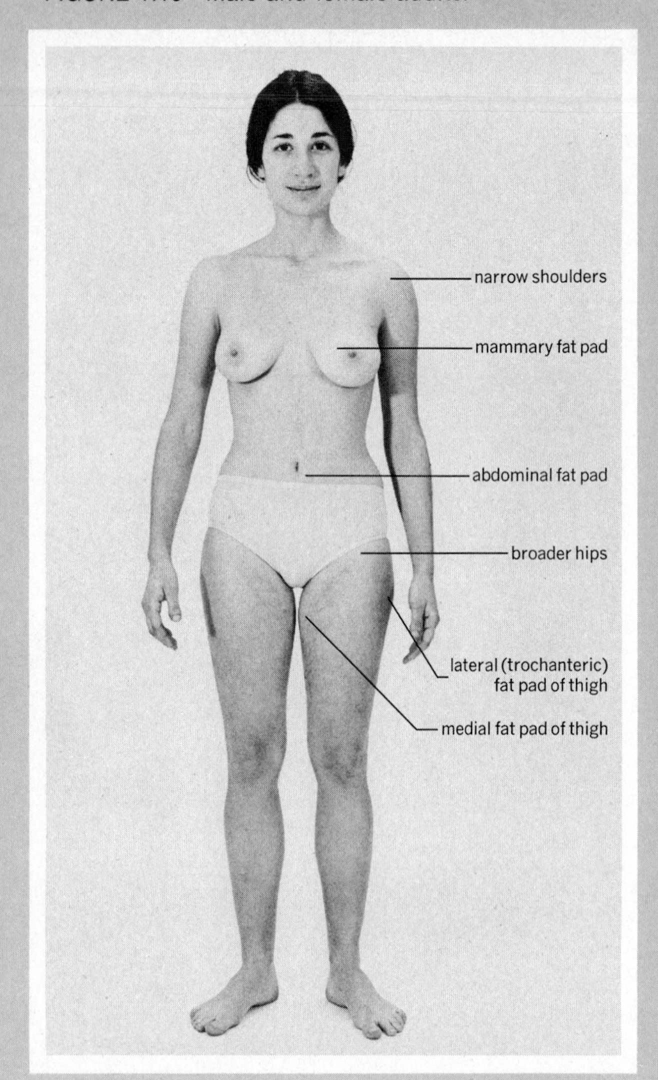

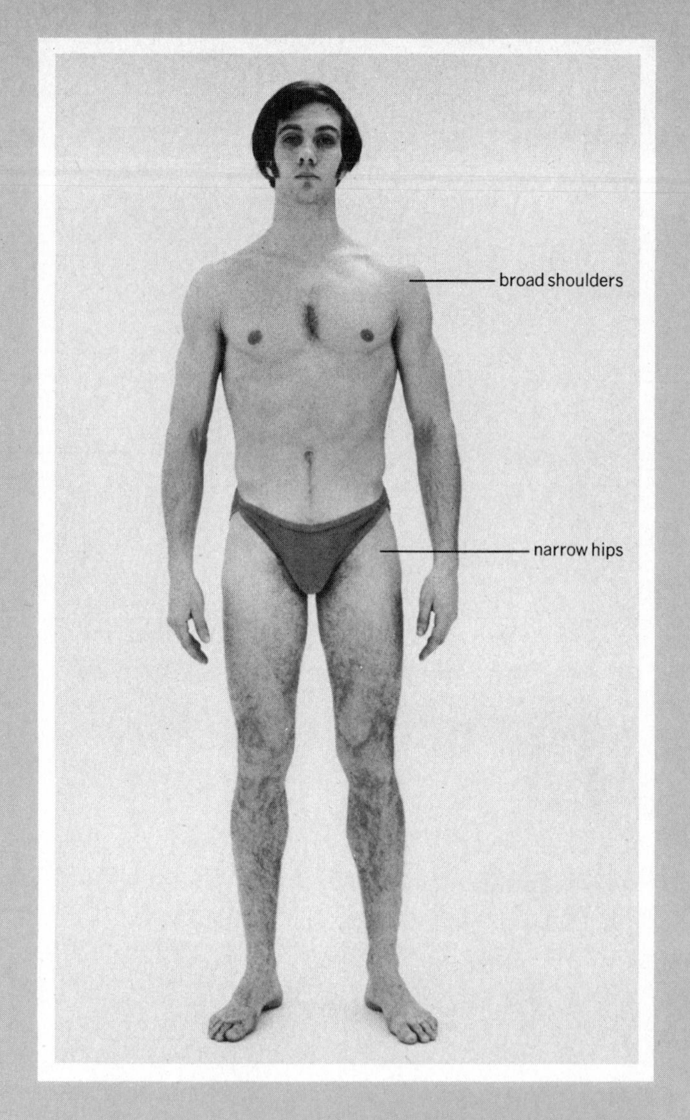

through the various planes of the body. A section dividing the body into equal right and left halves is one made in the MEDIAN OF MIDSAGITTAL PLANE. Any plane parallel to the median plane is PARASAGITTAL. A section dividing the body into anterior *(front)* and posterior *(back)* parts is made in the CORONAL *(frontal)* PLANE.

TRANSVERSE OF HORIZONTAL PLANES *(cross sections)* are at right angles to the other two and divide the body into superior and inferior portions. The terms longitudinal and cross section are used most commonly to describe the sections of organs themselves. A LONGITUDINAL SECTION is one cut parallel to the long dimension of an organ. A CROSS SECTION is one following the shorter dimension of an organ and is perpendicular to a longitudinal section.

General surface anatomy (Figs. 1.9 to 1.15)

It is not our purpose to pursue this subject in detail at this point but to provide certain general information and to encourage the student to continue this study as it applies to each of the systems studied. It is valuable not only to identify surface features but also to understand how they can be used to locate deeper structures of the body. For example, health personnel, in listening to heart sounds, need to know the surface points on the thorax where these sounds can be best heard. The bones which can be seen or palpated at the body surface give the position of underlying soft structures. Outlines of muscles can be seen and felt, their tendons followed, and their actions observed. Joint actions can be seen and blood vessels palpated to feel the pulsing of the blood. The valves of superficial veins can be located. Features of the skin such as thickness or thinness, looseness or tightness, and its relationship to underlying structures can be studied.

The body appearance. The external appearance of the body varies according to age, sex, and state of nutrition. Premature infants typically lack the thick subcutaneous layer of adipose tissue (fat pads) found in full-term infants, and the skin appears to hang loosely on the body. Full-term infants have a chubby appearance due to overall deposition of fat. The buccal fat pad in the cheeks gives the face the typical infant appearance. As growth occurs, the distribution of fat becomes more even on the body and is nowhere excessively abundant. At the time of adolescence and puberty, a sex difference becomes apparent. The female accumulates adipose tissue in restricted areas of the body, (fat pads), including the mammary glands, over the shoulders, buttocks, inner and outer sides of the thighs, lower abdomen, and over the symphysis pubis. In the male, the pads are thinner and more evenly distributed over the body, unless the individual is grossly overweight. In the latter case, the abdomen assumes the greatest role as a fat storage area. In middle age, obesity again becomes more predominant, while in old age, disappearance of fat and loss of elasticity of the skin causes the skin to hang loosely on the body.

Figure 1.10 shows adult male and female bodies.

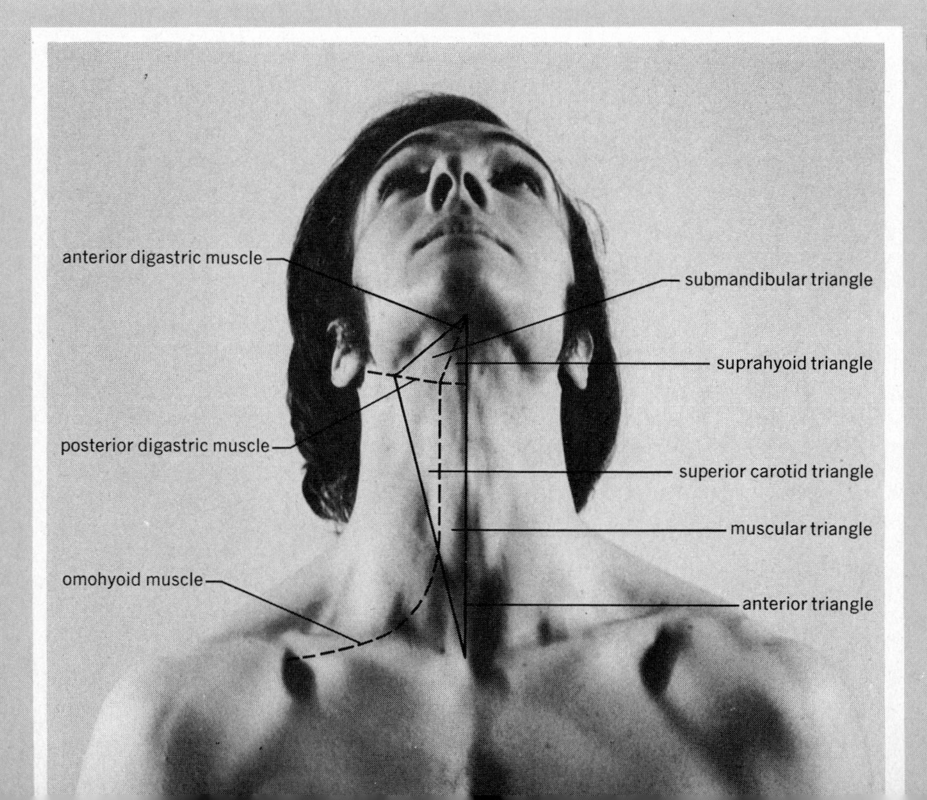

FIGURE 1.11 Anterior triangle.

FIGURE 1.12 Posterior cervical triangle.

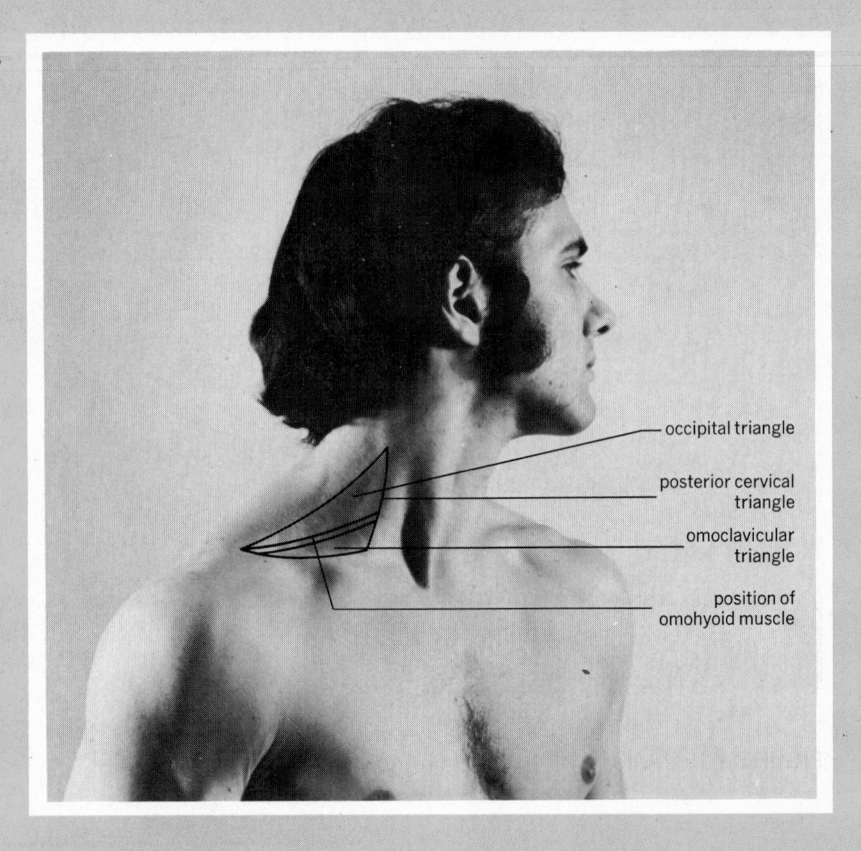

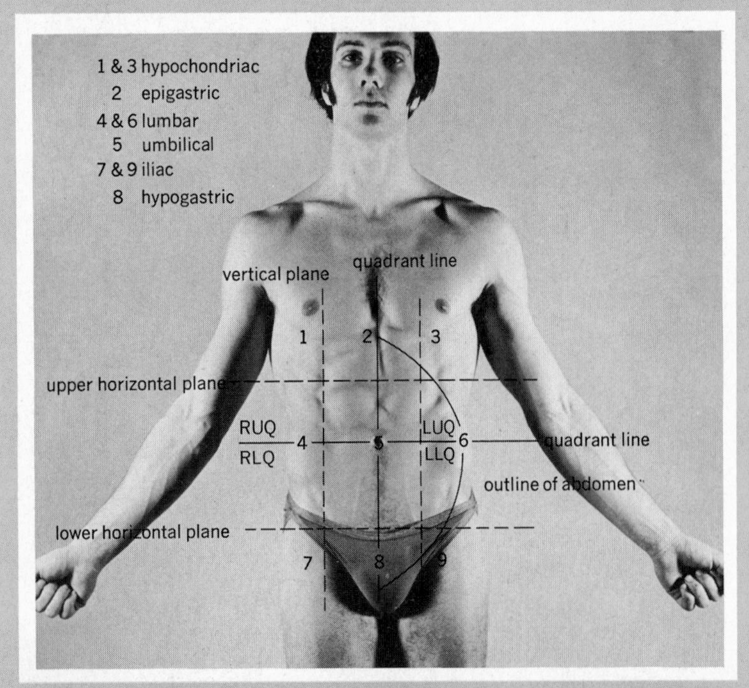

FIGURE 1.13 Abdominal reference lines.

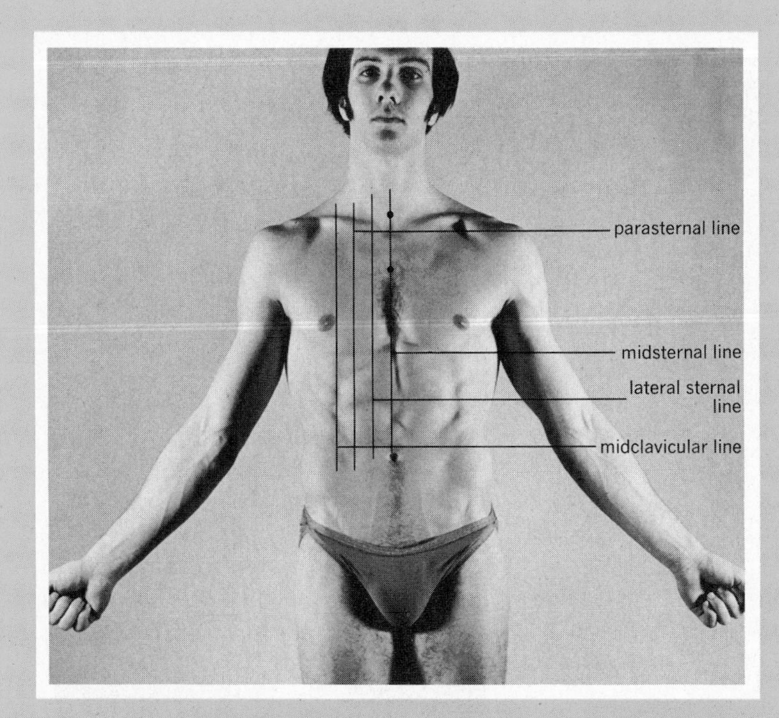

FIGURE 1.14 Thoracic reference lines.

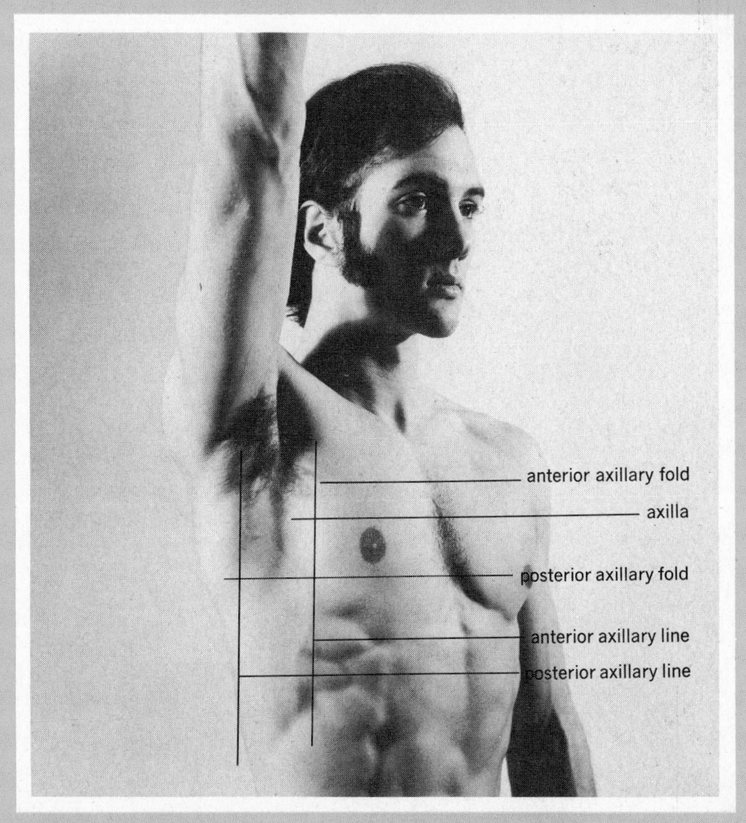

FIGURE 1.15 Axillary reference lines.

The male figure may be described as being triangular in shape with broad shoulders and narrow hips. The body is angular and lacks the rounded contours seen in the female. Muscular development is greater, and the general posture is more erect. The female body is more diamond-shaped. The shoulders are narrower, the hips broader. The body is rounded and softer in appearance due to the lesser development of muscles and a thicker layer of adipose tissue under the skin. The posture appears less erect, due primarily to a more pronounced curvature of the lower back.

Certain descriptive terms may be applied to the gross body areas (Fig. 1.9). These terms should be learned as an introduction to the descriptive anatomy that follows.

The head, supported by bones of the cranial walls, and the face provide many landmarks. These should be learned as one studies the skull, but can be previewed now by reference to Figs. 7.2 and 7.3.

The neck surfaces are not marked so much by skeletal as by muscular features (Figs. 1.11 and 1.12). The spinous process of the seventh cervical vertebra can be palpated at the base of the neck posteriorly, and the thyroid cartilage *(Adam's apple)* is the most prominent feature anteriorly. By pressing above the thyroid cartilage, the body of the hyoid can be felt.

The trunk region, made up of the chest or thorax, the back, and the abdomen, is illustrated here for general features of surface anatomy, but will be elaborated in the skeletal, muscular, and other systems. As the student will notice, the surface areas are marked off by vertical and horizontal lines appropriately named. These lines and the areas enclosed will take on additional meaning as indicators of the positions of underlying structures (Figs. 1.13 to 1.15).

Summary

1. The human being is an animal organism.

2. The human is classified as a member of the Phylum Chordata on the basis of having:
   - *a.* A notochord (embryologically).
   - *b.* Gill pouches (embryologically).
   - *c.* A dorsal hollow nerve cord.

3. The human is a vertebrate animal and has true celomic cavities as follows:
   - *a.* A thoracic cavity subdivided into a pericardial cavity and two pleural cavities.
   - *b.* An abdominopelvic cavity divided arbitrarily into abdominal and pelvic portions.

4. Other cavities possessed by the human are:
   - *a.* A dorsal cavity divided into cranial and vertebral portions, the former housing the brain, the latter the spinal cord.
   - *b.* A mediastinum between the pleural cavities.

5. The human's relationships to other animals are summarized on page 6.

6. Humans have their own unique characteristics:
   - *a.* Most highly developed cerebrum of the brain.
   - *b.* Bipedal locomotion.
   - *c.* Upright posture.
   - *d.* Most adaptable of all animals.
   - *e.* Capable of modifying their external environment more than other animals.
   - *f.* Highest reproductive efficiency, particularly in postnatal care.
   - *g.* Reproductively active throughout the year.
   - *h.* Intelligence, flexibility, individualization, and socialization are carried to a higher degree in humans than any other animals.
   - *i.* Humans are products of organic evolution but are unique in a new kind of evolution called *cultural.*
   - *j.* Humans are, or are capable of being, ethical animals.

7. The human's most serious problems relate to:
   - *a.* Overpopulation.
   - *b.* Destruction of the external environment.
   - *c.* The threat of and the act of war.

8. Levels of organization in humans (and other organisms) are considered in terms of:
   - *a.* Cells.
   - *b.* Tissues.
   - *c.* Organs.
   - *d.* Systems.

9. Parallels in ontogeny and phylogeny are noted.

10. Humans are creatures of dynamic balance, maintaining a near constancy of the internal environment.

a. The near constancy of the internal environment is referred to as homeostasis or, as some physiologists prefer, homeokinesis or homeodynamics.
b. Examples are: body temperature, acid-base balance, and osmotic balance.

11. Homeostasis involves the action of control systems.
a. The study of control systems is called cybernetics.
b. Physiology is largely a study of the control systems of living organisms.

12. Feedback, in which the output of a system is monitored and a signal sent back to the system, is a basic feature of control systems of the human body.
a. Negative feedback systems in humans are the kind most commonly involved where the result is an inhibition or slowing of action.
b. Positive feedback systems are those in which the feedback signal reinforces the input signal and output is increased.

13. Control in organisms goes back basically to the gene. Life began with the appearance of the first nucleic acids, of which genes are the modern descendants.
a. A gene is physically a unit or portion of a chromosome.
b. A gene is chemically the DNA fraction of a nucleoprotein.
c. A gene is operationally a fraction of a chromosome that controls one reaction in metabolism.
d. Genes account for the conservatism resulting in the continuity of species, for the variation within species, and for change, which is the basis for evolution.

14. As multicellular animals evolved, communication was at first cell-to-cell; their stimulus – modulator – response mechanism was essentially that within cells. Later, transportation systems appeared, enabling cells to extend their influence – largely chemical – to other parts of the body. Finally, the appearance of nervous systems brought control and integration to their present level as seen in humans.

15. To describe the human body a universal language became important. Such descriptions are all based on the body in the anatomical position – standing erect, eyes forward, arms at sides, palms forward, and with the heels approximated.

16. To gain additional information about human anatomy – the positioning of organs, for example – sections are made through the various planes of the body, namely:
a. Transverse (horizontal).
b. Coronal (frontal).
c. Median (midsagittal).
d. Parasagittal (sections parallel to median plane).

17. A study of surface anatomy provides knowledge not only of those surface areas of head, neck, trunk, and appendages, but also of the underlying soft structures.

The Cellular Level of Organization

Determinants of cellular morphology

A living organism may be defined as one that is capable of carrying out those activities necessary to ensure its survival. Such activities include: INTAKE of materials necessary for nutrition of the cell; ELIMINATION of wastes of activity that might prove toxic; the utilization or METABOLISM of materials, including energy-releasing reactions *(catabolism)* and synthetic reactions *(anabolism)*; EXCITABILITY, or the power to respond to environmental change; REPRODUCTION; and MOVEMENT.

This list of activities carried out by living organisms presumes that energy-yielding and synthetic enzyme systems are present to enable carrying out the activities. Viruses, for example, may be considered to be intracellular parasites, for they lack such enzyme systems and must depend on their host to provide the basic necessities for existence.

In those forms that do meet the "criteria of life" listed above, one or more units, known as CELLS, are usually present and are organized internally to permit the various activities to be carried out without interference between them.

Unicellular organisms must be capable of carrying out all of these activities within the limits of that single cell, and thus such an organism is more versatile, less specialized, and independent of other cells. Additionally, each cell looks like every other cell and is about the same size. Multicellular organisms, because of greater size and complexity, reserve for all cells only those processes necessary for survival (intake, metabolism, elimination) and show specialization of cells, division of labor between cells, and a relationship between structure and function. Thus, some cells become modified into coverings; others, such as muscle cells, become elongated and specialize in shortening or contraction; still others develop bizarre shapes or surface modifications to afford a large surface for contact (nerve cells), or to insure a large surface for intake and output of materials (intestinal epithelial cells).

With cellular specialization and division of labor comes an interdependency between the various cell types. For example, a skeletal muscle cell it not adapted to transport oxygen, but requires tremendous volumes of oxygen for its activity; it depends on red blood cells to supply the needed oxygen, and on the heart muscle cells to circulate the red blood cells. Of course, many other examples of interdependency are to be found in the body, and these will be discussed in the chapters that follow.

An additional consideration should be mentioned. Unicellular organisms usually live in a fluid environment over which they exert little or no control in terms of temperature, acidity, and other factors. Thus, any change in the environment must and will immediately affect the life-sustaining processes within the cell. Multicellular organisms, particularly the so-called "higher organisms," may have achieved a degree of independence from the external environment, at least as far as cell survival is concerned. This independence is achieved by the separation of an INTERNAL ENVIRONMENT from the external environment, and by the development of devices for maintaining the constancy of that internal environment within narrow limits. Cells are thus freer to concentrate on or specialize in particular activities without being concerned about such things as their nutrition, temperature regulation, and waste removal.

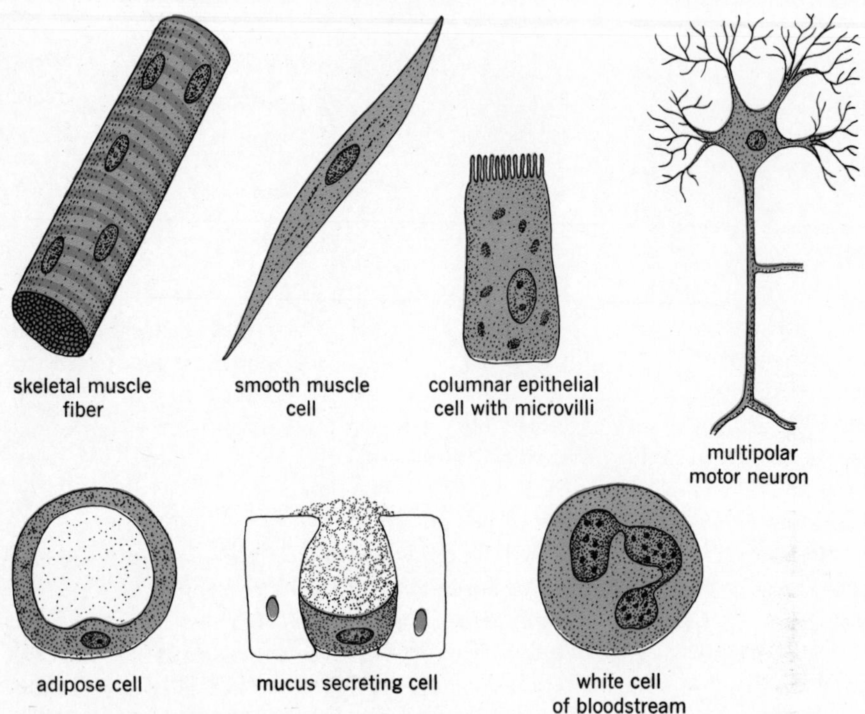

FIGURE 2.1 Diverse forms of mammalian cells (not to the same scale).

Cell size

Individual cells may vary from about 200 nm* in diameter to the size of an ostrich egg, about 20 cm (8 inches) in length. The lower limit is seen in the bacteria, and is set by the size required to contain the chemicals (e.g., nucleic acids and proteins) necessary to control metabolism and sustain life. Viruses may be as small as 10 nm in diameter, but are not considered to be true cells. The upper limit usually is influenced by the necessity for including in the cell large amounts of reserve nutrients for the development of offspring (as in eggs) and by the size of the offspring. Most cells lie in the size range of 0.5–20 μm.

As cell size increases, other problems are created that must be solved. If the cell has a single nucleus, increasing volumes of cytoplasm must be "controlled" or "given instructions" by the nucleus, and difficulties may arise in terms of circulating or transmitting that control to all parts of the cell. The cell may develop several nuclei, each of which governs or controls a portion of the cell, or the cell may develop methods such as cytoplasmic streaming to circulate the controlling chemicals within the cell. Also, as cell size becomes greater, surface area increases as the square of the increase ($12.57r^2$), while volume of the cell increases as the cube of the increase ($4.189r^3$). In other words, the amount of substance demanding nutrients and producing wastes goes up more

*m = meter
mm = millimeter = 1/1000 m or (10^{-3}m)
μm = micrometer or micron = 1/1,000,000 m or (10^{-6}m)
nm = nanometer or millimicron (mμ) = 1/1,000,000,000 m or (10^{-9}m)
Å = angstrom = 1/10 nm or (10^{-10}m)
pm = picometer (10^{-12}m)

rapidly than the surface through which the materials may pass. At some point, then, surface area becomes inadequate to service the active material; at this point the cell may develop devices (e.g., microvilli) for increasing its surface area, or it may become folded, elongated, or flattened to aid in solving the supply problems.

Cell shape

Shape of cells is variable (Fig. 2.1) and bears a great relationship to function and specialization of cells. Cells that must shorten (muscle) or ones that carry messages (nerve) obviously must have great length; fat (adipose) cells are "hollow" to store a fat droplet; mucus-secreting cells are "cup-like" to hold their secretion; red blood cells are flattened to provide a large surface-volume ratio for rapid diffusion of gases across their membranes. Many examples of the relationships or structure and function will be discussed in later chapters.

Cell structure and function

Most cells carry out a series of chemical reactions and activities to ensure their survival independent of special functions the cell may have. Hundreds of chemical reactions may be occurring simultaneously within the limits of a single cell. The compartmentalization of related reactions is necessary to avoid interference between particular metabolic schemes. Particular materials may be required at particular regions within the cell, and some materials may need to be kept out of or restricted to a particular cell site. These functions of compartmentalization, directing, and selectivity are carried out by a series of MEMBRANE SYSTEMS within the cell.

Membrane systems

STRUCTURE. A variety of membrane systems are found associated with cells (see Fig. 2.3). The outer limit of a cell is formed by the CELL or PLASMA MEMBRANE. Membranes also line or surround most of the CELLULAR ORGANELLES, to be discussed later. All membranes, regardless of where they are found, appear to be composed of two major components: lipids, substances insoluble in water, but soluble in substances such as alcohol, ether, and chloroform; and proteins, complex fibrous or globular units composed of smaller amino acids. What is different in different membranes are the ratios of the two components and their arrangement. Table 2.1 indicates the types and ratios of lipids and proteins in several representative membranes. These components are believed to be arranged either as a protein-lipid-protein "sandwich" in the membrane or as globular protein units that "float in a sea of lipids."

These two general theories of arrangement of components are diagrammed in Fig. 2.2. The arrangement of the components may result in the formation of channels or "pores" through the membrane. These channels are believed to have diameters of 4–8 Å and are thought to have positively charged ions (such as Na^+ or Ca^{++}) or ionized proteins lining their inner surfaces. In any event, due to size and/or charged linings, the pores do not represent areas where free passage of materials through the membrane can occur.

FUNCTION. The nature of the membrane allows it to control the movement of substances across the membrane. Most membranes are SELECTIVE, in that they allow some materials to cross the membrane while restricting the passage of others. Some factors determining membrane selectivity are:

TABLE 2.1 Composition of representative membranes

Membrane	Protein-lipid ratio (number : 1)	Lipids predominating*
Erythrocyte (plasma membrane)	1.5	Phospholipids (55%) Cholesterol (25%)
Mitochondrial (inner membrane)	3.0	Phospholipids (95%) Cholesterol (5%)
Myelin (nerve fibers)	0.25	Phospholipids (32%) Cholesterol (25%) Sphingolipids (31%)

* Phospholipids are a combination of a lipid substance and a phosphoric acid radical ($HPO_4^{=}$). Cholesterol is a complex lipid containing four ringlike structures. Sphingolipids contain a fatty acid, phosphoric acid, a substance called choline, and the alcohol sphingosine.

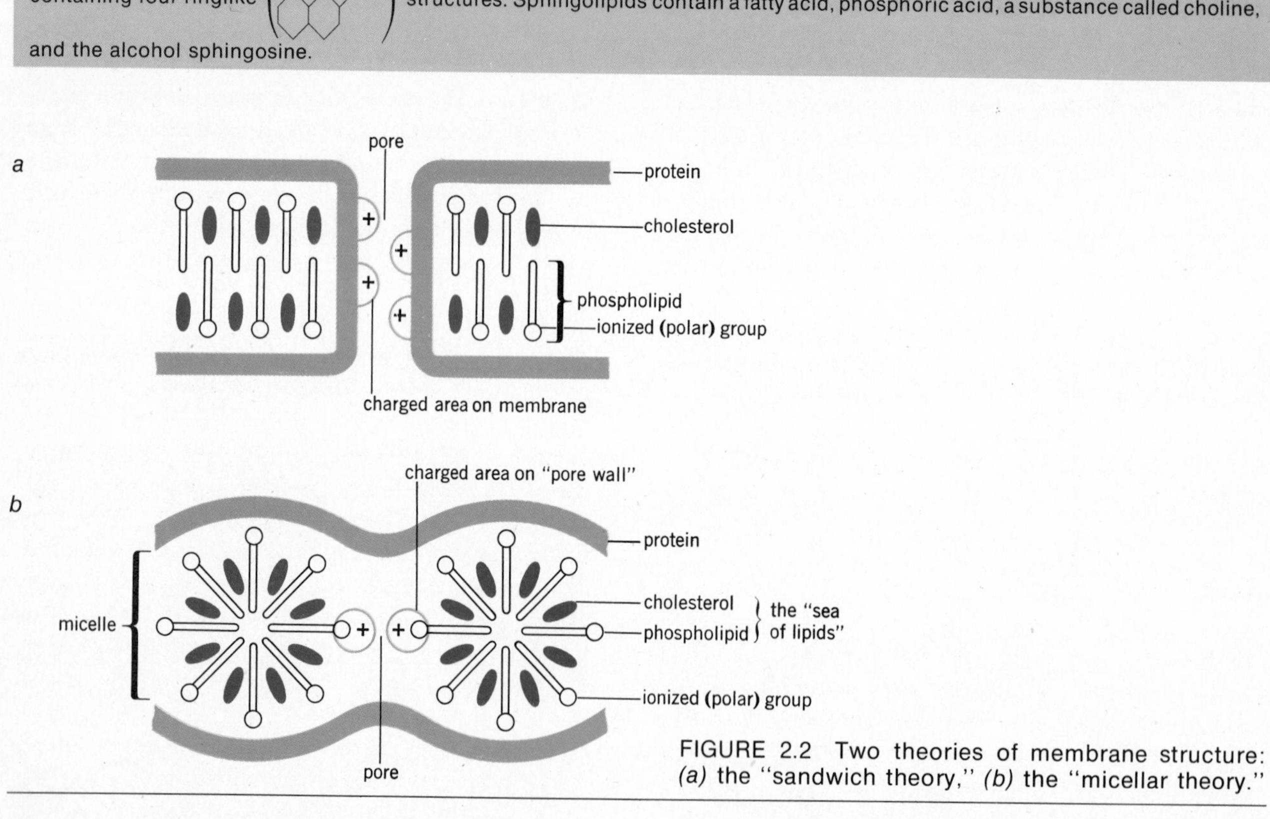

FIGURE 2.2 Two theories of membrane structure: *(a)* the "sandwich theory," *(b)* the "micellar theory."

Membrane thickness. Because of degree of hydration (water content), and ratios and orientation of the lipid and protein components, some membranes are thicker than others. A thicker membrane means a longer distance to traverse and usually a slower rate of passage of a substance through the membrane.

Size of substance passing through the membrane. Membrane structure, including pores, and the "meshes" of the protein strands composing it, creates "holes" of various sizes through which substances may try to pass. If the substance is smaller than the hole or pore, it will pass more easily than if it equals or is larger than the size of the hole or pore. Substances larger than the size of a glucose molecule (8.6 Å) pass with difficulty through most membranes.

Lipid solubility of the substance attempting to cross the membrane. Since membranes contain much lipid material, a substance that dissolves easily in the lipid should, and does, pass more easily through the membrane. For example, steroids, vitamins A, D, E, and K, and triglycerides all have high fat solubility.

Electrical charge of the substance passing through the membrane as related to charge on the membrane itself. Most membranes have negatively charged proteins and lipids at normal body pH, and their pores are usually lined by positively charged ions. If a substance trying to cross the membrane is itself charged, it may be repelled or attracted to the membrane (like charges repel, unlike charges attract). Charged substances generally pass less easily through membranes than do uncharged ones.

Active processes in or by the membrane. Membranes may contain chemicals that can actually attach to and carry substances across membranes (active transport) or, the membrane may respond by "sinking in" (pinocytosis) or surrounding (phagocytosis) certain materials that are on or near the membrane. Such processes enable the membrane to exert a very fine control over what ultimately crosses the membrane.

Membranes also participate in enzymatically controlled reactions. Cell membranes may contain digestive enzymes that break down chemicals; mitochondrial membranes contain enzymes for metabolism of glucose and other substances; membranes of other cellular organelles may aid in the synthesis of lipids and other complex materials. The "universal energy source" for cellular activity, called ATP (adenosine triphosphate) is split in many membranes.

Membranes appear to possess binding sites for materials that may be entering the cell, or that are influencing cell activity. For example, insulin, epinephrine, and other hormones may bind to cell membranes to exert their effects. Antibodies,

TABLE 2.2 Summary of functions of cell membrane

Function	Comments
Surrounding the cell	Separates the cell contents from the surrounding environment
Controlling passage of materials, such as	
Water	Freely permeable
Charged substances	Like charges repel, opposites attract. Relationship is between charge on membrane and charge of particle attempting to cross membrane
Large molecules	Large molecules pass more slowly unless transport system is present
Lipid soluble substances	Pass more rapidly than materials having low lipid solubilities
Digestion	Protein-, carbohydrate-, and lipid-digesting enzymes are present in most membranes
Transport	"Carriers" for active transport are found in membrane
Binding	Sites are described for hormones, antibodies, and other substances
Energy transformations	Adenosine triphosphate (ATP) is split in membrane

chemicals involved in the body's immune responses, also appear to bind to cell surfaces.

These activities of the cell membrane indicate that it is much more than a limiting structure. Its activities are summarized in Table 2.2.

Cellular organelles. The cellular organelles (Fig. 2.3) are described as being permanent, membrane surrounded (in most cases), metabolically active, self-reproducing, formed structures found in the protoplasm of the cell. The area be-

FIGURE 2.3 The central picture represents the structure of a cell as it might appear under light microscopy. The peripheral pictures illustrate the fine structure of the organelles as they would appear in electron micrographs.

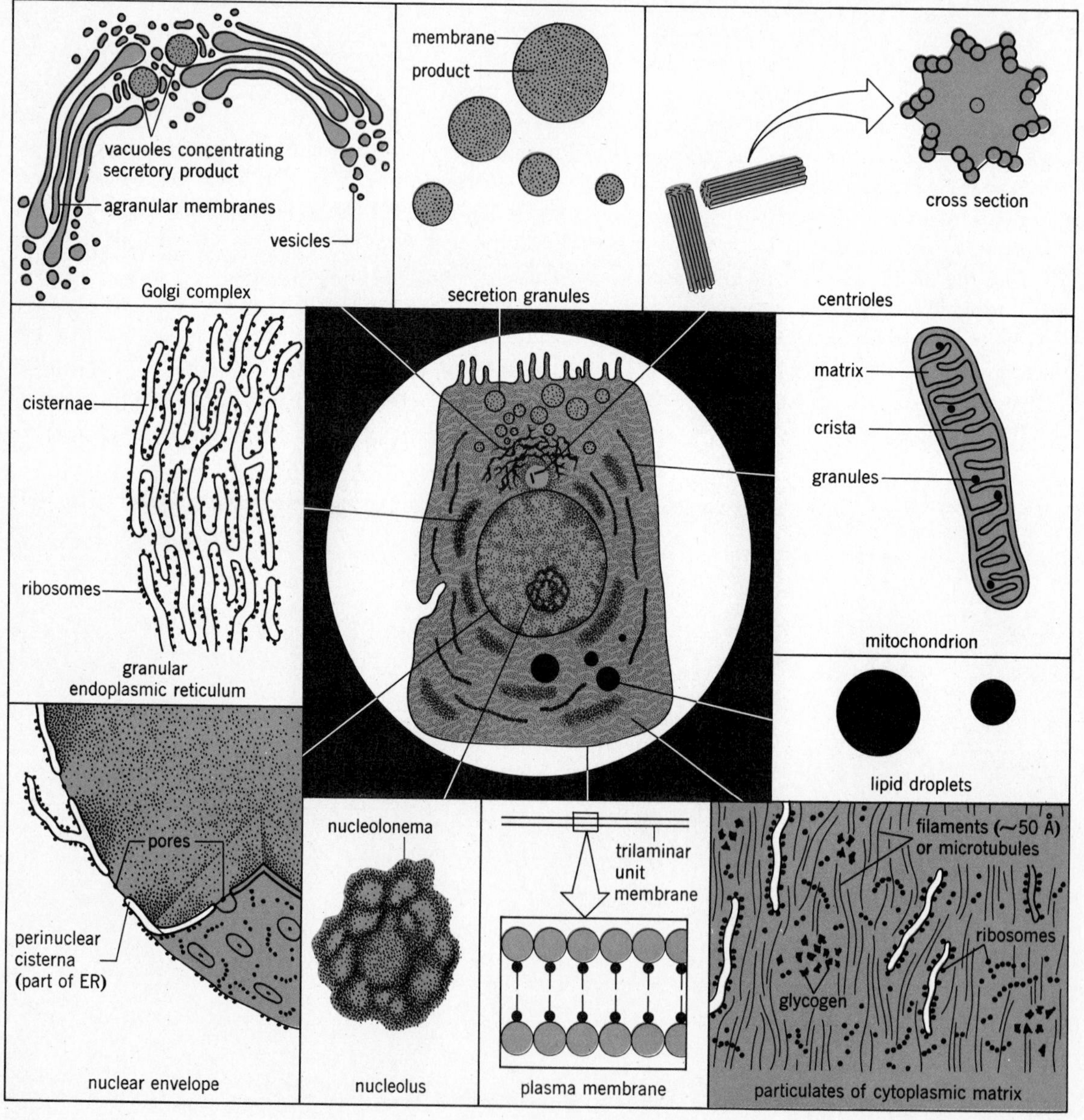

tween the cell and nuclear membrane is generally known as the cytoplasm and is described as being the region in which the bulk of the cellular energy-releasing and synthetic reactions occur. The organelles provide a means of isolating certain chemical reactions and/or preventing powerful enzyme systems from destroying cellular substance itself.

Endoplasmic reticulum. The endoplasmic reticulum (ER) is a membrane-lined netlike system of fluid-filled, tubular channels permeating all parts of the cytoplasm. The diameter of the tubes varies between different cell types; in some the tubules are 50–100 nm in diameter and appear as flattened vesicles; in other cells parts of the tubules are greatly enlarged to form vesicles and sacs known as CISTERNAE. The tubules have been shown to connect with, and may actually form, the tubes of the Golgi apparatus (see below). Also, the ER has been shown to have connections with the cell membrane and to form the outer of the two membranes surrounding the nucleus of the cell. The membranes of the ER have been shown to be selective, and may contain enzymes involved in the synthesis of several types of lipids. Two varieties of ER are described: ROUGH ER has ribosomes attached to its outer surface; SMOOTH ER lacks ribosomes and serves as the site of the lipid synthesis mentioned above. As a whole, the channels of the ER may also serve as a system for distribution of materials, acquired through the cell and nuclear membranes, throughout the entire cytoplasm (a "circulatory system" for the cell). MICROSOMES are believed to be vacuolelike fragments of rough ER that most often result when a cell is disrupted by experimental procedures.

Ribosomes. Ribosomes are small (100–150 Å) granules of a chemical called RIBONUCLEIC ACID (RNA) attached to a protein to give a combination known as RIBONUCLEOPROTEIN (RNP). Ribosomes have been shown to consist of a large and a smaller part; the large portion is involved in attracting amino acids during PROTEIN SYNTHESIS, while the smaller portion "locks" the ribosome to another molecule that dictates the placement of the amino acid in the growing protein (see Chapter 4). Ribosomes are commonly found linked in "chains" to form polyribosomes that appear to be involved in synthesis of very large protein molecules such as the immune substances (antibodies).

Golgi apparatus. The Golgi apparatus (also *complex* or *body*) is a system of canals and saclike vesicle stained selectively by the compound osmium tetroxide and silver salts. The organelle is usually localized within the cell, typically near the nucleus. The channels of the Golgi apparatus have been shown to connect with those of the ER, and the Golgi may receive its materials in this manner. Also, some evidence indicates a connection of the Golgi channels to the region around the nucleus. Functionally, the apparatus has been shown to contain lipid droplets, mucus, and protein polysaccharides that are believed to be synthesized by the organelle. Also, the apparatus appears to be the site where secretions (such as inactive forms of enzymes or hormones) are surrounded by a membrane ("packaged") before being released from the cell. In mucus-secreting cells, the Golgi apparatus is cuplike in shape and "contains" the mucus produced by or on the organelle.

Lysosomes. Lysosomes are particles ranging in size from 0.25–0.8 μm. Under the electron microscope, they appear as membrane-bound sacs or vacuoles of granular material. Analysis of the granules in the organelle shows them to be composed of powerful hydrolytic enzymes capable of breaking large molecules into smaller ones. In a sense, then, the lysosome acts as a "digestive system" for the cell. Some enzymes isolated from lysosomes are presented in Table 2.3. Lysosomes have also been implicated in causing cell death ("suicide packets") when cells are abnormal or injured. Such a mechanism "clears the way" for new cells to replace the damaged units, in a sort of self-destruct mechanism. The inflammation that occurs in rheumatoid arthritis may also be due partly to lysosome release of enzymes.

Mitochondria. Mitochondria are organelles usually between 0.5–1.0 μm in diameter, and up to 7 μm in length. They may be granular or filamentous in shape, and the shape may alter accord-

TABLE 2.3 Some lysosomal enzymes

Enzyme	Substrate (material acted on)
Collagenase	Collagen, connective tissue protein
Cathepsins	Cartilage, protein polysaccharides
Elastase	Elastin, connective tissue protein
Hyaluronidase	Connective tissue, protein polysaccharides
Protease	Protein polysaccharides
Acid peptidase	Fibrin

ing to the cell's physiological condition. The organelles may be derived from the cell membrane or ER, and show active movement in certain cells. Some investigators have suggested that mitochondria may be a separate organism (like a virus, perhaps) that has invaded a cell for sustenance. Structurally, a mitochondrion shows a two-layered membrane, the external one being regular in outline, while the inner one is thrown into shelflike or tubular structures called CRISTAE. The membranes and cristae have been shown to contain enzyme systems necessary for the degradation of glucose and other substances, and for the synthesis of ATP, the "universal energy source" for many cellular reactions.

Centrioles. Centrioles typically are paired structures that are concerned with cell division. They have an average diameter of about 0.2 μm, and a length of about 0.6 μm. They are not always visible, but may become so at the time of cell division. In most cells they are oriented at right angles to one another and have a structure consisting of nine "triplets" of fibrils arranged around their periphery. The centrioles and their surrounding areas of differentiated (lacks ribosomes, ER) cytoplasm constitute the CELL CENTER OF CENTRAL BODY.

Fibrils. Fibrils, or elongated filaments of protein, are common constituents of muscle cells, where they shorten, and ciliated cells. In cilia, the filaments have the appearance of being hollow (the proteins may form a spiral) and are arranged in a characteristic pattern of nine pairs of peripheral tubules around a central pair. Cilia "wave" and cause the movement of materials across the surface of the cells possessing them.

Microtubules. Microtubules are tiny hollow elongated tubules that appear to permeate all parts of the cytoplasm. They may give a form and structure or cytoskeleton to the cell; the fibers forming the "spindle" during cell division have been shown to be microtubules.

The term INCLUSIONS is usually given to masses of organic substances that are not metabolically active, but that may serve as fuel sources for the cell or represent wastes of cellular metabolism. Thus, lipid droplets, sugar stored in the form of glycogen, or crystals may be demonstrated in many cells. VACUOLES are often formed to contain such materials.

The nucleus

Overall direction of cellular activity is provided by the cell's NUCLEUS (Fig. 2.4). Two MEMBRANES limit the nucleus. The outer membrane is derived from the ER, and the inner membrane is a product of the nucleus. Nuclear pores, actually thin areas where the inner and outer membranes appear to be fused, allow entry and exit of large molecules to and from the nucleus. A NUCLEOLUS, composed chiefly of ribonucleic acid (RNA), floats in the NUCLEAR FLUID. Deoxyribonucleic acid (DNA) is present in the nucleus in the form of CHROMATIN MATERIAL. The chromatin material forms the chromosomes during cell division.

DNA contains the genes that determine the individuality of the organism. The genes influence cell function primarily by the synthesis of RNA in the process known as TRANSCRIPTION. A DNA nucleic acid chain (a gene?) causes the synthesis of an RNA chain that leaves the nucleus through

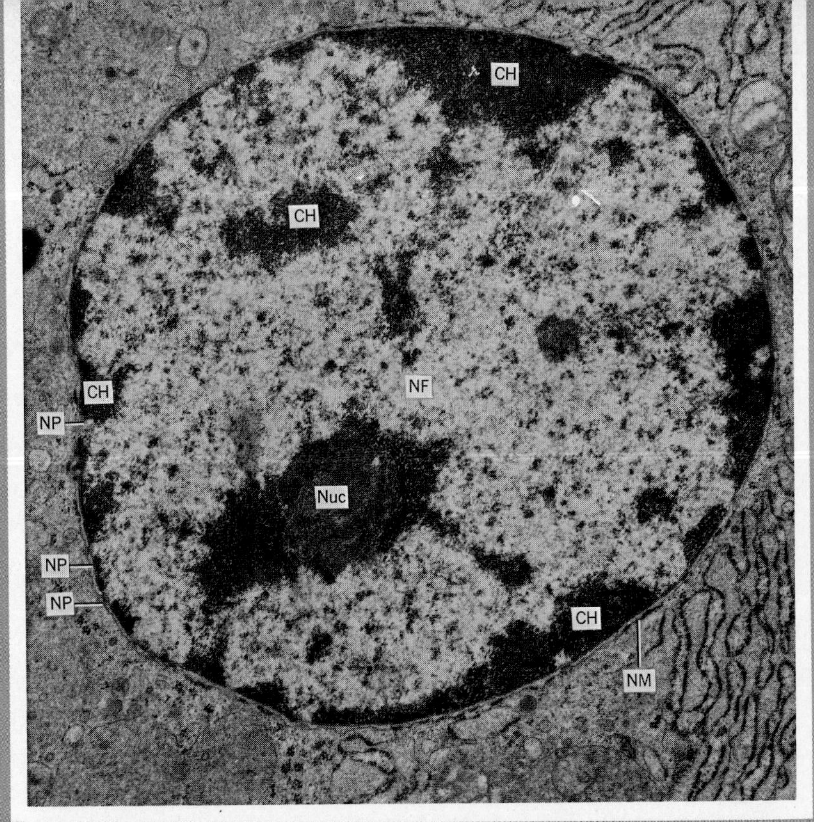

FIGURE 2.4 Nucleus from a mucous cell of the trachea of a laboratory mouse (×15,000). NM = nuclear membrane, NP = nuclear pores, CH = chromatin, NF = nuclear fluid (light areas). (Courtesy Norton B. Gilula, The Rockfeller University.)

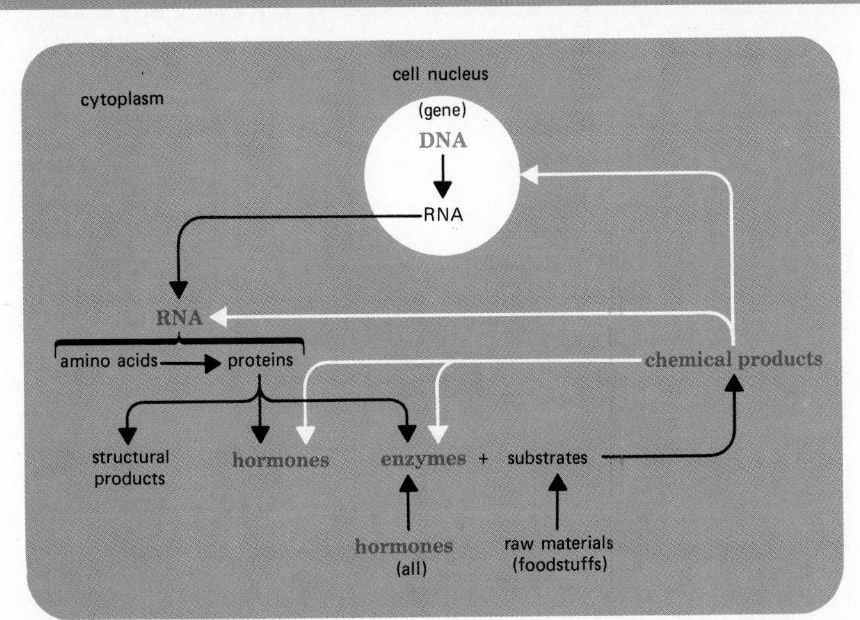

FIGURE 2.5 Control of cellular activity by genes and hormones. White lines indicate primary homeostatic mechanisms that may stimulate or inhibit activity.

TABLE 2.4 Summary of cell structure and function

Cell part	Structure	Function	Comments
Plasma membrane	Three-layered sandwich of lipid and protein, or units (micelles)	Protects, limits, controls entry and exit of materials	Selectively permeable
Cytoplasm	Contains many organelles and inclusions	Factory area; synthesizes, metabolizes, packages	Bulk of cell
ER	Hollow channels in whole cytoplasm; filled with fluid	Transport	"Circulatory system" of cell
Ribosome	Granules of nucleic acid and protein	Protein synthesis	Free or attached to ER
Mitochondria	Double layered with cristae	Production of ATP	"Powerhouse" of cell
Golgi apparatus	Flattened sacs with vacuoles	Secretion; condensation membrane	Packaging of materials
Central body	Centrioles and centrosome	Cell division	Lacking or non-functional in mature nerve cells
Lysosome	Membrane-surrounded sacs of enzymes	Reduces large molecules to smaller units	"Digestive system" of cell
Fibrils	Protein strands	Support and movement	Extreme development in muscle cells
Microtubules	Tiny hollow tubes	Support and conduction	Universal occurrence
Inclusions	Fats, sugar, wastes	Fuels and wastes	Not active metabolically
Vacuoles	Membranous sacs of fluid	Storage and excretion	Universal occurrence
Nucleus	Membrane, nucleolus, chromatin (DNA), and fluid	Directs cell activity	"Brains" of cell

the large pores and enters the cytoplasm. This RNA is known as *messenger-RNA (m-RNA)* because it carries a series of coded instructions for protein synthesis. The m-RNA attaches to the ribosomes. The cytoplasm contains *transfer-RNA (t-RNA),* which brings amino acids to the ribosomes where they are synthesized into proteins (enzymes, hormones) that control body chemical reactions. Thus, nuclear DNA controls activity primarily through the production of regulatory proteins.

Regulation of enzyme activity, once the enzymes are formed, may occur through substrate concentration, the accumulation of inhibitory metabolites, or by the action of hormones. The interrelationships described above are diagrammed in Fig. 2.5. A summary of cell structure and function is present in Table 2.4.

Methods of acquiring materials

Both unicellular and multicellular organisms utilize passive processes and active processes to acquire the materials necessary for their activity. PASSIVE PROCESSES occur according to physical laws and involve concentration and pressure differences as the driving forces for these processes. They require no energy output on the part of the cell. ACTIVE PROCESSES do require the participation of the cell as a supplier of energy and compounds necessary for the process.

Passive processes

DIFFUSION. Diffusion (Fig. 2.6) is defined as the tendency of molecules or ions to move from an area of higher concentration to one of lower concentration regardless of whether a membrane is present. The process eventually results in uniform distribution of the diffusing substance *(solute)* in the diffusing medium *(solvent).* Diffusion depends on the movement *(kinetic energy)* of the solute and solvent molecules and the collisions between these molecules that result in their separation. Rate of penetration of a substance is called the *permeability constant* (K), and is influenced by temperature, area of the cell membrane through which passage is occurring, size of diffusing molecules, and the magnitude of the diffusion gradient (the concentration difference for the diffusing material on the two sides of the membrane). Higher temperature increases kinetic energy and rate of diffusion; larger molecules diffuse more slowly because they are moved a lesser distance when collisions occur; and a larger gradient means a faster rate. Rate of diffusion slows as uniformity of distribution is approached and the gradient lessens. If a substance can diffuse through a membrane at all, it can obviously pass in both directions at once. A "net diffusion" of material will occur as a greater flow *from* the side with the greatest concentration of material. Examples of diffusion as it occurs in the human body include the absorption of prod-

FIGURE 2.6 The diffusion of dye molecules from an area of higher to lower concentration of dye.

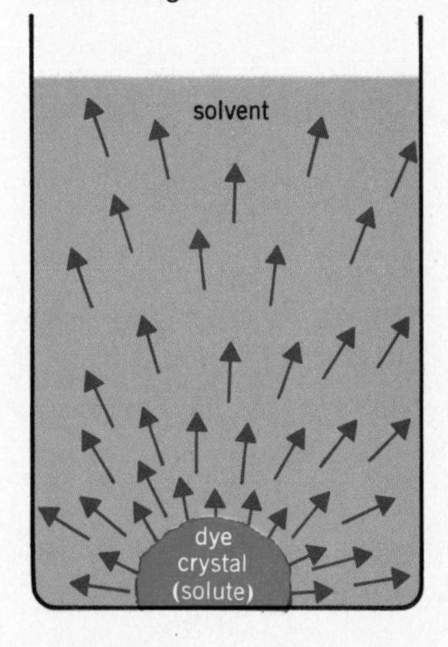

ucts of lipid digestion from the intestine into the lymph stream, the passage of gases across the lung surfaces, and the movement of any solute through the body fluids. The dispersion of milk or sugar in unstirred coffee or the distribution of an odor in a room are also examples of diffusion. Thus, diffusion may occur with or without the presence of a membrane.

OSMOSIS. Osmosis is an example of diffusion, but the term is restricted to mean the passage of only water through a selective or semipermeable membrane from an area of higher to lower water concentration. In order for only water to diffuse, there must be a selective membrane, which permits the passage of water in both directions but restricts the passage of solutes, and a difference in water concentration on the two sides of the membrane. Water concentration is inversely related to solute concentration on a given side of the membrane. Thus, the net water movement may be said to occur from the area of higher water concentration (lower solute concentration) to the area of lower water concentration (higher solute concentration). A measurable pressure, designated as the osmotic pressure, develops as the result of the separation of unequal solute concentrations. It varies directly with temperature and concentration. The formula used to calculate the osmotic pressure appears in the Appendix.

To illustrate the operation of the process, consider the placing of cells having selective membranes in solutions containing varying concentrations of solute (Fig. 2.7). If the cell is placed in a solution containing a solute concentration equal to that inside the cell (e.g., red cells in a physiological NaCl—0.9 percent solution), water concentrations are also equal, and there will be equal movement of water molecules into and out of the cell. The cell has been placed in an ISOTONIC (*iso*, equal) SOLUTION and will neither swell nor shrink. If the cell is placed in a solution that has less solute in it than in the cell, water concentration is relatively greater outside the cell and a net flow of water into the cell will occur. The cell will swell. In this situation, the cell has been placed in a HYPOTONIC (*hypo*, less) SOLUTION. If the cell is placed in a solution containing more solute than

FIGURE 2.7 Changes in cell size in *(a)* isotonic, *(b)* hypotonic, and *(c)* hypertonic solutions as osmosis occurs. Arrow length indicates direction of greatest water flow.

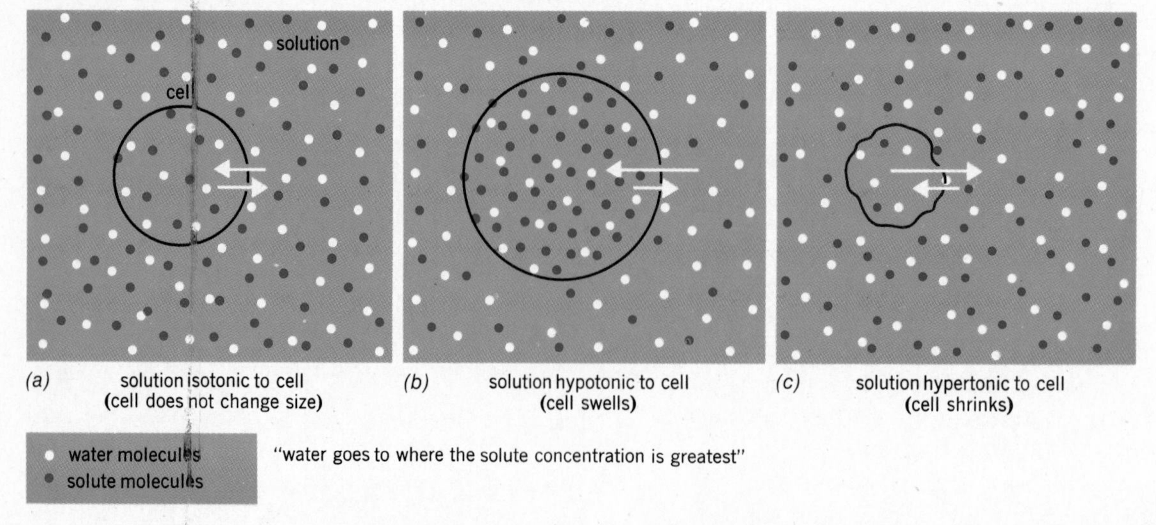

in the cell, concentration of water is relatively higher in the cell, and a net flow of water out of the cell will occur. The cell will shrink. In this situation, the cell is said to have been placed in a HYPERTONIC (*hyper*, greater) SOLUTION. Examples of osmosis occurring in the body include water movement through the walls of kidney tubules during urine formation; movement of tissue water into capillaries, and absorption of water through the walls of the alimentary tract as solutes are absorbed.

DIALYSIS. Dialysis occurs when there are two or more solutes on the same side of a membrane, and the membrane is not permeable to all of the solutes. The solutes to which the membrane is permeable will diffuse through the membrane according to their concentration gradients. Those to which the membrane is not permeable will not diffuse through the membrane. Thus, a separation of diffusible and nondiffusible solutes will occur. Dialysis is used to separate and purify solutes and is the basis of the operation of the artificial kidney (see Chapter 29).

FILTRATION. Filtration occurs because of a pressure differential on two sides of a membrane. Materials to which the membrane is permeable are forced through by the pressure. What passes thus depends mainly on the characteristics of the membrane, and the rate of passage depends on the magnitude of the pressure gradient. The passage of materials through capillary walls due to the blood pressure is an example of filtration as it occurs in the body.

FIGURE 2.8 A theory of active transport involving a diglyceride carrier. (After Hokin and Hokin, *J. Gen. Physiol.*, Vol. 44, 1960.)

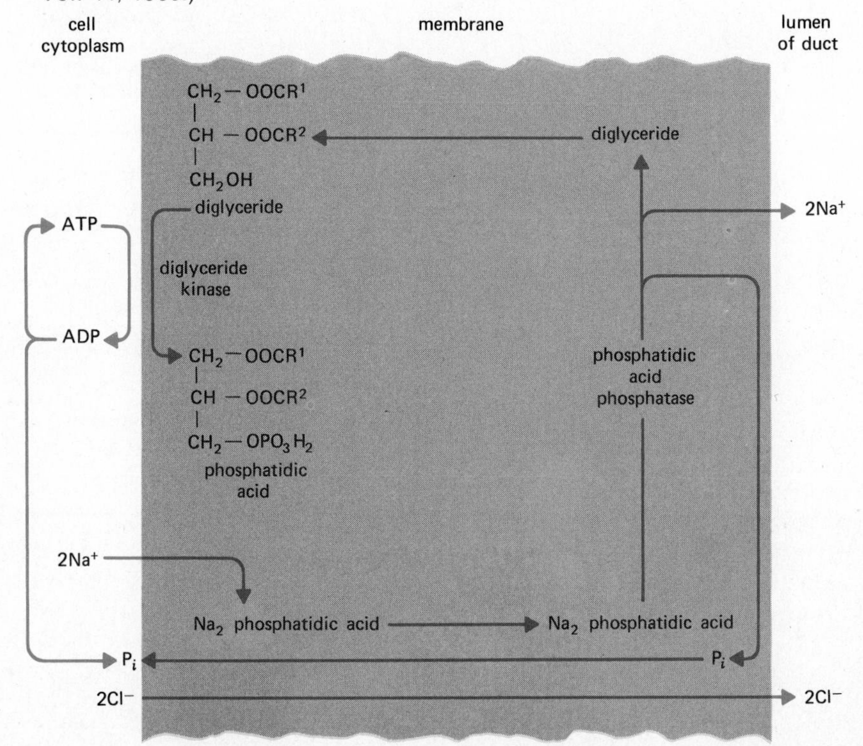

Active processes

ACTIVE TRANSPORT (Fig. 2.8) requires the presence of an *energy source,* a *carrier molecule,* and *enzymes,* all supplied by cellular activity. The energy source is usually adenosine triphosphate (ATP) and is used to activate the carrier enabling it to transport a substance. The carrier transports the substance across the membrane, and the enzymes attach and remove the activating substance to the carrier. Active transport carries substances at more or less constant rates, provided that temperature does not change; it can transport materials faster than they would have moved by simple diffusion, and it can transport against concentration gradients. It may be inhibited by the presence of metabolic poisons or by substances that may share a given carrier with another substance. Glucose, amino acids, sodium, potassium, hydrogen ions, vitamins, and small proteins are actively

FIGURE 2.9 Pinocytosis. *(a)* Particle enters a cleft in cell and becomes enclosed in a vesicle. *(b)* Particle is adsorbed on cell membrane and is enclosed in a vesicle. *(c)* Electron micrograph of pinocytic vesicle formation in a skeletal muscle capillary (× 22,000). (Electron micrograph courtesy Norton B. Gilula, The Rockefeller University.)

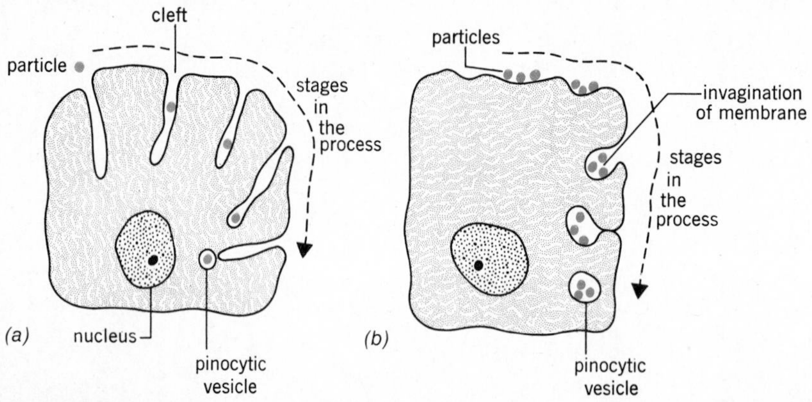

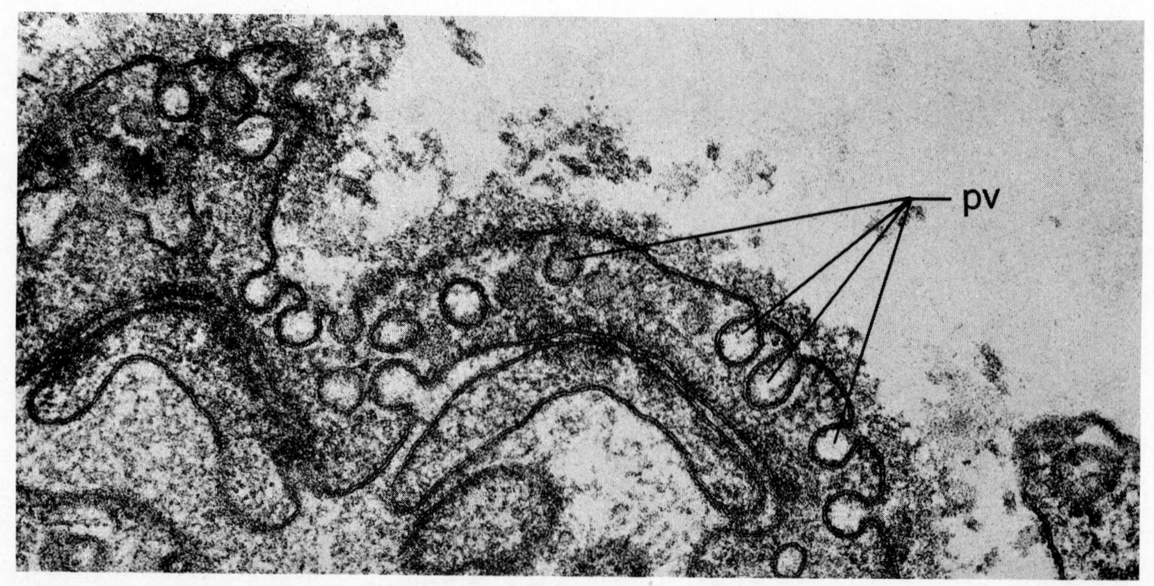

(c)

transported across most membranes in the human body.

PINOCYTOSIS (Fig. 2.9) involves the entry of large molecules into slitlike clefts in the cell's membrane, or their attachment to the cell surface. Both are followed by a "sinking" of the membrane, enclosing the molecule in an intracellular vesicle. Molecules too large to be transported or diffused may thus enter the cell.

Reverse pinocytosis, or EMEIOCYTOSIS, may occur when an intracellular vesicle moves to the cell membrane, fuses with it, and discharges the contents of the vesicle to the outside of the cell.

PHAGOCYTOSIS involves engulfing of a particle by a cell through the formation of pseudopods (Fig. 2.10). In this process, several steps are usually involved:

Attraction of phagocytes either to the site of injury or inflammation, or where there is material to be engulfed.

Coating of the particle by gamma globulin (opsoni-

FIGURE 2.10 A diagrammatic representation of phagocytosis by formation of pseudopods.

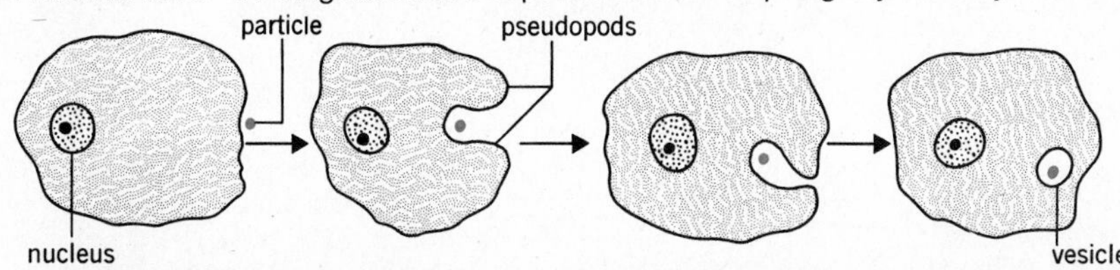

TABLE 2.5 A Summary of Methods of Acquiring Materials

Process	Type of material moved	Comments
PASSIVE PROCESSES		
Diffusion	Any solute	Most efficient for small particles. Rate decreases as size increases. No membrane necessary.
Osmosis (diffusion of water)	Water	Rate and direction determined by solute concentration. A selective membrane is required. One nonpermeable particle is required.
Filtration	All substances small enough to pass through the membrane	Force causing it is pressure. Membrane acts as a sieve.
ACTIVE PROCESSES		
Active transport	Ions, larger molecules, sugars, amino acids	Require carrier, energy, enzymes, all from cell. A membrane is required. A threshold or gradient is overcome.
Pinocytosis	Water and large molecules such as proteins, and lipids	"Sinking in" of membrane. Only method for intake of very large molecules.
Phagocytosis	Particulate materials are engulfed	Protects against bacterial invasion and rids the body of cellular debris.

zation) to increase its attractiveness to the phagocyte.

Ingestion by the phagocyte.

Enzymatic digestion of the particle, or generation of hydrogen peroxide by the phagocyte with destruction of the particle.

Phagocytosis is exhibited by motile white blood cells and the more or less fixed (nonmotile) macrophages in the tissues. Such activity is an important defense against bacterial invasion. A summary chart comparing the various methods of acquiring materials is presented in Table 2.5.

Summary

1. Living organisms are composed of one or more cells.

a. The cell carries out one or all of those activities that ensure survival, depending on whether the organism is uni- or multicellular.

b. Cells carry on activities such as intake and output of materials and metabolism; they are adaptable and irritable; they reproduce and move.

c. In multicellular organisms, cell specialization and division of labor occurs, and an interdependence of cells, tissues, organs, and systems results.

2. Cell size depends on nuclear-cytoplasm relationships and ability to transmit instructions to the cytoplasm from the nucleus. The ability of the cell surface to supply cytoplasmic volume with nutrients and remove its wastes is also a factor in determining ultimate cell size.

3. Cell shape is determined mainly by function and requirements of the cell for nutrients. Muscle cells are elongated so that they may shorten; nerve cells are elongated to transmit impulses to muscle and glands.

4. Cells consist of membrane systems, cytoplasm, organelles, and a nucleus.

a. Membrane systems govern entry and exit of materials into and out of the cell and its organelles, aid in enzymatic reactions, contain transport systems, and bind chemicals.

b. The cytoplasm contains organelles that carry out synthetic and degradative chemical reactions. Some organelles are common to all cells: the tubular endoplasmic reticulum (distribution), granular ribosomes (protein synthesis), the flattened sacs of the Golgi apparatus (synthesis of lipids, mucus), membrane-surrounded lysosomes (digestion), and granular mitochondria (ATP synthesis). Other organelles are found in only certain cells: centrioles (dividing cells), myofibrils (muscle), neurofibrils (nerve cells), and microtubules (ciliated cells). Inclusions are nonfunctioning and consist of fuel sources, wastes, and storage areas.

c. The nucleus directs cell activity by way of the nucleic acids DNA and RNA.

5. Cells acquire materials by passive and active processes.

a. Diffusion causes equal distribution of materials in a solution by molecular movement.

b. Osmosis is water movement (only) through a selectively permeable membrane due to a concentration difference of water. Cells show no change in size in an isotonic solution, swell in a hypotonic solution, and shrink in a hypertonic solution.

c. Dialysis involves selective membrane permeability to several solutes. They may be separated by this process.

d. Filtration occurs when substances are forced through membranes by pressure.

e. Active transport utilizes energy, carrier molecules, and enzymes to move substances that are not lipid soluble, have "wrong" electrical charges, are too large to pass by passive processes, or have a greater concentration on the side to which the substance is being moved through membranes.

f. Pinocytosis takes large molecules into the cell and encloses them in vesicles.

g. Phagocytosis occurs when cells engulf particles and eventually destroy or digest them. It is basically protective in function.

Cellular Composition: Membrane Potentials

Chemical organization of the cell

Living material is composed of free elements, ions, radicals, molecules, and compounds. ELEMENTS, numbering over 100, are substances in atomic form and are the simplest chemical units that retain the properties ascribed to them. Elements, alone or in combinations known as radicals, are usually capable of gaining or losing electrons, and thus may become ions or electrically charged units. A MOLECULE is formed of two or more atoms or ions to create a substance having properties different from the substances composing it. The term COMPOUND is used to refer to large complex molecules or the combination of molecules into a unit having properties different from the molecules composing it.

These materials may exist within the cell in a SOLUTION, in which the materials are found in the dissolved or unaggregated state. They may also exist as a COLLOIDAL SUSPENSION, in which the particles are large and form aggregates that do not settle, and that do not diffuse easily through membranes.

The living material within a cell is thus a heterogenous mixture of chemical substances in various states exhibiting different properties.

Composition of the cell

Four elements compose over 95 percent of the body cells, and thus of the body itself. These elements and their percent of body weight are: HYDROGEN (H) 10 percent, OXYGEN (O) 65 percent, CARBON (C) 18 percent, and NITROGEN (N) 3 percent.

Additional elements common to the body include calcium (Ca), sodium (Na), potassium (K), phosphorus (P), sulfur (S), and magnesium (Mg).

Hydrogen, oxygen, carbon, nitrogen, phosphorus, and sulfur are usually found linked into larger molecules and compounds to form the basis of body structure and function. These six elements form water, carbohydrates, lipids, proteins (some of which are enzymes), nucleic acids, hormones, and other substances essential for life. Analysis of the body for major *molecular* constituents indicates:

Material	Percent of body weight (average)
Water	60
Protein	17
Lipid	15
Carbohydrates	1
Other (nucleic acids, radicals)	5

Molecules and substances found in the cell

WATER composes 55–60 percent of the cell substance. It is a good solvent, is not toxic to the cell when isotonic, and is an excellent medium for heat transfer. It causes many substances to ionize

TABLE 3.1 Main electrolytes of the body

Substance	Charge	Element	or Radical	Anion	or Cation
Sodium (Na^+)	+1	X			X
Potassium (K^+)	+1	X			X
Calcium (Ca^{++})	+2	X			X
Magnesium (Mg^{++})	+2	X			X
Chloride (Cl^-)	−1	X		X	
Phosphate ($HPO_4^=$)	−2		X	X	
Sulfate ($SO_4^=$)	−2		X	X	
Bicarbonate (HCO_3^-)	−1		X	X	

or assume electrical charges, by loss or gain of electrons, making them more chemically reactive. All water in the body is available for movement by osmosis between fluid compartments. About 4 percent of the total body water appears to be associated with cell membranes and other materials as HYDRATION LAYERS. Such layers consist of water molecules attached by bonding forces to proteins, inorganic ions such as Na^+, and K^+, and other substances.

ELECTROLYTES are substances that can dissociate into electrically charged particles called ions. Ions may be elements or radicals (two or more atoms behaving as a single unit). Positively charged materials are attracted to a negatively charged electrical pole (cathode) and are thus called *cations;* negatively charged materials are attracted to a positively charged electrical pole (anode) and are thus called *anions.* A solution of electrolytes will conduct electrical current. The chief body electrolytes are presented in Table 3.1. These electrolytes are necessary for establishing osmotic gradients, functioning as components in buffer systems that resist change in the body's acidity, and aiding in the establishment of excitability (membrane potentials) in all cells.

PROTEINS are complex molecules composed of C (50–55 percent), H (6–8 percent), O (20–23 percent), N (15–18 percent), and S (0–4 percent). The basic building units of proteins are some 20 AMINO ACIDS (Fig. 3.1) linked together by peptide bonds (Fig. 3.2). Intermediate in complexity between amino acids and proteins are the PEPTIDES. A peptide containing less than 10 amino acids is known as an *oligopeptide;* one with more than 10 amino acids is known as a *polypeptide.* CONJUGATED PROTEINS are a combination of a protein and a nonprotein substance (e.g., nucleoproteins, lipoproteins).

Two broad categories of proteins are described. FIBROUS PROTEINS consist of elongated filamentous chains that are stable and relatively insoluble. Fibrous proteins are found as contractile elements in muscle (actin filaments), as the collagen molecules in connective tissues, and as the keratin molecules in the epidermis of the skin. GLOBULAR PROTEINS are more soluble and exhibit folding, which makes them more compact than the fibrous proteins. Hemoglobin (the pigment within red blood cells), myoglobin (a pigmented substance in muscle cells), cytochrome (a protein serving to transport electrons within cells), and the albumins

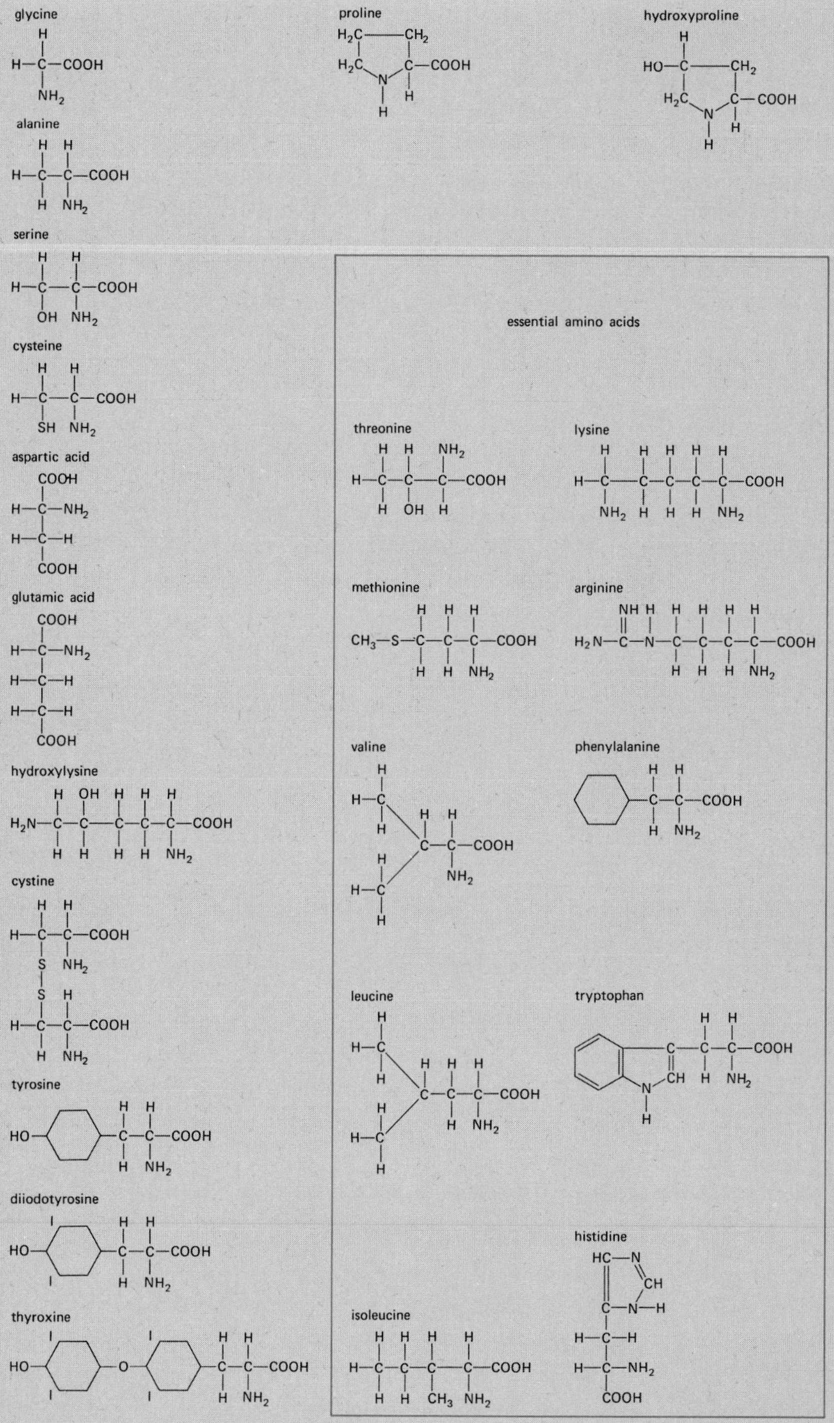

glycine

alanine

serine

cysteine

aspartic acid

glutamic acid

hydroxylysine

cystine

tyrosine

diiodotyrosine

thyroxine

proline

hydroxyproline

essential amino acids

threonine

lysine

methionine

arginine

valine

phenylalanine

leucine

tryptophan

histidine

isoleucine

FIGURE 3.1 Amino acids. *Essential* amino acids cannot be synthesized by body cells and must be ingested as such in the diet. The remaining amino acids *(nonessential)* are produced by body cells. Arginine and histidine are synthesized, but not in amounts sufficient to meet body needs.

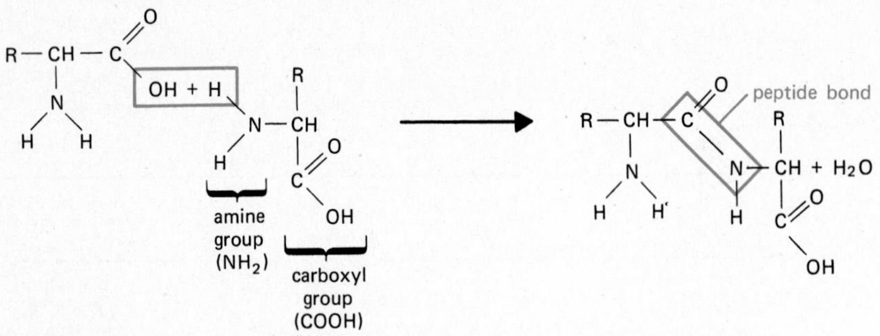

FIGURE 3.2 The linking of two amino acids by a peptide bond.

of the plasma are all globular proteins. Proteins may change from fibrous to globular states, as in the conversion of the soluble globular protein fibrinogen, to the insoluble fibrous protein fibrin, as blood clots.

ENZYMES permit the rapid progression of chemical reactions under mild temperature and pH conditions. They thus change the rates of chemical reactions and act as catalysts. Without enzymes, these reactions would proceed too slowly to be useful to the body. An enzyme is a protein synthesized by the cell that may require vitamins or metals as cofactors, and is inactivated or destroyed by heat, strong acids or bases, and organic solvents. Enzymes are SPECIFIC in that they usually catalyze only a single chemical reaction. *Group specificity* also exists for some enzymes in that a group of compounds that are similar in structure (e.g., hexoses) may be attacked by one enzyme. Specificity may depend on a "lock-and-key" type of arrangement between the enzyme and its substrate (the material with which the enzyme reacts). The relationship between an enzyme and its substrate is depicted in Fig. 3.3.

Enzymes may be INHIBITED by other compounds that are not the enzyme's primary substrate. The inhibition is usually the result of combination of the compound with the enzyme, reducing its activity.

Enzymes are named in various ways, including the substrate with which they react, and the type of reaction catalyzed. The suffix *-ase* is commonly employed to designate the enzyme. Some important categories of enzymes are presented in Table 3.2.

LIPIDS are insoluble in water, but are soluble in fat solvents such as ether, alcohol, and chloroform. Lipids are composed of C, H, and O and

FIGURE 3.3 The reaction of an enzyme with a substrate and the release of products.

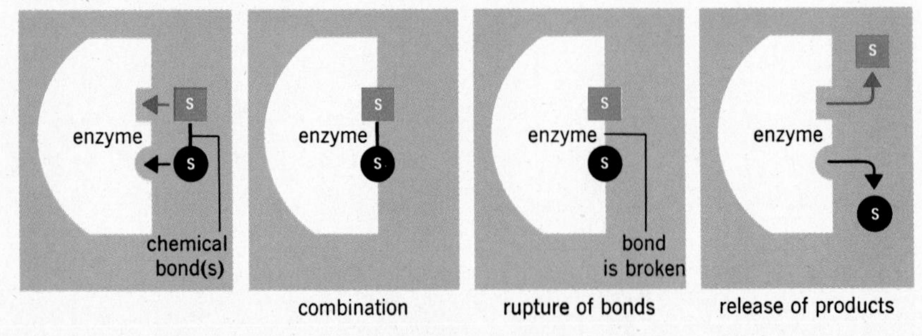

TABLE 3.2 Partial classification of enzymes

Enzyme	Substrate (material with which enzyme reacts)	Action	Examples
Hydrolases	None specific. Fats, phosphate-containing compounds, nucleic acids, many others	By adding water (hydrolysis), splits larger molecules into smaller units	Lipases, phosphatases, nucleases
Carbohydrases	Carbohydrates, starches, disaccharides	Fragment carbohydrates by hydrolysis to smaller units	Salivary amylase, pancreatic amylase, maltase, lactase, sucrase
Proteases	Proteins, smaller units containing fewer amino acids	Fragment proteins by hydrolysis to smaller units	Pepsin, trypsin, erepsin
Phosphorylases	Many molecules	Split or activate molecule by addition of phosphate group	Muscle phosphorylase, glucokinase, fructokinase
Dehydrogenases	Any compound capable of losing 2 H^+	Oxidizes compound by removal of 2 H^+	Succinic dehydrogenase, fumaric dehydrogenase
Oxidases	Any compound that may add oxygen	Oxidizes compound by addition of oxygen	Peroxidase
Aminases	Amine-containing compounds and keto acids	Addition or removal of amine groups	Transaminase, deaminase
Decarboxylases	Acids containing carboxyl group	Remove CO_2 from carboxyl group	Ketoglutaric decarboxylase
Isomerases	Chiefly carbohydrates	Moves a radical to different part of molecule	Isomerase

contain relatively few atoms of oxygen. Three categories of lipids are recognized.

Simple lipids (Fig. 3.4) include the fats, waxes, and oils. The triglycerides, compounds of three fatty acids with glycerol, are widespread in the body. They are the most common storage form of fat in adipose tissue and are the primary lipid metabolized for energy by the body.

Compound lipids (Fig. 3.5) consist of a lipid portion combined with a nonlipid portion. Examples are:

FIGURE 3.4 The structure of a triglyceride (tripalmatin).

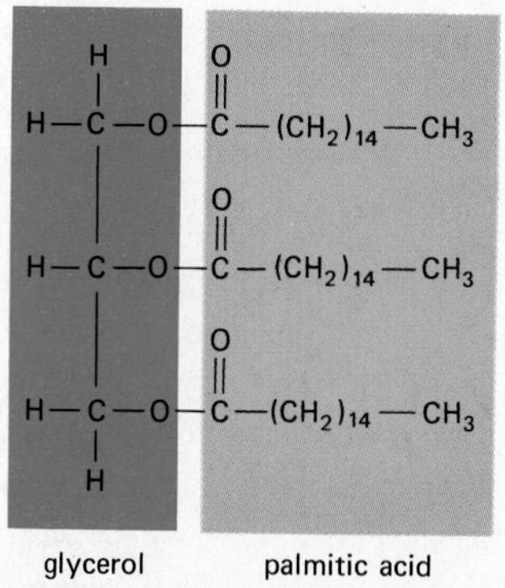

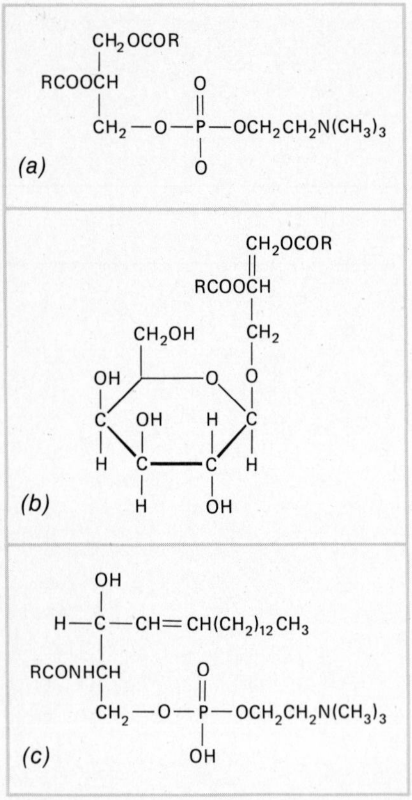

FIGURE 3.5 Some representative compound lipids. *(a)* Lecithin, a membrane phospholipid *(b)* Glycolipid of photosynthetic tissue. *(c)* Sphingolipid of nervous tissue. R = unspecified chemical groups.

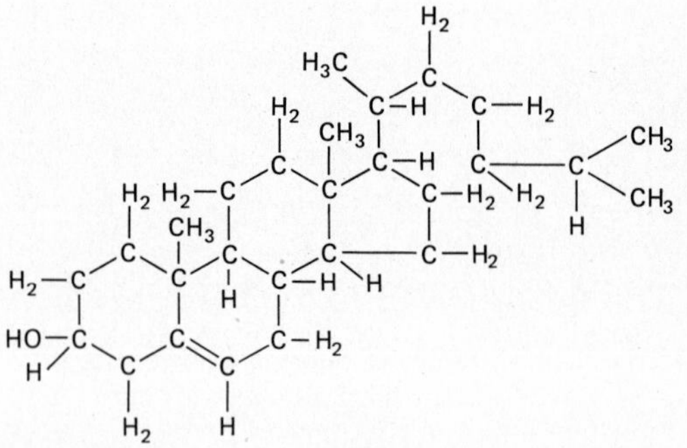

FIGURE 3.6 The structure of cholesterol, a derived lipid.

FIGURE 3.7 The structures of some representative monosaccharides.

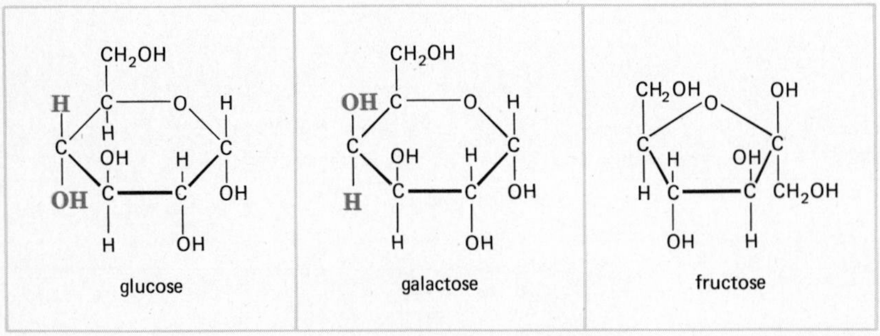

glucose galactose fructose

phospholipids, found in cell membranes; glycolipids, found in photosynthetic membranes, and sphingolipids, found in nervous tissue.

Derived lipids (Fig. 3.6) include the sterols and intermediate compounds in lipid metabolism. Sterols are derived from cholesterol, and include several steroid hormones (testosterone, estrogen, progesterone, adrenal cortical hormones). Cholesterol has been implicated as a lipid deposited in the walls of blood vessels leading to atherosclerosis and heart attack. Synthesis of cholesterol in the body is believed to be inversely related to dietary intake, and the consumption of low-cholesterol diets is thought to retard disposition of the compound in blood vessel walls, at

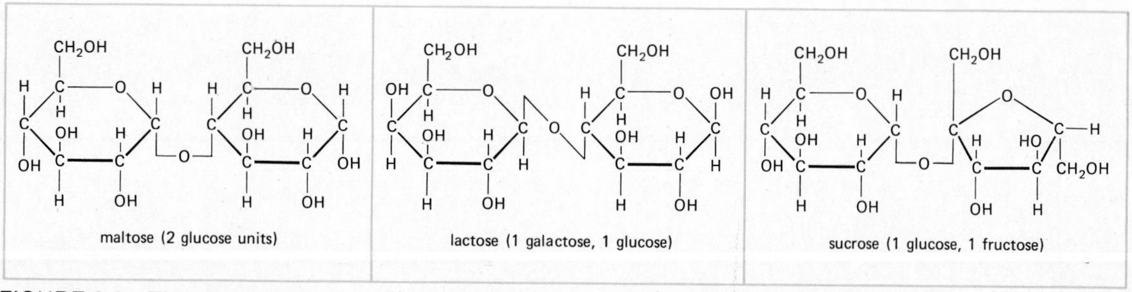

maltose (2 glucose units)

lactose (1 galactose, 1 glucose)

sucrose (1 glucose, 1 fructose)

FIGURE 3.8 The structures of some important disaccharides. Maltose is malt sugar, lactose is milk sugar, sucrose is cane or beet sugar.

least until the low dietary intake is compensated for by an increase in synthesis by the body.

Derived lipids include intermediates in the metabolism (especially hydrolysis) of lipids; for example, glycerol (derived from splitting of a triglyceride), short-chain fatty acids (produced during degradation of longer acids), and sterols (resulting from degradation of cholesterol).

Fatty acids form important sources of energy, releasing on combustion over twice the calories of heat energy per gram as do carbohydrates. In their storage depots in skin, they insulate and are thus important in control of body temperature. They form, as phospholipids, an important constituent of cell membranes. Sterols are found as components of many hormones.

CARBOHYDRATES include the starches and sugars. They are composed of carbon, hydrogen, and oxygen, with hydrogen and oxygen usually in a 2:1 ratio. Three categories of carbohydrates are recognized:

Monosaccharides. Simple sugars (Fig. 3.7). Pentoses (ribose) with five carbons, and hexoses (glucose, fructose, galactose) with six carbons are the most common examples. Simple sugars are the cell's preferred source of energy for ATP synthesis and other chemical reactions requiring energy.

Disaccharides. Double sugars (Fig. 3.8). These are two simple sugar units linked together. They usually represent intermediate stages in the synthesis or breakdown of other carbohydrates.

FIGURE 3.9 The glycogen molecule. Each circle represents a glucose molecule.

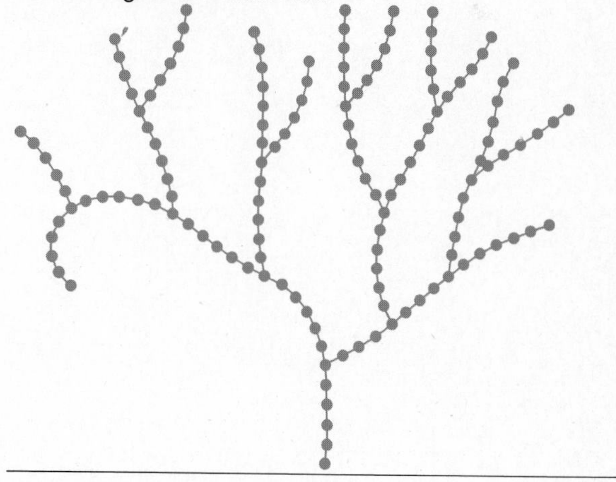

Polysaccharides. (Fig. 3.9). These are hundreds or thousands of simple sugar units linked into storage forms such as glycogen (animal starch), and plant starch. Polysaccharides may also be joined with other substances to form structural materials such as the protein polysaccharides.

NUCLEIC ACIDS (Fig. 3.10) are composed of nitrogenous ring-form bases called purines and pyrimidines (see Fig. 4.2) in combination with a sugar and phosphoric acid. According to the bases present and the sugar present, two types of nucleic acids are distinguished. DEOXYRIBONUCLEIC ACID (DNA) contains the bases adenine, guanine, cytosine, and thymine, plus the sugar deoxyribose; RIBONUCLEIC ACID (RNA) contains adenine, guanine, cytosine, and uracil, plus the sugar ribose. Both compounds also contain phosphoric

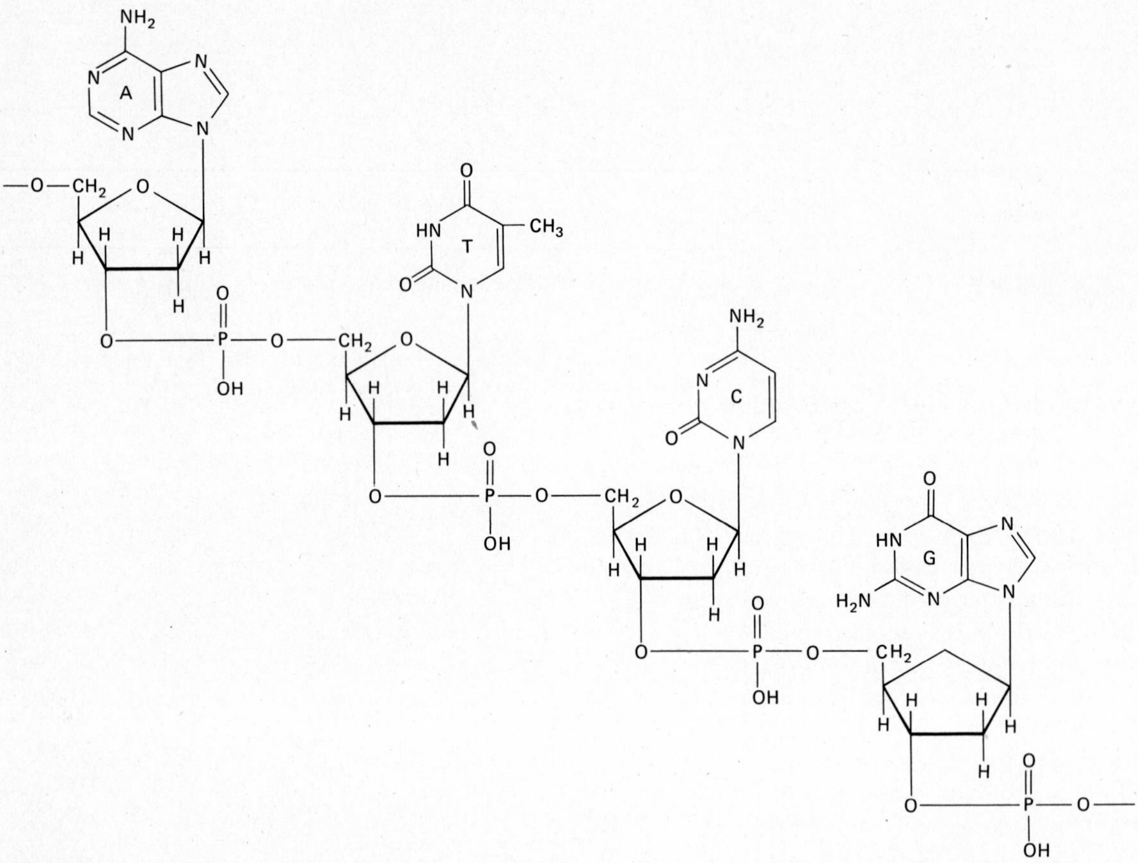

FIGURE 3.10 The general structure of a portion of a nucleic acid molecule containing four bases. A = adenine, T = thymine, C = cytosine, G = guanine.

acid. A *nucleoside* is a unit containing only the base and sugar components, while a *nucleotide* consists of the base, sugar, and acid. The nucleic acids control cellular activity through protein synthesis.

TRACE MATERIALS are essential for cell function but are required in only very small amounts. This category includes vitamins, certain metals, and hormones. The actions of such materials will be considered in greater detail in appropriate chapters later in the text.

A summary of the chemicals of the cell is presented in Table 3.3.

Membrane potentials

Having studied the chemicals composing living material, including the electrolytes, and the methods by which materials cross membranes of cells, it is appropriate at this time to examine the phenomenon of excitability in cells. The excitable state is one that all cells appear to exhibit and is

TABLE 3.3 Summary of the chemicals of living material

Substance	Location in cell	Function
Water	Throughout	Dissolve, suspend, and regulate other materials; regulate temperature
Inorganic salts	Throughout	Establish forces to govern water movement, pH, buffer capacity
Carbohydrates	Inclusions (non-living cell parts)	Preferred fuel for activity
Lipids	Membranes, Golgi apparatus, inclusions	Reserve energy source; give form and shape; protection; insulation
Proteins	Membranes, cytoskeleton, ribosomes, enzymes	Give form, strength, contractility, catalysts, buffering
Nucleic acids		
DNA	Nucleus, in chromosomes and genes	Direct cell activity
RNA	Nucleolus, cytoplasm	Carry instructions; transport amino acids
Trace Materials		
Vitamins	Cytoplasm	Work with enzymes
Hormones	Cytoplasm	Work with enzymes to activate or deactivate enzymes
Metals	Cytoplasm, nucleus	Specific functions in various synthetic schemes, e.g.: synthesis of insulin (zinc), maturation of red cells (cobalt, copper, iron)

the result of the creation of a TRANSMEMBRANE ELECTRICAL POTENTIAL difference. Such a potential difference between the inside and the outside of the cell enables various alterations in the internal or external environments of the cell (stimuli) to alter the potential and result in responses that will maintain or restore the homeostasis of the organism. Introduction of the concept at this time will provide an appreciation of the importance and universality of the phenomenon. The establishment of a transmembrane potential depends on unequal separation of ions on the two sides of the membrane. Methods by which ionic imbalances are created are discussed below.

LIMITING FREE DIFFUSION of ions is created by the presence of a selective membrane. Because of size, electrical charge, and other factors, rates of diffusion or transport may be influenced by the membrane. The relative sizes of the hydrated (having layers of water molecules around them) ions appear to be a major factor in membrane permeability to a particular ion. Cell membranes are usually more permeable to K^+ than Na^+, because the hydrated K ion is smaller than the hydrated Na ion.

The DONNAN-GIBBS EQUILIBRIUM is a special

type of diffusion in which a large, ionized, nondiffusible substance is present on one side of the membrane. Such a substance, in combination with a concentration gradient for diffusible ions, can create, on its own, an imbalance of ions when the process reaches an equilibrium state. At equilibrium, three criteria must be met:

The *product* of the diffusible ions must be equal on the two sides of the membrane.

The side containing the nondiffusible substance will contain a greater concentration of diffusible *cations* than the other side.

There must be electrical neutrality between ions on one side of the membrane.

The entire process may be illustrated by the following diagram:

membrane			membrane	
$5\ Na^+$	$10\ Na^+$		$5\ Na^+$ $5\ A^-$	$6\ Na^+$
$5\ A^-$	$10\ Cl^-$		$4\ Na^+$ $4\ Cl^-$	$6\ Cl^-$
start			equilibrium	

A^- = nondiffusible anion

Products of the *diffusible* ions at equilibrium

Left

$5\ Na^+ + 4\ Na^+ = 9\ Na^+$

$4\ Cl^- = 4\ Cl^-$

$9\ Na^+ \times 4\ Cl^- = \underline{\underline{36}}$

Right

$10\ Na^+ - 4\ Na^+ = 6\ Na^+$

$10\ Cl^- - 4\ Cl^- = 6\ Cl^-$

$6\ Na^+ \times 6\ Cl^- = \underline{\underline{36}}$

Note that positive and negative charges are balanced on a given side of the membrane.

ACTIVE PROCESSES, such as active transport, account for the greatest part of any membrane potential because separation of ions may be maintained at any level, depending on cellular supplies of the energy sources, enzymes, and carriers required in the process.

The ions involved in the creation of membrane potentials are the most abundant in the body fluid compartments. These ions are sodium, potassium, chloride, and ionized nondiffusible substances (e.g., protein) in the cell. Other ions contribute such a small fraction of the membrane potential that their effects can be largely ignored. For sodium, potassium, and chloride ions, determination of their concentrations in the extracellular and intracellular fluids shows the average values presented in Table 3.4. Note that sodium and chloride are concentrated outside the cell, and that potassium is concentrated inside the cell. Thus, by achieving an unequal separation of these ions, cells create an electrical potential difference and a state of excitability across their membranes.

The degree of separation of particular ions (the difference of concentration on the two sides of the membrane) determines the ease with which a stimulus will cause a response—that is, the threshold of response. Not all cells require the same strength of stimulus to cause an alteration in the membrane potential, and the strength required to

TABLE 3.4 Average concentrations of electrolytes in body fluids

Electrolytes	Concentration in extracellular fluid (meq/L[a])	Concentration in intracellular fluid (meq/L)
Sodium	145	12
Potassium	4	155
Chloride	120	4

[a] One equivalent weight of a substance is its atomic weight divided by its valance. It represents the number of grams of an element that will combine with 8 grams of oxygen or 1.008 grams of hydrogen. A milliequivalent is 1/1000 of an equivalent weight.

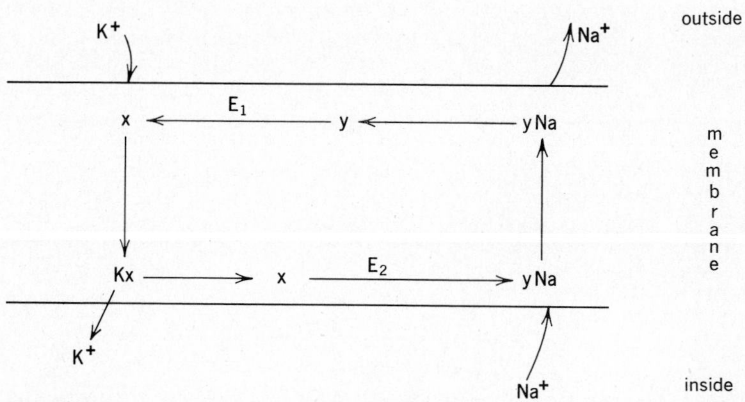

FIGURE 3.11 Sodium combines with a specific carrier (y) and is transported to the interstitial fluid. Carrier y is enzymatically transformed (E_1) into carrier x, which brings potassium in. Again, at the inside of the membrane, enzymatic transformation (E_2) changes x to y and the process is repeated.

cause an alteration may change according to the factors in the cell environment.

The basis of excitability in cells

Most cell membranes are quite permeable to K^+ and Cl^-, possess an active transport mechanism or "sodium-potassium pump" that keeps Na^+ largely on the outside of the cell and K^+ on the inside, and are impermeable to ionized proteins. The sodium-potassium pump mechanism is depicted in Fig. 3.11. Thus, Na^+ accumulates on the outside of the cell, and Cl^- remains with Na^+ by electrostatic attraction. K^+ tends to remain within the cell, placed there by the pump and retained by the negatively charged protein ions.

Analysis of the concentrations of these ions inside and outside the cell when it is not being stimulated (resting state) is shown in Fig. 3.12. Note that the distribution of diffusible ions is the same as shown in Table 3.4 for the body fluids. The total of the negative and positive charges inside and outside the cell indicates that the outside of the cell has a greater positive charge than the inside (29 as compared to 8). Thus, the outside is electrically positive to the inside. The inside,

FIGURE 3.12 The concentrations of ions in and around an excitable cell.

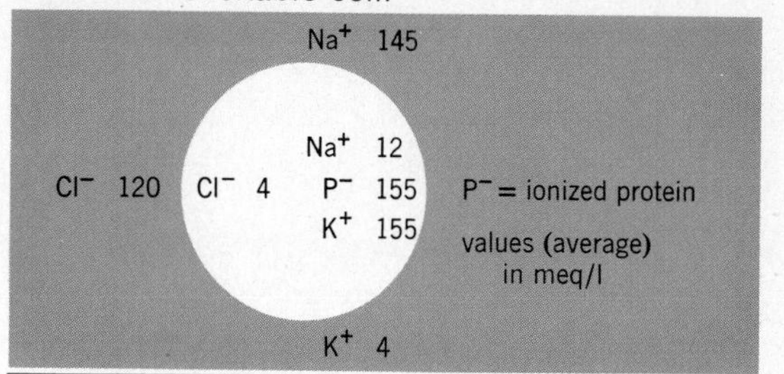

since it is less positive than the outside, is said to be electrically negative to the outside. This situation may be represented as shown in Fig. 3.13. An instrument for measuring electrical differences would then show an electrical difference if electrodes were placed inside the cell and on its surface. The membrane is said to be POLARIZED.

Measurement of the magnitude of the potential difference in various types of cells shows it to average 70 millivolts (mv). A negative sign is assigned to the value when the inside of the cell is negative to the outside; thus, the potential is represented as −70 mv. This potential is referred to as the RESTING POTENTIAL, since the cell has achieved it and is not reacting to stimuli.

Knowing the concentrations of ions inside and outside the cell, and applying the Nernst equation

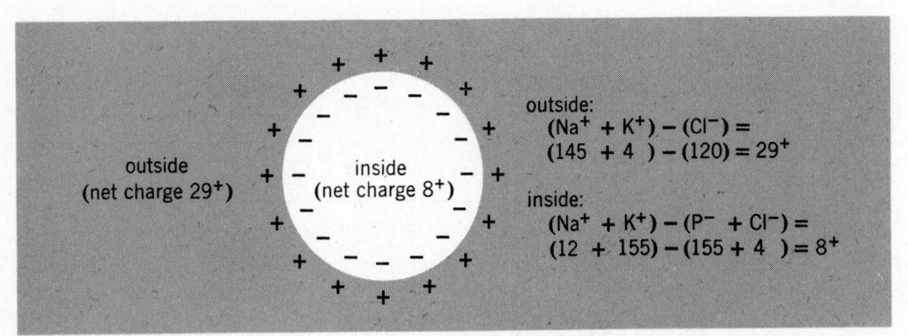

FIGURE 3.13 The charge relationships in an excitable nerve cell.

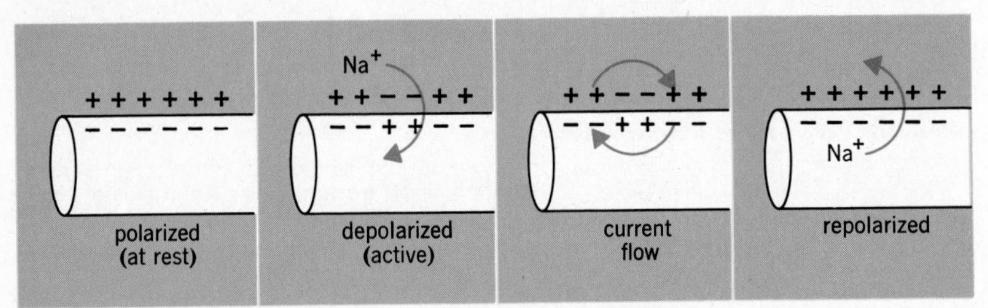

FIGURE 3.14 Depolarization, current flow, and repolarization in a nerve fiber.

(see Appendix), it is possible to calculate the resting potential as if it were due entirely to one ion. The potential due to sodium would be +60 mv; that due to potassium would be −90 mv; that due to chloride would be −70 mv. Remembering that the membrane permits relatively free diffusion of Cl^-, and that Cl^- follows Na^+ by electrostatic attraction, it is concluded that Cl^- distribution is the result of the potential and not the cause of it. Chloride may therefore be ignored as contributing to the potential. Thus, the resting potential is the result of permeability to both Na^+ and K^+ but with most of this potential due to membrane permeability to K^+ (remember that the smaller size of the hydrated K^+ allows it to diffuse more easily). The resting potential is thus described as being due to a "potassium diffusion potential."

The formation of transmissible impulses

Using the nerve fiber as the best example of a cell capable of forming a transmissible impulse, it is stated that impulse formation is associated with a sudden, localized, and transient increase in membrane permeability to Na^+, movement of Na^+ into the cell, and the creation of a recordable electrical change, the ACTION POTENTIAL. The alteration in permeability to Na^+ is believed to be the result of the removal of calcium ions from their positions lining pores in the cell membrane. In the resting state, Ca^{++} bound in the pores repels the like-charged Na ions. A stimulus may dislodge some Ca^{++} and thus permit an increasing inflow of Na^+. At any rate, the sodium pump is overwhelmed, and sodium moves into the fiber according to its electrical and chemical gradients. Potassium tends to leave the cell under *its* gradient and by the electrostatic repulsion of like charges, as Na^+ enters and drives K^+ from the fiber. However, K^+ leaves at a slower rate than sodium enters, since there is no great change in membrane permeability to K^+. The active region is thus said to be a "sodium membrane." As a result of Na^+ movement into the cell, the potential of the fiber is reversed and the inside of the fiber becomes positive to the outside. The fiber is thus said to have become depolarized. The presence of a depolarized area adjacent to a polarized area creates a "battery effect," and current flows between the two areas, utilizing the electrolyte-laden extracellular fluid and intracellular fluid as the medium of conductance of current. The current flow represents the action potential. These events are shown in Fig. 3.14. After sodium inflow, repolarization occurs as the sodium-potassium pump again achieves the upper hand, and Na^+ is pumped to the outside of the fiber, while K^+ is returned inside. Correlated electrical changes are illustrated in Fig. 3.15.

Propagation (conduction) of the action potential (Fig. 3.16)

The current flowing between polarized and depolarized regions of the nerve fiber may be shown to decrease exponentially with distance from the original center of the disturbance. It is nevertheless strong enough at a short distance from the center to depolarize that next short segment of the membrane. A new depolarized area farther down the membrane is created, current flows, another section is depolarized, and the disturbance is advanced or propagated down the fiber. Mean-

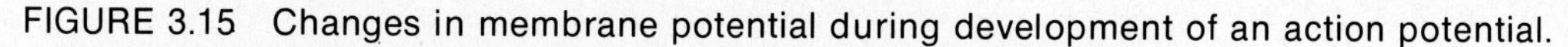

FIGURE 3.15 Changes in membrane potential during development of an action potential.

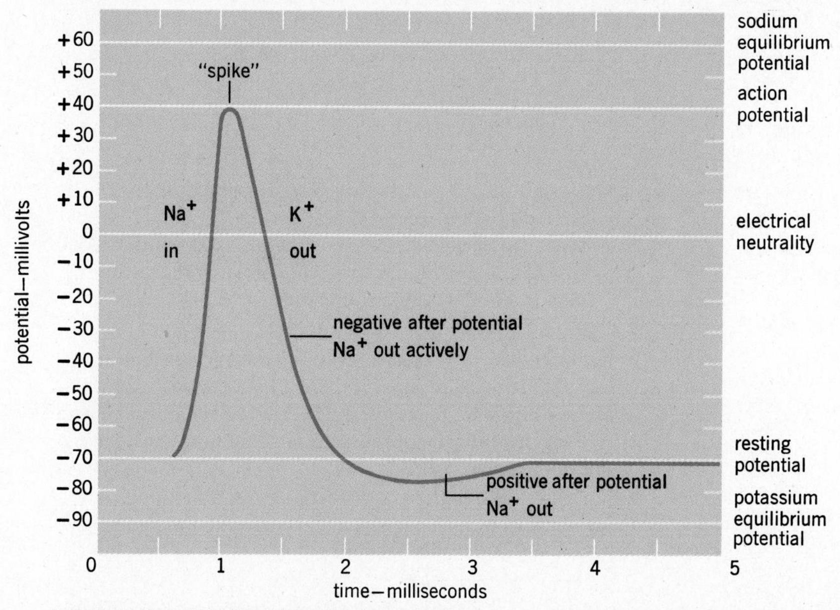

FIGURE 3.16 The events during propagation of a nerve impulse along an unmyelinated nerve fiber.

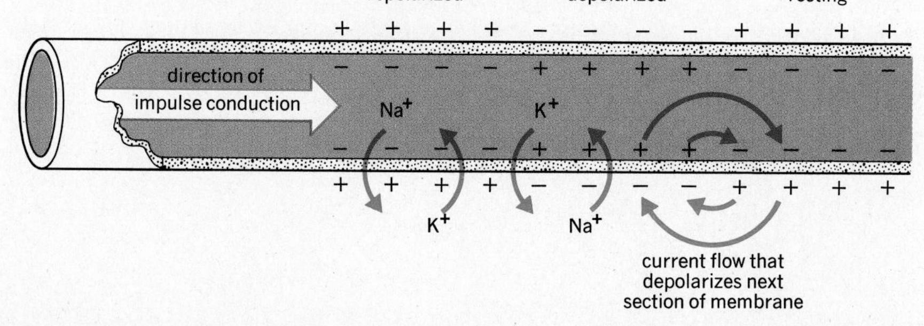

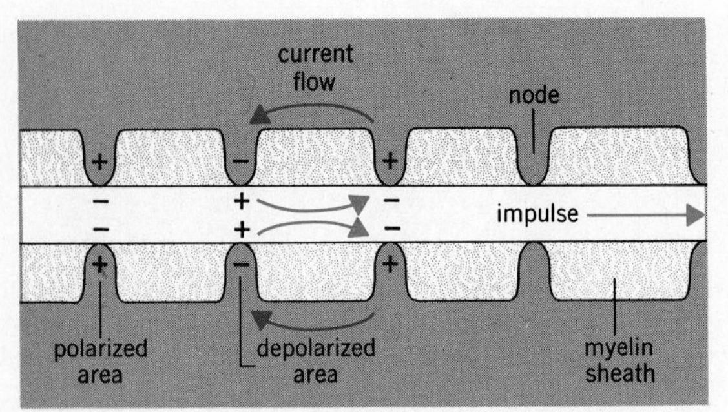

FIGURE 3.17 A diagrammatic representation of current flow in saltatory conduction.

while, repolarization is occurring behind the conducted impulse, and the fiber may respond to another stimulus. In a fiber possessing a segmented myelin sheath (myelinated or medullated fibers), current flow can occur only at the segments (nodes) in the myelin sheath (Fig. 3.17), since myelin is an efficient insulator against current loss. Since the nodes lie 2–3 mm apart, depolarization "jumps" from node to node, and conduction is more rapid than in the absence of a myelin sheath. Conduction in this manner is termed saltatory conduction.

It is important, as stated before, to realize that the events leading to the creation of an excitable state are common to nearly all body cells. While not all cells can form transmissible impulses, changes in the membrane potential can occur in any body cell that is polarized.

Summary

1. Cells are composed of elements, ions, radicals, molecules, and compounds.
   a. Substances may exist in the cell as solutions or in the form of colloidal suspensions. The latter consist of large molecules that do not settle out of suspension.
   b. Water serves as a solvent and as a medium in which chemical reactions occur.
   c. Electrolytes are electrically charged elements or radicals and are known as ions. They aid in establishing and maintaining the osmotic and pH homeostasis of the body fluids. They are also necessary for the establishment of the excitable state in cells.
   d. Proteins are structural materials, form enzymes, and are composed of amino acids linked by peptide bonds.
   e. Enzymes are catalysts for chemical reactions and are specific as to the materials on which they work. They require rather specific conditions of pH and temperature to exhibit their optimum activity.
   f. Lipids are fuel sources, form some hormones, and give form and insulation to the body.
   g. Carbohydrates are sugars and starches and are the preferred sources of energy for cellular activity.
   h. Nucleic acids provide direction of cellular activity, particularly protein synthesis.
   i. Trace substances include vitamins and metals. Such substances are required in only small amounts but are vital to normal cellular function.
2. Cells, by achieving a separation of ions, may create an excitable state and an electrical potential across their membranes. The main ions responsible for these potentials are sodium and potassium. The resting membrane has a −70 mv potential, with outside positive to the inside, and the cell is polarized.
3. Impulse formation occurs when membrane permeability increases to sodium ions. Sodium enters the cell, the potential is reversed, and the fiber is depolarized. The action potential is developed.
4. Impulses may be propagated down cells, such as nerve cells. A flow of electrical current occurs between a depolarized and a polarized area. The current depolarizes the next portion of the cell and is transmitted along the cell. Removal of sodium occurs by active processes, and the cell repolarizes.

4

Cellular Activity

Many of the processes described in the preceding chapters depend on the production of chemical compounds that exert control over the processes in terms of its rate and efficiency. Two important types of control mechanisms are available for governing these reactions: GENETIC CONTROL, exerted by way of DNA and RNA molecules, and CHEMICAL CONTROL, exerted by way of hormones. Such activity requires ENERGY that is provided by the metabolism of the basic foodstuff groups: carbohydrates, lipids, and proteins.

Genetic control

Nucleic acids

The nucleus of the cell contains molecules of deoxyribonucleic acid (DNA) and ribonucleic acid (RNA) that serve as the means of genetic expression. During cell division, chromosomes, whose number and morphology are characteristic for the species (Fig. 4.1), become visible.

The nucleic acid molecules are composed of subunits of structure known as NUCLEOTIDES that are composed of nitrogeneous bases (purines and pyrimidines), a pentose sugar (ribose or deoxyribose), and phosphate (Fig. 4.2). Phosphate-to-sugar bonds link nucleotides into chains.

In DNA, four common bases are found: ADENINE (A), GUANINE (G), CYTOSINE (C), and THYMINE (T). In 1953 Watson and Crick proposed that DNA consisted of two nucleotide chains, coiled as a double helix, and bound to each other by base-base linkages (Fig. 4.3). Only certain bases will combine with one another to form the linkages: A-T and C-G. The sugar is deoxyribose.

In RNA the sugar is ribose but URACIL (U) is substituted for thymine. Thus linkages may be made as A-U and C-G. Additionally, RNA consists of a single chain of nucleotides that may be folded upon itself to give the appearance of a double helix (Fig. 4.4). RNA exists in several forms designated as messenger RNA (m-RNA), ribosomal RNA (r-RNA), and transfer RNA (t-RNA).

The genetic code

Watson and Crick further proposed that the four bases composing DNA carried information that was coded according to the base sequence in the molecule. It was suggested that a sequence of three bases (a "triplet") corresponds to a particular amino acid. The placement of a triplet in the nucleic acid molecules will determine the position of the amino acid in a protein chain. Proteins, then, as hormones, enzymes, and structural materials, determine body function and structure. A sequence of bases responsible for the production of a protein that controls a particular body reaction or characteristic is believed to be a gene.

A three-base sequence coding for a given amino acid is termed a CODON. With four bases available to form codons, it is possible to form 64 different "three-letter words," as follows:

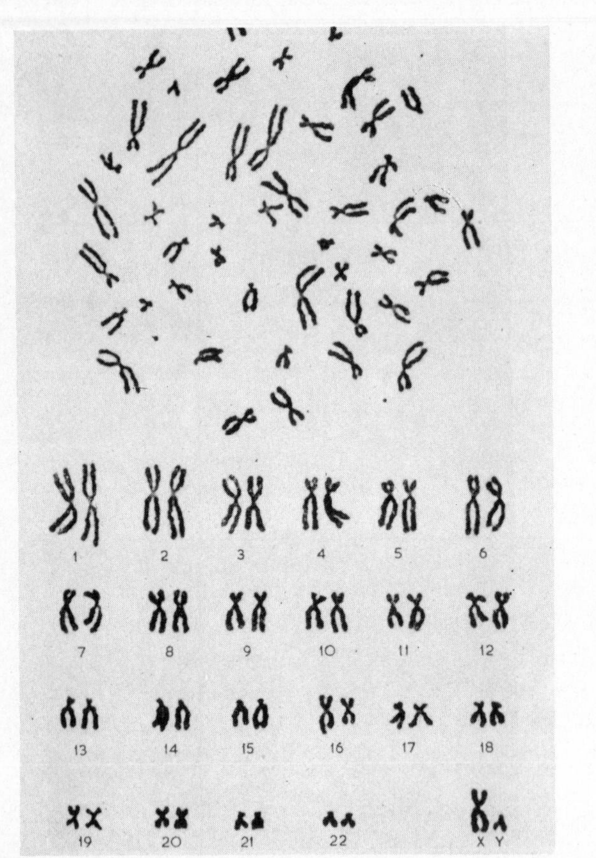

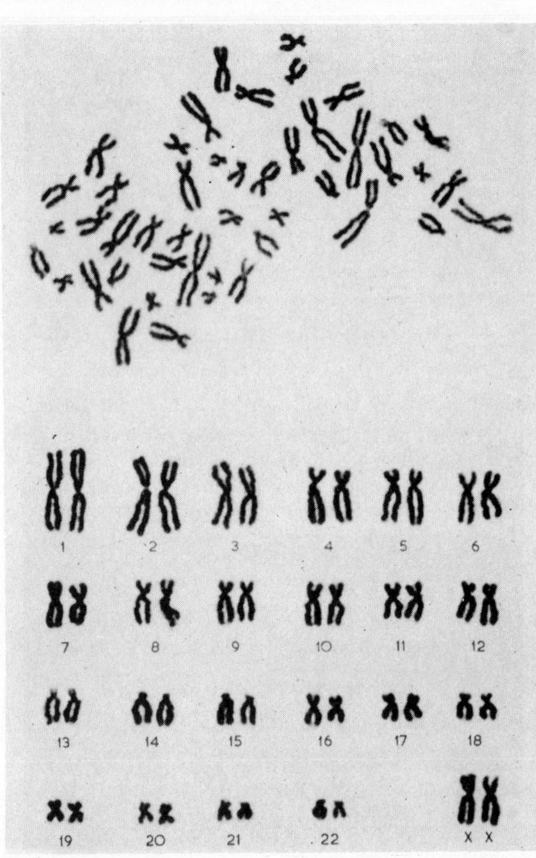

FIGURE 4.1 Normal human karyotype. Male (left), female (right). (Courtesy Little, Brown & Co. From L. S. Penrose, *Recent Advances in Human Genetics.*)

AAA AAG AAC AAT
AGA AGG AGC AGT
ACA ACG ACC ACT
ATA ATG ATC ATT

GAA GAG GAC GAT
GGA GGG GGC GGT
GCA GCG GCC GCT
GTA GTG GTC GTT

CAA CAG CAC CAT
CGA CGG CGC CGT
CCA CCG CCC CCT
CTA CTG CTC CTT

TAA TAG TAC TAT
TGA TGG TGC TGT
TCA TCG TCC TCT
TTA TTG TTC TTT

Since there are only about 20 different kinds of amino acids available for synthesis into proteins, but 64 codons, some codons must specify the same amino acid, or some codons may act as "punctuation marks" in the nucleic acid molecules to signal the beginning and end of genes. Codons that do not specify amino acids are termed "nonsense" codons.

By synthesizing particular triplet codons in the laboratory and then presenting them to a ribosome "soup," it is possible to achieve the syn-

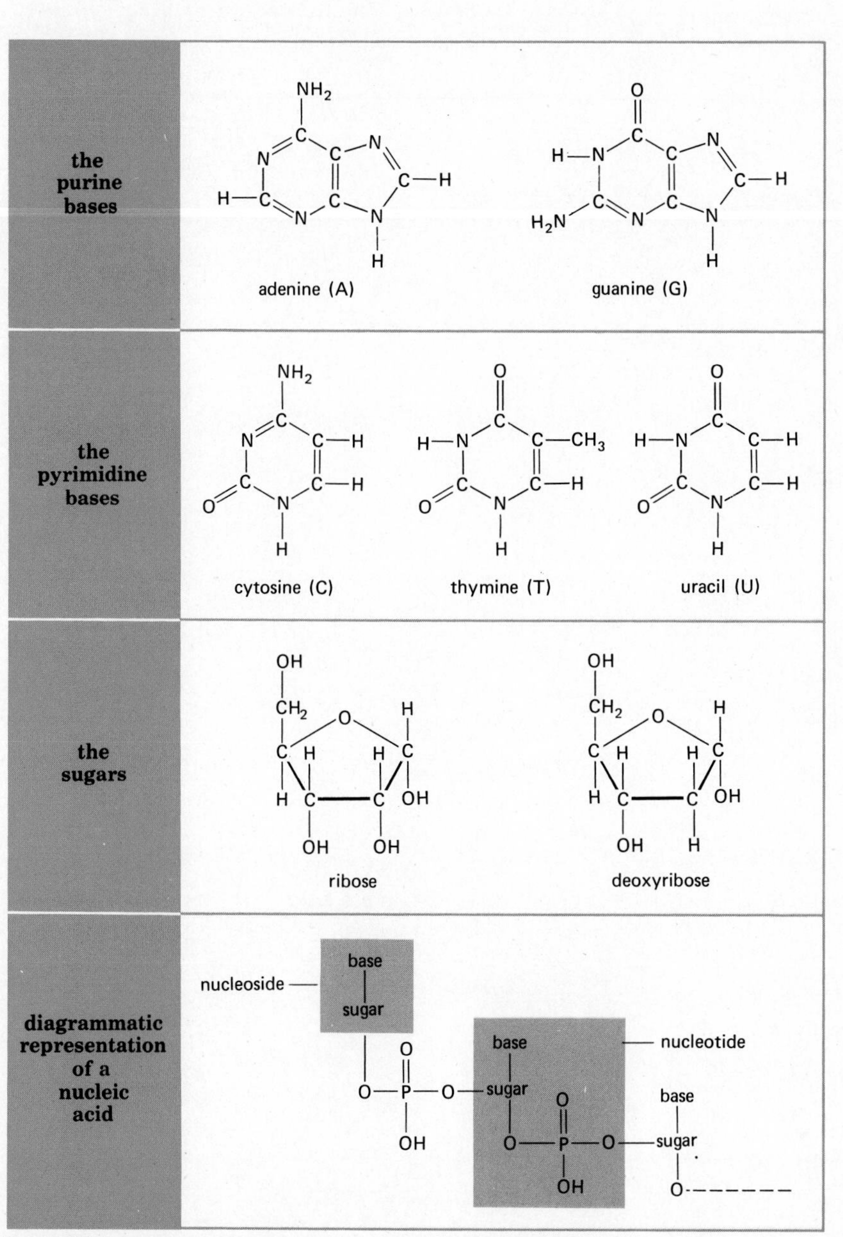

FIGURE 4.2 The constituents of the nucleic acids.

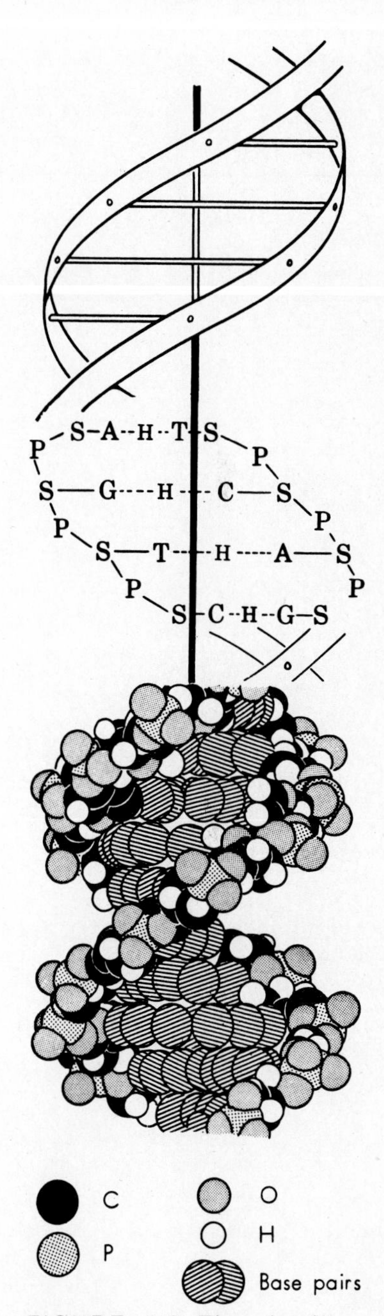

FIGURE 4.3 The double-strand helix of DNA represented three ways.

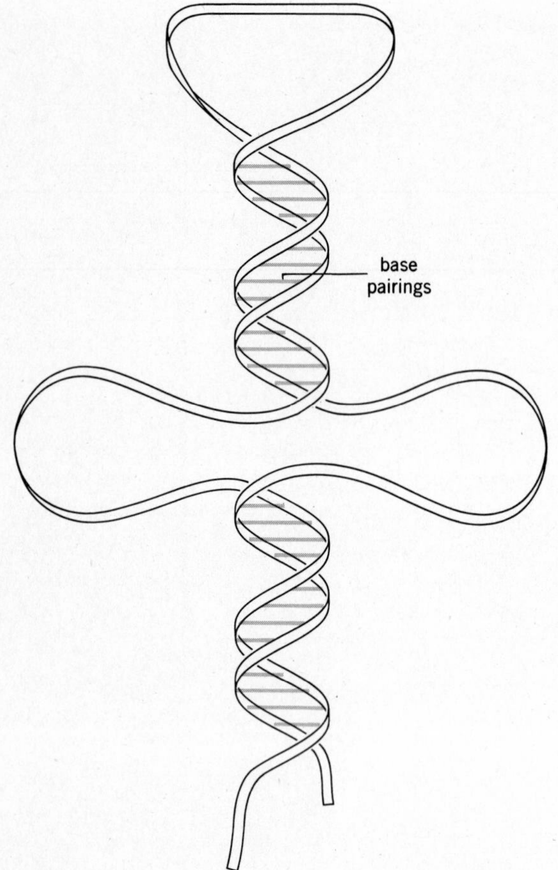

FIGURE 4.4 The single-strand "cloverleaf" helix of t-RNA.

thesis of proteins containing only one or a particular combination of amino acids. By such experiments, it has been determined that the DNA codons specify amino acids, as indicated below.

Isoleucine	Valine	Leucine	Phenylalanine
ATT	GTT	CTT	TTT
ATC	GTC	CTC	TTC
ATA	GTA	CTA	
		CTG	

Methionine	Valine/ Methionine	Proline	Leucine
ATG	GTG	CCT	TTA
		CCC	TTG
		CCA	
		CCG	

Threonine	Alanine	Histidine	Serine
ACT	GCT	CAT	TCT
ACC	GCC	CAC	TCC
AGA	GCA		TCA
ACG	GCG		TCG

Asparagine	Aspartic acid	Glutamine	Tyrosine
AAT	GAT	CAA	TAT
AAC	GAC	CAG	TAC

Lysine	Glutamic acid	Arginine	Cysteine
AAA	GAA	CGT	TGT
AAG	GAG	CGC	TGC
		CGA	
		CGG	

Serine	Glycine	"punctuation marks"
AGT	GGT	TAA
AGC	GGC	TAG
	GGA	TGA
	GGG	TGG

Arginine
AGA
AGG

In RNA, uracil replaces thymine in the triplet codes, specifying the same amino acids. DNA participates in protein synthesis by causing the production of m-RNA in the process known as TRANSCRIPTION. The process requires an enzyme known as RNA polymerase; it "reads" the DNA strand and simultaneously attracts and locks amino acids in proper sequence into m-RNA, which then separates from the DNA template and leaves the nucleus. The m-RNA moves to the cytoplasm.

Protein synthesis

To synthesize a specific protein in the process of TRANSLATION, an additional type of RNA is required. Transfer RNA (t-RNA) is collected on the nucleolus after being formed by DNA on a chromosome. It then moves into the cytoplasm. Transfer RNA (Fig. 4.5) consists of a twisted and looped single chain of nucleotides containing a series of three bases at the top of the middle loop. This three-base sequence is known as the ANTICODON

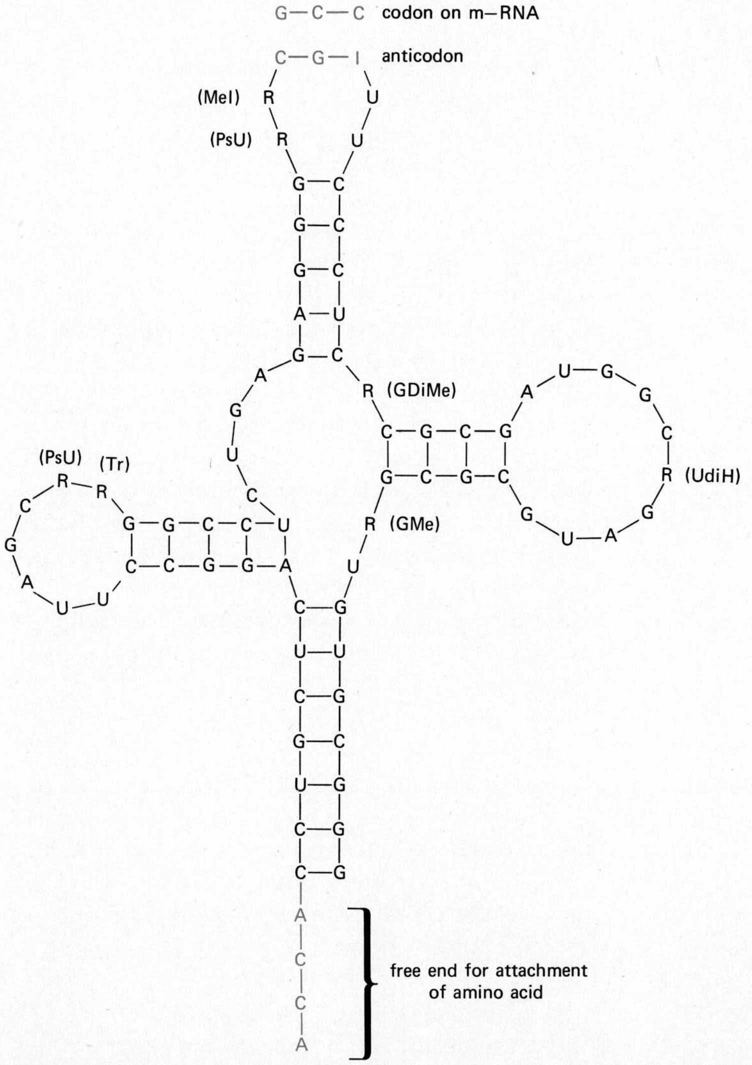

FIGURE 4.5 Diagram for alanine t-RNA. Each letter represents a nucleotide containing a nitrogenous base. R indicates the presence of one of the rare or unusual bases whose formulas are not given in this text. The rare nucleotides are thus indicated: MeI = methylinosine, PsU = pseudouridylic acid, Tr = ribothmidylic acid, Me + base letter = methylated derivative of common base (di Me = two methyl groups), DiH + base letter = di hydro (2–OH) derivative of common base.

and is associated with a particular amino acid, as is the codon. (Inosine, an additional base, is found in t-RNA. It has pairing properties similar to those of guanine.) The free end of the molecule, CCA, serves as the point of attachment for an amino acid associated with the anticodon. Since the anticodon will attach only to a particular codon on m-RNA, determined by base pairing, the t-RNA brings a particular amino acid to a ribosome where it is caused to take a particular position in the growing protein. This is the process designated as translation. Specifically:

The amino acid is activated by ATP and an activating enzyme.

Activated amino acids attach to their particular t-RNA molecules.

Ribosomes, consisting of ribosomal RNA and protein in roughly equal amounts, and the m-RNA molecule move relative to one another.

The amino acids are joined to the growing protein chain by the ribosome, and the t-RNA molecules carrying them are freed into the cytoplasm to pick up more acids.

The action of the ribosome is compared to the lock on a zipper, bringing together t-RNA and m-RNA.

The relations described are illustrated in Fig. 4.6.

Mutations

Alterations in base sequences—addition or removal of bases or triplets in the genetic code—result in scrambled codes and abnormal proteins. Such alterations constitute mutations. J. C. Kendrew illustrates the manner of production of mutations by the following passages:

> A mutation, interpreted in the light of the genetic code, may be compared to a misprint in a line of newspaper type:
>
> "Say it with glowers." In this example, there has been a SUBSTITUTION of an incorrect for a correct letter.
>
> "The prime minister spent the weekend in the country, shooting p_easants." In this case a letter is missing; a DELETION has occurred.
>
> "The treasury controls the public monkeys." Here, an additional letter is present in what is termed an INSERTION.
>
> "We put our trust in the Untied Nations." Two letters have been reversed in what is termed an INVERSION.
>
> "Each gene consists of xgmdonrsd ad zbqpt ytrs." Suddenly, a sequence becomes gibberish, forming NONSENSE WORDS.

It is easy to envision the havoc wrought in the code by these types of errors that result in abnormal proteins being produced. Most mutations usually affect only one representative of a pair of alleles controlling expression of a given characteristic. ALLELES are genes that occupy the same locus on a specific pair of chromosomes and control the heredity of a particular characteristic. If both members of a pair of alleles are identical, the individual is HOMOZYGOUS for the characteristic; if they are different, the individual is HETEROZYGOUS for the characteristic. An allele that is always expressed, whether the individual is homozygous or heterozygous, is said to be DOMINANT; one that is expressed only when homozygous is RECESSIVE. Thus, the normal allele may exert its effect more strongly *(dominance)*, and the mutation may not be expressed. The defect is carried, however, and may express itself at some future time.

Control of protein synthesis

In the schemes leading to synthesis of proteins, there are at least four points at which the process may be altered, resulting in slowing, blocking, or speeding of reactions:

At the synthesis of m-RNA from DNA.

The association of ribosome and m-RNA.

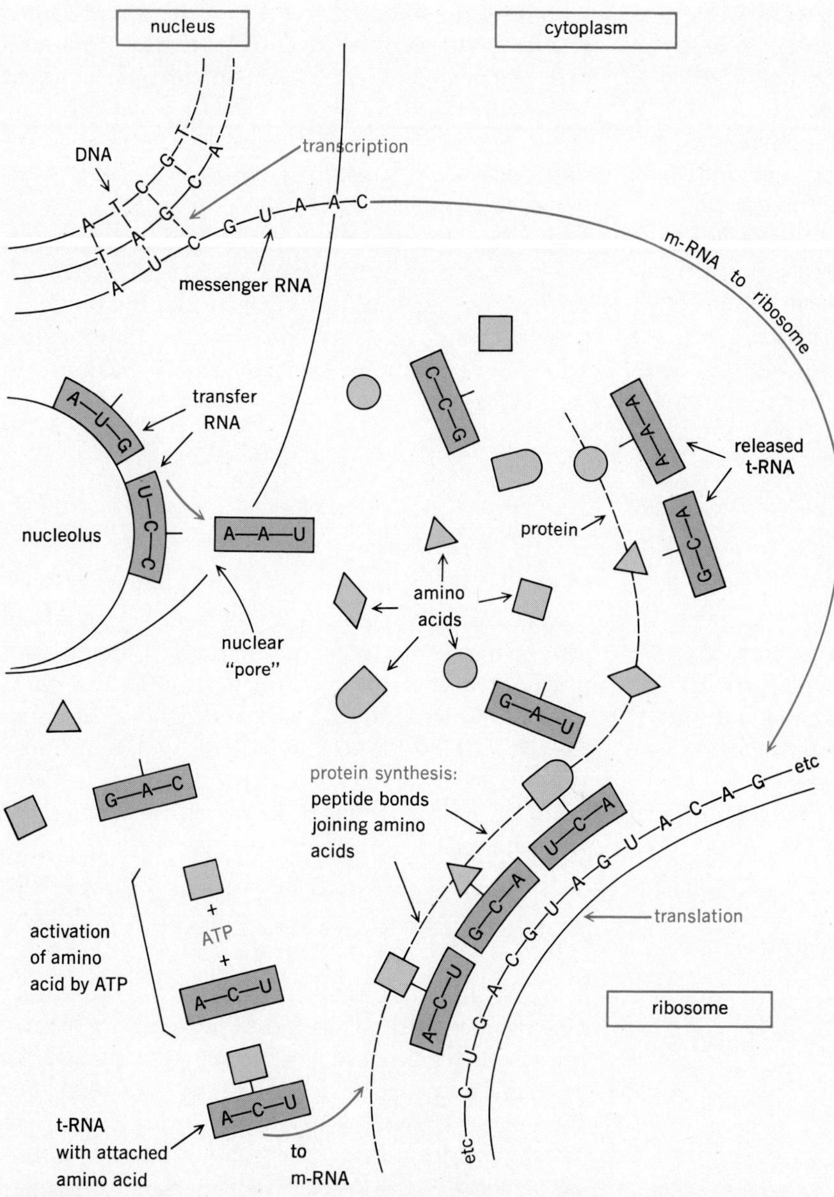

FIGURE 4.6 Transcription, translation, and protein synthesis in a cell.

Destruction of m-RNA in the cytoplasm.
Synthesis of t-RNA.

While the exact point at which the effect is exerted is not known, it is possible to state that control is exerted over protein synthesis. The following mechanisms are postulated to be methods by which control may be exerted.

INDUCTION is a process by which the presence of a chemical (e.g., lactose) may cause *(induce)* the synthesis of enzymes to metabolize the chemical. Obviously, the capacity to form the enzymes is already present in the cell and is "turned on" by the presence of the chemical. This mechanism has the advantage of conserving cellular resources until the need for the enzyme has arisen.

REPRESSION involves the ability of a chemical (e.g., histidine) to *inhibit* the synthesis of enzymes necessary to create the chemical. Accumulation of end products thus slows synthesis until the end products are utilized. Repression may also occur when a specific protein, called a repressor, is synthesized under the direction of a regulator gene (see below).

Many genes are apparently affected in the manners described above. These theories of action are incorporated into the OPERON CONCEPT. The concept further suggests that STRUCTURAL GENES, OPERATOR GENES and REGULATOR GENES are present in the genetic material. The structural genes direct the synthesis of specific enzymes and proteins via the m-RNA mechanism. An operator gene "turns on or off" the structural genes. The combination of operator and structural genes is the operon. A REGULATOR GENE may in turn control the operator gene through the synthesis of proteins called *repressors.* A scheme of checks and balances is thus created.

Chemical control

A wide variety of chemical substances are known to exert effects on cellular processes and on the composition of the cellular environment. Among the chemical substances most directly involved in control of cellular activity are the HORMONES, the products of activity of the endocrine glands. Four general theories have been proposed as explanations of how hormones exert their control:

Hormones influence membranes.
Hormones directly influence enzyme activity.
Hormones influence gene activity.
Hormones control the production and/or release or ions or other small molecules that may then exert an influence on cellular activity.

Hormonal effects on membranes

Membrane systems surround the cell and many of its organelles. Permeability of these membranes may be altered by the presence of hormones, as described in the following examples.

Insulin, a product of the pancreatic endocrine tissue, is postulated to increase cellular uptake of glucose through increasing plasma membrane permeability to glucose. Parathyroid hormone has been shown to alter mitochondrial membrane permeability to magnesium and potassium, and thus stimulates mitochondrial activity. This may lead to alterations in the rate of oxidative phosphorylation, the activity of the Krebs cycle, and shifts in ion concentrations on either side of the membrane.

Hormonal effects on enzymes

Though not definitely proved to have direct effects on enzymes *in vivo* (in the living organism), hormones have been shown to alter the activities of many enzymes *in vitro* (in the test tube). For example, estrogens (female sex hormones) may be shown to influence certain enzymes of the Krebs cycle.

Hormonal effects on genes

Many hormones appear to act as gene activators to alter rates of DNA and RNA synthesis or to act as inducers or repressors of gene activity. In the sequence of events occurring in the genetic control of cellular activity, there appear to be four points at which hormones could influence the gene:

By combining with DNA molecules or by controlling gene activators or inactivators within the nucleus.

By combining with the products of the regulator genes and thus acting as inducers or repressors of genes.

By altering the activity of an enzyme required for synthesis of an inducer or repressor.

By altering membrane transport of a particular substance necessary for synthesis of an inducer or repressor.

Examples of hormones acting on or within the nucleus include: the effect of *cortisone* (a hormone of the adrenal cortex) on stimulation of formation of glucose from amino acids (gluconeogenesis) by enhancing the rate of synthesis of enzymes required in the conversion; and the effect of *aldosterone,* an adrenal cortical hormone, on sodium and potassium transport through cell membranes

FIGURE 4.7 Genetic and hormonal controls on cellular activity (+ = stimulation; − = inhibition.)

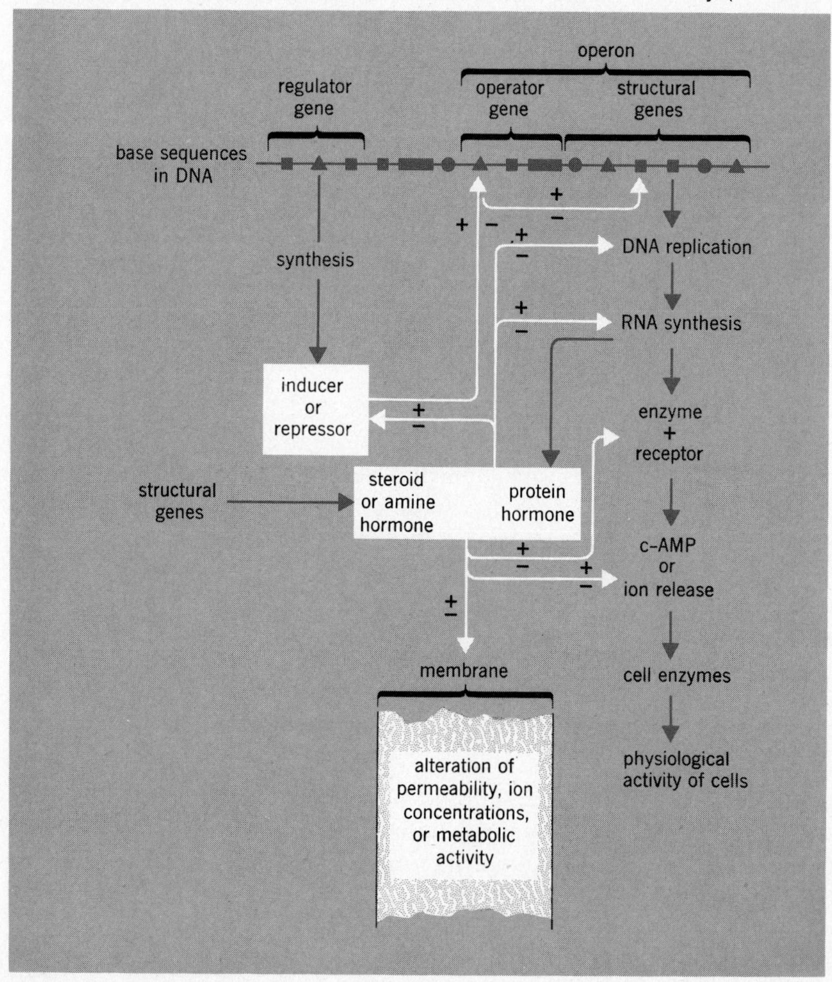

by gene activation and induction of formation of enzymes necessary to the transport process.

While exact mechanisms may not be available to specify the effects of hormones on genes, many (not all) hormones directly or indirectly influence gene activity; resultant changes in enzyme and protein concentrations may be the basis for the physiological effects demonstrated on hormone administration or deficiency.

Hormonal effects on small molecules and ions

The effects on a cell, or cell activity, of a hormone may not be through direct action of the hormone on a particular gene or enzyme, but on the production or destruction of a molecule that may act as an activator or inhibitor of the gene or enzyme. CYCLIC ADENOSINE MONOPHOSPHATE *(c-AMP)* is produced by the action of several hormones on the enzyme adenylcyclase. This enzyme causes the production of c-AMP from ATP (see Chapter 28). C-AMP results in the activation of enzymes. For example, epinephrine (from the adrenal medulla) results in increased c-AMP production, which activates the enzyme phosphorylase, which causes glycogen breakdown to glucose. Similarly, estrogens cause histamine release and influence activity of the uterus.

Ions, particularly calcium and sodium, may be released by hormones from areas of binding or storage. Oxytocin (a posterior pituitary hormone) stimulates calcium ion release, which then causes uterine contraction; aldosterone releases sodium ion that is required for active transport of amino acids into cell nuclei.

The combined action of genetic and chemical factors would thus seem to be necessary for control of cellular reactions. The interrelationships of these various control mechanisms are presented in Fig. 4.7.

Energy sources

To support the chemical reactions necessary to continuance of cellular activity, sources of energy are required. Energy is typically provided by the degradation or combustion of basic foodstuffs such as glucose and fatty acids. The ladderlike degradation of such molecules to form specific end products is carried out by specific metabolic pathways, schemes, or cycles. Most degradative metabolic pathways produce energy that is utilized to synthesize adenosine triphosphate (ATP). This compound serves as a means of transferring energy from the foodstuff to the physiological mechanism such as muscular contraction, membrane transport systems, or secretion of products by a gland. Among the more important and interrelated cycles that provide energy for the synthesis of ATP are glycolysis, Krebs cycle, beta-oxidation, and oxidative phosphorylation, which synthesizes the bulk of the cell's ATP. Other cycles are available for reactions of other compounds. The basic reactions in each of the schemes listed above are presented in this chapter, and the complete schemes may be found in Chapter 28. Many of the schemes are carried out in the cytoplasmic organelles, particularly the mitochondria.

Glycolysis

The anaerobic pathway of GLYCOLYSIS (Fig. 4.8) commences with glucose and results in the production of two molecules of pyruvic acid and two new ATP molecules. In the course of the reactions, four hydrogens are liberated and attach to large molecules known as hydrogen acceptors. In glycolysis, a total of 56,000 calories* per mole of glucose are evolved, of which 27.5 percent is captured in two new ATP molecules produced by the cycle. Further degradation of the pyruvic acid requires the removal of CO_2 and two H from each pyruvic acid resulting from glycolysis. Acetic acid is produced which is directed to the Krebs cycle.

* A calorie (small c) is the amount of heat required to raise the temperature of 1 gram of water 1°C. A Calorie (large C) or kilocalorie is 1000 times larger than a calorie, and raises the temperature of one kilogram of water 1°C.

The Krebs cycle

This cycle (Fig. 4.9) takes the two carbon acetic acid molecules produced from decarboxylation (removal of CO_2) of the pyruvic acid and produces two CO_2 molecules, one ATP, and eight H for each acetic acid channeled through the cycle. The cycle

FIGURE 4.8 The general reactions of glycolysis.

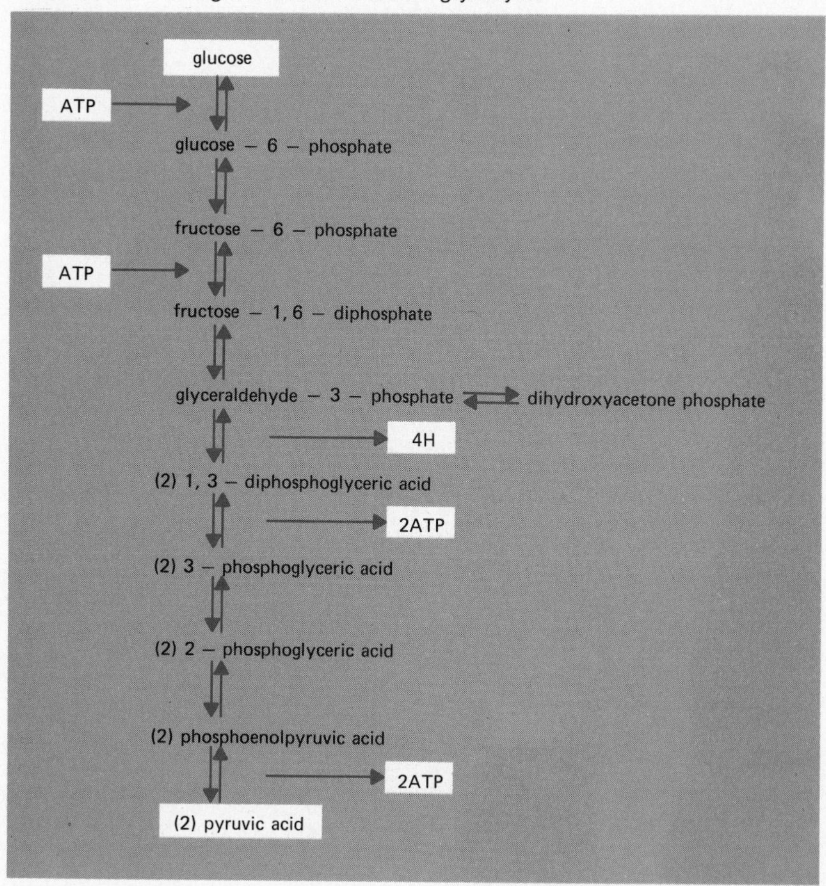

net reaction:

glucose → 2 pyruvic acid + 2ATP + 4H + 56 kcal/mole

is aerobic and releases 629,000 calories of heat energy per mole of pyruvic acid. Hydrogens released in the conversion of pyruvic acid to acetic acid and in the Krebs cycle are also bound to hydrogen acceptors.

Beta oxidation

Beta oxidation (Fig. 4.10) reduces long-chain fatty acids to two carbon units of acetic acid. Each cleavage of a two-carbon unit from the fatty acid also releases four H ions that, again, go to hydrogen acceptors. The acetic acid molecules then enter the Krebs cycle.

Oxidative phosphorylation

The hydrogens collected from the various cycles are channeled to yet another scheme designated *oxidative phosphorylation* (Fig. 4.11). This scheme,

FIGURE 4.9 The general reactions of the Krebs cycle.

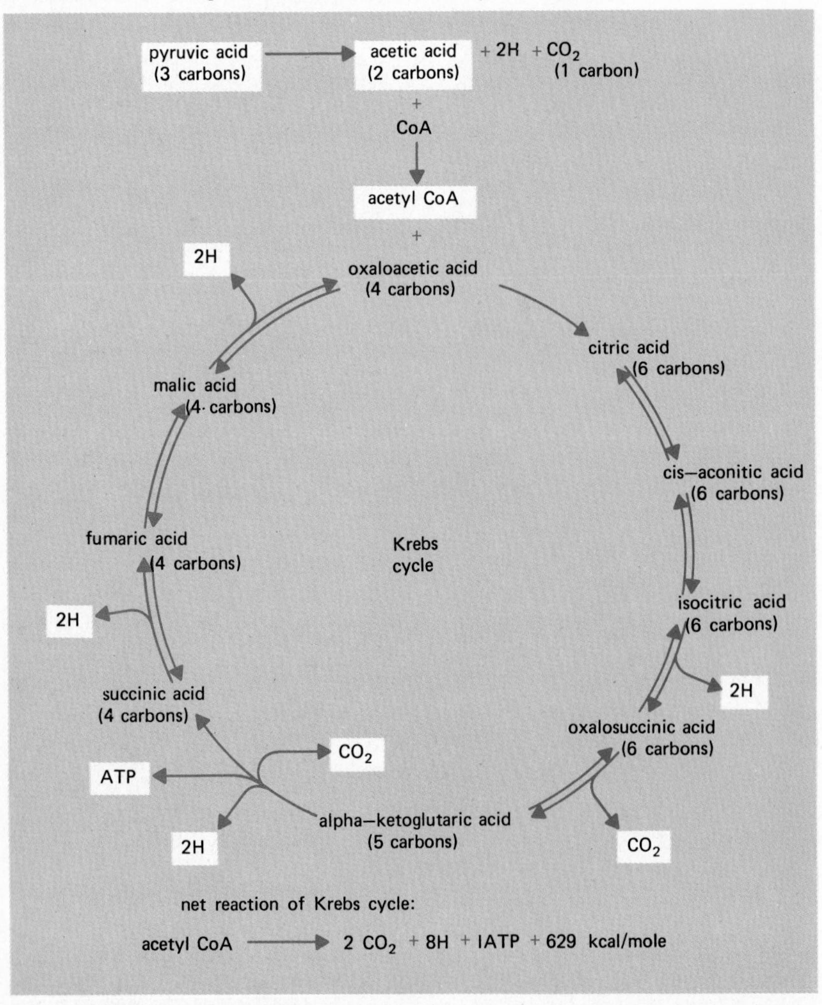

utilizing compounds known as *cytochromes,* transfers the hydrogen ions to oxygen to form water and liberates sufficient energy to synthesize two or three ATP molecules for each pair of hydrogens brought to the scheme on particular hydrogen acceptors. This scheme produces the bulk of the ATP synthesized from the degradation of a given foodstuff.

Other metabolic cycles

Glycolysis and beta oxidation are reversible if energy is supplied to the reactions. Thus, glucose or fatty acids may be synthesized from acetic acid. Additionally, by polymerization of glucose units, glycogen, a storage form of glucose, may be synthesized. Production of glycogen from glucose is termed GLYCOGENESIS (Fig. 4.12). Recovery of glucose from glycogen constitutes GLYCOGENOLYSIS (Fig. 4.13).

FIGURE 4.10 The general reactions of beta oxidation.

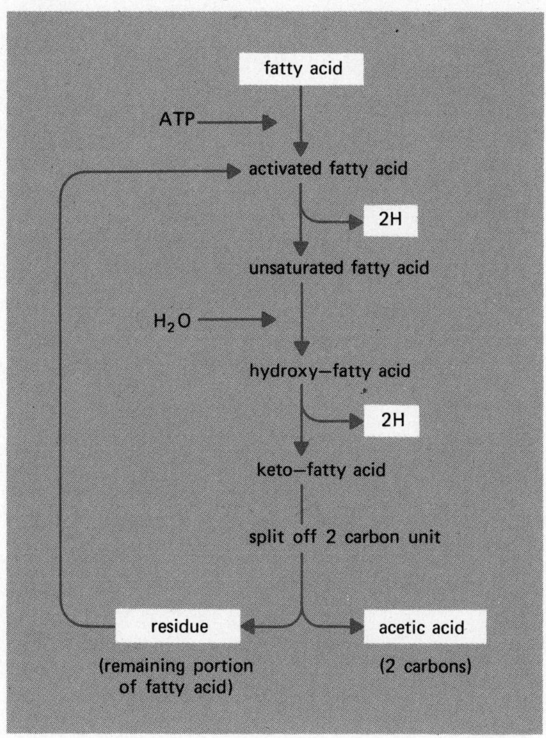

net reaction per split:

fatty acid ⟶ acetic acid + 4H + residue to other cycles (e.g. Krebs)

FIGURE 4.11 The general reactions of biological oxidation.

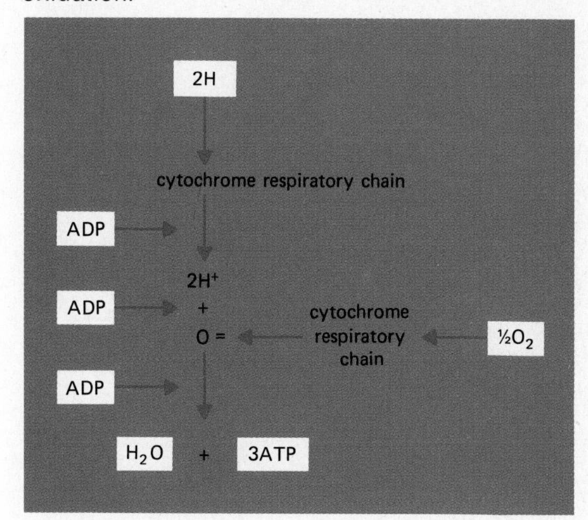

net reaction:

2H + ½O_2 + 3ADP ⟶ H_2O + 3ATP

FIGURE 4.12 The net reaction of glycogenesis.

glucose $\xrightarrow{\text{insulin}}$ glycogen

FIGURE 4.13 The net reaction of glycogenolysis.

glycogen $\xrightarrow[\text{glucagon}]{\text{epinephrine}}$ glucose

Amino acids may become energy sources for cellular activity if they are DEAMINATED (Fig. 4.14). This reaction produces *ammonia* and what is termed a *keto acid.* Certain keto acids are found in the schemes of glycolysis and beta oxidation and may thus suffer the same fate as compounds resulting from metabolism of glucose. In short, a given cycle treats the same compound similarly regardless of its source. Synthesis of glucose from any non-carbohydrate precursor (lipids and/or proteins) is termed GLUCONEOGENESIS.

Conversely, TRANSAMINATION of a keto acid (Fig. 4.15) produces amino acids by transferral of an amine group from an amine donor.

Finally, the three basic foodstuffs represented by glucose, fatty acids, and the nonessential amino acids are, to a large degree, interconvertible. Thus, relatively independent of dietary intake, essential materials for synthesis of body substances are assured. The "METABOLIC MILL" (Fig. 4.16) illustrates the interrelationships in the basic metabolism of the three basic foodstuffs.

FIGURE 4.14 The net reaction of deamination.

amino acid ⟶ keto–acid + ammonia

FIGURE 4.15 The net reaction of transamination.

keto–acid + amine group (from amine donor) ⟶ amino acid

FIGURE 4.16 The "metabolic mill" shows interrelationships of protein, fatty acids, and glucose metabolism.

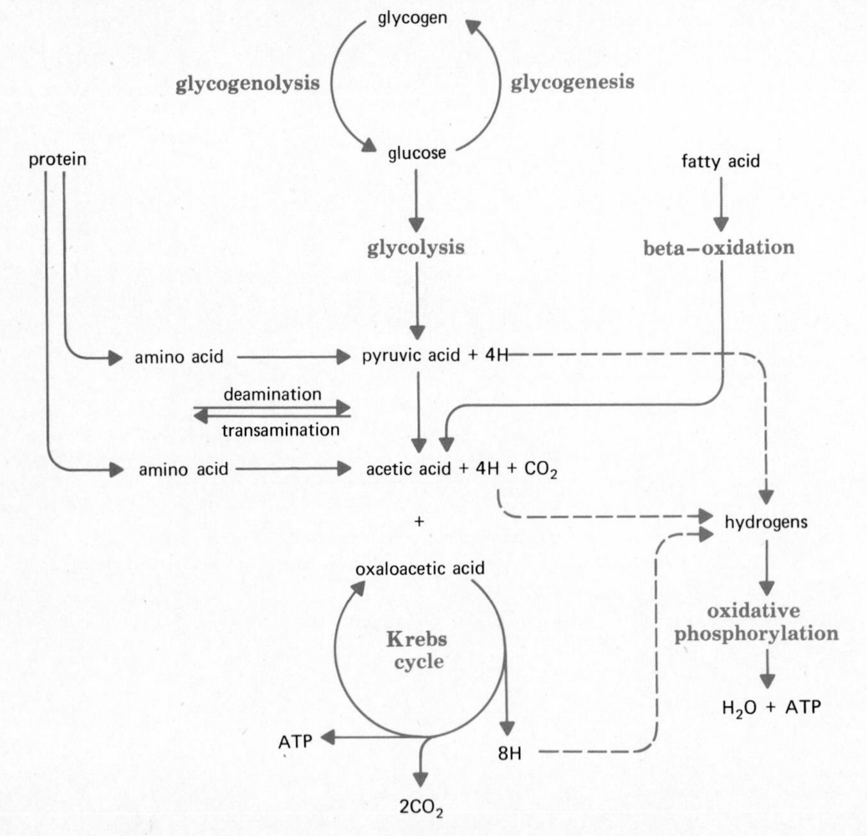

Summary

1. Control of cellular activity takes two forms: genetic and chemical.
   a. Genetic control is exerted by way of nucleic acids, the genetic code, and the synthesis of specific proteins. Mutations, or alterations in the code, create abnormal proteins and influence cell structure or activity.
   b. Chemical control is by way of hormones that influence membranes, enzymes, genes, and the production of small molecules.
2. Energy for cellular activity is provided by degradation of foodstuffs and the use of the energy in the foodstuffs for ATP synthesis. Several metabolic schemes are involved in ATP synthesis.
   a. Glycolysis reduces glucose to pyruvic acid.
   b. The Krebs cycle combusts pyruvic acid to CO_2.
   c. Beta oxidation degrades fatty acids.
   d. Oxidative phosphorylation produces water and most of the ATP molecules.
   e. Glycogenesis forms glycogen from glucose.
   f. Glycogenolysis releases glucose from glycogen.
   g. Deamination is the initial step in amino acid breakdown.
   h. Synthesis of glucose from amino acids is termed gluconeogenesis.
   i. Transamination synthesizes nonessential amino acids.
   j. All three foodstuffs (carbohydrates, fats, and protein) are interconvertible by way of the metabolic mill.

Cell Reproduction and Body Development

Cell reproduction

The body does not spend its life with the same cells with which it was born. Cells constantly die; the body is injured and must be repaired. Replacement of worn-out cells or the production of new cells, for purposes of repairing injury and continuance of the species, is common in most cell types within organisms. Two processes occur in the body that result in the production of new cells by division of preexisting cells.

Mitosis

Mitosis (Fig. 5.1) is the type of cell division that the body utilizes as a method of growth and to replace dead or lost cells. In this type of division, the nucleus undergoes a complicated series of changes to ensure that the daughter cells will receive the same complement of DNA (chromosomes) as was possessed by the parent cell. The

FIGURE 5.1 The major events occurring in mitosis as depicted in a cell having four chromosomes.

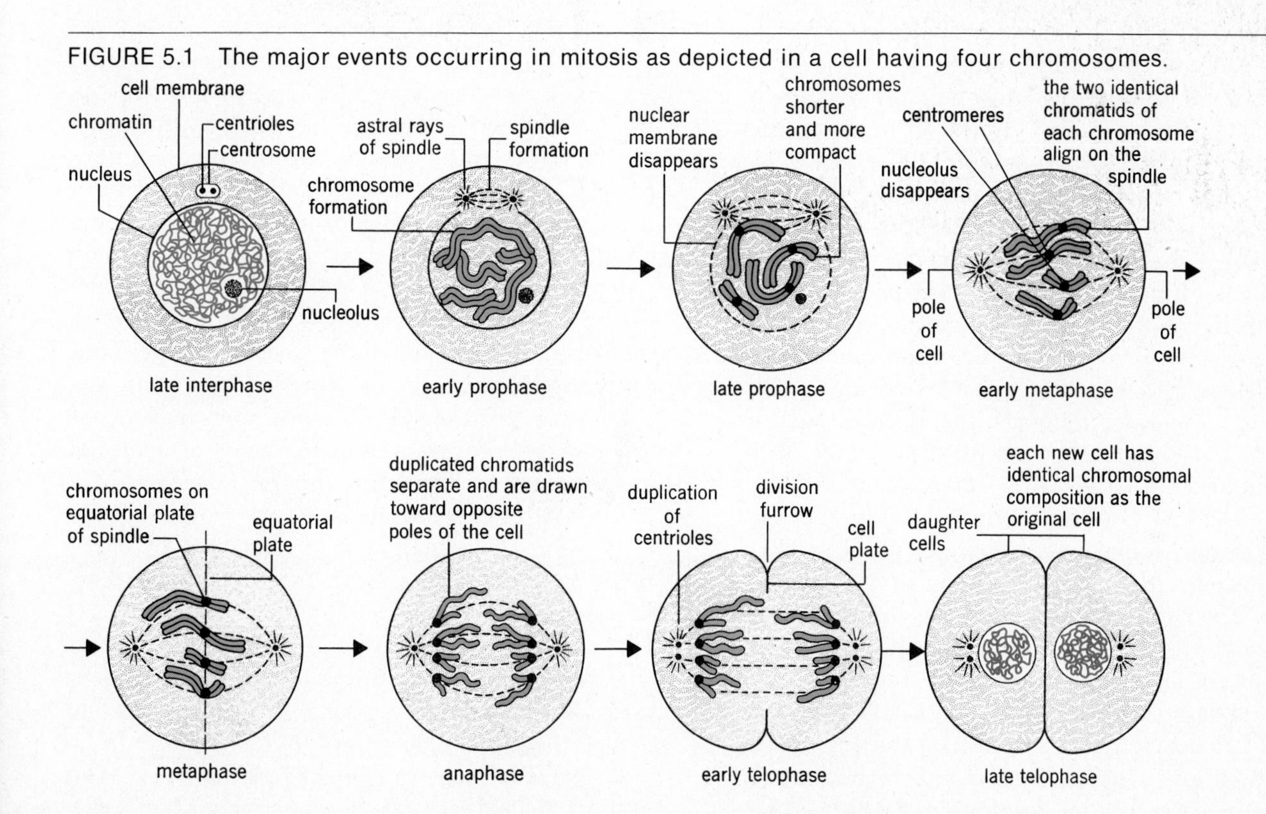

cytoplasm and its organelles undergo a more-or-less equal division between the daughter cells by simply cleaving in half. Mitosis is described as occurring in several stages, each designated by events that are characteristic for that stage.

In INTERPHASE, the cell appears typical for its type and is not undergoing division. It is engaged in accumulating chemicals for its own use, synthesis of proteins, metabolism of fuels, and replication of DNA and RNA. The chromatin material, and therefore each chromosome, is duplicated in this stage. (Geneticists "count" centromeres, as described below, in order to determine numbers of chromosomes. One centromere is equivalent to one chromosome, regardless of how many strands of DNA are attached to it.)

As the cell prepares to divide, the chromatin begins to condense and coil and becomes visible as deeply staining bodies that are the chromosomes. When the chromosomes become visible but show no pattern of arrangement, the cell may be said to have entered prophase.

In PROPHASE, each chromosome may be shown to consist of long, paired strands of material, the *chromatids*, connected at one point, the *centromere*. The paired centrioles separate and migrate to opposite poles (ends) of the cell, the nuclear membrane disappears, and a *spindle* of microtubules is formed between the centrioles.

Further condensation and shortening of the chromosomes occurs, and the chromosomes align themselves between the centrioles on an imaginary line, the *equatorial plate*, that divides the cell in half between the two centrioles. When this alignment occurs, the cell is in METAPHASE.

In ANAPHASE, the centromeres divide and the duplicated chromosomes separate, possibly due to centromere repulsion or pulling by the spindle fibers. The centromeres "lead" the chromosomes as they move, and the trailing chromosome arms assume a V or J shape during this stage.

In TELOPHASE, *cytoplasmic division* is initiated by a furrow that appears at or near the equatorial plane of the cell. Eventually, a complete membrane will form across the cell resulting in a nearly equal division of the cytoplasm and its constituents.

Simultaneously, the chromosomes are unwinding and becoming less deeply stained; they will return to the granular state characteristic of the interphase cell. The nuclear membrane reorganizes around the area where the chromosomes are located. Since the nuclear material of the daughter cells was duplicated in the original cell, and then separated by the division, chromosome number, DNA pattern, and identity of a given cell line are maintained.

Studies of the timing of mitosis in humans indicates that the whole process, in most cells that do divide, has a cyclical nature 12–24 hours in duration from interphase to interphase. Within this timespan, the actual mitotic division itself takes only about an hour.

Meiosis

Meiosis (Fig. 5.2) is a special type of cell division by which GAMETES (sex cells—sperm and ova) are produced. Each daughter cell formed by this process contains one-half the chromosome number present in the parent cell, and is said to be a HAPLOID cell. The chromosomes present in the daughter cells represent one of each chromosome pair possessed by the parent cell. The basic differences between this process and mitosis are: in meiosis, *two* cell divisions occur, one without previous duplication of chromosomes, and exchange of portions of chromosomes may occur during the process. This exchange of material creates genetic variability in the members of a given pair of chromosomes.

In MEIOTIC INTERPHASE, duplication of chromatin material occurs. In FIRST MEIOTIC PROPHASE, five stages are recognized:

1. *Leptotene* stage is characterized by the appearance of visible chromosomes that appear single but actually have a double set of DNA helices.

2. *Zygotene* stage is characterized by the pairing of homologous chromosomes (a pair of chromo-

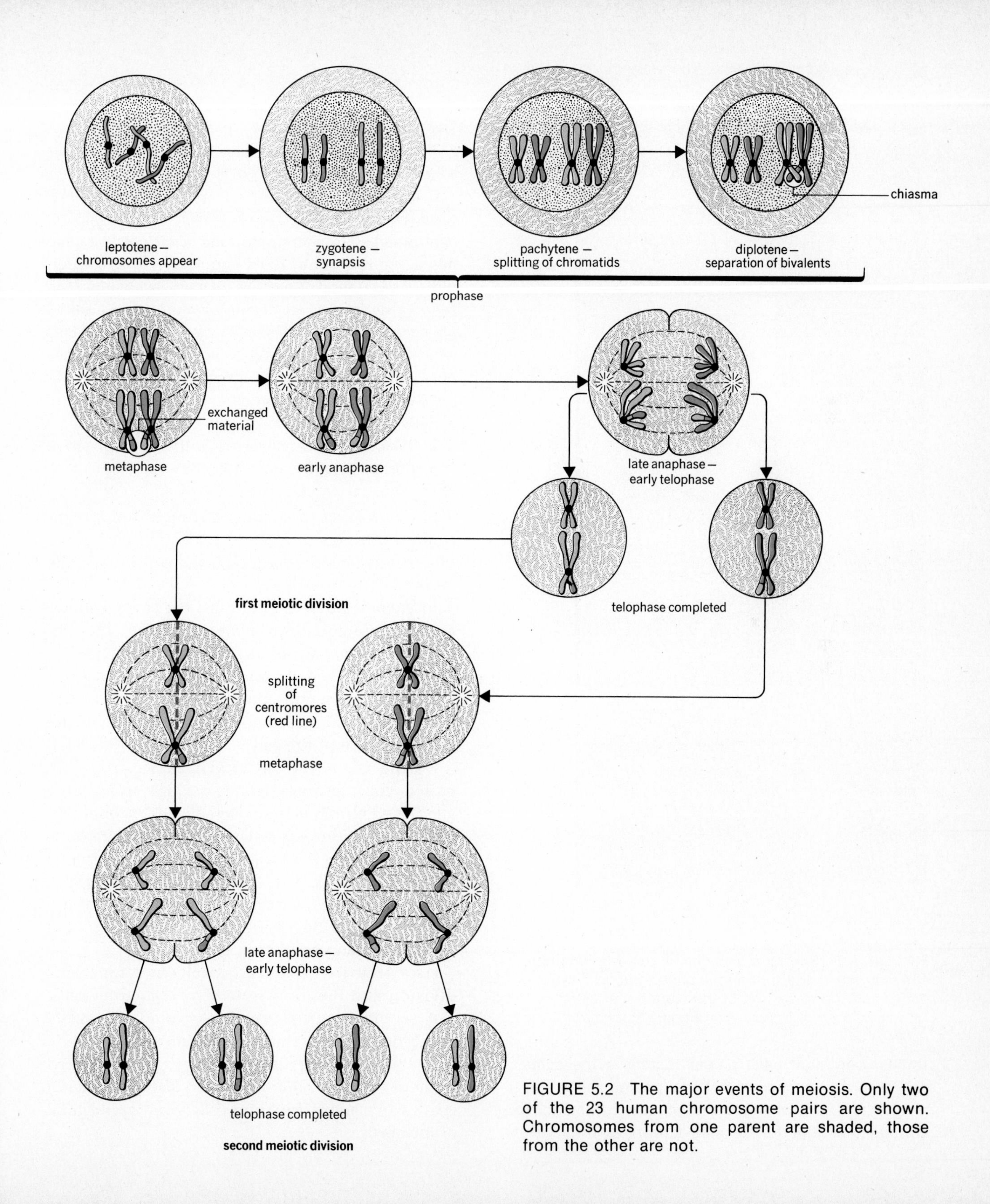

FIGURE 5.2 The major events of meiosis. Only two of the 23 human chromosome pairs are shown. Chromosomes from one parent are shaded, those from the other are not.

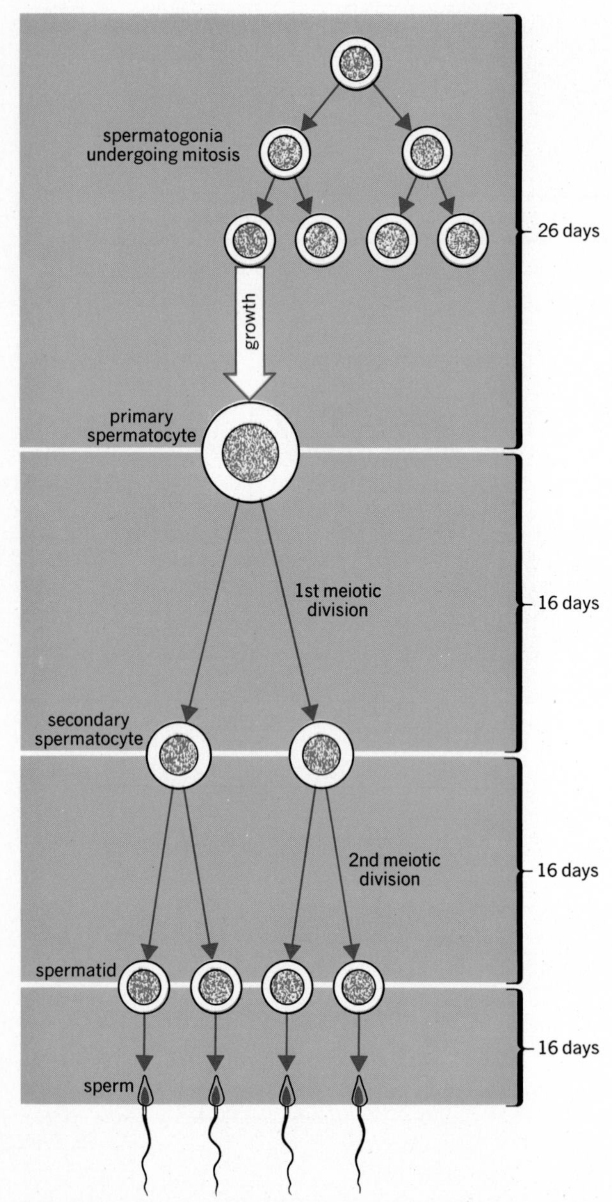

FIGURE 5.3 Spermatogenesis in the human male showing when the first and second meiotic divisions occur and the formation of four functional cells.

somes, one from each parent that have the same genes in the same order.) The pairing *(synapsis)* results in a parallel alignment with a point-for-point arrangement that forms units known as *bivalents. This pairing does not occur in mitosis.*

3. *Pachytene* stage is characterized by further shortening and thickening of the chromosomes.

4. *Diplotene* stage begins when a longitudinal separation of the two strands of each bivalent occurs; centromeres do not divide, and each chromosome has a "double set of arms" on it. During this longitudinal separation, the halves of each bivalent may be seen to be in contact at one or more places called *chiasmata.* These chiasmata may represent places where portions of homologous chromosomes have been exchanged in the process of *crossing-over.* As indicated earlier, this process creates variability in members of a chromosome pair.

5. *Diakinesis* stage involves further condensation and shortening of the chromosomes.

Final changes occurring during prophase include: centrioles migrate to opposite poles of the cell; the nuclear membrane disappears; the spindle is formed, as in mitosis.

In FIRST MEIOTIC METAPHASE, the chromosomes align on the equatorial plate of the cell. *The centromeres do not divide* at this time. The chromosomes have lined up on the equatorial plate in random fashion, with maternal and paternal numbers of each pair *not* aligned so that each is facing the same pole of the cell.

During FIRST MEIOTIC ANAPHASE, separation of paired chromosomes results in two things: the *number* of chromosomes is reduced by one-half, since separation occurs without centromere division—*this stage is thus a reductional division;* one member of each pair, some maternal in origin, some paternal, is drawn to opposite poles of the cell. Thus, further variability of daughter cell genetic makeup results.

FIRST MEIOTIC TELOPHASE results in cytoplasmic division and the formation of two daughter cells.

A SECOND MEIOTIC DIVISION occurs next—really a mitotic division; that is, the centromeres divide and the resulting chromosomes are separated without further change. Cytoplasmic division creates two more daughter cells, for a grand total of four cells.

As meiosis proceeds in the testes during formation of sperm *(spermatogenesis)*, four cells of equal size are formed. The first meiotic division occurs between the stages of development known as the *primary* and *secondary spermatocyte*, and the second meiotic division occurs between the stages of secondary spermatocyte and the *spermatid* (Fig. 5.3). The cycle of spermatogenesis in man has been estimated to be about 74 days.

In egg (ovum) production in the ovary *(oögen-*

FIGURE 5.4 Oogenesis in the human female showing when the first and second meiotic divisions occur and the formation of one functional cell and the polar bodies.

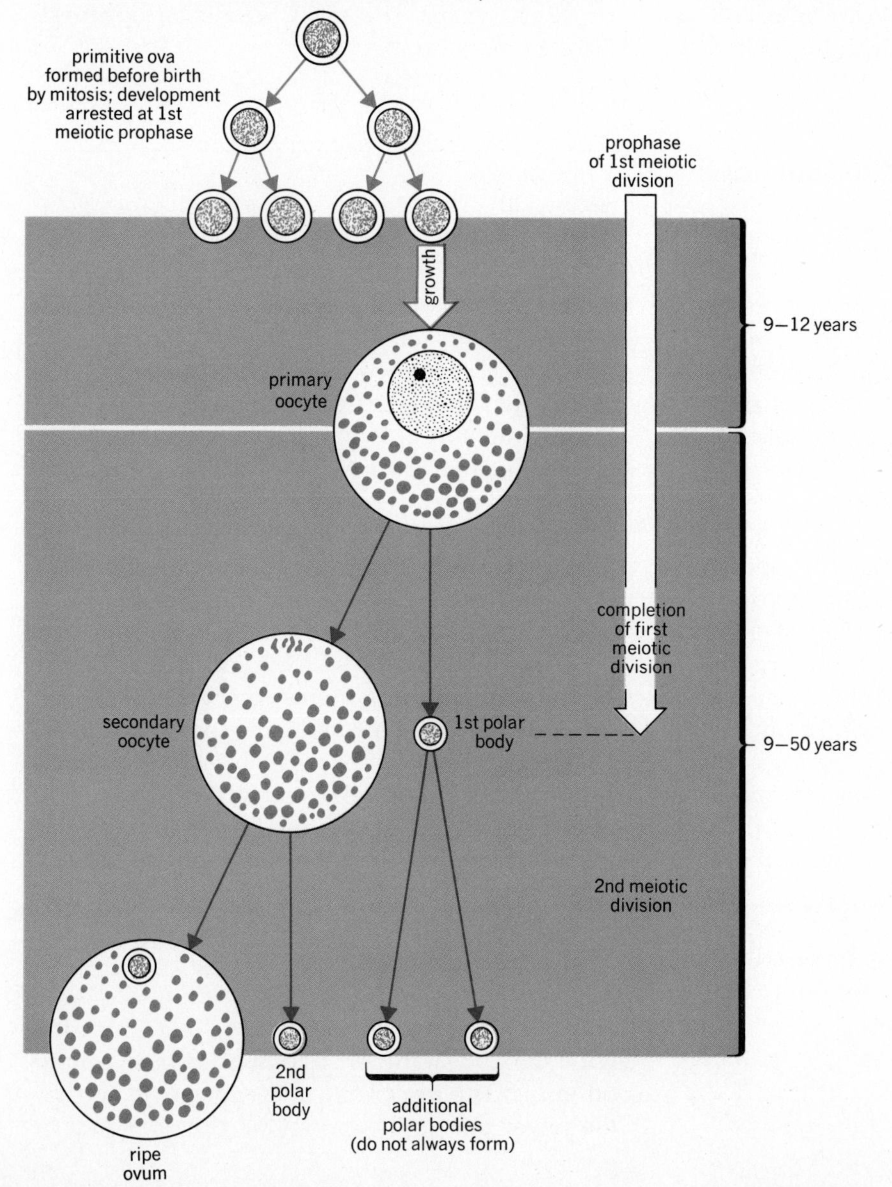

esis), embryonic and fetal formation of eggs has produced, by birth, some 400,000 *primordial follicles* that have reached first meiotic prophase. They remain "suspended" at this stage until puberty, when an average of one cell per month matures. Resumption of the first meiotic division occurs, and is usually completed by the time of ovulation (egg release from the ovary). During the first meiotic division, one of the two daughter cells receives nearly all of the cytoplasmic material, and is known as the *oocyte.* The other daughter cell is called a *first polar body.* The second meiotic division occurs as the ovum passes down the uterine tube, usually is completed after fertilization (if it occurs), and gives rise to the *ovum,* or mature egg, and a *second polar body.* Thus, only *one* functional cell results from ovarian meiosis, the polar bodies usually degenerating. Since usually only one ovum matures each month during the reproductive life of a female, the time for oogenesis may be 12–50 years. The major events in oogenesis are shown in Fig. 5.4.

The basic development of the individual

Human sperm and ova, produced by meiosis, each contain one-half the chromosome number characteristic of the species. This halved number of chromosomes is termed the haploid number and is 23 for the human. The haploid number of chromosomes consists of 22 AUTOSOMES, and an X or Y (SEX) CHROMOSOME. The meioses producing sperm create cells with either a 22 + X or 22 + Y content; those providing ova result in a 22 + X content. Restoring normal chromosome number (46) depends on the joining of an egg and sperm to form a zygote in the process of FERTILIZATION. Fertilization results in a 44 + XY (male) or 44 + XX (female) chromosome constitution in the zygote.

In about 85 percent of female body cells, the extra X chromosome may be visualized in the nucleus as the BARR BODY, or an extra mass of chromatin material, or as satellites on the nuclei of certain cells, such as blood cells. It is thus possible to specify the sex of the person if about a hundred cells can be examined.

Fertilization is the time of CONCEPTION. Development from conception onward may be described by the week in terms of the major processes concerned.

The first week

After fertilization, the zygote undergoes a series of mitotic divisions known as CLEAVAGE. A solid mass of cells, known as a MORULA, results. (The morula is about the size of a period in this text.) Little time for growth of cells occurs between cleavages, and the morula is only a little larger than the zygote. Cleavage results in a doubling of cell number for the first 8–10 divisions, and thereafter becomes irregular. As cleavage is occurring, the zygote is passing down the uterine (Fallopian) tube from ovary to uterus, a journey that takes about three days. Arriving in the uterine cavity, the morula adheres to the lining of the uterus (endometrium) and undergoes a REORGANIZATION into a BLASTOCYST. The blastocyst contains a cavity (the blastocele) and an INNER MASS of cells. This reorganization takes 3–4 days. The events of the first week are shown in Fig. 5.5.

The second week

The blastocyst undergoes further reorganization with the appearance of two cavities in the inner cell mass. The upper one is the AMNIOTIC CAVITY; the lower one is the CAVITY OF THE YOLK SAC. A

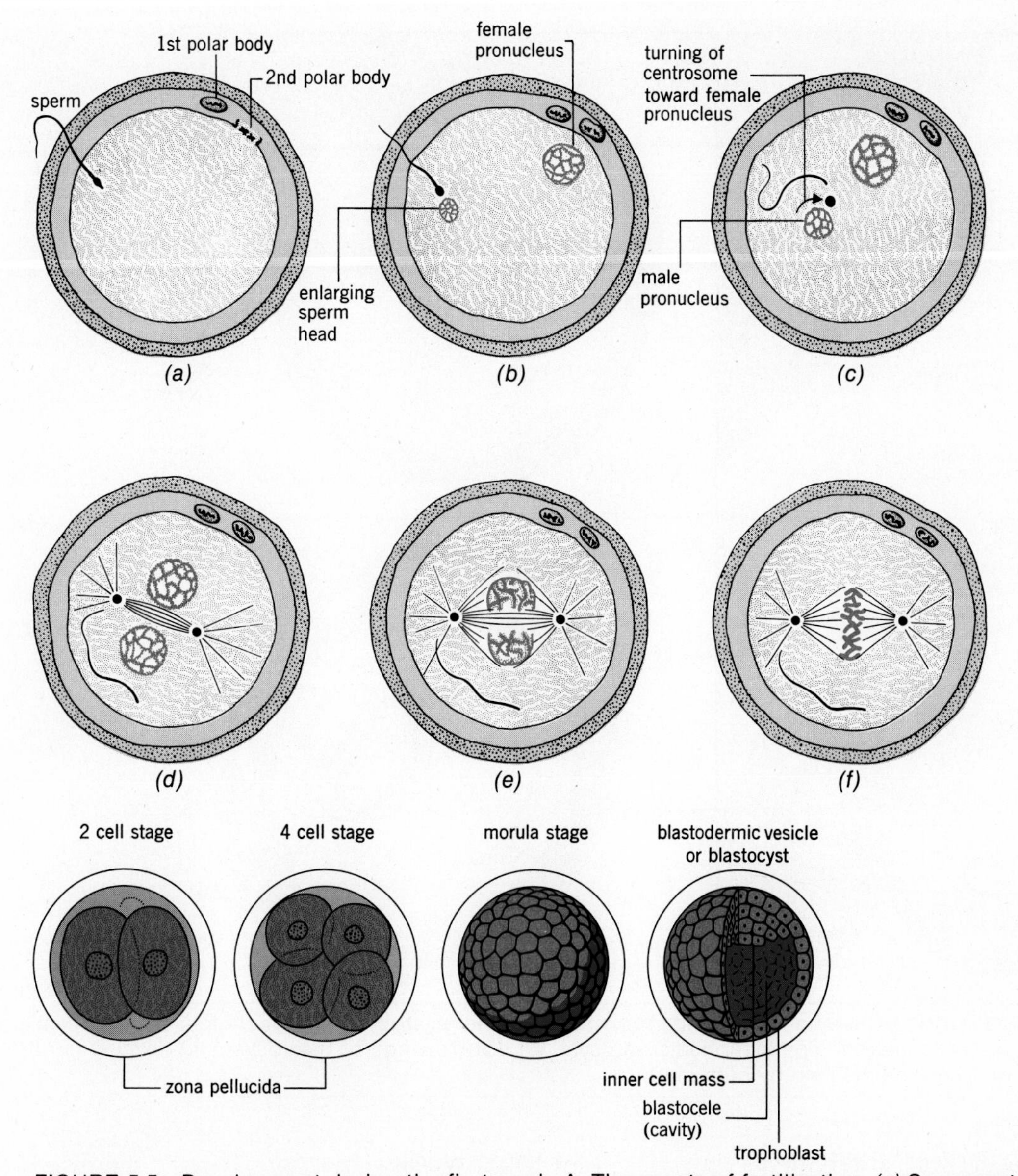

FIGURE 5.5 Development during the first week. A. The events of fertilization. *(a)* Sperm enters egg, *(b)* sperm head enlarges to form pronucleus, (*c*) rotation of centriole, (*d*) centriole divides and separates—spindle is formed, (*e*) pooling of chromosomes from pronuclei, (*f*) mitosis, metaphase. B. Cleavage and cell segregation.

two-layered plate of cells separates the two cavities and is known as the EMBRYONIC PLATE or DISC. Development to this stage is said to result in the formation of a *two-layered embryo.* The upper layer of cells in this embryo is termed ECTODERM, the lower layer ENDODERM. These layers form two of the three "germ layers" from which all body structures will develop. The embryo then under-

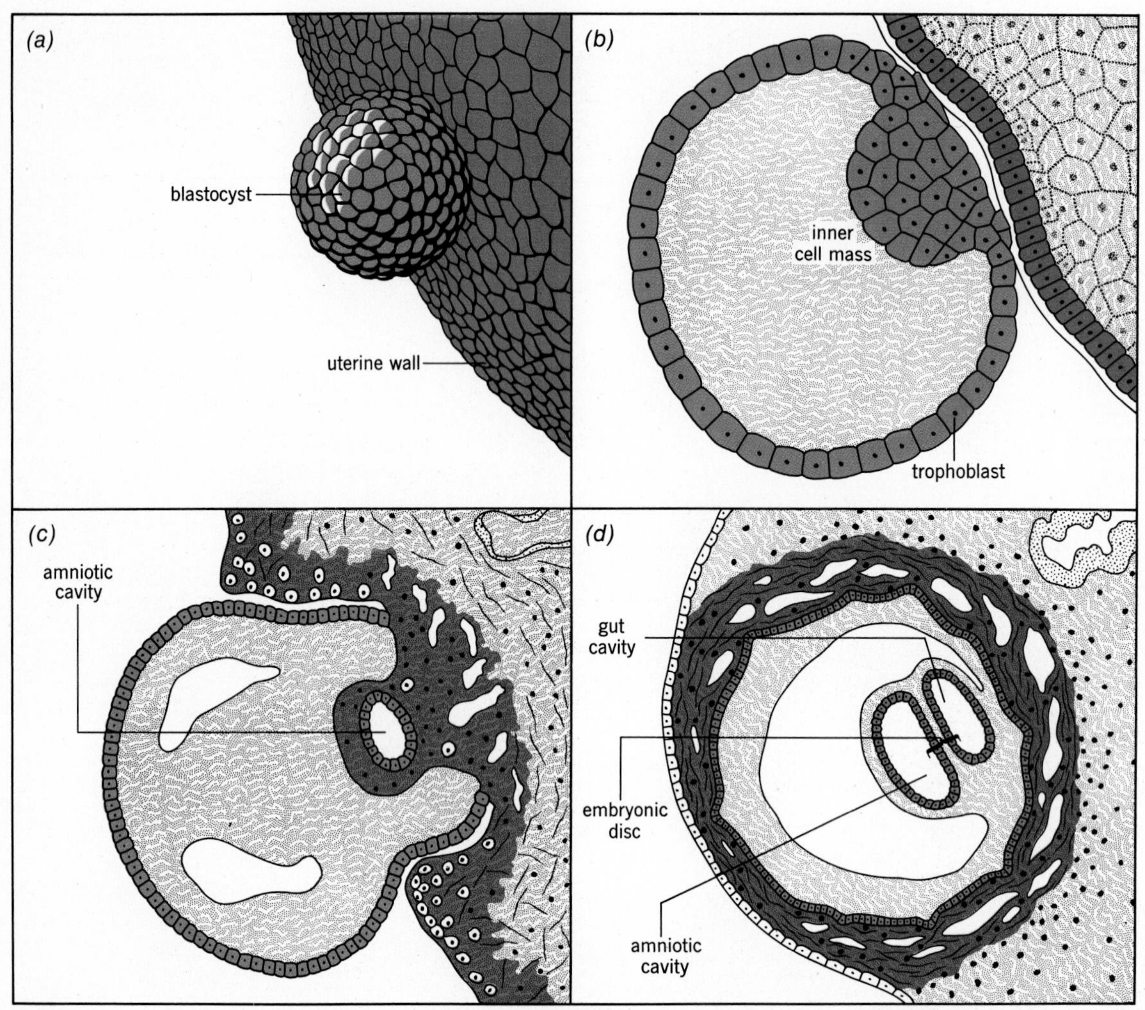

FIGURE 5.6 Development during the second week and implantation of the blastocyst. *(a)* Attachment of blastocyst to endometrium of uterus. (*b*) Appearance of sectioned blastocyst. (*c*) "Burrowing" of blastocyst and development of two-layered embryo. (*d*) Embryo enclosed by endometrium.

goes IMPLANTATION in the wall of the uterus. In this process, the blastocyst "digests" its way into the vascular and glandular endometrium of the uterus in order to establish a relationship with the mother's blood supply. This relationship assures the nutrients necessary for further growth and development. These processes are shown in Fig. 5.6.

The third week

The major event occurring at this time is the formation of the third basic germ layer. By mitotic cell division, a layer of cells termed the MESODERM is formed between the ectoderm and endoderm of the two-layered embryo. Table 5.1 shows the ultimate derivatives of these three germ layers.

TABLE 5.1 Tissues derived from germ layers in the human

Ectoderm	Mesoderm	Endoderm
1. Outer layer	1. Muscle:	1. Epithelium of:
(epidermis)	Skeletal,	Pharynx
of skin:	Cardiac,	Auditory (ear) tube
Skin glands	Smooth	Tonsils
Hair and nails		Thyroid
Lens of eye	2. Supporting	Parathyroid
	(connective)	Thymus
2. Lining tissue	tissue:	Larynx
(epithelium)	Cartilage	Trachea
of:	Bone	Lungs
Nasal cavities	Blood	Digestive tube
Sinuses		and its glands
Mouth:	3. Bone marrow	Bladder
Oral glands		Vagina and vestibule
Tooth enamel	4. Lymphoid tissue	Urethra and glands
Sense organs		
Anal canal	5. Epithelium of:	2. Pituitary
	Blood vessels	(anterior and
3. Nervous tissues	Lymphatics	middle lobes)
	Celomic cavity	
4. Pituitary	Kidney and ureters	
(posterior lobe)	Gonads and ducts	
	Adrenal cortex	
5. Adrenal medulla	Joint cavities	

The fourth week

The fourth week is associated with the formation of the NEURAL TUBE, the beginning of the nervous system, and of SOMITES or blocks of mesoderm along the backbone of the embryo. These somites are the forerunners of the bones and muscles of the back. The embryo appears as shown in Fig. 5.7. It should be noted that development occurs in a head-to-tail *(cephalocaudal)* direction, as does maturation of function, once an organ or system is formed.

FIGURE 5.7 The human embryo at four weeks of development.

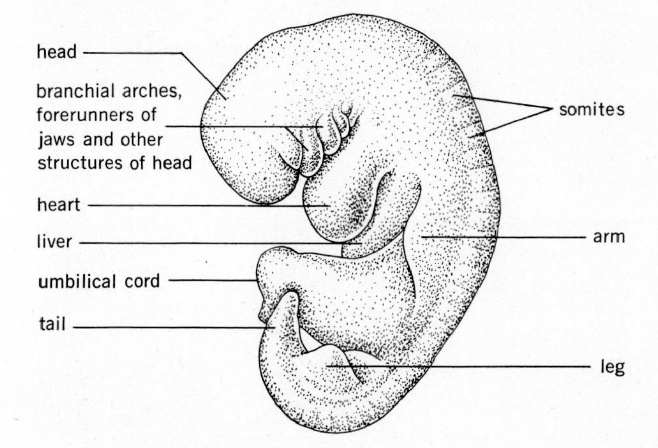

TABLE 5.2 Reference table of correlated human development (from Arey)

Age in Weeks	Size (C R) in Mm	Body form	Mouth	Pharynx and derivatives	Digestive tube and glands	Respiratory system	Coelom and mesenteries
2.5	1.5	Embryonic disc flat. Primitive streak prominent. Neural groove indicated.	–	–	Gut not distinct from yolk sac.	–	Extra-embryonic coelom present. Embryonic coelom about to appear.
3.5	2.5	Neural groove deepens and closes (except ends). Somites 1 – 16± present. Cylindrical body constricting from yolk sac. Branchial arches 1 and 2 indicated.	Mandibular arch prominent. Stomodeum a definite pit. Oral membrane ruptures.	Pharynx broad and flat. Pharyngeal pouches forming. Thyroid indicated.	Fore- and hind-gut present. Yolk sac broadly attached at mid-gut. Liver bud present. Cloaca and cloacal membrane present.	Respiratory primordium appearing as a groove on floor of pharynx.	Embryonic coelom a U-shaped canal, with a large pericardial cavity. Septum transversum indicated. Mesenteries forming. Mesocardium atrophying.
4	5.0	Branchial arches completed. Flexed heart prominent. Yolk stalk slender. All somites present (40). Limb buds indicated. Eye and otocyst present. Body flexed; C-shape.	Maxillary and mandibular processes prominent. Tongue primordia present. Rathke's pouch indicated.	Five pharyngeal pouches present. Pouches 1-4 have closing plates. Primary tympanic cavity indicated. Thyroid a stalked sac.	Esophagus short. Stomach spindle-shaped. Intestine a simple tube. Liver cords, ducts and gall bladder forming. Both pancreatic buds appear. Cloaca at height.	Trachea and paired lung buds become prominent. Laryngeal opening a simple slit.	Coelom still a continuous system of cavities. Dorsal mesentery a complete median curtain. Omental bursa indicated.
5	8.0	Nasal pits present. Tail prominent. Heart, liver and mesonephros protuberant. Umbilical cord organizes.	Jaws outlined. Rathke's pouch a stalked sac.	Phar. pouches gain dors. and vent. diverticula. Thyroid bilobed. Thyro-glossal duct atrophies	Tail-gut atrophies. Yolk stalk detaches. Intestine elongates into a loop. Caecum indicated.	Bronchial buds presage future lung lobes. Arytenoid swellings and epiglottis indicated.	Pleuro-pericardial and pleuro-peritoneal membranes forming. Ventral mesogastrium draws away from septum.
6	12.0	Upper jaw components prominent but separate. Lower jaw-halves fused. Head becomes dominant in size. Cervical flexure marked. External ear appearing. Limbs recognizable as such.	Lingual primordia fusing. Foramen caecum established. Labio-dental laminae appearing. Parotid and submaxillary buds indicated.	Thymic sacs, ultimobranchial sacs and solid parathyroids are conspicuous and ready to detach. Thyroid becomes solid and converts into plates.	Stomach rotating. Intestinal loop undergoes torsion. Hepatic lobes identifiable. Cloaca subdividing.	Definitive pulmonary lobes indicated. Bronchi sub-branching. Laryngeal cavity temporarily obliterated.	Pleuro-pericardial communications close. Mesentery expands as intestine forms loop.
7	17.0	Branchial arches lost. Cervical sinus obliterates. Face and neck forming. Digits indicated. Back straightens. Heart and liver determine shape of body ventrally. Tail regressing.	Lingual primordia merge into single tongue. Separate labial and dental laminae distinguishable. Jaws formed and begin to ossify. Palate folds present and separated by tongue:	Thymi elongating and losing lumina. Parathyroids become trabeculate and associate with thyroid. Ultimobranchial bodies fuse with thyroid. Thyroid becoming crescentic.	Stomach attaining final shape and position. Duodenum temporarily occluded. Intestinal loops herniate into cord. Rectum separates from bladder-urethra. Anal membrane ruptures. Dorsal and ventral pancreatic primordia fuse.	Larynx and epiglottis well outlined; orifice T-shaped. Laryngeal and tracheal cartilages foreshadowed. Conchae appearing. Primary choanae rupturing.	Pericardium extended by splitting from body wall. Mesentery expanding rapidly as intestine coils. Ligaments of liver prominent.
8	23.0	Nose flat; eyes far apart. Digits well formed. Growth of gut makes body evenly rotund. Head elevating. Fetal state attained.	Tongue muscles well differentiated. Earliest taste buds indicated. Rathke's pouch detaches from mouth. Sublingual gland appearing.	Auditory tube and tympanic cavity distinguishable. Sites of tonsil and its fossae indicated. Thymic halves unite and become solid. Thyroid follicles forming.	Small intestine coiling within cord. Intestinal villi developing. Liver very large in relative size.	Lung becoming gland-like by branching of bronchioles. Nostrils closed by epithelial plugs.	Pleuro-peritoneal communications close. Pericardium a voluminous sac. Diaphragm completed, including musculature. Diaphragm finishes its 'descent.'

Urogenital system	Vascular system	Skeletal system	Muscular system	Integumentary system	Nervous system	Sense organs	Age in weeks
Allantois present.	Blood islands appear on chorion and yolk sac. Cardiogenic plate reversing.	Head process (or notochordal plate) present.	–	Ectoderm a single layer.	Neural groove indicated.	–	2.5
All pronephric tubules formed. Pronephric duct growing caudad as a blind tube. Cloaca and cloacal membrane present.	Primitive blood cells and vessels present. Embryonic blood vessels a paired symmetrical system. Heart tubes fuse, bend S-shape and beat begins.	Mesodermal segments appearing (1 – 16 ±). Older somites begin to show sclerotomes. Notochord a cellular rod.	Mesodermal segments appearing (1 – 16 ±). Older somites show myotome plates.	–	Neural groove prominent; rapidly closing. Neural crest a continuous band.	Optic vesicle and auditory placode present. Acoustic ganglia appearing.	3.5
Pronephros degenerated. Pronephric (mesonephric) duct reaches cloaca. Mesonephric tubules differentiating rapidly. Metanephric bud pushes into secretory primordium.	Hemopoiesis on yolk sac. Paired aortae fuse. Aortic arches and cardinal veins completed. Dilated heart shows sinus, atrium, ventricle, and bulbus.	All somites present (40). Sclerotomes massed as primitive vertebrae about notochord.	All somites present (40).	–	Neural tube closed. Three primary vesicles of brain represented. Nerves and ganglia forming. Ependymal, mantle and marginal layers present.	Optic cup and lens pit forming. Auditory pit becomes closed, detached otocyst. Olfactory placodes arise and differentiate nerve cells.	4
Mesonephros reaches its caudal limit. Ureteric and pelvic primordia distinct. Genital ridge bulges.	Primitive vessels extend into head and limbs. Vitelline and umbilical veins transforming. Myocardium condensing. Cardiac septa appearing. Spleen indicated.	Condensations of mesenchyme presage many future bones.	Premuscle masses in head, trunk and limbs.	Epidermis gaining a second layer (periderm).	Five brain vesicles. Cerebral hemispheres bulging. Nerves and ganglia better represented. [Suprarenal cortex accumulating.]	Chorioid fissure prominent. Lens vesicle free. Vitreous anlage appearing. Octocyst elongates and buds endolymph duct. Olfactory pits deepen.	5
Cloaca subdividing. Pelvic anlage sprouts pole tubules. Sexless gonad and genital tubercle prominent. Müllerian duct appearing.	Hemopoiesis in liver. Aortic arches transforming. L. umbil. vein and d. venosus become important. Bulbus absorbed into right ventricle. Heart acquires its general definitive form.	First appearance of chondrification centers. Desmocranium.	Myotomes, fused into a continuous column, spread ventrad. Muscle segmentation largely lost.	Milk line present.	Three primary flexures of brain represented. Diencephalon large. Nerve plexuses present. Epiphysis recognizable. Sympathetic ganglia forming segmental masses. Meninges indicated.	Optic cup shows nervous and pigment layers. Lens vesicle thickens. Eyes set at 160° Naso-lacrimal duct. Modeling of ext., mid. and int. ear under way. Vomero-nasal organ.	6
Mesonephros at height of its differentiation. Metanephric collecting tubules begin branching. Earliest metanephric secretory tubules differentiating. Bladder-urethra separates from rectum. Urethral membrane rupturing.	Cardinal veins transforming. Inf. vena cava outlined. Atrium, ventricle and bulbus partitioned. Cardiac valves present. Stem of pulm. vein absorbed into l. atrium. Spleen anlage prominent.	Chondrification more general. Chondrocranium.	Muscles differentiating rapidly throughout body and assuming final shapes and relations.	Mammary thickening lens-shaped.	Cerebral hemispheres becoming large. Corpus striatum and thalamus prominent. Infundibulum and Rathke's pouch in contact. Chorioid plexuses appearing. Suprarenal medulla begins invading cortex.	Chorioid fissure closes, enclosing central artery. Nerve fibers invade optic stalk. Lens loses cavity by elongating lens fibers. Eyelids forming. Fibrous and vascular coats of eye indicated. Olfactory sacs open into mouth cavity.	7
Testis and ovary distinguishable as such. Müllerian ducts, nearing urogenital sinus, are ready to unite as uterovaginal primordium. Genital ligaments indicated.	Main blood vessels assume final plan. Primitive lymph sacs present. Sinus venosus absorbed into right atrium. Atrio-ventricular bundle represented.	First indications of ossification.	Definitive muscles of trunk, limbs and head well represented and fetus capable of some movement.	Mammary primordium a globular thickening.	Cerebral cortex begins to acquire typical cells. Olfactory lobes visible. Dura and pia-arachnoid distinct. Chomaffin bodies appearing.	Eyes converging rapidly. Ext., mid. and int. ear assuming final form. Taste buds indicated. External nares plugged.	8

TABLE 5.2 continued Reference table of correlated human development (from Arey)

Age in Weeks	Size (C R) in Mm	Body form	Mouth	Pharynx and derivatives	Digestive tube and glands	Respiratory system	Coelom and mesenteries
10	40.0	Head erect. Limbs nicely modeled. Nail folds indicated. Umbilical hernia reduced.	Fungiform and vallate papillae differentiating. Lips separate from jaws. Enamel organs and dental papillae forming. Palate folds fusing.	Thymic epithelium transforming into reticulum and thymic corpuscles. Ultimobranchial bodies disappear as such.	Intestines withdraw from cord and assume characteristic positions. Anal canal formed. Pancreatic alveoli present.	Nasal passages partitioned by fusion of septum and palate. Nose cartilaginous. Laryngeal cavity reopened; vocal folds appear.	Processus (saccus) vaginales forming. Intestine and its mesentery withdrawn from cord.
12	56.0	Head still dominant. Nose gains bridge. Sex readily determined by external inspection.	Filiform and foliate papillae elevating. Tooth primordia form prominent cups. Cheeks represented. Palate fusion complete.	Tonsillar crypts begin to invaginate. Thymus forming medulla and becoming increasingly lymphoid. Thyroid attains typical structure.	Muscle layers of gut represented. Pancreatic islands appearing. Bile secreted.	Conchae prominent. Nasal glands forming. Lungs acquire definitive shape.	Omentum an expansive apron partly fused with dorsal body wall. Mesenteries free but exhibit typical relations. Coelomic extension into umbilical cord obliterated.
16	112.0	Face looks 'human.' Hair of head appearing. Muscles become spontaneously active. Body outgrowing head.	Hard and soft palates differentiating. Hypophysis acquiring definitive structure.	Lymphocytes accumulate in tonsils. Pharyngeal tonsil begins development.	Gastric and intestinal glands developing. Duodenum and colon affixing to body wall. Meconium collecting.	Accessory nasal sinuses developing. Tracheal glands appear. Mesoderm still abundant between pulmonary alveoli. Elastic fibers appearing in lungs.	Greater omentum fusing with transverse mesocolon and colon. Mesoduodenum and ascending and descending mesocolon attaching to body wall.
20-40 (5-10 mo.)	160.0-350.0	Lanugo hair appears (5). Vernix caseosa collects (5). Body lean but better proportioned (6). Fetus lean, wrinkled and red; eyelids reopen (7). Testes invading scrotum (8). Fat collecting, wrinkles smoothing, body rounding (8-10).	Enamel and dentine depositing (5). Lingual tonsil forming (5). Permanent tooth primordia indicated (6-8). Milk teeth unerupted at birth.	Tonsil structurally typical (5).	Lymph nodules and muscularis mucosae of gut present (5). Ascending colon becomes recognizable (6). Appendix lags behind caecum in growth (6). Deep esophageal glands indicated (7). Plicae circulares represented (8).	Nose begins ossifying (5). Nostrils reopen (6). Cuboidal pulmonary epithelium disappearing from alveoli (6). Pulmonary branching only two-thirds completed (10). Frontal and sphenoidal sinuses still very incomplete (10).	Mesenterial attachments completed (5). Vaginal sacs passing into scrotum (7-9).

The fifth through eighth weeks

Completion of the embryo occurs during this period, and it assumes a clearly human form. The brain is enclosed, the digestive system forms, a heart is formed, and circulation of blood in vessels is established. The limbs, eyes, ears, and other features are evident (Fig. 5.8).

The ninth week to birth

This period constitutes the period of the FETUS, a creature with definite human form and all basic body systems. The fetus is, in some cases, capable of independent survival after about 26 weeks of development, a time which normally sees lung development far enough advanced to support life. Table 5.2 indicates that the fetal period is one

Urogenital system	Vascular system	Skeletal system	Muscular system	Integumentary system	Nervous system	Sense organs	Age in weeks
Kidney able to secrete. Bladder expands as sac. Genital duct opposite sex degenerating. Bulbo-urethral and vestibular glands appearing. Vaginal sacs forming.	Thoracic duct and peripheral lymphatics developed. Early lymph glands appearing. Enucleated red cells predominate in blood.	Ossification centers more common. Chondrocranium at its height.	Perineal muscles developing tardily.	Epidermis adds intermediate cells. Periderm cells prominent. Nail field indicated. Earliest hair follicles begin developing on face.	Spinal cord attains definitive internal structure.	Iris and ciliary body organizing. Eyelids fused. Lacrimal glands budding. Spiral organ begins differentiating.	10
Uterine horns absorbed. External genitalia attain distinctive features. Meson. and rete tubules complete male duct. Prostate and seminal vesicle appearing. Hollow viscera gaining muscular walls.	Blood formation beginning in bone marrow. Blood vessels acquire accessory coats.	Notochord degenerating rapidly. Ossification spreading. Some bones well outlined.	Smooth muscle layers indicated in hollow viscera.	Epidermis three-layered. Corium and subcutaneous now distinct.	Brain attains its general structural features. Cord shows cervical and lumbar enlargements. Cauda equina and filum terminale appearing. Neuroglial types begin to differentiate.	Characteristic organization of eye attained. Retina becoming layered. Nasal septum and plate fusions completed.	12
Kidney attains typical shape and plan. Testis in position for later descent into scrotum. Uterus and vagina recognizable as such. Mesonephros involuted.	Blood formation active in spleen. Heart musculature much condensed.	Most bones distinctly indicated throughout body. Joint cavities appear.	Cardiac muscle appearing in earlier weeks, now much condensed. Muscular movements in utero can be detected.	Epidermis begins adding other layers. Body hair starts developing. Sweat glands appear. First sebaceous glands differentiating.	Hemispheres conceal much of brain. Cerebral lobes delimited. Corpora quadrigemina appear. Cerebellum assumes some prominence.	Eye, ear and nose grossly approach typical appearance. General sense organs differentiating.	16
Female urogenital sinus becoming a shallow vestibule (5). Vagina regains lumen (5). Uterine glands appear (7). Scrotum solid until sacs and testes descend (7-9). Kidney tubules cease forming at birth.	Blood formation increasing in bone marrow and decreasing in liver. (5-10). Spleen acquires typical structure (7). Some fetal blood passages discontinue (10).	Carpal, tarsal and sternal bones ossify late; some after birth. Most epiphyseal centers appear after birth; many during adolescence.	Perineal muscles finish development (6).	Vernix caseosa seen (5). Epidermis cornifies (5). Nail plate begins (5). Hairs emerge (6). Mammary primordia budding (5); buds branch and hollow (8). Nail reaches finger tip (9). Lanugo hair prominent (7); sheds (10).	Commissures completed (5). Myelinization of cord begins (5). Cerebral cortex layered typically (6). Cerebral fissures and convolutions appearing rapidly (7). Myelinization of brain begins (10).	Nose and ear ossify (5). Vascular tunic of lens at height (7). Retinal layers completed and light perceptive (7). Taste sense present (8). Eyelids reopen (7-8). Mastoid cells unformed (10). Ear deaf at birth.	20-40 (5-10 mo.)

mainly of growth of the body; by 8–12 weeks, all body systems are present and need only to grow and undergo final refinement. The first 12 weeks of development are thus of critical importance in terms of establishing normal organs. Drugs used or diseases (e.g., rubella) contracted during this period may lead to malformations in the development of body organs and systems. CONGENITAL *(born with)* ANOMALIES or malformations are the result of developmental errors. They may be evident at birth or cause disorders later in life. Such malformations are the third leading cause of death in the United States between birth through four years of age, and the fourth leading cause of death between the ages of five through fourteen years. Since several organs and systems develop at the same time, a process affecting one system may result in changes in a simultaneously developing

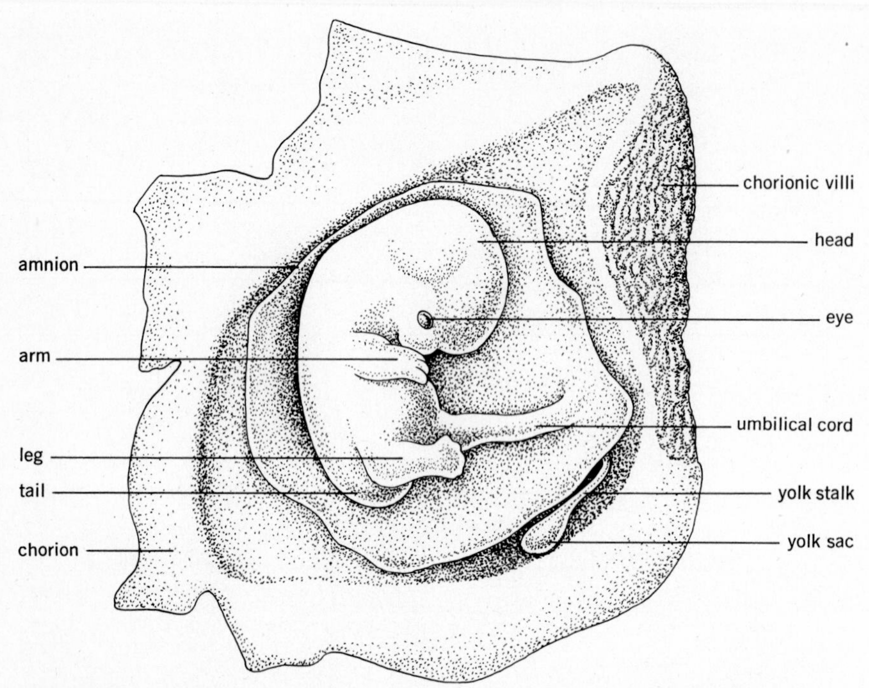

FIGURE 5.8 The human embryo at about eight weeks of development (twice natural size).

FIGURE 5.9 Changes in body proportions from before birth to adulthood.

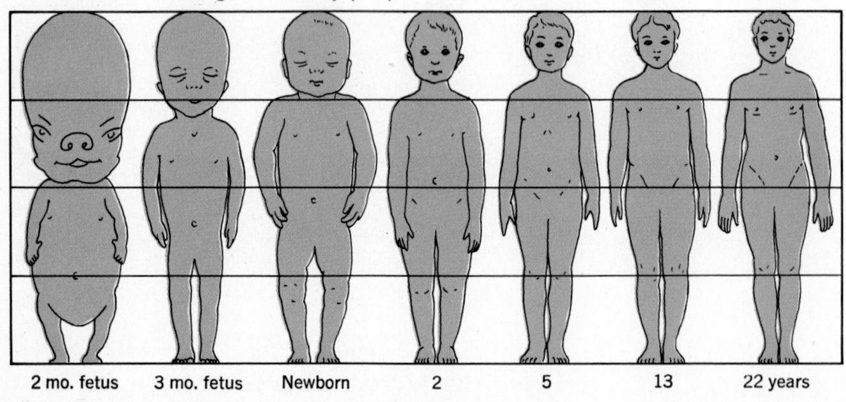

system, so that malformations are usually seen in more than one system.

During fetal growth, and after birth, body proportions change (Fig. 5.9). Thus, a newborn has a relatively larger head and shorter legs than does the child and adult. Growth in size is the most obvious change (Fig. 5.10) occurring after about two months *in utero*.

These sections have given a general view of body formation. The basic development of each system will be found at the beginning of appropriate chapters in the text.

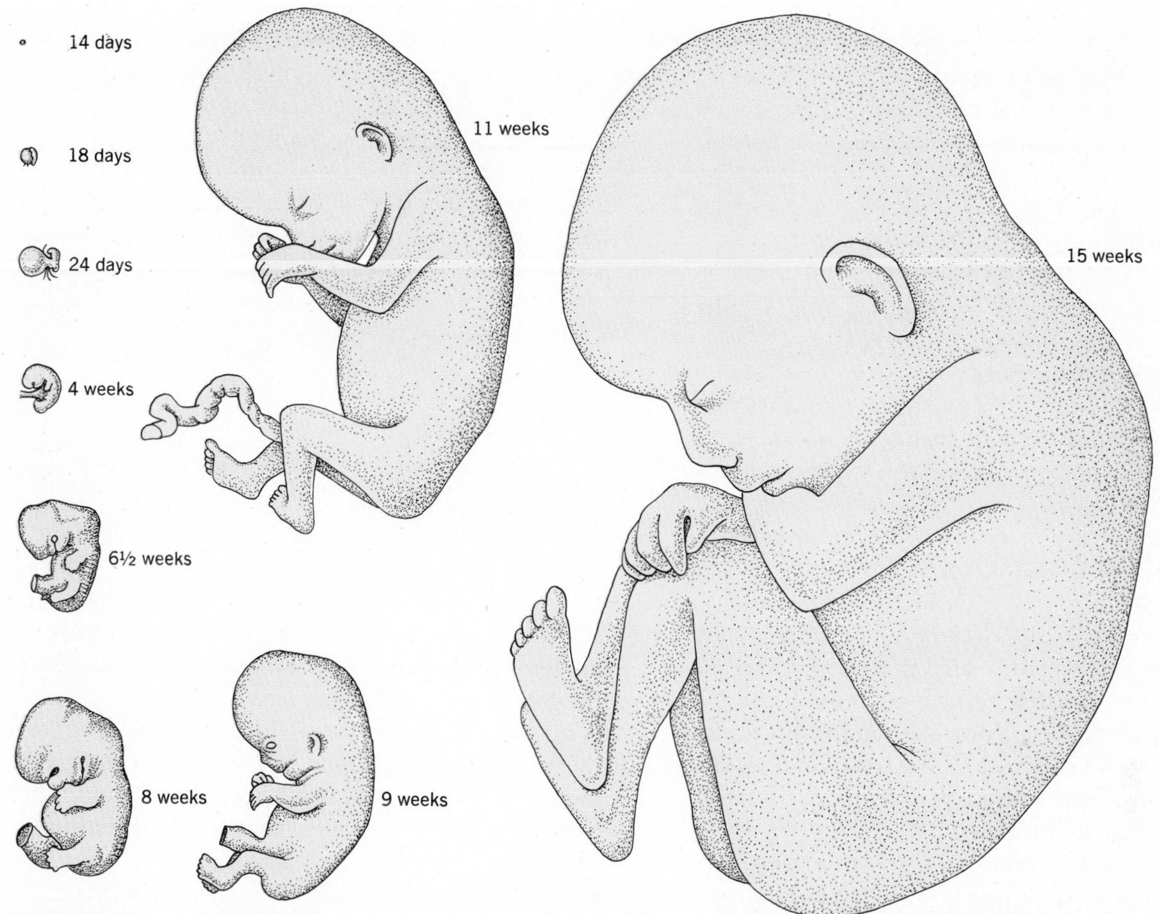

FIGURE 5.10 Changes in body size during development in the uterus (all figures natural size).

Neoplasms (cancers)

Cell division by mitosis is usually balanced to the needs of the body for cell replacement and repair. Sometimes, uncontrolled cell division may result in the development of NEOPLASMS or *cancer*.

Two major types of neoplasms are recognized. BENIGN NEOPLASMS usually grow slowly, are limited from surrounding normal tissue by connective tissue capsules, and are not considered grave threats to the body unless they are in an area that permits their undetected growth to large sizes. In such areas (e.g., brain), mechanical pressure forms the major threat the neoplasm poses as it grows. MALIGNANT NEOPLASMS grow rapidly, are not usually limited by capsules, and metastasize easily; that is, they shed cells into lymphatic or blood vessels. Such cells may be carried throughout the body to create neoplasms in many other body areas. Another method by which cancers may be spread is by "seeding" from a surgeon's gloves. While the causes of most neoplasms are not yet definitely known, many agents, including viruses, chemicals, and radiation have been shown to be *carcinogenic,* or capable of causing neoplasms. The American Cancer Society suggests seven "danger signals" that may indicate the presence of a neoplastic process in the body:

Any unusual discharge or bleeding from a body opening (e.g., anus, vagina, mouth, nose).

A sore that does not heal.

A change in bowel habits (including frequency and consistency) persisting more than three weeks.

Chronic indigestion or difficulty in swallowing.

Any change in size or color of a wart or mole.

A lump or thickening in the breast or elsewhere.

Persistent hoarseness or coughing.

Cancer is the *second leading cause* of death in the United States, and early detection utilizing awareness of the danger signals could reduce cancer deaths by one-third. Although cancers as a group cannot yet be cured, the treatments available are much more effective when begun early; thus prognosis is improved.

Prognosis in an individual with cancer depends on the degree of abnormality between normal and neoplastic cells, and on the length of time the process has proceeded before being discovered and treated. Once discovered, the neoplasm may be GRADED from I–IV according to the degree the neoplastic cells differ in organization and structure from normal cells. A grade I rating implies minimal differences from normal; a grade IV rating indicates maximal differentiation. The neoplasm may also be described as to the extent of its spread by the term STAGING. Four stages (I-IV) are employed, with stage I indicating limitation of the neoplasm to the site of origin, and stage IV indicating widespread metastasis. In neoplasms of the colon, the lesion may also be TYPED (A, B, C) according to the depth of penetration of the wall of the organ. Type A indicates involvement of mucosal and muscular layers; in type B the outer or serosal coat is involved; in type C the malignancy has spread to the lymph nodes in the area.

Since a neoplasm requires nourishment as do any cells, a neoplasm sustains itself at the expense of normal cells and the host in general. Thus emaciation, loss of weight, anemia, and disfigurement may result. Other systemic effects of neoplasia may be divided into four major groups:

Dermatological (skin)
Vascular
Hormonal
Neuromuscular

Certain carcinomas of the stomach are associated with the development of EXCESSIVE SKIN PIGMENTATION, particularly on the neck, groin, axillae, and around the nipples and umbilicus. It has been suggested that the pigmentation is the result of an immune response to breakdown products of the tumor.

INFLAMMATION OF VEINS (phlebitis) may suggest the presence of a tumor in the lung or pancreas long before the tumor itself becomes evident. The inflammation commonly results in thrombosis (plugging of a vessel by a blood clot).

STIMULATION OF ENDOCRINE SECRETION may result from the production by nonendocrine tumors of chemicals having hormonelike actions. Bronchial carcinomas secrete a substance having the effect of stimulating the secretion of the adrenal cortex. This results in the development of features mimicking the effects of a tumor of the cortex.

DEGENERATION OF NERVE CELLS in the cerebrum and/or cerebellum with loss of muscular control is seen in association with carcinoma. No explanation is available as to the mechanism for the production of the symptoms.

Treatment may involve surgical intervention, the use of chemotherapy, or ionizing radiation. The aim of treatment is removal of the neoplasm or inhibiting its growth, while promoting the patient's comfort and emotional well-being.

Summary

1. Cell reproduction produces new cells for maintenance of cell lines, repair, and reproduction of the species.
   a. Mitosis perpetuates a cell line, causes growth in size, and is used in repair of injury. It results in two cells that are genetically the same as the parent cell. Mitosis occurs in five stages:
   (1) Interphase
   (2) Prophase
   (3) Metaphase
   (4) Anaphase
   (5) Telophase
   b. Meiosis occurs in ovaries and testes, producing four sex cells that are haploid (one-half normal chromosome number) and different in genetic makeup from the parent cell. Meiosis involves two divisions and several recognized stages in the first meiotic prophase.
2. Basic human development may be regarded as occurring in weekly subdivisions.
   a. The first week of life is concerned with fertilization (union of egg and sperm), cleavage (mitotic division), and formation of a blastocyst.
   b. The second week is concerned with implantation (burrowing of blastocyst into uterus wall) and formation of a two-layered embryo composed of ectoderm and endoderm.
   c. Mesoderm formation occurs during the third week, and differentiation of cells into lines that will form specific tissues and organs takes place.
   d. The nervous system and somites begin formation during the fourth week.
   e. The laying down of the body systems and assumption of human form (fetus) occurs during the fifth to eighth weeks.
   f. Growth, and refinement of the body systems, occurs from the ninth week to birth. Body proportions and size change.
3. Neoplasms (cancers) are the result of uncontrolled mitotic cell division. They may be benign or malignant. Seven "danger signals" may suggest the presence of a neoplastic process in the body.
4. Neoplasms may spread via blood, lymph, or "seeding," and are thought to be due to carcinogenic agents and hereditary factors. Neoplasms are classified according to structural alterations, degree of spread, and tissue layers involved in an organ. They commonly produce nutritional, skin, vascular, hormonal, and neuromuscular symptoms, some before the lesion itself becomes evident. Treatment is aimed at removal, or inhibition of growth of the neoplasm.

Epithelia, Connective Tissues, and the Skin

Chapter 1 indicated that cells similar in structure and function, together with their associated intercellular material, form TISSUES. Four tissue groups are recognized: *epithelial, connective* (including blood and blood-forming tissues), *muscular,* and *nervous.* This chapter concentrates its discussion on the epithelial and connective tissue groups, and discusses the skin as the most obvious example of the combination of an epithelium and connective tissue.

Epithelial tissues

General characteristics

Epithelia COVER AND LINE actual or potential free surfaces of the body. They are therefore in direct contact with both the external and internal environments of the body, and are important in aiding maintenance of the homeostasis of the internal environment. The free surfaces of the skin, digestive system, respiratory system, excretory system, reproductive systems, and body cavities have surfaces that normally do not have another permanent tissue in direct contact with them, and these represent actual free surfaces. In a blood vessel, which is normally filled with blood, the free surface becomes apparent when the blood is drained from the vessel. Such a surface is an example of a potential free surface. Since epithelia exist on surfaces and form barriers between the environment and the underlying tissues, they must form as dense a covering as possible. They consist, therefore, of very CLOSELY PACKED CELLS with minimal amounts of material between the cells *(intercellular material).* Epithelia have NO BLOOD VESSELS of their own; the cells are nourished by diffusion from vessels in the underlying connective tissue. It would be impractical to have blood vessels so close to the surface that they are exposed constantly to danger of damage. Epithelia are commonly provided with NERVES, which pass into and between the cells. Epithelia usually rest on a BASEMENT MEMBRANE, a product of the epithelium and the underlying connective tissue. The membrane aids in affixing the epithelium to the connective tissue. Epithelia may originate from all three embryonic germ layers: ectoderm, endoderm and mesoderm.

The terms ENDOTHELIUM and MESOTHELIUM are applied to the mesodermally derived linings of blood vessels and true body cavities, respectively. One type of epithelium may develop from more than one germ layer, or one layer may give rise to several types.

Types of epithelia

Most epithelia may be named according to three general criteria: the NUMBER OF LAYERS OF CELLS in the epithelium, the SHAPE of the surface cells in the epithelium, and the presence of certain MODIFICATIONS on the free surfaces of the cells. The epithelia which cannot be named by applying these criteria are sometimes designated as *aberrant epithelia.*

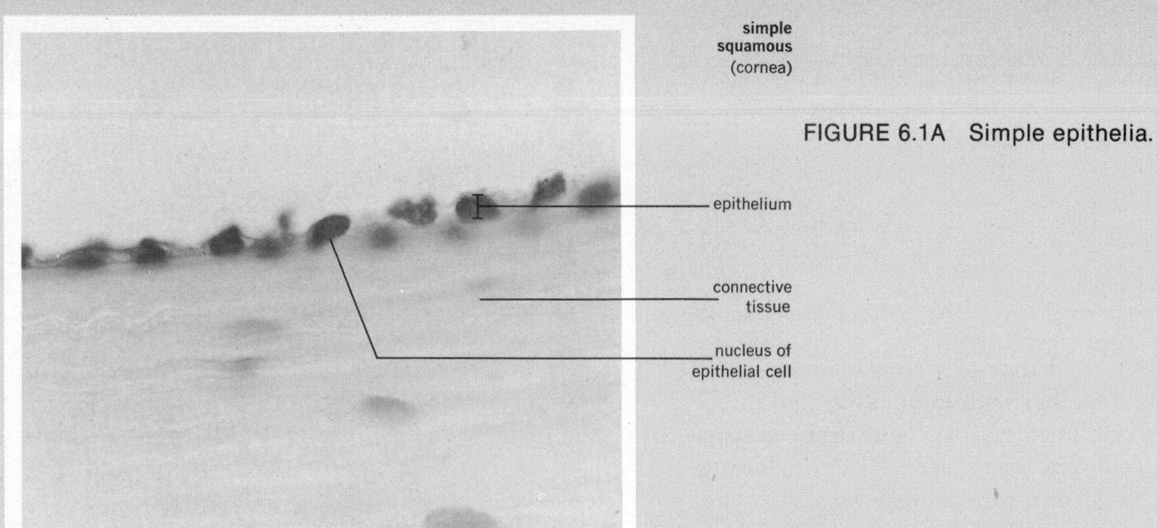
simple
squamous
(cornea)
epithelium
connective
tissue
nucleus of
epithelial cell

FIGURE 6.1A Simple epithelia.

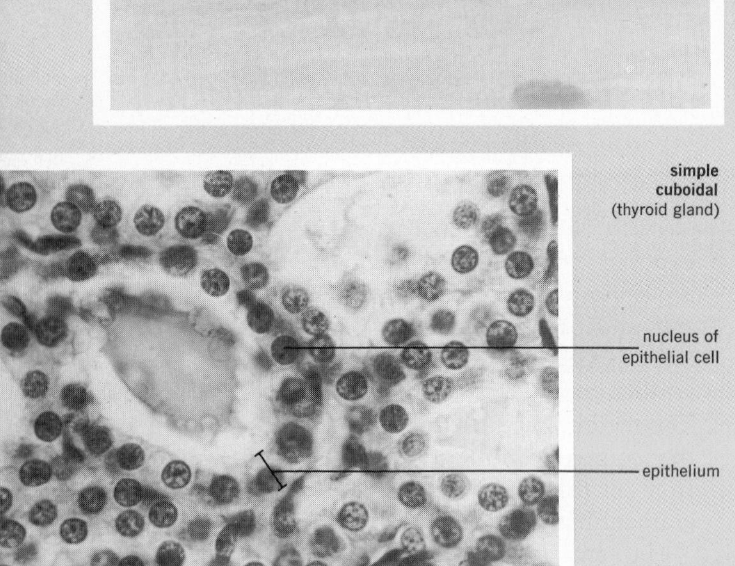
simple
cuboidal
(thyroid gland)
nucleus of
epithelial cell
epithelium

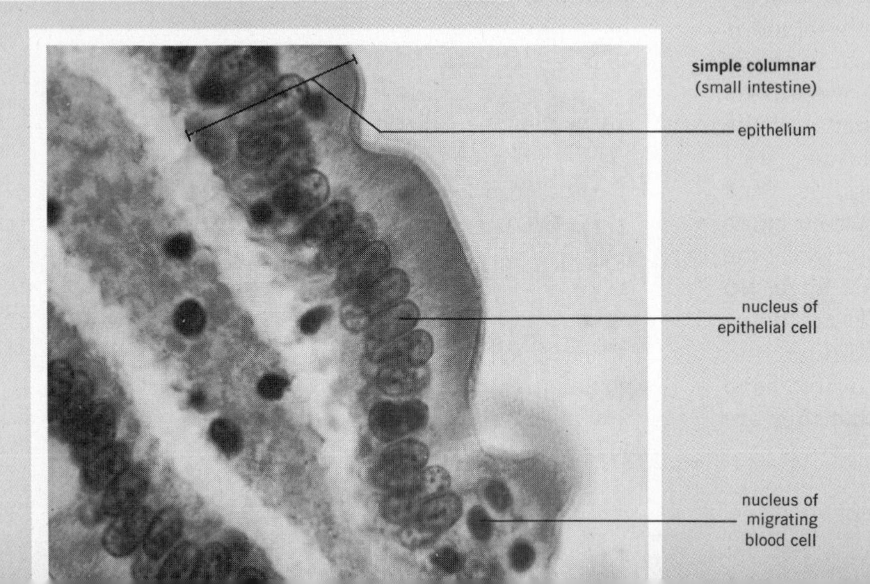
simple columnar
(small intestine)
epithelium
nucleus of
epithelial cell
nucleus of
migrating
blood cell

FIGURE 6.1B Stratified epithelia.

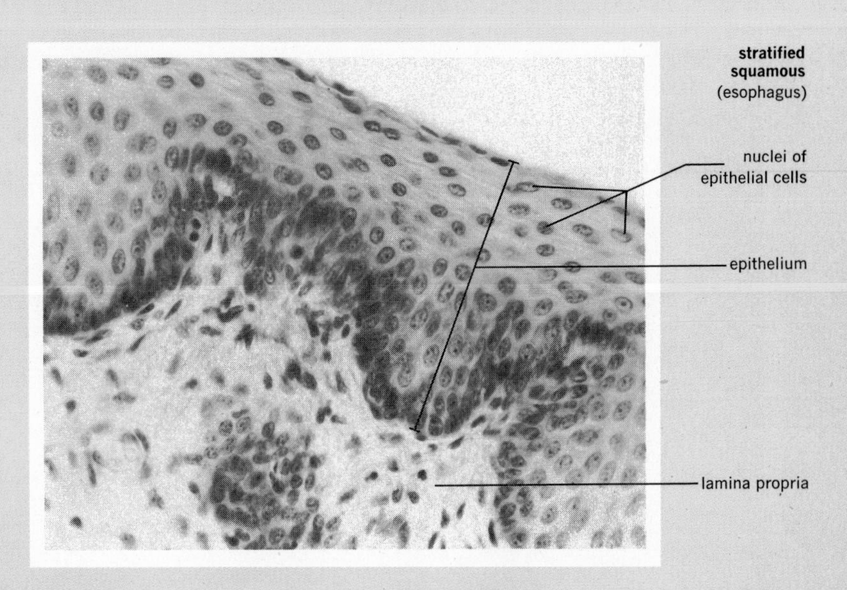

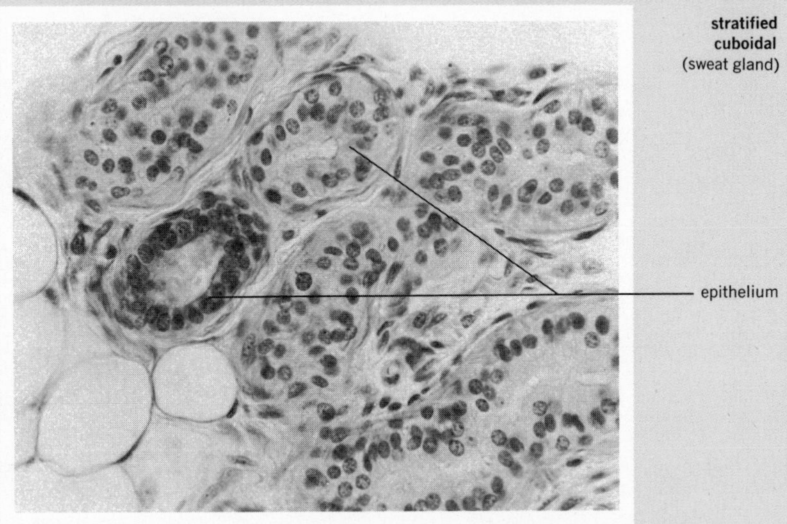

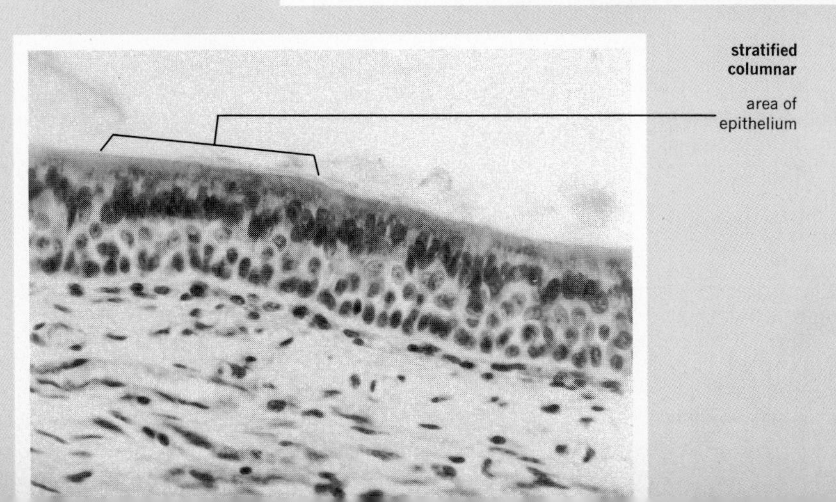

FIGURE 6.2 "Aberrant" epithelia

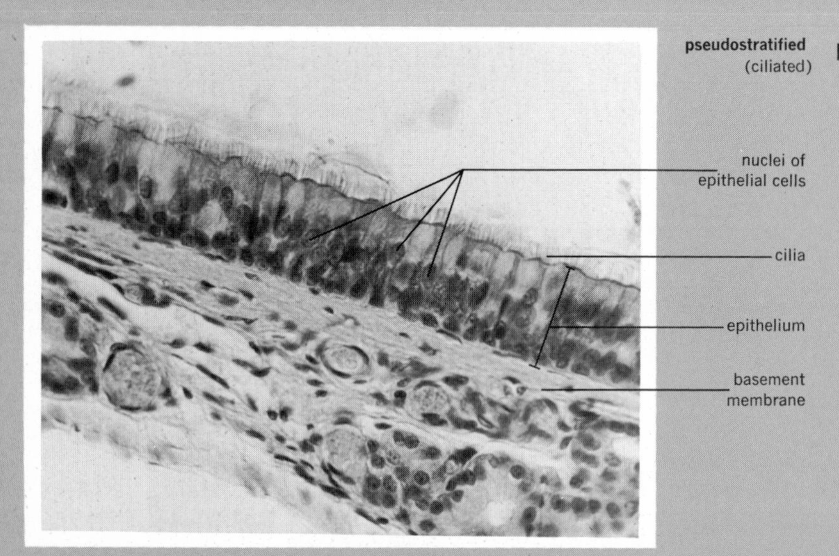

pseudostratified (ciliated)

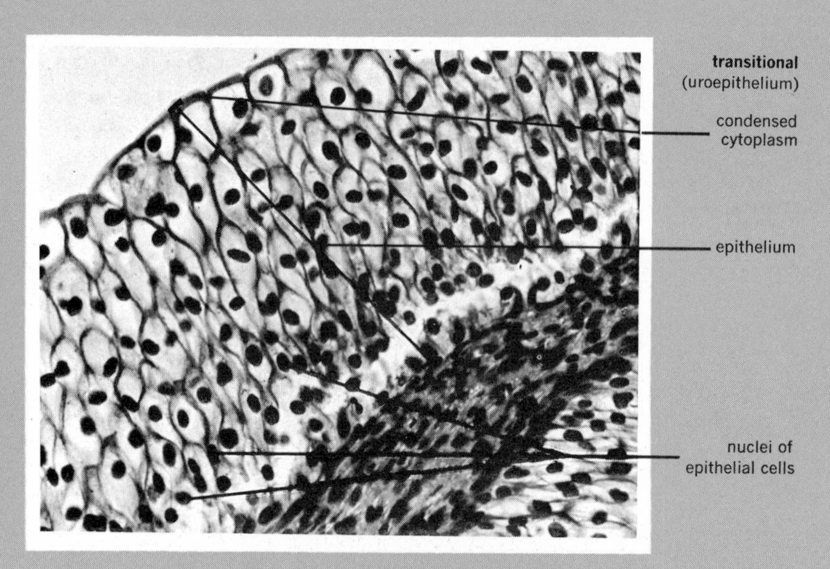

transitional (uroepithelium)

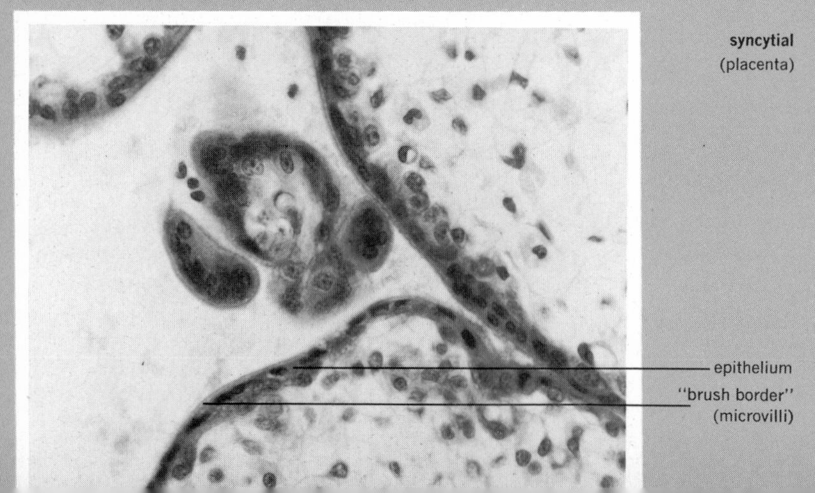

syncytial (placenta)

NUMBER OF LAYERS OF CELLS. A SIMPLE epithelium has only one layer of cells, all of which reach both the free surface and the basement membrane. The nuclei in the cells of a simple epithelium are all at the same level. A STRATIFIED epithelium consists of two or more layers of cells, of which only the top layer reaches the free surface and only the lowest layer reaches the basement membrane. In such a tissue, nuclei may be seen at many different levels, and the deeper-lying cells tend to become more round regardless of the shape of the surface layers. A stratified epithelium is therefore named according to the shape of the topmost layer or two of cells.

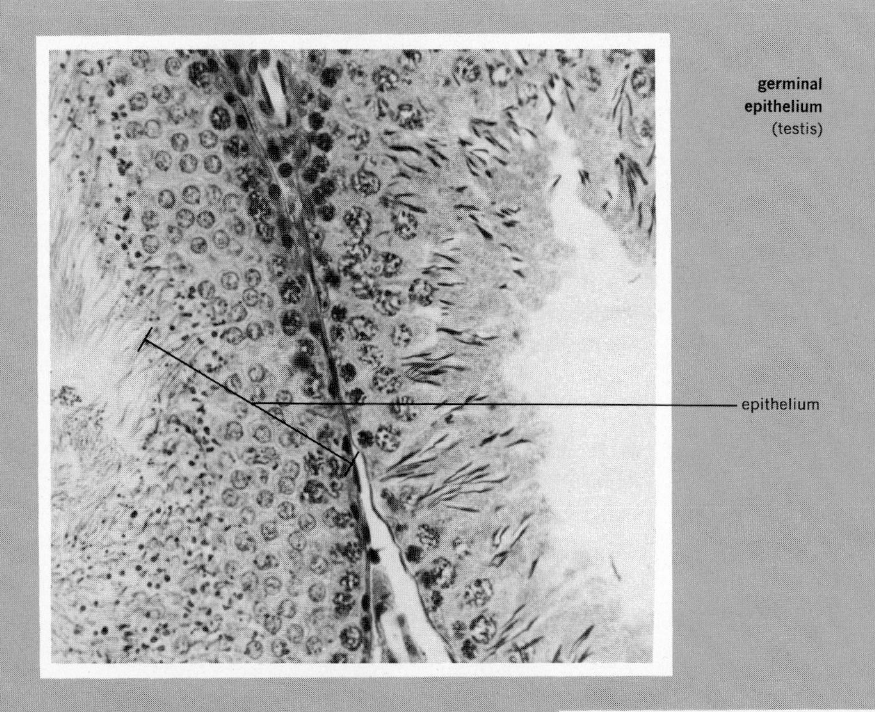

germinal epithelium (testis)

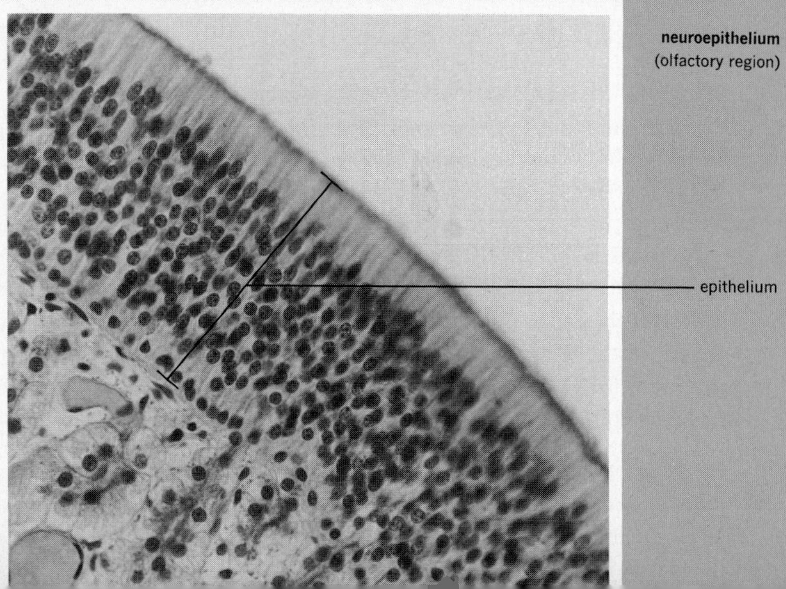

neuroepithelium (olfactory region)

SHAPE OF CELLS IN THE EPITHELIUM. There are three basic shapes in which epithelial cells may occur:

Squamous. The cells are flattened and length and width greatly exceeds height. Nuclei are flattened parallel to the surface and may form a bulge at the free surface.

Cuboidal. The cells are approximately equal in length, width, and height, and may be said to be *isodiametric* (equal). Nuclei are round, or nearly so, and the cell will appear square (if on a flat surface) or pyramidal (if lining a tubular structure).

Columnar. The cells are tall, and height exceeds length and width. The nuclei are compressed in a direction perpendicular to the free surface.

Intermediate forms between the three basic shapes are common. A rule of thumb that may be helpful when in doubt about the shape of any given cell is to look at the nucleus and determine its orientation and shape.

By combining these first two criteria, six different types of epithelia may be defined (Fig. 6.1 A, B):

SIMPLE SQUAMOUS. One row of flattened cells, nuclei flattened parallel to the surface. Simple squamous epithelium lines the alveoli of the lungs, glomerular capsule, and eardrum. The thinness of the tissue permits easy passage of materials through it.

Three tissues having an appearance identical to that of simple squamous epithelium, but differing in origin and potency are:

Endothelium. Mesodermal in origin, endothelium lines the internal surfaces of the heart, blood vessels, and lymph vessels.

Mesothelium. Mesodermal in origin, mesothelium lines body cavities not opening on the body surface. Specifically, it lines the abdominal, pleural, and pericardial cavities.

Mesenchymal epithelium. Originating from mesoderm, this epithelium lines the subdural and subarachnoid spaces of the central nervous system and the chambers of the eye and inner ear.

SIMPLE CUBOIDAL. One row of isodiametric cells, with round nuclei. It covers the ovary, lines kidney tubules and the smaller ducts of many glands. It is very active in secretion and absorption.

SIMPLE COLUMNAR. One row of tall cells, with nuclei elongated perpendicularly to the surface. The linings of the stomach, small intestine, gall bladder, and large ducts of many externally secreting glands are of this type of epithelium. It is also active in absorptive processes.

STRATIFIED SQUAMOUS. Several rows of cells, top layer(s) flattened. It is found in the epidermis of the skin, and lining the mouth, esophagus, anal canal, vagina, and cornea. It may be cornified, and is the toughest of all epithelia.

STRATIFIED CUBOIDAL. Several rows of cells, top layer cuboidal. A rare tissue, it is found in the secretory portions of sweat glands.

STRATIFIED COLUMNAR. Several rows of cells, top layer columnar. This tissue is found in the larynx, as a transitional form between stratified squamous and pseudostratified.

ABERRANT EPITHELIA. (Fig. 6.2). The following five types of epithelium cannot be named according to the criteria given above.

Pseudostratified epithelium appears to be stratified and has nuclei at many different levels. However, all cells may be shown to reach the basement membrane, but not all reach the surface. The epithelium is therefore simple. It is the typical epithelium of the respiratory system and portions of the male reproductive system.

Transitional or *uroepithelium* is a stratified tissue, but the top layer of cells is not uniform in shape. The tissue has the ability to stretch greatly under tension, and the cells may thus show many "transitional" forms. It lines the renal pelvis, ureters, and urinary bladder.

Syncytial epithelium lacks membranes between cells and is one continuous multinucleated mass. It covers the villi of the placenta.

Germinal epithelium is found in the testis and is a stratified tissue containing specialized cells leading to the formation of spermatozoa.

Neuroepithelium is found in the retina, olfactory area, taste buds, and inner ear, and is highly specialized for sensory perception. The epithelium is composed of nerve cells and supporting elements. Such epithelium is usually named according to its function, for example, "olfactory epithelium."

Surface modifications of epithelia

Since one side of the epithelium has a free surface, special structures or modifications of that surface may be present. Such modifications help to enable the epithelium to carry out its particular functions more efficiently.

CILIA AND FLAGELLAE. Cilia (Figs. 6.3 and 6.4) are multiple tiny hairlike projections arising from small granules located in the free ends of epithelial cells. According to some investigators, cilia are

FIGURE 6.3 Cross sections of cilia (c) from a mouse tracheal epithelial cell. Note the 9 + 2 arrangement of paired microtubules. (Courtesy Norton B. Gilula. The Rockefeller University.)

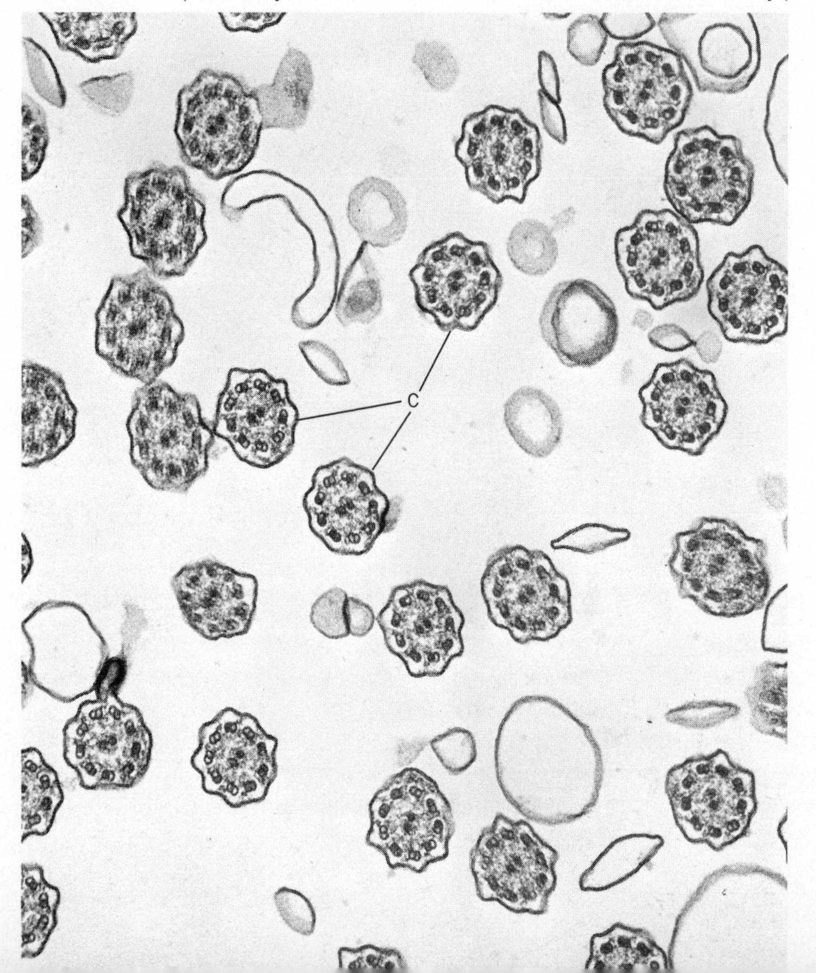

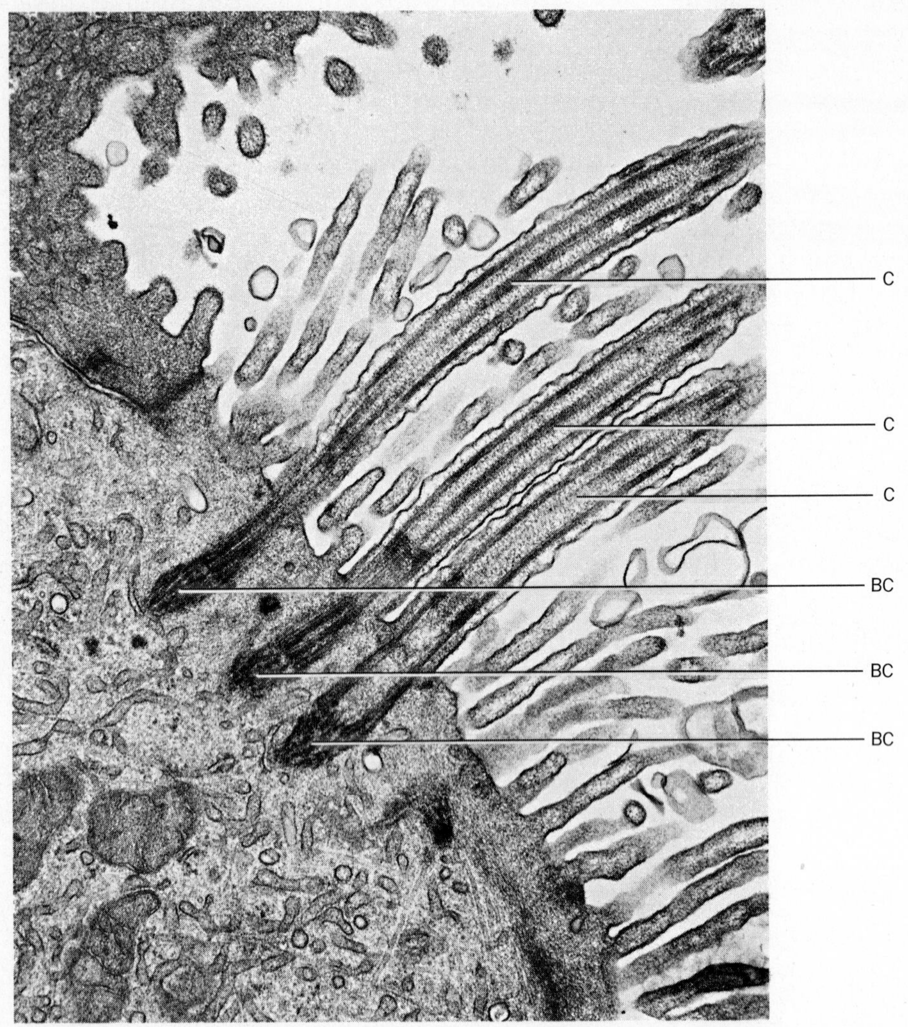

FIGURE 6.4 Longitudinal sections of cilia from a mouse tracheal epithelial cell. BC = basal corpuscle, C = cilium. (Courtesy Norton B. Gilula, The Rockefeller University.)

derived from centrioles. Flagellae are single, extremely long processes conferring independent movement to the cell possessing them, as in sperm. Both cilia and flagellae have a distinctive ultrastructure of smaller fibrils (as described in Chapter 2). Cilia may be motile *(kinocilia)* and aid the movement of materials over the cell surface. Motile cilia are the only surface modification normally included as part of the name of an epithelium. The adjective *ciliated* is used, and we may speak, for example, of a simple columnar ciliated epithelium.

MICROVILLI. Microvilli (Figs. 6.5 and 6.6) are tiny fingerlike projections of the cell cytoplasm at the free surface. They are commonly found on

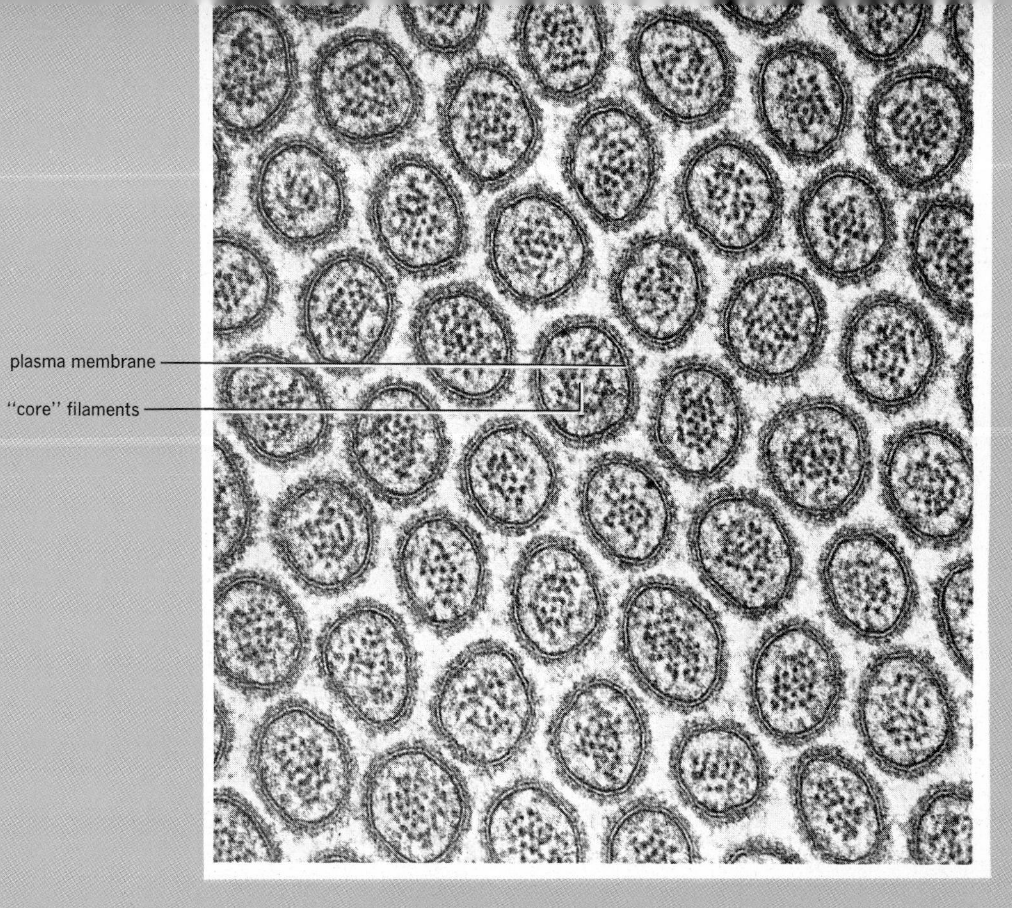

FIGURE 6.5 Cross sections of microvilli from the small intestine epithelial cells of a mouse (× 102,000). Note the trilaminar cell membranes. (Courtesy Norton B. Gilula, The Rockefeller University.)

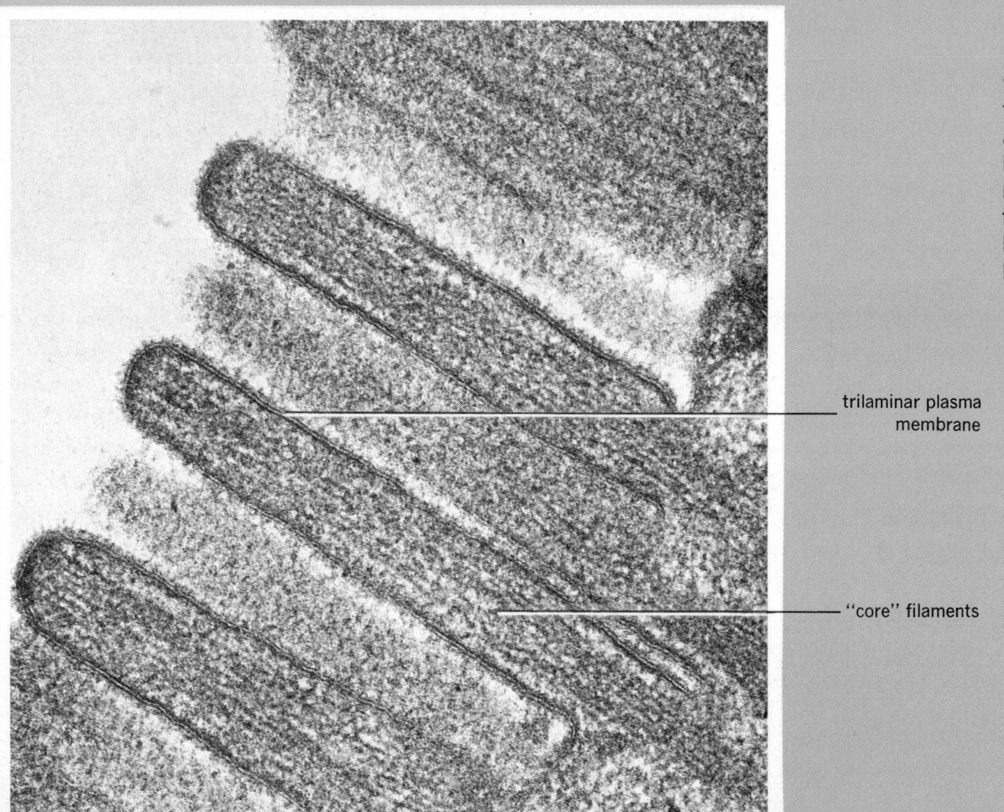

FIGURE 6.6 Longitudinal sections of microvilli from the small intestine epithelial cells of a mouse (× 82,000). Note trilaminar cell membranes and "core filaments." (Courtesy Norton B. Gilula, The Rockefeller University.)

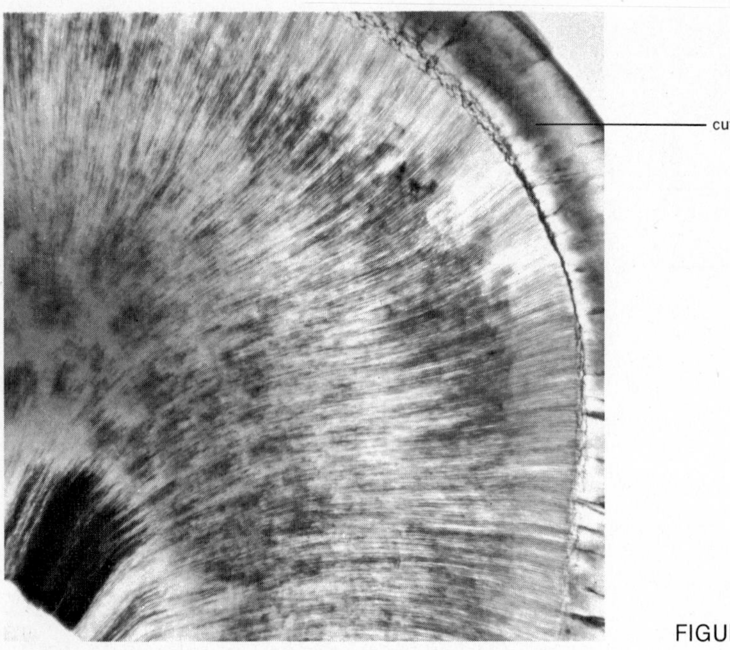

FIGURE 6.7 An example of a cuticle (tooth enamel).

those epithelia active in absorptive and secretory processes. They serve to increase the surface area through which such transport may occur. On the intestinal epithelium, the microvilli are short and very regular in arrangement and form the "striated border" of that organ. In the kidney tubules, gall bladder, placenta, and lining of the abdominal cavity, the microvilli are much longer, forming "brush borders." *Stereocilia* are microvilli that appear to be attached at their free ends. They are found in the epididymis and serve to channel nutrients to the sperm stored in that organ.

The cells of transitional epithelium have a thicker, denser layer of cytoplasm at the free surface of the top layer of cells, designated by some histologists as *condensed cytoplasm.* The layer appears, under the electron microscope, to be composed of many layers of membranes. It is believed to help protect the cells from the acidic urine and to provide the surface necessary when the epithelium is stretched.

CUTICLES. Layers of material that are produced at the free surface by cellular activity and that may be separated from that surface without damage to the cells are called cuticles (Fig. 6.7). True cuticles are rare in the body. The enamel of the tooth and the lens capsule are examples. The "cuticle" of the fingernails is merely the surface layers of the skin that are pulled out as the nail grows.

MODIFICATIONS OF THE INTERCELLULAR SURFACE. Various devices (Fig. 6.8) may be employed to hold cells together and to seal the free surface between cells. The ZONULA OCCLUDENS (the "terminal bar" of older literature) occurs at the free surface and involves an apparent fusion of membranes. In certain cells this type of junction forms what is called a "tight junction." The ZONULA ADHERENS is found between cells and involves a separation of membranes with filaments obvious in the adjacent cytoplasm. The MACULA ADHERENS *(desmosome)* also involves separated membranes,

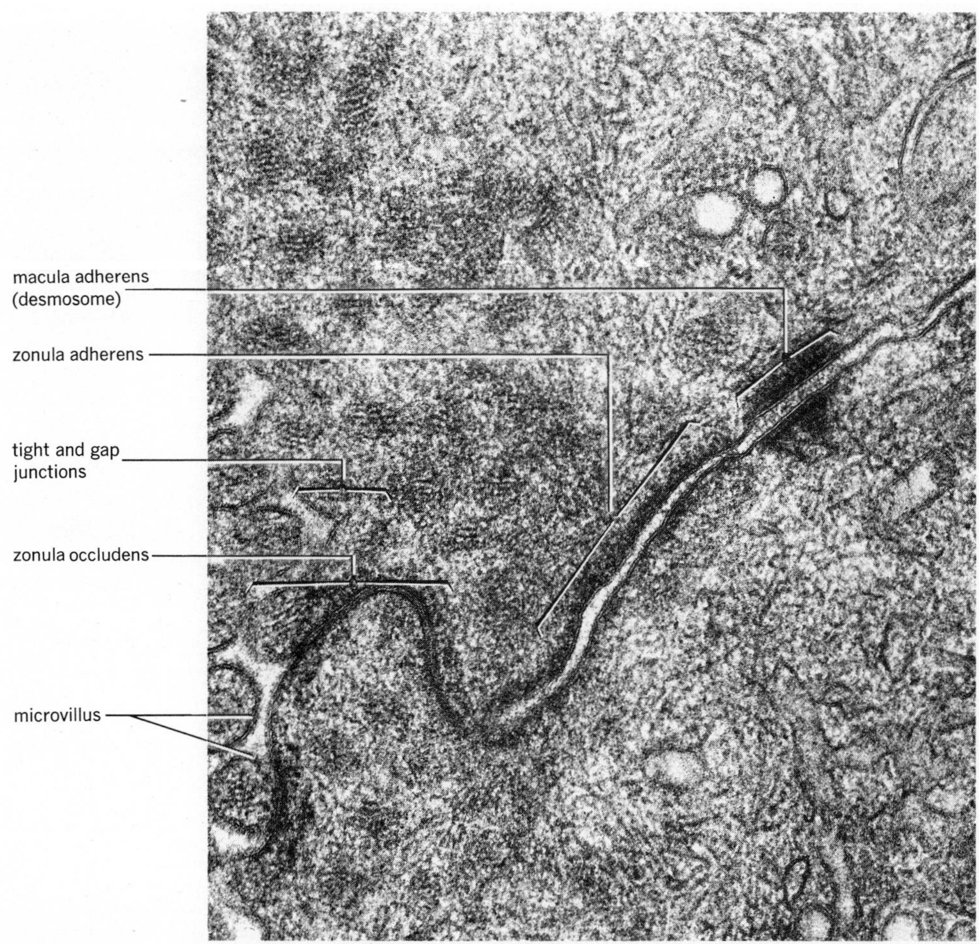

FIGURE 6.8 Junctional complexes from the small intestine of a mouse (× 65,000). (Courtesy Norton B. Gilula, The Rockefeller University.)

but with a dense plaque of cytoplasmic substance adjacent to the area of attachment.

Table 6.1 summarizes some basic facts about epithelia.

Epithelial membranes

The combination of an epithelium and an underlying connective tissue layer forms an epithelial membrane. Two types of such membranes are recognized in the body.

SEROUS MEMBRANES. Serous membranes (Fig. 6.9) form the outer coat of visceral organs, line the true body cavities, and form omenta and mesenteries. Omenta lie between certain organs in the abdominal cavity; mesenteries suspend many organs in the abdominal cavity. The epithelial component is a simple squamous mesothelium applied to a submesothelial connective tissue layer. Serous membranes are moistened by secretion of a watery fluid. Because they are always moist, they allow the free movement of materials over their surfaces or the movement

TABLE 6.1 Summary of Epithelium

Epithelial type	Characteristics	Examples of locations	Surface modification commonly present	Functions and comments
Simple squamous	One row flat cells, nuclei flattened and parallel to surface	Glomerular capsule, endothelium, mesothelium, Henle's loop of kidney	Microvilli (on mesothelium)	Exchange, because of thinness
Simple cuboidal	One row isodiametric cells, nuclei rounded	Kidney tubules, thyroid, surface of ovary	Microvilli (in kidney)	Secretion and absorption
Simple columnar	One row tall cells nuclei elongated perpendicular to surface	Stomach, small and large intestines, gall bladder, ducts, bronchioles, uterus	Microvilli (in gut) Cilia (in respiratory system)	Secretion and absorption (gut) movement of substances across surface if ciliated
Stratified squamous	Several layers of cells, top layer flattened	Skin (fully cornified), vagina (partially cornified), mouth, esophagus (slight or no cornification)	None	Protection, since cells are easily removed from the surface and replaced rapidly from below
Stratified cuboidal	Several layers of cells, top layer cuboidal	Sweat glands	None	Secretory

between an organ and the lining of its cavity. In the chest cavity, this movement is of particular importance because of the change in shape and size of the heart and lungs as they work. The serous membranes forming the omenta and mesenteries may become a reservoir for storage of fat.

MUCOUS MEMBRANES. Mucous membranes (Fig. 6.10) form the linings of the hollow body organs and systems that connect to the external body surface. They include the membranes lining the cavities of the mouth, digestive tract, and the reproductive, respiratory, and urinary systems. The epithelial type may vary, but the epithelium is usually moistened by mucus. The underlying connective tissue is known as the lamina *(tunica)* propria.

Epithelial type	Characteristics	Examples of locations	Surface modification commonly present	Functions and comments
Stratified columnar	Several layers of cells, top layer columnar	Larynx, upper pharynx	Cilia	Transition form between stratified squamous and pseudo-stratified
Pseudo-stratified	All cells reach basement membrane, not all reach surface. Nuclei at many levels	Nasal cavity, trachea, bronchi, male and female reproductive systems	Cilia	Movement of materials across surface
Transitional	Stratified, top layer not uniform in shape	Kidney pelvis, ureter, bladder	Condensed cytoplasm	Stretches and protects cells from acidic urine
Syncytial	Simple, no membranes between cells	Placental villus	Microvilli	Secretion and protects
Germinal	Several layers showing stages of sperm formation	Tubules of testis	None	Produces sperm
Neuro-epithelium	Nerve cells form part of epithelium	Taste buds, olfactory area, retina, cochlea	None	Sensory, usually as receptors of stimuli

Glands

Secretion forms one of the primary activities of epithelia. Glands are specialized epithelial cells or groups of epithelial cells producing substances differing in nature from the fluids otherwise available in the body. To produce a secretion different from the blood or tissue fluid, the cells must engage in metabolic activity over and above that required for their own maintenance. In short, production of a secretion involves the utilization of synthetic processes and active methods of transport.

The cells of a gland may remain in the epithelium *(intraepithelial)* or push into the underlying connective tissue, the latter forming a support for the gland. If a connection is retained to the epithelial surface, the gland is said to be an ex-

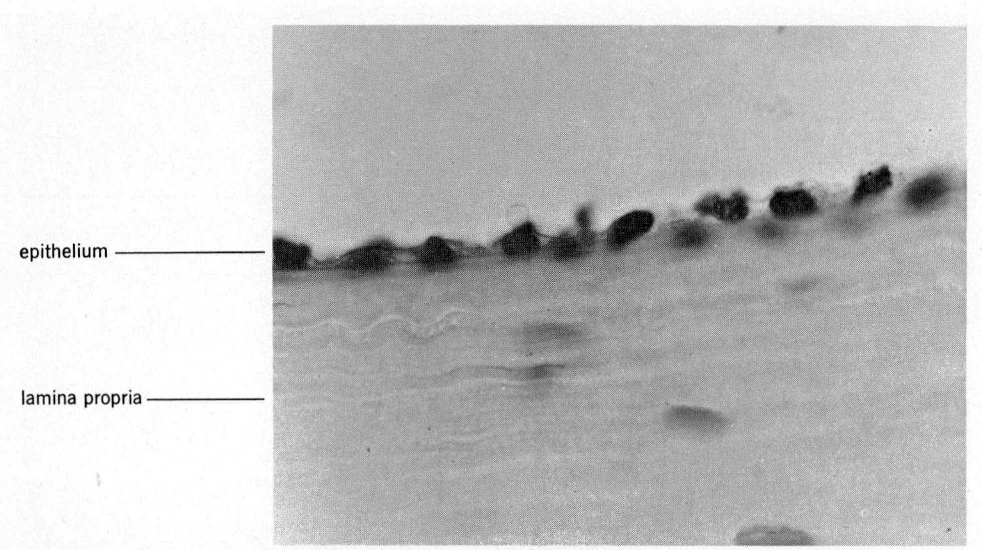

FIGURE 6.9 A serous membrane, composed of simple squamous epithelium and underlying connective tissue.

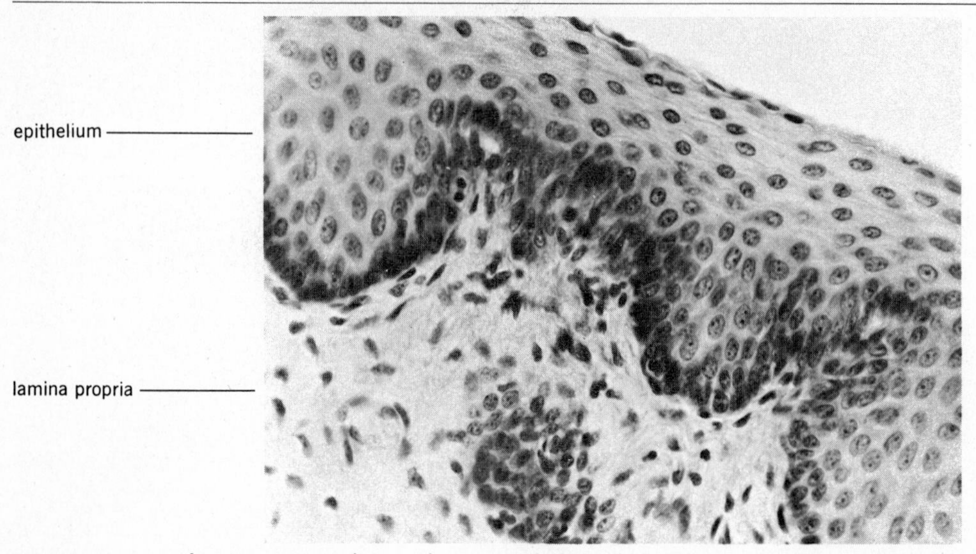

FIGURE 6.10 A mucous membrane, here consisting of stratified squamous epithelium and underlying connective tissue.

ternally secreting or EXOCRINE GLAND, and is possessed of *ducts* to convey the secretion to the exterior. If the surface connection is lost, the gland is *ductless,* and empties its secretion into the only other available channel, the blood stream. Such a gland is an internally secreting or ENDOCRINE GLAND. The structure of the endocrine glands is discussed in Chapter 31.

Exocrine glands may consist of only one cell *(unicellular)* or of many cells *(multicellular).* Mucus-secreting goblet cells of the intestine and respiratory passageways are examples of unicellular

TABLE 6.2 Summary of exocrine glands

Type of gland*	Diagram*	Characteristics	Examples
UNICELLULAR		One celled, mucus secreting	Goblet cells of resp. and diges. system
MULTICELLULAR Simple tubular	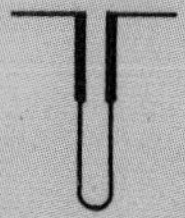	One duct—secretory portion straight tube	Crypts of Lieberkuhn of intestines
Simple branched tubular	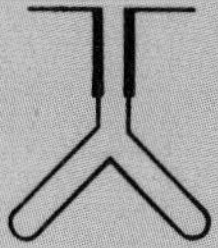	One duct—secretory portion branched tube	Gastric glands, uterine glands
Simple coiled tubular		One duct—secretory portion coiled	Sweat glands
Simple alveolar		One duct—secretory portion saclike	Sebaceous glands
Simple branched alveolar	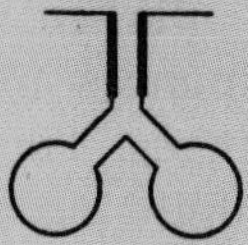	One duct—secretory portion branched and saclike	Sebaceous glands
Compound tubular	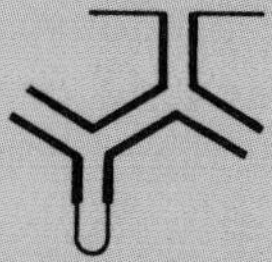	System of ducts—secretory portion tubular	Testes, liver
Compound alveolar	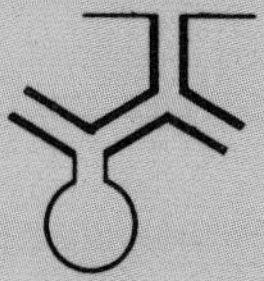	System of ducts—secretory portion alveolar	Pancreas, salivary glands, mammary
Compound tubulo-alveolar	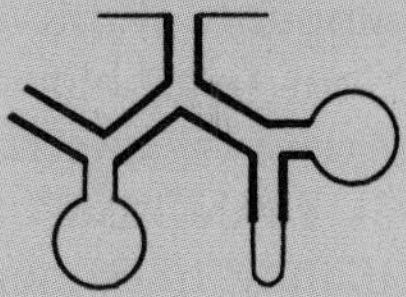	System of ducts—secretory portion both tubular and alveolar	Salivary glands

*Simple implies one duct in the gland; compound implies a series of branching ducts in the gland.

TABLE 6.3 Manner of production of secretion by exocrine glands

Type of gland	Examples
Merocrine—synthesized product independent of basic cell structure	Pancreas, salivary glands
Apocrine—some portion of the cell issued as part of the secretion	Mammary gland
Holocrine—product of gland is a cell, or cell as a whole is shed in the secretion	Testis, sebaceous glands

glands. All other glands are multicellular. Multicellular exocrine glands may be further characterized by the number of ducts conveying secretions from the secretory portion of the gland and by the shape of the secretory portion.

A SIMPLE GLAND has only one duct in the entire gland. A COMPOUND GLAND has many ducts within it, although only one may empty on the epithelial surface. The secretory portion may be *tubular* or *alveolar* (saccular) in shape. Combining these criteria leads to the designation of a variety of glands. Further descriptive terms, such as *branched* or *coiled*, may be applied to the glands.

The exocrine glands may further be characterized by the manner in which they produce their secretions. A MEROCRINE gland produces a secretion in which no part of the cell itself is found. An APOCRINE gland loses some portion of its cells in producing the secretion. A HOLOCRINE gland either produces a cell as its secretion (testis) or the entire cell is shed into the secretion. Tables 6.2 and 6.3 summarize facts about exocrine glands.

Clinical considerations

Certain of the epithelial types discussed above are particularly susceptible to environmental influences and may form the basis of neoplastic changes. For example, exposure of the skin to sunlight or other types of radiation (x-ray) may cause epitheliomas. Heavy smoking may cause the pseudostratified epithelium of the respiratory system to change to a stratified squamous type that may become malignant.

Connective tissues

General characteristics

With the exception of the tissues lining joints (synovial membranes) and the tissues forming the bursae and tendon sheaths, connective tissues DO NOT POSSESS FREE SURFACES. Unlike epithelia, they contain WIDELY SPACED CELLS and large amounts of intercellular material. With few exceptions (e.g., blood, mesenchyme), the INTERCELLULAR MATERIAL IS FIBROUS. Connective tissues are, as a rule, VASCULAR. They transmit blood vessels if they have none of their own. As their name suggests, connective tissues support and protect other tissues or organs within the body and form the primary structural tissues of the organism.

Types of connective tissues

The many types of connective tissues may be grouped, and named individually, by using several basic criteria.

If the tissue is present only in the embryo or fetus it is designated as an EMBRYONAL TISSUE. If the tissue is present without change after birth, it is designated as an ADULT TISSUE.

Adult tissues are grossly separated into three groups according to the overall consistency of the intercellular material, which may be more or less fluid, semisolid, or hard.

A "classification tree" that subdivides the connective tissues is presented in Fig. 6.11.

EMBRYONAL CONNECTIVE TISSUE (Fig. 6.12)

Mesenchyme. Mesenchyme is the primitive mesodermal undifferentiated tissue from which all other connective tissues arise. The cells are stellate in shape and typically form networks. While the cells may appear to be connected, they behave independently and may show ameboid motion. The cells lie in a histologically homogeneous ground substance composed of water, inor-

FIGURE 6.11 A classification of connective tissues. Separation criteria are shown in red; major groups are shown in blue; individual types are shown in black.

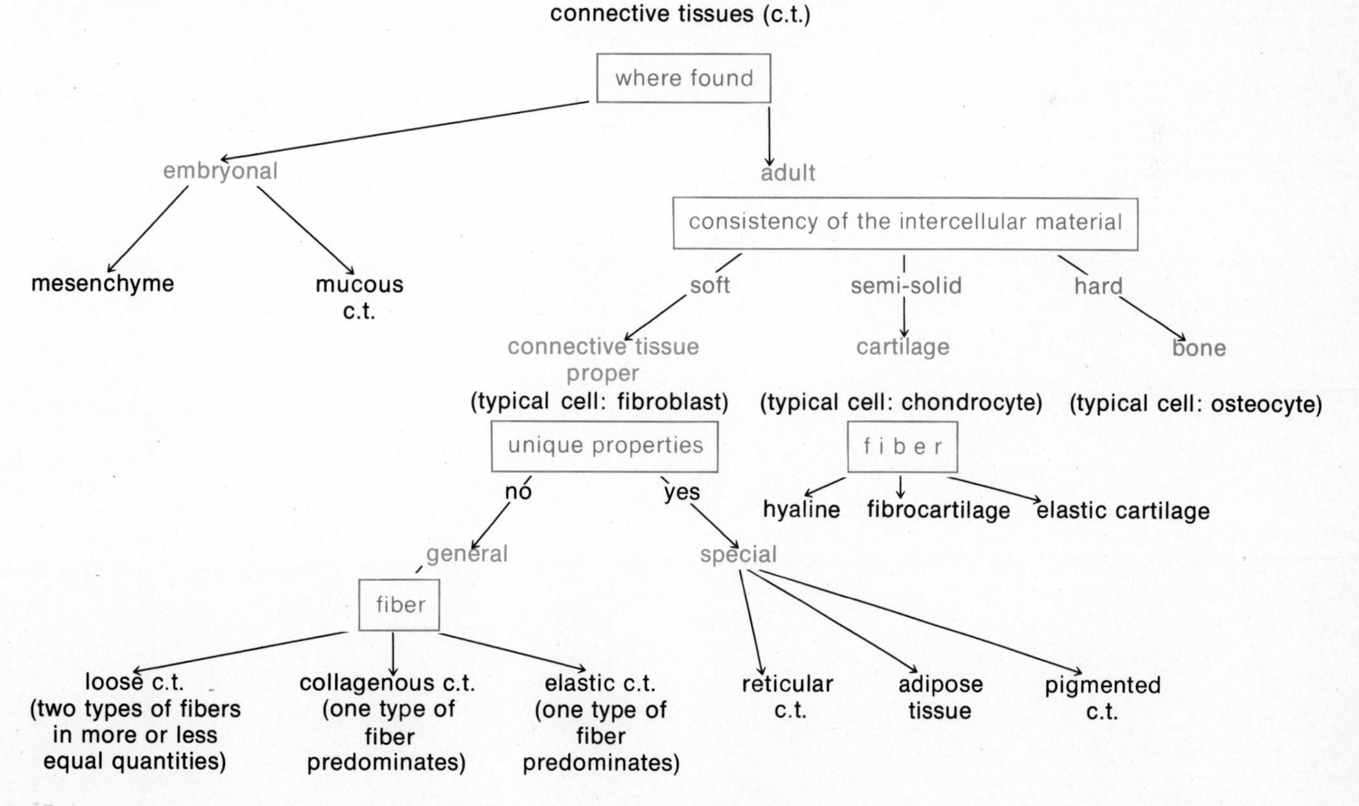

FIGURE 6.12 Embryonal connective tissues.

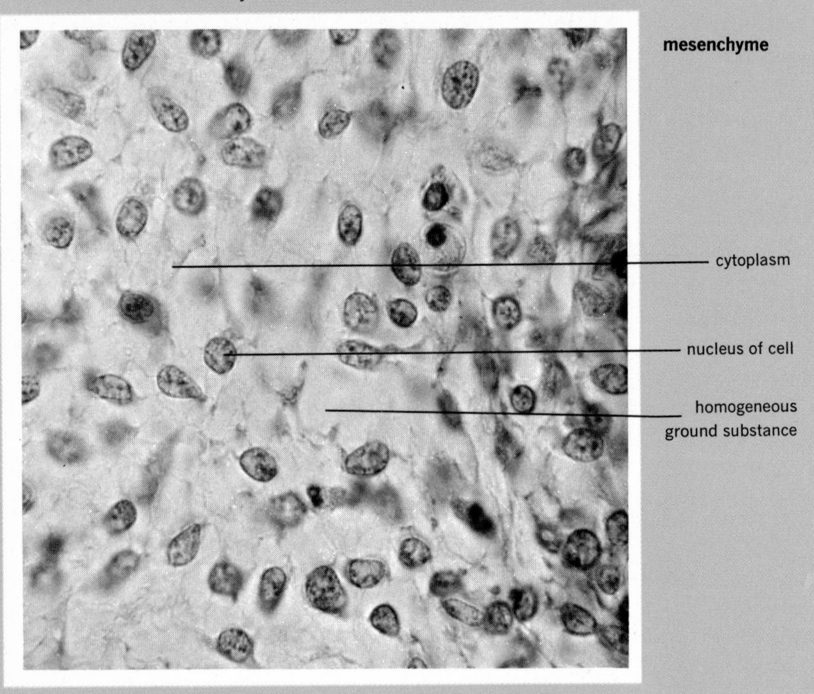

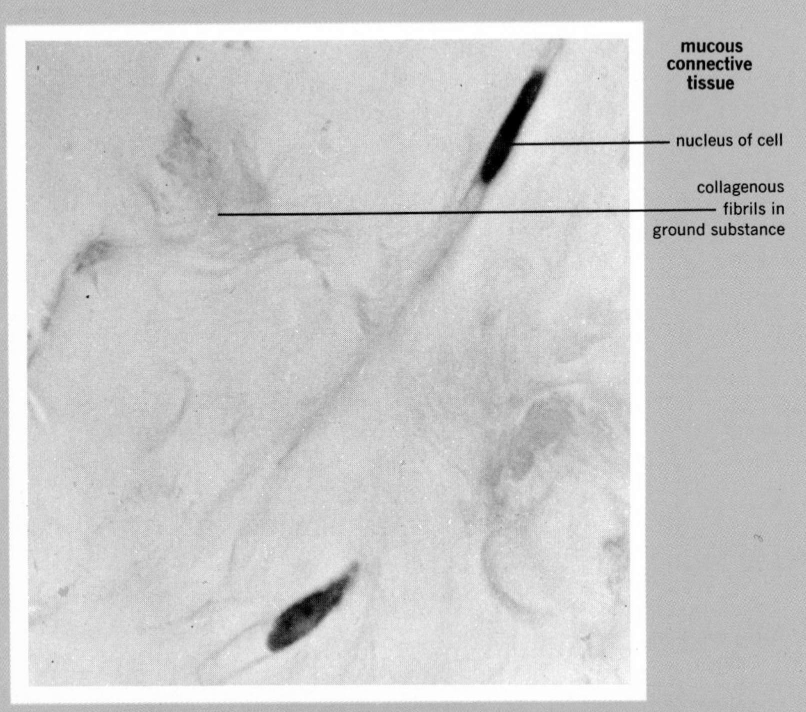

ganic ions, and protein polysaccharides (*mucopolysaccharides*). The viscosity of the ground substance varies with the amount of an enzyme (*hyaluronidase*) present in the ground substance, becoming more liquid as amount of enzyme increases. Mesenchyme is seen in sections of vertebrate embryos, under the skin, and along developing bones.

Mucous connective tissue. Mucous tissue is a differentiated tissue found only in the fetus. The cells are flattened or spindle-shaped and are set in a mucoid (mucuslike) ground substance. The ground substance has many fine fibrils in it (collagenous fibrils). Also known as Wharton's jelly, mucous tissue is found in the umbilical cord of the fetus.

ADULT CONNECTIVE TISSUES. Connective tissues proper (Fig. 6.13) are characterized by having a more or less fluid intercellular material, with the fibroblast as the typical cell. Five types are distinguished.

Loose connective tissue contains all of the structural elements found in other members of the group. Loose connective tissue is a nonoriented tissue. Two types of fibers, collagenous and elastic, are distributed in roughly equal quantities in the form of a three-dimensional "feltwork." At least nine types of cells are sprinkled randomly through the fibrillar mass, and all elements are suspended in a homogeneous ground substance.

COLLAGENOUS FIBERS are 1–12 μm in diameter; they consist of subunits termed *fibrils*, and appear as wavy refringent bundles in the tissue. The fibers may branch, but the fibrils do not. The fibers are composed of a protein called *collagen*, which in turn is made up of *tropocollagen* molecules secreted by the fibroblasts. The molecules are apparently polymerized into collagen outside the fibroblast. Collagenous fibers are extremely strong, but show little or no elasticity.

ELASTIC FIBERS are also 1–12 μm in diameter, but they are always smaller in caliber than the collagenous fibers when the two occur together in any given tissue. They are highly refractile, show no subunit of structure (are therefore homogeneous), and freely branch and rejoin with one another. They consist of a protein called *elastin* and have little strength but much elasticity.

The three most common cellular elements of loose connective tissue are:

Fibroblasts. The cell cytoplasm is usually not visible in the typical slide, but it gives the whole cell a flat or spindle-shaped outline if visible. Nuclei are large, oval, and typically possess two to three prominent nucleoli. The cells give rise to the fibers of the tissue.

Histiocytes or macrophages. These are cells capable of engulfing (phagocytosis) particulate matter, and thus form an important line of defense against bacterial invasion. They also aid in "cleaning up" the tissue if it has been damaged. The cells are of irregular outline, have a dense nucleus with a heavy nuclear membrane, and typically show irregular engulfed particles in the cytoplasm.

Blood cells. Because of their ameboid capacity, white blood cells are able to pass (diapedesis) through the walls of capillaries and enter the connective tissues. *Lymphocytes*, *eosinophils*, and tissue *basophils* (mast cells) may be found. These cells may also be important in defense against diseases. The tissue basophil has granules containing a heparinlike substance that may function to control the viscosity of the ground material. When tissue inflammation occurs, many neutrophils may also be found.

Other cell types found in the tissue include mesenchymal cells, typically located around blood vessels (pericytes), fat cells, pigment cells (melanocytes), and plasma cells. Plasma cells are an example of cells capable of forming antibodies.

The GROUND SUBSTANCE of loose connective tissue is a complex mixture of water, collagen, glycoproteins, and lipids, whose viscosity may change under conditions of inflammation and injury. Such changes in viscosity may aid the penetration of blood cells or wall off an area and prevent spread of the abnormal process.

FIGURE 6.13 Adult regular connective tissues.

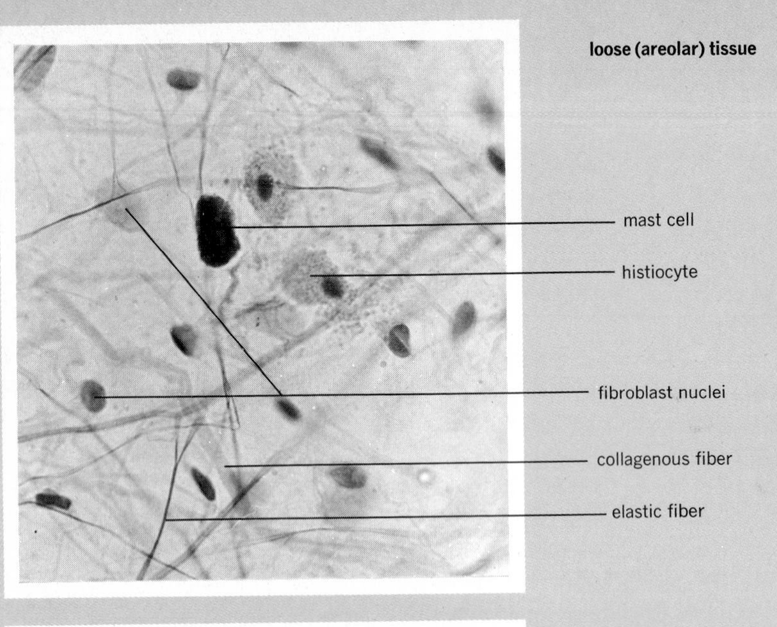

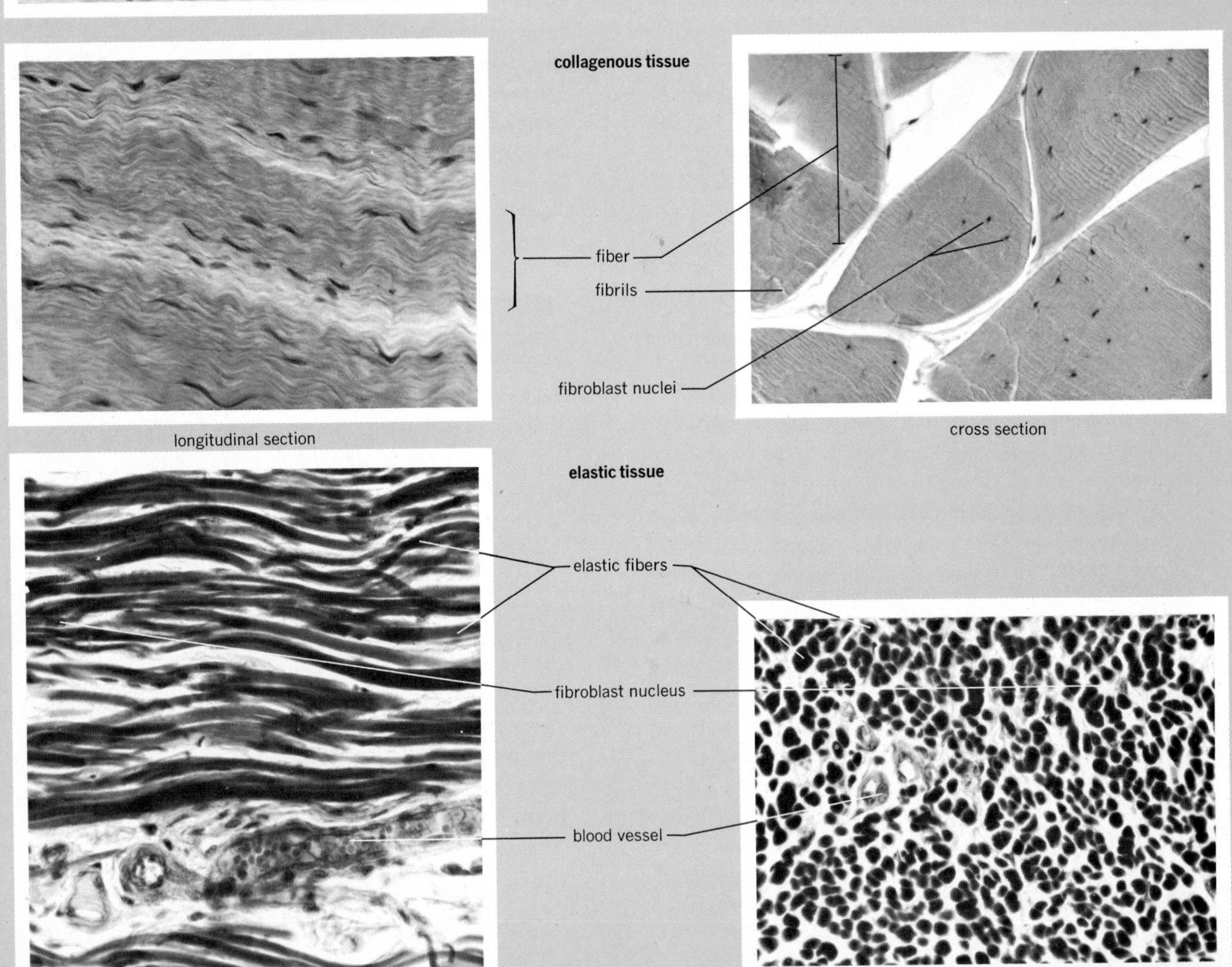

Loose connective tissue forms the subcutaneous layer of the skin (injections may be made into this tissue). It occurs as a "packing material" around body organs in the nooks and crannies of its cavities, and it forms layers around organs such as the large arteries. Its fibers confer both strength and elasticity while permitting relatively free movement.

A variant of loose connective tissue, termed DENSE IRREGULAR CONNECTIVE TISSUE, contains both collagenous and elastic fibers in which the fibrillar elements are somewhat larger than in loose connective tissue, and are compacted or more closely packed. The tissue forms the dermis of the skin and submucosa of the digestive tract.

Loose connective tissue may contain large numbers of pigment cells, forming pigmented connective tissue. Such tissue is found in the retina, choroid, and iris of the eye.

Collagenous connective tissue has a predominance of collagenous fibers arranged in an oriented manner. Flattened fibroblasts are found in rows between the fibers. The tissue is found in ligaments and tendons and in the sclera of the eye.

Elastic connective tissue has a predominance of elastic fibers also arranged in an oriented fashion. Fibroblasts are present in the spaces between the meshes of the tissue. Elastic connective tissue is found as the major tissue in the middle layer (media) of the aorta and pulmonary artery, and is found in the ligamentum nuchae and ligamenta flava of the vertebral column.

Reticular connective tissue (Fig. 6.14) is composed of small, irregular-diameter, interlacing fibers and forms the framework of many body organs, such as the spleen, liver, and lymph nodes. The tissue also holds smooth muscle cells into sheets. Reticular fibers are antecedent to collagenous fibers and may be demonstrated by silver-containing stains.

Adipose tissue (Fig. 6.14) consists essentially of loose connective tissue, wherein the primitive fibroblasts have become specialized for the storage of fat. The adipose cells typically show a "signet ring" shape, with the cytoplasm pushed peripherally in the cell by a large fat droplet. (In most slides, the fat has been dissolved out of the cell by the alcohols used in preparing the slide, so the cell presents an open cavity.) Adipose tissue insulates, acts as a shock absorber, is a storage form of energy, and gives form and strength to the body as a whole. The subcutaneous layer of the skin is a good place to find it.

CARTILAGES (Fig. 6.15). Cartilages have a semisolid intercellular substance termed the INTERTERRITORIAL MATRIX. The characteristic cells, CHONDROCYTES, typically lie singly or in groups in cavities in the matrix known as LACUNAE. Matrix substance between cells in a lacuna forms the TERRITORIAL MATRIX, and a darker staining region of newly formed matrix, the CAPSULE, surrounds the lacuna. The surface of a cartilage is covered by a membranous PERICHONDRIUM. Three varieties of cartilage are recognized.

Hyaline cartilage (gristle) is, in fresh section, a translucent firm mass. On a slide, in addition to the features described above, the interterritorial matrix *appears* devoid of fibers. Fibers are present, but not visible. Hyaline cartilage forms most of the embryonic skeleton, covers the end of bones forming freely movable joints, and is found in the nasal septum, costal, tracheal, and certain laryngeal cartilages.

Fibrocartilage has large numbers of visible collagenous fibers in the matrix. Other features are the same as previously described. Fibrocartilage forms the intervertebral discs and the symphysis pubis.

Elastic cartilage has many visible elastic fibers in the matrix as the differentiating characteristic. It may be found in the epiglottis and external ear.

Growth of cartilage occurs by two methods. APPOSITIONAL growth is the transformation of primitive cells in the perichondrium into chondrocytes and the formation of new cartilage at the surface of the mass. INTERSTITIAL GROWTH is formation of new cartilage on the walls of the

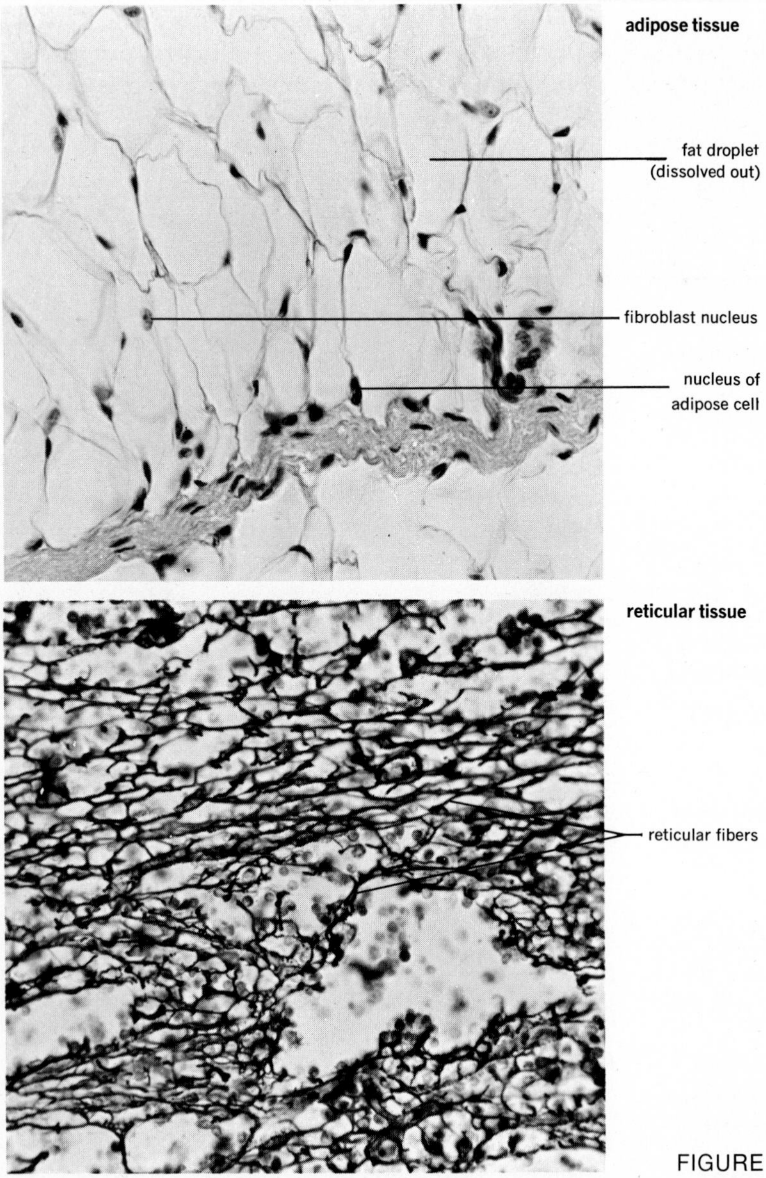

adipose tissue

fat droplet (dissolved out)

fibroblast nucleus

nucleus of adipose cell

reticular tissue

reticular fibers

FIGURE 6.14 Adult special connective tissues.

lacunae within the mass. Obviously, the greatest amount of cartilage growth is appositional growth, for to reduce the size of the lacuna is to sow the seeds of self-destruction. If damaged, cartilage heals slowly. Most cartilages have a poor blood supply, and regeneration is both slow and limited.

BONE. Bone has a HARD INTERCELLULAR MATERIAL, due to the deposition of inorganic salts in the matrix. The typical cells, OSTEOCYTES, lie in lacunae in the hard intercellular material. CANALICULI (tiny canals) connect lacunae with one another and give access to nearby blood vessels. Two basic

FIGURE 6.15 Cartilages.

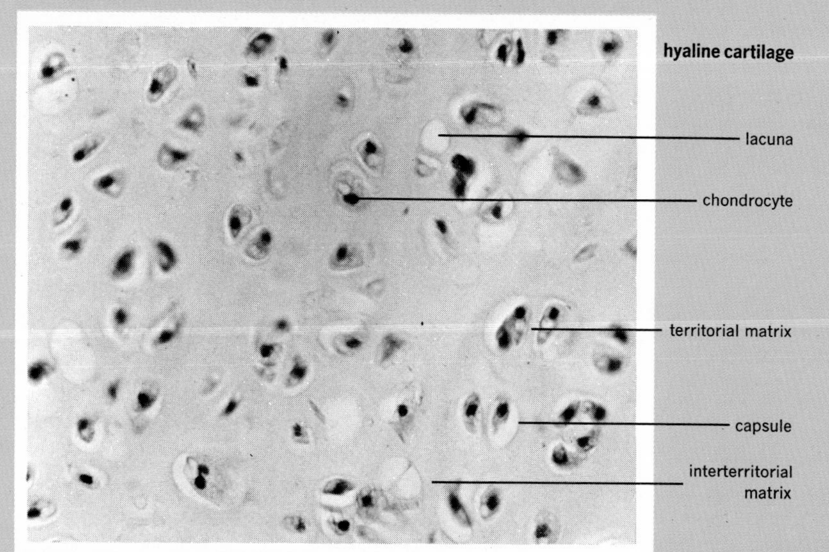

hyaline cartilage
lacuna
chondrocyte
territorial matrix
capsule
interterritorial matrix

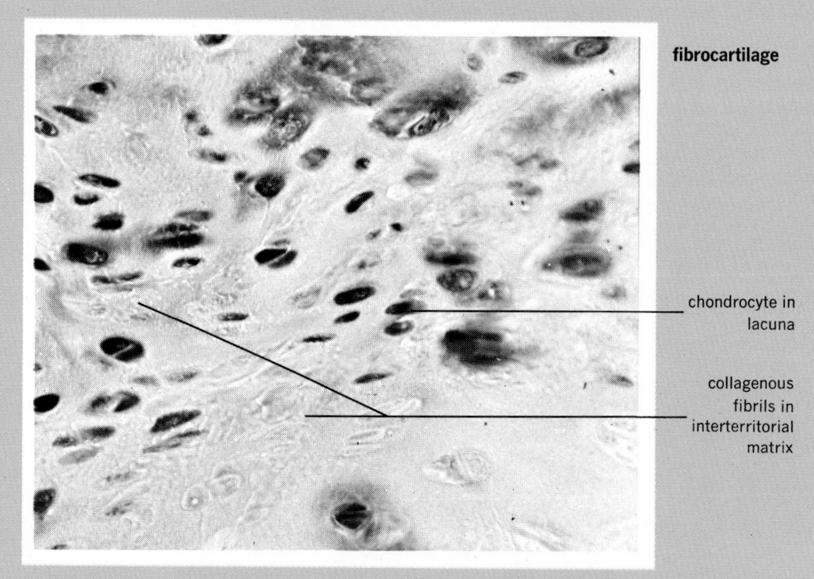

fibrocartilage
chondrocyte in lacuna
collagenous fibrils in interterritorial matrix

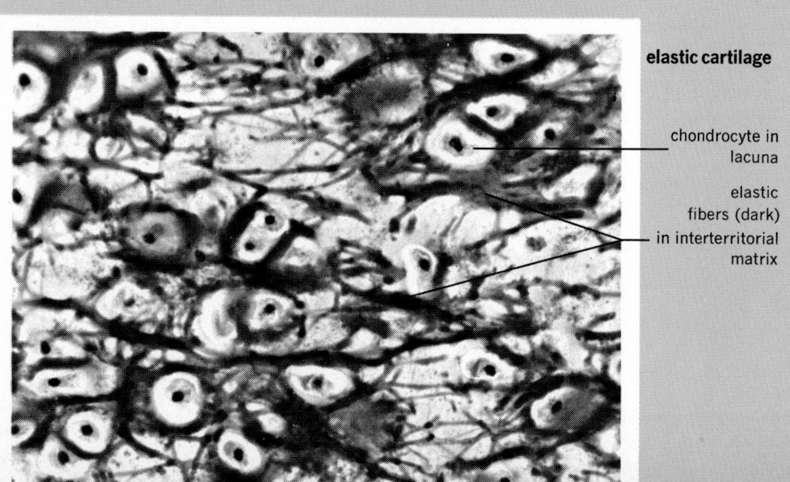

elastic cartilage
chondrocyte in lacuna
elastic fibers (dark) in interterritorial matrix

types of bone are recognized. They differ, not in chemical composition, but in arrangement. Both types are about 35 percent organic substance, consisting of bone collagen fibers and cells, and 65 percent inorganic substance, chiefly $[Ca_3(PO_4)_2]_3 \cdot Ca(OH)_2$.

CANCELLOUS or spongy bone consists of interlacing plates and bars of bony material with many spaces between. This type of bone is found in the ends of long bones and inside flat and irregularly shaped bones.

COMPACT BONE (Fig. 6.16) is extremely dense and consists of longitudinally oriented subunits of structure, the OSTEONS (Haversian systems). A canal in the center of the system carries blood vessels that communicate with the outside of the bone through obliquely oriented Volkmann canals.

Bone is constantly remodeled, particularly during youth, and it is possible to find incomplete systems in compact bone. These represent remodeled osteons and are known as INTERSTITIAL LAMELLAE. The outer and inner surfaces of the bone are covered by CIRCUMFERENTIAL LAMELLAE, lacking osteons.

A membranous PERIOSTEUM covers bone, except on articular surfaces, and is two-layered. The inner or *osteogenic layer* contains cells capable of differentiating into osteoblasts and forming new bone. The outer or *vascular layer* contains fewer cells and more blood vessels. The endosteum, a condensed layer of bone marrow stroma, lines the internal surface (marrow cavity) of long bones.

Other cells of bone. OSTEOBLASTS, recognizable by their position in rows on the surface of newly forming bone and by their deeply basophilic cytoplasm, are bone-forming cells. OSTEOCLASTS, giant multinucleated cells, also may be found on bone surfaces, and the function of bone destruction and bone remodeling are attributed to them. Both cells may be formed from mesenchyme or fibroblast cells. BONE MARROW CELLS fill the spaces

FIGURE 6.16 A cross section of a ground preparation of compact bone.

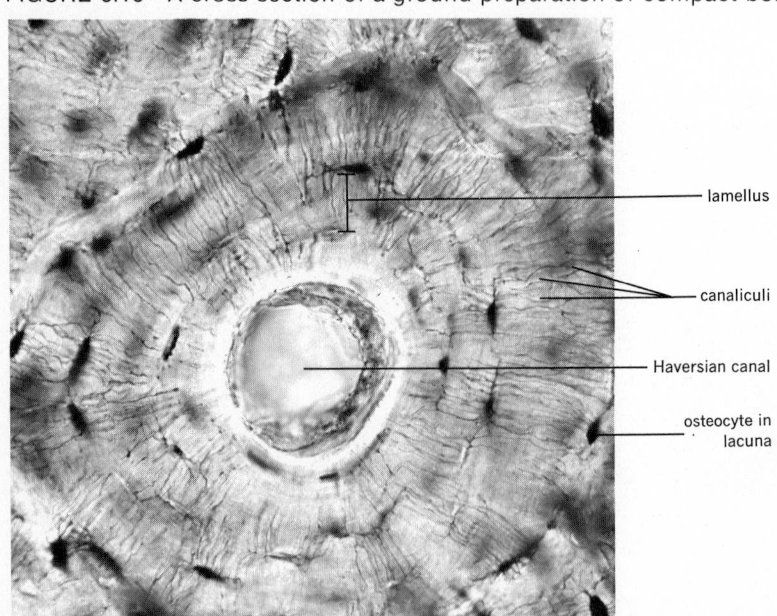

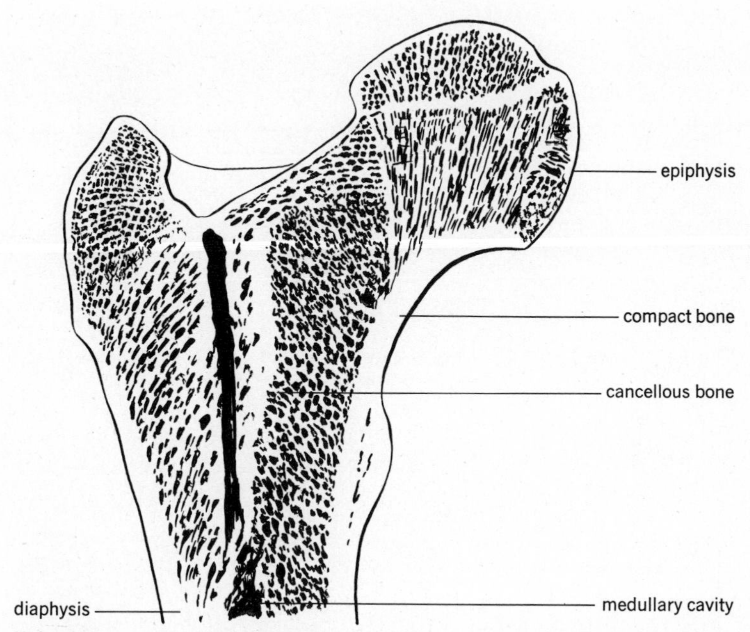

FIGURE 6.17 A diagram to show the arrangement of spongy bone in the proximal end of the femur.

between spicules of spongy bone. As bone is remodeled, particularly in young people, the spongy component tends to orient itself to resist the direction of force imposed on the bone. Figure 6.17 shows the vaults and bridgelike pattern of the spongy bone. This arrangement confers much strength with lightness on the bone as a whole.

Table 6.4 summarizes facts about the connective tissues.

Formation and physiology

Since the primary component of a connective tissue is the intercellular material, production of that material becomes the primary concern in the development and maintenance of the tissue. In loose connective tissue, and other fibrous tissues, the fibroblast is the cell that produces both the fibers and semiliquid components (ground substance) of the intercellular material. The ribosomes and Golgi apparatus of the fibroblasts synthesize the basic molecules (collagen, elastin, protein polysaccharides) necessary for intercellular materials, and these are then secreted from the cell and formed into fibers or ground substance. The fibroblasts of these tissues are stimulated to greater activity by foreign substances and by the products of cell injury, so that they become very important in healing of injuries. They also are stimulated by adrenal cortical hormones, and they play a role in the inflammatory process.

Cartilage relies on its chondrocytes to produce matrix and fibers. Again, the ribosomes and Golgi apparatus appear to be the sites of production of the necessary substances. When damaged, cartilage heals by forming new cartilage at the surface (appositional growth) through activity of cells in the perichondrium or by formation within the mass (interstitial growth). Formation of new cartilage is greatly influenced by several vitamins and hormones.

Vitamins A and C are necessary for formation

TABLE 6.4 Summary of the connective tissues

Adult or embryonic	Tissue group; consistency of intercellular material (ICM)	Typical cell	Tissue type	Other cells commonly found	Characteristics	Examples of locations
Embryonic	–	Mesenchyme	Mesenchyme	None	Stellate ameboid cells in networks	In embryo around forming bones; under skin
		Modified mesenchyme	Mucous c.t.	None	Cells flattened collagenous fibrils in ground substance	Umbilical cord
Adult	ICM Fluid c.t. proper	Fibroblast				
			Loose c.t.	Histiocyte, white blood cells, plasma cells, fat cells	Feltwork of collag. and elastic fibers	Around organs, subcutaneous tissue
			Dense irregular c.t.	Histiocyte, white blood cells, plasma cells, fat cells	Compacted loose c.t.	Submucosa of digestive tract, dermis of skin
			Collagenous	None	Wavy bundle of fibers, fibrils present	Ligaments, tendons, sclera of eye
			Elastic c.t.	None	Branching, homogeneous flbers	Aorta, ligamentum nuchae ligamenta flava
			Reticular c.t.	Reticular cell	Fine, irregular branching fibers	Interior of liver, spleen, lymph nodes

of fibers and ground substance. Vitamin D is necessary for the absorption of calcium for formation of bone from cartilage.

Growth hormone (from the pituitary gland) and thyroxin (from the thyroid gland) cause stimulation of production of cartilage during body growth.

Adult or embryonic	Tissue group; consistency of intercellular material (ICM)	Typical cell	Tissue type	Other cells commonly found	Characteristics	Examples of locations
			Adipose tissue	Fibroblasts	"Signet ring" cells with fat drops inside	Anywhere (almost); subcutaneous tissue of skin
			Pigmented	–	Masses of cells containing pigment	Iris, retina of eye; choroid of eye
Adult	Cartilage, ICM semisolid	Chondrocyte				
			Hyaline	None	Clear matrix (fibers not visible) cells in lacunae; capsule around lacunae	Embryonic skeleton, costal cartilages, nasal cartilages, articular surfaces
			Fibrocartilage	None	Visible collagenous fibers in matrix; other features as in hyaline	Intervertebral discs, symphyses
			Elastic cartilage	None	Visible elastic fibers in matrix; other features same as in hyaline	Epiglottis, external ear
	Bone, ICM hard	Osteocyte	Spongy bone	None *in* tissue	Interlacing plates and bars	Ends and interiors of bones
			Compact bone	None *in* tissue	Osteons present	Shafts of bones

Clinical considerations

In view of the many requisites for formation of ground substance and fibers of connective tissue and of the many things that may influence the course of development of the tissue, perhaps it is

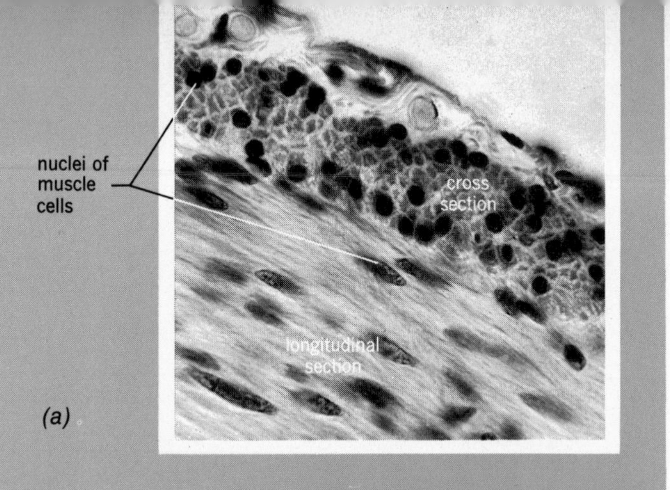

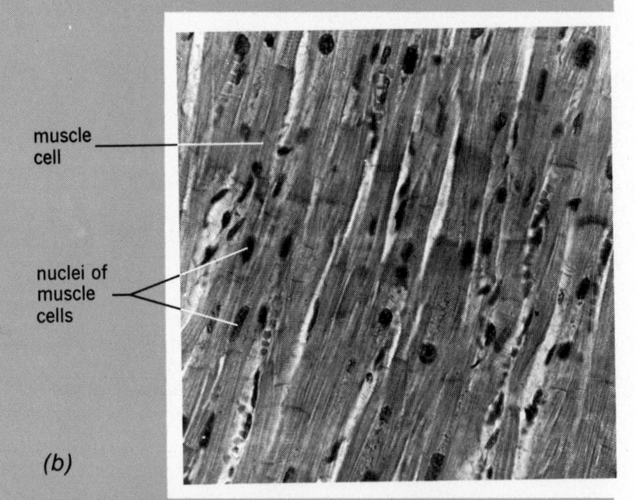

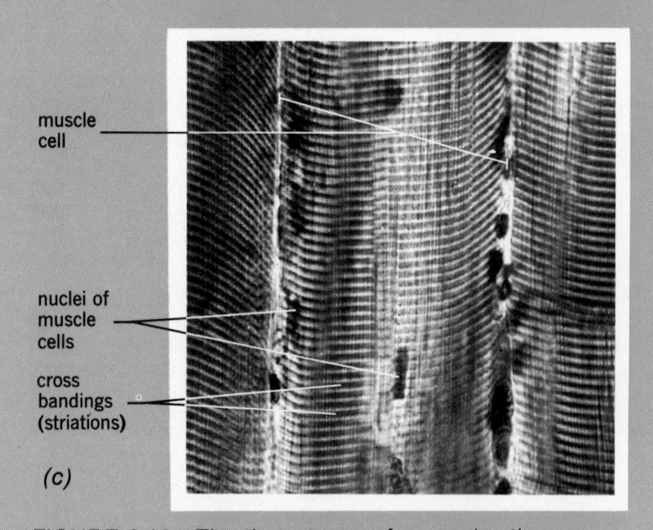

FIGURE 6.18 The three types of muscular tissue. *(a)* Smooth muscle in cross and longitudinal section. *(b)* Cardiac muscle in longitudinal section. *(c)* skeletal muscle in longitudinal section.

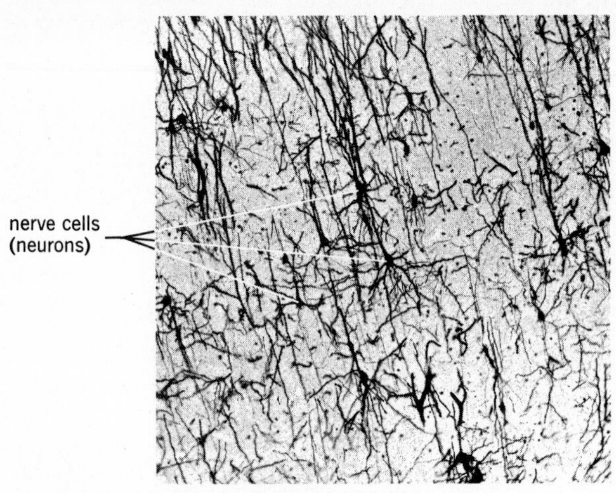

FIGURE 6.19 Nervous tissue, cerebrum.

FIGURE 6.20 The development of the skin. *(a)* 4 weeks, *(b)* 7 weeks, *(c)* 12 weeks, *(d)* newborn.

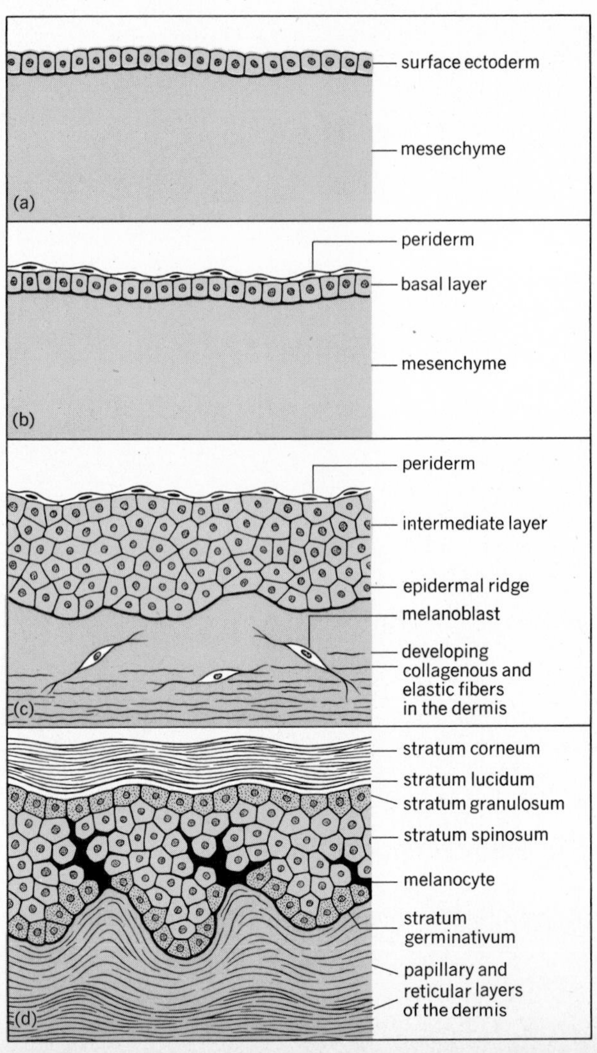

not surprising that the possibility of malfunctions of formation exist. It is perhaps more surprising that disorders do not occur more frequently.

The term *collagen diseases* is applied to those conditions resulting in pathological alteration in the connective tissue elements of the body, chiefly the fibrous portions of these tissues.

Collagen diseases appear to have some hereditary basis in the sense that in many cases the production of collagen proceeds at a faster rate than it can be utilized. This occurs without parallel increases in the numbers of fibroblasts, the supposed producers of collagen. It is therefore suspected that a basic metabolic defect of the synthetic processes within the fibroblast may be attributed to altered gene-controlled enzymatic mechanisms. A second factor in collagen diseases appears to be the formation of defective collagen, which is chemically different enough from normal collagen as not to be recognized by the antibody-producing sites of the body as "self." The collagen is thus treated as a foreign protein, antibodies are manufactured against it, and inflammation and degeneration result. This pattern is one of AUTOIMMUNITY and is more common in disease states than formerly recognized.

The collagen diseases are more common in the elements of connective tissue proper. Cartilages are more susceptible to disorders affecting the production of calcifiable matrix in the steps leading to bone formation. *Achondroplasia* results from "mutation" or abnormalities in the chromosomes and leads to defective endochondral bone formation. Intramembranous formation is usually unaffected or may be increased. As a result, the bones are short and wide, may be deformed (particularly in the appendages), and the general appearance presented is that of a normal head and trunk with dwarfing of the appendages. Mental ability usually is not affected, and achondroplastics may survive until old age.

Disorders primarily associated with bone are described in Chapter 7.

Muscular Tissues

Muscular tissues (Fig. 6.18) are the contractile tissues of the body. By shortening, they cause the movement of the body through space and the movement of materials through the body. Three types of muscular tissue are found in the body. Detailed descriptions of each are provided in later chapters.

Smooth muscle (Chapter 9) occurs in the body viscera and around blood vessels.

Cardiac muscle (Chapter 24) is found only in the heart.

Skeletal muscle (Chapter 9) attaches to and moves the skeleton.

Nervous tissue

Nervous tissue (Fig. 6.19) is the irritable (excitable) and conductile tissue of the body, originating and transmitting nerve impulses to body organs. Detailed descriptions of the nervous tissue are found in Chapter 11.

The integumentary system

The external surface of the body is formed by the integumentary system. The integumentary system includes: the SKIN, derivatives of the skin such as HAIR, NAILS, and GLANDULAR STRUCTURES, and several specialized types of RECEPTORS.

Development of the integument (Fig. 6.20)

The skin is derived from ectodermal and mesodermal germ layers. The outer layers that form the EPIDERMIS are derivatives of the ectoderm forming the outer layer of the embryo. Until about seven weeks of development the epidermis is composed of only one layer of cells, by eleven weeks two layers of cells are evident, and by birth the typical five-layered structure has been achieved.

The DERMIS, or connective tissue underlying the epidermis, is of mesodermal origin, and develops visible fibrillar elements by about eleven weeks of development. The appendages of the skin, hair, nails, and glands develop at about ten weeks (nails), sixteen weeks (glands), and twenty weeks (hairs).

The skin; structure

The skin itself is the largest organ of the body, is composed of epithelial and connective tissue components, and forms a pliable PROTECTIVE COVERING over the external body surface. It accounts for about 7 percent of the body weight and receives about 30 percent of the left ventricular output of blood. The term *protective,* as used here, includes not only resistance to bacterial invasion or attack from the outside, but also protection against large changes in the internal environment. Control of body temperature, prevention of excessive water loss, and prevention of excessive loss of organic and inorganic materials are necessary to the maintenance of internal homeostasis and continued normal activity of individual cells. In addition, the skin acts as an important area of storage, receives a variety of stimuli, and synthesizes several important substances used in the overall body economy.

An adult has an average skin surface area of 1.75 m^2 (about 3000 $in.^2$). Grossly, the skin presents several unique features. FLEXION CREASES appear where the skin folds during the movement of joints. The hand shows many such creases. FLEXION LINES occur where the skin must stretch, but to limited degrees. The back of the hand and fingers show such lines well. FRICTION RIDGES occur on those parts of the body involved in grasping. These ridges occur on the finger and toe tips and on the sole of the foot and palm. On the fingers, they are used for identification purposes because the pattern of the ridges in these areas is characteristic for the individual. Firmness of attachment varies from loose (as over the elbow) to tight (as on the scalp). Generally, the greater the degree of movement required, the looser the attachment will be. Skin varies in thickness in different areas of the body. In general, the thickness depends on the degree of mechanical attrition to which the area is subjected. The thickest skin (5–6 mm) is found on the hands and feet. Calluses and bunions represent even greater thickening in areas of extreme pressure or wear and tear. The eyelids, eardrum, and penis have the thinnest (½ mm) skin covering. The skin of other areas of the body is 1-2 mm thick, intermediate between the areas mentioned above. Skin may be pigmented to various degrees due to the presence of pigment or pigment cells. Various shades of redness depend on the caliber of the blood vessels in the skin and on the degree of oxygenation of the blood in these vessels.

Histologically, skin is classed as being THICK or THIN. The terms not only refer to total depth

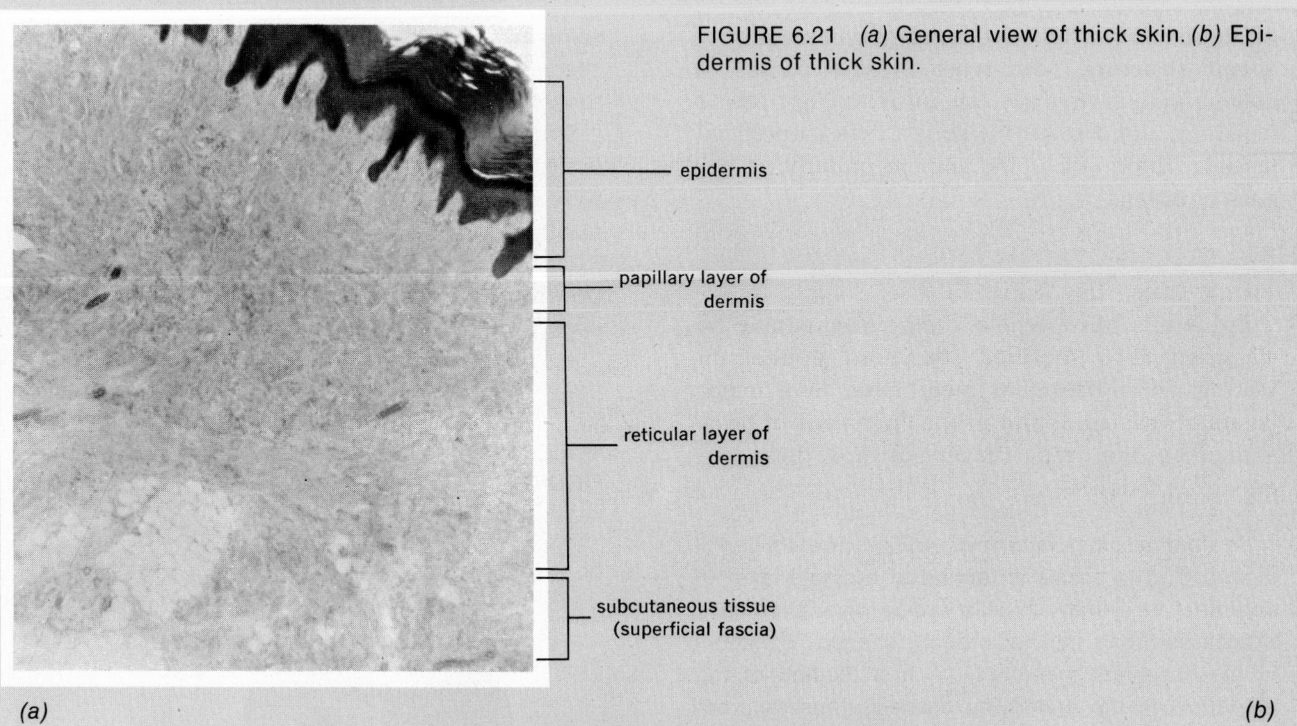

FIGURE 6.21 *(a)* General view of thick skin. *(b)* Epidermis of thick skin.

(a)

(b)

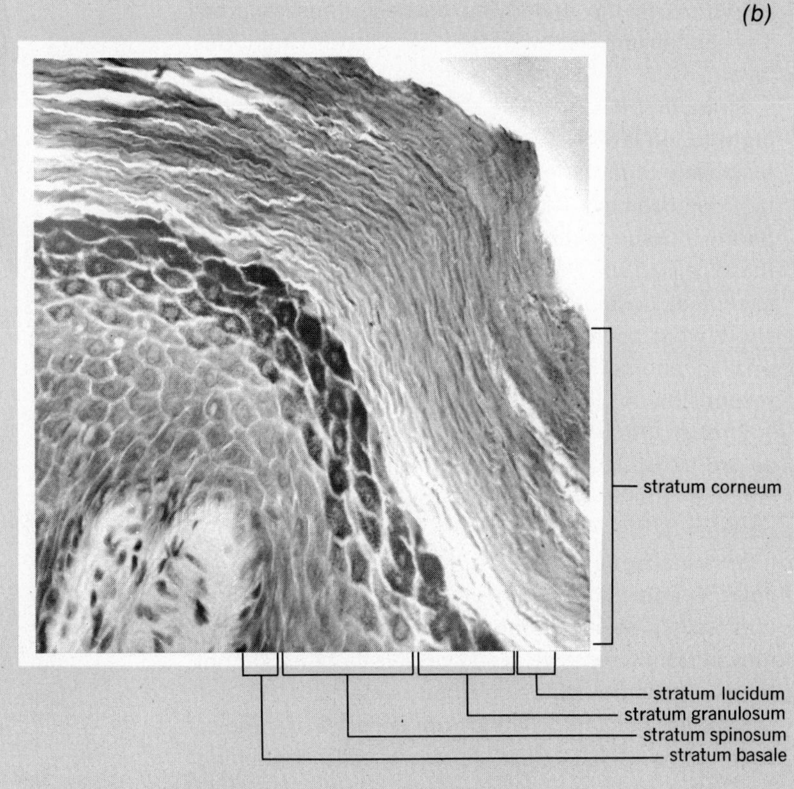

of the covering, but also imply different microscopic structure. Both types of skin possess a superficial EPIDERMIS, an underlying DERMIS (corium), and a SUBCUTANEOUS LAYER (superficial fascia), which affixes the skin to underlying tissues or organs.

THICK SKIN. (Fig. 6.21).

Epidermis. As a whole, the epidermis may be classified as a stratified squamous epithelium resting on a basement membrane. Five major layers of cells are found in the epidermis of thick skin. From the deepest layer outward, these layers are as follows:

Stratum basale (stratum cylindricum, stratum germinativum). The basale is formed of a single layer of columnar or cylindrical cells whose lower ends have processes fitting into reciprocally shaped "pockets" in the basement membrane. A firm fixation of the epidermis to the underlying tissue is thus assured. The epithelium renews itself by mitosis from this layer.

Stratum spinosum. The spinosum is composed of eight to ten layers of polygonal cells. In life, the cells fit closely together and are attached to one another by desmosomes. The techniques employed in preparing a tissue section for microscopic viewing cause these cells to shrink. They pull away from one another, except at desmosomal attachments. The cells thus show what appear to be processes extending from one to another. These processes or "spines" are responsible for the name of the layer.

Stratum granulosum. The granulosum is composed of two to five layers of rhombic cells containing dark staining granules of keratohyaline. This material represents the first step in the formation of keratin, a proteinaceous substance that will fill the surface cells. Granulosum cells usually show nuclei in various stages of degeneration. The cells are so far removed from the dermal blood vessels that they cannot be maintained by diffusion of nutrients.

Stratum lucidum. The lucidum consists of three to four layers of flat, clear, refractile cells containing semiliquid droplets of eleidin. Eleidin represents the second stage in keratin formation. No cellular structure is evident in this layer.

Stratum corneum. The corneum consists of twenty-five to thirty layers of flat, dead, scalelike cells filled with keratin. The most superficial layers are constantly being shed and form the stratum disjunction.

The lower surface of the entire epidermis is folded. Downward projections of epidermis are termed *rete pegs.*

FIGURE 6.22 The cleavage lines of the skin.

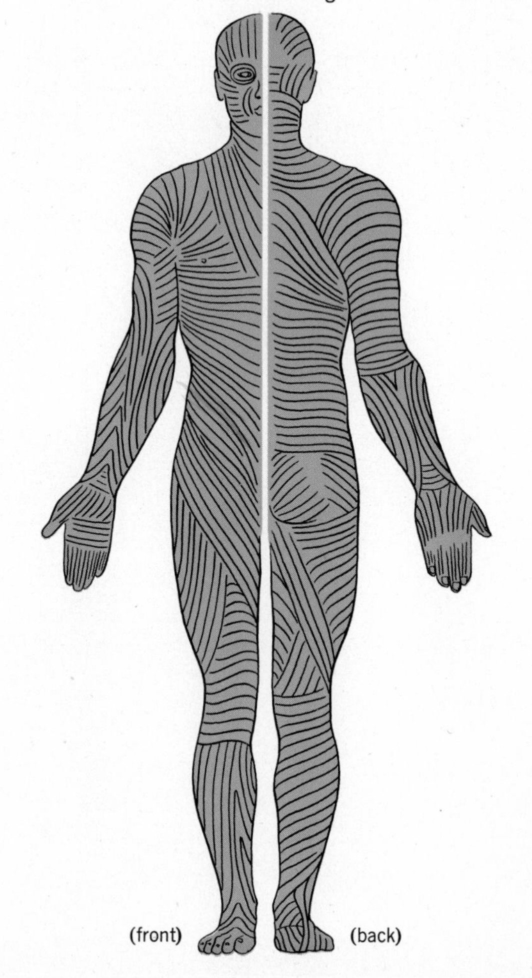

The dermis. Normal dermis is a tough, flexible, elastic layer of dense, irregularly arranged connective tissue. Its elasticity creates CLEAVAGE LINES (Fig. 6.22) for the skin as a whole. Skin, if cut, draws away from the wound in certain well-defined directions. These are the cleavage lines. (Surgeons generally make incisions parallel to these lines to promote more rapid healing.) The upper layer of the dermis is termed the PAPILLARY LAYER. It is about one-fifth of the total dermis and possesses papillary pegs, upfolds which interdigitate with the downfolds of the epidermis. Connective tissue fibers in the papillary layer are smaller than in the remainder of the dermis. The RETICULAR LAYER forms the remainder of the dermis. The dermis contains many blood vessels arranged in two layers. In the deep portion of the reticular layer, many sinuous arteries are found. Branches ascend to the lower border of the papillary layer to form a subpapillary arterial plexus. From this plexus, capillaries form loops within the papillary pegs (Fig. 6.23). The arrangement of the capillary loops and their connecting vessels allows more or less blood to flow through the dermis as a whole, a fact of importance in temperature regulation.

The subcutaneous layer. The subcutaneous layer is composed of loose connective tissue typically containing much adipose tissue *(panniculus adiposus)*. There is no clear line of demarcation between it and the dermis, other than fatty infiltration. The tissue is loose enough to accommodate significant volumes of fluid, and it is into this layer that subcutaneous injections are made.

THIN SKIN. In thin skin (Fig. 6.24), all layers are reduced in depth. In addition, a granulosum and lucidum are usually lacking. If subjected to increased attrition, thin skin may assume the appearance of thick skin.

FIGURE 6.23 The arrangement of blood vessels in the skin.

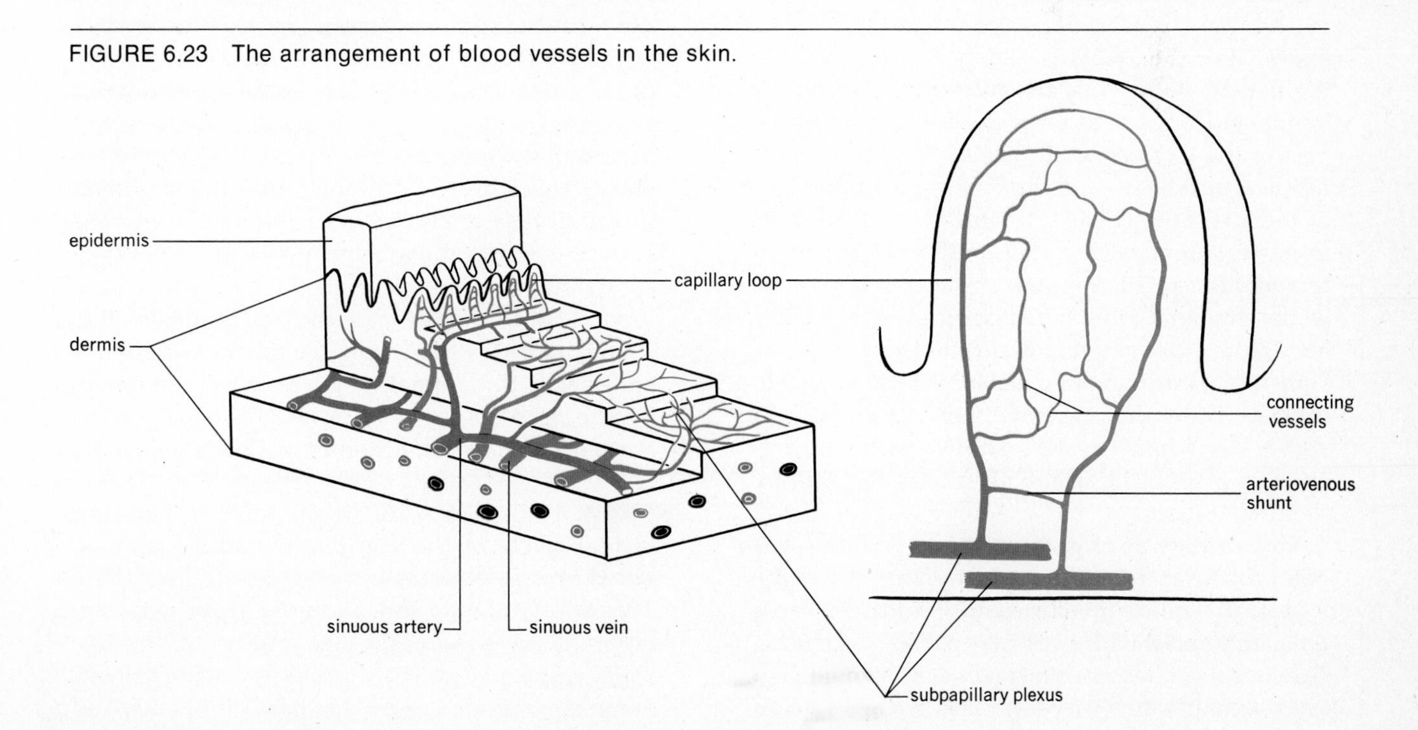

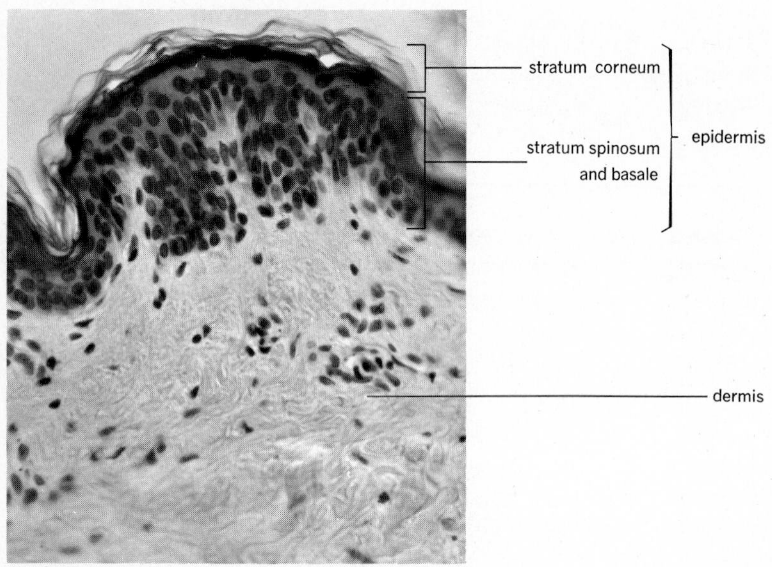

FIGURE 6.24 The appearance of thin skin.

Appendages of the skin

HAIR. Hair (Fig. 6.25) is widely distributed over the human body. Only the anterior surface of the hands and fingers, soles of the feet, posterior aspect of the last two phalanges of the fingers, lips, nipples, umbilicus, and the skin-covered portions of the male and female genitalia are truly hairless. In lower animals, the hair performs important temperature-regulating and protective functions. In the human, the total hair covering is not dense enough to be of any importance in these respects. Hair color is provided by melanins located within the hair, or by the presence of air within the hair shaft. "Gray hair has air." Lanugo or down hair appears first on the fetal body and is replaced by terminal hairs.

A downgrowth of epidermal cells into the dermis forms a HAIR FOLLICLE. Germinal cells at the base of the follicle give rise to the hair itself. These cells are nourished by vascular DERMAL PAPILLAE. The hair itself has a central MEDULLA, surrounding CORTEX, and a covering CUTICLE. The follicle consists of INNER and OUTER ROOT SHEATHS and a CONNECTIVE TISSUE SHEATH. The hair has a BULB on its lower end. The SHAFT is the visible portion of the hair. Hairs always have SEBACEOUS GLANDS associated with them. These glands secrete an oily SEBUM that keeps the hairs pliable and the skin moist. Hairs are set obliquely into the skin, and in the obtuse angle between the follicle and the surface is found the ARRECTOR PILI MUSCLE. When contracting, the muscle pulls the hair into a more vertical position. In hair-covered animals, this action increases the insulative capacity of the fur and makes the animal look larger. In the human, "goose flesh" is the only result; this occurs as the erected hair pushes a mound of skin to one side.

NAILS. Nails (Fig. 6.26) are modified corneal and lucidal layers of the digits. The nail has a ROOT, BODY, and FREE EDGE and rests on the NAIL BED. The EPONYCHIUM and HYPONYCHIUM are skin layers at the base of the nail and under the free edge, respectively. The LUNULA is a whitish half-moon shaped area under the base of the nail, and

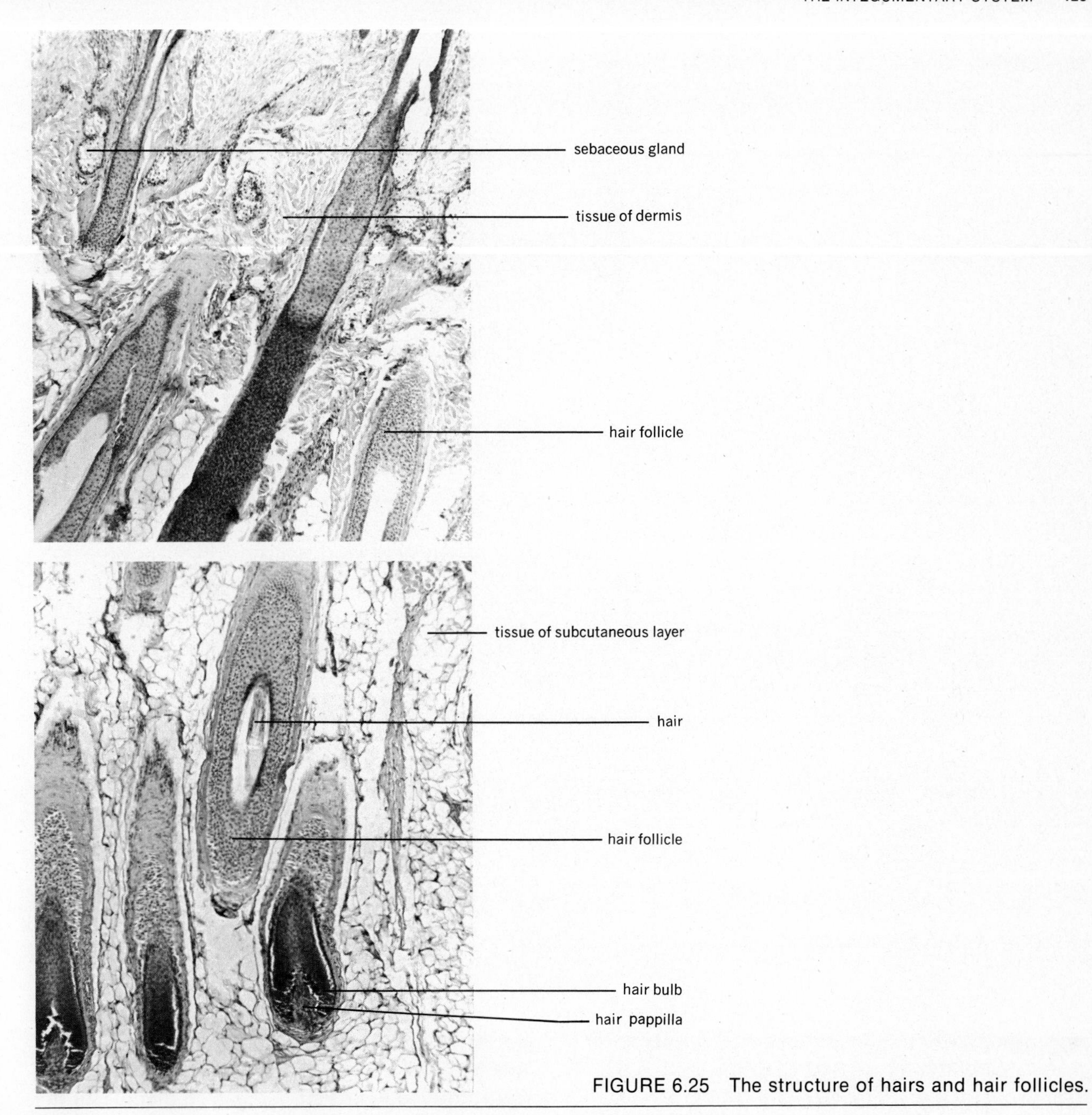

FIGURE 6.25 The structure of hairs and hair follicles.

it represents the active growing region of the nail. The nail is bordered proximally and laterally by NAIL FOLDS. In both longitudinal and cross sections the nails are curved, giving much strength for the thinness.

FIGURE 6.26 Nails.

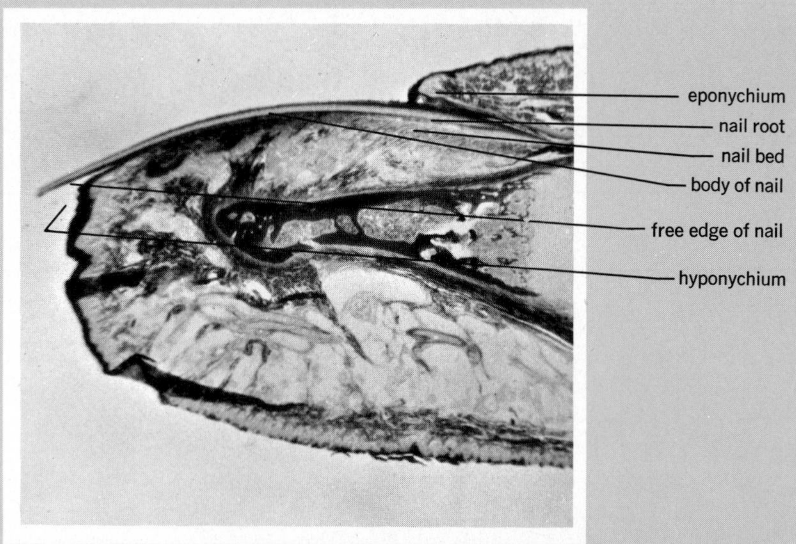

A. longitudinal section

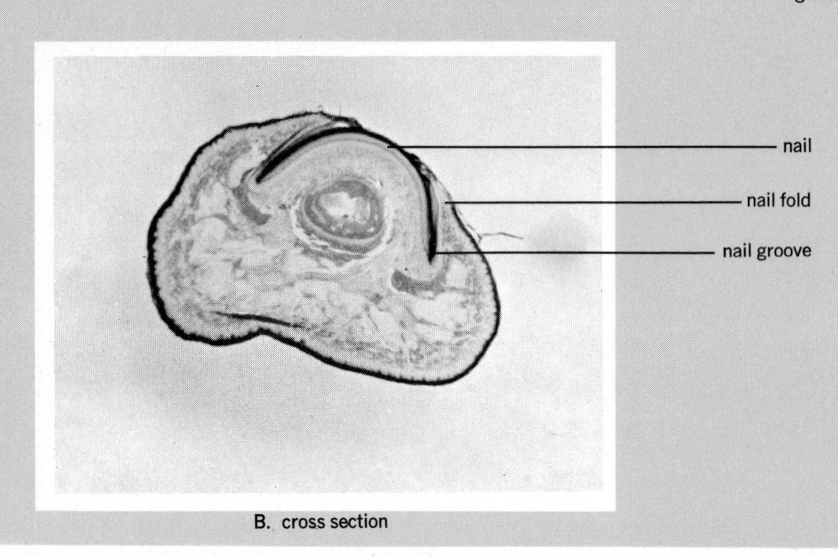

B. cross section

GLANDS. Two important types of glands are associated with and derived from the epidermis. SEBACEOUS GLANDS (Fig. 6.27) are simple alveolar holocrine glands. Indifferent cells line the gland and transform into sebaceous cells, which then disintegrate to form the secretion (sebum).

SUDORIFEROUS *(sweat)* GLANDS (Fig. 6.27) are simple coiled tubular structures. The so-called *eccrine* sweat glands are widely distributed over the body and produce a watery merocrine secretion important in temperature regulation. In the axilla, anal, genital, and ear lobe regions of the body are found *apocrine* sweat glands. Their secretion is whitish and cloudy and contains much organic substance.

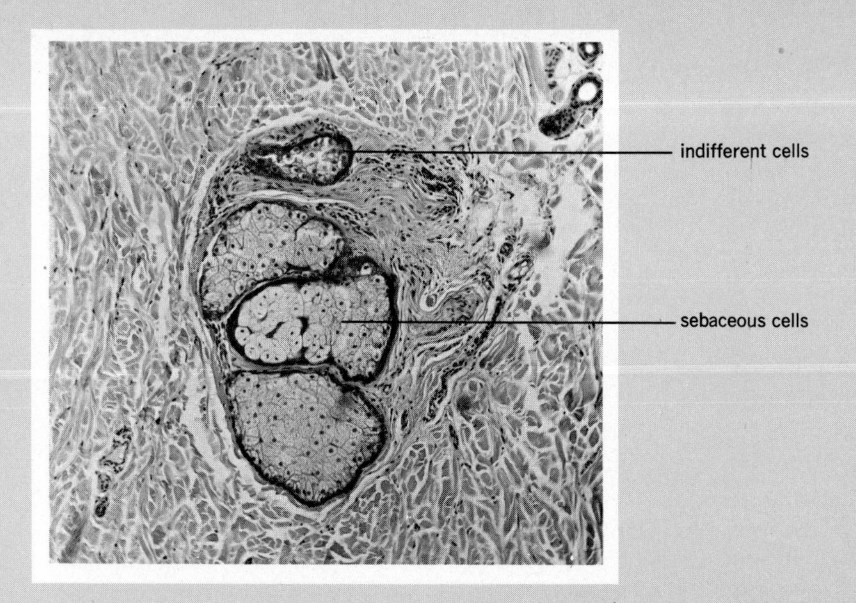

FIGURE 6.27 Sebaceous (top) and sweat (bottom) glands.

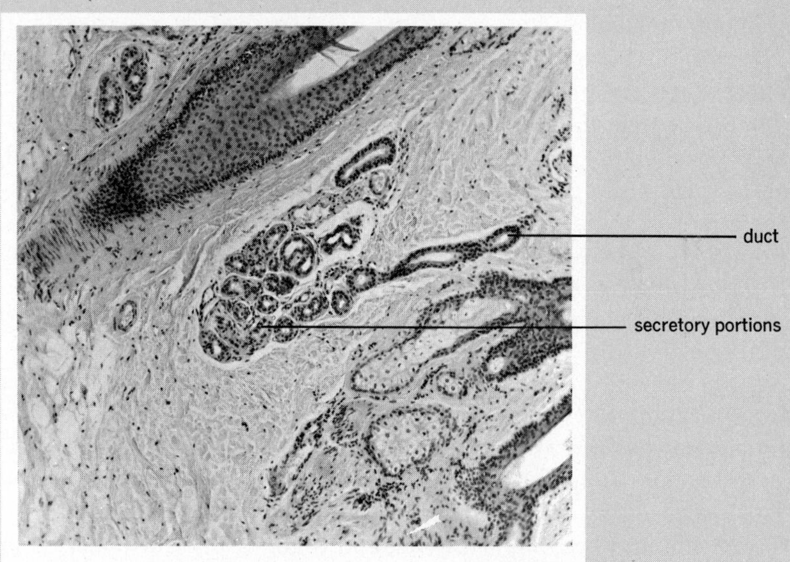

Characteristics and functions of the skin

MECHANICAL PROPERTIES. Skin on the body is normally under tension and will retract if it is cut. This phenomenon indicates that the elastic elements of the dermis are normally slightly stretched. With loss of elastic fibers (for example, in aging) the skin loses its ELASTICITY and tends to sag.

RESILIENCY of the skin may be defined as the tendency or ability to return to its original shape when stretched. This property varies with age and again depends on the presence of elastic fibers in the dermis. Young skin averages a 92 percent resiliency; aged skin, 81 percent. It is of interest to note that application of estrogenic hormone preparations to nonresilient skin can lead to redevelopment of elastic fibers and an increase in blood supply of the dermis. Some gain in resiliency may be realized. The TENSILE STRENGTH of the skin refers to the force required to elongate it. In general, the younger the skin, the more easily it stretches. Older skin is tougher.

ELECTRICAL BEHAVIOR. The skin behaves as though it were a negatively charged membrane. It may absorb anions, but does not pass them to the inside. Basic dyes penetrate easily, acidic ones do not. An "electrical double layer," consisting of H^+ externally and OH^- internally, exists at the line of junction between the cornified layers of the skin. This layer tends to limit the passage of charged substances.

PROTECTION. The skin presents several lines of defense against the environment.

A SURFACE FILM composed of water, lipids, amino acids, and polypeptides is derived from the secretions of the sweat and sebaceous glands and from the breakdown of the cornified surface cells. It has a pH of 4–6.8 and forms an effective antiseptic layer, retarding the growth of bacteria and fungi on the skin surface. Its water content *moistens* the skin, and the lipids *lubricate* and aid in waterproofing the surface. The *waterproofing* function is not perfect, as evidenced by the wrinkles skin develops (fingers) on long exposure to water. This is caused by water entering the upper layers of the skin and causing it to swell and fold.

The presence of the horny layer of the skin, with its KERATINIZED CELLS, acts as a physical barrier to the penetration of most chemicals, bacteria, parasites, and a good part of the environmental radiation (e.g., sunlight). At the junction between the cornified and noncornified layers of the skin is a region of positively and negatively charged ions that will repel charged substances trying to enter the body.

These devices, which prevent entry of substances into the body from outside, also act to prevent the loss of essential body constituents from the inside. Thus the body does not "dry out," or lose essential proteins, salts, and other chemicals.

PERCUTANEOUS ABSORPTION. This term refers to the penetration of substances through the skin and into the bloodstream. Historically, the skin was considered to be impermeable to all substances. We know today that gases and lipid soluble substances pass relatively easily through the skin, but that the skin remains essentially impermeable to electrolytes and water. Such results imply a barrier to the penetration of materials. The barrier does not lie in the corneum. The corneum behaves like a gross sieve. An initial barrier is imposed by the electrical double layer at the cornified-noncornified epidermal junction. The electrical properties of this area probably limit the passage of electrolytes. A second barrier exists at the basement membrane. Most noncharged substances appear to be held up at this level. If a substance passes these two barriers, no restriction is imposed on its entry into the blood stream.

Fat solubility increases penetration, chiefly because the skin surface is covered by a waxy layer composed of cholesterol, cholesterol esters, and waxes. This layer hinders the penetration of aqueous solutions, as well as facilitating penetration of solutions of oils. Substances that dissolve keratin (keratolytics), including salicylic acid (aspirin), enhance penetration of fat-soluble substances.

A second possible route of passage through the skin is afforded by the hair follicles and sweat glands. These might be suspected of affording an easier route of entry, since no penetration of a primary barrier is involved. Some absorption takes place through hair follicles, but apparently none occurs through sweat glands.

Some specific substances and their routes and manner of passage through the skin are summarized in Table 6.5.

These data suggest that application of substances

TABLE 6.5 Percutaneous absorption by the skin

Substance	Route	Reason for penetration
Phenols	Skin surface	Coagulates skin proteins and destroys barrier
Hormones (testosterone, estrogen)	Skin surface	Fat soluble
Vitamins (A,D,K)	Skin surface	Fat soluble
Organic bases	Skin surface	Fat soluble
Aspirin	Skin surface	Keratolytic
Gases (O_2, CO_2)	Skin surface	Are not changed; small molecules which diffuse rapidly
Animal and vegetable fats	Hair follicle	

directly to the skin has practical value in the treatment of certain conditions. If the object is to achieve a higher local concentration of substance, it is of value. Remember, however, that the vehicle employed to dissolve the substance is important, as is the depth to which any material may penetrate the skin.

SENSORY FUNCTIONS. The skin mediates four main modalities of sensation:

Pain, including itch and tickle
Touch, light and heavy
Cold
Warmth

The distribution of the "sensory spots" serving the four major modalities is in the form of a mosaic and differs in density in different areas of the body. The sense of light touch is attributed to the *Meissner's corpuscles* (Fig. 6.28), irregular areas about 5 mm in diameter. They are most numerous on the fingertips and on the lips. The sensation of touch may also be aroused by bending of hairs. Heavy touch is assigned to the *Pacinian corpuscles* (Fig. 6.28). These receptors lie deeper in the dermis than the Meissner's corpuscles and require a heavier pressure to be stimulated.

Pain elicited from the skin has two qualities. It may be bright and pricking (temporary and easily localized), or burning and longer lasting.

Thermal sensations are attributed to *Ruffini corpuscles* (warmth) and *Krause corpuscles* (cold). Excitation of thermal receptors occurs when the normal temperature gradients are changed by thermal stimuli.

THERMOREGULATION. The body loses heat by four methods:

Radiation involves the transfer of heat from a warmer to a cooler area by electromagnetic waves (infrared). Under normal conditions, about 60 percent of the body's heat loss is by this method.

Conduction is transfer of heat from a warmer to a cooler area if they are in contact. Normal clothing plays little role in changing physical heat loss. Heat stasis may result, with change of total body temperature if insulating materials are employed.

Convection is transfer of heat to a moving medium,

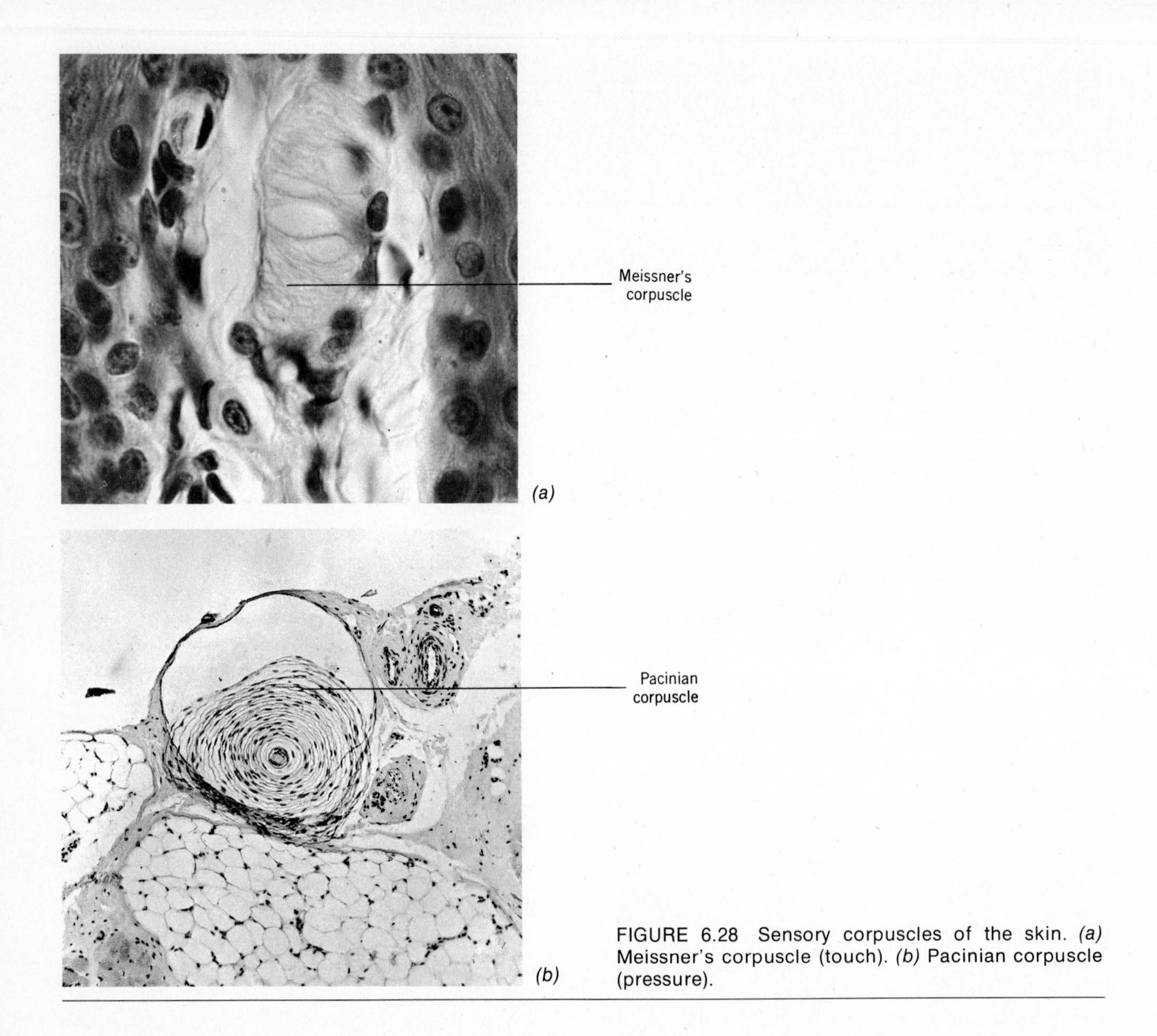

FIGURE 6.28 Sensory corpuscles of the skin. *(a)* Meissner's corpuscle (touch). *(b)* Pacinian corpuscle (pressure).

such as air or water. Convection carries off a great deal of heat as long as the moving medium has a lower than body temperature.

Evaporation causes heat loss because of the heat required to vaporize water on the skin surface.

The role of the skin in heat loss revolves primarily around its ability to act as a radiant surface and transfer heat from the cutaneous blood vessels to the exterior. Obviously, the amount of heat available for transfer depends on the diameter of the cutaneous vessels and the amount of blood therefore circulated through the skin. Vessel diameter is nervously controlled (Fig. 6.29). Within the hypothalamus are "heat-gain" and "heat-loss" centers. These centers receive input information from superficial and deep-lying thermal receptors, and they also monitor the temperature of the blood. The heat-gain center sends nerve fibers to cutaneous vessels, which constrict them, and to skeletal muscles to increase their tone. Heat radiation through the skin is thus decreased, while heat production is increased. The heat-loss center sends fibers to cutaneous vessels to dilate them and to sweat glands to stimulate secretion. Radiation and evaporation are increased.

Sweating forms a second mechanism of thermoregulation directly involving the skin. Eccrine sweat glands produce the clear sweat responsible for heat regulation. Eccrine sweat is composed of 99–99.5 percent water and 0.5–1 percent solids. Half of the solids are inorganic, chiefly NaCl; the other half are organic, including urea, lactic acid, amino acids, glucose, and B vitamins. Freshly secreted eccrine sweat has a pH of 5.73–6.49. Control of eccrine secretion is nervous to those glands located on the palms, soles, head, and trunk. Glands in other areas appear to respond directly to local application of heat. The term *hyperidrosis* means excessive sweating. Apocrine gland secretion contains protein, sugars, ammonia, ferric

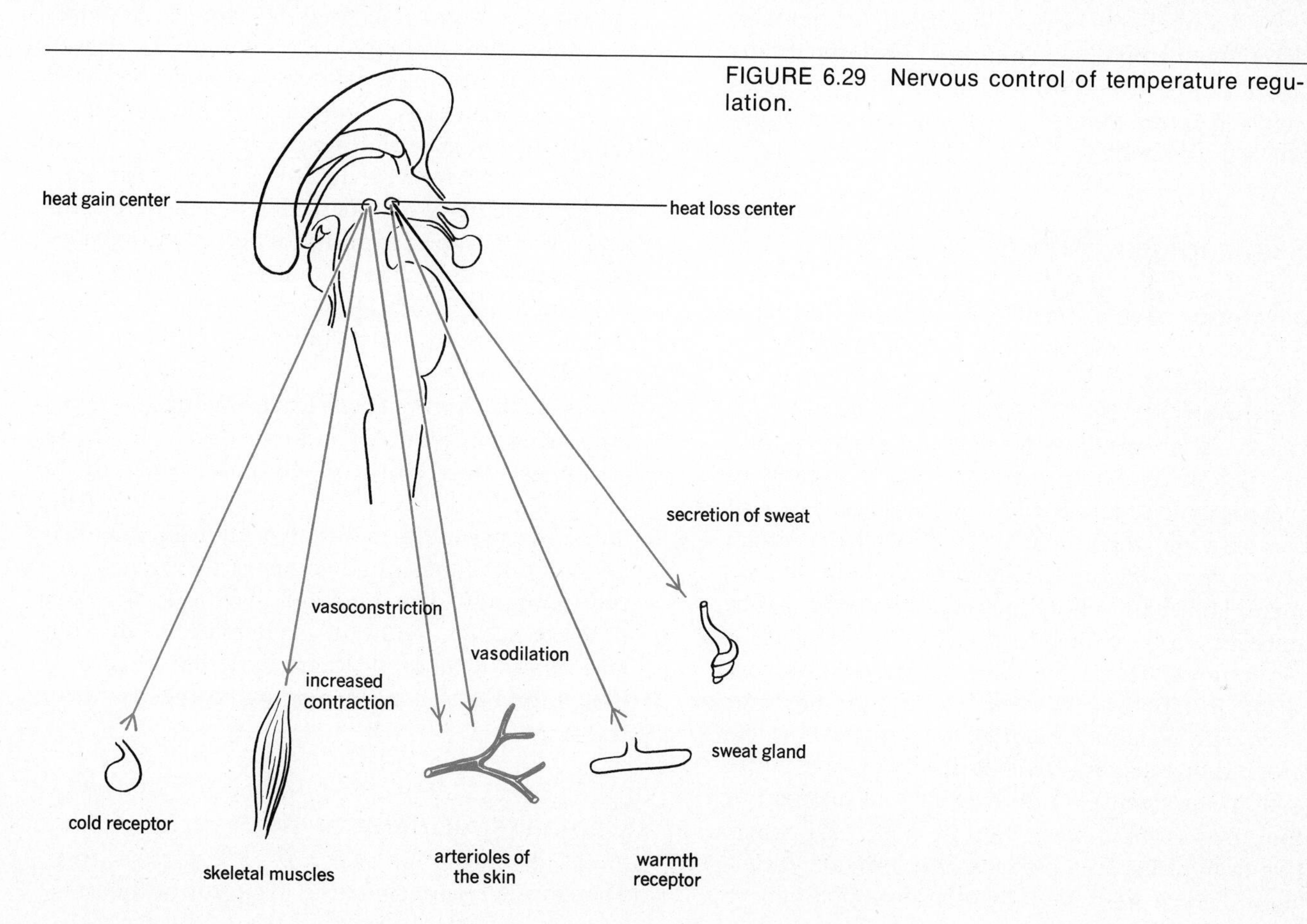

FIGURE 6.29 Nervous control of temperature regulation.

ion, and fatty acids. The secretion is cloudy, may contain odoriferous substances (sex attractants in some animals), or be colored (hippopotamus, "sweating blood"). Bacterial action on the constituents of apocrine sweat may increase odor, but is not necessarily required to create an odor. Apocrine glands receive a dual innervation and respond primarily to mental rather than thermal changes.

SYNTHETIC FUNCTIONS. Although the skin manufactures many substances used in its own economy (for example, KERATIN), one substance is manufactured that exerts effects on areas far removed from the skin. VITAMIN D is manufactured from dehydrocholesterol, a sterol found in skin. Through the action of ultraviolet light on the above compound, vitamin D_3 is formed, absorbed into the blood stream, and becomes involved in the metabolism of calcium and phosphate in the body.

Skin and hair color

Most of us have some color to our skin and hair. The color itself and the intensity of the color are determined by several factors.

PIGMENTS in the skin and hair cells are produced by a variety of biochemical reactions that are genetically determined. *Melanin* is a yellow to black pigment found only in the basal layer of the skin of white people, but in all epidermal layers in the skin of black people. Amount of pigment, and therefore darkness of skin color, is not determined by a single gene, but by as many as eight genes. Thus, the range of skin color is very great. In albinos (*albus,* white) a gene has mutated, and melanin cannot be synthesized because of a missing enzyme. Such individuals have white skin (no pigment), white hair, and no pigment in the irises of their eyes. Their eyes are very sensitive to light because the lack of iris pigmentation allows much light to enter. *Carotene* is a yellow pigment found in the horny layer of the skin. It is more abundant in the skin of certain Asian races. Mixtures of the two pigments may give yellows and reds of various intensities.

Pigments in the skin afford protection against the damaging effects of solar radiation on vascular elements and other structures beneath the epidermis.

Hair color is determined by the same pigments as are found in the skin, and intensity of color in both skin and hair are correlated. As we age, pigment synthesis in the hair follicles diminishes, and many of the hair cells contain air; this causes graying or whitening of the hair.

The AMOUNT OF BLOOD circulating through the skin and the thickness of the skin determines its "pinkness." Thinner skin allows more color of the blood to show through. The amount of HEMOGLOBIN WITH OXYGEN ON IT also determines skin color. Richly oxygenated hemoglobin is bright red; hemoglobin with little oxygen on it is bluish in color. If sufficient "blue hemoglobin" is present in the capillary networks of the skin, a blue color *(cyanosis)* is imparted to the whole skin. *Jaundice* is a yellow color imparted to the skin by excessive red blood cell destruction, or liver malfunction. It results from excessive bilirubin (a pigment) in the blood stream.

EXPOSURE TO ENVIRONMENTAL FACTORS, particularly radiation, can also influence skin color. The ultraviolet radiation of sunlight is damaging to the blood vessels and other tissues of the skin. One response to the radiation is tanning, in which the amount of melanin is temporarily increased in the epidermis. The pigment then absorbs more of the radiation. Sunlight also dehydrates the skin and makes it more "leathery," so that a tan may look nice but can be damaging to the health of the skin.

Clinical considerations

The skin is a very sensitive indicator of the presence of abnormal processes at the skin surface, and within the body itself. Rashes or eruptions of the skin may occur as a result of infection or other processes.

LESIONS. Skin eruptions ("breaking out" with a visible change in the skin surface) accompany many diseases. The changes that occur follow a similar series of events regardless of the agent causing them. A given condition may stop in any one of the stages, or progress through all of them. The terms below are given in order of their appearance. A good example on which to observe the follow-through of each stage is the smallpox vaccination. The various stages through which skin lesions progress are:

Macule. A discolored area on the skin, neither raised nor depressed.

Papule. A red, elevated area on the skin.

Vesicular Stage

Vesicle. A pinhead to split-pea-sized elevation filled with fluid.

Pustule. A vesicle containing pus.

Bulla. A large vesicle or blister.

Crust. Dry exudate (weeping of fluid and/or pus) adhering to the skin.

Lichinification. Thickening and hardening of the skin.

Scar. Replacement of cells by fibrous tissue as healing progresses.

Keloid. A large, elevated scar.

COMMON DISORDERS OF THE SKIN. INFECTIONS are perhaps the most common disorders affecting the skin. *Bacteria* (mostly staphylococcus and streptococcus) are the most common infectious agents. Boils are a result of coccal infection. *Viruses* may produce warts or herpes (itching vesicles). *Fungi* produce cracks or fissuring as in athletes foot and scalp infections (ringworm). The fungi thrive in warm, moist environments, making for easy transfer from a tennis shoe to a locker room and showers or swim areas. Insects such as mites, lice, bees, and spiders may bite or sting, causing eruptions on the skin.

ACNE or "pimples" is a physiological disturbance of puberty and early adulthood affecting more than 80 percent of teenagers. It is caused by the effect of androgenic hormones (male sex hormones) on the sebaceous glands and hair follicles. It occurs in both sexes, because androgens are produced by the adrenal glands, testes, and ovaries. Acne occurs in four grades, depending on severity.

Grade I acne consists of the formation of a fatty-keratin plug, a blackhead, in the opening of a follicle on the skin.

Grade II acne results when the plugged duct ruptures and spills sebum into the surrounding tissues. The sebum irritates the tissue and a pustule forms. Bacteria may also be trapped in the tissue.

Grade III acne occurs when there is tissue destruction by the inflammation, and scarring results.

Grade IV acne results in lesions extending to the shoulders, arms, and trunk, with severe scarring occurring.

Treatment involves washing with a mild soap and warm water as needed, and avoiding foods that may aggravate the condition (e.g., chocolate, nuts, cola drinks). Sun may be helpful, and some of the newer topical medications (e.g., vitamin A preparations) are promising. Squeezing or "popping" the pustules should not be attempted since this may increase chances of scarring. Too frequent washing also may traumatize the skin. If the face is disfigured by acne, psychologically sensitive individuals may withdraw from social contact with their fellows, with resultant need for counseling. Dermabrasion, in which the skin is "ground down" to the level of the dermal upfolds, may be utilized to remove scars. This method of treatment is done only with the advice of a skin specialist, because not all types of acne scarring will be improved by the procedure.

Eczema, seborrhea, and psoriasis are other disorders of the skin whose names are commonly read or heard in the mass media.

ECZEMA refers to a chronic irritation of the skin in which there is redness, eruptions, watery discharge, and formation of crusts and scabs. Its cause is not known, and it probably represents a symptom rather than being a disorder in itself.

SEBORRHEA is a disorder of the sebaceous glands in which there is excessive production of sebum that may accumulate as crusts and scales on the head, face, or trunk.

PSORIASIS results in the formation of silvery scales on scalp, elbows, knees, or the body generally. The cause is unknown, and the disease can occur at any age in both sexes. Treatment is directed toward relief of symptoms, as there appears to be no cure.

FIGURE 6.30 The rule of nines, used to calculate the extent of burns.

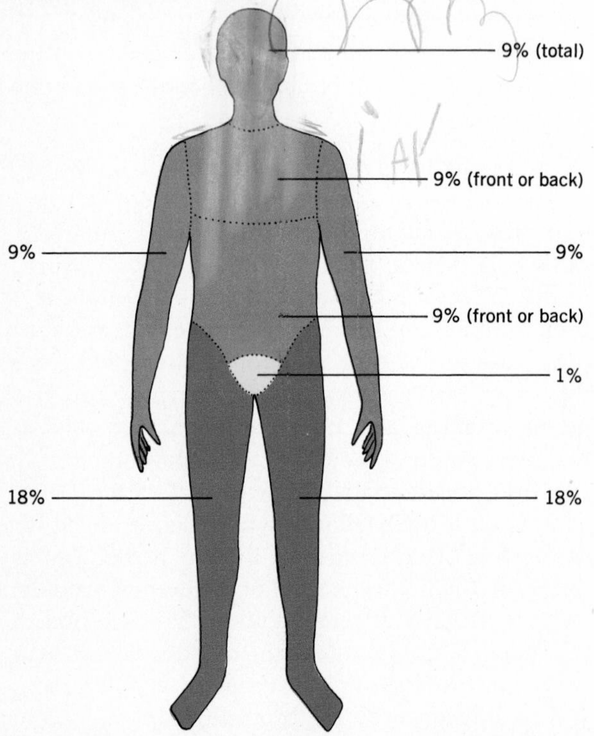

BURNS. One of the first hazards humans recognized in their efforts to change the quality of their survival was burns. Today nearly one million people are burned severely enough to be hospitalized annually, and about 7000 of these people expire. Burns are one of the major categories included under accidents, the first leading cause of death between 1–44 years of age.

EXPOSURE TO FLAME, scalding, hot objects, or some chemicals destroys the skin, and its protective functions are lost. INFECTION and loss of fluids, electrolytes, and blood proteins follow as a result of skin destruction. A burn is described in two ways: first, by the extent of destruction of surface area [the "rule of nines," (Fig. 6.30.) is a convenient method of estimating the extent of destruction]; second, the depth of the burn is described by "degrees."

FIRST-DEGREE BURN involves only the surface layers of the epidermis. The skin is reddened and tender to the touch. A sunburn is a good example of a first-degree burn.

SECOND-DEGREE BURN is one in which much of the epidermis is destroyed, but some epidermal remnants are present. It extends into the dermis. There is redness, and blisters are usually present. A burn resulting from briefly touching a hot object is usually a second-degree burn. This type of burn is usually very painful because of irritation of the nerves of the dermis by products of cell destruction.

THIRD-DEGREE BURN involves all skin layers, with no epidermal remnants present in the burned area. Charring of the skin is common, and destruction may involve muscles, tendons, and bones. The skin is insensitive to stimuli because of destruction of nerves in the dermis and hypodermis. On healing, fibrous masses of dead tissue *(eschars)* may form, which limit movement at joints. Third-degree burns may result from prolonged contact with hot objects, from open flame, steam (car radiators), chemicals (strong acids or alkalies), and hot liquids (paraffin, oil).

There are three considerations of primary importance in the treatment of burns:

Prevent, as best as possible, secondary infection in the burned area.

Maintain water and electrolyte homeostasis by administration of appropriate solutions. Urine output is a convenient measure of adequacy of fluid intake.

Encourage adequate nutrition. A burned individual is one who is not interested in eating. Intravenous *(IV)* feeding may be required. Mobilizing the patient as early as the burns permit encourages recovery and feeding.

Summary

1. Cells that are similar form tissues. Four tissue groups compose the body: epithelial, connective, muscular, and nervous tissues.
2. Epithelia have special characteristics.
   a. They cover and line free surfaces of the body.
   b. They are composed of closely packed cells.
   c. They are active as secretory and absorptive tissue.
   d. They rest on a basement membrane and are nourished from blood vessels beneath the epithelium.
   e. They show great powers of regeneration.
3. Epithelia are either one-layered (simple) or many-layered (stratified) and contain cells that are flat (squamous), cuboidal (cube-shaped), or columnar (tall).
   a. Simple squamous epithelium is one layer of flat cells. It is adapted for exchange of materials.
   b. Simple cuboidal epithlium is one layer of cube-shaped cells. It is adapted for absorption and secretion.
   c. Simple columnar epithelium is one layer of tall cells. It is an absorptive and secretory tissue.
   d. Stratified epithelium is several layers of cells, with the top layer flat. It is adapted to resist mechanical wear and tear.
   e. Pseudostratified epithelium lines the respiratory and parts of the reproductive systems. It generally has cilia.
   f. Transitional epithelium lines the urinary system and stretches as the organs expand.
   g. Syncytial epithelium has no membranes between cells; it covers the placenta.
   h. Germinal epithelium lines the testis tubules and covers the ovaries. It produces sperm in the testes.
   i. Neuroepithelium is specialized for reception of stimuli. It is found in the organs of special sense (eye, ear, nose, tongue).
4. Epithelia and their underlying connective tissue form epithelial membranes.
   a. Mucous membranes line cavities that open on a body surface. They are moistened by mucus.
   b. Serous membranes line internal body cavities. They are moistened by a watery secretion.
5. Glands are derivatives of epithelia.
   a. Exocrine glands secrete onto a body surface.
   b. Endocrine glands secrete into the blood stream.
6. Connective tissues have special characteristics.
   a. They connect, support, and protect body structures.
   b. They have large amounts of intercellular material that is usually fibrous.
   c. They are vascular.
   d. They contain scattered cells.
7. There are several important types of connective tissues.
   a. Loose connective tissue contains strong white fibers and stretchy elastic fibers. The cells called fibroblasts produce fibers. Loose connective tissue occurs under the skin.
   b. Adipose tissue contains many fat cells. It stores energy, insulates, and gives form to the body.
   c. Reticular tissue forms the internal framework of many body organs (e.g., spleen, liver, lymph nodes).
   d. Fibrous connective tissue is dense, forms tendons and ligaments, and is found in blood vessel walls.
   e. Cartilage is semisolid and forms part of the skeleton (hyaline), discs between vertebrae (fibrous), and outer ear (elastic).
   f. Bone is hard, forms the skeleton, and protects body organs.
8. The components of the intercellular substance of a connective tissue are produced by the characteristic cells of that tissue. Connective tissues reflect nutritional or hormonal deficiencies very quickly.
9. Connective tissues may form abnormally and result in malformed skeletal structures.
10. Muscular tissue is contractile and occurs in three varieties.
    a. Skeletal muscle attaches to and moves the skeleton.
    b. Cardiac muscle is found in the heart and circulates the blood.
    c. Smooth muscle is found in internal organs and controls blood pressure and movement through hollow organs.
11. Nervous tissue is excitable and conductile and forms sensory, interpretive, and motor pathways.
12. The skin is the major organ of the integumentary system and has three main layers.
    a. Five layers occur in the epidermis.
    b. The dermis contains fibrous connective tissue, sensory corpuscles, blood vessels, and glands.
    c. The subcutaneous tissue contains much fat.

13. Hairs, nails, and glands are appendages (derivatives) of the skin.
   a. Hairs lie in a follicle and grow from the follicle.
   b. Nails are hardened structures useful in grasping small objects.
   c. Sweat glands help cool the body; sebaceous glands lubricate the hair and skin.

14. The functions of the skin include:
   a. Protection. A surface film of chemicals acts as an antiseptic and reacts with toxic materials. The horny layer forms a physical barrier to entry of microorganisms and radiation. Loss of body components is also prevented.
   b. Absorption. Gases and lipid soluble materials pass through the skin, as do those substances that dissolve the keratin in the epidermis.
   c. Reception of stimuli. Sensory receptors sensitive to heat, cold, touch, pressure, and pain are found in the skin.
   d. Temperature regulation. Heat radiates from the blood passing through the skin. Sweat glands produce a fluid that evaporates and cools the skin.
   e. Synthesis of materials. Keratin and vitamin D are produced by skin cells.
   f. Excretion. Water, salts, urea, and ammonia are excreted by the skin glands.

15. Skin color depends on:
   a. Pigment cells
   b. Blood flow
   c. Oxygen levels of the blood
   d. Bilirubin levels
   e. Exposure to sunlight

16. Hair color is determined by pigments or air in the hair cells.

17. The skin is a sensitive indicator of whole-body physiology.
   a. It may develop lesions; a common series of phases occurs.
   b. It may develop disorders including infections, acne, seborrhea, and psoriasis.
   c. It may be burned.

18. Burns are described as to the extent of burning and depth.
   a. Extent is referred to by percent.
   b. Depth is referred to as first, second, and third degree. Reddening and tenderness characterizes a first-degree burn; destruction, but some epithelial remnants, characterizes a second-degree burn; existence of no epithelial remnants characterizes a third-degree burn.

19. Treatment of burns is centered around prevention of secondary infection, maintenance of fluid and electrolyte balance, and maintenance of adequate nutrition.

7

The Skeletal System

The skeletal system

Definition

The skeletal system of the human is made up predominantly of organs called BONES, and of minor components of cartilage. The bones are joined at the articulations or joints, enabling them to be moved in meaningful relationship one to another. The skeletal muscles provide the energy source for movement, and are capable of converting stored chemical energy into mechanical energy, that is, energy of action.

Kinds of skeletons

The skeleton of humans, as of other vertebrates, is a living internal or ENDOSKELETON. As such, it grows as the body grows; it adapts itself to the condition of life of the individual. It is capable of self-repair following disease or injury. This is in contrast to the external or EXOSKELETON so well demonstrated in the insects and other Arthropods. The exoskeleton is nonliving, and is produced by underlying living tissues. The organism, to grow, must shed the exoskeleton and then replace it after growth is accomplished. It does not have the adaptability of an endoskeleton. It is, however, a good protection for the animal. The exoskeleton in vertebrates is limited to scales, shells *(as in turtles),* feathers, and *(in humans)* to nails and hair. In this chapter we are concerned only with the endoskeleton.

Functions

The skeletal system is a FRAMEWORK OF SUPPORT for the soft tissues of the body. It is basic to the FORM OF THE BODY, and, in the human, its ERECT POSTURE. The skeletal muscles attach to the bones and, in the appendicular skeleton especially, the BONES ARE USED AS LEVERS, with the articulations acting as fulcra around which movement takes place. The skeleton therefore plays a PASSIVE, but essential, role in MOVEMENT.

The skeleton provides PROTECTION for vital organs, such as the central nervous system, which is housed in the cranial cavity *(formed by the bones of the skull)* and vertebral canal *(formed by the vertebrae).* The heart, lungs, and some major blood vessels are enclosed in the thoracic cavity with its protective framework of vertebrae, ribs, sternum, and costal cartilages.

The skeleton is a RESERVOIR OF MINERALS such as calcium, phosphorus, magnesium, and citrate, and the bones are involved in the metabolism of these substances. Certain bones also serve as centers for blood cell formation or HEMOPOIESIS. Blood formation takes place mostly in red bone marrow in the proximal epiphyses of the femur and humerus, in the ribs, sternum, clavicle, os coxae, vertebrae, and in the diploë (spongy bone) of cranial bones. Yellow bone marrow, found mostly in the shafts of long bones, consists primarily of fat cells and may also become active in the formation of red cells, granulocytes, and platelets. It is an emergency reserve for blood cell formation.

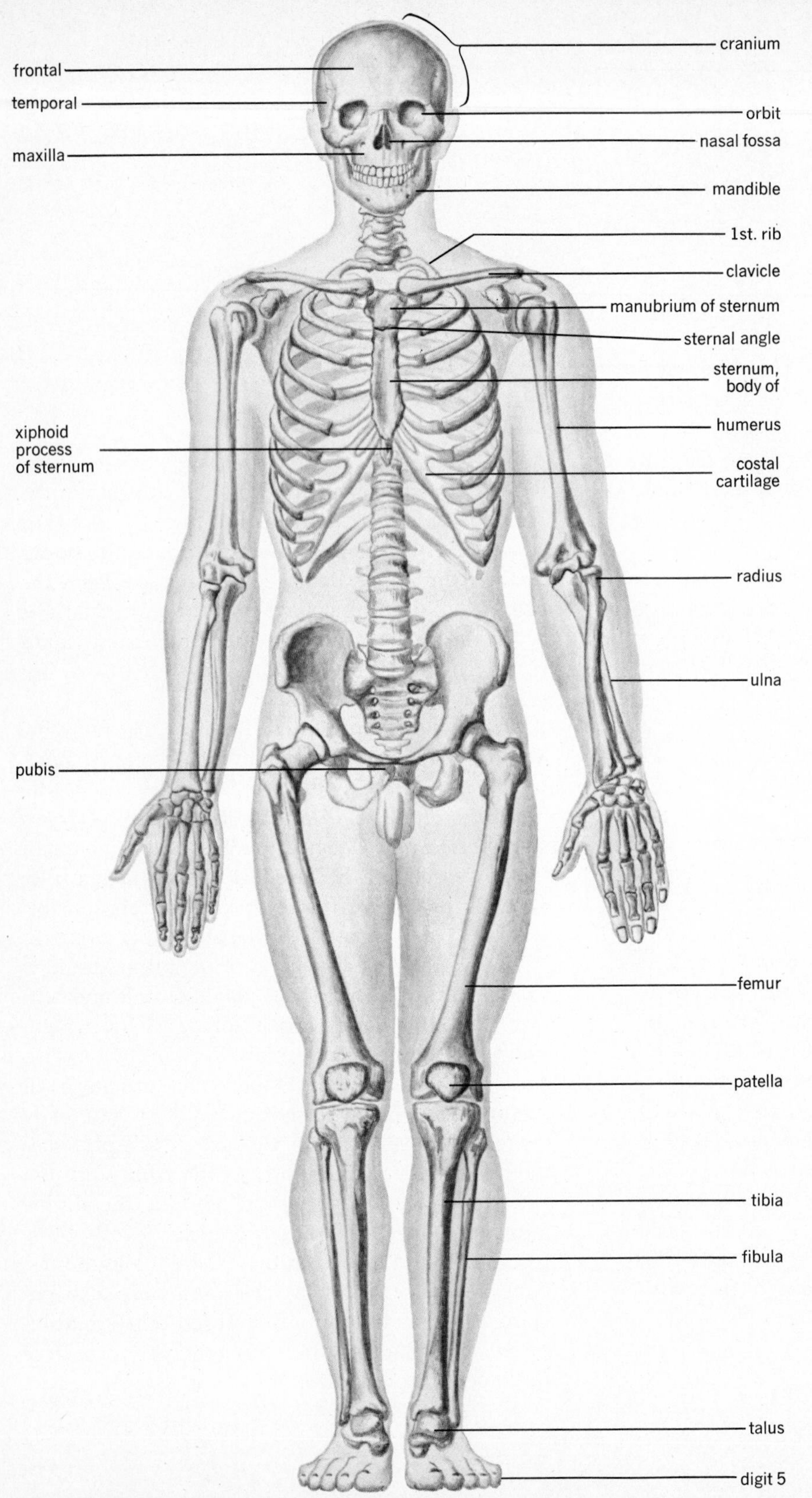

FIGURE 7.1 Anterior and posterior view of the skeletal system.

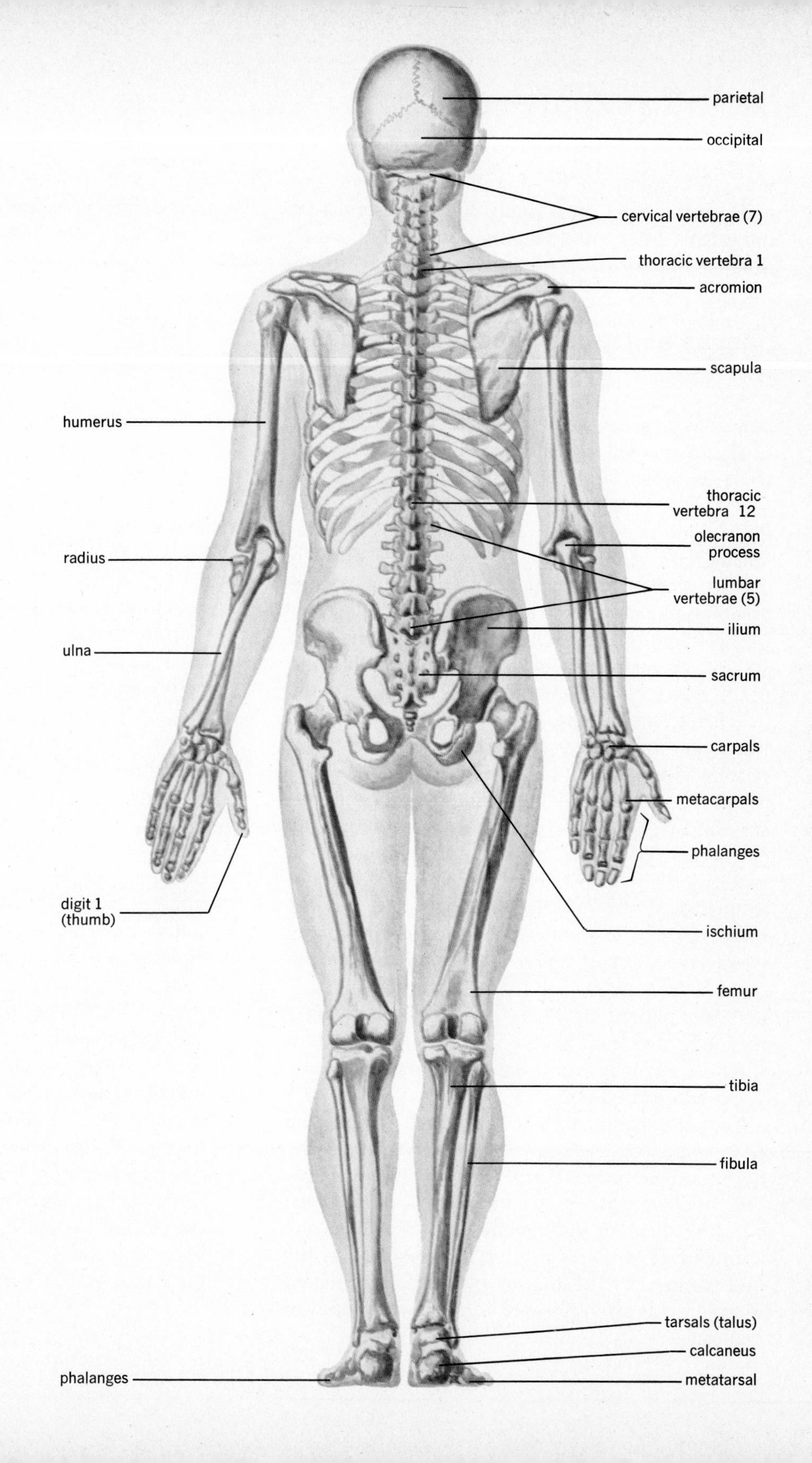

parietal
occipital
cervical vertebrae (7)
thoracic vertebra 1
acromion
scapula
humerus
thoracic
vertebra 12
olecranon
process
radius
lumbar
vertebrae (5)
ilium
ulna
sacrum
carpals
metacarpals
phalanges
digit 1
(thumb)
ischium
femur
tibia
fibula
tarsals (talus)
calcaneus
phalanges
metatarsal

From another point of view, bones, because they are among the most stable of body organs and are the ones most likely to be preserved in the earth's crust, are a valuable source of information for EVOLUTIONARY STUDIES. They have been the objects of search for centuries by archeologists and paleontologists in all parts of the world, and have contributed much to our understanding of the history of life and of humans.

Shapes and gross structure of bones

Bones, as organs, usually are classified according to shape and fall logically into four groups: long, short, irregular, and flat (Fig. 7.1).

LONG BONES are those whose length is greater than their width. The humerus, femur, tibia, radius, and ulna are obvious examples. Though not so obviously, the bones of the fingers and toes, the phalanges, are also long bones.

The femur, as an example of a long bone (Fig. 7.2), is characterized by a long shaft called the DIAPHYSIS and two enlarged and modified ends, the EPIPHYSES. If the femur is cut lengthwise, two types of bony substance are apparent: the COMPACT BONE forming the thick wall of the diaphysis and a thinner covering over the epiphyses, and SPONGY OR CANCELLOUS BONE in the epiphyses. The cancellous bone is composed of a pattern of latticework, bars, and spicules of bone-enclosing spaces which in life contain the red bone marrow. The pattern of the cancellous bone, if observed superficially, may seem to have no organization. However, critical studies indicate that the pattern develops gradually as a young person grows and matures and is arranged in such a way as best to resist the stresses normally imposed on the bone.

The diaphysis has an internal space, the MEDULLARY OR MARROW CAVITY, which in the living subject is filled with yellow bone marrow. The hollow shaft is so constructed as to resist bending due not only to the load carried on the femur head but also to other stresses such as the pull of muscles. The bone as a whole is engineered to give great strength with a minimal amount of material.

The articulating surfaces of long bones are covered with smooth, shiny ARTICULAR CARTILAGE, of the hyaline type. The rest of the living bone has an external covering of a fibrous membrane, the PERIOSTEUM. A condensation of the stroma of the bone marrow forms a lining of the internal surfaces of long bones. This is called the ENDOSTEUM.

SHORT BONES are those whose length and width are roughly equal and which are closely joined by ligamentous structures. They are found in the carpus *(wrist)* and tarsus *(ankle)* regions of the skeleton. They consist of an inner cancellous structure covered externally by a thin layer of compact bone. They contain red bone marrow in life.

IRREGULAR BONES are those having complex shapes and structures, such as the vertebrae and some of the bones of the skull—the ethmoid and sphenoid. They have a cancellous structure internally and are covered by compact bone. Red bone marrow occupies the internal spaces.

FLAT BONES are composed of two more-or-less parallel plates of compact bone, the OUTER AND INNER TABLES, separated by cancellous bone, the DIPLOË. They present large surfaces for muscle attachment. Red bone marrow occupies the spaces in the diploë. The scapulae and bones of the cranial vault are good examples of flat bones.

Short, irregular, and flat bones, like the long bones, are provided with periosteum except on their articulating surfaces. Endosteum also lines their spongy bone spaces.

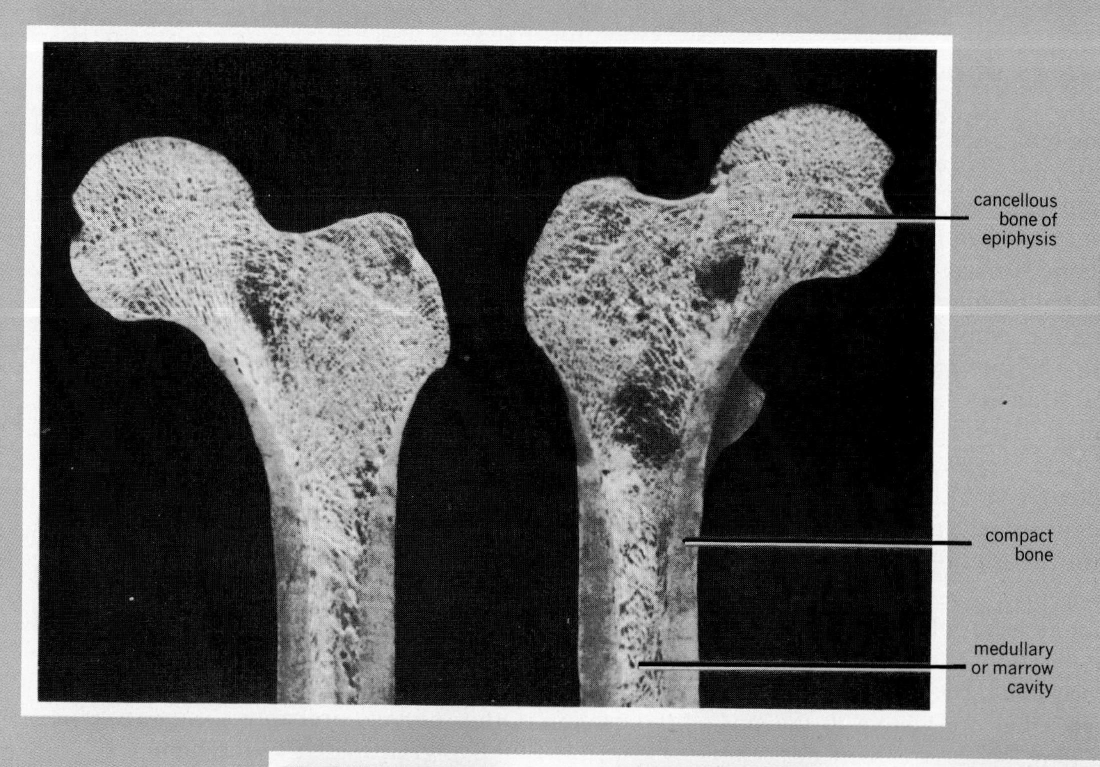

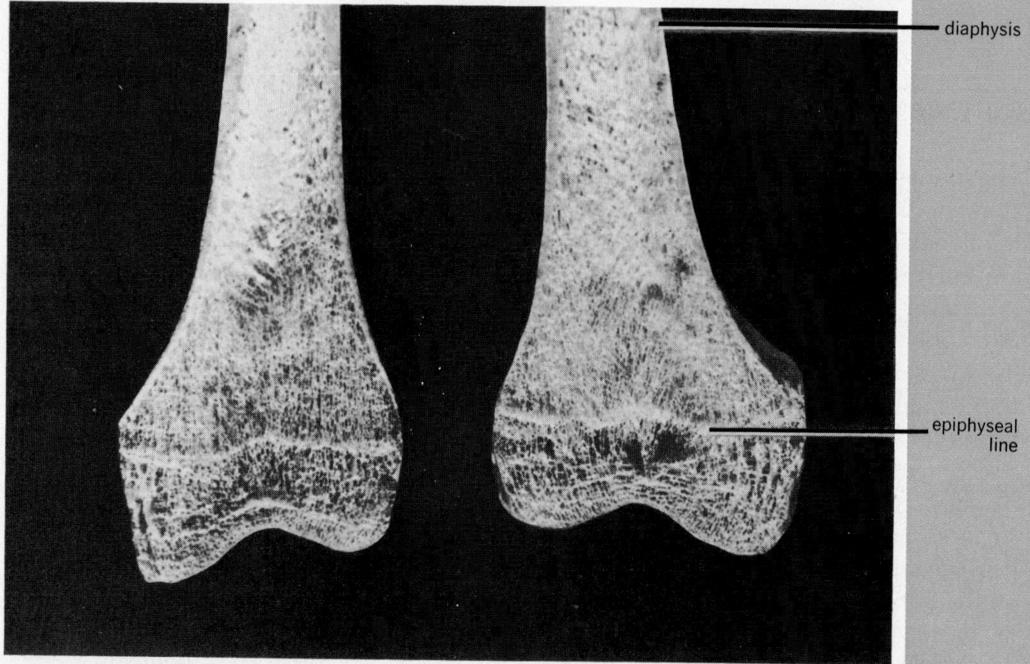

FIGURE 7.2 Photographs of the proximal and distal ends of a split femur.

Vessels and nerves of bones

Bone as living tissue requires a blood supply, and because it is a site for blood cell formation a particularly rich supply is necessary. Some evidence of such a supply is seen in the macroscopic FORAMINA on the bone surface; many more openings, not visible to the unaided eye, are present.

Blood vessels form a compact network in the periosteum of compact bone and from here send tiny branches, the arterioles, into minute openings in the compact tissue. The vessels then ramify through the canal system, Volkmann and Haversian (see below), of the compact tissue and reach the marrow. Some small vessels, the venules, retrace the course of the arterioles to return to the periosteum.

The cancellous bone is supplied by larger and less numerous vessels from the periosteum, that penetrate the compact bone. The foramina through which these vessels enter are visible to the unaided eye and in long bones are seen near the ends of the bones. Other large foramina in this area are for the return vessels, the veins.

Besides the vessels mentioned above, there are NUTRIENT ARTERIES and veins. In the long bones one or more prominent foramina are present midway in the diaphysis into which nutrient arteries enter and pass directly to the marrow cavity. Here they usually divide into proximal and distal branches to carry blood to the marrow and also to send branches to anastomose with vessels of the compact and cancellous bone. The veins from this system emerge in the company of the artery just described, through large foramina at the ends of long bones or through the tiny openings in the compact bone.

LYMPHATIC VESSELS also penetrate the canal system accompanying the vessels described above.

Many SENSORY NERVES are present in the periosteum. Other nerves, probably VASOMOTOR, accompany the arterial vessels in the bone.

Surface features of bones

The bones present many surface markings that suggest the function in which the bones are involved. The ends of weight-bearing long bones are enlarged, hard, smooth, and covered with a thin layer of hyaline cartilage. Articulating surfaces of bones vary in shape and partly determine the kind of action possible at joints. Bones have roughened areas and prominences of different forms for muscular, tendinous, and ligamentous attachments. Some bones have grooves for the passage of blood vessels and nerves over their surfaces and perforations through the bone for passage of blood vessels or nerves. Depressions on the bone surface may serve also for articulation or for attachment of ligaments or muscles.

Surface features of bone may be described as follows:

Articular surfaces

Condyle. A relatively large, convex prominence; for example, the occipital condyles or the prominences seen at the distal ends of a long bone such as the femur.

Facet. A smooth, flat, or shallow surface as seen on the thoracic vertebrae for the attachment of the ribs.

Head. A rounded surface often set off from the rest of the bone by a neck, or narrowed region, such as at the proximal ends of the femur and humerus.

Nonarticular surfaces

Crest. A prominent, often roughened border or ridge, such as the crest of the ilium.

Epicondyle. A prominence above or on a condyle, such as the epicondyles of the femur.

Fissure. A narrow, cleftlike opening, such as in the orbital fissure of the sphenoid bone.

Foramen. An opening through a bone for the passage of nerves and/or blood vessels.

Fossa. Generally, a deeper depression; for example, the olecranon and coronoid fossae at the distal end of the humerus.

Fovea. A shallow depression, such as the fovea capitis of the head of the femur.

Incisura. A notch, usually in the border of a bone, such as the sciatic incisura or notch.

Line. A slight ridge, such as seen in the linea aspera of the femur.

Meatus. A canal running within a bone, such as the external auditory meatus.

Notch. An indentation, usually in the border of a bone, such as the sciatic notch or incisura.

Process. The generic term for any prominent, roughened projection from a bone.

Sinus. Air spaces within certain bones of the skull connecting into the nasal passageways.

Spine. An abrupt projection from the bone surface that may be blunt or sharply pointed, such as the spine of the scapula.

Sulcus. A groove, such as the sagittal sulcus of the skull.

Trochanter. A large, blunt process found only on the femur.

Tubercle. A smaller, rounded eminence such as seen on the humerus and on most ribs.

Tuberosity. A large, often rough eminence such as seen on the ischium.

Histology of bone

The architecture of bone tissue can be studied either by cutting sections of dried bone and grinding them down to sufficient thinness to study by transmitted light under the microscope, or by steeping the bone in dilute acid to dissolve out the lime salts, a process called decalcification. This leaves the bone soft so that it can be bent, even tied in a knot, if the bone is slender enough, or sections can be cut and prepared for microscopic study.

A cross section of compact bone examined under a low power of the compound microscope reveals a number of areas, each with a pattern of concentric rings around a central canal (Fig. 7.3). These areas are called HAVERSIAN SYSTEMS OR OSTEONS. The central canal of an osteon is called the HAVERSIAN CANAL and in living bone contains blood vessels, nerve fibers, and, in some of the larger ones, lymphatic vessels. The concentric rings around the Haversian canal are called LAMELLAE and are composed of inorganic impregnated intercellular material and collagenous fibrils grouped into bundles following a parallel course. Between the lamellae and also arranged concentrically are small dark spots, actually spaces, the LACUNAE, which house bone cells. Radiating out from the lacunae through the lamellae are fine, dark lines that anastomose and actually represent tiny canals, the CANALICULI, which contain the processes of bone cells. Those from the lacunae closest to the Haversian canal enter that canal; others open onto the free surfaces of the bone.

Among the circular osteons are other lamellae with their lacunae and canaliculi running in various directions. They are called INTERSTITIAL LAMELLAE and represent remodeled or incomplete osteons. The outer part of the compact bone has lamellae running parallel to its circumference and hence are called CIRCUMFERENTIAL LAMELLAE. Interstitial and circumferential lamellae lack osteons.

A longitudinal section of compact bone reveals other important structural details. The Haversian canals can be seen to run more or less parallel to the long axis of the bone. They interconnect at

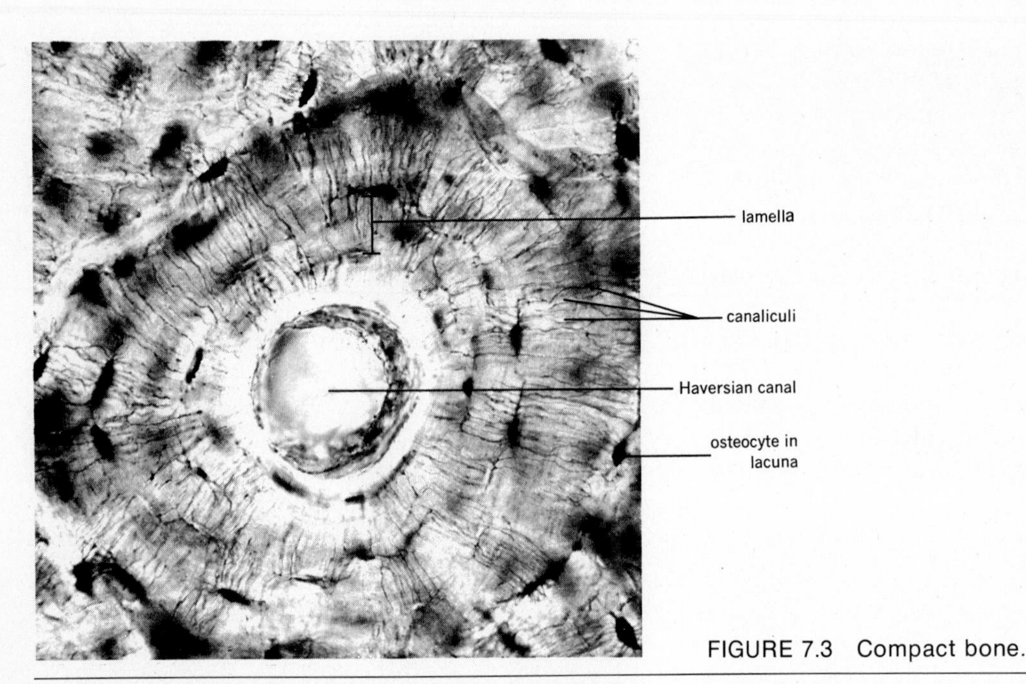

FIGURE 7.3 Compact bone.

FIGURE 7.4 A diagram of a longitudinal section of a bone to show Volkmann canals.

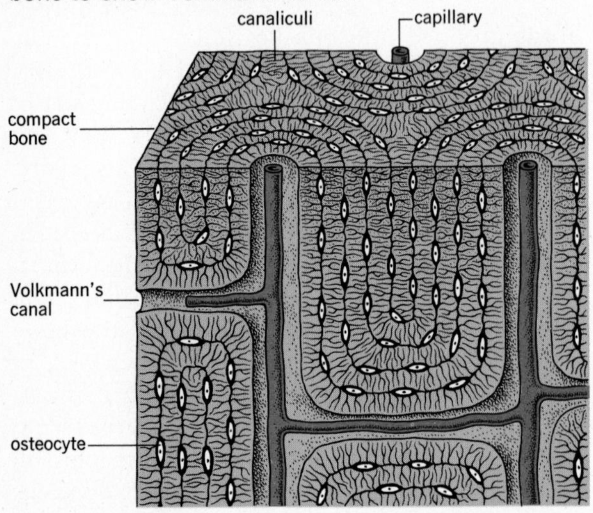

intervals and also send oblique canals to open on the external and internal surfaces of the bone. These are the VOLKMANN CANALS (Fig. 7.4). Since all of these canals carry blood vessels and nerves and many of them carry lymphatic vessels, they provide an effective system of circulation and communication within the living bone.

The lacunae of living bone house the bone cells or OSTEOCYTES, which with their long processes reach out through the canaliculi. Through the canaliculi, they are also able to receive essential nutrients from the blood supply and to return to it the waste products of metabolism.

In cancellous bone the spicules, plates, and bridges are made up of lamellae, canaliculi, and lacunae with osteocytes but there are no osteons. The spaces within cancellous bone carry blood vessels, nerves, and lymphatics and serve in a sense as do the Haversian canals in making nutrients and other materials available to the osteocytes.

Periosteum and endosteum

The PERIOSTEUM is a fibrous membrane covering the bone externally except on the articular surfaces. It varies in the tightness of its attachment to underlying bone, being more closely and firmly attached at the epiphyses of long bones than on the diaphysis. It is more firmly attached also at the points of origin and insertions of tendons and of ligaments. It is more firmly attached to short bones.

The periosteum is made up of two layers: the outer one is dense, fibrous, and vascular, and is made up mostly of collagenous fibers; the inner layer is more cellular, looser, and more elastic. The two layers are more clearly demarcated in developing bone when the bone-forming cells or osteoblasts are an obvious feature of the inner layer. While this layer does not contain osteoblasts in the adult, some cells may be converted to osteoblasts when a bone is stimulated, as in a fracture. The inner layer then becomes once again well defined. Vessels and nerves pass through the inner layer to reach the bone and its canal system.

It is by means of collagenous fibers of this inner layer of the periosteum, SHARPEY'S FIBERS, that the periosteum is attached to the underlying bone. These fibers pass through the circumferential *(periosteal)* lamellae like nails and hence sometimes are called PERFORATING FIBERS. They may be seen in some of the interstitial lamellae but not in those of the osteons.

ENDOSTEUM lines the marrow cavities, covers the surfaces of spongy bone plates and spicules, and extends as a lining into the canal system of compact bone. It is a delicate layer compared to the periosteum. It is composed of condensed connective tissue and serves in bone-forming *(osteogenic)* processes and in the resorption of bone, thus aiding in the modeling of developing bone.

OSTEOCLASTS are giant multinucleated cells found on bone surfaces where they are believed to be involved in bone destruction and remodeling. They sometimes are called bone-destroying cells. They are mildly phagocytic but likely produce the enzyme acid phosphatase to destroy bone. Bone resorption can and does take place in their absence.

Bone formation *(osteogenesis)*

Bone tissue formation, wherever it occurs, initially has a common pattern. It appears first in the form of spicules, either in fibrous membranes or in cartilage.

OSTEOBLASTS, or bone-forming cells, are differentiated from mesenchyme. They are medium-sized cells with delicate processes that often touch the processes of neighboring cells. These cells synthesize OSTEOCOLLAGENOUS FIBERS and an AMORPHOUS GROUND SUBSTANCE—a soft material called OSTEOID TISSUE.

The formation of osteoid is followed by deposition in it of lime salts—calcification. As these calcified areas form and enlarge to become spicules, osteoblasts form an epithelioid layer around them and continue to add to their size.

Osteoblasts themselves become imprisoned in their own matrix and become the bone cells or OSTEOCYTES. The spaces in which they are imprisoned are called LACUNAE, and the canals extending from the lacunae to enclose the cytoplasmic processes of the osteocytes are the CANALICULI.

As spicules elongate and branch they tend to interconnect, thus forming a network constituting spongy bone. Continued deposition on this spongy or cancellous network makes the tissue increasingly dense. This kind of bone, which is spongy and has lamellae not laid down in layers but at random, is characteristic of most bone until about the time of birth. After birth the nonlamellar fetal skeleton is replaced by one in which both spongy and compact bone are lamellar.

Other bone is laid down by osteoblasts in the periosteum to form the BONE COLLAR. The ENDOSTEUM also is involved in bone formation as well as in the resorption of bone.

Organogenesis of bones

Organogenesis involves the consideration of how specific bones develop and how they are influenced in form, strength, and effectiveness by mechanical and metabolic requirements.

Two types of bone are described on the basis of the tissues in which they develop. One, INTRAMEMBRANOUS BONE, is formed in a layer of primitive connective tissue, the mesenchyme, and when completed is often called MEMBRANE *(dermal)* BONE; the second, INTRACARTILAGENOUS OR ENDOCHONDRAL BONE, is preceded by a model of HYALINE CARTILAGE which must be replaced as the bone develops. These sometimes are called CARTILAGE or REPLACING BONES.

Both types of bone formation are involved in most bones, such as the temporal, sphenoid, and occipital. Every cartilage bone ends up quite largely a membrane bone through secondary processes.

Intramembranous ossification

This process (Fig. 7.5) is seen in its purest form in the bones of the cranial vault and the irregular facial bones. In other bones such as the sphenoid, temporal, and occipital, part of the bone is intramembranous, part endochondral. The surface layers of all bones are intramembranous.

An area where bone is to develop is called an OSSIFICATION CENTER. The earliest of such centers appears at about the eighth week of fetal life and is characterized by cell growth and proliferation and increased vascularity. The process that follows has been described above under osteogenesis.

At birth intramembranous ossification in bones has spread parallel to the flat surfaces of the membrane, and bones have met other bones along most of their margins. They are covered by a peri-

FIGURE 7.5 Intramembranous bone formation.

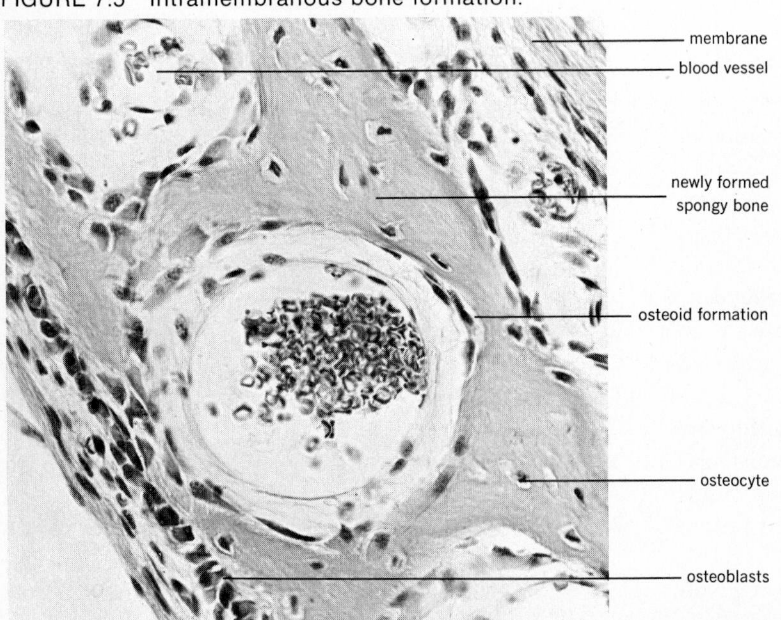

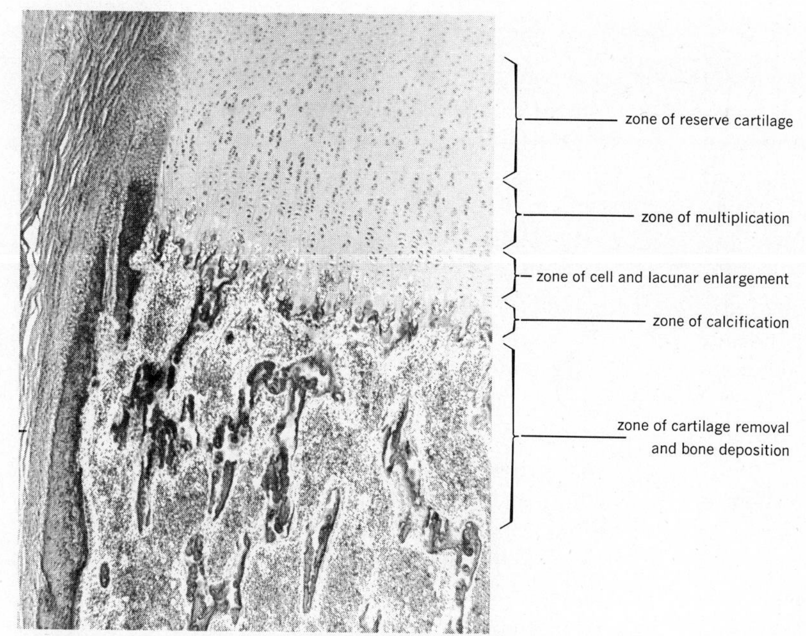

FIGURE 7.6 Intracartilagenous bone formation.

osteum, on the inner surface of which are osteoblasts that have helped to thicken the bone. At this stage the bone is a compact mass of cancellous bone, but now the osteoblasts of the periosteum begin to lay down parallel lamellae, and the tables characteristic of flat bones are formed. At the same time there is a thinning of the cancellous bone forming the larger spaces, the DIPLOË.

The curvature and increased size of the bones of the cranial wall necessitated by the growth of the brain are accomplished by differential growth and resorption of the bone surfaces. As one table of a flat bone grows, the other table is in part resorbed. At the same time, Haversian systems (osteons) develop within the compact bone of the tables (see page 146).

Intracartilagenous (endochondral) ossification

Since most of the bones of the skeleton are preceded by models of hyaline cartilage, the story of their development is more complicated than that of intramembranous bones. It is particularly complicated in irregular bones because these have many ossification centers of both primary and secondary types. The process is simplest in short bones for they are mostly without epiphyses. Long bones are intermediate in complexity of development, usually having one primary center for the diaphysis and one secondary center for each epiphysis.

The following description is of the development of a long bone such as a femur or humerus. Although the description is organized into discrete steps, one should remember that it is in fact a continuous process (Fig. 7.6).

1. The cartilage model is of the hyaline type and has a peripheral sheath of fibrous membrane, the PERICHONDRIUM.

2. Primary ossification centers appear as early as the second fetal month in some bones, and as late as childhood in short bones of the wrist *(carpus)* and ankle *(tarsus).*

3. A PRIMARY CENTER OF OSSIFICATION appears in about the center of the diaphysis of the long bone, and hence is often called the DIAPHYSEAL CENTER. It is marked by an enlargement and an increase in number of CARTILAGE CELLS *(chondrocytes)* and consequently a squeezing of the matrix into thin bars and plates between the cells.

4. Lime salts are deposited in the cartilage matrix to form a CALCIFIED CARTILAGE, which stains heavily with basic dyes. It is called a PROVISIONAL CALCIFICATION because most of it will later dissolve.

5. The enlarged cartilage cells that have become walled off from nutrients diffusing through the matrix die, and the matrix also becomes unstable and begins to dissolve, leaving irregular cavities in the matrix and a few undissolved calcified spicules.

6. At the same time the above processes are occurring in the center of the diaphysis, the inner layer of the perichondrium around the center of the diaphysis becomes active, forming osteoblasts. These begin to lay down a COLLAR of spongy bone. Therefore, the perichondrium is now called the PERIOSTEUM. It should be noted that the bone collar is formed intramembranously.

7. The vascular connective tissue of the periosteum sends PERIOSTEAL BUDS through the bone collar to invade the calcified cartilage at the primary ossification center.

8. The bud advances and the thin partitions between dying cartilage cells are dissolved. The bud tissues now proliferate rapidly, filling the spaces left by the dissolved matrix and the destroyed cartilage cells.

9. Some of the mesenchyme cells of the bud give rise to BONE MARROW CELLS and others to OSTEOBLASTS.

10. Osteoblasts now line up on the undissolved spicules and plates of calcified cartilage and begin to lay down osteoid tissue, which becomes calcified to form a spongy bone. ENDOCHONDRAL BONE FORMATION THEREFORE INVOLVES TWO CALCIFICATIONS.

11. Further ossification takes place from the primary center toward the ends of the bone following the same sequence of events described above. Lateral branches from the periosteal bud move toward the ends to complete the process.

12. The various stages in intracartilagenous bone formation can be seen by examining a longitudinal section of a developing bone (see Fig. 7.6).

13. Meanwhile, the bone collar has continued to develop and enlarge, not only in thickness but also extending further toward the ends of the developing bone.

14. Internally, as endochondral bone development extends toward the ends of the bone, there is a resorption at the midpoint of the primary center, leaving a cavity—THE MARROW CAVITY. This cavity increases in length as bone development extends toward the ends. Since this reduction of endochondral bone tends to weaken the structure, thickening by growth of the bone collar provides the necessary compensating strength.

15. The marrow cavity thus formed fills with differentiating tissue—the secondary marrow or DEFINITIVE RED MARROW.

16. The secondary ossification centers are much later to develop, generally starting after birth. They appear in the cartilage remaining at the ends of a developing long bone and are called EPIPHYSEAL CENTERS. They will form the bony epiphyses.

17. The sequence of secondary ossification is much like that outlined above except that the cartilage cells form irregular clusters rather than rows. The vascular buds erupt from the perichondrium and move to the center of the developing epiphysis where osteoblasts lay down bone on the surface of exposed calcified cartilage. From this center ossification spreads peripherally.

18. The cartilage in the epiphysis is not all replaced by the spongy bone. Some remains on the free surface as ARTICULAR CARTILAGE. There is also a plate of hyaline cartilage left between the diaphysis and epiphysis; it is a GROWTH OR EPIPHYSEAL CARTILAGE OR DISK. It allows for the continued growth in length of the bone.

19. As new bone is added at the diaphyseal side of

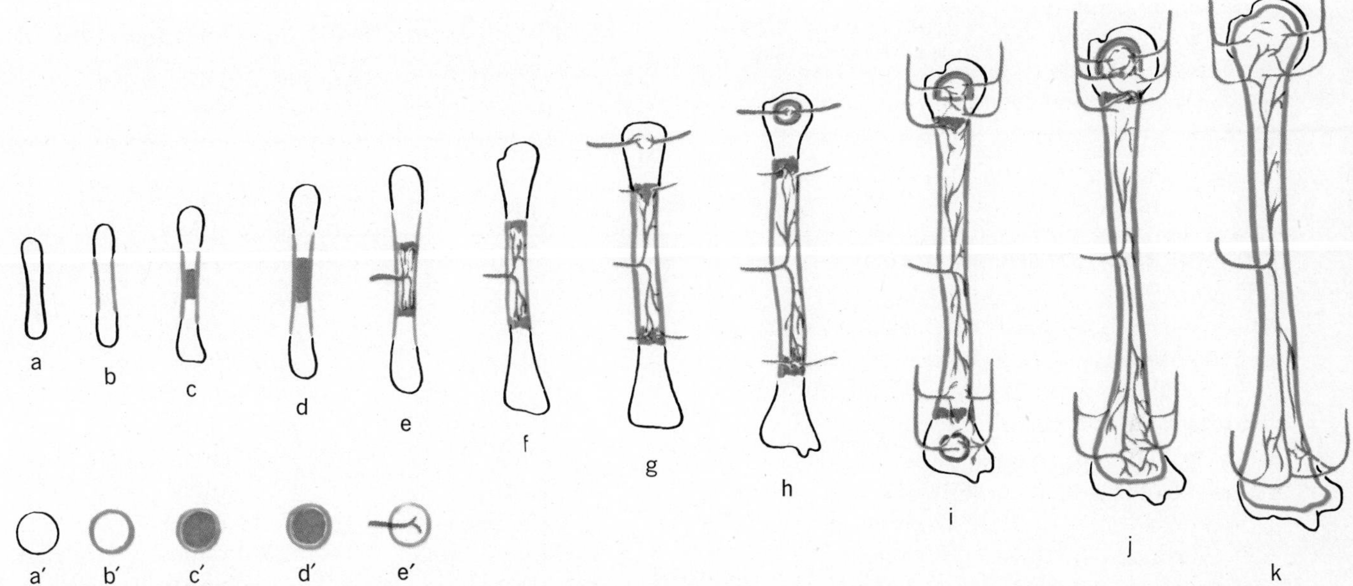

FIGURE 7.7 Diagram of the development of a typical long bone and its blood supply. *Green,* bone; *blue,* calcified cartilage; *red,* arteries. a′, b′, c′, d′, e′, are cross sections through the centers of a, b, c, d, e, respectively. a, cartilage model; b, development of the bone collar; c, development of calcified cartilage in primary center; d, extension of the bone collar; e, f, the invasion of the cartilage by vascular sprouts and mesenchyme with formation of two areas of bone formation toward the bone ends; g, h, i, secondary centers develop in the bone ends as the central area expands; j, k, epiphyseal plates disappear and the blood vessels of the diaphysis and epiphyses intercommunicate. (Bloom and Fawcett, *A Textbook of Histology,* courtesy of W. B. Saunders Co.)

the epiphyseal disk, there is a compensating addition of cartilage on the epiphyseal side. This continues until the bone has gained its full length, at which time the epiphyseal disk itself becomes ossified and can be recognized in adult bone as the EPIPHYSEAL LINE. This union by bone of diaphysis and epiphysis is called SYNOSTOSIS.

20. Growth in diameter of long bones is accomplished by deposition of bone by the periosteum.

21. As the diameter of the bone increases there is an erosion of the periosteal bone internally to widen the marrow cavity. The marrow cavity began originally by the destruction of endochondral bone; it is continued by destruction of periosteal bone.

22. The result of the above processes is a bone in which the marrow cavity fills most of the diaphysis except at the ends, where some spongy bone remains and continues into the epiphyses.

23. As noted above, the bone first formed by either intramembranous or endochondral ossification is nearly all spongy or cancellous bone and is nonlamellar. After birth the periosteal and endosteal bone laid down is compact and lamellar.

24. Formation of OSTEONS *(Haversian systems)* takes place in a variety of ways, one being the filling in of the lateral branches of the periosteal buds as they parallel the length of the bone. This is accomplished by the osteoblasts laying down concentric lamellae on the walls of the canals from the outside inward and enclosing the blood vessel or vessels of the bud, which then lie in the CENTRAL HAVERSIAN CANAL. The spaces where the periosteal buds invaded the bone collar remains as the VOLKMANN CANALS. These canals also reach the marrow cavity and constitute an important part of the system of circulation within living bone (Figs. 7.4 and 7.7).

25. Bones continue to grow and to be remodeled until the twenty-fifth year of life, and even after that remodeling continues, depending on forces from the environment and metabolic influences operating on the bone.

26. Bone development and growth are controlled by hormones from the pituitary gland and also by sex hormones.

Organization of the skeleton

The skeleton is generally conceded to be made up of 206 BONES (Fig. 7.1). They are usually organized in the following logical manner for purposes of learning and understanding:

Axial skeleton	80 bones
Skull	29 bones
Cranium	8 bones
Face	14 bones
Hyoid	1 bone
Ossicles (ear bones)	6 bones
Malleus	2 bones
Incus	2 bones
Stapes	2 bones
Vertebral column	26 bones
Cervical vertebrae	7 bones
Thoracic vertebrae	12 bones
Lumbar vertebrae	5 bones
Sacrum	1 bone (5 fused)
Coccyx	1 bone (3 to 5 usually fused)
Thorax	25 bones
Sternum	1 bone
Ribs	24 bones
Appendicular skeleton	126 bones
Shoulder girdle	4 bones
Clavicle	2 bones
Scapula	2 bones
Upper extremity	60 bones
Humerus	2 bones
Ulna	2 bones
Radius	2 bones
Carpals	16 bones
Metacarpals	10 bones
Phalanges	28 bones
Pelvic girdle	2 bones
Os coxae	2 bones
Lower extremity	60 bones
Femur	2 bones
Fibula	2 bones
Tibia	2 bones
Patella	2 bones
Tarsus	14 bones
Metatarsus	10 bones
Phalanges	28 bones
Total	206 bones

If one were to count the two SESAMOID BONES commonly occurring under the head of the first metatarsal bone of each foot, the total number of bones would be 210.

The skull

The skull rests on the superior end of the vertebral column which, because of its special modifications, allows freedom of movement of the skull and hence of the head. This versatility of movement of the skull increases the effectiveness of special sense organs, such as the eyes, ears, and nose, which are housed in and protected by the skull bones, by allowing them to be directed to-

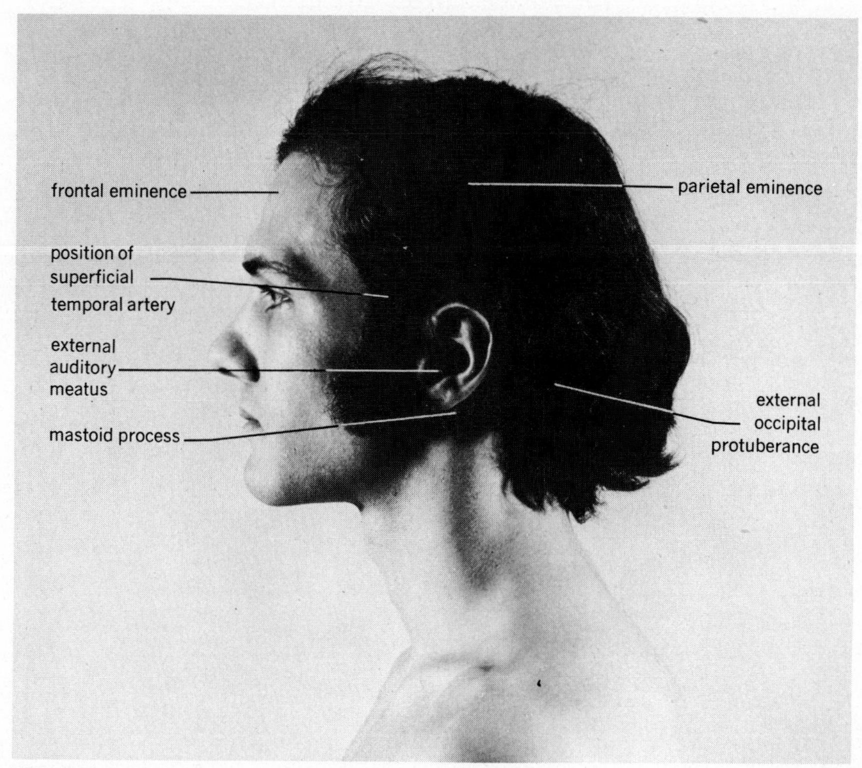

FIGURE 7.8 Side view of cranium.

ward the source of stimulation. The skull forms the CRANIAL CAVITY, housing and protecting the brain. Physiologically, the skull not only serves for support and protection, but also, through the freely movable mandible, contributes to mastication and speech.

Many features of the skull are easily palpated through the skin surface of the head. These should be identified by reference to Figs. 7.8 and 7.9 as the student progresses with the study of the skull.

Classification of skull bones

For purposes of description and organized learning, the bones of the skull may be divided into two groups, cranial bones and facial bones.

- Cranial bones (8)
  - Single bones
    - Frontal
    - Occipital
    - Ethmoid
    - Sphenoid
  - Paired bones
    - Temporal
    - Parietal
- Facial bones (14)
  - Single bones
    - Mandible
    - Vomer
  - Paired bones
    - Maxillae
    - Zygomatics
    - Nasals
    - Lacrimals

Palatines
Inferior nasal conchae

In addition to the 22 BONES listed above, there are three tiny bones referred to as the OSSICLES in the middle ear cavity of each temporal bone. They are the MALLEUS, INCUS, and STAPES. There is also a HYOID BONE in the base of the tongue that rightfully belongs to the skull. These bones bring the total number in the human skull to 29.

General description

The bones of the cranium and face are immovably joined by fibrous articulations, with the exception of the mandible. The mandible forms with the temporal bone a synovial, free-moving joint, enabling this bone to function in chewing, talking, and other related actions. The fibrous articulations of the skull are mostly of the suture type with interdigitating, overlapping, or abutting articular surfaces or edges.

CRANIAL VAULT SUTURES. The cranial vault of the skull (Fig. 7.10) shows clearly some of the sutures: the CORONAL SUTURE, which joins the frontal and parietals, the SAGITTAL SUTURE between the two parietal bones, and the LAMBDOIDAL SUTURE between the parietals and the occipital. Laterally, the parietals are joined with the temporal bones at the SQUAMOSAL SUTURE (Figs. 7.11

FIGURE 7.9 Front view of face.

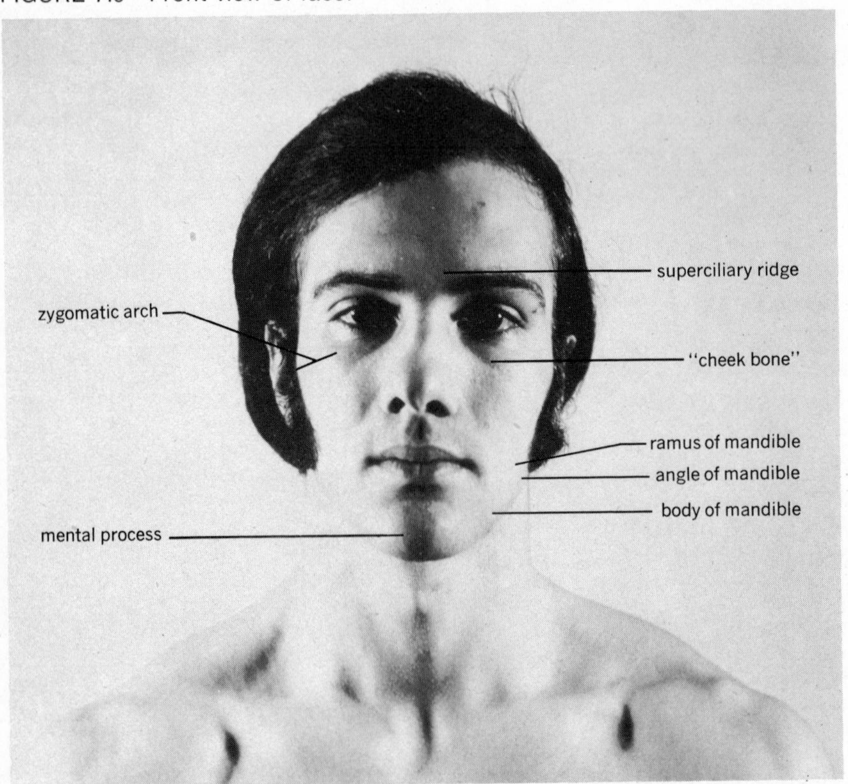

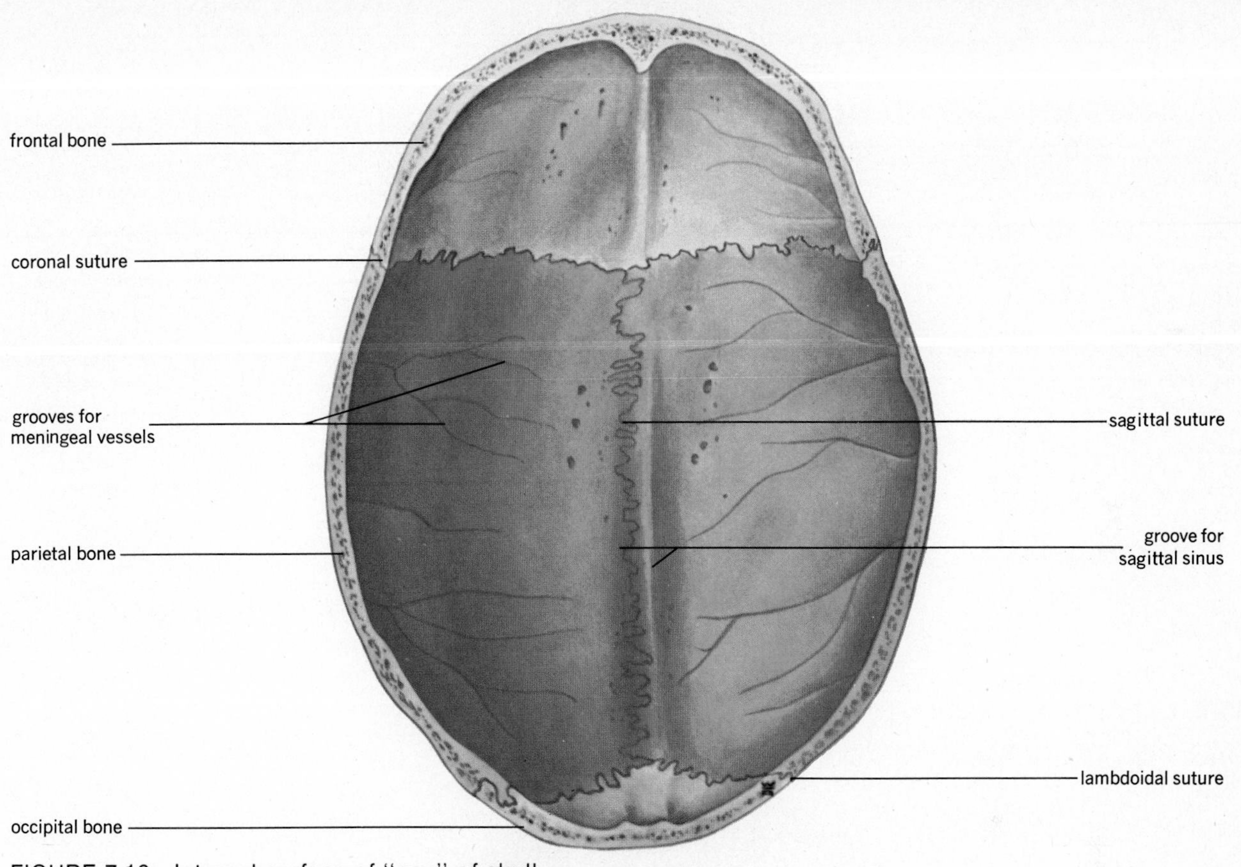

FIGURE 7.10 Internal surface of "cap" of skull.

FIGURE 7.11 Lateral view of the skull of a newborn.

parietal
frontal fontanel
frontal
occipital fontanel
area of the squamosal suture
area of the lambdoidal suture
squamous of temporal
mastoid of temporal
mastoidal fontanel
supraoccipital
exoccipital
sphenoidal fontanel
lacrimal
sphenoid
zygomatic
maxilla
mandible
external acoustic meatus

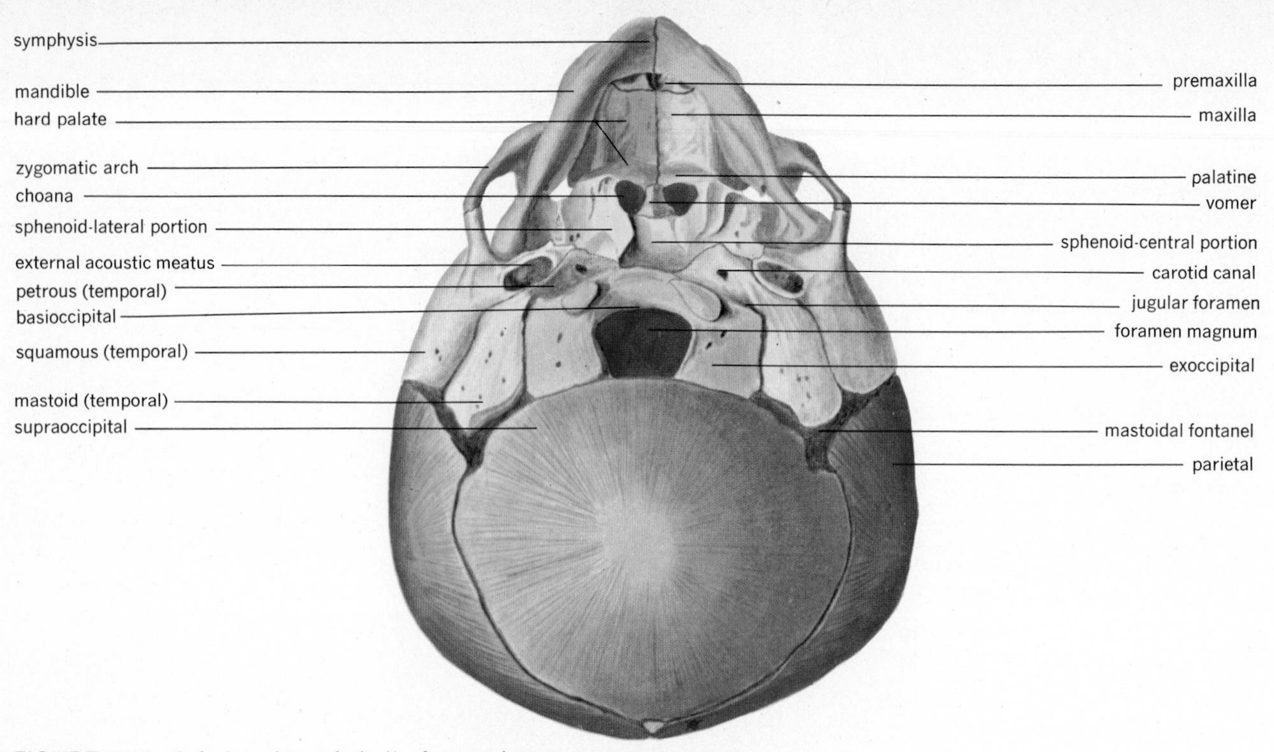

FIGURE 7.12 Inferior view of skull of a newborn.

FIGURE 7.13 Anterior view of the skull.

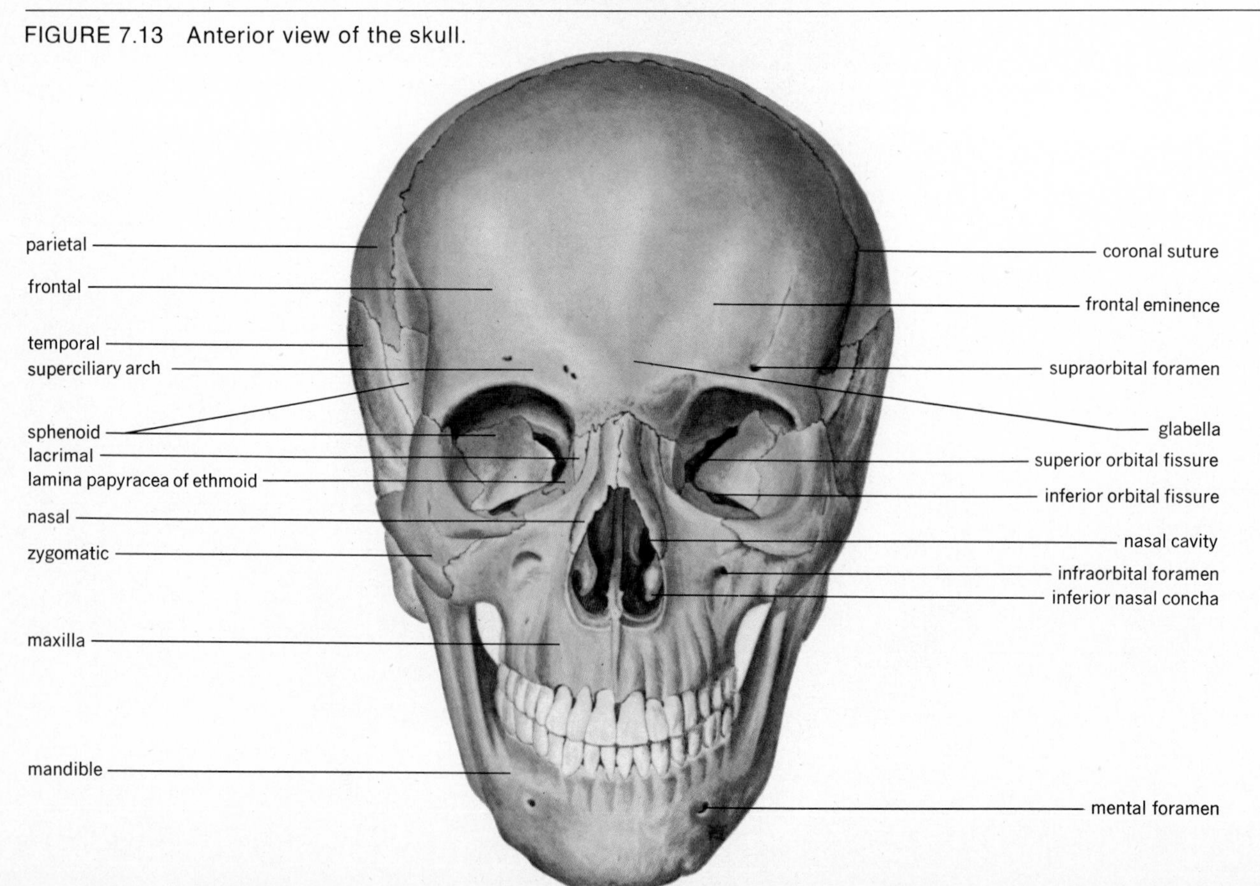

and 7.15). Small bones of irregular occurrence may appear along these sutures and are called SUTURAL or WORMIAN BONES (Fig. 7.28).

Numerous other sutures appear between the bones of the face and in the base of the skull, usually named according to the bones involved.

FONTANELS. In the skull of the newborn (Figs. 7.11 and 7.12), membranous areas—the fontanels—persist at the points of joining of the above-named sutures. The bones of the cranial vault are intramembranous bones, and the fontanels are fibrous membranes in which ossification has not yet taken place. Also, the two frontal bones have not yet joined on the midline, but later in development a FRONTAL *(metopic)* SUTURE grows together and in the adult is obliterated. Between parietals and frontal bones at the midline of the skull is the largest fontanel, the FRONTAL OF ANTERIOR FONTANEL. This was used prior to x-ray as a landmark in obstetrical diagnosis, because it can be palpated through the mother's rectum and the position of the fetus determined. The frontal fontanel closes at about 18 months of age. The OCCIPITAL OF POSTERIOR FONTANEL is at the midline where the lambdoidal and sagittal sutures meet. It closes about two months after birth. Two pairs of LATERAL FONTANELS are also found: the ANTERO-

FIGURE 7.14 View of a median section of the skull.

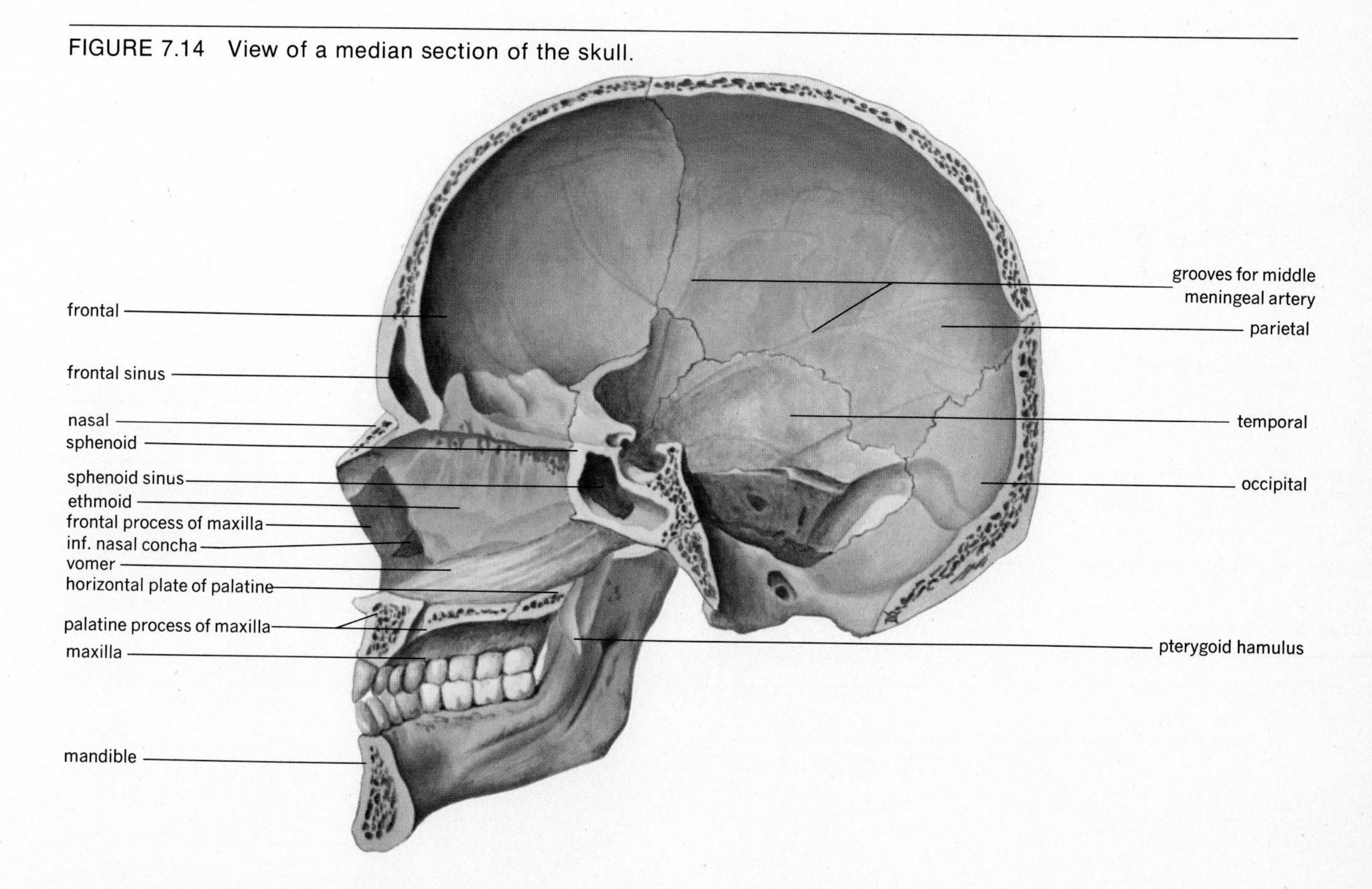

LATERAL or SPHENOIDAL FONTANELS, which lie at the juncture of parietal, frontal, sphenoid, and temporal bones; and the POSTEROLATERAL or MASTOIDAL FONTANELS, which are where occipital, parietal, and temporal bones join. These fontanels and the fibrous membranes that persist between the bones of the cranial vault ease the process of childbirth because they allow the bones to override each other to help to accommodate the skull to the size of the birth canal.

ANTERIOR VIEW. In the anterior view (Fig. 7.13), the bones underlying the forehead and face are shown. Conspicuous features are the ORBITS *(orbital fossae)* housing the eyes, the NASAL FOSSAE housing the nose, and the jaws marking the entrance to the digestive system and supporting the teeth.

The SQUAMA or vertical plate of the frontal bone underlies the forehead and forms the anterior part of the cranial vault. It slopes gradually forward from the coronal suture and then turns abruptly downward. To either side of the midline it bulges slightly to form the FRONTAL EMINENCES. Below each eminence, a flattened ridge, the SUPERCILIARY ARCH, curves transversely across the bone at about the level of the eyebrow. Between the frontal eminences and the superciliary arches is a flattened area, the GLABELLA, which extends downward to where the frontal and nasal bones articu-

FIGURE 7.15 Right lateral view of the skull.

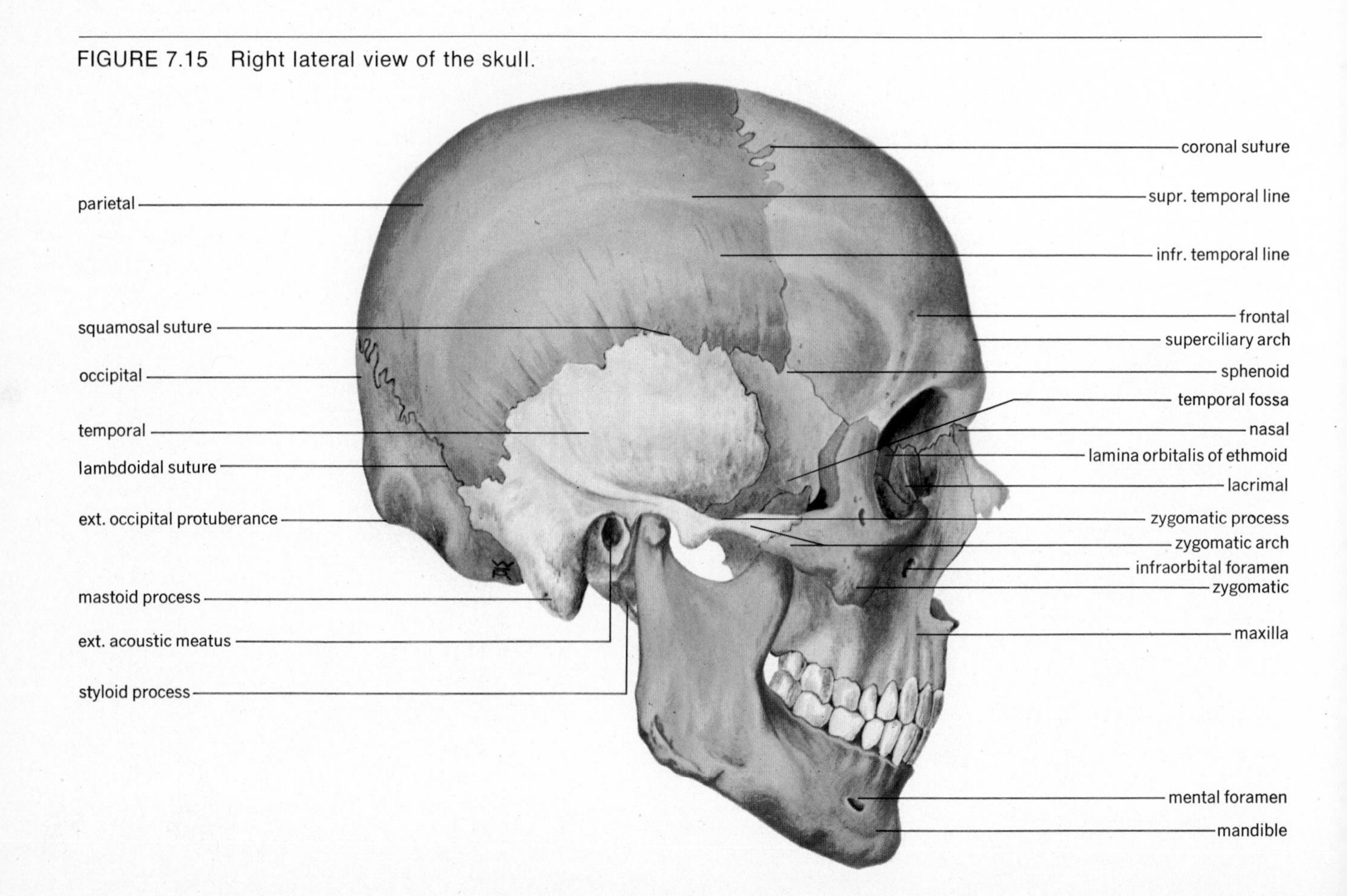

late. The frontal bone thickens below the superciliary arches and turns under and backward at the SUPRAORBITAL MARGINS as a HORIZONTAL or ORBITAL PLATE to form the roof of the orbits and part of the floor of the cranial cavity. In the supraorbital margin is a notch or foramen for passage of the supraorbital nerves and blood vessels, the SUPRAORBITAL NOTCH or FORAMEN. The FRONTAL SINUSES lie medially behind the superciliary arches and empty into the nasal cavities (Fig. 7.14).

The ORBIT houses the bulb or eyeball, and the muscles, nerves, and blood vessels associated with it, as well as nerves and blood vessels passing to the face. The orbit is pyramidal in shape with the apex pointing inward. The base of the pyramid forms the orbital aperture. Its superior margin is formed, as indicated above, by the supraorbital margin of the frontal bone. The medial margin of the orbit is formed mostly by the frontal process of the maxillary bone, while the inferior margin is formed by the body of the maxilla and a part of the zygomatic bone. The zygomatic also forms most of the lateral margin of the orbit. The medial wall of the orbit is formed by the frontal process of the maxilla, the lacrimal, and the lamina orbitalis *(orbital plate)* of the ethmoid bone (Figs. 7.15 and 7.16). The roof is formed by the orbital plate of the frontal bone, the floor is formed by the maxilla and zygomatic bones, and the lateral wall is formed anteriorly by the zygomatic and posteriorly by the great wing of the sphenoid.

Running along the floor of the orbit toward the apex and between the maxilla and great wing of the sphenoid is an irregular crevice, the INFERIOR ORBITAL FISSURE, which in life carries nerves and blood vessels to the deep structures of the face in the INFRATEMPORAL FOSSA. At the apex of the orbit is the OPTIC FORAMEN and canal, which carries the optic nerve and ophthalmic artery between the cranial cavity and the orbit. Just lateral to the optic canal is the SUPERIOR ORBITAL FISSURE, which transmits the oculomotor, trochlear, opthalmic division of the trigeminal, the abducens nerves, a branch of the middle meningeal artery, and the opthalmic veins into the orbit (Fig. 7.13).

The LACRIMAL BONE (Fig. 7.16) in the anterior medial wall of the orbit is the smallest bone of the face. It is best identified by a vertical ridge, the POSTERIOR LACRIMAL CREST, which divides its lateral surface into two areas. In front of the crest is a groove that joins with a similar groove on the

FIGURE 7.16 The right lacrimal bone in situ (left), and in lateral and medial view.

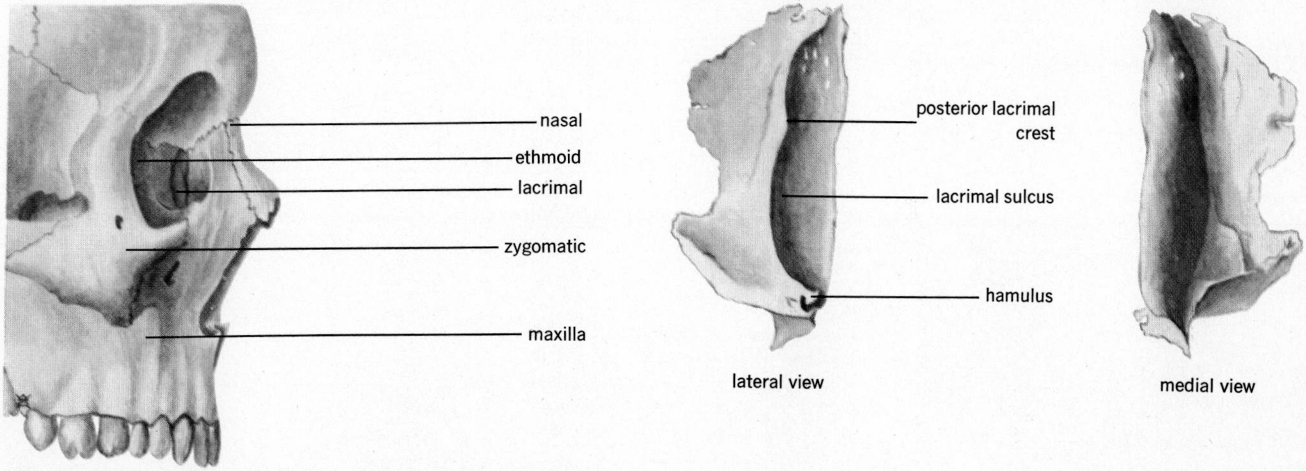

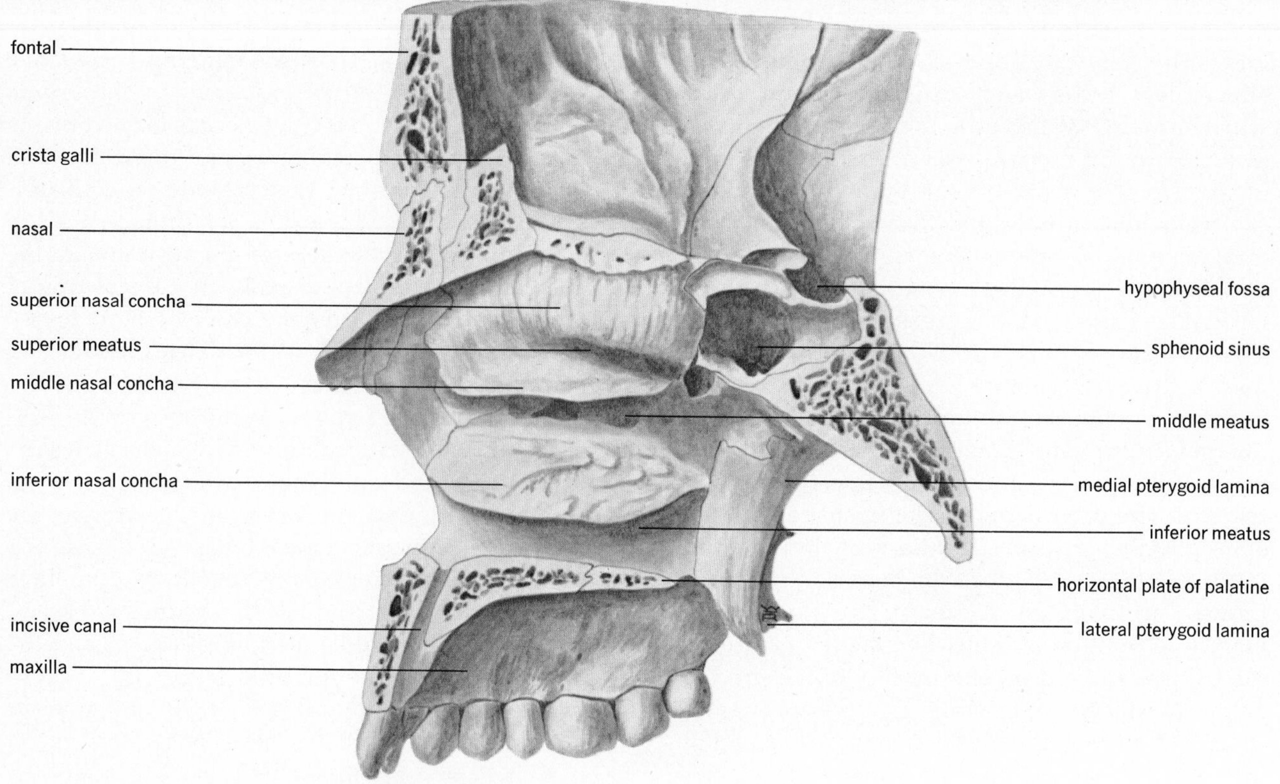

FIGURE 7.17 Medial view of the right lateral wall of the nasal cavity.

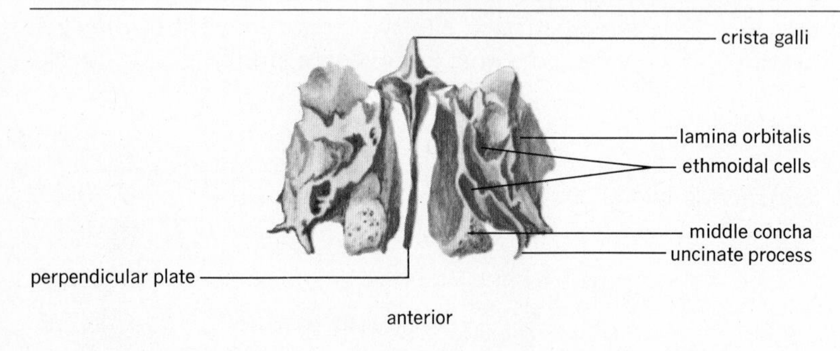

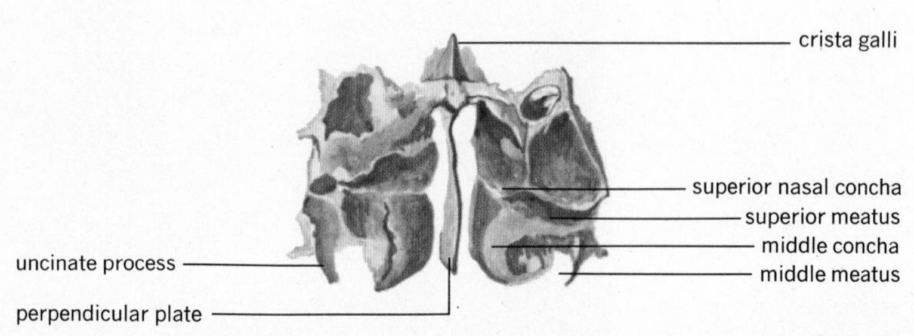

FIGURE 7.18 Anterior and posterior views of the ethmoid.

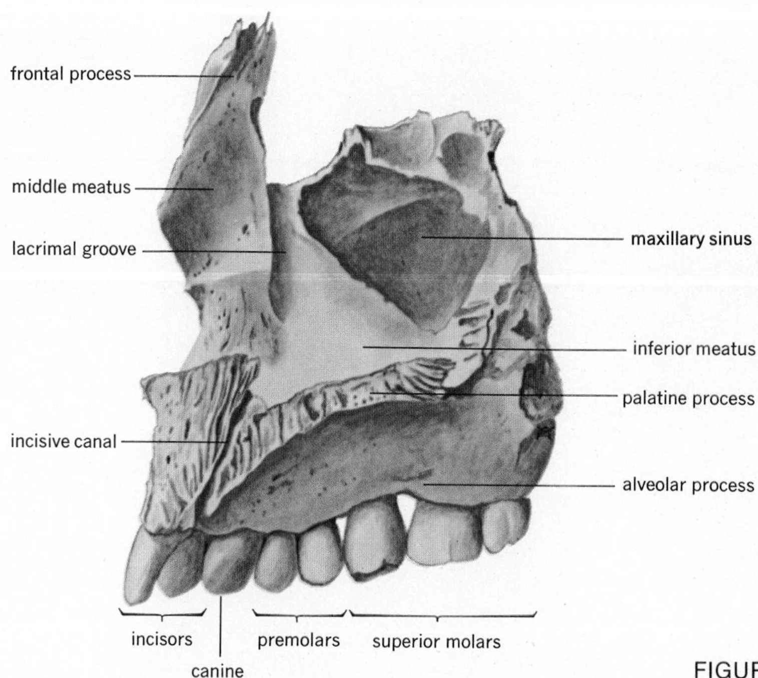

FIGURE 7.22 Medial view of the right maxilla.

MAXILLARY SINUSES or ANTRA OF HIGHMORE, within their bodies. These communicate with the nasal cavities. The maxillaries also bear teeth in sockets on their ALVEOLAR PROCESSES. Below the orbit is the INFRAORBITAL FORAMEN for the passage of intraorbital vessels and nerves. The *zygomatic process* projects laterally from the body of the maxilla to join the zygomatic bone (Figs. 7.11 and 7.15).

The MANDIBLE (Figs. 7.14, 7.15, and 7.23), which is separate and free-moving, completes the skeleton of the face. It articulates, by means of the only SYNOVIAL JOINT in the skull, with the temporal bone to form the TEMPOROMANDIBULAR JOINT. It consists of a horseshoe-shaped body that arches backward and RAMI that rise vertically on each side to articulate with the skull. The body has an ALVEOLAR PROCESS superiorly with 16 sockets to house the lower teeth. The inferior border of the body is rounded and supports the musculomembranous structures forming the floor of the mouth. The two halves of the mandible develop separately and join on the anterior medial line to form the mental or MANDIBULAR SYMPHYSIS. A MENTAL FORAMEN appears on the lateral surface to each side of the symphysis, transmitting mental nerves and vessels. Medially and to either side of the symphysis is a depression for the sublingual salivary gland, the SUBLINGUAL FOSSA, and posterior to that and extending to the ramus is a SUBMANDIBULAR FOSSA for the salivary gland of the same name. A prominent ridge, the MYLOHYOID LINE, extends upward and backward from near the symphysis to the ramus. At about the center of the medial side of the ramus is a MANDIBULAR FORAMEN, which carries inferior dental (alveolar) nerves and vessels into the lower jaw. In front of and above the foramen is a flap or tongue of bone, the LINGULA.

LATERAL VIEW. The study of the lateral view of the skull (Fig. 7.15) allows a review of the bones

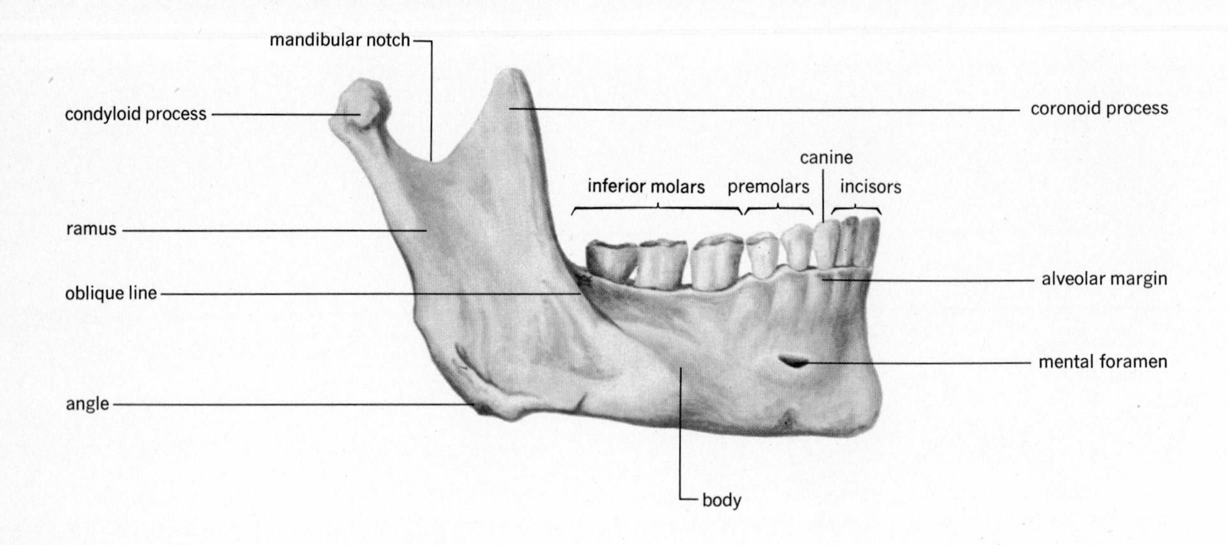

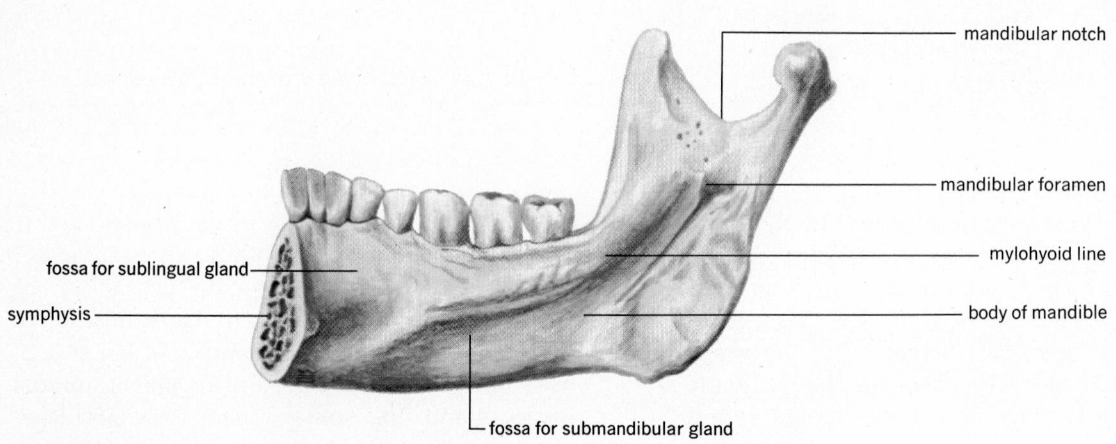

FIGURE 7.23 The lateral (top) and medial (bottom) view of the mandible.

of the anterior aspect as seen in profile and clarifies their relationships. The frontal bone is seen as a part of the cranial vault. It articulates at the coronal suture with the parietal bones, with the great wing of the sphenoid, and, by its zygomatic process, to the frontal process of the zygomatic bone. The zygomatic bone in turn articulates, as described above, with the zygomatic process of the maxilla anteriorly and forms a part of the margin and lateral wall of the orbit. Posteriorly, by its temporal process, it articulates with a long process of the temporal bone, the zygomatic process. Together, these structures form the ZYGOMATIC ARCH, which bridges a depressed area on the lateral surface of the skull, the TEMPORAL FOSSA. The temporal fossa houses the temporalis muscle, an important muscle of the mandible. This aspect also reveals to better advantage the relationship of the body to the ramus of the MANDIBLE and the angle. The superior side of the ramus has two processes: the thin, triangular CORONOID PROCESS anteriorly and the thick CONDYLOID PROCESS posteriorly. These are separated from each other by the prominent MANDIBULAR NOTCH. The head of the condyloid process fits transversely into the MANDIBULAR FOSSA of the temporal bone to form the TEMPOROMANDIBULAR JOINT (Fig. 7.15). This joint contains an articular disc and has a synovial membrane that forms separate cavities on each side of the disc. A loose, common, fibrous, articular capsule encloses the whole joint, and added support and strength are given on the anterior and lateral surfaces by a temporomandibular ligament. Opening and closing the jaws is primarily a hinge movement, the head of the mandible moving on the concave inferior surface of the articular disc. When the mandible is moved forward, the disc and mandibular head move as one onto the articular eminence lying in front of the mandibular fossa. When moved backward, the disc and mandibular head slip back into the mandibular fossa. In side-to-side movements, the disc and head move as a unit across the mandibular fossa. Chewing, biting, and grinding may combine all of the above movements.

The remaining bones of the lateral aspect of the skull are the parietal, the temporal, and the occipital.

Note in Fig. 7.15 the SUPERIOR TEMPORAL LINE that arises from the zygomatic process of the frontal bone, arches backward over the side of the frontal and parietal, and then curves forward along the supramastoid crest to the base of the zygomatic process. A less conspicuous inferior temporal line parallels it. Inferior to the temporal lines is the SQUAMOUS SUTURE, which articulates the squamous portion of the temporal with the parietal bone.

The TEMPORAL FOSSA, more clearly shown, is shallow posteriorly, but dips anteriorly to its deepest point behind the frontal process of the zygomatic bone. It houses the temporal muscle, which inserts on the coronoid process of the mandible and, when contracted, elevates the mandible to close the mouth.

The other part of the TEMPORAL BONE seen in the lateral aspect is the AUDITORY-MASTOID AREA (Fig. 7.24). The EXTERNAL ACOUSTIC MEATUS is the conspicuous opening leading into the middle ear cavity. It lies just behind the mandibular fossa. Projecting forward and downward from the external auditory meatus is the slender, long STYLOID PROCESS, which serves for the attachment of certain tongue, pharynx, and neck muscles. Posterior to the meatus is the conspicuous, blunt, triangular MASTOID PROCESS, the interior of which is hollowed out by the MASTOID AIR CELLS that communicate with the middle ear cavity. At the posterior side of the lateral aspect, the confluence of the squamosal and lambdoidal sutures is seen, as well as the occipital bone. The latter is better described from the basal aspect of the skull.

THE BASE OF THE SKULL. The base of the skull is best seen after removal of the mandible (Fig. 7.25). As seen in this aspect, it is oval in outline. The MAXILLARY BONES are seen anteriorly bearing the teeth in the sockets of their ALVEOLAR MARGINS. Within the horseshoe formed by the teeth is the HARD PALATE, which is composed of the horizontal palatine processes of the maxillae and the horizontal portion of the PALATINE BONES. In the midline anteriorly is the INCISIVE FOSSA; posteriorly, the hard palate forms part of the border of the CHOANAE or INTERNAL NARES. The VOMER divides the choanae into right and left portions and forms a part of the bony nasal septum. The body of the SPHENOID and its MEDIAL PTERYGOID LAMINA complete the walls of the choanae. Lateral to the medial pterygoid lamina is the PTERYGOID FOSSA, bordered laterally by the LATERAL PTERYGOID LAMINA. The infratemporal fossa lies between the lateral pterygoid lamina and the zygomatic arch and is occupied in life by muscles of mastication. Lateral and superior to the infratemporal fossa is the temporal fossa. The GREAT WING of the SPHENOID spreads laterally and superiorly to form a part of the wall of the temporal fossa and of the orbit.

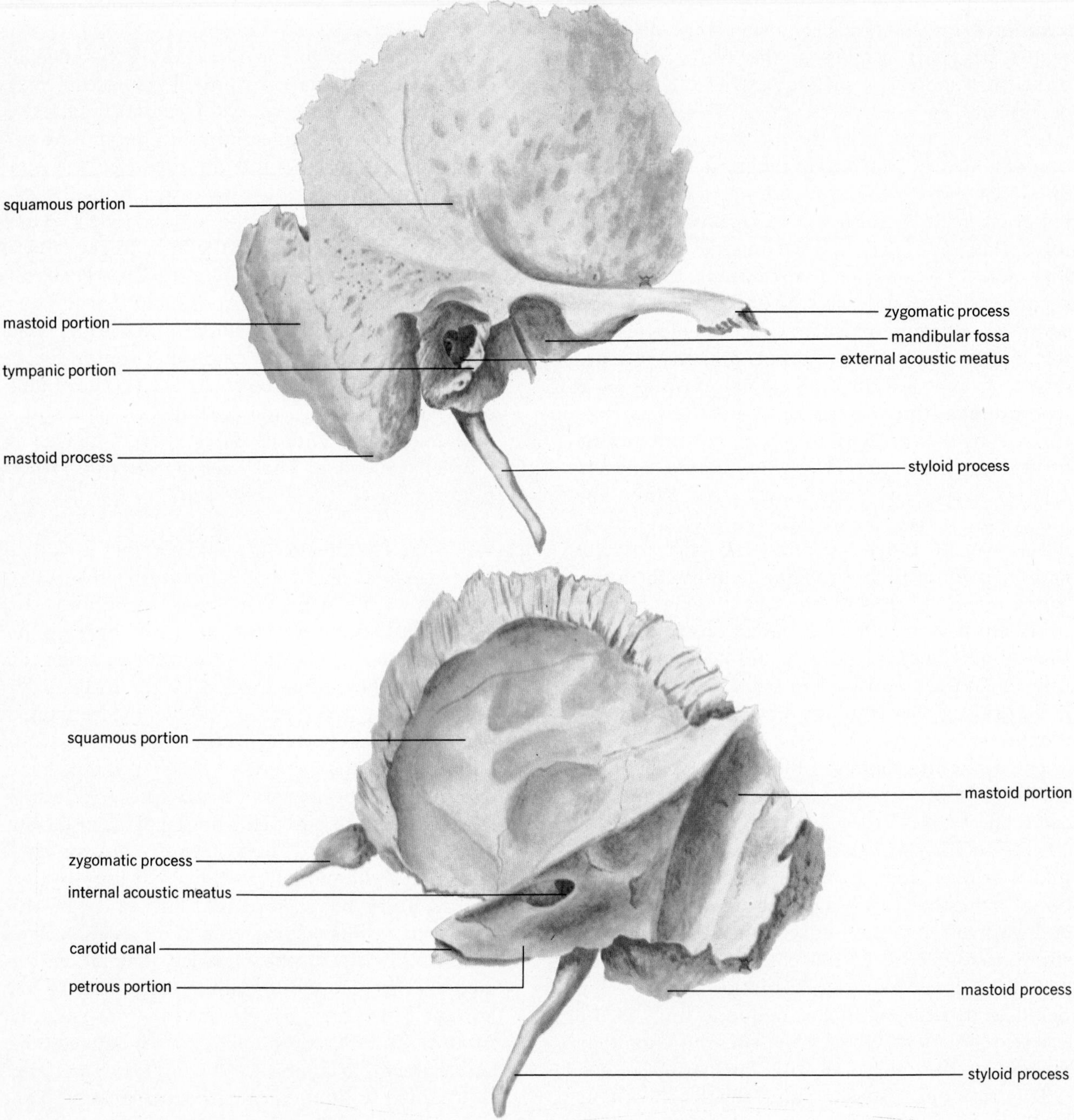

FIGURE 7.24 The right temporal bone. Lateral view (top), medial view (bottom).

FIGURE 7.25 The skull as viewed from below.

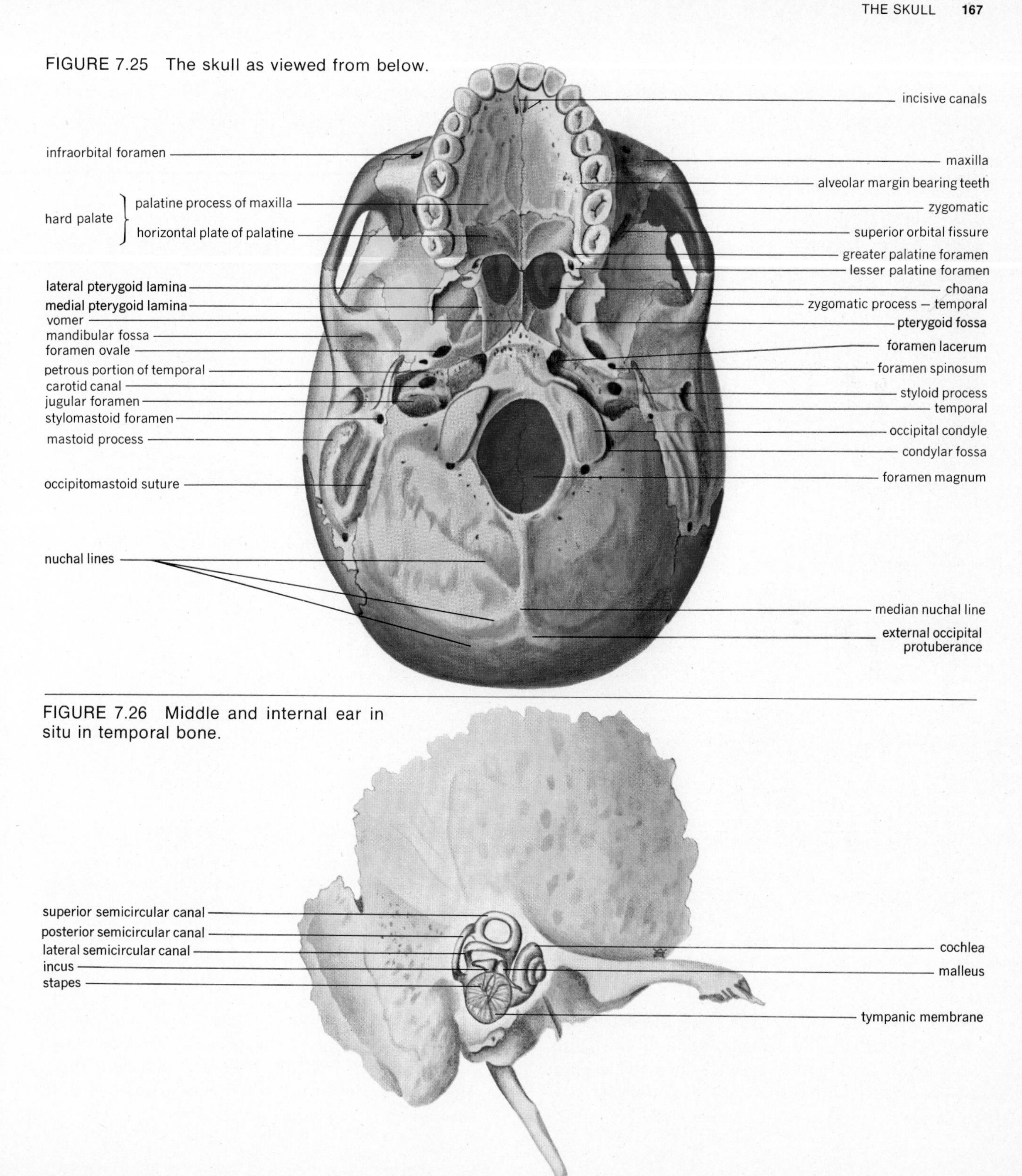

FIGURE 7.26 Middle and internal ear in situ in temporal bone.

TABLE 7.1 Foramina of the skull of humans

Foramina of facial bones	Location	Structures passing through
1. Incisive (Stensen; Scarpa, sometimes)	Posterior to incisor teeth in hard palate	Stensen, anterior branches of descending palatine vessels; Scarpa, nasopalatine nerves
2. Greater palatine	Palatine bones at posterior-lateral angle of hard palate	Greater palatine nerve
3. Lesser palatine	Palatine bones at posterior-lateral angle of hard palate	Lesser palatine nerves
4. Supraorbital (or notch)	Frontal bone, above orbit	Supraorbital nerve and vessels
5. Infraorbital	Maxilla, body	Infraorbital nerve and vessels
6. Zygomaticofacial	Zygomatic bone, malar surface	Zygomaticofacial nerve and vessels
7. Zygomaticotemporal	Zygomatic bone, temporal surface	Zygomaticotemporal nerve
8. Zygomaticoorbital	Zygomatic bone, orbital surface	Zygomaticotemporal and zygomaticoorbital nerves
9. Mental	Mandible, lateral anterior surface, inferior to second premolar	Mental nerve and vessels
10. Mandibular	Mandible, about center on medial side of ramus	Inferior alveolar vessels and nerves
Foramina of cranial bones	**Location**	**Structures passing through**
1. Olfactory	Ethmoid, cribriform plate	Olfactory nerves (I)
2. Optic	Sphenoid, superior surface	Optic nerves (II); ophthalmic arteries
3. Superior orbital fissure	Sphenoid, between small and great wings	Oculomotor (III); trochlear (IV); trigeminal (V) ophthalmic branch; abducens (VI); orbital branches of middle meningeal artery; superior ophthalmic vein; branch of lacrimal artery

Posteriorly, the body of the sphenoid articulates with the body of the occipital bone. In the living subject the area between the choanae and in front of the occipital bone is occupied by the pharynx.

Laterally, the sphenoid and occipital bones articulate with the temporal bone. The mandibular fossa can be seen at the base of the zygomatic process and in front of it the articular eminence. The squamous portion of the TEMPORAL is best studied in the lateral view of the skull, but the

Foramina of cranial bones	Location	Structures passing through
4. Inferior orbital fissure	Sphenoid; zygomatic; maxilla; palatine	Trigeminal (V), maxillary nerve; infraorbital vessels
5. Rotundum	Sphenoid, greater wing	Maxillary nerve (V)
6. Ovale	Sphenoid, greater wing	Mandibular nerve (V); accessory meningeal artery; lesser petrosal nerve
7. Spinosum	Sphenoid, greater wing	Mandibular nerve, recurrent branch (V); middle meningeal vessels
8. Lacerum	Sphenoid, greater wing and body; apex of petrous temporal; base of occipital	Meningeal branch of the ascending pharyngeal artery; internal carotid artery
9. Internal acoustic meatus	Temporal, petrous portion	Facial nerve (VII); acoustic nerve (VIII); internal auditory artery; nervus intermedius
10. Jugular	Temporal, petrous portion; occipital	Glossopharyngeal (IX); vagus (X); accessory (XI); internal jugular vein
11. Hypoglossal canal	Occipital	Hypoglossal nerve (XII); meningeal artery
12. Condyloid canal	Occipital	Vein from transverse sinus
13. Carotid canal	Temporal, petrous portion	Internal carotid artery
14. Stylomastoid	Temporal, between mastoid and styloid processes	Facial nerve (VII)
15. Mastoid	Temporal, mastoid portion	An emissary vein
16. Foramen magnum	Occipital	Medulla oblongata and its meninges; accessory nerves (XI); vertebral arteries

PETROUS PORTION pushes mediad in between the sphenoid and the occipital. It is the hardest bone in the body and houses the essential parts of the ear (Fig. 7.26). The CAROTID CANAL opens on its inferior surface. A long slender spine, the styloid process, projects downward from the petrous temporal. It is frequently broken off in preserved skulls. Posterior and lateral to the styloid process is the mastoid portion of the temporal with its mastoid process containing the mastoid air cells.

The OCCIPITAL BONE forms the most posterior part of the base of the skull (Fig. 7.27). From its articulation with the sphenoid its spreads posteriorly and laterally, articulating with the petrous and mastoid portions of the temporal bones. It has a large opening, the FORAMEN MAGNUM, through which the brainstem passes to continue into the spinal cord. The large OCCIPITAL CONDYLES lie to either side of the anterior border of the foramen magnum. From the foramen magnum the occipital turns gradually and then abruptly upward, articulating, as noted previously, along the lambdoidal suture with the parietal bones. Medially this portion has a MEDIAN NUCHAL LINE that ends superiorly in the EXTERNAL OCCIPITAL PROTUBERANCE. The external surface of the occipital carries several additional nuchal lines.

A number of FORAMINA are easily recognized in this view of the base of the skull. They provide for the transmission of blood vessels and nerves. These should be noted on the drawings of the skull and studied in Table 7.1.

THE CRANIAL CAVITY (Figs. 7.10 and 7.28). The internal surface of the cranial vault is concave in all directions and its markings reflect structures on the surface of the brain. The SUPERIOR SAGITTAL SINUS leaves a groove along the midline. MIDDLE MENINGEAL ARTERIES leave an impression of themselves as they course upward across the parietal bones. Impressions of lesser vessels may also be observed as can the coronal and sagittal sutures.

Looking into the floor of the cranial cavity one sees three large fossae arranged like steps. The highest step, the anterior cranial fossa, houses the large frontal lobes of the cerebrum and lies above the orbits. The central step, the middle cranial fossa, receives the temporal lobes of the cerebrum. The lowest step, the posterior cranial fossa, contains in life the medulla oblongata and the cerebellum. Each of these fossae reflects by the markings on its walls the form of the brain and the blood vessels that in life rest against these walls.

The walls of the ANTERIOR CRANIAL FOSSA are

FIGURE 7.27 A view of the occipital bone from below.

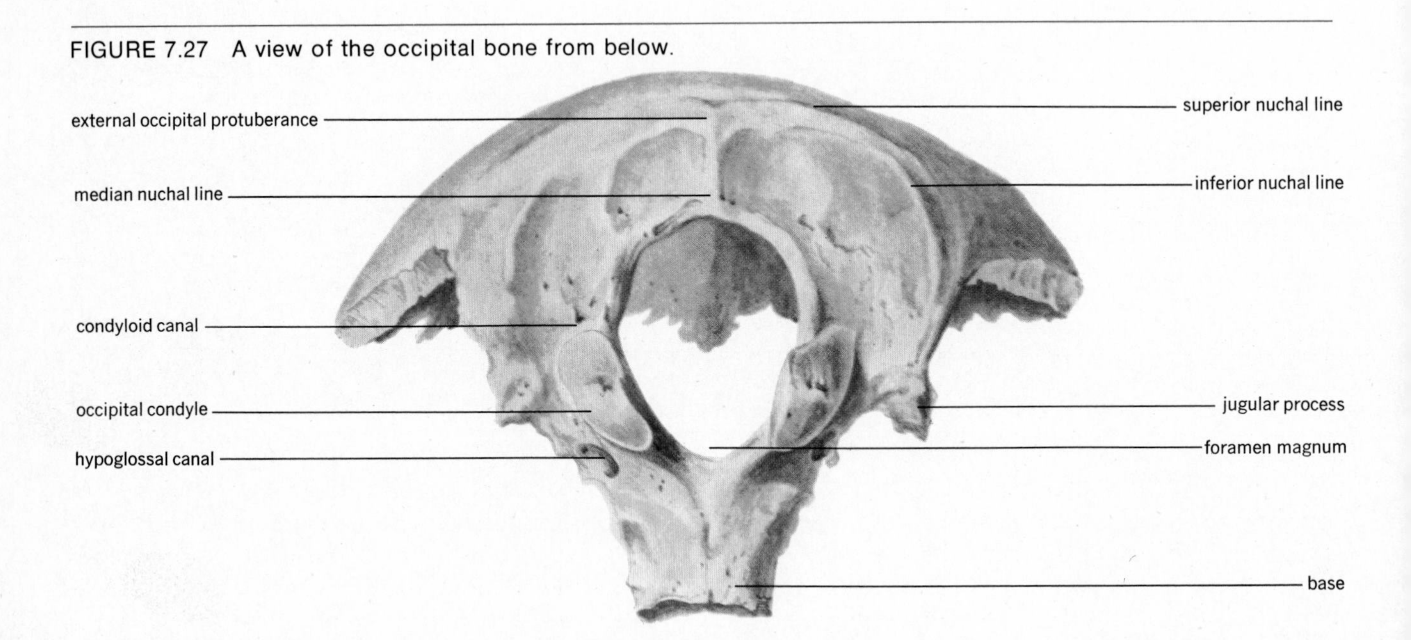

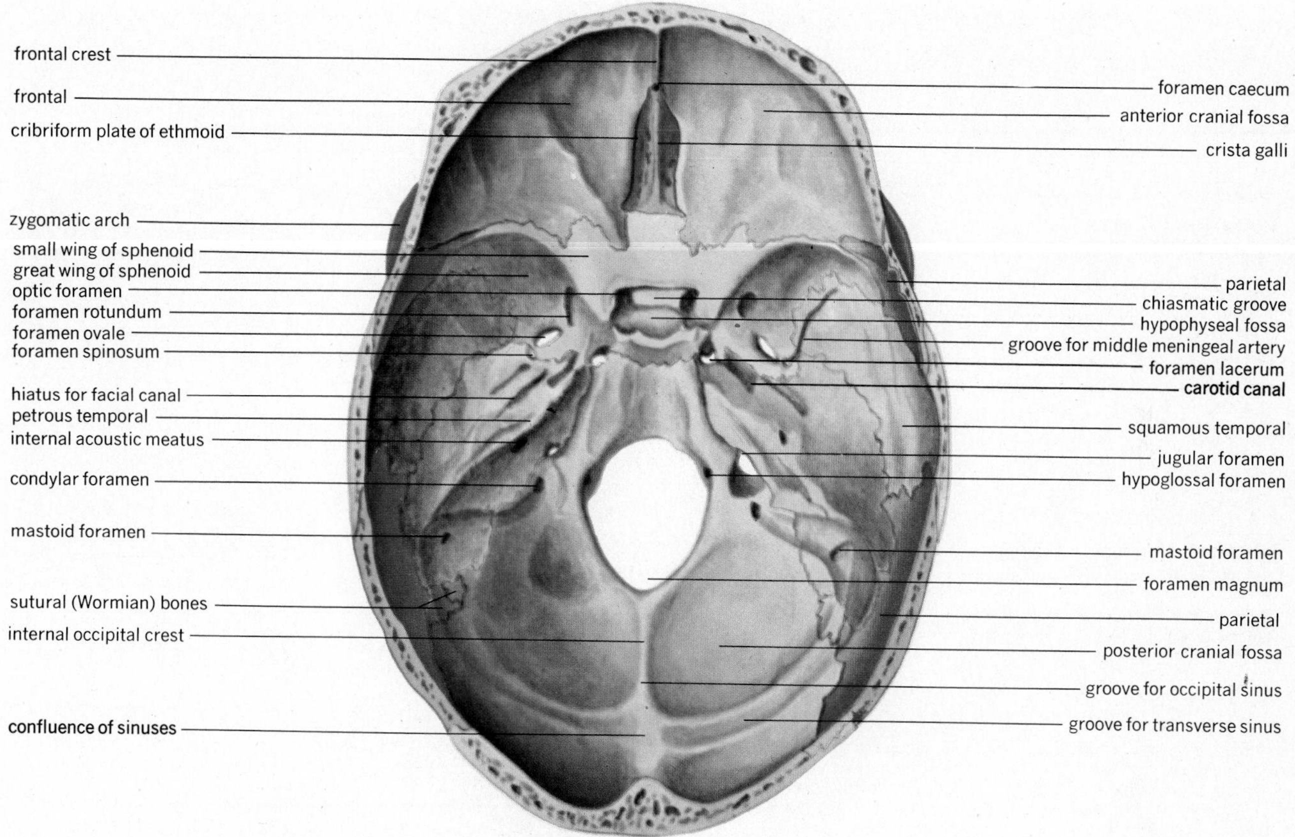

FIGURE 7.28 Floor of the cranial cavity as viewed from above.

composed largely of the frontal bone, its ORBITAL PLATES forming a large part of the floor of the fossa. Between the orbital plates and fitting into the ethmoid notch of the frontal bone is the CRIBRIFORM PLATE of the ETHMOID. These plates are perforated by numerous FORAMINA that carry the OLFACTORY NERVE FILAMENTS from the olfactory cavities. Rising between the troughs of the cribriform plates is a sharp upward projection, the CRISTA GALLI, which serves as a point of attachment for the dura mater of the brain (Figs. 7.18 and 7.28). On the midline of the anterior vertical wall of the anterior cranial fossa is a sharp FRONTAL CREST that affords attachment for the FALX CEREBRI, a dural membrane lying between right and left cerebral hemispheres of the brain.

The posterior part of the floor of the anterior cranial fossa and its posterior sharp margin is formed by the BODY and LESSER WINGS of the SPHENOID bones. To either side of the midline this margin of the sphenoid projects backward, overhanging the middle cranial fossa as the ANTERIOR CLINOID PROCESSES (Fig. 7.29). The OPTIC CANALS that transmit the optic nerves emerge from under these processes. They lead into the CHIASMATIC GROOVE where a partial crossing over of the optic nerves takes place.

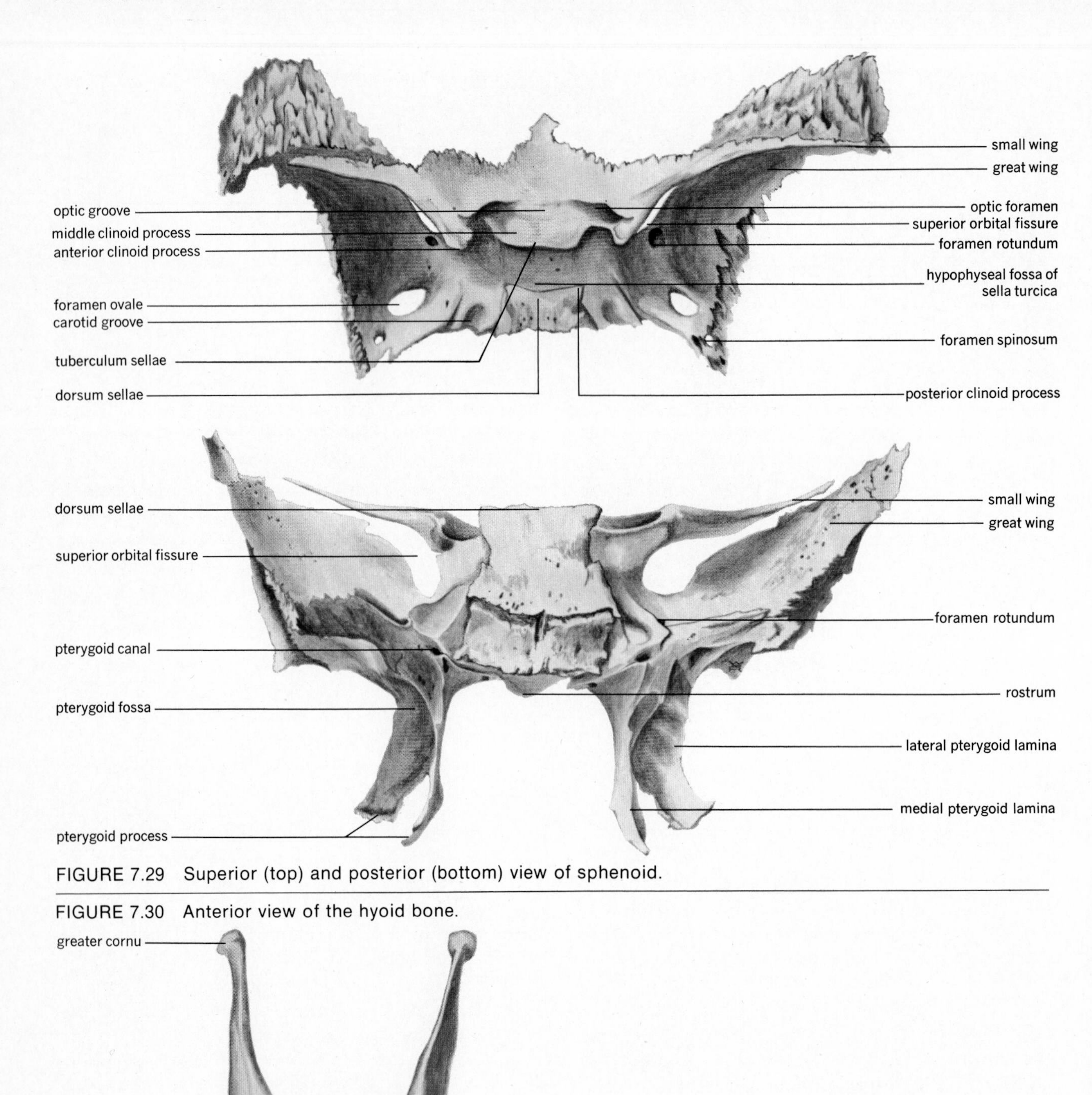

FIGURE 7.29 Superior (top) and posterior (bottom) view of sphenoid.

FIGURE 7.30 Anterior view of the hyoid bone.

The anterior cranial fossa terminates abruptly at its posterior margin. Behind and below it is the MIDDLE CRANIAL FOSSA. The medial portion of the middle cranial fossa is formed by the elevated body of the SPHENOID BONE (Fig. 7.29). This is often called the SELLA TURCICA OR TURK'S SADDLE. A deep fossa in its superior surface is the HYPOPHYSEAL *(pituitary)* FOSSA, which in life houses the HYPOPHYSIS OR PITUITARY GLAND. The hypophyseal fossa is bounded anteriorly by the TUBERCULUM SELLAE, from the posterior border of which project small MEDIAL CLINOID PROCESSES. The posterior wall of the fossa is the DORSUM SELLAE bearing POSTERIOR CLINOID PROCESSES.

The medial part of the fossa drops off laterally into deep cups bounded anteriorly by the great wings of the sphenoid, laterally by the squamous portion of the temporal, and posteriorly by the petrous temporal. Five prominent foramina or fissures are found in the floor and walls of the middle cranial fossa. Anteriorly and medially and under the anterior clinoid processes are the SUPERIOR ORBITAL FISSURES that transmit the oculomotor (III), trochlear (IV), ophthalmic division of the trigeminal (V), and the abducens (VI) nerves into the orbits. Below the orbital fissures are the FORAMINA ROTUNDA that transmit the maxillary (V) nerves and, posterior and lateral to these, the FORAMINA OVALE for the passage of the mandibular (V) nerve. The last foramina in this row are the small FORAMINA SPINOSAE, which carry the middle meningeal arteries and the recurrent branches of the mandibular nerves. Medial to this row of foramina and to each side of the sella turcica are shallow carotid grooves and below them the FORAMINA LACERUM that transmits the internal carotid arteries and the cavernous sinuses (Table 7.1).

The POSTERIOR CRANIAL FOSSA is the largest and deepest of the three cranial fossae. It is limited anteriorly by the dorsum sellae of the sphenoid bone, the base of the occipital, and the petrous portions of the temporal bones. The mastoid portions of the temporals, the mastoid angles of the parietals, and the occipitals limit the fossa laterally while the occipital completes the posterior wall. The large FORAMEN MAGNUM occupies the central portion of the floor of the fossa. From its posterior border an INTERNAL OCCIPITAL CREST extends along the midline of the occipital bone and gives attachment to the FALX CEREBRI. At the end of the occipital crest is the INTERNAL OCCIPITAL PROTUBERANCE. Grooves for venous sinuses are prominent in the walls of the posterior cerebral fossa. The groove for the OCCIPITAL SINUSES follows the internal occipital crest. At the internal occipital protuberance the grooves for OCCIPITAL, SAGITTAL, and TRANSVERSE SINUSES come together. The transverse sinuses extend laterally from the protuberance and then downward and medially to end at the JUGULAR FORAMINA, which lie between the base of the occipital and the petrous portion of the temporal. MASTOID FORAMINA and CONDYLOID CANALS may be seen entering the lower part of the grooves for the transverse sinuses. A large opening is seen in the posterior part of the PETROUS TEMPORAL—the INTERNAL ACOUSTIC MEATUS, which transmits the FACIAL and VESTIBULOCOCHLEAR NERVES and AUDITORY ARTERIES.

The HYOID (Fig. 7.30) or tongue bone lies above the larynx and is suspended from the tips of the STYLOID PROCESSES of the temporal bone by the STYLOHYOID LIGAMENT. It is composed of a transverse body and greater and lesser cornua. It serves for the attachment of a number of muscles of the tongue, the neck, and pharyngeal regions.

The vertebral column

Definition and function

The vertebral column or spine (Fig. 7.31) is a part of the axial skeleton. It is a strong, flexible rod that protects the spinal cord and gives support to the head, to the thorax (and through it to the supe-

FIGURE 7.31 Lateral view of the vertebral column.

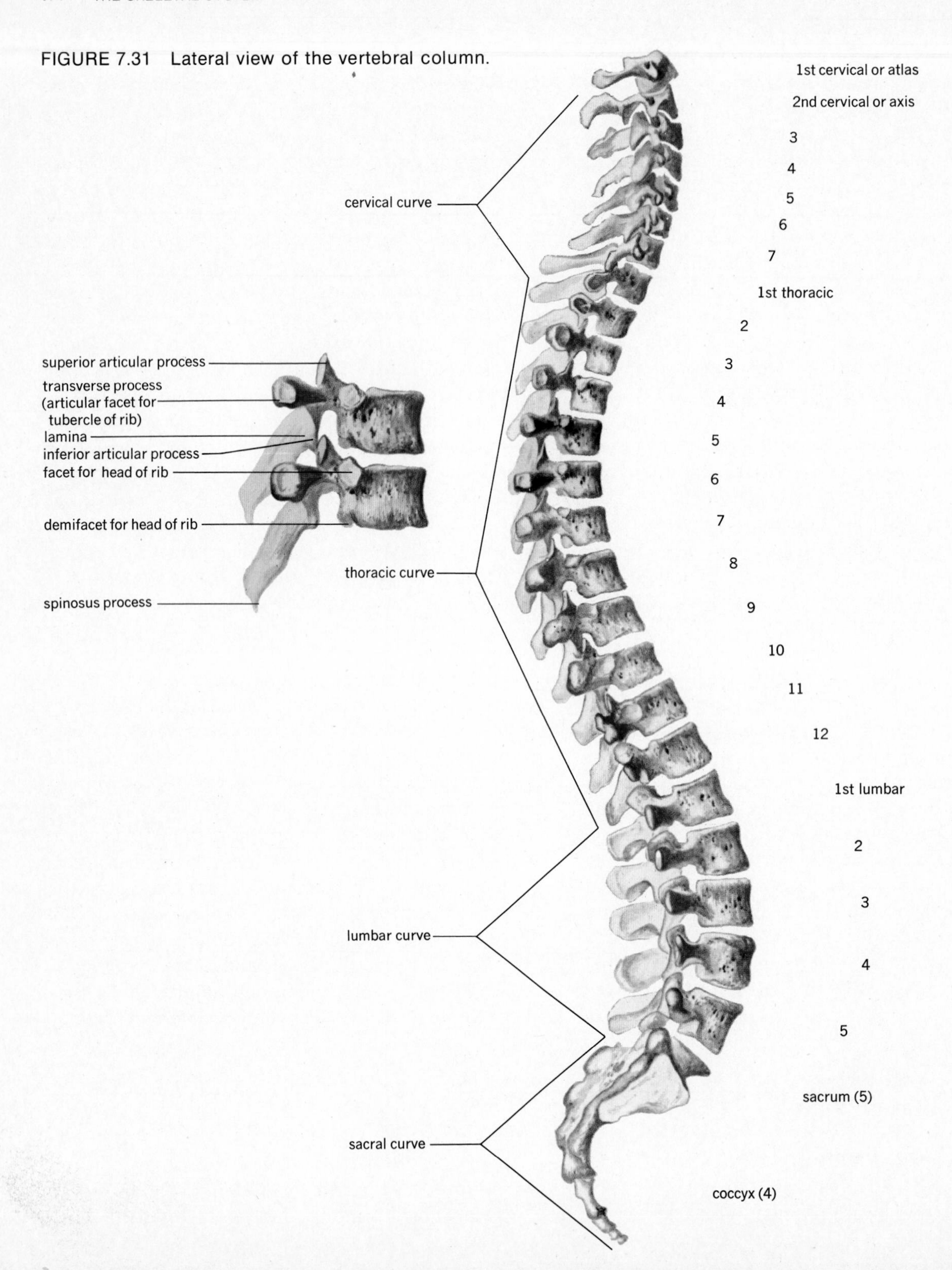

rior limb), and to the pelvic girdle and inferior limb. It provides protection for the vital organs in the thorax and for structures within the pelvic cavity. It serves as a base for muscle attachment and is a key to the posture of the individual. Its intervetebral foramina transmit the spinal nerves.

The vertebral column is composed of a series of 33, or sometimes 34, vertebrae arranged in the following groups: 7 CERVICAL, 12 THORACIC, 5 LUMBAR, 5 SACRAL, and 4 to 5 COCCYGEAL. In the adult, the sacral vertebrae are fused into one bone, the SACRUM, and the coccygeal into one bone, the COCCYX. Counted on this basis the vertebral column contains 26 individual bones. Also important to the column are the fibrocartilaginous INTERVERTEBRAL DISCS that alternate with the vertebrae and lie between their bodies.

Curvatures of the spine

Viewed from the lateral side, the vertebral column shows four curves, alternately convex and concave anteriorly. They serve to absorb shock. In the fetus these curves do not exist; instead, there is one large curve that is concave anteriorly. As a child learns to hold his head up (at about the third month), the cervical vertebrae are lifted upward and a CERVICAL CURVE gradually develops. As the child pulls itself up to the erect position and learns to walk, an anterior convexity develops in the lumbar region, the LUMBAR CURVE. In the thoracic and sacrococcygeal regions the curvatures retain the anterior concavity of the fetus. The THORACIC and SACRAL CURVES are for this reason referred to as PRIMARY CURVES; the CERVICAL and LUMBAR CURVES, having been modified from the fetal condition, are SECONDARY CURVES. If the vertebral column is viewed from the anterior or posterior aspect it is normally without curves, although slight curvatures to the right are quite common.

Abnormal curvatures of the spine are sometimes seen. A marked curvature to the right or left is called SCOLIOSIS. It may be caused by muscular paralysis on one side of the spine, by poor posture, or by disease processes that deteriorate the bone. An excessive posterior convexity of the spine is called KYPHOSIS and usually involves the thoracic region, giving a humpbacked appearance. There may be an excessive lumbar curvature or swayback condition known as LORDOSIS. These abnormal curves may be caused by TUBERCULOSIS of the vertebrae, causing a softening of the bodies of the vertebrae so that they crush under the influence of the weight of the body. The column then tends to telescope or fold, exaggerating otherwise normal curves. Other disease processes affecting bone may cause the same problems; POOR POSTURE, MALNUTRITION *(rickets)*, and wearing of improper shoes may be causative factors.

Components of a typical vertebra

The vertebrae have a common pattern of structure, although they differ in size, shape, and finer details. Each has evolved adaptive features that determine its function as a part of the vertebral column.

The parts of a typical vertebra are the BODY *(centrum)*, VERTEBRAL ARCH, and a variety of processes, articulating surfaces, and foramina. The body and vertebral arch form the walls of the VERTEBRAL FORAMEN through which the spinal cord passes.

The body forms the anterior and thickest part of the vertebra. Its flattened, rough superior and inferior surfaces give attachment to the fibrocartilaginous INTERVERTEBRAL DISCS. The circumferences of these surfaces are slightly raised to form a rim. Nutrient foramina are common in the anterior and lateral surfaces, and on the posterior surface facing the vertebral canal are one or two irregular apertures for the passage of the basivertebral veins.

The VERTEBRAL ARCH extends posteriorly from

the body of the vertebra. It is composed of a stout PEDICLE on each side and flat LAMINAE extending from the pedicles mediad, where they meet on the midline to complete the arch.

Seven processes rise from the vertebral arch. At the junction of the pedicle and lamina on each side a TRANSVERSE PROCESS extends laterally. A single SPINOUS PROCESS extends posteriorly and inferiorly from the posterior midline of the vertebral arch. These are easily felt along the midline of the neck, thorax, and back. Two pairs of articular processes extend from the junction of pedicles and laminae, one pair SUPERIORLY, the other INFERIORLY. A layer of hyaline cartilage covers their articular surfaces.

Regional variations in the vertebral column

While all vertebrae have a common pattern of structure, the vertebrae of each region have their own identifying features. The sizes of the BODIES vary, becoming larger and stronger in progressing from the superior end of the column to the lower lumbar region. This is a logical arrangement, since the more inferior vertebrae carry more of the body weight.

The INTERVERTEBRAL DISCS also become gradually thicker and are thickest in the lumbar region. The VERTEBRAL CANAL, although fairly uniform in size throughout its length, enlarges slightly and becomes more triangular in the cervical and lumbar regions where it accommodates the cervical and lumbar enlargements of the spinal cord. In the thoracic region it is rounder and smaller.

The SPINOUS PROCESSES are short and BIFID in most of the cervical region. They become longer and quite narrow in the thoracic region and slant downward so that each process overlaps the one inferior to it like shingles on a roof. In the lower thoracic, and especially in the lumbar region, they become more and more massive and are square as seen from the lateral aspect.

The TRANSVERSE PROCESSES of the cervical region are most distinctive in that they have a TRANSVERSE FORAMEN for the passage of vertebral arteries and veins. In the thoracic region they possess

FIGURE 7.32 Posterior view of the atlas and axis.

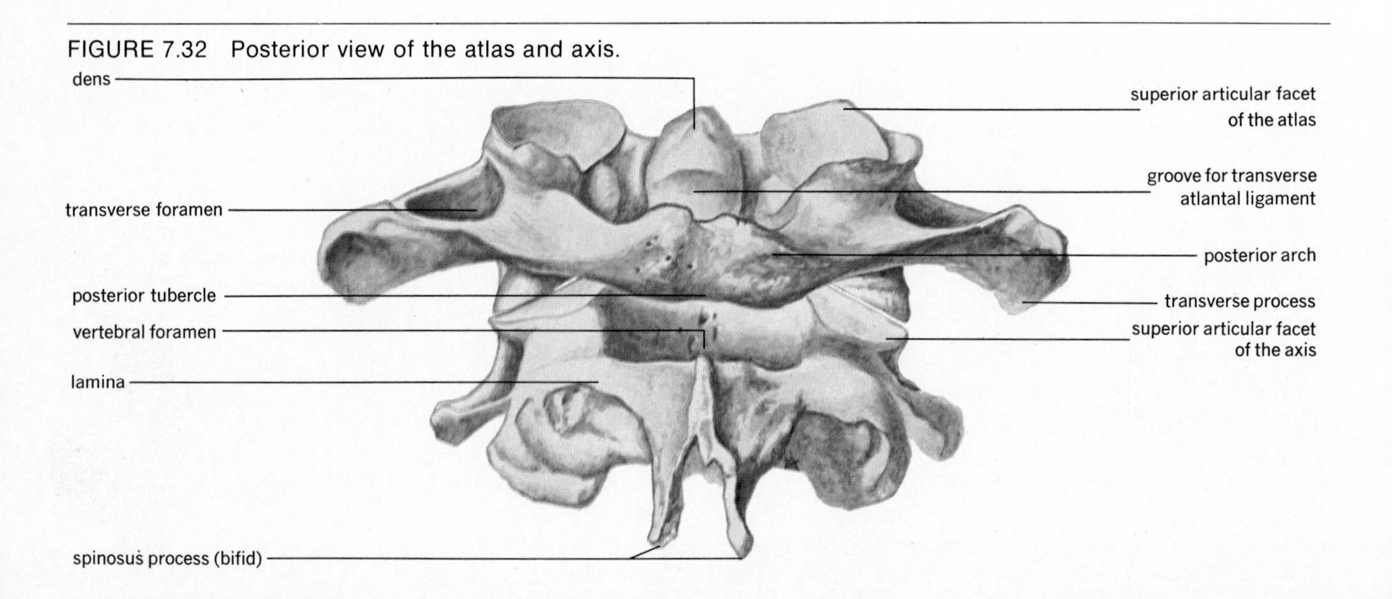

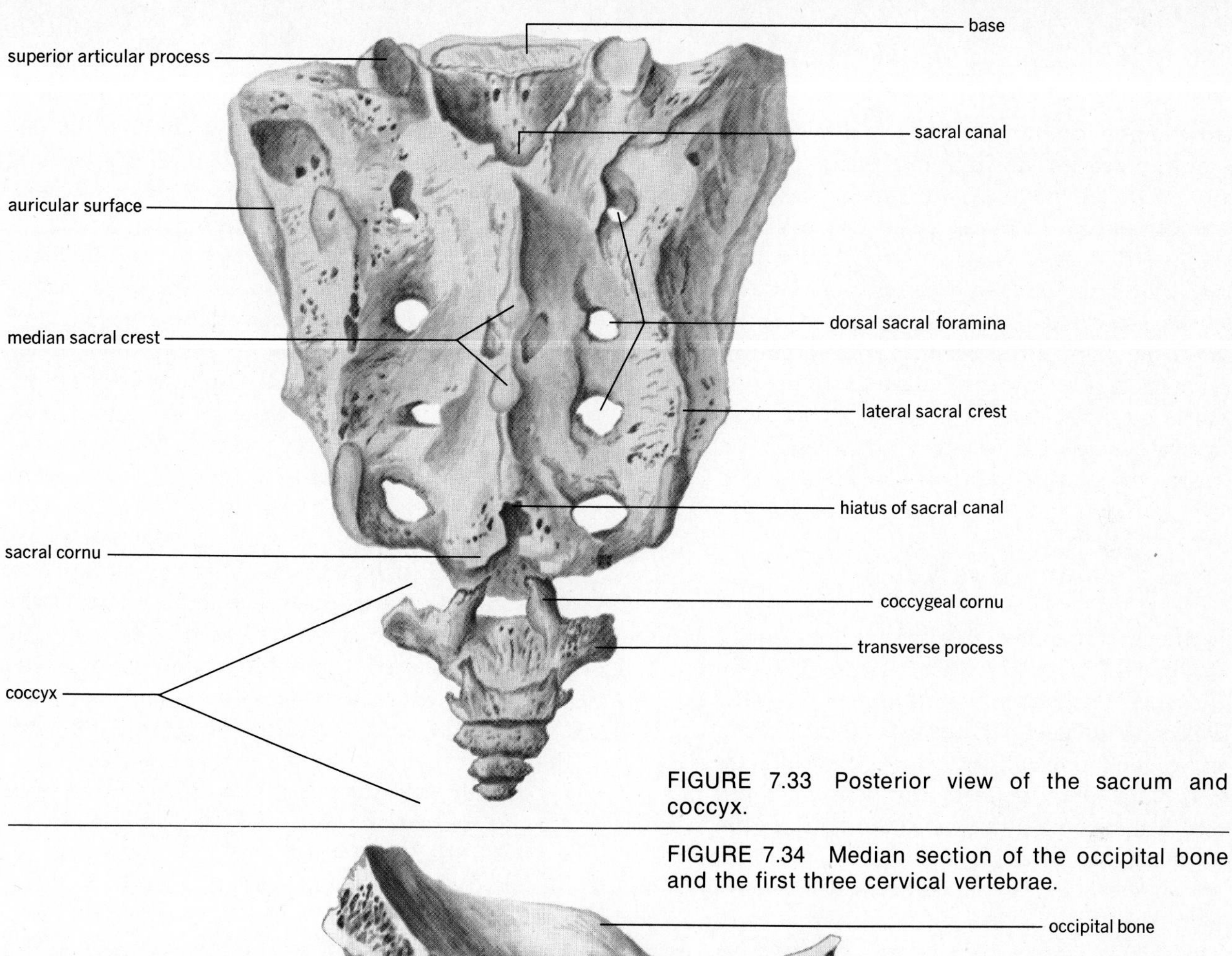

FIGURE 7.33 Posterior view of the sacrum and coccyx.

FIGURE 7.34 Median section of the occipital bone and the first three cervical vertebrae.

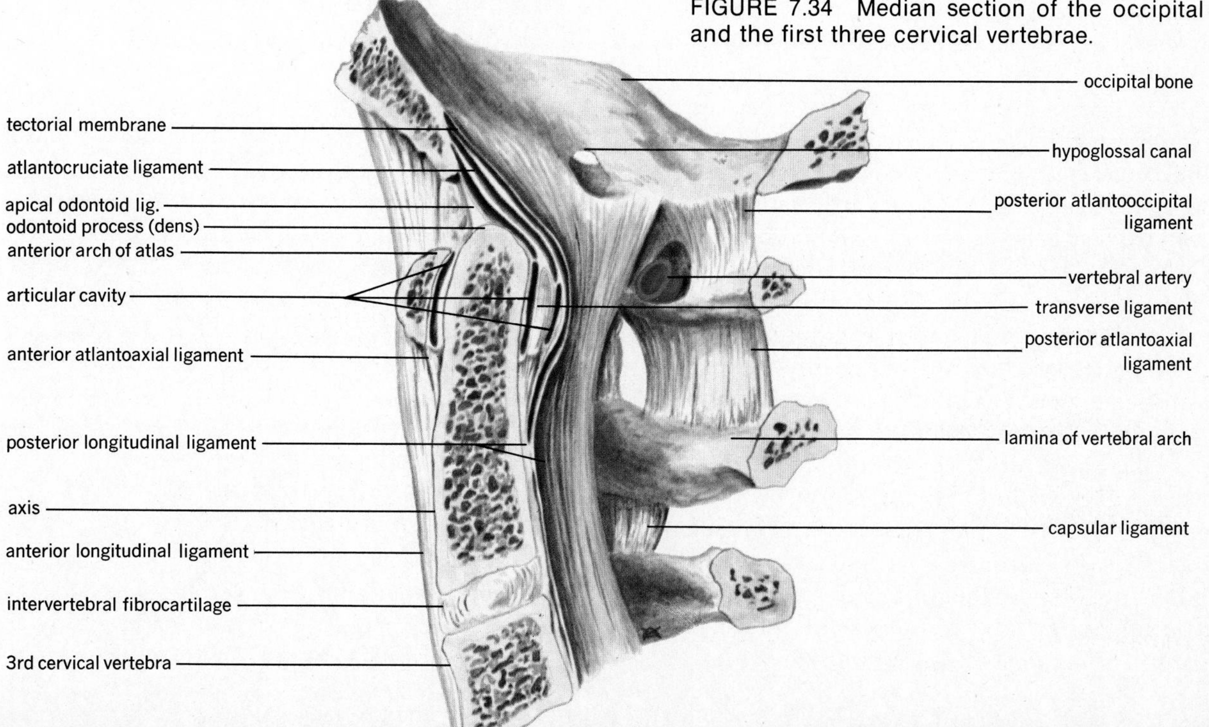

shallow articular FACETS or DEMIFACETS for the articulation of the tubercles of the ribs. In the lumbar region neither facets nor transverse foramina are found and the processes come straight out to the sides.

The articular processes vary greatly from the superior to the inferior ends of the spine. They are farthest apart in the cervical region, come quite close together in the thoracic region, and, in the lumbar region, the articulating surfaces of the SUPERIOR PROCESSES turn inward and grasp the outward-turning INFERIOR PROCESSES. This PREVENTS ROTATION of the spine in this region.

Specialization of vertebra

Some vertebrae have become highly adapted and serve special functions (Fig. 7.32). Notable among these are the first and second cervical vertebrae, called the atlas and axis, respectively. The ATLAS is a ring of bone with anterior and posterior arches and large lateral masses. The body has been lost and has become attached to the body of the axis as the dens, where it serves as a pivot around which the atlas can rotate and allows the movements involved in shaking the head to mean "no." The lateral masses have large concave facets on their superior surfaces for articulation with the occipital condyles of the occipital bone (THE ATLANTOOCCIPITAL JOINT). At this articulation the head can be flexed and extended as in nodding the head to mean "yes." Some lateral movement is also allowed. Circular, flat (or slightly concave) articular facets are found on the inferior surfaces of the lateral masses that articulate with the axis. The transverse processes of the atlas are wide and each contains a large transverse foramen.

The AXIS is characterized by the prominent dens mentioned above, and by a prominent BIFID SPINOUS process and small transverse processes. Its body is prolonged downward anteriorly where it overlaps the body of the third cervical vertebra.

The SEVENTH CERVICAL VERTEBRA, OR VERTEBRA PROMINENS, is marked by a large, nonbifid spinous process that serves for attachment of the LIGAMENTUM NUCHAE. It is easily palpated at the base of the neck and serves as an important surface feature and LANDMARK for estimating the positions of underlying structures such as nerves.

The SACRUM is composed of five modified and fused sacral vertebrae and serves as a strong foundation for the attachment of the pelvic girdle (Fig. 7.33). It is triangular in shape, and on its concave pelvic surface are four transverse lines that mark the joining of the bodies of the vertebrae. Four pairs of large foramina, the ANTERIOR SACRAL FORAMINA, are seen here. The posterior surface is convex and has along its midline the MIDDLE SACRAL CREST and three or four tubercles representing reduced spinous processes of the vertebrae. Where the laminae and spinous processes do not form, a SACRAL HIATUS or opening is left. This occurs, for example, in the region of the fourth and fifth sacral vertebrae. Four POSTERIOR SACRAL FORAMINA are seen, and near the apex of the sacrum are the SACRAL CORNUA.

Laterally, the sacrum is broad above where it has a large AURICULAR SURFACE for articulation with the ilium to form the SACROILIAC JOINT. Below this it narrows into little more than a ridge.

The TRIANGULAR COCCYX is the most rudimentary part of the vertebral column and is a vestige of a tail (Fig. 7.33). It is made up of four vertebral components, all of which are hardly recognizable as vertebrae. Only the first one has small transverse processes and coccygeal cornua that project upward to join the SACRAL CORNUA. It also has an intervertebral disc and is joined to the sacrum. The remaining coccygeal vertebrae, each smaller than the one above, are little more than reduced bodies or centra.

Articulations and movements of the vertebral column

The articulations within the vertebral column fall into two categories, cartilaginous and synovial. The CARTILAGINOUS ARTICULATIONS are of the

SYMPHYSIS TYPE. They consist of pads of fibrocartilage, the intervertebral discs, interposed between the bodies of adjacent vertebrae. The cartilages vary in thickness and therefore in resilience; they are thickest in the lumbar region, thinnest in the thoracic. This is one of the factors influencing movement in the vertebral column, the greatest flexion and extension being in the LUMBAR REGION, the least in the THORACIC (Fig. 7.34).

INTERVERTEBRAL DISCS are frequently ruptured, causing the soft inner NUCLEUS PULPOSUS to break through its tough covering, the ANNULUS FIBROSUS. In some cases, rather than breaking the annulus, the nucleus pulposus may cause it to protrude. In either situation, posterior or posterolateral protrusions are likely to impinge on the spinal cord or spinal nerves, causing pain in the back and legs. The condition is most common in the lumbar region, sometimes in the cervical, and seldom in the thoracic. In any case, one should see a physician for diagnosis and treatment. Sometimes traction and bed rest are necessary, or one may have to resort to surgery, which may involve removal of the disc and fusion of the vertebrae.

The synovial articulations lie between the superior and inferior articulating surfaces of the vertebrae, and, as indicated earlier, the relationships of these processes are another determining factor in movement. Since these processes are directed more or less upward and downward in the cervical region they allow, along with the symphyses, a variety of movements: flexion, extension, rotation, abduction, and adduction. In the thoracic region the articular processes face anteriorly and posteriorly, and all of the above movements are possible, but more restricted. In the lumbar region the articular processes face outward and inward, one grasping the other in such a way as to prevent ROTATION. Yet, because the intervertebral discs are so thick and compressible, flexion and extension are greatest in the lumbar region, and some abduction and adduction are possible.

The joint between the atlas and axis, the ATLANTOAXIAL JOINT, is a special case based on the modifications of the vertebrae described earlier (Fig. 7.34).

Articulations between the vertebral column and other skeletal structures also concern us here. The relationship of the atlas to the skull at the atlanto-occipital joint has been described. Some of the ligaments that maintain it are shown in Fig. 7.34.

The sacrum joins the ilium to form the SACROILIAC JOINT. This is partly a fibrous joint and partly a synovial joint, the former giving strength and and the latter slight movement. Since the whole weight of the trunk, head, and neck rests on this joint, it is under stress at all times, except when the individual is lying down. Strong ligaments support it, as seen in Fig. 7.52.

The thoracic vertebrae give attachment to the ribs at articular facets on their bodies and transverse processes. These are SYNOVIAL JOINTS allowing the ribs to swing forward and upward.

The thorax

The thorax (Fig. 7.35) consists of 12 pairs of RIBS, the STERNUM, and the COSTAL CARTILAGES. The thoracic vertebrae complete the thorax posteriorily. The narrow INLET of the conical-shaped thorax is superior; its broad OUTLET, which is closed by the DIAPHRAGM, is inferior. It is flattened from front to back and in cross section is kidney-shaped since the ribs swing posteriorly from their attachment to the thoracic vertebrae and then forward.

Ribs and costal cartilages

The first seven, or TRUE RIBS, attach through their costal cartilages to the sternum. The eighth

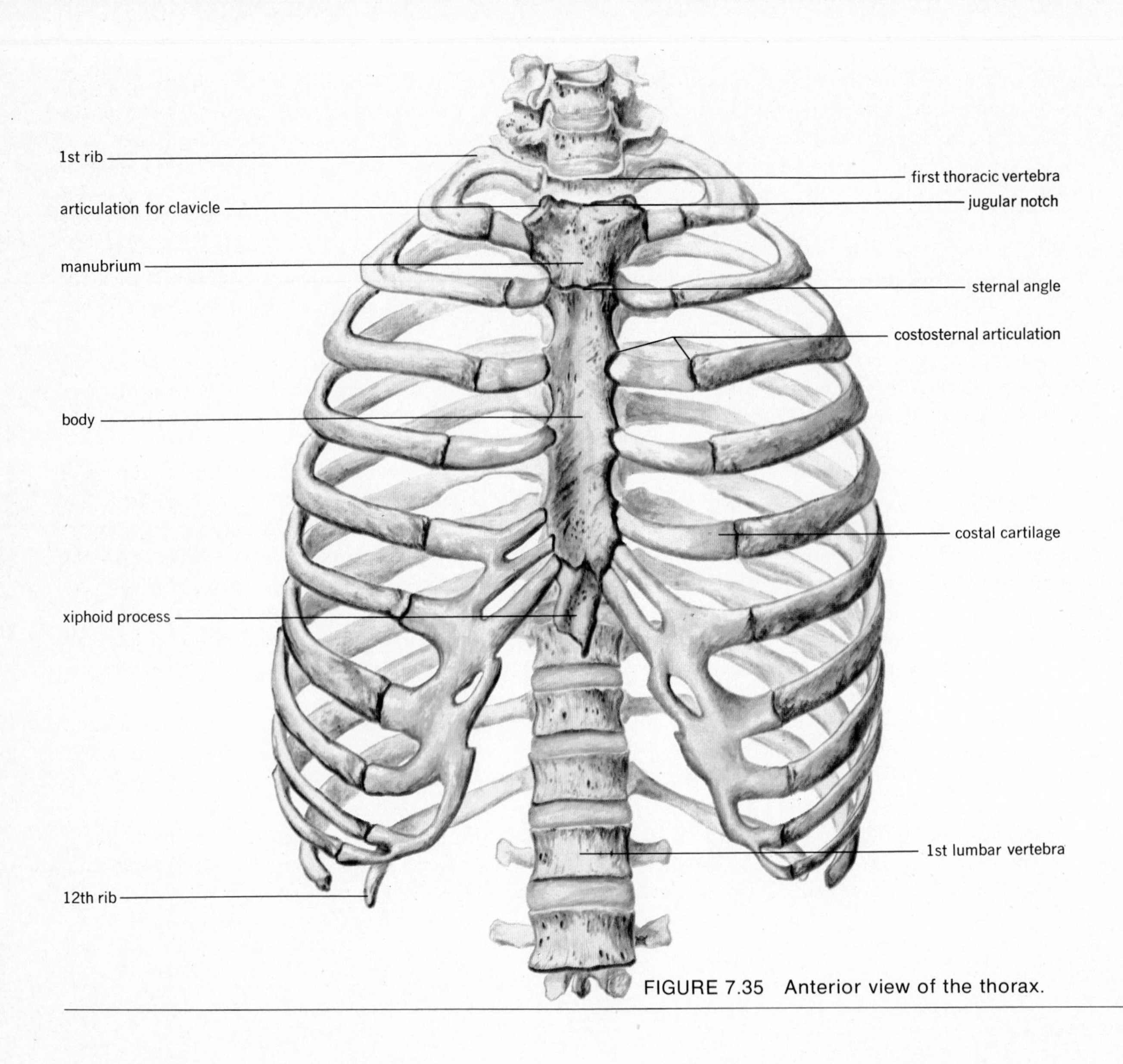

FIGURE 7.35 Anterior view of the thorax.

through the twelfth are FALSE RIBS, so-called because their costal cartilages do not attach directly to the sternum. The costal cartilages of the eighth through the tenth ribs attach to each other and then to the cartilage of the seventh rib. The eleventh and twelfth ribs are called FLOATING RIBS because their anterior ends terminate freely in the body wall.

From number one to seven (Fig. 7.36) the ribs increase in length; they then taper off to the small twelfth rib. A TYPICAL RIB, such as number six (Fig. 7.37 and Fig. 7.38), consists of a HEAD with TWO ARTICULAR FACETS separated by an INTERARTICULAR CREST. The facets articulate with the demifacets of the bodies of adjacent thoracic vertebrae. Next to the head is a short NECK that ter-

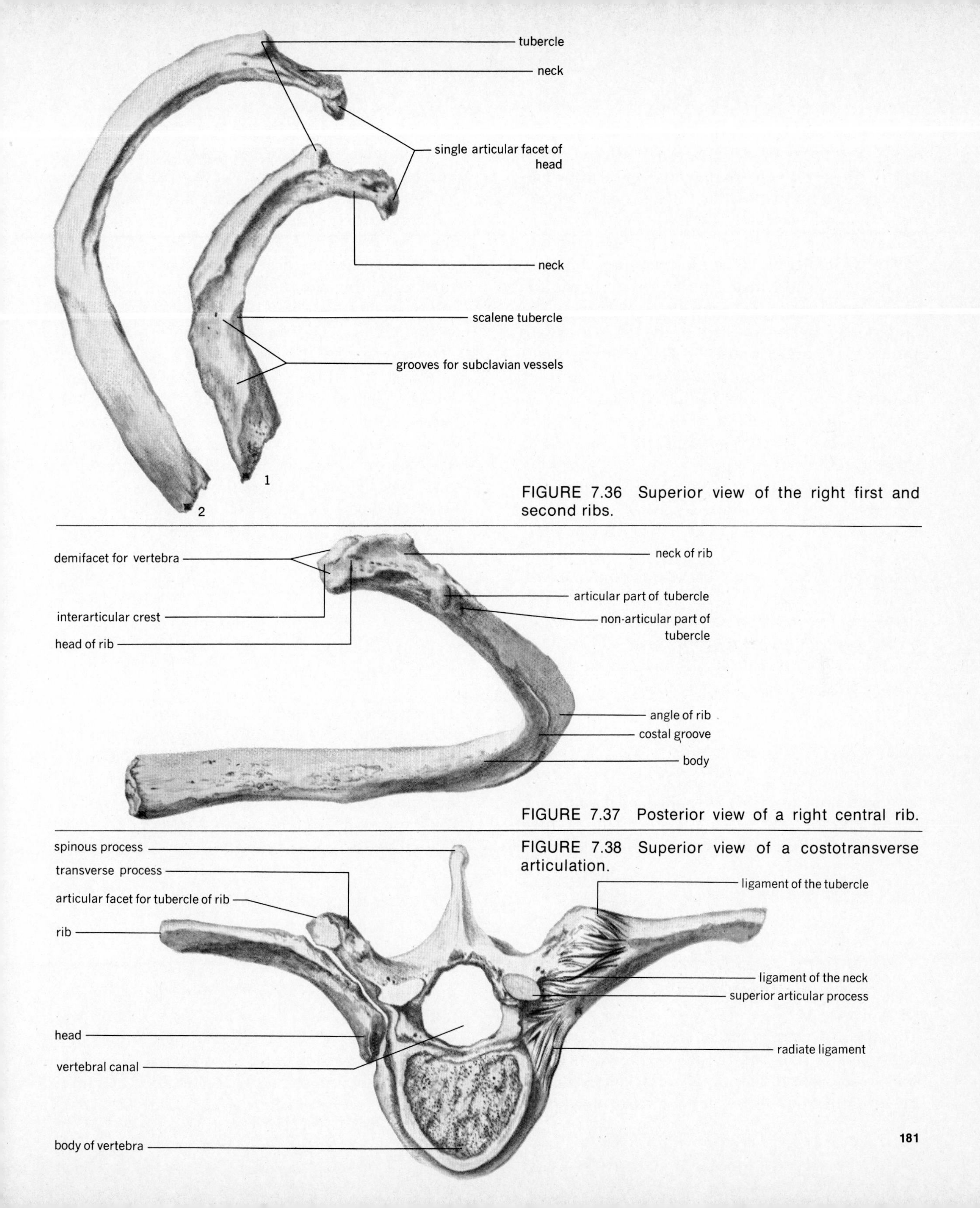

FIGURE 7.36 Superior view of the right first and second ribs.

FIGURE 7.37 Posterior view of a right central rib.

FIGURE 7.38 Superior view of a costotransverse articulation.

minates at the TUBERCLE. The tubercle is divided into ARTICULATING and NONARTICULATING PORTIONS. The lower and more medial of the tubercles articulates with the facet on the transverse process of a thoracic vertebra. The nonarticular tubercle serves for the attachment of a ligament. Beyond the tubercle the rest of the rib constitutes the BODY or SHAFT. A short distance from the tubercle the shaft bends abruptly to form the ANGLE. The superior border of the shaft of the rib is rounded; the inferior border is sharper. The external surface of the rib is convex; the inner surface has a COSTAL GROOVE for the passage of the INTERCOSTAL VESSELS and NERVES. The anterior end of the rib is flattened and has a depression in it for attachment of the costal cartilage.

Some ribs deviate from the above descriptions. The FIRST RIB is usually the shortest and is flattened in the horizontal plane. It has the greatest curvature, but NO ANGLE, and the head has a SINGLE FACET for articulation with the first thoracic vertebra. The tenth, eleventh, and twelfth ribs also have only ONE ARTICULAR FACET on their heads, and the elventh and twelfth have NO NECK or TUBERCLE. The twelfth rib may have NO ANGLE and is sometimes shorter than the first rib.

Sternum

The sternum (Fig. 7.35) is located in the midline of the anterior thoracic wall. It is made up of three parts: a superior manubrium, an elongated body, and an inferior xiphoid process. The MANUBRIUM is triangular in shape and has on its superior surface a depression, the SUPRASTERNAL *(jugular)* NOTCH, to each side of which are oval articular surfaces for the proximal ends of the clavicles. On the lateral border, superiorly, there is a facet for articulation of the costal cartilage of the first rib; inferiorly, there is a demifacet for the cartilage of the second rib. Where the manubrium joins the body there is a raised transverse line called the STERNAL ANGLE. The body of the sternum is flat, broadest at its lower end, and laterally has demifacets at the superior and inferior ends with four full facets between them. These are for the attachment of the costal cartilages of ribs two to seven. The XIPHOID varies in form and is cartilaginous in early age. It gradually ossifies with accumulating years. It has a demifacet laterally at its superior end that, with the demifacet of the body, serves for attachment of the seventh costal cartilage.

Functions

The thorax provides PROTECTION for vital organs such as the heart and lungs. It gives SUPPORT to the pectoral (shoulder) girdle. It is essential in the BREATHING process because the inspiration of air depends on the enlargement of the thorax by muscle action. It gives ORIGIN TO MANY MUSCLES that move the upper limb.

Appendicular skeleton

Definition and functions

The appendicular skeleton, composed of pectoral and pelvic girdles and upper and lower extremities, serves the locomotor and manipulatory needs of the body. When vertebrates evolved from the aquatic to the terrestrial environment, paired limbs changed from balancing organs in an environment of water that gave support to the body to organs of SUPPORT AND LOCOMOTION. It became necessary to LIFT THE BODY above the substratum in order to move effectively. Many terrestrial vertebrates, such as salamanders and alligators, still have paired limbs out to the side; in higher vertebrates, such as mammals, limbs are placed under the torso and so lift it well above the ground for more efficient LOMOTION. In humans, the body became erect and the upper extremities were made available for a variety of uses in the MANIPULATION and MODIFICATION OF THE ENVIRONMENT. The human HAND is an extraordinary tool that, coupled with a SUPERIOR BRAIN, has enabled humans to DOMINATE THE WORLD.

As this diversity of functions of the girdles and the upper and lower extremities suggests, there is a corresponding variation in their structure. The pectoral or shoulder girdle has no articulation with the vertebral column. Through its anterior structure, the clavicle, it articulates with the sternum. Its posterior component, the scapula, rides freely in a complex musculature. The shoulder joint itself is freely movable. The whole complex is relatively weak, but highly versatile in movement. This freedom of movement serves as a good basis for the upper extremity, enabling it to be used in a multitude of ways and positions.

The pelvic (hip) girdle, in contrast, is articulated securely to a special fused part of the vertebral column, the sacrum. It gives a strong and stable support to the lower extremity on which the weight of the body is carried and balanced in locomotion (walking, running, and jumping) or just in standing. The hip joint where the lower extremity joins the pelvic girdle, although it is a ball-and-socket or universal joint, is still a strong one, the head of the femur fitting the deep bony acetabulum of the girdle.

Pectoral girdle

The PECTORAL GIRDLE consists of two pairs of bones, the POSTERIOR SCAPULAE *(shoulder blades)* and the ANTERIOR CLAVICLES *(collar bones).* This

FIGURE 7.39 Anterior view of sternoclavicular and sternocostal articulations.

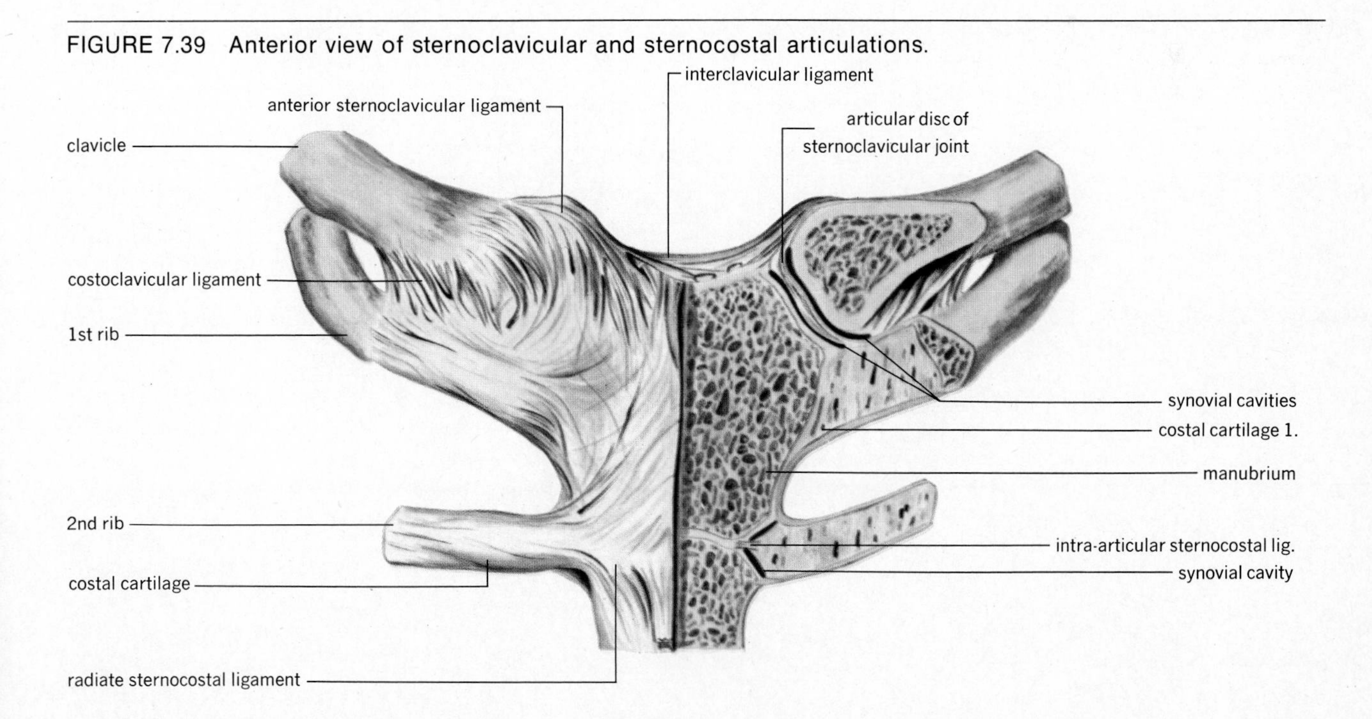

girdle rests on the thorax, but articulates with it only anteriorly where the clavicle joins the manubrium of the sternum *(the sternoclavicular joint)* (Fig. 7.39).

CLAVICLES. As seen from the superior or inferior aspect, the clavicles (Fig. 7.40) are shaped like a shallow S. Their medial ends are rounded and articulate with the sternum. Their lateral ends are flattened and broad and articulate with the acromion processes of the scapulae—the ACROMIOCLAVICULAR JOINTS (Fig. 7.41). On the inferior surface of the bone laterally is a CORACOID TUBEROSITY for the attachment of the CORACOCLAVICULAR LIGAMENT, which crosses over to the CORACOID PROCESS of the scapula to form a fibrous joint. The clavicle holds the shoulder joint out to the side so that the arm can swing freely; when it is broken the shoulder collapses. Since the clavicle is subcutaneous, it is subject to fracture by blows to the shoulder. Also, because it is the only means by which the pectoral girdle–upper extremity complex is attached to the axial skeleton, it is often broken by falling on the outstretched arm; for example, when one uses the arm to break a fall.

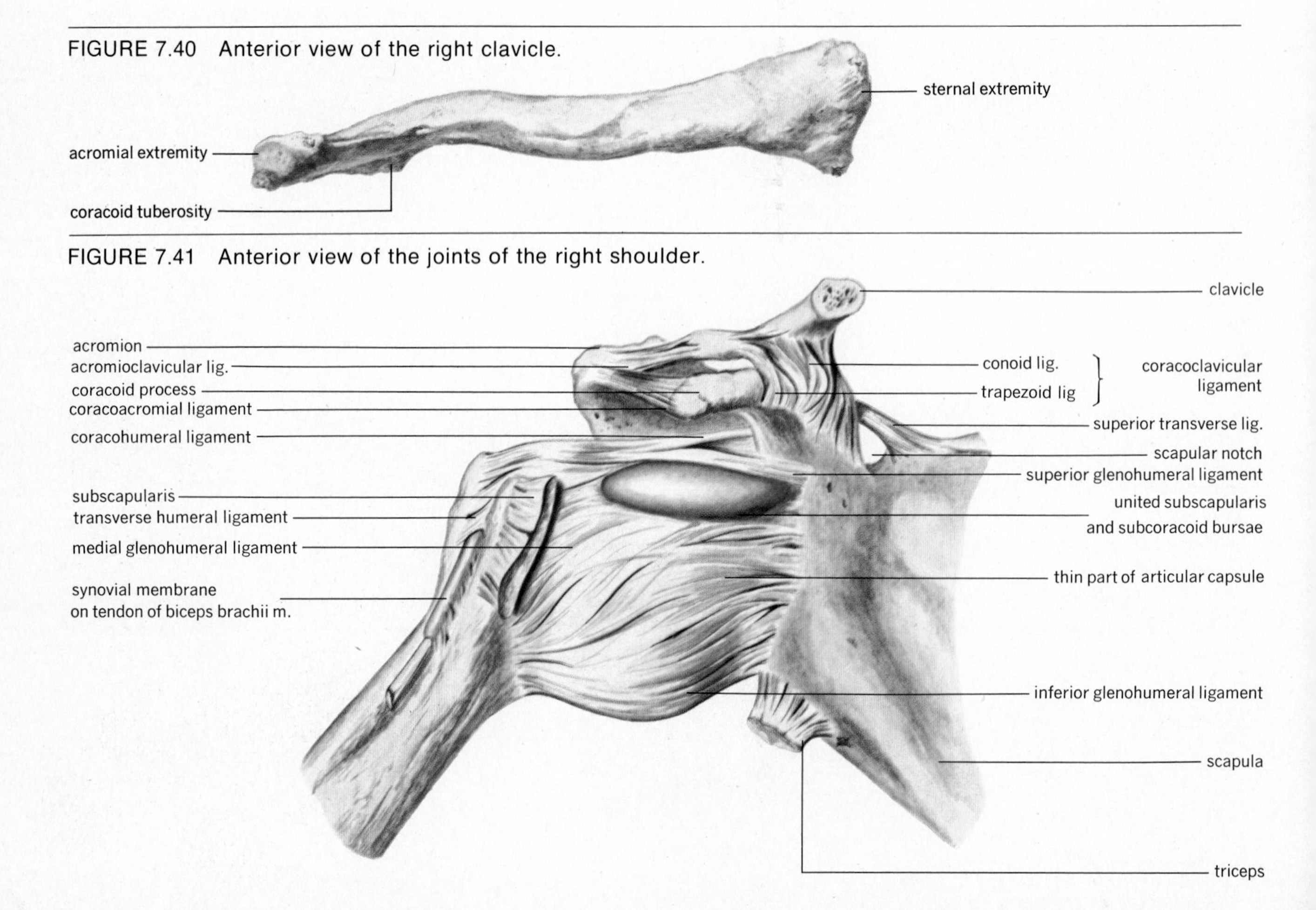

FIGURE 7.40 Anterior view of the right clavicle.

FIGURE 7.41 Anterior view of the joints of the right shoulder.

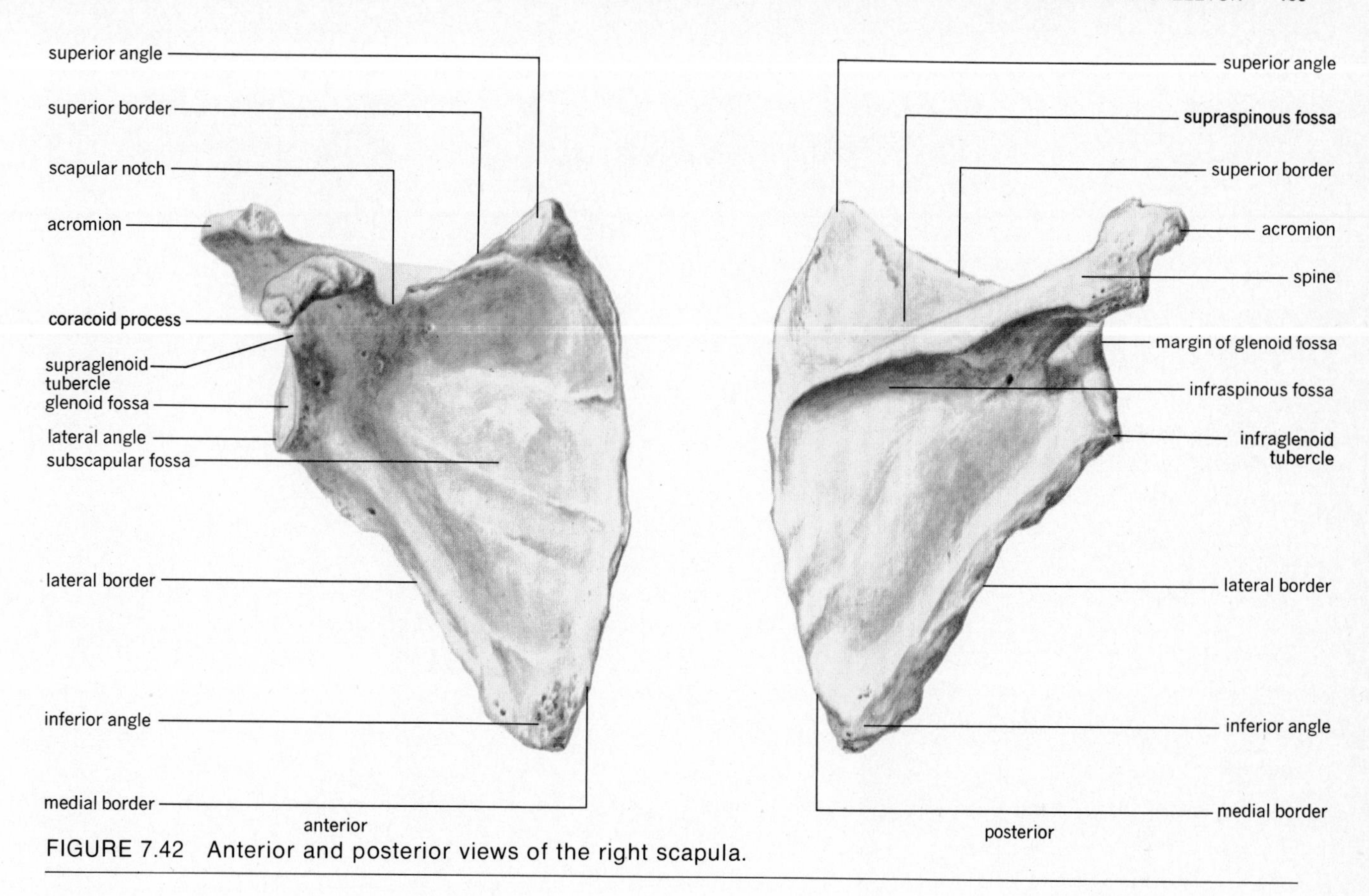

FIGURE 7.42 Anterior and posterior views of the right scapula.

SCAPULAE. The SCAPULAE (Fig. 7.42) lie over the posterior wall of the rib cage between ribs two and seven and about two inches from the vertebral column. They consist of a triangular, flattened body from which rises posteriorly a SPINE, the lateral end of which is expanded and flattened to form the ACROMION. The body has a thin medial or VERTEBRAL BORDER, a thick LATERAL or AXILLARY BORDER, and a SUPERIOR BORDER. It has three angles: SUPERIOR, INFERIOR, and LATERAL. At the lateral angle is a smooth, ovoid-shaped depression, the GLENOID FOSSA, which in the living subject is deepened by a rim of cartilage, the GLENOID LABRUM, and which serves for the articulation of the humerus. A SUPRAGLENOID TUBERCLE is found above the rim of the fossa, and an INFRAGLENOID TUBERCLE lies below. At the beginning of the superior border is a strong, curved CORACOID PROCESS. Just beyond the coracoid process on the superior border is the SCAPULAR NOTCH. The superior border, the shortest of the three borders, ends at the superior angle. The ANTERIOR or COSTAL SURFACE of the BODY of the scapula is concave, the SUBSCAPULAR FOSSA, and lies over the rib cage. The posterior surface is convex and is divided by the spine into two unequal parts: a smaller SUPRASPINOUS FOSSA above the spine and a larger INFRASPINOUS FOSSA below the spine. These fossae are the surfaces of origin for some of the major muscles of the shoulder.

FIGURE 7.43 Anterior and posterior view of the right humerus.

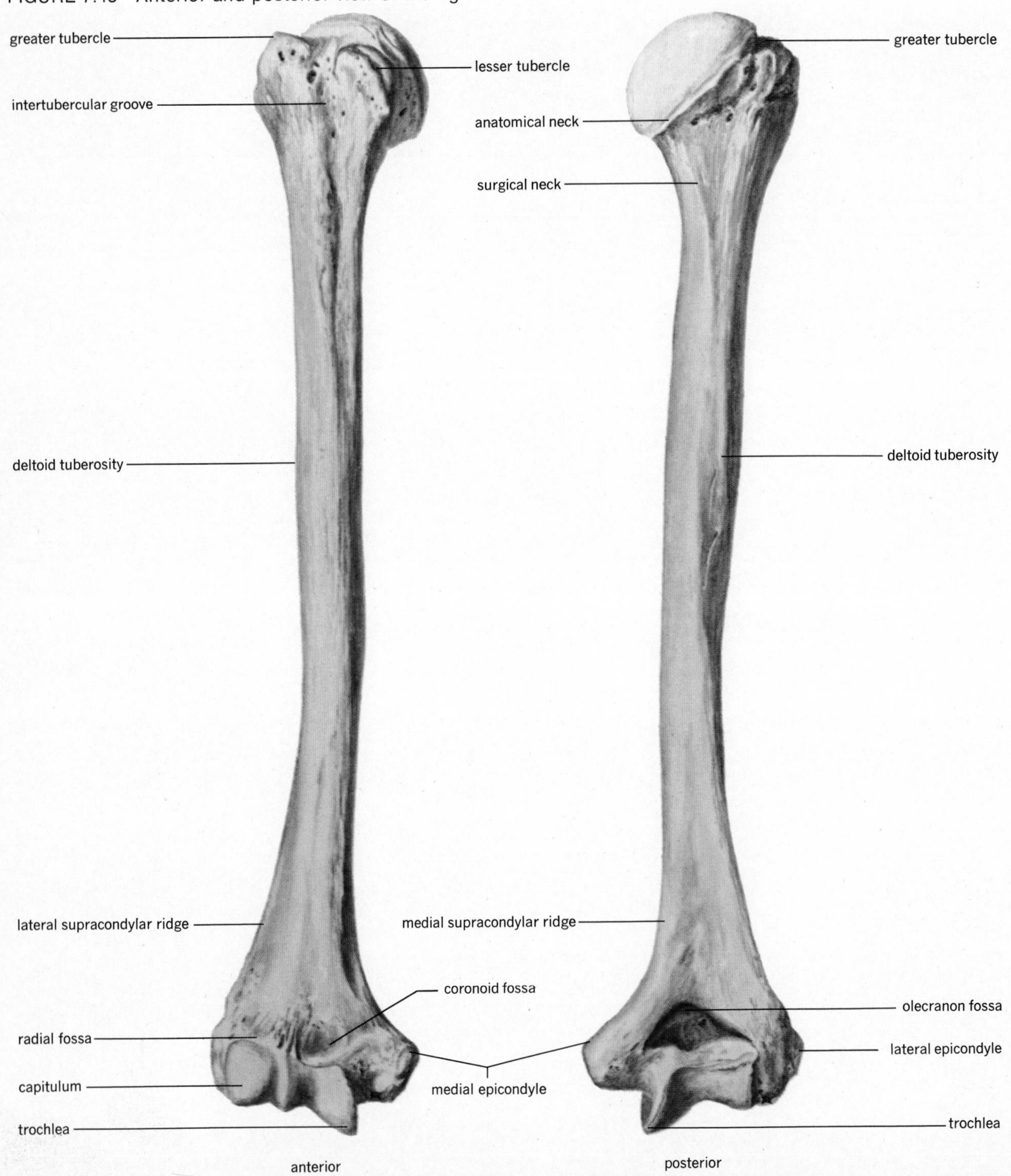

Upper extremity

The bones of the upper extremity consist of the humerus in the arm, the radius and ulna in the forearm, and the carpals (wrist), metacarpals (palm), and phalanges (fingers or digits) in the hand—a total of 30 bones.

HUMERUS. The humerus (Fig. 7.43) is the longest and largest bone of the upper extremity. It articulates at its proximal end with the glenoid fossa of the scapula to form the shoulder joint. At its distal end it joins the radius and ulna at the elbow joint. Its proximal end consists of a HEAD, NECK, GREATER and LESSER TUBERCLES, and the INTERTUBERCULAR SULCUS or GROOVE. The head fits into the glenoid fossa. The neck separates the head of the humerus from the rest of the bone. This is the ANATOMICAL NECK. A second neck, the SURGICAL NECK, is the narrow part of the humerus just below the tubercles and is frequently the site of fracture. The intertubercular groove (bicipital groove) is fairly deep and lies on the upper anterior surface of the humerus between the two tubercles. In the intact body, the tendon of the long head of the biceps brachii muscle lies in this groove.

The SHAFT or BODY of the humerus is almost cylindrical at its upper end. It gradually becomes triangular, and is flattened at its lower end and broad where it is continuous with the distal extremity of the bone. At about the middle of the shaft, on the anterolateral surface, is a slightly raised, roughened area, the DELTOID TUBEROSITY, which serves for the attachment or insertion of the deltoideus muscle.

The distal extremity of the humerus is flattened anteroposteriorly and curved slightly forward. It has a double articulating surface consisting of the lateral CAPITULUM and the adjoining medial TROCHLEA. The RADIAL, CORONOID, and OLECRANON FOSSAE and the LATERAL and MEDIAL EPICONDYLES are the other features on the distal end of the humerus. The capitulum articules with the head of the radius; the trochlea articulates with the semilunar notch of the ulna to form the hinge type of elbow joint. The radial and coronoid fossae are on the anterior surface of the bone; the former receives the head of the radius and the latter receives the coronoid process of the ulna when the forearm is flexed. The olecranon fossa, on the posterior side of the bone, receives the olecranon process of the ulna when the forearm is extended and limits that movement. The lateral and medial epicondyles are eminences lying beside and above the condyles.

ULNA. The ulna (Fig. 7.44) is the medial bone of the forearm and is larger and longer than the radius, to which it lies parallel. Its proximal end is large and highly adapted to form a part of the elbow joint. The bone tapers off through its shaft to the smaller distal end, which articulates with the radius.

The proximal end of the ulna is dominated by the OLECRANON PROCESS, which forms the tip of the elbow. The deep SEMILUNAR NOTCH of the ulna fits around the trochlea of the humerus. Below the semilunar notch is the CORONOID PROCESS, and in front of it is the TUBEROSITY of the ulna, a roughened area for the insertion of the brachialis muscle. The RADIAL NOTCH is an articular depression on the upper part of the ulna lateral to the coronoid process. The round head of the radius fits into this notch. The SHAFT or BODY of the ulna has a sharp lateral INTEROSSEOUS BORDER for the attachment of the interosseous ligament, which connects with a similar border on the medial side of the shaft of the radius. The distal end of the ulna is small and round with a blunt projection, the STYLOID PROCESS, on its posterior side.

RADIUS. The radius (Fig. 7.44) is the lateral bone of the forearm. Its proximal end is round and its circumference fits into the radial notch of the ulna; its superior surface, which is slightly concave, articulates with the capitulum of the humerus. On the medial side just below the proximal end

FIGURE 7.44 Anterior view of proximal, intermediate, and distal articulations of the right forearm.

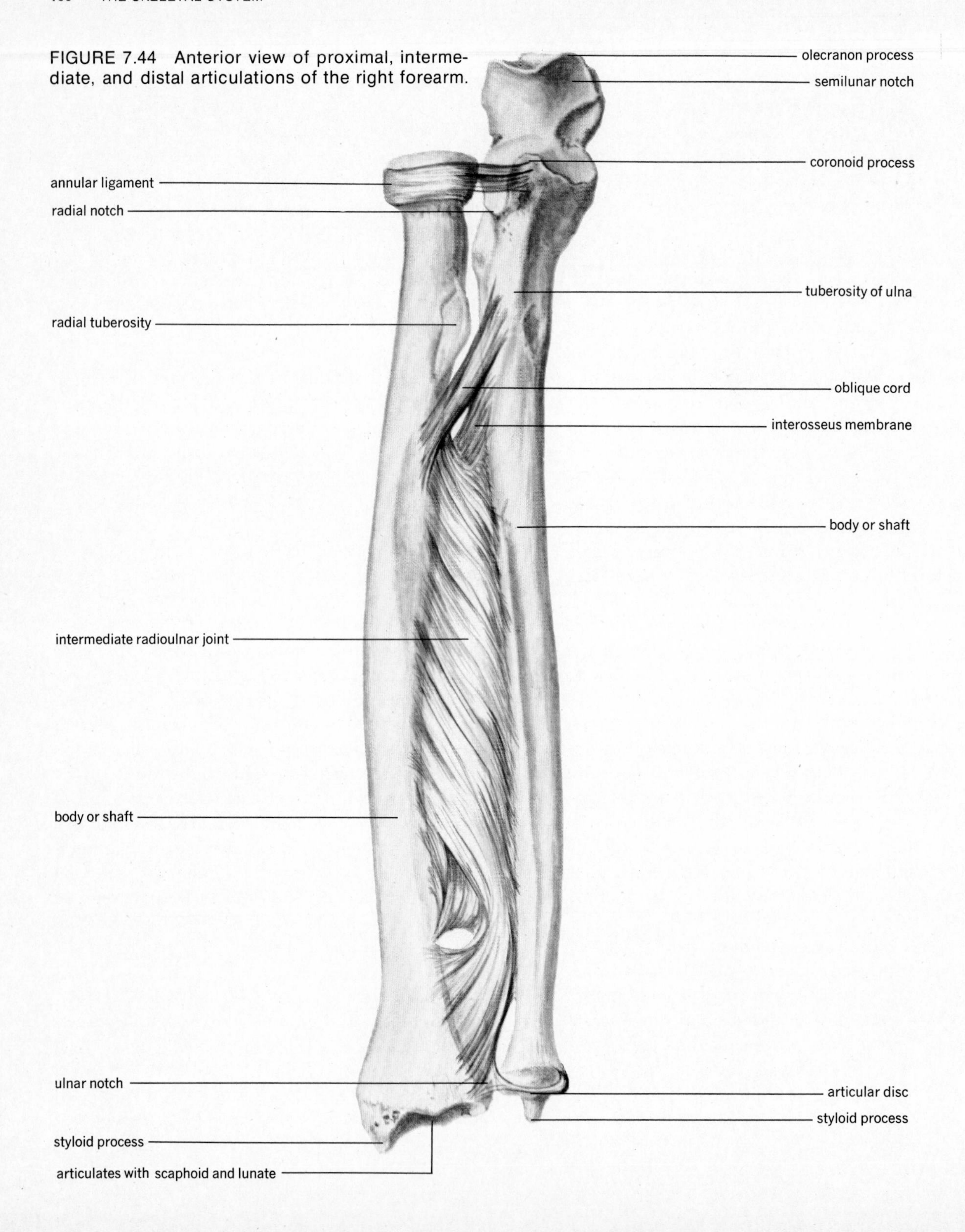

is a raised, roughened area, the RADIAL TUBEROSITY, for the insertion of the biceps brachii muscle. The SHAFT of the radius, almost uniform in diameter through most of its length, widens distally to form a broad, concave inferior surface for articulation with the scaphoid and lunate of the carpus or wrist. The distal end has vertical grooves on its convex posterior surface for the passage of tendons. A STYLOID PROCESS is on the lateral side, and a concave ULNAR NOTCH for articulation with the distal end of the ulna is medial.

CARPUS. The carpus (Fig. 7.45) consists of eight bones arranged in two transverse rows, each with four bones. In the proximal row, from lateral to medial side, are the scaphoid *(navicular)*, lunate, triquetrum *(triangular)*, and pisiform; in the distal row, lateral to medial, are the trapezium *(greater multangular)*, trapezoid *(lesser multangular)*, capitate, and hamate. The names of the bones are somewhat suggestive of their shapes. They are bound closely together, but the joints between them are synovial. Little movement is allowed,

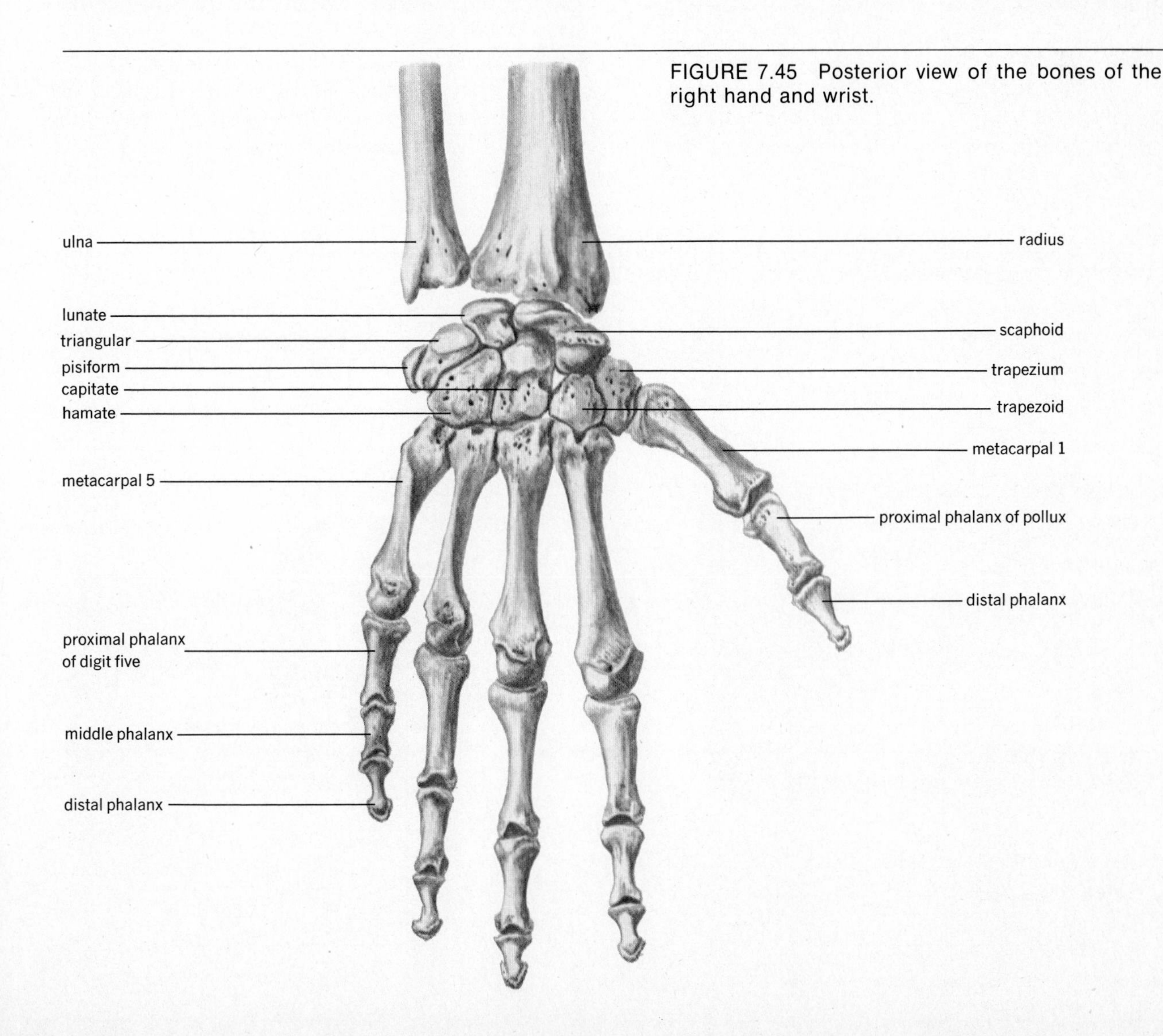

FIGURE 7.45 Posterior view of the bones of the right hand and wrist.

however, the greatest being in the midcarpal or transverse intercarpal joint, which lies between proximal and distal rows of carpals.

METACARPUS. The metacarpus (Fig. 7.45) consists of five bones, numbered from one to five, starting with the lateral bone. Each has a proximal base, a shaft, and a distal head. The first metacarpal is more freely movable than the other four, which articulate with one another and with carpal bones. The heads of the metacarpals articulate with the digits.

PHALANGES. The phalanges (Fig. 7.45) are the bones of the digits. There are two phalanges in digit one (the thumb), and three in each of the other four digits (fingers). The phalanges of the fingers are called proximal, middle, and distal phalanges.

ARTICULATIONS OF THE UPPER EXTREMITY. The principal joints to be considered here are the shoulder, elbow, and wrist joints. Others will be mentioned, and some understanding of them can be gained by careful study of the illustrations.

The SHOULDER JOINT (Fig. 7.41) brings the upper extremity into functional relationship to the shoulder girdle. The rather expansive, rounded head of the humerus fits into the shallow glenoid fossa of the scapula. Because of the great discrepancy in the surface areas of these two articulating surfaces, only a small part of the head of the humerus is at any time in contact with the glenoid fossa. Its surface is covered in part by the

FIGURE 7.46 Medial and lateral view of the elbow and distal radioulnar joints.

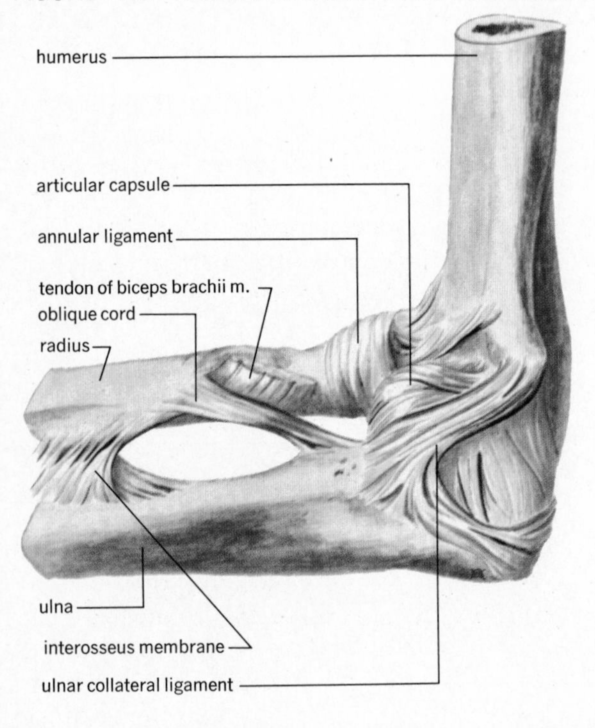

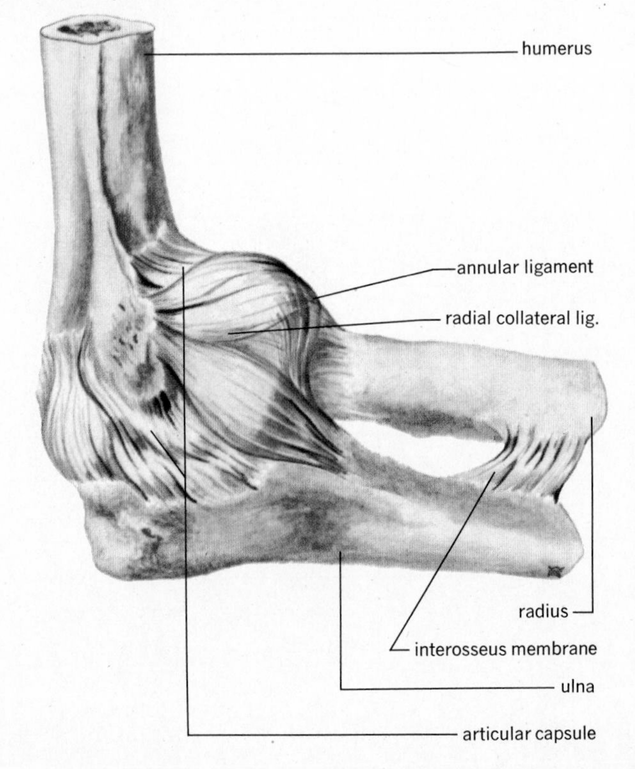

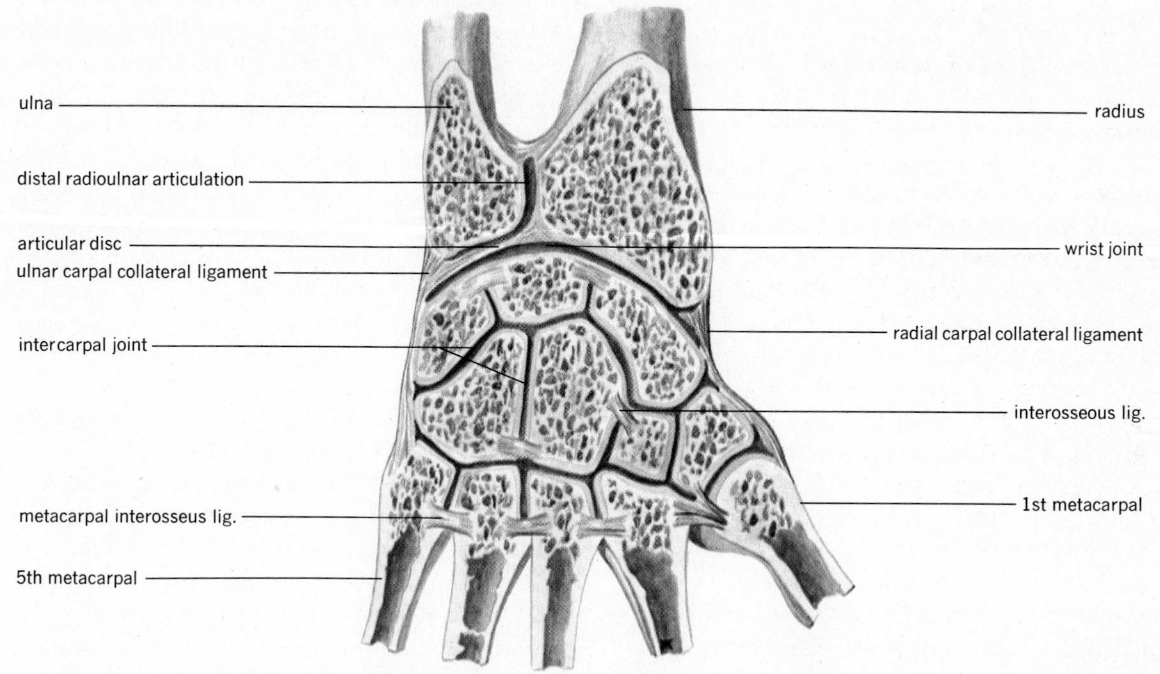

FIGURE 7.47 Synovial articulations of the right hand and wrist as shown in section.

cartilaginous GLENOID LABRUM, which deepens the glenoid fossa in the living subject, and by the ARTICULAR CAPSULE, which encloses the joint. The articular capsule is also a loose-fitting structure that attaches around the border of the bony glenoid fossa and around the anatomical neck of the humerus, allowing a separation of as much as 2.5 cm between the bones. This laxity of the joint, along with the relatively loose attachment of the scapula, allows for the great freedom of movement characterizing the shoulder and hence the arm. The fibrous articular capsule has one opening for the passage of the TENDON OF THE LONG HEAD OF THE BICEPS BRACHII. The tendon does not penetrate the synovial membrane, however, but pushes against it and becomes surrounded by it. The shoulder joint is further supported by the CORACOHUMERAL and GLENOHUMERAL LIGAMENTS, and by muscles such as: the supraspinatus above, the tendons of the teres minor and infraspinatus posteriorly, the subscapularis anteriorly, and the long head of the triceps below. FLEXION, EXTENSION, ABDUCTION, ADDUCTION, ROTATION, and CIRCUMDUCTION ARE MOVEMENTS POSSIBLE IN THE SHOULDER JOINT.

The ELBOW JOINT (Fig. 7.46), broadly considered, includes the joining of the semilunar notch of the ulna with the trochlea of the humerus, the relation of the head of the radius with the capitulum, and the relation of the circumference of the radial head with the radial notch of the ulna. All of these are enclosed under a COMMON CAPSULE and have a COMMON SYNOVIAL CAVITY. The RADIOULNAR JOINT may also be considered separately (Figs. 7.44 and 7.47). The capsule is reinforced by RADIAL and ULNAR COLLATERAL LIGAMENTS.

"TENNIS ELBOW," an EPICONDYLITIS, often affects the lower end of the humerus on the lateral side. It is painful to the touch and even more painful if the elbow is twisted. An EPICONDYLITIS ON THE MEDIAL SIDE of the lower arm is called "GOLFER'S ELBOW" and manifests similar symptoms to "tennis elbow."

The HUMEROULNAR ARTICULATION is a typical hinge joint. The semilunar notch fits closely over the trochlea, allowing only flexion and extension of the forearm. The HUMERORADIAL ARTICULATION, where the concave superior surface of the radius fits the humeral capitulum, follows the hinge action of the elbow. It also allows the rotation of the radius at the PROXIMAL RADIOULNAR ARTICULATION where the circular head of the radius, held in position by an ANNULAR LIGAMENT, turns against the radial notch of the ulna. The rotation allows SUPINATION and PRONATION of the hand. An INTERMEDIATE RADIOULNAR JOINT consists of an INTEROSSEOUS MEMBRANE extending between the adjacent borders of the shafts of the radius and ulna. It is a FIBROUS JOINT (Fig. 7.44). The DISTAL RADIOULNAR JOINT is held by the DORSAL and PALMAR RADIOULNAR LIGAMENTS. The movement here is pivotlike—the distal end of the radius, at its ulnar notch, riding around the ulna to its anterior and medial border in pronation and back to the anatomical position in supination.

The RADIOCARPAL or WRIST ARTICULATION is largely between the radius and the scaphoid and lunate, and between the disc at the distal end of the ulna and the triquetrum (Fig. 7.47). It is a CONDYLOID JOINT allowing flexion, extension, abduction, adduction, and circumduction of the hand. The joint is enclosed in a CAPSULE made up of DORSAL and PALMAR RADIOCARPAL and RADIAL and ULNAR COLLATERAL LIGAMENTS.

The INTERCARPAL and CARPOMETACARPAL JOINTS have one continuous SYNOVIAL CAVITY, with the exception of the joint between the pisiform and triquetrum, and the trapezium and first metacarpal. There is little movement between carpals, except at the MIDCARPAL JOINT between proximal

FIGURE 7.48 The third digit of the right hand showing metacarpophalangeal and interphalangeal joints.

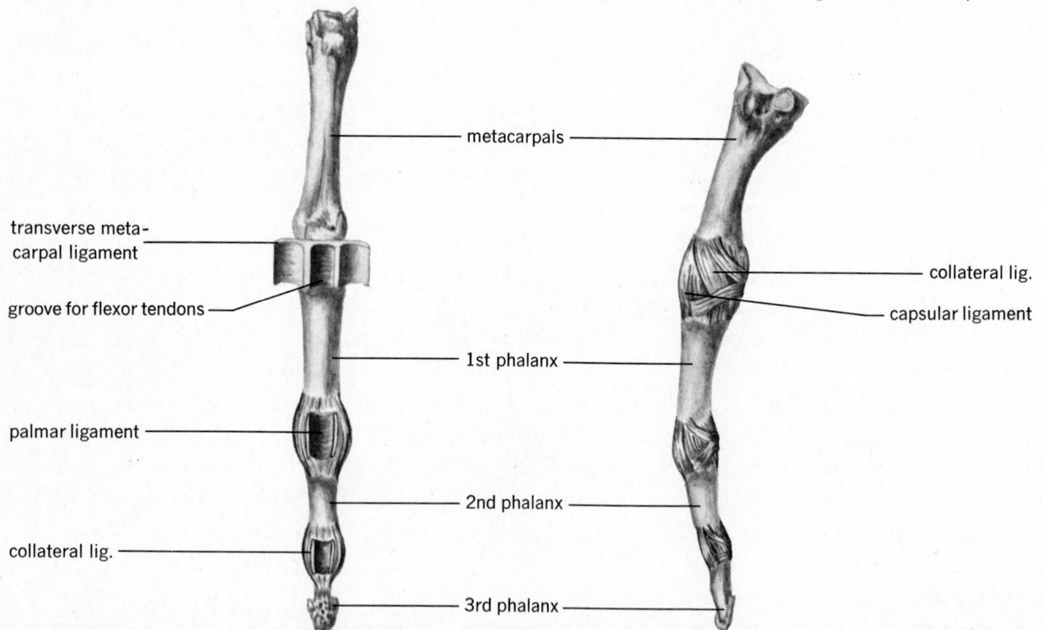

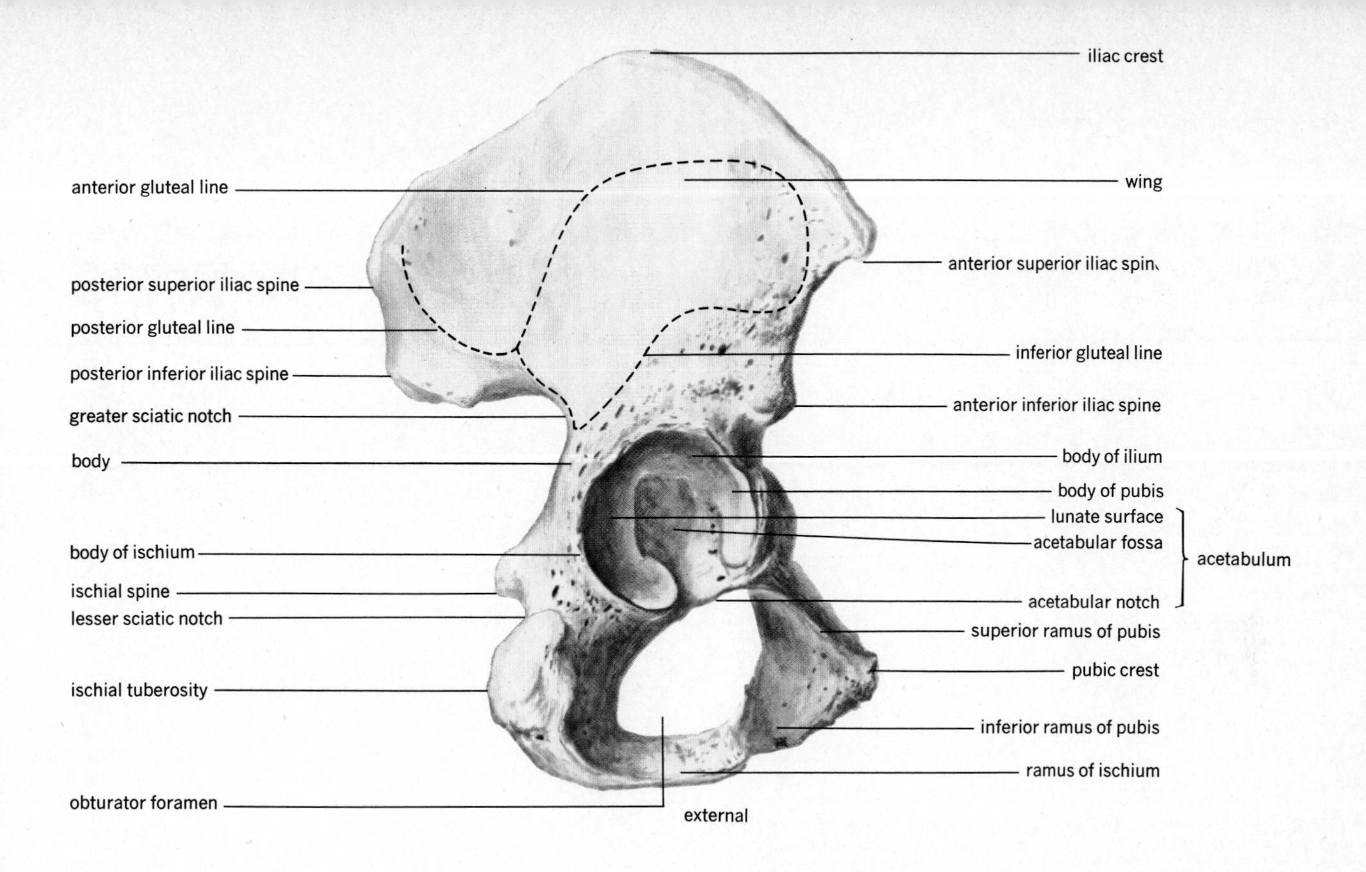

FIGURE 7.49 The right os coxae viewed from the external and internal surfaces.

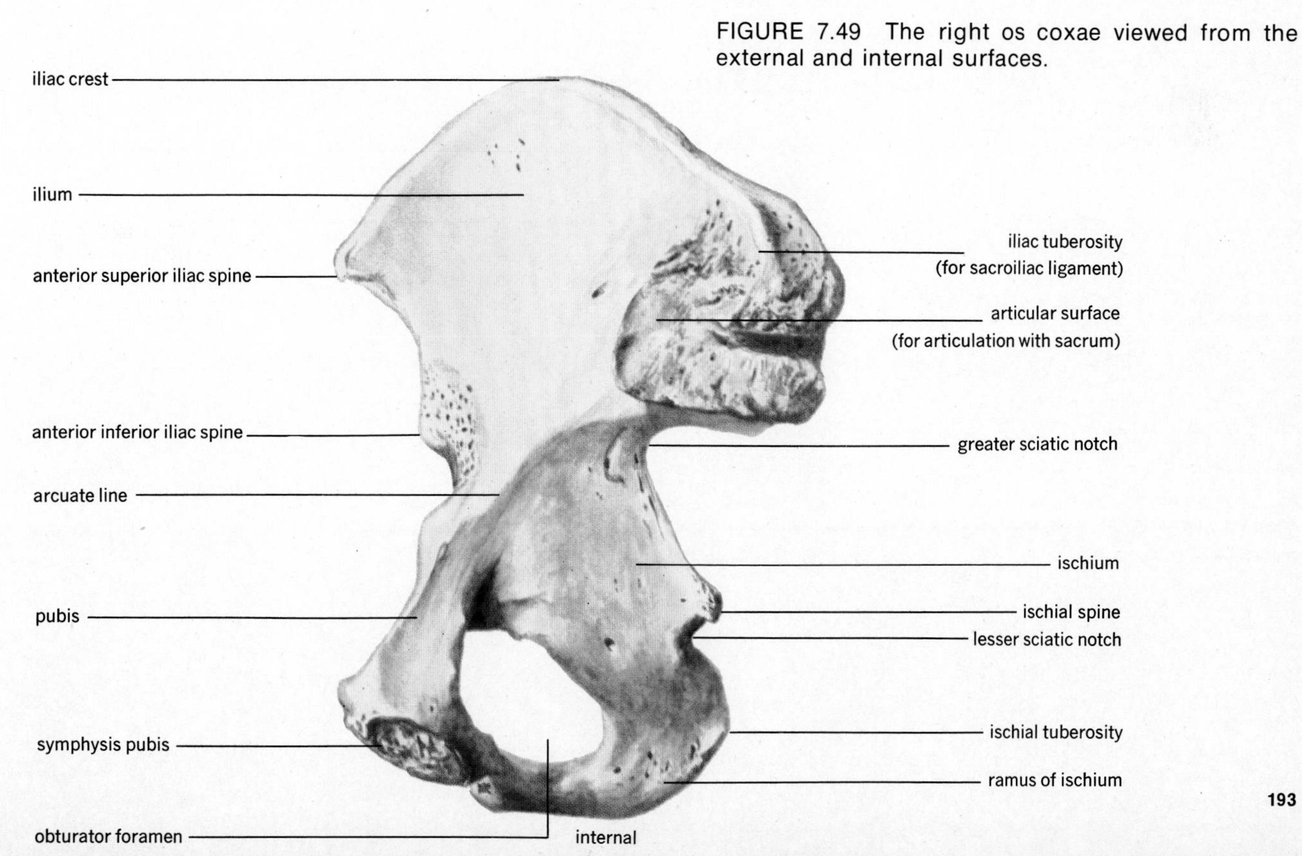

and distal rows, where the principal actions are flexion and extension. DORSAL, PALMAR, and COLLATERAL LIGAMENTS are among those supporting these articulations.

The CARPOMETACARPAL JOINTS TWO TO FIVE allow a minimum of gliding movements (Fig. 7.47). The first metacarpal is also positioned in such a way as to cause the thumb to face medially where it can be used with the digits opposite it. It is said to be OPPOSABLE, and forms a SADDLE JOINT. The TRANSVERSE METACARPAL LIGAMENT binds the heads of the metacarpals, except for metacarpal one which remains free.

The metacarpals, with the exception of the first, have rounded heads that fit into shallow concavities on the bases of the proximal phalanges forming the METACARPOPHALANGEAL JOINTS, which are CONDYLOID. The joint between the first metacarpal and thumb is hingelike or GINGLYMOID and allows only flexion and extension. PALMAR and TWO COLLATERAL LIGAMENTS are provided for each joint (Fig. 7.48).

INTERPHALANGEAL JOINTS are of the hinge type, each having a palmar and two collateral ligaments (Fig. 7.48). The tendons of the extensor muscles serve as posterior ligaments. Their action is flexion and extension.

Pelvic girdle

The pelvic girdle consists of the two OS COXAE

FIGURE 7.50 Right os coxae of a child showing cartilage.

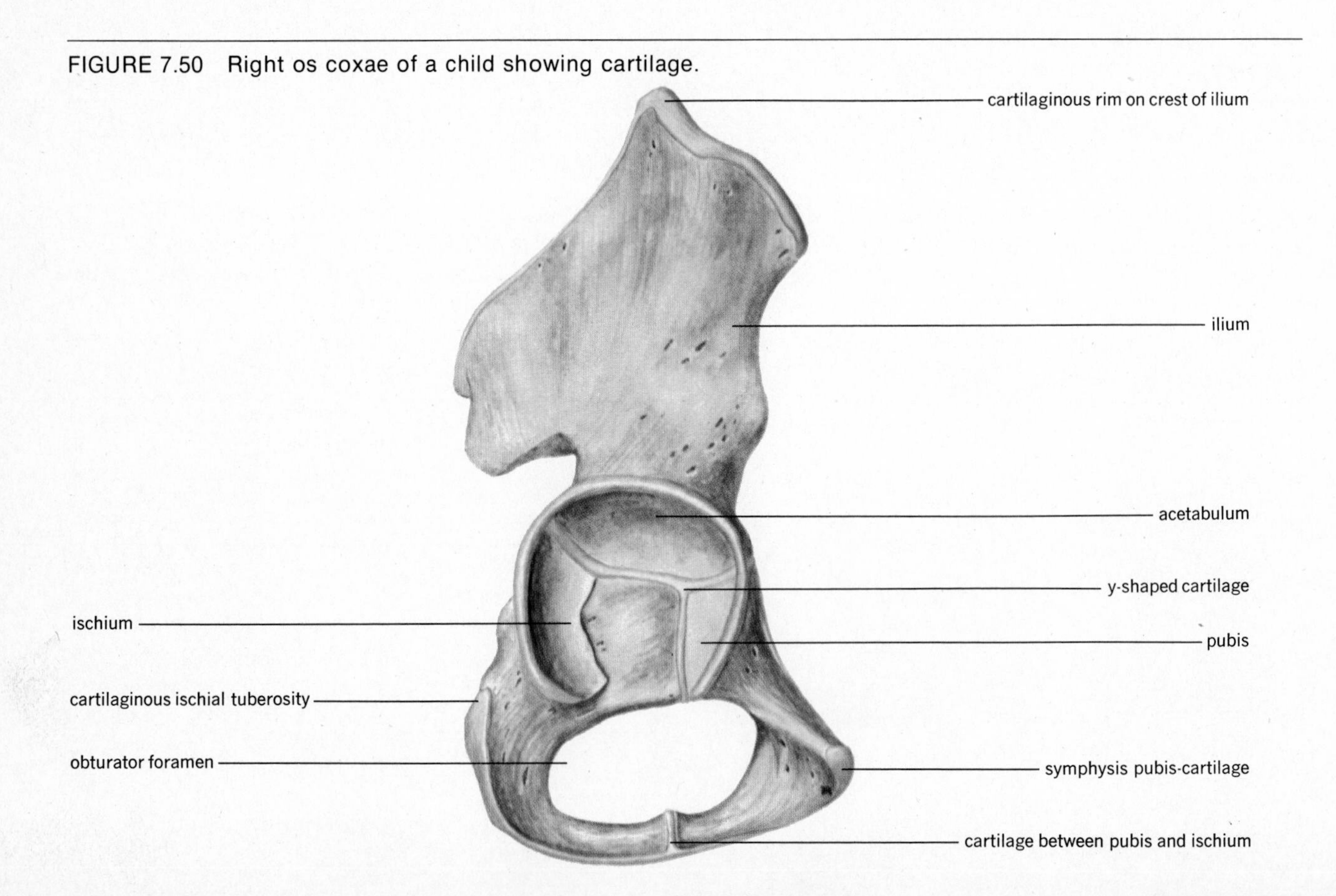

(Fig. 7.49) or INNOMINATE BONES that form, with the SACRUM and the COCCYX, the BONY PELVIS. The os coxae join each other anteriorly at the SYMPHYSIS PUBIS. The articulation between the sacrum and the os coxae on each side is the SACROILIAC JOINT. The os coxae of the newborn are made up of three parts: a superior ILIUM, an inferior and anterior PUBIS, and an inferior and posterior ISCHIUM (Fig. 7.50). These three components come together in a deep fossa laterally, the ACETABULUM, for the articulation of the lower extremity. The girdle is constructed in such a way as to give strength, support, and protection, as indicated in the previous comparison of pectoral and pelvic girdles.

ILIUM. The ilium (Fig. 7.49 and 7.50) is the largest component of the os coxae. It is superiorly placed, its upper portion, the ALA (wing), being broad and expansive; its lower portion, the BODY, narrows, but contributes about two-fifths of the wall of the acetabulum. Its superior border is called the ILIAC CREST, which ends anteriorly in the ANTERIOR SUPERIOR ILIAC SPINE. Below the POSTERIOR SUPERIOR ILIAC SPINE is a notch and below it, in turn, a POSTERIOR INFERIOR ILIAC SPINE. Below this spine is the GREATER SCIATIC NOTCH, which in the living state is closed by the SACROSPINOUS LIGAMENTS to form a FORAMEN. Anteriorly, a notch occurs below the anterior superior iliac spine, followed by an ANTERIOR INFERIOR ILIAC SPINE.

The internal surface of the ilium has a smooth concave surface called the ILIAC FOSSA where the iliacus muscle has its origin. Also, posterior to the fossa is a divided roughened area, the superior portion of which is the ILIAC TUBEROSITY and the inferior portion of which is the AURICULAR SURFACE, for articulation with the sacrum.

The external surface of the ilium shows three arched lines: the POSTERIOR, ANTERIOR, and INFERIOR GLUTEAL LINES. Between them the three gluteal muscles have their origin.

ISCHIUM. The ischium (Figs. 7.49 and 7.50), the lower and posterior part of the pelvic girdle, contributes about two-fifths of the acetabular wall. Posteriorly, on the body of the ischium, is a prominent ISCHIAL SPINE below which is the LESSER SCIATIC NOTCH and below it, in turn, the ISCHIAL TUBEROSITY. The remaining portion of the ischium is the RAMUS. It extends anteriorly and slightly upward and joins the inferior ramus of the pubis and, with the pubis, forms the OBTURATOR FORAMEN.

PUBIS. The pubis (Figs. 7.49 and 7.50) is the anterior and inferior part of the pelvis. Its BODY forms the remaining one-fifth of the acetabular wall. From the body the SUPERIOR RAMUS of the pubis extends downward to the SYMPHYSIS PUBIS where the pubic bones of opposite sides are articulated. The INFERIOR RAMUS extends downward and backward from the symphysis to join the ramus of the ischium.

THE BONY PELVIS. As indicated above, the bony pelvis (Fig. 7.51) is composed of the two os coxae or hip bones anteriorly and laterally, and the sacrum and coccyx posteriorly. The pelvis is divided into a GREATER *(false)* PELVIS above and a LESSER *(true)* PELVIS below by a plane passing through the PROMONTORY of the SACRUM, the ARCUATE *(iliopectineal)* LINES, and the SUPERIOR MARGIN of the PUBIC BONES and the SYMPHYSIS PUBIS.

The GREATER or FALSE PELVIS is formed laterally by the expansive ilia; posteriorly it is formed by the upper portion of the sacrum, and anteriorly there is no bony component, only the abdominal wall. It contains parts of the intestines, gives them support, and also throws some of their weight onto the abdominal wall.

The LESSER or TRUE PELVIS is well protected by its walls, which are composed of a part of the ilium, the ischium, pubis, sacrum, and coccyx. It has openings above and below, called the inlet and outlet of the pelvis, respectively. It contains the urinary bladder anteriorly, the rectum posteriorly, and between these organs, in the female, the vagina and part of the uterus.

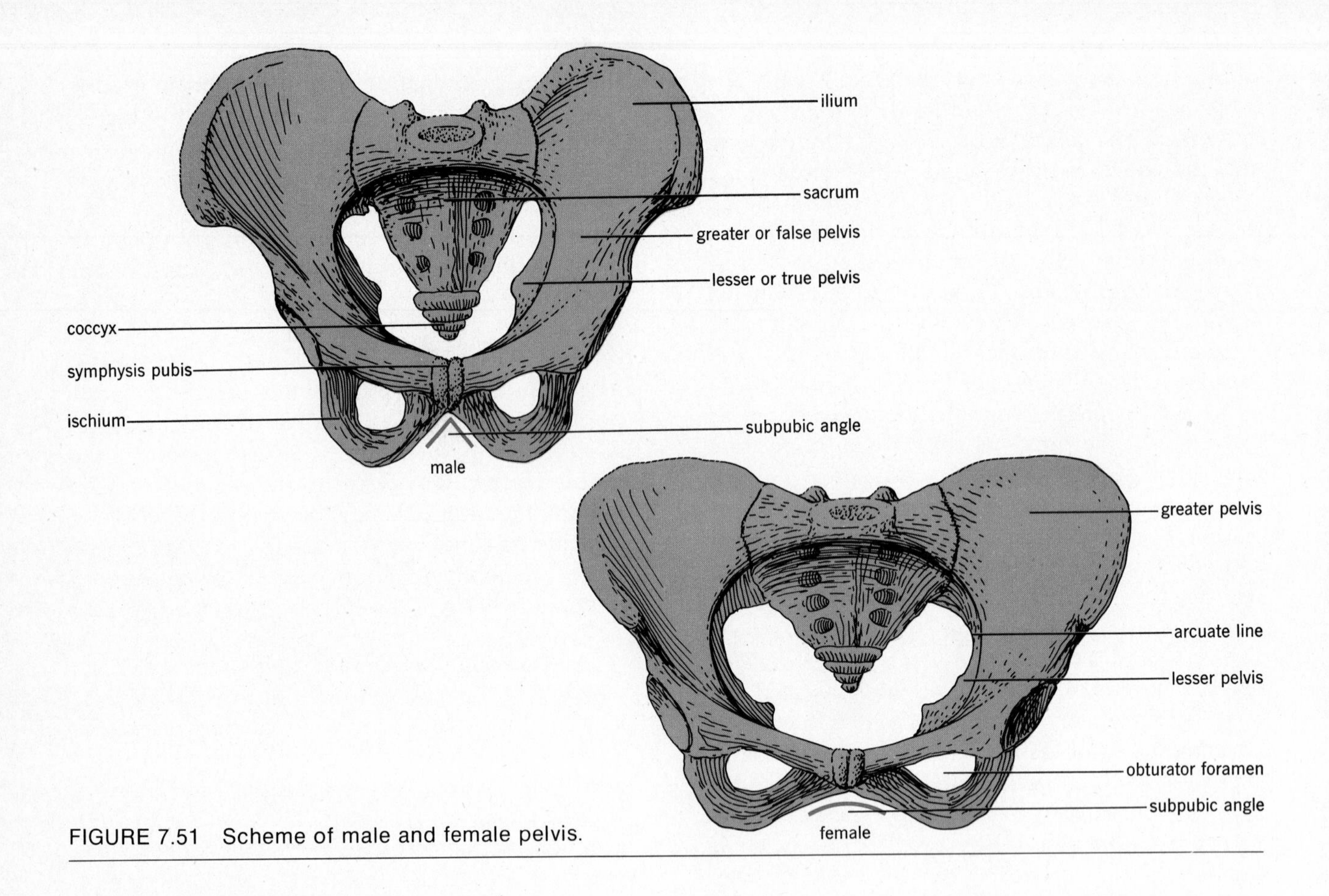

FIGURE 7.51 Scheme of male and female pelvis.

DIFFERENCES BETWEEN MALE AND FEMALE PELVES. The female pelvis is characterized by adaptations to the childbearing function. It is wider, more spacious, and lighter in construction than that of the male. It is shallower than that of the male and its diameters are greater. The ilia flare out more to the sides than in the male, which makes for the broader hips of the female. The inlet of the lesser or true pelvis of the female is almost round, that of the male is heart-shaped. The cavity of the true pelvis is shallower and wider than in the male. The sacrum is shorter, wider, and less curved than in the male; the coccyx is more flexible; the ischial spines and tuberosities turn outward and hence are farther apart; the pubic arch forms an obtuse angle rather than the acute angle of the male. These features all contribute to the wider outlet of the true pelvis in the female.

ARTICULATIONS OF THE PELVIS. The articulations of the pelvis (Fig. 7.52) are the SACROILIAC JOINT, the SYMPHYSIS PUBIS, and the SACROCOCCYGEAL SYMPHYSIS. The sacroiliac joint is a double structure in that the upper part is a fibrous articulation, the lower part synovial. The joint, however, is almost immovable because of the roughened interlocking surfaces of the articulating bones. It does give resilience to this weight-bearing structure. ILIOLUMBAR, ANTERIOR SACRO-

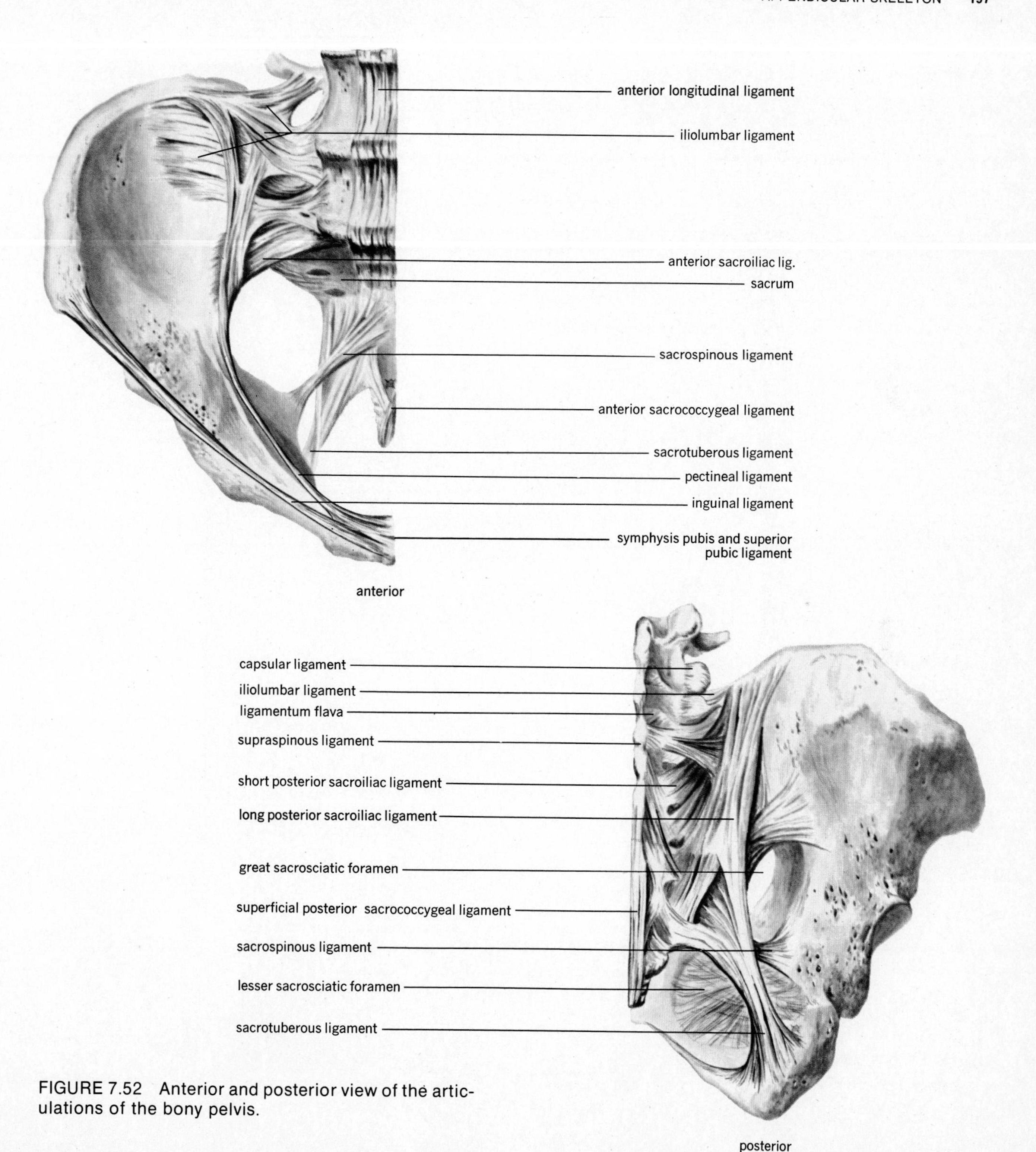

FIGURE 7.52 Anterior and posterior view of the articulations of the bony pelvis.

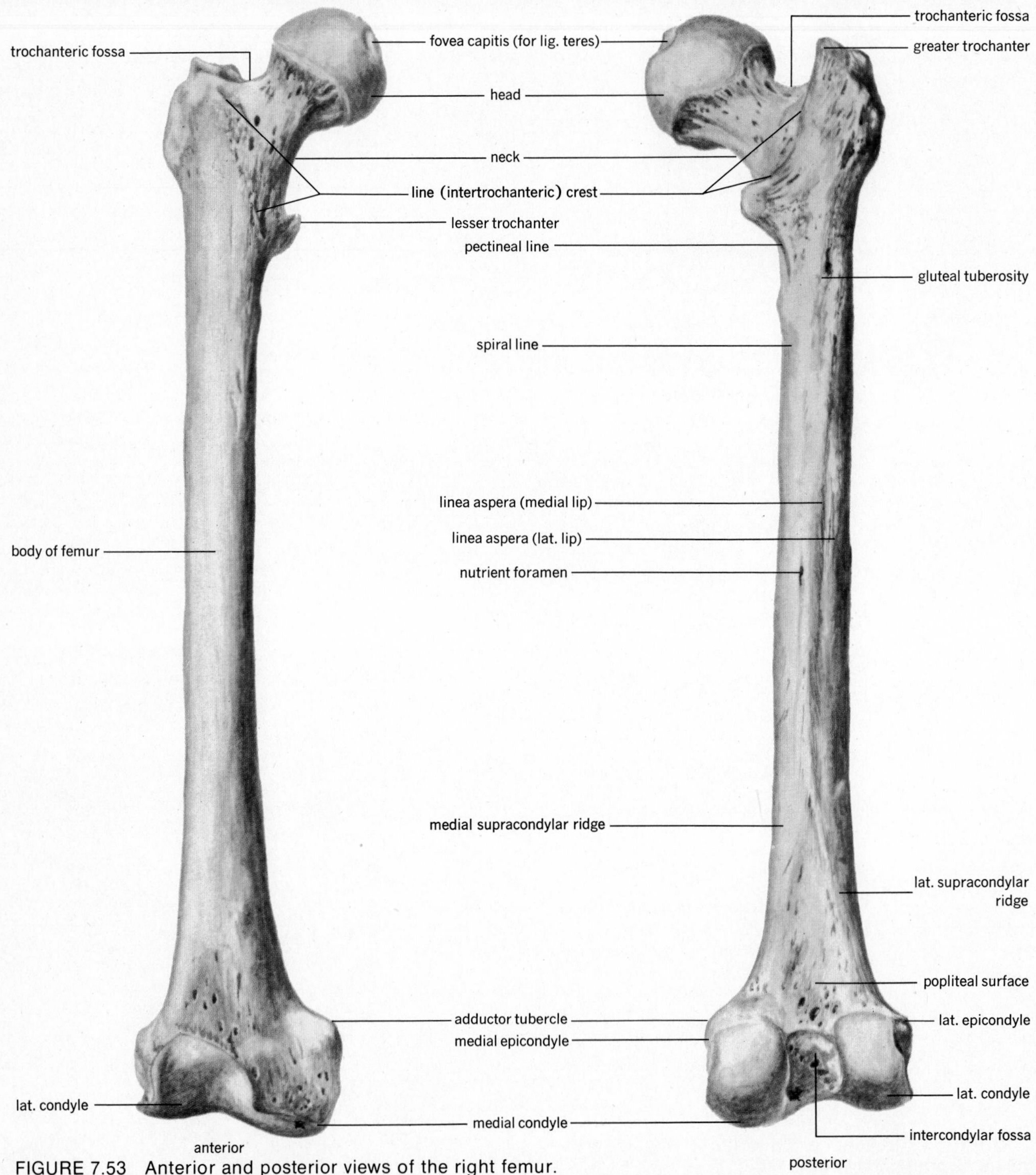

FIGURE 7.53 Anterior and posterior views of the right femur.

ILIAC, LONG, and SHORT POSTERIOR SACROILIAC LIGAMENTS support this articulation directly while the SACROSPINOUS and SACROTUBEROUS LIGAMENTS help by preventing any abnormal tipping of the pelvis.

The SYMPHYSIS PUBIS, between the two pubic bones anteriorly, is a cartilaginous joint similar to those between the bodies of the vertebrae. It is held by the SUPERIOR PUBIC LIGAMENT, which lies between the pubic tubercles, and the ARCUATE PUBIC LIGAMENT, which arches inferior to the cartilage and between the inferior pubic rami.

The SACROCOCCYGEAL SYMPHYSIS is similar to those joints between the bodies of the vertebrae. Its ligaments, such as the dorsal, ventral, and lateral SACROCOCCYGEAL LIGAMENTS and the INTERARTICULAR LIGAMENT, are also similar (Fig. 7.52).

All three of the articulations of the pelvis are influenced by hormones during pregnancy. The ligaments are softened, as are the fibrocartilages of the symphysis joints, resulting in slight looseness and movement of the joints—an aid to childbirth.

The lower extremity

The lower extremity consists of the femur in the thigh, the patella, the tibia and fibula in the leg, and the tarsals (ankle), metatarsals, and phalanges in the foot—a total of 30 bones.

FEMUR. The femur (Fig. 7.53) is the largest and longest bone of the skeleton. Its HEAD articulates with the acetabulum, and the SHAFT of the bone is directed downward and mediad to approach the femur of the opposite side. This convergence of the femurs brings the knee joints nearer the line of gravity of the body. The female pelvis, being broader than the male, causes the bones to converge even more, and thus many women appear to be knock-kneed. True KNOCK-KNEE *(genu valgum)* is a deformity in which the legs are bent outward below the knee. When the knees are together the ankles cannot be approximated. BOW LEG *(genu varum)*, in contrast, is an outward curving of one or both legs at or below the knee.

The femur consists of a body or shaft and proximal (upper) and distal (lower) ends. The proximal end consists of a rounded HEAD, NECK, and GREATER and LESSER TROCHANTERS. The head is nearly spherical with a smooth surface, except for the OVAL FOVEA CAPITIS near its center where the LIGAMENTUM TERES is attached. The NECK extends between the head and the shaft of the femur. It angles upward, mediad, and slightly forward. The angle between the neck and the shaft is slightly less acute in the male than in the female due to the greater width of the female pelvis. The greater and lesser trochanters are prominences at the junction of the neck and the shaft. Between them on the anterior surface is a narrow, roughened INTERTROCHANTERIC LINE, while on the posterior surface of the bone they are connected by a raised INTERTROCHANTERIC CREST.

The shaft is cylindrical in cross section and smooth except for a vertical, roughened ridge on the posterior surface. This ridge is divided superiorly into three, and inferiorly into two, diverging lines. The undivided ridge is the LINEA ASPERA, which serves for the attachment of a number of thigh muscles. One of the three upper diverging lines runs toward the greater trochanter and is the GLUTEAL TUBEROSITY. The second, directed toward the lesser trochanter, is the PECTINEAL LINE. The third line twists around beneath the lesser trochanter and is the SPIRAL LINE. The two lower diverging lines are the LATERAL and MEDIAL SUPRACONDYLAR RIDGES, which enclose a triangular space, the POPLITEAL SURFACE.

The distal or lower expanded end of the femur includes: the MEDIAL and LATERAL CONDYLES for articulation with the tibia; a depressed area between them posteriorly, the INTERCONDYLAR FOSSA, which is limited above by the INTERCONDYLAR LINE; and the LATERAL and MEDIAL

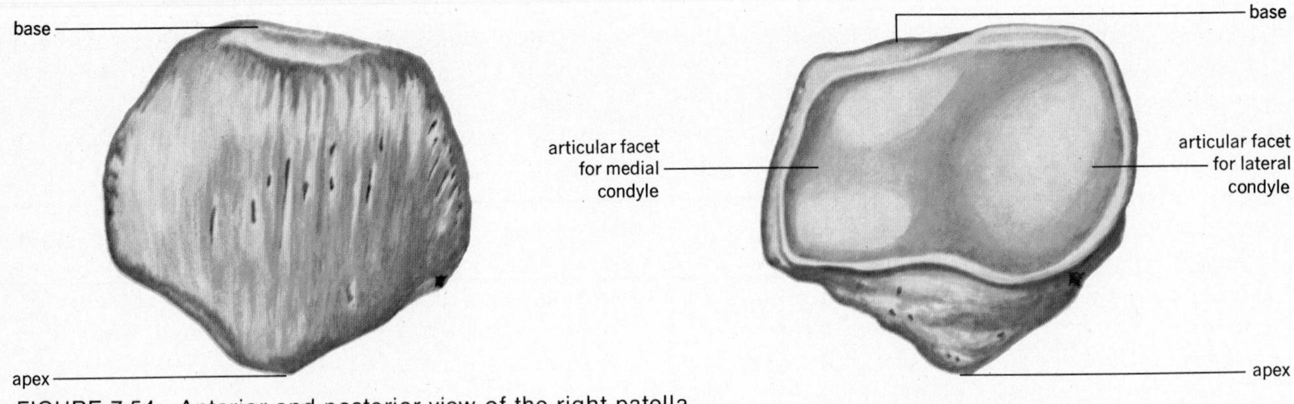

FIGURE 7.54 Anterior and posterior view of the right patella.

EPICONDYLES lying above the condyles. A small tubercle, the ADDUCTOR TUBERCLE, is found where the medial supracondylar ridge joins the epicondyle. The femur articulates above with the acetabulum of the pelvis, below with the tibia, and anteriorly and distally with the patella.

PATELLA. The patella (Fig. 7.54) lies in the tendon of the quadriceps femoris muscle. Bones of this type, formed in tendons, are called SESAMOID BONES. The patella articulates with only one bone, the femur.

TIBIA. The tibia (Fig. 7.55) is the medial and larger of the two bones of the leg. It is the weight-bearing bone of the leg, carrying weight to the ankle of the foot. Its entire length is subcutaneous and can be felt along the anterior and medial surface of the leg. The tibia is often called the shinbone.

The proximal end of the tibia is expanded into LATERAL and MEDIAL CONDYLES to form a broad surface for articulation with the large condyles of the femur. The two shallow articular areas of the tibia are separated by the INTERCONDYLAR EMINENCE and are deepened by flat, semicircular wedges of cartilage, the LATERAL and MEDIAL SEMILUNAR CARTILAGES (MENISCI). On the posterolateral aspect of the lateral condyle is the FIBULAR FACET for articulation with the fibula. A TIBIAL TUBEROSITY is seen on the anterior surface of the upper end of the bone for attachment of the PATELLAR LIGAMENTS.

The shaft or body of the tibia tapers away from the proximal end and widens again distally. It has a raised line on its lateral border, the INTEROSSEOUS BORDER, for the attachment of the INTEROSSEOUS MEMBRANE that extends between the tibia and fibula.

The distal end of the tibia is shaped to articulate with the talus bone at the ankle joint. It is broad and has a downward-projecting MEDIAL MALLEOLUS that grasps the medial side of the talus. On the lateral side of the lower end of the tibia is a FIBULAR NOTCH for articulation with the fibula.

FIBULA. The fibula (Fig. 7.55) is a slender bone, triangular in cross section, that articulates laterally and posteriorly with the tibia. Its upper end is rounded and has an articular surface medially for articulating with the tibia. Its lower end is triangular and, as the LATERAL MALLEOLUS, extends beyond the tibia to form the lateral part of the ankle joint. The lateral and medial malleoli can be easily palpated. An INTEROSSEOUS MEMBRANE connects the twisted shaft of the fibula with the tibia.

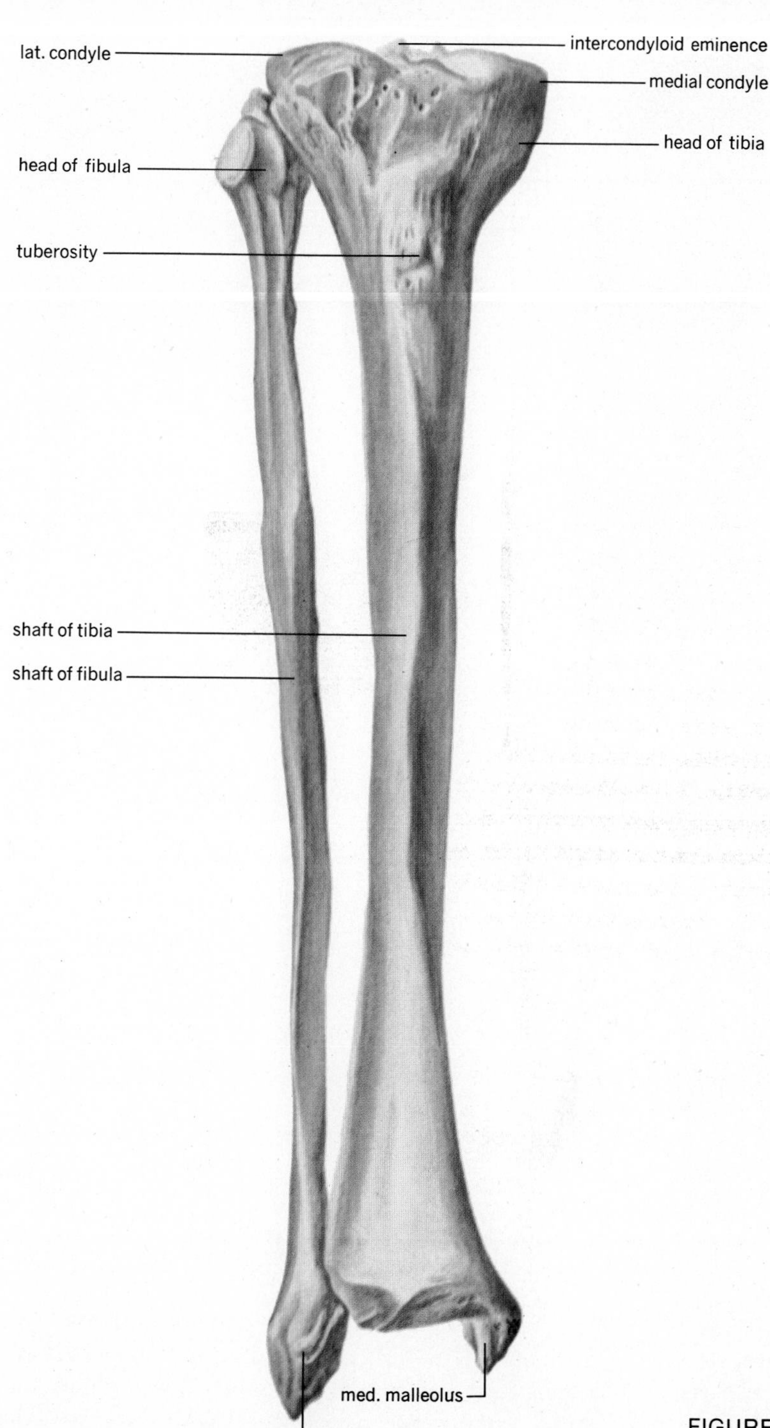

FIGURE 7.55 Anterior view of the right tibia and fibula.

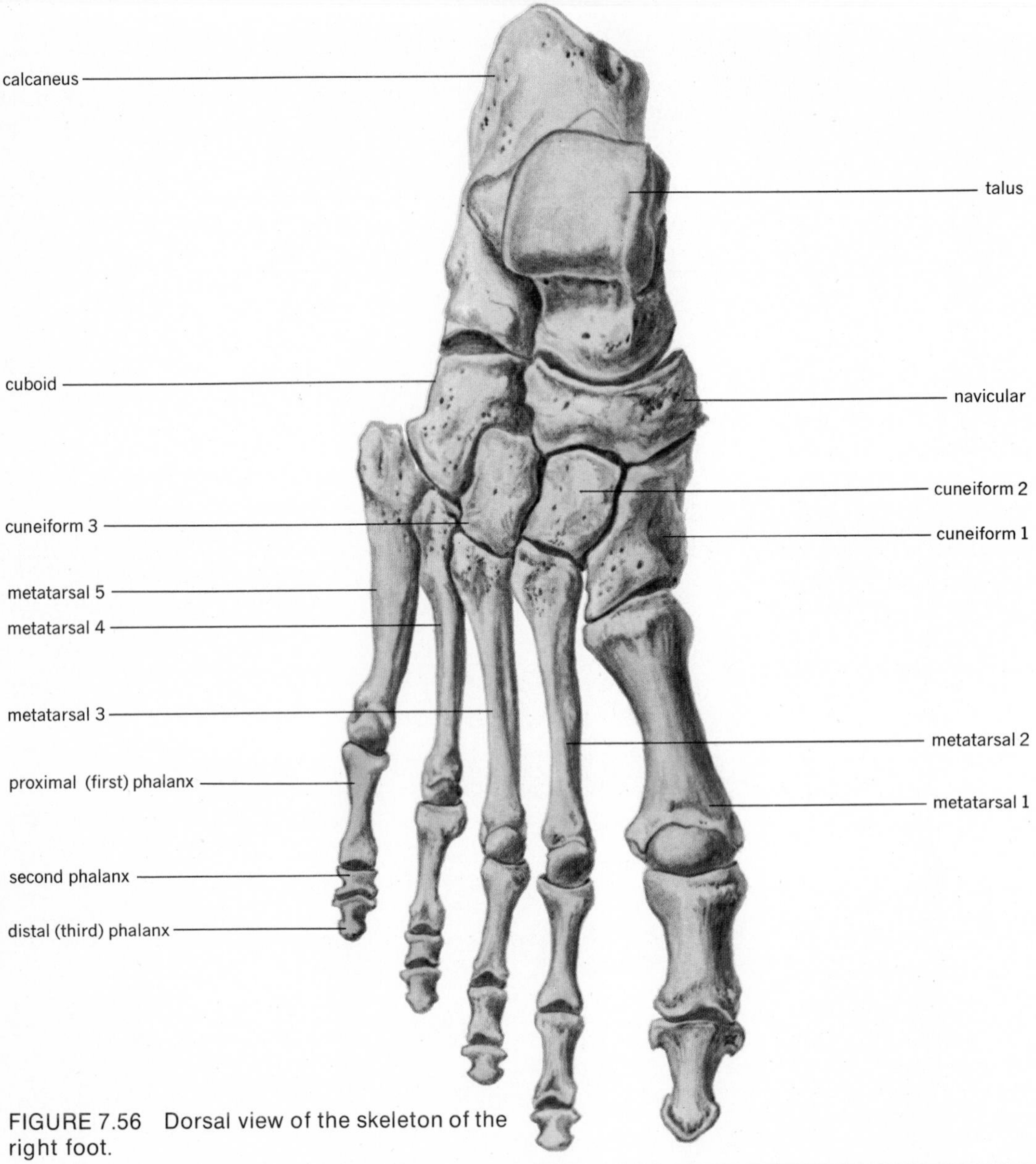

FIGURE 7.56 Dorsal view of the skeleton of the right foot.

TARSUS. The seven bones of the tarsus (Figs. 7.56 and 7.57) can be divided into two groups: the TALUS and CALCANEUS on the posterior part of the foot; and the CUBOID, NAVICULAR, and three CUNEIFORMS anteriorly. The TALUS, formerly called the astragalus, is the only bone of the foot to articulate with the tibia and fibula. These three bones form the hingelike ankle joint at which the entire foot can be DORSIFLEXED or PLANTAR FLEXED. The talus bears the entire weight carried

FIGURE 7.57 Medial view of the skeleton of the right foot.

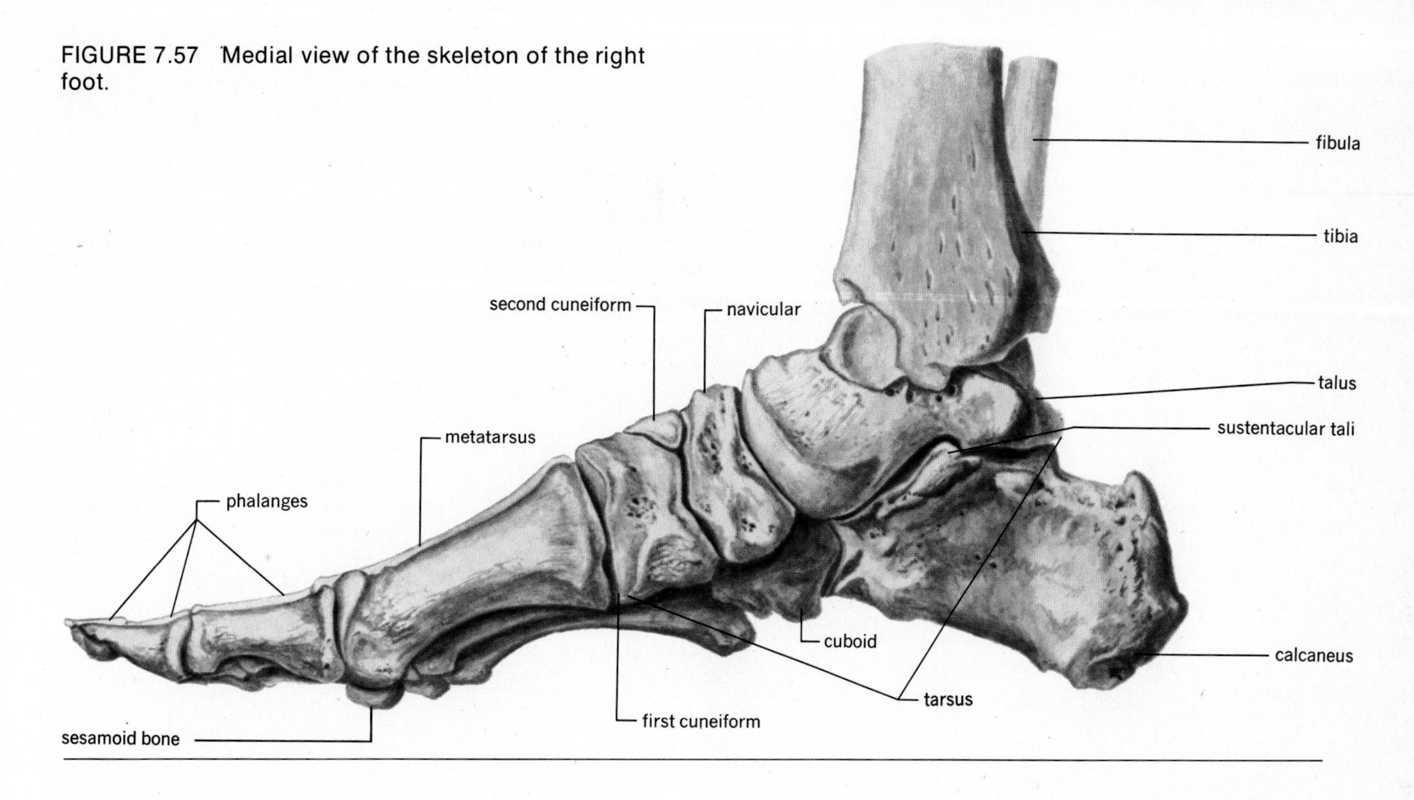

by its extremity. One half of the weight is transferred downward from the talus to the heel, the other half forward to the tarsals that form the keystones of the arch of the foot. They increase the flexibility of the foot, especially its twisting movements.

METATARSUS. The five metatarsals (Figs. 7.56 and 7.57) articulate proximally with the cuboid and with the middle, intermediate, and lateral cuneiforms. Their distal articulation is with the proximal phalanges of the toes. They form the forward pillar of the longitudinal arch. The first metatarsal is thicker and shorter than the rest and has two small sesamoid bones under its head.

PHALANGES. The arrangement of phalanges (Figs. 7.56 and 7.57) is essentially like that of the hand. The first or big toe has two phalanges, quite large and heavy, and the other four toes have three each. The proximal phalanges are the largest; the middle ones are short; and, except for the first toe, the distal ones are short and have expanded ends.

ARTICULATIONS OF THE LOWER EXTREMITY. The hip joint (Fig. 7.58) is a good example of a synovial joint serving three important functions. It supports one half of the body weight above the pelvis; it is involved in the transmission of weight; it is responsible for movement. It involves the OS COXAE and the FEMUR, the former presenting a deep bony socket, the ACETABULUM, and the latter a rounded head that fits into the acetabulum. The acetabulum is deepened in the living subject by a rim of cartilage, the ACETABULAR LABRUM, and a TRANSVERSE LIGAMENT bridges the gap formed by the acetabular notch. The acetabulum, thus constituted, virtually grasps the head of the femur beyond its greatest diameter and

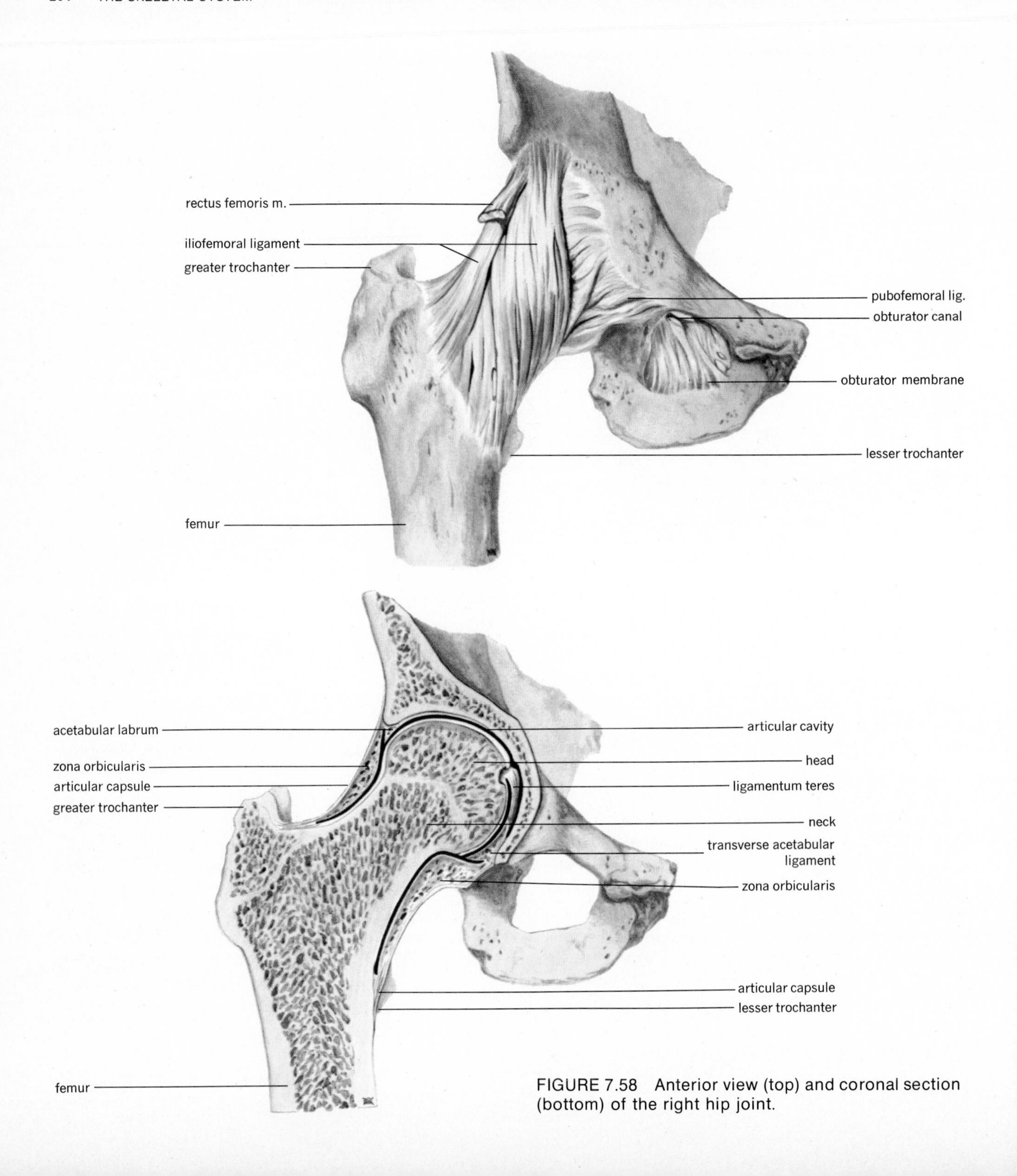

FIGURE 7.58 Anterior view (top) and coronal section (bottom) of the right hip joint.

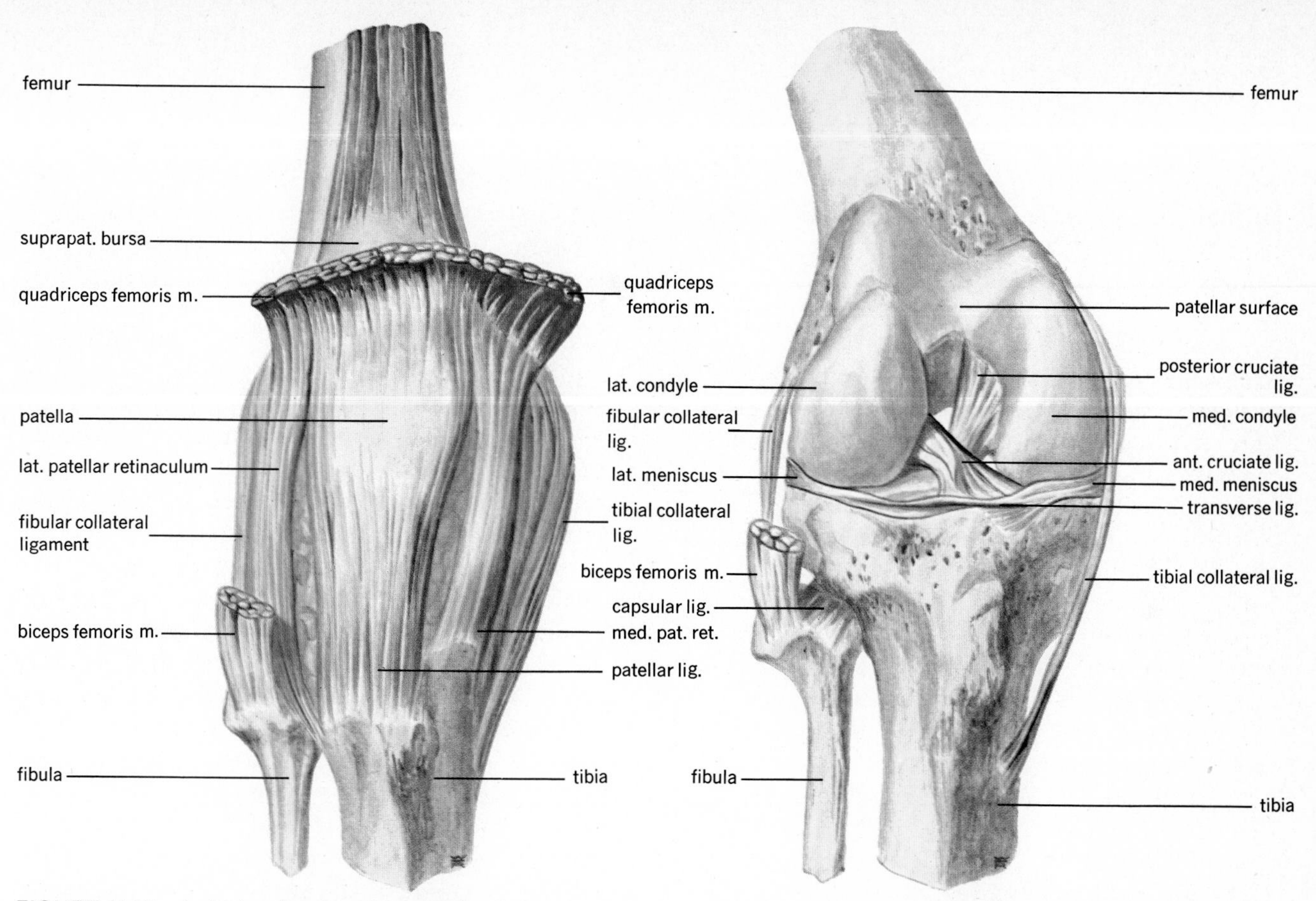

FIGURE 7.59 Anterior (top) and posterior (bottom) views of the superficial and deep structures of the right knee.

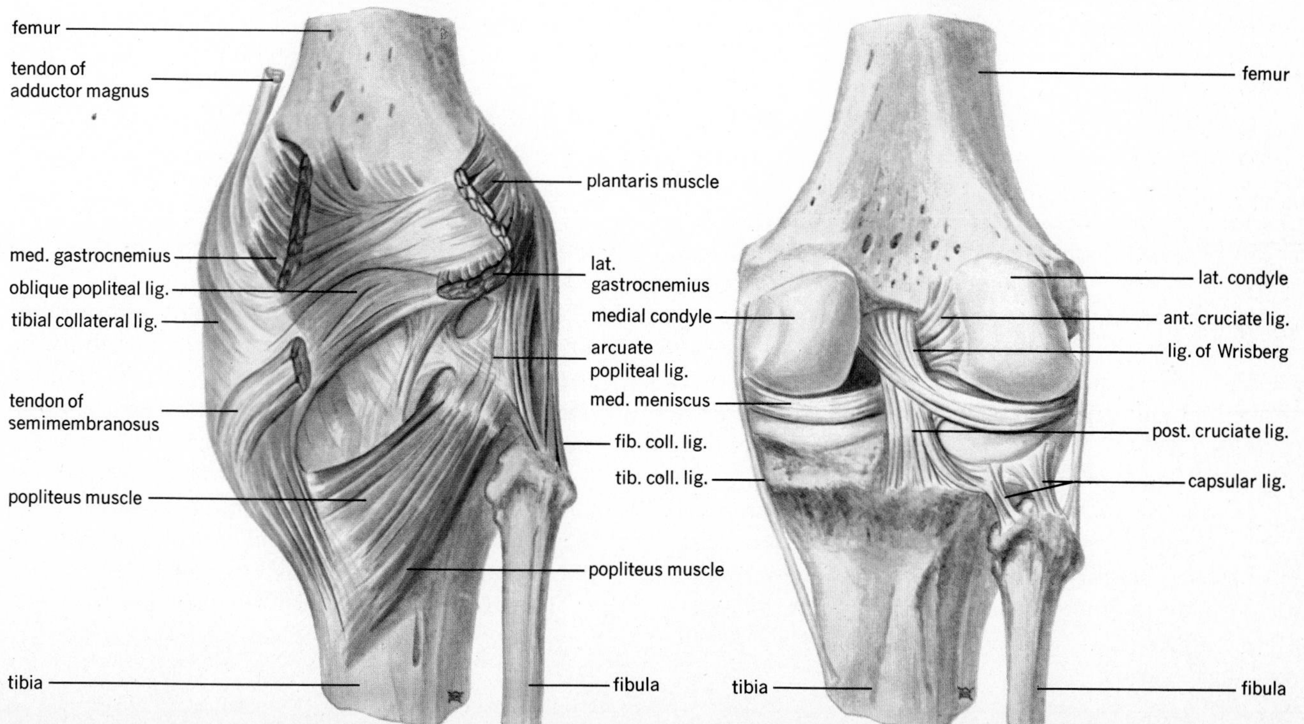

thus serves to strengthen the joint. The FIBROUS CAPSULE is attached to the rim of the acetabular labrum and encloses most of the neck of the femur. The SYNOVIAL MEMBRANE follows the fibrous capsule, is extremely lax, and is reflected on the neck of the femur. It becomes an acetabular fat pad in the acetabular fossa.

The ligaments of the hip joint are thickenings of the fibrous capsule and are named for the parts they connect; they are the ILIOFEMORAL, PUBOFEMORAL, and ISCHIOFEMORAL LIGAMENTS. These ligaments serve to check the range of movement of the pelvis on the femurs more than to strengthen a joint, which is already sturdy because of its bony components. A ligament of the femoral head *(ligamentum teres femoris)* runs from the FOVEA of the femoral head into the acetabulum to carry blood vessels into the head.

The movements of the hip joint are flexion, extension, abduction, adduction, and medial and lateral rotation. While the movements at this joint are relatively free, they do not equal the range of those of the shoulder joint where the scapula is also free to move.

The KNEE JOINT is the most complicated joint in the body and is inherently unstable (Fig. 7.59). Yet the two knee joints together carry the total weight of the body superior to them and are involved in the important functions of walking, running, standing, and kicking. The large medial and lateral condyles of the femur rest on the smaller and shallow corresponding condyles of the tibia. The condyles of the tibia are deepened by crescentic, wedge-shaped rims of fibrocartilage, the SEMILUNAR CARTILAGES or MENISCI.

The FIBROUS ARTICULAR CAPSULE encloses not only the articulations between the femur and tibia, but also the one between patella and femur. The capsule is thin posteriorly, is strengthened laterally and medially by the FIBULAR and TIBIAL COLLATERAL LIGAMENTS, and is replaced anteriorly by the TENDON of the QUADRICEPS FEMORIS MUSCLE AND THE PATELLA. The collateral ligaments prevent side-to-side movements of the joint. As the joint becomes straight at full extension the collateral ligaments tighten, which adds stability to the joint. A thickening in the posterior part of the capsule, the OBLIQUE POPLITEAL LIGAMENT, also prevents overextension. A TRANSVERSE LIGAMENT runs between the menisci anteriorly.

The CRUCIATE LIGAMENTS are intraarticular, arising from the nonarticular portion of the superior surface of the tibia. One passes anterior and the other posterior to the intercondylar eminence. They cross one another as they rise through the intercondylar fossa. The anterior cruciate ligament passes backward and attaches to the internal surface of the lateral condyle of the femur. The posterior cruciate ligament crosses the anterior ligament and passes forward to attach to the internal surface of the medial condyle of the femur. These ligaments are stabilizers of the knee joint and prevent anteroposterior displacement.

The SYNOVIAL CAVITY is large and has many ramifications due to the large intercondylar fossa and the many peculiarities of this complex joint. It extends several inches upward under the quadriceps tendon and muscle, forming a synovial pouch filled with synovial fluid and acting as a BURSA. The synovial membrane lines all of the synovial cavity except the bearing surfaces of the bones and the menisci.

The knee joint is a hinge allowing extensive flexion and extension and slight rotation when the knee is in the flexed condition.

There are three TIBIOFIBULAR JOINTS: the superior tibiofibular, the interosseous membrane, and the inferior tibiofibular (Figs. 7.59, 7.60, and 7.61). At none of these joints is there any appreciable amount of movement. They allow the fibula to spring a bit as an accommodation to ankle movement without losing the stability necessary for foot action.

The SUPERIOR TIBIOFIBULAR joint is a synovial joint and allows some gliding movement between the tibia and fibula. It is held by anterior and posterior ligaments of the fibular head (Fig. 7.59).

The INTEROSSEOUS MEMBRANE stretches between

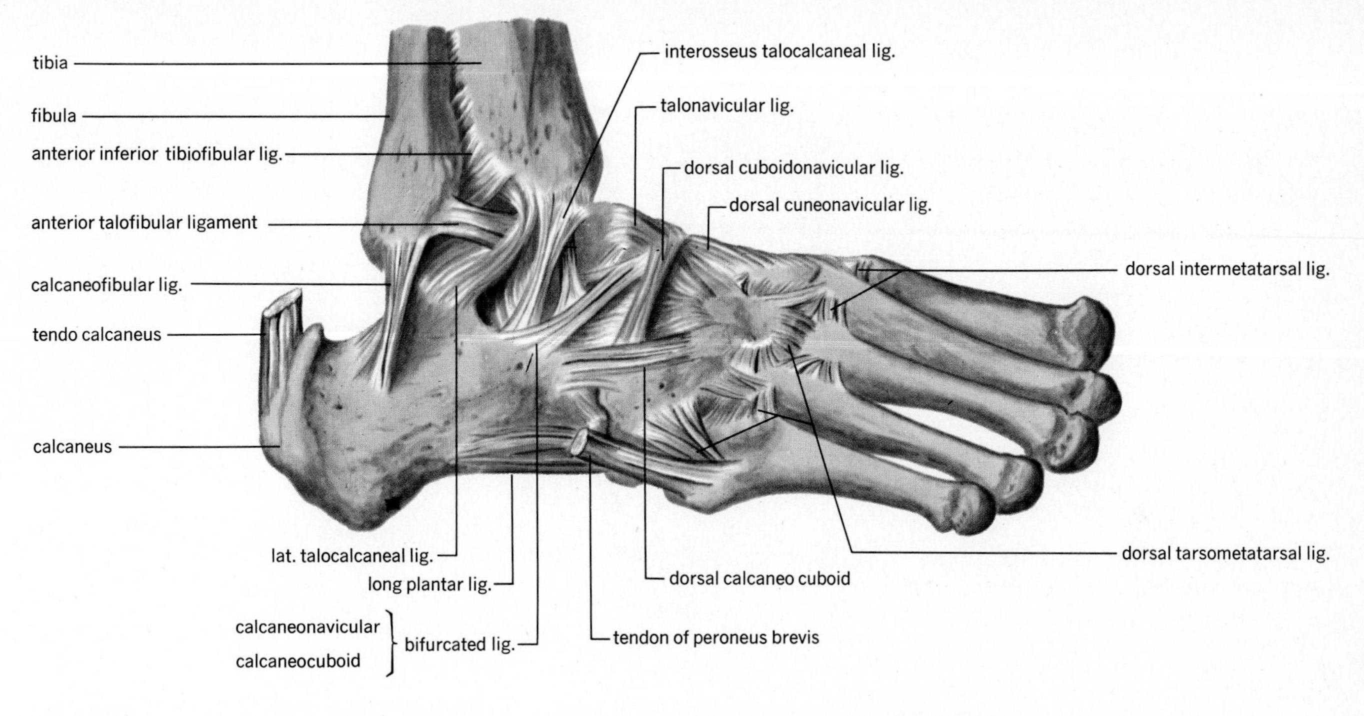

tibia
fibula
anterior inferior tibiofibular lig.
anterior talofibular ligament
calcaneofibular lig.
tendo calcaneus
calcaneus
interosseus talocalcaneal lig.
talonavicular lig.
dorsal cuboidonavicular lig.
dorsal cuneonavicular lig.
dorsal intermetatarsal lig.
dorsal tarsometatarsal lig.
lat. talocalcaneal lig.
long plantar lig.
dorsal calcaneo cuboid
calcaneonavicular
calcaneocuboid
bifurcated lig.
tendon of peroneus brevis

FIGURE 7.60 Ligaments and tendons of the lateral (top) and medial (bottom) views of the right foot.

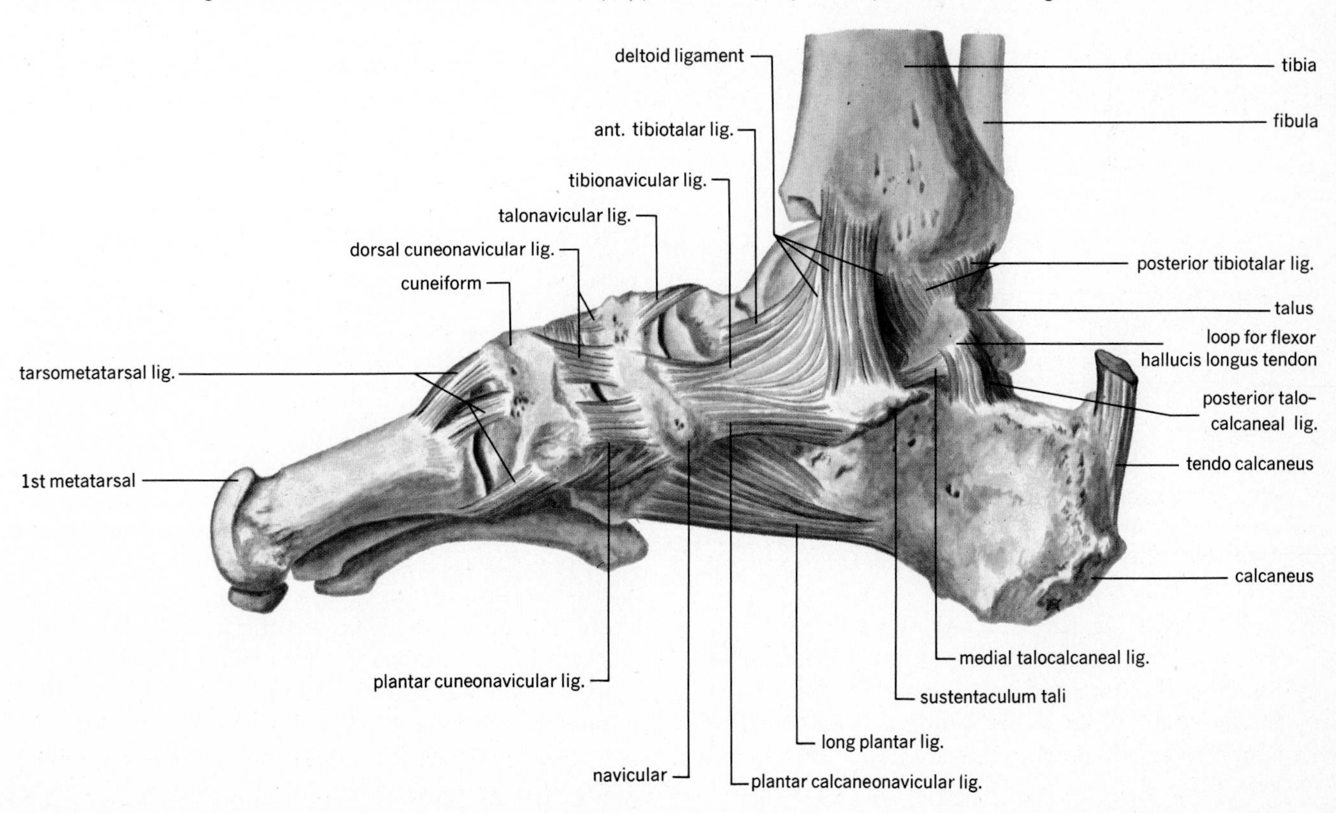

deltoid ligament
ant. tibiotalar lig.
tibionavicular lig.
talonavicular lig.
dorsal cuneonavicular lig.
cuneiform
tarsometatarsal lig.
1st metatarsal
tibia
fibula
posterior tibiotalar lig.
talus
loop for flexor hallucis longus tendon
posterior talo-calcaneal lig.
tendo calcaneus
calcaneus
medial talocalcaneal lig.
sustentaculum tali
long plantar lig.
plantar calcaneonavicular lig.
navicular
plantar cuneonavicular lig.

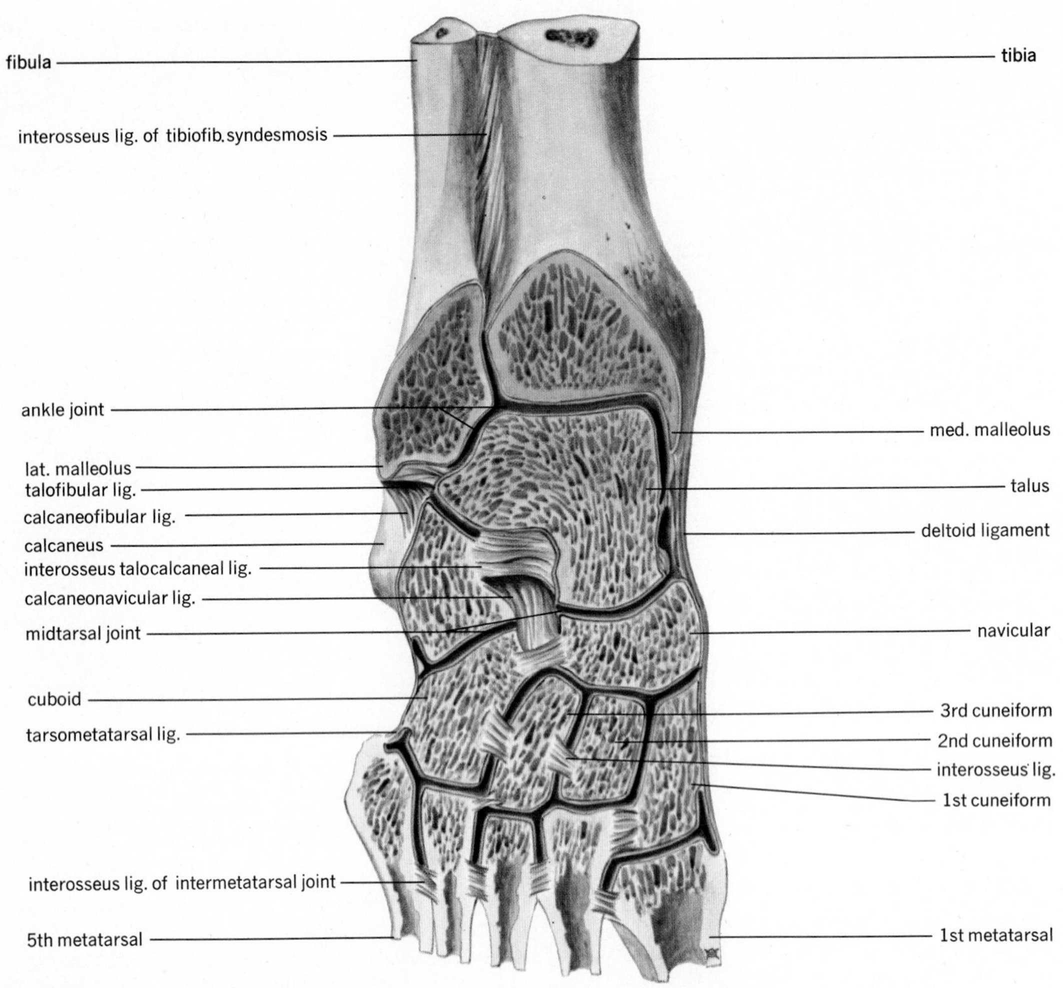

FIGURE 7.61 Synovial articulations of right foot as shown in section.

the interosseous borders of tibia and fibula to form a fibrous joint. The interosseous membrane receives muscle attachments that extend onto it from the adjacent bones.

The INFERIOR TIBIOFIBULAR joint is a fibrous joint, a SYNDESMOSIS. An interosseous tibiofibular ligament holds the fibula firmly in the fibular notch of the tibia, while anterior and posterior tibiofibular ligaments strengthen the joint externally.

The TALOCRURAL OR ANKLE JOINT is of the synovial hinge variety (Figs. 7.60 and 7.61). The convex upper surface of the talus fits into the concave inferior end of the tibia and is grasped medially by the medial malleolus of the tibia and laterally by the lateral malleolus of the fibula. This arrangement allows dorsiflexion and plantar flexion, but lateral movements are largely prevented by the presence of reinforcing ligaments in the fibrous capsule of the joint.

A number of collateral ligaments extend from the tibial and fibular malleoli to the talus and from the malleoli to other tarsal bones (Fig. 7.60). Among these are the ANTERIOR and POSTERIOR TALOFIBULAR LIGAMENTS, and the CALCANEOFIBULAR LIGAMENT on the lateral side. On the medial side are found the ANTERIOR and POSTERIOR TIBIOTALAR LIGAMENTS, the TIBIOCALCANEAL LIGAMENT, and the TIBIONAVICULAR LIGAMENTS. These medial ligaments are quite close together and spread out like a fan. They constitute what is commonly called the DELTOID LIGAMENT. The lateral ligaments are the ones injured when the foot is forcefully twisted inward; the deltoid ligament is injured on violent eversion. In these instances we say we have sprained the ankle.

Reference to Fig. 7.61 will give a good idea of the complexity of joints in the foot. They fall roughly into five classifications: intertarsal, tarsometatarsal, intermetatarsal, metatarsophalangeal, and interphalangeal joints. All are synovial joints.

Among the intertarsal joints the TRANSVERSE TARSAL OR MIDTARSAL JOINT is of particular importance. It lies between the talus and navicular and the calcaneus and cuboid. At this point movements of abduction, inversion, and eversion of the foot take place. The TALOCALCANEAL JOINT is also involved with these movements.

The tarsometatarsal joints are between the three cuneiforms, the cuboid, and the bases of the metatarsals (Fig. 7.61). Note that the medial cuneiform is longer than the others and that the second metatarsal, articulating with the middle cuneiform, is wedged in between the medial and lateral cuneiforms and metatarsals one and three. Movement at these joints is limited to a slight gliding of one bone in relation to another because dorsal, plantar, and interosseous tarsometatarsal ligaments bind them closely together.

Intermetatarsal joints involve both proximal and distal ends of these bones. DORSAL, PLANTAR, and INTEROSSEOUS METATARSAL LIGAMENTS bind these bones proximally, allowing only slight gliding movements. Distally, the heads of all of the metatarsals are joined by a TRANSVERSE METATARSAL LIGAMENT. Recall that in the hand the corresponding transverse metacarpal ligament did not join the first metacarpal.

The metatarsophalangeal joints allow flexion, extension, abduction, and adduction. They are condyloid joints and are held by plantar ligaments ventrally, while dorsally the tendons of extensor muscles take the place of ligaments. Strong, rounded collateral ligaments are found on the lateral and medial sides of each joint.

INTERPHALANGEAL JOINTS are GINGLYMOID or hinge joints and hence allow only flexion and extension. Their ligaments are essentially like those of the metatarsophalangeal joints—namely, plantars and collaterals.

ARCHES OF THE FOOT. The arches of the foot are the lateral and medial longitudinal arches and the transverse arch. The former have been briefly described above. The MEDIAL LONGITUDINAL ARCH is higher than the lateral, as can be seen by a footprint of a normal foot, in which the medial side does not leave a print. The medial arch starts in the calcaneus and rises to its sustentaculum tali and the head of the talus. It then descends forward through the navicular, the three cuneiforms, and the first three metatarsals, whose distal ends or heads form three of the six contacts of the arched foot. The TALUS is the KEYSTONE OF THIS ARCH, the CALCANEUS the POSTERIOR PILLAR, and the remaining bones the ANTERIOR PILLAR. The lateral arch, like the medial one, begins at the calcaneus as its posterior pillar; it rises to its high point and KEYSTONE in the CUBOID, and its anterior pillar consists of the fourth and fifth metatarsals whose heads complete the six points of contact of the longitudinal arch. The points of contact, in summary, are the calcaneus and the heads of the five metatarsals (Figs. 7.57 and 7.60).

The TRANSVERSE ARCH involves the concave inferior surfaces of the navicular and cuboid and the interlocking cuneiforms. The latter hold the articulating metatarsals in the same domed position.

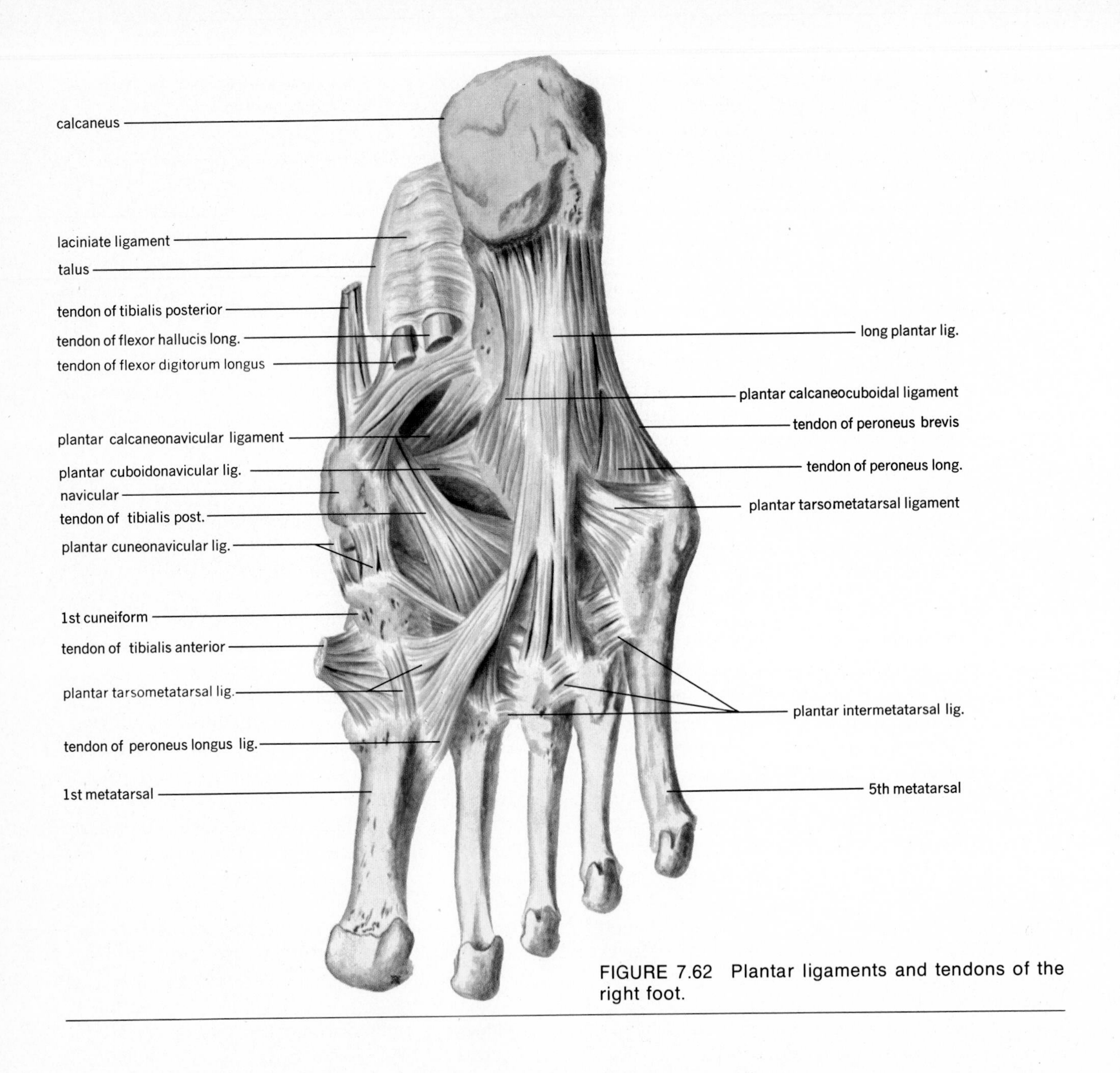

FIGURE 7.62 Plantar ligaments and tendons of the right foot.

The arches are held primarily by ligaments and in part by muscles such as the peroneus longus and the tibialis posterior.

While many ligaments have been described in the account of the foot, all contributing to maintaining its integrity, three are of special importance in relationship to the arches of the foot. The medial longitudinal arch is supported by a strong, elastic ligament that runs longitudinally under the talus from the calcaneus to the navicular, the PLANTAR CALCANEONAVICULAR LIGAMENT. It is often called the SPRING LIGAMENT because of its elasticity and the resilience it gives to the foot as it prevents the flattening of the arch. This ligament is coated with hyaline cartilage that enables the bones to move more freely on it (Fig. 7.62).

The much lower lateral longitudinal arch is supported and strengthened by the LONG PLANTAR and the PLANTAR CALCANEOCUBOID LIGAMENTS (Fig. 7.62).

These arches not only give resilience, strength, and stability to the foot, but they also spread the superimposed weight of the erect body about equally between the calcaneus and the heads of the metatarsals. In walking, one comes down on the calcaneus (heel) and then the weight is shifted forward along the lateral side of the foot to the heads of the metatarsals. The head of the first metatarsal takes about 50 percent of the weight as one comes up on the front of the foot and is stronger in construction to carry this greater burden.

Disorders of bones

Bones are subject to a variety of disorders. OSTEOPOROSIS, which is a loss of mineral and protein matrix, may be caused by a number of factors such as excess of glucocorticoid, an estrogen deficit, liver disease, immobilization, or a failure in development.

OSTEOMALACIA *(rickets)* results from failure to calcify developing bone matrix because of a lack of adequate amounts of calcium and phosphorus in the body fluids. This in turn may be caused by a VITAMIN D DEFICIENCY, which results in inadequate calcium absorption. It may be due also to excessive kidney excretion. The condition is usually corrected by increasing the intake of calcium orally or by intravenous injection and by providing vitamin D.

INFECTION is a common cause of bone or skeletal problems. OSTEOMYELITIS is an example, one often caused by *Staphylococcus aureus.* TUBERCULOSIS of the bone is still seen, but is not as common as it was when it was a formidable crippler.

FRACTURES are the main concern, for they are very common in modern life. Experimental studies indicate that bone is more readily broken by tension and torque forces than by compression forces. Fractures obviously involve the soft tissues around the bones such as muscles, blood vessels, nerves, joints, and skin. This makes the handling of fractures a very critical matter. Too often more harm than good is done by people who are eager to help but are uninformed.

The most familiar classification of fractures is SIMPLE or CLOSED where the skin is intact and COMPOUND or OPEN where the bone has broken through the skin, leaving the patient very subject to infection. A more sophisticated and meaningful classification is as follows (Fig. 7.63):

1. *Compression.* Fractures occur mostly in softer or cancellous bones as epiphyses of long bones, vertebrae, and pelvis.
2. *Avulsion.* Ligaments remain intact but the bone to which they attach is pulled apart.
3. *Transverse.* Fracture is straight across a bone, usually due to a heavy blow to the bone.
4. *Oblique.* Caused by a heavy blow to the bone.
5. *Spiral.* Due to a twisting or rotary force.
6. *Comminuted.* Bone is fragmented into a number of pieces.
7. *Pathological.* Caused by weakening of bone by disease processes.

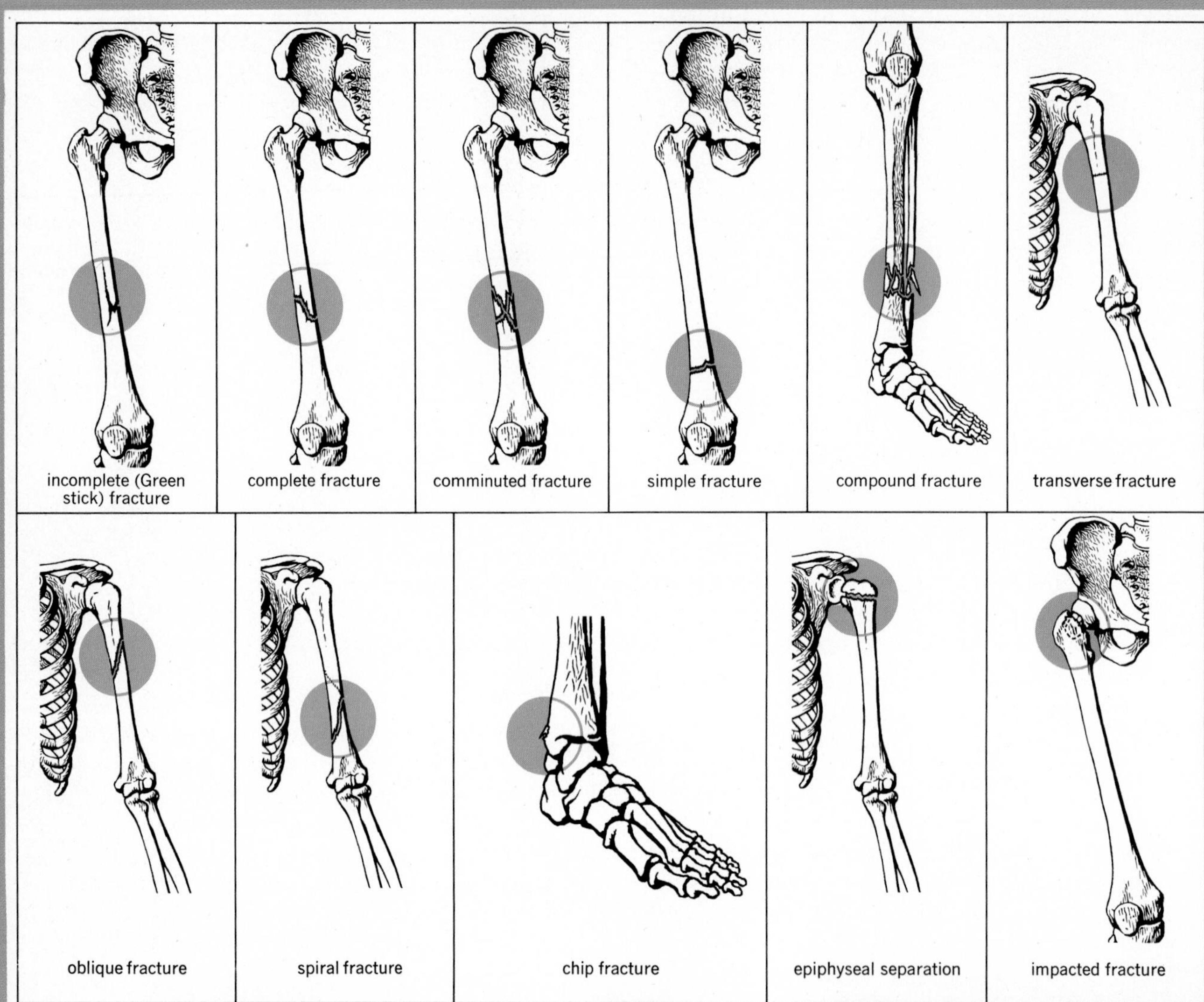

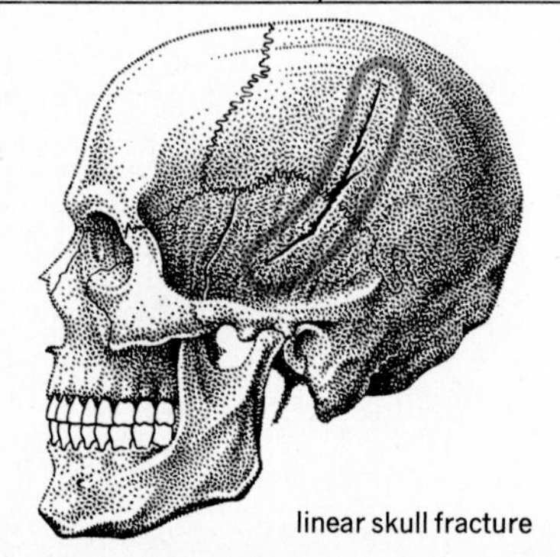

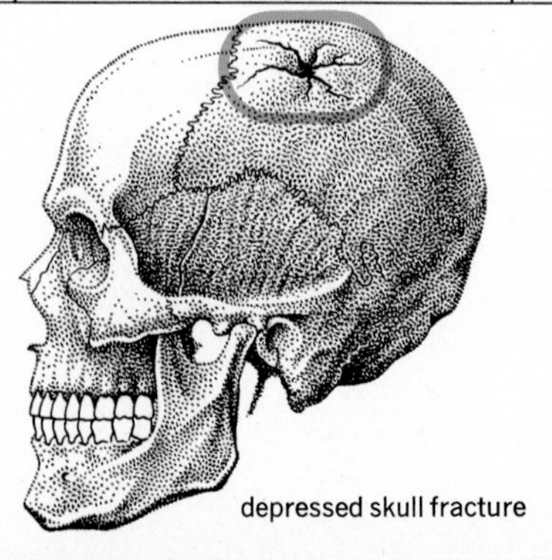

FIGURE 7.63
Some types of fractures

The diagnosis of fractures is not always easy. As with many injuries, there is EDEMA and HEMORRHAGE, and pain when the area is moved or touched. Often only an x-ray will reveal a cracked or chipped bone. In other cases the diagnosis can be made simply if there is a gross deformity or shortening of the part, if there is a grating of the bone when moved *(crepitus),* or if there is definite movement along a long bone. Even in these cases of certain fracture one should have x-rays of the bone and the joints above and below it to see if there are additional fractures.

HEALING OF FRACTURES is a unique phenomenon only in the sense that when the process is completed the continuity of the bone is reestablished. This is not so with many important structures such as muscle, brain, and kidney, where only fibrous tissue is formed by the healing process.

Healing depends, as we have indicated above, on the kind of care received at the time of the fracture and during the period required to reach adequate facilities and personnel for care and treatment. Healing involves the following sequence of events:

1. Blood clot forms. The clot is a fibrin network in which many red and white blood cells are enmeshed. Phagocytic cells—the neutrophils and monocytes—cause breakdown and removal of fibrin and cell debris.

2. Connective tissue cells appear that become chondroblasts and osteoblasts. Endosteum and periosteum furnish additional osteoblasts.

3. Fibroblasts and osteoblasts synthesize and secrete collagen, which takes the form of fibrils within the wound site.

4. The connective tissue matrix begins to calcify. Intramembranous bone formation moves ahead to complete the formation of a bony callus.

The speed with which the healing process takes place depends on the character of the fracture itself and a good vascular supply, which in turn depends on proper immobilization of the fractured bone. Early weight-bearing on the factured limb will improve circulation and speed healing. Bone grafts, which are a source of bone cells and serve as a scaffolding, thereby hasten healing.

GREEN-STICK FRACTURES are common in children. In these a long bone breaks much like a live willow twig. It breaks on the convex side and the periosteum splits open, but on the concave side the bone is only bent and the periosteum remains intact. Setting must be carefully done but healing is very fast.

Slippage of the epiphyseal plate and its fracture occur easily in children and can result in arrested growth and severe deformity. Care should be exercised care when jerking children around by their appendages.

COLLE'S FRACTURE, named for an Irish physician, is a break at the distal end of the radius in which the hand is displaced posteriorly by the pull of the muscles of the forearm.

POTT'S FRACTURE, named for the London surgeon who experienced one and wrote a description of it, involves the lower end of the fibula and the medial malleolus of the tibia. The ligaments here are so strong that in twisting the ankle, as one might in a skiing accident, the bones fracture and the ligaments remain intact. There are variations of this fracture and the causes may vary. The fracture results in the displacement of the talus. Since this is a weight-bearing joint and essential in walking, great care must be taken in restoring proper alignment of parts. Pins or other holding devices may be necessary in some cases.

Bone marrow (myeloid tissue) and hemopoiesis

Bone marrow, found in the medullary cavities of long bones and in the spaces in spongy bone, is the site of hemopoiesis or the development of blood. Bone marrow consists of a supporting and spongelike framework of reticular tissue, the stroma, in which are located blood vessels, the sinusoids, and various stages in the development of blood cells. Marrow is divided into active red marrow and yellow marrow. The marrow in the embryo and newborn is all red, but as age progresses some of it is gradullay changed to yellow marrow, until in the adult there are about equal amounts of each. Red marrow in the adult is found in the vertebrae, sternum, ribs, the diploë of skull bones, and the epiphyses of long bones. Yellow marrow is in the marrow cavities of the diaphyses of long bones.

Contained in the stroma of the bone marrow are primitive and PHAGOCYTIC RETICULAR CELLS. These are attached to the reticular fibers. The SINUSOIDS are lined with flattened, FIXED MACROPHAGE CELLS like those found in the sinuses of lymph nodes. Through the walls of the sinusoids, cells from the marrow constantly pass into the blood stream. FAT CELLS are always present in small numbers in red bone marrow or myeloid tissue but are so numerous in yellow marrow that they obscure most other cells. LYMPH NODULES are found in bone marrow but no lymph vessels have been demonstrated.

Free cells in bone marrow are varied in form and consist of: the mature erythrocytes and granular leucocytes, just as found in the blood stream; and immature cells, the hemocytoblasts, which, through a series of developmental stages, give rise to the various blood cells. LYMPHOID TISSUES such as lymph nodes and spleen give rise to lymphocytes and monocytes. Since, as indicated above, mature blood cells are found in the myeloid or marrow tissue, it serves as a ready or emergency source of these cells as well as a place for their manufacture. The spleen and liver are also areas of blood cell formation in the embryo.

No attempt has been made to summarize this chapter since it consists mainly of anatomical terms.

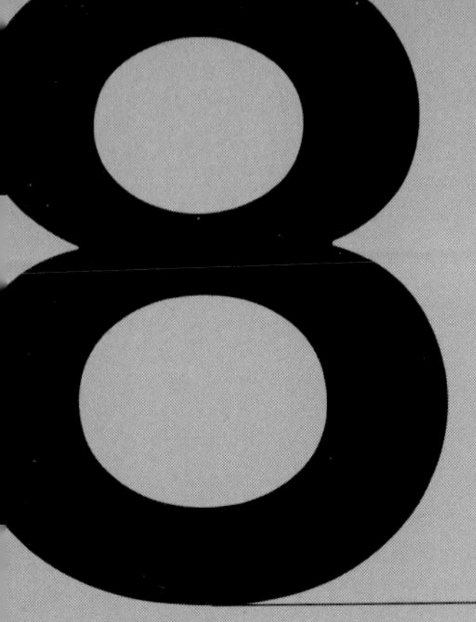

Articulations

A joint or articulation is formed where a bone joins another bone, or where a cartilage joins a bone. The structure of the joint depends largely on the function it must serve. Accordingly, the union may be rigid or may permit variable degrees of motion. If permitted, motion may occur in one, two, or three planes of movement, and the joint may be termed uniaxial, biaxial, or triaxial.

Joints depend for their security on closely fitting bony parts, ligaments, or muscles. The closer the fit of the bones, the stronger the joint, but the greater may be the restriction on axes of movement.

Classification of joints

Utilizing the criteria of structure, degree, and type of motion permitted, three categories of joints are recognized.

Fibrous joints

Fibrous joints (Fig. 8.1), or *juncturae fibrosae,* are generally immovable joints. No joint cavity is present and the bones involved are held together by thin fibrous membranes or by surrounding connective tissue. The category includes several subgroups.

SYNDESMOSIS. Two bones are held tightly by connective tissue between them or surrounding the area. The junction between the distal ends of the tibia and fibula is a syndesmosis.

SUTURES. Found only in the skull, sutures involve a nearly bone-to-bone union with only minimal amounts of connective tissue between the bones.

Some types of suture have interlocking projections that increase the strength of the union. According to the shape of the interlocking portions,

FIGURE 8.1 Examples of fibrous joints.

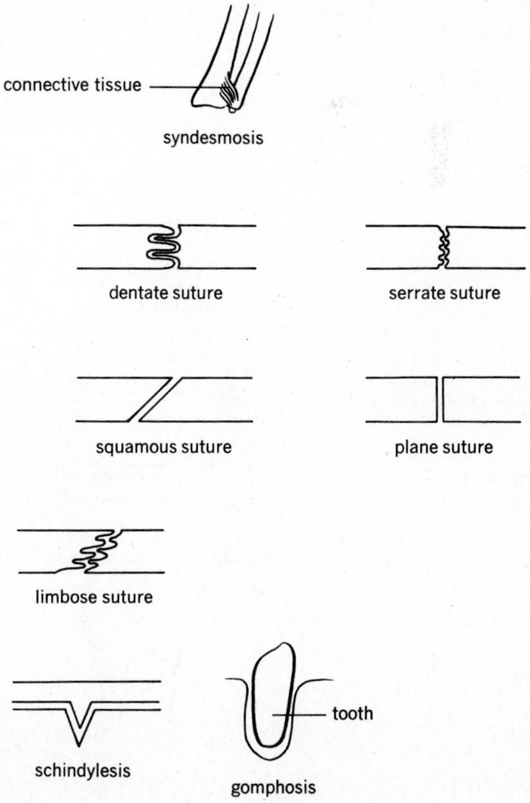

sutures may be called *dentate* (toothed or toothlike, such as the sagittal suture), *serrate* (finer sawtooth-like parts, such as between the two parts of the frontal bone in the fetus), or *limbose* (suture is beveled and interlocked, such as the coronal suture). If no interlocking of parts occurs, the suture may be called *squamous* (if the parts are beveled, such as the squamosal suture), a *plane* suture (flat parts opposed, such as between the nasal bones), a *schindylesis* (a blade fits into a cleft, such as the sphenoid-ethmoid joint), or a *gomphosis* (teeth in the sockets of the maxillae and mandible).

Cartilagenous joints

Cartilagenous joints (Fig. 8.2) *(juncturae cartilaginea* or *amphiarthroses),* are slightly movable joints. No joint cavity is present and a pad of cartilage joins the two bones. Fibrocartilage or hyaline cartilage is the joining material. Since both types of cartilage have only a little flexibility, the joints formed have but slight movement. According to the type of cartilage present, two subtypes of cartilagenous joints are recognized.

SYNCHONDROSIS. The connecting material is

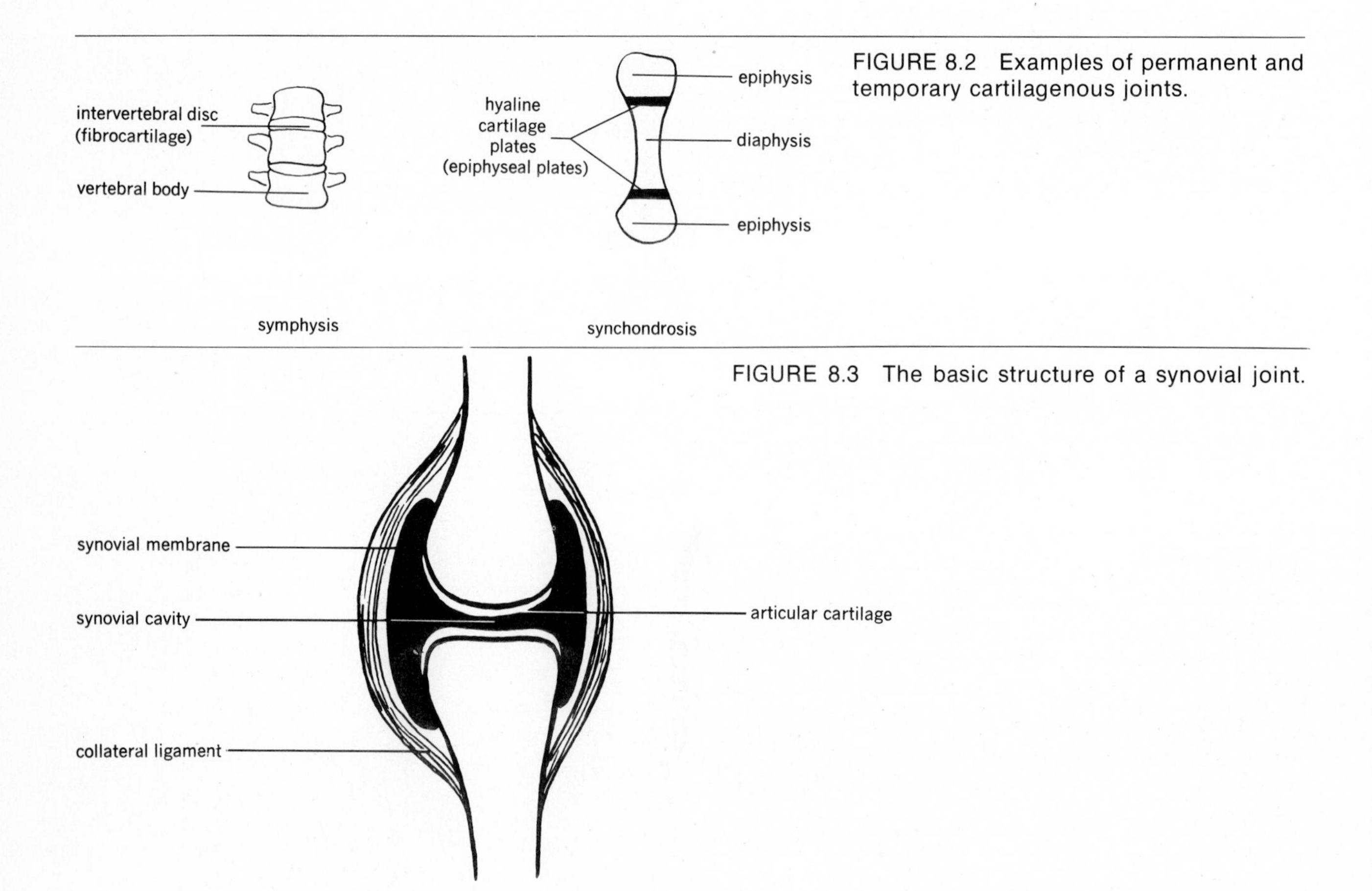

FIGURE 8.2 Examples of permanent and temporary cartilagenous joints.

FIGURE 8.3 The basic structure of a synovial joint.

hyaline cartilage. Such joints occur between the ends (epiphyses) of a growing bone and its shaft (diaphysis). The cartilage will ultimately be replaced by bone; the joint is therefore only temporary. Permanent synchondroses occur between the ends of the ribs and the sternum.

SYMPHYSIS. The connecting material is fibrocartilage, as in the intervertebral joints and symphysis pubis. Movement in this type of joint may be summated, or is the result of adding together many small movements to create great flexibility; this is demonstrated in the spine.

Synovial joints

Synovial joints (Fig. 8.3) *(diarthroses* or *juncturae synoviale),* are *freely movable joints.* The typical synovial joint has a JOINT CAVITY *(synovial cavity),* a SYNOVIAL MEMBRANE that lines the cavity on all but the articular surfaces and secretes a SYNOVIAL FLUID to lubricate the joint, and surrounding CAPSULAR *(collateral)* LIGAMENTS to hold the joint together. The two bones involved in the joint have their articular surface covered with hyaline cartilage (articular cartilage) for smooth action. Such joints may have a pad of cartilage between

FIGURE 8.4 The six basic types of synovial joints, illustrated by joints of the body.

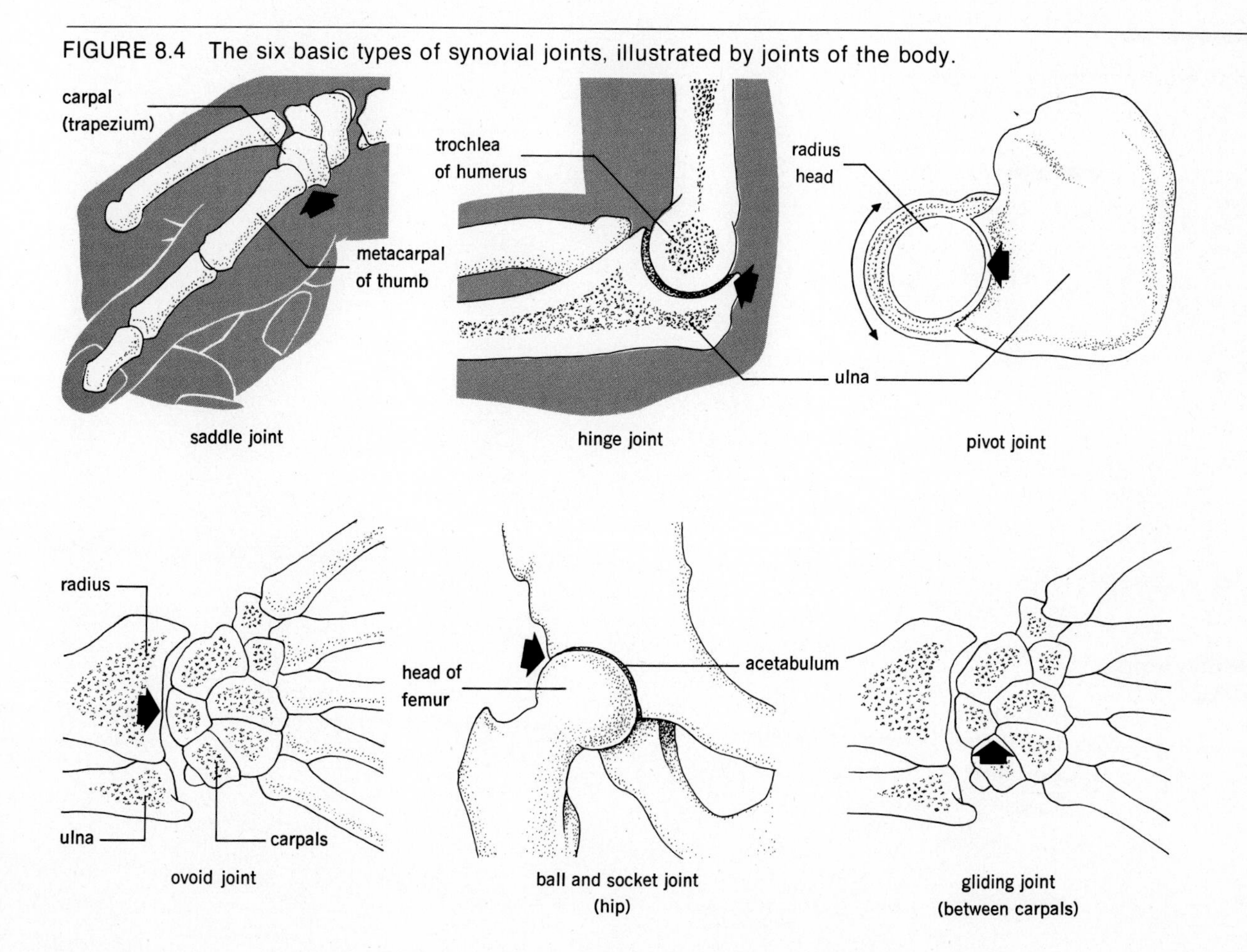

the two bones for shock-absorbing or security purposes. Six varieties of diarthroid joints (Fig. 8.4) are recognized, based on the structure of the bones involved and the degrees or numbers of direction of movement permitted.

HINGE *(ginglymus).* A spool-like surface on one bone fits into a half-moon surface on the other bone. Only a single plane of motion is permitted, usually flexion-extension, and the joint is uniaxial. The elbow and knee joints are examples of this type.

PIVOT *(trochoid).* A conical, pointed, or rounded

TABLE 8.1 Summary of articulations

Major category	Type	Characteristics	Axes of motion permitted	Examples
Fibrous joints		No cavity, immovable	None	Sutures, syndesmoses
	Syndesmosis	Bones held together by connective tissue	None	Tibia-fibula
	Suture	Bone-bone joint, minimal c.t. between	None	Skull
	Dentate	Toothed interlocking parts	None	Parietal bones
	Serrate	Finely toothed interlocking parts	None	Frontal
	Limbose	Interlocking and beveled	None	Frontal-parietal
	Squamous	Beveled	None	Parietal-temporal
	Plane	Flat surfaces opposed	None	Nasal bones
	Schindylesis	Blade in cleft	None	Spenoid-ethmoid
	Gomphosis	Cone in socket	None	Teeth in jawbones
Cartilagenous joints		No cavity; slightly movable; a bone, cartilage, bone joint	Compression	Synchondroses, symphyses
	Synchondrosis	Bone, hyaline cartilage bone	None	Shaft and ends of long bones (temporary) Ribs to sternum (permanent)
	Symphysis	Bone, fibrocartilage, bone	None	Symphysis pubis, vertebral column

surface articulates with a shallow depression on the other bone. Only rotation is permitted and the joint is again uniaxial. The joint between the radius and humerus is a pivot.

OVOID *(condyloid).* An egg-shaped surface on one bone fits into a reciprocally shaped surface on the other. The joint permits side-to-side and back-and-forth motions and is biaxial. The wrist joint between the carpals and the ends of the radius and ulna, is of this type.

SADDLE. Imagine a saddle on the back of a horse, the saddle representing one bone, the horse the

Major category	Type	Characteristics	Axes of motion permitted	Examples
Synovial joints		Cavity, freely movable	One to three	Six types
	Hinge	Spool in halfmoon	Uniaxial; flexion and extension	Knee, elbow, ankle
	Pivot	Cone in depression	Uniaxial; rotation	Radio-humeral, atlas-axis
	Ovoid	Egg in depression	Biaxial; flexion-extension, adduction	Wrist
	Saddle	"Saddle on horse"	Biaxial; flexion-extension, abduction-adduction	Carpo-metacarpal joint of thumb
	Ball and Socket	Ball in cup	Triaxial; flexion-extension, adduction-abduction, rotation (circumduction)	Hip, shoulder
	Gliding	Nearly flat surfaces opposed	Biaxial; flexion-extension, abduction-adduction	Intercarpal joints, intertarsal joints

other. Such a joint permits two directions of movement, side-to-side and back-and-forth, and is biaxial. The joint between the trapezium of the carpals and the metacarpal of the thumb is a saddle joint.

BALL-AND-SOCKET *(spheroidal).* A ball-like surface fits into a cuplike depression on the other bone. All planes of motion are permitted and the joint is triaxial. The shoulder and hip joints are of this type.

GLIDING *(arthrodial).* The bony surfaces are nearly flat, and side-to-side and back-and-forth motion is permitted. The joint is biaxial. Intercarpal joints are of this type.

A summary of the articulations is presented in Table 8.1.

Motions permitted in freely movable joints are of four general types.

GLIDING MOTIONS occur in a side-to-side and back-and-forth direction. A twisting or rotational motion is generally not permitted because of ligaments or the proximity of other bones. Such a joint would be classed as biaxial.

ROTATION occurs as a turning about a central point or around the long axis of a bone without any other motion being permitted. If permitting only one type of motion, a joint would be termed uniaxial.

ANGULAR movements alter the angle between two bones. *Flexion* decreases the angle; *extension* increases it. *Abduction* moves a part away from the midline of the body or midline of the appendage. *Adduction* moves the part toward the midline of body or appendage. Joints usually permit the corresponding pairs of actions, singly or together. Uniaxial or biaxial joints may permit angular movements.

CIRCUMDUCTION occurs when the distal end of an appendage is caused to describe a circle, while the proximal end remains essentially stable within the joint capsule. Circumduction involves an orderly progression of flexion, abduction, adduction, and extension, and generally involves rotation as well. Since movement occurs in all three planes, a joint permitting circumduction is usually triaxial.

Synovial fluid

Synovial fluid is the fluid filling synovial joints. It is secreted by the synovial membrane and has a total volume in adult humans of about 100 ml. Its function is to provide a thin film of fluid over the articulating surfaces for smooth action as the bones move on one another. Its chemical and physical properties resemble those of interstitial fluid, from which it is derived. Analysis of knee joint synovial fluid in humans shows it to have the following characteristics:

Total solids (percent)	3.4
pH	7.39
Viscosity (relative to water)	235
Specific gravity (average)	1.012
Na (meq)	135
Protein (g/100 ml)	2.8
Mucin (g/100 ml)	0.85
Uric acid (g/100 ml)	3.9
Hyaluronic acid (mg/100 ml)	155

The high viscosity of the fluid may be attributed to its mucin and protein content.

If excess amounts of fluid are produced (for example, after injury to the joint), movement of the joint may become restricted.

Clinical considerations

Some reflections on specific joints

The structure and functions of the joints of the body have been considered in Chapter 7. The following discussion is presented to emphasize certain joints commonly injured as a result of sports activities or everyday life.

The SCAPULOHUMERAL *(shoulder)* JOINT is formed by the head of the humerus articulating with the glenoid fossa of the scapula. The fossa has a low rim of cartilage (the glenoid labrum) to deepen it and is surrounded by several ligaments. The muscles around the shoulder also contribute to its security. A "shoulder separation," or dislocation of this joint, occurs commonly as a result of falling on the outstretched arm (football and basketball players). In this disorder, the humerus head is driven out of the glenoid fossa, usually in an upward (superior) direction. Yanking a child upward by the arm may also result in damage to the joint.

The SACROILIAC JOINT, or union between the sacrum and ilium, is a partially fibrous, partially synovial joint. The joint surfaces are rough and the bones are firmly bound by many ligaments; thus, no movement is normally permitted. A "sacroiliac slip" involves stretching of the ligaments holding the joint, and slight movement of the bones may occur. Stretching or tension applied to the sciatic nerve may then result in intense leg pain. "Slips" may occur if heavy loads are carried that put downward pressure on the joint, or in landing on the feet without bending the knees. This tends to push the lower appendages strongly into the hip bones and thus forces upward pressure on the sacroiliac joints.

The HIP JOINT, between the acetabulum and femoral head, is constructed for weight bearing as well as movement. The socket *(acetabulum)* is deep, and is surrounded by many capsular ligaments. The supporting ligaments are numerous and heavy and seem to serve to limit motion more than to secure the joint. The joint is rarely dislocated if normal. In congenital malformation of the hip, the acetabular rim is low, and the femoral

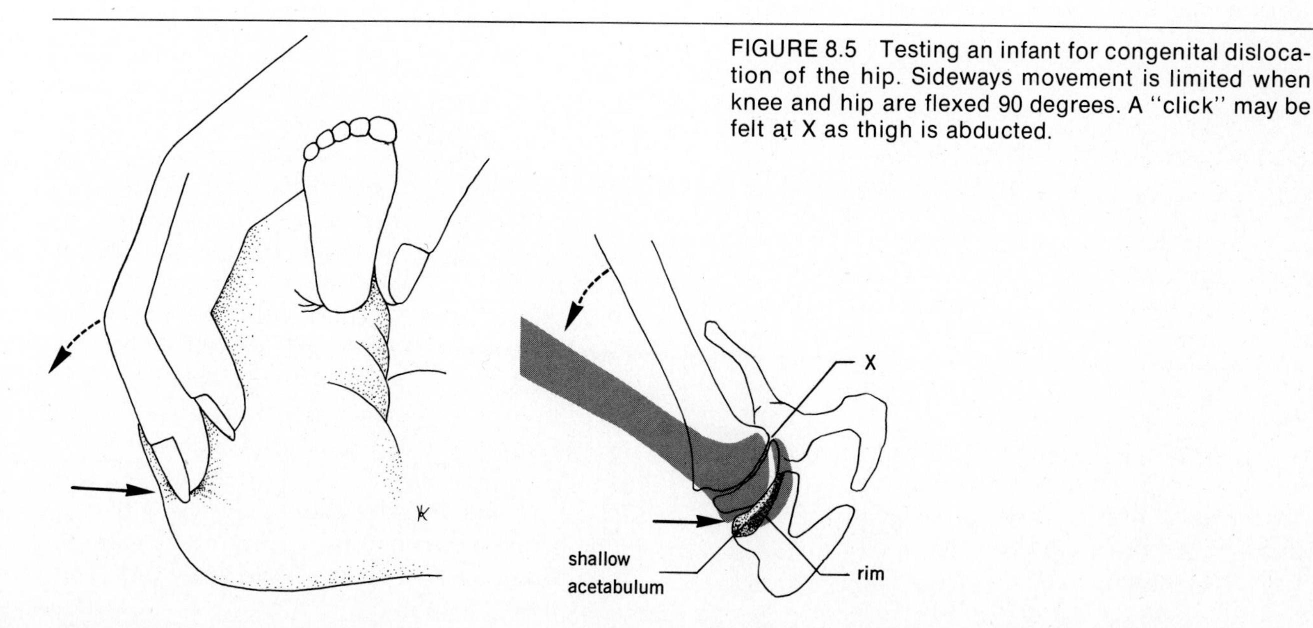

FIGURE 8.5 Testing an infant for congenital dislocation of the hip. Sideways movement is limited when knee and hip are flexed 90 degrees. A "click" may be felt at X as thigh is abducted.

heads slips easily over or out of it. Limited sideward movement of the limb (Fig. 8.5) when hip and knee are flexed (frog position) is usually indicative of congenital malformation. The disorder should be diagnosed in infancy.

The KNEE JOINT is regarded by many as one of nature's mistakes, because it is basically unstable. Two nearly flat surfaces are held together by a number of ligaments. The cruciate ligaments tend to limit anterior-posterior movement, while the collateral ligaments, with the menisci, tend to limit sideward motion. The joint may be disrupted by strong rotation, by blows to either side, and by blows from the front that tend to overextend it. Such traumas are inherent in the game of football. Crushing of menisci and tearing of ligaments is common and often requires surgery to repair the damage. Figure 8.6 shows which structures will be disrupted according to the direction of the force imposed on the joint.

Traumatic damage to a joint

A DISLOCATION is defined as loss of continuity of the joint structures. It may occur as the result of a blow, or because of malformation. A SPRAIN occurs when ligaments and muscles are torn and tendons are stretched as the result of trauma. Pain is common, is often intense, and rapid development of edema (swelling) is usual. Hemorrhages into the skin (ecchymoses), or into the joint itself are initially black or bluish and change to greenish-brown or yellow with time. The bruise changes color with the time because of the breakdown of blood pigments to bilirubin (yellow) and biliverdin (green). A STRAIN is caused by overstretching or pulling of muscles and tendons, without tearing.

Inflammation of joints

ARTHRITIS is a term used to designate joint inflammation. There are many types of arthritis. A few of the more common types are as follows.

RHEUMATOID ARTHRITIS is a disease for which no specific cause is known. It may occur at any age. The synovial membranes become inflamed and infiltrated with lymphocytes from the bloodstream and antibody-producing plasma cells. These cells produce a *rheumatoid factor,* an immune globulin that represents an antibody to the chemicals produced by the inflamed membrane; the factor further increases the inflammation of the joint. As cells die, their lysosomes release their enzymes into the joint tissues, and ligaments, bone, and articular cartilage are eroded. The condition may develop suddenly or slowly, and may involve one joint or several in a progressive fashion. Symmetrical involvement of joints of the hands, feet, wrists, and elbows is common. Hip and knee joint involvement may occur later, making locomotion extremely difficult. The joints are very tender, and the patient often complains of depression, fatigue, and morning stiffness of the joints. Deformities such as enlargement of the affected joint are common and develop rapidly. Treatment of the disorder may involve resting the affected joints, the use of salicylates (aspirin) for relief of pain, and the use of steroids (cortisone) to reduce inflammation. Surgery to remove the affected synovial membranes sometimes offers relief, and severe cases, where joint mobility is lost, may be treated by replacement of the entire joint with a plastic or stainless-steel prosthesis (artificial structure).

OSTEOARTHRITIS is a chronic degenerative disease, especially of weight-bearing joints. It is characterized by: fissuring and destruction of articular cartilage, without great inflammation; overgrowth of bone at the margins of the joint, often with twisting and enlargement of the affected appendage (such as fingers). Pain is minimal, but may increase with hard joint usage. This type of arthritis rarely occurs before 40 years of age, is slow to develop, and usually does not cripple the subject unless it occurs in the hip joints. If advanced, the disorder may be treated with anti-inflammatory substances or surgical intervention to remove excess bone or replace the joint.

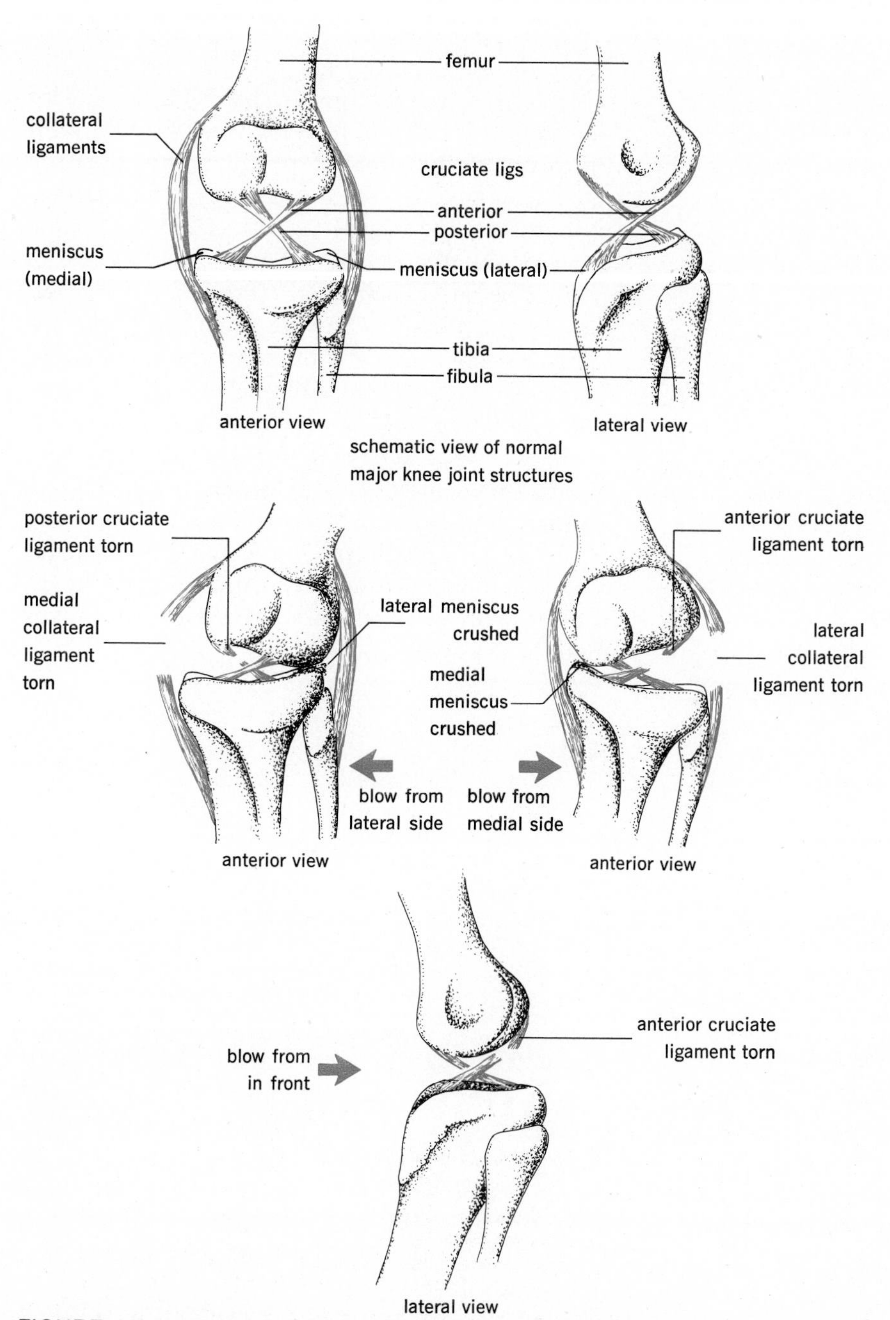

FIGURE 8.6 A diagrammatic representation of the structures of the left knee that will most likely be injured by trauma from various directions.

GOUTY ARTHRITIS is the result of a metabolic error involving the degradation of nitrogenous bases. Uric acid crystals are deposited in certain joints (such as ankle, knee, hands) with development of intense pain and swelling. Relief may be afforded by decreasing purine intake in the diet, and by the use of drugs (such as colchicine) that reduce inflammation and deposition of crystals in the joints.

SEPTIC ARTHRITIS arises as a result of the presence of bacteria or their products in the bloodstream, with involvement of the joints. Gonococcal (gonorrhea), nonhemolytic streptococcal (strep), and staphylococcal (staph) infections are most commonly involved as causative agents.

BURSAE are simple synovial sacs filled with synovial fluid placed between two frictional surfaces. They serve to cushion structures or increase motility between two moving parts. They may or may not communicate with a joint and may be placed beneath the skin (such as the bursa in front of the patella), beneath muscles (as in the shoulder region where the deltoid overlies the shoulder joint), or beneath tendons (as beneath the patellar tendon). The large bursae found associated with the major body joints (hip, knee, shoulder) are usually named by their positions. Small bursae without specific names are associated with most other synovial joints.

BURSITIS refers to acute or chronic inflammation of a bursa. Bursitis may result from injury to a joint that has bursae around it, rheumatoid inflammation, gout, and bacterial infection (syphilis, tuberculosis). The most common forms include: *olecranon bursitis* (tennis elbow), *prepatellar bursitis* (housemaid's knee), *ischial bursitis* (tailor's bottom), and *subdeltoid bursitis* (found in quarterbacks and baseball pitchers). Chronic bursitis may result in eventual deposition of calcium in the sac with extreme pain resulting from movement of the area.

Summary

1. An articulation is a junction between bones, or between a bone and a cartilage.
   a. Joints may permit no movement, slight movement, or free movement.
   b. Joints are secured by bony parts, ligaments, muscles, or combinations of these.
2. Three categories of joints are recognized.
   a. Fibrous joints are immovable and are formed by collagenous tissue or nearly bone-to-bone junctions (sutures of skull). Fibrous joints may become bone-to-bone junctions as they age.
   b. Cartilagenous joints are slightly movable and are joined by hyaline cartilage (synchondrosis) or fibrocartilage (symphysis). Synchondroses may be temporary or permanent.
   c. Synovial joints are freely movable and have a joint cavity, with synovial membranes and fluid, and are of six types.
   (1) Hinge. Allows back-and-forth motion (fingers, knee, elbow).
   (2) Pivot. Permits turning or rotation (head, radius on humerus).
   (3) Ovoid. Allows side-to-side and back-and-forth (wrist).
   (4) Saddle. Allows side-to-side and back-and-forth (thumb).
   (5) Gliding. Allows side-to-side and back-and-forth (intercarpal, intertarsal).
   (6) Ball-and-socket. Permits side-to-side, back-and-forth, and rotation (shoulder, hip).
3. Synovial fluid is produced by the synovial membrane. It is thick and lubricates synovial joints.
4. Specific joints are briefly discussed: shoulder, sacroiliac, hip, knee.
5. Disorders of joints include:
   a. Dislocations. Loss of continuity of joint structures.
   b. Sprains. Tearing of ligaments and muscles with pain and edema.
   c. Strains. Overstretching of ligaments and muscles without tearing.
   d. Arthritis. Inflammation of a joint. It may be due to reactions secondary to infection, may be a disease of aging, or may be the result of metabolic disorders.
6. Bursae are synovial sacs that cushion or allow easier movement between two frictional surfaces. Bursitis is inflammation of a bursa.

The Structure and Properties of Skeletal and Smooth Muscle

Types of muscular tissue

Muscular tissue is one of several excitable tissues and is *the* contractile tissue of the body. Its cells are elongated and capable of shortening to cause movement, propulsion of something through an organ, or change in size and shape of an organ in which it is found. Three types of muscular tissue are described.

SKELETAL MUSCLE attaches to and moves the skeleton. It is a *voluntary* type of muscle; that is, it contracts under the influence of the will, and is normally stimulated by outside nerves.

CARDIAC MUSCLE is found only in the heart, is *involuntary*, and contracts because of inherent stimulation. It causes the circulation of blood through the body. Its properties are described in Chapter 24.

SMOOTH *(visceral)* MUSCLE occurs primarily in internal body organs. It is usually involuntary—that is, possessing the ability to contract without direct nervous stimulation. It is slow acting, requiring from about 2–20 seconds to complete a series of contraction and relaxation. It propels materials through the digestive, reproductive, and urinary systems. It also controls the diameter of blood vessels and the tubes of the respiratory system.

The three types of muscle are estimated to form about 50 percent of the body weight.

Skeletal muscle

Development and growth

Skeletal muscle develops from mesoderm, and the somites (see Fig. 5.7) first give evidence of the formation of the tissue. The cells giving rise to skeletal muscle tissue are termed MYOBLASTS. The mature muscle cell (or fiber) is a multinucleate structure thought to result from embryonic fusion of many uninucleate myoblasts. By about the fifth month of development, the number of fibers is fixed, and any increase in size (growth) of the muscle from this time onward is the result of increase in size of fibers.

Smooth muscle also develops from myoblasts, but does not acquire the stripes (striations) of skeletal muscle.

Structure

SKELETAL MUSCLE (Fig. 9.1) is composed of multinucleated, long cylindrical FIBERS or cells. Size varies from 10–100 μm in diameter, and from 2–3 mm to about 7.5 cm in length. Each fiber is anatomically distinct from other fibers and carries characteristic cross-bandings or STRIATIONS on smaller units known as FIBRILS *(myofibrils)*. The various subdivisions of the striations receive designations and contain specific and different chemical substances. The distance between two **Z** lines is known as a SARCOMERE and is considered to be the unit that shortens when the muscle contracts. Four major proteins may be isolated from muscle. They form the FILAMENTS. MYOSIN,

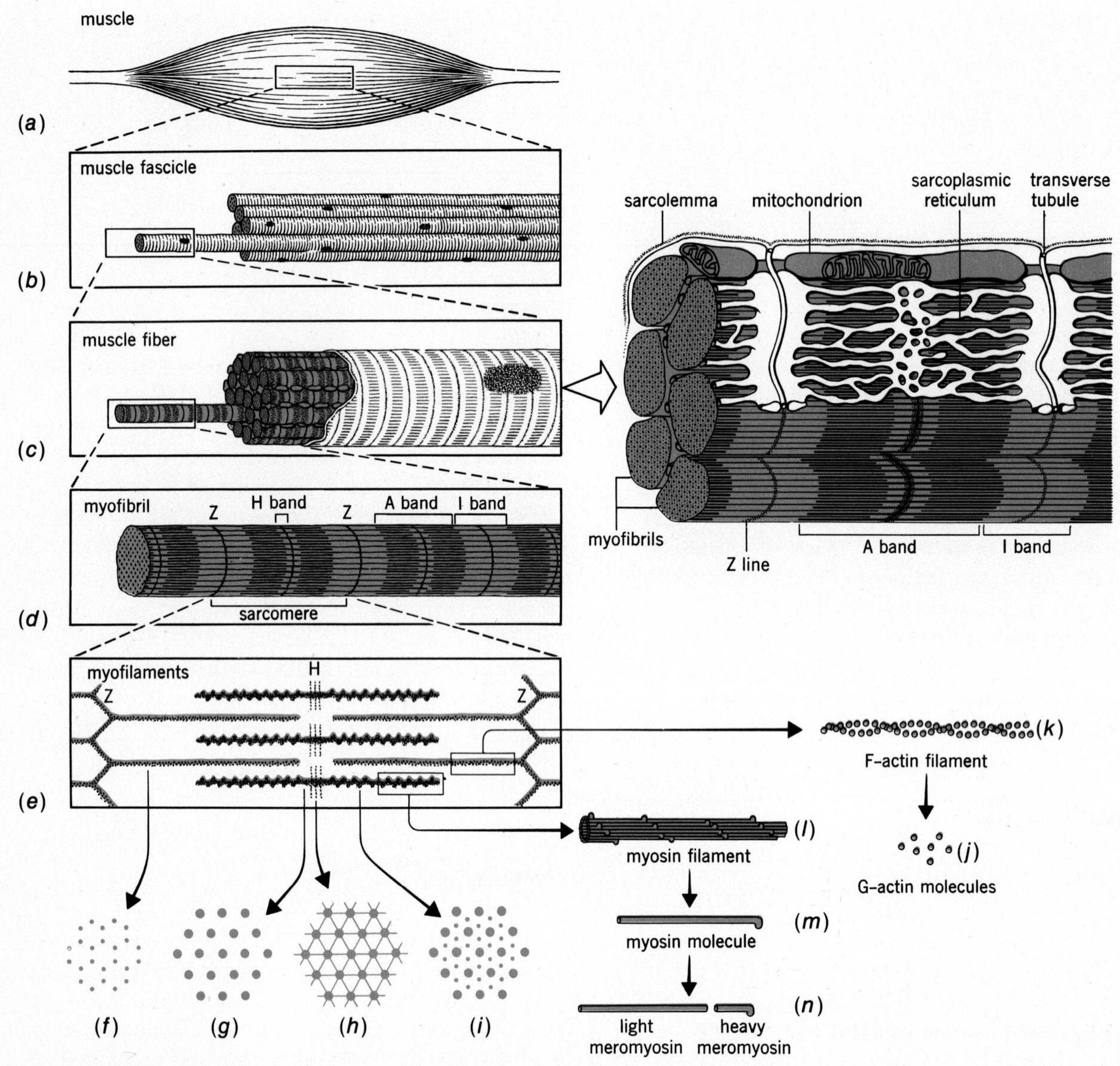

FIGURE 9.1 The structure of skeletal muscle. (F), (G), (H), and (I) are cross sections of the filaments at the points indicated.

a large (MW 220,000–600,000) protein possessing enzymelike properties and having clublike ends on its molecules, lies betwen **Z** lines. ACTIN is a smaller (MW about 60,000) protein thought to exist as a coiled double helix that passes through or is attached to the **Z** lines. Actin appears to be the contractile protein of muscle and exists in two forms: *globular* or *G-actin* exists in rounded globular units; *fibrous* or *F-actin* is composed of G-actin units in a polymerized form. F-actin composes the helices of the protein. TROPOMYOSIN (MW about 50,000) is believed to be attached to actin. TROPONIN is bound to actin by tropomyosin and may inhibit the contractile process or ensure relaxation. Other chemicals present in muscle include *creatine phosphate, glucose, ATP,* Ca^{++}, the usual intra- and extracellular ions K^+, Na^+, Cl^-, and myoglobin. The latter substance is composed of a heme molecule and a single globin chain and combines with oxygen much as does hemoglobin.

A system of T-TUBULES (see Fig. 9.1) reaches from the external surface of the fiber to the fibrils inside, and carries extracellular fluid to the inside. A second system of tubules abuts against but does not directly connect with the T-tubules and surrounds each fibril within the fiber. This system is known as the SARCOPLASMIC RETICULUM and contains or binds most of the muscle's Ca^{++} when the muscle is at rest. Terminal cisternae of the sarcoplasmic reticulum lie in pairs associated with the T-tubules. The three structures constitute the *triads.*

Skeletal muscle is normally stimulated to contract by nerves that reach it from the brain and spinal cord. The connection between the nerve and the muscle fiber is made by way of the MYONEURAL JUNCTION (Fig. 9.2). The junction is made by a nerve fiber that has lost all coverings or sheaths it may have possessed, and that has divided into a number of *end feet* or terminal nerve branches. The end feet contain many small vesicles (synaptic vesicles) of the chemical *acetylcholine;* the latter functions as the transmitter to get the nerve impulse across the junction to the muscle fibrils. The end feet fit into depressions (the synaptic troughs) in a thickened portion of the muscle membrane of the junction known as the MOTOR END PLATE. The motor end plate exhib-

FIGURE 9.2 The myoneural junction as seen by light and electron microscopy. *(a)* Section of the junction as seen by electron microscopy. *(b)* Surface view of the junction as seen by light microscopy. *(c)* Details of the junction as seen by electron microscopy.

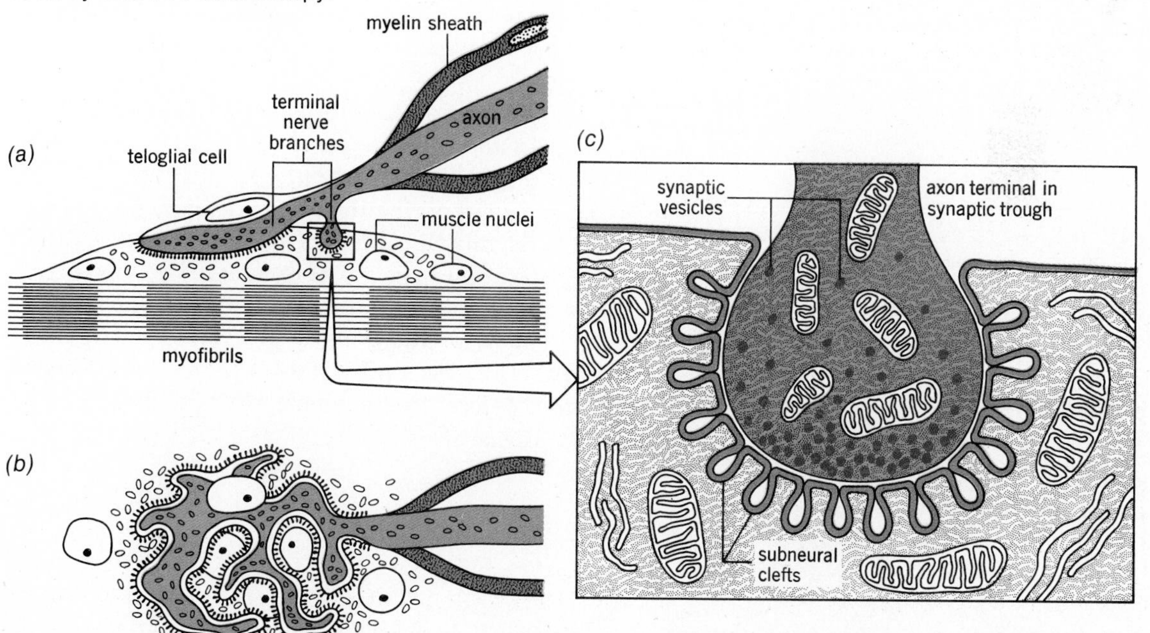

TABLE 9.1 Distribution of ions in resting muscle (average values)

Ion	Interstitial fluid (meq/L)	Intracellular fluid (meq/L)
Na^+	145	12
K^+	4	155
Cl^-	120	3.8
HCO_3^-	27	8
Ca^{++}	10	–
$^aA^-$	Trace	155

[a] Ionized protein and other organic substances.

FIGURE 9.3 Changes in lengths of the sarcomere during muscle contraction.

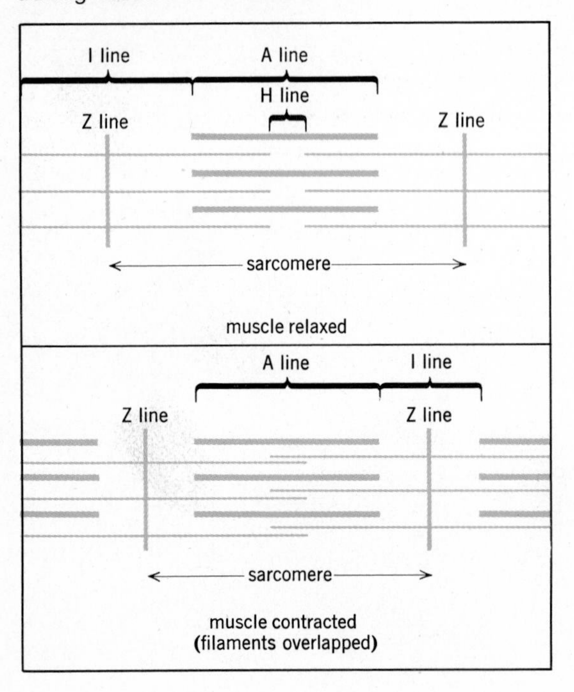

its many folds known as the *subneural clefts* or *palisades.* The myoneural junction is responsive to chemicals in its environment. For example, nicotine increases transmission across the junction; curare prevents transmission and can lead to muscular paralysis.

Two types of skeletal muscle are recognized. RED MUSCLE is composed of larger, slower-acting muscle fibers, which have less-evident striations and which contain more myoglobin, a pigment resembling hemoglobin. They are darker in color than the other type of muscle, because of their greater content of myoglobin. This type of muscle appears to be more resistant to fatigue and is capable of holding sustained contractions without rapid development of fatigue. The muscles that sustain the standing position of the body against gravity are of this type. WHITE MUSCLE, or *phasic muscle,* contains less myoglobin, is faster reacting, and tires more rapidly than the red muscle. White muscle is found in those body areas where fine, skilled movements are carried out, such as the eye muscles and hand muscles.

Muscles grow larger (hypertrophy) when they are regularly exercised and when there is an increase in the amount of sarcoplasm, not by an increase in number of fibers or fibrils. Denervation of a muscle results in atrophy, or decrease in size, as does disuse (for example, when a bone is broken and the limb must be placed in a cast until the fracture heals). Atrophy is attributed to a decrease in the amount of sarcoplasm. Damage to skeletal muscle fibers, if not too great, leads to regeneration from existing fibers if nerve connections are still present. This process occurs by enlargement of the ends of the damaged fibers to form "muscle buds" that become new fibers. Large defects in the muscle are filled by a connective tissue scar.

Physiology

THE ESTABLISHMENT OF THE EXCITABLE STATE IN MUSCLE (see also *Chapter 3*). Resting muscle exhibits a resting potential between the intra- and extracellular compartments of about 90 mv. This potential is due to the separation of ions by active and passive mechanisms and results from the ionic distributions shown in Table 9.1. Contraction of muscle is brought about by depolarization of the fibers and a series of electrochemical changes that culminate in the shortening of the F-actin filaments.

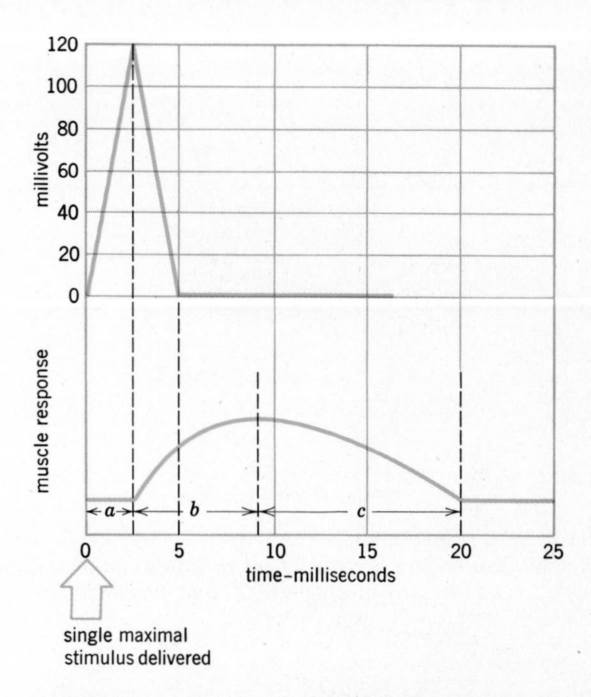

FIGURE 9.4 Electrical and mechanical events correlated for skeletal muscle. Note asynchrony of events. *a*, Latent period, *b*, contraction, *c*, relaxation.

FIGURE 9.5 A length-tension curve. Length is measured in arbitrary units, with 100 representing the resting length of the muscle in the body. Maximum tension is developed when muscle is at its normal body length.

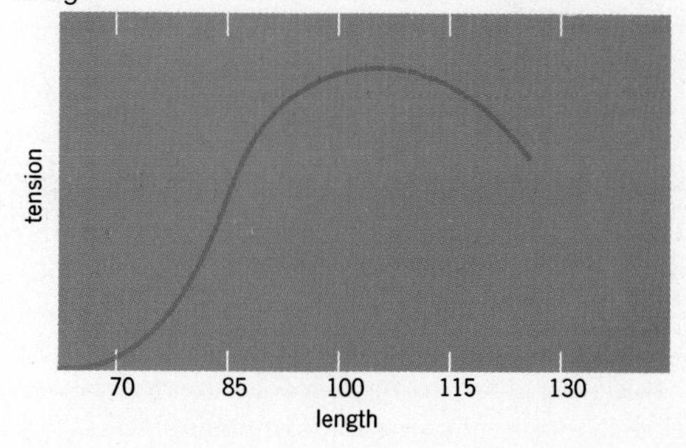

THE EVENTS OF MUSCULAR ACTIVITY; CONTRACTION AND RELAXATION. In the resting state, actin and myosin are separate molecules. Actin contains bound ATP attached to it as well as the tropomyosin-troponin complex. Muscle calcium is restricted to (and bound to) the membranes of the sarcoplasmic reticulum, and the motor nerve end feet of the myoneural junction contain vesicles of a transmitter substance known as acetylcholine. The membranes of the T-tubules and reticulum are polarized due to the ionic separation. A nerve impulse reaching the nerve end feet causes release of acetylcholine. The acetylcholine increases the permeability of the muscle membrane to sodium ions, and the membrane potential rises a few millivolts to create the *end plate potential*. The end plate potential then excites (depolarizes) the motor end plate. The depolarization wave is transmitted across the surface to the membranes of the reticulum that also depolarize. The bound Ca^{++} is released and moves from the reticulum to the myosin. Myosin is activated by the Ca^{++} and exhibits an ATPase activity, splitting a phosphate from ATP. The tropomyosin-troponin complex is split from actin, which then forms a cross bridge (possibly via ADP) with the enlarged end on the myosin molecule. The cross bridges then oscillate back and forth, attaching to the active filaments and drawing them together, then detaching and forming a new bond farther down the filament (Fig. 9.3). The actin molecules slide past one another and the **Z** lines are drawn closer together.

Relaxation is accomplished by active removal of Ca^{++} to its storage place in the reticulum. This

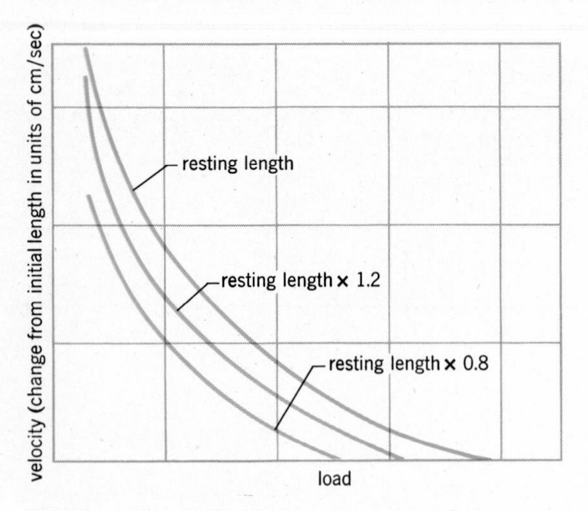

FIGURE 9.6 Several force-velocity curves for skeletal muscle. Shortening is most rapid when the muscle is loaded by a weight equivalent to that normally imposed by body structures.

halts ATPase activity of myosin, the cross bridges break, ADP is resynthesized to ATP, the tropomyosin-troponin complex is reattached to the actin ("relaxing protein"?), and the actin lengthens, possibly due to elastic recoil or by the attachment of the relaxing protein.

The foregoing description of muscular activity presents the so-called SLIDING FILAMENT MODEL, which embodies the current notions of how muscles contract and relax.

OTHER EVENTS OCCURRING DURING MUSCLE ACTIVITY

Electrical phenomena. An action potential always accompanies the depolarization of the muscle membranes. It has the form shown in Fig. 3.15 and may be recorded as an *electromyogram*. In a single mammalian skeletal muscle fiber, the potential lasts about 2 ms. Muscular contraction, however, does not begin synchronously with the action potential (Fig. 9.4). Approximately 2 ms are required for the chemical reactions that lead to cross-bridge formation to occur. The elapsed time constitutes the latent period of the muscle.

Length-tension relationships. If a muscle or a fiber at rest is passively stretched, it exhibits *elasticity* in much the same manner as a rubber band. That is, to a point, stretching causes an increase in tension. If the muscle or fiber is fixed at the length it normally has in the body (resting length) and is stimulated to contract maximally, it will develop its greatest contractile tension. In other words, it appears to be more efficient when placed under slight tension by the skeleton than it is at its resting length in the body. These relationships are shown in Fig. 9.5.

Force-velocity relationships. Velocity or speed of muscular contraction varies inversely with the weight the muscle must move and becomes zero when the load is too heavy to move (Fig. 9.6). Contractions that result in development of tension *and* shortening of the muscle fibers are termed ISOTONIC CONTRACTIONS, while those in which tension develops but no shortening occurs are termed ISOMETRIC CONTRACTIONS.

The term *isometric* is probably familiar to those who have undertaken training programs with a view to increasing muscular strength. Isometric exercises, in which one muscle or group of muscles is pitted against another with no shortening of the muscles allowed, have been shown to cause a more rapid increase in muscular strength than isotonic exercises, such as lifting barbells. While a complete physiological explanation for this phenomenon is lacking, several things occur more rapidly under an isometric program: muscle fibers increase in size more rapidly; fibers that are smaller or not normally active are brought into play more rapidly; faster increase in capillary networks occurs; and motor end plates increase in size and chemical content, bringing stronger stimuli to the muscle.

Heat production. Nearly all the chemical processes in resting muscle release heat *(resting heat)* to the environment. Among the processes causing heat release during rest are normal metabolic reactions and maintenance of muscle tone. Isometric and isotonic contractions release increased amounts of heat. Heat produced during one complete cycle of activity is known as ACTIVATION HEAT and may be subdivided into several categories.

INITIAL HEAT is produced synchronously with stimulation and contraction of the muscle and represents the heat released by the processes that get the contraction started—that is, activation of myosin, splitting of ATP, and formation of cross bridges. Oxygen is not required for these reactions.

RELAXATION HEAT is produced when stimulation ceases and the muscle relaxes. It correlates with the load on the muscle and does not appear if the load is removed from the muscle before it relaxes.

RECOVERY HEAT is produced for some time after relaxation has occurred and represents oxidative metabolism designed to replenish energy sources and rid the muscle of metabolites. It does not appear in the absence of oxygen.

Utilizing heat production and knowledge of the energy available for contraction, it can be calculated that efficiency of muscular activity is about 25 percent. The formula usually employed to calculate efficiency is:

$$\text{efficiency} = \frac{W}{A + 0.16\, P_0 x + 1.18\, W}$$

where

W = work (load × distance)
A = initial heat, constant at about 0.1 g cm
P_0 = maximum tension developed
x = distance shortened

ENERGY SOURCES FOR MUSCULAR ACTIVITY (see also Chapter 4). Resting metabolism in muscle is largely dependent on glycolytic decomposition of glucose and beta-oxidation of fatty acids. The RQ (respiratory quotient or ratio of CO_2 released to O_2 required in combustion of a foodstuff) is about 0.7, indicating primary reliance on fats. Ninety percent of the energy produced during rest comes from fat, 10 percent from glucose. Energy for the initial phases of contraction is provided by the breakdown (hydrolysis) of ATP. This reaction does not require oxygen, and the muscle may continue activity anaerobically as long as ATP concentrations are adequate. If contraction must be maintained for long periods of time, resynthesis of ATP is required. The energy for ATP synthesis is provided by several mechanisms:

Breakdown of phosphocreatine: phosphocreatine + ADP ⇆ creatine + ATP

Anaerobic combustion of glucose (glycolysis): glucose + 2 ATP → 2 lactic acid + 4 ATP

Aerobic combustion of glucose (Krebs cycle and oxidative phosphorylation): glucose + 2 ATP → $6CO_2$ + $6H_2O$ + 38 ATP

Aerobic combustion of fatty acids (β-oxidation, Krebs cycle, and oxidative phosphorylation): fatty acid → CO_2 + H_2O + ATP (amount depends on length of chain)

The last two processes require oxygen, and if oxygen is not available in sufficient amount to ensure these reactions, the muscle may proceed anaerobically and incur an OXYGEN DEBT. The debt represents the amount of oxygen required to combust the lactic acid formed during anaerobic metabolism. The energy produced during recovery from oxygen debt and during relaxation is correlated with the relaxation and recovery heat production.

PHYSIOLOGICAL PROPERTIES AND REACTIONS OF SKELETAL MUSCLE. Skeletal muscle is organized into MOTOR UNITS (Fig. 9.7), each unit consisting of a single nerve fiber and the muscle fibers it serves. Many motor units compose a given muscle. Stimuli delivered to a muscle may be SUBTHRESHOLD *(subliminal)* or of insufficient strength to obtain a response. A MAXIMAL stimulus is strong enough to cause contraction of all muscle fibers in the muscle.

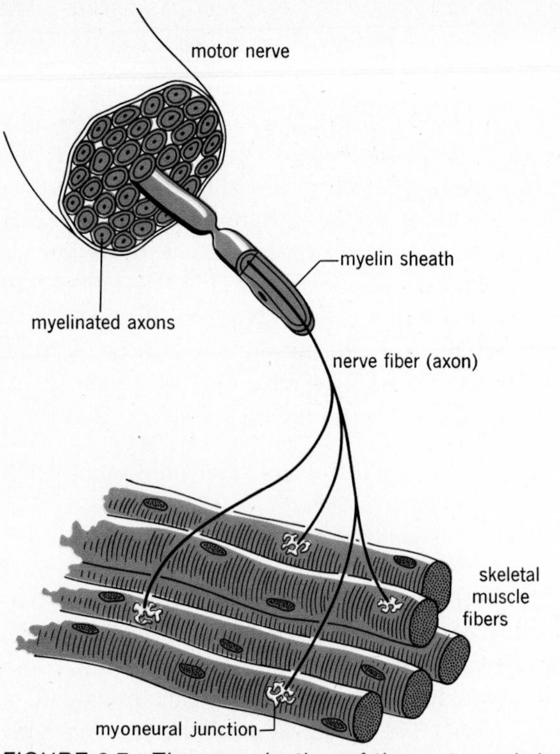

FIGURE 9.7 The organization of the motor unit (one motor nerve fiber and the muscle fibers it supplies).

GRADING OF TENSION is a matter of exciting more motor units. Excitation of increasing numbers of motor units involves stronger stimuli delivered to the nerves that cause depolarization of more nerve fibers and contraction of more motor units.

Each motor unit follows the ALL-OR-NONE LAW: a stimulus that exceeds the threshold required to depolarize the motor unit will result in a maximal contraction. Conversely, if the depolarization threshold is not exceeded, no contraction will occur. According to the conditions around and in the muscle at the time of stimulation, such as pH, temperature, and chemical factors, the strength of the response may vary.

The muscle fibers exhibit INDEPENDENT IRRITABILITY; that is, they may be caused to contract by direct application of stimuli. This property makes it possible to maintain healthy muscle status when the muscle has been denervated, if stimulation is applied periodically by external electrode.

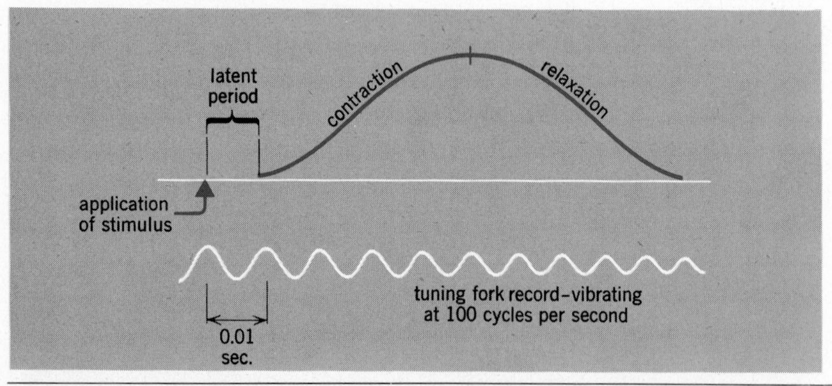

FIGURE 9.8
A single muscle twitch.

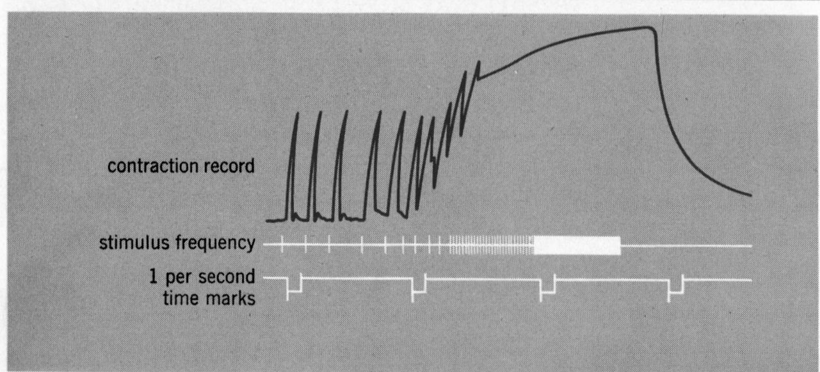

FIGURE 9.9 The genesis of tetanus in a skeletal muscle with increasing frequency of stimulation.

A single stimulus to a muscle is followed by a single response known as a TWITCH. Normally not occurring in the body, the twitch may be reproduced in the laboratory and gives information as to the phases of muscular activity (Fig. 9.8).

Skeletal muscle exhibits a very SHORT REFRACTORY PERIOD and is capable of responding to a second stimulus about 1/500 sec after the first. The short refractory period enables fusion and summation of twitches (wave summation) to produce the sustained maximal (tetanic) or near-maximal (subtetanic) contractions (Fig. 9.9). Wave summation occurs when frequency of stimulation is rapid enough to cause a second response before the muscle has completed a previous response. Shortening is greater with each succeeding stimulus. In subtetanic contractions different motor units contract asynchronously; that is, some units are contracting while others are relaxing. Development of fatigue is thus minimized. Subtetanic contractions occur in the body to maintain posture and in response to volleys of impulses reaching the muscles over their motor nerves.

Skeletal muscle exhibits TREPPE (Fig. 9.10), or a small increase in strength of contraction, even when a maximal stimulus is delivered to the muscle. Treppe is described as a result of lowered resistance to movement of myofibrils as a muscle "warms up" with repeated stimuli. More energy may be directed toward shortening and less to overcoming internal resistance.

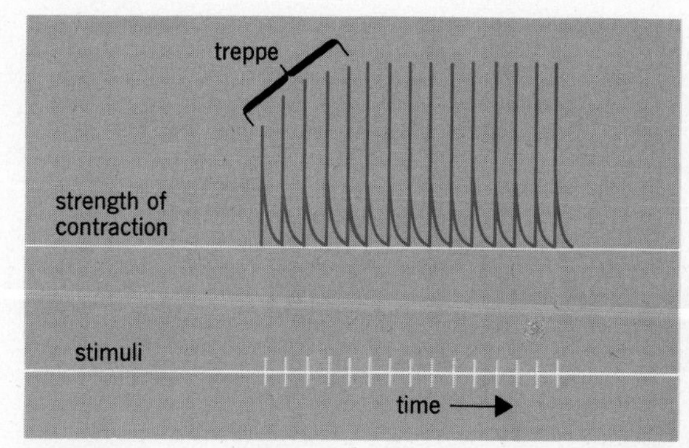

FIGURE 9.10. Treppe.

The problem of TONUS in skeletal muscle remains to be resolved. The muscle is said to maintain a subtetanic contraction or state of continuous slight contraction that keeps it in readiness for greater activity or that maintains posture. In the sense that an innervated muscle offers resistance to extension and a denervated one does not (that is, it has *no* tone), it is obvious that the muscle is partially contracted all the time, and that the contraction depends on an intact nerve supply. The so-called "relay theory" suggests that the contraction is maintained without fatigue by asynchronous contraction of alternate motor units.

The structure and properties of skeletal muscle are summarized in Table 9.2.

Smooth muscle

Structure

Smooth muscle (Fig. 9.11) consists of spindle-shaped fibers 2–5 μm in diameter and 50–100 μm in length. Each fiber possesses a single nucleus located near its center in the widest portion of the cell. A sarcolemma surrounds the fiber. Fibrils are few and NONSTRIATED. Most smooth muscle is INVOLUNTARY; that is, it can contract without outside stimulation, although its activity may be modified by nervous stimulation. The fibers are capable of maintaining a more-or-less constant contraction (tone) regardless of length. Concentrations of actin and myosin are

TABLE 9.2 Summary of structure and properties of skeletal muscle

Item	Function or structure	Comments
Fiber	The basic unit or "cell" of skeletal muscle	Unit of structure of all named muscles. Size varies according to muscle size.
Sarcolemma	Membrane around fiber	Provides limit to fiber, some control of entry.
Sarcoplasm	The cytoplasm of the fiber	Contains mitochondria, reticulum, myofibrils.
Myofibrils	Longitudinally arranged units of the fiber	Cross banded (striated)
Myofilaments	Protein strands longitudinally arranged inside myofibrils	Contractile units of the muscle. 2 types: thick (contains myosin); thin (contains actin).
ATP, CP, K^+, Ca^{++}, $PO_4^{\equiv}$, glucose	Chemicals, within the fiber	Necessary for contraction and nutrition.
Sarcoplasmic reticulum	The "endoplasmic reticulum" of the fiber	Houses Ca^{++} until required for contraction.
T-tubules	Tubules separate from sarcoplasmic reticulum which communicate to fiber exterior	Avenue for passage of substances into fiber.
Endomysium Perimysium Epimysium	Connective tissue binding fibers into a muscle	Endo—binds fibers together. Peri—surrounds fasciculi. Epi—surrounds muscle.
Interdigitating filament model	Theory of arrangement of filaments in myofibrils	Allows explanation of muscle contraction.
Muscle contraction	Shortening of fibers	Ultimate cause of movement.
Creatine phosphate	Compound in muscle releasing energy for ATP synthesis	Immediate source of energy for ATP synthesis.
Glucose, Lactic acid, Fatty acid.	Combusted to provide energy for ATP synthesis	Provide energy to sustain activity.
Isometric contraction	Tension developed but no shortening	Posture.
Isotonic contraction	Contraction with shortening	Movement.
Twitch	Illustrates phases of muscular activity. 1 response to 1 stimulus.	0.1 second duration (frog muscle). In body, does not normally occur.
Tetanus	Fusion of twitches; sustained maximal contraction	Result of short refractory period allowing twitch fusion.
Tonus	Sustained partial contraction	Depends on intact motor areas in brain, intact upper and lower motor neurons.
Treppe	Increasing strength of contraction with repeated strong stimuli	"Warming up" allows greater contraction.

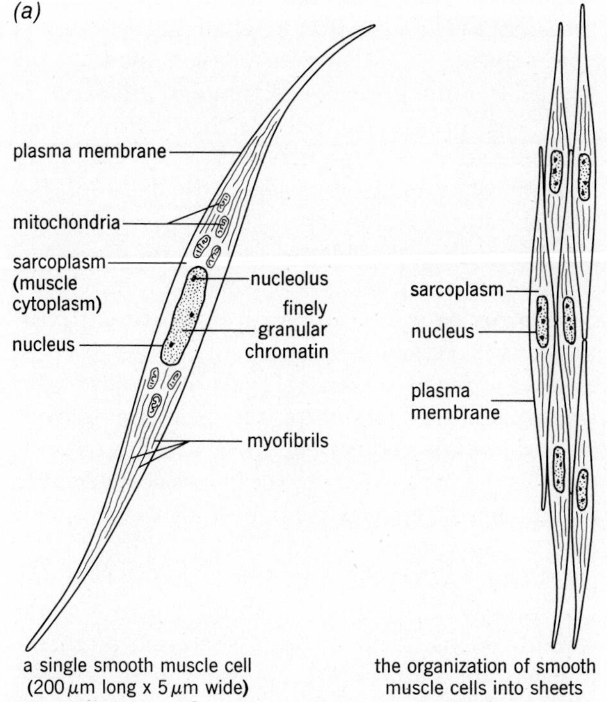

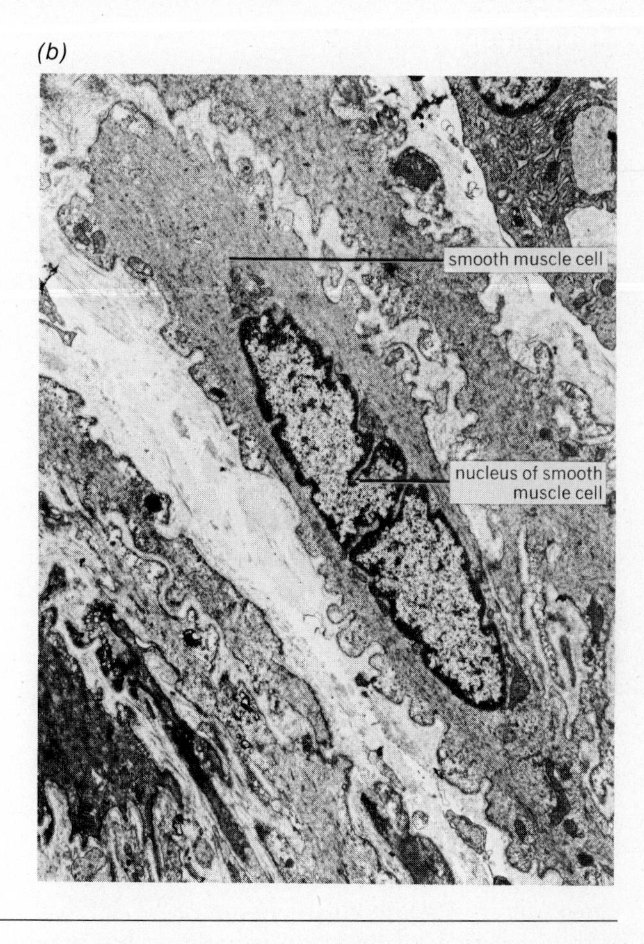

FIGURE 9.11 The morphology of smooth muscle. *(a)* Diagrammatic representation of the structure of smooth muscle. *(b)* Smooth muscle cell as seen under the electron microscope (× 8,000). (Electron micrograph courtesy Norton B. Gilula, The Rockefeller University.)

about seven times lower in smooth muscle than in skeletal, as are concentrations of ATP and phosphocreatine. The muscle fibers appear to lack T-tubules and have a poorly developed sarcoplasmic reticulum.

Types of smooth muscle

Two categories of smooth muscle may be distinguished.

UNITARY SMOOTH MUSCLE exhibits spontaneous activity and conducts impulses from one muscle cell to another as though the entire muscle mass were a single cell. This type of muscle is found in the alimentary tract, uterus, ureters, and small blood vessels.

MULTIUNIT SMOOTH MUSCLE does not exhibit spontaneous contraction, usually requires stimulation by nerves, and can grade its strength of contraction. This type of muscle is found in the ciliary muscles and iris of the eye and in larger blood vessels.

Some smooth muscle does not fall clearly into either group. For example, in the bladder, stretch and/or nerve stimulation can cause contraction. Unitary smooth muscle has been studied most extensively and the following comments are based primarily on investigation of this type.

Spontaneous activity

Spontaneous activity depends on LOCAL DEPOLARIZATION similar to that occurring in cardiac nodal tissue. The membrane potential is about two-thirds that of skeletal muscle and nerve (about 60 mv) and decays so that spontaneous depolarization occurs. This depolarization does not arise in a constant area in the muscle but "migrates" from one place to another. The waves or areas of depolarization are then conducted from cell to cell by way of low-resistance "tight junctions" between adjacent cell membranes. Actual contraction time is long, presumably because of the time required for the few myofibrils to develop tension. Contraction of smooth muscle is believed to occur by a process similar to that described for skeletal muscle. Strength of smooth muscle contraction depends on the frequency of spontaneous discharge from individual cells and the number of cells contracting together. Smooth muscle responds to several influences.

Stretching usually serves as an adequate stimulus for depolarization and is followed by contraction.

CHEMICAL FACTORS. ACETYLCHOLINE stimulates most unitary smooth muscle to increased activity. The mechanism of its action is not clear but may be to increase membrane permeability to Ca^{++} or Na^{+}, similar to its effect on skeletal muscle. Epinephrine effects depend on type of muscle and animal species. EPINEPHRINE usually inhibits contraction of gut musculature and stimulates activity of uterine and vascular muscle. This chemical probably exerts its action by changing membrane permeability to Ca^{++} or Na^{+}. NOREPINEPHRINE is similar to epinephrine in action and effect but is about 100 times less effective than epinephrine. ESTROGENIC HORMONES and OXYTOCIN also modify activity of smooth muscle, particularly of the uterus. Estrogen stimulates activity, as does oxytocin. Progesterone tends to quiet or inhibit activity. Other chemicals affecting smooth muscle are reserpine, ephedrine, and botulinus toxin. RESERPINE, a chemical employed as an antihypertensive agent, causes relaxation of arteriolar smooth muscle. EPHEDRINE, a substance employed in nasal sprays, constricts vessels in the respiratory system. BOTULINUS TOXIN blocks acetylcholine release and leads to muscular paralysis.

NERVOUS FACTORS. Autonomic nerves from the "involuntary nervous system" supply most smooth muscle and, by their secretion of chemicals at their endings, exert effects similar to those produced by acetylcholine and norepinephrine. Most smooth muscle receives parasympathetic, (energy-conserving and "normal activity") acetylcholine-secreting nerves and sympathetic (energy-utilizing and "emergency activity") norepinephrine-secreting nerves. No specialized end plate exists on the muscle to transmit the effects of nerve stimulation to the muscle. The axons themselves contain vesicles of chemicals that presumably diffuse directly to the muscle cells to influence muscle activity.

Clinical considerations

Since skeletal muscle normally depends on stimulation through nerves for contraction and maintenance of functional status, denervation is typically followed by ATROPHY or diminished size of the muscle. Even without denervation, prevention of movement of parts of the body, such as when a limb is placed in a cast, is associated with atrophy. Denervation also increases the sensitivity of the muscle to acetylcholine and may result in irregular contractions of the muscle *(fibrillation).*

MUSCULAR DYSTROPHY refers to a group of hereditary muscular diseases characterized by degeneration of muscle proteins. *Duchenne dystrophy,* the most common type, is sex-linked and recessive and affects only young males. The legs are affected initially and cause a waddling gait, frequent falls, and eventual confinement to a wheelchair. *Facioscapulohumeral dystrophy* affects both sexes and is inherited as a dominant characteristic. Progression of the disease is slower than in the Duchenne type and involves the upper limbs more than the lower.

MYASTHENIA GRAVIS is a disorder characterized by muscle weakness, "sagging" of facial muscles, and fatigue on minor muscular effort. This disorder appears to be one of neuromuscular transmission. Investigation of the disorder discloses failing neuromuscular transmission, small size of acetylcholine vesicles in myoneural junctions, and decreased number of vesicles. An additional possibility for abnormal function is hyperactive cholinesterase, an enzyme that normally destroys acetylcholine in neuromuscular junctions. Physostigmine, a cholinesterase inhibitor, increases muscular contractility. The exact mechanism involved in production of myasthenia is not known. Genetic and immunologic causes have been advanced to explain the disorder.

PARALYSIS refers to inability to contract a muscle voluntarily. The muscle may be in a contracted *(spastic)* state, or in a relaxed *(flaccid)* state. Spastic paralysis occurs if the motor tracts of the brain and spinal cord are damaged. Flaccid paralysis occurs if the motor nerves from the brain or cord to the muscles are damaged. These differences enable determination of where the damage exists when a muscle is paralyzed.

FASCICULATIONS, FIBRILLATIONS, and TREMOR refer to involuntary, repetitive contractions of muscles. The usual cause is nerve damage.

In MUSCLE SPASM, a forcible, often painful contraction of a muscle occurs, usually of appendage muscles. Chemical causes are most common, such as electrolyte imbalances (blood levels of Ca, Na, K) or chemical toxins (tetanus). Massage often relieves the spasm by increasing blood flow to the affected part.

MUSCLE CRAMPS, although similar to spasm, may have a different explanation. Exercise causes intense afferent nerve stimulation, through accumulation of chemicals or other causes. The spinal cord responds to this increased input of signals by an increased output of impulses that cause reflex contraction of the muscle. This may, in turn, lead to greater stimulation of the afferent nerves, and a "vicious cycle" is established. The cramp may often be promptly relieved by voluntary contraction of the opposing muscle group, while preventing movement of the body part. This causes a reciprocal inhibition of the contraction. A "charley horse" or "pulled muscle" may be painful and cause spasm and cramps. It usually involves tearing of muscle fibers.

MUSCLE SORENESS, especially after exercise, has not been satisfactorily explained. Metabolites of activity (for example, lactic acid) may cause edema (swelling), putting tension or pressure on muscle nerves and creating pain. Damage to the muscle fibers (rupture) or connective tissue (tearing) or sustained contraction have all been advanced as causes of soreness.

FATIGUE is a term used in several ways. It usually refers to the failure of a muscle to contract when stimulated. It may be due to failure of a metabolic process to maintain ATP levels, or exhaustion of acetylcholine vesicles in the myoneural junction. The term also refers to the subjective sensation of "tiredness" that may occur after exercise, emotional upheaval, and during physical illness.

HERNIA (commonly called "rupture") is a term used to describe the protrusion of abdominal viscera through a defect in the muscle or connective tissue of the abdominal wall, diaphragm, or pelvic floor. Hernias are the result of abdominal weakness, and an intraabdominal pressure that causes tearing or separation of the components of the abdominal wall. Hernias are more common in obese persons, older individuals, those whose livelihood requires heavy lifting, and in those whose style of life results in lack of use and weakening of abdominal muscles.

According to where the hernia occurs, several types are described.

INGUINAL HERNIAS account for 80 percent of all hernias. The defect is in the inguinal canal, which carries the spermatic cord from the scrotum in the male and the round ligament of the uterus to the labia majora in the female. Increased intraabdominal pressure, as in coughing, sneezing, lifting, straining during defecation, or distension caused by fluid or gas, may result in a separation or tearing of the inguinal muscle or abdominal fascia. Abdominal viscera, primarily the small intestine, may then protrude into the scrotum (in the male) or labia majora (in the female). Inguinal hernia in the male is recognized best by the presence of a soft mass (the visceral protrusion) through the inguinal ring and by the inability to palpate (feel) the structures of the spermatic cord above the mass. In the female, a labium will be enlarged as the viscera protrude through the canal.

FEMORAL HERNIA occurs when there is an enlargement of the femoral ring that normally passes the blood vessels to and from the thigh. A bulging of the skin in the groin is common as intestines push into the area.

UMBILICAL HERNIA occurs around the umbilicus and is more common in infants.

HIATUS HERNIA is protrusion of the stomach into the chest cavity as a result of a weakness in the opening (hiatus) passing the esophagus through the diaphragm.

Surgical repair is the preferred treatment for those whose activities may lead to difficulties with the hernia. Surgery repairs the defect in the muscle or fascia. After surgery, any type of strenuous activity or lifting should be avoided for several weeks to allow for proper healing and to prevent adhesions from forming. Common sites of herniation are shown in Fig. 9.12.

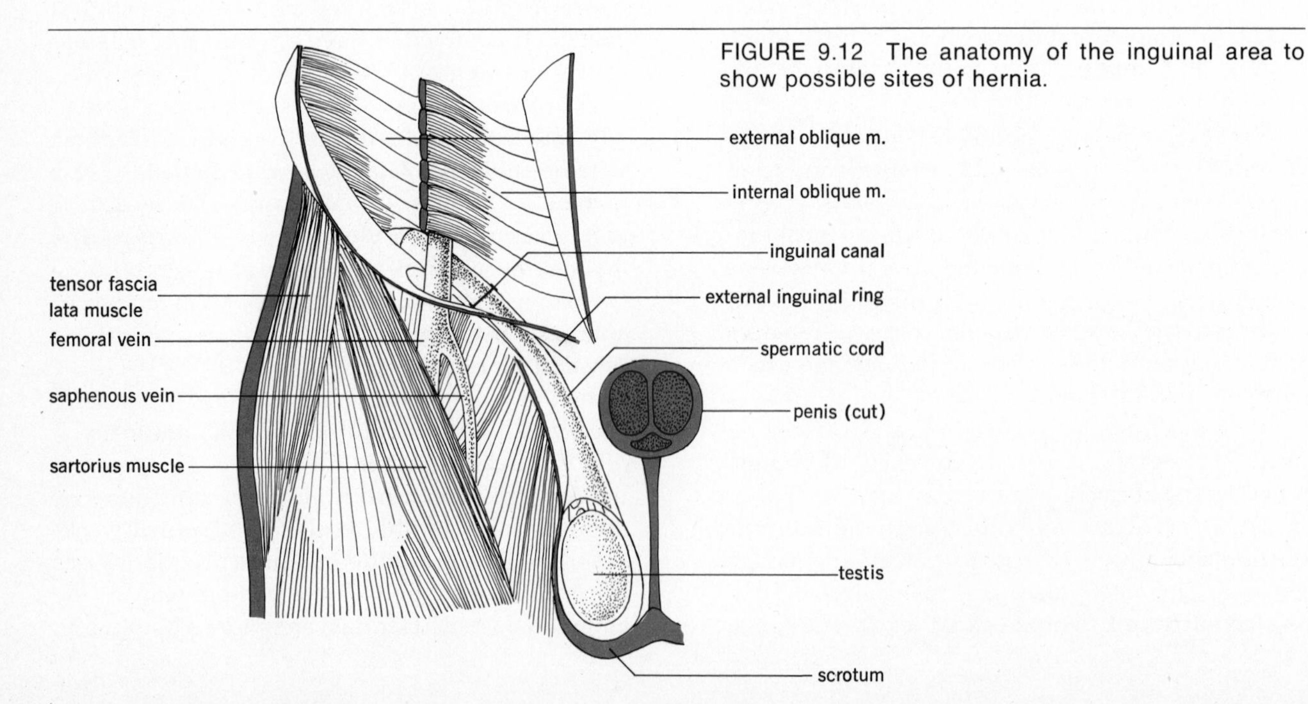

FIGURE 9.12 The anatomy of the inguinal area to show possible sites of hernia.

Summary

1. Muscular tissue is characterized by its ability to contract or shorten on stimulation. Three types of muscular tissue are found in the body: skeletal, smooth, and cardiac.
2. Skeletal muscle exists as individual fibers, with smaller fibrils and filaments inside.
   a. Myosin and actin are muscle proteins composing the filaments and are involved in contraction. Tropomyosin and troponin are proteins associated with relaxation.
   b. Systems of T-tubules and sarcoplasmic reticulum tubules provide communication for the extracellular fluid to the inside of the fiber and contain the Ca^{++} necessary for muscular contraction.
   c. Skeletal muscle is voluntary, normally stimulated to contraction by nerves. A myoneural junction makes the connection between nerve and muscle.
   d. Red and white muscle types are distinguished. Red is slower contracting, is less susceptible to fatigue, contains more myoglobin, and is found in postural muscles. White is faster reacting, is more susceptible to fatigue, and is found in the appendages.
   e. Increases and decreases in size of skeletal muscles depend on changes in amount of sarcoplasm rather than in numbers of fibers or fibrils.
   f. Regeneration of muscle occurs if lesions are small. Scar tissue formation occurs with large lesions.
3. Muscle is an excitable tissue and is polarized by the processes discussed in Chapter 3.
4. Muscular contraction and relaxation occurs in several steps.
   a. A nerve impulse arrives at the end of a motor nerve and increases its permeability to Ca^{++}. Ca^{++} enters and ruptures the vesicles of acetylcholine.
   b. The ends of the nerve release acetylcholine.
   c. Depolarization of T-tubule and sarcoplasmic reticulum membranes allows Ca^{++} to move to the filaments.
   d. Myosin is activated and splits a $PO_4^{=}$ from ATP.
   e. Cross bridges between myosin and actin are formed.
   f. The actin filaments slide past one another and the muscle shortens.
   g. Relaxation occurs as myosin is deactivated and the cross bridges break.
5. Other activities occur during muscle activity.
   a. An action potential may be recorded.
   b. The muscle develops tension according to its initial length.
   c. Speed of movement is inversely related to weight or load moved.
   d. Heat is produced during contraction (initial heat), during relaxation (relaxation heat), and after activity has ceased (recovery heat).
6. Energy for muscular activity is supplied from several sources. ATP is utilized first.
   a. Phosphocreatine breakdown supplies energy for ATP resynthesis.
   b. Anaerobic combustion of glucose and oxidation of fatty acids provides much ATP.
   c. Aerobic combustion of glucose and oxidation of fatty acids provides much ATP.
   d. Oxygen debt may result if anaerobic sources are required to supply energy faster than the products of their activity can be removed by the aerobic metabolic cycles.
7. Skeletal muscle exhibits other properties and reactions.
   a. It can grade its strength of contraction according to how many fibers are activated.
   b. Individual fibers contract maximally or not at all when stimulated (all-or-none law).
   c. Muscles may be stimulated directly and caused to contract (independent irritability).
   d. In the laboratory a muscle can exhibit a type of contraction called a twitch. A twitch shows the phases of activity (latent period, contraction, relaxation).
   e. It shows a short refractory period, which enables it to be tetanized (sustained maximal contraction).
   f. It shows treppe (increased contraction with maximal stimulation).
   g. It shows tonus (a sustained subtetanic contraction).
8. Smooth muscle is nonstriated and slow reacting. It exists as two types.
   a. Unitary smooth muscle is involuntary and is found in gut and small blood vessels.
   b. Multiunit smooth muscle requires nervous stimulation. It is found in the eye and large blood vessels.
9. Spontaneous activity is exhibited by unitary smooth muscle and is influenced by several factors.
   a. Stretching invokes a reflex contraction.
   b. Acetylcholine and norepinephrine stimulate or inhibit smooth muscle. Response depends on muscle characteristics in different body areas. Acetylcholine usually stimulates gut muscle; norepinephrine inhibits it.
   c. Nervous factors are involved through their production of acetylcholine and/or norepinephrine. No motor end plates are found on smooth muscle.
10. Clinical conditions are associated with skeletal muscle.
   a. Dystrophy is when the muscle atrophies or shrinks.
   b. Myasthenia gravis is a disorder of neuromuscular transmission.
   c. Paralysis is the usual result of motor nerve damage.
   d. A variety of involuntary contractions may follow nerve damage.
   e. Spasm is a painful state of contraction most commonly due to chemical irritants or ionic imbalances.

f. Soreness is due to nerve irritation as a result of edema, rupture of fibers, or tearing of connective tissues.

g. Muscle cramps occur when sensory nerve stimulation causes reflex contractions.

h. Fatigue refers to loss of contraction of muscle on stimulation. It may be caused by loss of energy sources, failure to transmit an impulse across the neuromuscular junction, or emotions, or circulatory failure.

i. Hernia or rupture refers to protrusion of abdominal viscera through a defect in the muscular walls of the abdominopelvic cavity. Inguinal hernias are the most common type. Hernias may be repaired surgically.

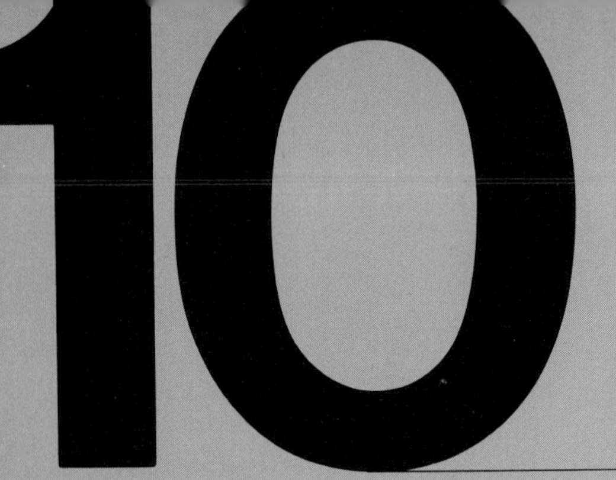

The Skeletal Muscles

General principles of myology

Definitions

Muscles can only actively shorten, and they cause movement by pulling on the skeleton. Since joints may allow several directions of motion, it follows that there must be separate muscles or groups of muscles to achieve each type of motion. A muscle causing a desired action is said to be an AGONIST or prime mover. Usually, a muscle having an opposite effect on the bone must relax to permit the movement to occur. This second muscle is termed the ANTAGONIST. According to the actions involved, a given muscle may be an agonist or an antagonist at different times. Most joints are operated by antagonistic groups of muscles. Still another type of muscle occasions little movement of a joint, but may steady the joint or eliminate undesirable movements. Such a muscle is termed a SYNERGIST (*syn*, together + *ergos*, work), or *fixator*. Fixators take no part in the movement, but steady the body to provide a stable platform against which motion may occur. For example, the subclavius steadies the clavicle during arm movements. A muscle attaches, usually via a tendon, to one bone or a connective tissue sheath, which moves to a lesser degree than the bone or connective tissue to which the other end of the muscle attaches. The lesser moving part serves as the ORIGIN of the muscle and the greater moving part serves as the INSERTION. The ACTION is the particular type of motion occurring as the muscle pulls on the insertion.

FIGURE 10.1 The three classes of levers.

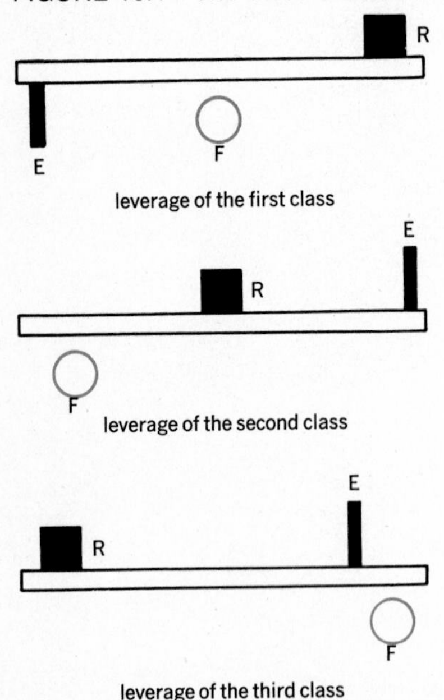

Levers

The muscles utilize the bones and joints as LEVERS to achieve movement. Any lever has several basic parts.

The fulcrum (F). This is the point about which the lever moves. In the body, the fulcrum is provided by a joint.

The point of effort (E). This is where the muscle inserts; the point at which the pull of the muscle is applied.

The resistance (R). This is the weight that must be moved by the muscle's pull. It is usually considered to be concentrated at some point on the lever.

The effort arm (EA). This is the distance from fulcrum to point of effort. It is usually some part of the length of the bone being moved.

The resistance arm (RA). This is the distance from the fulcrum to the area where the resistance is concentrated. It also is a distance along the length of the bone.

According to the placement of fulcrum, effort, and resistance relative to one another, three classes of levers are recognized (Fig. 10.1).

Class I. The fulcrum is placed between effort and resistance. Seesaws, post-hole diggers, and pry bars are everyday examples of first-class levers. Not too many levers of this class are found in the body, for it is not common to have a joint in the middle portion of the bone. The action of the triceps in extension of the forearm is a good example of a first-class lever.

Class II. The fulcrum is toward the end of the lever, and resistance is placed between fulcrum and effort. A wheelbarrow is an example of a second-class lever. This class requires less effort to move a given weight than the other classes. It is not agreed that any second-class levers exist in the body. Raising oneself on one's toes, the fulcrum being the ball of the foot, is considered by some to be an example.

Class III. The fulcrum is toward the end, and effort is between fulcrum and resistance. This type is the most common in the body, for it provides a joint at the end of a bone, and much space for muscle attachment along the effort arm.

Efficiency of a given type of lever may be calculated by the simple formula

$$E \times EA = R \times RA$$

FIGURE 10.2 The various arrangements of fibers in the muscles of the body.

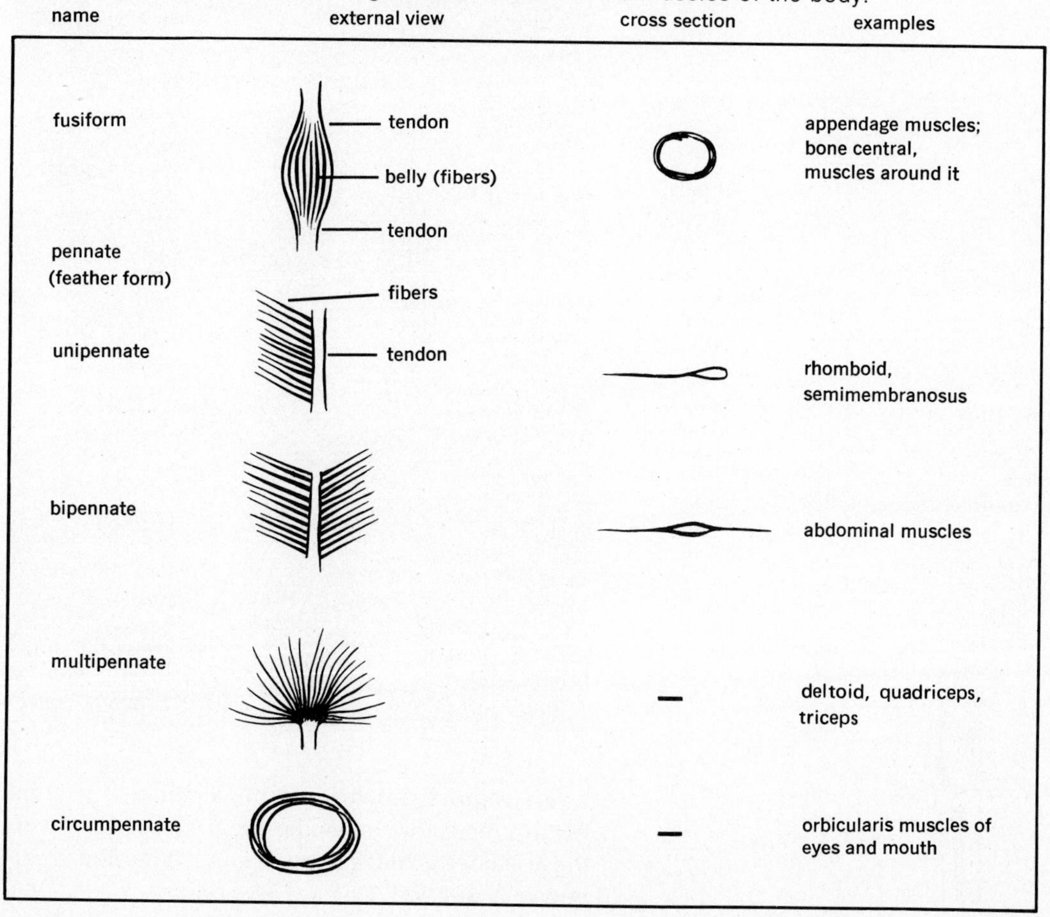

name	external view	cross section	examples
fusiform	tendon; belly (fibers); tendon		appendage muscles; bone central, muscles around it
pennate (feather form)	fibers		
unipennate	tendon		rhomboid, semimembranosus
bipennate			abdominal muscles
multipennate			deltoid, quadriceps, triceps
circumpennate			orbicularis muscles of eyes and mouth

For example, typical measurements for the triceps action on the elbow might be:

$E = ?$ Thus $E \times 1 = 10 \times 10$

$EA = 1$ in. or $E = \dfrac{10 \times 10}{1}$

$R = 10$ lb

$RA = 10$ in. $E = 100$ lb to move a 10-lb weight

For the second class

$E = ?$ Thus $E = \dfrac{10 \times 12}{10}$

$EA = 10$ in.

$E = 12$ lb

$R = 10$ lb

$RA = 12$ in.

For the third class

$E = ?$ $E = \dfrac{10 \times 12}{2}$

$EA = 2$ in.

$E = 60$ lb

$R = 10$ lb

$RA = 12$ in.

It becomes obvious that first- and third-class levers require much more power than is exerted by the resistance to make them work. However, these classes achieve a great range of motion when power is applied.

FIGURE 10.3 The relationships of bursae to tendons and bones. Bursae facilitate movement of tendons over bones.

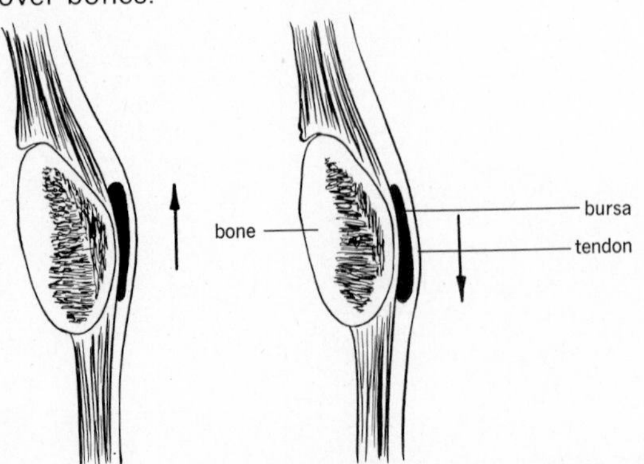

TABLE 10.1 Muscle actions

Action	Definition
Flexion	Decrease of angle between two bones
Extension	Increase of angle between two bones
Abduction	Movement away from the midline (of body or part)
Adduction	Movement toward midline (of body or part)
Elevation	Upward or superior movement
Depression	Downward or inferior movement
Rotation	Turning about the longitudinal axis of the bone
Medial	Toward midline of body ("inward")
Lateral	Away from midline of body ("outward")
Supination	To turn the palm up or anterior
Pronation	To turn the palm down or posterior
Inversion	To face the soles of the feet toward each other
Eversion	To face the soles of the feet away from each other
Dorsiflexion (=flexion)	At the ankle, to move the top of the foot toward the shin
Plantar flexion (=extension)	At the ankle, to move the sole of the foot downward, as in standing on the toes

Arrangement of fibers in a muscle

According to placement on the body and the area to which the effort of a muscle is applied, muscles may vary in the arrangement of fibers. A simple classification of fiber arrangement is shown in Fig. 10.2. It may be noted that muscle shape tends to be related to where the muscle occurs on the body. Fusiform muscles are found mainly on the appendages, where the muscles are arranged around the centrally placed bone; pennate muscles are found on broad surfaces such as abdomen or thorax; circumpennate muscles are around orifices to open or close them.

Tendons and bursae

Tendons usually make the attachment of a muscle to the bone for several reasons. The muscular portion may not be long enough to reach between the two bones. A tendon can be any length and also is small in size. Tendons may pass over bony prominences that would destroy a muscle fiber. Because of their small size, many more tendons can pass over a joint than could accommodate the fleshy portion.

Tendons are subjected to wear where they pass over bony prominences. Small fluid-filled sacs known as bursae (see Chapter 8) are placed between the tendon and the bone (Fig. 10.3) to cushion the tendon.

Actions

As indicated previously, muscle contraction results in the motion of a bone, the motion caused being termed the action of the muscle. Table 10.1 defines the basic types of actions. Note that the actions are paired in antagonistic movements.

The muscles

A general description of the major skeletal muscles follows in the next sections. Muscles are organized regionally into groups that act differently on particular body joints. As an aid to remembering the muscles, it is noted that muscles are generally named by their shape, location, origin, and insertion, action, or combinations of these features. Origins, insertions, actions, and innervations of individual muscles are given in Table 10.2, at the end of the chapter. Figure 10.4 shows the muscular system of the human as it would appear with the skin removed.

Muscles of the head and neck

Facial muscles

Facial muscles (Fig. 10.5) are the muscles of facial expression. Their contractions move the fleshy parts of the face for speech and cause the various expressions associated with emotions and feelings. The facial muscles include the following.

Orbicularis oculi encircles the eye and closes or winks the eye.

Orbicularis oris encircles the mouth and closes and protrudes ("puckers") the lips.

Levator labii superioris elevates the upper lip to give an "expression of contempt" (or as in saying "yech!" or "yuck!").

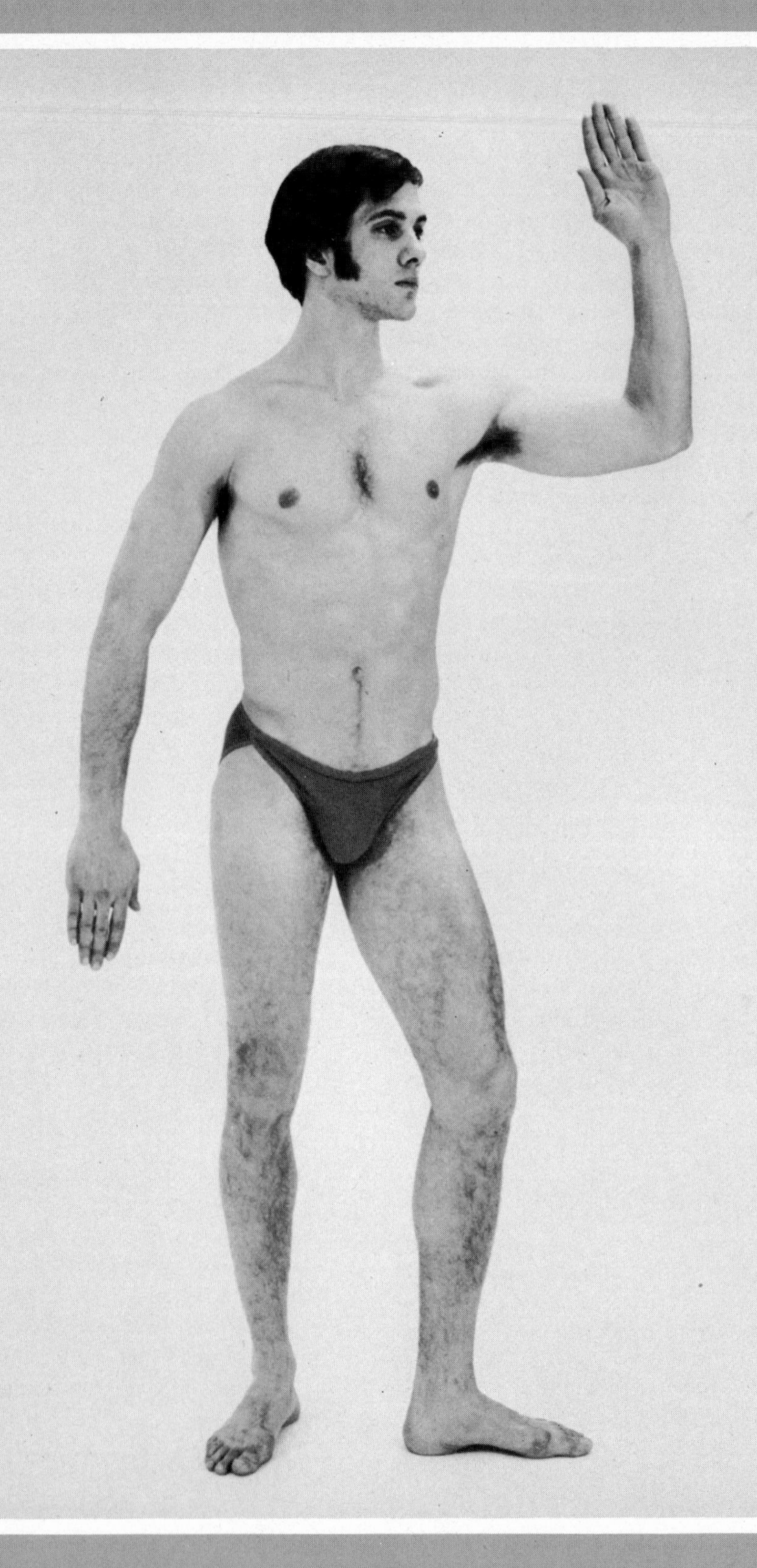
(a)

FIGURE 10.4 Views of the skeletal muscles of the human. *(a)* Anterior, *(b)* posterior.

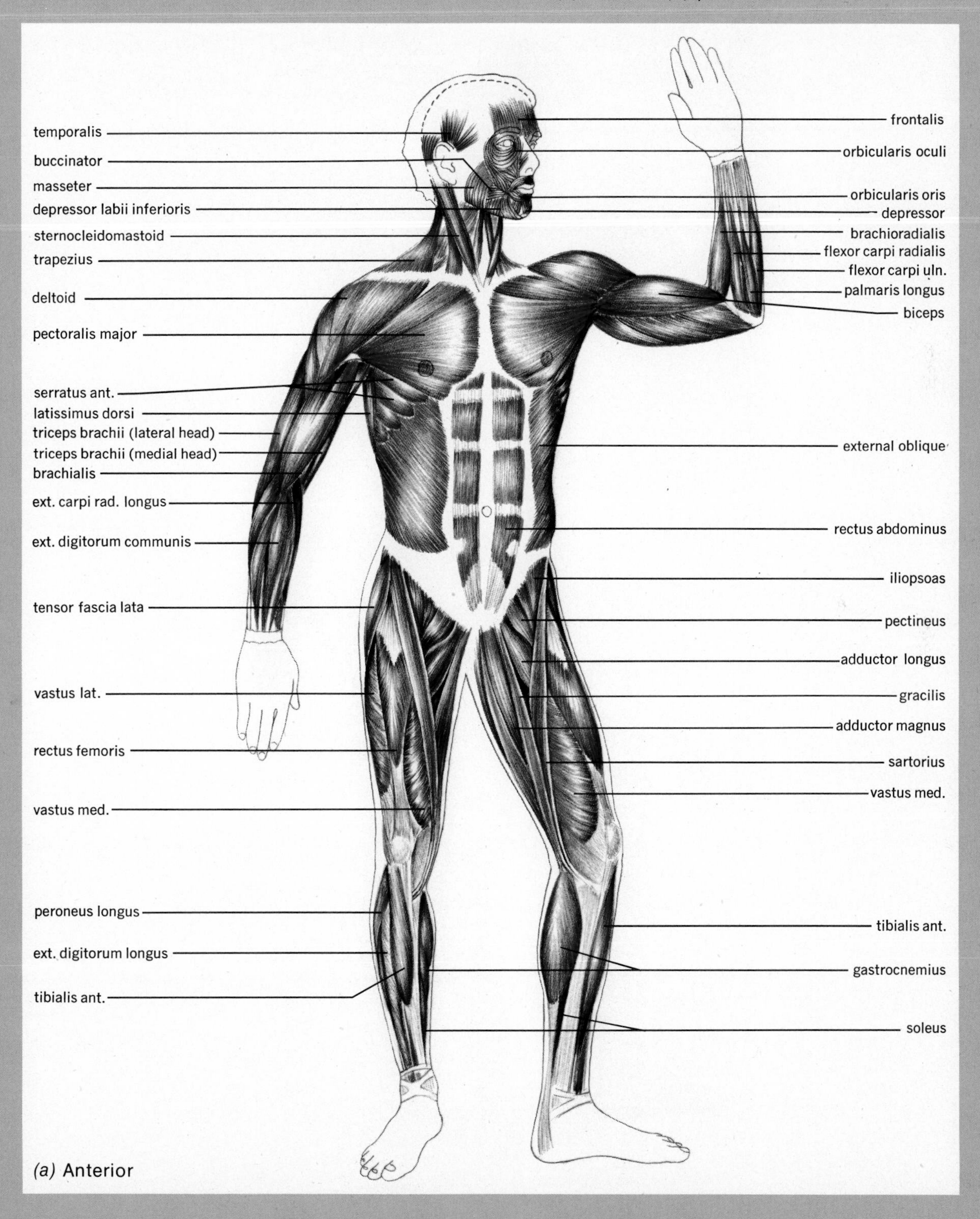

(a) Anterior

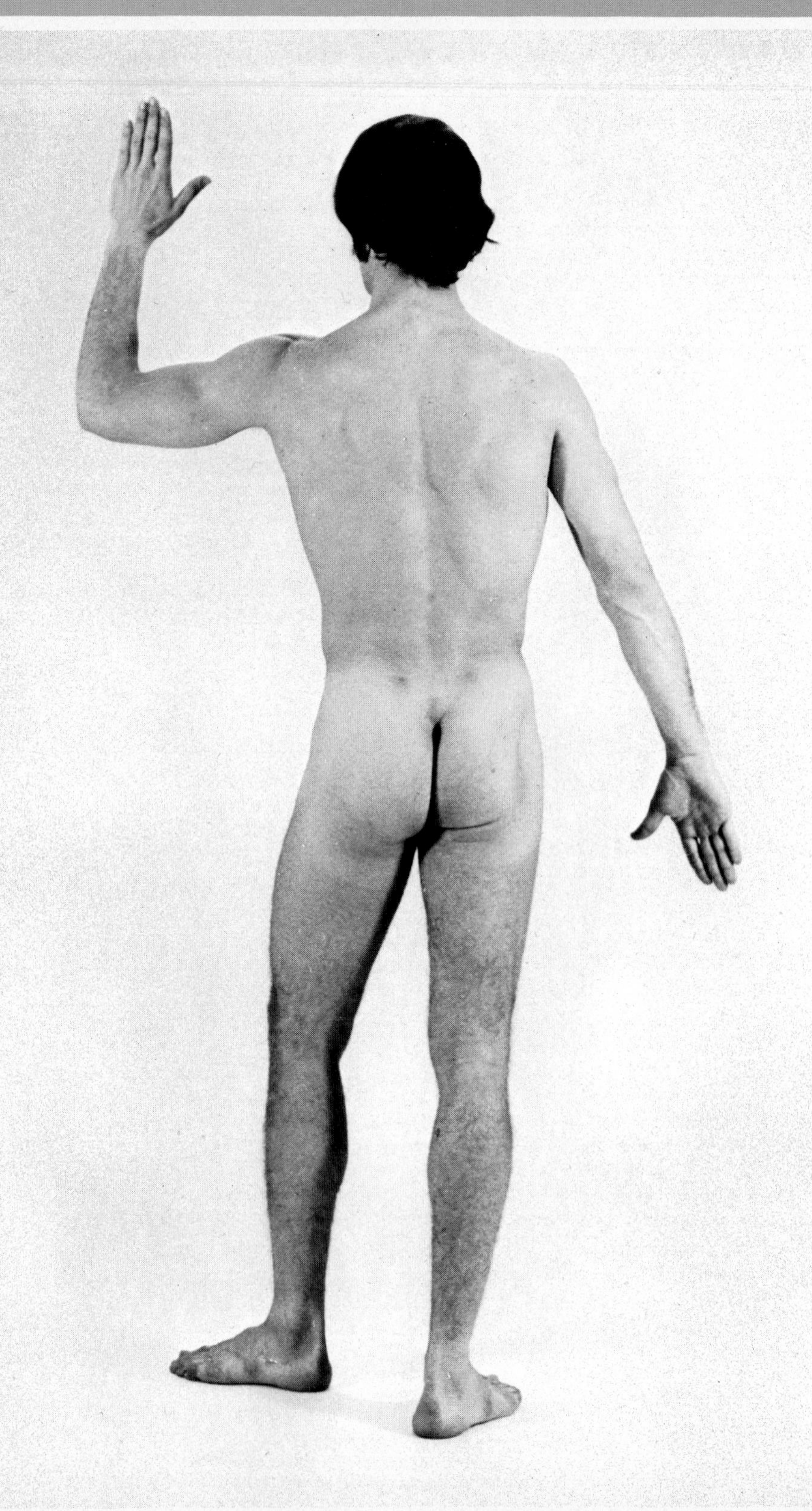

(b)

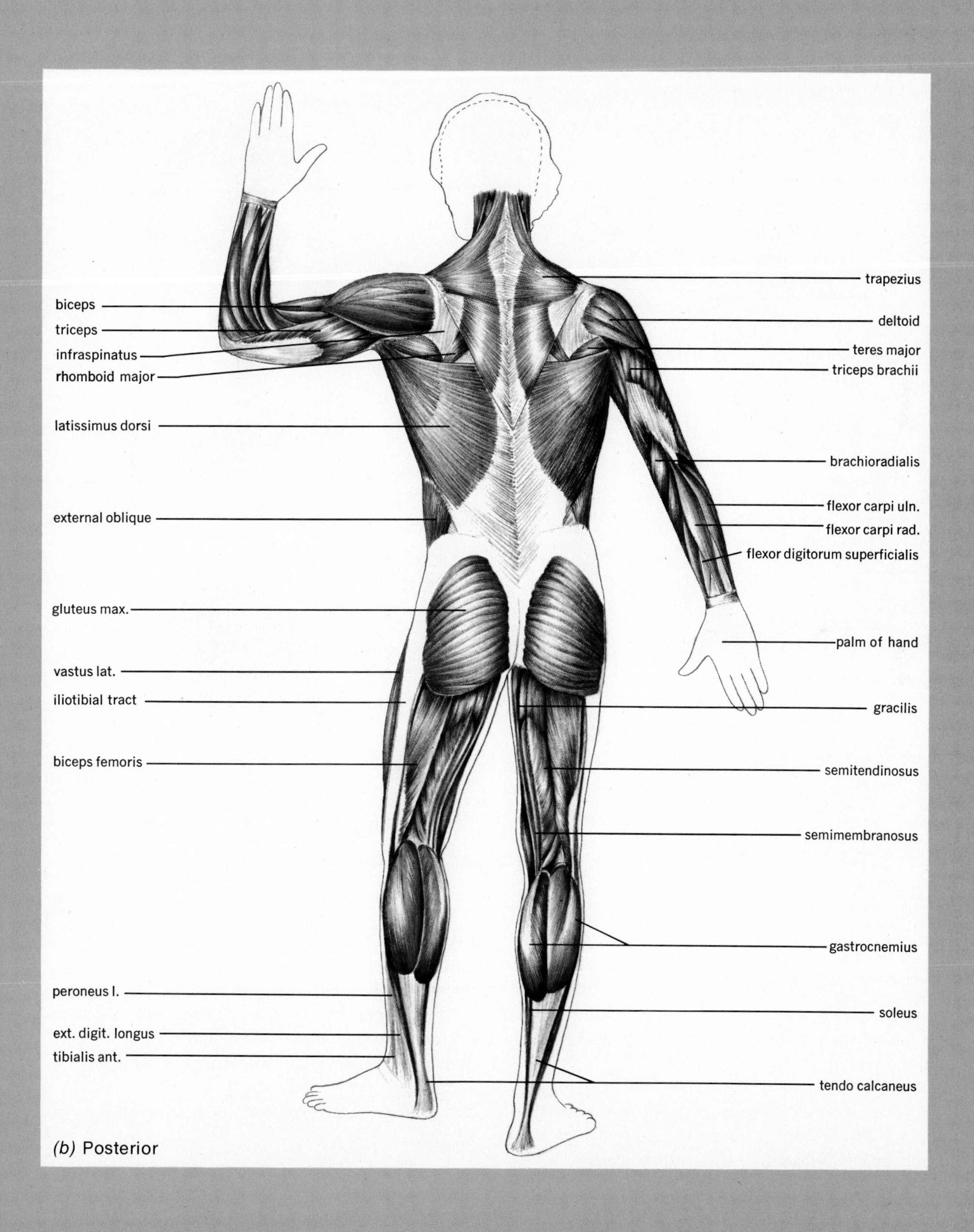
trapezius
biceps
deltoid
triceps
infraspinatus
teres major
rhomboid major
triceps brachii
latissimus dorsi
brachioradialis
flexor carpi uln.
external oblique
flexor carpi rad.
flexor digitorum superficialis
gluteus max.
palm of hand
vastus lat.
iliotibial tract
gracilis
biceps femoris
semitendinosus
semimembranosus
gastrocnemius
peroneus l.
soleus
ext. digit. longus
tibialis ant.
tendo calcaneus

(b) Posterior

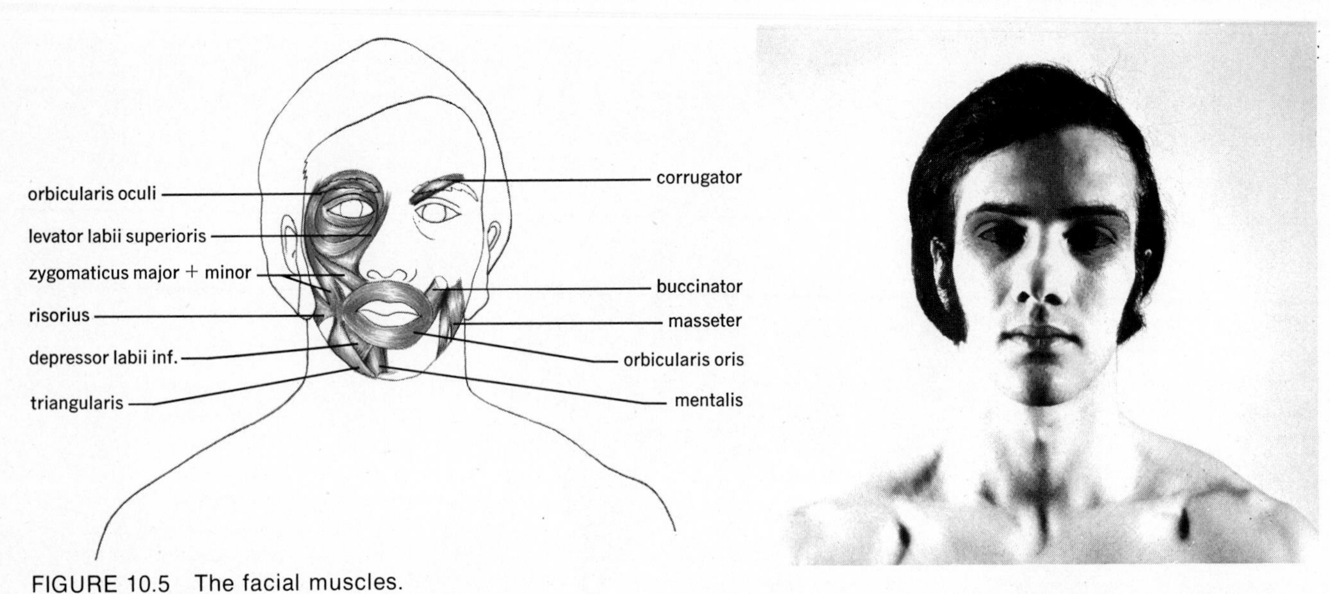

FIGURE 10.5 The facial muscles.

FIGURE 10.6 The cranial muscles.

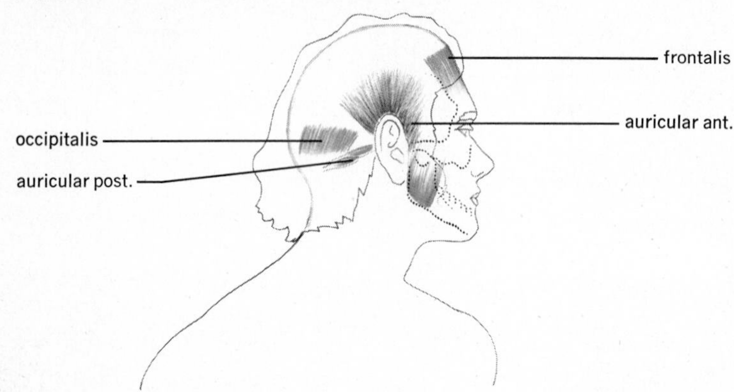

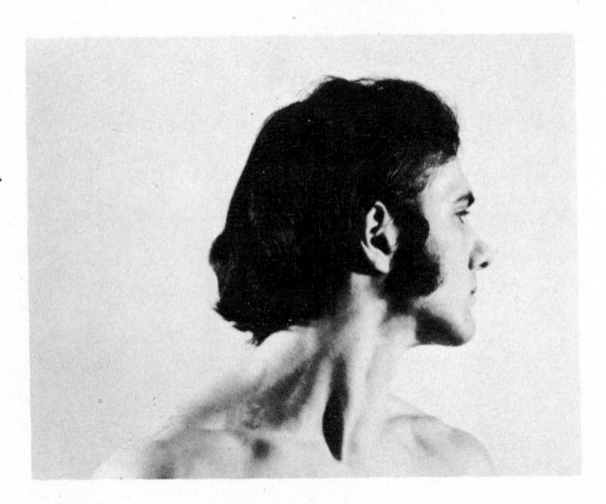

Zygomaticus draws the corners of the mouth up and back, as in smiling.

Risorius draws the corners of the mouth directly sideways as in a grimace. (Children often put a finger in each corner of their mouth and pull; this imitates the action of the risorius.)

Triangularis draws the corners of the mouth down and back to create an expression of sadness ("drooping mouth").

Depressor labii inferioris depresses the lower lip.

Mentalis elevates the lower lip and protrudes it, as in pouting.

Buccinator lies beneath the muscles listed above and compresses the cheek. It is sometimes called the "trumpeter's muscle," because it is used strongly in playing a brass musical instrument such as the trumpet.

Corrugator produces "frown lines" in the center forehead.

Cranial muscles

The cranial muscles (Fig. 10.6) lie on the forehead, back of the head, and around the ears.

The EPICRANIUS is the only cranial muscle. It is divided into an *occipitofrontal group* consisting of an anterior frontalis and posterior occipitalis, and a *temporoparietal group* consisting of the auricular muscles attaching to the pinna (flap) of the ear. The occipitalis and frontalis attach to a broad flat tendon *(galea aponeurotica)* over the top of the skull. The frontalis pulls the scalp forward and produces transverse wrinkles in the forehead ("expression of surprise"). Occipitalis draws the scalp backward.

The auricular muscles are rudimentary in the human. A superior muscle draws the pinna upward; a posterior muscle draws it backward, and an anterior muscle draws it forward.

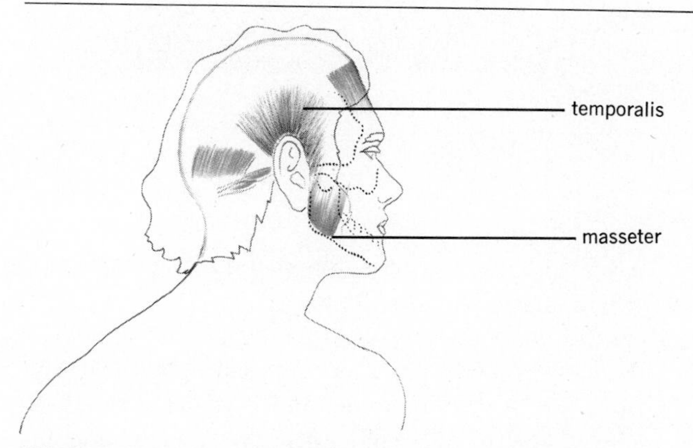

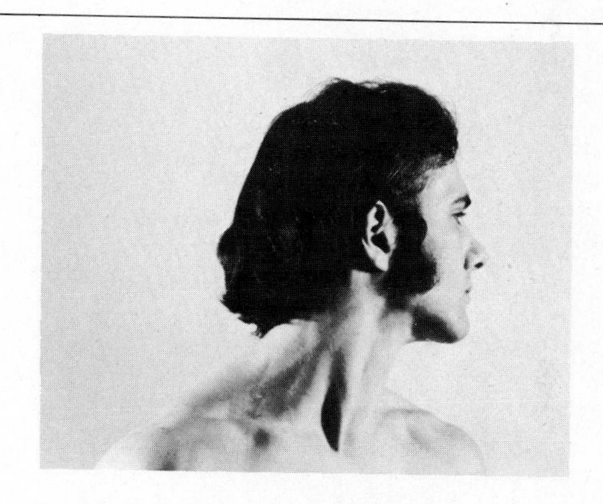

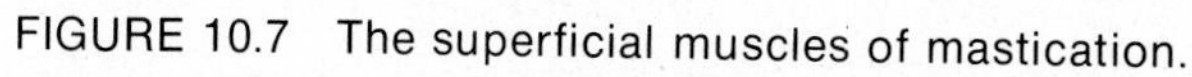
FIGURE 10.7 The superficial muscles of mastication.

FIGURE 10.8 The deep muscles of mastication.

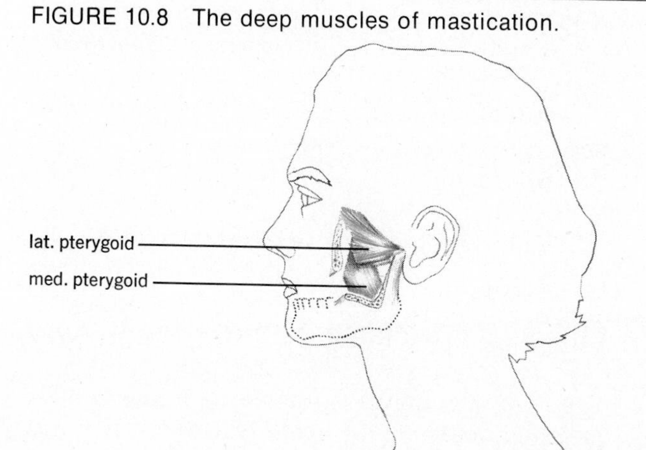

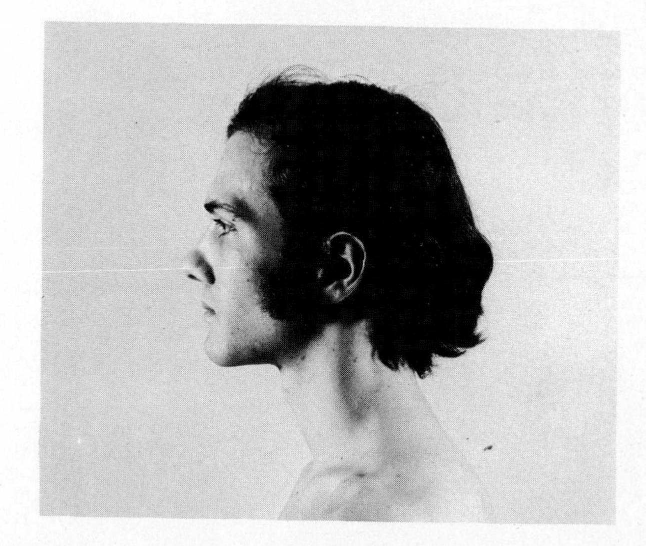

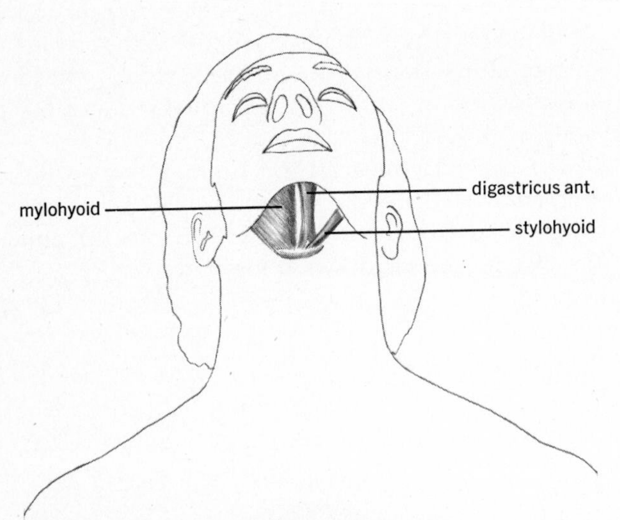

FIGURE 10.9 The suprahyoid muscles.

FIGURE 10.10 The infrahyoid muscles.

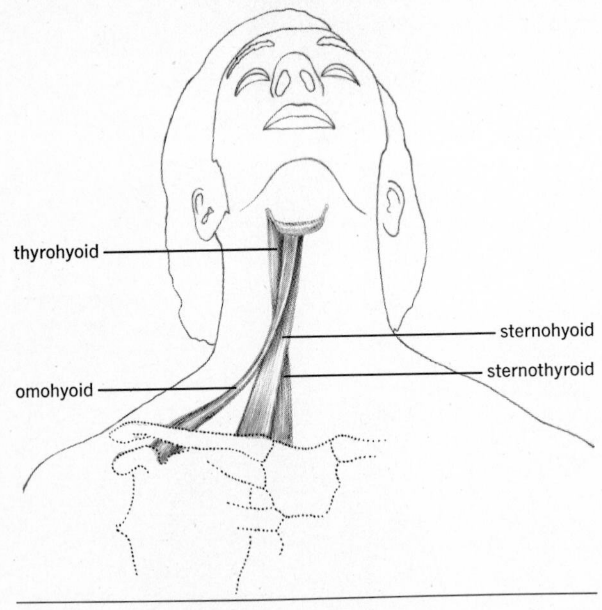

Muscles of mastication *(chewing)*

Four muscles are the most important in mastication (Figs. 10.7 and 10.8).

The *temporalis* fills a depression on the side of the skull and elevates the mandible, closing the mouth, and clenching the teeth.

Masseter, covering the ramus of the mandible, is the most powerful jaw closing muscle; it also clenches the teeth. Both temporalis and masseter can be felt contracting by clenching the jaws and feeling the skull in the appropriate area.

Medial pterygoid is internal to the mandibular ramus and elevates the jaw, clenching the teeth only weakly.

Lateral pterygoid lies between the medial pterygoid and the ramus, and depresses the mandible, opening the jaw (mouth).

The pterygoids also move the mandible laterally.

Muscles associated with the hyoid bone

These muscles may be grouped according to their position relative to the hyoid bone (Figs. 10.9 and 10.10). Most are named by their origin and insertion, and contribute to the movements that occur when we swallow.

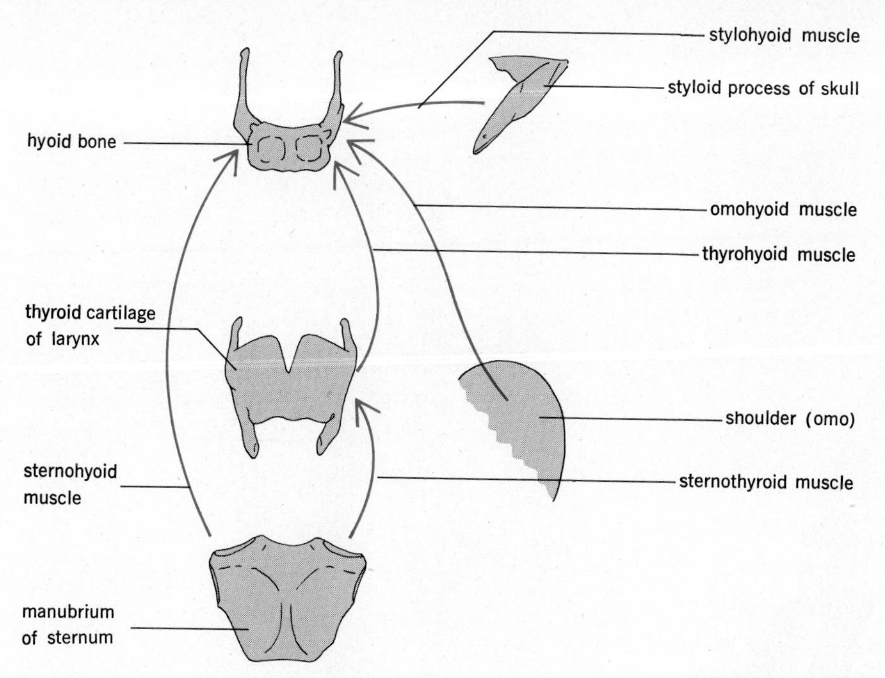

FIGURE 10.11 A diagrammatic representation of how the muscles of the hyoid area are named.

FIGURE 10.12 The anterior neck muscles.

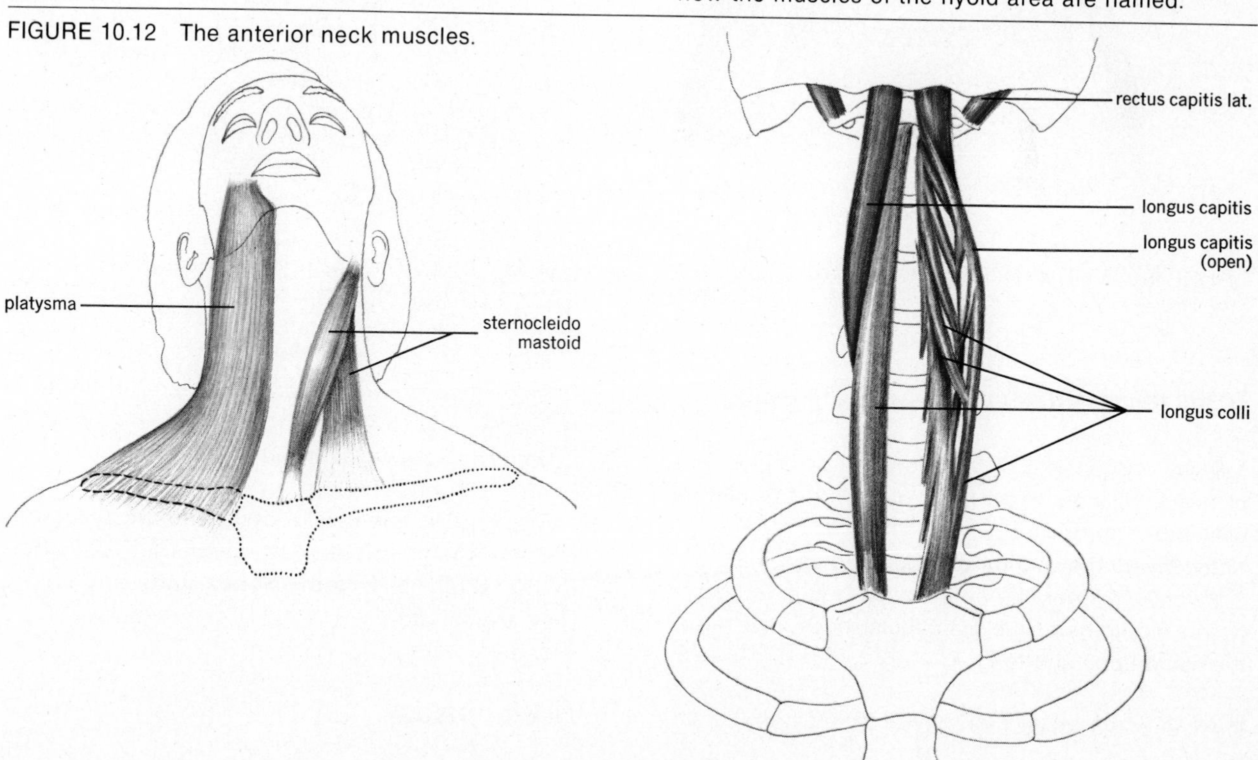

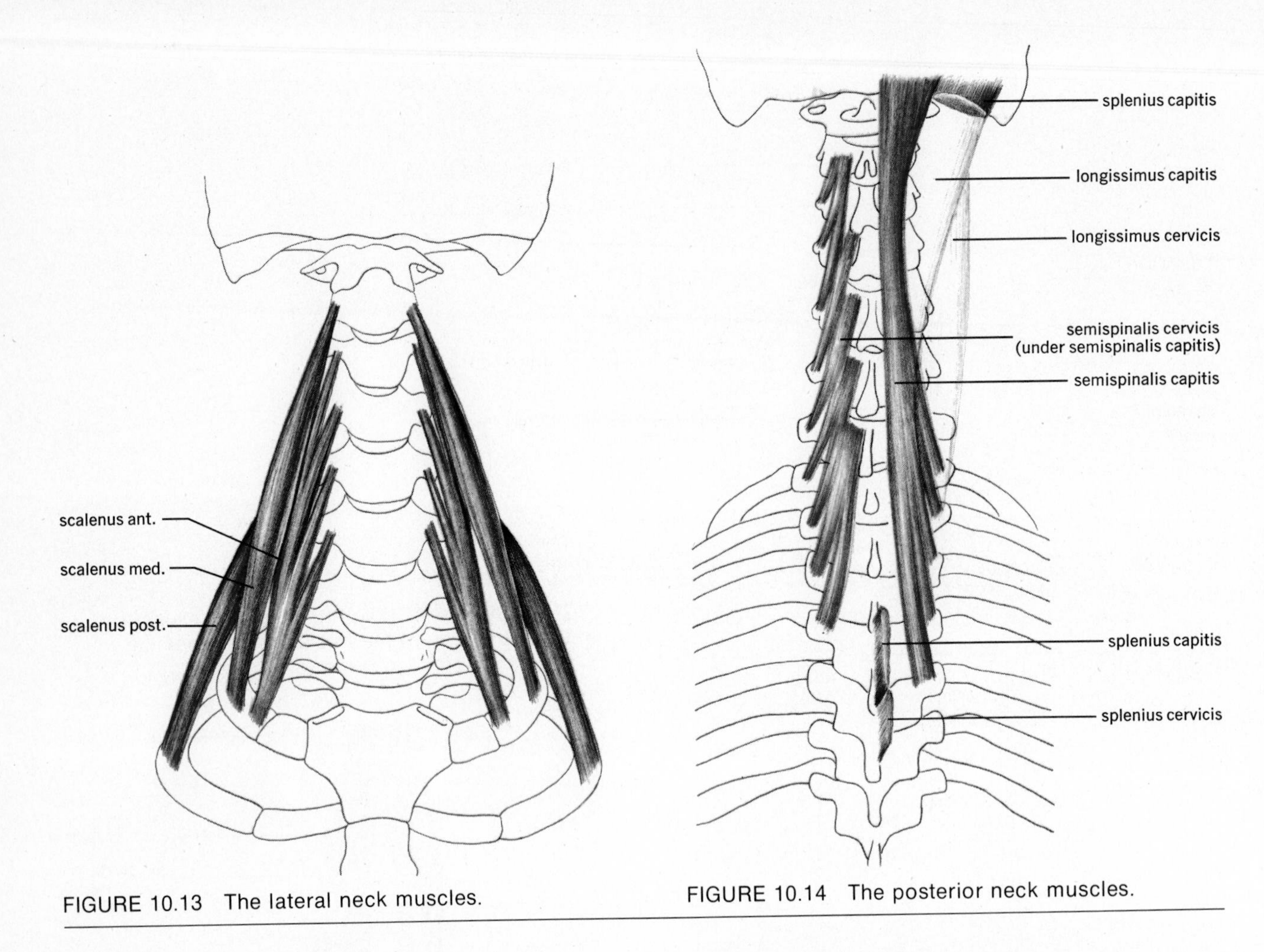

FIGURE 10.13 The lateral neck muscles.

FIGURE 10.14 The posterior neck muscles.

SUPRAHYOID MUSCLES (above hyoid bone).

Digastric consists of two portions or bellies. The posterior belly draws the hyoid backward, the anterior belly draws it forward.

Stylohyoid draws the hyoid bone up and back.

Mylohyoid forms the floor of the mouth. It lies between the limbs of the mandibular body, and raises the hyoid bone and tongue.

INFRAHYOID MUSCLES (below hyoid bone).

Sternohyoid draws the hyoid downward.

Sternothyroid draws the larynx downward.

Thyrohyoid draws the larynx upward, if the hyoid is fixed.

Omohyoid draws the hyoid downward.

Figure 10.11 presents a diagrammatic representation of the positions of these muscles in relation to the skull, neck, and thorax, and shows how they are named.

Neck muscles

The muscles of the neck (Figs. 10.12, 10.13, and 10.14) consist of two superficial muscles on the anterior neck and a number of deep muscles that attach to the vertebrae and skull.

SUPERFICIAL MUSCLES. The PLATYSMA is a broad sheet covering the inner side of the shoulder and lateral neck and mandible. It draws the corners of the mouth down and sideward, as in screaming, or in an expression of horror.

The STERNOCLEIDOMASTOID muscles flex the neck if the muscles are operating together and, singly, rotate the head to the opposite side while pointing the chin upward. The primary action of these muscles is on the head, and only secondarily on the neck.

DEEP MUSCLES. These may be divided into anterior, lateral, and posterior groups.

The anterior group includes LONGUS COLLI, LONGUS CAPITUS, and RECTUS CAPITUS, all of which flex the neck on the chest.

The lateral group includes the three SCALENES, which bend the neck to the side and help to elevate the ribs if the neck is fixed.

The posterior group includes the SEMISPINALIS, LONGISSIMUS, and SPLENIUS, which extend the neck and raise the chin.

Muscles operating the vertebral column

Movements possible

The column may be flexed or bent forward, extended or bent backward, abducted or bent to the side, adducted or returned to the midline, and rotated or twisted (Fig. 10.15). Some of the muscles responsible for these actions have other functions, and their names may appear in other sections.

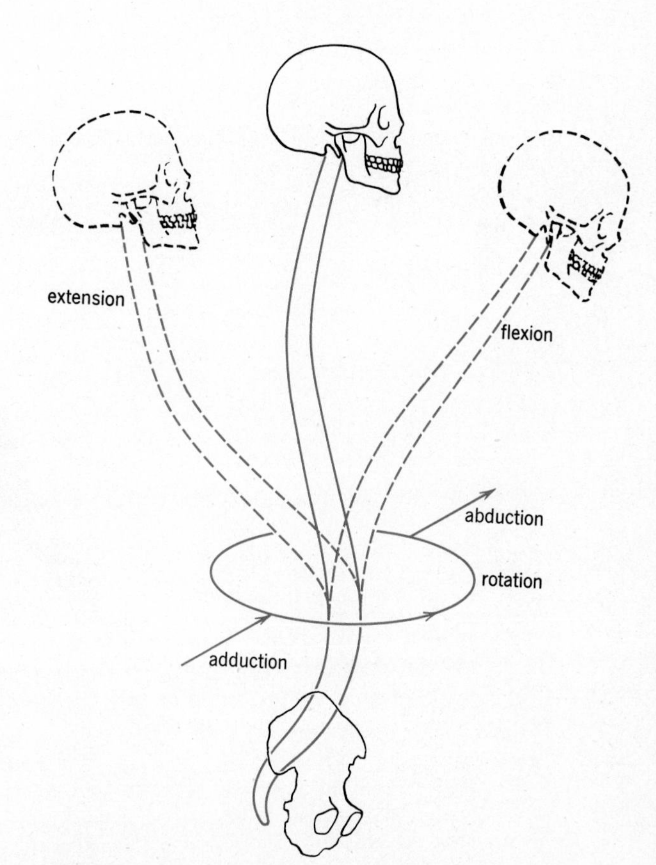

FIGURE 10.15 Movements of the spinal column.

Flexors of the spine

Rectus abdominus is the strongest flexor of the spine (Fig. 10.16). It runs down the midline of the belly. One uses it strongly in a sit-up.

Psoas and *iliacus* flex the spine if the thigh is fixed.

Extensors of the spine

Three muscles of the back, known collectively as the ERECTOR SPINAE or sacrospinalis, are the most important extensors of the spine (Fig. 10.17), and form the major postural muscles holding our backs straight against gravity when we sit or stand.

Iliocostalis, *longissimus*, and *spinalis* are the three muscle groups, listed laterally to medially.

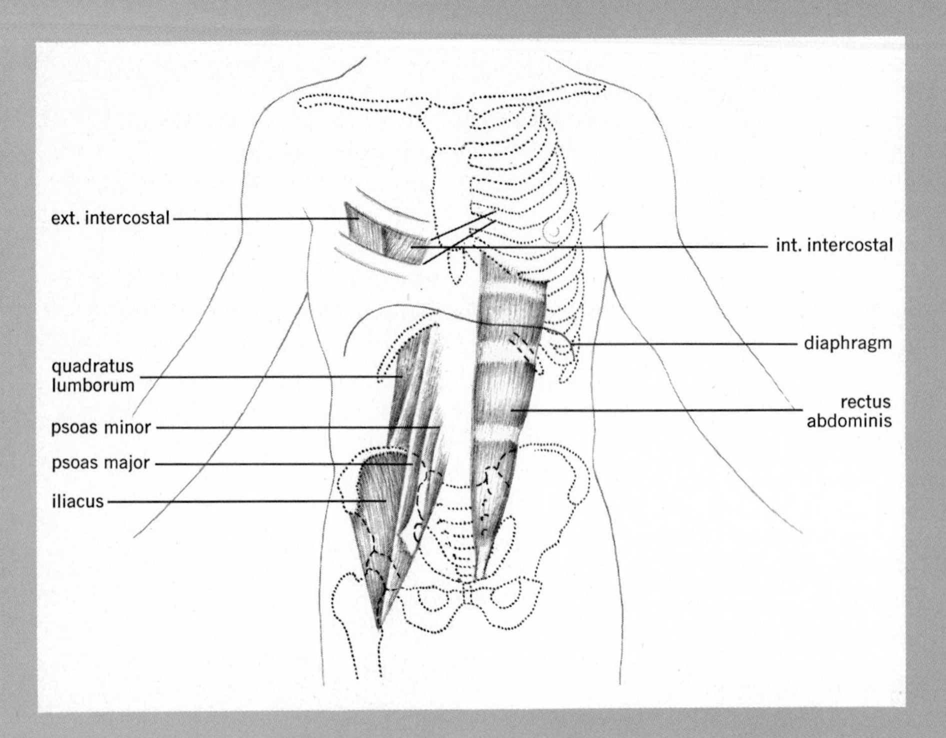

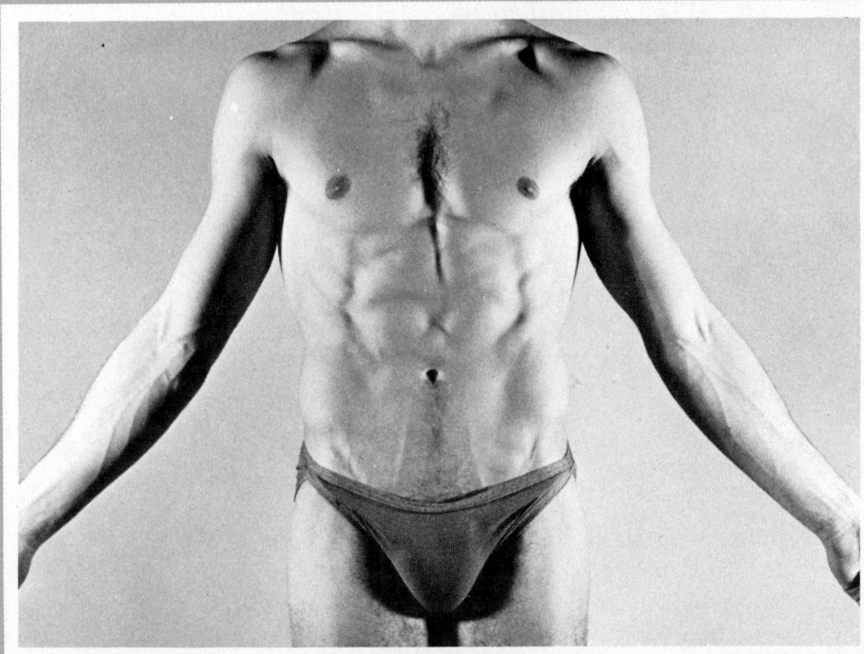

FIGURE 10.16 Flexors of the spinal column and muscles of respiration.

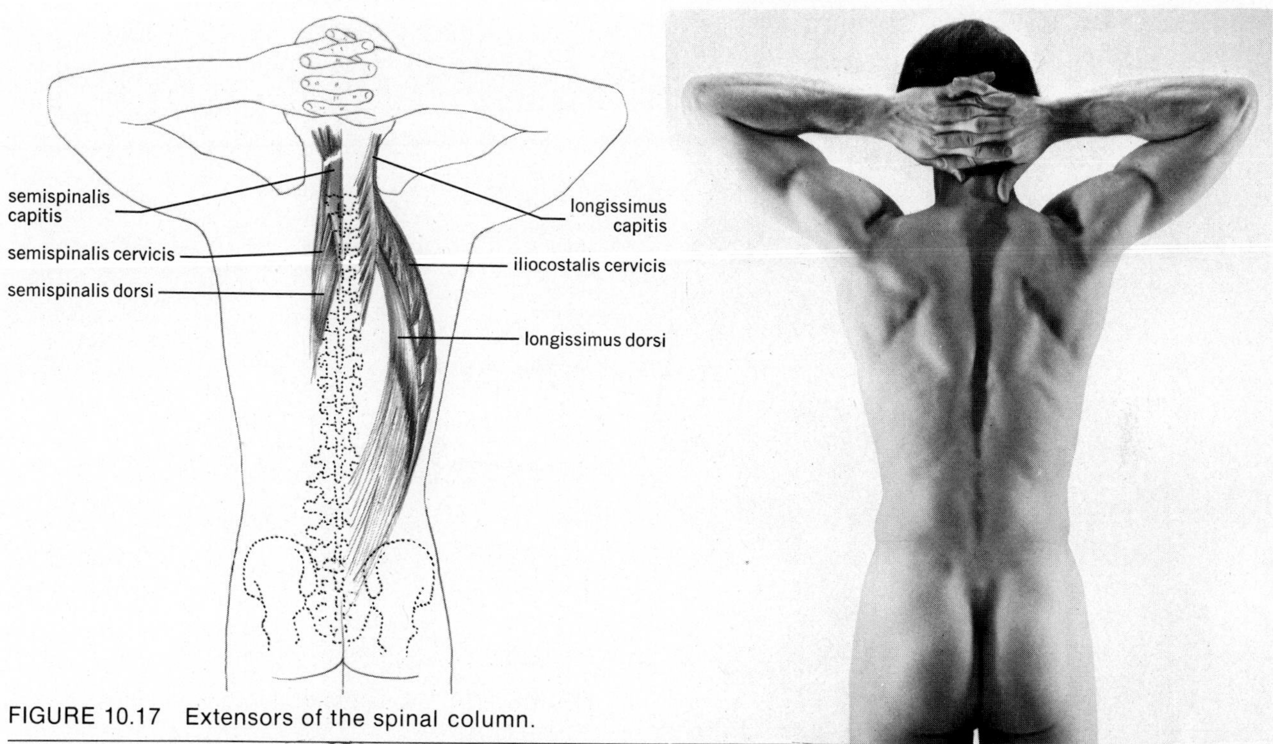

FIGURE 10.17 Extensors of the spinal column.

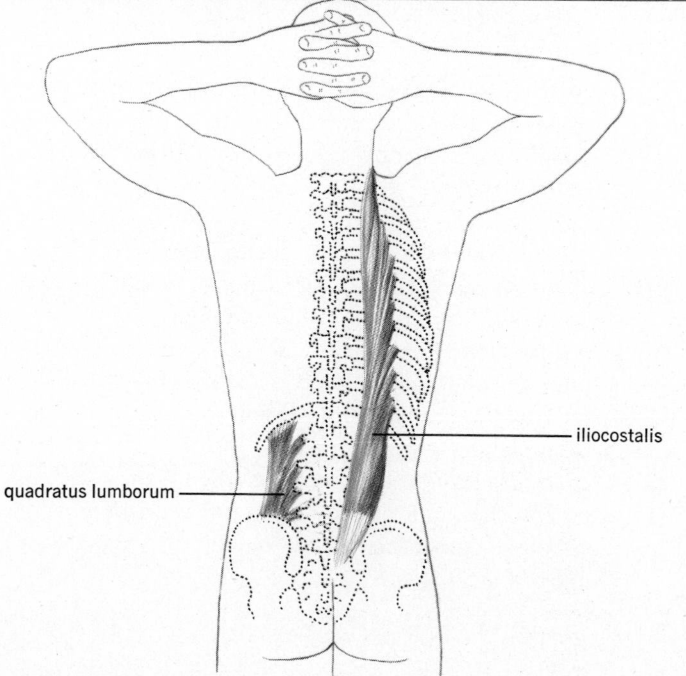

FIGURE 10.18 Abductors and adductors of the spinal column.

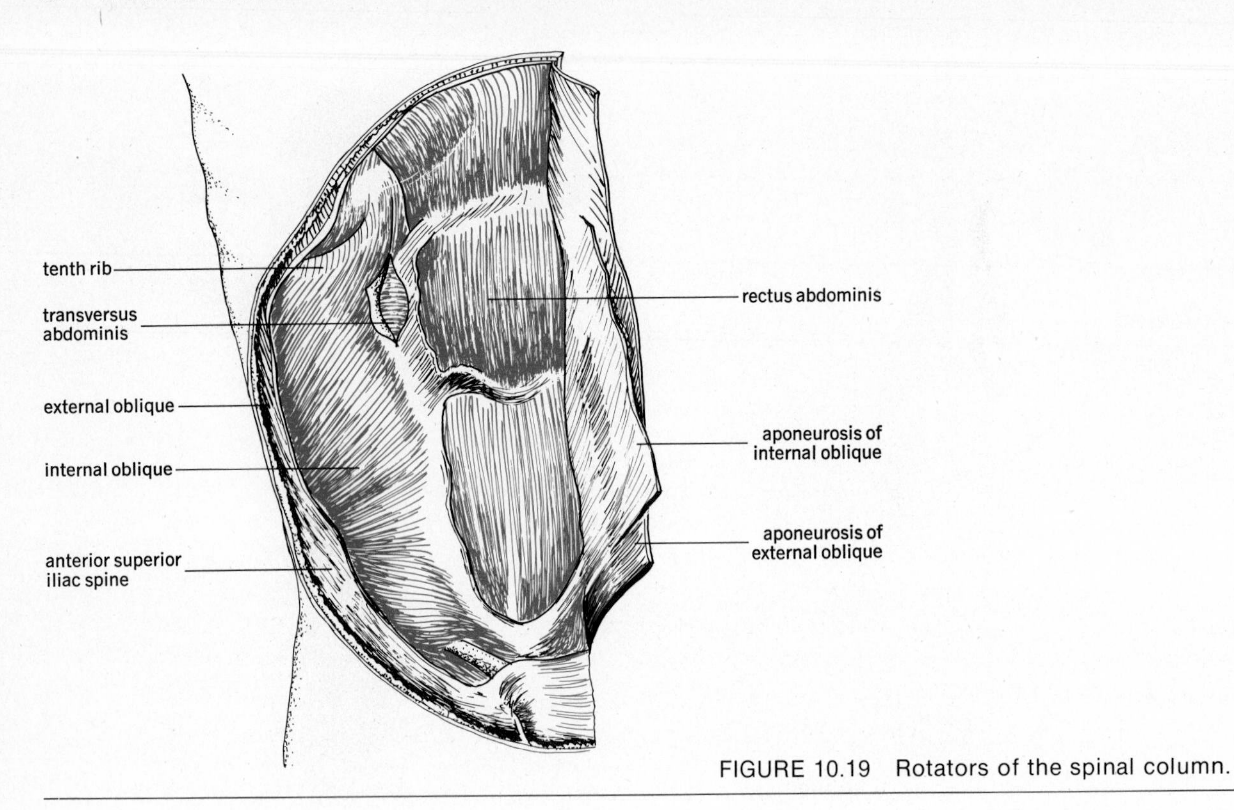

FIGURE 10.19 Rotators of the spinal column.

Abductors and adductors of the spine

ILIOCOSTALIS is lateral enough in placement to bend or to abduct the column (Fig. 10.18). The same muscle on the opposite side of the vertebral column adducts the column. For example, the left iliocostalis abducts the column to the left; the right iliocostalis adducts the column to the midline.

QUADRATUS LUMBORUM pulls the lumbar vertebrae and bends the column to the same side as the muscle contracting. The same muscle on the opposite side returns the column to the midline position.

Rotators of the spine

Three abdominal muscles may help to rotate the spine (Fig. 10.19) and also are important in compressing the abdominal contents. Weakness of these muscles allows protrusion of the abdomen (a "pot belly").

Each *external oblique,* acting alone, rotates the spine so as to bring the same shoulder forward. Together, they compress the abdomen.

Each *internal oblique,* acting alone, brings the opposite shoulder forward. Together, they compress the abdomen. (Figure out which muscles, and which sides, would be used to twist the trunk to the left.)

The *transverse abdominal* is mainly an abdominal compressor, utilized strongly in urination, defecation, vomiting, assisting during childbirth, and forced expiration.

Muscles of the thorax

The muscles of the thorax are those associated with the ribs and are concerned with breathing.

Movements possible

The ribs may be elevated, which expands the thorax in the front-to-back and side-to-side directions, and depressed, which decreases these dimensions (Fig. 10.20). The vertical dimension of the thorax may be increased.

Elevators of the ribs

The EXTERNAL INTERCOSTALS lie superficially between the ribs (Fig. 10.21). They elevate the ribs and create what is called *costal breathing.* Costal breathing provides about two-thirds of the air exchanged during normal breathing.

SERRATUS POSTERIOR INFERIOR also elevates the ribs.

With forced inspiration, the scalenes (see above) and the pectoralis minor (see below) aid in rib elevation.

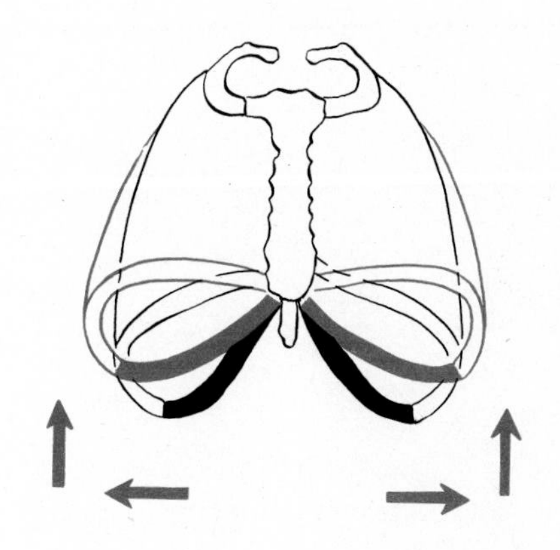

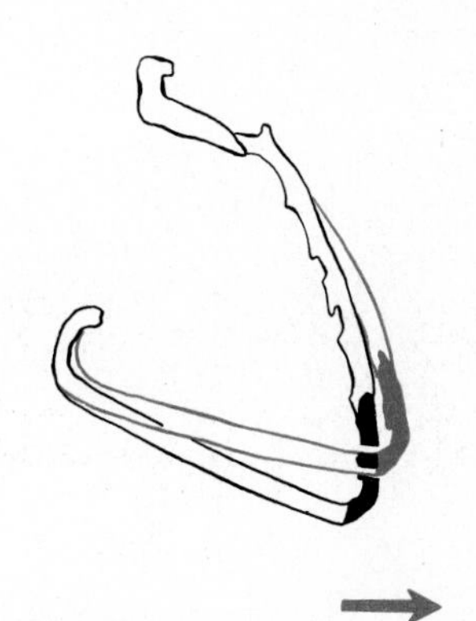

FIGURE 10.20 Movements of the thorax.

FIGURE 10.21 Elevators of the ribs.

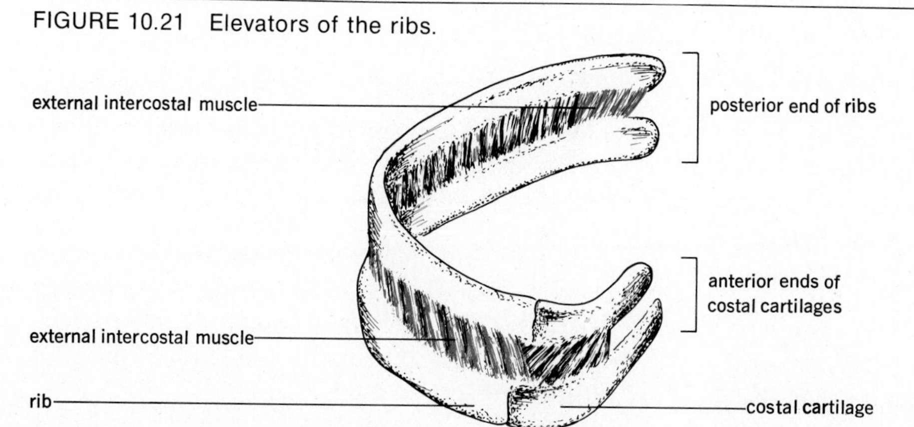

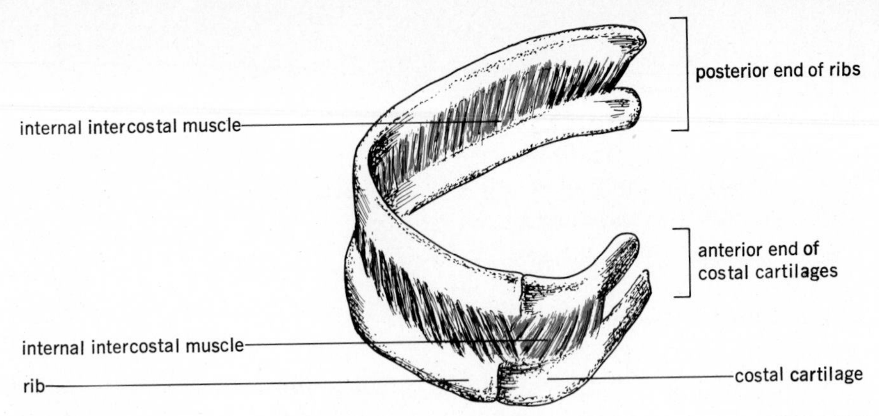

FIGURE 10.22
Depressor of the ribs.

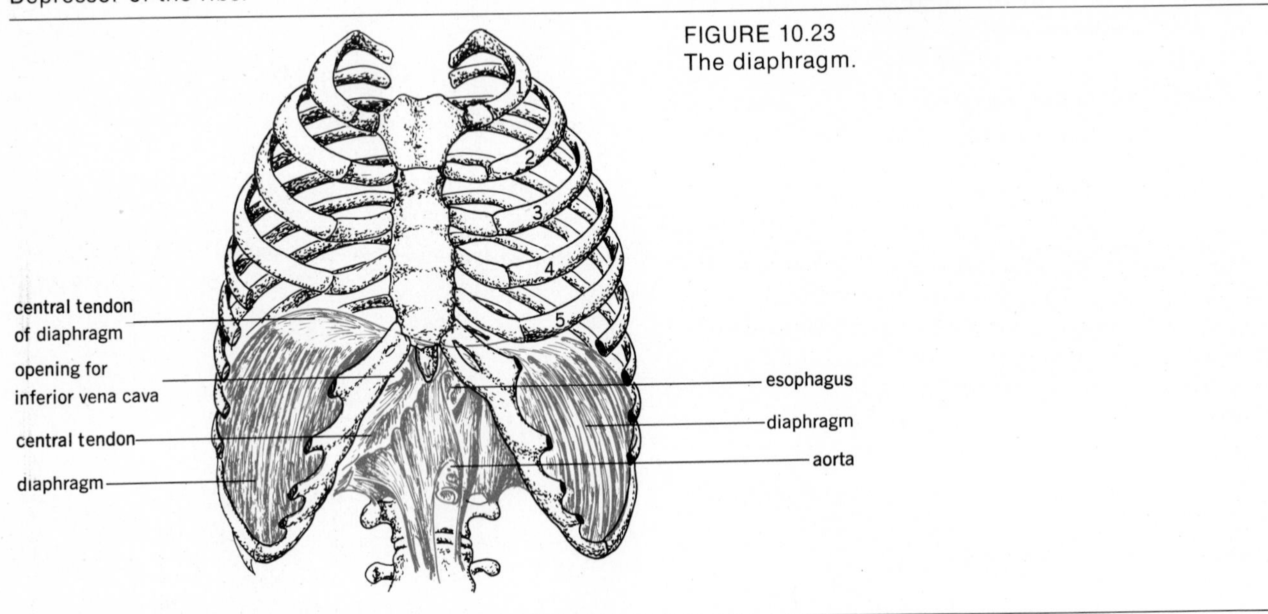

FIGURE 10.23
The diaphragm.

Depressors of the ribs

The INTERNAL INTERCOSTALS lie deep to the externals, and depress the ribs (Fig. 10.22).

Increasing the vertical dimension of the thorax

This task is handled by the DIAPHRAGM (Fig. 10.23), a dome-shaped muscle lying between the thoracic and abdominal cavities. Its fibers insert into a central tendon that is drawn downward when the muscle contracts. Its action accounts for about one-third of the air exchanged during normal breathing. It becomes the most important breathing muscle when increased depth of breathing is required.

The abdominal muscles (rectus, obliques, transverse) are brought into play during forced expiration. Their contraction presses the abdominal viscera against the diaphragm and hastens its return to its original position.

The muscles forming the floor of the pelvic cavity are shown in Fig. 30.23.

Muscles of the shoulder (pectoral) girdle

Movements possible

The scapulae may be elevated, depressed, abducted, adducted, and rotated (Fig. 10.24). The clavicles may be elevated and depressed. The muscles involved may be divided into an anterior group (Fig. 10.25) and a posterior group (Fig. 10.26).

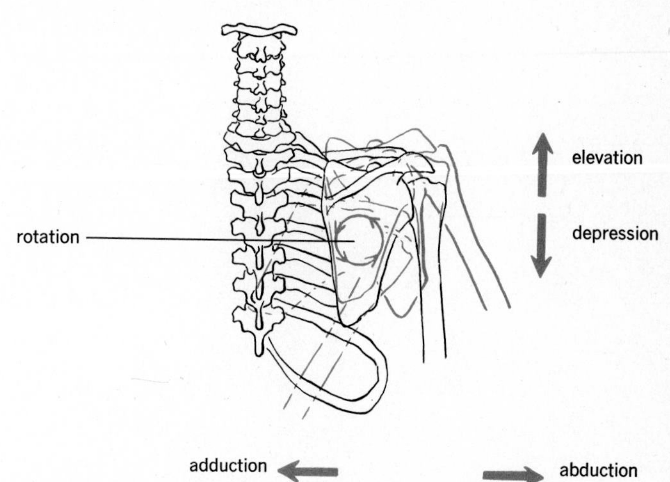

FIGURE 10.24 Movements of the shoulder girdle.

Anterior group

Pectoralis minor draws the scapula forward and down, and if the scapula is fixed, helps to elevate the ribs for breathing.

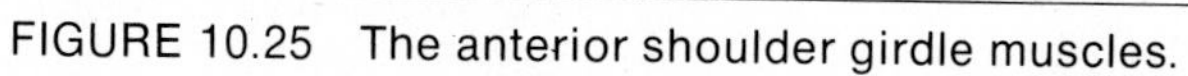

FIGURE 10.25 The anterior shoulder girdle muscles.

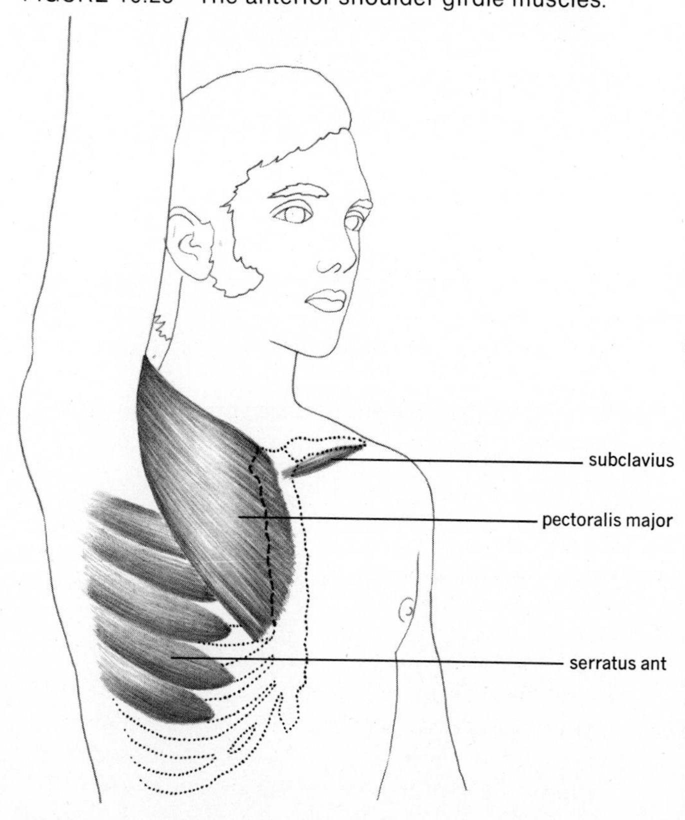

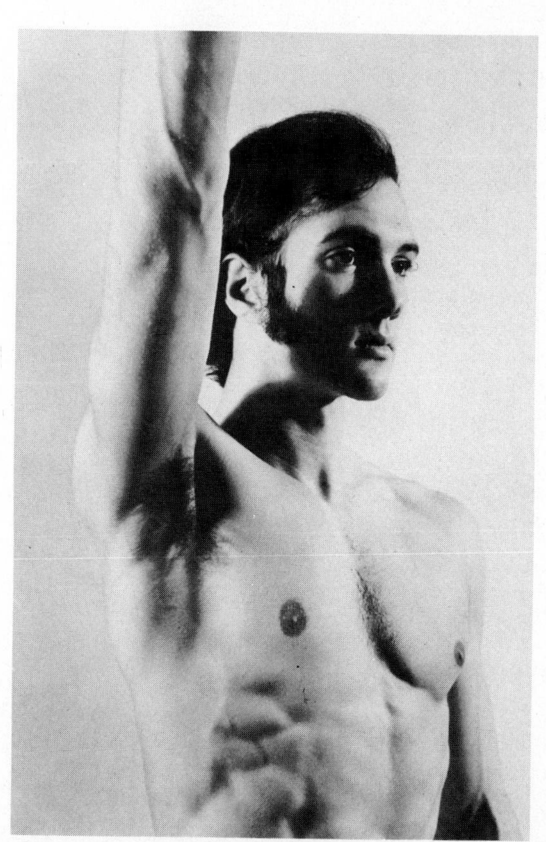

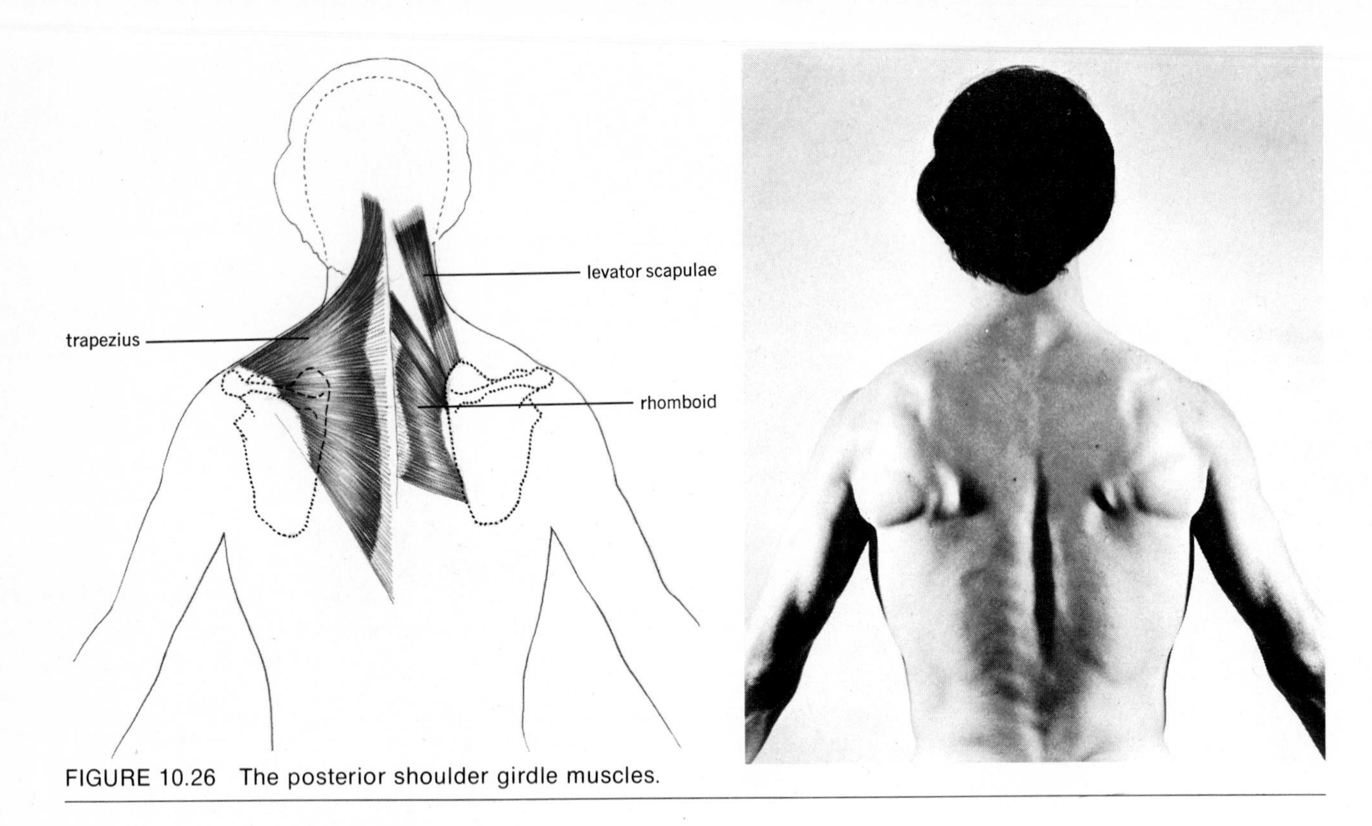

FIGURE 10.26 The posterior shoulder girdle muscles.

Serratus anterior abducts the scapula and acts strongly in pushing movements.

Subclavius is a synergistic muscle that depresses and holds the clavicle firmly.

Posterior group

Trapezius has three parts: the upper portion elevates the scapula; the middle part adducts the scapula; the lower part depresses the scapula.

Rhomboid is a strong adductor of the scapula and also elevates it. It is made up of two parts, a small superior *rhomboid minor*, and a larger inferior *rhomboid major.*

Levator scapulae elevates the scapula, and "shrugs" the shoulder (along with the upper trapezius).

Muscles moving the shoulder joint

Movements possible

As a ball-and-socket joint secured mainly by muscles, the shoulder joint is the most freely movable joint in the body (Fig. 10.27). The humerus thus shows a wide range of motion. It may be flexed, extended, abducted, adducted, and rotated medially or laterally. Circumduction is a circular movement of the entire upper limb, composed of the preceding motions occurring in sequence.

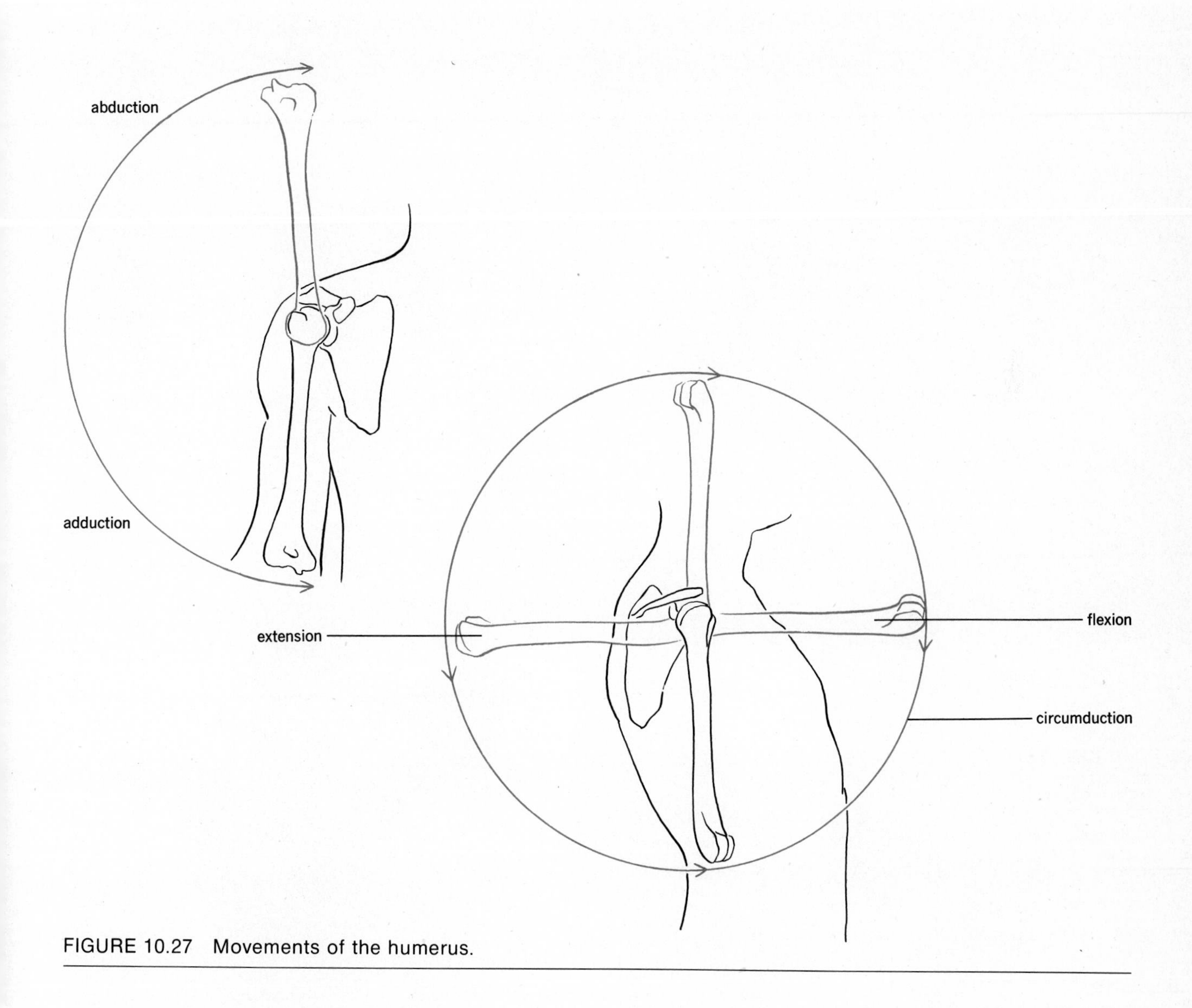

FIGURE 10.27 Movements of the humerus.

Flexors of the humerus

Pectoralis major (the "pects" or pectorals) flexes and medially rotates the humerus, as in throwing or delivering a jab.

Coracobrachialis is a weak flexor (Fig. 10.28).

Extensors of the humerus

Latissimus dorsi (the "lats") is a powerful extensor of the arm, and with the pectoralis major, adducts the humerus.

Teres major acts with the latissimus (Fig. 10.29).

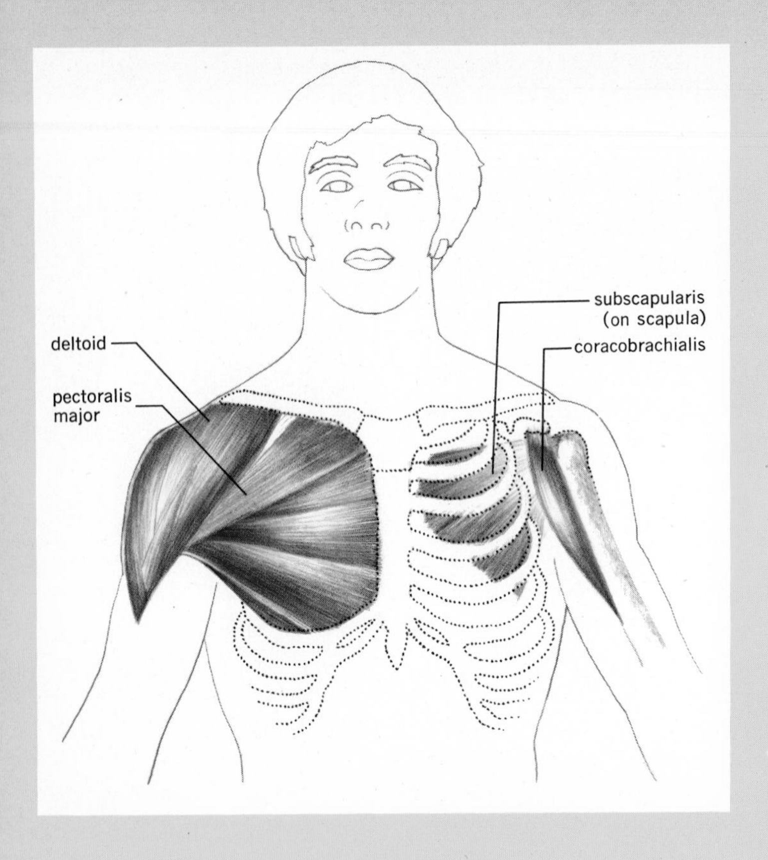

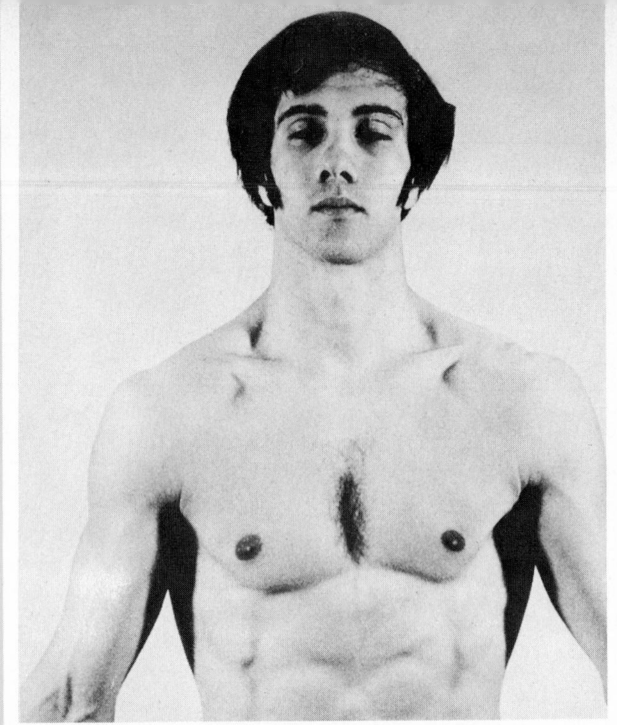

FIGURE 10.28 Flexors of the humerus.

FIGURE 10.29 Extensors of the humerus.

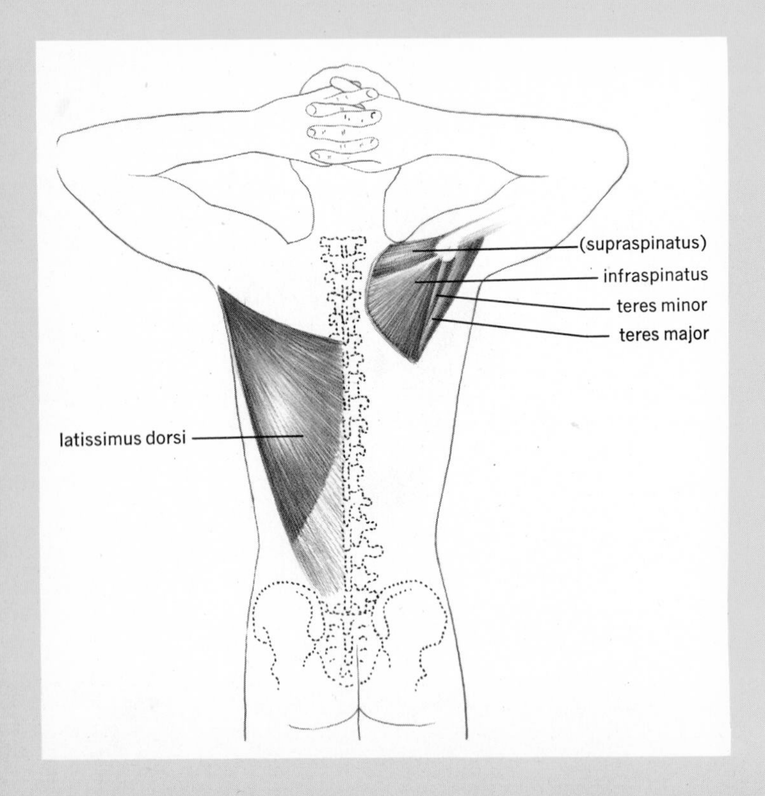

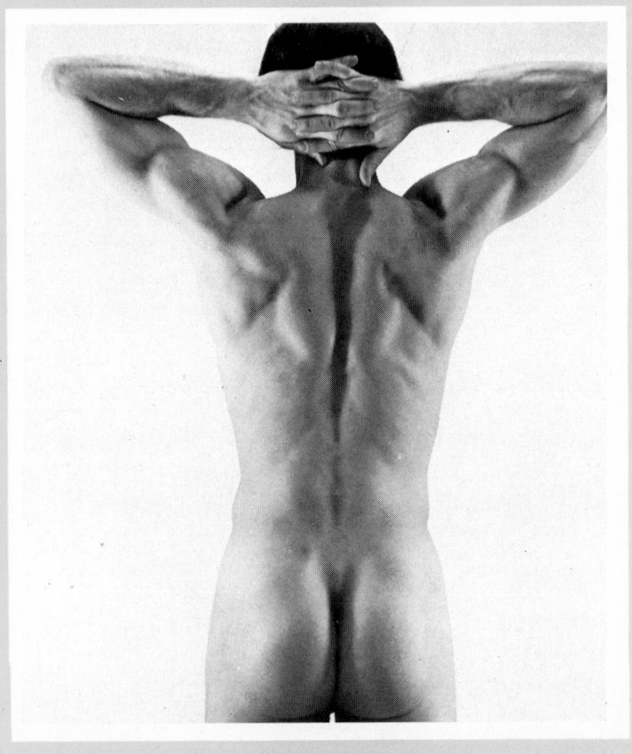

Abductors of the humerus (see Figs. 10.28 and 10.29)

The deltoid, acting as a whole, abducts the shoulder as in holding the arms straight out to the side.

Supraspinatus acts with the deltoid.

Rotators of the humerus (see Figs. 10.28 and 10.29)

Subscapularis medially rotates the humerus.

Infraspinatus and *teres minor* laterally rotate the humerus.

Muscles moving the forearm

Movements possible

The term *elbow joint* refers to three joints between the humerus, ulna, and radius that are enclosed in a common capsule. The three joints are specifically named humeroulnar, humeroradial, and radioulnar. The humeroulnar joint may only be flexed and extended. The humeroradial joint occurs between the humerus and head of the radius and may be rotated medially (pronation) or laterally (supination). The radioulnar joint occurs between the proximal ends of radius and ulna and allows the rotation mentioned above (Fig. 10.30).

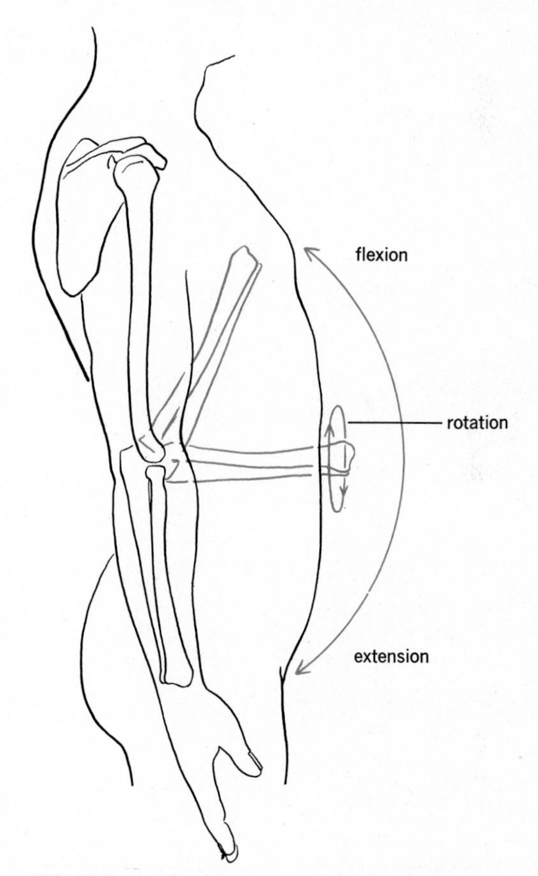

FIGURE 10.30 Movements of the forearm.

Flexors of the forearm

Biceps brachii is a two-headed muscle strongly flexing and laterally rotating the forearm (Fig. 10.31). It is used strongly in a "pull up" (chinning), and is often displayed by small boys to impress others with their "muscle."

Brachialis is a pure flexor, and lies deep to the biceps.

Brachioradialis lies in line with the thumb on the lateral forearm. It flexes the forearm and returns the forearm to the midpoint position from either full pronation or full supination.

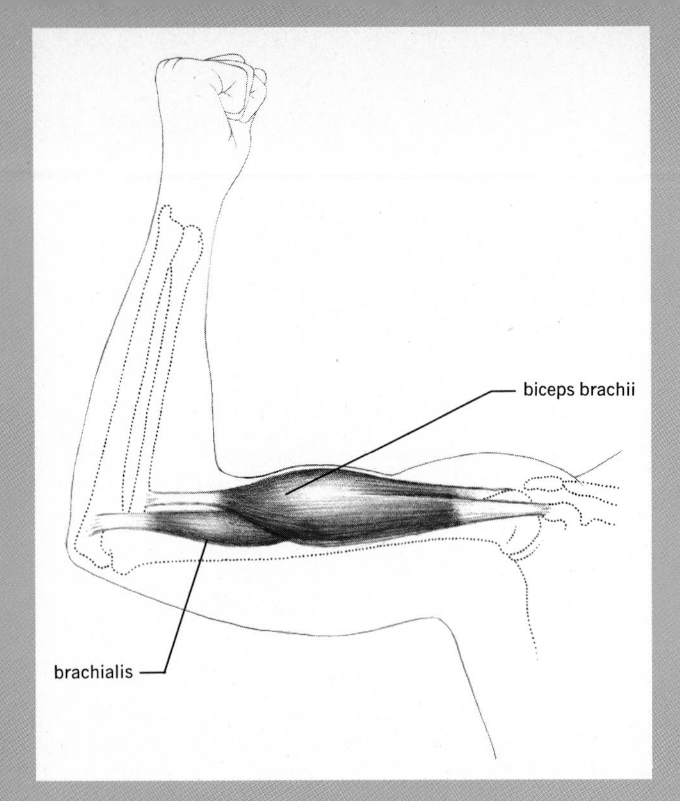

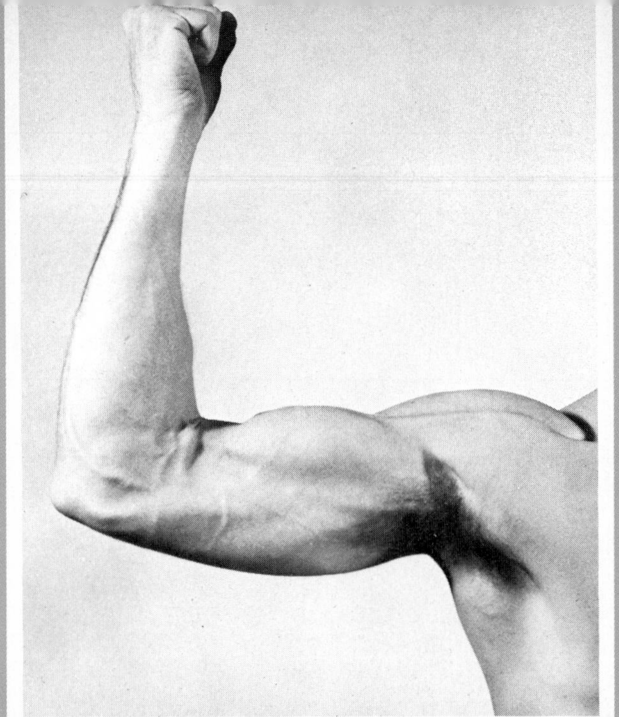

FIGURE 10.31 Flexors of the forearm.

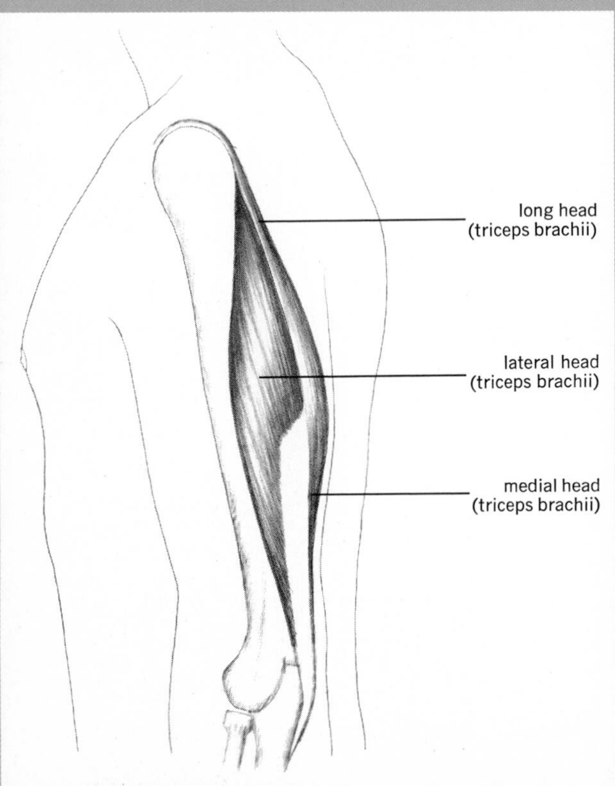

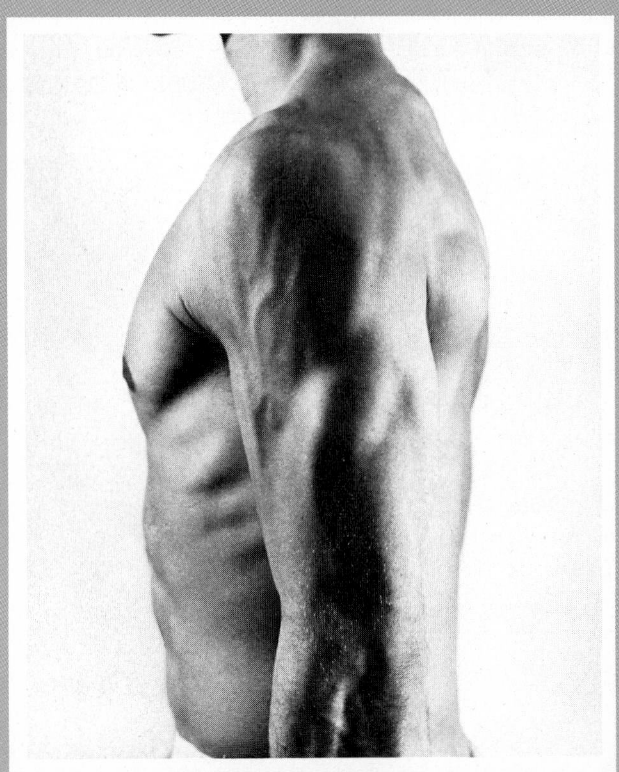

FIGURE 10.32 The extensor of the forearm.

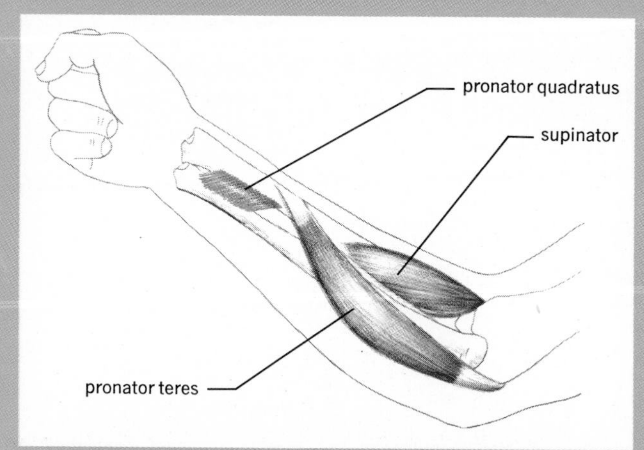

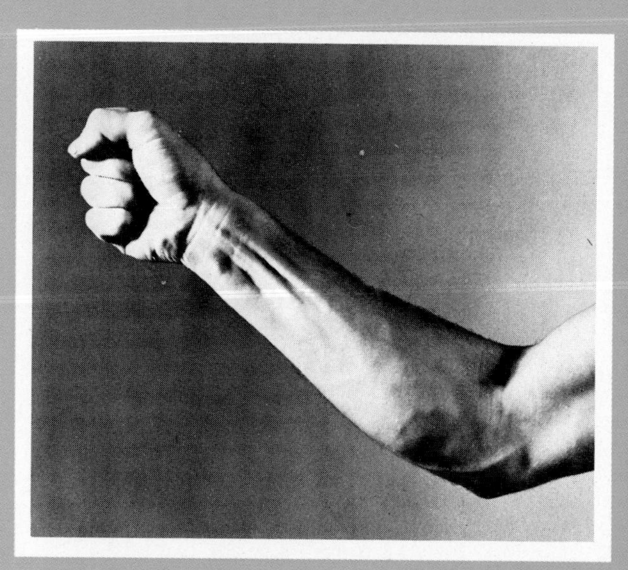

FIGURE 10.33 The rotators of the forearm.

Extensor of the forearm

The TRICEPS BRACHII is the only extensor of the forearm (Fig. 10.32). It has three heads, designated *lateral, long,* and *medial,* from lateral to medial on the posterior upper arm. All have a common tendon of insertion on the point of the elbow (olecranon process).

Rotators of the forearm

In addition to the brachioradialis and biceps, the supinator supinates (laterally rotates) the forearm (Fig. 10.33).

The PRONATOR TERES and PRONATOR QUADRATUS pronate (medially rotate) the forearm.

Muscles moving the wrist and fingers

Movements possible

The wrist and fingers may be flexed, extended, abducted, and adducted (Fig. 10.34).

Flexors of the wrist

Wrist flexors (Fig. 10.35) form a superficial group of muscles on the anterior distal humerus and the anterior forearm. Their tendons cross the anterior aspect of the wrist to attach to the carpal bones. From lateral to medial, there are three muscles.

Flexor carpi radialis flexes the wrist and slightly abducts it.

Palmaris longus flexes the wrist as its only action.

Flexor carpi ulnaris flexes the wrist and slightly adducts it.

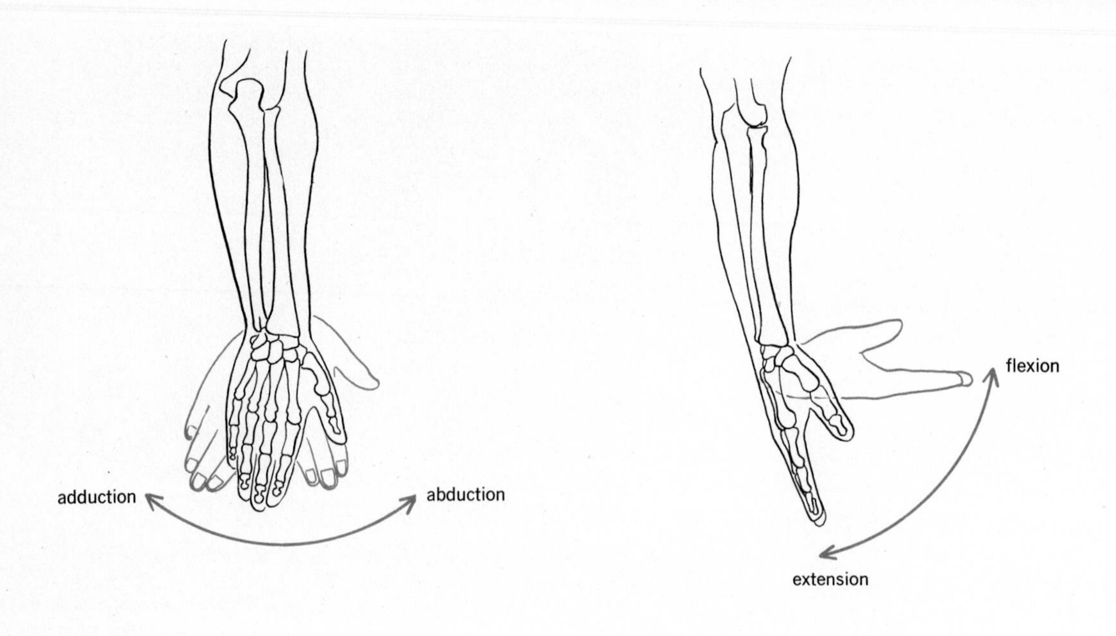

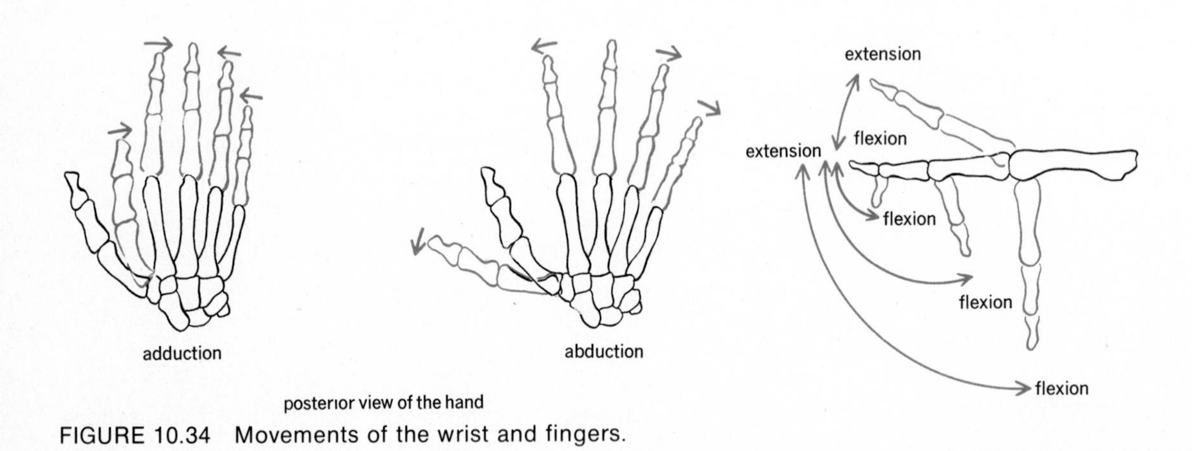

FIGURE 10.34 Movements of the wrist and fingers.

Flexors of the fingers (see Fig. 10.35)

Beneath the three muscles named above are two layers of muscles that attach to the phalanges of the fingers and flex them (Fig. 10.35). Each muscle has four parts to it, each of which sends a tendon to each finger. Although named as one muscle, the parts operate as single muscles, giving humans the ability to move one finger separately from the other fingers.

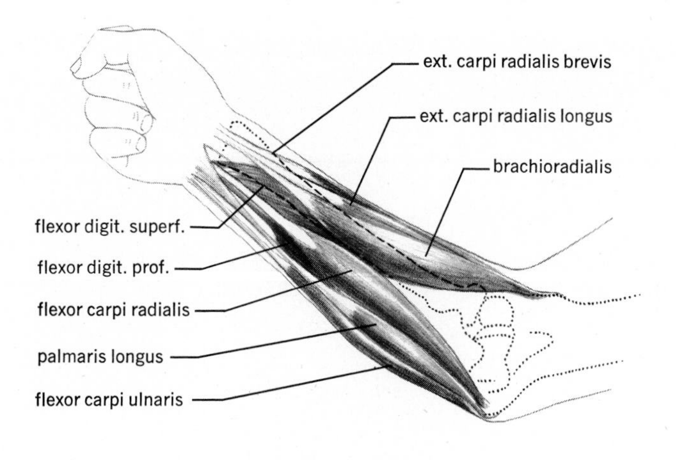

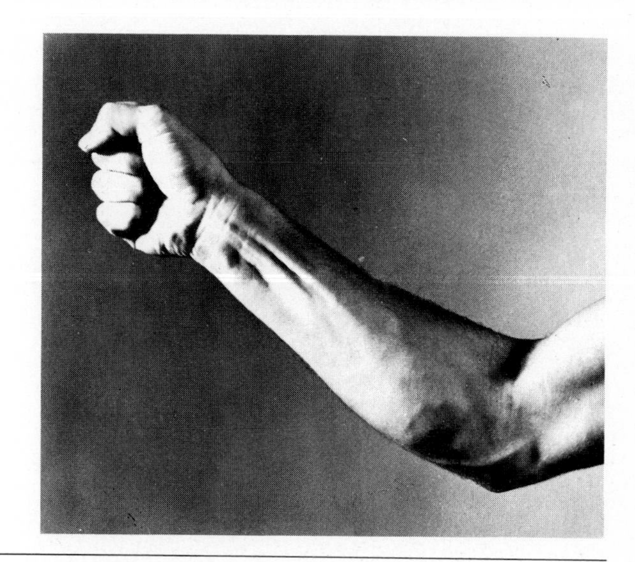

FIGURE 10.35 Flexors of the wrist and fingers.

Flexor digitorum superficialis is the first layer, and its tendons reach the second phalanx of each finger.

Flexor digitorum profundus is the second layer, and its tendons pass to the distal phalanges.

Both muscles are required to make a fist.

Extensors of the wrist and fingers

Five muscles in a single layer across the posterior aspect of the forearm are the extensors of wrist and fingers (Fig. 10.36). From lateral to medial, they are as follows:

Extensor carpi radialis longus extends and slightly abducts the wrist.

Extensor carpi radialis brevis acts as does the longus.

Extensor digitorum communis is a three-headed muscle sending tendons to all phalanges of the index, middle, and ring fingers, to extend them separately or together.

Extensor digiti minimi is a separate muscle sending tendons to all phalanges of the little finger to extend it.

Extensor carpi ulnaris extends and slightly adducts the wrist.

Abduction and adduction of the wrist

Strong abduction of the wrist occurs through the cooperation of flexor carpi radialis and extensor carpi radialis longus and brevis.

Strong adduction occurs as the result of cooperation between flexor and extensor carpi ulnaris. Figure 10.37 diagrammatically shows the positions of all the muscles described above with their direction of pull (arrows), and illustrates the manner in which strong adduction and abduction of the wrist is produced by muscle cooperation.

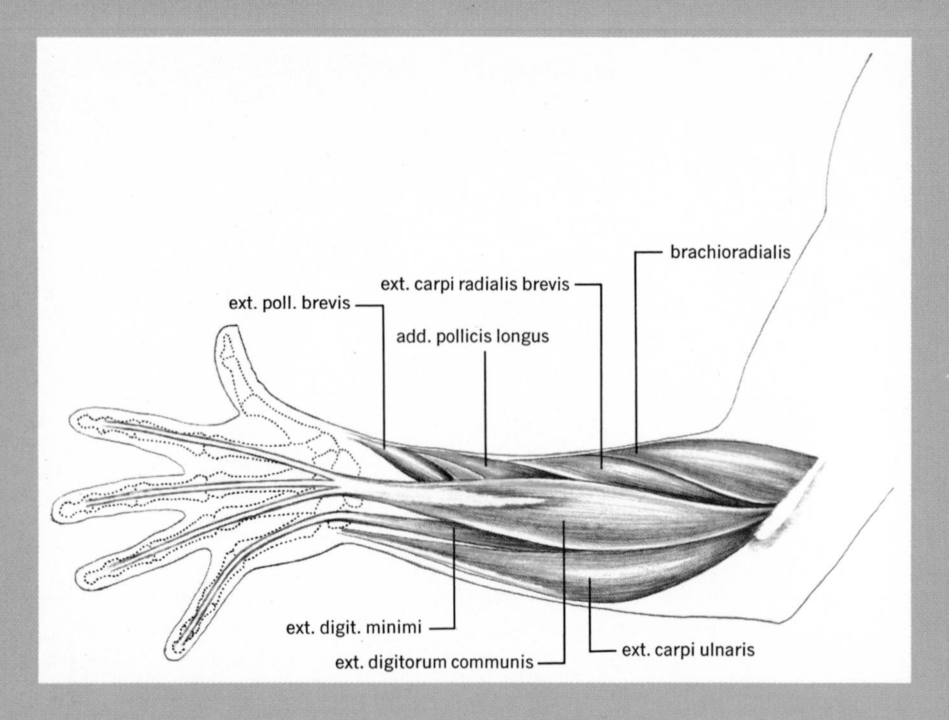

FIGURE 10.36 Extensors of the wrist and fingers.

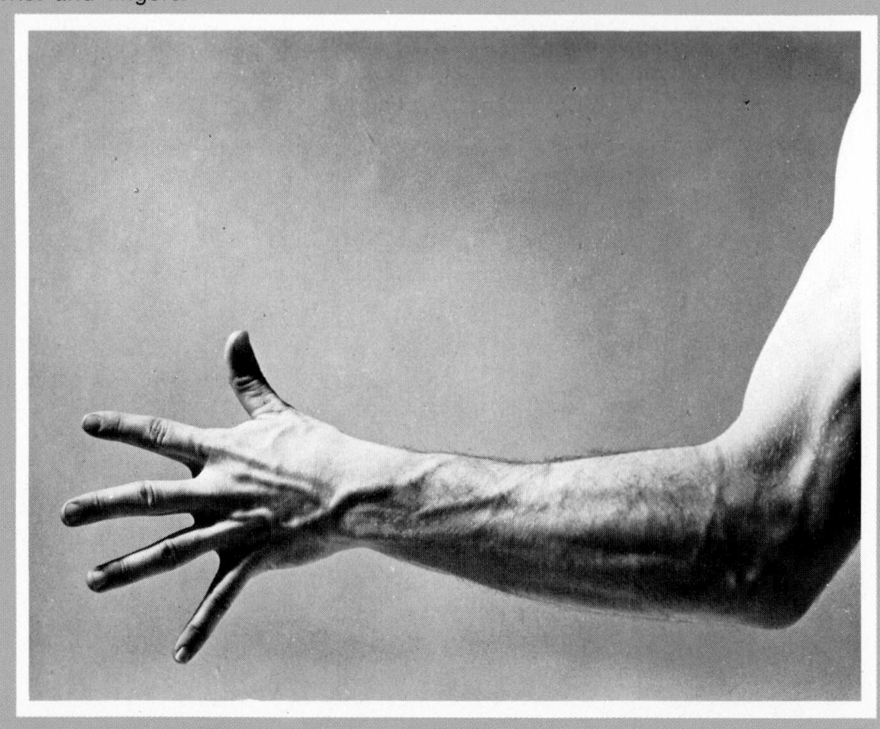

FIGURE 10.37 A diagram illustrating the origin of wrist abduction and adduction.

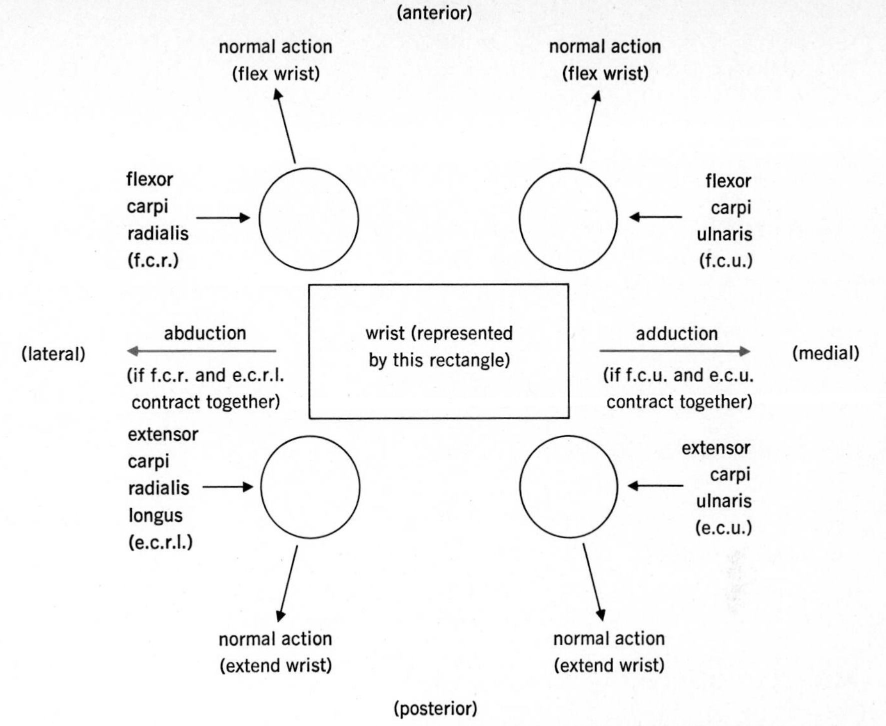

FIGURE 10.38 Muscles of the thumb and hand.

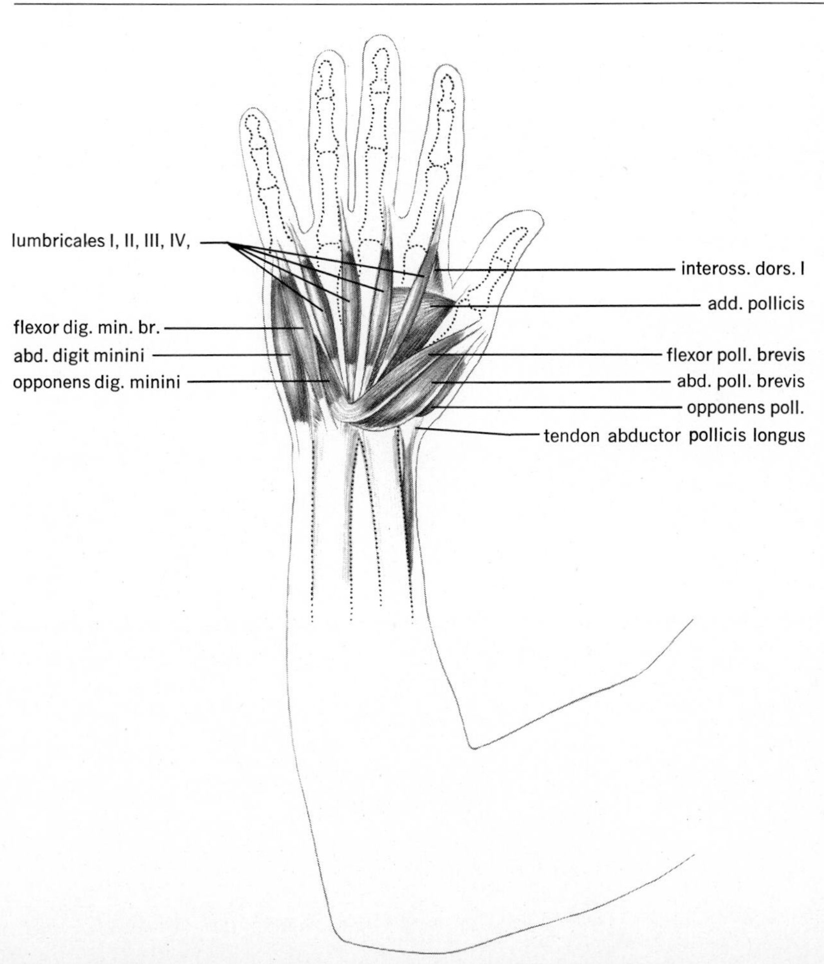

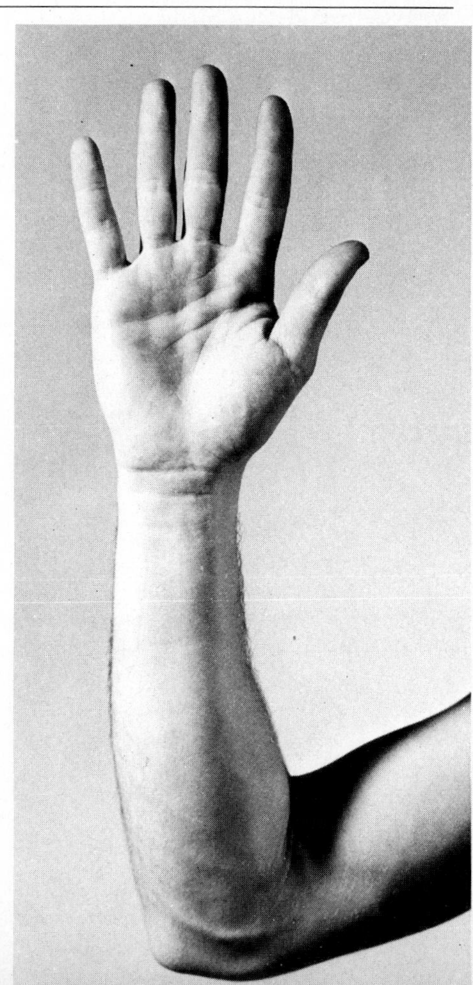

Muscles of the thumb

The thumb has its own supply of muscles independent of those of the fingers. This allows the thumb to be placed in opposition to the fingers for grasping and manipulation of objects. Some of these muscles are shown in Fig. 10.38. All thumb muscles carry the action produced, and the word *pollicis* (*pollex,* thumb) in their names.

Miscellaneous hand muscles

Fig. 10.38 also shows some of the muscles within the hand that permit abduction and adduction of the fingers. The INTEROSSEI and LUMBRICALES are the two main groups.

Muscles operating the hip and knee joints

Movements possible

The hip joint is a ball-and-socket joint that allows the femur (thigh) to be flexed, extended, abducted, adducted, and medially and laterally rotated (Fig. 10.39). The leg may only be flexed and extended. Some of the muscles operate both joints at once. Locating the muscles by position forms a convenient method of classifying them.

FIGURE 10.39 Movements of the thigh and leg.

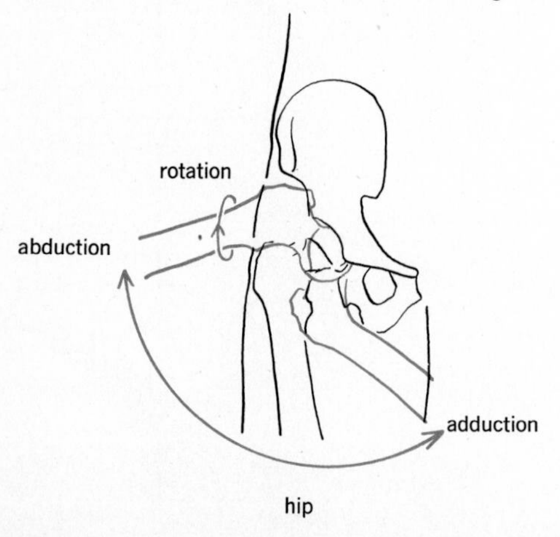

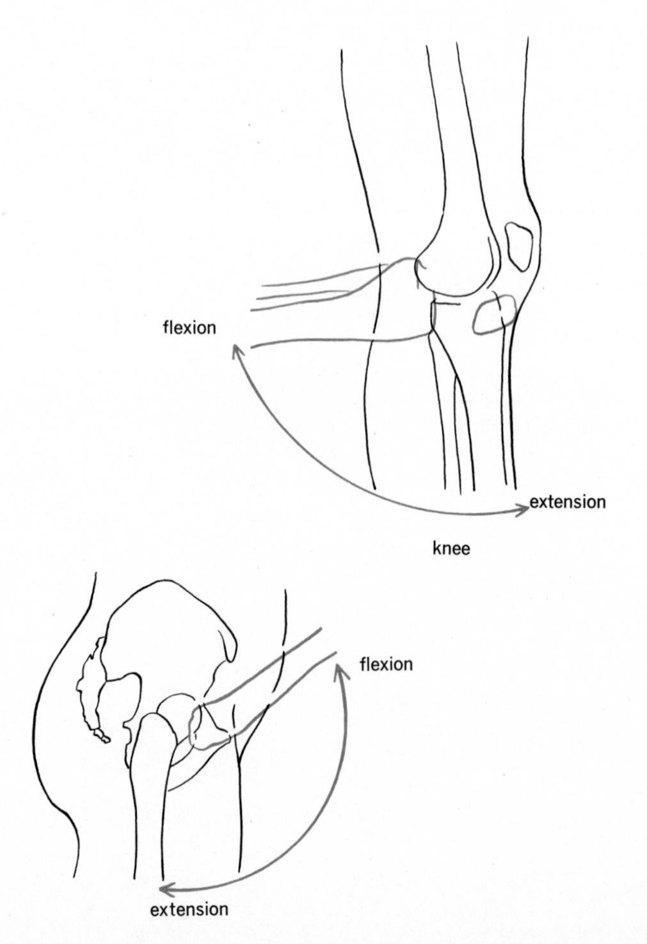

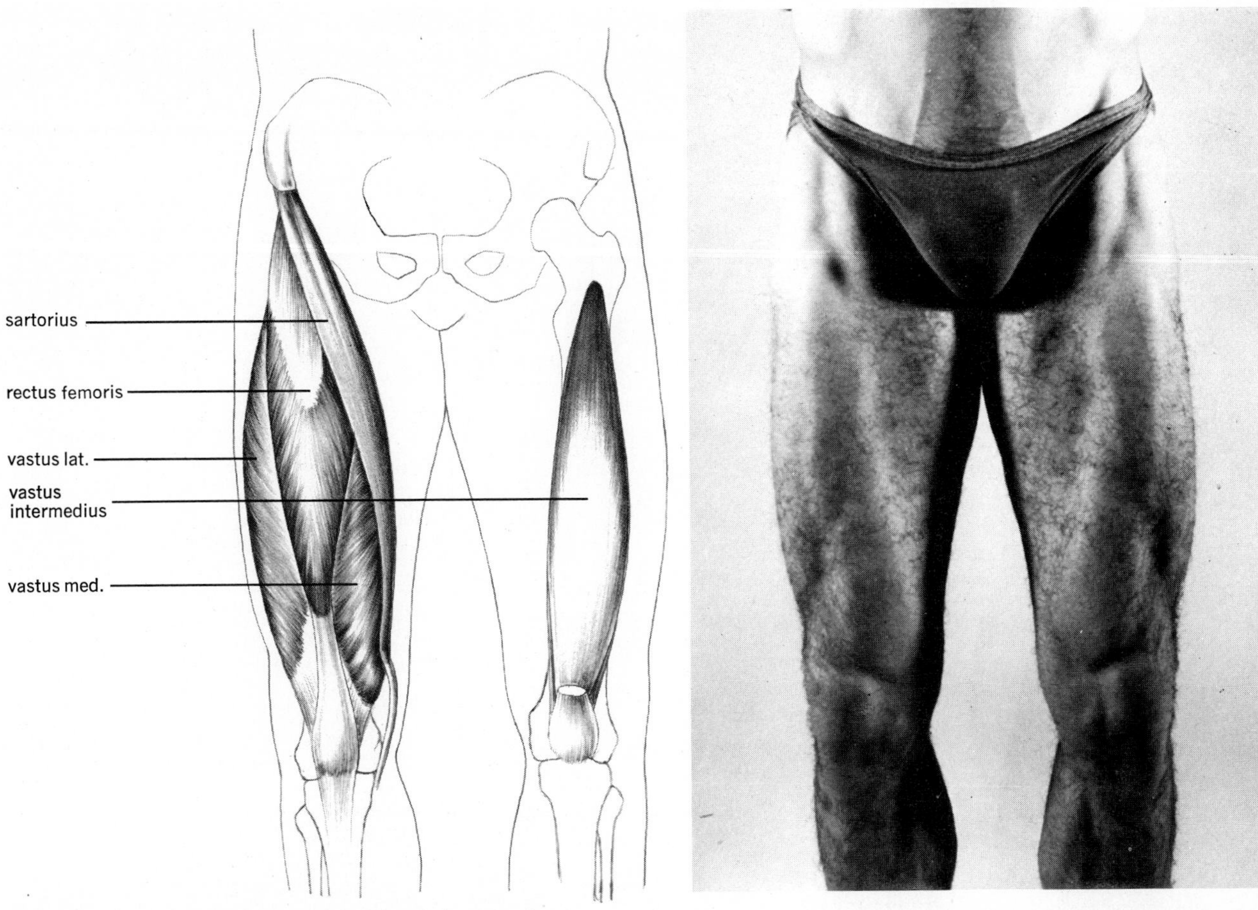

FIGURE 10.40 Anterior thigh and pelvic muscles.

Anterior thigh and pelvic muscles

These muscles are flexors of the thigh and/or extensors of the leg (Fig. 10.40).

ILIACUS and PSOAS, sometimes considered as a single muscle named *iliopsoas,* flex the thigh and medially rotate it. The muscles are strongly used in climbing, kicking, walking, and running.

RECTUS FEMORIS flexes the thigh and extends the leg.

VASTUS MEDIALIS, VASTUS INTERMEDIUS, and VASTUS LATERALIS extend the leg as their only action. These three and the rectus femoris form the *quadriceps femoris,* an important muscle group in locomotion.

SARTORIUS flexes both thigh and leg. It is the main muscle involved in crossing the knees or putting an ankle on the opposite knee.

Posterior thigh and pelvic muscles

These muscles are extensors of the thigh and/or flexors of the leg (Fig. 10.41).

Gluteus maximus (buttock muscle) extends the

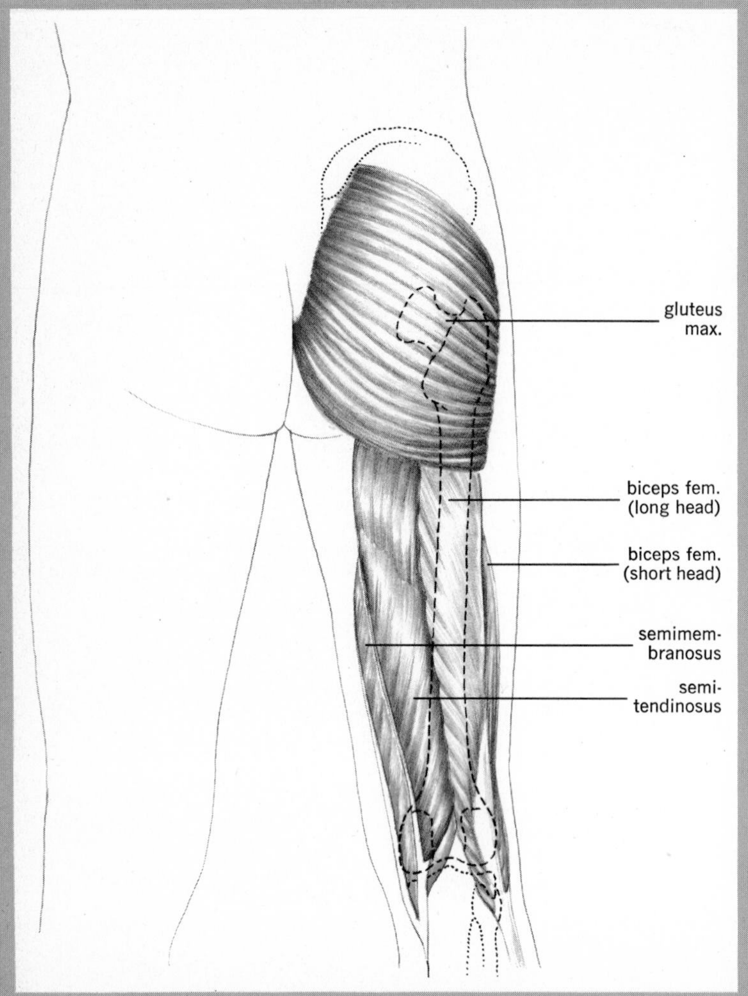

FIGURE 10.41 Posterior thigh and pelvic muscles.

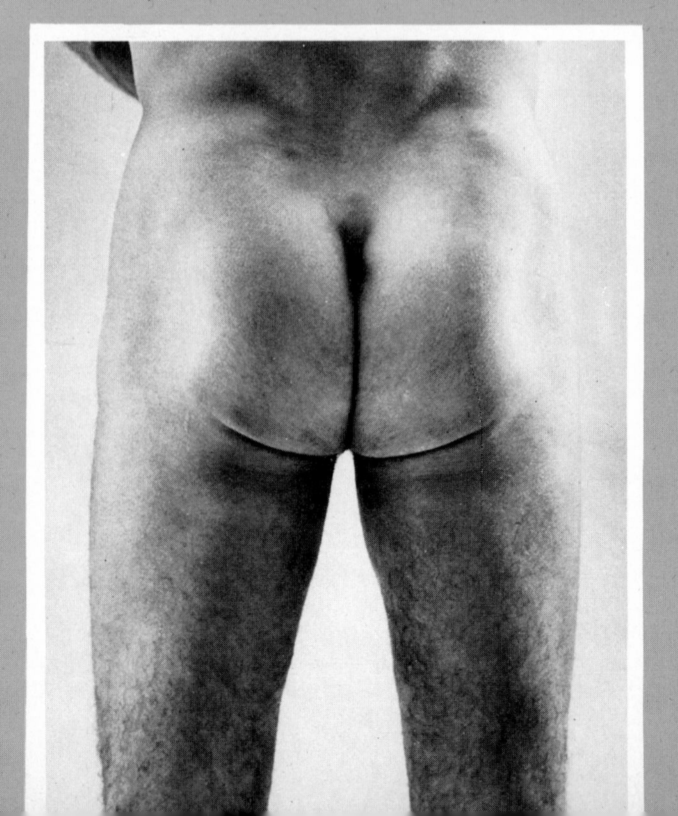

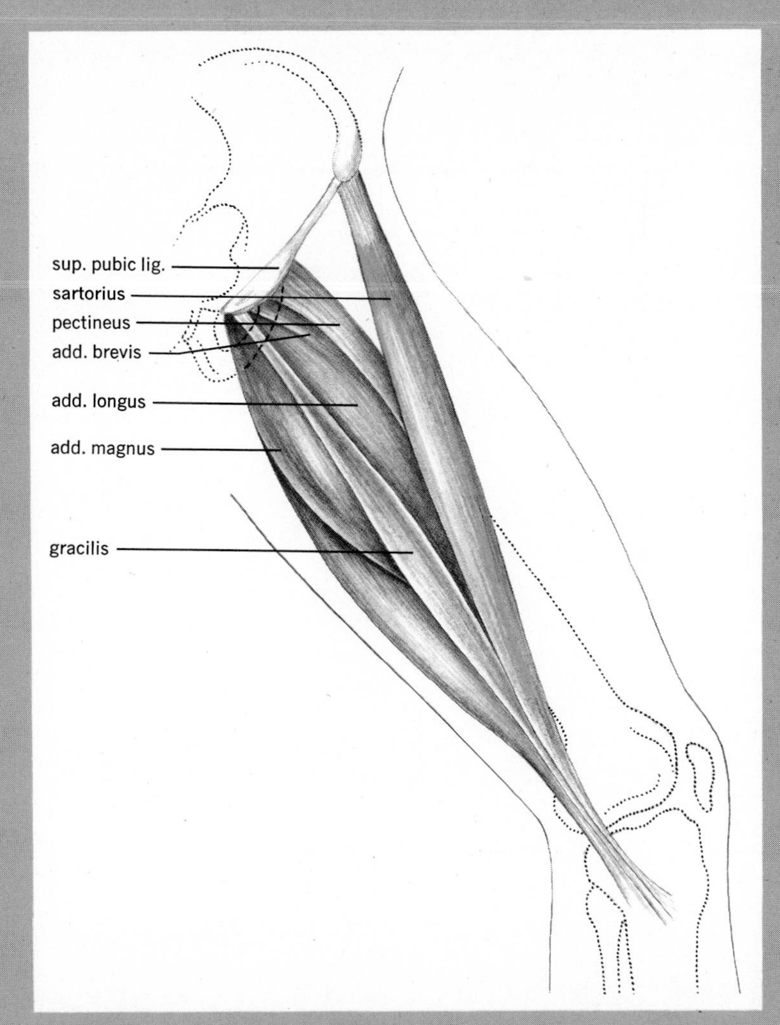

FIGURE 10.42 Medial thigh muscles.

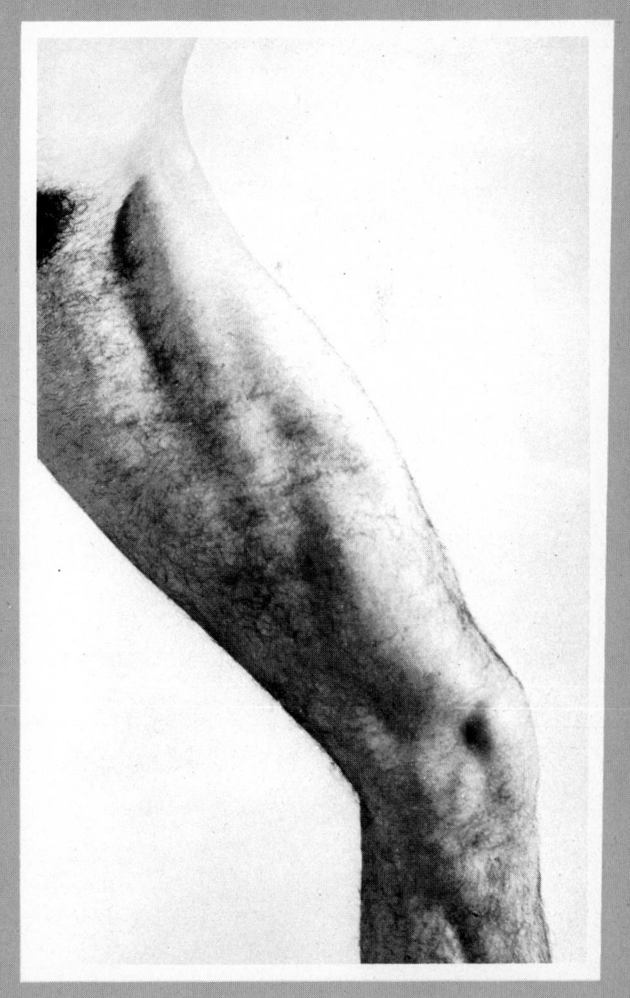

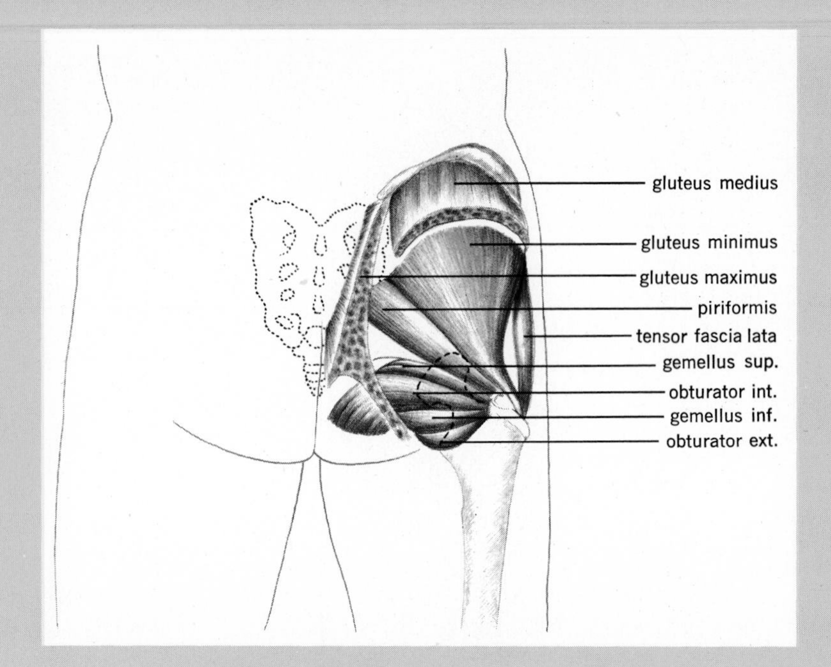

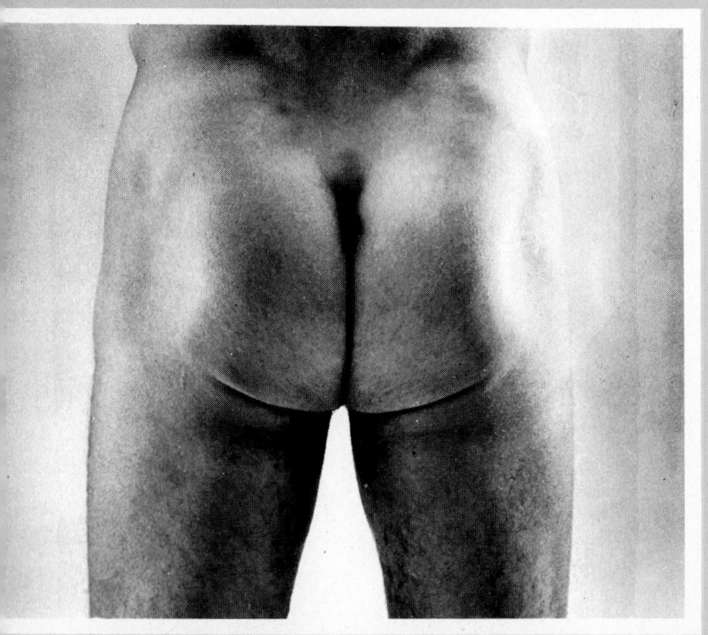

FIGURE 10.43 Posterior iliac muscles.

thigh and rotates it laterally. It is strongly used in bicycling and stair climbing to straighten the thigh on the hip.

Biceps femoris is the most lateral posterior thigh muscle; it extends the thigh and flexes the leg.

Semimembranosus and *semitendinosus* lie centrally and medially on the posterior thigh, and extend the thigh and flex the leg. The last three muscles together constitute the "hamstrings."

Medial thigh muscles

These muscles are adductors of the thigh or femur and several carry that action as part of their name (Fig. 10.42).

Adductor magnus, adductor longus, and *adductor brevis* are the strongest adductors of the thigh, and give the rounded contour to the inner side of the thigh.

Pectineus also adducts, and lies above the adductor longus.

Gracilis is a straplike muscle on the medial side of the thigh, also acting to adduct the thigh.

Posterior iliac muscles

These muscles abduct the thigh (Fig. 10.43).

Gluteus medius and *gluteus minimus* are the middle and deep muscle layers beneath the gluteus maximus.

Tensor fascia lata actually lies more on the lateral side of the ilium, and also abducts the thigh.

Posterior sacral muscles

This group includes six muscles in four groups that laterally rotate the thigh. From the superior to inferior direction the muscles are: PIRIFORMIS, GEMELLUS (Superior, Inferior), OBTURATOR (Internus, Externus), QUADRATUS FEMORIS.

Muscles operating the ankle and foot

Movements possible

The ankle may be flexed (dorsiflexion) and extended (plantar flexion). Inversion and eversion occur within the foot. The toes may be flexed, extended, abducted, and adducted. The muscles are grouped into anterior, posterior, and lateral crural (*cruralis,* leg) groups (Fig. 10.44).

Anterior crural muscles

The anterior crural muscles (Fig. 10.45) are on the anterolateral aspect of the leg; they dorsiflex the ankle, invert the foot, and extend the toes.

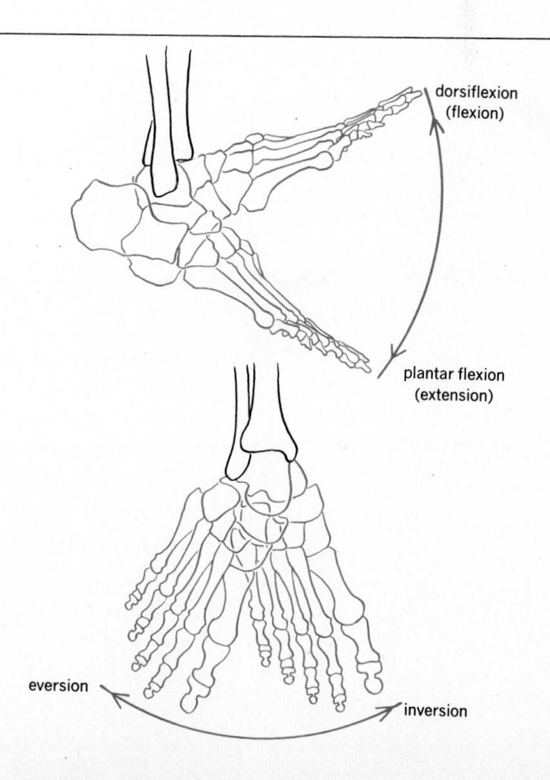

FIGURE 10.44 Movements of the ankle and foot.

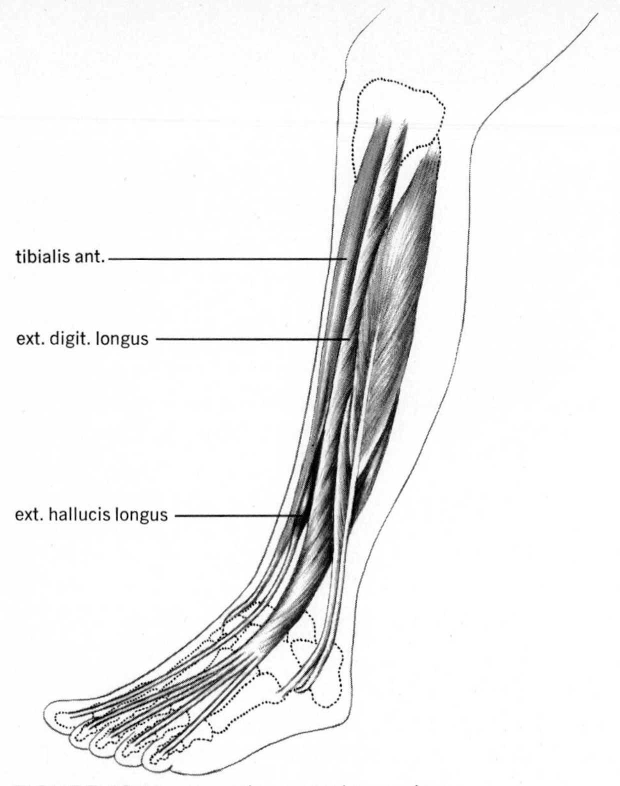

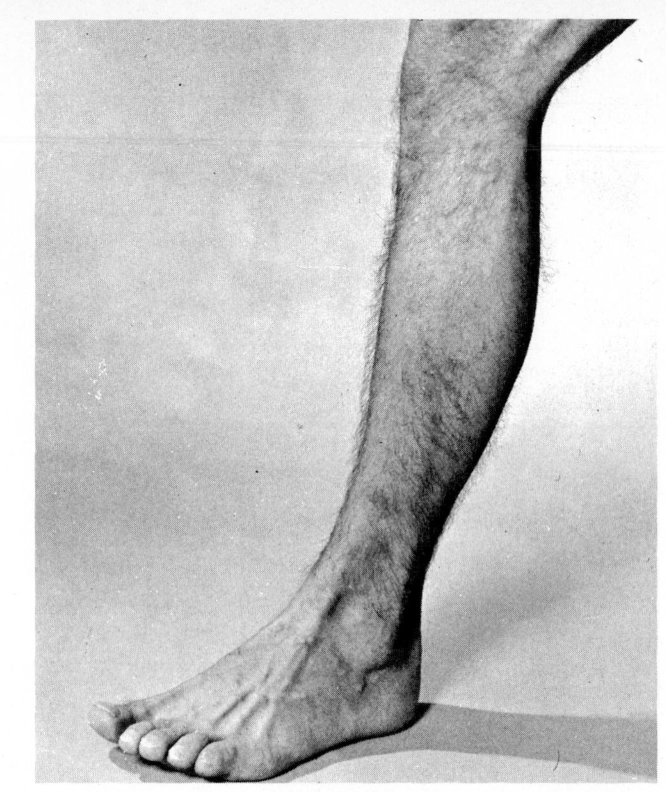

FIGURE 10.45 Anterior crural muscles.

FIGURE 10.46 Posterior crural muscles.

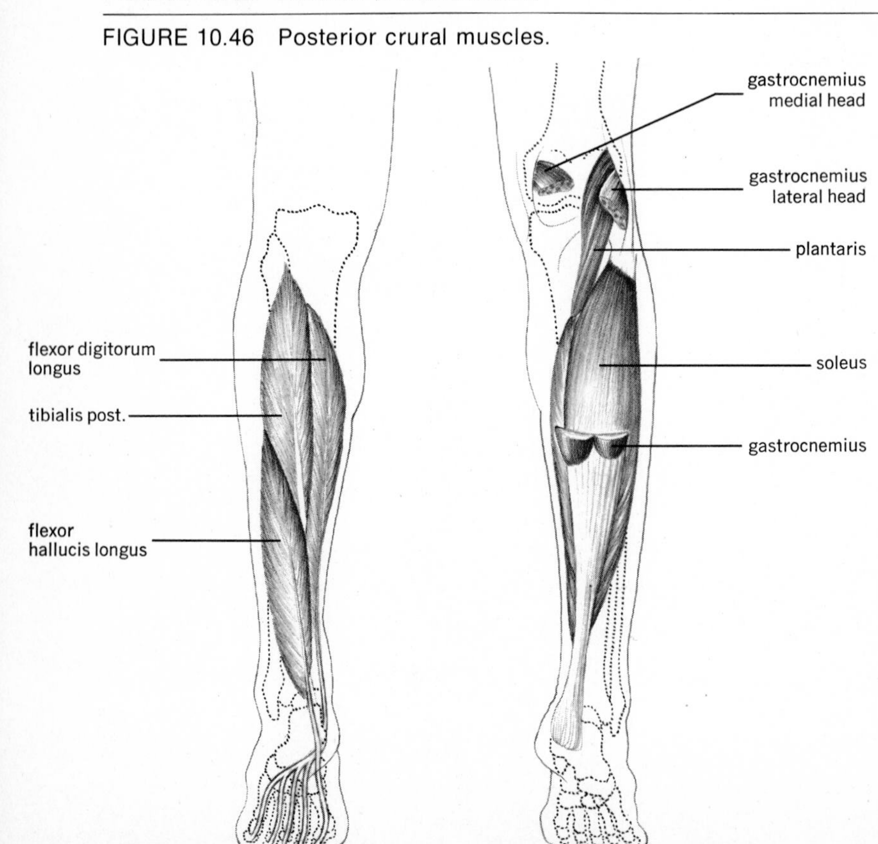

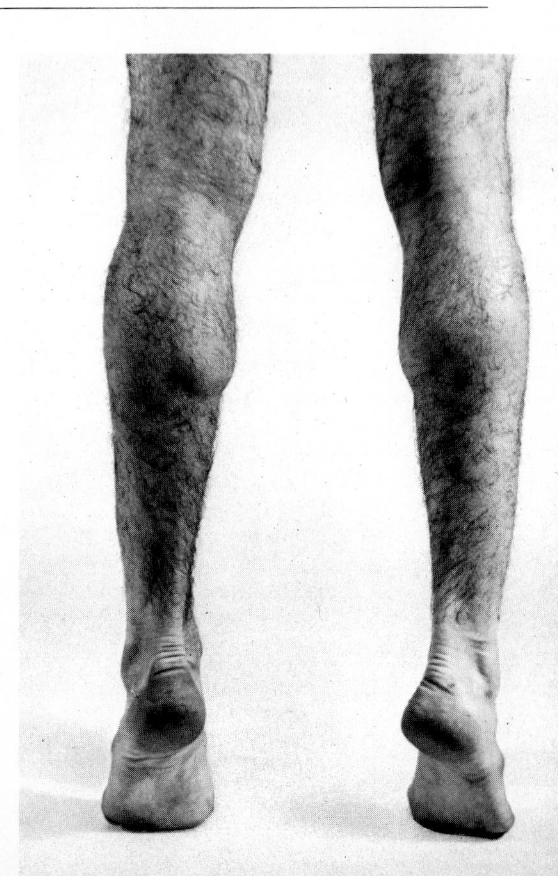

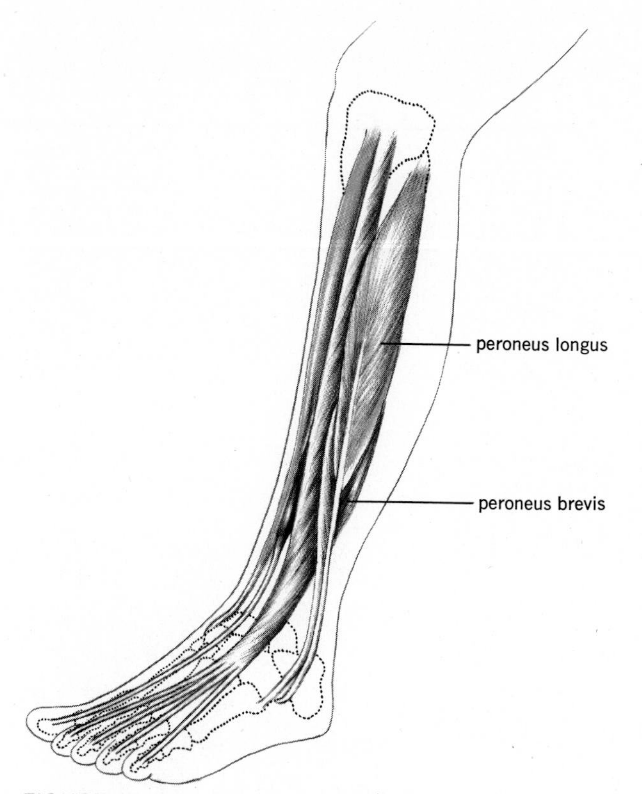

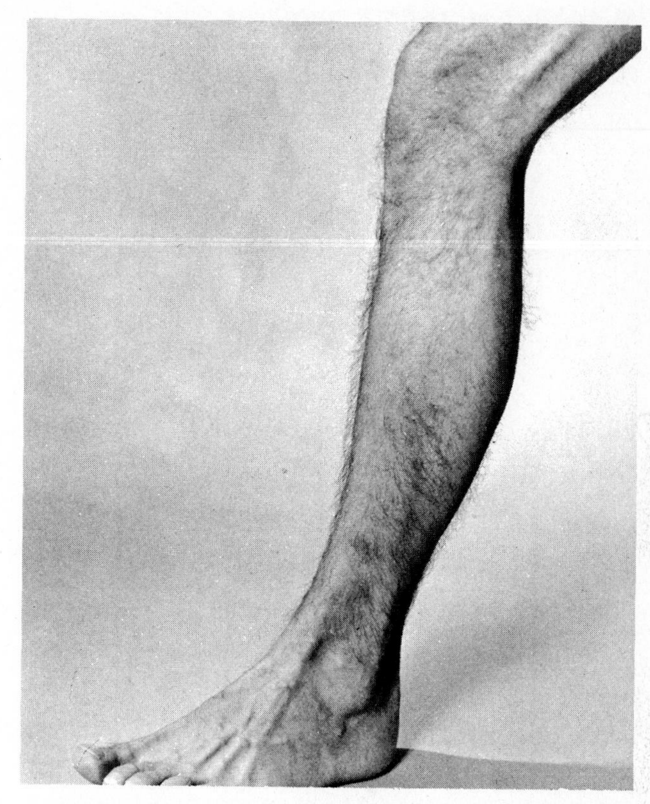

FIGURE 10.47 Lateral crural muscles.

Tibialis anterior dorsiflexes and inverts the foot. It is the site of a muscle cramp known as "shin splints."

Extensor digitorum longus extends all four outer toes at the same time, because it consists of one muscle that gives rise to four tendons.

Extensor hallucis longus extends the big toe.

Peroneus tertius dorsiflexes the foot.

Posterior crural muscles

The posterior crural muscles (Fig. 10.46) are on the posterior aspect of the leg; they plantar flex the foot and flex the toes.

Gastrocnemius (calf muscle) is an important muscle in walking and running, because it provides the "push" for locomotion. If its tendon, the tendon of Achilles at the back of the ankle, is severed, these activities cannot be performed. It plantar flexes the foot and enables humans to rise on their toes.

Soleus lies beneath gastrocnemius and has the same action.

Plantaris plantar flexes the foot.

Tibialis posterior plantar flexes the foot.

Flexor digitorum longus flexes the four outer toes at the same time.

Flexor hallucis longus flexes the big toe.

Lateral crural muscles

PERONEUS LONGUS and BREVIS plantar flex and evert the foot (Fig. 10.47).

Miscellaneous foot muscles

Within the foot, as in the hand, there are intrinsic muscles that hold the bones firm so as to provide a stable platform for locomotion. These are shown in Fig. 10.48. The INTEROSSEI and LUMBRICALES are the most important groups.

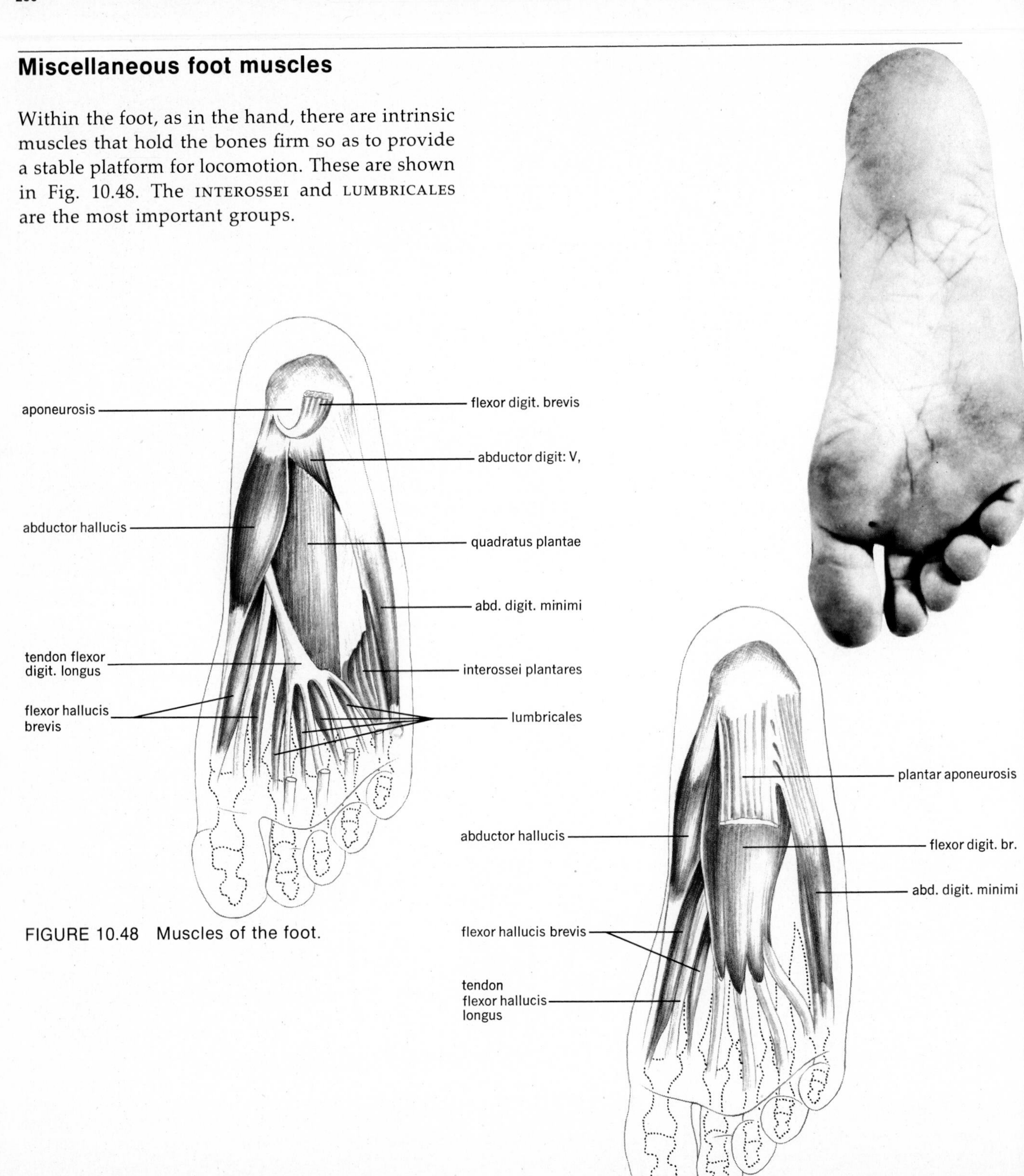

FIGURE 10.48 Muscles of the foot.

TABLE 10.2 Muscles of the body arranged alphabetically (includes those mentioned or diagrammed in the text) (italics indicate major muscles)

Muscle	General location	Origin	Insertion	Action	Innervation
Abductor digiti minimi (foot)	Sole of foot	Calcaneus	First phalanx 5th toe	Abducts 5th toe	Greater sciatic
Abductor hallucis	Sole of foot	Calcaneus	First phalanx great toe	Abducts 1st toe	Greater sciatic
Abductor pollicis brevis	Base of thumb on palm	Scaphoid multangular	First phalanx thumb	Abducts thumb	Median nerve
Adductor brevis	Medial thigh	Inferior pubic ramus	Upper third linea aspera	Adducts thigh (some flexion and medial rotation)	Obturator nerve
Adductor hallucis	Sole of foot	Metatarsals	First phalanx great toe	Adduction great toe	Greater sciatic
Adductor longus	Medial thigh	Crest and symphysis of pubis	Middle third linea aspera	Adducts thigh (some flexion and medial rotation)	Obturator nerve
Adductor magnus	Medial thigh	Ramus of pubis ischium	Lower third linea aspera	Adducts, flexes, medially rotates thigh	Obturator and sciatic
Adductor pollicis	Between thumb and index finger palm of hand	Capitate, multangulars, metacarpals 2 to 4	First phalanx thumb	Adducts thumb	Ulnar nerve
Auricularis					
Anterior	Anterior to pinna	Fascia of scalp	Pinna	Moves pinna forward	7th cranial nerve
Posterior	Posterior to pinna	Fascia of scalp	Pinna	Moves pinna backward	
Superior	Superior to pinna	Fascia of scalp	Pinna	Elevates pinna	
Biceps brachii	Anterior upper arm	Long head; above glenoid fossa Short head; coracoid process	Radial tuberosity	Flexes and supinates forearm	Musculo-cutaneous nerve
Biceps femoris	Posterior thigh	Long head; ischial tuberosity Short head; linea aspera	Lateral head fibula	Extends thigh, flexes knee	Sciatic nerve
Brachialis	Anterior upper arm	Lower half anterior humerus	Coronoid process (ulna)	Flexes forearm	Musculo-cutaneous nerve
Brachioradialis	Lateral forearm	Lateral upper ⅔ of humerus	Styloid process (radius)	Flexes forearm (semipronates semisupinates)	Radial nerve
Buccinator	Cheek of face	Alveolar processes of maxillae and mandible	Orbicularis oris fibers	Compresses cheek	7th cranial nerve

Table 10.2 continued Muscles of the body arranged alphabetically (includes those mentioned or diagrammed in the text) (italics indicate major muscles)

Muscle	General location	Origin	Insertion	Action	Innervation
Coraco-brachialis	Medial upper arm	Coracoid process	Midhumerus	Flexes and adducts humerus	Musculo-cutaneous nerve
Corrugator	Forehead	Inner end superciliary ridge	Skin above orbit	Produces frown	7th cranial nerve
Deltoid	Top of shoulder	Clavicle, acromion and spine of scapula	Deltoid tuberosity	Abducts, flexes extends humerus	Axillary nerve
Depressor labii inferioris	Inferior to lower lip	Mental process	Skin of lower lip	Depresses lower lip	7th cranial nerve
Diaphragm	Between thoracic and abdominal cavities	Xiphoid process, last 6 ribs, lumbar vertebrae	Central tendon	Increases vertical dimension of thorax	Phrenic nerve
Digastricus	Runs from mastoid process to hyoid	Posterior belly: mastoid notch of temporal	Greater horn, hyoid	Elevates hyoid, opens mouth	7th cranial nerve
		Anterior belly: mandible	Body of hyoid	Draws hyoid forward	5th cranial nerve
Epicranius *Frontalis*	Forehead	Muscles above orbit	Galea	Raises eyebrows, draws scalp forward	7th cranial nerve
Occipitalis	Base of skull	Superior nuchal line and mastoid temporal	Aponeurotica	Pulls scalp backwards	7th cranial nerve
Extensor carpi radialis brevis	Posterior forearm	Lateral epicondyle humerus	Top of 3rd metacarpal	Extends and abducts wrist	Radial nerve
Extensor carpi radialis longus	Posterior forearm	Lateral epicondyle humerus	Top of 2nd metacarpal	Extends and abducts wrist	Radial nerve
Extensor carpi ulnaris	Posterior forearm	Lateral epicondyle humerus	Base of 5th metacarpal	Extends and adducts wrist	Radial nerve
Extensor digiti minimi	Posterior forearm	Lateral epicondyle, humerus	Top of 1st phalanx, 4th finger	Extends 4th finger	Radial nerve
Extensor digitorium communis	Posterior forearm	Lateral epicondyle, humerus	2nd and 3rd phalanges of fingers	Extends fingers	Radial nerve
Extensor digitorum longus	Anterolateral leg	Lateral tibial condyle, distal fibula	2nd and 3rd phalanges, four outer toes	Extends four outer toes	Common peroneal nerve

Muscle	General location	Origin	Insertion	Action	Innervation
Extensor hallucis longus	Anterior fibula	Anterior surface of fibula	Top of great toe	Extends great toe	Common peroneal nerve
Extensor pollicis brevis	Posterior radius	Dorsal surface of radius	1st phalanx of thumb	Extends thumb	Radial nerve
Extensor pollicis longus	Posterior ulna	Dorsal surface of ulna	2nd phalanx of thumb	Extends thumb	Radial nerve
External intercostals	Between ribs—superficial	Inferior border of rib	Superior border of rib below	Elevates ribs	Intercostal nerves
External oblique	Abdomen—superficial	Anterior inferior surface, lower 8 ribs	Linea alba, pubis, iliac crest	Flexion, rotation of spine	Intercostal nerves
Flexor carpi radialis	Anterior forearm superficial	Medial epicondyle, humerus	Base 2nd metacarpal	Flexes and abducts wrist	Median nerve
Flexor carpi ulnaris	Anterior forearm superficial	Medial epicondyle of humerus, distal ulna	Base 5th metacarpal pisiform	Flexes and adducts wrist	Ulnar nerve
Flexor digitorum longus	Posterior tibia	Posterior surface tibial shaft	Distal phalanges 4 outer toes	Flexes toes	Tibial nerve
Flexor digitorum profundus	Anterior forearm deep	Proximal three-fourths ulna	Base of distal phalanges	Flexes fingers	Median and ulnar nerves
Flexor digitorum superficialis	Anterior forearm middle	Medial epicondyle, humerus; coronoid process, ulna; radial shaft	Second phalanges of fingers	Flexes fingers	Median nerve
Flexor hallucis longus	Posterior fibula	Lower two-thirds fibula	Undersurface of phalanges of great toe	Flexes great toe	Tibial nerve
Flexor pollicis brevis	Base of thumb on palm	Multangulars	Base of 1st phalanx of thumb	Flexes and adducts thumb	Median nerve
Flexor pollicis longus	Base of thumb on palm	Anterior surface, radius; coronoid process, ulna	Base of distal phalanx of thumb	Flexes and adducts thumb	Median nerve
Gastrocnemius	Calf of leg, superficial	Posterior surface, condyles of femur	Calcaneus	Plantar flexion flexes knee	Tibial nerve

Table 10.2 continued Muscles of the body arranged alphabetically (includes those mentioned or diagrammed in the text) (italics indicate major muscles)

Muscle	General location	Origin	Insertion	Action	Innervation
Gemellus	Posterior pelvis	Inferior: Ischial tuberosity Superior: Ischial spine	Greater trochanter	Rotates thigh laterally	Obturator nerve
Gluteus maximus	Buttocks, superficial	Gluteal line of ilium	Gluteal tuberosity, femur	Extends and laterally rotates thigh	Sciatic
Gluteus medius	Buttock, intermediate	Outer surface of ilium	Greater trochanter	Abducts and medially rotates thigh	Sciatic
Gluteus minimus	Buttock, deep	Outer surface of ilium	Greater trochanter	Abducts and medially rotates thigh	Sciatic
Gracilis	Medial thigh	Symphysis pubis	Medial head tibia	Adducts thigh	Obturator nerve
Iliacus	Iliac fossa	Iliac fossa and crest, sacrum	Lesser trochanter	Flexes and medially rotates thigh	Femoral nerve
Iliocostalis	Midback	Iliac crest, ribs	Ribs and cervical transverse processes	Extends spine	Posterior branches of spinal nerves
Infraspinatus	Scapula below spine	Infraspinous fossa	Greater tubercle, humerus	Lateral rotator of humerus	Axillary nerve
Internal intercostals	Between ribs, deep	Inner surface of rib	Superior surface of rib below	Depresses ribs	Intercostal nerves
Internal oblique	Abdomen, middle layer	Iliac crest, fascia of back	Lower 3 ribs, linea alba, xiphoid process	Flexes spine, compresses abdomen	Intercostal nerves
Interossei					
Foot	Between metatarsals	Adjacent sides of metatarsals	Bases 1st phalanges	Abducts toes flexes "knuckles"	Common peroneal nerve
Hand	Between metacarpals	Adjacent sides of metacarpals	Bases 1st phalanges	Abducts fingers flexes "knuckles"	Ulnar nerve
Latissimus dorsi	Lower back, superficial	Spinous processes T6-L5, iliac crest last 3 ribs	Bicipital groove, humerus	Extension, adduction, medial rotation humerus	Radial nerve
Levator ani	Forms part of pelvic floor	Pubis, spine of ischium	Coccyx, central tendon of perineum	Supports and raises pelvic floor	Pudendal nerve
Levator labii superioris	Above upper lip	Lower rim of orbit	Skin of upper lip	Elevates upper lip	7th cranial nerve

Muscle	General location	Origin	Insertion	Action	Innervation
Levator scapulae	Posterior neck	Transverse processes C1-4	Superior angle, scapula	Elevates scapula	Cervical nerves
Longissimus	Midback	Transverse processes C5-T5	Mastoid process, skull	Extends spine, bends it to side	Posterior branches of spinal nerves
Longus capitus	Lateral neck	Transverse processes C3-6	Occipital bone	Flexes neck	Cervical nerves
Longus colli	Anterior neck, deep	Transverse processes and bodies C3-T7	Atlas, cervical bodies	Flexes and rotates neck	Cervical nerves
Lumbricales Foot	Plantar surface metatarsals	From flexor digitorum longus	Tendons of extensor digitorum longus	Flexes toes	Common peroneal nerve
Hand	Anterior surface metacarpals	Tendons of flexor digitorum profundus	Tendons of extensor digitorum communis	Flexes knuckles	Median and ulnar nerves
Masseter	Side of mandibular ramus	Zygomatic process of maxilla, zygomatic arch	Ramus, angle and coronoid process of mandible	Elevates mandible	5th cranial nerve
Mentalis	Chin	Mental symphysis	Skin of chin and lower lip	Depresses lip, wrinkles chin	7th cranial nerves
Mylohyoid	Floor of mouth	Mylohyoid lines	Hyoid body	Elevates hyoid and tongue	5th cranial nerve
Obturator	Posterior pelvis	Externus: Rim of obturator foramen	Greater trochanter	Laterally rotates thigh	Obturator nerve
		Internus: Rim of obturator foramen, pubis, ischium			Obturator nerve
Omohyoid	Shoulder to hyoid	Scapula, superior border	Body of hyoid	Depresses hyoid	Cervical nerves
Orbicularis oculi	Around orbit	Medial surface of orbit	Skin of eyelid	Closes eye (wink, blink)	7th cranial nerve
Orbicularis oris	Around mouth	Skin of lips, fibers of other facial muscles	Corners of mouth	Closes and puckers lips	7th cranial nerve

Table 10.2 continued Muscles of the body arranged alphabetically (includes those mentioned or diagrammed in the text) (italics indicate major muscles)

Muscle	General location	Origin	Insertion	Action	Innervation
Palmaris longus	Anterior forearm	Medial epicondyle humerus	Palm of hand	Flexes wrist	Median nerve
Pectineus	Median thigh	Iliopectineal line, pubis	Base of lesser trochanter	Flexes and adducts thigh	Femoral nerve
Pectoralis major	Chest, superficial	Clavicle, sternum, costal cartilages true ribs	Bicipital groove	Flexes, adducts, medially rotates humerus	Pectoral nerve
Pectoralis minor	Chest, deep	Ribs 3 to 5	Coracoid process	Draws scapula forward and down	Pectoral nerve
Peroneus brevis	Lateral leg	Lower two-thirds fibula	Base 5th metatarsal	Plantar flexes and everts foot	Common peroneal
Peroneus longus	Lateral leg	Upper two-thirds fibula	Base 1st metatarsal, 1st cuneiform	Plantar flexes and everts foot	Common peroneal
Peroneus tertius	Anterior ankle	Anterior fibula	Dorsal surface 5th metatarsal	Dorsiflexion of foot	Common peroneal
Piriformis	Posterior pelvis	Anterior sacrum	Greater trochanter	Lateral rotation thigh	Sacral nerves
Plantaris	Posterior lower thigh	Linea aspera	Calcaneus	Plantar flexion	Tibial nerve
Platysma	Anterior neck	Fascia of deltoid and pectoralis major	Skin of lower face	Depresses lower lip	7th cranial nerve
Pronator teres	Anterior upper forearm	Medial epicondyle, humerus; and coronoid of ulna	Midradius	Pronation	Median nerve
Pronator quadratus	Anterior lower forearm	Body of ulna	Lower 1/4 radius	Pronation	Median nerve
Psoas	Posterior wall of pelvic cavity	Transverse processes lumbar vertebrae	Lesser trochanter	Flexes thigh	Femoral nerve
Pterygoid					
Lateralis	Medial to ramus of mandible	Great wing of sphenoid, lateral pterygoid process	Condyloid process	Protrudes and opens jaw, moves jaw side to side	5th cranial nerve
Medialis	Medial to lateralis	Lateral pterygoid process, palatine, maxilla	Ramus and angle of mandible	Closes jaw	5th cranial nerve

Muscle	General location	Origin	Insertion	Action	Innervation
Quadratus femoris	Posterior pelvis	Ischial tuberosity	Femur	Laterally rotates thigh	Sacral nerve
Quadratus lumborum	Between last rib and iliac crest	Iliac crest, lower lumbar vertebrae	Rib 12 and upper lumbar vertebrae	Flexes lumbar spine	Lumbar nerves
Quadratus plantae	Sole of foot in front of heel	Calcaneus	Tendons of flexor digitorum longus	Flexes 4 outer toes	Sciatic nerve
Rectus abdominus	Midabdomen	Pubic crest	Cartilages of ribs 5 to 7	Flexes spine	Intercostal nerves
Rectus capitus	Anterior neck, deep	Atlas	Base of occipital bone	Flexes neck	Cervical nerves
Rectus femoris	Anterior thigh, superficial	Anterior inferior iliac spine, acetabulum	Patella	Flexes hips, extends knee	Femoral nerve
Rhomboid	Back, deep	Spinous processes T2–5	Lower third vertebral border of scapula	Adducts and elevates scapula	Cervical nerve
Risorius	Lateral to mouth	Buccinator fascia (masseter)	Skin of corners of mouth	Pulls mouth laterally	7th cranial nerve
Sartorius	Anterior thigh	Anterior superior iliac spine	Medial head, tibia	Flexes hip and knee	Femoral nerve
Scalenes	Lateral neck	Transverse processes cervical vertebrae	1st and 2nd ribs	Elevate ribs and rotates and bends neck	Cervical nerves
Semimembranosus	Posterior thigh	Ischial tuberosity	Medial condyle of tibia	Extends hip, flexes knee	Sciatic nerve
Semispinalis	Midback	Transverse processes lower cervical and thoracic vertebrae	Spinous processes cervical vertebrae and occipital	Extends and rotates spine	Posterior branches of spinal nerves
Semitendinosus	Posterior thigh	Ischial tuberosity	Medial tibial shaft	Extends hip,flexes knee	Sciatic nerve
Serratus anterior	Lateral thorax	Upper 8 or 9 ribs	Vertebral border, scapula	Abducts scapula	Cervical nerves
Serratus posterior	Back, deep	Spinous processes thoracic and lumbar vertebrae	Ribs	Increases lateral dimensions of thorax	Intercostal nerves

Table 10.2 continued Muscles of the body arranged alphabetically (includes those mentioned or diagrammed in the text) (italics indicate major muscles)

Muscle	General location	Origin	Insertion	Action	Innervation
Soleus	Calf, deep	Head of fibula, upper tibia	Calcaneus	Plantar-flexion	Tibial nerve
Spinalis	Posterior neck	Same as semispinalis	Same as semispinalis	Extends spine	Posterior branches spinal nerves
Splenius	Posterior neck	Spinous processes lower cervical and upper thoracic vertebrae	Nuchal lines and cervical transverse processes	Extension of neck, rotation	Cervical nerves
Sternocleido-mastoid	Lateral neck	Sternum and clavicle	Mastoid process of temporal bone	Rotates head, flexes neck	11th cranial and cervical nerves
Sternohyoid	Anterior neck	Sternum	Body of hyoid	Depresses hyoid	Cervical nerves
Sterno-thyroid	Anterior neck	Sternum	Thyroid carti-lage of larynx	Depresses larynx	Cervical nerves
Stylohyoid	Runs from styloid process of skull to hyoid bone	Styloid process	Body of hyoid	Elevates and retracts hyoid	7th cranial nerve
Subclavius	Beneath clavicle	Middle of 1st rib	Clavicle	Depresses clavicle	Cervical nerves
Subscapularis	Scapula, anterior	Subscapular fossa	Lesser tubercle, humerus	Medially rotates humerus	Axillary nerve
Supinator	Anterior upper forearm	Lateral epicondyle humerus	Radial shaft	Supination	Radial nerve
Supra-spinatus	Scapula, above spine	Supraspinous fossa	Greater tubercle, humerus	Abducts humerus	Cervical nerve
Temporalis	Lateral skull	Temporal fossa	Coronoid and ramus, mandible	Closes jaws	5th cranial nerve
Tensor fascia lata	Lateral hip	Anterior superior iliac spine	Fascia lata	Abducts and flexes thigh	Sciatic nerve
Teres major	Inferior angle of scapula to humerus	Inferior angle of scapula	Lesser tubercle, humerus	Adduction, extension, medial rotation humerus	Cervical nerves
Teres minor	Above teres major	Axillary border, scapula	Greater tubercle, humerus	Lateral rotation of humerus	Axillary nerve

Muscle	General location	Origin	Insertion	Action	Innervation
Thyrohyoid	Larynx to hyoid	Thyroid cartilage of larynx	Greater horn of hyoid	Elevates larynx	12th cranial and cervical nerves
Tibialis anterior	Anterolateral leg	Upper two-thirds tibia	First cuneiform and 1st metatarsal	Dorsiflex and invert foot	Common peroneal nerve
Tibialis posterior	Posterior leg	Shaft of tibia and fibula	Navicular, calcaneus, all cuneiforms	Plantar flexion and inversion of foot	Tibial nerve
Transverse abdominal	Abdomen, deep	Iliac crest, lumbar fascia, last 6 ribs	Xiphoid process linea alba, pubis	Compresses abdomen	Intercostal nerves
Trapezius	Upper back, superficial	Superior nuchal line, ligamentum nuchae, spines C7-T12	Clavicle, spine and acromion of scapula	Elevates, adducts depresses scapula	11th cranial and cervical nerves
Triangularis	Below corners of mouth	Body of mandible	Skin of lower lip	Depresses lower lip	7th cranial nerve
Triceps	Posterior humerus	Lateral head, humerus Long head, scapula Medial head, humerus	Olecranon process	Extends elbow	Radial nerve
Vastus intermedius	Anterior thigh, deep	Anterior and lateral femur shaft	Patella	Extends knee	Femoral nerve
Vastus lateralis	Lateral thigh	Upper half, lateral linea aspera	Patella	Extends knee	Femoral nerve
Vastus medialis	Medial thigh	Upper half, medial linea aspera	Patella	Extends knee	Femoral nerve
Zygomaticus	Above corners of mouth	Zygomatic bone	Skin of corners of mouth	Draws mouth up and back	7th cranial nerve

11 The Basic Organization and Properties of the Nervous System

The nervous system provides a means of DETECTING CHANGES (stimuli) in the external and internal environments of the body, and for INTERPRETING or ANALYZING the nerve impulses resulting from those changes. It may initiate and CONTROL RESPONSES to stimuli so that homeostasis is maintained. Control is rather specific in that organs and not body processes are usually caused to alter their activity.

Development of the nervous system

The nervous system (Fig. 11.1) is the first body system to develop, and it develops from the ectodermal germ layer. A NEURAL PLATE, an area of thickened ectoderm, appears by 18 days of embryonic development. A groove, the NEURAL GROOVE, develops in the plate, and the lips of the groove form two NEURAL FOLDS. The folds enlarge, meet in the midline, and create a hollow NEURAL TUBE. Neural crest material separates from the fused neural folds and comes to lie above (posterior to) the neural tube. It will ultimately form the neurons of the sensory portion of the peripheral nervous system, autonomic ganglia, the medulla of the adrenal gland, and melanocytes. The anterior end of the neural tube forms three enlargements that are subdivided as development proceeds (Table 11.1). Two general types of cells differentiate as the system forms: NEUROBLASTS will give rise to neurons, cells exhibiting the ability to form and conduct nerve impulses, and that

TABLE 11.1 Derivatives of the primary neural tube enlargements

Primary enlargement	Subdivision(s)	Structure(s) arising from subdivision
Forebrain (Prosencephalon)	Telencephalon	Cerebrum Basal ganglia
	Diencephalon	Thalamus Hypothalamus
Midbrain (Mesencephalon)	None	Midbrain
Hindbrain (Rhombencephalon)	Metencephalon	Pons Cerebellum
	Myelencephalon	Medulla

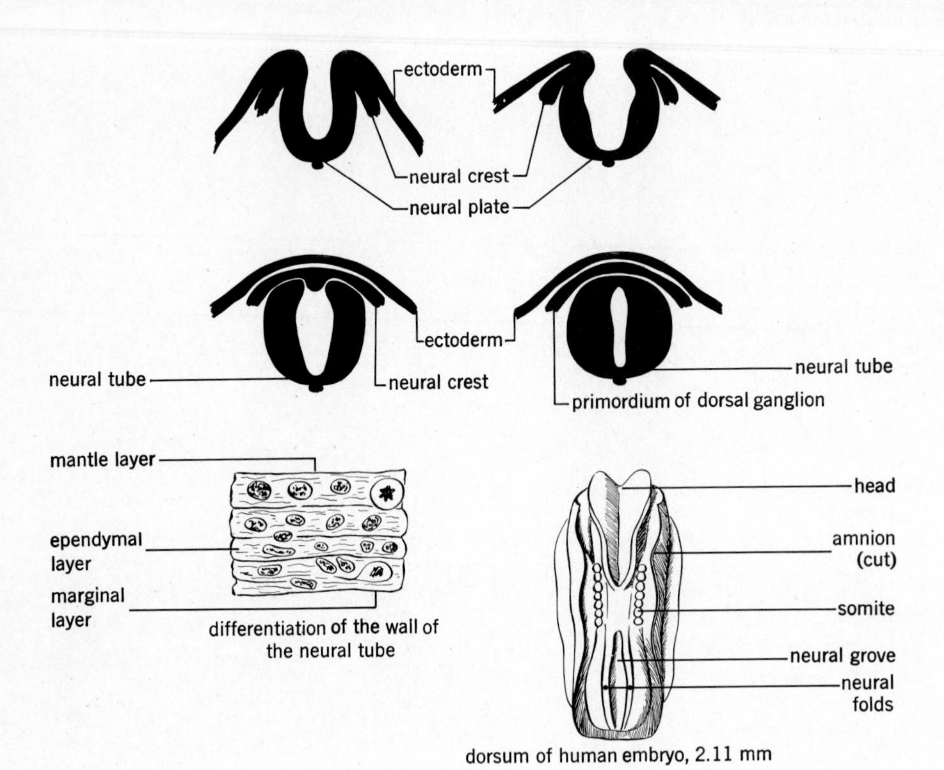

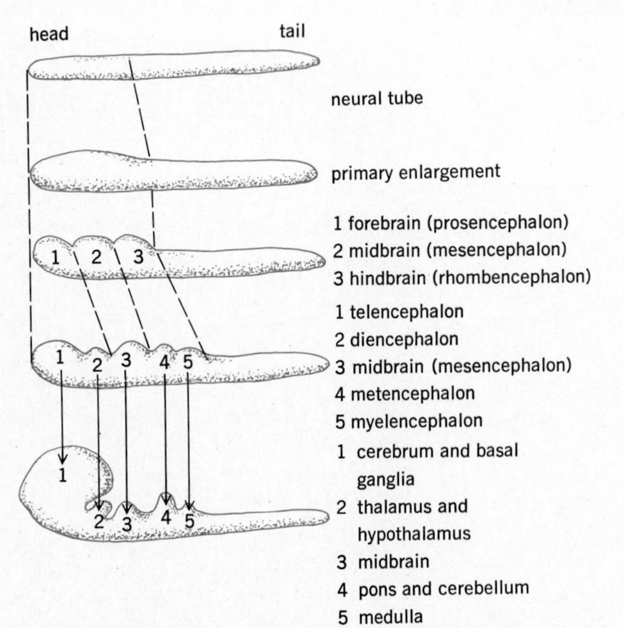

FIGURE 11.1 Early development of the nervous system.

will become the structural and functional units of the nervous system; SPONGIOBLASTS will form the glial cells, cells that connect, support, and perform nutritive and synthetic functions in the nervous system. Glia far outnumber neurons in the nervous system. Peripheral nerves grow from the central nervous system to supply their organs.

Organization of the nervous system

The nervous system has several divisions or parts, distinguished on the basis of location and function. A useful system of classification is presented below.

The CENTRAL NERVOUS SYSTEM (CNS) lies enclosed within the skull and vertebral column, and consists of two organs: the brain and the spinal cord.

The PERIPHERAL NERVOUS SYSTEM consists of all nervous tissue outside of the skull and vertebral column; it includes the cranial and spinal nerves. The divisions of the peripheral nervous system include:

The SOMATIC SYSTEM supplies motor and sensory fibers to the skin and skeletal muscles.

The AUTONOMIC SYSTEM supplies smooth muscle, cardiac muscle, and glands in the body viscera. Most body organs receive nerves from two divisions of the autonomic system to control their activity: the PARASYMPATHETIC DIVISION generally maintains normal function and conserves body resources; the SYMPATHETIC DIVISION generally causes organ changes that increase the body's ability to resist stress.

The SPECIAL SENSORY SYSTEM supplies motor and sensory fibers to the organs of special sense (eye, ear, olfactory area, taste buds).

Two other terms may well be learned at this time. SENSORY implies dealing with sensation and the carrying of impulses toward (afferent) the CNS from the periphery; MOTOR implies dealing with muscular or secretory activity, and the carrying of impulses away from the CNS (efferent) to the peripherally located smooth, skeletal and cardiac muscle, and glandular tissue. The muscle and glands form EFFECTORS that aid in maintaining homeostasis.

Cells of the nervous system

The structure of neurons

Neurons may be quite varied in size and shape (Fig. 11.2), but all possess certain features in common, as is illustrated by the sensory and motor neurons of the peripheral nervous system (Fig. 11.3). The cell body of the neuron is surrounded by a typical cell membrane, and contains the NEUROPLASM (cell cytoplasm) and the NUCLEUS. The nucleus is separated from the cytoplasm by a nearly invisible NUCLEAR MEMBRANE, and usually exhibits a prominent NUCLEOLUS and granular CHROMATIN MATERIAL. Collectively, the nuclear contents are called the KARYOPLASM. The neuroplasm contains the usual cellular organelles (mitochondria, ER, Golgi apparatus), but appears to lack or have a nonfunctioning cell center (centrioles and their surrounding cytoplasm) after about 16 years of age. Mitosis of nerve cells is believed to cease at about 3–4 years of age. Thus, "adult" or "mature" neurons do not have the capability of undergoing mitosis to replace destroyed or damaged cells. Additionally, the neuroplasm contains NISSL BODIES, masses of

FIGURE 11.2 Some different forms of neurons from the human nervous system.

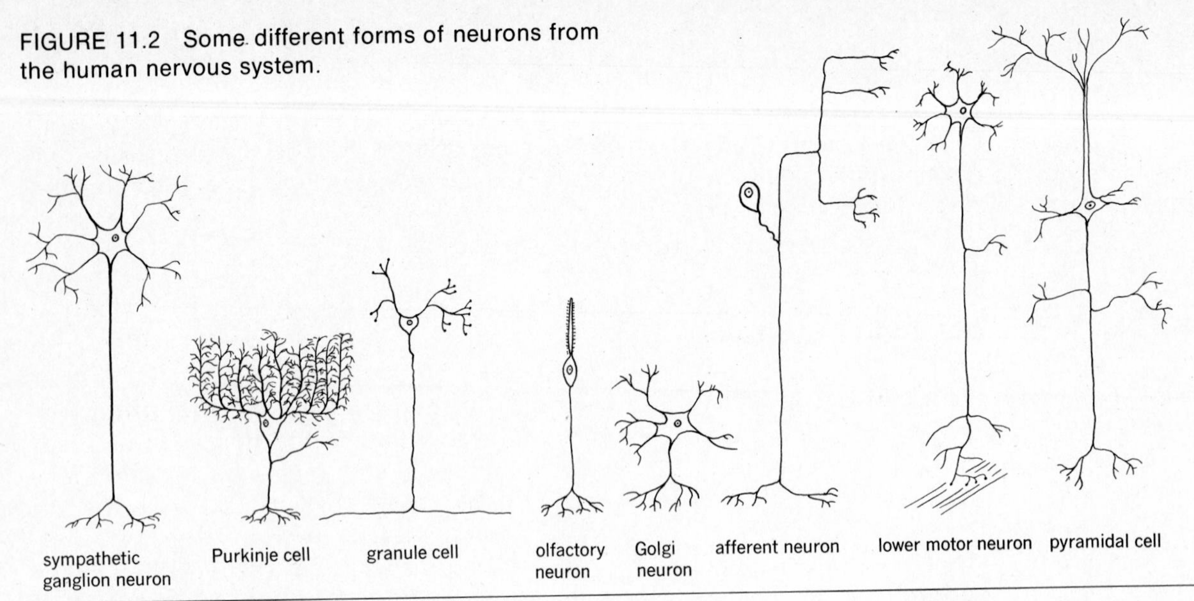

FIGURE 11.3 The two major types of neurons in the body. *(a)* Motor neuron of spinal cord. *(b)* Sensory neuron.

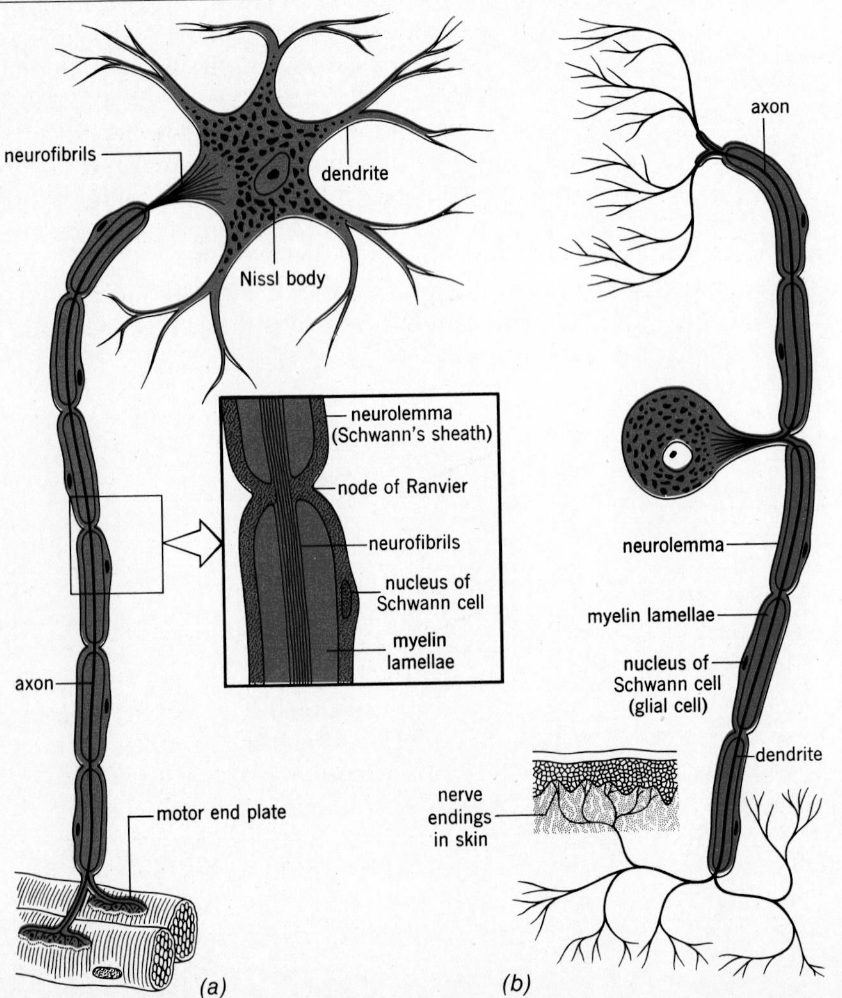

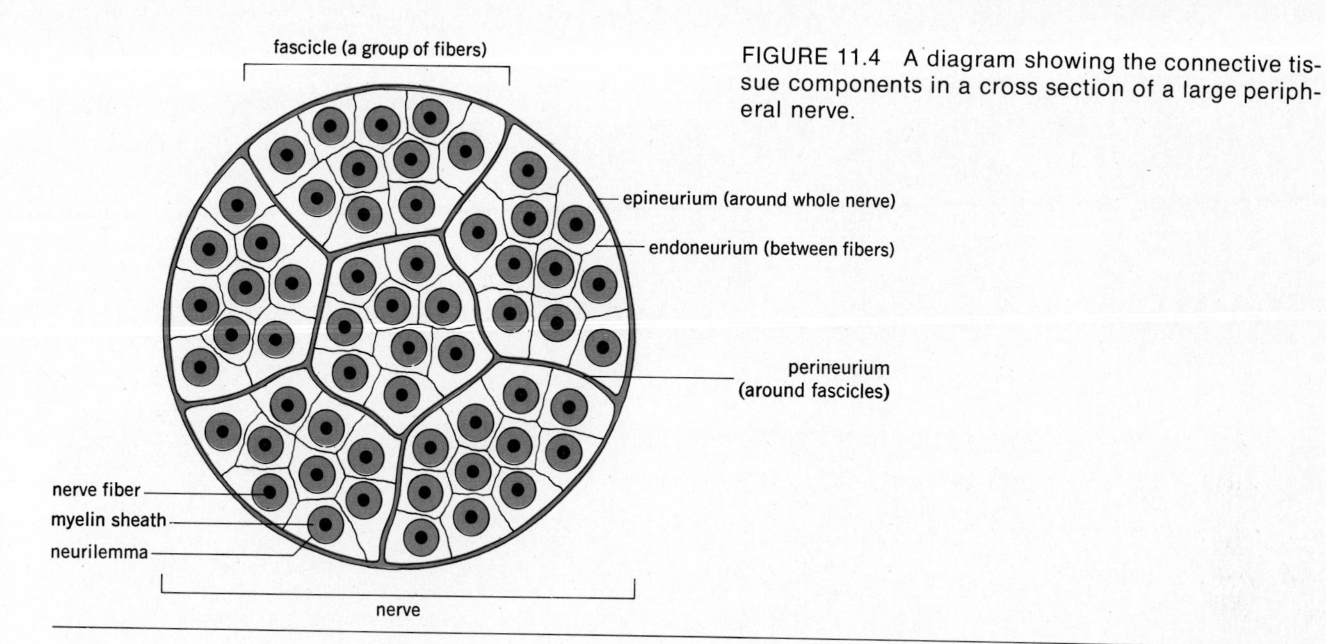

FIGURE 11.4 A diagram showing the connective tissue components in a cross section of a large peripheral nerve.

granular endoplasmic reticulum concerned with protein synthesis, and hollow microtubules known as NEUROFIBRILS. The latter may be concerned with support or the spreading of nutrients or depolarization waves through the cell.

The cell body gives rise to one or more PROCESSES that extend from the cell body, and that are involved in receiving and transmitting nerve impulses. The processes are structurally and/or functionally of two types: AXONS are long, usually single, sparsely branched efferent processes of regular diameter; DENDRITES are usually shorter, highly branched, multiple, and afferent processes of irregular diameter. In a sensory neuron (see Fig. 11.3), the axon and dendrite are constructed similarly, but one process acts as a dendrite, conducting toward the cell body, and the other acts as an axon, conducting away from the cell body. The term NERVE FIBER refers to the neuronal processes, and these are the units that are bound into the large structures called the nerves of the body (Fig. 11.4). Both types of processes may develop sheaths or remain "naked." A MYELIN SHEATH (Fig. 11.3) is a segmented, yellowish-white, fatty covering acting to insulate the fiber and to speed the conduction of nerve impulses; a NEURILEMMA (SCHWANN SHEATH) is a thin living membrane surrounding the fiber or its myelin sheath. The neurilemma is essential for regeneration of nerve fibers, and aids in guiding a regenerating fiber to something approximating its original connections. Peripheral myelin sheaths are believed to be produced by the wrapping of Schwann cells around the nerve fiber and by the synthesis of myelin as the wrapping proceeds (Fig. 11.5). Both coverings (myelin and neurolemma) may be found in various combinations, as shown in Fig. 11.6.

Another factor, the fiber diameter, influences the speed at which a nerve fiber will conduct an impulse. About 1 meter per second increase in conduction speed is seen for each micrometer increase of overall fiber diameter. Larger fibers conduct impulses more rapidly because current flow is easier through the neuroplasm (internal resistance is less). In a sense, the fibers behave like copper wires; a larger wire has a lower resistance. The names given to fibers of various sizes and

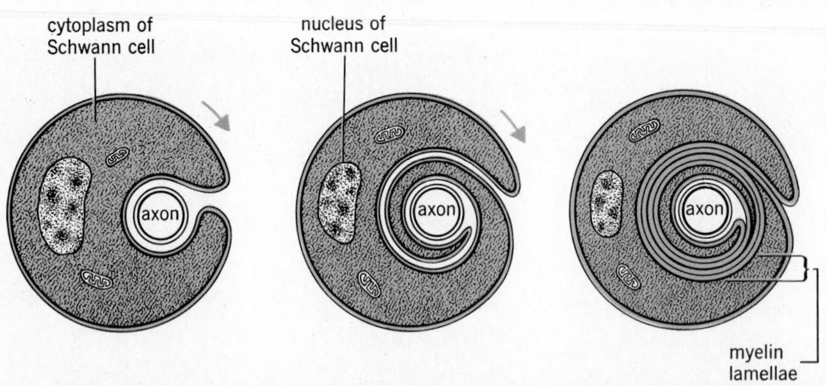

FIGURE 11.5 The formation of myelin lamellae around a nerve fiber. The Schwann cell is believed to "wrap around" the fiber.

TABLE 11.2 Characteristics and types of nerve fibers according to diameter

Type	Diameter (μm)	Myelinated	Conduction velocity (m/sec)	Occurrence
A fibers	1–20	Yes	5–100	Motor and sensory nerves
B fibers	1–3	Some	3–14	Autonomic system
C fibers	<1	No	<3	Skin, viscera

their conduction speeds are presented in Table 11.2.

Properties of neurons

Neurons are excitable and conductile cells. The basis of excitability and conductivity was presented in Chapter 3, and may be profitably reviewed at this time. The steps in the formation and conduction are summarized in Fig. 11.7. Additional neuronal properties are presented below.

THRESHOLD. Neuronal membranes typically have a voltage that must be exceeded before depolarization will occur. To be effective, a stimulus must possess a certain strength. The term *threshold* is used to describe both quantities. Threshold of the fiber is influenced by pH, temperature, and ions in the environment of the fiber; thus, the neuron may respond to stimuli of given strengths at one time but not at another. This variability of threshold may act to prevent extraneous or unimportant stimuli from causing a response, and gives a degree of control of response to the neuron itself.

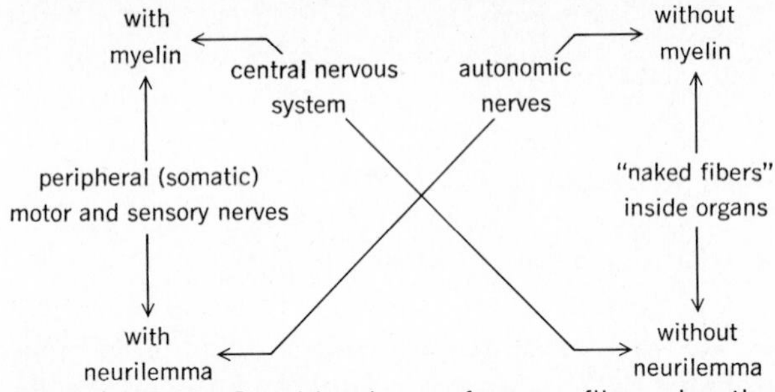

FIGURE 11.6 Combinations of nerve fiber sheaths in the nervous system.

SUMMATION. A stimulus not strong enough to cause depolarization is termed a *subthreshold* (subliminal) *stimulus.* Although not strong enough

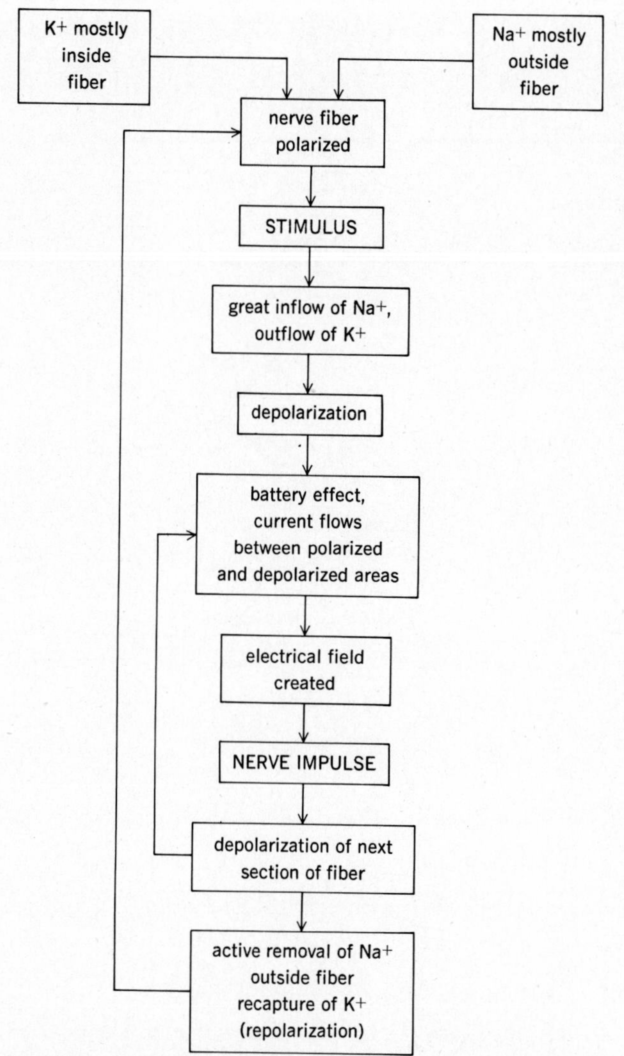

FIGURE 11.7 The events in formation and conduction of a nerve impulse.

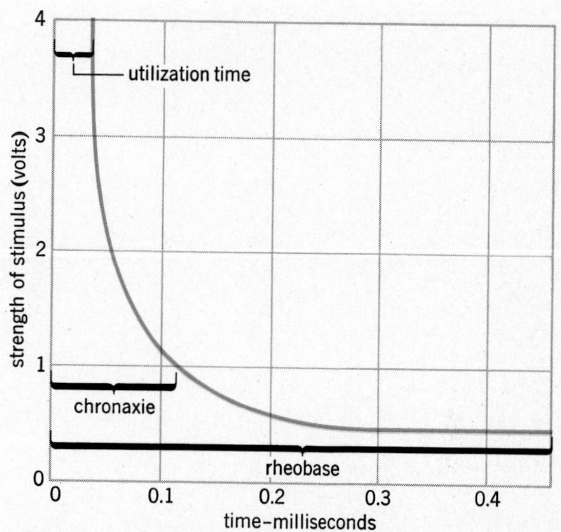

FIGURE 11.8 A strength-duration curve for the stimulation of a large myelinated axon. The stronger the stimulus, the shorter time it must act.

by itself to produce depolarization, it does alter the resting potential of the nerve fiber by lowering it slightly. If a series of closely spaced subthreshold stimuli are delivered to the fiber, each reduces the potential slightly and eventually the threshold for stimulation may be reached. This type of summation (adding) of stimulus effect is termed TEMPORAL SUMMATION, or summation in time. SPATIAL SUMMATION also exists, whereby subliminal stimuli are carried over several neurons to a final common neuron that responds to each weak stimulus by a slight lowering of the resting potential. The additive effect may result in depolarization of the "final common path" neuron, because the effect of each preceding stimulus lasts about 15 ms.

STRENGTH-DURATION RELATIONSHIPS. Even if a stimulus is of threshold strength it must act a certain length of time to be effective. The higher the strength of the current, the shorter is the time required for the stimulus to act. These relationships are depicted in a strength-duration curve, shown in Fig. 11.8. The term RHEOBASE (threshold) is used to define the minimum current strength required to excite, and CHRONAXIE is the duration required for a 2× rheobase strength stimulus to excite the neuron. The latter quantity provides a convenient measure of rapidity of response in neurons. Chronaxie is preferable to rheobase in

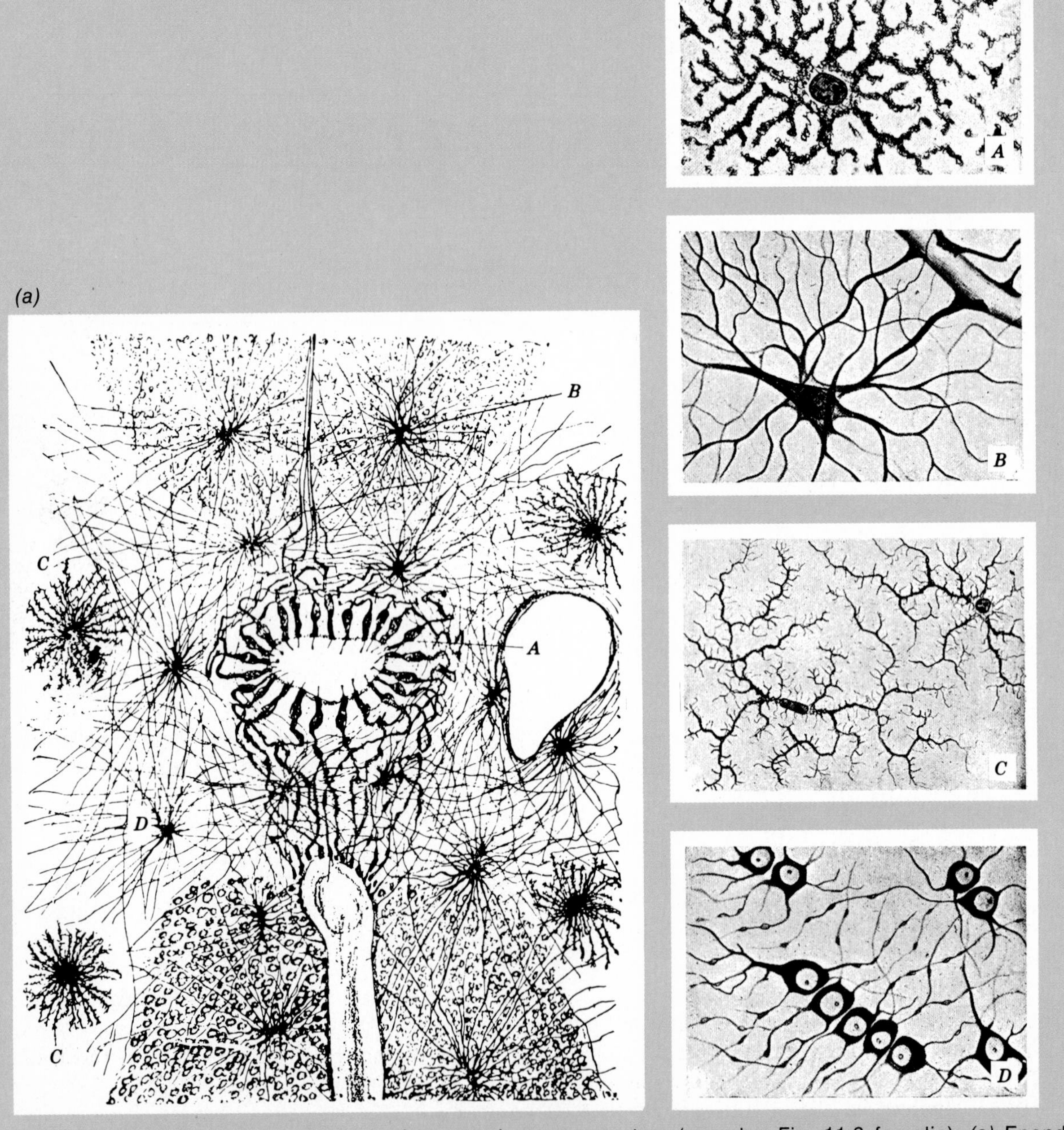

FIGURE 11.9 Some types of glia in the central nervous system (see also Fig. 11.3 for glia). *(a)* Ependyma and neuroglia in the region of the central canal of a child's spinal cord: A, Ependymal cells; B and D, fibrous astrocytes; C, protoplasmic astrocytes. Golgi method. *(b)* Interstitial cells of the central nervous system: A, Protoplasmic neuroglia; B, fibrous neuroglia; C, Microglia; D, oligodendroglia. (Courtesy W. B. Saunders, from J. W. Ranson and S. L. Clark, *The Anatomy of the Nervous System,* Tenth Edition, Saunders, Philadelphia, 1964.)

determinations of excitability because rheobases are not constant.

ACCOMMODATION. The application of a *continued* or chronic rheobasic current to a nerve fiber results in an increase in threshold termed *accommodation.* It is believed that the current allows ions to build up on the outside of the fibers at the point of stimulus application and makes the fiber more difficult to excite. This property implies that continued stimuli of given strength will eventually be ignored as threshold rises. This phenomenon is the antithesis of summation.

ALL-OR-NONE RESPONSES. For conducted impulses, the law of a maximal response or no response to an adequate stimulus is followed. A stimulus is either too weak to cause a propagated impulse or it is not. Subthreshold stimuli can, however, cause local, nonconducted alterations in membrane potential.

REFRACTORY PERIOD. For a short time after the development of an action potential, it is not possible to cause another impulse to be generated, no matter what strength of stimulus is employed. This period of time is designated as the ABSOLUTE REFRACTORY PERIOD; it is related to the time that the neuron is in a depolarized condition. As repolarization progresses, a stronger-than-normal stimulus may cause another impulse to be generated, even though a full return to original status has not occurred. This period of time is designated as the RELATIVE REFRACTORY PERIOD. Absolute refractory periods in the largest axons last only about 1 ms and theoretically such an axon can be driven at a frequency of 1000 stimuli per second.

The structure and function of glia

The glial elements (Fig. 11.9) are the nonconductile cells of the nervous system and are of several types.

EPENDYMA are of ectodermal origin; they line the cavities of the central nervous system. They are wedge-shaped, ciliated cells, and appear to act only as an epithelium for the cavities of the central nervous system.

ASTROCYTES, also of ectodermal origin, form supporting networks in the central nervous system and commonly attach to blood vessels. They are believed to transfer nutrients from blood vessels to neurons.

OLIGODENDROCYTES, of ectodermal origin, form myelin in the central nervous system.

MICROGLIA are small glial cells in the central nervous system that are of mesodermal origin, and that may become phagocytic under certain conditions. They are stimulated to show their phagocytic properties in the case of injury to the central nervous system.

SCHWANN CELLS, of ectodermal origin, have already been described.

The synapse

A SYNAPSE is defined as an area of functional but not of anatomical continuity between two neurons. The axon of one neuron *(presynaptic)* and the dendrites and/or soma of another neuron *(postsynaptic)* are approximated to one another. It is an area where transmission of information may be blocked, passed, or altered, and is thus an area where control over the impulse is exerted. It is possessed of properties different from those of the neurons forming it. Mammalian synapses transmit chemically, although electrical synapses have been discovered in invertebrates and lower vertebrates.

Chemical transmission across the synapse

The terminal ends of axons may be enlarged to form end feet, end knobs, boutons terminaux (end buttons), or other specialized structures. Within the enlargements are found small VESICLES or granules of chemical substances. Only one chemical will be found in the vesicles of a given axon. A nerve impulse arriving at an enlargement is believed to alter its membrane permeability and allow the entry from the extracellular fluids of a material (Ca^{++}?) that disrupts the vesicles, causing release of the chemical. The chemical then diffuses across the synaptic cleft to influence the membrane of the next neuron. In order that the effect of the chemical not persist, an ENZYME that destroys the transmitter agent is present. An additional method of removing the transmitter from the synapse is uptake by the postsynaptic neurons. A variety of possible and actual transmitter agents with corresponding enzymes are known to exist. These are summarized in Table 11.3. A chemical method of transmitting impulses across the synapse endows it with properties different than those of the neurons themselves.

Physiological properties of synapses

ONE-WAY CONDUCTION is explained by the fact that only axonal endings contain the transmitter

TABLE 11.3 Synaptic transmitters

Substance or agent	Enzyme or process destroying the agent	Where found, highest concentration	Comments
Acetylcholine	Cholinesterase	Neuromuscular junctions; peripheral synapses between neurons; brain stem, thalamus, cortex, retina	In CNS, not distributed evenly. Perhaps serves specific functions in each area
Norepinephrine	Monamine oxidase (MAO) or catechol-o-methyl transferase (COMT)	Postganglionic sympathetic neurons; midbrain, pons, medulla	Low concentration in cerebrum
Serotonin	MAO	Hypothalamus, basal ganglia, spinal cord	May be involved in expression of behavior
Histamine	Imidazole-N-methyl transferase, histaminase	Hypothalamus	In mast cells (tissue basophils)
Amino acids, for example GABA (gamma amino-butyric acid)	Oxidation or conversion	Cortex, visual area	GABA is inhibitory to most synapses

vesicles and therefore only an axon-dendrite transmission can occur.

FACILITATION (increasing the ease of synaptic passage) of the impulse occurs when the chemical lowers the potential (hypopolarization) of the postsynaptic neuron and creates an *excitatory postsynaptic potential* (EPSP). Summation (see below) may occur in this manner. The EPSP enables easier depolarization of the neuron. Both acetylcholine and norepinephrine appear to produce EPSPs.

INHIBITION occurs if the chemical transmitter raises the potential (hyperpolarization) of the postsynaptic neuron and it is rendered extremely difficult to excite. In this case an *inhibitory postsynaptic potential* (IPSP) is created. Gamma aminobutyric acid (GABA) is a chemical producing IPSP.

SYNAPTIC DELAY, or, the time for an impulse to cross the synapse, results from the time required for vesicle breakdown, diffusion of chemical, and excitation of the postsynaptic neuron. The length typically amounts to 0.5–1.0 ms.

FATIGUE may occur at the synapse if a high frequency of stimulation releases the transmitter chemicals from the vesicles at faster rates than they may be resynthesized. Synaptic fatigue is evidenced by failure of impulses to cross the synapse and demonstration of ability of both pre- and postsynaptic neurons still to carry impulses if stimulated independently. Neurons apparently do not fatigue.

OCCLUSION occurs when stimulation of two separate but closely associated neurons give a greater total postsynaptic potential if stimulated separately than if stimulated together. The phenomenon is explained by the fact that areas served by each fiber overlap and, if stimulated simultaneously, excite some of the same neurons (Fig. 11.10).

TEMPORAL SUMMATION of subthreshold stimuli occurs through persistence of chemical in the synapse for about 15 ms. If a small amount of chemical is released, a slight decrease in postsynaptic membrane potential occurs. Another subthreshold stimulus adds another small amount of chemical, and eventually sufficient chemical may accumulate to trigger depolarization of the postsynaptic neuron.

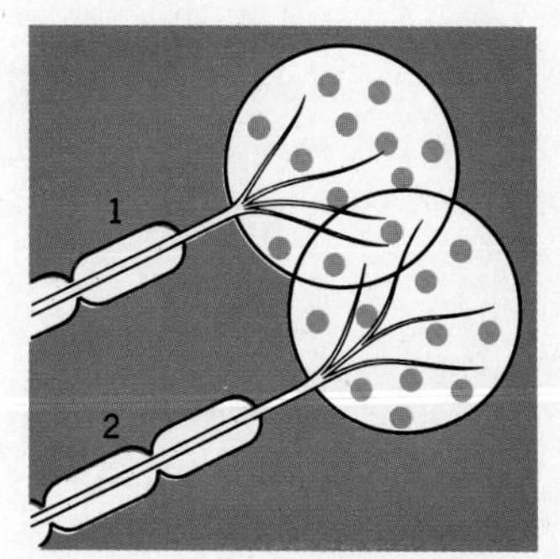

FIGURE 11.10 The production of occlusion. Each large circle contains 12 "responding units." If (1) is stimulated, 12 units respond; if (2) is stimulated, 12 units respond; if (1) and (2) are stimulated, 22 units respond.

SPATIAL SUMMATION occurs when subthreshold amounts of chemical are released by several neurons impinging on a common neuron, with the total amount of chemical exceeding the threshold of the postsynaptic neuron and causing its depolarization.

REPETITIVE DISCHARGE in the postsynaptic neuron may occur when a single stimulus is applied to the presynaptic neuron. This is explainable by assuming persistence of chemical in the synapse coupled with a short postsynaptic neuron refractory period. The postsynaptic neuron, on achieving repolarization, still faces enough chemical to cause depolarization again. This will continue until the enzyme brings chemical levels to below threshold levels.

ENLARGEMENT OF SYNAPTIC FEET has been observed in certain synapses as learning and conditioning occur. A larger knob is regarded by some as one that contains more chemical and that can release more upon stimulation. Such a synapse could then form a preferred pathway for impulse conduction, a requisite for learning or habituation of response.

DRUG EFFECTS may be exerted on synapses. Caffeine is a common drug that increases the ease of stimulation of postsynaptic neurons. Hypnotics, analgesics, and anesthetics (bromides, aspirin, morphine, opium) increase thresholds for excitation and thereby decrease synaptic activity. Curare blocks myoneural junction transmission by preventing acetylcholine from engaging the receptor sites on the motor end plate, and no depolarization of the membrane can occur. Physostigmine, parathion, malathion, and strychnine inactivate cholinesterase and thus allow continuance of the effect of acetylcholine, by preventing its hydrolysis. This can result in muscular paralysis.

CHANGES IN pH alter synaptic activity. Alkalosis increases excitability, while acidosis depresses it. Epileptic seizures may be precipitated by hyperventilation that lowers H_2CO_3 and increases pH in an individual who has an epileptic locus in the brain.

HYPOXIA produces cessation of synaptic activity. As one peruses this list of synaptic properties, the number implying alteration or control exerted on the impulse should be noted. A great degree of control, by inhibition, facilitation, summation, and other processes, is therefore exerted at the synaptic level, particularly in terms of screening out irrelevant information and amplifying those of importance.

The reflex arc; properties of reflexes

The REFLEX ARC (Fig. 11.11) is the simplest functional unit of the nervous system capable of detecting change and causing a response to that change. Many body activities (for example, heart and breathing rates, gut motility) are controlled reflexly. This term implies an AUTOMATIC ADJUSTMENT to maintain homeostasis without conscious effort. A reflex arc has five basic parts:

A *receptor* to detect change.

An *afferent neuron* to conduct the impulse resulting from stimulation of the receptor, to the central nervous system.

A *center* or synapse where a junction is made between neurons. An *internuncial neuron* is commonly found at this point.

An *efferent neuron* to conduct impulses for appropriate response to an organ.

An *effector*, or organ that does something to maintain homeostasis.

Functionally, reflexes may be classed as *extero-*

FIGURE 11.11 The components of a reflex arc.

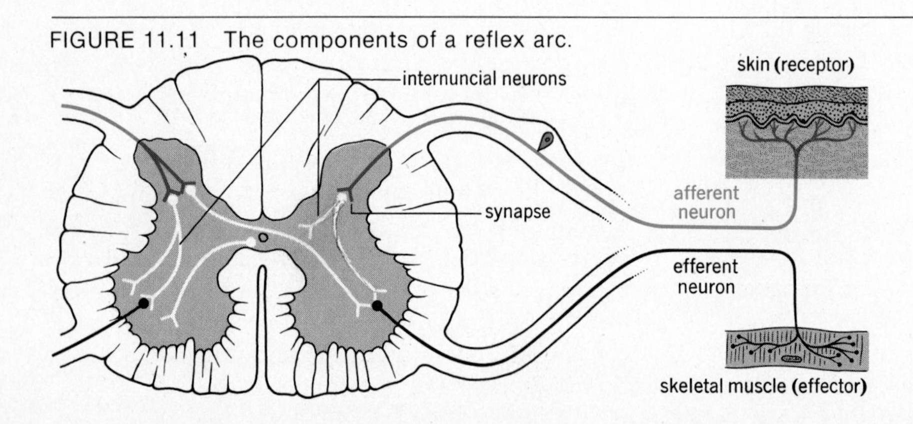

ceptive if the receptor is at or near a body surface; *interoceptive* if the receptor is within a visceral organ; *proprioceptive* if the receptor is in a muscle or tendon.

Activity that results from passage of impulses over a reflex arc is called the REFLEX ACT or the REFLEX. Reflex acts, regardless of what the stimulus and response are, have several characteristics in common.

Activity of the effector depends on stimulation of a receptor. That stimulation may or not rise to the conscious level. The arc serving the act must be complete; if any part of the arc is not working, the effector will not respond.

Reflex activity is *involuntary;* it cannot be stopped by an act of will. Reflex activity is stereotyped (repetitious); stimulation of a given receptor always causes the same effector response. Thus, if the patellar tendon is tapped, the knee always extends. It is thus *predictable.*

Reflex activity is *specific* in terms of the stimulus and the response. A particular stimulus causes a particular response. Response is usually *purposeful.*

The value of these reflex properties to the body is obvious. A particular stimulus causes a predictable and appropriate response, which will usually result in or aid in maintenance of homeostasis. If responses were different to the same stimulus, the body would have a difficult time maintaining its posture, balance, and overall functions.

Clinical considerations

Degeneration and regeneration in the nervous system

Injury to neuron cell bodies is usually followed by death of the entire neuron. Damage to an axon results in changes in the cell body (the axon reaction) and in the axon distal to the point of injury (Wallerian degeneration).

In the AXON REACTION, the cell body swells, and the Nissl substance fragments into fine granules. In WALLERIAN DEGENERATION, the distal segment of the axon swells, the myelin sheath degenerates, the axon itself degenerates, and only the outer layer of the neurilemma or glial cells remains. The products of degeneration are removed by glial cells.

In peripheral axons, ameboid, bulbous ends develop on the injured axon, and these grow through the neurilemmal tubes at about 2 mm per day. Reestablishment of connections, if it occurs, is more widespread than originally, and control is less precise than before.

In the central nervous system, regeneration is more limited than peripherally. The damaged region is filled with a special type of scar tissue by astrocyte proliferation that tends to block passage of any regenerating axons. Since central axons lack neurilemmal sheaths, guiding of the regenerating axon is lacking, and reestablishment of normal connections is haphazard and is complicated by the fact that millions of axons are present in the cord and brain. Thus, chances of making proper connections are low.

Disorders

MULTIPLE SCLEROSIS is a disease of unknown origin and has no cure. It is characterized by degeneration of myelin sheaths in the brain and spinal cord. Changes in function that commonly occur include: lack of ability to concentrate, speech disturbances, visual changes, and increased vigor of knee and ankle jerk reflexes.

POLIOMYELITIS (infantile paralysis, acute anterior poliomyelitis) is an acute viral infection that may result in destruction of the cell bodies of the neurons to skeletal muscle (paralysis will result), or of certain of the neurons, responsible for breathing, in the medulla of the brainstem ("bulbar" polio). The viruses are apparently in the system all the time, and resistance to the disease depends on the development of immunity against the organisms. Immunity is easily acquired by vaccination of the individual by dead virus (Salk vaccine) given parenterally (by any route other than the alimentary tract), or by the oral ingestion of live, weakened virus (Sabin vaccine). Many states in the United States require that a child present evidence of vaccination against polio before starting school; many children are not vaccinated until this time, and thus the greatest danger of infection is in the preschool population. Statistics indicate a slow increase in polio cases, probably as a result of failure to immunize preschool children against the disease.

NEURITIS is a term referring to a syndrome of reflex, sensory, motor, and vasomotor symptoms that occurs as a result of a lesion in a nerve root to or from the central nervous system, or as a result of a lesion of a peripheral nerve. Mechanical damage (trauma), infections (bacteria or viruses), or toxins (poisons) are most commonly involved as etiologic agents in neuritis. Pins-and-needles sensations, pain, numbness, muscular weakness, and sweating of the affected body part are some of the more common symptoms that develop.

NEURALGIA is a term referring to pain that develops within the distribution of a peripheral sensory nerve when no apparent cause for the pain can be demonstrated. The pain is commonly described as intense, stabbing, or shooting in quality, and is difficult to relieve. Analgesics usually afford only temporary relief.

SHINGLES *(herpes zoster)* is an acute infection of sensory nerves as they enter the spinal cord and is due to a virus. Recovery is usual as antibodies are produced to the virus.

Summary

1. The nervous system provides a means of controlling activities in the body.
2. The system is primarily ectodermal in origin.
   a. The brain and spinal cord develop from the neural tube.
   b. Peripheral nerves are outgrowths of the tube.
   c. The forebrain forms the cerebrum, thalamus, and hypothalamus; the midbrain remains as such; the hindbrain forms the pons, cerebellum, and medulla.
3. The system is organized into central and peripheral portions.
   a. The peripheral is subdivided twice more, into somatic and autonomic; the autonomic into parasympathetic and sympathetic.
4. Neurons and glia form the nervous system.
   a. Neurons are excitable and conductile and have a structure similar to other cells.
   b. The characteristic thing about many neurons is their length.
   c. Glia are connective, supportive, and nutritive cells; some also are phagocytic and form myelin.
5. Excitability depends on separation of ions, chiefly Na^+, and K^+.
   a. A fiber is polarized when ions are separated.
   b. A stimulus causes loss of ion separation, allowing Na^+ to enter the fiber.
   c. Repolarization occurs as K^+ diffuses out and the $Na^+ - K^+$ pump restores resting concentrations by active transport.
6. Conductivity depends on current flow between polarized and depolarized areas, and movement of that current along the fiber.
   a. Speed of conduction depends on whether the fiber has a myelin sheath (faster) or is larger in diameter (faster).
7. Neurons show other physiological properties.
   a. They depolarize completely or do not at all when stimulated (all-or-none law).
   b. They repolarize rapidly (short refractory period).
   c. They have stimulus thresholds.
   d. They must be stimulated for a certain time before they depolarize.
   e. They require nutrients to live.
8. A synapse is a junction between two neurons.
   a. No anatomical connections exist.
   b. Conduction is chemical.
   c. Synapses control passage across themselves.
9. Synaptic properties differ from those of neurons.
   a. Conduction is one way.
   b. Impulses may pass more easily or be blocked.
   c. It takes longer to cross a synapse.
   d. Drugs affect synapses easily.

10. Neurons form reflex arcs as the simplest multineuron controlling device.
    a. Reflex arcs always have five parts: receptor, afferent nerve, synapse (center), efferent nerve, and effector.

11. Peripheral nerve fibers may regenerate; central ones show little regeneration. Neuron cell body damage usually causes death of the entire neuron. Demyelinating diseases result in loss of function.

12 The Spinal Cord and Spinal Nerves

The spinal cord

The location and structure of the cord

The spinal cord is the portion of the central nervous system located within the vertebral or spinal canal formed by the arches of the vertebrae. It, like the brain, is surrounded by three membranes collectively known as the MENINGES. An outer, tough *dura mater* ("hard mother") serves as the

FIGURE 12.1 Three views of the gross anatomy of the spinal cord. *(a)* Lateral view, *(b)* anterior view, *(c)* posterior view.

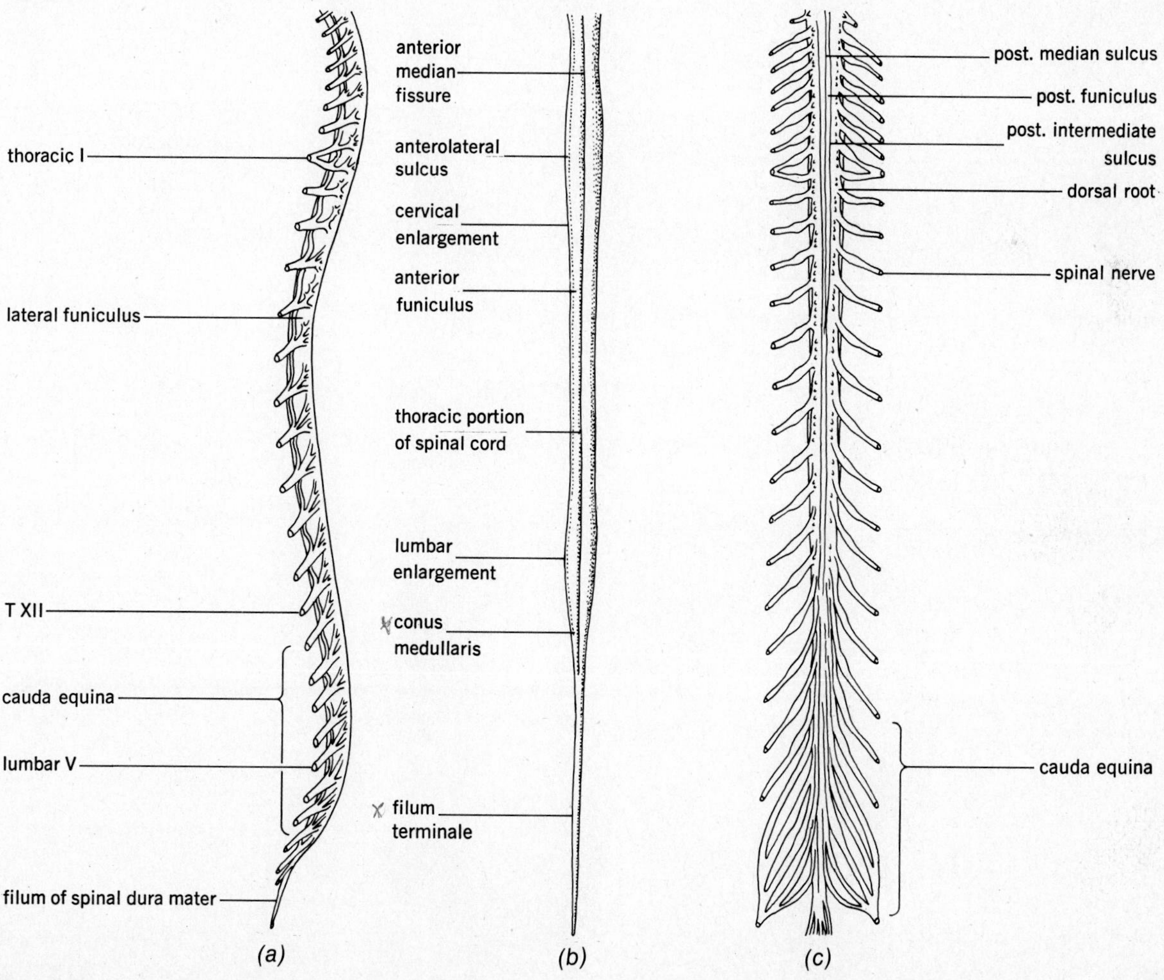

major protective membrane. A middle *arachnoid* (like a web of a spider) is a delicate membrane enclosing a subarachnoid space filled with the *cerebrospinal fluid.* An inner *pia mater* ("tender mother") is a thin vascular membrane carrying blood vessels into the cord. The diameter of the cord itself does not exceed three-fourths of an inch, and thus it does not completely fill the spinal canal. The meninges and loose connective tissue fill the space of the canal not occupied by the cord itself.

In the infant, the cord has nearly the same length as the spinal column. As growth occurs, the column grows at a faster rate than does the

FIGURE 12.2 A cross section of the spinal cord.

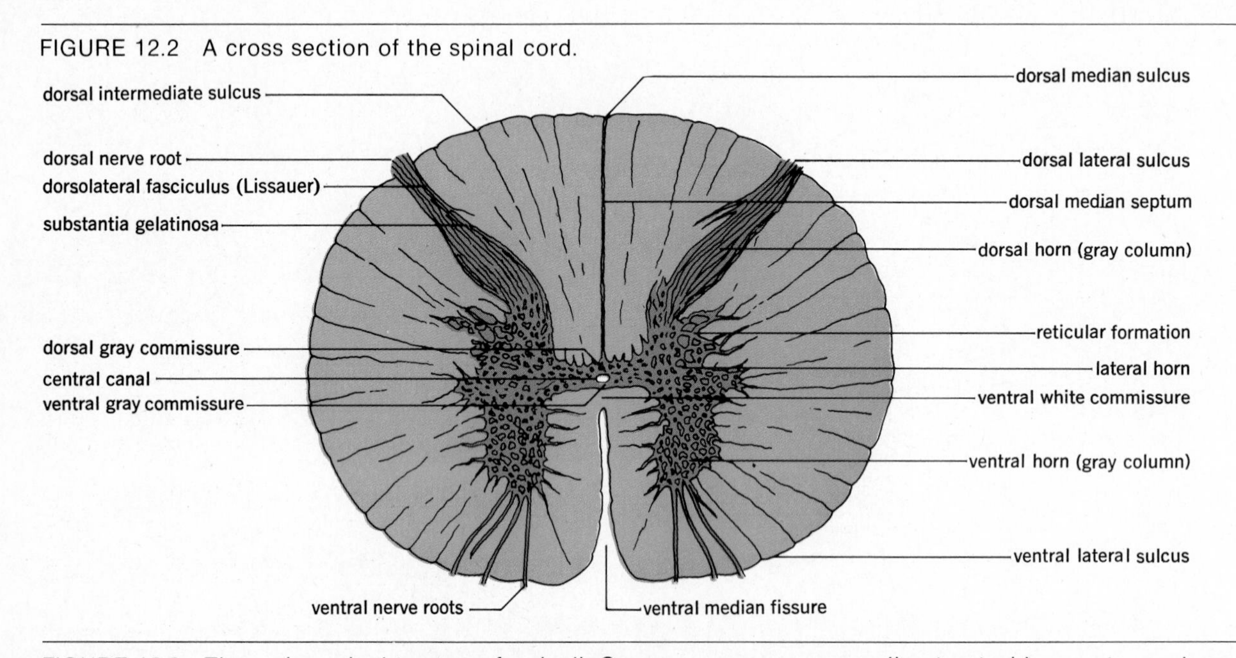

FIGURE 12.3 The major spinal tracts or fasciculi. Orange: sensory or ascending tracts, blue: motor or descending tracts.

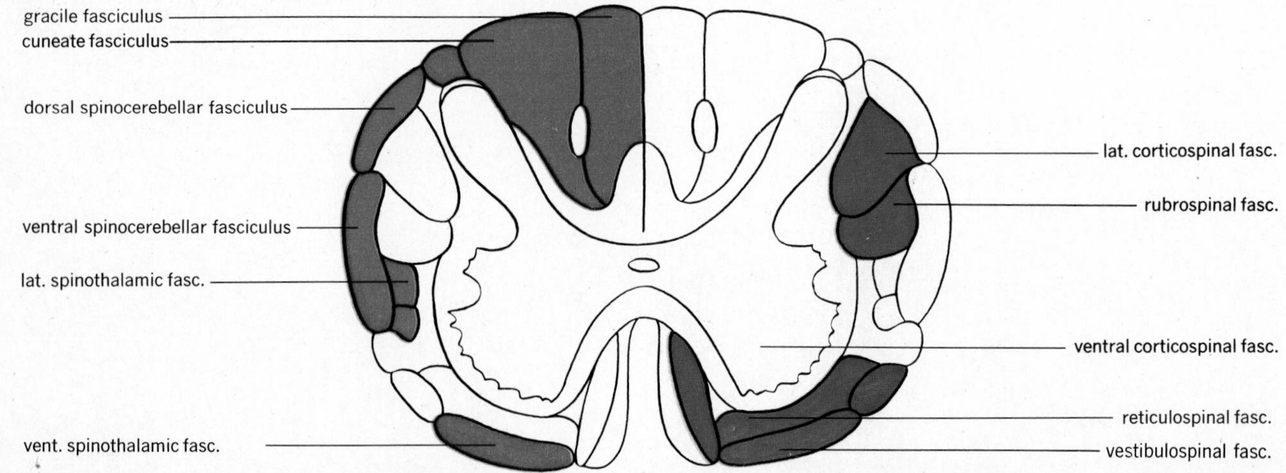

TABLE 12.1 Major spinal tracts

	Name of tract	Area of origin, or region from which tract receives fibers	Area of termination	Crossed or uncrossed	Function or impulse(s) carried
DESCENDING	Corticospinal				
	Lateral	Motor areas of cerebral cortex	Synapses in cord	Crossed in medulla, 80% of fibers	Carry voluntary impulses to skeletal muscles
	Ventral			Uncrossed, 20% of fibers	
	Rubrospinal	Red nucleus of basal ganglia in cerebrum	Synapses in cord	Crossed in brainstem	Involuntary impulses to skeletal muscles concerned with tone, posture.
	Reticulospinal	Reticular formation of brainstem	Synapses in cord	Crossed in brainstem	Increases skeletal muscle tone and motor neuron activity
	Vestibulospinal	Vestibular nuclei of brainstem	Synapses on motor neurons of cord	Uncrossed	regulates muscle tone to maintain balance and equilibrium
ASCENDING	Spinothalamic				
	Lateral	Skin	Thalamus, relays to cerebral cortex	Crossed in cord	Pain and temperature
	Ventral	Skin	Thalamus, relays to cerebral cortex	Crossed in cord	Crude touch
	Spinocerebellar				
	Dorsal	Muscles and tendons	Cerebellum	Uncrossed	Unconscious muscle sense for controlling muscle tone, posture
	Ventral		Cerebellum	Uncrossed	
	Gracile	Skin and muscles	Medulla, relays to cerebral cortex	Uncrossed	Touch, pressure, two-point discrimination, conscious muscle sense concerned with appreciation of body position
	Cuneate			Uncrossed	

cord, and thus the cord, in the adult, does not occupy the full length of the spinal canal. In the adult, the cord is about 18 inches in length; it stops at the level of the second lumbar vertebra, a fact of importance when performing a spinal tap or puncture.

An external view of the cord (Fig. 12.1) shows it to be grooved by several SULCI or shallow indentations, and by a deeper anteriorly placed FISSURE. It may be divided, on the basis of what portions give rise to spinal nerves that exit between the four regions of vertebrae, into CERVICAL (C), THORACIC (T), LUMBAR (L), and SACRAL (S) regions. An upper CERVICAL ENLARGEMENT marks the entry into the cord of nerves from the upper limb, and a lower LUMBAR ENLARGEMENT marks the entry into the cord of nerves from the lower limb. The tapering end of the cord is the CONUS MEDULLARIS, from which the FILUM TERMINALE, a fibrous derivative of the pia mater, extends through the rest of the vertebral canal and aids in anchoring the cord.

A cross section of the cord (Fig. 12.2) shows again the sulci and fissure mentioned above and the organization of the substance of the cord into an inner H-shaped mass of GRAY MATTER, surrounded by WHITE MATTER. The gray matter consists of neuron cell bodies, synapses, and many blood vessels, all of which contribute to its grayish-red color in fresh section. There are several COLUMNS or horns of gray matter and COMMISSURES connecting the two sides of the cord. These subdivisions are labeled in Fig. 12.2. The white matter is divided, by the gray columns, into three regions or FUNICULI, also labeled in Fig. 12.2. These white-matter areas consist primarily of myelinated nerve fibers traveling up and down the cord. Myelin, in fresh section, has a yellowish-white color, hence the general name of these regions.

The white matter contains functional areas known as TRACTS (Fig. 12.3). The tracts are not visible as separate anatomical structures, but their presence may be demonstrated by experimental procedures or by the type of loss that occurs when the cord had been damaged. The names and functions of the major spinal tracts are presented in Table 12.1. The fibers of the *descending tracts* constitute the upper motor neurons of the system supplying skeletal muscle with fibers. *Ascending tracts* are sensory in function, conveying impulses from sensory receptors to the brainstem nuclei, cerebellum, and cerebrum.

The cord and reflex activity

In the human, most of the controlling and analytic functions of the cord that appear in the lower animals have been moved into the brain. The human cord thus serves as an area for sensory input, and houses the cell bodies of somatic and autonomic neurons that supply muscle and glands in the body. Since synapses may occur within the cord between sensory and motor neurons, reflex arcs are possible, and *most of the activity the cord itself controls is reflex in nature.* Some spinally controlled reflex activities are described below.

Reflex control appears to be exerted primarily on muscular activity, requiring the presence of receptors known as MUSCLE SPINDLES (Fig. 12.4) or receptors (pain, heat) in the skin. Muscle spindles contain several tiny skeletal muscle fibers known as INTRAFUSAL FIBERS, in contrast to the EXTRAFUSAL FIBERS of the muscle itself. Two types of intrafusal fibers, distinguished by arrangement of their nuclei, are surrounded by centrally placed spiral sensory endings of what are called *Ia afferent neurons.* The intrafusal fibers receive motor neurons called *gamma efferent neurons,* and these fibers control the contraction of the outer ends of intrafusal fibers and thus the tension or stretch applied to the central area in which the sensory endings are found. The spindle as a whole responds to stretching of the muscle, but the gamma fibers control when the spindle "fires" by adjusting the tension on the intrafusal fibers. In general, the greater the contraction of the intrafusal

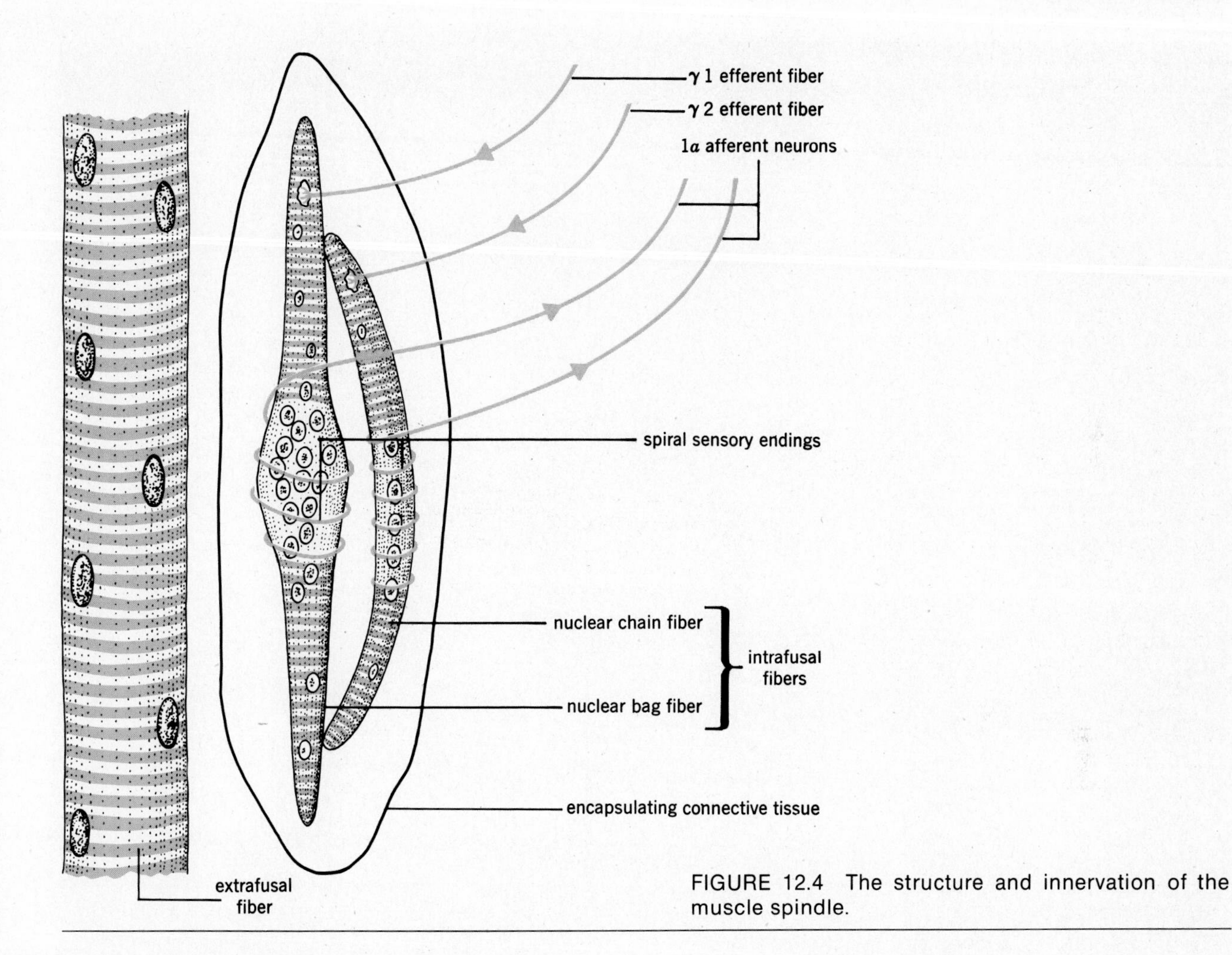

FIGURE 12.4 The structure and innervation of the muscle spindle.

fibers, the less stretch required of the muscle to cause the spindle to form a new impulse.

Specific reflex activities of muscle mediated by the spindles include the following:

EXTENSOR THRUST REFLEX. If pressure is applied to the sole of the foot, the leg extensors are stretched and contract reflexly. This tends to maintain posture.

STRETCH (MYOTATIC) REFLEXES. A sudden tension such as tapping, applied to a muscle or its tendon, momentarily stretches the muscle receptors and causes a reflex contraction of the same muscle. A "jerk" results. The knee jerk, ankle jerk, and abdominal reflexes are examples of myotatic reflexes. These "simple" reflexes are usually two-neuron reflexes and are controlled by somewhat restricted or localized levels of the cord. They are commonly used by the neurologist to assess cord function at various levels, as shown in Table 12.2.

WITHDRAWAL REFLEX. A reflex involving skin

TABLE 12.2 Some reflexes used to assess cord function

Name of reflex	Level of cord involved	How elicited	Interpretation
Jaw jerk	C3	Tap chin with percussion hammer; jaw closes slightly.	Marked reaction indicates upper neuron lesion.
Biceps jerk	C5,6	Tap biceps tendon with percussion hammer; biceps contracts.	If reflex is absent or greatly exaggerated, cord damage, or damage to sensory or motor nerves is probable.
Triceps jerk	C7,8	Tap triceps tendon with percussion hammer; triceps contracts.	
Abdominal reflexes	T9-L2	Draw key or similar object across abdomen at different levels; muscles contract.	
Knee jerk	L2,3	Tap patellar tendon with percussion hammer; quadriceps contract to extend knee.	
Ankle jerk	L5	Tap Achilles tendon; gastrocnemius contracts to plantar flex foot.	
Plantar flexion	S1	Firmly stroke sole of foot from heel to big toe; toes should flex.	If toes fan and big toe extends (Babinski sign), upper neuron lesion is present.

receptors is the withdrawal reflex. Painful stimulation of skin receptors (e.g., touching a hot object) typically causes withdrawal of the body part from the stimulus in what is termed a FLEXION REFLEX. In many cases, if one limb is flexed, the other must extend to maintain balance and posture and to aid in pushing the body away from the stimulus. This latter response is termed a CROSSED EXTENSION REFLEX. The nervous pathways for withdrawal reflexes are diagrammed in Fig. 12.5.

The cord controls visceral reflexes concerned with bladder and bowel evacuation. These reflexes will be discussed in greater detail in chapters to follow.

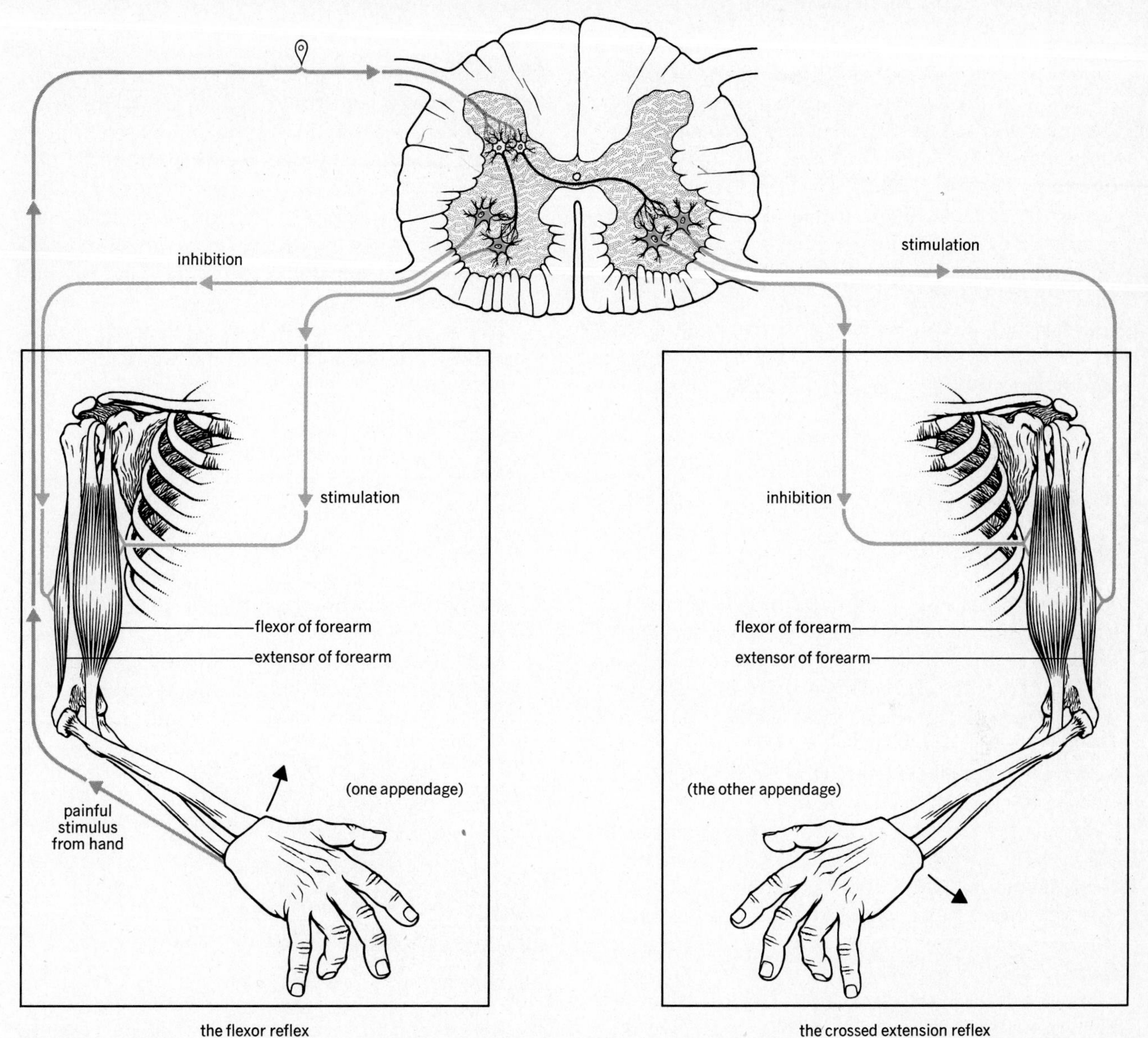

FIGURE 12.5 The interrelationships of the flexor reflex, the crossed extension reflex, and the phenomenon of reciprocal inhibition.

The pyramidal and extrapyramidal pathways

The cord also contains two functionally grouped "systems" that are responsible for the various activities skeletal muscle shows. The term PYRAMIDAL SYSTEM includes the corticospinal tract, mentioned earlier, and the corticobulbar tract related to certain cranial nerve nuclei. The neurons involved in this system are regarded as carrying the nerve impulses that result in precise, non-stereotyped voluntary movements of the skeletal muscles. The connection to skeletal muscles is made by the tract fibers (constituting the UPPER MOTOR NEURON) to motor neurons in the ventral gray column of the spinal cord. These latter neurons carry impulses peripherally to the skeletal muscles and constitute the LOWER MOTOR NEURONS.

The EXTRAPYRAMIDAL SYSTEM includes the spinal motor tracts other than corticospinal and corticobulbar (e.g., vestibulospinal, rubrospinal, reticulospinal) and conveys impulses to the lower motor neurons that are responsible for controlling muscle tone and contraction of the intrafusal fibers of the muscle spindles. In short, this "system" controls the involuntary activity of skeletal muscles resulting in maintenance of posture, reflex response, and control of muscular coordination during movements.

In damage to the cord, both systems are usually involved, and deficits in voluntary movement and tone are seen.

The spinal nerves

Number and location

The cord gives rise to 31 pairs of spinal nerves, divided into 8 cervical (C), 12 thoracic (T), 5 lumbar (L), 5 sacral (S), and 1 coccygeal. They are distributed to the skin and muscles of particular body regions (Fig. 12.6) with overlap of about 30 percent by neighboring nerves. Thus, if a given nerve is cut or damaged, complete function in a body area will not usually be lost.

The spinal nerves exit from the vertebral canal through the *intervertebral foramina* between vertebral pedicles. In the upper portions of the cord, the foramina of exit are nearly opposite the cord segment of nerve origin, and the nerves exit immediately to the side. In the lower segments, since the cord is shorter than the spinal column, nerves travel downward within the canal until they find their foramina of exit. Large numbers of these nerves form the CAUDA EQUINA (horse's tail) of the cord (see Fig. 12.1).

Segmental distribution

Segmental loss of function is more easily understood if one considers the distribution of the spinal nerves (Fig. 12.7). Each spinal nerve has a component that supplies a skin area (dermatome) with sensory fibers. Thus, testing for cutaneous sensitivity at various points in the body may reveal the presence of lesions in the spinal nerves. If a single spinal root is interrupted, however, loss is minimal, since the dermatomes supplied by adjacent spinal nerves overlap, as indicated below.

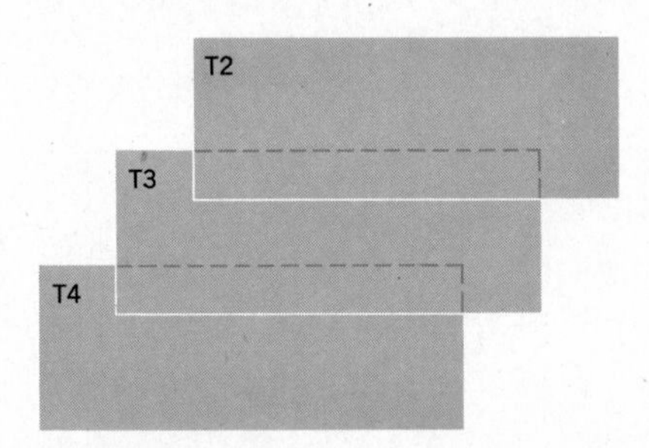

Motor components of the spinal nerves also tend to be distributed in segmental fashion, and damage to the cord is followed by motor loss in areas supplied by motor nerves originating below the level of the lesion.

Connections with the spinal cord

Each spinal nerve (Fig. 12.6) has a DORSAL ROOT *(sensory)* that conveys impulses to the cord.

FIGURE 12.6 (Top) The relationships of the spinal nerves to the spinal cord. *(a)* The cervical region, *(b)* the lumbar region, *(c)* the origin of the 31 pairs of spinal nerves shown on one side only. (Bottom) The connections of one pair of spinal nerves to the cord.

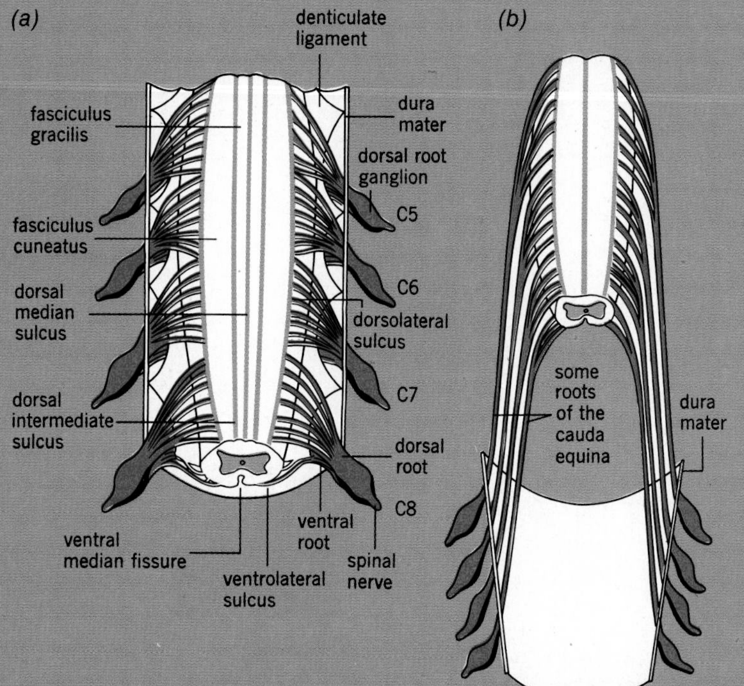

(a)

denticulate ligament
dura mater
fasciculus gracilis
dorsal root ganglion
C5
fasciculus cuneatus
C6
dorsal median sulcus
dorsolateral sulcus
C7
dorsal intermediate sulcus
dorsal root
C8
ventral root
ventral median fissure
spinal nerve
ventrolateral sulcus

(b)

some roots of the cauda equina
dura mater

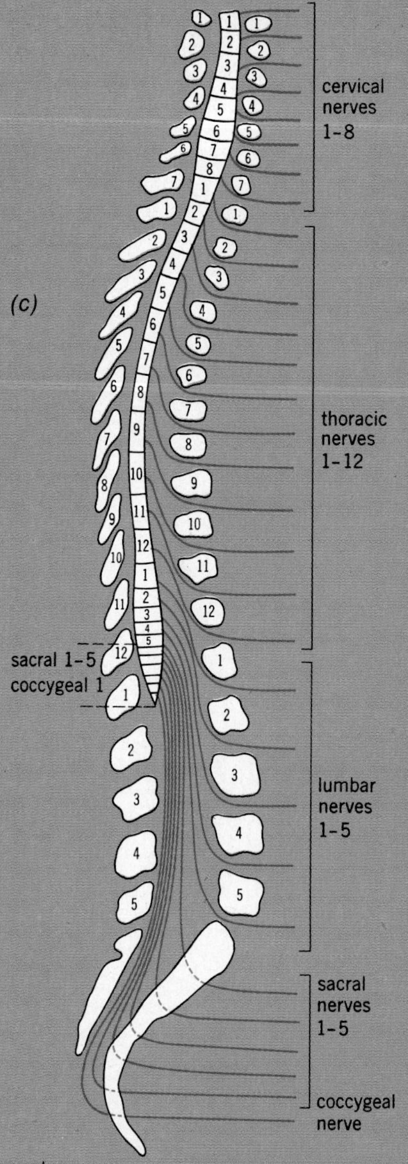

(c)

cervical nerves 1-8
thoracic nerves 1-12
sacral 1-5
coccygeal 1
lumbar nerves 1-5
sacral nerves 1-5
coccygeal nerve

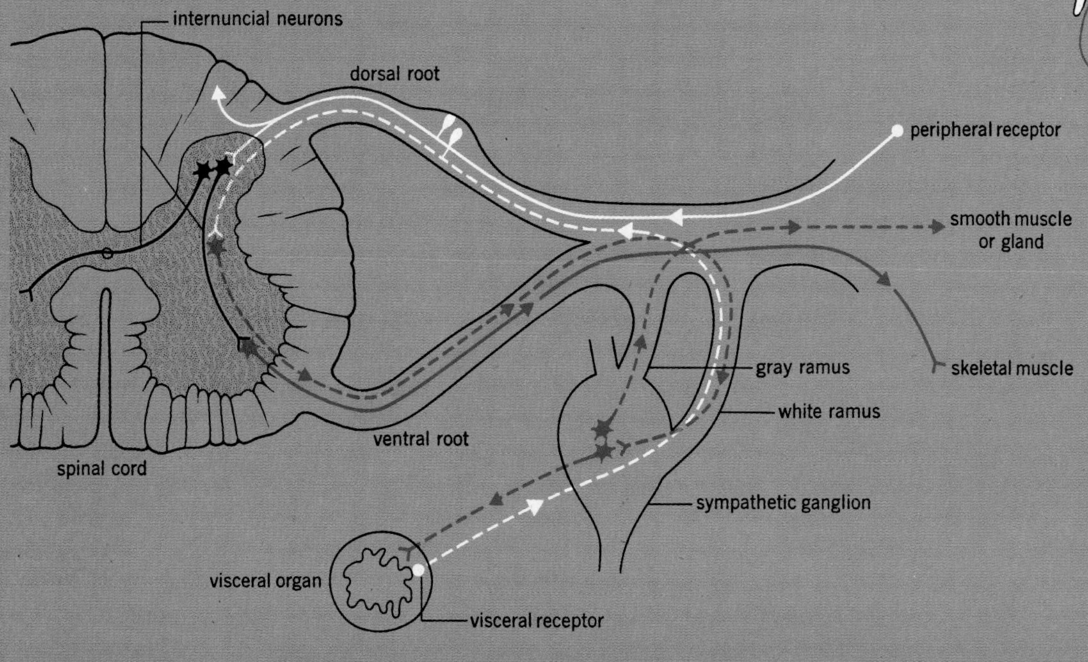

internuncial neurons
dorsal root
peripheral receptor
smooth muscle or gland
gray ramus
skeletal muscle
white ramus
ventral root
spinal cord
sympathetic ganglion
visceral organ
visceral receptor

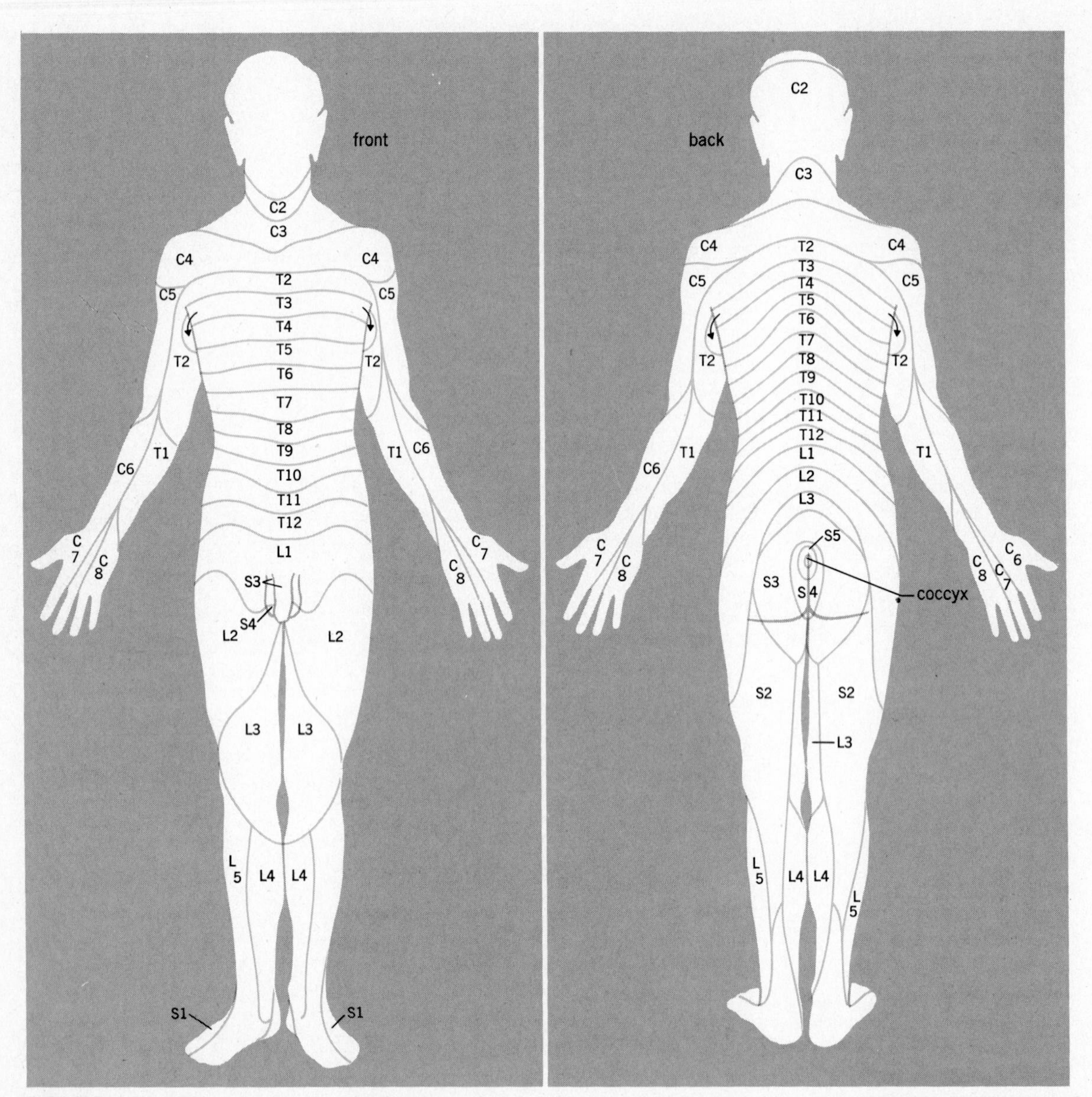

FIGURE 12.7 The segmental distribution of the spinal nerves to the skin segments (dermatomes) of the body. C = cervical, T = thoracic, L = lumbar, S = sacral.

Cell bodies of these neurons are located in a DORSAL ROOT GANGLION of each nerve. A VENTRAL ROOT *(motor)* conveys impulses from cell bodies in the cord to muscles and glands. COMMUNICATING RAMI allow connections for autonomic fibers. Peripherally, somatic nerves form anterior and posterior branches to skin and skeletal muscles. The ventral somatic roots to the skeletal muscles constitute the lower motor neurons of the motor system.

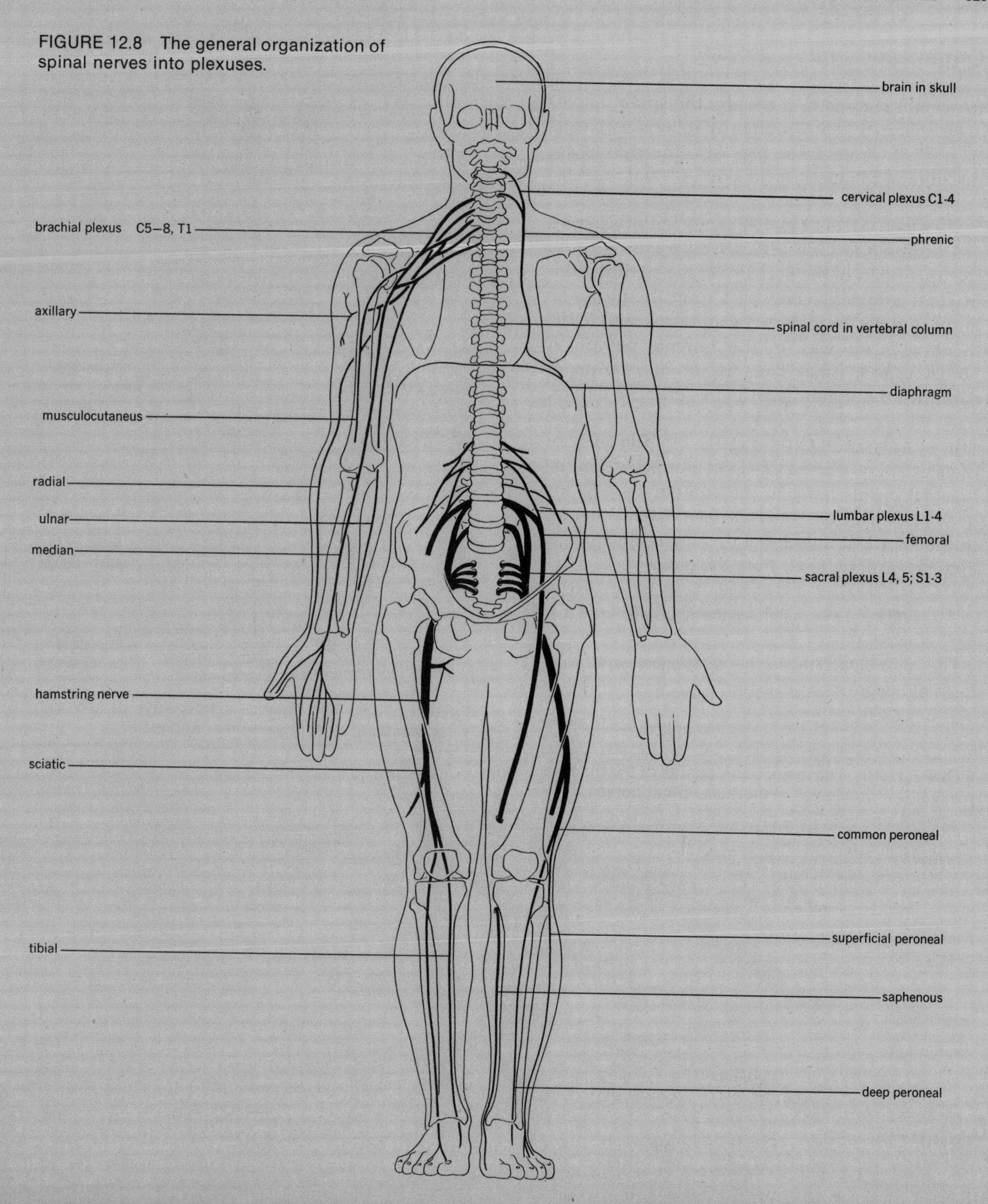

FIGURE 12.8 The general organization of spinal nerves into plexuses.

Organization into plexuses

In several regions of the body, the spinal nerves are "collected" into groups that form a plexus (Fig. 12.8). From these plexuses, the nerves are distributed to muscles and skin.

THE CERVICAL PLEXUS (Fig. 12.9). Cervical nerves 1–4 combine to form the cervical plexus lying in the upper part of the neck under the sternocleidomastoid muscle. Nerves pass from the plexus to muscles and skin of the scalp and neck, and to the diaphragm by way of the phrenic nerves. A "broken neck" may thus result in respiratory paralysis through injury to the phrenic nerves.

THE BRACHIAL PLEXUS (Fig. 12.10). The anterior roots of spinal nerves C5–T1 combine to form the brachial plexus lying in the anterior base of the neck just superior to the clavicle. It sends nerves to the skin and muscles of the upper appendage. Five major nerves exit from the plexus.

The *axillary nerve* supplies the shoulder joint and deltoid muscle.

The *musculocutaneous nerve* supplies the structures of the anterior upper arm.

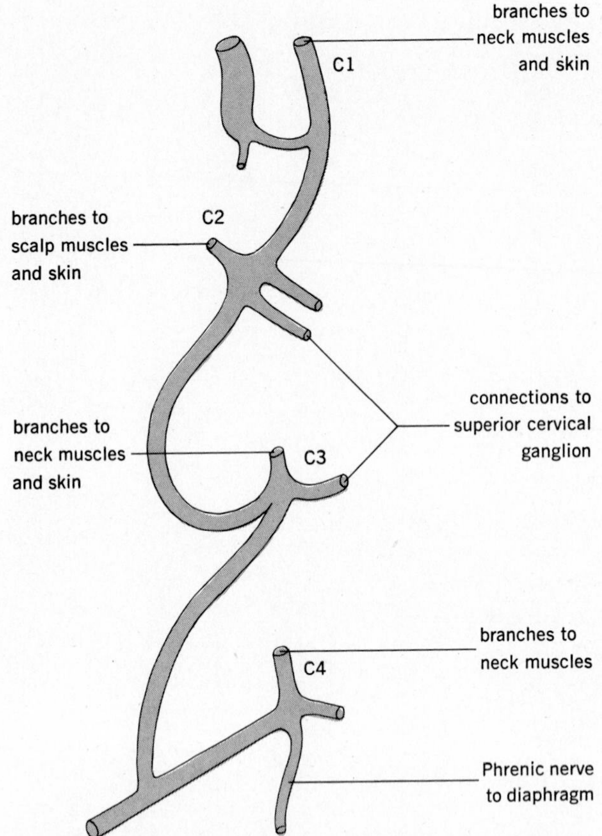

FIGURE 12.9 The major components of the cervical plexus. C = cervical spinal nerves.

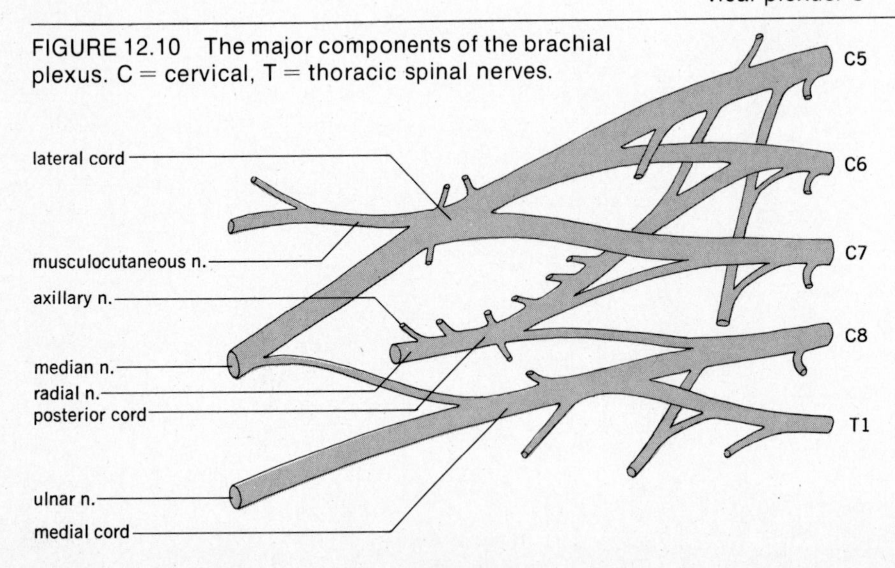

FIGURE 12.10 The major components of the brachial plexus. C = cervical, T = thoracic spinal nerves.

The *median nerve* supplies the anterior forearm.

The *ulnar nerve* serves the same areas as the median nerve.

The *radial nerve* supplies the posterior upper arm, forearm, and hand.

The nerves are superficial at the axilla. Injury here may result in paralysis. The "crazy" or "funny bone" is the ulnar nerve as it passes over the medial epicondyle of the humerus.

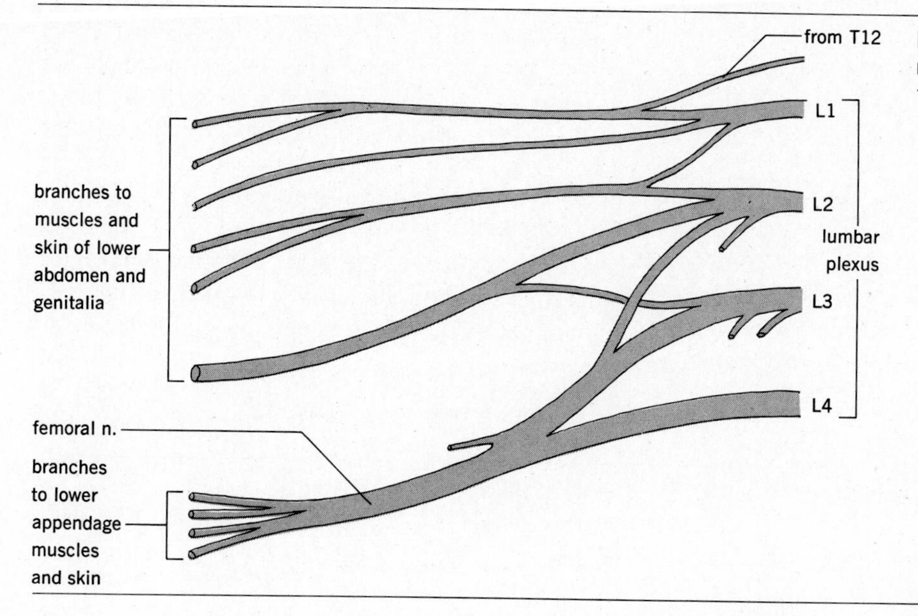

FIGURE 12.11 The major components of the lumbar plexus. T = thoracic, L = lumbar spinal nerves.

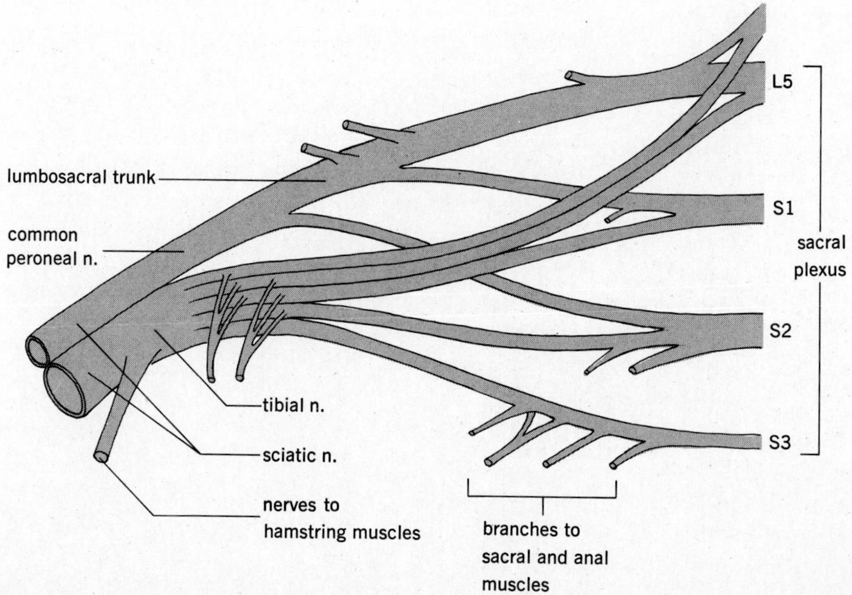

FIGURE 12.12 The major components of the sacral plexus. L = lumbar, S = sacral spinal nerves.

THE INTERCOSTAL NERVES. Thoracic nerves 2–12 form the intercostal nerves supplying the intercostal and abdominal muscles. A named plexus is not formed.

THE LUMBAR PLEXUS (Fig. 12.11). Lumbar nerves 1–4 combine to form the lumbar plexus lying within the posterior part of the psoas muscle. Three main nerves exit from the plexus.

The *femoral nerve* supplies the skin and muscles of the anterior thigh.

The *lateral cutaneous nerve* supplies the skin and muscles of the upper lateral thigh.

The *obturator nerve* supplies the skin and muscles of the medial thigh.

THE SACRAL PLEXUS (Fig. 12.12). Lumbar nerves 4 and 5, and sacral nerves 1–3, combine to form the sacral plexus lying in the middle portion of the false pelvis. Some anatomists consider the lumbar and sacral plexuses to be one large plexus, which they call the lumbosacral plexus. There is some variation in opinion as to which nerves contribute to the two plexuses, particularly the lower lumbar nerves. The major nerve of the sacral plexus is the SCIATIC NERVE, which forms several important branches.

The *nerve to the hamstring muscles* ("hamstring nerve") rises from the sciatic nerve in the thigh and supplies the skin and muscles of the posterior thigh.

The *tibial nerve* rises as one branch from the sciatic just above the knee and supplies the skin and muscles of the knee joint and posterior leg.

The *common peroneal nerve* is the other branch of the sciatic just above the knee. It continues down the leg, giving branches that supply the skin and muscles of the knee, anterior and lateral leg, and foot.

Clinical considerations

Cord and nerve injury

Dislocation of vertebrae, stab wounds, or bullet wounds may result in damage to the cord with consequent loss of function. Specific loss depends on what particular cord area and tract(s) are involved. If the cord is severed (complete section), all sensory and motor function will be lost below the cut. Paralysis is of the spastic type (muscle is paralyzed in a contracted condition), characteristic of an upper motor neuron lesion.

Damage to the lower motor neurons results in a flaccid paralysis (muscle paralyzed in a relaxed condition), restricted to the muscles served by the damaged nerve. Regeneration of upper motor neurons occurs, but the chances that normal connections will be reestablished are minimal; thus, restoration of function is poor. Peripheral nerves may regenerate, but the new connections established are more widely distributed and less precise than originally, so that function may return but will be less exact than before.

Hemisection of the cord (cutting halfway through the cord) is usually produced by surgical division of the cord and results in the Brown-Sequard syndrome. This is characterized by loss of voluntary muscle movement, touch, and pressure on the same side of the cut; and loss of pain, heat, and cold on the opposite side of the cut.

By remembering the spinal tracts and the areas supplied by the spinal nerves, it is easy to determine the spinal level of damage and the side of damage.

Spinal shock

If the cord is completely cut or is severely traumatized (injured) the following losses or alterations of function will result.

Permanent loss of voluntary muscular activity and sensation will occur in the body segments below the level of the cut, because the damaged tracts will not undergo regeneration and reestablishment of original connections.

The *development of spinal shock* will occur. This is a temporary synaptic depression that results in loss of reflex activity below the level of the cut. The length of time the state lasts depends on the position of the animal in the phylogenetic scale, extending from minutes for an animal such as a frog to months in humans. As recovery from shock proceeds, reappearance of reflex activity occurs in a typical pattern:

Knee jerk and other simple reflexes reappear.

Flexion and crossed extension reflexes reappear.

Visceral reflexes, including bladder-evacuating reflexes, and sexual reflexes reappear (erection may occur but rarely ejaculation).

The human is subject to a postural hypotension (low blood pressure) when the body position changes from lying to sitting or standing. The individual whose cord has been damaged no longer has connections with the vasoconstrictor centers located in the lower part of the brainstem *(medulla oblongata)*. Postural changes are not met by vasoconstriction (narrowing of blood vessels) to maintain blood pressure. Since there is no sensation from the body parts below the level of section, the individual may also develop skin ulcerations from pressure and ischemia and not be aware of them.

Spinal puncture *(tap)*

Since the cord stops at about the second lumbar vertebra, it is possible to insert a needle into the subarachnoid space and withdraw spinal fluid. The puncture is usually done at the level of the fourth lumbar vertebra. The fluid may be withdrawn for analysis, or may be replaced with an equivalent volume of anesthetic to create spinal anesthesia. Puncture must be done carefully to avoid damaging nerves of the cauda equina.

Intramuscular injections

Intramuscular injections involve giving medications by insertion of a needle into a muscle. Large nerves may lie in the area into which the injection is to be made, and their path and size must be recalled to avoid injuring them. Branches of the axillary nerve lie beneath the deltoid muscle; the sciatic nerve lies beneath the buttock and lateral thigh muscles; the femoral nerve lies under the anteromedial thigh muscles. Paralysis or sensory loss may result if the nerve is damaged.

Summary

1. The spinal cord integrates reflex activity and transmits impulses to and from the brain. The spinal nerves carry impulses to and from the periphery.

2. The cord is located within the vertebral canal.
   a. It is surrounded by fluid, meninges, and connective tissue.
   b. It is shorter than its canal.
   c. It has four major regions.
   d. A cross section shows gray and white matter.
   e. White matter contains the spinal tracts, areas of particular motor and sensory function.

3. The cord is responsible for much reflex activity.
   a. Stretch, flexion, crossed extension, and withdrawal reflexes are controlled by the cord.
   b. Visceral reflexes such as bladder emptying and defecation are controlled by the cord.

4. The pyramidal pathways are generally responsible for control of voluntary muscular movement. The extrapyramidal pathways deal with tone and reflex activity. Both are usually involved in cord damage.

5. Thirty-one pairs of spinal nerves originate from the cord.

a. They connect to the cord by dorsal and ventral roots.
b. They form four major plexuses from which nerves pass to skin and skeletal muscles. These plexuses and the major nerves originating from each area are:
   (1) Cervical plexus—nerves to neck and shoulder muscles; phrenic nerves to diaphragm.
   (2) Brachial plexus—axillary, musculocutaneous, median, ulnar, and radial nerves to upper appendage.
   (3) Lumbar plexus—femoral, lateral cutaneous, and obturator nerves to upper thigh.
   (4) Sacral plexus—sciatic branches to give hamstring nerve; forms tibial nerve, and common peroneal nerve to lower appendage.

6. Thoracic nerves 2–12 form the intercostal nerves that supply the intercostal and abdominal muscles; plexus is not formed.

7. Cord injury produces motor and sensory loss, depending on degree of damage and area of damage.

8. Spinal shock, characterized by loss of function, is the result of damage to the cord. Return of function is in a definite order.

9. Spinal puncture is possible because of the fact that the cord does not extend through the entire vertebral canal.

The Brain and Cranial Nerves

The term *brain* refers to the large nervous structures contained within the cranial cavities of the skull. Three major parts (Fig. 13.1) are recognized as composing the brain.

The BRAINSTEM forms the more or less central portion of the brain; it consists, from inferior to superior, of the *medulla* (oblongata), *pons, midbrain,* and *diencephalon.* The latter includes the thalamus and hypothalamus. Many of the cranial nerves that supply fibers to head, neck, and visceral organs have their points of origin and termination within the brainstem.

The CEREBELLUM fills the posterior cranial cavity of the skull and lies posteriorly to the medulla and pons. It is concerned with control of the involuntary aspects of skeletal muscle activity, such as control of tone and muscular coordination during movement.

The CEREBRUM, composed of two hemispheres, is the "crowning glory" of the human. In it are lodged areas for voluntary movement of skeletal muscles, interpretation of sensation, and the so-called "silent areas" from which no specific movement or sensations may be elicited by stimulation. These latter areas appear to be concerned with behavior, social and moral sense, thought, and those other aspects of human endeavor that set us above the "lower animals."

The human brain has, rightly or wrongly, been compared to a computer (or is it the other way around?!). In an infant, the machine is largely unprogrammed, and experience fills it with the information that will be used throughout life.

The brainstem

The medulla oblongata

STRUCTURE. The medulla forms the inferior 3 cm or so of the brainstem. It is slightly wedge-shaped, being wider at its upper end. It continues inferiorly with the spinal cord, from which it is marked externally by the first cervical spinal nerve outlet; the spinal nerve origin also corresponds to the level of the foramen magnum. Internally, the separation of medulla and cord is marked by an extensive rearrangement of gray and white matter. The gray and white components are "all mixed together" within the medulla, as opposed to the internal H-shaped gray, surrounded by white organization of the cord. Groups of nerve cell bodies known as NUCLEI are scattered throughout the medulla. Some of these nuclei form the so-called VITAL and NONVITAL CENTERS of the medulla; other nuclei serve cranial nerves IX through XII, whose fibers emerge from the medulla. On the anterior external surface of the medulla are two elongated elevations known as the PYRAMIDS. The OLIVES are two rounded elevations on either side of the upper part of each pyramid.

FUNCTION. The VITAL CENTERS of the medulla include centers necessary for survival of the organism.

The CARDIAC CENTERS include cardioacceleratory and cardioinhibitory centers that receive sensory fibers from blood vessels, heart, lungs, cerebrum, and other body areas. These centers provide for the reflex control of heart rate according to body activity levels and such factors as the amount of carbon dioxide and oxygen in the bloodstream. This reflex control of heart rate is described in greater detail in Chapter 24.

Two of three paired RESPIRATORY CENTERS (discussed more fully in Chapter 26), including the inspiratory and expiratory centers, are in the medulla. The inspiratory centers provide stimuli to the muscles (diaphragm, external intercostals) responsible for the changes that draw air into the lungs, while the expiratory center provides impulses that tend to interrupt the inspiratory cen-

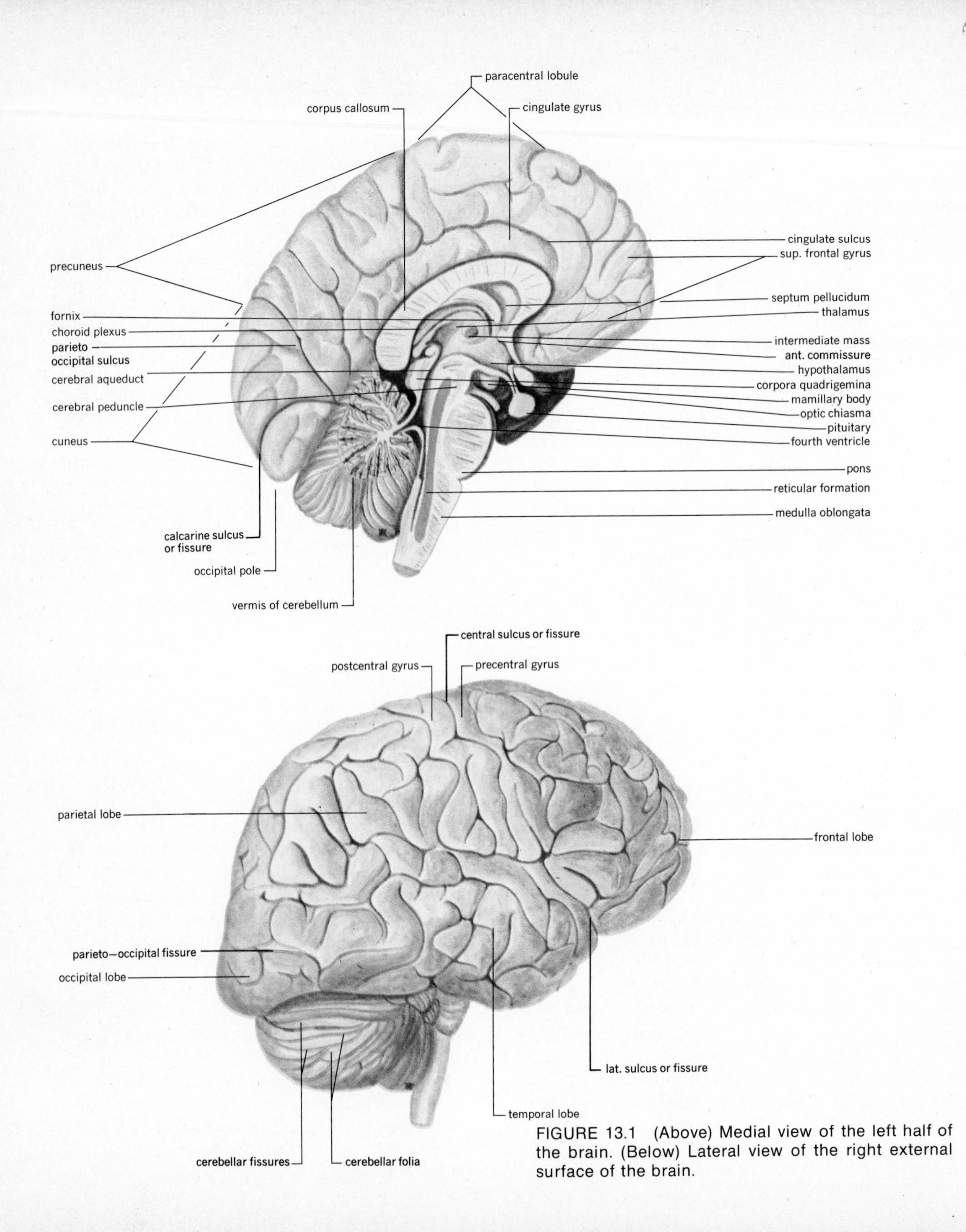

FIGURE 13.1 (Above) Medial view of the left half of the brain. (Below) Lateral view of the right external surface of the brain.

ter's activity and allow expiration of air from the lungs.

The VASOMOTOR CENTERS include vasoconstrictor and vasodilator centers (described in more detail in Chapter 25); these centers reflexly control the diameter of muscular blood vessels in the body and therefore aid in controlling blood pressure.

Interconnections are made between all of these vital centers so that responses of breathing, heart action, and blood pressure are coordinated. For example, activity speeds heart rate, raises blood pressure, and accelerates breathing to achieve the desired end result of speeding the flow of oxygenated blood to active areas.

NONVITAL CENTERS of the medulla include those for *swallowing, vomiting, sneezing,* and *coughing.*

The major voluntary motor pathways from the cerebrum (pyramidal or corticospinal tracts) undergo an 80 percent crossing in the medulla, and these crossed fibers form the pyramids.

The olives contain synapses for fibers rising in the equilibrium structures of the inner ear (semicircular canals and maculae) and send fibers to the cerebellum to ensure the accuracy and efficiency of postural adjustments.

The paired gracile and cuneate tracts of the spinal cord synapse within the medulla, in nuclei of the same names; from these nuclei, fibers pass to the opposite sides of the medulla and ascend to the thalamus in the *medial lemniscus.* The decussation (crossing) of the medial lemniscus thus occurs in the medulla.

CLINICAL CONSIDERATIONS. Damage to the medulla may occur in blows to the back of the head or upper neck; such blows are particularly dangerous because of the possibility of damage to the vital centers. Involvement of the respiratory centers may lead to respiratory paralysis and death, unless a positive- or negative-pressure breathing apparatus is immediately available. Before the advent of vaccines for polio virus, the respiratory centers were a favorite target for the virus, and bulbar polio, with its damage to the inspiratory centers, resulted in the victim being confined to an iron lung (negative-pressure apparatus).

The pons

STRUCTURE. The pons is about 2.5 cm long and forms a conspicuous bulge on the anterior surface of the brainstem superior to the medulla. It is named for the appearance of the bridgelike fibers that run laterally across the bulging surface. A BASAL OR ANTERIOR PORTION of the pons receives fibers from a cerebral hemisphere and sends fibers to the opposite cerebellar hemisphere over the CEREBELLAR PEDUNCLES. The dorsal (posterior) portion or TEGMENTUM of the pons contains ascending and descending tracts and the nuclei of cranial nerves V to VIII. A respiratory nucleus called the PNEUMOTAXIC CENTER is also located in the dorsal pons.

FUNCTION. The basal portion of the pons acts as a synaptic or relay station for motor fibers from the cerebrum to the cerebellum. The cranial nerve nuclei provide for sensory and motor fibers to the skin of the head and certain muscles, for audition, taste, and eye movement. The pneumotaxic center provides an additional means of inhibiting inspiratory center activity to allow for "breathing out."

CLINICAL CONSIDERATIONS. Damage to the pons may produce motor deficits through involvement of the peduncles. Movements are less precise, and posture is difficult to maintain. If the pneumotaxic center is involved, apneustic breathing, characterized by long sustained inspirations, may result.

The midbrain

STRUCTURE. The midbrain is a wedge-shaped

portion of the stem, about 1.5 cm long, located superior to the pons. Its anterior portion is composed chiefly of two large bundles of fibers called the CEREBRAL PEDUNCLES. These structures carry most of the efferent cerebral motor fibers to the cerebellum and spinal cord. The posterior portion contains the SUPERIOR and INFERIOR COLLICULI, two pairs of rounded elevations, and the nuclei of cranial nerves III and IV.

FUNCTION. The midbrain, by way of the peduncles, serves as a motor relay station for fibers passing from the cerebrum to the cord and cerebellum. The colliculi integrate a variety of visual and auditory reflexes, including those concerned with avoiding objects that are seen and muscular responses (turning the head) to sounds to achieve greatest benefit from a sound. Righting reflexes also appear to be controlled (in part) by the midbrain. The latter are reflex responses that maintain the body in proper orientation to the environment.

CLINICAL CONSIDERATIONS. Damage to the midbrain is associated with loss of motor function if the peduncles are involved, and disturbances of visual, auditory, and righting reflexes if the colliculi are affected.

The diencephalon

The diencephalon forms the superior end of the brainstem and is part of the original forebrain. It develops into larger, superiorly placed, paired THALAMI, and an inferiorly placed, single HYPOTHALAMUS.

THE THALAMUS

Structure. Each thalamus (Fig. 13.2) measures about 3 cm front to back and 1.5 cm in height and width, and together they compose about four-fifths of the diencephalon. The thalami are connected across the midline by the *intermediate mass.* Developmentally, three portions differentiate within the thalamus primordia to give rise to an upper *epithalamus,* a central *dorsal thalamus,* and a lower *ventral thalamus.* Several nuclei, or groups of cell bodies, are found in each area (Fig. 13.3). Fibers entering the thalamus are mainly sensory in nature, and outgoing fibers from the dorsal and ventral thalamus pass primarily to the sensory and motor regions of the cerebral cortex.

Function. The epithalamus consists of the PINEAL GLAND (epiphysis) and the paired HABENULAR NUCLEI, located anterior to the pineal gland. The epithalamus receives fibers from the olfactory regions of the cerebrum and from the limbic system. The latter is concerned with emotions and their expression. It sends fibers to the midbrain that then pass to the autonomic nuclei of the brainstem and to the reticular formation (these are discussed in greater detail in Chapters 15, 19, and 20). These connections provide a basis by which olfactory stimuli can influence emotional behavior, as in mating, and the activity of the viscera. The pineal gland is a cone-shaped structure measuring about 7 by 5 mm; in the human it appears to be concerned with the onset of puberty and the establishment of circadian rhythms (see Chapters 31 and 32).

The dorsal and ventral thalamus (commonly referred to simply as *thalamus*) is the largest portion and contains the following nuclei:

The reticular nucleus receives fibers from the reticular formation and sends fibers to the cerebral cortex. It participates in those changes that aid in maintaining the waking state of the organism.

The midline nuclei receive sensory information from receptors in the body viscera and from taste buds, and are involved in autonomic reactions to such stimuli.

Specific thalamic nuclei include:

The MEDIAL GENICULATE BODY (or nucleus) that receives fibers from the hearing structures of the

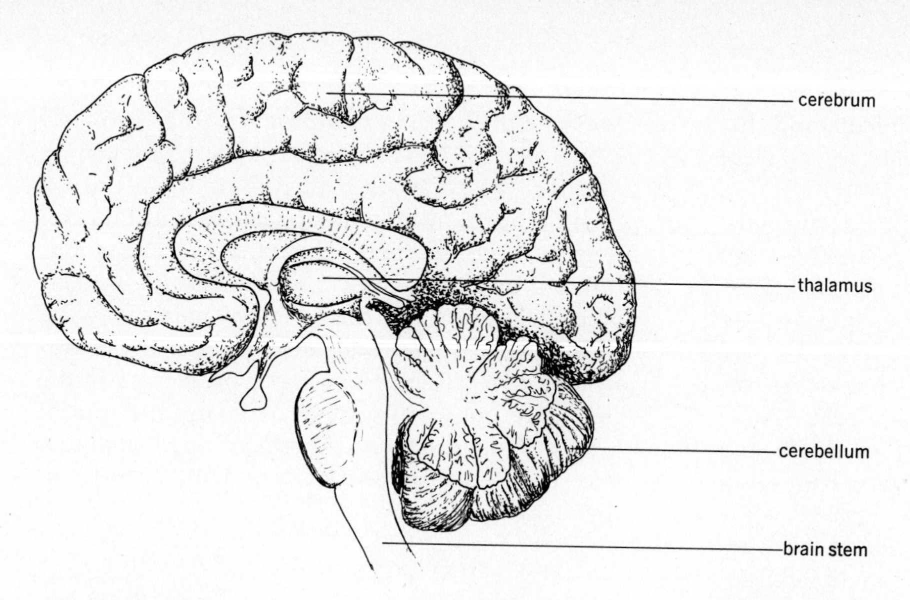

FIGURE 13.2 The location and internal structure of the thalamus.

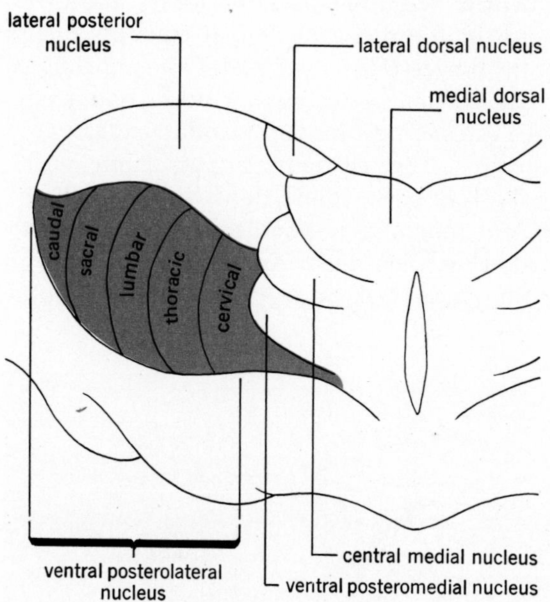

FIGURE 13.3 The major thalamic nuclei. *(a)* View from the dorsolateral direction. *(b)* Posterior view.

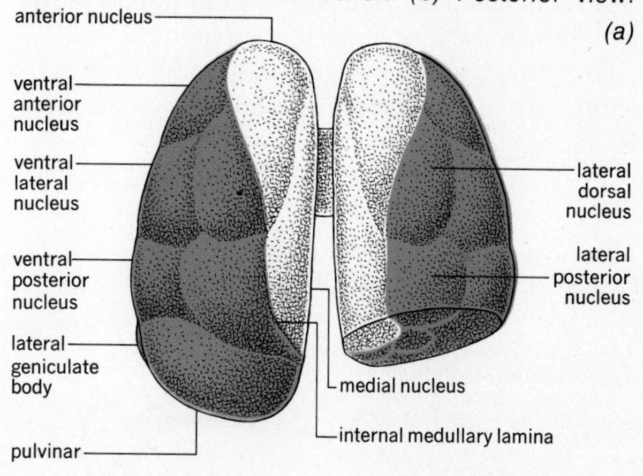

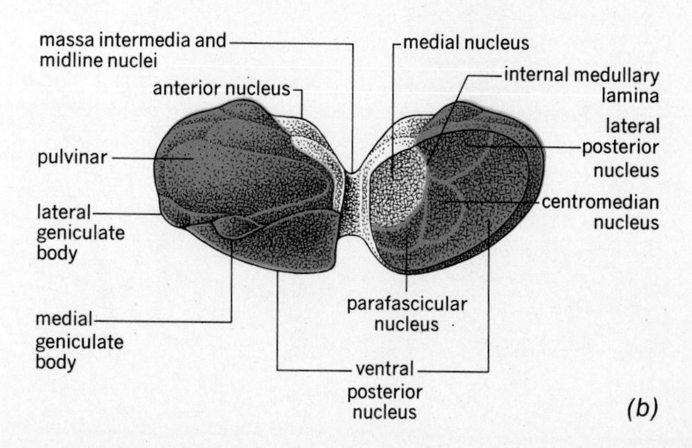

inner ear and from the inferior colliculi, and that sends fibers to the auditory cortex for relay of auditory information.

The LATERAL GENICULATE BODY (or nucleus) serves functions similar to the medial body except it operates in the visual system. It receives fibers from the eye and superior colliculus and projects to the visual cortex.

The VENTRAL POSTERIOR NUCLEUS receives the spinothalamic tracts (pain, heat, cold, crude touch), the medial lemniscus (fine touch and pressure), and fibers from the face conveying sensory information over cranial nerve V. The body is represented in this nucleus in a topographical fashion, with the lower body laterally, and the head medially. Outgoing fibers terminate in the sensory cortex for discrimination and localization.

The VENTRAL LATERAL and VENTRAL ANTERIOR NUCLEI have motor functions and form part of the extrapyramidal system described earlier.

In summary, the thalamus is primarily a sensory relay station, connecting sensory pathways to the cerebral cortex.

Clinical considerations. Damage to the sensory portions of the thalamus produces the THALAMIC SYNDROME. A lesion here is usually vascular in origin (a vessel blocked or ruptured) and results in little or no awareness of sensation until a certain point is reached. Then the sensation may "burst" upon the conscious and become intolerable, especially the sensation of pain. In older individuals, tremor and muscular rigidity (Parkinson's disease) may develop with lesions in the thalamic motor regions. Administration of L-DOPA (*l*evorotatory *d*ihydroxy *p*henyl*a*lanine) may be used to relieve or lessen the severity of the symptoms.

THE HYPOTHALAMUS

Structure. The hypothalamus forms the lower one-fifth of the diencephalon. It contains many separate nuclei (Fig. 13.4) and has connections with the thalamus, cerebral cortex, brainstem, limbic system, and pituitary gland.

Functions. The cells of the hypothalamic nuclei not only have nerve connections with many body areas, but they also respond to the properties of the blood reaching them. Among the blood properties to which the area responds are pH, osmotic

FIGURE 13.4 The nuclei of the hypothalamus.

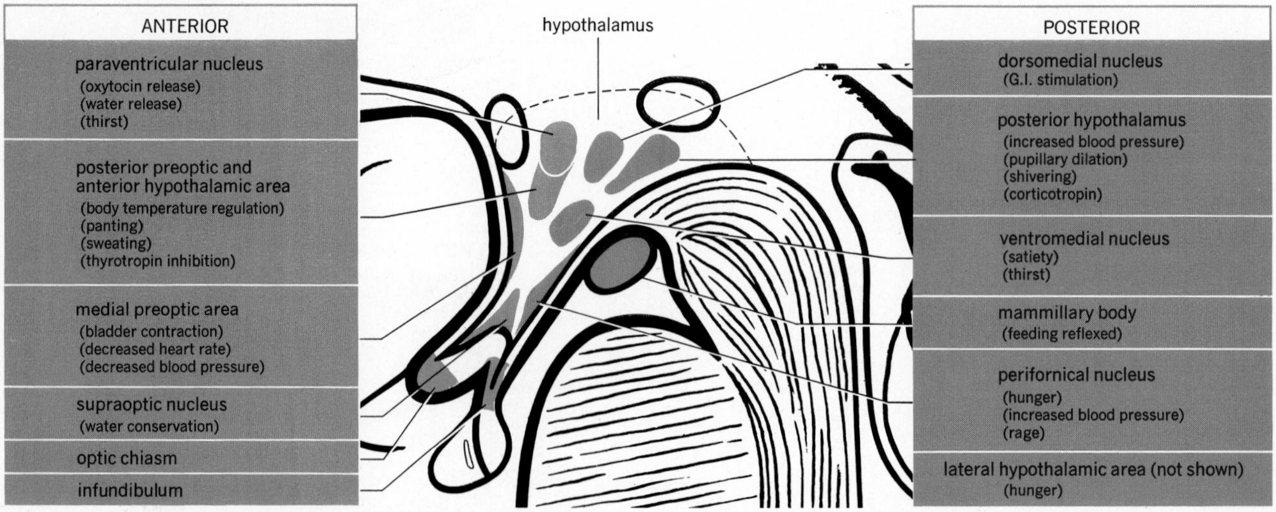

ANTERIOR
paraventricular nucleus
(oxytocin release)
(water release)
(thirst)
posterior preoptic and anterior hypothalamic area
(body temperature regulation)
(panting)
(sweating)
(thyrotropin inhibition)
medial preoptic area
(bladder contraction)
(decreased heart rate)
(decreased blood pressure)
supraoptic nucleus
(water conservation)
optic chiasm
infundibulum
hypothalamus
POSTERIOR
dorsomedial nucleus
(G.I. stimulation)
posterior hypothalamus
(increased blood pressure)
(pupillary dilation)
(shivering)
(corticotropin)
ventromedial nucleus
(satiety)
(thirst)
mammillary body
(feeding reflexed)
perifornical nucleus
(hunger)
(increased blood pressure)
(rage)
lateral hypothalamic area (not shown)
(hunger)

pressure, and glucose levels. Output from the hypothalamus is designed to control many body processes concerned with maintenance of homeostasis.

Functions of the hypothalamus include the following:

Temperature regulation. Human body temperature is normally maintained within a degree or two of 98.6°F or 37°C. A proper balance between heat production and loss is maintained by HEAT LOSS and HEAT GAIN CENTERS in the hypothalamus. The heat loss center causes dilation of skin blood vessels, sweating, and decreased muscle tone, all of which increase heat loss or reduce heat production. The heat gain center causes skin vessel constriction, shivering, and cessation of marked sweating, all of which conserve body heat or increase heat production.

Regulation of water balance. Hypothalamic neurons continually monitor blood osmotic pressure and adjust the tonicity of the body fluids by ADH *(antidiuretic hormone)* that is passed to the posterior lobe of the pituitary gland for storage and release. The hormone permits greater water reabsorption from the kidney tubules.

Control of pituitary function. At last count, nine chemicals influencing pituitary function have been isolated from the hypothalamus. Properly termed PITUITARY REGULATING FACTORS, these chemicals are produced in the hypothalamus as a result of bloodborne and nervous stimuli, are passed by blood vessels to the anterior lobe of the pituitary gland, and there stimulate or inhibit production and release of hormones.

Control of food intake. Initiation and cessation of feeding is controlled by hypothalamic FEEDING and SATIETY CENTERS.

Regulation of gastric secretion. The amount of gastric juice produced by the stomach is increased by hypothalamic stimulation. Thus, emotions may trigger release of gastric juice when there is no food in the stomach and may lead to ulcer development.

Emotional expression. The hypothalamus is one part of a system necessary for expression of reactions of rage and anger. It also appears necessary for sexual behavior.

Clinical considerations. Depending on where damage occurs in the hypothalamus, and its extent, a wide variety of symptoms may develop. Those associated with disturbances of water balance, temperature regulation, and emotions are the most common. *Diabetes insipidus* is the production of a large volume of very dilute urine as a result of diminished ADH secretion; inability to maintain near-constant body temperature indicates hypothalamic damage.

The cerebellum

The cerebellum is the second largest portion of the brain and is an important component of the motor system of the body.

STRUCTURE. The cerebellum (Fig. 13.5) lies posteriorly on the medulla and pons. It has a small, centrally placed VERMIS and two larger CEREBELLAR HEMISPHERES. Both vermis and hemispheres are thrown into small folds known as FOLIA. Attaching the cerebellum to the brainstem are paired bundles of fibers, the CEREBELLAR PEDUNCLES. These bundles carry afferent and efferent impulses to and from the organ. Afferent fibers come from muscles, tendons, skin, inner ear, midbrain, and cerebral cortex. Efferent impulses pass to thalamus, basal ganglia, and spinal cord.

The cerebellum has an outer CORTEX of gray matter and an inner MEDULLARY BODY of white matter. The latter has a treelike arrangement and is known as the ARBOR VITAE. The characteristic cell of the cerebellum is the Purkinje cell (see Fig. 11.2).

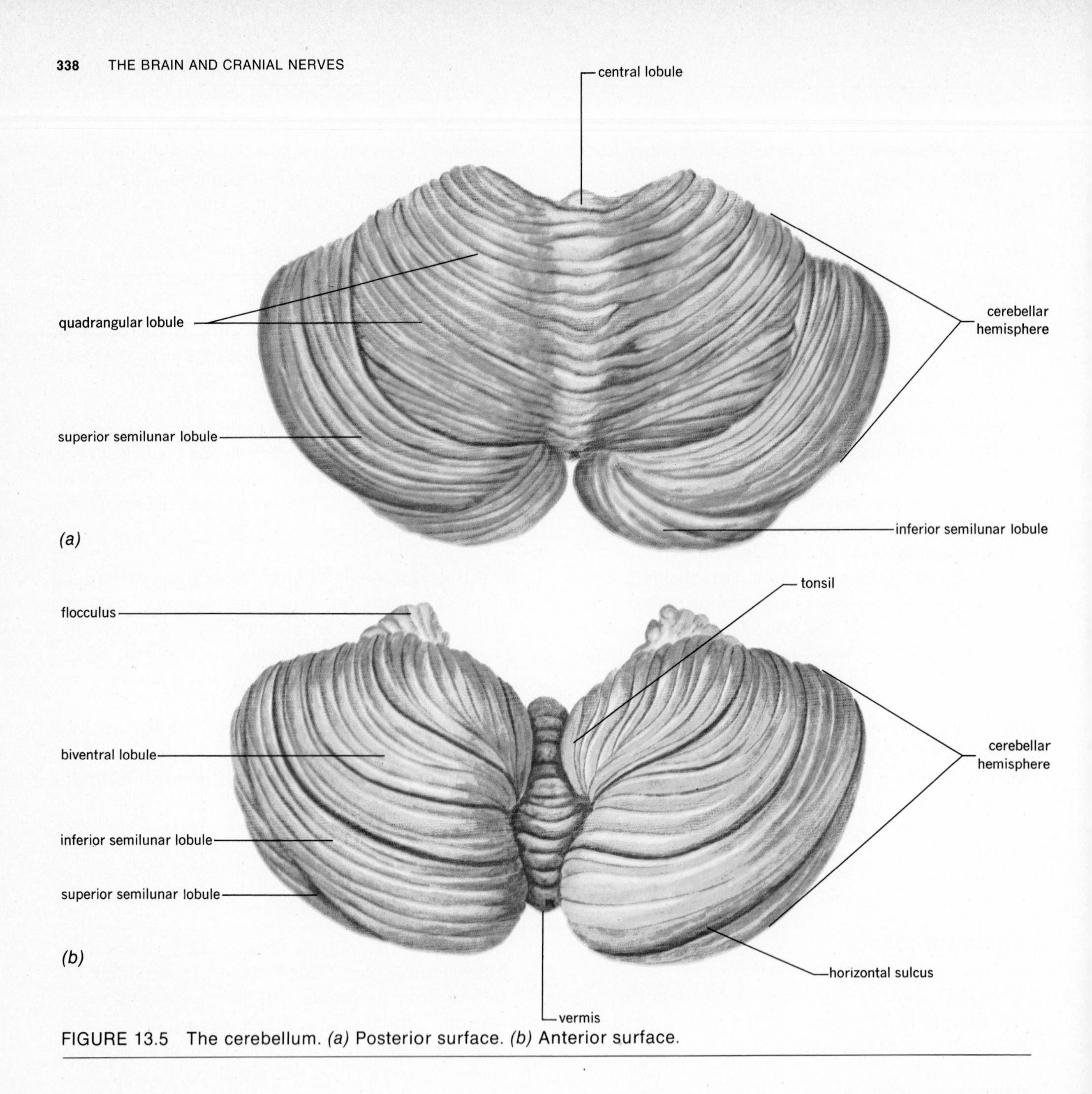

FIGURE 13.5 The cerebellum. *(a)* Posterior surface. *(b)* Anterior surface.

FUNCTION. The cerebellum operates completely at the subconscious level. It COORDINATES ALL MUSCULAR ACTIVITY, INTEGRATES MUSCULAR MOVEMENT, and PREDICTS when to stop movements, so that precise direction of body movement may result.

Additionally, it COORDINATES REFLEXES that serve to maintain posture and equilibrium.

CLINICAL CONSIDERATIONS. Damage to the cerebellum is followed by TREMOR, ASYNERGIA (lack of muscular coordination), puppetlike or jerky movements, PAST POINTING (overshooting the end point of a movement), FAILURE TO DIRECT MOVEMENTS PROPERLY, cerebellar ATAXIA (a reeling walk), and loss of muscle tone.

The severity of the loss of function depends on the amount of tissue damaged, not the location. If lesions are small, compensation occurs readily, and little evidence of damage may appear.

The reticular formation

The reticular formation is a diffuse network of gray matter placed centrally throughout the brainstem. Its main functions appear to be to alert or arouse the organism and to coordinate reflex and voluntary movements. The reticular formation and its cortical fibers are sometimes designated as the RETICULAR ACTIVATING SYSTEM (RAS) (see also Chapter 19). Drugs that keep a person alert or that depress consciousness act, in part, on the RAS.

The cerebrum

The cerebrum develops from the telencephalon, the second subdivision of the original forebrain. It is the largest portion of the brain in humans.

The cerebral hemispheres

STRUCTURE. The cerebrum (Fig. 13.6) is divided into two lateral halves or HEMISPHERES by the LONGITUDINAL FISSURE. Each hemisphere is further separated by fissures into LOBES that are located and named according to the cranial bones they lie beneath.

A *central fissure* separates a frontal lobe from a parietal lobe.

The *parietooccipital fissure* separates the parietal lobe from an occipital lobe.

The *lateral fissure* separates a temporal lobe from the frontal and parietal lobes.

The cerebrum, throughout its surface, is thrown into upfolds or GYRI, separated by shallow depressions known as SULCI. Each gyrus has a name; two should be emphasized. The PRECENTRAL GYRUS lies anterior to the central fissure; the POSTCENTRAL GYRUS lies posterior to the central fissure.

The arrangement of gray and white matter in the cerebrum is reversed as compared to that of the cord. An outer layer of gray matter 2–5 mm thick forms the CEREBRAL CORTEX. This layer contains an estimated 12–15 billion neuron cell bodies. The inner white matter, or MEDULLARY BODY, contains a large number of fiber tracts connecting the two halves of the cerebrum *(commissural fibers),* tracts coming into or leaving the hemispheres *(projection fibers),* and fibers connecting one part of a hemisphere with another *(association fibers).*

THE BASAL GANGLIA (Fig. 13.7). The basal gan-

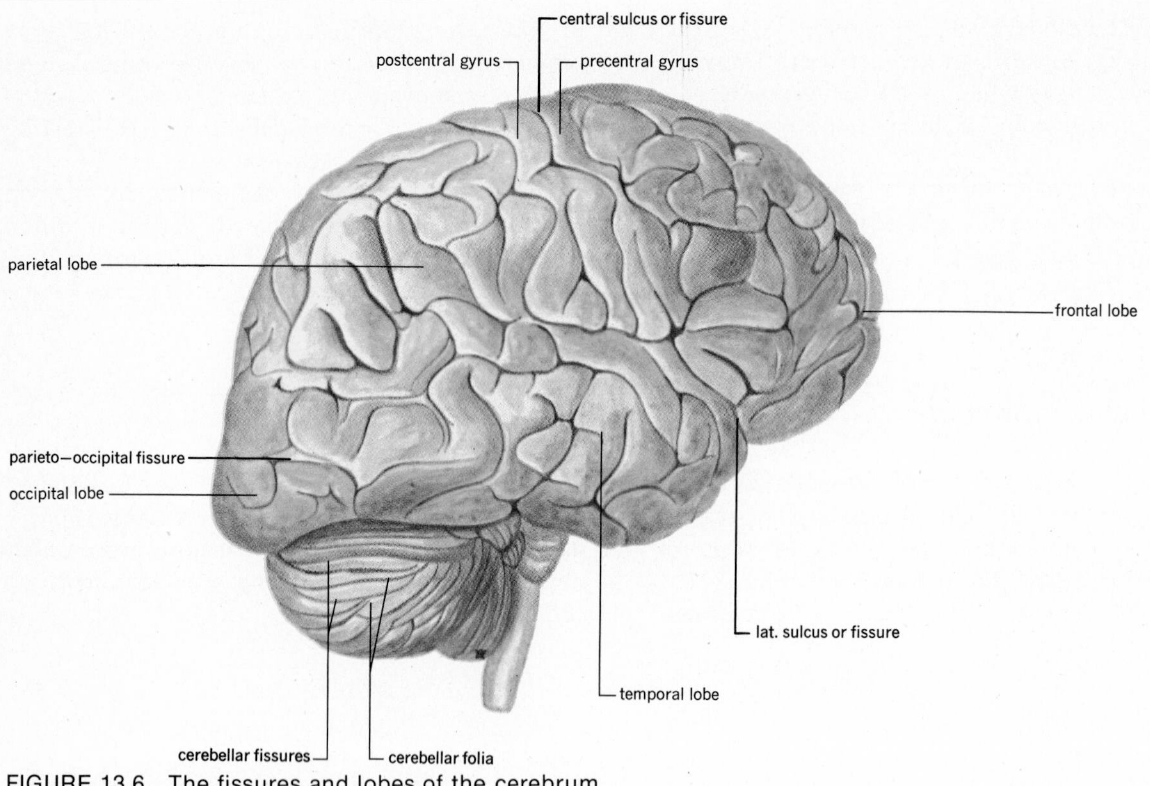

FIGURE 13.6 The fissures and lobes of the cerebrum.

glia are nuclei buried deep within the hemispheres. They include the CAUDATE NUCLEUS, PUTAMEN, GLOBUS PALLIDUS, SUBTHALAMIC NUCLEI, SUBSTANTIA NIGRA, and RED NUCLEUS.

THE LIMBIC SYSTEM (Fig. 13.8). The limbic system is composed of certain cerebral and brainstem structures (see below).

FUNCTION. Specific functional areas within the cerebral cortex exist in the different lobes. A convenient method of designating the areas is to number them according to the scheme devised by Brodmann (Fig. 13.9).

A *primary somatic motor area* (area 4) lies in the precentral gyrus. The voluntary movements of skeletal muscles are controlled from this region. Representation of body parts is unequal and "upside down" in this region, with more area given to those body parts where skilled or complex movement is required (Fig 13.10).

A *premotor area* (areas 6 and 8) lies anterior to the motor area, and coordinates eye and head movements. The region also creates a postural background (synergistic action) against which skilled movements are performed.

The *primary somatic sensory region* (areas 3, 1, and 2) lies in the postcentral gyrus. As in area 4, the body is represented unequally and upside down in this region (Fig. 13.11). This area interprets most of the skin sensations (heat, cold, touch, pressure, some localization of pain).

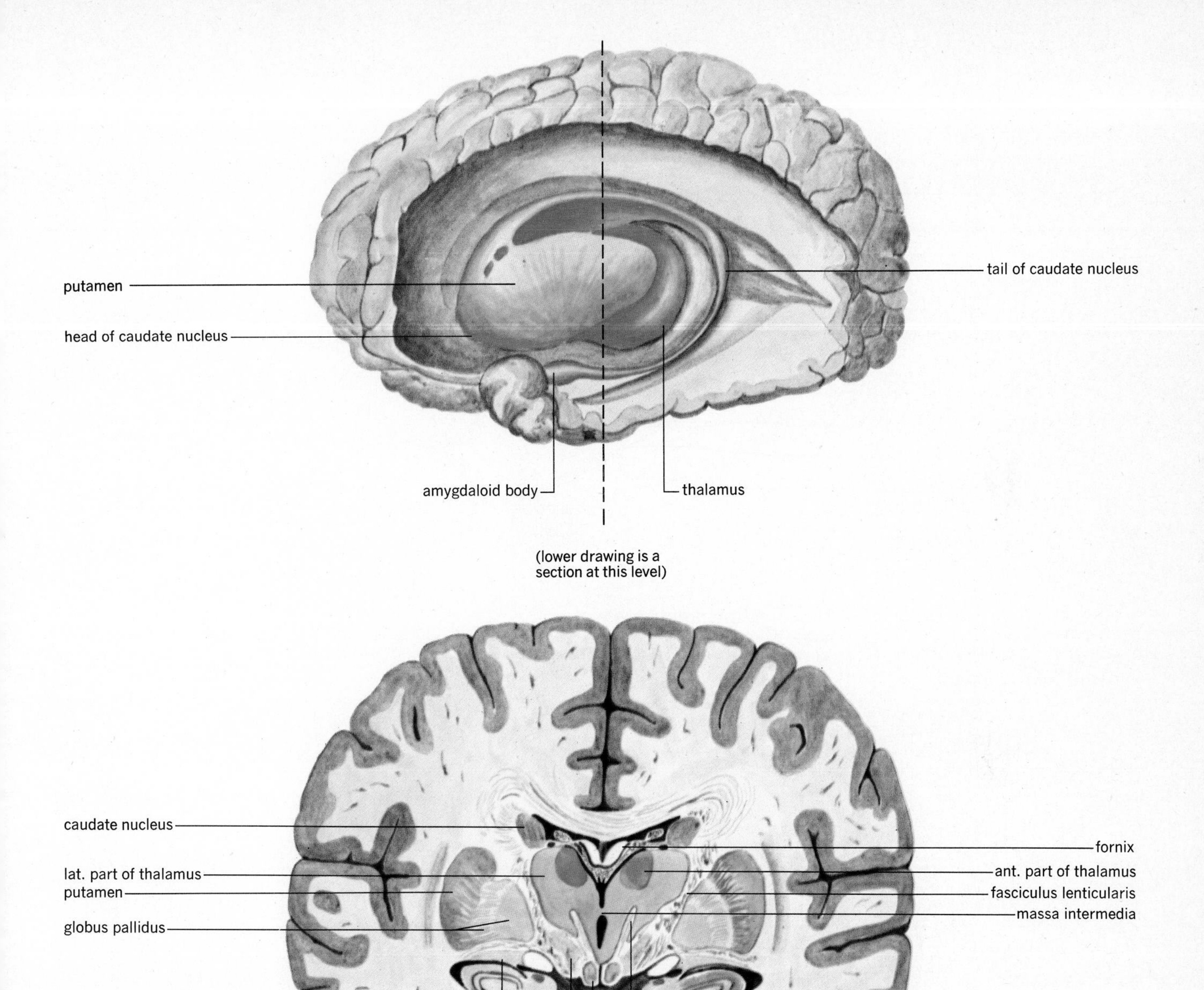

FIGURE 13.7 The basal ganglia. (Above) Several major ganglia projected on the cerebral hemisphere. (Below) Frontal section of the cerebrum to show positions of the ganglia.

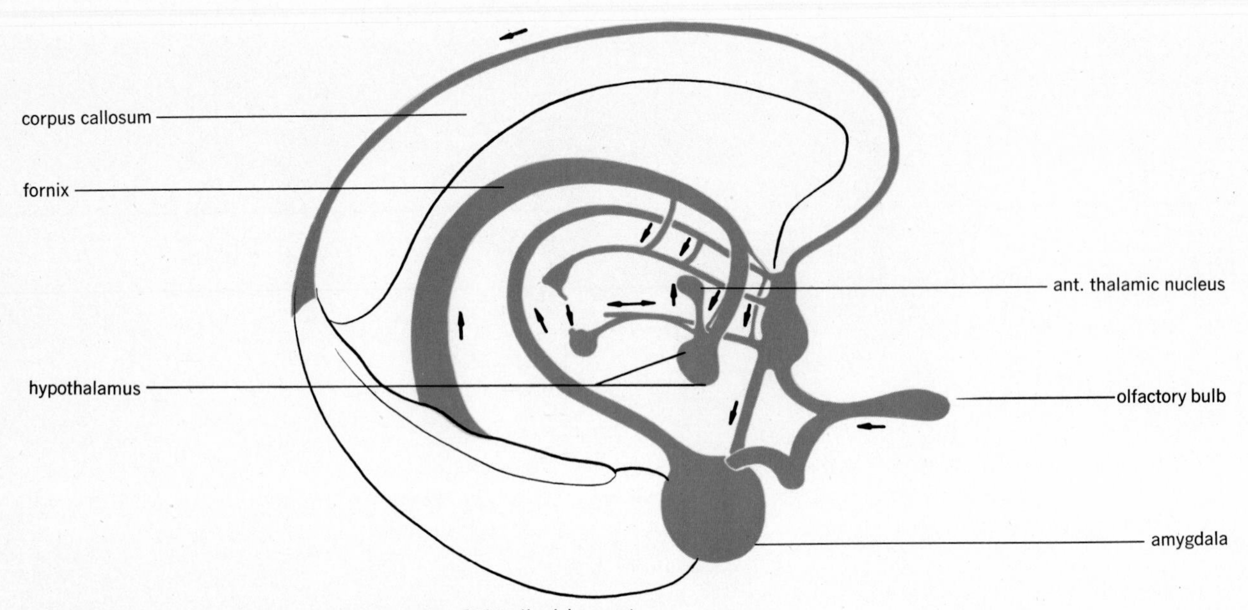

FIGURE 13.8 The major components of the limbic system.

FIGURE 13.9 Some functional areas of the cerebrum as numbered by Brodmann.

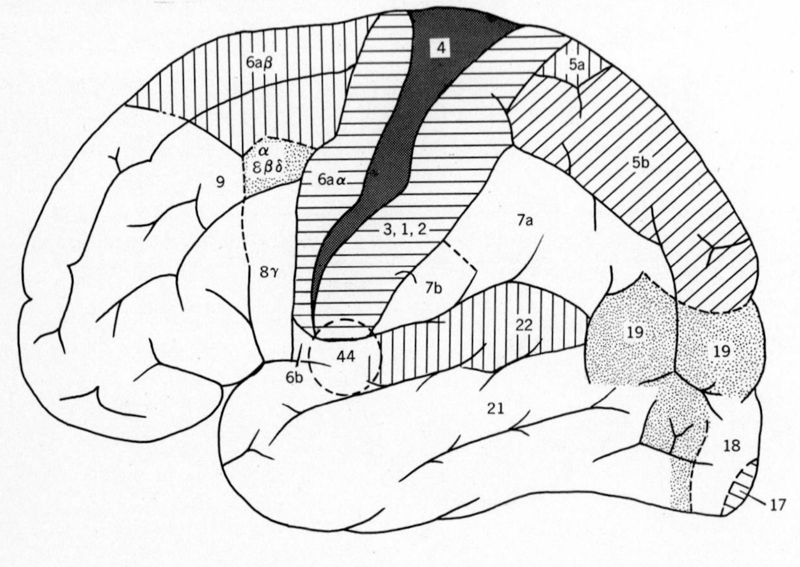

FIGURE 13.10 Cortical location of motor functions in the primary motor area.

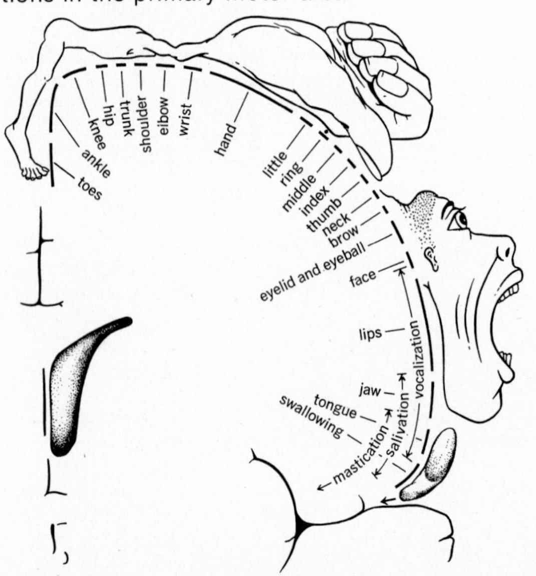

The *visual area* (area 17) lies in the occipital lobe and receives impulses from the eyes by way of the optic nerves, tracts, and radiations.

The *auditory area* (areas 41 and 42) lies in the temporal lobe and receives impulses from the cochleas of the inner ear over the auditory pathways.

Broca's speech area (area 44) includes parts of the motor, sensory, and auditory areas and is concerned with formation of words.

Association areas of the cerebrum are those from which specific motor responses or sensations cannot be elicited when they are stimulated.

The *prefrontal areas* (areas 8–12) control behavior, particularly in inhibition of rate reactions.

The *parietal areas* (areas 5, 7, 19, 39, and 40) are concerned with interpretation of size, shape, texture, degree of heat, and other qualities of objects touched.

The *temporal areas* (areas 20–22) are concerned with learning and memory of things seen and heard.

The *basal ganglia*, as a group, control muscle tone, inhibit movement, and control tremor.

The *caudate nucleus* and *putamen* regulate gross intentional movements of the body, such as movement of a whole limb through space.

The *globus pallidus* provides positioning of the body so that more specific or discrete movements may occur properly.

The *subthalamic nuclei*, *substantia nigra*, and *red nucleus* are part of the *extrapyramidal motor system* (see below) concerned primarily with muscle tone and damping of tremor.

The LIMBIC SYSTEM is composed of hypothalamus, thalamus, the amygdaloid nuclei, and several fiber tracts connecting these areas (see Fig. 13.8). The entire system is concerned with emotions and motivation. Expression of rage and sexual behavior are among the functions these structures serve. "Pleasure centers," which create positive drives toward stimuli, are also found in these areas.

Functionally, two more groupings of efferent fibers from the brain may be made.

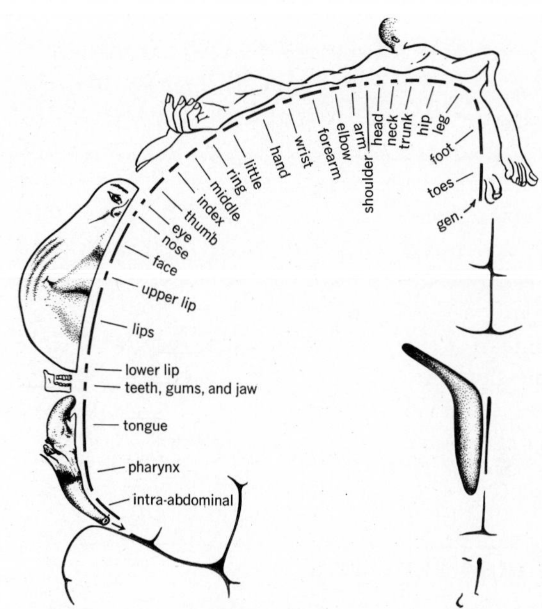

FIGURE 13.11 Cortical location of sensory functions in the primary sensory areas.

The PYRAMIDAL SYSTEM refers to motor fibers originating from the cerebral cortex and passing through the pyramids. About 40 percent of the fibers in this system come from the primary somatic motor area (area 4); the remainder come from other cortical regions.

The EXTRAPYRAMIDAL SYSTEM consists of motor fibers from basal ganglia, brainstem, cerebellum, and other noncortical areas. Although the pyramidal system has traditionally been considered the "voluntary motor system" of the body, this is not strictly true, since only about 40 percent of its fibers come from area 4. The extrapyramidal system, however, is "involuntary" in the sense that it operates without acts of will and controls tone and reflex movement.

CLINICAL CONSIDERATIONS. Damage to the primary somatic areas (motor, sensory, visual, auditory) of the cerebrum will produce specific losses of motor or sensory function. Paralysis, partial deafness, or blindness are typical signs of damage to these regions.

Damage in the association areas produces signs correlated with the area involved. Frontal lobe lesions produce behavioral alterations; parietal lobe lesions result in inability to perceive the body correctly as a result of sensory loss; temporal lobe lesions produce *agnosia* (failure to recognize familiar objects), *aprexia* (inability to perform voluntary movements, especially of speech), and *aphasia* (inability to understand written or spoken words).

Damage to the basal ganglia results in loss of muscle tone and in inability or difficulty in making precise movements. Also, Parkinson's disease, characterized by involuntary tremors, is most consistently associated with damage to the caudate, pallidus, or substantia nigra.

Damage to limbic structures produces exaggerated responses of rage and hypersexuality.

Damage to pyramidal or extrapyramidal systems or both produces motor losses characterized by spastic paralysis. If the damage occurs above the level of the brainstem, the opposite side of the body is usually affected; below the brainstem the same side is usually affected. This occurs because crossing of motor fibers (e.g., lateral corticospinal tract, rubrospinal tract) occurs in the brainstem.

The cranial nerves

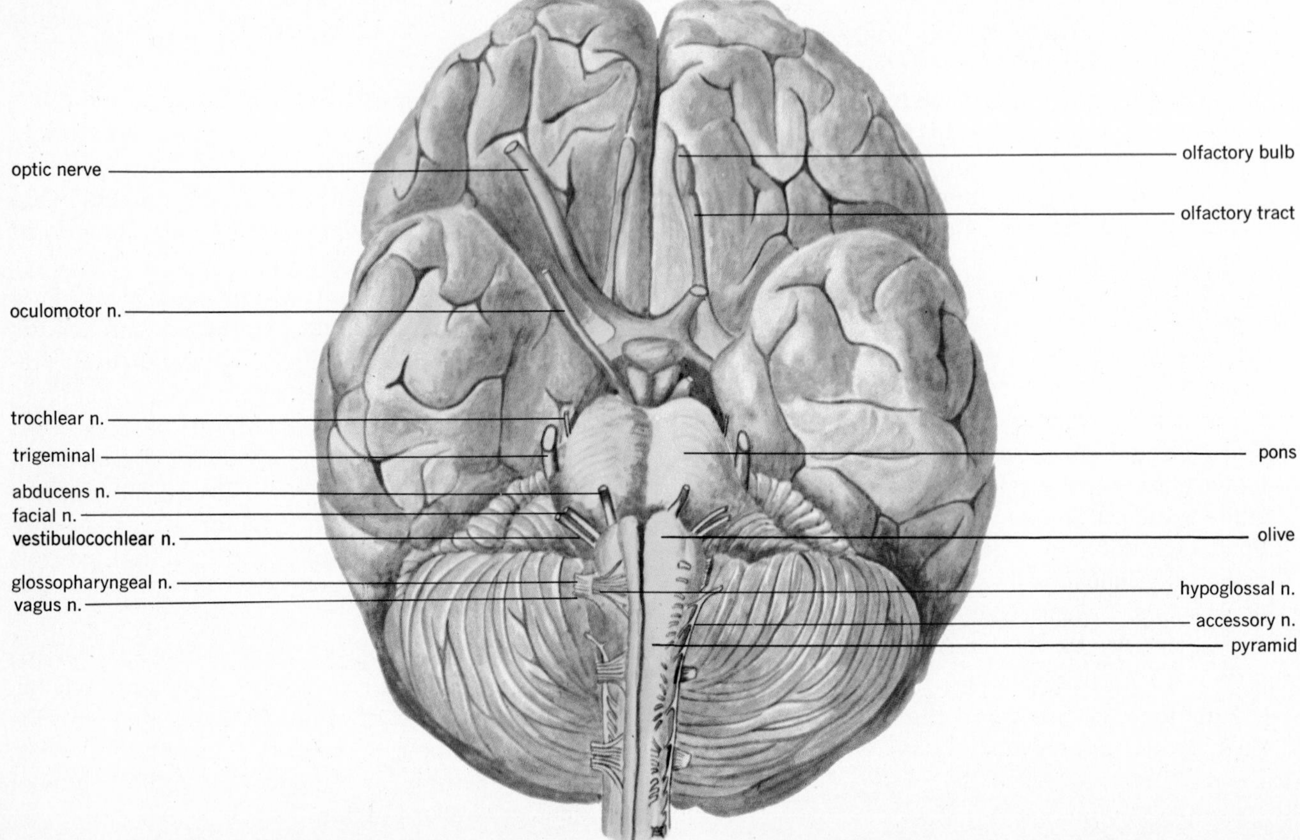

FIGURE 13.12 A basal view of the brain to illustrate the origins of the cranial nerves.

The brain gives rise to 12 pairs of cranial nerves that supply motor and sensory fibers to structures in the head, neck, and shoulder regions; one pair supplies fibers to body viscera. These fibers form, with the spinal nerves, the peripheral nervous system.

The relationships of the nerves to the brain are shown in Fig. 13.12 and their nuclei in the brainstem are shown in Fig. 13.13. A general description of each nerve follows, and a chart of the nerves is shown in Table 13.1.

I. *Olfactory nerve.* The nerve of smell, the olfactory nerve passes from nasal cavities to the olfactory bulb of the cerebrum.

II. *Optic nerve.* The nerve of sight, the optic nerve conveys impulses from retina to brainstem. These impulses are then relayed to the occipital lobe over the optic radiation.

III. *Oculomotor nerve.* Motor fibers in this nerve control four of the six eye muscles that turn the eyeball (medial, inferior, and superior rectus, and inferior oblique), cause pupillary size changes, and aid in

FIGURE 13.13 Diagram illustrating the positions of the cranial nerve nuclei in the brainstem.

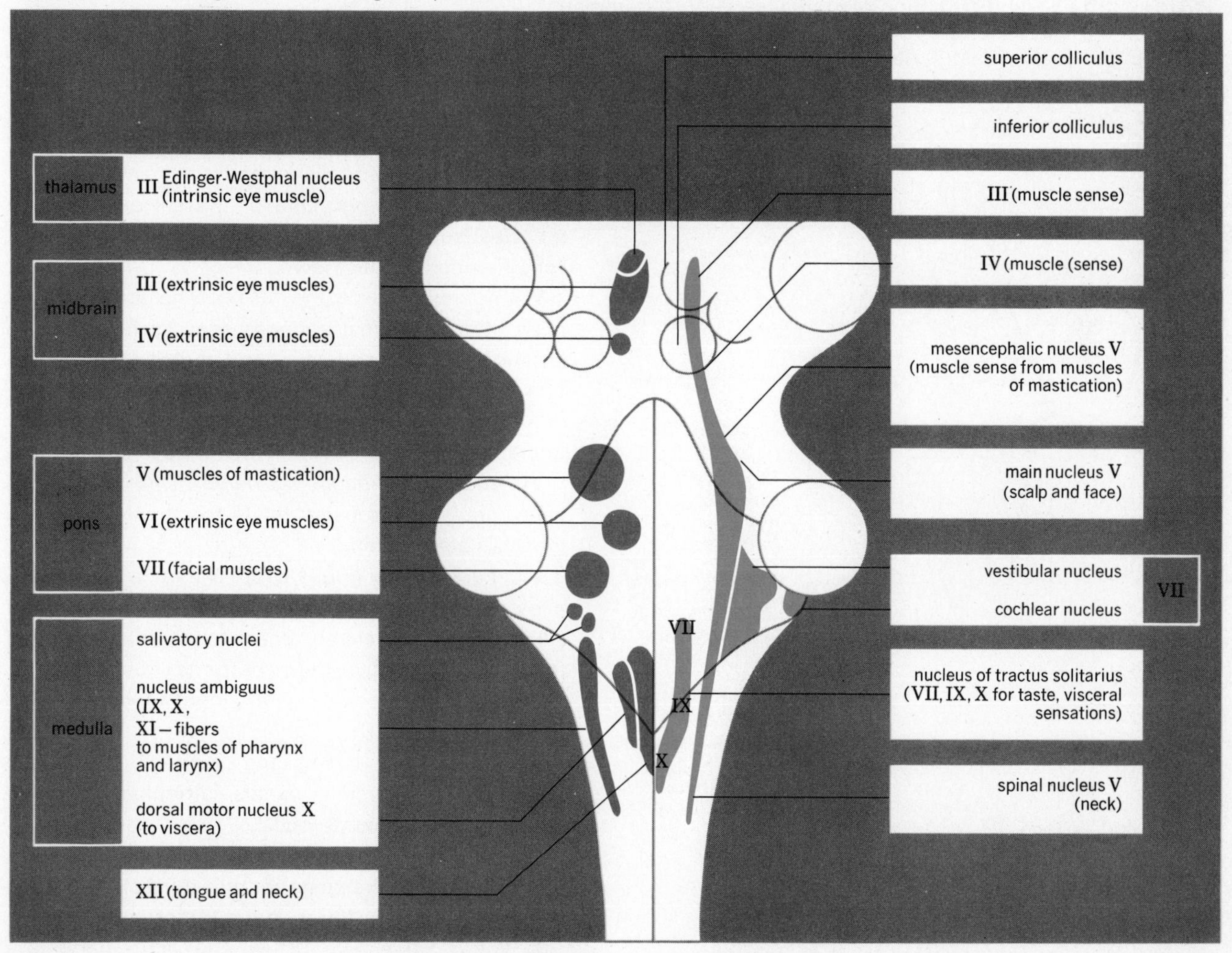

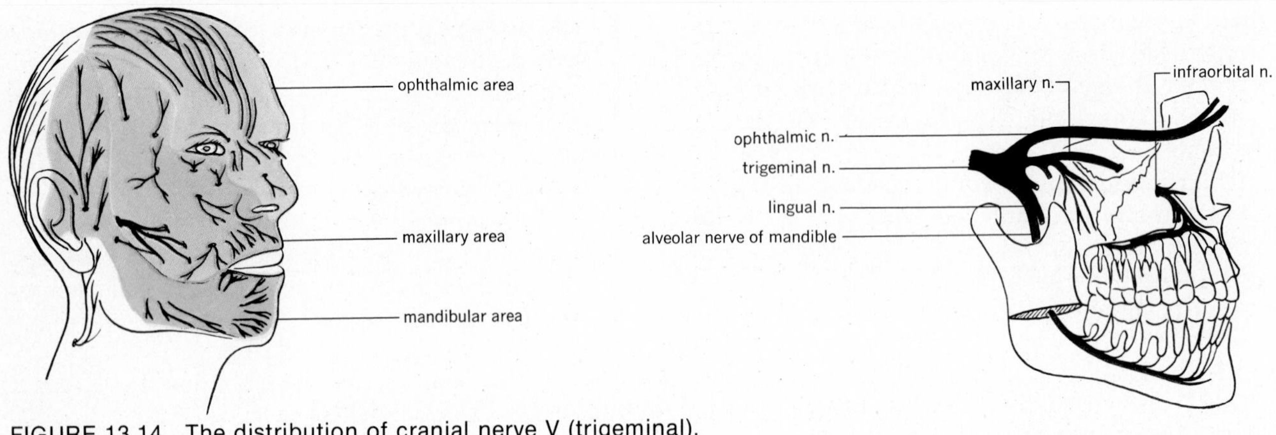

FIGURE 13.14 The distribution of cranial nerve V (trigeminal).

FIGURE 13.15 The distribution of cranial nerve VII (facial).

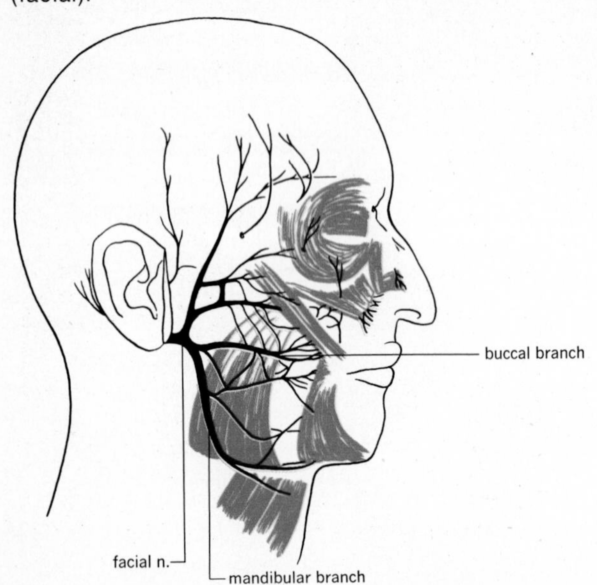

focusing. A sensory component relays impulses from the muscles, iris, and ciliary body.

IV. *Trochlear nerve.* One of the six extrinsic eye muscles (superior oblique) is supplied by motor and sensory fibers of this nerve.

V. *Trigeminal nerve* (Fig. 13.14). The largest cranial nerve, the trigeminal supplies sensory fibers to the anterior cranium and face, and motor fibers to the chewing muscles. *Ophthalmic, maxillary,* and *mandibular* branches form the nerve and supply sensory fibers to all structures within their area of distribution.

VI. *Abducent (abducens) nerve.* This nerve supplies motor and sensory fibers to one extrinsic eye muscle (lateral rectus).

VII. *Facial nerve* (Fig. 13.15). The facial nerve supplies motor fibers to the muscles of the face and the salivary glands. Sensory fibers for taste are derived from the anterior two-thirds of the tongue.

VIII. *Vestibulocochlear (statoacoustic, acoustic,* or *auditory) nerve.* This nerve carries sensory fibers from the cochlea and organs of equilibrium of the inner ear (semicircular canals, utriculus, and sacculus), to the temporal lobe.

IX. *Glossopharyngeal nerve.* This nerve supplies sensory fibers to the posterior one-third of the tongue for taste and motor fibers to the throat muscles.

X. *Vagus nerve* (Fig. 13.16). The vagus nerve is a

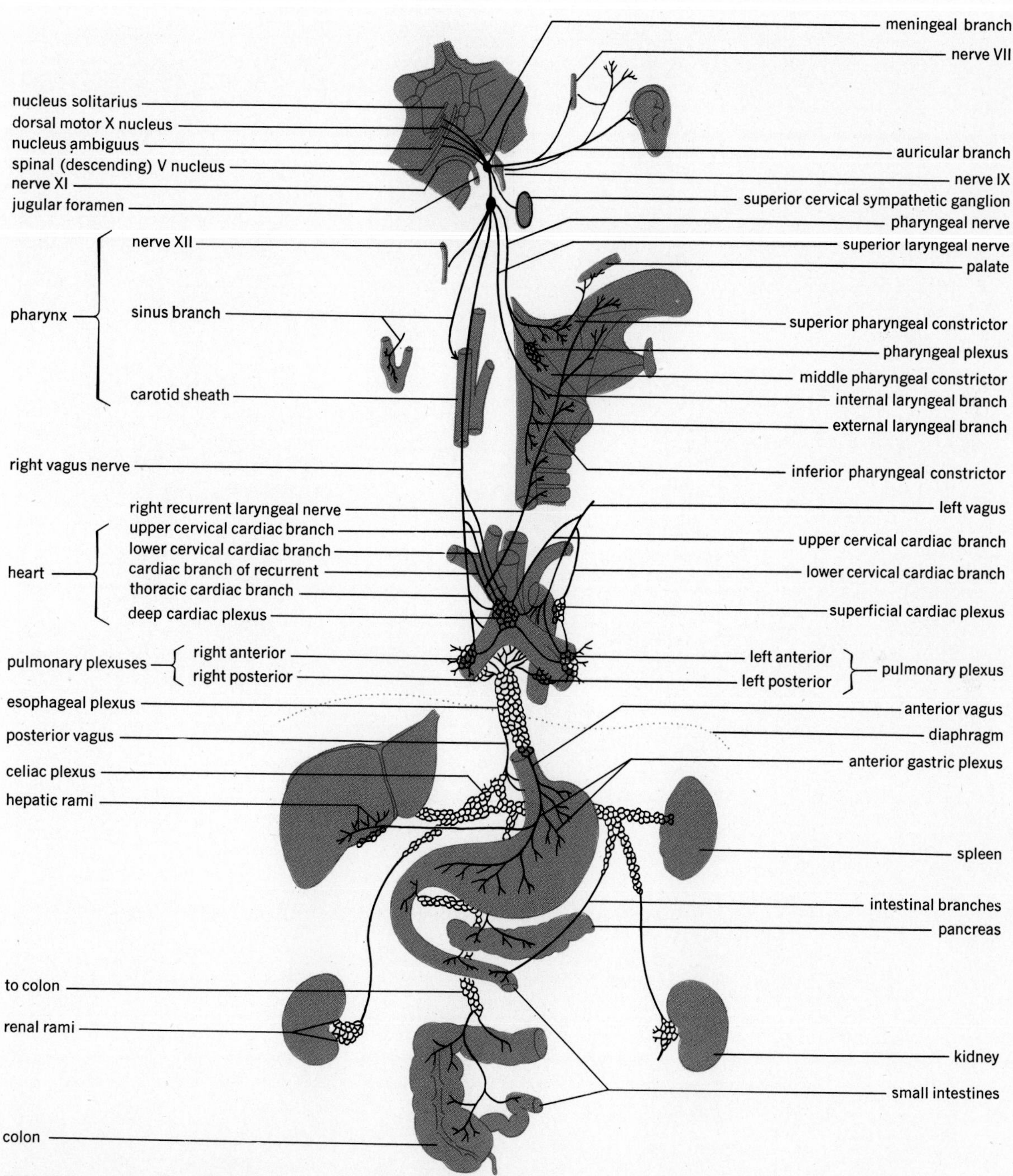

FIGURE 13.16 The distribution of cranial nerve X (vagus).

TABLE 13.1 Summary of the cranial nerves

Nerve	Composition M = motor; S = sensory	Origin	Connection with brain or peripheral distribution	Function
I. Olfactory	S	Nasal olfactory area	Olfactory bulb	Smell
II. Optic	S	Ganglionic layer of retina	Optic tract	Sight
III. Oculomotor	MS	M—Midbrain	4 of 6 extrinsic eye muscles (superior rectus, medial and inferior rectus, inferior oblique)	Eye movement
		S—Ciliary body of eye	Nucleus of nerve in midbrain	Focusing, pupil changes, muscle sense
IV. Trochlear	MS	M—Midbrain	1 extrinsic eye muscle (superior oblique)	Eye movement
		S—Eye muscle	Nucleus of nerve in midbrain	Muscle sense
V. Trigeminal	MS	M—Pons	Muscles of mastication	Chewing
		S—Scalp and face	Nucleus of nerve in pons	Sensation from head
VI. Abducent	MS	M—Nucleus of nerve in pons	1 extrinsic eye muscle (lateral rectus)	Eye movement
		S—1 extrinsic eye muscle	Nucleus of nerve in pons	Muscle sense
VII. Facial	MS	M—Nucleus of nerve in lower pons	Muscles of facial expression	Facial expression
		S—Tongue (ant. 2/3)	Nucleus of nerve in lower pons	Taste
VIII. Vestibulocochlear (Statoacoustic, acoustic, auditory)	S	Internal ear: balance organs, cochlea	Vestibular nucleus, cochlear nucleus	Posture, hearing
IX. Glossopharyngeal	MS	M—Nucleus of nerve in lower pons	Muscles of pharynx	Swallowing
		S—Tongue (post 1/3), pharynx	Nucleus of nerve in lower pons	Taste, general sensation
X. Vagus	MS	M—Nucleus of nerve in medulla	Viscera	Visceral muscle movement
		S—Viscera	Nucleus of nerve in medulla	Visceral sensation
XI. Accessory	M	Nucleus of nerve in medulla	Muscles of throat, larynx, soft palate, sternocleidomastoid, trapezius	Swallowing, head movement
XII. Hypoglossal	M	Nucleus of nerve in medulla	Muscles of tongue and infrahyoid area	Speech, swallowing

TABLE 13.2 Methods of assessing cranial nerve function

Nerve being tested	Method	Comments
Olfactory	Ask patient to differentiate odors of coffee, bananas, tea, oil of cloves, etc., in each nostril.	Test may be voided if patient has a cold or other nasal irritation.
Optic	Ask patient to read newspaper with each eye; examine fundus with ophthalmoscope.	Perform test with lenses on, if patient wears them.
Oculomotor	Shine light in each eye separately and observe pupillary changes; look for movement by moving finger up, down, right, and left; have subject's eyes follow finger movements.	Look for defects in both eyes at same time. Light in one eye causes other one to constrict to lesser degree.
Trochlear		Requires special devices. If eye movement is normal, nerves are probably all right.
Trigeminal	Motor portion: have patient clench teeth and feel firmness of masseter; have patient open jaw against pressure.	Subject should exhibit firmness and strength in both tests.
	Sensory portion: test sensations over entire face with cotton (light touch) or pin (pain).	Deficits in particular areas indicate problems in one of the three branches.
Abducent	Perform test for lateral movement of eyes.	Must be very careful to observe any deficit in lateral movement.

very important component of the autonomic system. It supplies motor and sensory fibers to nearly all thoracic and abdominal viscera.

XI. *Accessory (spinal accessory) nerve.* This nerve supplies motor fibers to muscles of the throat, larynx, and soft palate and to the trapezius and sternocleidomastoid muscles.

XII. *Hypoglossal nerve.* This nerve supplies some of the tongue muscles and the infrahyoid muscles.

CLINICAL CONSIDERATIONS. Functions of the cranial nerves are easily assessed by simple procedures, and these form an important part of a neurological examination. The various procedures are presented in Table 13.2.

TABLE 13.2 continued Methods of assessing cranial nerve function

Nerve being tested	Method	Comments
Facial	Have subject wrinkle forehead, scowl, puff out his cheeks, whistle, smile. Note nasolabial fold (groove from nose to corners of mouth).	Nerve supplies all facial muscles. Fold is maintained by facial muscles and may disappear in nerve damage.
Vestibulocochlear (statoacoustic, acoustic, auditory)	Cochlear portion: test auditory acuity with ticking watch; repeat a whispered sentence; use tuning fork. Compare ears. Examine with otoscope.	Ears may not be equal in acuity; note.
	Vestibular portion: not routinely tested.	If subject appears to walk normally and keep his balance, nerve is usually all right.
Glossopharyngeal and Vagus	Note disturbances in swallowing, talking, movements of palate.	
Accessory	Test trapezius by having subject raise shoulder against resistance; have subject try to turn head against resistance.	Strength is index here.
Hypoglossal	Have subject stick out tongue. Push tongue against tongue blade.	Tongue should protrude straight; deviation indicates same side nerve damage. Pushing indicates strength.

Summary

1. The brain includes nerve structures in the cranium. It is divided into:
   a. Brainstem: medulla, pons, midbrain, diencephalon: thalamus, hypothalamus.
   b. Cerebellum.
   c. Cerebrum.
2. The medulla is the lowest part of the brainstem.
   a. It contains many important reflex centers for respiration, heart rate, and blood vessel size.
   b. Motor tracts from the cerebrum cross here.
   c. Damage to the medulla may interfere with breathing.

3. The pons lies above the medulla.
   a. It contains one of several respiratory centers (pneumotaxic).
   b. It connects to the cerebellum by way of the cerebellar peduncles.
   c. Damage here produces motor and respiratory deficits.
4. The midbrain lies above the pons.
   a. It shows the cerebral peduncles, receiving fibers from the cerebrum.
   b. It contains the corpora quadrigemina that connect eye and ear with motor fibers for righting and adjustments to visual and auditory stimuli.
   c. Damage here produces deficits in maintenance of balance and equilibrium.
5. The diencephalon includes the thalamus and hypothalamus.
   a. The thalamus has a number of nuclei that are sensory and motor in function. Sensory nuclei provide a means of organizing sensory impulses and may provide crude awareness of sensation. Motor nuclei are concerned with motor activity of emotions. Damage results in the thalamic syndrome.
   b. The hypothalamus contains many nuclei. These nuclei control body temperature, water balance, the pituitary gland, food intake, gastric secretion, and emotional response. Damage produces deficits in the above functions.
6. The cerebellum lies posterior to the medulla and pons.
   a. It has a vermis and two hemispheres, an outer gray cortex and inner white matter.
   b. It coordinates and controls muscular movement and tone.
7. The reticular formation is concerned with arousal and alerting of the organism. It is part of the reticular activating system.
8. The cerebrum is the largest part of the brain.
   a. Each hemisphere contains several lobes, separated by fissures. The hemispheres are folded.
   b. The basal ganglia are nuclei deep within the cerebral hemispheres.
   c. The limbic system is composed of cerebral and lower level structures.
9. The cerebrum contains motor areas, sensory areas, visual and auditory areas, and association areas concerned with sensory interpretation and behavior. Damage interferes with movement, sensory reception, and understanding.
10. The basal ganglia control muscle tone, inhibit movement, and control muscle tremor. Damage produces tremor and loss of tone.
11. The limbic system contains centers for expression of emotion and pleasure centers. Damage produces exaggerated display of emotions and sexual behavior.
12. The pyramidal system includes nerve fibers causing and modifying voluntary movement; the extrapyramidal system controls muscle reflexes and tone. Damage to either system produces spastic paralysis.
13. Twelve pairs of cranial nerves arise from the brain. They supply head and neck with motor and sensory fibers.

Blood Supply of the Central Nervous System; Ventricles and Cerebrospinal Fluid

Blood supply

The central nervous system is completely dependent on its blood supply to provide the glucose and oxygen it requires for continuation of activity. Interruption of the blood supply to these organs for more than a few minutes is invariably associated with the development of alterations in sensory, motor, and intellectual functions.

Brain

The paired INTERNAL CAROTID ARTERIES, derived from the common carotid arteries of the neck, and the paired VERTEBRAL ARTERIES, rising from the subclavian arteries beneath the clavicles, form the two systems of arterial vessels to the brain (Fig. 14.1). The internal carotids, or their branches, form three pairs of CEREBRAL ARTERIES supplying the cerebrum. The vertebral arteries join on the anterior surface of the brainstem to form the single BASILAR ARTERY that gives rise to vessels supplying the cerebellum and brainstem. The carotid and vertebral vessels are joined at the CIRCLE OF WILLIS. The circle consists of the posterior communicating arteries, the carotids, part of the anterior cerebral arteries, and the anterior communicating artery; it permits blood from one system to flow into the area of the other system, in case one vessel of a system is narrowed or blocked. The veins draining the brain are largely unnamed except for the DURAL SINUSES (Fig. 14.2).

Spinal cord

A single ANTERIOR SPINAL ARTERY (see Fig. 14.1) rises in a Y-shaped manner from the paired vertebral arteries. Rising as separate vessels from the vertebral or posterior cerebellar arteries are the paired POSTERIOR SPINAL ARTERIES. These three vessels run the entire length of the cord. SPINAL BRANCHES rise from the vertebral, cervical, lumbar, and sacral arteries and from the abdominal aorta, pass to the cord, and contribute additional blood flow to the cord. These vessels enter through the intervertebral foramina and branch to supply the cord.

THE BLOOD-BRAIN BARRIER. A variety of substances that pass easily through capillary walls elsewhere in the body are slowed or stopped as they attempt to pass through the walls of cerebral capillaries. Protein molecules, antibiotics, urea, chloride, and sucrose are some materials that pass with difficulty through cerebral capillaries. Such observations suggest that there is a blood-brain barrier restricting solute passage. Electron microscope studies of the anatomy of the cerebral vessels show the major features depicted in Fig. 14.3. Several facts should be emphasized.

Cerebral capillaries seem to lack the "pores" found in capillaries in other body areas.

Endothelial cells of cerebral capillaries have overlapping ends sealed by impervious tight junctions. This arrangement creates four membranes through which a solute must pass; this limits passage through the intercellular space.

A continuous basement membrane surrounds the entire circumference of each capillary.

About 85 percent of the outer surface of each capillary is covered by glial cells. To get to a neuron, a substance may have to pass through the capillary cell and the glial cells, slowing its arrival at the neuron.

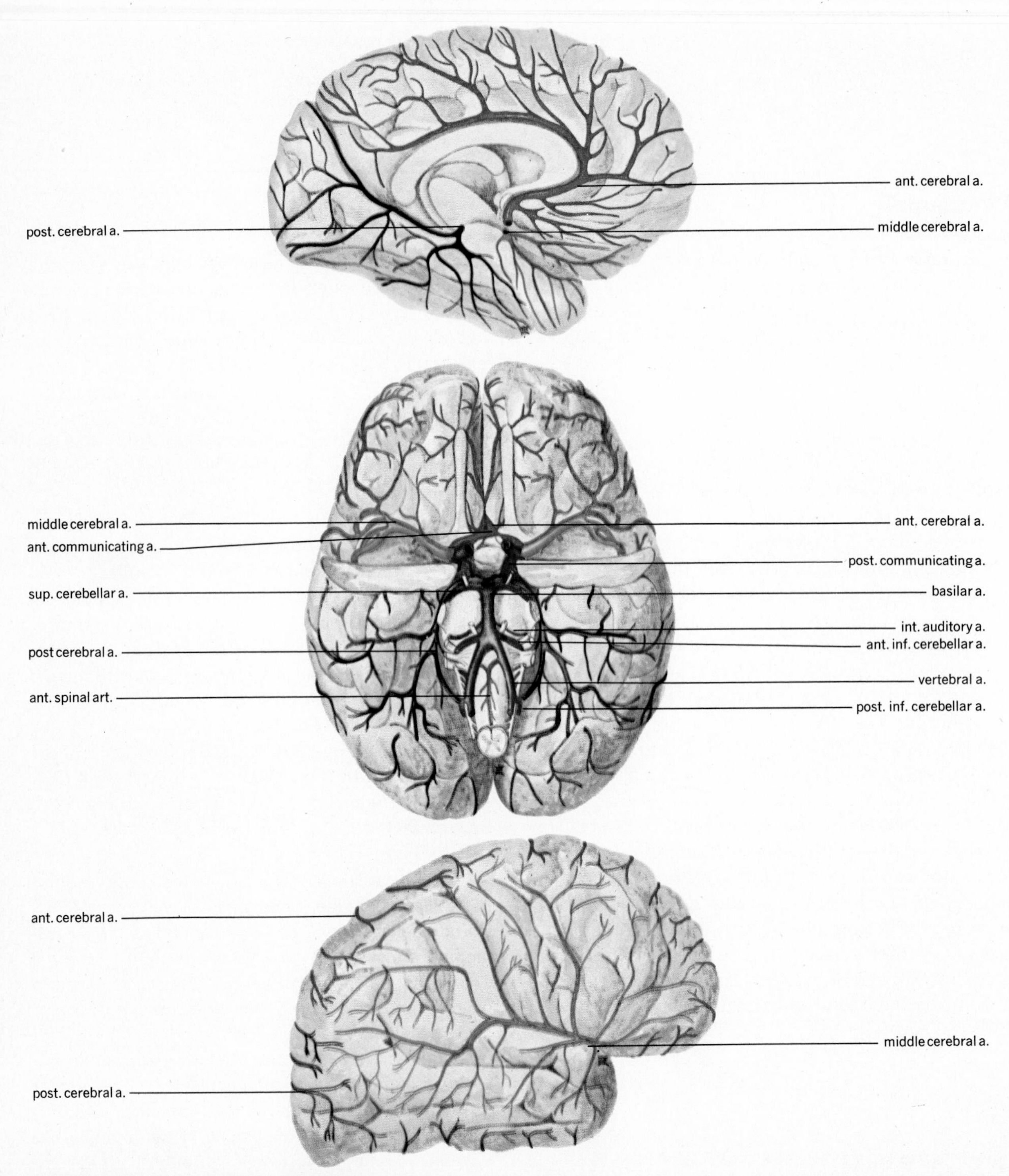

FIGURE 14.1 The arteries of the brain.

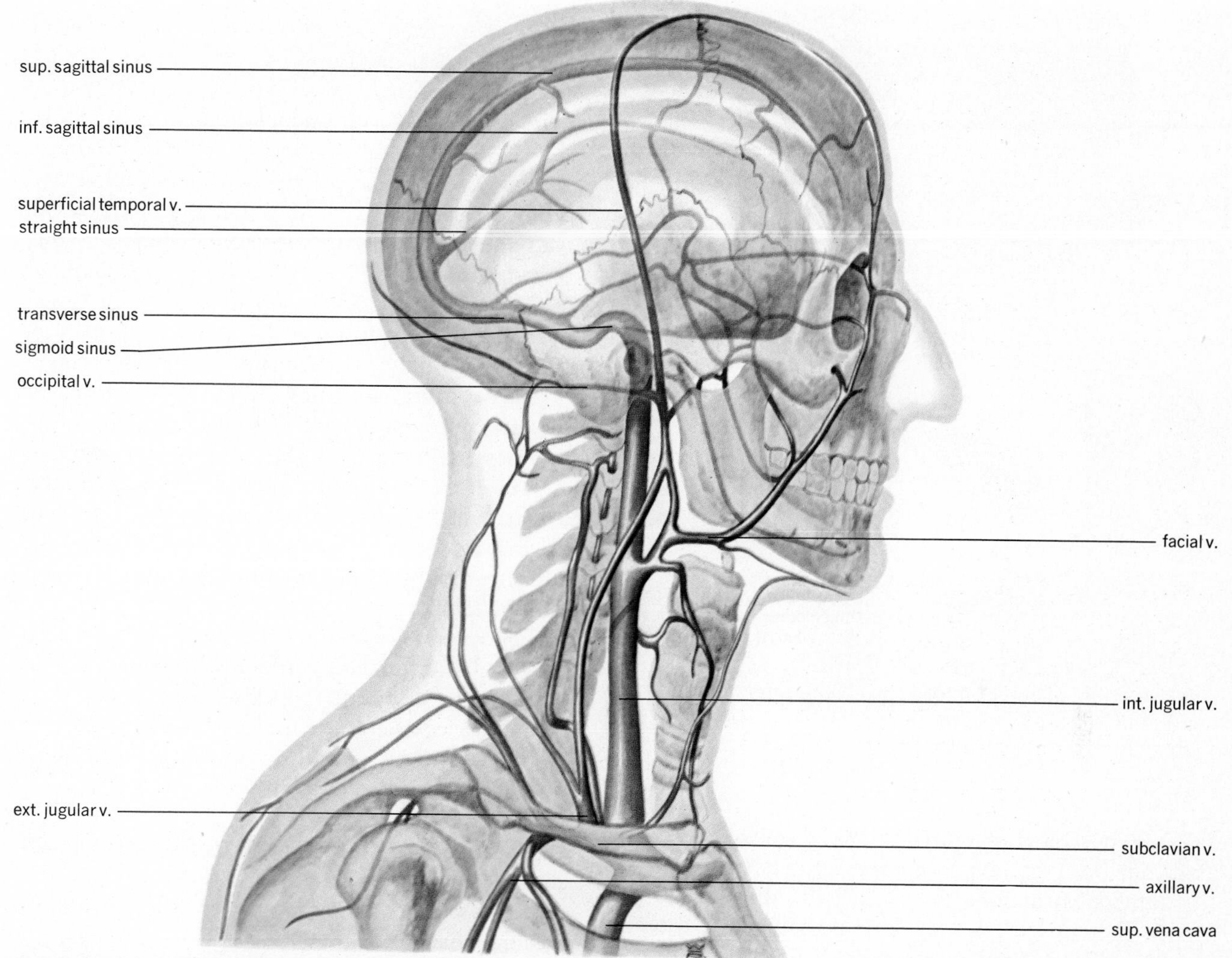

FIGURE 14.2 The dural sinuses and some facial and cervical veins.

While this "barrier" may act to protect the neurons from potentially harmful materials, it makes antibiotic treatment of brain inflammation extremely difficult.

Clinical considerations

CEREBRAL VASCULAR LESIONS [*"strokes," cerebral vascular accidents (CVA)*] account for more neurological disorders than any other category of pathlogical processes. No attempt will be made to present all possible disorders involving CNS blood vessels; rather, several broad causative categories of vascular disorders, and their common symptomology, will be discussed. CVA is the third leading cause of death in the United States.

CLASSIFICATION OF VASCULAR DISORDERS OF THE BRAIN. In vascular disorders of the brain,

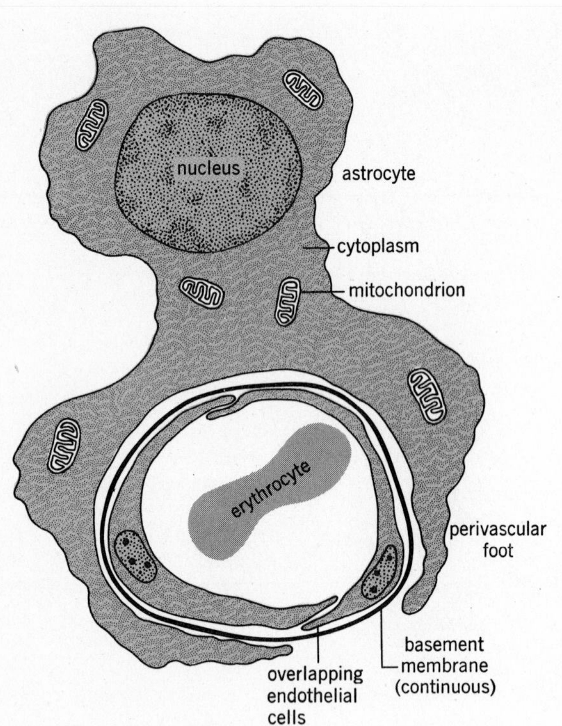

FIGURE 14.3 The structures involved in the creation of the blood-brain barrier.

normal nervous function is impaired in two ways: FAILURE OF CIRCULATION, because of clots, plaques, or other obstructions to blood flow; RUPTURE OF VESSELS, with development of pressure (by the clot) or dissolution of tissue (by deprivation of nutrients).

Based on cause, failure of circulation or rupture of vessels falls into three major categories:

Narrowing or occlusive disorders, in which vessel diameter is reduced or flow is blocked and cells starve.

Embolic disorders, in which floating masses (clots, air bubbles, fatty plaques) plug a vessel(s).

Weakening of vessel walls, because of failure of tissues to develop or their destruction, is usually followed by *aneurysm* (a blisterlike dilation of the wall) and vessel rupture.

ATHEROSCLEROSIS, or deposition of fatty substances (cholesterol, lipoid material) in the inner wall of a vessel, is a prime cause of occlusive disorders. Also, SYPHILIS may thicken a blood vessel wall and reduce its diameter. Since blood flow is proportional to the fourth power of the radius of the vessel, small decreases in size mean great decreases in blood flow.

EMBOLISM may occur by sudden decompression (a scuba diver surfacing without exhaling) or by nitrogen coming out of solution and forming bubbles in the tissues (the bends), by entry of air into the bloodstream after cardiac, pulmonary, or cerebral surgery, or by fatty plaques or clots tearing loose from their place of formation (as after heart surgery or "reaming out" of arteries to remove fats).

WEAKNESS is usually the result of infection, genetic disorders that influence development of tissue coats, or toxins (poisons).

In all cases, the gradual or sudden reduction of blood flow to brain neurons produces a more or less common series of symptoms.

Headache that is vague as to location and usually short-lived.

Dizziness and *faintness* that is exaggerated by postural changes.

Nausea and *vomiting*.

Unequal reflexes (e.g., pupillary reflex) on the two sides of the body.

Muscular weakness, often one-sided.

Mental confusion and difficulties in speaking.

These symptoms do not lead to the diagnosis of specific disorders, but suggest that a CVA has occurred.

Ventricles and cerebrospinal fluid

The nervous system develops as a hollow tube, and the cavity of this tube has been modified into a series of brain cavities known as the VENTRICLES; the central canal of the spinal cord represents the remnant of the hollow cavity in that organ. The ventricles are filled with CEREBROSPINAL FLUID, and the fluid also lies around the brain and cord, enclosed by the meninges.

TABLE 14.1 Plasma and CSF compared for some major constituents

Constituent or property	Plasma	CSF
Protein	6400–8400 mg%	15–40 mg%
Cholesterol	100–150 mg%	0.06–0.20 mg%
Urea	20–40 mg%	5–40 mg%
Glucose	70–120 mg%	40–80 mg%
NaCl	550–630 mg%	720–750 mg%
Magnesium	1–3 mg%	3–4 mg%
Bicarbonate (as vol % of CO_2)	40–60 mg%	40–60 mg%
pH	7.35–7.4	7.35–7.4
Volume	3–4 liters	200 ml
Pressure	0–130 mm Hg	110–175 mm CSF

Ventricles (Fig. 14.4)

Each cerebral hemisphere contains one LATERAL VENTRICLE (these are the first and second ventricles), and each communicates through an INTERVENTRICULAR FORAMEN with the THIRD VENTRICLE. The third ventricle is a single, narrow, slitlike structure lying between the two halves of the diencephalon. The CEREBRAL AQUEDUCT is a small channel running through the midbrain and pons

FIGURE 14.4 The ventricles of the brain.

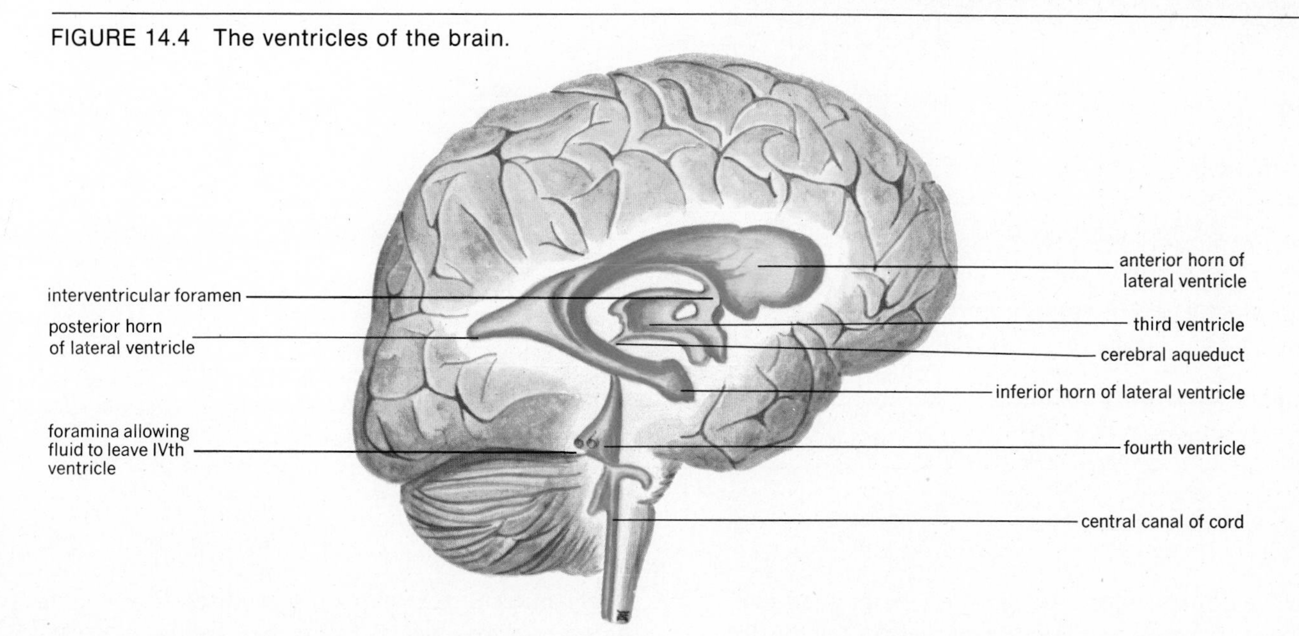

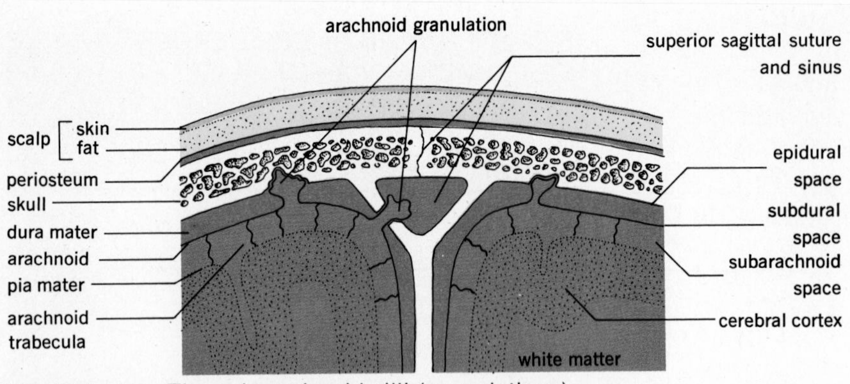

FIGURE 14.5 The subarachnoid villi (granulations).

FIGURE 14.6 The surgical treatment of hydrocephalus.

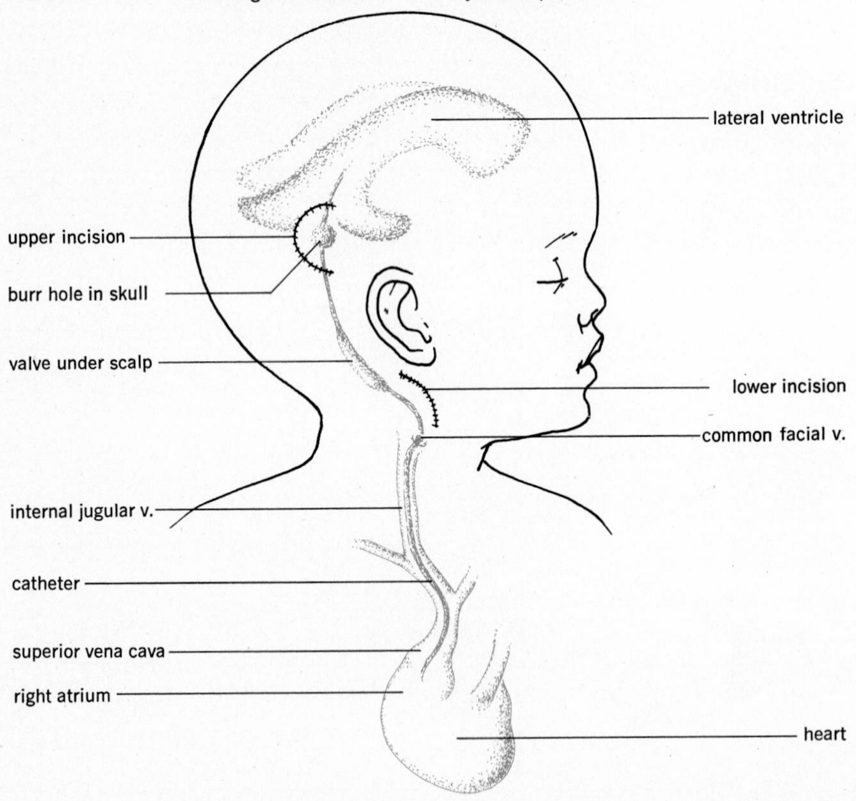

to the FOURTH VENTRICLE located in the medulla. Three openings are found in the roof of the fourth ventricle: two lateral foramina of Luschka, and a single midline foramen of Magendie. These foramina communicate with the subarachnoid space of the cord and brain.

K. WELLEIN

Cerebrospinal fluid

Vascular structures known as CHOROID PLEXUSES, located in the lateral, third, and fourth ventricles, produce the cerebrospinal fluid (CSF) from the blood plasma. CSF differs from plasma in enough ways to suggest that it is formed by active processes and not filtration (Table 14.1).

The fluid circulates in a top to bottom direction—that is, from lateral ventricles, to third ventricle, to aqueduct, to fourth ventricle, to subarachnoid space. Absorption of the fluid into blood vessels occurs through spinal venous vessels and SUBARACHNOID VILLI (Fig. 14.5) in the skull.

The fluid acts as a shock absorber for the brain and cord, is easily removed by spinal puncture for analysis, and compensates for changes in blood volume, keeping cranial volume constant.

Clinical considerations

The term HYDROCEPHALUS (*ydor*, water + *kephale*, head) means an increased accumulation of CSF in the ventricles of the brain. It may be due to increased production of CSF, to decreased absorption, or blockage of the flow of CSF.

CHOROID PLEXUS TUMOR may increase production of CSF. A pneumoencephalogram, in which the CSF in the ventricles is replaced by air, usually demonstrates the presence of such a tumor.

DECREASED ABSORPTION is usually due to failure of the subarachnoid villi (see Fig. 14.5) of the meninges to develop.

BLOCKAGE, by tumor or failure of the ventricular system to develop, accounts for about one-third of the cases. The cerebral aqueduct, because of its small size, is a common site of blockage of CSF flow.

As a result of excessive fluid filling the ventricle(s), ballooning of the ventricles may occur, and the head enlarges because the sutures between the cranial bones have not yet grown together. Diagnosis of the condition is made in infants by repeated measurements of head circumference to demonstrate the abnormal rate of increase of head size.

The treatment is usually surgical—bypassing the obstruction (if present) with a tube and a one-way valve. The CSF may be delivered to the right atrium of the heart or the peritoneal (abdominal) cavity (Fig. 14.6).

Summary

1. The central nervous system has blood vessels designed to provide it with nutrients and to remove its wastes of activity.
   a. The brain is supplied by the internal carotid and vertebral arteries.
   b. The spinal cord is supplied by anterior and posterior spinal arteries and by spinal arteries rising from the aorta.
   c. Capillaries in the cerebrum are less permeable to solute passage than elsewhere in the body. A blood-brain barrier exists. The barrier has an anatomical basis.
2. Vascular disorders of the central nervous system deprive nerve cells of their nutrients and lead to cell death. They are of three types.
   a. Occlusive disorders result from narrowing of vessels and deprivation of neurons of oxygen and fuels.
   b. Embolic disorders result when floating masses or bubbles lodge in a cerebral or brain vessel.
   c. Weakening of vessel walls may result in rupture.
3. Symptoms of vascular disorders are basically the same regardless of cause.
   a. Headache, dizziness, nausea, unequal reflexes on both sides of the body, muscular weakness, and mental confusion are the most common symptoms.
4. A series of cavities or ventricles, filled with cerebrospinal fluid, are found within the brain. The fluid also fills the subarachnoid space of brain and cord.
   a. There are four ventricles.
   b. The fluid is formed from plasma in choroid plexuses. It is removed by absorption from the subarachnoid spaces.
   c. Excessive cerebrospinal fluid, from whatever cause, produces hydrocephalus.

The Autonomic Nervous System

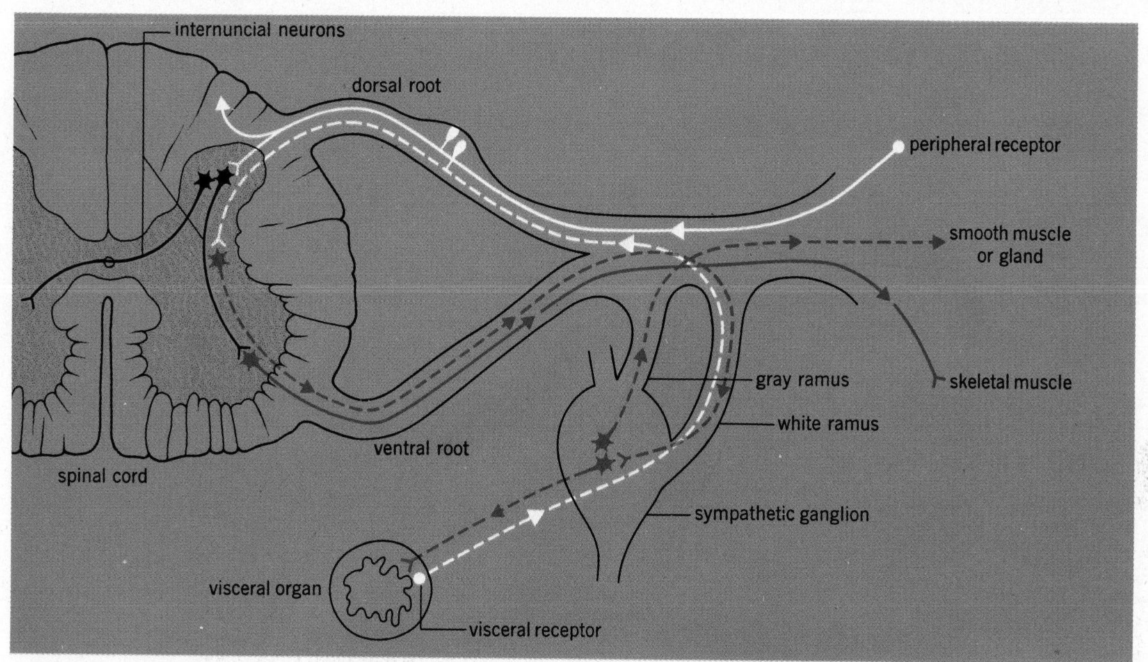

FIGURE 15.1 The connections of the autonomic nervous system fibers to the spinal cord.

The autonomic nervous system has been classically defined as that part of the peripheral nervous system supplying motor fibers to smooth and cardiac muscles and glands, operating "automatically" at the reflex and subconscious levels. Such a definition is too limited, for it is obvious that there are fibers of sensory nature that pass from organs to the central nervous system (Fig. 15.1) and that form the afferent limbs of many autonomic reflex arcs. Today, we recognize that there are VISCERAL AFFERENT FIBERS, carrying sensory input from organs, and VISCERAL EFFERENT FIBERS to organs.

Connections of the system with the central nervous system

Afferent pathways

Visceral afferent fibers enter the spinal cord through the posterior (dorsal) roots, in company with somatic afferents from the skin and skeletal muscles. One neuron reaches from the periphery to the central nervous system, and its cell bodies are found in the dorsal root ganglia of the spinal nerves. Synapses usually occur in the dorsal gray column. Visceral afferents may also enter in various cranial nerves, and in this case the cell bodies are found with the sensory nuclei of those nerves located in the brainstem.

Efferent pathways

Visceral efferent fibers have their cell bodies in

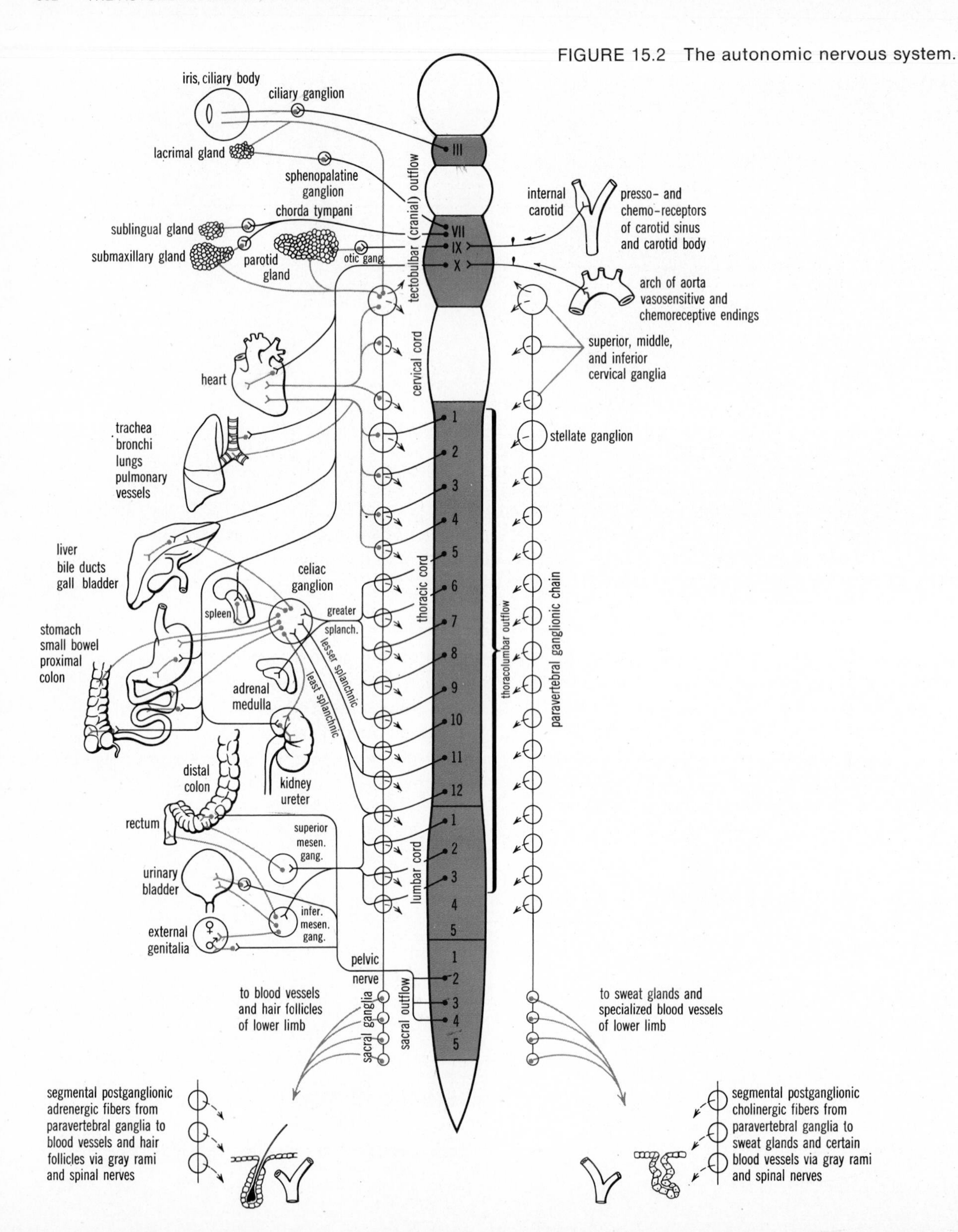

FIGURE 15.2 The autonomic nervous system.

the lateral gray columns of the spinal cord, or in the motor nuclei of certain cranial nerves. Those that leave by way of anterior spinal nerve roots separate from the somatic fibers by way of the WHITE RAMI. They may undergo synapses in ganglia (see below) close to the cord or pass through the ganglia to synapse in ganglia closer to the organs innervated. If a synapse does occur in the ganglia close to the cord, the nerve fibers usually rejoin the spinal nerve by way of the GRAY RAMI, and are then distributed to organs located within the skin and skeletal muscles (mostly blood vessels and skin glands). Two autonomic neurons usually extend from the CNS to the organ innervated. The one passing from the CNS to a synapse in the periphery is called the PREGANGLIONIC NEURON; the one passing from the synapse to the organ supplied is called the POSTGANGLIONIC NEURON. Transmission from pre- to postganglionic neurons is chemical, utilizing acetylcholine. Postganglionic neurons produce either acetylcholine or norepinephrine at their endings, and this accounts for the different effects exerted on the organs by the two divisions of the autonomic system.

Autonomic ganglia and plexuses

The term GANGLION is usually used to refer to masses of nervous tissue that contain nerve cell bodies and/or synapses and that are located in the periphery. The term *plexus* usually refers to a network of nerve fibers or blood or lymphatic vessels. In describing the nervous system, the two terms are *sometimes* used interchangeably, with the term ganglion understood to mean a synaptic region.

Synapses between pre- and postganglionic neurons occur in one of three areas in the periphery.

The LATERAL OR VERTEBRAL GANGLIA, also called the *sympathetic ganglia,* form paired chains of 22 ganglia that lie alongside the vertebral bodies in the thoracic and abdominal cavities. They receive preganglionic fibers from the thoracic and lumbar spinal nerves.

The COLLATERAL OR PREVERTEBRAL GANGLIA form several groups of nerves and synapses in association with visceral organs or large blood vessels. The *cardiac, celiac* (solar), and *mesenteric plexuses* are three important ones located near the heart, spleen, and urinary bladder, respectively. These plexuses receive preganglionic fibers from the cranial, thoracic, lumbar, and sacral portions of the central nervous system.

TERMINAL GANGLIA or plexuses lie close to or within the organ supplied. The nerve plexuses in the wall of the intestine are examples of terminal ganglia. These ganglia receive preganglionic fibers from the cranial and sacral portions of the central nervous system.

Divisions of the system: structure and function

Preganglionic autonomic fibers form three outgoing groups (outflows). The CRANIAL OUTFLOW is composed of the motor fibers of cranial nerves III, VII, IX, and X. The THORACOLUMBAR OUTFLOW consists of the visceral efferents derived from all thoracic and lumbar spinal nerves. The SACRAL OUTFLOW consists of the visceral efferents derived from the second through fourth sacral spinal nerves. Functionally, the fibers are grouped into a PARASYMPATHETIC *(craniosacral)* DIVISION and a

SYMPATHETIC *(thoracolumbar)* DIVISION. The two divisions provide a double innervation for most (but not all) body organs (Fig. 15.2).

The parasympathetic division

The parasympathetic division, formed from the cranial and sacral outflows, supplies fibers to all

TABLE 15.1 Effects of autonomic stimulation

Organ affected	Parasympathetic effects	Sympathetic effects
Iris	Contraction of sphincter pupillae; pupil size decreases	Contraction of dilator pupillae; pupil size increases
Ciliary muscle	Contraction; accommodation for near vision	Relaxation; accommodation for distant vision
Lacrimal gland	Secretion	Excessive secretion
Salivary glands	Secretion of watery saliva in copious amounts	Scanty secretion of mucus rich saliva
Respiratory system:		
Conducting division	Contraction of smooth muscle; decreased diameters and volumes	Relaxation of smooth muscle; increased diameter and volumes
Respiratory division	Effect same as on conducting division	Effect same as on conducting division
Blood vessels	Constriction	Dilation
Heart:		
Stroke volume	Decreased	Increased
Stroke rate	Decreased	Increased
Cardiac output and blood pressure	Decreased	Increased
Coronary vessels	Constriction	Dilation
Peripheral blood vessels:		
Skeletal muscle	No innervation	Dilation
Skin	No innervation	Constriction
Visceral organs (except heart and lungs)	Dilation	Constriction

autonomic effector organs except the adrenal medulla, sweat glands, splenic smooth muscle, and blood vessels of skin and the skeletal muscles. The preganglionic fibers are relatively long; they synapse in collateral or terminal ganglia or both. The postganglionic fibers are shorter, secrete acetylcholine at their endings, and are thus called *cholinergic* fibers. The cranial portion of this divi-

Organ affected	Parasympathetic effects	Sympathetic effects
Stomach:		
Wall	Increased motility	Decreased motility
Sphincters	Inhibited	Stimulated
Glands	Secretion stimulated	Secretion inhibited
Intestines:		
Wall	Increased motility	Decreased motility
Sphincters:		
Pyloric, iliocecal	Inhibited	Stimulated
Internal anal	Inhibited	Stimulated
Liver	Promotes glycogenesis; promotes bile secretion	Promotes glycogenolysis; decreases bile secretion
Pancreas (exocrine and endocrine)	Stimulates secretion	Inhibits secretion
Spleen	No innervation	Contraction and emptying of stored blood into circulation
Adrenal medulla	No innervation	Epinephrine secretion
Urinary bladder	Stimulates wall, inhibits sphincter	Inhibits wall, stimulates sphincter
Uterus	Little effect	Inhibits motility of non-pregnant organ; stimulates pregnant organ
Sweat glands	No innervation	Stimulates secretion (produces "cold sweat" when combined with cutaneous vasoconstriction)

sion tends to protect and conserve body resources and to preserve normal resting functions. The sacral portion provides a group of mechanisms for evacuation of the colon, rectum, and urinary bladder.

The sympathetic division

The sympathetic division is formed by the thoracic and lumbar outflows. Preganglionic fibers are relatively short; they synapse in vertebral and/or collateral ganglia. The postganglionic fibers are longer and most of them secrete norepinephrine at their endings. These fibers are thus called *adrenergic* (from noradrenalin, another name for norepinephrine). This division supplies the same organs as does the parasympathetic division, and is the only nerve supply to the adrenal medulla, sweat glands, skin and skeletal muscle blood vessels, and spleen. In the latter group of organs, activity is controlled only by increase or decrease in sympathetic stimulation. Activity of the sympathetic division tends to increase utilization of body resources. Massive sympathetic discharge is associated with liberation of epinephrine (adrenalin) from the adrenal medulla and with preparation of the entire body for a "fight-or-flight" posture against the agent causing the great stimulation.

Operation of both divisions is TONIC or continuous, so that many body functions, even at resting levels, reflect a balance between the activity of both parts. Response to the parasympathetic system is more specific, with individual organs being influenced; this is because the parasympathetic preganglionic fibers connect with fewer "individual" postganglionic fibers. Response to sympathetic stimulation is more diffuse because preganglionic fibers connect to larger numbers of postganglionic neurons, and because norepinephrine is destroyed more slowly than acetylcholine. Table 15.1 summarizes the effects on body organs of stimulation of the two divisions.

Norepinephrine and epinephrine cause different sympathetic effects within the body. To explain these differential effects it is assumed that two different kinds of receptor substances or sites are found in effector cells—that is, substances or regions that may be affected by epinephrine, norepinephrine, or both. The two substances or sites are designated ALPHA RECEPTORS and BETA RECEPTORS. In theory, epinephrine can affect only cells containing alpha sites. Alpha receptors promote responses such as vasoconstriction, dilation of the pupil, and relaxation of intestinal muscle. Beta receptors promote responses such as vasodilation, cardiac acceleration and increased strength of heart contraction, vasodilation of coronary and skeletal blood vessels, and relaxation of uterine muscle.

Drugs exist that may compete for the receptor sites on the effector organs, and may thus block the expression of a certain sympathetic effect. For example: *phenoxybenzamine* blocks alpha receptors and leads to vasodilation, an effect useful in controlling hypertension; *propanolol* blocks beta receptors and leads to cardiac slowing and regularization of heart beat.

Higher autonomic centers

The medulla of the brainstem contains the vital centers (vasomotor, cardiac, respiratory) that are involved in controlling blood pressure and breathing. The cerebral cortex contains, in the frontal lobes, areas that produce both sympathetic and parasympathetic effects. These regions may be tied in with expression of emotions, such as anger and rage.

The hypothalamus appears to be involved to a great degree in the control of what are sometimes

TABLE 15.2 Drugs and the autonomic system

Drug	How acts	Use	Comments
Reserpine	Depletes norepinephrine from postganglionic endings by increasing release. Causes decrease in heart action.	To alleviate hypertension	Is one of, and the most potent of a series of alkaloids derived from the Rauwolfia plant. Norepinephrine loss results in vasodilation and fall in blood pressure, and decrease in heart action.
Guanethidine	As above, but exerts no effect on heart	As above	Limited side effects make it a "clinically advantageous drug"
Methyldopa	Lowers brain and heart content of norepinephrine by interfering with synthesis.	As above	May produce toxicity of liver.
Hydralazine	Depress vasoconstrictor center and inhibit sympathetic stimulation.	As above	Many side effects are common.
Veratrum (a series of plant alkaloids)	Slow heart action, stimulate vagus (parasympathetic) activity.	As above	Some side effects.
Hexamethonium	Blocks pre- to post-ganglionic transmission, Anticholinergic.	As above	
Acetylcholine	Increase or mimics parasympathetic stimulation. Causes some vasodilation.	As above	Not routinely employed. It is rapidly destroyed and affects many organs other than blood vessels.
Atropine	Inhibits acetylcholine and parasympathetic effects	Eye examinations; before general anaesthesia to dry respiratory secretions.	Sympathomimetic
Pilocarpine	Mimics parasympathetic stimulation.	Treatment of glaucoma.	Parasympathomimetic

called the vegetative (involuntary) functions of the body. Many of the functions of the hypothalamus discussed in the preceding chapter are autonomic in nature. (See Fig. 13.4 for a review of the involvement of the hypothalamus in autonomic function.)

Clinical considerations

Some clinical conditions are life-threatening and are believed to be the result of excessive autonomic activity. One of these is HYPERTENSION *(high blood pressure).* Surgical or chemical intervention designed to relieve the excessive autonomic activity (if this is the cause) is sometimes employed.

Surgery and the autonomic system

Since the effect of sympathetic stimulation on most visceral blood vessels is to constrict them, and since constriction of large numbers of blood vessels raises the blood pressure, removal of the sympathetic effect should lower the blood pressure. A SYMPATHECTOMY, which removes the sympathetic ganglia from the levels of T10–L2, may be performed to alleviate hypertension. If the procedure is effective, blood pressure will fall, but since the ganglia are gone, the individual has lost the ability to raise blood pressure in stress or in changes of position.

Drugs and the autonomic system

An alternative to surgical intervention in disorders of autonomic function is to employ drugs that either block or depress the effect of one of the divisions of the system, increase the activity of a division, or mimic the effects of stimulation of a division.

Drugs that mimic effects of stimulation of a given division are said to be *parasympathomimetic* or *sympathomimetic.* Those that prevent synaptic transmission are ganglionic blocking agents. Table 15.2 presents a summary of some drugs that affect the autonomic system, with notes as to how they act. Most are employed to alleviate hypertension by causing vasodilation.

Summary

1. The autonomic system controls visceral activity and glandular secretion.
2. Visceral afferent fibers convey sensory impulses from viscera to CNS. Visceral efferent fibers convey motor impulses from CNS to organs.
3. The efferent pathway consists of two neurons and one synapse. The neurons are designated as pre- and postganglionic neurons. Transmission across the synapse is by acetylcholine. Postganglionic fibers may produce either acetylcholine or norepinephrine.
4. Autonomic ganglia and plexuses provide areas for synapse between pre- and postganglionic fibers.
5. Two divisions of the autonomic system are designated.
   a. The parasympathetic (craniosacral) division is composed of cranial and sacral outflows and conserves body resources.
   b. The sympathetic (thoracolumbar) division consists of thoracic and lumbar outflows and increases utilization of body resources.
   c. Most body viscera receive fibers from both divisions. Sympathetic fibers are the only nerve supply to sweat glands, adrenal medulla, spleen, and blood vessels of skin and skeletal muscle.
   d. Effects of both systems are tonic (continuous).

6. Higher centers in the brainstem, hypothalamus, and cerebral cortex are responsible for several autonomic effects.
7. In treatment of hypertension, or high blood pressure, surgical or chemical intervention is often necessary.
   a. Sympathectomy is a surgical procedure involving removal of sympathetic ganglia to remove constricting nerve impulses to visceral blood vessels.
   b. Drugs may block the constricting effect of the sympathetic nervous system on blood vessels.

Sensation

Maintenance of homeostasis by reaction to change and adjustment of function would seem to suggest the presence of structures capable of detecting change and initiating the appropriate homeostasis-maintaining response. In general, a structure capable of detecting an alteration in some internal or external condition is termed a RECEPTOR. Receptors may not only be able to detect a change, but may also be able to provide signals related to the intensity or degree of change and may provide the ability to "ignore" a continuing stimulus once it has been recognized and acted upon.

Receptors

Definition and organization

A receptor may be defined as the peripheral end of a sensory neuron, and it may possess a complicated or simple structure. A given afferent nerve fiber (neuron) and all of its peripheral and central branches constitute a SENSORY UNIT that responds best to a particular type of stimulus (a modality) delivered within that unit's PERIPHERAL RECEPTIVE FIELD, or area supplied. Sensitivity to particular types of stimuli may depend on the density (numbers of receptors in a given area), or the threshold of a given receptor may change depending on a variety of factors. Thus, humans achieve an ability to restrict or limit certain sensory experiences to those body areas where it is most appropriate to have them, and an ability to "ignore" some stimuli if they are not essential to body functioning.

Properties

A receptor usually responds best, although not exclusively, to one particular type of stimulus. This property is embodied in the LAW OF ADEQUATE STIMULUS. For example, pressure on the eyeball or light entering it both create a sensation of seeing a light, although the stimuli are different. In a sense, then, each receptor is specific to a given stimulus, and this accounts for the many different types of receptors the body possesses. When stimulation of a receptor results in the formation of a nerve impulse, that impulse does not differ in its characteristics from the impulse rising from another receptor. What difference in sensation humans are able to appreciate is due to the central connection ultimately made by the fibers from the receptor. This is called the LAW OF SPECIFIC NERVE ENERGIES. Thus, we might see thunder and hear lightning if the central connections of eye and ear could be reversed. INTENSITY of a stimulus is conveyed either by the stimulation of more receptors by a stronger stimulus or by an increased frequency (number per second) of nerve impulse formation by a receptor when it is more strongly stimulated. All receptors appear capable of TRANSDUCING *(changing)* a physical or chemical change into a nerve impulse. This is believed to occur as a result of permeability changes in the receptor when it is stimulated, the creation of a GENERATOR POTENTIAL by the receptor, and depolarization of the nerve fiber by the generator potential. ACCOMMODATION by receptors is evidenced by a decreasing frequency of receptor discharge as a stimulus of constant strength is continued.

Two types of receptors may be distinguished on the basis of their ability to accommodate. SLOWLY ADAPTING RECEPTORS do not accommodate or do so very slowly; they therefore continue to transmit information to the CNS as long as they are stimulated. Examples include: muscle, tendon, and joint receptors, that continue to inform us of limb position; pain fibers that convey continuing "hurts" of the body; the chemical (O_2, CO_2) sensitive receptors in the aorta and carotid arteries that convey information about low blood oxygen, or high blood CO_2 levels. RAPIDLY ADAPTING RECEPTORS react strongly when a stimulus is first presented, but slow or stop firing as a stimulus continues unchanged. Touch and pressure receptors and olfactory receptors are examples of this type. Their continued function is not required for survival of the organism.

Classification

Receptors are classified or named according to where they are located or what their adequate stimulus is. EXTEROCEPTIVE receptors are located at or near a body surface (eye, touch, pressure); ENTEROCEPTIVE receptors are located within body organs other than muscles or tendons (stretch receptors of blood vessels, lungs); PROPRIOCEPTIVE receptors are found in muscles and tendons (muscle spindles, Golgi organs). Physiologically, a more meaningful classification of receptors is to name them according to the type of stimulus to which they best react, as shown in Table 16.1.

TABLE 16.1 Classification of receptors by stimulus

Adequate stimulus	Example(s)
Mechanical (mechanoreceptors) pressure, bending, tension	Touch and pressure receptors in skin. Pressure or stretch sensitive receptors (baroreceptors) in blood vessels, lungs, gut. Equilibrium and balance receptors in inner ear (semicircular canals, maculae). Organ of hearing (hair cells of cochlea). Kinesthetic receptors in muscles, tendons, joints.
Chemical (chemoreceptors) substances in solution	Taste Smell
Stimulation by type or concentration of chemical	Carotid and aortic bodies (CO_2 and O_2 sensitive) Osmoreceptors in hypothalamus
Light (photoreceptors)	Eye
Thermal change (thermoreceptors)	Heat Cold
Extremes of most any stimuli (nociceptors)	Pain

Somesthesia; the study of specific sensations

The term *somesthesia* refers to those sensations elicited from the "common" receptors other than those for sight, hearing, equilibrium, taste, and smell. They are mostly cutaneous and visceral sensations and those from muscles and tendons.

Tactile sensations: touch and pressure

The same basic stimulus, mechanical distortion of a skin surface, is effective in stimulating the touch receptors (MEISSNER'S CORPUSCLES) and pressure receptors (PACINIAN CORPUSCLES) (Fig. 16.1). The Meissner's corpuscles are located in the top layers of the dermis, while the Pacinian corpuscles are deeper in the dermis and have more layers of tissue around them. Therefore, a lighter stimulus (touch) is effective on the Meissner's receptor. A sense of touch may also be felt when body hairs are bent, because of the presence of unspecialized fibers located around hair follicles. The sense of touch is most acute where density of receptors is greatest (lips, tongue, fingertips); where density is greatest, stimulation thresholds are lowest. Thus, blind persons are easily able to read the raised dots of the braille alphabet with the fingers. Touch, combined with the mental concepts of shape and size, may be utilized to determine shapes of objects and their size, if size differs (in either direction) by 10 percent or more

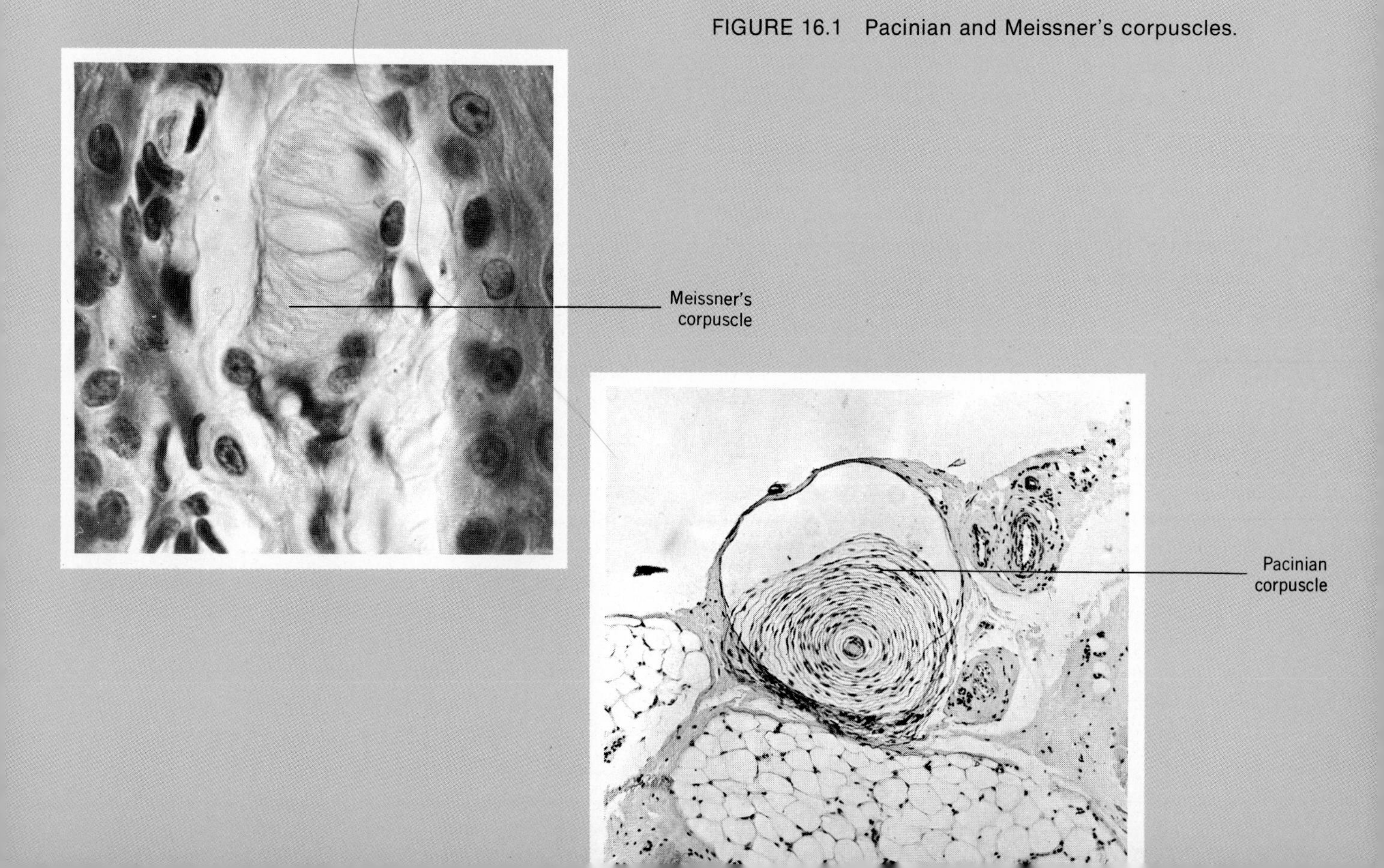

FIGURE 16.1 Pacinian and Meissner's corpuscles.

(Weber-Fechner law). Touch sensations are conveyed over the gracile and cuneate tracts of the spinal cord (fine discrimination), or the ventral spinothalamic tracts (crude touch), and pass to the thalamus and parietal sensory areas (areas 3, 1, and 2).

Thermal sensations: heat and cold

The temperature sensations of the skin are served by RUFFINI CORPUSCLES *(warmth)*, and KRAUSE'S CORPUSCLES *(cold)*. In some places on the body where these sensations may be appreciated, there are no discrete receptors, but "sensory end bulbs" of nonspecific structure may be found. In any event, sensitivity to heat and cold appears in sporadic spots, with cold spots outnumbering warmth spots by four or ten to one. Density of both types of receptors is greatest on hands and face. Warmth receptors fire between 20°C and 45°C, with a peak at 37.5–40°C. Cold receptors fire most strongly at 15–20°C and again at 46–50°C. The latter range of discharge accounts for the feelings of being cold when stepping into a hot shower *(paradoxical cold)*. The lateral spinothalamic tract serves as the afferent pathway up the cord for these sensations, and the thalamus and parietal cortex are areas for relay and interpretation for the sensations.

Pain

Pain is a protective sensation for the body in that it indicates overstimulation of any sensory nerve and damage or potential damage to body tissues. It is served by naked nerve endings in skin and organs. Pain is said to occur in three varieties.

"Bright" or pricking pain is the type experienced when the skin is cut or jabbed with a sharp object. It may be intense but is relatively short-lived and is easily localized.

Burning pain, of the sort experienced when the skin is burned, is slower to develop, lasts longer, and is less accurately localized. It commonly stimulates cardiac and respiratory activity and is difficult to endure.

Aching pain develops when visceral organs are stimulated. It is persistent, often produces feelings of nausea, is poorly localized, and is often *referred* to areas of the body distant from the actual area where damage is occurring. It is of the utmost importance to the physician, for it may signal life-threatening disorder of vital organs.

There appears to be no particular adequate stimulus for pain; it may be evoked by mechanical, electrical, thermal, or chemical stimuli. All appear to result in permeability changes leading to generation of pain impulses. Adaptation or accommodation does not appear to occur; a person may be distracted from a painful stimulus or ignore it, but if attention is paid to it, it appears to persist unchanged. Ischemia (lowered blood flow) of an area may, by retaining metabolites or other chemicals in an area, contribute to development of pain.

Afferent neurons enter via the dorsal roots and synapse in the dorsal gray column. Second-order neurons cross to the opposite lateral spinothalamic tract and ascend to the thalamus. It is believed that most of the interpretation of pain occurs here. Evidence for thalamic interpretation of pain revolves mainly around the fact that removal of the sensory cortex does not eliminate the ability to appreciate pain. The cortex may contain areas of pain appreciation and certainly aids in localization of painful stimuli, particularly if touch receptors are stimulated simultaneously with pain fibers.

REFERRED PAIN. Sensory impulses from visceral organs and skin often impinge on the same or closely spaced neurons. Intense stimulation of visceral pain neurons may cause irradiation or spread to neurons serving the skin, and the sen-

TABLE 16.2 Referred pain

True point of origin of pain	Common cause	Area to which pain is referred
Heart	Ischemia secondary to infarction	Base of neck, shoulders, pectoral area, arms
Esophagus	Spasm, dilation, chemical (acid) irritation	Pharynx, lower neck, arms, over heart ("heart burn")
Stomach	Inflammation, ulceration, chemicals	Below xiphoid process (epigastric area)
Gall bladder	Spasm, distention by stones	Epigastric area
Pancreas	Enzymatic destruction, inflammation	Back
Small intestine	Spasm, distention, chemical irritation, inflammation	Around umbilicus
Large intestine	As above	Between umbilicus and pubis
Kidneys and ureters	Stones, spasm of muscles	Directly behind organ or groin and testicles
Bladder	Stones, inflammation, spasm, distension	Directly over organ
Uterus, uterine tubes	Cramps (spasm)	Lower abdomen or lower back

sation feels as if it were coming from the skin itself (it is referred to the skin). Visceral pain commonly results from spasm, distention, ischemia, or chemical irritation of an organ. The list presented in Table 16.2 indicates to what body area pain originating in various viscera is referred.

HEADACHE. Perhaps the most common pain complaint is headache. From whatever cause, headache pain is nearly always appreciated as deep, diffuse, and aching in quality; it nearly always causes reflex contractions of cranial muscle that may aggravate the pain. Its causes are diverse.

Tension applied to blood vessels (arteries and veins) within the skull or to the membranes around the brain causes headache. Tumors are a possible cause of tension or pressure being applied to cranial structures.

Dilation of the arteries to the brain, resulting from vasodilation or high blood pressure (increased intracranial pressure) produces pain. Spasm, producing ischemia, may be the cause of migraine headache.

Inflammation of paranasal sinuses or any pain-sensitive structure within the skull may produce headache.

Spasm of cranial muscles produces so-called "tension headaches" and results most commonly from emotional involvement or fatigue.

Eye disorders, such as spasm of the extraocular muscles or irritation of the conjunctivae, produces pain behind the eyes.

The role of aspirin in the relief of pain (analgesic action) revolves about its ability to depress nervous conduction or raise the threshold of conduction of impulses in or near the hypothalamus.

DISORDERS OF PAIN PERCEPTION. PARESTHESIA *(abnormal sensation)* rises from irritation or damage to peripheral nerves or cord tracts that results in impulses being transmitted to the brain. The brain then localizes the pain in the area from which that nerve impulse would normally have come.

HYPERESTHESIA involves a greater than normal sensitivity to pain—that is, a lower pain threshold. The receptor itself may be hypersensitive (primary hyperesthesia), or conduction somewhere along the pathway may be facilitated (secondary hyperesthesia).

Cord damage will produce localized or generalized loss of specific sensations depending on extent, area, and side of damage. HEMISECTION of the cord, such as might be carried out experimentally, produces loss of touch and pressure on the same side (ipsilateral) as the damage and loss of pain, heat, and cold on the opposite side (contralateral) as the injury. Most cord damage is not as "neat" as a hemisection, and thus variable deficits in sensory loss are the rule.

The THALAMIC SYNDROME rises from ischemia or damage to the posteroventral portion of the thalamus and results in loss of all sensations from the opposite side of the body, loss of ability to control muscular movements precisely (ataxia), and continuous pain, which resists relief by normal methods of control.

Kinesthetic sense

The sense of body position and movement of joints embraces that body sense known as KINESTHESIA. It originates peripherally in sensory organs located in muscles, joints, and tendons (see Fig. 12.4). Movement of joints and shortening or stretching of muscles and tendons elicit impulses that pass to the cord and enter the dorsal columns. Conscious muscle sense, or appreciation of body position, is carried primarily in the gracile and cuneate tracts, while impulses related to maintenance of muscle tone, unconscious muscle sense, are transmitted primarily over the spinocerebellar tracts to the cerebellum (see Fig. 12.3). Further discussion of the role of the muscle spindles is provided in the section on postural maintenance.

Synthetic senses

The body is capable of appreciating such sensations as itch, tickle, and vibration. No specific receptors exist for such sensations; they appear to be a result of simultaneous stimulation of one or more of the basic sensations previously described. ITCH appears to rise from chemical irritation of pain fibers in the skin and mucous membranes of the nose. If the stimulus moves across the skin, TICKLE results. VIBRATIONAL SENSITIVITY appears to be due to rhythmic and repetitive stimulation of pressure receptors. The pressure receptors respond maximally to repetitive stimuli at 250–300 cps.

TABLE 16.3 Summary of somesthetic senses

Sensation	Receptor	Adequate stimulus	First order neuron	Second order neuron	Third order neuron	Comments
Touch	Meissner's corpuscle	Mechanical pressure to skin	Spinal or cranial afferent	Gracile and cuneate tracts	Thalamus to cortex	Light pressure causes change in receptor shape
Pressure	Pacinian corpuscle	As above	As above	As above	As above	Heavier pressure causes change in receptor shape
Heat	Sensory ending	Rise of tempera- ture, especially between 37–40°C	As above	Spinothalamic tracts	As above	–
Cold	Sensory ending	Fall of temperature, especially between 15–20° C	As above	As above	As above	–
Pain	Naked nerve endings	Overstimu- lation of any nerve; strong stimulation of naked nerves as by ten- sion, pressure	As above	As above	As above	Three types of pain: bright, burning, visceral
Kinesthesis	Muscle, joint and tendon organs	Tension, stretch or move- ment	As above	Conscious: gracile and cuneate tracts.	Thalamus to cortex	Conscious is body position
				Unconscious: spino- cerebellar tracts	Cord to cere- bellum	Unconscious is tone.
Synthetic (itch, tickle, vibration)	None specific	Chemical and mechanical stimulation	–	–	–	Itch: chemical stimulation of pain fibers Tickle: moving stimulation Vibration: touch organs
Visceral (hunger, thirst)	None specific	Mechanical and drying	–	–	–	Most visceral afferents are carried in the vagus nerve.

Visceral sensations

Visceral or organic sensations result from stimulation of internal organs and include hunger and thirst. Other sensations (bladder and rectal fullness) will be discussed in appropriate chapters that follow.

HUNGER is associated with powerful contractions of the stomach, which may be due to lowering of the blood glucose levels. The compression of the gastric nerves by the powerful contractions may create the unpleasant and often near-painful quality of the sensation.

THIRST occurs when the oral membranes dry as a result of decreased salivary secretion. The decreased secretion is, in turn, due to dehydration, and it signals a need for fluid intake.

A summary of somesthetic senses appears in Table 16.3.

Summary

1. Receptors sense change internally and/or externally to the body, and their nerves convey impulses to the central nervous system for interpretation.

2. All receptors show several basic properties.
   a. They respond best to one particular form of change (law of adequate stimulus).
   b. All impulses from different receptors are alike. Differences depend on central connections (law of specific nerve energies).
   c. Intensity of stimulation may be signalled either by increased frequency of receptor discharge or by stimulation of more receptors.
   d. Receptors accommodate or slow their rate of impulse discharge with continued stimuli of the same strength.

3. Receptors are classified by their location and their adequate stimulus.
   a. Receptors located at or near body surfaces are exteroceptive; in viscera, enteroceptive; in muscles and tendons, proprioceptive.
   b. By stimulus, receptors may be chemical, pressure, light, thermal, or pain sensitive.

4. Somesthesia refers to touch, pressure, heat, cold, and pain receptors.
   a. Touch and pressure are served by Meissner's and Pacinian corpuscles. The impulses pass through ventral spinothalamic, gracile, and cuneate tracts to the thalamus and then to the sensory cortex.
   b. Heat and cold may have specialized corpuscles. They are conveyed over the lateral spinothalamic tracts to the thalamus and then to the sensory cortex.
   c. Pain is served by naked nerve endings, is conveyed over lateral spinothalamic tracts, and is interpreted in the thalamus. Localization of pain occurs in the sensory cortex. Pain is bright, burning, or aching in character and may be referred to areas other than its actual point of origin.

5. Headache is a type of pain resulting from tension on internal skull structures, increased intracranial pressure, inflammation, cranial muscle spasm, and eye disorders.

6. Paresthesia (abnormal sensation) and hyperesthesia (increased sensitivity) are abnormalities of pain perception.

7. Kinesthesia refers to the sensation of body position and joint movement. It originates from receptors in muscles and tendons.
   a. Conscious muscle sense (of body position) is carried in gracile and cuneate tracts to the cerebrum.
   b. Unconscious muscle sense (muscle tone reflex) is carried over the spinocerebellar tracts to the cerebellum.

8. Synthetic senses result from simultaneous stimulation of other receptors.
   a. Itch results from chemical irritation of pain fibers.
   b. Tickle results from movement across the skin.
   c. Vibration sensitivity is due to rhythmic stimulation of pressure receptors.

9. The thalamus interprets pain and acts as the relay station to the cerebrum for all other somesthetic sensations.

17 The Eye and Vision

The structure of the eye

Light and lenses; image formation
- Properties of light and lenses
- Image formation by the eye

Disorders of the eye
- Disorders of image formation
- Other disorders

Retinal function
- Structure and functions of rods and cones

The visual pathways
- Efferent pathways from the eye
- Other functions of the brain in vision

Summary

The eye is a complex peripheral exteroceptor that is specialized to respond to quanta of light energy. It also contains within itself mechanisms that normally enable a clear and undistorted image of the external world to be focused on the photosensitive portion of the eyeball.

The structure of the eye (Fig. 17.1)

Each EYEBALL is an organ approximately 25 mm (1 in.) in diameter that lies within the orbit of the skull. Anteriorly, it is protected by the EYELIDS; these structures guard against the entry of light, afford some protection against mechanical forces, and spread the tears over the eye when they are blinked. Each eyelid is lined with about 100 eyelashes that act as a trigger, when bent, for the blink reflex. Also, when the lids are nearly closed, the lashes act to scatter light, which contributes to diminishing light entering the eye. Blinking is protective in nature and is triggered by additional stimuli such as a bright light flashed suddenly into the eye (dazzle reflex), rapidly approaching objects (menace reflex), or touching the cornea. Large sebaceous glands (Meibomian glands) asso-

FIGURE 17.1 The eye and eyelids in section.

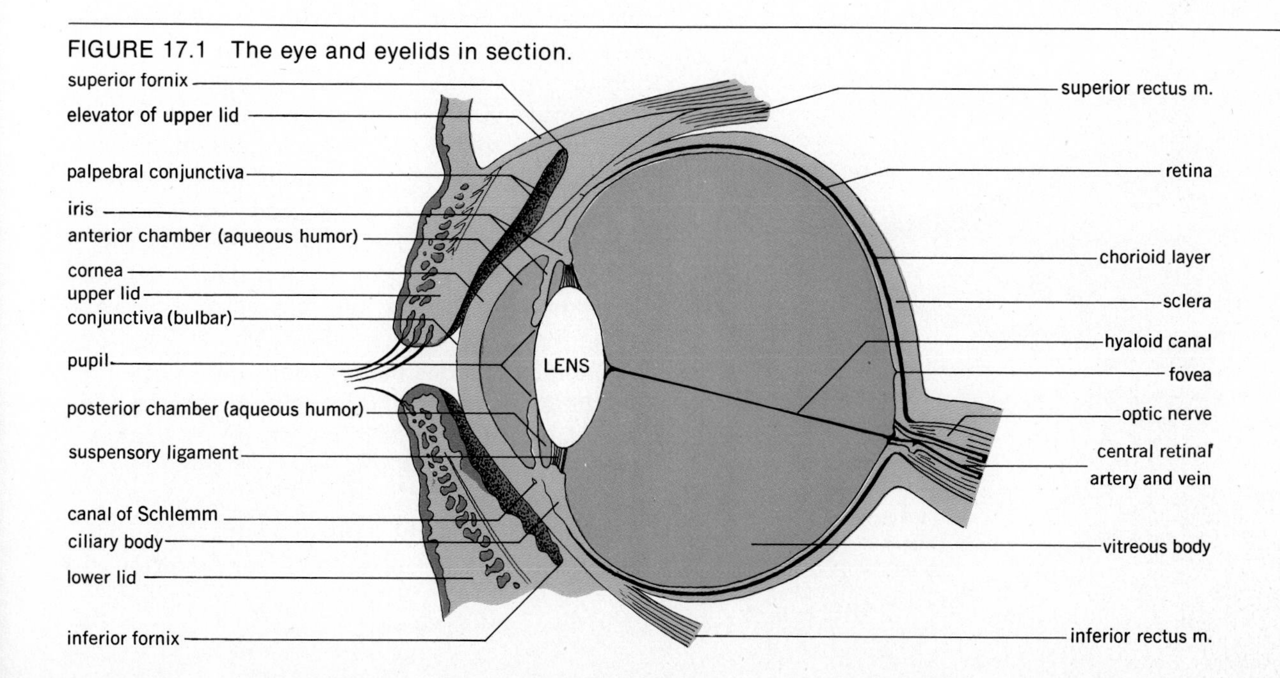

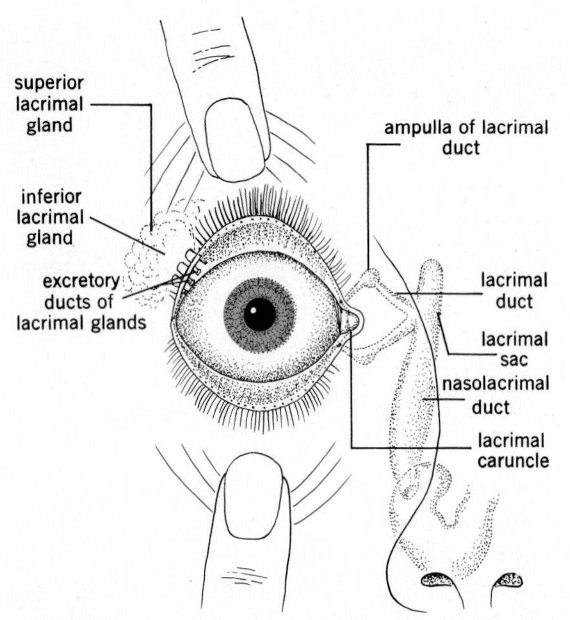

FIGURE 17.2 The lacrimal apparatus of the right eye.

FIGURE 17.3 The six extrinsic muscles of the left eye, viewed from the lateral side.

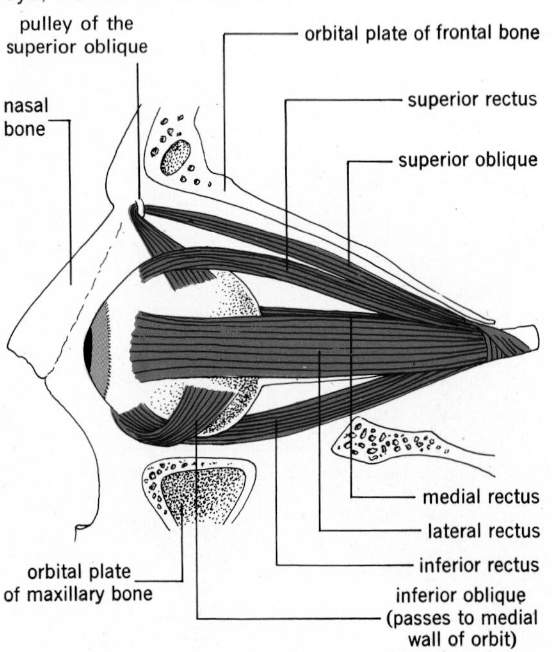

ciated with the lashes secrete the material often found in the eye when humans awake.

A two-lobed LACRIMAL GLAND (Fig. 17.2) is located on the upper outer aspect of each eyeball. Several small ducts empty LACRIMAL FLUID *(tears)* onto the eyeball at the upper outer corner of the lids. Blinking spreads the fluid over the eyeball to prevent its drying, lubricates the eye and lids for free movement, and destroys bacteria through its content of the enzyme *lysozyme.* About 1 ml of fluid is secreted each day by each lacrimal gland. The fluid is drained into the nasal cavities by a system of ducts (see Fig. 17.2).

Six EXTRINSIC EYE MUSCLES (Fig. 17.3) attach to and move each eyeball. Their names and actions are presented in Table 17.1.

Strabismus (squint) is failure of one of the eyes to be directed at an object when looking at it. The affected eye may be deviated away from the object in any direction. Causes of strabismus include paralysis of an ocular muscle, a too-short or too-long muscle, or disuse of the eye because of a severe defect of vision.

Three tunics or coats of tissue form the eyeball. The outer coat is called the FIBROUS TUNIC and is composed of the *sclera* or "white" of the eye, and the anterior transparent *cornea.* Both parts of the fibrous tunic are composed of collagenous tissue. The cornea is not vascular and has a low antigenicity, so that it is rarely rejected when transplanted from one person to another. The middle tunic is designated the UVEA and consists of the posterior vascular *chorioid* and the anterior *ciliary body.* The latter structure is subdivided into the ciliary muscle, the iris, the suspensory ligaments of the lens, and the lens itself. The ciliary body is generally responsible for control of light entering the eye and for fine focusing of images. The inner layer of the eye is the RETINA, the photosensitive layer containing the *rods* and *cones,* or visual receptors.

A watery fluid, the AQUEOUS HUMOR, secreted by the ciliary body fills the spaces (anterior and

TABLE 17.1 Extrinsic eye muscles

Name	Innervation (cranial nerve)	Action on eyeball
Lateral rectus	VI Abducent	Laterally
Medial rectus	III Oculomotor	Medially
Superior rectus	III Oculomotor	Superiorly and medially
Inferior rectus	III Oculomotor	Inferiorly and medially
Superior oblique	IV Trochlear	Inferiorly and laterally
Inferior oblique	III Oculomotor	Superiorly and laterally

posterior chambers) in front of and behind the iris. The fluid is drained from these spaces by a venous ring, the CANAL OF SCHLEMM. Blockage of the canal causes a rise in intraocular pressure and development of the disorder GLAUCOMA. The gel-like VITREOUS BODY fills the eyeball behind the lens, aiding in maintaining the shape of the eyeball.

The OPTIC NERVE exits from each eyeball slightly to the inner or nasal side of the posterior pole of the eye. At the point of exit of the nerve, no visual receptors are present, and a "blind spot" is created.

Light and lenses; image formation

Properties of light and lenses

Light is a form of energy that behaves like a wave composed of individual particles or quanta. That portion of the entire spectrum of electromagnetic radiation constituting the visible portion utilized by visual processes lies between the wavelengths of 400 nm (blue) and 800 nm (red).

Light is refracted or bent as it passes from one medium to another, as from air to water, or air to tissues to water. Light rays entering the eye come in more or less straight lines from distant objects or as diverging lines from near objects; in any case, refraction is necessary in order to focus images sharply on the retina. White light may also be separated into its spectral colors, as by a prism. Only three pure light colors (red, blue, and green) are required to create all other colors.

The ability of a lens to bend or refract light rays (its "strength") depends on its curvature and the medium through which the light passes before it strikes the lens. Its focal length refers to the distance from the lens at which it forms a sharp primary image. In general, the greater the curvature of a lens, the greater its strength, and the shorter its focal length will be. The term DIOPTER is used as a unit by which the strength of a lens can be measured. A convex lens focusing light rays sharply at a distance of 1 m is said to have a strength of +1 diopter. Convex lenses focusing an image closer than 1 m are stronger and have higher diopter numbers (a 50 cm focal length =

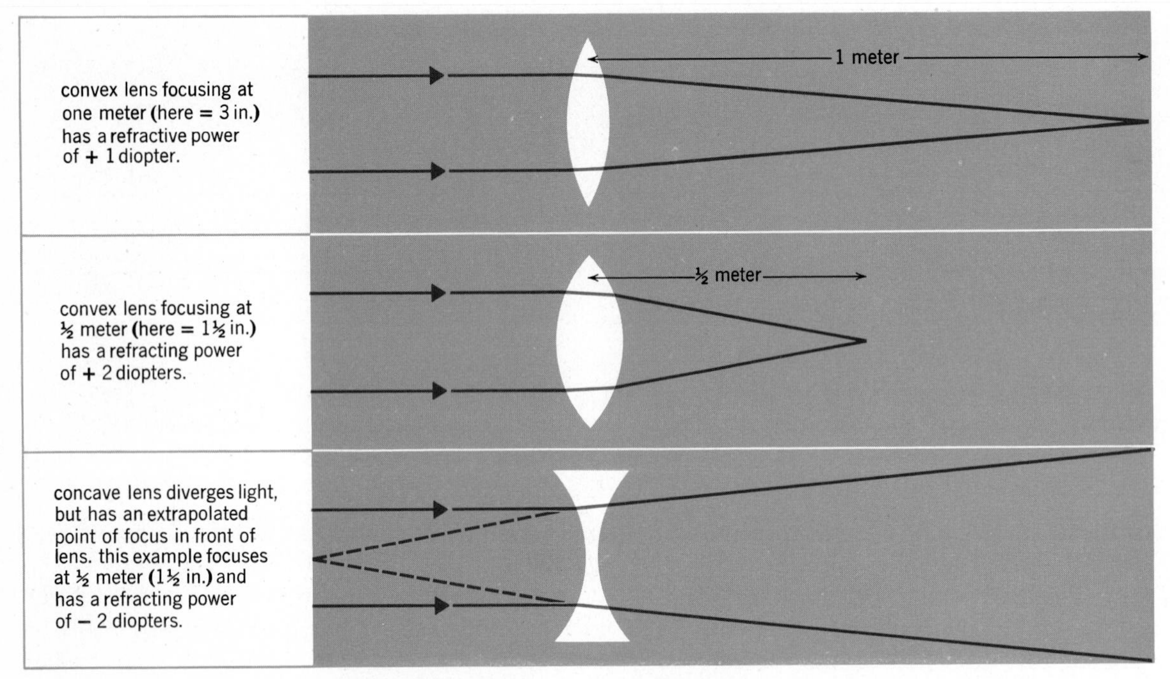

convex lens focusing at one meter (here = 3 in.) has a refractive power of + 1 diopter.
1 meter
convex lens focusing at ½ meter (here = 1½ in.) has a refracting power of + 2 diopters.
½ meter
concave lens diverges light, but has an extrapolated point of focus in front of lens. this example focuses at ½ meter (1½ in.) and has a refracting power of − 2 diopters.

FIGURE 17.4 Determination of the refractive power of lenses.

+2 diopters; a 20 cm focal length = +5 diopters). A concave lens diverges light rays, but its focal length can be extrapolated to a point on the same side as the light source, and diopter numbers are assigned a negative value (Fig. 17.4), but on the same basis as for convex lenses.

Image formation by the eye

As light rays enter the eye, they are refracted by four structures: the anterior surface of the cornea, the posterior surface of the cornea, the anterior surface of the lens, and the posterior surface of the lens. The anterior corneal surface refracts with a strength of +48.2 diopters, and the posterior surface with −5.9 diopters. The anterior lens surface has a lesser curvature (+5.0 diopters) than the posterior surface (+8.3 diopters). With the listed diopters, the human eye can focus a sharp image at a distance as close as 18 mm; the eyeball is 25 mm long, so that size and focal distance are nicely matched with a "reserve" created. In other words, the system is stronger than it need be, and focusing is possible.

As images of progressively closer objects are focused on the retina, a series of three basic changes occur, collectively called ACCOMMODATION.

The *ciliary muscle*, attached to the chorioid, pulls the chorioid anteriorly, relieving tension on the lens ligaments and allowing the natural elasticity of the lens to cause it to assume a greater curvature and become a stronger lens.

The *pupil constricts* due to activity of a circularly arranged mass of muscle in the iris called the sphincter pupillae. This action corrects two aberrations: spherical aberration results from the peripheral portions of the lens acting to focus light rays to a different point than does the central part; chromatic aberration is the result of the peripheral lens regions acting as a prism and splitting light rays into their component spectral colors. The constricting action of the iris blocks these peripheral rays.

The *eyes converge,* or are directed at the approaching object.

An emmetropic or normal eye carries out these changes so that focal power matches eyeball length, and a sharp image is created on the retina.

Disorders of the eye

Disorders of image formation

If the eyeball is too long for the focusing power of the lens system, the image will be focused ahead of the retina, and blurring of vision will occur. Such an eye is said to be MYOPIC *(hypometropic* or *nearsighted).* If the eyeball is too short for the focusing power, images will be focused behind the retina, and the eye is HYPERMETROPIC *(farsighted).* The placement of appropriately shaped glass or plastic lenses ahead of the eyeball will correct both situations (Fig. 17.5).

ASTIGMIA or ASTIGMATISM results when the corneal or lens surface does not possess equal curvatures in the lateral and vertical directions. The cornea or lens thus forms two images, one for each curvature. Double vision or blurring of vision usually results.

PRESBYOPIA *(old eye)* results from hardening of the lens with age. It usually assumes a more flattened shape and cannot round up, even though the ciliary muscle is still contracting. The subject with presbyopia moves reading matter farther away from the eyes in an attempt to maintain proper focus. Convex glass or plastic lenses placed ahead of the cornea will correct this problem.

Other disorders

CATARACT is opacity of the lens or its capsule, or both. Aging may produce cataract, as may trauma, infection, or diabetes. The cataract develops in stages, beginning with spokelike opacities and proceeding to swelling, shrinkage, and loss of transparency. Removal of the lens (extraction) is the treatment employed when the lens becomes so opaque that it interferes with the pursuit of normal activities.

GLAUCOMA is an increased intraocular pressure due to blockage of the canal of Schlemm. Blockage may result from eye infections, congenital malformations, or trauma. Treatment may be attempted by medication, usually miotics, drugs that constrict the pupil and reduce aqueous humor production by carbonic anhydrase inhibition. If medication cannot control the pressure, surgical intervention is indicated, with creation of an outflow canal to the subconjunctival space.

The CONJUNCTIVA is a membrane lining the lids and covering the anterior surface of the eyeball. CONJUNCTIVITIS is inflammation of the conjunctiva. TRACHOMA is bilateral viral conjunctivitis and is estimated to affect some 15 percent of the world's population. Untreated, it can lead to blindness through corneal ulceration and destruction.

KERATITIS is corneal inflammation usually of bacterial etiology. Corneal transplant may be performed for conditions that badly damage the cornea. The cornea is an avascular structure that derives its nutrients from and places its wastes into the aqueous humor. It is minimally antigenic and does not have to be tissue typed before transplanting.

HORDEOLUM *(sty)* involves inflammation of the hair follicles of the lids. It is most commonly due to staphylococcal organisms.

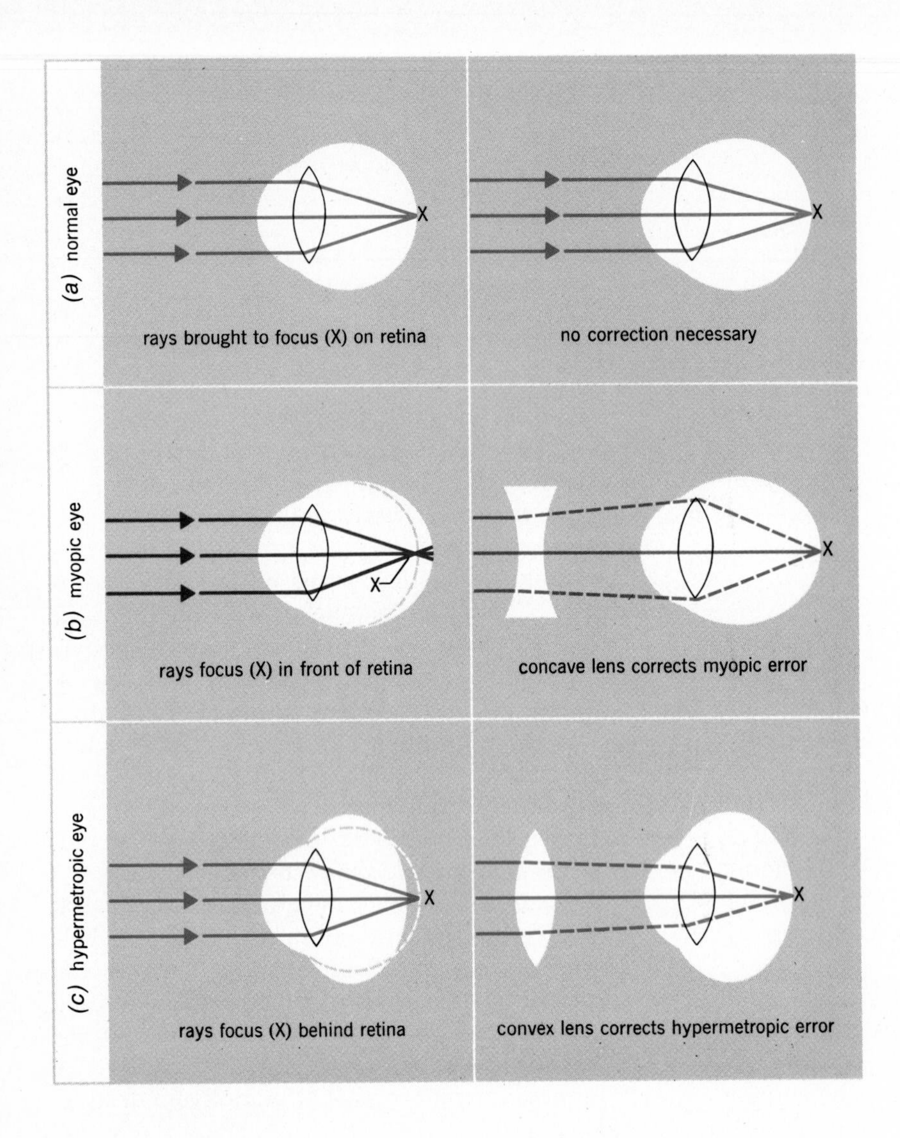

FIGURE 17.5 *(a)* Emmetropic, *(b)* myopic, and *(c)* hypermetropic eyes, and type of lens required to correct abnormalities.

Retinal function

Structure and function of rods and cones

The RETINA is composed of 10 layers of neurons and their processes (Fig. 17.6). The second layer, the BACILLARY LAYER, contains the light-sensitive portions of the visual receptors, the RODS and CONES (Fig. 17.7). The rods and cones are found in all parts of the retina except in the blind spot (no receptors) and in the fovea (where cones alone are found).

FUNCTIONS OF THE RODS. Rods are receptors for low-intensity light and are the night vision

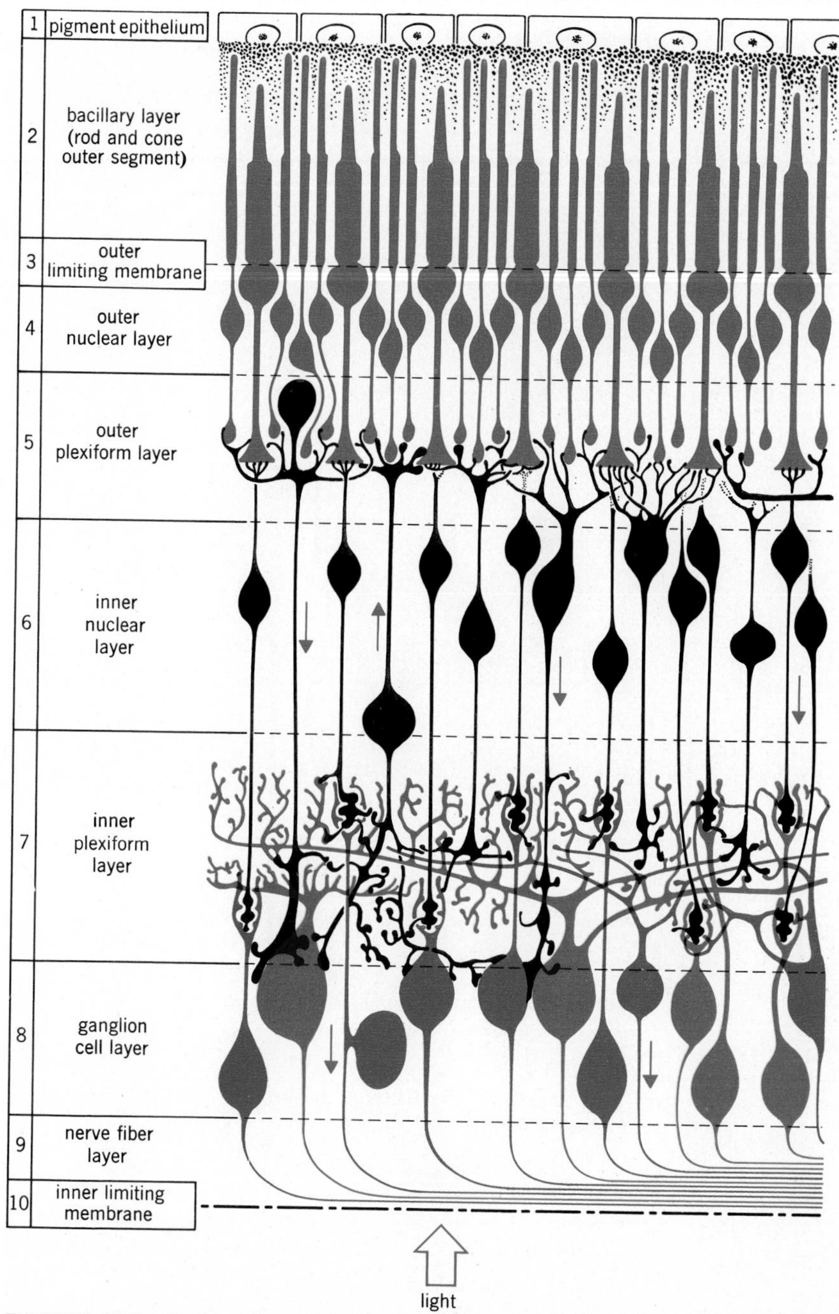

FIGURE 17.6 A diagram illustrating the cellular organization of the retina.

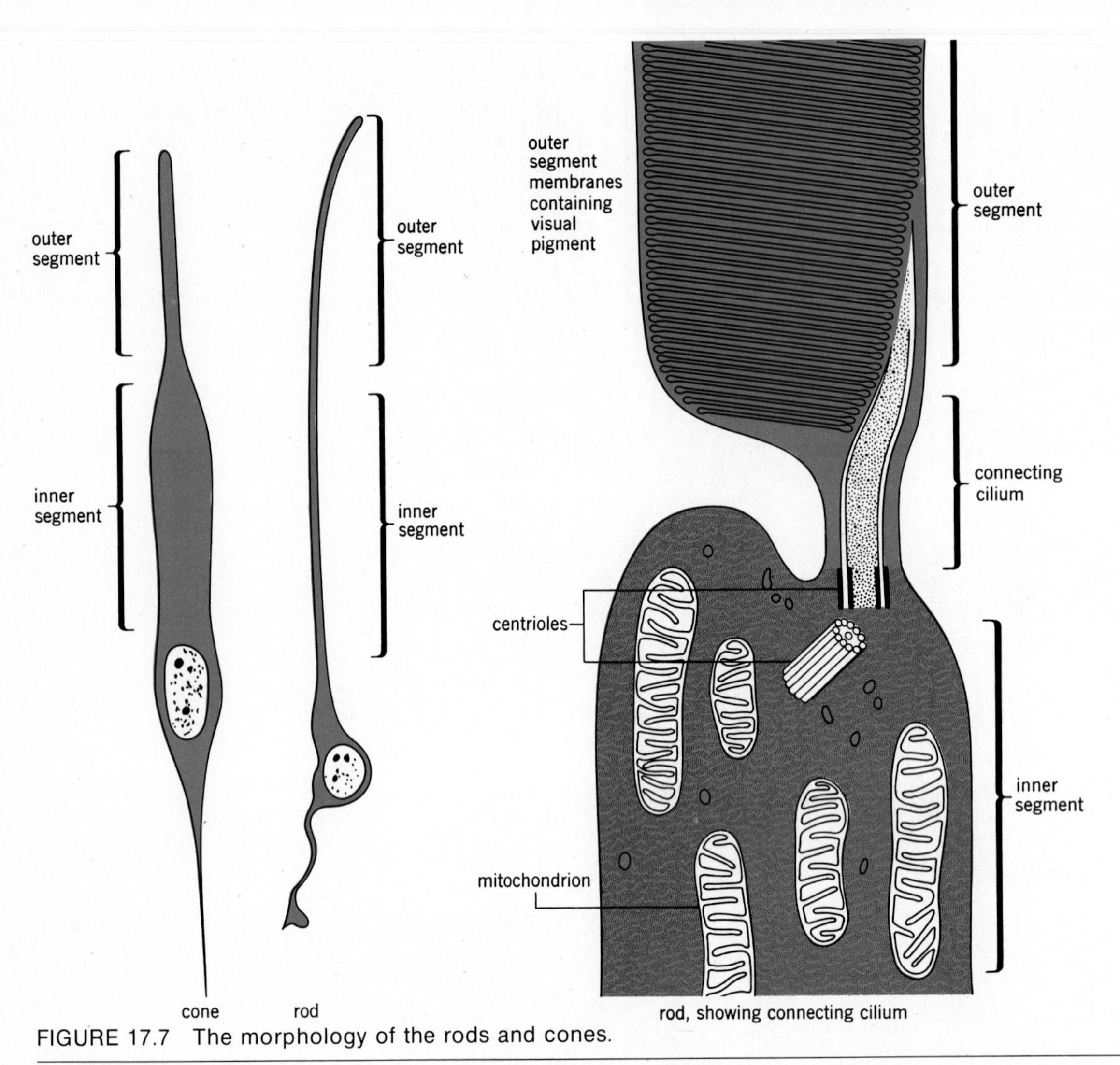

FIGURE 17.7 The morphology of the rods and cones.

FIGURE 17.8 A diagram illustrating the breakdown and resynthesis of rhodopsin in the rod cells of the retina.

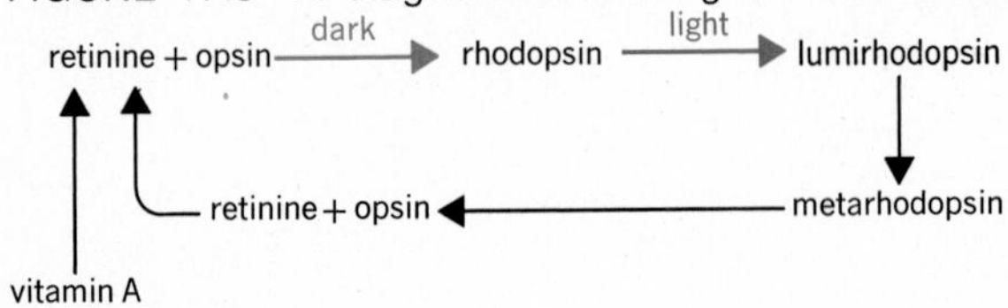

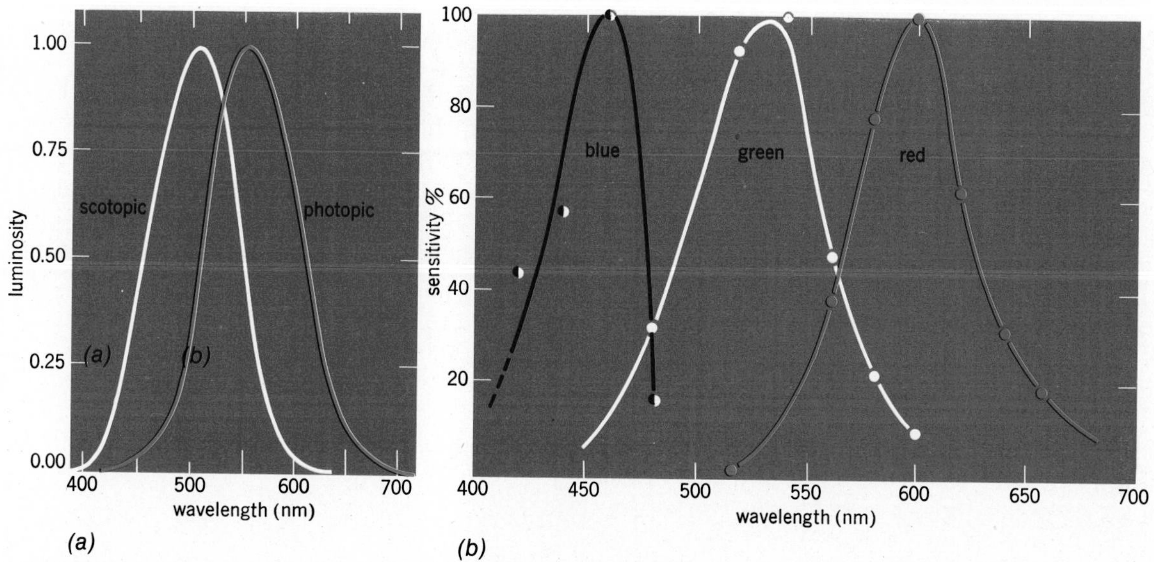

FIGURE 17.9 Absorption curves for *(a)* rods versus cones, and *(b)* the three types of cones. The curves are used to demonstrate the presence of two types of receptors (rods and cones) and three types of cones.

(scotopic) receptors. They have no color function. They contain a deep purple visual pigment called RHODOPSIN. Rhodopsin consists of a derivative of vitamin A designated as retinine and a protein called an opsin. As light strikes the rhodopsin, the series of changes presented in Fig. 17.8 is postulated to occur. Splitting of the rhodopsin molecule is believed to cause depolarization of the rod cell. A single quantum of light is believed to cause the splitting of one molecule of the pigment in the outer segments.

FUNCTIONS OF THE CONES. There appear to be three types of cones, each with a different visual pigment in it. Each cone has been shown to contain a pigment with a different maximum absorption of light of different color (Fig. 17.9). Current theories suggest that differential stimulation of blue-, green-, and red-sensitive cones create all other color by "mixing" colors. It is presumed that the cone pigments *(iodopsin* for blue, *chlorocruorin* for green, *erythrocruorin* for red) are split in the same general manner as described for rhodopsin. Color blindness is thus easily explained on the basis of a missing cone or cone pigment. Color blindness is a sex-linked condition, with **X** chromosomes carrying an abnormal gene, and the male usually showing the disorder because he lacks a normal gene on his **Y** chromosome to mask the expression of the abnormal gene.

In addition to reacting to light, the retina is involved in several other visual phenomena.

VISUAL ACUITY, as determined by ability to discriminate two separate lines, depends on the presence of an unlighted cone between two lighted ones. The area of sharpest vision is in the fovea, where only cones are present and where cone diameter is narrowest, and the image can become smallest before falling on adjacent cones.

The ability to FUSE SEPARATE IMAGES into continuous motion, as in viewing a motion picture film, is due to persistence of the products of pigment splitting and to the adding of the next stimulus to the first.

AFTER-IMAGES, or the continuing view of a scene after the scene is no longer being viewed, are also explained by persistence of the products of pigment splitting.

The visual pathways (Fig. 17.10)

Efferent pathways from the eye

Impulses originating in the visual receptors are ultimately gathered into the OPTIC NERVE that leaves the eyeball. At the OPTIC CHIASMA, nerve fibers from the nasal portion of each retina undergo a crossing. Fibers proceed to the LATERAL GENICULATE BODY (a part of the thalamus) where a synapse occurs. Fibers from the geniculate body form the OPTIC RADIATION to the *occipital* LOBES of the brain (area 17). Areas 18 and 19, lying anterior to 17, provide integration, interpretation, and some storage of visual information. The effects of lesions in various parts of the visual pathways is also shown in Fig. 17.10.

FIGURE 17.10 The central visual pathways and the types of visual loss that will occur with lesions at different points in the pathways. Shaded areas indicate loss or deficit of vision.

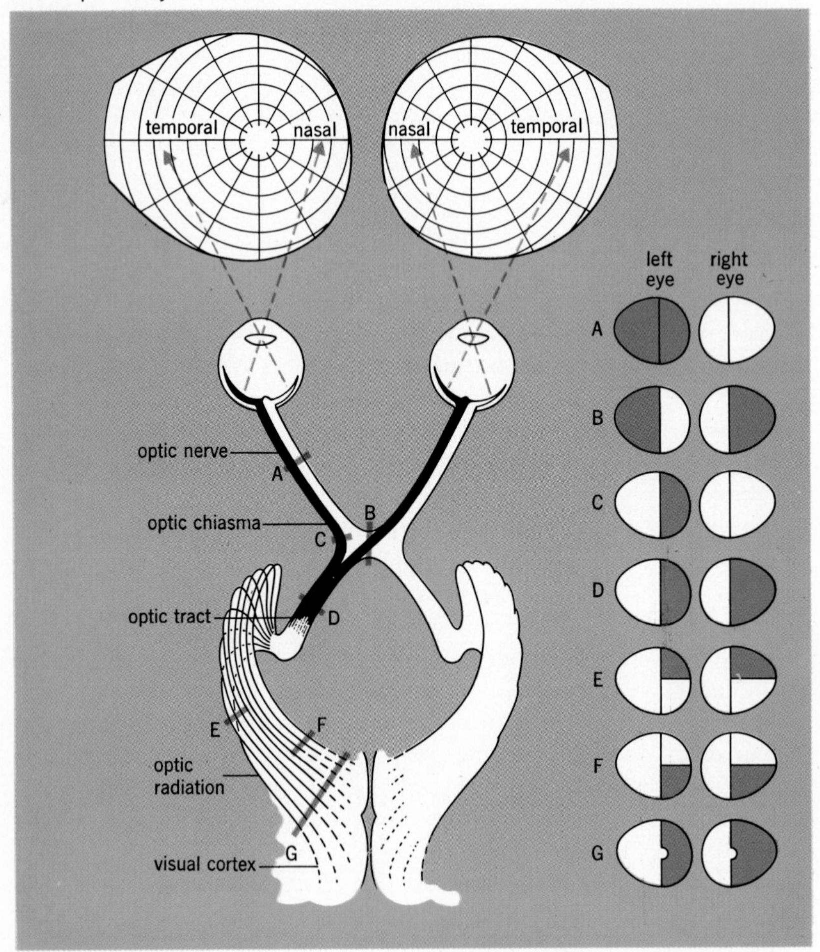

Other functions of the brain in vision

DEPTH and DISTANCE PERCEPTION is provided by the fact that humans have two eyes, each of which views an object from slightly different angles. The brain thus integrates these views into an appreciation of depth and distance.

Summary

1. The eye is the organ of vision. It is located in the orbit, is protected by the eyelid, and is lubricated by lacrimal fluid. Six eye muscles turn the eye within the orbit.
2. Three tunics surround the eyeball.
   a. The outer fibrous tunic includes the sclera and cornea.
   b. The middle tunic is the uvea and includes the ciliary body and vascular chorioid.
   c. The inner retina contains the visual cells, rods, and cones.
3. Image formation occurs when light enters the cornea and is refracted by the cornea and lens to focus on the retina.
   a. Focusing occurs by accommodation, which involves change in lens curvature, constriction of the pupil, and movements of the eyeballs.
   b. The iris, by enlarging or constricting the pupil, corrects for spherical and chromatic aberration and controls light entering the eye.
4. Disorders of the eye may involve the cornea, the whole eyeball, the lens, and ocular fluid. There are several common disorders.
   a. Corneal or lens imperfections cause astigmatism.
   b. Eyeballs too long or short for the lens system produce near- and farsightedness, respectively.
   c. Presbyopia is associated with hardening of the lens and the inability to focus near objects.
   d. Cataract is lens opacity.
   e. Glaucoma is increased intraocular pressure as a result of excess production or decreased drainage of aqueous humor.
   f. Conjunctivitis is inflammation of the membrane covering the anterior eyeball.
5. The retina contains rods and cones.
   a. Rods respond to low light intensity and give night vision.
   b. The cones respond to higher intensity of light and give daylight and color vision.
   c. Both types of receptors contain pigments that are broken by light to give stimuli to depolarize the cell.
6. The retina determines several other visual phenomena.
   a. Visual acuity is determined by density of cones in the retina.
   b. Visual persistence in the retina enables fusion of individual events into a continuous sequence.
   c. After-images are explained by persistence of products of retinal pigment destruction.
7. The optic nerves, optic tract, and optic radiation convey visual impulses to the occipital lobe for processing.
   a. Lesions in each area produce characteristic types of visual loss.
8. The possession of two eyes enables binocular vision and its attendant abilities to perceive depth.

Hearing, Equilibrium, Taste and Smell

The four senses described in this chapter may be said to constitute the senses of the ear, nose, and tongue. The ear contains structures responsible for the ability to hear and for equilibrial responses to motion and head position. Taste and smell are complementary sensations that give an appreciation of the fine qualities of foods and may also give protection against the intake of toxic substances.

The ear and hearing

The ear is divided into three portions designated as the OUTER, MIDDLE, and INNER EARS (Fig. 18.1). That portion of the inner ear concerned with hearing is the cochlea.

The nature of sound waves

Sound waves require molecules to be present so that the wave can produce alternate compression

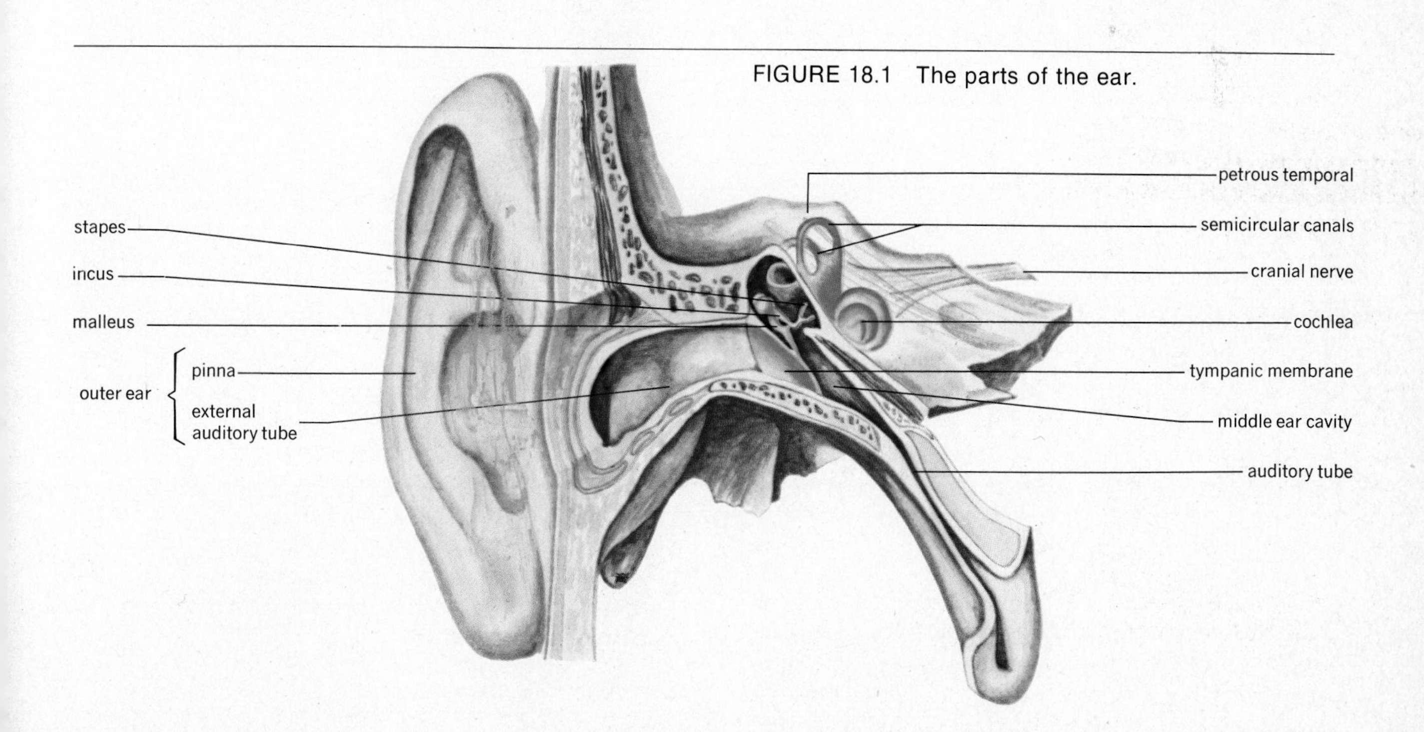

FIGURE 18.1 The parts of the ear.

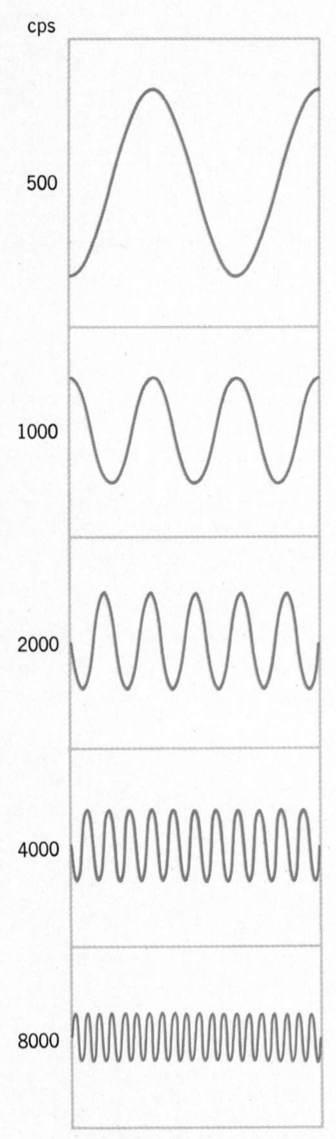

FIGURE 18.2 The relationships of frequency (cycles per second) of a sound wave to pitch.

and decompression of those molecules. A sound wave thus transmits itself like a ripple in a pond. Transmission of sound waves can occur through air, fluids, or solid material such as bone. A sound has three important properties:

Pitch (frequency) is determined by the number of cycles (or vibrations) per second (Fig. 18.2), with higher frequencies meaning higher pitches.

Intensity refers to the *loudness* of the sound, and is measured in decibels.

Timbre or *quality* of a sound is due to the presence in a vibrating object of harmonics, higher or lower tones that accompany the fundamental tone produced. Thus, the same note sounds different when played by a piano, violin, or reed instrument.

Table 18.1 gives an idea of the decibel rating system for some common sounds.

Each 10 decibel change reflects a tenfold change in power or intensity of a sound; thus, the ear can discriminate as much as a 10^{13} change in intensity. Above about 100 decibels, sounds are so loud as

TABLE 18.1 Decibel ratings of sounds

Sound	Rating—decibels
Absolute silence	0
Watch ticking	20
Residential street, no traffic	40
Stream	50
Automobile at 30 ft	60
Conversation at 3 ft	70
Loud radio	80
Truck at 15 ft	90
Car horn at 15 ft	100
Pneumatic hammer at 3 ft	120
Amplified rock music	130
Propeller airplane at 15 ft	130
Jet aircraft at takeoff	150^{+}

to become painful, and may damage the ear structures if allowed to continue. To some degree, intensity of the waves reaching the cochlea is controlled by muscles associated with the auditory structures (see below).

The outer ear

The outer ear is composed of the fleshy appendage attached to the side of the skull, the PINNA or AURICLE, and the EXTERNAL AUDITORY MEATUS, the 2.5-cm long canal that leads from the outside to the eardrum. The pinna (Fig. 18.3) contains a supporting framework of elastic cartilage that is covered with skin. The normal pinna has several parts, labeled in Fig. 18.3. Deformities of the pinna may indicate malformations elsewhere in the body, particularly of the kidneys (the two structures develop at about the same time); recently, the presence of a groove in the earlobe has been correlated with a greater incidence of heart attack in those possessing the groove. A "cauliflower ear" results when trauma to the pinna produces bleeding into and/or rupture of the cartilage and its membrane; the healing process results in fibrosis and excessive production of new cartilage that gives an abnormal shape to the pinna.

The thin skin lining the canal is lubricated and protected by the secretion of cerumen, the brown and bitter-tasting "ear wax" produced by modified sweat glands known as ceruminous glands. It is supposed to discourage the entry of insects into the canal. Cerumen sometimes accumulates to the point where sound waves may be blocked or greatly reduced, with decrease of ability to hear (a type of transmission deafness).

FIGURE 18.3 The pinna.

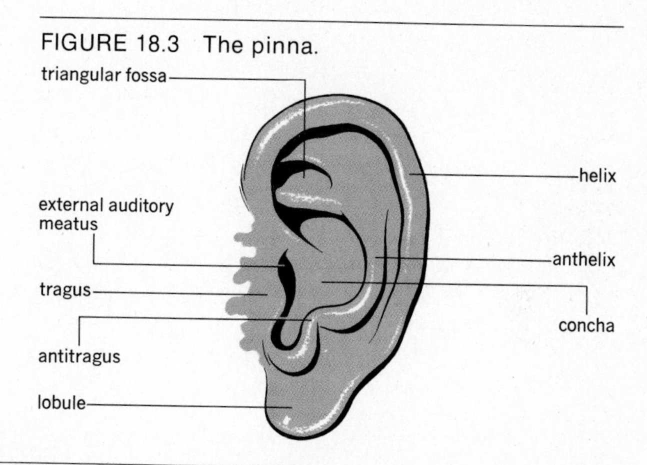

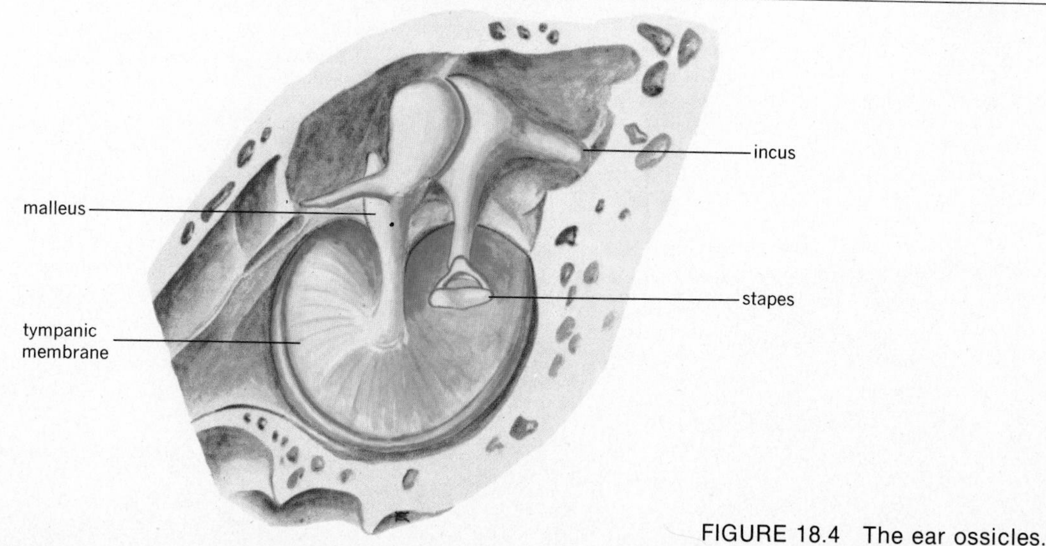

FIGURE 18.4 The ear ossicles.

The middle ear

The middle ear consists of an irregular air-filled cavity (middle ear or TYMPANIC CAVITY), the contents of that cavity, and the PHARYNGOTYMPANIC *(Eustachian)* TUBE that communicates with the throat. The tube permits equalization of air pressure on the two sides of the eardrum to prevent rupture and can also act as a means of distributing throat infections to the middle ear. The lateral wall of the cavity is formed by the EARDRUM *(tympanum* or *tympanic membrane).* This membrane is caused to vibrate by sound waves; it sets into motion the three EAR OSSICLES. The ossicles (Fig. 18.4) consist of the MALLEUS *(hammer),* INCUS *(anvil),* and STAPES *(stirrup)* that act as a system of levers to transmit vibrations of the eardrum to the cochlea and to reduce the movement and increase the pressure of the stapes in the OVAL WINDOW. This is necessary because the eardrum vibrates in air, while the organ of hearing must vibrate in the fluid of the cochlea, and it takes less force to cause a structure to vibrate in air than in fluid. Another opening, the ROUND WINDOW, is found in the medial wall of the middle ear cavity. It is closed by the mucous membrane lining the cavity.

The tendons of two small muscles lie within the middle ear cavity. The tendon of the TENSOR TYMPANI muscle (innervated by cranial nerve V) attaches to the handle of the malleus and limits malleus movement when it contracts. The tendon of the STAPEDIUS muscle (innervated by cranial nerve VII) attaches to the stapes and, when it contracts, limits the movement of the stapes in the oval window. The muscles form the effectors for the AUDITORY REFLEX that protects the ear structures from damage by loud sounds.

The middle ear cavity communicates with the mastoid air cells in the mastoid process of the temporal bone. Infections (e.g., otitis media) of the middle ear cavity may thus cause infections of the air cells (mastoiditis). Antibiotics are generally employed to treat such infections.

The inner ear and hearing

The inner ear consists of the OSSEOUS LABYRINTH, a series of channels and spaces within the petrous

FIGURE 18.5 The membranous labyrinth of the inner ear.

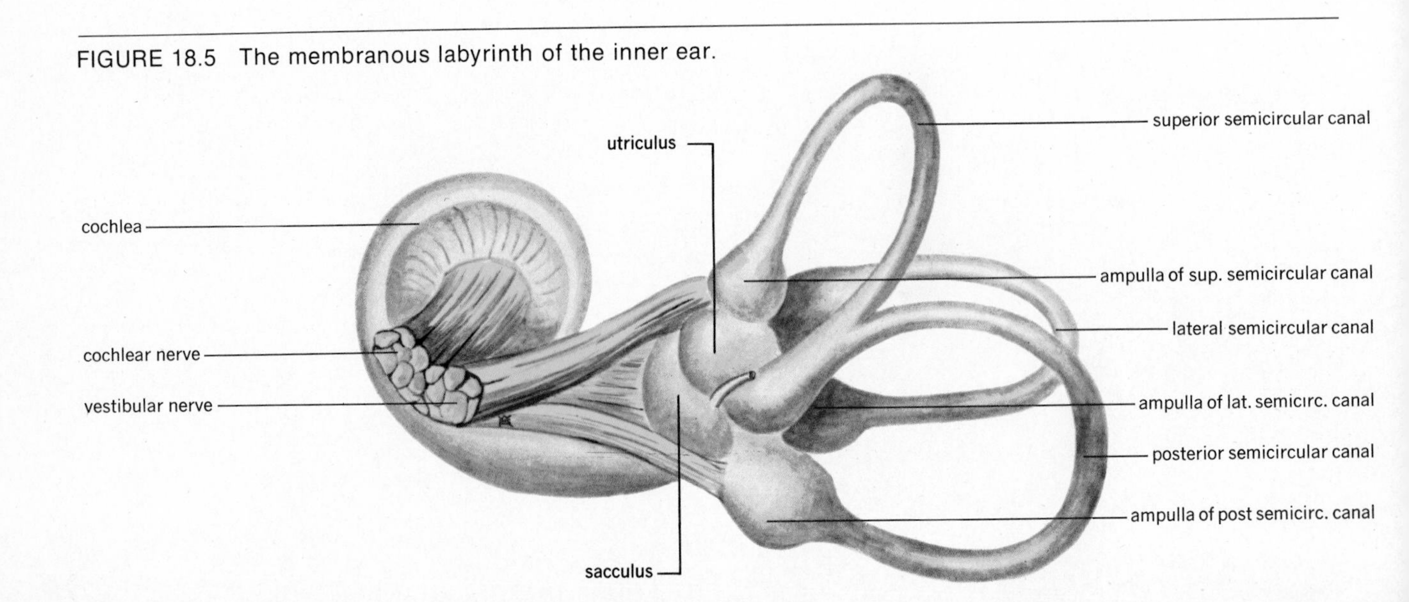

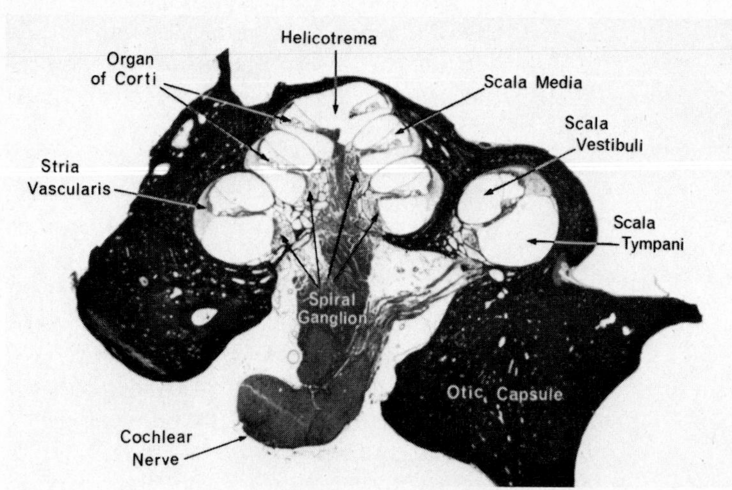

FIGURE 18.6 A photograph showing the basic structure of the cochlea.

FIGURE 18.7 A diagram showing the structure of one of the cochlear turns.

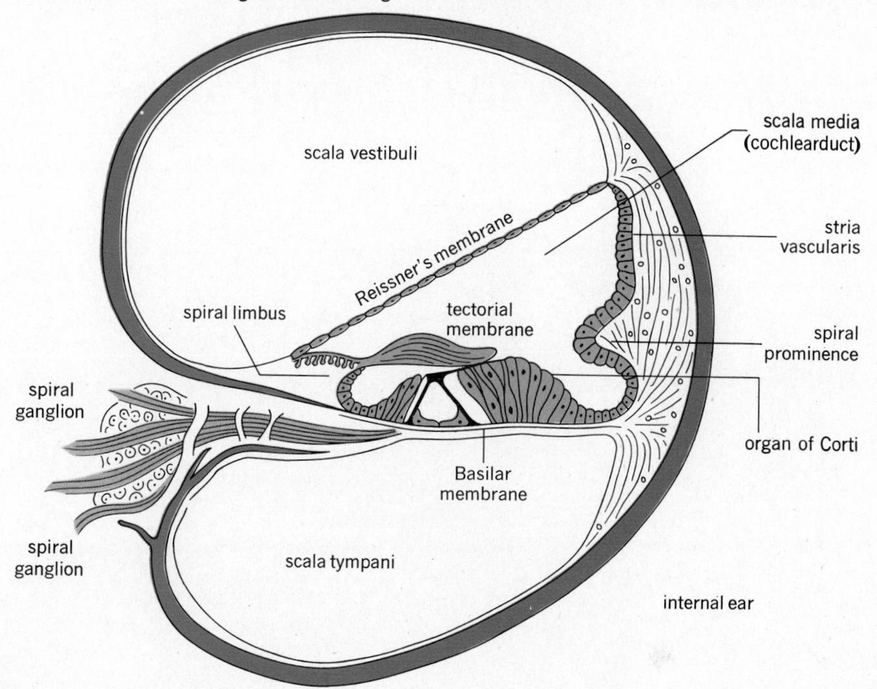

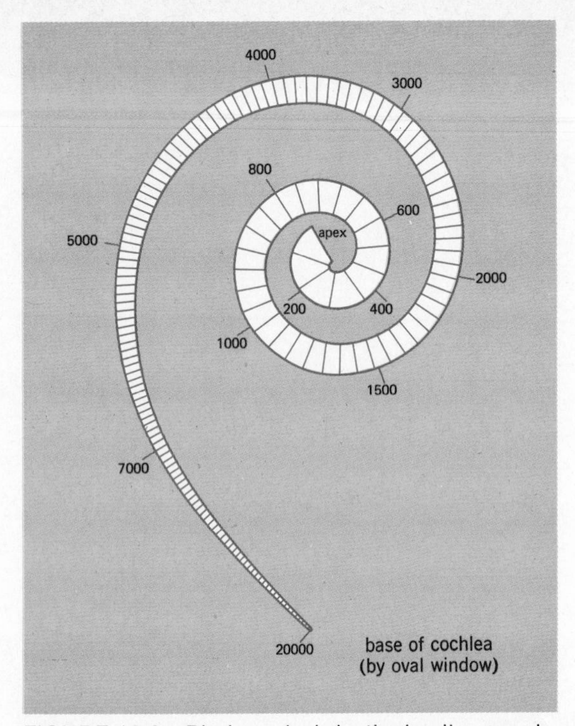

FIGURE 18.8 Pitch analysis by the basilar membrane. Numbers indicate where a sound of a given frequency is analyzed. The membrane averages 32 mm long, 0.04 mm wide at the base and 0.5 mm wide at the apex.

portion of the temporal bone, and the MEMBRANOUS LABYRINTH (Fig. 18.5) the continuous series of living tissues lining the bony channels. ENDOLYMPH fills the membranous labyrinth, while PERILYMPH fills the spaces between the membranous labyrinth and the walls of the bony channels. The inner ear contains the COCHLEA, the organ for hearing, and the UTRICULUS, SACCULUS, and SEMICIRCULAR CANALS that are responsible for reactions essential to maintenance of posture and equilibrium.

The portion of the inner ear concerned with hearing is a coiled snailshell-like structure known as the COCHLEA (Figs. 18.6 and 18.7). It wraps two and one-half times around a central "core" of bone known as the MODIOLUS. The bony labyrinth of the cochlea contains a series of membranous structures (the membranous labyrinth) that divide it into three channels. The VESTIBULAR and

FIGURE 18.9 The organ of Corti. The sensory elements are the hair cells.

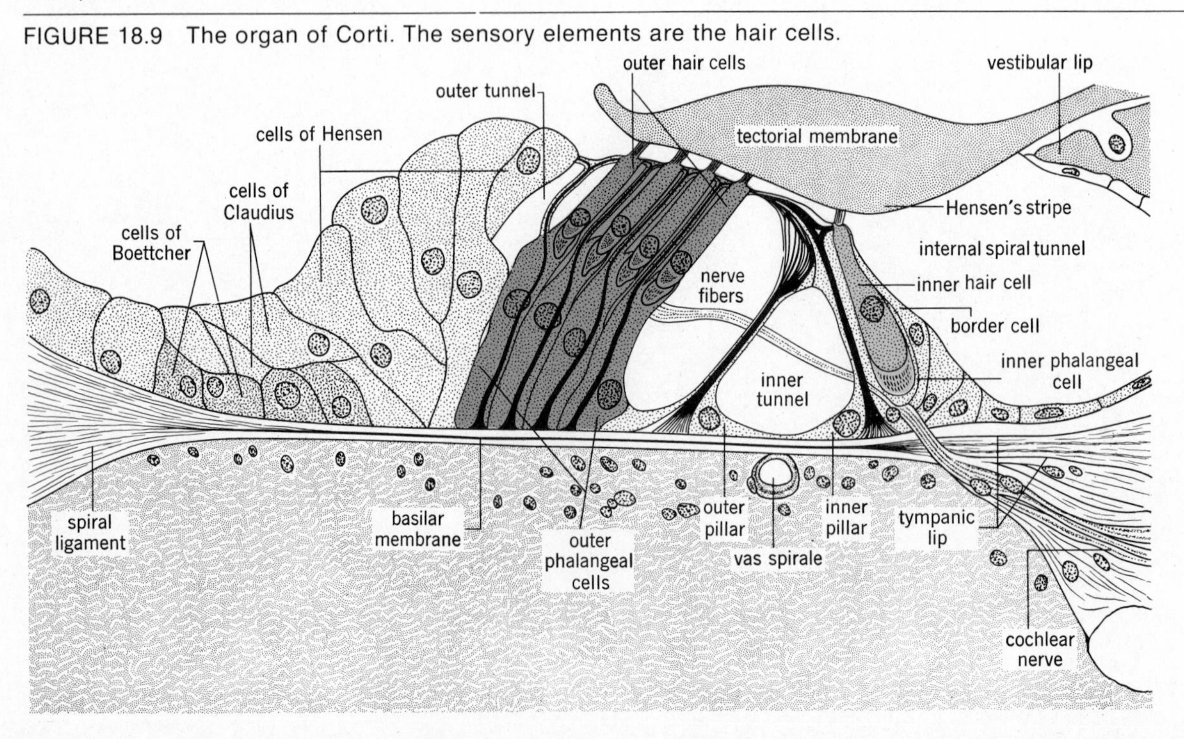

BASILAR MEMBRANES form a SCALA MEDIA *(cochlear duct)* between them; it is filled with endolymph. A SCALA VESTIBULI lies between the vestibular membrane and the bony cochlear wall, while a SCALA TYMPANI lies between the basilar membrane and the bony cochlear wall. These canals are filled with perilymph and communicate with one another at the apex of the cochlea through the HELICOTREMA. The scala vestibuli communicates with the oval window, and the scala tympani with the round window at the base of the cochlea. The ORGAN OF CORTI, responsible for hearing, rests on the basilar membrane in the scala media.

Activation of the hair cells of the organ of Corti is achieved by shock waves set up in the cochlear fluid by the vibrations of the stapes in the oval window. The BASILAR MEMBRANE of the organ is caused to move up and down by these shock waves. The basilar membrane (Fig. 18.8) consists of some 20,000–30,000 separate strands of tissue that lengthen as the membrane follows the two and one-half turns of the cochlea. The membrane is believed to vibrate selectively, in different regions, according to the frequency (pitch) of the incoming sound waves. Thus, pitch analysis may be a function of the cochlea. The transduction of basilar membrane movement into nerve impulses occurs by the HAIR CELLS of the organ of Corti (Fig. 18.9), whose processes are bent against the tectorial membrane as the basilar membrane vibrates. The events leading to nerve impulse formation are summarized in Fig. 18.10.

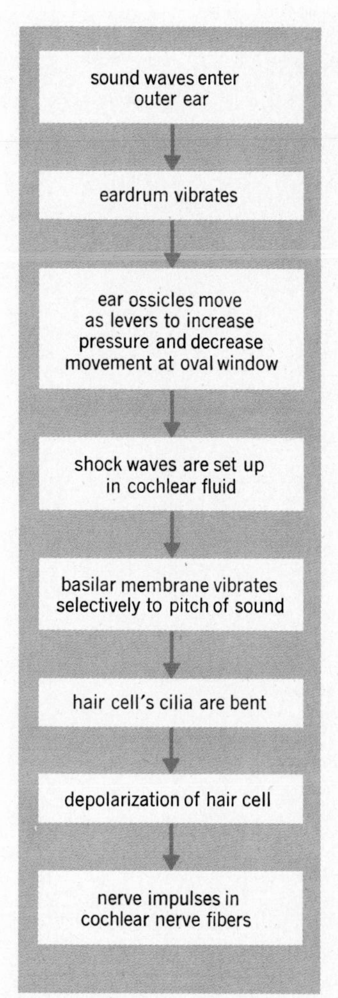

FIGURE 18.10 The events that occur when a sound wave strikes the eardrum.

Auditory pathways (Fig. 18.11)

The nerve fibers that leave the hair cells leave the cochlea over the cochlear portion (COCHLEAR NERVE) of cranial nerve VIII (vestibulocochlear). They pass to COCHLEAR NUCLEI in the medulla where synapses occur. The cochlear nuclei show activity in response to changing sound intensity, so that some analysis of intensity may occur here. Connections of the cochlear nuclei are made to the INFERIOR COLLICULI (for auditory-motor reflexes) and to the MEDIAL GENICULATE BODY (a part of the thalamus), and are then relayed to the AUDITORY CORTEX in the temporal lobes of the cerebrum (areas 41 and 42). The temporal lobes also permit discrimination of sound intensity and

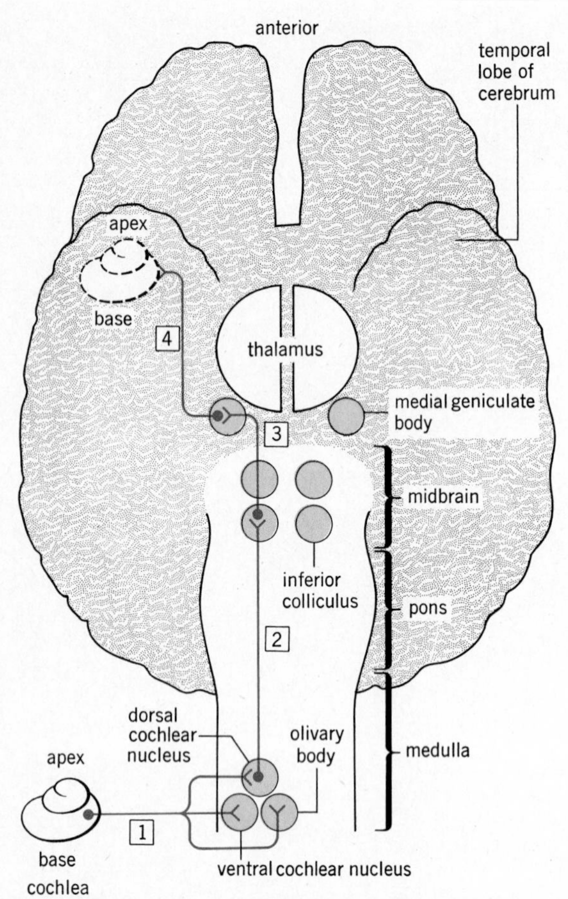

FIGURE 18.11 A diagram illustrating the auditory pathways. Four neurons are involved.

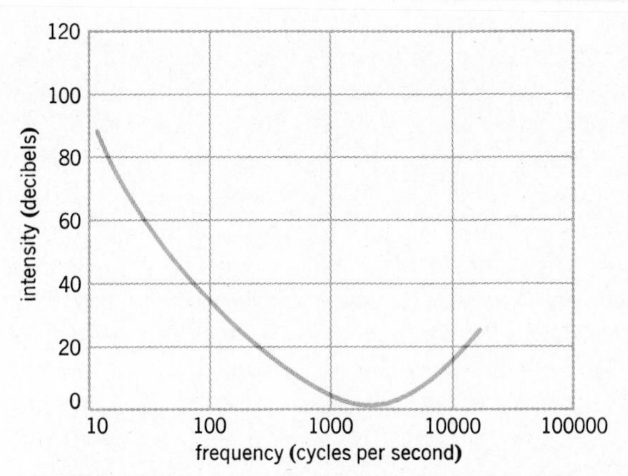

FIGURE 18.12 An audiogram of a person with normal hearing. Range is about 20–20,000 cycles per second, and acuity is inversely proportional to frequency.

tonal patterns, aid in localization of the sound, and are required for comprehension of the spoken word.

Clinical considerations

Deafness is defined as loss of ability to hear and may be partial or complete.

TRANSMISSION DEAFNESS results from failure of sound waves to reach the eardrum (plugging of the canal by wax) or by failure of the ossicles to transmit vibrations to the cochlea (ossicle fusion, otitis media, fixation of stapes in oval window).

NERVE DEAFNESS results from damage to the cochlea, auditory pathways, and/or the brain. Some occupations (foundry workers, aircraft mechanics) are associated with high-intensity sound that can cause degeneration of parts of the organ of Corti.

Testing for hearing and its disorders is usually done by using an AUDIOMETER. The record is called an AUDIOGRAM (Fig. 18.12); it relates both intensity and pitch to the ability to hear.

The inner ear and equilibrium

The inner ear includes organs for determination of body position and changes of acceleration, leading to maintenance of posture and equilibrium. These are the semicircular canals and the maculae (Fig. 18.13).

The SEMICIRCULAR CANALS are three fluid-filled

channels in each ear. An enlarged AMPULLA contains a CRISTA. The crista is bent by movement of the fluid as changes in acceleration and direction of motion occur. The three canals are placed in three mutually perpendicular planes to one another so that any head movement will cause excitation of one or more cristae. The MACULAE are organs of position, not movement, and are activated by gravity pulling on the OTOLITHS ("ear stones") of the organ. This force bends the cilia of the macula. Activation of either of these receptors results in reflex responses *(labyrinthine reflexes)* tending to maintain the body properly in space. Labyrinthine reflexes are categorized into two groups.

ACCELERATORY REFLEXES, arising from stimulation of the semicircular canals, produce response of eye, neck, trunk, and appendages to starting, stopping, or turning motions. NYSTAGMUS, a horizontal, vertical, or rotary movement of the eyes, occurs on angular acceleration. Limb responses are typically those of extension on the side away from an angular acceleration and toward the force if linear. Thus, a person leans toward a curve, forward on acceleration, and backward on deceleration.

FIGURE 18.13 Diagrams illustrating the location and morphology of the equilibrium structures of the inner ear. *(a)* Crista ampullaris of semicircular canal. *(b)* Macula of the sacculus or utriculus.

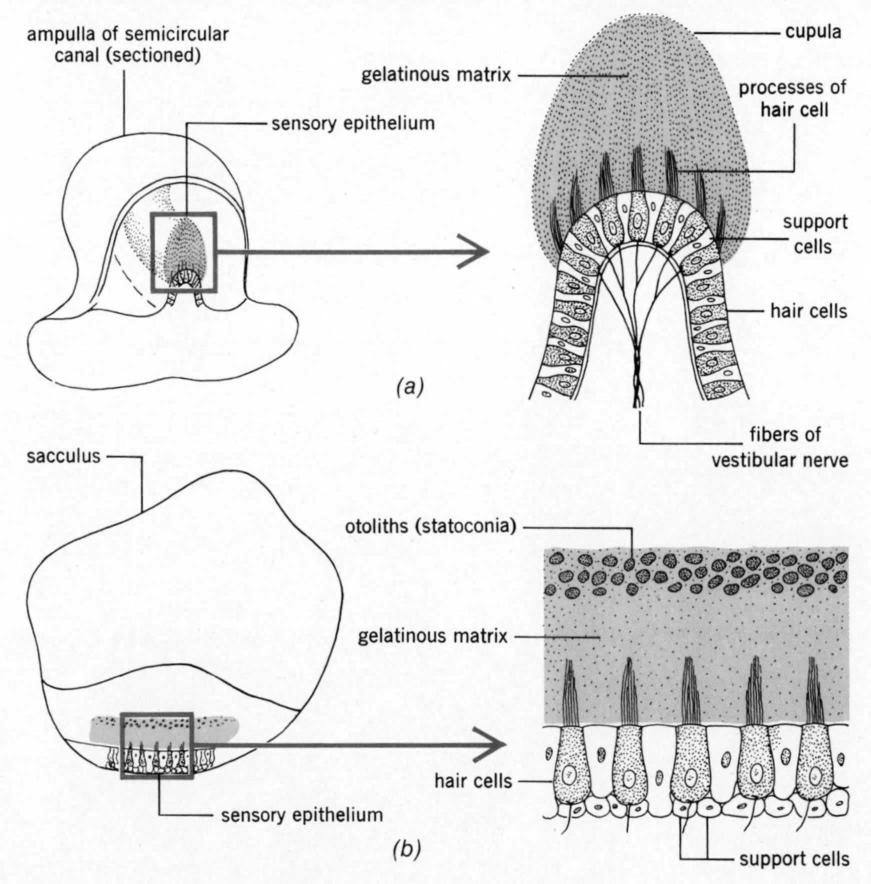

POSITIONAL REFLXES, resulting from stimulation of the maculae, involve righting reflexes and notification of head position. The pathway for communication of information from the canals and maculae is the vestibular portion (vestibular nerve) of cranial nerve VIII. The fibers pass to brainstem vestibular nuclei, located in the lower lateral pons, and thence to the cerebellum, where appropriate muscular response is provided.

Disorders and testing of labyrinthine function

Although not strictly a disorder, MOTION SICKNESS involves repetitive changes in angular and linear acceleration affecting the receptors. Nausea and vomiting may result. Infections of the vestibular nerve may result in loss of equilibrium, as may trauma injuring the nerve. Indeed, any basic loss of equilibrium without change of muscle tone is indicative of altered vestibular function. TESTING function typically involves spinning the subject in a Barany chair with the head in various positions to excite the different canals. If nystagmus, dizziness, and loss of equilibrium are not produced, a lesion is presumed to exist somewhere in the system.

Taste and smell

Taste and smell are chemical senses in that the adequate stimuli for the receptors are chemicals in solution in water or air. The senses reinforce one another; as anyone with a cold knows, food can become tasteless. Taste may protect against the intake of toxic substances such as alkaloids and, as a component of appetite, plays a role in food selection. It has also been established that animals deficient in certain nutritional elements may exhibit an increased interest in those foods containing the missing elements and that taste provides the sensory clue to proper selection of foods. Smell provides humans with the ability to detect odors that may represent something harmful to them (e.g., spoiled food, insecticides, and gases).

FIGURE 18.14 The taste buds.

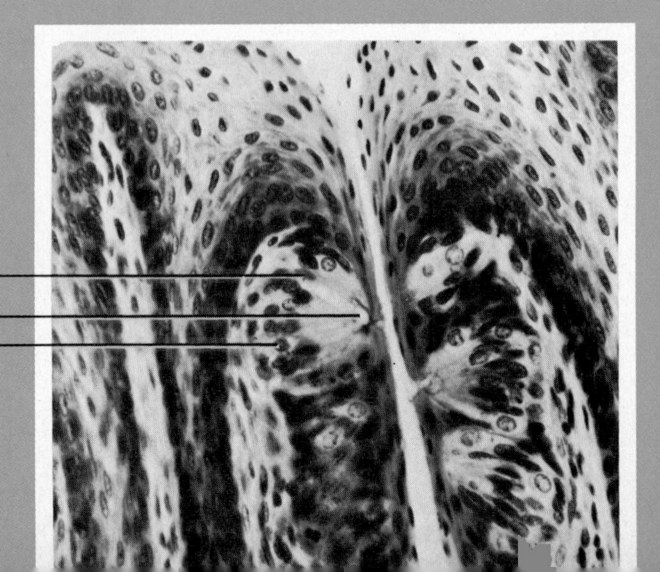

Taste

The receptors for the sense of taste are the TASTE BUDS (Fig. 18.14) located primarily on the tongue. A few buds may be found on the soft palate, epiglottis, and pharynx. According to the stimulus that causes the maximum response by the taste buds, four basic types of buds and thus four basic taste sensations are described.

SOUR TASTE is produced by hydrogen ion (acids). Strong acids, because of their greater liberation of hydrogen ion, produce a stronger taste than weak acids.

SALTY TASTE is due to cations of ionized salts; for example, Na^+, NH_4^+, Ca^{++}, Li^+, K^+. Sodium compounds and ammonium compounds are most effective in stimulating salty taste.

SWEET TASTE requires the presence of compounds with hydroxyl (—OH) groups in the molecule. Many substances meet this criterion, including sugars, alcohol, amino acids, ketones, and lead salts. In terms of sweetness, sugars are second to saccharine (Fig. 18.15) (500 times as sweet as glucose) and chloroform (40 times as sweet as glucose).

BITTER TASTE is produced by alkaloids and long-

saccharine — sodium saccharine $\cdot\, 2H_2O$

FIGURE 18.15 The formulae of saccharine and sodium saccharine. Saccharine is less soluble (1 g/290 ml. of water) than is sodium saccharine (1 g/1.5 ml. of water). Both are about 500 times sweeter than sugar.

FIGURE 18.16 The olfactory epithelium. *(a)* Light microscope view. *(b)* Structure as revealed under the electron microscope.

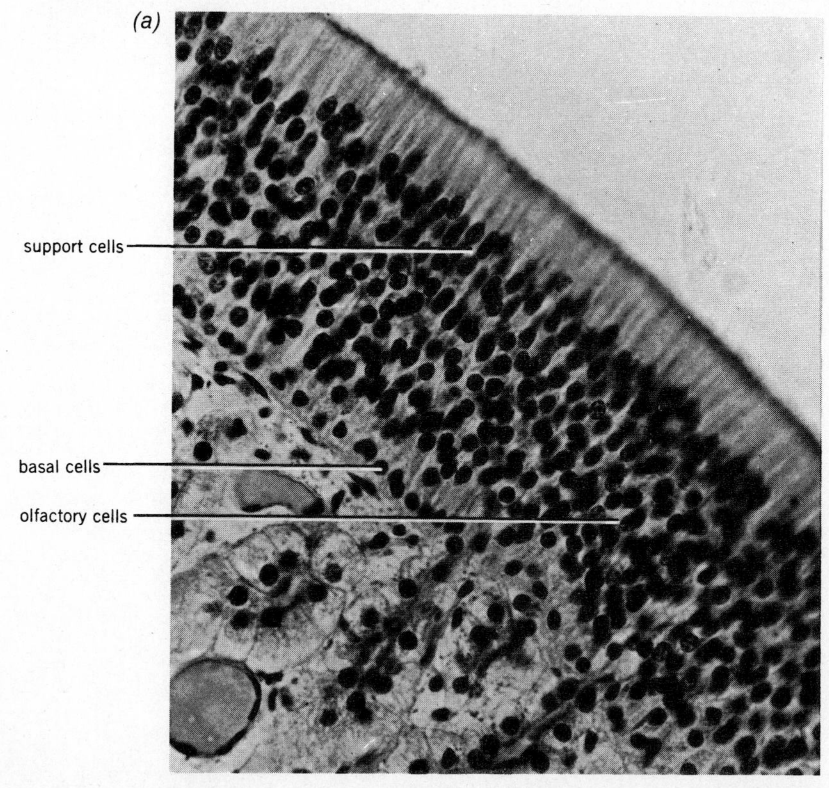

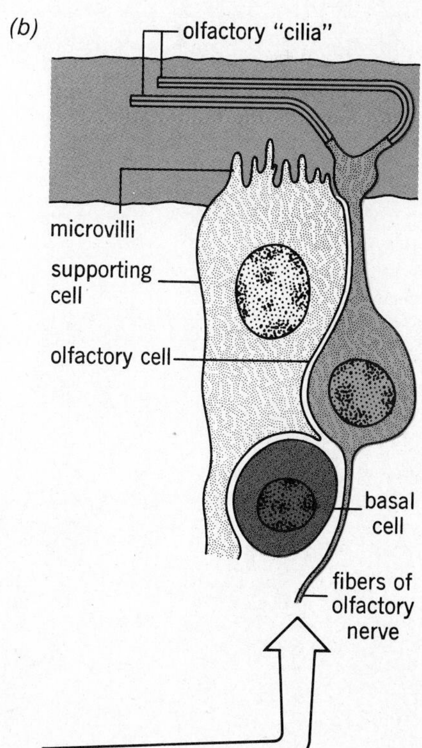

chain organic molecules. Thus, strychnine, quinine, caffeine, and nicotine all taste bitter.

The buds serving the four basic taste sensations do not have the same threshold for stimulation. The sensation with the lowest threshold (greatest sensitivity) is bitter (0.000008 M solution), followed by sour (0.0009 M solution), salt (0.01 M solution), and sweet (0.02 M solution).

While the manner in which the substance to be tasted causes depolarization of the taste cells is unknown, it has been suggested that the material must fit a depression (a receptor site) on the microvilli of the taste cells in the manner of a lock and key. A "fit" is followed by depolarization of the membrane and generation of a nerve impulse.

Buds located on the anterior third of the tongue pass impulses to cranial nerve VII (FACIAL). Buds on the posterior two-thirds of the tongue send impulses over branches of cranial nerve IX (GLOSSOPHARYNGEAL). Buds on palate, pharynx, and epiglottis are innervated by cranial nerve X (VAGUS). The afferent nerves pass to the tractus solitarious in the BRAINSTEM. From here impulses pass to the THALAMUS and thence to the PARIETAL CORTEX. Fibers also pass to the salivary nuclei and vagus nuclei to serve the salivary reflex and the cephalic phase of gastric secretion.

Smell

The receptors for the sense of smell are located in the OLFACTORY EPITHELIUM located in the apices of the nasal cavities (Fig. 18.16). The classification of odors is highly subjective and is based on the assumption that molecular shape of the substance and depression of the microvilli must achieve a match before a depolarization can occur. Knowledge of molecular shape can lead to prediction of what the material will smell like. According to this theory, there are seven basic ODORS: camphoraceous, musky, floral, ethereal, pungent, putrid, and pepperminty.

Thresholds of smell are lower than those for taste. On the average, smell is 25,000 times more sensitive than taste. Concentrations of substances only 10–50 times above threshold levels trigger

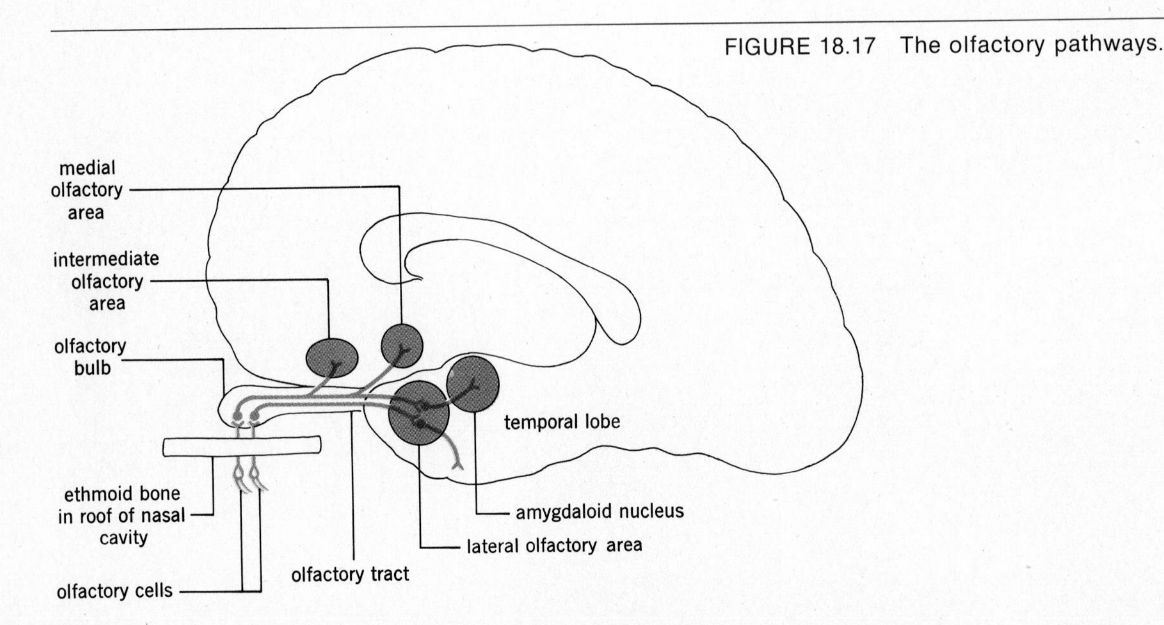

FIGURE 18.17 The olfactory pathways.

maximum firing by olfactory cells. In other words, relatively low concentrations of materials result in maximum response. This suggests that the olfactory receptors are designed for qualitative rather than quantitative detection of odors.

The neural pathways for smell (Fig. 18.17) include the axons of the OLFACTORY CELLS that pass through the roof of the nasal cavity to synapse with mitral cells in the OLFACTORY BULB of the brain. Fibers from the mitral cells form the OLFACTORY TRACT that leads to the OLFACTORY AREAS of the brain. Three olfactory areas are described. A *lateral olfactory area* includes a portion of the temporal lobe and a part of the amygdaloid nuclei. The area is termed the primary olfactory area because it is the principal area for olfactory awareness. The *intermediate olfactory area* lies at the bases of the olfactory tracts in the lower frontal cortex. It appears to be primarily a region for communication and interconnection between the different parts of the olfactory system. A *medial olfactory area* lies on the medial aspect of the frontal lobe beneath the anterior portion of the corpus callosum. This area appears to be involved with the limbic system in expression of emotion. In rats, a "pleasure center" has been shown to be located here; if the center is implanted with an electrode connected to a self-stimulating device, the rat seeks stimulation for its pleasure even to the exclusion of eating.

The olfactory regions send fibers to autonomic centers that can produce reflex behavior in response to olfactory stimuli. Salivation, sexual behavior, and emotional expression are examples of reflex behavior induced by olfactory cues.

DISORDERS of the olfactory system are rare but have considerable importance when present. Interference with the sense of smell may be produced by a *tumor* pressing on or invading the olfactory tract or bulb. Temporal lobe lesions may produce olfactory *hallucinations,* usually of disagreeable odors. Such disagreeable odors often precede the development of an epileptic seizure.

Summary

1. Sound waves are vibrations of air molecules. Sound has three qualities.
   a. Pitch depends on frequency of the sound vibrations.
   b. Intensity refers to the loudness of a sound; it is measured in decibels.
   c. Timbre refers to the quality of the sound.
2. The outer ear gathers and directs sound waves to the eardrum.
3. The middle ear contains a series of ear bones (ossicles) that reduce the amplitude and increase the force of eardrum movements. This is required because the organ of hearing floats in fluid.
4. The organ of hearing is contained within the cochlea. It is known as the organ of Corti.
   a. The organ contains hair cells that convert movement into electrical impulses.
   b. The organ determines pitch by selective vibration according to frequency of the incoming sound wave.
5. The auditory pathways conduct impulses from the cochlea to the brainstem cochlear nuclei, then to the auditory area in the temporal lobe of the cerebrum.
6. Disorders of hearing may result from plugging of the outer ear, inability of the ossicles to move, or damage to the nervous structures of cochlea and/or auditory pathways.
7. The semicircular canals and maculae of the inner ear respond to movement and position. Muscular responses to maintain posture and equilibrium result.
8. Taste is a chemical sense. Four tastes exist.
   a. Sour is triggered by H^+.
   b. Sweet is triggered by OH.
   c. Salty is triggered by metallic ions.
   d. Bitter is triggered by alkaloids.
9. The taste buds are the receptors. Taste pathways include cranial nerves VII and IX and the parietal lobe.
10. Smell is a chemical sense. Seven subjective senses exist.
    a. The olfactory pathways lead from the nasal cavities to the temporal lobes and frontal lobes.

19 Electrical Activity in the Brain; Wakefulness and Sleep; Association Areas

It is possible to record electrical impulses from the living brain. Such impulses represent action potentials in the billions of neurons forming the brain. The patterns of electrical discharge change according to the level of consciousness of the individual and may give clues as to the health of the brain. When stimulated, large regions of the brain give no obvious motor response or sensory impression; such regions are termed association areas, and they appear to be concerned with the "higher" mental functions exhibited by the brain.

Electrical activity; the EEG

Continuous activity

Several basic types of normal brain waves, or electrical activity, may be recorded from the brain, using a device known as an electroencephalograph. The record the machine draws is known as an ELECTROENCEPHALOGRAM or EEG; it is obtained from needlelike electrodes placed in the skin of the head. The basic wave patterns are described below and illustrated in Fig. 19.1. The origin of such waves is not known, but is believed by some investigators to be the result of impulses "circling" over neuronal chains in the brain.

ALPHA WAVES occur at frequencies of 8–13 per second and have voltages of about 50 microvolts (μv). They are most easily obtained from the frontal and occipital regions of the brain and show their most typical form when the subject is awake but relaxed with the eyes closed. When the eyes are open or attention is directed at something, the alpha waves tend to lose the typical pattern and to become much lower in voltage. The change in pattern with fixing of attention is called "desynchronization of the alpha rhythm." Recently, claims have been made that it is possible to train an individual to "bring up" the alpha pattern at will—i.e., to put the brain into a relaxed state voluntarily.

BETA WAVES have frequencies of 15–60 per second, with voltages of 5–10 μv. They are usually recorded best from the frontal and parietal regions of the scalp and are characteristic of the attentive or stimulated brain.

DELTA WAVES have a frequency of 1–5 per second, with voltages of 20–200 μv; they occur during deep sleep, in severe brain disease, and in infants. They usually indicate a brain that is "unconscious" or damaged and can be recorded from most any part of the cortex.

THETA WAVES have frequencies of 5–8 cycles per second and voltages of about 10 μv. They may be recorded from the temporal lobes, and are best seen in children or during emotional stress.

Evoked activity

In addition to the electrical activity that may be continuously recorded from the brain, there may be recorded activity associated with sensory receptor stimulation. Such activity is called *evoked* ("called forth") *activity,* because it is always associated in time with sensory stimulation. For example, a click in the ear, a light shone in the eye, or a pinprick is followed by a burst of electrical activity in the brain. Such activity usually appears

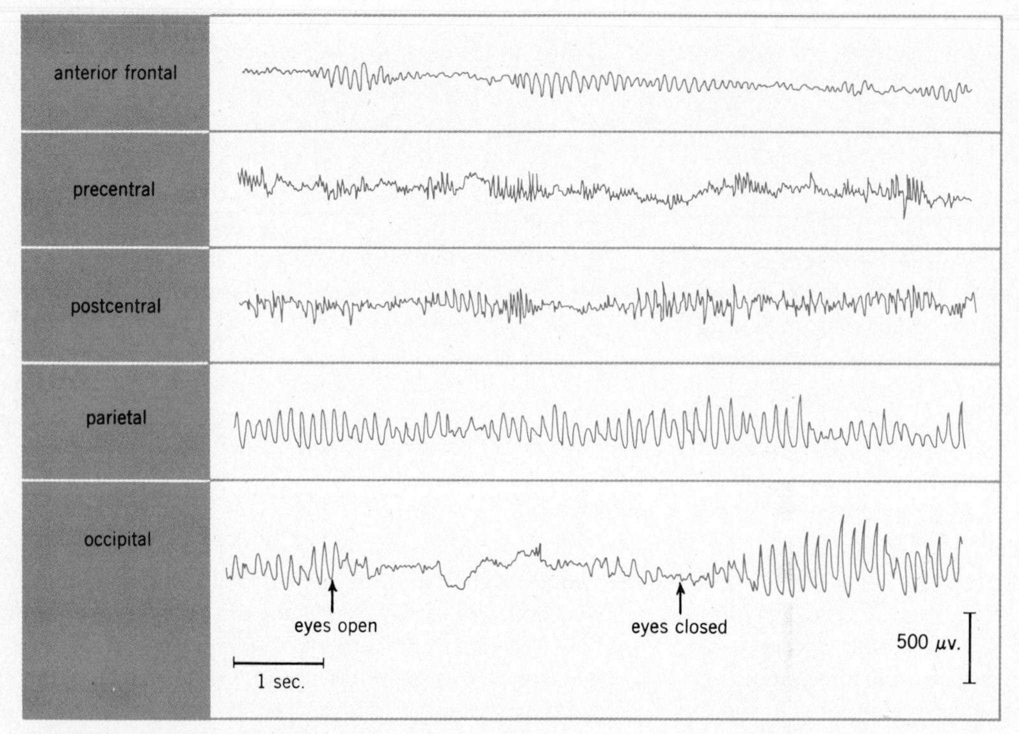

FIGURE 19.1 Normal EEG patterns from different regions of the cerebral cortex. Alpha waves are most prominent in parietal and occipital areas, and are blocked (desynchronized) when the eyes are opened or attention is directed. Beta waves predominate in the prefrontal areas.

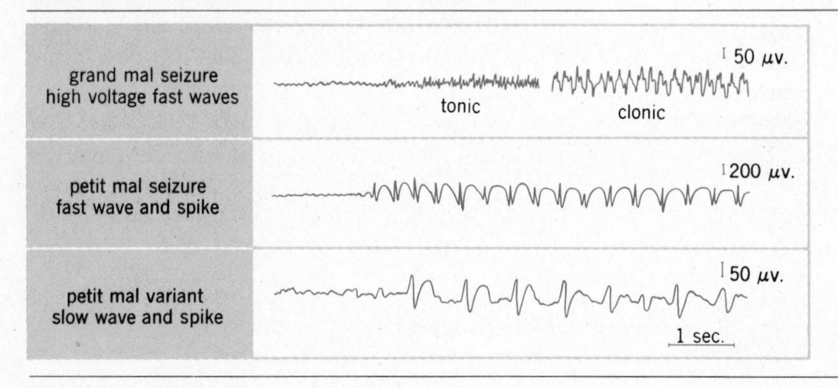

FIGURE 19.2 Some abnormal EEGs in epilepsy.

in the brain region where the fibers concerned with the specific sense terminate, such as the temporal area for sound, occipital for vision, and parietal for most other sensations. Evoked activity can lead to learning and memory storage.

Abnormal activity; epilepsy

In some individuals, bursts of abnormal activity (Fig. 19.2) may occur in response to fever, infections, lack of oxygen to the brain, trauma to the brain, and taking drugs. In other individuals, abnormal activity has no obvious cause or association. The abnormal electrical activity is often associated with disturbances of consciousness, motor and sensory activity, or mental confusion, and a convulsion or SEIZURE has occurred. Fever, infections, lack of oxygen to the brain, tumors, allergic reactions, trauma to the brain, and drugs have all been implicated as causes of convulsive seizures. In other cases there is no obvious cause of the seizure.

If there is a recurring pattern to the seizures the subject is usually said to have epilepsy. EPILEPSY affects about 0.5 percent of the population. It is divided into two general types:

SYMPTOMATIC *(Jacksonian)* EPILEPSY may be demonstrated to have a cause, such as a brain lesion.

IDIOPATHIC EPILEPSY occurs without any demonstrable cause.

According to the type of symptoms the subject shows, several kinds of epilepsy are described.

Grand mal epilepsy results in loss of consciousness, falling, and spasm of the muscles, often preceded by a hallucination of a disagreeable odor.

Petit mal epilepsy clouds the consciousness and is not associated with loss of consciousness or muscular spasms. It produces a "blank" for 1–30 seconds in the subject's behavior, because the subject stops whatever he is doing when the attack begins, and then resumes it when the attack is over.

Psychomotor epilepsy lasts 1–2 minutes and is characterized by loss of contact with the environment, staggering, muttering, and mental confusion for several minutes after the attack is over.

Infantile spasm occurs in the first three years of life and may be replaced by other types of seizures later in life. It is characterized by flexion of trunk and arms and extension of the legs.

If a lesion such as a tumor or scar may be shown to cause the seizure, surgery may be indicated. Management of seizures is often achieved by the administration of anticonvulsant drugs alone or in combinations (e.g., phenobarbital, diphenylhydantoin, bromides).

Wakefulness and sleep

Wakefulness

Located in the brainstem extending from the medulla to the midbrain is a column of neuronal nuclei (excluding cranial nerve nuclei) called the RETICULAR FORMATION. The formation receives impulses from many motor regions of the brain and is part of the motor system (extrapyramidal) controlling muscle activity. It also receives fibers from the spinal sensory tracts and sends fibers to the thalamus and ultimately to the cortex in general. Stimulation of the upper portions of the formation causes immediate and marked activation of the cerebral cortex, a state of alertness and attention and, if the subject is sleeping, immediate awakening. The upper part of the formation and its pathways to the thalamus and cortex have been designated the RETICULAR ACTIVATING SYSTEM (RAS) for the alerting function (Fig. 19.3), and its activity is believed to be responsible for the waking state of the brain and therefore the organism. Supporting such a contention is the fact that coma develops when the brainstem reticular formation is affected by tumors, hemorrhage, trauma, or drugs.

Maintenance of the waking state is achieved by a positive feedback mechanism in which RAS stimulation of the cortex causes the cortex to raise the activity of the RAS, stimulate visceral activity, and cause release of epinephrine from the adrenal medulla. These effects all cause increased input to the reticular formation that increases cortical acti-

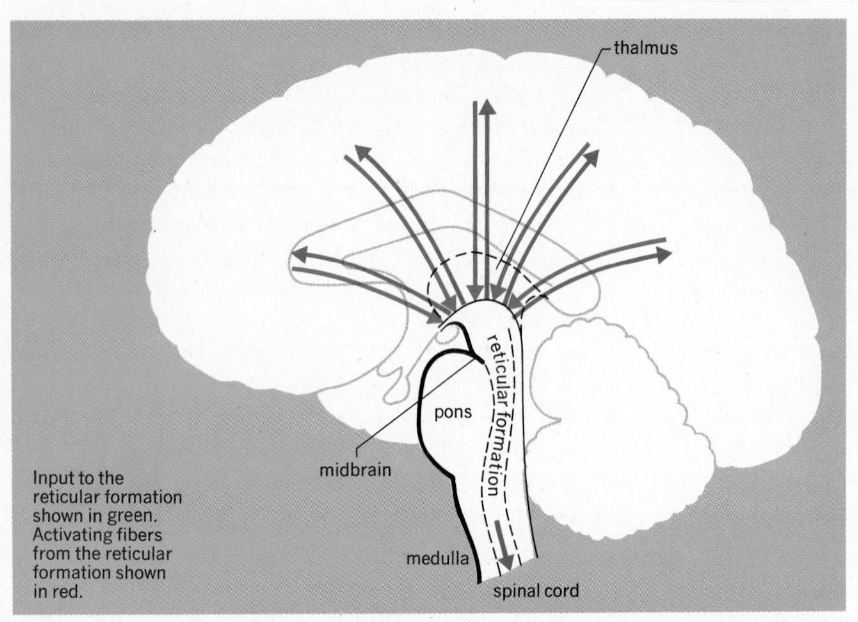

FIGURE 19.3 A diagram representing the pathways of the reticular activating system (RAS). Blue: input to the reticular formation. Red: activating fibers from the reticular formation.

vation. Presumably, the achievement of a non-waking state requires depression of activity in the RAS.

Sleep

Sleep is an interruption of consciousness and the mental activity associated with the waking state. Sleep may be one of the most basic needs of higher organisms. Length of sleep is a highly individual matter, and may be much of the day for infants or only a few hours for adults.

There are several theories as to how sleep occurs. Reduction of external stimuli, and therefore input to the RAS, is claimed to be a requisite for sleep; production of certain chemicals or "hypnotoxins" [serotonin, gamma aminobutyric acid (GABA), dihydroxyphenylalanine (DOPA)] has been claimed to be necessary for sleep to occur. At any rate, loose-fitting clothing, a relaxed, warm environment, and decrease of metabolism, body temperature, and adrenal steroid production all contribute to a sense of drowsiness and desire for sleep. Sleep progresses in four stages.

STAGE 1 SLEEP is signaled by slowing and decrease in amplitude of the EEG alpha rhythm. A person experiences a drifting or floating sensation; if awakened during this stage of sleep, humans will usually assert that they were not asleep, but merely closed the eyes.

STAGE 2 SLEEP is characterized by the appearance of "*sleep spindles*" in the EEG record. These are bursts of 14–15 cps waves lasting several seconds. During this stage the sleeper may be easily awakened.

STAGE 3 SLEEP is associated with the appearance of *delta waves* in the EEG. Breathing is slow and even, and pulse rate is slow (about 60); temperature and blood pressure continue to decline. This is a stage of intermediate-depth sleep.

STAGE 4 SLEEP is deep or *oblivious sleep*. External stimuli, such as clicks, may be recorded in the EEG, so the brain is still receptive. However, the sleeper is not aroused by such stimuli. The brain responds to but does not record such stimuli.

A sleeper appears to descend into deep sleep, then emerge into a modified stage 1 sleep. This occurs throughout the sleep period in cycles 80–120 minutes long. During the emergent phase, the eyes may be seen to undergo rapid movements (REM, rapid eye movement). REM sleep is associated with dreaming and often with movement, accelerated respiration, and accelerated heart rate.

Deprivation of sleep is associated with the development of bizarre behavior and temporary neurosis and/or psychosis. After four to five days of sleep deprivation, there is a decrease in energy mobilization by the body. ATP stores are decreased, there are elevated levels of adrenal cortical "stress hormones" in the bloodstream, and a chemical similar to serotonin and LSD-25 appears in the bloodstream. Memory fails, normal mental agility is impaired, and attention span is severely limited. Hallucinations are common, perhaps brought on by the LSD-like chemical mentioned above. Hallucinations and reality may merge into a continuum from which the only release is sleep. Release is accomplished by allowing sleep to occur, and the subject may sleep 12–14 or more hours, when the time spent in REM sleep typically increases. REM sleep may be essential both for release of psychological tension and for "exercising" the brain. Up to 10 days may be required for completely normal return of all body functions; the necessity of sleep may be concerned with prevention of such disorders as noted above.

Coma

Coma is defined as a state of unconsciousness from which even the most powerful external stimuli cannot arouse the subject. It seems to require suppression of neuronal function or damage to neurons such as by compression (cerebrovascular accident with clot formation; tumor), deprivation of nutrients ("stroke," rupture of vessels), poisoning (hepatic coma), hyperglycemia (diabetic coma), and infections (meningitis).

Emergence from a coma depends on removal of the cause (if possible) and on the nature and degree of damage to neurons that has occurred. If cell destruction has taken place, chances for recovery are less than if the cells have been merely compressed. The brain area most commonly involved in production of coma is the brainstem between the anterior end of the third ventricle and the medulla. In other words, cerebral injuries do not in and of themselves produce coma; there must be brainstem involvement as a result of the action of the causative agent.

Association areas

Those areas of the cortex from which specific sensory or motor response cannot be elicited constitute the association areas. They include the PREFRONTAL REGIONS (areas 8–12), the frontal lobes anterior to the motor region (areas 44–47), the TEMPORAL LOBE (areas 20–22), and the PARIETAL LOBE (areas 5, 7a, 7b, 19, 39, and 40). (See Fig. 13.9)

The prefrontal areas

Removal of portions of the prefrontal areas does not greatly affect reflexes, movement, or general intelligence but does cause marked disturbances of behavior.

HYPERACTIVITY, characterized by aimless, incessant pacing, may occur.

DELAYED RESPONSE REACTIONS are virtually eliminated; these are tested for as follows. If a normal animal is allowed to see food or another reward being placed in a given position beneath a cup or other masking device, then has it hidden from view and subsequently is offered the substance still hidden, it will, with training, remember and successfully choose the proper cup and receive the reward after delays of up to 90 seconds. In a prefrontally lesioned animal, 5 seconds reduces the choice to mere chance.

EMOTIONAL ALTERATIONS, including excessive rage reaction and easily elicited reactions, result. This suggests involvement of the prefrontal areas in inhibition of excessive emotional display.

Reactions typical of neurosis may be prevented by prefrontal ablation. An animal exhibiting neurotic behavior, particularly that associated with emotional outbursts, rage reactions, and anxiety, may, after surgical isolation of the prefrontal regions, exhibit a more calm type of behavior. These reactions are used to justify psychosurgery (frontal lobotomy) to relieve such states in the human.

In humans there is much more variability and diversity in the symptoms displayed as a result of frontal lobe damage by tumor or trauma and by surgical intervention. As with lower animals, the greatest changes are evidenced in personality, with little effect on general intelligence. The sorts of changes resulting are illustrated by the "crowbar case" of Phineas P. Gage, "an efficient and capable foreman" who had a tamping iron blown through the frontal regions of his brain. The changes resulting were described by his attending physician: "He is fitful, irreverant, indulging in the grossest profanity (which was not previously his custom), manifesting but little deference to his fellows, impatient of restraint or advice when it conflicts with his desires, at times ...obstinate, capricious, and vacillating.... His mind was radically changed, so that his friends said he was no longer Gage." Such a patient also may show lack of foresight, refuse to accept the responsibilities of life, and fail to discern the seriousness of a situation. The only consistent conclusion to be drawn from such injuries is that some alteration of personality will result, but its direction, duration, and severity are not predictable from one individual to the next. For these reasons, surgical isolation (prefrontal lobotomy) of the prefrontal areas for relief of anxiety and other neuroses is discouraged.

The temporal lobes

The temporal association areas appear to be concerned with the LEARNING and MEMORY of visual discriminations and with comprehension and memory of visual and auditory inputs. As evidence of damage to the area, loss of memory of things learned usually occurs, and may occur in one or more of three forms.

Agnosia, or loss of ability to recognize objects that are usually familiar. Failure may manifest itself through lack of ability to distinguish things touched *(astereognosis)*, heard *(auditory agnosia)*, or seen *(visual agnosia)*, or the lack of ability to distinguish right from left *(autotopognosis)*.

Apraxia, or an inability to perform voluntary movements, particularly of speech, when no true paralysis is present.

Aphasia, or any inability to use or comprehend ideas communicated by spoken or written symbols. Aphasia may be motor, resulting in inability to express oneself, or sensory, resulting in a deficit in understanding.

Alexia, a form of aphasia, refers specifically to inability to comprehend the written word (word blindness) or to speak words correctly. It is usually first discovered in reading defects of schoolchildren.

The parietal lobes

Lesions in the parietal lobes may also produce agnosia if the lesion is on the "dominant" side of the cerebrum (left side in right-handed persons, and vice versa). Lesions on the nondominant side produce AMORPHOSYNTHESIS, characterized by defective perception of one side of the body, to the extent that it is ignored or denied to exist.

Summary

1. Electrical activity in the brain is the result of "spontaneous" activity or evoked activity.
   a. Electrical activity may be recorded as an EEG (electroencephalogram).
   b. Continuous electrical activity is seen as the basic alpha, beta, delta, and theta waves.
   c. Evoked activity is the result of specific sensory input to the brain and is reflected as changes in the EEG.
   d. Abnormal electrical activity is associated with convulsive disorders such as epilepsy.
2. The waking state of the organism is maintained by input from sensory receptors acting through the reticular formation of the brainstem. The reticular activiting system (RAS) consists of the brainstem reticular formation and its efferent pathways to the thalamus and cortex. It maintains the waking state of the organism by positive feedback.
3. Sleep is an interruption of the waking state.
   a. Sleep occurs usually once in a 24-hour period and is associated with depression of body temperature, heart rate, blood pressure, respiration, and metabolism.
   b. Sleep occurs in four stages, in cycles of 80–120 minutes. Emergence to REM sleep occurs after stage 4 sleep and is associated with dreaming. Dreaming appears to serve psychological and physiological needs.
   c. Sleep deprivation is associated with disorders of personality and behavior. Chemicals produced during deprivation may account for these symptoms.
4. Coma is a state of unconsciousness from which the subject cannot be aroused.
   a. Coma is usually associated with damage to brain cells or deprivation of glucose and oxygen.
   b. Recovery from coma depends on the extent and duration of damage.
5. The cerebrum contains areas from which no specific motor or sensory response can be elicited. These are association areas concerned with behavior, intelligence, memory, and emotions.
   a. Personality derangements are characteristic of prefrontal damage.
   b. Disorders of memory and understanding are most common with temporal lobe lesions.
   c. Defective sensory perception on one side of the body may be produced by parietal lobe lesions.

Emotions, Learning, and Memory

Emotion, learning, and memory are interwoven in that emotion requires previous experience by which to measure the need for, and degree of, expression of emotion.

Emotional expression

Aspects of emotion

Emotional expression appears to be associated with the structures of the LIMBIC SYSTEM (Fig. 20.1). These structures are involved primarily in expression of emotions. Emotion, however, has four aspects.

Cognition, or perception and evaluation of a particular situation before a reaction is made.

FIGURE 20.1 Some structures of the limbic system. M = mammillary body, H = habenula.

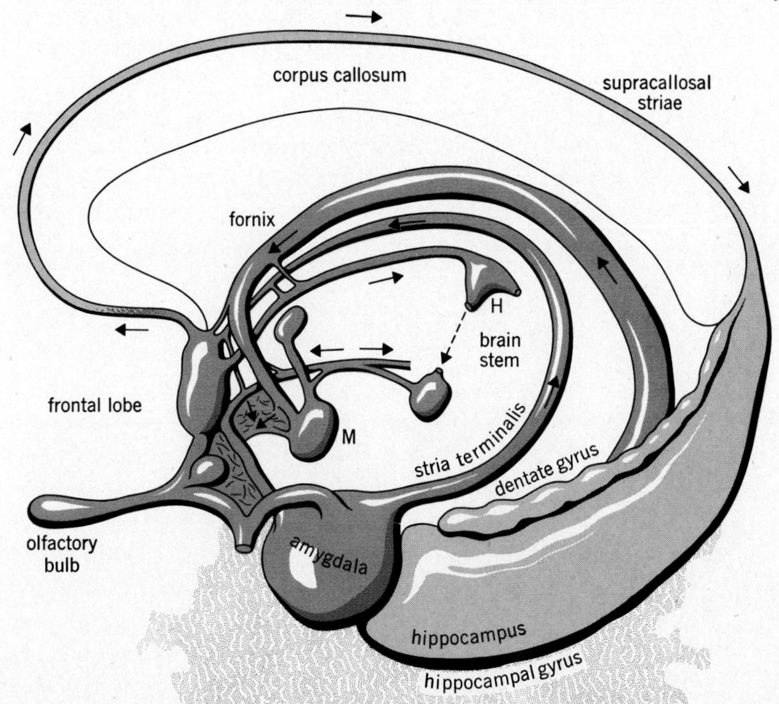

Expression, or external sign of emotion.

Experience, or the subjective evaluation of an incident as compared to previous incidents.

Excitement, or degree of "vividness" of the emotion.

Most, if not all, of the aspects of emotion except expression are provinces of the cerebrum.

Excitement is theorized to result from simultaneous activation of the reticular system and the cortex.

Of interest is the development of emotions. While there is no definite agreement as to whether a newborn exhibits emotions or not, several general statements may be made.

A newborn appears to have a comfort-discomfort reaction. It may cry if something makes it hurt, cold, or lonely. It may turn up the corners of the mouth (a smile?) in response to internal stimuli (need for defecation, gas). By age two to eight weeks it may exhibit a true smile in response to comfort, satisfaction of needs, or the appearance of a parent in the field of view. If there is a hereditary factor in emotions, it appears to lie in the area of setting basic neural pathways that can be "filled" with experience as life progresses. In a sense, the situation is like a newly built computer; the circuits are there but the instrument is as yet unprogrammed. The infant learns very rapidly that certain activities on its part cause reactions on the part of its caretakers, and so positive or negative reinforcement of behavior and emotional expression occurs. By two years of age, the child has probably achieved a full set of emotional expressions. It may have shown frustration at the time of weaning; it may have developed guilt about its toilet training; it shows curiosity as new items appear in its world; it may show anxiety at the anticipation of danger. Fear does not seem to be inherent in the child but is learned.

The limbic system

Included as parts of this system are the AMYGDALOID NUCLEI, the HYPOTHALAMUS, FORNIX, and several SUBCORTICAL NUCLEI, whose general locations are shown in Fig. 20.1. Stimulation of the amygdala produces aggressive responses, while their destruction causes violent animals to become passive, exhibit no fear of objects that formerly produced fear, and exhibit hypersexuality. Integrity of the hypothalamus is required for outward expression of emotion, particularly the visceral components. Directed emotional expression, such as attack on a specific object, requires the involvement of the cortex. Thus, all structures are integrated to produce the full range of emotional expression.

Also included as part of the limbic system are areas that, when stimulated, produce emotional feelings characteristic of pleasure. The most effective regions are in the posterior hypothalamus and preoptic regions. Regions such as those just described can be characterized as areas of motivation; stimulation produces activity directed toward securing additional stimulation.

Learning and memory

Learning and memory are perhaps inseparable qualities of the nervous system, for one usually must have a base for comparison (provided by memory) against which to measure whether some new bit of information shall be permanently incorporated into the nervous system.

LEARNING may be produced through conditioning, as illustrated by Pavlov's dog experiments, or by reinforcing either negatively or positively a particular activity such as bar pressing to obtain food.

Several areas of the nervous system appear to

be involved in learning. The spinal cord may be involved in the learning of simple conditioned responses such as those involving withdrawal reactions to painful stimuli; the cortex appears to speed the rate of learning; the reticular formation may be "trained" to arouse the cortex, since less stimulation is required to produce the same evoked cortical potential as learning proceeds. In short, no specific area appears to be involved to the exclusion of others in the learning process. Learning may proceed through the establishment of "preferred pathways." If a particular circuit is used again and again, enlargement of synaptic endings, increase in numbers of vesicles, and increase in number of endings occur. Thus, the next time the same stimulus occurs, it tends to follow a given path, and learning has taken place.

MEMORY is another problem. Two questions are immediately important: *where* is the information stored, and *how* is it stored?

The question of *where* is not completely answerable today. Stimulation of points within the temporal lobes evokes recall of memories of long-past experiences. The memories themselves are probably not stored there; the stimulation merely activates pathways that cause retrieval from brainstem or frontal and parietal storage areas. The HIPPOCAMPAL AREA appears to be necessary in the storage of information as memories for recent events, since destruction of that area impairs recent memory.

The process of memory formation appears to require the involvement of three mechanisms: one for momentary retention (remembering a phone number only long enough to dial it); memory for events occurring minutes to hours before *(short-term memory);* memory for events occurring in the past *(long-term memory).* Both the momentary and short-term memories appear to depend on some electrical phenomenon such as impulses circling around neuronal chains in the cerebral association areas, because they may be abolished by electroshock. The latter procedure imposes an external electrical field on the entire brain and causes all neurons simultaneously to assume a refractory state. This is thought to result in blockage of electrical impulses circling the neuronal chains. Some four hours after the experience, electroshock does not affect the memory. Between about ten minutes and four hours, administration of an antibiotic (puromycin, which is also an inhibitor of protein synthesis) prevents the establishment of long-term memory. It prevents "consolidation of the memory trace" or the creation of a permanent record of the memory as a chemical of some sort. It is tempting, then, to suggest that long-term memory involves the production of a protein. Protein synthesis requires one or more forms of RNA. A theory has thus evolved to suggest that consolidation of memory traces results from production of specific RNA "templates" that cause production of specific proteins for each bit of information stored. Retrieval of the memory is associated with destruction of the protein but not of its template so that, once recalled, the memory is not lost forever. Supporting this theory are investigations that show increases of RNA in the brain as learning and experience increases, and by the fact that RNAase inhibits learning (memory?) in certain worms. Nevertheless, humans are a long way from a "smart pill." However, some drugs (see the Appendix) do increase rate of learning. For example, stimulants such as caffeine, physostigmine, amphetamines, nicotine, and strychnine improve learning by speeding consolidation of the trace. Pemoline (Cylert), a mild stimulant, is also a drug increasing RNA synthesis, and it also speeds learning.

Summary

1. Emotions have four aspects and apparently are learned.
   a. Cognition or perception of a situation.
   b. Expression or external sign.
   c. Experience or evaluation.
   d. Excitement or vividness of the emotion.
2. The limbic system involves the amygdaloid nuclei, hypothalamus, and several other nuclei and connecting tracts.
   a. The amygdaloid nuclei are concerned with aggression.
   b. The hypothalamus is concerned with outward expression of emotion.
   c. The hypothalamus is also the site of areas concerned with motivation.
3. Learning is the ability to acquire a bit of knowledge or a skill. It may be retained, in part, as a memory.
   a. Short-term memory appears to be served by electrical phenomena.
   b. Long-term memory appears to require the production of a chemical, possibly a protein. RNA may be involved in the protein synthesis.
4. Certain drugs accelerate learning, perhaps by stimulating RNA and/or protein synthesis.

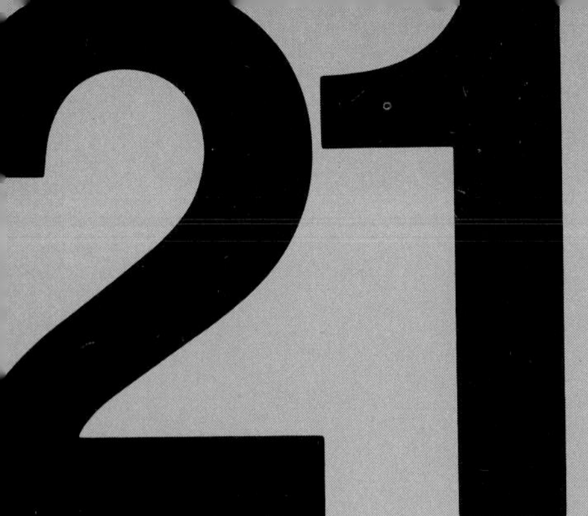

Body Fluids and Acid-Base Balance

Life is said to have originated in the sea—that is, in a watery environment. The original life forms thus existed in a medium that provided all their nutrient needs, and that carried away the wastes of their activity. This same principle of "bathing the cells" is followed by the human, but the fluid environment is separated from the external environment. The body fluids form an INTERNAL ENVIRONMENT for the cells, and the pH, osmotic pressure, temperature, and other characteristics of the fluids must be maintained within narrow limits to ensure continued normal cell function.

Fluid compartments

The total quantity of fluid in the body is influenced by several factors (Fig. 21.1 and Table 21.1). Younger individuals tend to have a higher body water content than older persons; adult females tend to have less body water than adult males; obese individuals have less body water than lean persons. Regardless of actual quantity, there are two major divisions recognized in total body water. The term EXTRACELLULAR FLUID refers to all fluids not contained within cells; the term INTRACELLULAR FLUID refers to all fluids contained within the cells. The figures mentioned in subsequent paragraphs are for an average adult.

Extracellular fluid

Extracellular fluid (ECF) accounts for about 37 percent of the total body water and has the following subdivisions.

PLASMA. This is the fluid portion of the blood

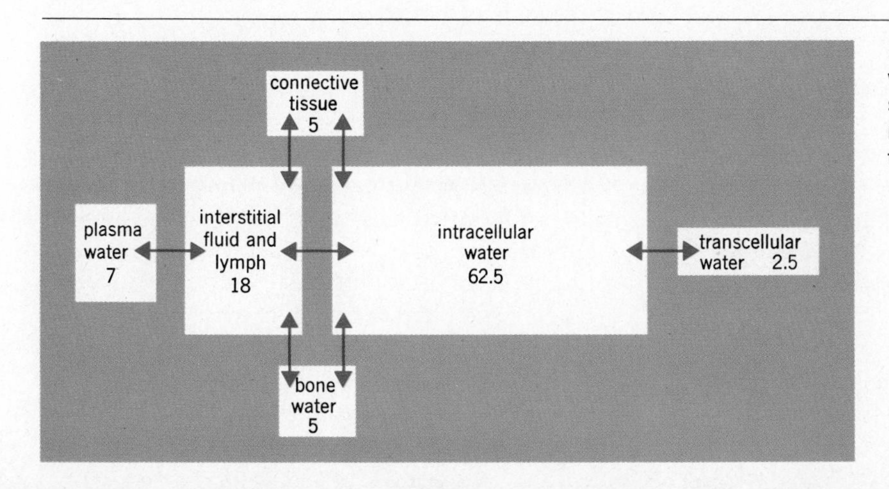

FIGURE 21.1 Relationships of body water compartments. Numbers represent the percent that each compartment comprises of total body water in the adult.

TABLE 21.1 Compartments of the body water

Compartment	Approximate % of body weight					Approximate % of body fluid				
	Premature	Newborn	1 yr.	2 yr.	Over 18 yr.	Premature	Newborn	1 yr.	2 yr.	Over 18 yr.
Extracellular fluid	**50**	**40**	**30**	**24**	**21.4**	**56**	**53**	**47**	**40**	**37.5**
Plasma	5	5	5	5	5	6	7	7	7	7
Interstitial fluid (including connective tissue and bone)	45	35	25	19	15	50	46	40	33	28
Transcellular fluid	—	—	—	—	1.4	—	—	—	—	2.5
Intracellular fluid	**40**	**35**	**34**	**36**	**38.6**	**44**	**47**	**53**	**60**	**62.5**
Total body water	**90**	**75**	**64**	**60**	**60**	**100**	**100**	**100**	**100**	**100**

contained within the blood vessels of the body. It amounts to about 7 percent of the body water.

INTERSTITIAL FLUID AND LYMPH. Interstitial fluid, or tissue fluid, is the fluid immediately around the cells. It is formed by filtration of plasma water and solutes and is not contained within any type of vessel. Lymph is tissue fluid that has entered lymphatic vessels. It has the same basic composition as tissue fluid. These two subdivisions together account for about 18 percent of the body water.

FLUIDS OF DENSE CONNECTIVE TISSUE AND BONE. These fluids are actually a part of the interstitial fluid, but the water in these tissues exchanges very slowly with the interstitial fluid and thus behaves like a separate compartment. These fluids compose about 10 percent of the body water.

TRANSCELLULAR FLUIDS. This category includes all fluids separated from other ECF by an epithelial membrane. The fluids of the eye, joints, and true body cavities, the cerebrospinal fluid, and the fluids of the hollow organs of the digestive, respiratory, and urinary systems are included as transcellular fluids. These fluids comprise about 2.5 percent of the body water.

Intracellular fluid (ICF)

Fluid contained within the body cells amounts to about 63 percent of total body water.

Water requirements

Maintenance of normal body water volumes and solute concentrations in the various compartments requires that loss be balanced by intake or production of fluids. Actual amounts involved are age dependent, as shown in Table 21.2, and are influenced by such factors as activity levels and levels of basic metabolic processes.

An infant, whose kidneys are not yet able to concentrate urine to adult levels, loses more water via the urine and thus requires a greater intake. Also, expressed as calories per kilogram of body weight, infant and child basal metabolic rate (BMR) is higher than that of an adult (60–80 C/kg to 25–30 C/kg), because of the greater heat production by the younger individual. This causes a greater vaporization of water at lung and body surfaces, and a greater loss that requires replacement. Additionally, infant breathing rates are higher than those of an older individual, and a greater water loss occurs despite the small volume of air exchange.

A normal adult adds approximately 2500 ml of water per day to the total fluid volume. Of this total, about 2300 ml enters the fluid compartments by gut absorption of water ingested in food and drink. About 200 ml of water is added per day to the body fluids by metabolic reactions. Loss of water from the body occurs through fixed or uncontrollable paths of loss and by variable or controllable pathways.

FIXED LOSS ROUTES include:

TABLE 21.2 Water requirements as related to age and weight

Age	Weight (kg)	Water required /kg body weight/day (normal) (ml)
3 days	3	80–100
10 days	3.2	125–150
6 months	7.3	130–155
12 months	9.5	120–135
2 years	12	115–125
6 years	20	90–100
10 years	29	70–85
14 years	45	50–60
Adult	55–70	21–43

The lungs—about 300 ml per day.

Perspiration—about 500 ml per day.

Feces and evaporation from the mouth—about 200 ml per day.

The VARIABLE ROUTE OF LOSS is through the kidney—about 1500 ml per day.

One may note that the route by which the greatest amount of fluid may be lost is also the one that is most subject to control; thus, body fluid content may be adjusted more easily. Also, loss via the kidney is inversely proportional to loss by other routes.

Water absorbed from the alimentary tract passes primarily into the plasma, from which it may be lost through the lungs, skin, and kidneys. Gut absorption is by *osmosis*. Passage of water from plasma to interstitial fluid is by *filtration* from blood capillaries. Return of water from the tissues to the bloodstream is by *osmosis* into blood capillaries (90 percent) and return via lymph vessels (10 percent). Water enters cells by osmosis, and lymphatics by the "pumping action" of muscular contraction and tissue pressure. Some of these relationships are shown in Fig. 21.2, which emphasizes the critical role of the plasma proteins (which are too large to filter in significant amounts from capillaries) in establishing the osmotic pressure that returns the bulk of tissue water to the capillaries.

Composition of the compartments

In general, the subdivisions of the ECF (except transcellular fluids) are nearly identical in content of electrolytes and small molecules, differing mainly in protein concentration. The composition of ICF is different from that of ECF in terms of quantity and type of solutes. The major ions of ECF are sodium and chloride; those of ICF are potassium, phosphate, and ionized protein. These differences are presented in Table 21.3 and Fig. 21.3.

The primary organ regulating the solute con-

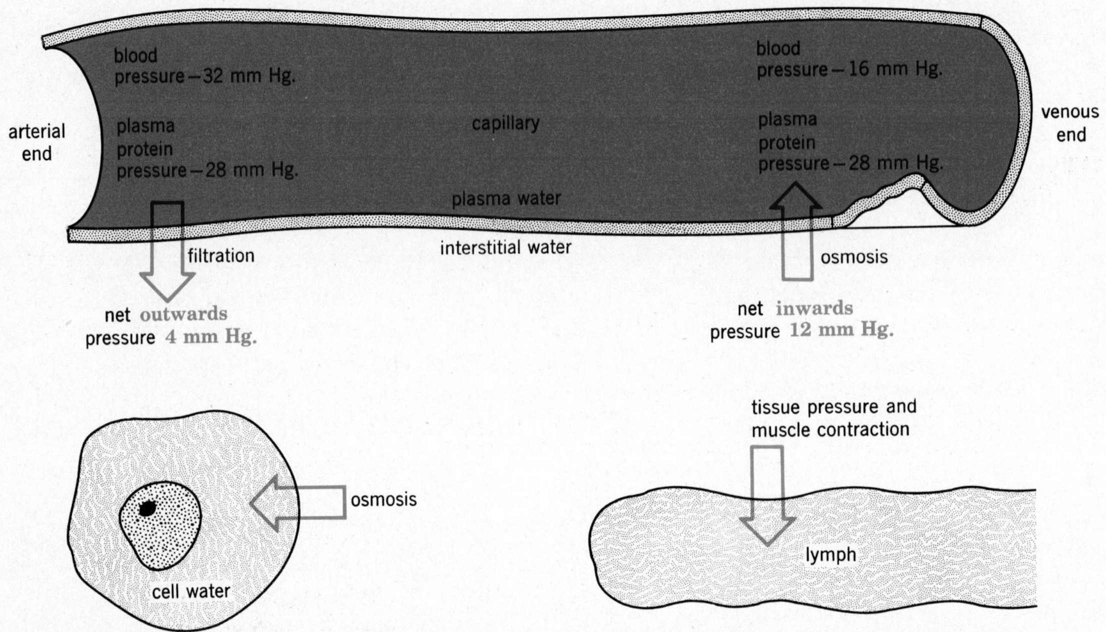

FIGURE 21.2 Some determinants of water movement between body water compartments.

centration of ECF, and therefore also of ICF, is the kidney. It can control the blood levels of several dozen materials more or less independently of one another for homeostasis of fluid-solute composition.

Homeostasis of volume and osmolarity of ECF

Maintaining the volume and solute content of the ECF compartment within normal limits involves controlling both the intake and output of fluids and solutes. Some of the mechanisms involved in this regulation are presented below.

THE THIRST MECHANISM. Increased loss of water over intake leads to slowing of salivary secretion and dryness of the mouth. Fluid is taken in to relieve the dry sensation. Reduction of ECF volume, loss of ICF, and a low cardiac output also create sensations of thirst. It has been claimed that integration of these causes into the sensation of thirst is mediated by a hypothalamic region close to the areas producing ADH (see below).

MECHANICAL FACTORS. Loss of fluid in amounts sufficient to reduce blood volume and lower cardiac output reduces kidney filtration of water. Burns, hemorrhage, and the decreased cardiac output associated with heart failure are examples of the operation of this factor. On the other hand, if ECF volume is increased to a point where cardiac output is increased, as by salt and water retention or high-salt diets, rise of blood pressure occurs, and greater kidney filtration eliminates the excess water.

HORMONAL FACTORS. The ALDOSTERONE MECHANISM (Fig. 21.4) is a device by which aldosterone, an adrenal cortical hormone, causes an increase in sodium reabsorption from the kidney tubules that eventually raises blood sodium levels. Sodium is a cation, and it causes anions (HCO_3^-,

TABLE 21.3 Composition of extracellular and intracellular fluid compartments

Constituents and Properties	Extracellular fluid: Plasma	Extracellular fluid: Interstitial fluid	Intracellular fluid
Sodium	142 meq/L	145 meq/L	10 meq/L
Potassium	4 meq/L	4 meq/L	160 meq/L
Calcium	5 meq/L	5 meq/L	2 meq/L
Magnesium	2 meq/L	2 meq/L	26 meq/L
Chloride	101 meq/L	114 meq/L	3 meq/L
Sulfate	1 meq/L	1 meq/L	20 meq/L
Bicarbonate	27 meq/L	31 meq/L	10 meq/L
Phosphate	2 meq/L	2 meq/L	100 meq/L
Organic acids	6 meq/L	7 meq/L	—
Proteins	16 meq/L	1 meq/L	65 meq/L
Glucose (av)	90 mg %	90 mg%	0-20 mg%
Lipids (av)	0.5 gm %	—	—
pH	7.4	7.4	6.7-7.0*

*Average value; difficult to measure.

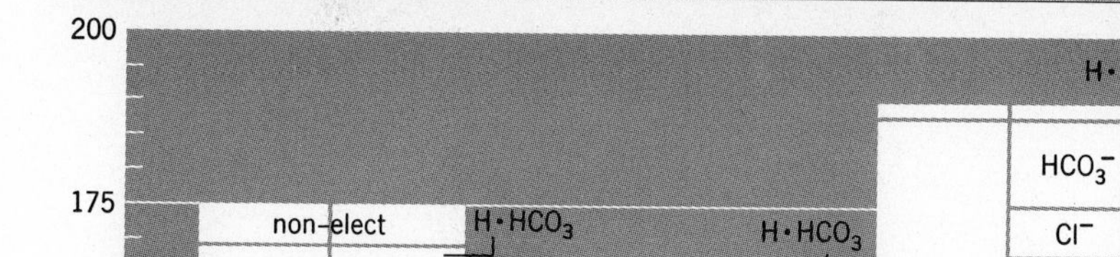

FIGURE 21.3 A comparison of the constituents of the three major compartments of the body water.

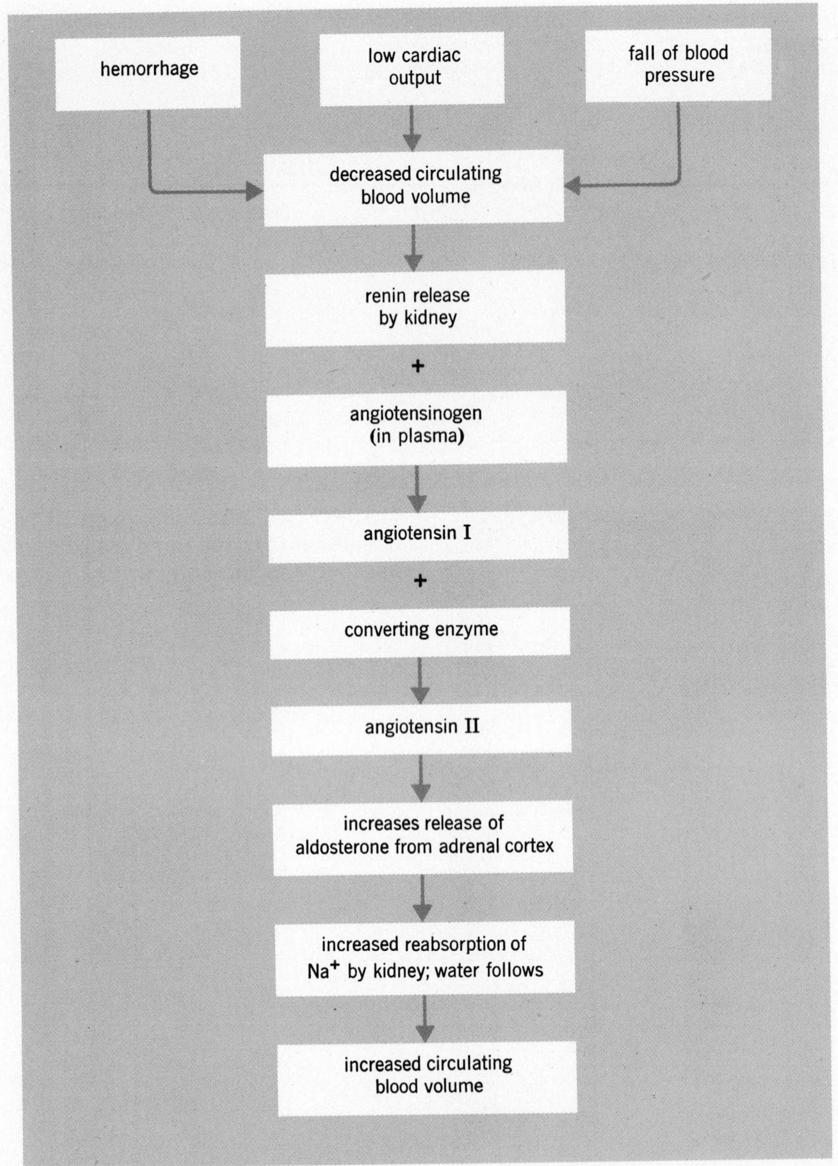

FIGURE 21.4 The mechanism controlling aldosterone secretion and the effects of its secretion.

Cl^-) to follow it. As these solutes move out of the tubules, water follows osmotically to maintain isotonicity. Thus, total blood volume is increased as fluids and electrolytes are balanced. However, this mechanism can achieve only about a 10 percent increase in volume before the mechanical system mentioned above causes kidney filtration to offset the aldosterone effect.

ANTIDIURETIC HORMONE (ADH) (Fig. 21.5) is a substance produced by the hypothalamus that is stored in and released from nerve fibers in the posterior lobe of the pituitary gland. Hypothalamic

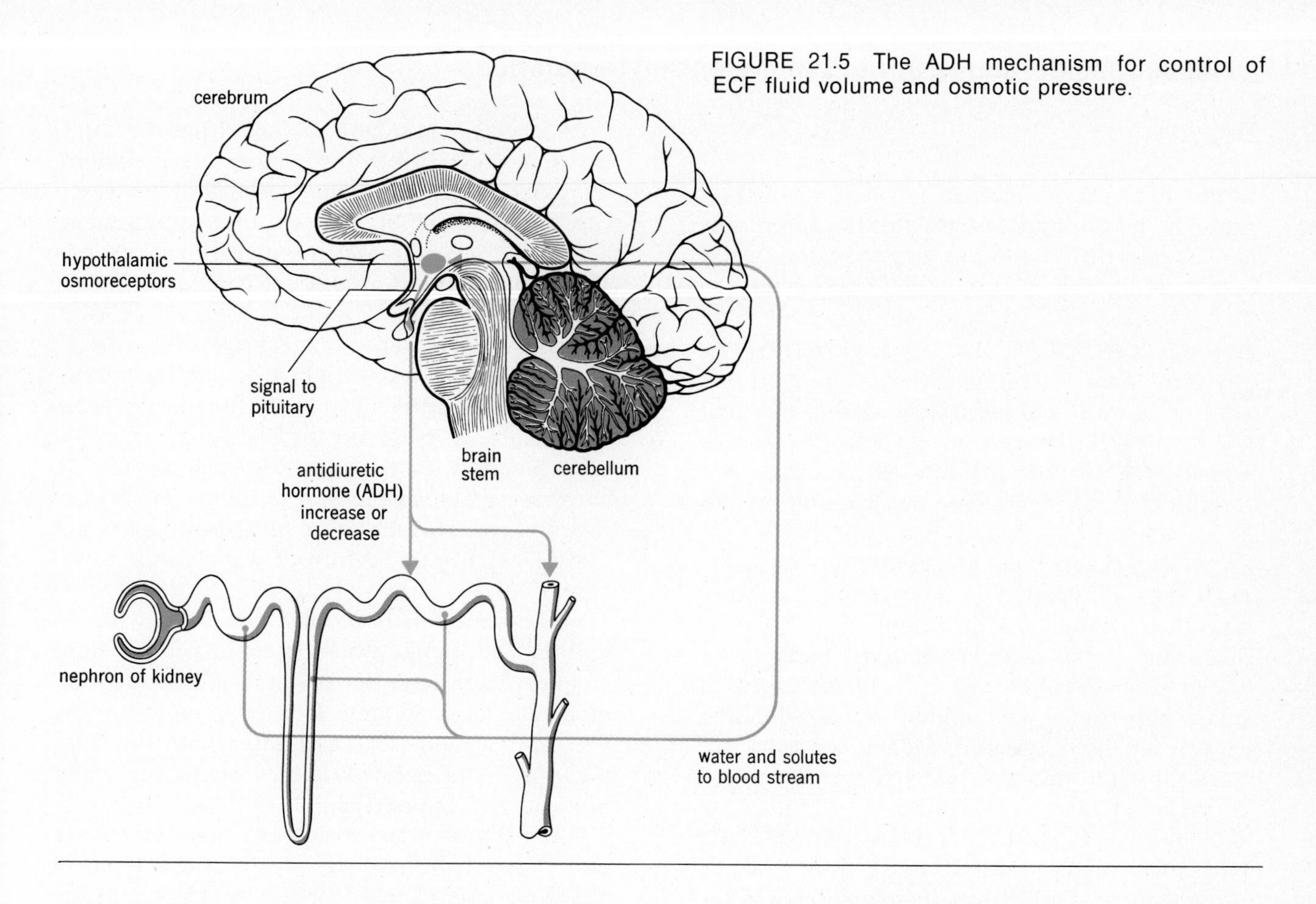

FIGURE 21.5 The ADH mechanism for control of ECF fluid volume and osmotic pressure.

production and release from the posterior lobe is directly proportional to blood osmostic pressure (osmolarity). The hormone permits or causes greater water absorption from the collecting tubules and the proximal portions of the papillary ducts as they pass through the medulla.

If blood water content is increased, the need is to get rid of the extra water and ADH release is slowed; less water is reabsorbed and more is excreted. Of course, the reverse is also true.

Some investigations suggest the presence of a NATRIURETIC HORMONE or "third factor" that is supposed to reduce the absorption of sodium by the kidney. Its source is unknown.

NERVOUS REFLEX FACTORS. Many circulatory reflexes, to be described in greater detail in subsequent chapters, affect ECF volume. For example, baroreceptors in several areas of the body cause constriction of the afferent arterioles of the kidney, thereby reducing filtration and conserving ECF. Increase of blood volume causes volume receptors in the thorax to cause decreased secretion of ADH and less reabsorption of water by the kidney.

Clinical considerations of fluid and electrolyte balance

Dehydration

If the homeostatic mechanisms fail to operate properly, loss of fluid, electrolytes (ions), or both may occur. Three conditions may arise depending on the relative losses of fluid and electrolytes.

EQUAL LOSS OF FLUID AND ELECTROLYTE ("isotonic dehydration"). There is a proportionate loss of fluid and electrolyte so that the total volume of ECF changes, but its osmotic pressure remains within normal limits. Hemorrhage is a condition in which isotonic dehydration occurs.

EXCESSIVE FLUID LOSS AS COMPARED TO ELECTROLYTES. ("hypertonic dehydration"). More fluid than electrolyte is lost, with resulting concentration of the ECF. Water thus tends to be drawn from cells. Burned individuals typically suffer hypertonic dehydration. Again, infants, because of their immature kidney function, are more subject to this type of dehydration.

EXCESSIVE ELECTROLYTE LOSS AS COMPARED TO FLUIDS ("hypotonic dehydration"). More solute than fluid is lost, and therefore the ECF becomes diluted. Water thus tends to enter cells. Severe diarrhea may cause this type of dehydration.

Edema

Edema is a condition caused by accumulation of fluid primarily in the interstitial compartment. Some factors involved in the production of edema are presented below.

INCREASED FILTRATION of fluid from the capillaries, as in high blood pressure (hypertension), increases the volume of interstitial fluid. Increased capillary permeability may permit loss of protein, as does kidney disease in which protein is lost. The osmotic return of water to capillaries is thus decreased.

DECREASED PUMPING EFFICIENCY of the right ventricle of the heart (right-side cardiac failure) causes an impediment to blood flow in the veins and results in body organs becoming engorged with blood. Venous pressure increases, creating an increased blood pressure at the venous end of the capillary. This increased pressure opposes and diminishes osmotic return of water to the capillaries.

SALT RETENTION is usually associated with retention of enough water to maintain the fluid isotonic. Thus, retention of salt results in retention of excess fluid volume. It is of interest to note that the "weight loss" associated with the commencement of a low-salt diet represents water loss and not that of tissue.

BLOCKAGE OF LYMPH DRAINAGE from the tissues will prevent the removal of about one-tenth of the fluid filtered from capillaries. The condition is called *lymphedema* and usually develops gradually. Causes include failure of vessels to develop, compression or blockage of vessels by scar tissue as after surgery, or infections by parasites or microorganisms.

Increased volume of fluid in the interstitial compartment may in turn cause entry of excessive water into the cell, as interstitial dilution occurs and pressure increases. Symptoms associated primarily with the central nervous system appear (disoriented behavior, convulsions, coma).

Measurement of the volumes of the fluid compartments

Volumes of certain compartments may be directly measured if a substance that will confine itself primarily to that compartment is available. Other compartment volumes must be calculated. A known

TABLE 21.4 Measurement of body water compartment volumes

Compartment	Measured or calculated	Substance(s) employed	Volume in 70 kg man (liters)	Comments
Total water	Measured	Heavy water (D_3O)	46	Mixes with water molecules in all compartments
ECF	Measured	Polysaccharides Disaccharides (sucrose) Radioactive sodium	17	Small enough to filter from blood capillaries. Do not enter cells or are actively kept out of cells
Plasma	Measured	Radioactive serum albumin	3	Generally too large to filter from blood capillaries.
Interstitial fluid	Calculated (ECF minus plasma)		14	Values of these depend on accuracy of measurements of above-listed compartment volumes
ICF	Calculated (total minus ECF)		29	

concentration of the substance is injected, time is allowed for distribution to occur, and then the concentration of the material is measured in the urine and in the compartment. The volume of distribution is calculated by:

$$\text{volume of distribution} = \frac{(\text{quantity administered}) - (\text{quantity excreted})}{\text{concentration of sample}}$$

Only total water, plasma volume, and ECF can be measured; the others must be calculated. Some typical measurements are shown in Table 21.4.

Acid-base balance

The body fluids normally have the ability to resist changes in their degree of acidity and alkalinity (pH). The term *buffering* is used to describe this ability to resist changes in pH. The primary force tending to alter body fluid pH is the addition of hydrogen ion (H^+) to the fluids.

Hydrogen ion is derived from the hydrogen atom by a loss of the single orbiting electron the atom possesses. The ion is extremely reactive and combines with water to form a hydronium ion (H_3O^+). The H_3O^+ is the form in which the H^+ actually exists in the body. Conventionally, however, it is referred to as H^+, and this terminology will be followed in the ensuing discussion.

Acids and bases

In modern chemical terminology, a material capable of losing H^+ is termed an *acid* while a substance capable of accepting H^+ is a *base.* "Strong acids" are substances that give up H^+ easily, and that usually have a high degree of ionization or dissociation. Hydrochloric acid, for example, is a strong acid because it is 100 percent dissociated in dilute solutions:

$$HCl \longrightarrow H^+ + Cl^-$$

A "weak acid" is a substance with a lesser tendency to release its H^+, and consequently has a low degree of dissociation. Acetic acid is a weak acid, being only slightly dissociated in solution:

$$HAc \rightleftharpoons H^+ + Ac^-$$

The concentration of hydrogen ion in a solution, represented by $[H^+]$, is conveniently described by the symbol pH. Strictly defined, pH, or degree of acidity of a solution, is given by the equation:

$$pH = \log\left(\frac{1}{[H^+]}\right) \text{ or } pH = -\log\,[H^+]$$

The pH scale runs from 0 to 14, or from strongly acid (low pH) to weakly acid (high pH) solutions. Conversely, it may be said that this scale represents a change from weakly basic (low pH) to strongly basic (high pH) solutions. At a pH of 7, a solution is neither acidic nor basic; it is neutral, because it contains equal concentrations of H^+ and base. A change of one pH unit (for example, from 6 to 5) represents a tenfold change in $[H^+]$. These basic facts indicate the severity of the problem faced by the body in its attempts to remove H^+.

Sources of hydrogen ion

Most of the H^+ added to the body fluids rises from three sources:

The COMPLETE COMBUSTION OF CARBON COMPOUNDS. Complete combustion of any fuel produces carbon dioxide. This reacts with water to produce carbonic acid, which, although classed as a weak acid, is produced in quantity in the body.

$$CO_2 + H_2O \rightarrow H_2CO_3 \rightarrow H^+ + HCO_3^-$$

To remove this source of H^+, one must eliminate the CO_2 leading to its formation.

PRODUCTION OF ORGANIC ACIDS from the incomplete combustion of carbohydrates, fats, and protein. Lactic acid, pyruvic acid, and fatty acids are examples of such weak acids, all of which can liberate H^+ from their carboxyl groups.

Although not a large source of H^+, MEDICATION with ammonium salts or other acidic materials may increase $[H^+]$. For instance, ammonium chloride is a common constituent of materials that increase mucus secretion in the respiratory system and therefore act as expectorants facilitating expulsion of sputum. After absorption, the ammonium portion of the molecule is converted to urea in the liver with production of H^+. This H^+ is normally buffered by the bicarbonate system. Ingestion of aspirin (salicylic acid) is another source of H^+.

Methods of controlling free hydrogen ion

BUFFERING. If a weak acid and its completely ionized salt, or conjugate base, are present in a solution, the addition of H^+ to the solution will result in little change of pH. Such a pair of substances constitutes a "buffer pair" or "buffer system," and it is capable of buffering or removing free H^+ from the solution. The operation of such a system may be illustrated by the buffer system of acetic acid (HAc) and sodium acetate (NaAc) in a solution:

$$HAc \rightarrow H^+ + Ac^- \text{ (weak acid)}$$

$$NaAc \rightarrow Na^+ + Ac^- \text{ (conjugate base)}$$

If H^+ is added in quantity to the solution, it reacts with Ac^- supplied in excess by NaAc to form the slightly dissociated HAc.

$$H^+ + Na^+ + Ac^- \rightarrow HAc + Na^+$$

The body has several buffer systems that operate in the fashion indicated; for example:

Carbonic acid and sodium bicarbonate.

$$H_2CO_3 \rightleftharpoons H^+ + HCO_3^- \text{ (weak acid)}$$

$$NaHCO_3 \longrightarrow Na^+ + HCO_3^- \text{ (conjugate base)}$$

Addition of H^+ to this system results in formation of H_2CO_3.

Proteins with H^+ and proteins with Na^+ or K^+. The carboxyl group of the amino acid can lose H^+ and acts as a weak acid.

$$R - COOH \rightleftharpoons R - COO^- + H^+ \text{ (weak acid)}$$

$$R - COONA \longrightarrow R - COO^- + Na^+ \text{ (conjugate base)}$$

Addition of H^+ to this system increases formation of $R - COOH$.

Phosphate systems, with K^+ or Na^+.

$$\text{(K) or } NaH_2PO_4 \rightleftharpoons NaHPO_4^- + H^+ \text{ (weak acid)}$$

$$\text{(K) or } Na_2HPO_4 \longrightarrow NaHPO_4^- + Na^+ \text{ (conjugate Base)}$$

As above, H^+ addition results in formation of more weak acid.

The carbonic acid and sodium bicarbonate system forms the primary buffer system of the extracellular fluid. The body normally maintains a ratio of 1 molecule of H_2CO_3 to 20 molecules of $NaHCO_3$. The extra HCO_3^- constitutes the "alkali reserve" of the body and represents the extra buffering capacity of the fluid. Protein and hemoglobin systems operate mainly within body cells.

DILUTION. As H^+ is produced by cells, it is dissipated through the ECF. Although it does not remove H^+, dilution prevents local buildup of H^+.

EXCRETION OF CO_2. The excretion of CO_2 is effected by the lungs. In the lungs, an enzyme designated carbonic anhydrase increases the rate of decomposition of carbonic acid into CO_2 and H_2O. The CO_2 is then exhaled.

$$H_2CO_3 \xrightarrow[\text{anhydrase}]{\text{carbonic}} H_2O + CO_2$$

To replenish the H_2CO_3, bicarbonate ion produced from H_2CO_3 dissociation or taken in the diet recombines with H^+, thus removing H^+.

$$H^+ + HCO_3^- \rightarrow H_2CO_3$$

SECRETION OF H^+. The secretion of H^+ is effected by the kidney. By active transport, kidney tubule cells (mainly proximal and distal tubule cells) are able to move H^+ into the tubule lumen where they will ultimately be excreted in the urine.

FIGURE 21.6 Mechanisms affecting body fluid pH.

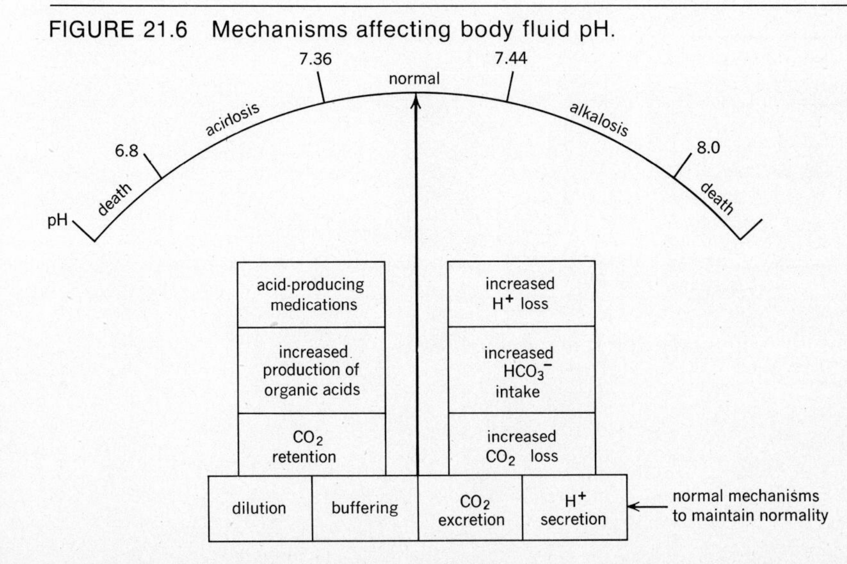

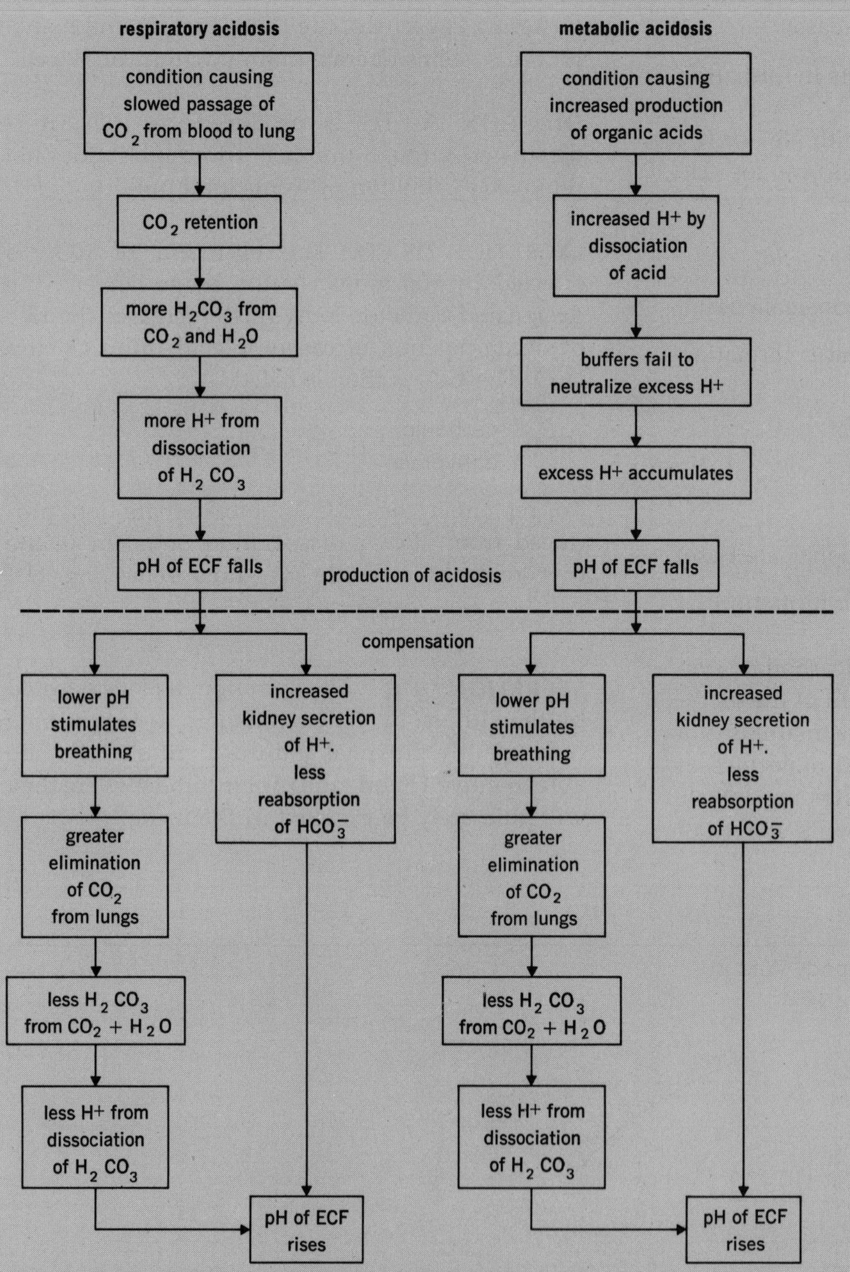

FIGURE 21.7 Production and compensation of respiratory and metabolic acidosis.

   (2) Production by metabolism.
   c. Routes of loss of water include:
   (1) "Fixed loss": lungs, perspiration, feces.
   (2) Variable loss: kidneys.

4. Most absorbed water enters the plasma.
   a. Plasma water is filtered into the interstitial fluid.
   b. Plasma water is lost through kidneys, skin, lungs, and feces.

5. Water is returned to the plasma by osmosis due to plasma proteins.

6. Interstitial water either passes osmotically into the plasma, is removed by lymph vessels, or enters cells.

7. Intracellular water passes into the interstitial fluid compartment.

8. All extracellular fluids are nearly identical in electrolyte types and concentrations but differ in protein concentration. Sodium and chloride are the primary extracellular electrolytes. Protein concentration is highest in the plasma compartment.

9. Intracellular fluid contains much more protein than extracellular fluid and contains potassium and phosphate as the primary electrolytes.

10. Regulation of volume and osmolarity of extracellular fluid is by:
   a. The thirst mechanism that results in increased water intake.
   b. Mechanical factors that adjust kidney filtration to blood volume.
   c. Hormonal factors (aldosterone, ADH, natriuretic factor) that control electrolyte and water reabsoprtion.
   d. Nervous reflexes that adjust ADH secretion to blood volume.

11. Dehydration results when fluid and electrolyte loss exceeds intake.
   a. Isotonic dehydration involves equivalent loss of fluid and electrolytes.
   b. Hypertonic dehydration involves greater fluid than electrolyte loss. Cell water is lost.
   c. Hypotonic dehydration involves greater electrolyte than fluid loss. Water enters cells.

12. Edema results from accumulation of fluid in the interstitial compartment. It occurs when the following factors are in effect:
   a. Filtration from blood vessels exceeds osmotic return.
   b. The heart fails to pump blood effectively and venous pressure increases.
   c. There is salt and water retention.
   d. Lymph vessels are blocked.

13. Measurement of certain fluid compartment volumes may be made by the dilution method.
   a. Plasma, extracellular fluid, and total water are measurable.
   b. Interstitial and intracellular fluids must be calculated.

14. Regulation of pH of body fluids ensures continuance of enzymatic reactions in the body.
   a. Hydrogen ion produced during metabolism, from the reaction of CO_2 and H_2O, tends to shift pH toward more acid values.
   b. Regulation of pH maintains pH at 7.4 ± 0.02.

15. Acids are substances that release hydrogen ions.
   a. Strong acids release much hydrogen ion.
   b. Weak acids release little hydrogen ion.

16. Bases are substances that accept free hydrogen ions and neutralize their effects.

17. Sources of hydrogen ion include:
   a. Reaction of CO_2 with H_2O to produce carbonic acid with release of hydrogen ion.
   b. Release of hydrogen ion from organic acids.
   c. Ingestion of acidic substances.

18. Methods of minimizing the effect of free hydrogen ion are:
   a. Buffering. A weak acid and its corresponding completely ionized salt are a buffer pair. Hydrogen ion added to the salt forms the weak acid.
   b. Dilution. Prevents local accumulation.
   c. CO_2 elimination by the lungs to prevent reaction with H_2O.
   d. Secretion of hydrogen ion by the kidney.
   e. Uptake by hydrogen acceptors.

19. The body has base in excess of normal need to neutralize hydrogen ion. The extra base is the alkali reserve or base excess.

20. Acid-base disturbances include acidosis ($pH < 7.38$) and alkalosis ($pH > 7.42$).
   a. Respiratory acidosis results from CO_2 retention.
   b. Metabolic acidosis results from loss of alkali, excessive intake of acids, or accumulation of organic acids.
   c. Respiratory alkalosis results from excessive CO_2 loss.
   d. Metabolic alkalosis results from loss of hydrogen ion.

21. Compensation of acidosis and alkalosis occurs through alterations in rate and depth of breathing and by changes in kidney secretion and reabsorption of H^+ and HCO_3^-.

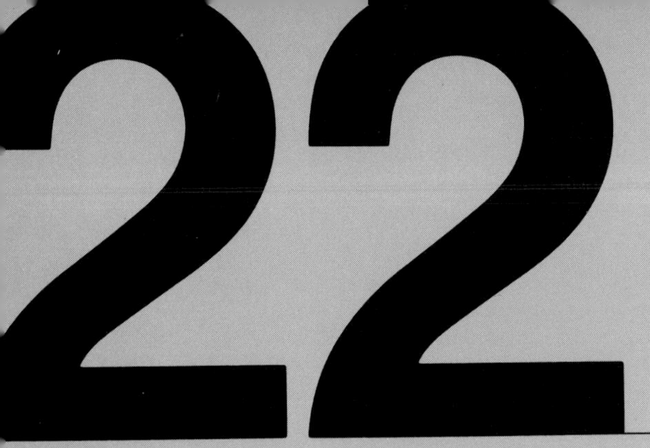

The Blood and Lymph

Blood is a connective tissue with a complex liquid intercellular material (the plasma) in which cells or cell-like structures are suspended (the formed elements). Furthermore, plasma is an extracellular fluid, the one contained within the blood vessels of the body. Blood comprises about 7 percent of the body weight, and, in the 70 kilogram (154 lb) male adult, amounts to 5–6 liters in volume. The average female adult has about 5 liters of blood. It is easy to obtain blood from the superficial veins of the body, and since it circulates to and from cells, its composition reflects cellular activity. Therefore, analysis of the blood may give a good idea of the status of body function.

Lymph is formed by filtration of the blood; it was originally interstitial or tissue fluid. When it enters lymphatic vessels, it may properly be called lymph.

Development of the blood

Development of the blood and blood vessels (angiogenesis) occurs simultaneously, beginning at 15–16 days of embryonic development. Early development is necessary because the human

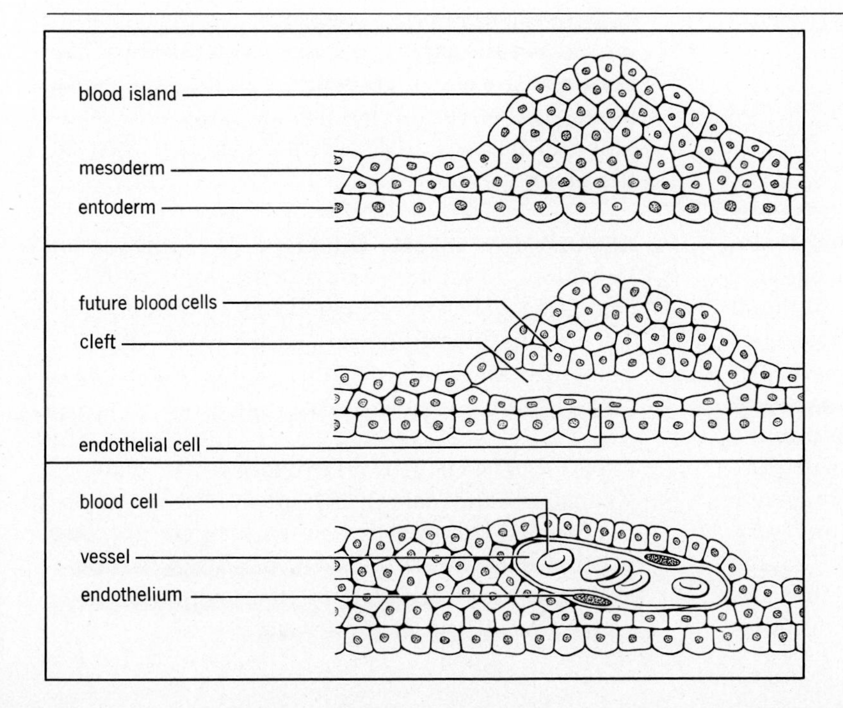

FIGURE 22.1 The development of blood islands.

zygote contains little yolk to sustain development, and there is a need to bring nutrients and oxygen from the placenta to the embryo to ensure continued differentiation and growth.

The first region to show development of vessels and blood is the extraembryonic (outside of the embryo) mesoderm of the yolk sac. Here, cells called ANGIOBLASTS come together to form masses known as BLOOD ISLANDS (Fig. 22.1). Spaces appear within these islands, and the cells at the periphery of the space flatten to form the endothelial lining of what is now a primitive blood vessel. Budding of the endothelial cells extends the network of vessels. As the cavities form and their linings develop, endothelial cell division creates primitive STEM CELLS within the vessels. These give rise first to cells resembling nucleated red blood cells that are capable of transporting oxygen and carbon dioxide. Plasma is secreted by these same endothelial cells. At about two months of age, the embryo begins to produce its own blood cells, first in the liver and later in the spleen, bone marrow, and lymph nodes. At birth, only the bone marrow and lymph organs remain as sites of blood production. Actual circulation of blood through the embryo has begun by the end of the third week of development.

Functions of the blood

Functions of the blood may be considered to center around two main activities: TRANSPORT and REGULATION OF BODY HOMEOSTASIS. The latter is in turn concerned with regulation of water content (particularly of the interstitial fluid compartment), pH (buffering), temperature, protection against blood loss, and protection from invasion by foreign chemicals and microorganisms.

Transport

Because it has a liquid intercellular material, the blood can DISSOLVE and/or SUSPEND many materials and carry them to and from the cells as it is circulated by heart action. The cells of the blood may BIND blood gases (red cells) and hormones (white cells) for transport. The plasma contains amino acids, simple sugars, fats, vitamins, and other nutrients that have been picked up from the gut and carried to body cells in the plasma. The plasma carries metabolic wastes to appropriate organs of excretion; it carries regulatory substances such as enzymes—some unique to the blood, others that are being carried from an organ of production to a site of action—and hormones that are not attached to blood cells.

Regulation of homeostasis

REGULATION OF TISSUE WATER CONTENT. Fluids in the interstitial and intracellular body water compartments are ultimately derived from the blood. The hydrostatic pressure of the blood causes FILTRATION of materials from the capillaries, and the plasma protein content causes the OSMOTIC RETURN of interstitial fluid to the capillaries. Therefore, the passive exchange and regulation of tissue water content depends in part on the blood composition.

REGULATION OF pH. The blood has been shown to contain a number of BUFFER SYSTEMS that resist change in pH. Arterial blood has a normal pH of 7.4; venous blood has a normal pH of 7.36. The inorganic salts and proteins of the plasma and cells play a major role in this regulation.

REGULATION OF BODY TEMPERATURE. Water is a substance that can absorb much heat with relatively small changes in its own temperature. The plasma water TAKES UP HEAT produced by metabolic processes and carries it to skin and lungs where excess heat is eliminated.

PROTECTION. If a blood vessel is damaged, blood loss (hemorrhage) may occur with potentially harmful consequences. By its ability to COAGULATE or clot, the blood aids in protecting the body against loss of blood. These reactions involve both the plasma and formed elements. Several of the white cells of the blood are good PHAGOCYTES, aiding in the removal of the debris of injury and engulfing microorganisms and particles. Some white cells also produce or contain chemicals known as ANTIBODIES that can neutralize foreign chemicals (antigens) that may enter the body.

All of these functions of the blood should emphasize its essential role in total body function.

Components of the blood

The liquid portion of the blood is known as the plasma, and the cells or cell-like units suspended within the plasma are the formed elements.

Plasma

Plasma is the straw-colored liquid portion of the blood composing 55–57 percent of the volume of normal adult blood. It contains an amazing variety of substances, and in general resembles cytoplasm in terms of the major categories of substances it contains. The major plasma constituents are presented in Table 22.1

It should be noted that plasma proteins are characteristic of the blood and are involved in control of plasma protein osmotic pressure and antibody activity.

SOURCES OF PLASMA COMPONENTS. Water and electrolytes of the plasma are derived almost entirely by ABSORPTION from the alimentary tract. The albumins, alpha and beta globulins, and fibrinogen are produced by the liver. Gamma globulins are PRODUCED BY PLASMA CELLS (see Chapter 23) and by the disintegration of white cells in the bloodstream. Large organic molecules such as carbohydrates, lipids, and enzymes are absorbed from the gut or produced by various body organs (many from the liver). Wastes are contributed by all body cells as they metabolize substances.

Formed elements

Three categories of formed elements are found in normal human peripheral blood:

ERYTHROCYTES, or red blood cells.
LEUCOCYTES, or white blood cells.
PLATELETS, or thrombocytes.

ERYTHROCYTES. Mature human erythrocytes (Fig. 22.2) are biconcave, nonnucleated disks averaging (in the bloodstream) 8.5 μm in diameter, 2 μm thick on the edges, and 1 μm thick in the center. Normal erythrocytes are remarkably constant in size and may be used to estimate the sizes of other blood cells. The shape of the cell permits the greatest possible surface-volume ratio, and rapid diffusion of gases through the membrane will occur. Although lacking a nucleus, the mature cells show a low level of metabolic activity (O_2, ATP and glucose consumption, CO_2 production) associated with membrane transport systems and catabolism of materials.

Numbers. Erythrocytes are the most numerous of the formed elements. Some values at different ages are presented in Table 22.2. It may be emphasized that a newborn has more cells than anyone else, having lived in a low-oxygen environment (the uterus). Adult males, because of their generally larger body size and muscular development, require more cells to carry the oxygen the body needs.

TABLE 22.1 Major constituents of plasma

Component	Amount	Comments
Water	90% of plasma	Dissolves, suspends, causes ionization, carries heat
Electrolytes (meq/l)	About 1%	Create osmotic pressure, buffer, irritability in all tissues
Sodium	136–145	
Potassium	3.5–5.0	
Calcium	4.5–5.0	
Magnesium	1.5–2.5	
Chloride	100–106	
Bicarbonate	26–28	
Phosphate	2	
Protein	17	
Other	6	
Nitrogenous substances (mg/100 ml blood)		Nutrients or wastes
Nonprotein nitrogen (NPN)	33	
Blood urea nitrogen (BUN)	8–25	
Creatinine	0.7–1.5	
Amino acids	0.13–3.0	Amount depends on each acid, 26 have been found in plasma
Plasma proteins	6–8% of plasma	As a group, contribute to blood viscosity, reserve of amino acids, clotting, antibodies
Albumins	4.5%, 4–5 g%	Oncotic pressure (plasma colloidal osmotic pressure)
Globulins	2%	
Alpha	510–980 mg%	Serve general functions of plasma proteins
Beta	550–1010 mg%	
Gamma	1200–2150 mg%	Antibodies
A	140–260 mg%	Most common in secretions, general lytic
M	70–130 mg%	First to appear
G	700–1450 mg%	Natural and acquired antibodies
D	300 mg%	Function unknown
E	100 mg%	Reagins, incomplete antibodies as those in allergies
Fibrinogen	0.3%; 0.15–0.30 g%	Clotting
Lipids (mg/100 ml blood)		
Fatty acids	190–420	Usually in transit or fuels
Cholesterol	150–280	
Triglycerides	About 20	
Glucose (mg/100 ml blood)	60–110	Preferred fuel source for cell activity
Enzymes	A wide variety acting on all categories of materials. Some increase with pathological conditions in the body, and are thus aids in diagnosis. Lactic dehydrogenase (LDH) and serum glutamic-oxaloacetic transaminase (SGOT) increase in heart damage	
Vitamins A, D, tocopherol, thiamine, riboflavin, B_6, nicotinic acid, B_{12}, folic acid, C, and others		

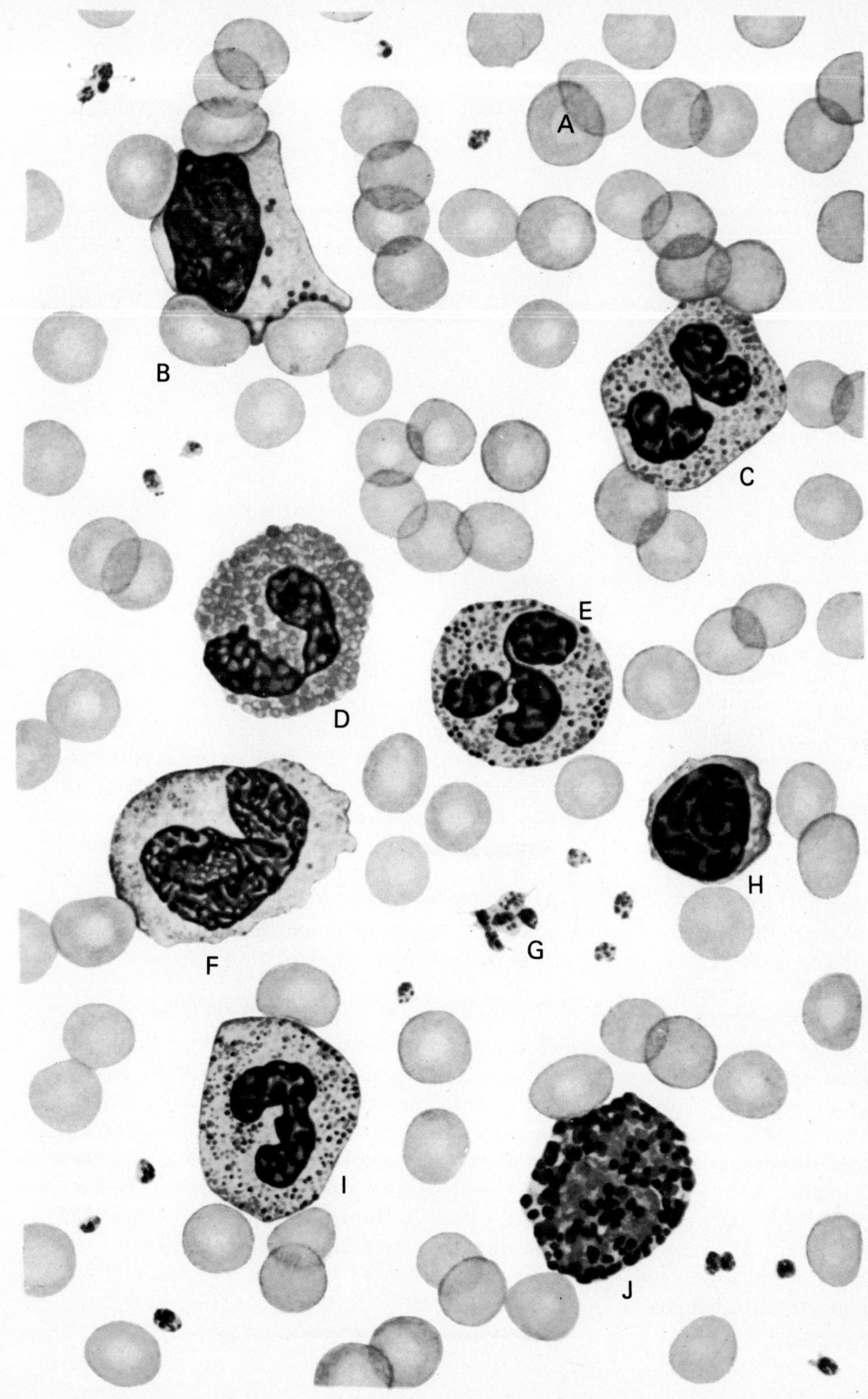

FIGURE 22.2 The morphology of the mature blood elements. (Courtesy L. W. Diggs, M.D. From L. W. Diggs, D. Strum, A. Bell, "Morphology of Blood Elements," Abbott Laboratories, 1954.)

Legend key.
CELL TYPES FOUND IN SMEARS OF PERIPHERAL BLOOD FROM NORMAL INDIVIDUALS. The arrangement is arbitrary and the number of leukocytes in relation to erythrocytes and thrombocytes is greater than would occur in an actual microscopic field.

A Erythrocytes
B Large lymphocyte with azurophilic granules and deeply indented by adjacent erythrocytes
C Neutrophilic segmented
D Eosinophil
E Neutrophilic segmented
F Monocyte with blue gray cytoplasm, coarse linear chromatin and blunt pseudopods
G Thrombocytes
H Lymphocyte
I Neutrophilic band
J Basophil

TABLE 22.2 Numbers of erythrocytes by ages

Age	Mean, million per/cu mm	Comments
Newborn	6.1	Life *in utero* is life at a low oxygen concentration, thus more red cells are present. At birth, a loss of red cells occurs with breathing of higher O_2 levels. An increase in number occurs at puberty. Males have more muscle than females, and thus have a greater oxygen demand, hence more cells.
1 day	5.6	
3 weeks	4.9	
2 months	4.5	
1 year	4.6	
10 years	4.8	
Adult male	5.4	
Adult female	4.8	

Production. As stated in the section on blood development, during intrauterine life production of red cells occurs in the yolk sac, spleen, liver, lymph nodes, and red bone marrow. At or shortly before birth, all areas except the red bone marrow cease production of red cells. Primitive cells remain in the liver and spleen, however, and may, in certain pathological states, resume limited production of red cells. The nature of the stem cell that gives rise to erythrocytes (and indeed all blood cells) is still disputed, but a series of stages is seen (Fig. 22.3) as the cells mature. The respiratory pigment, hemoglobin, increases in concentration as the cell matures. Rates of production have been estimated to be 5–10 million per second.

Life history. Mature elements are released from the marrow looking something like lumpy balls of dough. They assume their normal contours and enter blood vessels where they circulate for 90–120 days. As they circulate through the body, they are subjected to a lot of wear and tear, and they apparently undergo minor chemical changes as they age. "Old cells" are phagocytosed by cells of the liver, spleen, and marrow. Since the number of red cells per cubic millimeter of blood does not normally change very much, rates of destruction are balanced to production—that is, 5–10 million cells per second. The materials of the erythrocyte are largely conserved and are "recycled" into the production of new cells.

Hemoglobin (Fig. 22.4). Hemoglobin is a compound known as a respiratory pigment that is capable of combining with several gases. It forms a loose and reversible chemical combination with oxygen according to the equation

hemoglobin + oxygen ⇆ oxyhemoglobin

If hemoglobin were not present in the blood cells, 100 ml of blood could carry only about 1/3 ml of oxygen; with hemoglobin present, some 20 ml of oxygen can be carried per 100 ml of blood. While hemoglobin can combine with other gases, such as CO_2 and CO, its primary function is to bind and transport O_2 from the lungs to the body cells. The synthesis of hemoglobin (Fig. 22.5) requires a variety of substances, some of which are shown in Table 22.3.

IRON is the heavy metal that actually binds the

oxygen molecules so that they may be carried to cells. The normal male adult requires about 1/2 mg of iron in his diet per day to meet requirements for hemoglobin synthesis. A menstruating female may require 2 mg per day to meet her requirements. Absorption is active from the intestine, and the iron combines with a beta-globulin in the blood called transferrin, and in this form is carried to the liver. In the liver, iron is stored by being combined with a cellular protein called apoferritin, and a compound called ferritin is formed. The metabolism of iron is summarized in Fig. 22.6. If iron intake exceeds need, some tissues will form a substance called hemosiderin to "store" the excess iron. Its presence in the tissues causes hemochromatosis.

ERYTHROPOIETIN is a substance that is synthesized by the kidney in response to lowered oxygen levels in the bloodstream. It stimulates erythrocyte production and can cause an increase of about 2 percent of the total erythrocyte mass.

Destruction of the erythrocyte and the fate of its components. Aged erythrocytes are phagocytosed by macrophages in the liver, spleen, and bone marrow. The heme is split from the globin portion of the molecule. The globin is degraded to its constituent amino acids that are reutilized by body cells, and the heme is converted to biliverdin. The iron is removed and largely conserved for resynthesis into new hemoglobin. Biliverdin undergoes reduction to BILIRUBIN and is released into the plasma. Bilirubin combines with albumin, is absorbed by liver cells, and is conjugated with glucuronic acid to form a bile salt. It is next secreted into the small intestine, acted upon by bacteria, and converted to UROBILINOGEN. Some of the urobilinogen remains within the tract and imparts the orange-brown color to the feces. The remainder is reabsorbed by the intestine into the bloodstream and is executed by the kidney as the amber coloring matter of the urine. These relationships are summarized in Fig. 22.7.

Clinical tests for hepatic function often include a determination of plasma bilirubin levels. When total plasma bilirubin content is greater than 2 mg per 100 ml, the scleras, mucous membranes, and skin may become yellowed, and jaundice or icterus is said to have developed. Several causes for elevated bilirubin levels exist, including:

TABLE 22.3 Requirements for erythrocyte production

Substance	Description and/or use
Lipid	Cholesterol and phospholipids; incorporated in membrane and stroma
Protein	Incorporated into cell membrane
Iron	Incorporated into hemoglobin
Amino acids	Incorporated into hemoglobin
Erythropoietin	A glycoprotein, it is released from the kidney with hypoxia, hemorrhage, and excessive androgen secretion, and stimulates production of erythrocytes
Vitamin B_{12}	Used in the formation of DNA in nuclear maturation
Intrinsic factor	A mucopolysaccharide produced by the stomach. It combines with vitamin B_{12} and insures absorption of the vitamin from the gut
Pyridoxin	Increases the rate of cell division
Copper	Catalyst for hemoglobin formation
Cobalt	Aids synthesis of hemoglobin
Folic Acid	Promotes DNA synthesis

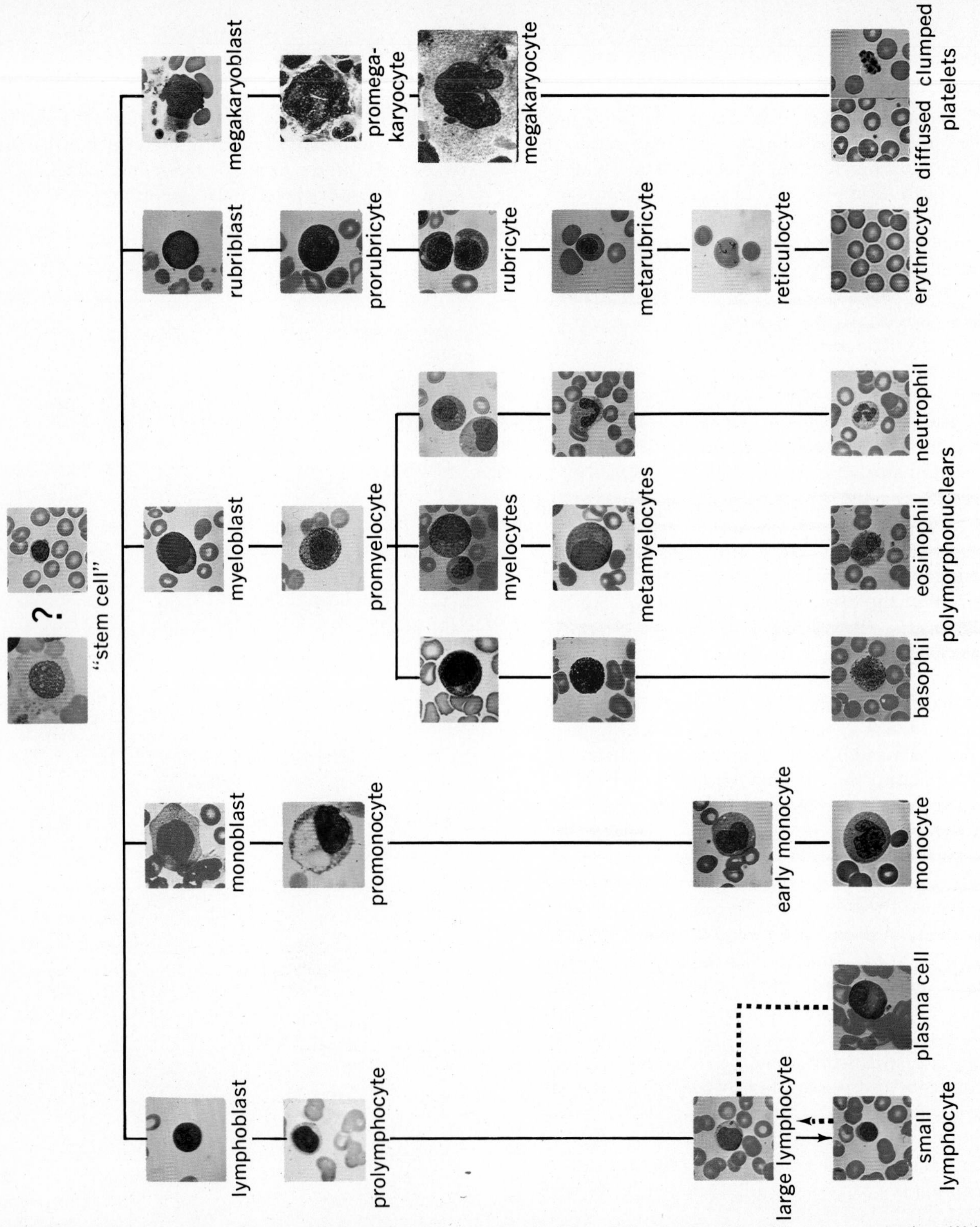

FIGURE 22.3 The maturation of blood cells. The exact morphology of the stem cell is still open to question. Wright stain, × 1200. (Courtesy American Society of Hematology National Slide Bank and Health Sciences Learning Resources Center, University of Washington, used with permission.)

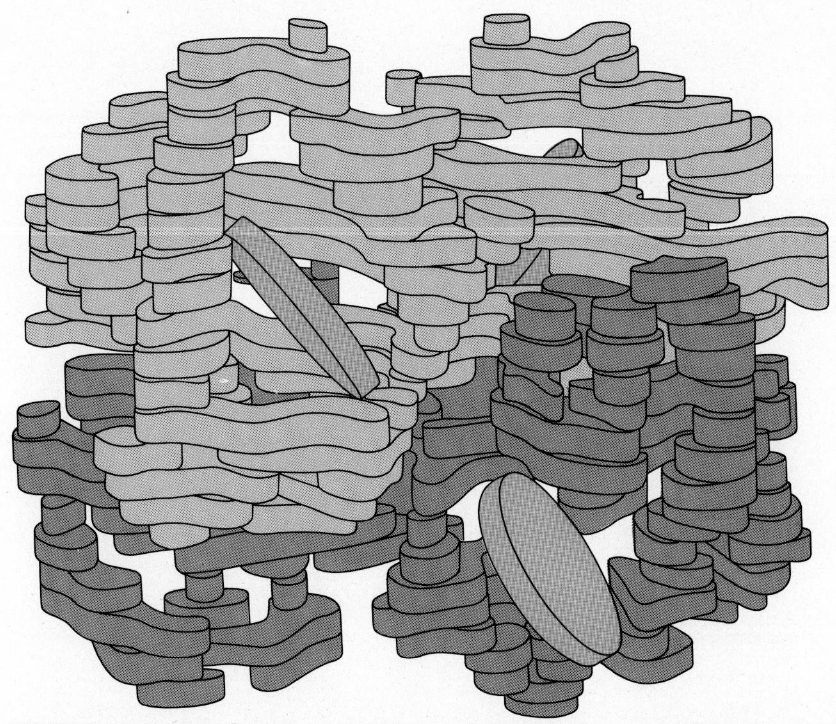

FIGURE 22.4 Hemoglobin molecule, as deduced from x-ray diffraction studies, is shown from the side. The irregular blocks represent electron-density patterns at various levels in the hemoglobin molecule. The molecule is built up from four subunits: two identical alpha chains (light gray blocks) and two identical beta chains (dark gray blocks). Each chain enfolds a heme group (colored disk), the iron-containing structure that binds oxygen to the molecule. (From "The Hemoglobin Molecule" by M. F. Perutz. Copyright © November 1964 by Scientific American, Inc. All rights reserved.)

EXCESSIVE PRODUCTION of bilirubin that overloads the liver cells, as in excessive destruction of erythrocytes (e.g., sickle cell anemia, malaria).

DEFECTIVE LIVER UPTAKE of the pigment (e.g., infectious hepatitis).

OBSTRUCTION of the common bile duct (e.g., stones) that blocks secretion into the small intestine and causes entry of excessive bilirubin into the bloodstream. Small stones may form in the gall bladder and leave it to become lodged in the common bile duct to the small intestine. Blockage causes a greater absorption of bilirubin (bile) into the bloodstream from the hepatic cells, since the normal route of excretion (bile ducts) is plugged.

Other functions served by the erythrocytes. Hemoglobin (a conjugated protein) is a good BUFFERING compound, and the erythrocytes also contain potassium bicarbonate for buffering purposes. The units also contain an enzyme, CARBONIC ANHYDRASE, that speeds the conversion of carbonic acid to carbon dioxide and water. Transport of CO_2 across membranes is thus speeded by the action of the enzyme that ensures a high-diffusion gradient.

Clinical considerations. Abnormalities in erythrocyte formation include the processes that result in ABNORMAL SHAPE of cells (poikilocytosis), ABNORMAL SIZE of cells (anisocytosis), or

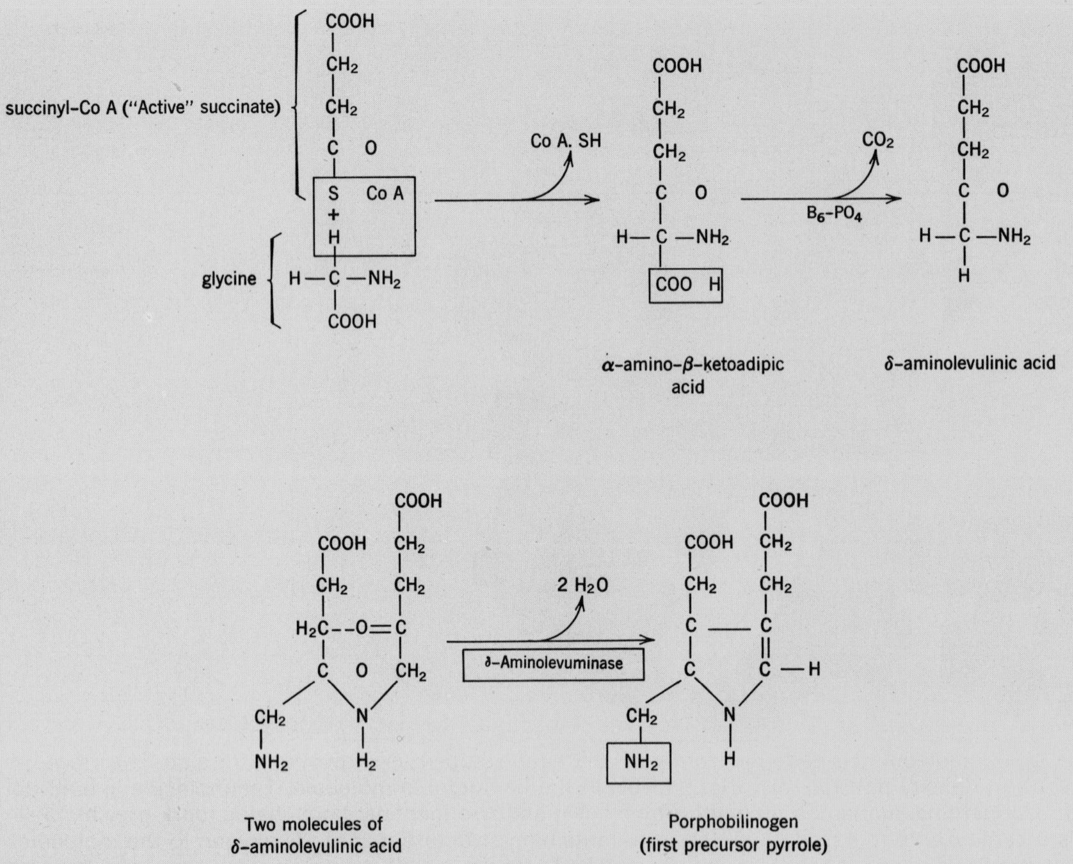

succinyl-Co A ("Active" succinate)

glycine

Co A. SH

CO_2

B_6-PO_4

α-amino-β-ketoadipic acid

δ-aminolevulinic acid

2 H_2O

δ-Aminolevuminase

Two molecules of δ-aminolevulinic acid

Porphobilinogen (first precursor pyrrole)

FIGURE 22.5 The steps in the synthesis of hemoglobin.

ABNORMAL NUMBERS of cells (polycythemia or anemia). Hemorrhage, heavy exercise, and exposure may result in a release into the bloodstream of immature erythrocytes that are larger than normal and that may contain nuclei or fragments of nuclei. Table 22.4 summarizes some of the common types and causes of anemia. Polycythemia may be a response to decreased oxygen content in the air that is breathed. This is caused by release of erythropoietin and is called a physiological polycythemia. Pathological polycythemia results from tumorlike conditions in the bone marrow that vastly overproduces erythrocytes. The blood flows very sluggishly because of the great increase in viscosity resulting from the presence of increased number of cells. Extraction of oxygen from the hemoglobin is thus greater, and the person typically shows a bluish color of the skin (cyanosis).

LEUCOCYTES. Leucocytes are a heterogeneous population of nucleated cells that do not contain hemoglobin. There are five varieties, as shown in Fig. 22.2. The morphology, characteristics, and

uroporphyrinogen III

6H

auto-oxidation in light

uroporphyrin III

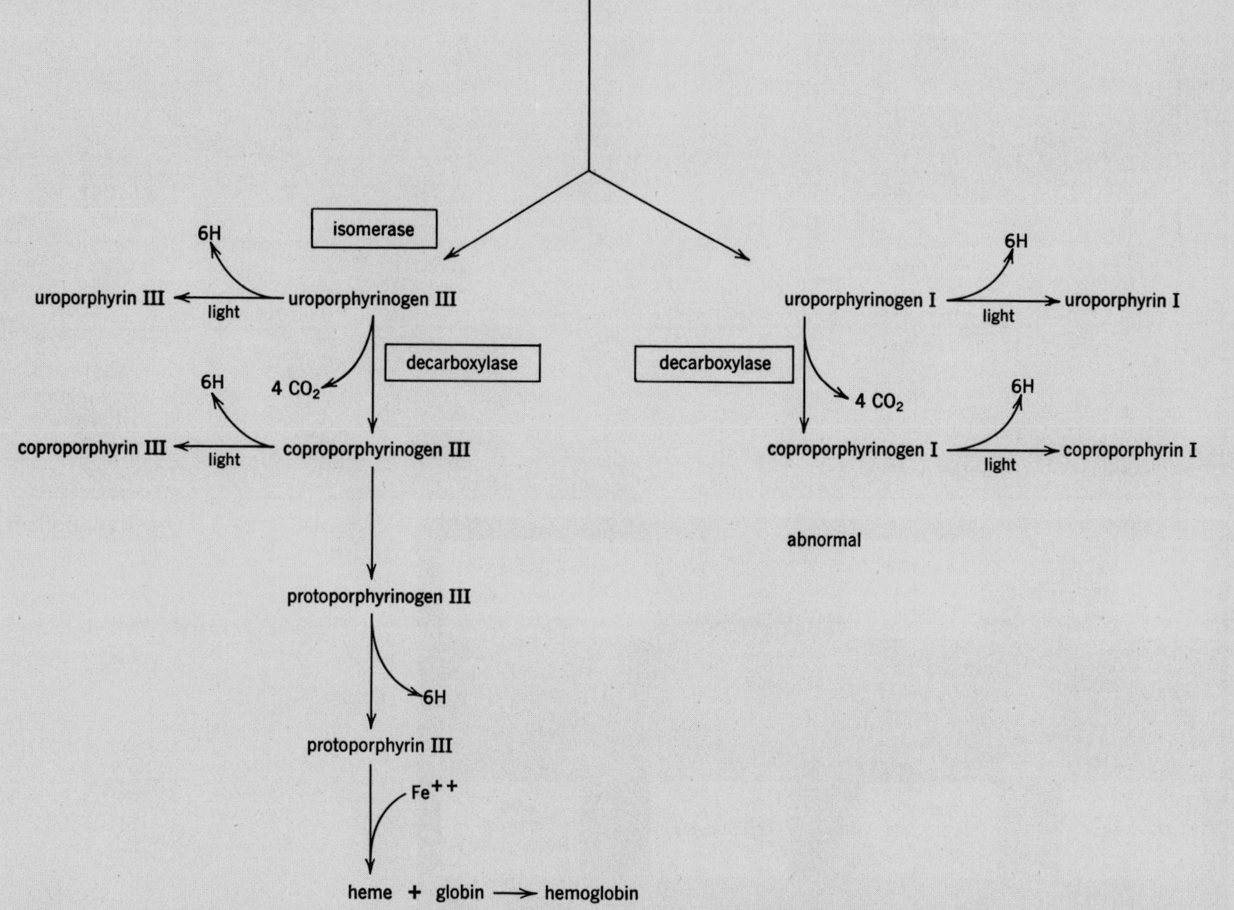

TABLE 22.4 Some common anemias

	Type of anemia	Causes	Characteristics	Symptoms	Treatment
INCREASED LOSS	*Hemorrhagic* Acute	Trauma Stomach ulcers Bleeding from wounds	Cells normal	Shock	Transfusion
	Chronic (iron deficiency)	Stomach ulcers Excessive menses	Cells small (microcytes), deficient in hemoglobin content	None or fatigue	Iron administration
	Hemolytic	Defective cells Destruction by parasites, toxins, antibodies	Young cells (reticulocytes) very prominent serum haptoglobin* reduced. Morphology of cells may be abnormal when detected.	None or fatigue	Dependent on cause
DECREASED FORMATION	*Deficiency states* Folic acid	Nutritional deficiency	Cells large (macrocytes), with normal hemoglobin content	None or fatigue	Folic acid
	Vitamin B_{12}	Lack of intrinsic factor in stomach (pernicious anemia)	Cells large with normal hemoglobin content		Administration of vitamin B_{12}
	Hypoplastic or *Aplastic*	Radiation Chemicals Medications	Bone marrow hypoplastic	None or fatigue	Transfusion Androgens Cortisone

*A protein strongly binding hemoglobin in plasma.

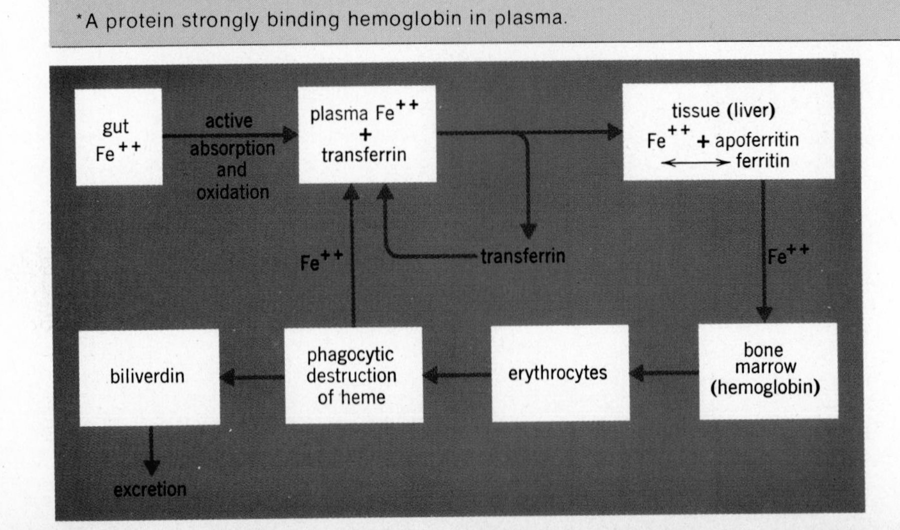

FIGURE 22.6 The metabolism of iron.

origins of these cells are described in Table 22.5.

Numbers. Leucocytes are the least numerous of the formed elements; numbers may vary physiologically with age and state of the body. Some values of the cells according to age are presented in Table 22.6. Leucocytes are counted in the differential count. In this procedure, 100 stained leucocytes are counted on a slide and numbers of each type are recorded. The final number in each category is then its percent of total leucocyte number.

Life history. Granular (myeloid or bone marrow origin) leucocytes include NEUTROPHILS, EOSINOPHILS, and BASOPHILS. They are also known as granulocytes or polymorphonuclear (PMN) leucocytes and are shed into the bloodstream and remain functional for 12 hours to 3 days. When liberated into the bloodstream, about 75 percent possess two or three lobes in their nucleus. As they age, the number of lobes increases. The classification of myeloid leucocytes may be tabulated in series on a page with the younger (fewer lobes) leucocytes on the left and the older leucocytes (more lobes) on the right. The term "shift to the left" implies a greater proportion of younger forms in the bloodstream, and a "shift to the right" implies a greater proportion of older cells. The relative proportions may suggest the presence of abnormal marrow activity.

Nongranular (lymphoid) leucocytes (that is, LYMPHOCYTES and MONOCYTES) have been estimated to remain functional for 100–300 days under normal body conditions. Monocytes circulate as such for about 3 days, then leave the bloodstream and become the macrophages (large, more or less fixed phagocytic cells) of the liver, spleen, loose connective tissue, and other smaller body areas.

The normal production of both categories of cells appears to depend primarily on the levels of adrenal steroid hormones and whether or not infection is present in the body. Steroid hormones increase the production of neutrophils, while products of infection tend to increase both neutrophils and lymphocytes. Additionally, a leucocyte-mobilizing factor is released by injured cells. The cells are lost from the circulation by migration through the walls of the alimentary tract, by destruction (pus formation) during infections, or by disintegration in the bloodstream. Mononucleosis, a debilitating infectious disease thought to be viral in origin, results in increase of abnormal mononuclear cells (lymphocytes and monocytes) in the bloodstream, fatigue, fever, and the appearance of an abnormal antibody in the bloodstream (Paul-Bunnel antibody).

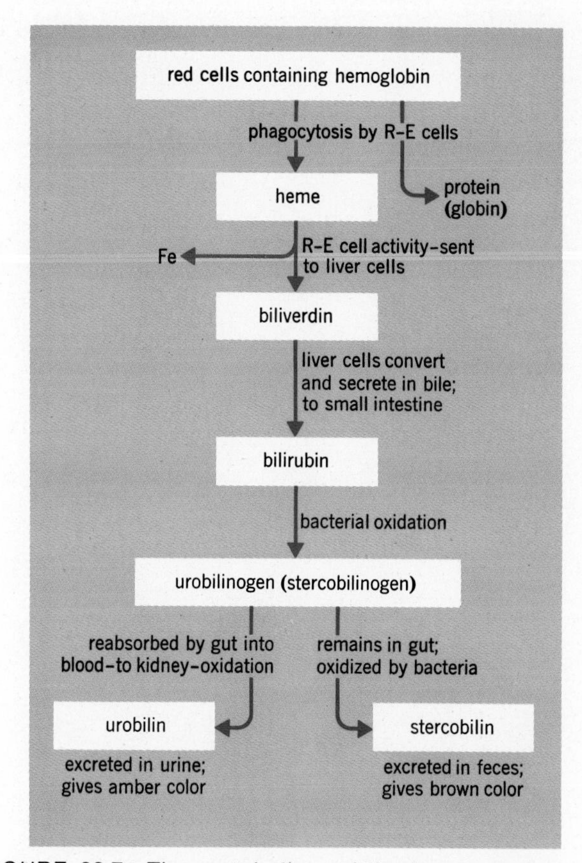

FIGURE 22.7 The metabolism of the heme portion of the hemoglobin molecule.

Requirements for production. Materials required for leucocyte production are the same as

TABLE 22.5 Summary of formed elements

Element	Normal numbers	Origin (area of)	Diameter (micron)	Morphology	Function(s)
Erythrocytes	4.5-5.5 million /mm	Myeloid (marrow)	8.5 (fresh) 7.5 (dry smear)	Biconcave, non-nucleated disc; flexible	Transports O_2 and CO_2 by presence of hemoglobin; buffering
Leucocytes	6000-9000/mm^3		9-25		
Neutrophil	60-70% of total	Myeloid	12-14	Lobed nucleus, fine heterophilic specific granules	Phagocytosis of particles, wound healing. Granules contain peroxidase for destruction of microorganisms.
Eosinophil	2-4% of total	Myeloid	12	Lobed nucleus; large, shiny red or yellow specific granules	Detoxification of foreign proteins? Granules contain peroxidases, oxidases, trypsin, phosphatases. Numbers increase in autoimmune states, allergy, and in parasitic infection (schistosomiasis, trichinosis, strongyloidiasis).
Basophil	0.15% of total	Myeloid	9	Obscure nucleus; large, dull, purple specific granules	Control viscosity of connective tissue ground substance? Granules contain heparin (liquefies ground substance) serotonin (vasoconstrictor), histamine (vasodilator)
Lymphocyte	20-25% of total	Lymphoid			
Small			9	Nearly round nucleus filling cell, cytoplasm clear staining	Phagocytosis of particles, globulin production
Large			12-14	Nucleus nearly round, more cytoplasm	
Monocyte	3-8% of total	Lymphoid	20-25	Nucleus kidney or horseshoe-shaped, cytoplasm looks dirty	Phagocytosis, globulin production
Platelets (Thrombocytes)	250,000-350,000 mm^3	Myeloid	2-4	Chromomere and hyalomere	Clotting

for cells in general, but especially important is folic acid. This substance is required for the normal production of the purine and pyrimidine components of nucleic acids as the cells mature.

Functions of the leucocytes. All leucocytes possess, to some degree, four basic properties that relate to their functions in the body.

AMEBOID MOTION. By production of cytoplasmic extensions or pseudopods, leucocytes are capable of independent movement through the tissues. Neutrophils, lymphocytes, and monocytes possess this property to the greatest degree.

CHEMOTAXIS. The cells are attracted to (positive chemotaxis) or repelled from (negative chemotaxis) areas of injury or inflammation. A specific polypeptide (leucotaxine), nucleic acids, and positively charged particles are strong positive chemotaxic agents. Negatively charged particles repel most leucocytes.

PHAGOCYTOSIS. The ability to engulf and digest or kill bacteria and products of cell death is best developed in the neutrophils, lymphocytes, and monocytes. The latter two cells may also migrate into the tissues and become macrophages, larger cells with an increased capacity for phagocytosis.

DIAPEDESIS. The ability of leucocytes to pass through the walls of capillaries to reach an area of inflammation of infection.

The functions of the cells thus relate to defense against bacteria, foreign particulate matter, and the removal of the debris resulting from injury. Functions are shown in Table 22.5.

Clinical considerations. LEUKEMIA is regarded as a neoplastic disease of the blood-forming tissues that causes unrestricted production of mature leucocytes and their precursor cells. Numbers may reach 250,000–500,000/mm^3. Unlimited production of white cells depresses erythrocyte and platelet formation and may lead to anemia and a tendency to bleed.

TABLE 22.6 Numbers of leucocytes by ages

Age	Number of cells (per cu mm)
Newborn	10,000–45,000
3 days-4 years	5,000–24,000
4-7 years	6,000–15,000
8-18 years	4,500–15,000
Adult	5,000– 9,000

Causes of leukemia are not definitely known, although viruses and radiation have been shown to cause the disease. The disease occurs in acute and chronic forms. ACUTE LEUKEMIA is sudden in onset, causes anemia, bleeding is usually present, and response to treatment is usually poor. The CHRONIC FORM occurs more often in adults and may be myelocytic (leucocytes of marrow origin are increased) or lymphoid (leucocytes of lymphoid origin are increased). General weakness, fatigue, and enlargement of the spleen are common symptoms.

PLATELETS. Platelets or thrombocytes (see Fig. 22.1) originate in the bone marrow from a giant cell known as a megakaryocyte. The units are 2–4 μm in diameter and consist of a granular structure (chromomere) surrounded by a light-colored area (hyalomere). They are not nucleated and carry one of several materials essential to the CLOTTING of the blood. Their numbers are normally 250,000–300,000/mm^3 of blood.

Hemostasis

When blood vessels are damaged, a series of reactions occurs that aids in preventing blood loss through the wound. These reactions constitute HEMOSTASIS (*aima*, blood + *statikos*, standing). Three types of reactions occur.

There is NARROWING OF THE VESSELS (vasocon-

TABLE 22.7 Factors definitely implicated in blood coagulation

International committee designation	Synonyms	Origin	Location
Factor I	Fibrinogen	Liver	A plasma protein
Factor II	Prothrombin	Liver	A plasma protein
Factor III	Thromboplastin	By series of reactions in blood; also found, as such, in cells	Produced in the clotting process or released into fluids by injured cells
Factor IV	Calcium	Food and drink; from bones	As Ca^{++} in plasma
Factor V	Labile factor (accelerator globulin)	Liver	Plasma protein
Factor VI*			
Factor VII	SPCA (serum prothrombin conversion accelerator)	Liver	Plasma
Factor VIII	AHF (antihemophilic factor) AHG (antihemophilic globulin)	Liver	Plasma
Factor IX	PTC (plasma thromboplastin component)	Liver	Plasma
Factor X	Stuart-Prower factor; develops full factor III power	Liver	Plasma
Factor XI	PTA (plasma thromboplastin antecedent)	Liver	Plasma
Factor XII	Hageman factor; contact factor; initiates reaction	?	Plasma
Factor XIII	Fibrin stabilizing factor; renders fibrin insoluble in urea. (Laki-Lorand factor)	?	Plasma
Platelet Factor	Cephalin	Marrow	Platelets

*No longer considered a separate entity; considered to be identical to Factor V.

striction) in the area of the wound, which presumably causes some reduction of the blood flow in the injured area.

A PLATELET PLUG is formed across the tear in the vessel.

The blood changes from its normal fluid consistency to a gelated or semisolid state in the process of COAGULATION or clotting.

Vascular responses and the platelet plug

Vasoconstriction occurring when blood vessels are traumatized is first due to nerve impulses arriving at the vessels because of the pain associated with the injury. A reflex involving the spinal cord is responsible, but this response lasts only a few minutes. Second, the vessel muscle undergoes an intense spasm because of the direct mechanical damage to the vessels. This reaction lasts for 20–30 minutes.

Injury to the vessels also creates a roughened surface to which the platelets adhere. Several layers of platelets accumulate across a tear in a vessel and may completely seal the injury. This platelet mechanism operates continually to seal the ruptures in the small vessels of the body that occur as a normal consequence of living. If platelets are low in numbers, many tiny pinpoint hemorrhages appear on the body surface as capillaries rupture and are not sealed by platelet plugs.

Coagulation

At least 12 factors or chemical substances are required for the blood to clot. These are present in the platelets and plasma or are produced during the clotting reactions. They are shown in Table 22.7.

Two "systems" of clotting are recognized in the body. The EXTRINSIC SYSTEM results in clotting of the blood when a vessel and body cells are damaged and thromboplastin is liberated. The thromboplastin, in combination with Ca^{++}, SPCA, labile factor (V), and Stuart-Prower factor (X), creates extrinsic prothrombin activator that converts prothrombin into thrombin. The thrombin, in the presence of several accessory factors, converts soluble fibrinogen to insoluble fibrin.

The INTRINSIC SYSTEM is initiated by release of platelet factor. A rupture of a blood vessel creates a roughened surface to which platelets adhere and partially plug the break; then development of intrinsic prothrombin activator commences. The creation of the activator requires four plasma factors and the presence of platelets. In the next reaction, generation of thrombin, the prothrombin activator in the presence of several accessory factors, converts prothrombin to thrombin. In the last reaction, generation of fibrin, thrombin converts fibrinogen to fibrin. The steps in both systems are presented in Fig. 22.8.

The clotting of blood may therefore be viewed as a series of autocatalytic reactions culminating in the formation of an insoluble protein network, fibrin. A "white clot" is first formed of fibrin, which subsequently entangles platelets, white cells, and red cells. The normal color of a clot is thus due to the trapped elements. The platelets have the additional function of causing the clot to shrink (syneresis), and an even tighter bond across an injured area is formed. An internal clot is eventually destroyed by the activity of neutrophils and by an enzyme known as plasmin.

Anticoagulants

Any procedure or chemical that reduces the capability of the blood to coagulate is an ANTICOAGULANT. If one of the necessary materials in the process is removed, the process will not go to completion. Of the various materials present, Ca^{++} is probably the easiest to remove. Mixing withdrawn blood with sodium or ammonium oxalate or citrate results in an exchange of Na^{+} or NH_4^{+} for Ca^{++}. The blood has been DECALCIFIED. HEPARIN is an organic anticoagulant that may be injected into the body and that acts to decrease thromboplastin generation. It also is antithrombic and interferes with the reactions of phase II. DICOUMAROL, a

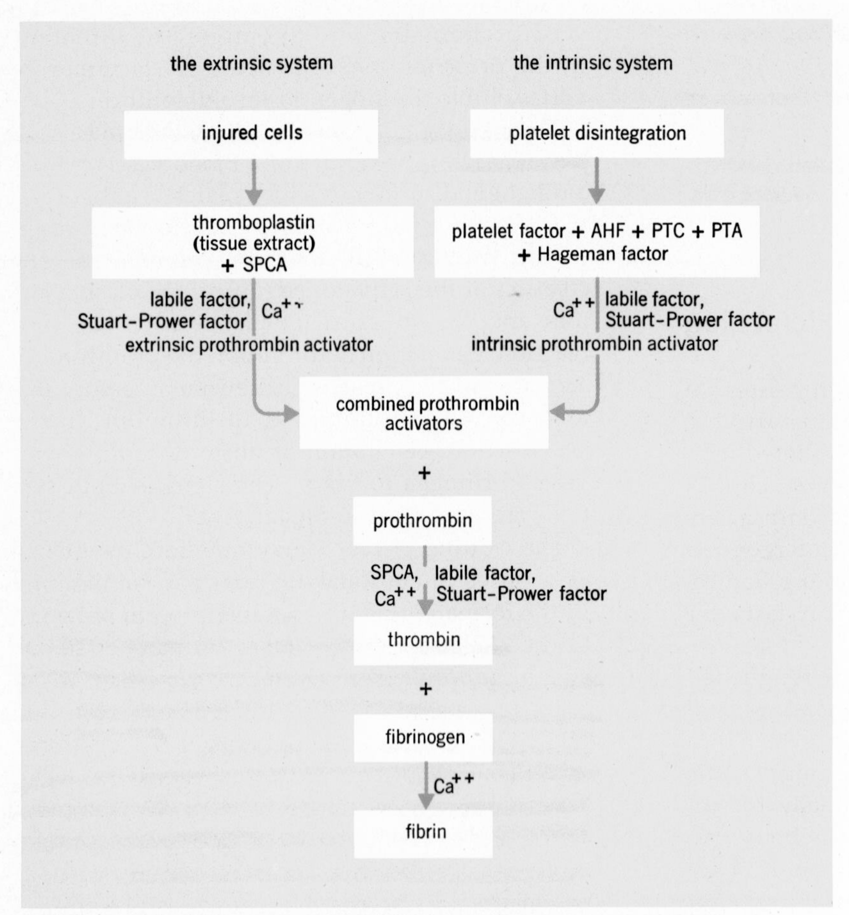

FIGURE 22.8 The extrinsic and intrinsic systems of blood clotting.

product first isolated from spoiled sweet clover, interferes with liver synthesis of prothrombin and factors V and VII. Its effect is not fully developed until 36–48 hours after intake but lasts longer than the other organic anticoagulants.

Clinical considerations

The HEMOPHILIAS are genetically determined conditions that result in failure of the blood to clot. Hemophilia A, or classical hemophilia, is mediated by a sex-linked recessive gene and results in failure to synthesize AHF. Hemophilia B, or Christmas disease, is also transmitted by a sex-linked recessive gene and results in failure of PTC synthesis. Hemophilia C is transmitted by an autosomal dominant gene and results in failure of PTA synthesis. Afibrinogenemia, or, more correctly, hypofibrinogenemia (deficiency of fibrinogen in the blood), may be either congenital or acquired following severe liver damage.

Blood may spontaneously clot within the vessels to give an INTRAVASCULAR CLOT or thrombus. Slow blood flow and the presence of roughened surfaces on a blood vessel (as in atherosclerosis) appear to predispose an individual to intravascular clotting. A thrombus may be quite serious if thrombosis of a vital vessel occurs; for example, thrombosis of coronary or cerebral vessels. The term *embolus* may refer to a floating clot that has detached from its site of formation. The danger here is that the clot may lodge in a vital vessel and result in embolism. Clots floating from the lower appendages appear particularly dangerous because they have a good chance of passing through the right side of the heart, and lodging in the lungs to create a pulmonary embolism. Surgery or trauma may result in the formation of clots in veins that may become emboli.

Lymph and lymph organs

Formation, composition, and flow of lymph

Tissue or interstitial fluid is formed by the filtration of materials from blood capillaries under the hydrostatic pressure of the heart action. All components of the blood except red cells and platelets may normally be found in tissue fluid. Upon entering the lymph vessels, the fluid is properly termed *lymph*. Concentration of all diffusible substances in lymph is nearly the same as that of plasma, while protein concentration is always less than that of plasma and is variable according to where the sample is taken. Protein concentration is always less than that of plasma since filtration of protein is slower than that of the other substances. Some representative figures for various components in blood and lymph are presented in Table 22.8.

Entry of tissue fluid, and any particles or materials within it, into lymphatic vessels is facilitated by several factors.

Lymphatics tend to run in the fibrous fascial planes of the body (Fig. 22.9). Relaxation of the tissues around vessels tends to expand them and to draw fluid into the vessels. Contraction of muscle compresses the vessels, moving the fluid onward.

Permeability of lymphatics is greater than that of blood capillaries, allowing easy penetration of protein and other large molecules. Particulate matter may enter by phagocytosis or pass directly through the walls as in wounds. The exact method of entry is unknown.

Classically, it has been taught that pressure gradients decrease from blood capillaries (15–0.7 mm Hg) to interstitial spaces (2.2–0.2 mm Hg) to lymphatic (1.2–0.2 mm Hg) vessels. Hydrostatic pressures thus force fluid into the lymphatics. Increase of venous pressure will increase lymph flow by providing more fluid for the lymphatics. Recently, it has been demonstrated that interstitial fluid pressure may be negative (−7 mm Hg). In this case, expansion and compression of the tissue "pumps" fluid into the lymph vessels.

Flow of lymph through the vessels varies, depending primarily on the level of muscular activity in the body. A typical amount of lymph drained from a medium-sized lymphatic vessel of a human would be about 1.5 ml/min at moderate activity levels. Direction of flow is controlled by valves that permit a movement of fluid only toward the subclavian veins.

Flow of lymph through the vessels depends on several factors.

MASSAGING ACTION of contracting and relaxing skeletal muscles that compress and release the lymphatics. Even during sleep, muscular movements occur to ensure lymph flow. On the other hand, standing still for long periods of time may lead to edema of the legs and feet.

HYDROSTATIC PRESSURE GRADIENT, which decreases from periphery to entry of the ducts into the subclavian veins, provides a means of causing lymph flow.

INSPIRATORY ACTIVITY expands the thorax and reduces the pressure on the ducts; they therefore

TABLE 22.8 Comparison of blood and lymph for various components (values in mg percent unless otherwise noted)

Constituent	Lymph (thoracic duct)	Plasma
Calcium	7.7	9.5
Chloride	335	335
Inorganic phosphate	3.9	4.0
Potassium	18.3	19.1
Sodium	290	290
Protein (total) g %	2.8–3.6	6.5
Albumin g %	1.6–2.4	3.5
Globulin g %	1.2	2.5
Fibrinogen g %	0.2	0.4
Amino acids	2.4	2.4
Nonprotein nitrogen	23.4	25.7
Urea	22	24
Glucose	136	135
Lipid g %	0.2–7.3 (intestinal)	0.3–0.6
Cholesterol	75	120
Leucocytes (lymph cells)/mm^3	3500–75,000	1500–1850

expand. Like a suction pump, lymph is drawn into the expanded vessel from more peripheral areas, both above and below the thorax.

The discussion so far has dealt with the function of the lymph in returning excess fluid and filtered protein to the blood-vascular system. Other functions served are associated with defense and leucocyte production.

Lymph nodes

The lymph nodes (Fig. 22.10) are ovoid or rounded structures varying in size from 1–25 mm in diameter. They are placed in the course of most medium-sized lymph vessels and thus have incoming (afferent) and outgoing (efferent) lymphatics. Fibrous tissue surrounds the node, and an internal stroma of reticular connective tissue forms the basic outline of the node. Lymph moves through the nodes by way of lymph sinuses. The sinuses contain two types of cells:

PRIMITIVE *(reticular)* CELLS that give rise to lymphocytes and/or plasma cells.

MACROPHAGES or phagocytes, members of the reticuloendothelial system, line the walls of the sinuses.

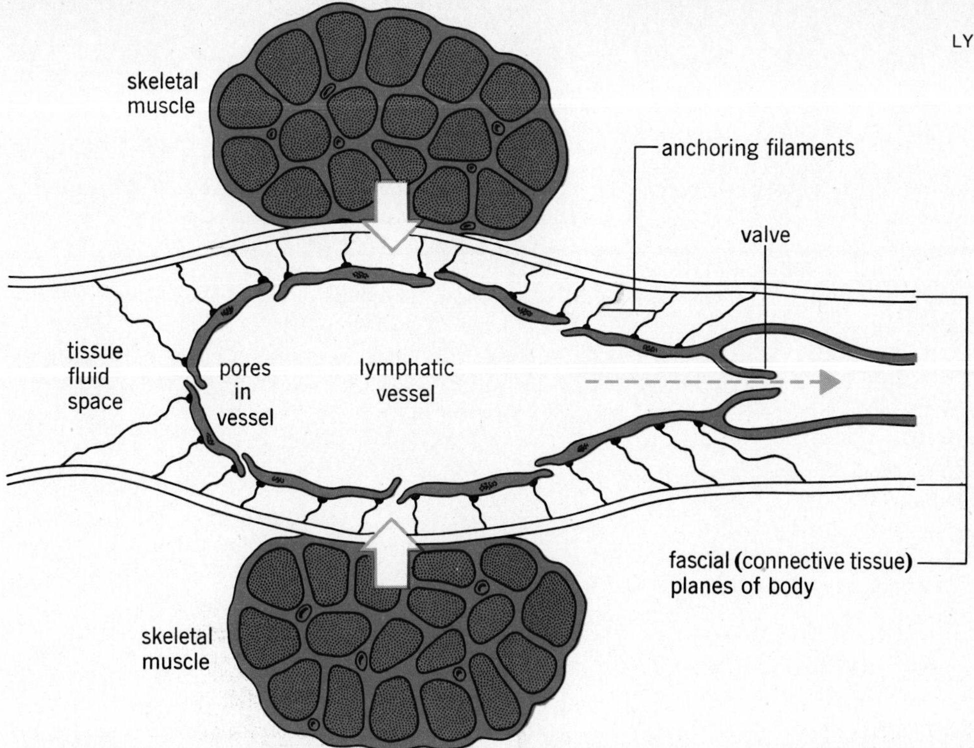

FIGURE 22.9 The relationships of lymphatic vessels to skeletal muscles and fascial planes of the body. The anchoring filaments tend to expand the vessel; muscle contractions tends to compress the vessels (↓). Lymph tends to move to the right (→) and is prevented from backflowing by the valve. A "bellows" or pumping action thus draws tissue fluid into the vessels and moves it onward.

FIGURE 22.10 A diagram of a lymph node.

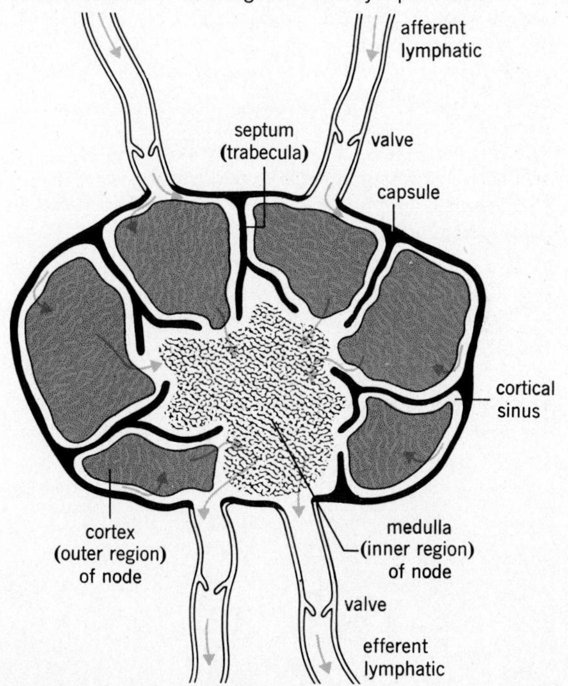

Lymphocytes are shed into the efferent lymph flow and ultimately enter the bloodstream. Plasma cells may remain in the node or may seed other areas of the body. They represent areas of antibody production. The macrophages engulf particulate matter entering the node and, in the case of a carcinoma, trap metastasizing cancer cells. Spleen, thymus, and tonsils share the lymphocyte-producing function with the nodes; the spleen is an important area of phagocytosis of aged erythrocytes.

Summary

1. The blood comprises 6–8 percent of the body weight; in an adult, the blood amounts to 5–6 liters in volume.
2. Functions of the blood include:
   a. Transport of nutrients, oxygen, wastes, hormones, and enzymes.
   b. Aiding in regulation of cell and tissue water content.
   c. Aiding in regulation of pH by buffering.
   d. Aiding in regulation of body temperature.
   e. Prevention of blood loss by clotting.
   f. Protection by cells and chemicals.
3. Blood consists of a liquid (plasma, about 55 percent) and cells (formed elements, about 45 percent).
4. The plasma has a composition similar to the cell's (Table 22.1).
   a. Plasma contains water, salts, and proteins as the materials present in greatest concentration.
   b. The proteins consist of albumins that create most of the plasma colloidal osmotic pressure, globulins, where most of the antibodies of the blood are found, and fibrinogen, an important clotting material.
   c. Most of the water and salts of the plasma are derived by absorption from the gut; the liver contributes most of the plasma proteins.
5. The formed elements are cells or parts of cells.
   a. Erythrocytes, or red blood cells, carry oxygen and carbon dioxide and aid in buffering. Erythrocytes are formed in bone marrow, live about three to four months in the bloodstream and then are phagocytosed and destroyed. The most potent stimulus to their production is reduced oxygen concentration to the kidney; this releases erythropoietin to stimulate formation. Their production requires many materials (Table 22.2). Their destruction produces bilirubin, an amber pigment that is excreted in urine and feces. They contain hemoglobin, a substance combining with oxygen and carbon dioxide. Hemoglobin contains iron as the metal enabling it to combine with gases. Anemia (Table 22.3) refers to insufficient concentration of hemoglobin in the blood.
   b. Leucocytes, or white blood cells, are ameboid, phagocytic, and aid in protecting the body against microorganisms. Leucocytes also aid in wound healing and in removing products of cellular destruction. There are five types, three originating from bone marrow, two from lymph organs. Each has a characteristic morphology (Table 22.4).
   c. Platelets are fragments of large bone marrow cells and are involved in blood clotting.
6. Hemostasis includes processes that operate protectively to prevent blood loss from injured vessels. Two processes are concerned.
   a. Vascular reactions, chiefly constriction, reduce blood flow to the injured area.
   b. Coagulation, or the change of the blood from a liquid to a gelated stage, bridges ruptures in injured vessels.
   c. Coagulation involves an intrinsic pathway that occurs within the blood and an extrinsic pathway involving injured cells.
   d. The blood may be rendered incapable of clotting by removal of an essential component of the scheme (e.g., Ca^{++}) or by organic anticoagulants that inhibit one of the steps.
   e. Genetic disorders that result in failure to produce one of the essential materials (e.g., hemophilia) may also render the blood incapable of clotting.
7. Lymph is interstitial fluid that has entered lymphatic vessels.
   a. Salts, glucose, lipids, and water are the principal constituents of lymph.
   b. Lymph enters lymphatic vessels as muscles contract and relax; lymph vessels are quite permeable to large molecules; tissue pressure also causes entry of interstitial fluid into lymph vessels.
   c. Flow of lymph through lymph vessels occurs as muscles contract and relax, "massaging" the lymph through the vessels by a pressure gradient that decreases as the veins are reached and by inspiratory activity that "draws" lymph into the thoracic cavity.
8. Lymph nodes are sites of production of lymphocytes and plasma cells. They filter or cleanse the lymph as it passes through the nodes. Phagocytes in the nodes carry out the cleansing process.

Tissue Response to Injury; Immunity; Blood Groups

Tissue response to injury

Cellular response to injury

- The reticuloendothelial system
- Antigens and antibodies
- Immunity
- Allergy
- Transplantation

The blood groups

- The ABO system
- The Rh system
- Other blood group substances
- Transfusions

Summary

The human body is provided with a variety of devices designed to protect it from injury or invasion by toxic substances or microorganisms. The skin and mucous membranes provide a first line of defense against invasions. This barrier is largely associated with the production of immunoglobulins (especially IgA) by the exocrine-secreting glands in certain of these surfaces (e.g., respiratory system). If this line is breached, the body tissues and cells respond to try to stop the infectious agent(s) at the site of penetration, and white blood cells provide the last line before body cells in general are "attacked" by the invader.

This chapter provides a discussion of the basis for the body's ability to react to a diverse series of challenges.

Tissue response to injury

If the skin or a mucous membrane is penetrated by some agent, the area around the point of penetration typically becomes tender and warm to the touch, reddened, and painful. These sensations and responses are part of a tissue response to injury known as the INFLAMMATORY REACTION. The purpose of such a reaction is twofold:

To DESTROY or neutralize the agent responsible for the reaction.

To FACILITATE REPAIR of the injury.

The types of agents that initiate the inflammatory response are diverse, including chemicals, thermal injury (burns by scalding water or other hot liquids), radiation, and mechanical agents (trauma such as crushing accidents). Regardless of the agent, the body response is more or less the same, following a series of steps, summarized in Fig. 23.1.

A short-lived CONSTRICTION of blood vessels in the injured area tends to localize the agent. This is followed by an OPENING of blood vessels to the area that presumably brings more blood cells to the region.

As damaged CELLS DIE (necrosis), lysosomes release their enzymes and this causes further cell death in the area.

Damaged mast cells (tissue basophils) release HISTAMINE, a potent vasodilator substance. Histamine creates an increased blood flow (hyperemia) to the injured area, and redness with local temperature elevation results.

VENULES AND CAPILLARIES DILATE because of histamine release and mechanical damage, and the injured area may have an increased fluid content (edema), with formation of what is called the "inflammatory exudate." Loss of clotting factors through these vessels may cause a "clotting" of the tissue fluid that "walls off" the injured area and slows passage of noxious agents from the original site of injury.

LEUCOCYTES MARGINATE (line up or adhere to) the inside of the capillaries in the injured area, and then pass through *(diapedesis)* the capillary walls to join the battle against the agents.

PHAGOCYTOSIS of agents and dead cells occurs, healing progresses, and replacement of lost cells occurs by mitosis.

How long the inflammatory process takes and

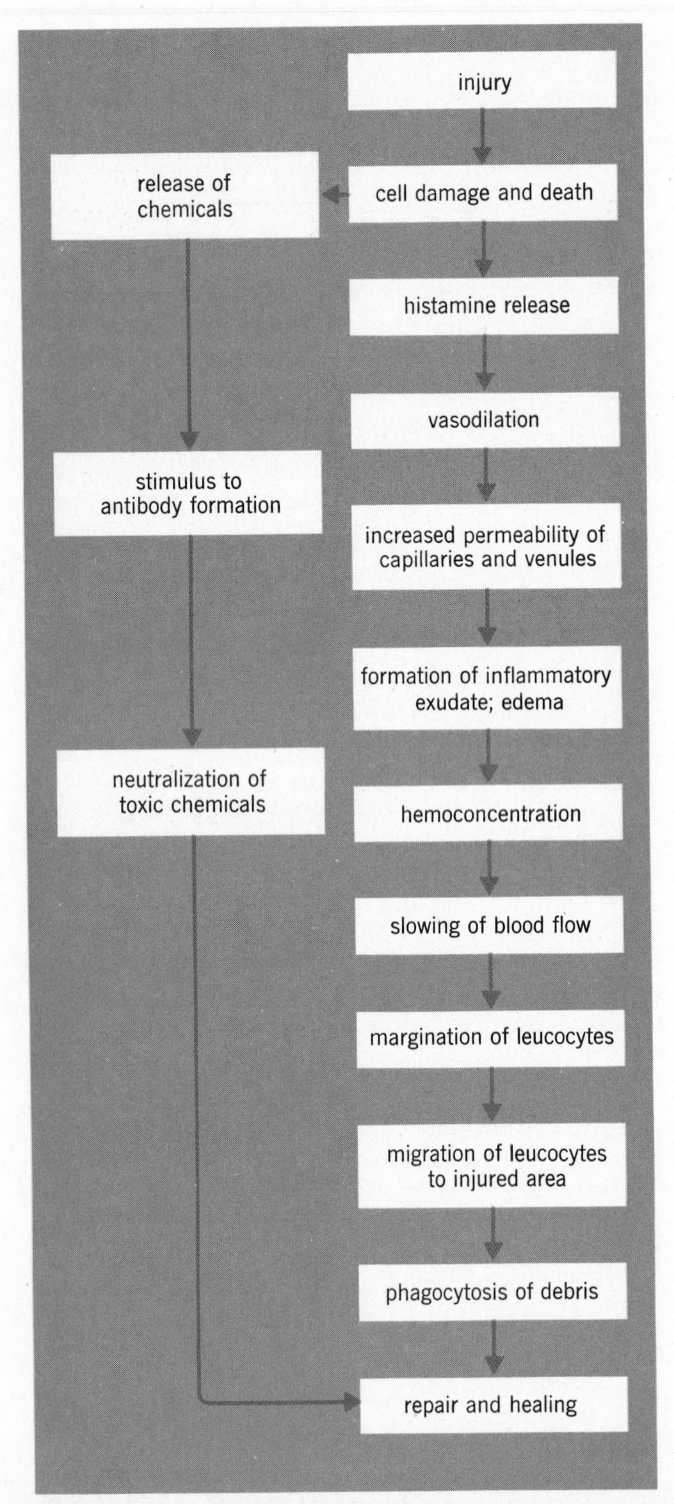

FIGURE 23.1 The steps in the inflammatory response.

its eventual outcome are influenced by several factors.

The nature and intensity of the irritating agent. If the invading agent is bacterial, it may produce an enzyme (hyaluronidase) that liquefies the gellike intercellular material and facilitates the spread of the agent. Recovery will obviously be slower if a larger body area is involved. Agents that cause deep or extensive injury will result in a longer healing process.

The duration of the irritation. Any given agent, if it lasts a long time, will cause more extensive destruction. BURNS are a good example of the effect of this factor.

Burns, either from thermal, chemical, or radiation sources, may be classed as first, second, or third degree in severity. In a FIRST-DEGREE BURN, only the epidermis is involved and reddening with no blistering occurs. In a SECOND-DEGREE BURN, dermis and deeper tissues are involved but some epidermal remnants are always present in the burned area. There is blistering. In a THIRD-DEGREE BURN, bones, muscles, and tendons are usually involved and no epidermal remnants are present in the burned area. Ulceration is common. In general, the deeper the burn, the greater the protein, fluid, and electrolyte loss, and the greater the danger of (and from) bacterial contamination.

The tissue affected. In general, the more vascular the tissue, the faster the recovery. Tissues of younger people recover more quickly than older ones, presumably because of greater rates of mitosis and replenishment of cells and intercellular substance.

The effectiveness of the defensive response. Implied here is the ability of the white blood cells to get to the injured area, the effectiveness of phagocytosis, the ability to wall off the infection, and the ability to produce antibodies.

Cellular response to injury

Specific body cells are involved in combating an invader or the damage resulting from the agent.

The reticuloendothelial system

The more-or-less fixed phagocytic cells of body tissues and organs are called MACROPHAGES. They and the PLASMA CELLS that produce antibodies comprise the reticuloendothelial (RE) system. The cells of this system can ingest particles such as microorganisms and cell debris, and can combat the chemicals produced by the inflammatory reaction. The plasma cells of the RE system are believed to be derived from lymphocytes.

Two types of lymphocytes are believed to be present in the body. T-CELLS, which originate from the thymus and bone marrow, require a chemical from the thymus to make them capable of responding to chemicals produced within the body or those introduced by tissue transplantation. Some T-cells become "killer cells" that reject tissue transplants; others become "primed cells," capable of responding to a second exposure of a given antigen. A first exposure merely sensitizes the cell and makes it capable of a reaction. B-CELLS originate from the bone marrow and form the lymphoid tissue of the spleen, lymph nodes, and lymph nodules of the intestine. They may require a thymus chemical to transform them into antibody-producing plasma cells. The relationship of these two types of lymphocytes is shown in Fig. 23.2. The two types of cells thus appear to create protection against all possible means of challenge by foreign cells and/or chemicals.

Antigens and antibodies

In addition to phagocytosis, the body possesses plasma cells that can form chemicals or antibodies in response to the presence of foreign chemicals, or antigens, that may enter the body.

ANTIGENS are usually large protein, polysaccharide, or nucleic acid molecules (as in viruses) that cause the plasma cells of the body to produce modified globulins in response to the presence of the antigen. The globulins are known as ANTIBODIES. Antigens and antibodies are usually specific in that they only react with one another and not with other antigens or antibodies. In reacting, the two materials may cause a variety of visible or nonvisible reactions including neutralization, in which the effect of the antigen is removed; precipitation, where an aggregate is formed, settles out, and is phagocytosed; agglutination, in which cells are clumped together; or lysis, which destroys the cells.

Another mechanism by which phagocytosis of microorganisms is enhanced is the phenomenon of *opsonization.* Opsonins are specific antibodies produced in response to surface antigens on viral or bacterial invaders. When the opsonin reacts with the antigen that caused its production, the reaction causes the surface of the invader to become more sticky, and this renders it more susceptible to phagocytosis by the body macrophages and blood cells.

The cells producing most of the antibodies, the plasma cells, are thought to be derived from lymphocytes originating in bone marrow or lymphatic tissues during early development. They migrate to various body areas, settle, and proliferate to form a colony or CLONE of cells. In order for such cells to become immunologically competent, or

FIGURE 23.2 Relationships of T and B lymphocytes.

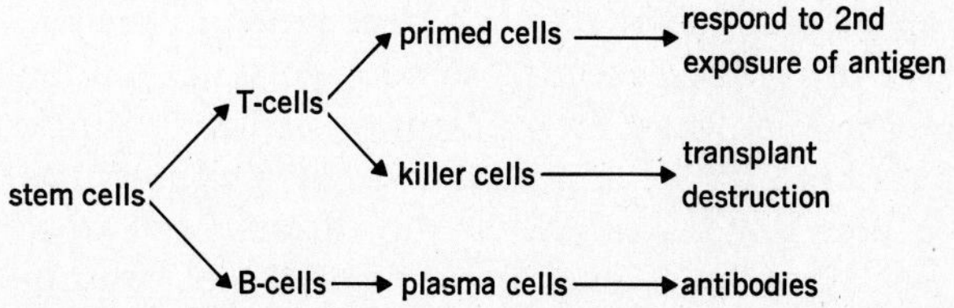

capable of antibody production, secretions from the thymus gland may be required to transform the lymphocytes into plasma cells. Removal of the thymus in newborn animals results in a deficit in cellular immunity but not in humoral immunity (see below), and therefore tissue transplants are not rejected and delayed hypersensitivty reactions do not occur.

Two major theories have been advanced to account for antibody production. According to the TEMPLATE THEORY of antibody production, the antigen must contact or enter the plasma cell, take over the DNA-replicating apparatus of the cell, and cause it to produce a specific protein that is the mirror image of the antigen. These two then fit together like a lock and key. The SELECTIVE HYPOTHESIS (or modified clonal theory), which appears to be emerging as the accepted theory, suggests that the ability to form antibodies already exists in the clone cells, and that all that is required is a stimulus (provided by the antigen) to "turn on" the plasma cell. This theory also suggests that there is a separate cell for each antibody (total number may reach millions) and that the diversity of cells is brought about by mutation during cell replication before any antigens enter the body. Thus, humans may possess cells to produce antibodies against any possible antigen that might be encountered during their lives. This theory explains the inability of the body to react to its own chemicals as being due to clone destruction during development. As new chemicals are produced (which act as antigens), they cause antibody production, and then an antigen-antibody reaction occurs that destroys the clone producing the antibody.

Viruses entering the body may enter cells and "take over" the DNA-replicating apparatus of the cell to make more viral molecules. Such virus-infected cells also produce a large protein molecule known as INTERFERON. By some as-yet-unexplained mechanism, interferon prevents viral replication, possibly by blocking viral m-RNA attachment to cellular ribosomes. When the infected cell dies, the interferon is released and may "coat" noninfected cells to prevent viral entry. Interferon has been suggested as a possible method for controlling the "common cold." The difficulty with the use of interferon as a "cold cure" lies in the fact that it is produced in very small quantities by human cells, and it appears to be species specific; that is, only human interferon is effective in humans.

Immunity

An AUTOIMMUNE DISEASE results from a change in a normal body chemical, rendering it antigenic and capable of stimulating antibody production.

The production of antibodies in response to foreign chemicals that enter the body is one basis of acquiring immunity to a variety of diseases. Methods used to induce immunity include the following.

The use of live, attenuated (weakened) organisms that retain the power to induce antibody formation but do not usually cause clinical disease (e.g., polio, measles).

The use of similar organisms to provoke the production of antibodies similar or identical to those occasioned by another organism (e.g., cowpox for smallpox).

The use of killed organisms that cause no disease but induce antibody formation (e.g., polio, typhoid).

The use of toxoids, which are toxins (harmful chemicals) that have been modified to remove their toxic effect but not their ability to stimulate the production of antibodies (e.g., diphtheria, tetanus).

The injection of pooled human gamma globulin, which may contain antibodies to a variety of antigens.

The use of antitoxins—antibodies to specific toxins or organisms (e.g., antitoxins to rattlesnake venom).

There are several types of immunity.

TISSUE *(cellular)* IMMUNITY. Some antibodies, once produced, remain within the plasma cell. Such antibodies may be retained for long periods of time and may confer lifelong immunity in some cases (e.g., measles).

HUMORAL IMMUNITY. Other antibodies, rather than remaining in the plasma cells, are secreted into the bloodstream, where they are subject to destruction or elimination, and thus do not usually confer immunity for extended periods of time. Examples include immunity to diphtheria and tetanus, which both require periodic "boosters" to restimulate antibody production.

ACTIVE IMMUNITY. This results when the individual contracts a disease or is given a mild case of it by vaccination and then responds by production of an antibody.

PASSIVE IMMUNITY. This occurs when an individual is given an antibody from another person or animal by injection. It provides immediate protection and provokes long-range protection.

ACTIVE-PASSIVE IMMUNITY. Antibodies are given at the time of exposure to reduce the severity of the disease, and contracting the disease itself confers immunity. An example is giving gamma globulin injections to a child who is incubating the measles virus.

VACCINATION is a method widely used today to induce immunity to a variety of diseases. Injection or the introduction into the body of dead antigenic bacteria, attenuated live organisms, or some of their chemical products is utilized to trigger antibody production with resultant immunity. Table 23.1 shows some of the diseases against which humans may be protected by vaccination.

In the development of active immunity, a specific time period must elapse between the arrival of the antigen and the production of sufficient

TABLE 23.1 Some diseases with methods and duration of protection

Disease	Method used	Duration of protection
Measles Rubella Rubeola	Attenuated organism	Expected to be life[a]
Poliomyelitis	Attenuated or killed organism	After a series of injections or oral adminstrations boosters are required about every 10 years
Pertussis (whooping cough)	Killed organism	About 10 years
Smallpox	Similar organism toxoid	3-10 years
Diphtheria	Toxoid	About 10 years
Tetanus	Toxoid Antitoxin (for exposure)	To 10 years after first booster or if penetrating injury of skin
Mumps	Attenuated organism	Expected to be life[a]
Typhoid	Killed organism	3 years (minimum)
Influenza	Attenuated organism	1 year for a given strain

[a] Vaccines have been available too short a time to assess their protection for life; they presently produce a lasting effect.

antibody to have an immunizing effect. It usually requires one to two weeks after the first presentation of an antigen for significant quantities of antibodies to appear. On a second exposure to the same antigen, antibodies appear within days, reach higher concentrations (titers), and last longer than on a first exposure. This "secondary response" is the basis for boosters and maintains immunity to specific diseases.

Allergy

Allergies represent an incomplete or abnormal reaction to an antigen. They are typically chronic and occur in individuals who have a hereditary predisposition to develop allergies. An antigen may trigger incomplete or partial antibody production, not enough to confer immunity. Also usually associated with allergy is some type of reaction that results in tissue damage and the liberation of histamine. To histamine release may be attributed many of the symptoms of allergy such as watery nasal discharge, hives, and edema. It also accounts for the waxing and waning nature of the symptoms and the shift from one organ or system to another. The use of antihistamines in relief of allergic symptoms is based on this consideration. Three types of allergies are usually recognized: precipitin, delayed-reaction, and atopic allergies.

PRECIPTIN ALLERGIES. These occur when enough antibody is produced to react with an antigen and cause precipitation of the antigen-antibody complex. Since most of these reactions occur between a tissue antibody and the antigen, severe tissue damage results and may cause the following conditions.

Anaphylactic shock. Very low blood pressures, weak heart action, edema, and circulatory collapse are the chief symptoms of anaphylactic shock. They are thought to be due to release of cellular substances (histamine, serotonin, bradykinin) when the antigen combines with tissue-fixed (cellular) antibodies. The condition develops very rapidly and gives little time for effective treatment. The "penicillin reaction" exhibited by some individuals is an example of an anaphylactic reaction.

Anaphylaxis may be passively transferred in the serum from the allergic individual to a normal person as in a transfusion. After a time lapse, contact by the normal individual with the antigen responsible for the shock in the allergic person may cause shock in the normal individual.

Serum sickness. Similar in symptoms to anaphylactic shock, this develops more slowly (one to two weeks) and is more amenable to treatment. It results when serum from a foreign animal is injected into a human.

Arthus reaction. In the Arthus reaction, the antigen-antibody complex is toxic to tissues and causes a local inflammatory reaction.

DELAYED-REACTION ALLERGIES. Exemplified by the reactions of the tuberculin skin test and the reactions to poison oak and ivy, these allergies cause initial sensitization to an antigen, but a second exposure is required to get a reaction.

ATOPIC ALLERGIES. These include hay fever, bronchial asthma, and dermatitis. They result from production of incomplete antibodies (reagins or IgE). Atopic allergies differe from most allergies in that they are inherited, the antibody is deposited in cutaneous (skin) tissues and may enter the bloodstream, and edema is the primary reaction that appears. Symptoms include bronchiolar MUSCLE SPASM, PRURITUS (severe itching), IRRITATION of the nasal mucosa with nasal discharge, and URTICARIA (hives).

ALLERGIC RESPIRATORY DISEASE represents one of the most common types of atopic allergies in children. Allergic rhinitis (nasal discharge), allergic bronchitis, and bronchial asthma are the most common. The interference with breathing in asthma is due to bronchiolar muscle spasm. Increased mucus secretion and edema may exaggerate the blockage of air movement.

Many types of allergies, particularly atopic allergies, may be treated by giving the individual regular doses of the antigen into the skin in an attempt to build up antibody production in the process of desensitization. Specific IgG or blocking antibodies (nonallergic, not tissue fixed) are increased, and symptoms are relieved by removing the allergen before it can combine with the tissue-fixed antibody.

Transplantation

The modern human's average lifespan is less than 75 years. Organs wear out, become diseased, or otherwise become incapable of maintaining normal function. Transplantation or replacement of defective organs with healthy organs from other individuals is a relatively new field of medicine. At the present time, only limited survival is possible with organ transplants, due to the REJECTION PHENOMENON.

The rejection phenomenon is the result of an antigen-antibody reaction, with the transplanted organ serving as the antigen to call forth the production of antibodies by the recipient's body. To minimize or reduce the production of antibodies and prolong the operation and life of the transplanted organ, several procedures may be employed either singly or in various combinations.

X-RAY OR GAMMA RADIATION kills the cells responsible for antibody production. However, radiation is indiscriminate in terms of cell destruction and may damage bone marrow and lymphoid organs and lower resistance to disease to a point where ordinary stressors may prove fatal to the recipient.

TISSUE MATCHING, utilizing lymphocytes in a type of cross match, determines how close the antigens of the donor match those of the prospective recipient. The closer the match, the less treatment will be required to combat the rejection.

DRUG THERAPY, using cortisonelike materials, is utilized to reduce inflammatory reactions resulting from the transplant.

ANTILYMPHOCYTE serum, prepared by injecting human lymphocytes into horses, is used to react with human lymphocytes in the recipient. If the cells are destroyed, they cannot become plasma cells and thus produce antibodies.

LYMPH, collected from the lymph ducts, may be treated to REMOVE WHITE CELLS from the fluid, and the fluid is returned to the recipient. The lymphocytes are thus not available as potential antibody producers.

DEVELOPMENT OF TOLERANCE to foreign organ antigens through the injection of small doses of these antigens over long periods of time may reduce the necessity for any external therapy and is one of the more promising avenues of current research.

CULTURING OF TISSUES before transplanting has recently been shown to result in a higher degree of "takes" by transplanted tissue. Skin, adrenals, and other organs have been cultured. It has been suggested that the tissues revert to an embryonic state on culturing and are not antigenic; thus, they do not stimulate antibody production.

Transplants that are AUTOLOGOUS, that is, from one part of a person's body to another or from one identical twin to another, are genetically identical and do not stimulate antibody production. HOMOLOGOUS TRANSPLANTS, from one individual to another member of the same species, are less successful because of rejection. HETEROLOGOUS TRANSPLANTS, from another animal to a human, are least successful.

The blood groups

On the surface of the erythrocytes are found genetically determined chemical substances that act as antigens. Because they are substances unique to the bloodstream, these materials are designated as ISOANTIGENS, isoagglutinogens, or agglutinogens. Two main categories of substances are recognized: one group is designated as the ABO SYSTEM, the other as the RH SYSTEM.

The ABO system

Two basic isoantigens are involved, designated A and B. With two antigens, four possible combinations may exist. The cell may have one or the other, both, or neither antigen on its surface. The particular antigen present determines the blood group. The relationships of antigens to blood group and frequency of occurrence are shown in Table 23.2.

Genetically, A and B inheritance is dominant to O, and each is the result of two allelic genes (one chromosome, occupy the same locus, and control the same characteristics). Therefore, the following genotypes could exist:

Blood group	Genotype
A	AA, Ai
B	BB, Bi
AB	AB
O	ii

where

A = group A } codominant and dominant
B = group B } to group O
AB = group AB
i = group O} recessive to both A and B

The plasma contains ANTIBODIES (*isoantibodies, isoagglutinins,* or *agglutinins*) that correspond to the antigens described above. They are designated by the terms a, anti-A, or alpha and b, anti-B, or beta. Agglutinins are not present at birth. They are gamma globulins whose production is believed to occur between two and eight months of age in response to A and/or B antigens taken in foods, particularly meats. The antibody present in the blood of any one given individual is the reciprocal of the antigen. Recall the clonal hypothesis, which states that the cells capable of forming corresponding antibodies to the blood antigens are destroyed embryonically, avoiding a reaction as new antigens are produced. The setup of both antigens and antibodies in the various blood groups would thus be as follows:

TABLE 23.2 Blood groups and frequency of occurrence

Antigen present	Blood group (same as antigen)	Frequency of population	
		Percentage of white	Percentage of black
A	A	40.8	27.2
B	B	10.0	19.8
AB	AB	3.7	7.2
Neither A nor B	O	45.5	45.8

Blood group	Dominant antigen(s)	Antibody
A	A	b
B	B	a
AB	AB	none
O	none	ab

The importance of the blood groups becomes apparent when one considers the necessity of TRANSFUSING BLOOD between one individual and another. It must be remembered that the mixing of red cells having a given antigen with plasma containing a corresponding antibody must result in an antigen-antibody reaction. The bloods must "match" or correspond as far as the antigens are concerned if such a reaction is to be avoided. The bloods may be TYPED and CROSS-MATCHED to determine the compatibility or "likeness" of blood given during transfusion.

Typing for the ABO system involves placing pure agglutinin (a or b, available commercially) on a microscope slide, mixing with it the unknown blood, and watching for a visible antigen-antibody reaction. It is important to remember that when typing the blood one looks for a reaction, while in the actual transfusion one avoids the possibility of a reaction. When inspecting the slide of blood and antibody, the question is: what had to be present on the erythrocytes of the unknown blood in order to give an antigen-antibody reaction with a known antibody? The following chart summarizes the possible combinations that may result (plus indicates a reaction; minus indicates no reaction) when *typing* the blood on a slide.

Antibody (on slide)		Group and antigen
a	b	present on red cells
+	−	A
−	+	B
+	−	AB
−	−	O

The Rh system

The Rh system is composed of three allelic genes for each type. The dominant genes are designated conventionally as CDE and the recessives as cde. Eight different genotypes are possible and are typed into two categories as Rh positive and Rh negative. The individual with a D gene will be RH POSITIVE (about 85 percent of Caucasian Americans, 88 percent of black Americans); the individual with no D gene will be RH NEGATIVE (about 15 percent of Caucasian Americans, 12 percent of black Americans). The specific combinations are:

CDE cDE cDe CDe	Rh^+	cdE Cde CdE cde	Rh^-

An individual who is Rh positive has the antigen on the erythrocyte and has no corresponding antibody in the plasma. An individual who is Rh negative has no antigen on the cell and, similarly, has no antibody in the plasma. However, the clones of the Rh negative person have never been challenged, and the person still possesses the ability to form antibodies if positive cells ever enter the body. There are two general sets of circumstances under which Rh positive cells might enter the body of an Rh negative individual.

The first is transfusion. Blood is typed for the Rh factor in the same manner as for ABO factors; that is, by placing serum with Rh antibody on a slide, adding blood to it, and watching for a reaction. A reaction means Rh positive and the antigen is present; no reaction means Rh negative, and the antigen is not carried by the blood typed. If the person typing the blood makes an interpretive error, positive cells could be transfused into a negative person.

The second situation involves a possible reaction between an Rh positive child *in utero* and an Rh negative mother. The inheritance of Rh factor depends on two genes, with Rh positive dominant to Rh negative. The child will inherit one gene for the condition from each parent. If the father if Rh

Genotypes of			
Father	Mother	Child (one gene from each parent)	Interpretation
RhRh	RhRh	RhRh	Child and mother are Rh^+. No reaction possible
Rhrh	Rhrh	RhRh or Rhrh or rhrh	Child is either Rh^+ (3 chances in 4) or Rh^- (1 chance in 4) relative to a positive mother. No reaction is possible between child and mother
RhRh or Rhrh	rhrh	Rhrh or rhrh	Child is Rh^+, mother is Rh^-. 50:50 possibility of reaction exists

negative, there is no possibility of a child-mother difficulty due to Rh factors.

The following examples will illustrate the conditions under which a reaction between child and mother might occur.

Normally, the blood supplies of the mother and child are separated by membranes in the placenta. If the membranes become permeable to the child's red cells, positive cells could enter the blood of the negative mother and cause her body to produce antibodies to the child's cells. The antibodies may then make their way back into the child through the placental membranes where an antigen-antibody reaction may occur. On the other hand, the antibodies may remain in the mother's bloodstream possibly to affect a future child.

ERYTHROBLASTOSIS FETALIS is a hemolytic disorder that results in the child when antibodies pass from mother to fetus during its sojourn *in utero*. If there is evidence that the fetus is developing erythroblastosis and is considerably less than 32 weeks of age, it may be too immature and anemic to survive a Cesarean section (surgical removal through the mother's abdomen) and an exchange transfusion (see below). In these circumstances, the fetus may be given transfusions of group O, Rh negative blood injected into its peritoneal cavity to sustain life until delivery and specific supportive care is possible (at about 32 weeks). If suspected severe erythroblastosis is confirmed after the child is born, an exchange transfusion may be performed. Group O, Rh negative or group specific (A, B, AB) Rh negative blood is used. Mild cases may require no treatment or only a transfusion of a small quantity of red blood cells that match the child's.

Research has shown that Rh antibody concentrations usually rise in the bloodstream of the Rh negative mother about 72 hours after the birth of her Rh positive child. This result is interpreted to mean that it is at placental separation (afterbirth) that fetal blood may enter the maternal circulation, presumably through wounds created by placental separation. To desensitize the mother (that is, to remove these antigens), massive doses of anti D gamma globulin (RhoGAM*, Rho-Immune**) may be given within 72 hours after each childbirth. The antibody reacts with fetal Rh antigens present in the maternal circulation to form a complex that is excreted through the kidney. The mother's blood is thus cleared of antigens, and the chance that a baby will be born with erythroblastosis is reduced.

An Rh positive mother is no threat to an Rh negative fetus *in utero*. Even if the mother were to pass antigens to the fetus, the immaturity of its immune system and the dilution of fetally pro-

* Ortho Chemical Co.
** Lederle Laboratories.

duced antibodies by the mother's great blood volume (seven to eight times that of the child) results in low antibody concentrations.

Other blood group substances

Genetic variants or mutants of common factors have been demonstrated in certain family groups. Their antigens have been designated M, N, S, s, P, Kell, Duffy, Lewis, Kidd, Diego, Lutheran, and others. While reactions involving these antigens are rarely seen, transfusion of such a blood into a recipient may cause sensitization in the recipient. That is, the recipient will produce antibodies to the antigen, and these antibodies will remain for long periods of time in the bloodstream.

The Hr (cde) system is another system of antigens and is supposedly the reciprocal of the Rh system. If Rh antigens are absent, Hr antigens are thought to be present. Several Hr factors have been demonstrated in this system, the most important of which are Hr (c), Hr′ (d), and Hr″ (e).

Transfusions

In any transfusion, the two bloods involved are typed and cross-matched. The latter procedure tells if the bloods are compatible for all factors, not just the ABO and Rh factors. In this test, erythrocytes from the donor are mixed with plasma of the recipient ("major side") and plasma of the donor is mixed with the erythrocytes of the recipient ("minor side"). Evidence of an antigen-antibody reaction in either mixture is sufficient grounds for rejecting the donor's blood as suitable for transfusion.

TRANSFUSION REACTIONS. These include:

HEMOLYSIS of red cells, with consequent anemia.

KIDNEY FAILURE as a result of chemicals released by the antigen-antibody reaction and resultant renal vasoconstriction.

PLUGGING OF KIDNEY TUBULES by hemoglobin released by the lysis of erythrocytes.

FEVER as a result of toxins released by the antigen-antibody reaction.

TRANSFUSION FLUIDS. BLOOD is obviously the ideal transfusion fluid, inasmuch as it supplies all needed blood components. Unfrozen blood may not be stored for more than two to three weeks, because the cellular elements begin to disintegrate. Methods that involve quick freezing indicate the possibility of storage for several years. PLASMA (blood with formed elements removed) is used to aid in maintenance of blood volume. It contains all the organic materials of the blood, including antibodies and protein, and thus it must be type-specific to avoid any antigen-antibody reaction. Plasma aids in maintaining blood volume because its proteins draw water osmotically into the capillaries from the tissues. Plasma may be stored indefinitely. Solutions of COLLOIDS or "plasma expanders," such as albumin and dextran, may be transfused to maintain blood volume. These are artificial solutions and also draw water from tissue spaces into the bloodstream. They thus increase the volume of the plasma at the expense of the interstitial fluid. Dextran, a synthetic polysaccharide, is the colloid most often used as a plasma expander. It consists of aggregates of molecules from 10,000–100,000 molecular weight. The small aggregates filter through the kidney; the larger ones remain in the vessels and are metabolized or broken down over the course of several days. The substance thus exerts a relatively long-lived effect and is not antigenic (as albumin could be). The packed cell volume (hematocrit) is monitored to determine the need for more or less expander to be added to the bloodstream. Solutions of CRYSTALLOIDS, such as NaCl, may be employed to increase the blood volume temporarily. Their effect is only temporary because the particles are too small to remain in the capillaries, and they are filtered rapidly into the tissues. Thus, the use of crystalloids may be associated with the development of edema.

Summary

1. The inflammatory response occurs when tissues respond to injury. It aids in destruction of the injurious agent and in repair of the injury. Several steps occur in the response.
   a. Vasoconstriction.
   b. Cell death.
   c. Histamine release and vasodilation.
   d. Increased venule, and later capillary permeability leads to formation of exudate.
   e. Leucocyte invasion of the injury and commencement of healing.

2. Intensity of injury, duration of irritation, type of tissue injured, and individual response to injury determine the course of the response. Vascular tissues heal more rapidly.

3. The reticuloendothelial (RE) system is important in body defense against disease. The definition of the RE system is: all fixed phagocytic cells of the body plus cells capable of producing antibodies.

4. Antibodies, modified globulins, are produced in response to antigens or foreign chemicals that enter the body.
   a. Plasma cells, derived from lymphocytes, are the cells producing most of the antibodies.
   b. Plasma cells produce antibodies only when the antigen enters an unspecialized plasma cell and directs synthesis of antibodies specific to that antigen (template theory). Or clones are genetically produced by mutation, are specialized as to the antigen to which they may react, and possess the ability to produce antibodies to any antigen that may enter the body and stimulate only a certain cell to activity (selective hypothesis).
   c. Antigen-antibody reactions occur if an antigen and its corresponding antibody are brought together. The reaction is usually protective in nature and removes the antigen as a threat to the body.

5. Antibodies may create immunity to many disease agents that act as antigens.
   a. Immunity may be achieved by exposing the body to killed or live organisms or by injections of chemicals (toxoids, antitoxins, or gamma globulins).
   b. Several types of immunity exist: cellular immunity (involving antibodies in cells), humoral immunity (involving antibodies in the bloodstream), and various combinations of active immunity (involving the disease agent and production of antibody by the individual) and passive immunity (involving administration of antibody from another person or animal).

6. Allergy is an incomplete response to an antigen, and lasting immunity is not achieved. Types of allergies include:
   a. Precipitin allergies, where antigen-antibody reactions occur within cells, and anaphylactic shock may result.
   b. Delayed-reaction allergies, where symptoms develop slowly and sensitize the individual to the antigen. Another exposure to the antigen is required to develop an antigen-antibody reaction.
   c. Atopic allergies are common in children and include hay fever and asthma. They appear to be inherited.

7. Transplantation of organs causes antigen-antibody reactions with the donor's organ(s) acting as the antigen(s) and the recipient's body producing the antibodies.

8. The blood groups are examples of blood antigens and antibodies.
   a. The ABO group consists of two antigens (A and B) and two antibodies (a and b) usually present in reciprocal relationship. For example, Ab, Ba, AB, and Oab.
   b. The Rh group consists of one antigen (Rh), and it may be present (Rh positive) or absent (Rh negative) in the blood.
   c. Transfusion of blood requires that the antigens are the same between donor and recipient bloods to avoid an antigen-antibody reaction.
   d. Erythroblastosis fetalis may result if a mother produces antibodies to her child's blood Rh antigens while the child is still *in utero*. Destruction of red cells and plugging of kidney tubules with hemoglobin are two of the results of such a reaction.

9. Replacement of lost blood may be made by transfusion of whole blood, plasma, colloids, or crystalloids. Whole blood provides all missing elements; the other solutions provide volume.

24

The Heart

As indicated in a previous chapter, the cardiovascular system is the first system to become functional in the embryo. By the end of the third week of development, blood is being circulated through a system of blood vessels. Early development of the cardiovascular system is correlated with the necessity for providing a source of nutrients for the rapidly growing and developing embryo; an early requisite for circulation is the development of a pump to circulate the blood. While, at this early stage, the pump or heart does not have the final structure it will acquire later, it is developed to the point where it can achieve a directed, pressurized flow of blood through the organism.

Development of the heart (Fig. 24.1)

Normal development

At about two weeks of age, a mass of mesoderm called the CARDIOGENIC PLATE develops in the anterior end of the embryo. Over the next two to four days, the mesoderm organizes into paired

FIGURE 24.1 Stages in the development of the heart (numbers 5–8 occur between 5 and 8 weeks).

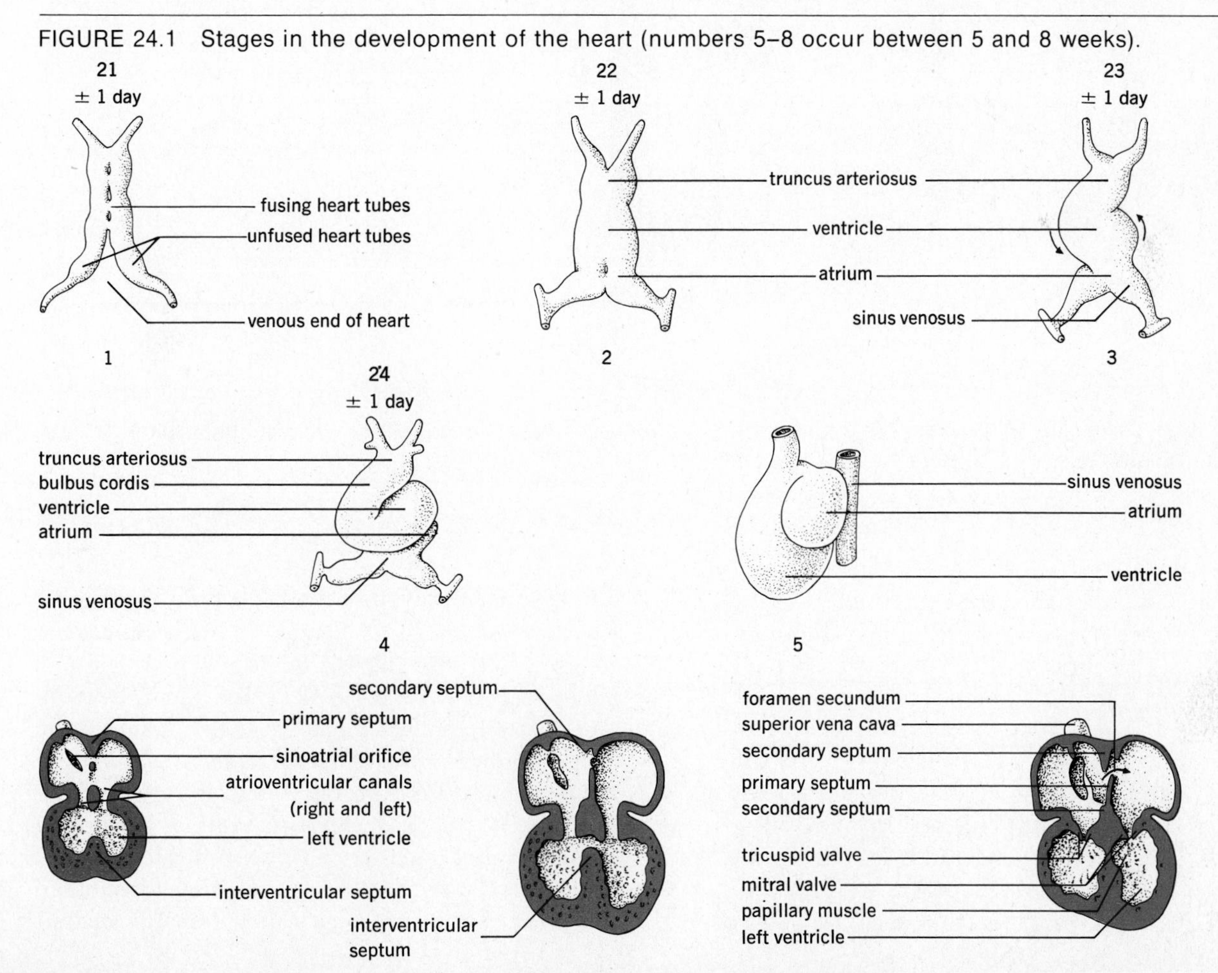

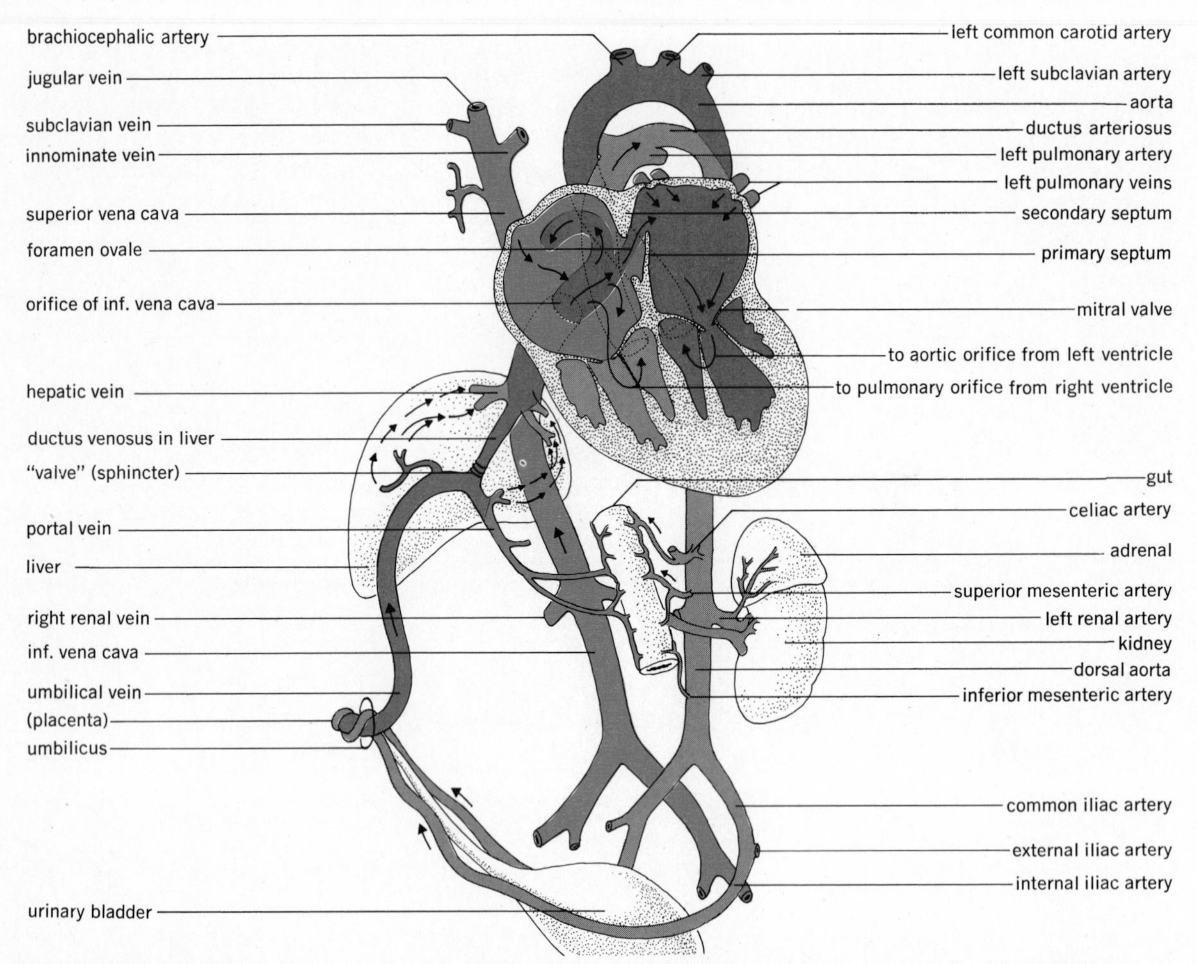

FIGURE 24.2 The plan of the fetal circulation. Colors indicate state of oxygenation (red, highest; blue, lowest; gray, intermediate).

elongated strands called HEART CORDS. Cavities develop within the cords, transforming them into hollow HEART TUBES. The tubes next fuse, beginning with the front or anterior end, and by the 22nd day a single tube is present. During the fourth week, CONSTRICTIONS develop in the tube that mark the positions of the bulbus arteriosus (bulbus cordis), ventricle, atrium, and sinus venosus, from anterior to posterior. It may be noted that the position of the chambers (from front to back) is reversed compared to that of the fetus or adult. Rapid GROWTH of the heart tube occurs, and the ventricle and sinus venosus fold over and come to lie posterior to the atrium, which is the normal fetal position. Between the fifth and seventh weeks, a series of changes occur that result in the four-chambered structure of the heart.

The common atrium is divided into two chambers by a PRIMARY SEPTUM that is perforated almost immediately by an opening called an ostium. Later, a SECONDARY SEPTUM develops just to the right of the first one, but it remains incom-

FIGURE 24.3 Some of the most common congenital anomalies of the heart.

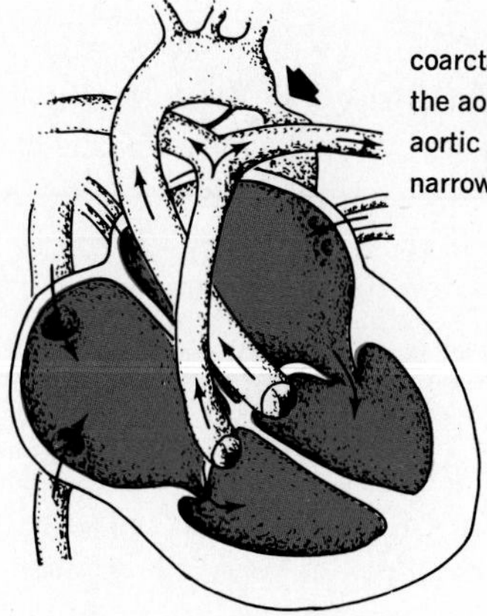

coarctation of the aorta. the aortic lumen is narrowed.

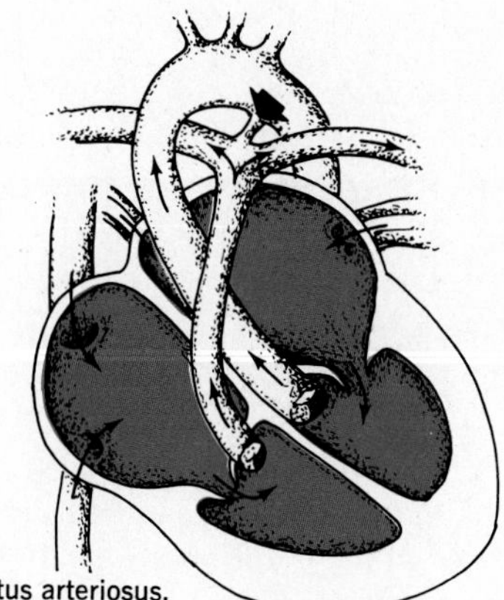

patent ductus arteriosus. the shunt between the pulmonary artery and aorta is open. blood bypasses the lungs.

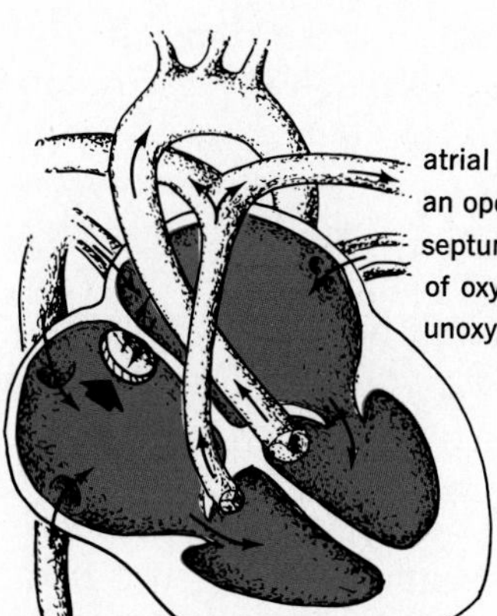

atrial septal defect. an opening in the atrial septum permits mixing of oxygenated and unoxygenated blood

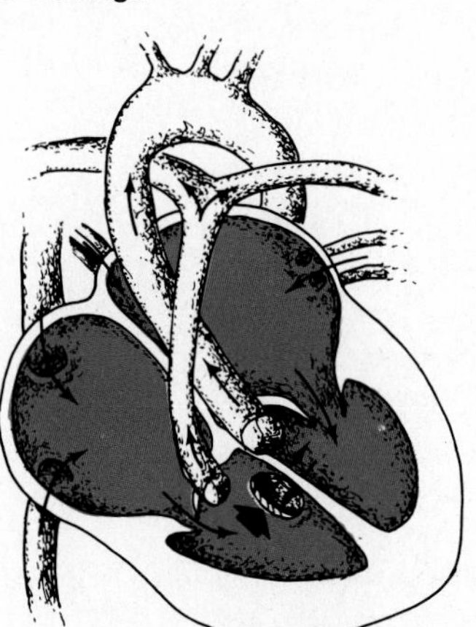

ventricular septal defect. an abnormal opening between the ventricles. blood passes from left to right ventricle.

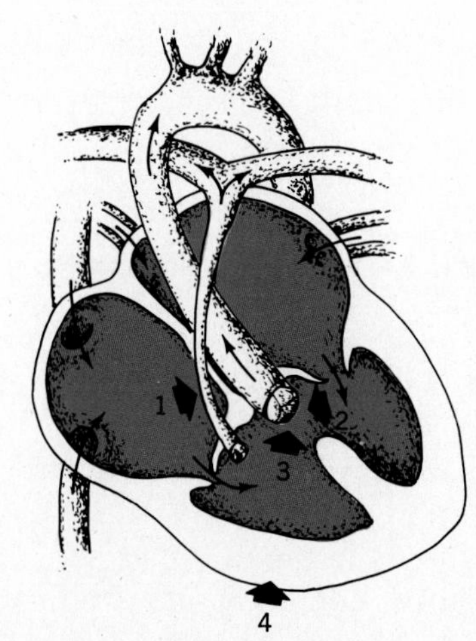

tetralogy of Fallot. four defects occur simultaneously. (1) narrowing of pulmonary artery or valve (2) ventricular septal defect, (3) aorta (overriding) receiving blood from both ventricles (4) hypertrophy of wall of right ventricle.

plete. The opening between the two portions of the secondary septum is the FORAMEN OVALE. The foramen allows a large portion of the blood entering the right atrium of the heart to go to the left atrium and bypass the nonfunctional lungs of the fetus. The lower part of the primary septum overlaps the foramen, and pressure differences between the two atria cause the septum to seal the foramen at birth to establish the adult type of flow through the heart.

The VENTRICLE IS PARTITIONED into two chambers by an interventricular septum, which grows upward from the apex of the heart.

The BULBUS ARTERIOSUS IS SUBDIVIDED into two vessels, the aorta and pulmonary arteries, which leave the left and right ventricles, respectively.

The three layers of the HEART WALL are formed, as is the special NODAL TISSUE that causes the heart to beat.

Simultaneously, connections with the placenta, for exchange of nutrients and wastes, are established. The heart and circulation show the features depicted in Fig. 24.2.

A critical time period in these changes occurs when the atria and ventricles are being divided. Interference with these processes, as by rubella or rubeola (measles) virus, may lead to VENTRICULAR SEPTAL DEFECTS (VSD), ATRIAL SEPTAL DEFECTS (ASD), or defects in the arteries. At birth, the shunts in the fetal circulation are closed, and the normal adult pattern of circulation is established. Two shunts are closed: the foramen ovale and the ductus arteriosus (which allows any blood entering the pulmonary artery to go to the aorta, bypassing the lungs). The ductus arteriosus is closed by contraction of the muscle in the vessel, possibly in response to the higher blood oxygen levels occurring when breathing begins.

Clinical considerations

Congenital anomalies of the heart include: failure of the ductus arteriosus to close at birth (patent ductus arteriosus); failure of the foramen ovale to close; and a variety of atrial and ventricular septal defects that may occur in various combinations. The most common of these congenital defects are shown in Fig. 24.3. Short descriptions of each condition are provided under each picture. Valves may also not develop properly, or may be damaged by disease and trauma (as in hypertension). If a valve fails to open completely, it is said to be stenosed; the condition is called STENOSIS. If the fibrous rings supporting the valves are stretched, the orifice may not seal properly when the valve closes. Such a valve is said to be INCOMPETENT, and it permits regurgitation of blood in a wrong-way flow. Rheumatic fever, the result of streptococcal bacterial infection, may result in stiffening and erosion of the mitral cusps causing stenosis and incompetence.

Basic relationships and structure of the heart

Size

At almost any age after birth, the size of the human heart approximates the size of the clenched fist of its owner. At birth, the heart measures about 25 mm (1 in.) in length, width, and depth and weighs about 20 g (about 3/4 oz). In the adult, the heart measures about 12.5 cm (5 in.) long, 9 cm (3 1/2 in.) wide, and 5 cm (2 in.) deep; it weighs about 300 g ((11 oz). Two periods of the greatest rate of growth of the heart are between 8 and 12 years of age and between 18 and 25 years.

Location and description

The adult human heart (Fig. 24.4) is located within the middle portion of the thorax (chest) known as the MEDIASTINUM. The broad upper end, or BASE, is directed toward the right shoulder, and the

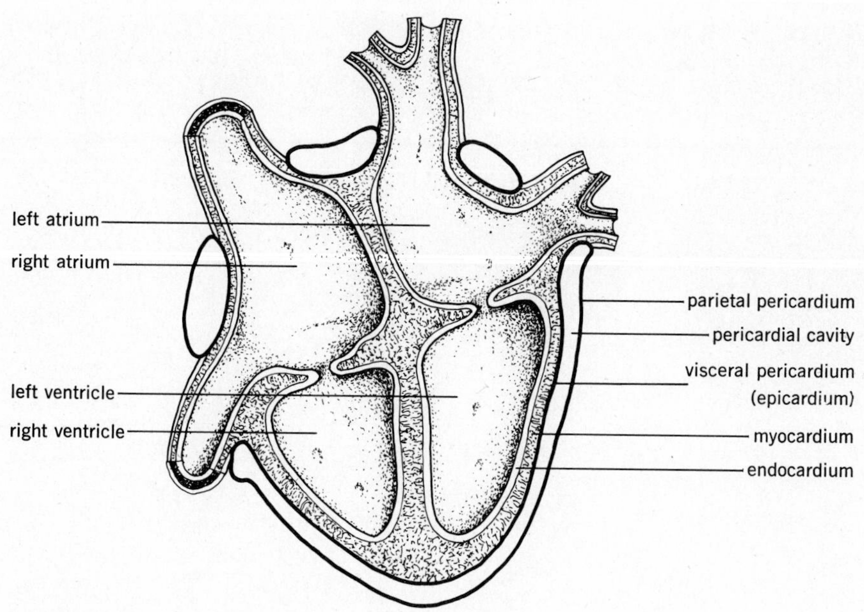

FIGURE 24.4 The relationships of the pericardium to the heart.

more pointed lower end, or APEX, is directed toward the left hip. If a line were drawn from the center of the base to the apex, it would define the AXIS of the heart and demonstrate that the heart is not set vertically within the chest. Neither is the heart centered within the chest; about one-third of it lies to the right of the midsternal line, two-thirds to the left. The organ lies, in the vertical plane, at the level of the fifth to eighth thoracic vertebrae (T5-8).

The heart is enclosed within a double-walled sac known as the PERICARDIUM (see Fig. 24.4). This sac is composed of two layers of tissue. An outer tough FIBROUS PERICARDIUM composed of collagenous tissue attaches to the bases of the pulmonary artery and aorta and to the diaphragm and sternum. This layer aids in maintaining the heart's position and protects it. The SEROUS PERICARDIUM is a thin, delicate membrane that lines the fibrous sac and that is reflected onto the outer surface of the heart at the bases of the great arteries. This membrane produces a film of fluid that lubricates the sac internally to allow nearly frictionless movement of the heart within the pericardium. The part of the serous pericardium lining the fibrous sac is called the PARIETAL PERICARDIUM, and the portion on the heart itself is called the VISCERAL PERICARDIUM, or epicardium. A space is found between the parietal and visceral layers called the PERICARDIAL CAVITY. In PERICARDITIS, or inflammation of the serous layers, fluid production is decreased, and adhesions (sticking together of the membranes) may occur with development of pain. In CARDIAC TAMPONADE, excessive fluid secretion or blood accumulates in the pericardial cavity, interfering with the heart's action.

Heart wall

As the heart develops, three basic layers of tissue form in its walls (Fig. 24.5). An inner ENDOCARDIUM is composed of a lining endothelial layer and several layers of connective tissue and smooth muscle. The valves of the heart (see below) are also formed from the endocardium. In some

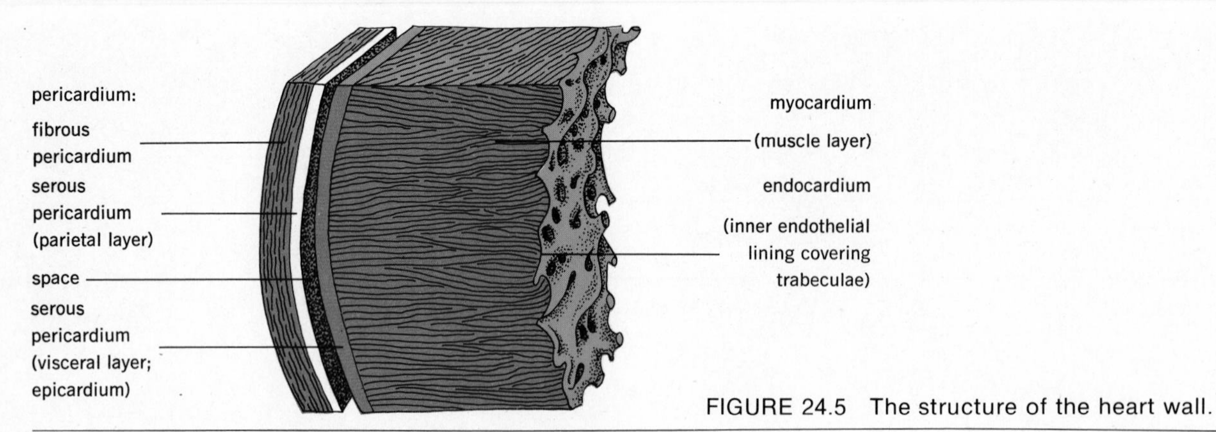

FIGURE 24.5 The structure of the heart wall.

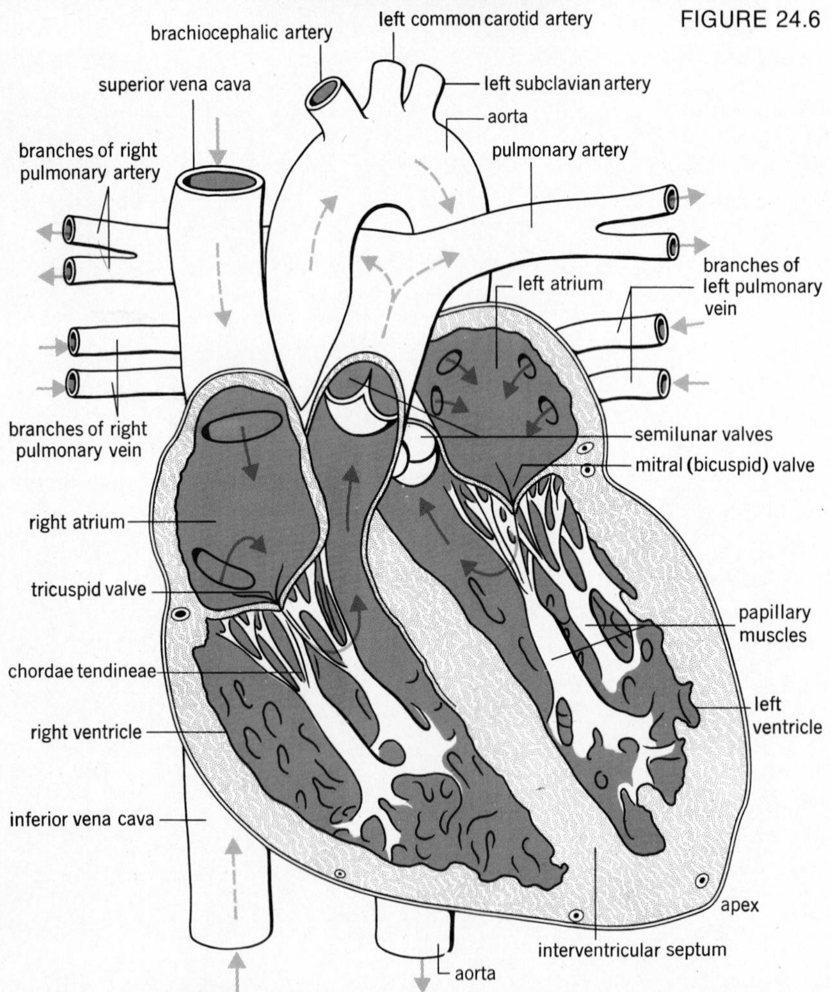

FIGURE 24.6 The heart in frontal section.

parts of the heart, a *subendocardial layer* is present that contains the specialized tissue responsible for contraction of the heart muscle. A central MYOCARDIUM, composed of cardiac muscle, forms about 75 percent of the wall thickness. The cardiac muscle is responsible for the ability of the heart to circulate blood. The outer layer of the heart wall is the EPICARDIUM or the layer of the serous pericardium described above. The epicardium carries the larger vessels of the coronary circulation and may also contain many fat cells.

Chambers and valves

The heart is actually two pumps in one (Fig. 24.6). The right side of the heart acts as a pump to receive blood from the body generally and to send it to the lungs for oxygenation and carbon dioxide elimination. This circuit constitutes the PULMONARY CIRCULATION. The left side of the heart receives blood from the lungs and pumps it to the body generally through the SYSTEMIC CIRCULATION.

The upper receiving chambers of the heart are called the ATRIA, and the lower pumping chambers are the VENTRICLES. The term *auricle* refers to the triangular appendage attached to the anterior portions of each atrium; they are not hollow. Thus *auricle* is not synonymous with *atrium*. The right atrium is a thin-walled chamber that receives blood from the "great veins" or venae cavae. It pumps its blood through the RIGHT ATRIOVENTRICULAR ORIFICE into the right ventricle. The orifice is guarded by the right atrioventricular or TRICUSPID VALVE. This valve consists of three flaps (cusps) of tissue attached by their bases to a fibrous ring around the orifice. The free edges are attached to strands of fibrous tissue called CHORDAE TENDINAE. The chordae attach to PAPILLARY MUSCLES of the ventricular wall and prevent the valve from reversing as the ventricle contracts. The right ventricle has a thicker muscular wall than the right atrium and pumps blood into the pulmonary artery for oxygenation in the lungs. In the base of the pulmonary artery is the PULMONARY VALVE *(pulmonary semilunar valve)*. It is composed of three pockets of tissue, without supporting chordae; the valve prevents return of blood to the right ventricle. After being oxygenated in the lungs, the blood returns to the thin-walled left atrium through four pulmonary veins. No valves are present in these vessels. The LEFT ATRIUM sends blood through the LEFT ATRIOVENTRICULAR ORIFICE into the LEFT VENTRICLE. This orifice is guarded by the left atrioventricular valve, also known as the MITRAL or BICUSPID VALVE. It has a structure similar to that of the tricuspid valve, except it has only two flaps of tissue. The very thick-walled left ventricle pumps blood into the aorta for distribution to the entire body. A valve, the AORTIC VALVE *(aortic semilunar valve)*, is found in the base of the aorta. It is constructed like the pulmonary valve and prevents return of blood to the left ventricle.

Thickness of chamber wall reflects the amount of work each chamber must do. Atria receive blood and move it to the adjacent ventricles; thus, little force is required, and the atria have thin walls. The right ventricle pumps to the lungs, a task requiring more force; thus, it has a thicker wall. The left ventricle pumps to the entire body and has a wall about three times thicker than the wall of the right ventricle.

Blood supply

Two CORONARY ARTERIES (Fig. 24.7), a right and a left, rise from the aorta just behind the pockets of the aortic valve. This origin permits the heart to receive blood with the highest oxygen level and under the highest possible pressure. The right coronary artery passes to the posterior side of the heart, giving off branches to the left ventricle, right ventricle, and right atrium. The left coronary artery supplies the left ventricle and left atrium. Anastomoses, or communications between the smaller branches of both coronary arteries, are common. About 48 percent of the United States population has what is called a right coronary predominance, in which the right coronary artery

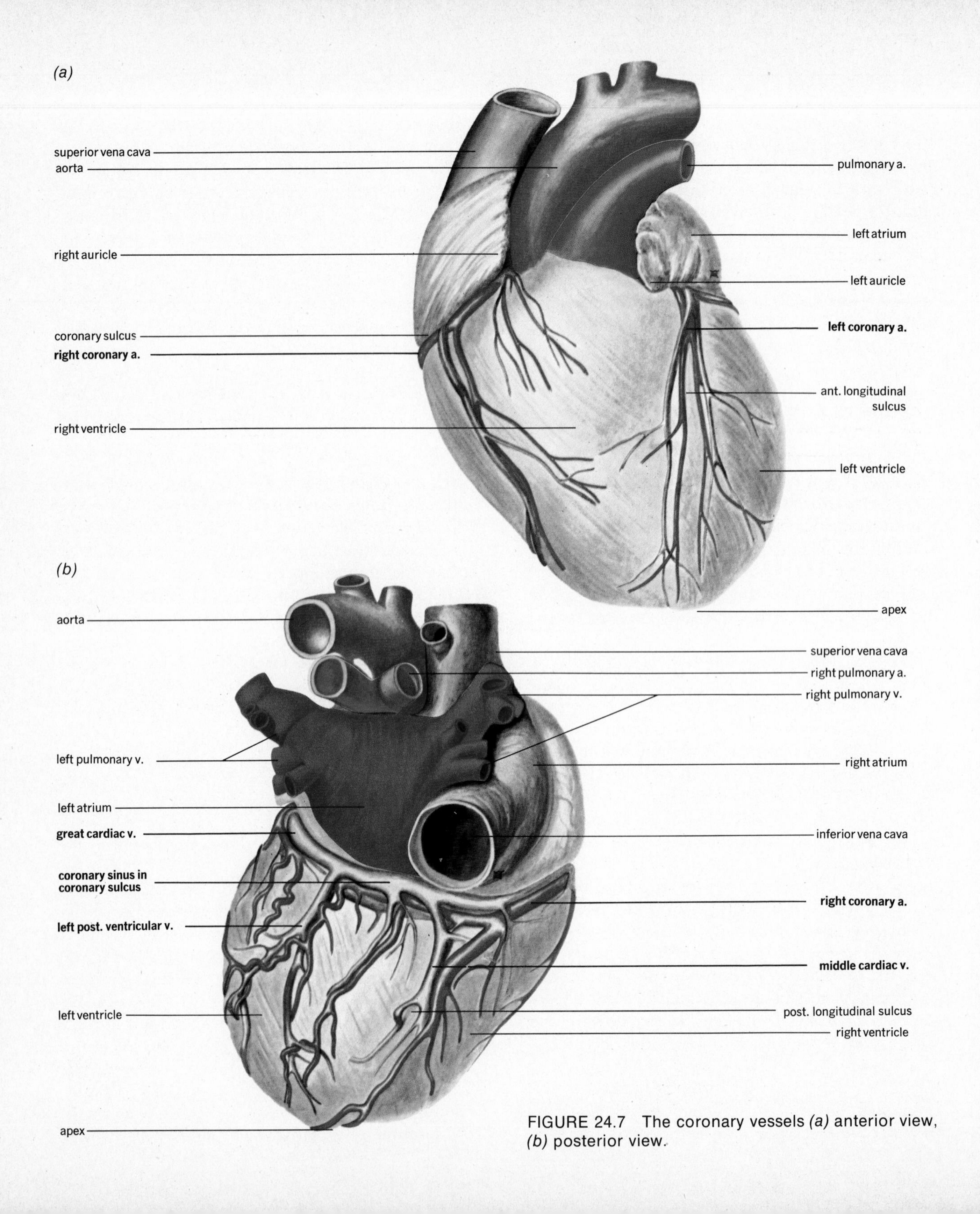

FIGURE 24.7 The coronary vessels *(a)* anterior view, *(b)* posterior view.

supplies more than just the right side of the heart. About 18 percent have left coronary predominance. The remainder, about 33 percent, have a balanced coronary supply.

The arteries form CAPILLARY and SINUSOIDAL (large irregular vessels) BEDS in the myocardium, and substances diffuse to and from these beds to the muscle and other layers of the heart wall. About 30 percent of the blood reaching the heart wall passes through the sinusoids directly into the ventricular cavities. The remaining 70 percent is collected by the CORONARY VEINS, which, by way of the CORONARY SINUS, empty into the right atrium.

Blood flows through the system of coronary arteries, capillaries, and veins only when the myocardium is not contracted. Contraction compresses the small vessels of the heart and stops flow. Also, when the ventricles are ejecting blood, the open semilunar valve in the aorta blocks the openings of the coronary arteries. Thus, the only time nutrients may be supplied to the heart is when the myocardium is relaxing or at rest and when the aortic valve is closed.

The coronary arteries bring nutrients and oxygen to the heart to sustain its activity. About 60 percent of the energy for heart activity is derived from the metabolism of fatty acids. About 35 percent comes from the metabolism of carbohydrates, and 5 percent from the metabolism of amino acids. Inorganic salts are also extremely important in the regulation of the beat and depolarization of the heart muscle. These are discussed in a later section.

CLINICAL CONSIDERATIONS. BLOCKAGE of the coronary vessels, as by a blood clot, results in coronary thrombosis; the myocardium is deprived of blood flow, and death of tissues beyond the block *(infarction)* may occur. If the thrombosis results in incomplete blockage of the vessel, a reduced blood flow to the area supplied *(ischemia)* may be the only result. Decreased flow of blood to the heart is usually associated with the development of the pain of ANGINA. In this condition, pain associated with myocardial ischemia is referred to the left chest and arm. If the vessel plugged is relatively small, the anastomoses between coronary arteries may permit sufficient blood flow to compensate for the blockage. If a large vessel is plugged, the amount of tissue that dies may be too great to permit continued heart action. The level of serum glutamic oxaloacetic transaminase (SGOT, an enzyme involved in heart metabolism) bears a relationship to severity of cardiac muscle death following a thrombosis. Normally found in a concentration of about 10–40 units per milliliter of serum, the enzyme increases in amount according to the severity of cardiac damage. Thus an SGOT of 90–100 units indicates mild heart muscle damage, and 200 or more units indicates severe damage. The SGOT may be used to indicate the severity of the initial attack and to follow recovery from the attack.

Nerves

The heart receives motor fibers from both divisions of the autonomic nervous system. PARASYMPATHETIC FIBERS are carried in the vagus nerves and act primarily to SLOW the rate of beat. SYMPATHETIC FIBERS, carried in the cardiac nerves, INCREASE both heart rate and strength of beat. Sensory fibers arise within the heart and are carried in the vagus nerves to the brainstem. These fibers carry impulses for the reflex control of heart rate and carry pain impulses to the central nervous system.

Physiology of the heart

Tissues of the heart

Four major types of tissue are concerned in setting the basic structure and functions of the heart. EPITHELIAL TISSUES line the inner and outer surfaces of the organ; CONNECTIVE TISSUES form a cardiac skeleton that supports the valves and affords attachment for the CARDIAC MUSCLE bundles. Finally, a specialized type of cardiac muscle called NODAL TISSUE is responsible for generating and distributing the electrical disturbances that cause muscular contraction. The nodal tissue and cardiac muscle are the subjects of this section.

NODAL TISSUE. Nodal tissue develops from the cardiac muscle as the heart is formed. It has largely lost its power of contractility, but has developed the properties of spontaneous depolarization and conductivity to a high degree. It gives the heart the ability to contract independently of any outside force. Hearts that beat because of inherent factors are called *myogenic,* while those that depend on extrinsic nerves to beat are called *neurogenic.* The tissue is organized into several discrete masses and bundles (Fig. 24.8).

The SINOATRIAL (SA) NODE measures about 18 mm in length by 6 mm wide (about ¾ × ¼ in.) and is located in the wall of the right atrium where it is joined by the superior vena cava. While all parts of the nodal tissue can spontaneously depolarize, the SA node does it most rapidly, and it thus dominates the rest of the system. Its rate sets the rate of heartbeat and it acts as the *pacemaker* of heartbeat with a basic rate of 75–80/min. The phenomenon of spontaneous depolarization is explained by assuming that an original membrane potential of about 60 mv "decays" or decreases because of a membrane that does not keep sodium completely out of the fiber but allows it to "leak" into the fiber. At some point, a threshold for depolarization is reached, and the formation of an impulse is triggered. K^+ flows out,

FIGURE 24.8 The locations of the nodal tissue in the human heart.

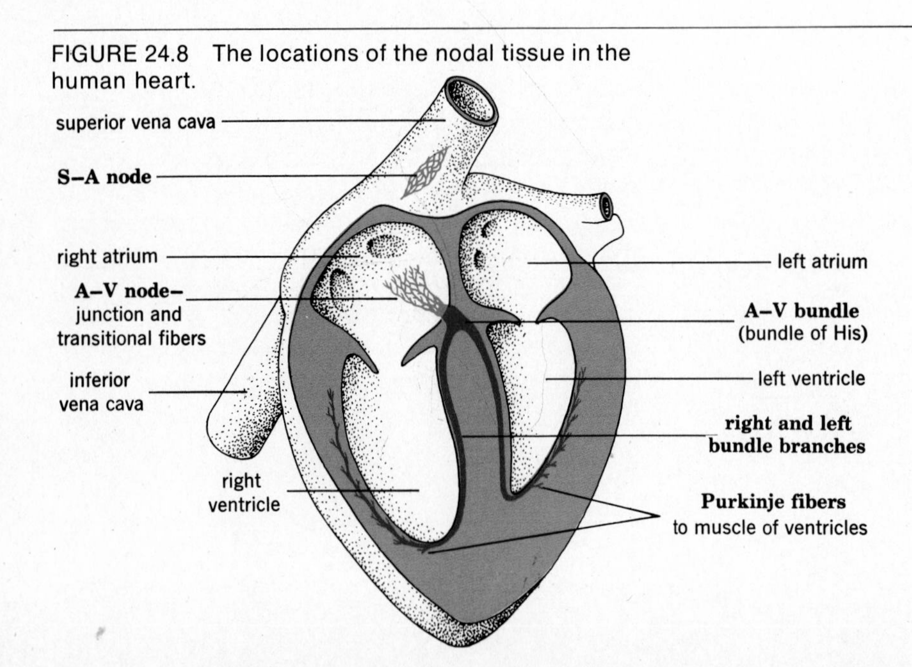

TABLE 24.1 Some characteristics of nodal tissue and cardiac muscle

Area	Velocity of conduction (m/sec)	Inherent rate of discharge (impulses/min)
SA node	0.05	70–80 (pacemaker)
Atrial muscle	0.3	–
AV node	0.05–0.1	40–60
AV bundle	2–4	35–40
Purkinje fibers	2–4	15–40
Ventricular muscle	0.3	–

slows, is recaptured, and Na^+ is pumped out of the nodal cell. The original polarized state is reestablished and the "decay process" starts again. The normal domination of the rest of the system by the SA node is explained by the fact that depolarization of the system caused by an SA node impulse is followed by repolarization. Then another impulse from the SA node causes depolarization before the depolarization that would be caused by the slower rhythm possessed by the other parts of the system.

The impulse from the SA node spreads through the atrial muscle, causing it to contract as the impulse passes through it. As the impulse passes through the muscles, it excites the AV NODE located in the right atrium near the base of the interatrial septum. Since the distance from SA to AV node is shorter than from the SA node to the farthest reaches of the left atrium, the impulse that excites the AV node cannot be immediately passed on to the ventricles; the ventricles must not contract before the atria have finished their activity. Specialized slow-conduction junctional and transitional tissues of the AV node accomplish this "AV delay." Then, once through the AV node, the object is to conduct the impulses to the ventricles as rapidly as possible. Impulses next pass through the AV BUNDLE, its BRANCHES, and the PURKINJE FIBERS to the ventricles for coordinated contraction of the ventricular muscle. These latter parts of the system have the fastest conduction rate. Some facts as to rates of impulse formation and conduction in nodal tissue and muscle are shown in Table 24.1.

CARDIAC MUSCLE. The cardiac muscle of the heart (Fig. 24.9) consists of branching and rejoining lightly striated fibers. The fibers are organized in two masses, one for the atria and one for the ventricles, and have a distinct pattern of distribution, particularly in the ventricles (Fig. 24.10). The two masses are electrically insulated from one another by the cardiac skeleton of the valves, so that the route for passage of impulses from atria to ventricles is over the nodal path. The individual cardiac MUSCLE CELLS are anatomically separated from one another by the membranes of the INTERCALATED DISCS, but these normally offer no resistance to the passage of depolarizing impulses along the fiber. Thus, each mass of cardiac muscle behaves as a functional SYNCYTIUM and contracts nearly simultaneously. The muscle follows the all-or-none law, whereby a threshold stimulus causes a maximal contraction according to the conditions (pH, temperature, chemicals), present at the time. The fibers follow the LAW OF

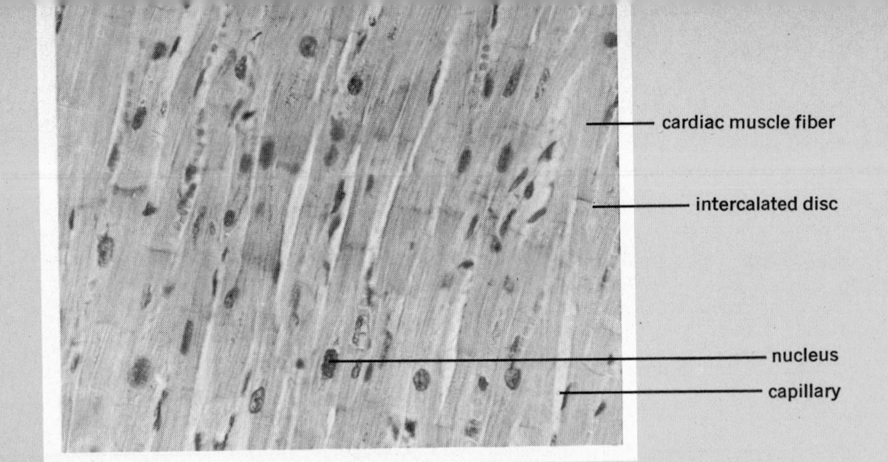

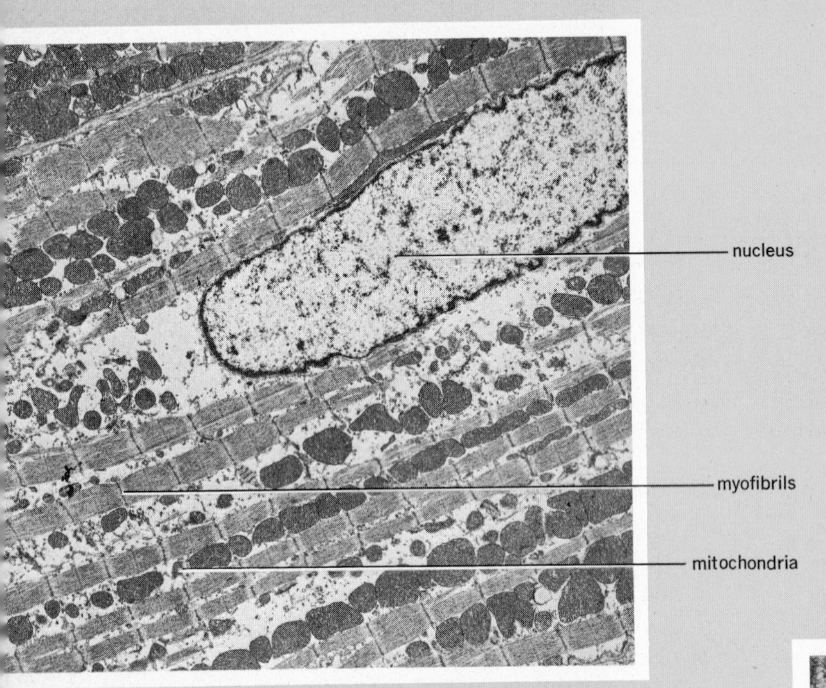

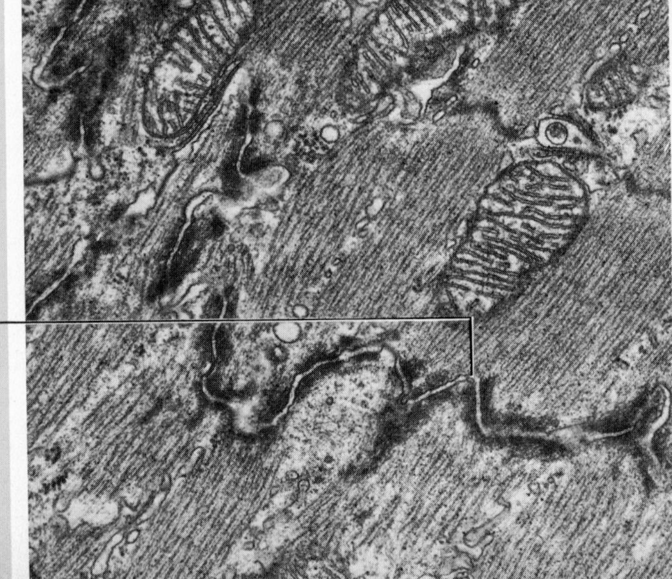

FIGURE 24.9 The histology of cardiac muscle. (Courtesy Norton B. Gilula, The Rockefeller University.)

THE HEART (Frank-Starling law), by which tension applied to the muscle will result in a stronger contraction, to the point where the tension begins to cause mechanical damage to the muscle. The tension applied to the muscle fibers with the heart *in situ* (in position in the body) is caused by the degree of filling of the ventricle just before ventricular contraction (the end diastolic volume). After being depolarized, the cardiac muscle repolarizes more slowly than skeletal muscle and passes through an ABSOLUTE REFRACTORY PERIOD during which no stimulus, no matter how strong, will cause a second response. Then comes a RELATIVE REFRACTORY PERIOD, during which the muscle is responsive to a stronger-than-normal stimulus. The absolute refractory period extends partway into the relaxation phase of the muscular activity, and thus it is not usually possible to cause the heart to undergo a sustained contraction.

After the relative refractory period has ended, a second normal-strength stimulus may cause a ventricular EXTRASYSTOLE (an additional contraction) that is of lesser strength than a normal contraction (because filling and tension is less and a weaker beat occurs). A COMPENSATORY PAUSE

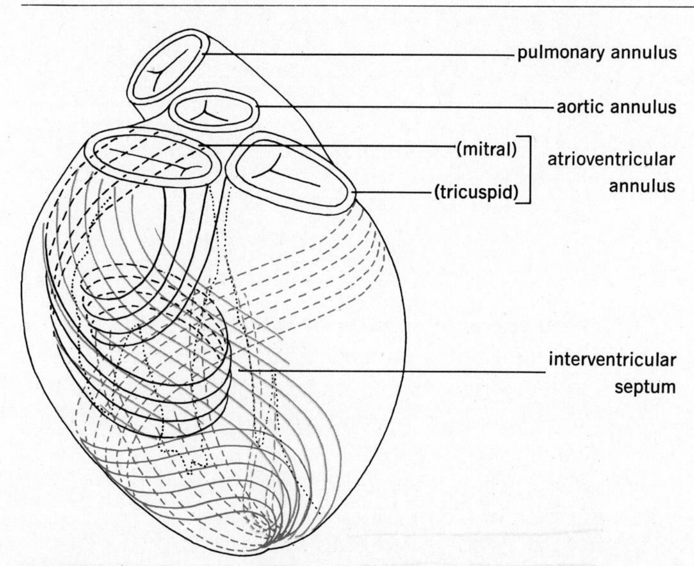

FIGURE 24.10 A diagram representing the orientation of the major bundles of cardiac muscle in the heart.

FIGURE 24.11 The genesis of an extrasystole and the compensatory pause. Stimulus applied at ⇧ causes extrasystole; next normal SA node impulse arriving at * finds muscle refractory from extrasystole. Muscle does not respond until SA node impulse arrives at normal time.

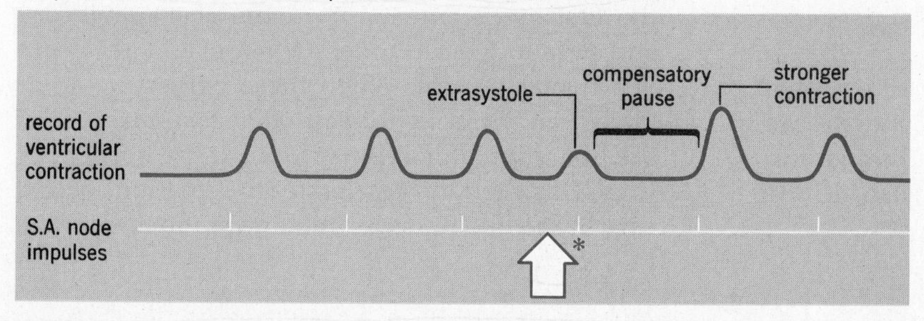

TABLE 24.2 Summary of properties of nodal tissue and cardiac muscle

Tissue	Property	Comments
Nodal	Autorhythmicity	Allows a basic rate of beat (70–80/min) to be established. Not dependent on outside factors
	Conductivity	Fastest in AV bundle onward to excite ventricular muscle nearly simultaneously. Slowest in AV node
Cardiac muscle	Syncytial arrangement of fibers	Allows rapid spread of depolarization through a mass of muscle. Gives near-simultaneous contraction for greatest pressure development
	Follows the all-or-none law	Maximal contraction occurs if stimulus is strong enough to cause depolarization. Strength varies according to environment and mechanical factors (tension)
	Follows the law of the heart	Increased tension results in stronger contraction. Allows variation in amount of blood pumped. Adjustability
	Has long refractory period	No tetanus. Some filling before contraction to maintain pumping action

(skipped beat) will follow the extrasystole, and it will not respond until yet another SA node impulse arrives (Fig. 24.11).

It may be noted how these properties of nodal tissue and cardiac muscle assure a continual, rhythmic, strong, and adjustable beat for the efficient circulation of blood through the body. Some of these properties are summarized in Table 24.2.

Factors controlling rate and rhythm of nodal tissue activity

It was indicated in the previous section that strength of the contraction of cardiac muscle can be changed according to body conditions and needs. Changes in *rate* and *rhythm* are usually due to the following factors:

The operation of nervous activity.

Changes in blood temperature with activity.

Alterations of ionic concentrations or ratios that alter depolarization and conductivity in nodal and muscular tissues.

Changes in the chemical environment of the heart because of factors other than ions.

NERVOUS ACTIVITY. The nerves to the heart (Fig. 24.12) are derived from the autonomic nervous system.

The CARDIAC NERVES *(sympathetic)* come from the first five segments of the thoracic spinal cord and pass to the SA node, AV node, and atrial and ventricular muscle. When these nerves are stimulated, an increase in heart rate occurs because of the liberation of norepinephrine at the nerve endings. This chemical accelerates the reactions occurring in the nodal tissue and lowers the membrane potential so that depolarization is easier and time for depolarization is decreased. Since the effect of stimulating these nerves is to accelerate heart rate, they are known as CARDIOACCELERATOR NERVES.

The VAGUS (X cranial) nerves *(parasympathetic)* arise from the medulla of the brain and supply the SA and AV nodes and the atrial musculature. Stimulation of these nerves causes a marked slowing of heart rate because of the liberation of acetylcholine at the nerve endings. This chemical inhibits the reactions in the nodal tissue and creates a state of hyperpolarization that requires a longer time to decay. Since these nerves cause a slowing of heart rate, they are called CARDIOINHIBITORY NERVES.

Both sets of nerves have connections with CARDIOACCELERATOR and CARDIOINHIBITORY CENTERS (two of several "vital centers") in the medulla of the brainstem, so that input from receptors in various body regions can cause reflex alterations in heart action (see Fig. 24.24). Both sets of nerves are tonically (continually) active, so that heart rate at any given moment reflects the result of a balance in the action of the two sets of nerves.

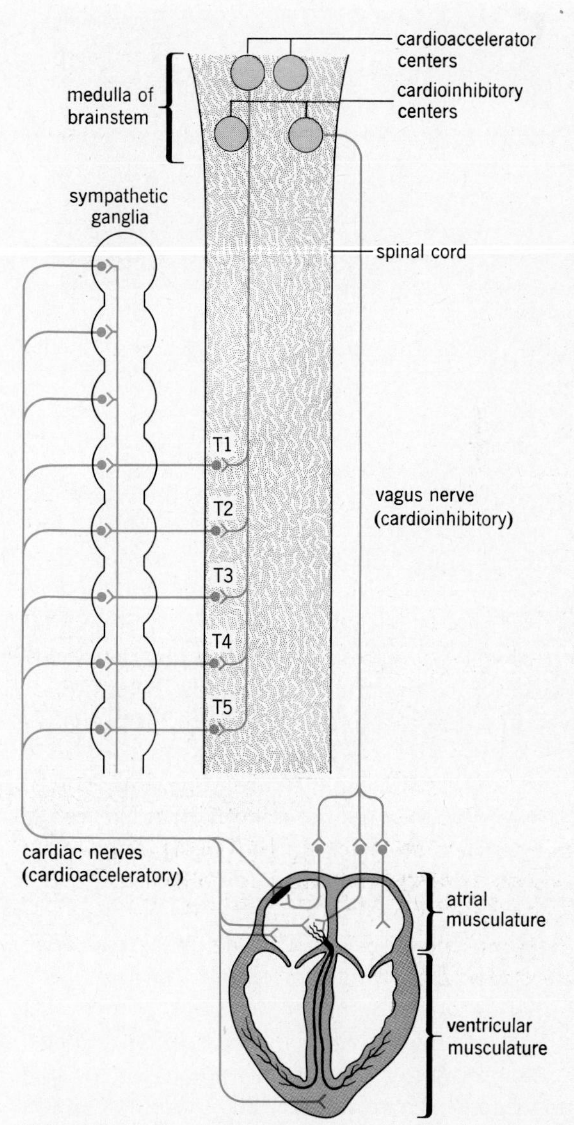

FIGURE 24.12 A diagram showing the arrangement of the autonomic nerves to the heart. Only one side of each set is represented for clarity (T = thoracic)

TEMPERATURE. As activity increases, more heat is produced, warming the blood that eventually returns to the right atrium and accelerating the reactions of the SA node. A fall of blood temperature has the opposite effect. Alterations of blood temperature within the body are of so small a degree as to have little physiological significance, but must be considered in the transfusion of large volumes of blood. The incidence of heart block has been shown to increase when cool or cold blood is transfused. The artificial lowering of body temperature (hypothermia), as carried out in certain types of surgical procedures, also slows heart rate.

IONIC EFFECTS. Potassium, sodium, and calcium ions must be present in the extracellular fluid in certain ratios to one another, and they must be maintained within rather narrow limits to assure normal heart action.

Potassium. Increase of potassium in ECF to two or three times the normal value causes a decrease

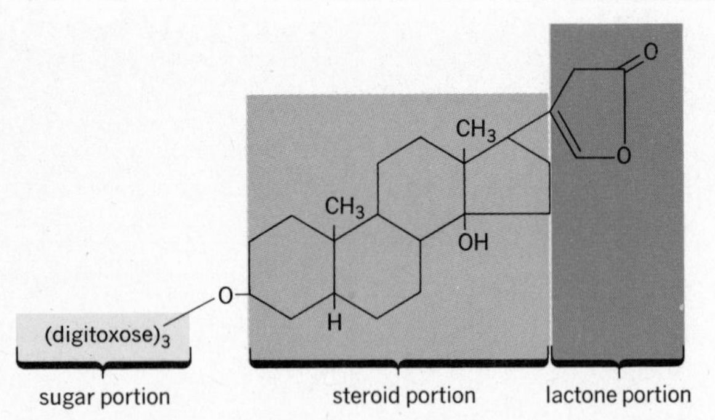

FIGURE 24.13 The chemical structure of digitalis.

in conduction velocity, especially in the AV node, and may lead to arrhythmias, heart block, and fibrillation (rates of 300–400/min). Ectopic foci (see below) also develop more easily with rise of ECF potassium levels. Decreases in ECF potassium levels decrease rate by causing a hyperpolarization of the muscle and nodal tissue.

Sodium. Little alteration in heart activity is seen with either increase or decrease of ECF sodium, unless the changes are quite large. Large increases (two to three times normal) of Na^+ cause hyperpolarization and slowed heart rate; large decreases (10–20 percent of normal) also cause a decreased rate.

Calcium. Increases in ECF calcium levels cause hyperpolarization and slowing of heart rate. Also, stimulation of strength of contraction occurs, and less and less relaxation of the muscle takes place. A state resembling tetanus may result, called "calcium rigor." Decrease of ECF calcium levels increase the development of ectopic foci (see below) and fibrillation.

EFFECTS OF OTHER CHEMICALS ON HEART RATE AND RHYTHM. Oxygen levels of the bloodstream, unless extremely low or prolonged, have little effect on heart action. Excessive CO_2 levels, by lowering pH, have a depressing effect on SA and AV node function, the muscle relaxes more, and a greater end diastolic volume is achieved (triggers the law of the heart), with an increase in strength of contraction. This effect is therefore a beneficial one in terms of increasing the circulation of blood. Lactic acid acts in the same manner. Carbon dioxide increase and pH decrease also have the effect of stimulating heart rate by exciting the medullary cardioaccelerator centers.

Norepinephrine, epinephrine, and serotonin are all catecholamines; that is, substances formed by hydroxylation (adding OH), decarboxylation (removal of CO_2) and methylation (adding $-CH_3$) to the amino acids phenylalanine and tyrosine. As a group, such substances usually increase strength *and* rate of heart contraction. Digitalis (digitoxin) is a "cardiac glycoside" often used to increase the efficiency of heart action. Glycosides consist of a sugar, a steroid, and a lactone (Fig. 24.13), the sugar differing in the several glycosides. Digitalis increases the strength of cardiac muscle contraction (inotropic effect) by causing a greater influx of Ca^{++} across the muscle membrane, increasing release of Ca^{++} bound in the sarcoplasmic reticulum, and slowing loss of Ca^{++} from the fiber after depolarization (see Chapter 9 for the role of Ca^{++} in muscle contraction). It also alters the electrical behavior of the muscle (electrophysiologic action) by increasing the rate of spontaneous depolarization, decreasing conduction velocity, and shortening of the refractory periods. Digitalis is a toxic substance, causing nausea and vomiting, diarrhea, headache, and increasing the chances for development of arrhythmias and heart block. The line between a therapeutic and a toxic dose is very fine and subject to individual variation, and its effects should be closely monitored in each and every person.

ABNORMAL RHYTHMS OF THE HEART. The term ARRHYTHMIA is applied to any variation from the normal rhythm and sequence of excitation of the heart. Arrhythmias result from the following factors:

Alterations in rate of SA node activity.
Interference with conduction.
Presence of ectopic foci.
Various combinations of the above.

Alterations in rate of SA node activity. BRADYCARDIA (decreased rate) or TACHYCARDIA (increased rate) of SA node activity, and thus heart rate, are nomally associated with breathing and sleep. Heart rate increases on inspiration, decreases on expiration, and decreases due to increased vagal stimulation during sleep.

Abnormal alterations may also be associated with DRUG INTAKE. Some drugs that increase heart rate are atropine, caffeine, epinephrine, and camphor. Some that decrease rate are amylnitrite, nitroglycerine, and alchohol in small quantities.

Interference with conduction. HEART BLOCK results from damage to various parts of the nodal conducting system, or from excessive vagal stimulation, and results in an obstruction to the passage of electrical impulses from atria to ventricles. Three degrees of block are recognized.

FIRST-DEGREE BLOCK causes a pause or delay in transmission of depolarization and thus slows ventricular rate without separation of atrial and ventricular beating rates.

SECOND-DEGREE BLOCK (partial or incomplete block) results in some sinoatrial impulses being missed by the ventricle. A slight separation of atrial and ventricular rhythm results (e.g., 75 beats/min of the atria, with 60 beats/min of the ventricles).

THIRD-DEGREE BLOCK (complete block) results in complete separation of atrial and ventricular beat, each area being controlled by that part of the nodal system closest to the muscular tissue. Thus, the atria may still be responding to sinoatrial stimulation (rate 80 beats/min) while the ventricles are responding to atrioventricular node or bundle stimulation (rate 40 beats/min). If cyanosis, faintness, and convulsions are seen, insufficient cardiac output is indicated. Third-degree block may require implantation of an artificial pacemaker if the rate falls to a point where adequate cardiac output cannot be maintained.

The pacemaker is usually implanted in the abdomen or beneath the muscle in the axilla. It provides a source of stimulations that are delivered to the heart by long wires or leads. The leads may be inserted through a vein (transvenous route) into the right ventricle for temporary or permanent pacing, or they may be surgically inserted in the ventricular wall (transpericardial route) for permanent pacing. The latter procedure involves opening the chest. The pacer may provide one of three types of pacing.

SET OR ASYNCHRONOUS PACING. This provides a constant rate of stimulation regardless of sinoatrial activity or alterations in muscular activity.

SYNCHRONOUS PACING. The pacer fires on detection of sinoatrial node activity and delivers ventricular stimulations that vary with SA node activity. This type of pacing thus varies with muscular activity levels.

DEMAND PACING. The pacer is set to fire only when it detects a rate of sinoatrial node discharge that is less than some preset value. Thus, this type only stimulates the ventricles when the heart itself fails to achieve an adequate rate of beat.

Some of the hazards associated with pacemaker installation include infections, thrombosis, power failure, and lead breakage. Isotope-powered pacemakers, and those which may be recharged from outside the body without removal of the pacemaker, may improve the reliability of these instruments.

Presence of ectopic foci. FLUTTER and FIBRILLATION are complete incoordination of contraction and the assumption of very rapid (to over 200 beats/min) heart rates. In most cases, they are due to areas of the heart, other than nodal tissue, assuming pacemaker activity. Such areas compete with the SA node to stimulate the muscle. The muscle may be divided into several separately contracting areas, each controlled by its own pacemaker. The abnormal areas of depolarization are the ectopic foci. The development of circular repetitive areas of depolarization constitute "circus movements" (Fig. 24.14). Different areas of the muscle contract at different times and coordination is lost. In most cases, throwing the heart into a simultaneous state of depolarization by the application of external shock *(defibrillation)* usually reestablishes the coordination necessary for efficient circulation of blood.

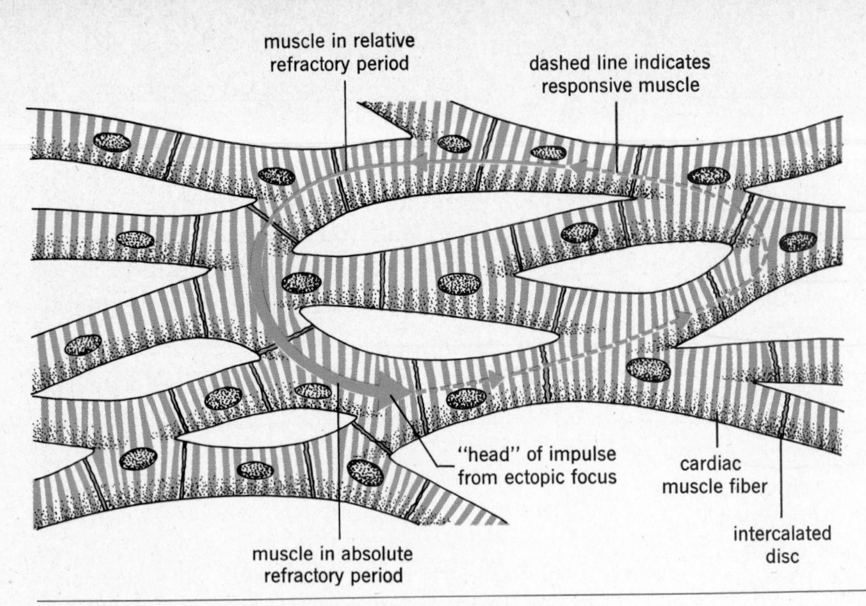

FIGURE 24.14 The development of a circular conduction pathway in the heart.

FIGURE 24.15 The timing of a cardiac cycle. Rate of beating is 70–72 beats/minute. 1 = atrial systole, 2 = atrial diastole, 3–8 atrial diastasis. A–C = ventricular systole, D = ventricular diastole, E–H = ventricular diastasis.

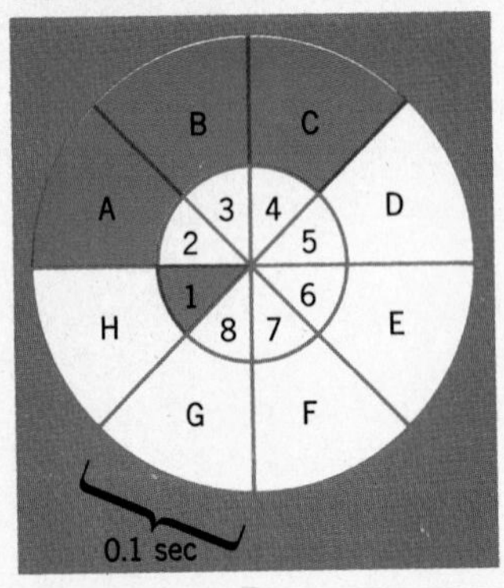

FIGURE 24.16 The normal electrocardiogram.

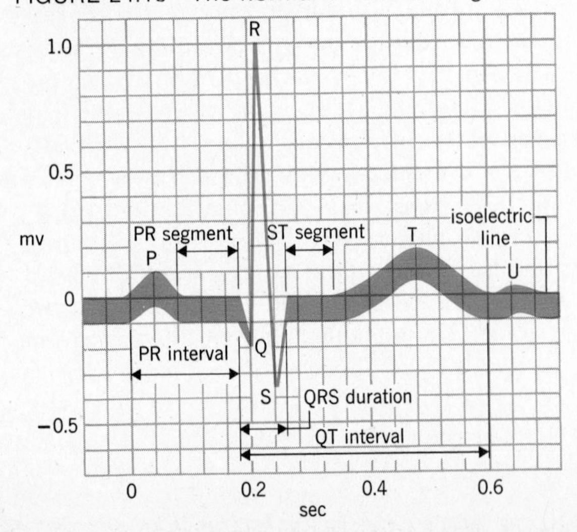

The cardiac cycle

One complete series of events during heart activity constitutes one cardiac cycle. Since electrical activity precedes all other events in a cycle, the beginning of a cycle is usually taken as the generation of an impulse by the SA node. The events of a single cycle give information as to the electrical and mechanical events that occur as the heart beats. Within the framework of a single cycle, three important types of events occur.

ELECTRICAL DISTURBANCES may be recorded as an electrocardiogram (ECG or EKG).

The contraction of the heart chambers produces PRESSURE and VOLUME CHANGES that cause blood to flow.

There are SOUNDS as the heart works.

TIMING OF THE CYCLE. Rate of heartbeat varies according to many factors. At rest, however, a normal adult heart rate is usually between 70 and 80 beats per minute. Figure 24.15 illustrates the phases of a cardiac cycle when the resting rate is 70–72 beats per minute. SYSTOLE is a term referring to contraction of a chamber, DIASTOLE refers to the relaxation phase of activity usually associated with a rapid filling of the chamber, and DIASTASIS refers to a period of slow filling, usually associated with no activity ("rest") on the part of the muscle.

TABLE 24.3 Phases of the electrocardiogram		
Item (designation)	Approximate time (sec)	Causes or events occurring
P wave	0.1	Excitation of the atria
PQRS interval	0.12-0.2	Wave travels through AV node, bundle, and Purkinje fibers
QRS complex	0.08	Excitation of the ventricles
T wave	0.1	Repolarization of the ventricles
QT interval	0.35	Time required for complete excitation and recovery of the ventricles
S-T segment	0.1	Represents time for complete excitation of ventricles
U wave	0.1	Thought to be due to slow repolarization of papillary muscles. Appears with hypokalemia. Not present in all ECGs

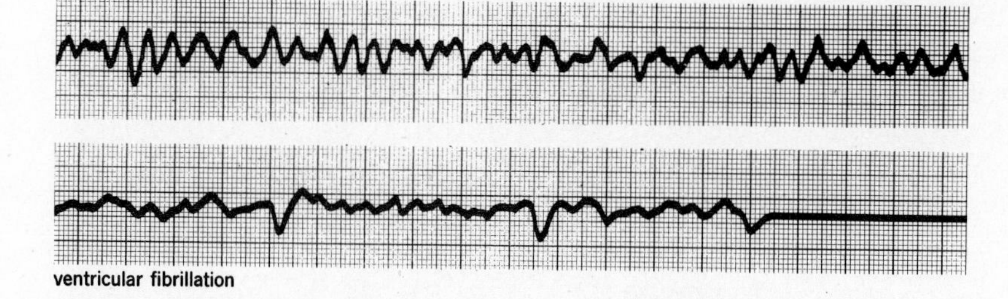
ventricular fibrillation

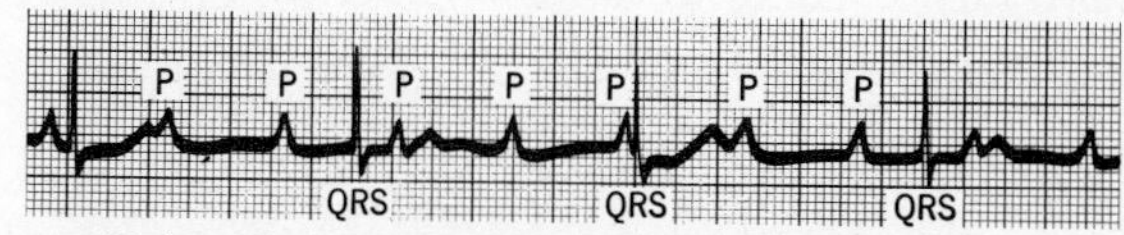

complete heart block (atrial rate, 107; ventricular rate, 43)

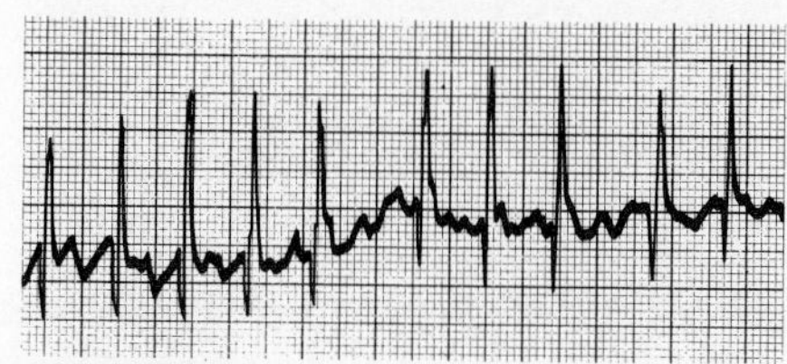
atrial fibrillation

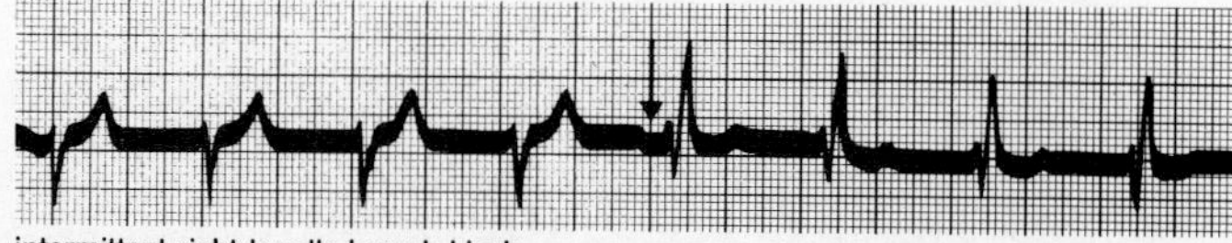
intermittent right bundle branch block

FIGURE 24.17 Some abnormal electrocardiograms in various conditions.

The last phase (diastasis) is followed by contraction of the chamber.

ELECTRICAL CHANGES; THE ELECTROCARDIOGRAM. As the depolarization wave initiated by the SA node sweeps over the heart, active tissue becomes electrically negative to inactive areas. This generates a cardiac electrical field that may be recorded as an electrocardiogram (ECG or EKG). A normal ECG is shown in Fig. 24.16. The various deflections are designated by letters; between letters are a variety of intervals or complexes. The deflections, intervals, and complexes are summarized with an explanation of their causes in Table 24.3.

The value of the ECG lies in the fact that any alterations of conduction and status of the nodal or cardiac tissue must and will be reflected as a deviation from the normal ECG. Alterations in polarity or voltage are seen in the vertical plane; alterations in timing are seen horizontally. The analysis of ECGs has proceeded to the point that definitive alterations are characteristic of specific disorders of the heart (Fig. 24.17). The ECG of a damaged heart may be followed as the organ recovers and this gives clues as to the rapidity and completeness of the recovery process.

In recording an ECG in a human subject, electrodes may be attached to any part of the body (Fig. 24.18). Conventionally, the so-called STANDARD LEADS are attached to the two wrists and to the left ankle. This type of placement creates a triangle within which the heart is enclosed. Lead I lies between the right and left wrists. Lead II lies between the right wrist and left ankle, and lead III between the left wrist and left ankle (Fig. 24.18*a*). The electrical deflections in these standard leads are shown in Fig. 24.19. The height of the deflections in these leads is determined by the amount of tissue depolarizing and by the status of the muscle and nodal tissue.

NINE ADDITIONAL LEADS may be employed to assess the function of particular areas of the myocardium. Six positions on the chest designated

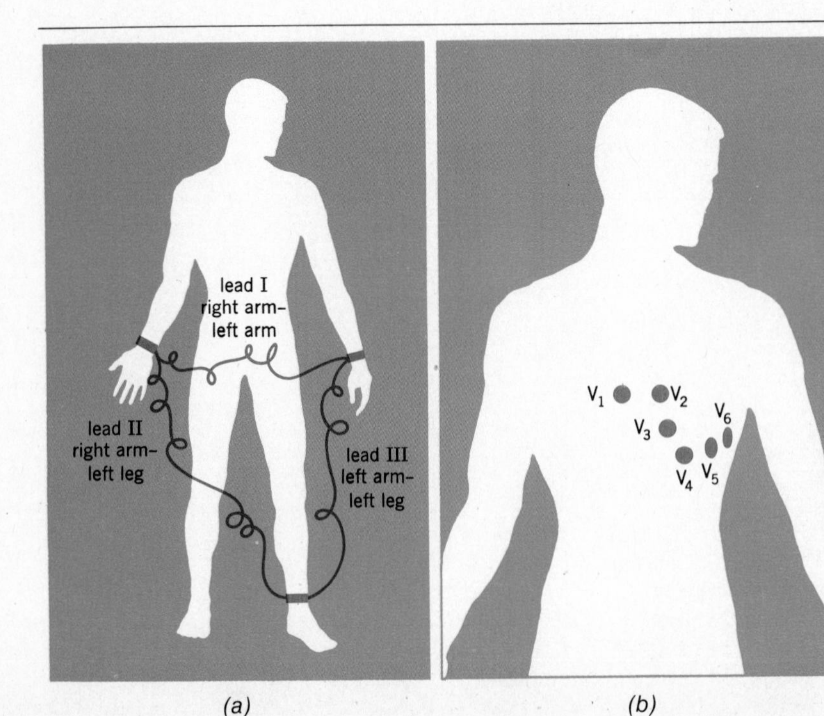

FIGURE 24.18 *(a)* The three standard electrocardiographic leads, and *(b)* the chest leads.

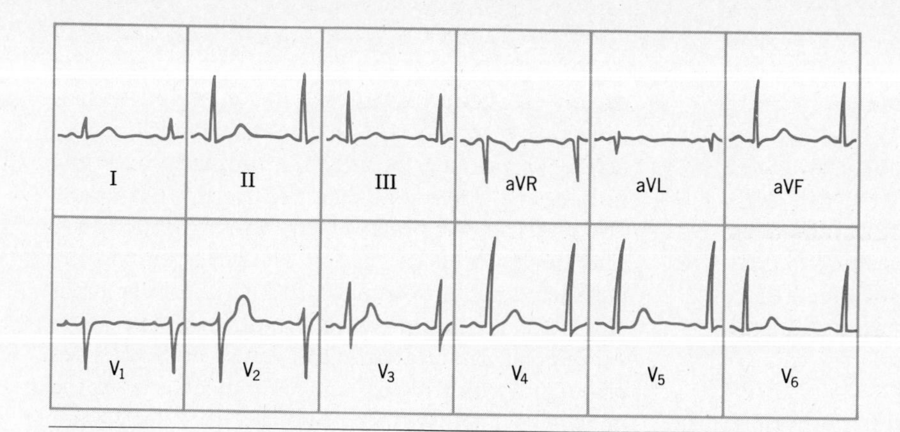

FIGURE 24.19 A normal 12-lead electrocardiogram.

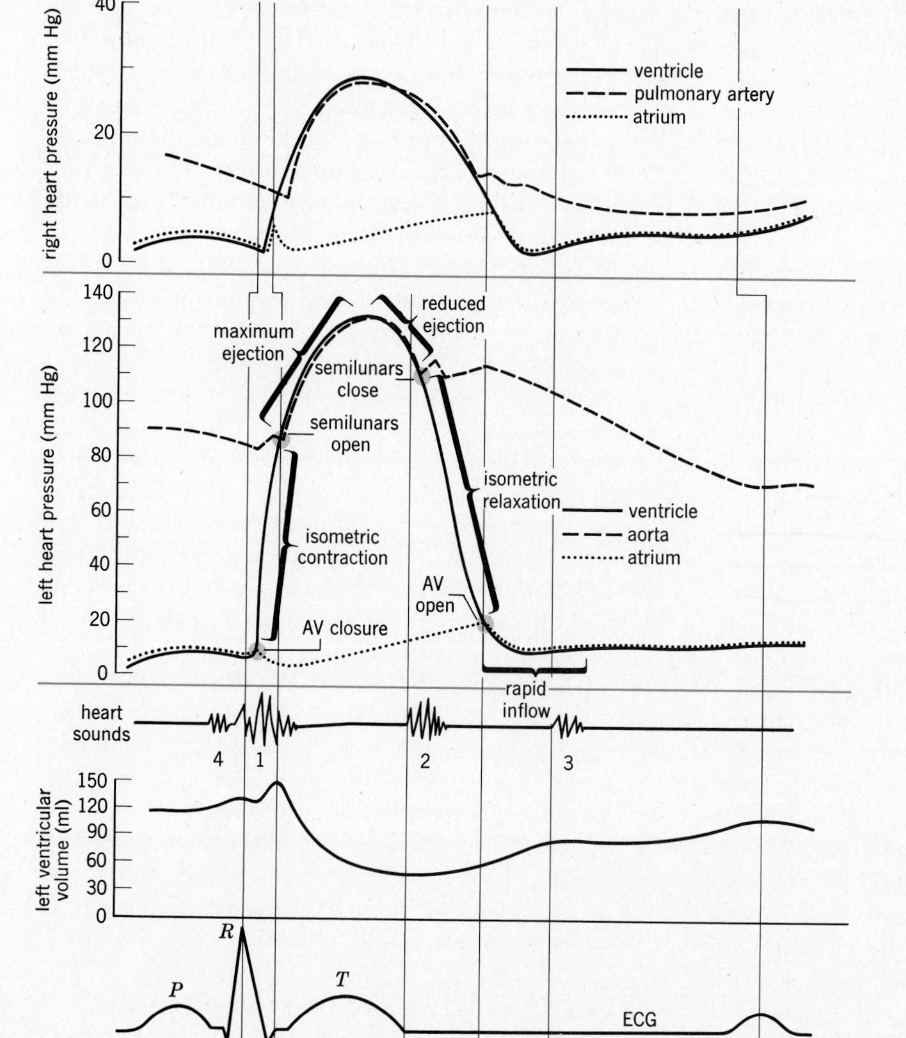

FIGURE 24.20 Correlation of events in a cardiac cycle.

V_I to V_{VI} (Fig. 24.18*b*) may be employed. Wiring the three standard leads so that the positive lead is alternately on the left ankle, right arm, or left arm produces three leads designated aVF, aVR, and aVL, respectively. This type of wiring "looks at the heart" from the particular lead that is positive and gives an electrical picture of that surface of the organ. Records from chest leads and the aV leads are shown in Fig. 24.19.

PRESSURE AND VOLUME CHANGES. Pressure changes occur within the atria, ventricles, and arteries leaving the ventricle (aorta and pulmonary artery). Significant volume changes occur only in the ventricles as they contract and relax. These changes and their correlation with ECG and heart sounds are shown in Fig. 24.20. The figure should also be correlated with the description that follows.

Atrial contraction commences immediately after the P wave on the ECG. The right atrial pressure rises from about 3 mm Hg diastolic pressure to a systolic pressure of 7 mm Hg; left atrial pressure rises from 3 mm Hg diastolic to about 10 mm Hg systolic. This force imparted to the blood by atrial contraction is of minimal importance to ventricular filling. Most of the filling of the ventricle occurs when the ventricle relaxes and creates a lowered pressure that "draws" blood into the expanding chamber. Filling continues during diastasis and terminates with initiation of ventricular systole. The atria thus act as receiving and storage chambers for blood returning from lungs and body. During atrial contraction, the tricuspid and mitral valves are open because the ventricular pressure is lower than that of the atria; blood flows from atrium to ventricle. Atrial excitation is followed by the QRS segment of the ECG and ventricular excitation; the ventricles contract. Ventricular pressure begins to rise. It rises above that of the atria, and at this point the tricuspid and mitral valves are firmly closed (first heart sound). There is no exit of blood from the ventricles, since they have not generated sufficient pressure to open the arterial valves, closed from the preceding cycle. The ventricles thus enter a period of isometric contraction. The right ventricle must achieve a pressure of only 18 mm Hg to open the pulmonary valve; the left ventricle must create about 80 mm Hg pressure to open the aortic valve. The thicker wall of the left ventricle renders the ventricle capable of generating this higher pressure. At the point when ventricular pressure exceeds arterial pressure, the valves of the arteries are opened and blood is rapidly ejected from the ventricles. Peak systolic pressures of about 30 mm and 130 mm Hg are created by the right and left ventricle, respectively. A momentum is imparted to the blood that causes continued ejection but at a slower rate, even though the ventricles are now starting to relax and pressure within them is falling. Pressure continues to decline within the ventricles until pressure falls below that within the arteries. The blood attempts to return to the ventricles and, in so doing, closes the arterial valves (second heart sound). An incisura (notch) in the arterial pressure curve occurs as the blood rebounds from the closed valve and sets up a secondary shock wave in the artery and pressure rises. The ventricles are again closed chambers, and no blood is moving into them; a period of isometric relaxation is entered. Fall of pressure in the ventricles continues until it drops below that of the atria. At this point, the tricuspid and mitral valves open and ventricular filling begins. Note that movement of blood, and valve action, is secondary to pressure changes.

The volume of the ventricles during diastasis is normally 100–150 ml per chamber. Under resting conditions, about two-thirds of the contained blood is ejected as the ventricles contract. This leaves a reserve of blood in the ventricle that may be ejected by a more forcible contraction. This ability to increase the amount of blood per beat is an important factor in changing the cardiac output to meet varying body demands.

HEART SOUNDS. Two distinct sounds are normally heard through a stethoscope as the heart works. The FIRST HEART SOUND occurs early in

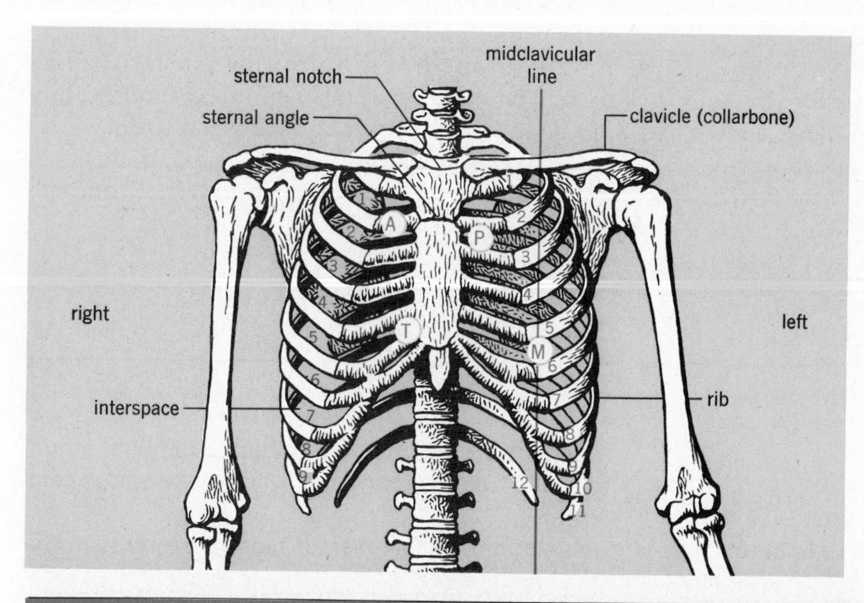

FIGURE 24.21 Chest areas where adult valve sounds may be heard most clearly. Interspaces are numbered according to the rib that lies above. The second rib is the one easiest to locate because it attaches at the sternal angle about two inches below the sternal notch. A = aortic valve, P = pulmonary valve, T = tricuspid valve, M = mitral (bicuspid) valve.

TABLE 24.4 Summary of events occurring during the cardiac cycle

ECG, electrocardiogram	Depolarization of nodal tissue and cardiac muscle	Characteristic pattern for normal and abnormal states. Has diagnostic value
Pressure changes	Contraction of cardiac muscle	Pressure changes eject blood from a chamber and operate the valves
Two sounds normally heard	Closure of valves Turbulence of blood flow Muscle and valve vibrations	First sound due to all three causes. Second sound due only to valve closure

ventricular contraction, is rather drawn out, is of low frequency (30-45 cps), and is phonetically represented by the word "lubb." Three factors contribute to its production.

VIBRATION OF THE ATRIOVENTRICULAR VALVES during and after closure.

TURBULENCE in the stream of blood as it is ejected through the arterial openings.

VIBRATIONS IN THE MUSCLE of the ventricles as they contract.

The SECOND HEART SOUND, represented phonetically by the word "dup," occurs as ventricular pressure begins to fall from peak values. It is shorter in duration than the first sound and is higher pitched (50–70 cps). CLOSURE OF THE ARTERIAL VALVES causes this sound.

With amplification, a THIRD and FOURTH HEART SOUND (Fig. 24.20) may be detected. The third heart sound occurs during or after atrial contraction and is attributed to the tensing of ventricular walls due to the rush of blood from atrium to ventricle. The fourth heart sound may be heard during atrial contraction and is attributed to the muscular contraction of these chambers.

Intensity of the sounds depends on the force with which the heart is beating, becoming louder with stronger activity. Splitting of the sounds may occur, particularly the second sound, since the valves do not close synchronously (together). The sites where the sounds created by the action of each valve may be best appreciated are shown in Fig. 24.21.

MURMURS may occur due to regurgitation (backflow) of blood when valves fail to close completely. *Functional murmurs* are no threat to the body since they usually occur only during strenuous exercise and are usually outgrown. *Resting murmurs* are a cause for concern, for they usually reflect the presence or result of an abnormal process in the body. Narrowing (stenosis) of an orifice or artery may also produce modification of a sound. The heart sounds thus have some diagnostic value to the physician. The major events of the cardiac cycle are shown in Fig. 24.20 and are summarized in Table 24.4.

The heart as a pump

Cardiac output

Contraction of the cardiac muscle creates a pressure on the blood within the heart chambers, causing it to circulate through the blood vessels. The magnitude (height) of the pressure, the speed with which the blood is circulated, and the efficiency with which the heart meets tissue demand for blood depends primarily on the CARDIAC OUTPUT (CO), defined as the amount of blood ejected per minute by each ventricle. CO is the product of two other factors: the rate of heartbeat (STROKE RATE or SR) and the volume of blood ejected per beat (STROKE VOLUME or SV).

$$\underset{\text{(ml blood/min)}}{\text{CO}} = \underset{\text{(beats/min)}}{\text{SR}} \times \underset{\text{(ml blood/beat)}}{\text{SV}}$$

The normal adult has a resting heart rate of 70–80 beats/min and a stroke volume of 60–70 ml of blood per beat. Thus, a resting cardiac output may be calculated as lying between 4200–5600 ml of blood per minute.

Alterations in cardiac output associated with activity or other changes are met by altering stroke rate, stroke volume, or both simultaneously.

Factors controlling changes in cardiac output

Five major mechanisms are normally responsible for modifying cardiac output.

The VOLUME OF BLOOD returned to the right atrium from the tissues. This is the venous inflow or venous return.

Alterations in the ease of blood flow through the small arteries of the body. This is termed the PERIPHERAL RESISTANCE.

ALTERATIONS OF TEMPERATURE.

CHEMICAL INFLUENCES.

NERVOUS FACTORS.

VENOUS INFLOW. During exercise, for example, the volume of blood returned to the right heart through the veins is increased by the massaging action of the active skeletal muscles. The right atrium, and next the right ventricle, are filled to a greater degree, and as a result are stretched to a greater degree than normal. This invokes the law of the heart, and contractions become stronger, ejecting more blood per beat; in short, stroke volume increases. More blood through the lungs increases venous return to the left atrium and ventricle, and left ventricular output into the aorta is increased. Thus, after a few beats, right heart inflow and left heart output are balanced. This mechanism is an example of autoregulation by the heart—that is, adjustment without involvement of outside factors. These factors are presented in Fig. 24.22.

PERIPHERAL RESISTANCE. The small arteries (arterioles) of the body have large amounts of

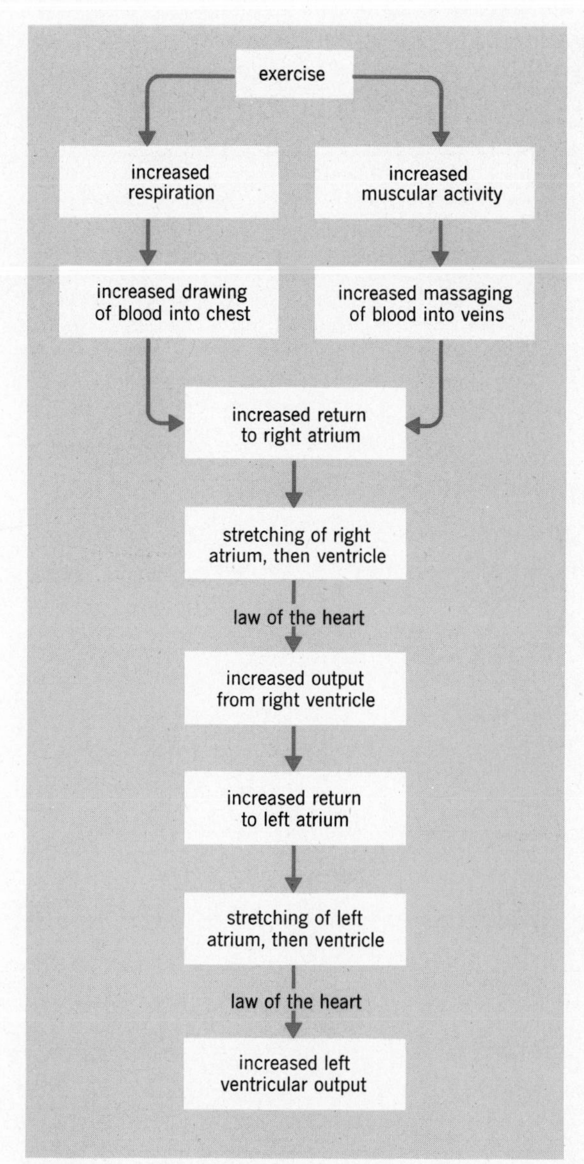

FIGURE 24.22 The effects of exercise on heart action.

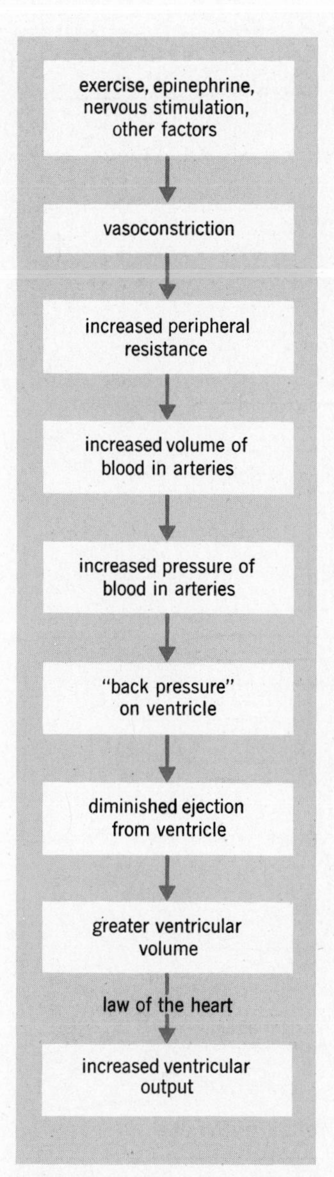

FIGURE 24.23 The effects of vasoconstriction on heart action.

circularly disposed smooth muscle around them. The vessels can thus change their diameter to a greater degree of their total diameter than any other vessel. A decrease of diameter (*vasoconstriction*) and an increase of diameter (*vasodilation*) increase and decrease peripheral resistance, respectively. Increased resistance calls for a greater pressure to maintain adequate flow through the narrowed channels. The ventricles increase their strength of contraction, emptying the chamber

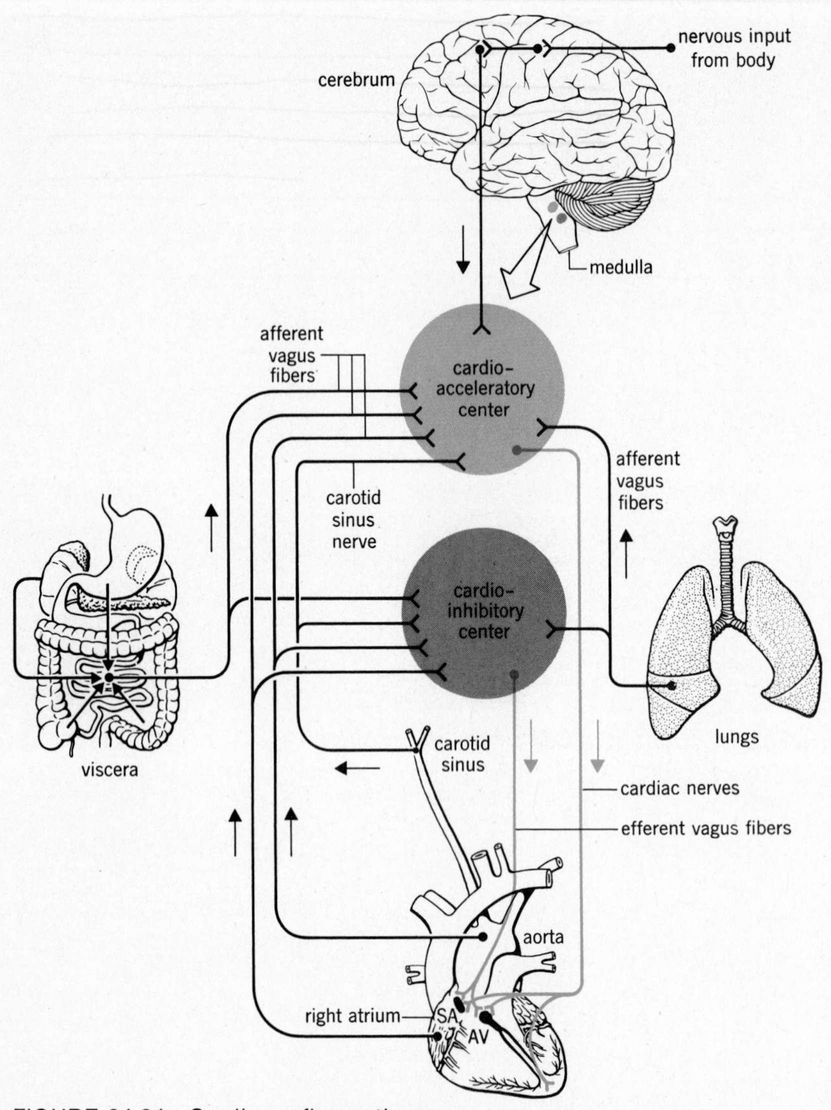

FIGURE 24.24 Cardiac reflex pathways.

more completely and thus raising stroke volume, cardiac output, and arterial blood pressure. These events are shown in Fig. 24.23.

TEMPERATURE. As indicated in an earlier section, a rise in temperature produces a faster rate primarily, with a small increase in volume. The effect on rate is apparently exerted directly on the SA node, accelerating the chemical reactions within it. The speed of transmission through the rest of the nodal tissue is also increased. The small rise in volume is accounted for by the elevated metabolism and increased strength of contraction associated with rise of temperature.

CHEMICALS. Epinephrine, a hormone of the

TABLE 24.5 Cardiac reflexes

Name of reflex	Input from	Stimulus triggering	*Effect on CIC[a]	*Effect on CAC	*Effect on SR	*Effect on SV
Aortic depressor	Baroreceptor (sensitive to stretch) in aorta	Stretch of aorta	+	−	↓	0
Carotid sinus	Baroreceptor in carotid sinus	Stretch of sinus	+	−	↓	0
—	Baroreceptors in pulmonary arteries	Stretch of arteries	+	−	↓	0
Bainbridge reflex (existence disputed)	Vena cava or right atrium	Increased filling of atrium (stretching)	−	0/+	↑	0
—	Cerebrum—anger/fear	Stress, thoughts	−	+	↑	0
—	Strong stimulation of any nerve	Pain	−	+	↑	↑

*+ (stimulate); − (inhibit); 0 (no change); ↑ (increase); ↓ (decrease).
[a] CIC = cardioinhibitory center
CAC = cardioacceleratory center
SR = stroke rate
SV = stroke volume

adrenal medulla, is liberated with exercise or stress on the body. The chemical enhances the rate of SA node activity and also increases contractile strength. Large rises in cardiac output are thus brought about. The hormone is rapidly deactivated by the liver so that its effects, while great, are short-lived. Elevated levels of CO_2 in the blood increase relaxation of the ventricles, causing a greater filling and therefore more tension on the muscle. A stronger contraction results, and stroke volume increases. Increased CO_2 levels decrease pH by the reaction of CO_2 and H_2O. This decrease of pH has the same effect as increased CO_2 levels.

NERVOUS REFLEXES. Nervous reflexes provide a means by which extrinsic factors may influence heart action (Fig. 24.24). It may be recalled that nerve fibers (cardioinhibitory nerves) rise from neuronal cell bodies in the medulla of the brainstem (cardioinhibitory centers), and that these fibers pass through the vagus nerves to terminate primarily on the SA and AV nodes. From the superior cervical ganglia, sympathetic fibers (cardioaccelerator nerves) derived from medullary centers (cardioaccelerator centers) terminate on the SA and AV nodes and ventricular muscle. The centers receive input from many body areas including the right atrium, aorta, carotid sinus, lungs, viscera, and cerebrum.

Vagal stimulation results in a decreased rate

TABLE 24.6 Mechanisms controlling cardiac output

Mechanism	Component(s) of CO affected most	Comments
Increased venous return to right atrium	Stroke volume	Brought about by law of the heart (increased tension causes stronger contraction and greater emptying)
Increased peripheral resistance	Stroke volume	As above
Temperature	Stroke rate	Effect on nodal tissue primarily
Chemicals		
Epinephrine	Both	Increases rate and strength
CO_2	Stroke volume	Greater filling causes more tension and stronger contraction
pH	Stroke volume	May act in same manner as CO_2
Acetylcholine	Both	These chemicals are liberated at nerve endings and cause change in rate. Acetylcholine slows, and norepinephrine speeds heart rate
Norepinephrine	Both	
Nervous reflexes	Stroke rate	Allows input from any body area: most necessary for continual adjustment to body requirements

FIGURE 24.25 Some factors responsible for increase in cardiac output.

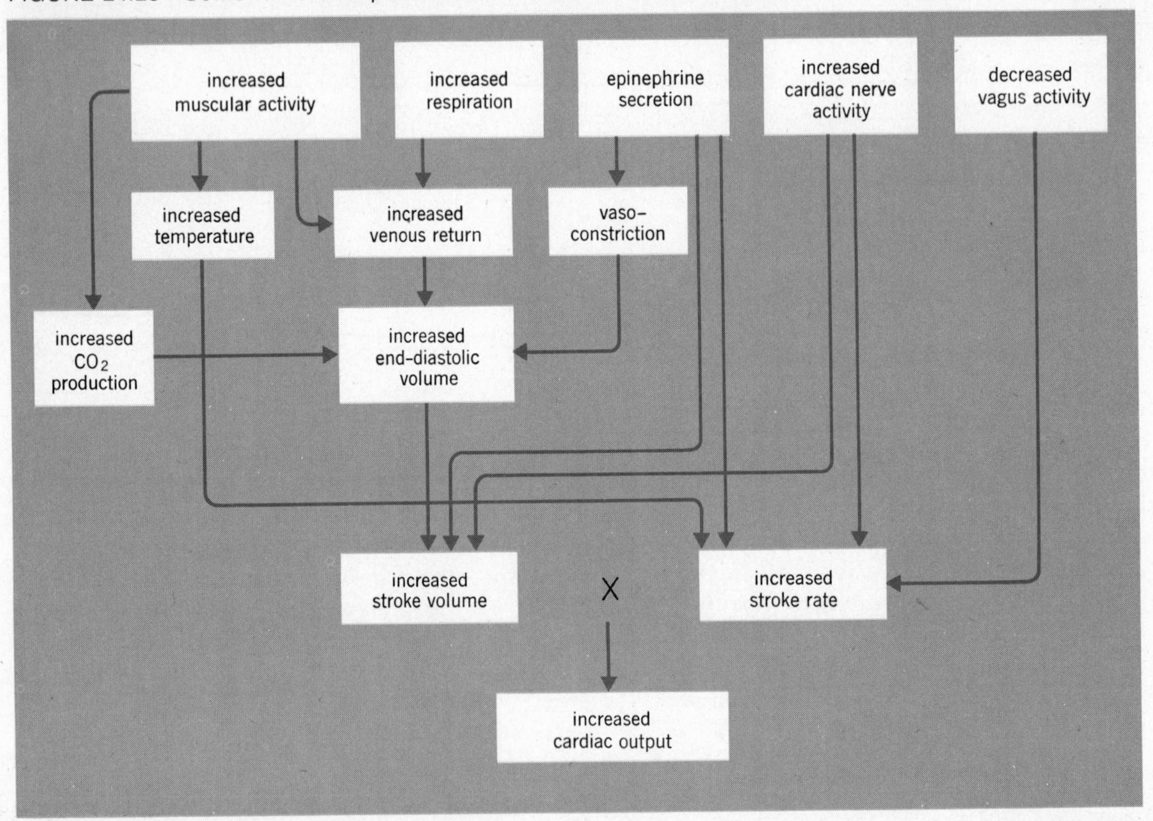

of heartbeat due primarily to the liberation of acetylcholine at the vagal terminations. Sympathetic stimulation results in increased heart rate and increased strength of ventricular contraction through the release of norepinephrine (similar to epinephrine) at the sympathetic terminals. Some nervous reflexes altering heart action are summarized in Table 24.5.

Nervous reflexes provide a continuous and critical control primarily over the stroke rate. A summary of mechanisms controlling cardiac output and thus the efficiency of the heart as a pump are presented in Table 24.6 and Fig. 24.25.

CARDIAC RESERVE. The foregoing discussion has indicated that the heart has a great capacity to increase the amount of blood it pumps per minute and, in so doing, is normally able to supply the needs of changing body demands. Table 24.7 presents data relating to distribution and total cardiac output at various activity levels. The ability to increase output constitutes the cardiac reserve. The magnitude of the reserve may be demonstrated by a ventricular function curve (Fig. 25.26). Note that there is a limit to output imposed by the anatomical and physiological properties of the heart. A further limit on output is placed by the heart rate. When rate rises above 140–150 beats/min, insufficient time is allowed for ventricular relaxation and filling. Consequently, stroke volume decreases, and output may decrease.

HEART FAILURE. Inability of cardiac output to keep pace with normal body demands for nutrients and waste removal constitutes heart failure. This usually occurs when myocardial contraction fails and output falls to 2–2.5 liters of blood per minute. Tissues are starved, suffocated, submerged in their own wastes, or poisoned by the

FIGURE 24.26 Ventricular function curves for the right and left ventricles. Output increases according to input to a maximum value.

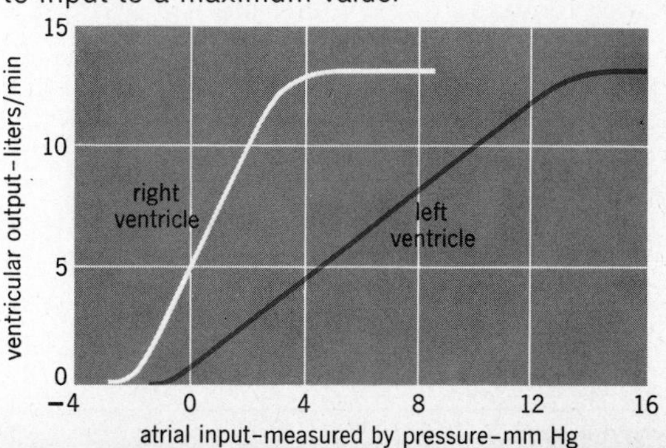

TABLE 24.7 Some figures showing distribution of the cardiac output to various body areas and total cardiac output of a normal adult during various activity levels

Area supplied	Activity level and blood flow (ml/min)			
	Rest	Light	Strenuous	Maximal
Visceral organs (gut, liver, pancreas, lungs)	1400	1100	600	300
Kidneys	1100	900	600	250
Brain	750	750	750	750
Heart (coronary)	250	350	750	1000
Skeletal muscles	1200	4500	12500	22000
Skin	500	1500	1900	600
Other	600	400	400	100
Total cardiac output	5800	9500	17500	25000

Note that blood is diverted away from "nonessential" areas (gut) to skeletal muscles and heart as activity increases. Note also that brain supply remains constant.

failure of the kidney to filter wastes from the blood. One ventricle or both may fail, creating such additional symptoms as pulmonary edema (left ventricle failure), low blood pressure (left ventricle failure), and increased venous pressure (right ventricle failure).

Measurement of cardiac output

It is difficult to measure cardiac output accurately since heart action is influenced by many factors. Two of the better methods employed to measure cardiac output are the Fick and the dilution methods.

THE FICK METHOD. The FICK METHOD compares the oxygen or carbon dioxide content of femoral vein and femoral artery blood, and—knowing the rate of O_2 absorbed or CO_2 eliminated per minute by the lungs—calculates cardiac output by the equation

$$\text{cardiac output (l blood/min)} = \frac{O_2 \text{ absorbed or } CO_2 \text{ eliminated (ml/min)}}{\text{AV } O_2 \text{ or } CO_2 \text{ difference (ml/l)}}$$

For example,

O_2 intake = 250 ml O_2/min

arterial O_2 = 19 ml O_2/100 ml blood

venous O_2 = 14.5 ml O_2/100 ml blood

AV O_2 difference = 4.5 ml O_2/100 ml blood or 45 ml O_2/l blood

$$\text{cardiac output} = \frac{250 \text{ ml } O_2/\text{min}}{45 \text{ ml } O_2/\text{l blood}} \text{ or } 5.5 \text{ l/min}$$

The Fick method is applicable to any organ that exhibits an AV O_2 difference and is amenable to sampling of arterial and venous blood; it may be used to calculate blood flow through organs.

DILUTION METHOD. Injection of a known amount of material into the circulation is followed by measurement of the dilution of the material during a known period of time. A curve is obtained, the area of which may be calculated and related to cardiac output since the rapidity of dilution of the material, and thus the slope and height of the curve, depends on the volume of blood circulated in a measured timespan.

Summary

1. The heart provides a pressure to circulate blood to the body to provide nutrients and remove wastes.
2. The heart develops early in embryonic life to assure circulation of substances necessary for cellular activity.
   a. At two weeks a cardiogenic plate is present.
   b. At three weeks paired heart tubes are formed.
   c. At four weeks fusion of the tube forms the heart.
   d. Between five and seven weeks the four chambers are established, the wall is developed, and circulation of blood is established.
3. The heart has certain structural and locational characteristics.
   a. It is about the size of a clenched fist (age-dependent).
   b. It is located in the chest, about one-third to the right of the midline, two-thirds to the left.
   c. It is enclosed in a pericardial sac having fibrous and serous portions.
   d. The wall is composed of an inner endocardium, a middle myocardium, and an outer epicardium.
   e. It has four chambers, with two atria and two ventricles.
   f. The right side of the heart pumps into the pulmonary circulation; the left to the systemic circulation.
   g. Valves are present between atria and ventricles, are in the great arteries, and control direction of blood flow.
4. Congenital anomalies may cause defects in formation of the heart; infections may damage it.
5. The heart is supplied with blood through the coronary arteries.
   a. Flow through the vessels occurs only when the heart muscle is not contracted.
   b. The coronary circulation provides nutrients to sustain heart activity and ions to govern rate and strength of beat.

c. Blockage of the arteries is associated with "heart attack," tissue death, and possible cessation of life.

6. Nerves to the heart include:
a. Parasympathetic nerve (vagus) to slow heart rate.
b. Sympathetic nerve (cardiac nerves) to speed heart rate and strength of contraction.

7. Two types of tissue are most important in heart activity.
a. Nodal tissue (SA node, AV node, AV bundle, bundle branches, Purkinje fibers) originates and distributes the stimuli necessary for heart muscle contraction. The SA node is the pacemaker.
b. Cardiac muscle contracts and creates pressure to circulate the blood. It is extensible, involuntary, inherently rhythmic, cannot be thrown into a sustained contraction if ionic concentrations are normal, and follows the law of the heart and the all-or-none law.

8. A cardiac cycle is one complete series of events during cardiac activity.
a. It normally lasts about 0.8 seconds.
b. It creates electrical changes (ECG) that reflect the status of the nodal and cardiac tissue. The ECG is used to determine injury to the heart and recovery from injury.
c. Pressure changes operate valves and move blood.
d. Sounds are heard as the heart works.
e. Murmurs during the cycle usually indicate improper valve functioning.

9. The heart, by its output of blood, establishes a basic pressure that circulates blood. Two factors determine output.
a. Stroke rate (beats/min) is controlled primarily by chemicals and nervous reflexes (see Table 24.5).
b. Stroke volume (ml/beat) is determined primarily by chemical and mechanical factors.

10. Heart failure occurs when output does not meet body needs for blood.

The Blood and Lymphatic Vessels; Regulation of Blood Pressure

Blood vessels of the body act as the "system of plumbing" to contain the blood as it is circulated through the body. The entire system of vessels is believed to be continuous—that is, a closed system of tubes. Such an arrangement allows for more development of pressure on the blood by heart action. The vessels are living structures that can actively and passively alter their diameter to control and alter blood pressure and blood flow through the body organs and tissues.

Development of the blood vessels

The basic steps in the formation of blood vessels from blood islands was considered in Chapter 22. It may be recalled that vessel formation occurs first in the yolk sac outside of the embryo proper at 15–16 days of age. The networks of vessels are extended by "budding" from these first vessels. About the 18th day, embryonic vessels begin to form, and by the 21st or 22nd day, the vessels have developed to the degree shown in Fig. 25.1. Tissues forming the walls of the blood vessels are derived from mesoderm. Discussion of further development and refinement of the vascular system is beyond the scope of this text. The reader is referred to the table of correlated human development (Chapter 5) for more detailed discussions of vessel development.

FIGURE 25.1 Development of the blood vessels by about 21 days of embryonic life.

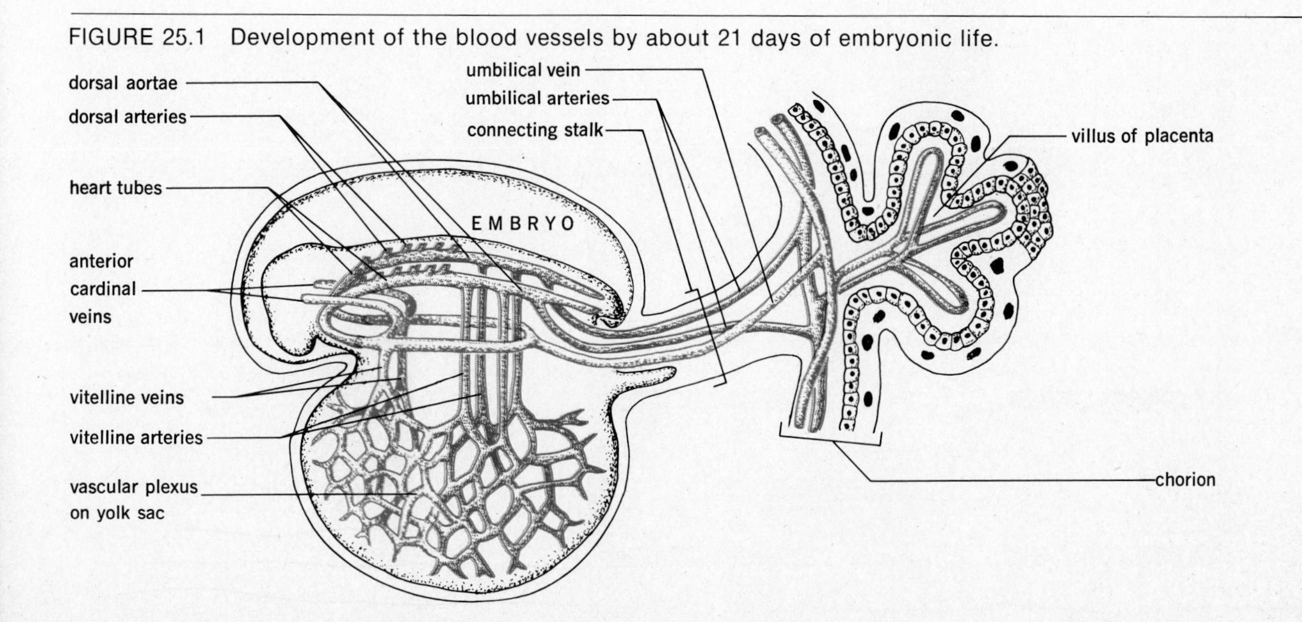

Types of blood vessels

The blood vessels that carry blood from the heart to body tissues and organs and back to the heart again fall into three major categories.

ARTERIES carry blood away from the heart regardless of the degree of oxygenation of the blood they contain. Most have elastic tissue and/or smooth muscle in their walls allowing passive and active change in diameter. According to structure and size, arteries may be classed as *large arteries* (elastic arteries), *medium-sized arteries, and small muscular arteries* (the latter are also known as arterioles). *Metarterioles* are very small vessels, differing in size and structure from arterioles, that guard the entry into capillary beds.

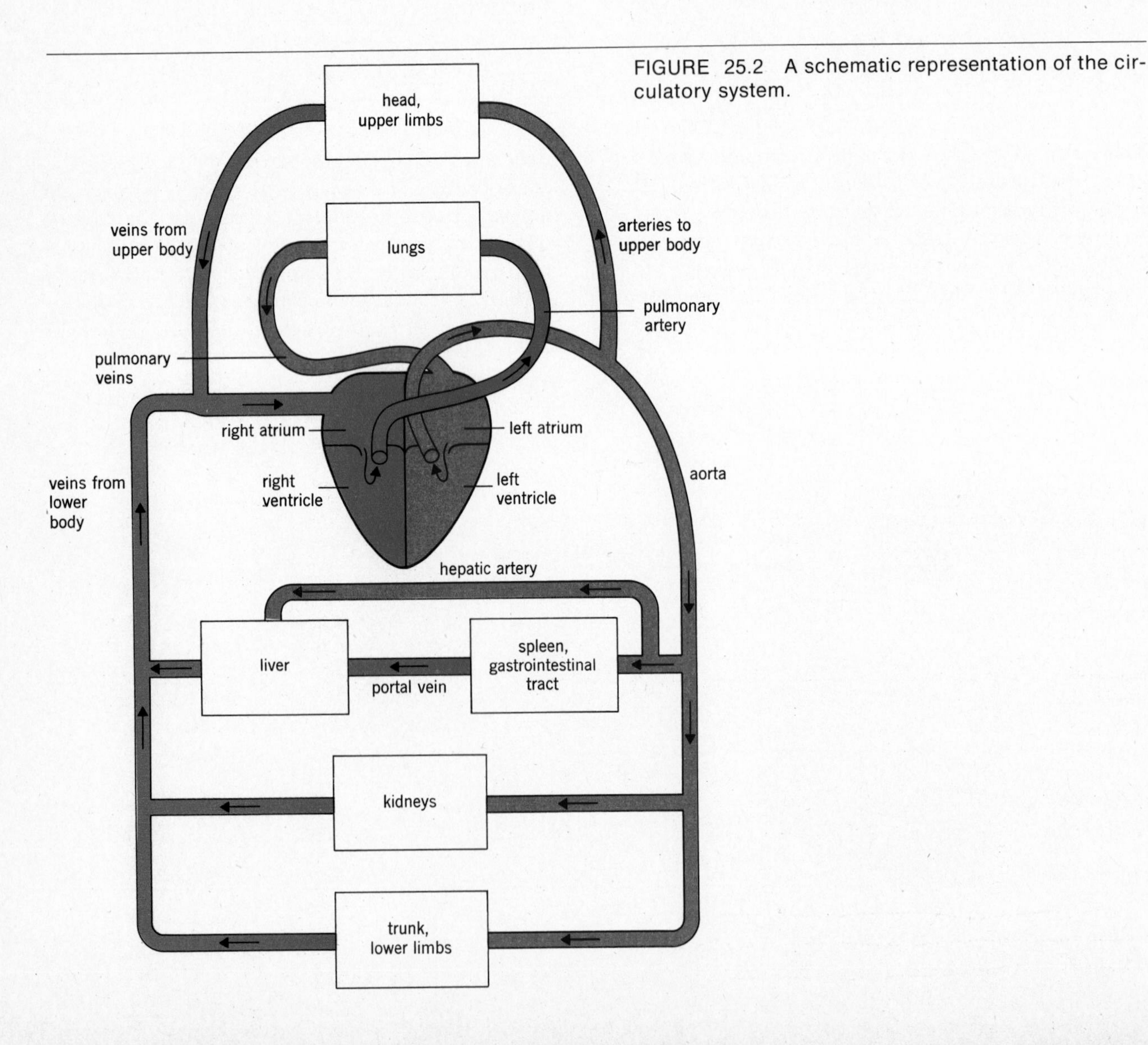

FIGURE 25.2 A schematic representation of the circulatory system.

CAPILLARIES are microscopic (about 7 μm) vessels permeating body tissues and organs, and serving as the area where exchange of materials takes place between cells and the bloodstream. They are the most numerous vessels (one estimate suggests there are over 60,000 miles of capillaries in the adult body).

VEINS carry blood toward the heart, regardless of the degree of oxygenation of the blood they contain. Connective tissue is the major component of vein walls, although some contain enough smooth muscle to allow active changes in diameter. Like arteries, veins may be subdivided into small (venules), medium, and large veins on structural and size bases.

The general arrangement of these vessels in the body is shown in Fig. 25.2.

Two main circulation pathways are evident:

The PULMONARY CIRCULATION receives blood from the right ventricle and returns it to the left atrium.

The SYSTEMIC CIRCULATION receives blood from the left ventricle and returns it to the right atrium.

Lymphatic vessels may be compared to the capillaries and veins of the blood vessel system. Lymph capillaries are blind-ended tubes in the tissues and organs. They form larger collecting vessels and lymph ducts that empty into blood veins of the upper chest region.

Structure of blood vessels

Arteries

As stated in a previous section, arteries (Fig. 25.3) are of four sizes or types. The large, medium, and small arteries all have THREE LAYERS OF TISSUE in their walls.

An inner TUNICA INTIMA is composed of the *endothelial* simple squamous lining and a thin subendothelial *connective tissue* layer. In medium and large arteries, an elastic membrane called the *internal elastic lamina* develops as a third layer of the intima. In atherosclerosis, deposition of lipids occurs within the tissues of the tunica intima.

A middle TUNICA MEDIA contains many elastic laminae (large arteries) or layers of smooth muscle (medium arteries and arterioles). Metarterioles have an endothelial tube and one layer or one smooth muscle cell around the tube; thus, they are two-layered. The media is the best-developed layer of arteries.

An outer TUNICA ADVENTITIA, usually one-third to one-half the thickness of the media, forms the third layer of tissue in large, medium, and small arteries. Metarterioles lack this layer. Small blood vessels that nourish the tissues of the vessel wall may be found in the adventitia. They form what is called the *vasa vasorum* (blood vessels to blood vessels).

A summary of the structure of arteries is presented in Table 25.1.

CLINICAL CONSIDERATIONS. ARTERIOSCLEROSIS is a generalized term referring to any vascular condition that results in thickening and loss of elasticity and resiliency of the artery wall.

ATHEROSCLEROSIS refers specifically to the changes that occur as fats accumulate in the intima and subintimal layers of the vessel wall. Accumulation of fats in the walls of arteries leads to narrowing of their size and to decrease of blood flow to the tissues supplied by the affected vessel.

Atherosclerosis is present in almost all animal species, and its changes can be detected at any age. It is considered to be an inevitable result of aging. Although the exact cause of the disorder is not known, circumstantial evidence suggests that a high-fat (cholesterol) diet, cigarette smoking, and lack of exercise accelerate the changes associated with the atherosclerotic process.

The changes that occur have been shown to occur in the following order.

TABLE 25.1 Summary of artery structure

Type of vessel	Average size (mm)	Example(s) or location	Tissue in: Tunica intima	Tunica media	Tunica adventitia	Comments
Large artery	20–25	Aorta, pulmonary artery	Endothelium and c.t. (connective tissue)	Elastic c.t.	Loose c.t.	Elastic vessels expand to contain cardiac output; recoil to push blood onward.
Medium artery	2–10	Named vessels of arms, legs, viscera	As above	25–40 layers of smooth muscle	As above	Muscle makes them important in terms of changing size; not too numerous
Small artery (arteriole)	2 mm to about 20 μ	Unnamed; close to or in tissues and organs	As above	5–10 layers of smooth muscle	As above	Large numbers make these the primary controllers of blood flow and pressure.
Metarteriole	7–10 μm	Unnamed; found at entrances to capillary beds	As above	1 layer or 1 cell of smooth muscle	None	Acts as sphincters to control flow in capillary beds

Fatty streaks appear in the arteries in the first days, months, or years of life, and deposition continues into the second decade of life.

Fibrous and fatty plaques appear beginning with the second decade of life and increase until some individuals show the signs of vessel narrowing and diminished blood flow to body organs (cardiac infarction, stroke, gangrene).

Once a lesion is established, it acts as an irritant to the artery wall, causing an inflammatory reaction that further thickens and stiffens the vessel wall. Ulceration may then cause rupture of the wall, or ischemia may cause tissue death.

Treatment of the disorder is centered about reducing dietary intake of lipids, which is thought to reduce the rate of accumulation of fats in the vessel walls. If an artery is of such a size and is in a favorable location, THROMBOENDARTERECTOMY may be performed. This procedure in effect "reams out" the vessel, removing its lipid deposits and increasing blood flow. Bypass grafting of blood vessels to renew the supply to an organ or tissue that has been deprived of blood is sometimes successful.

In some individuals, the tunics of the vessel walls do not develop properly, and the wall thins out. Under the pressure of the blood, the artery develops a thin blisterlike ANEURYSM that may rupture, leading to hemorrhage, shock, and death.

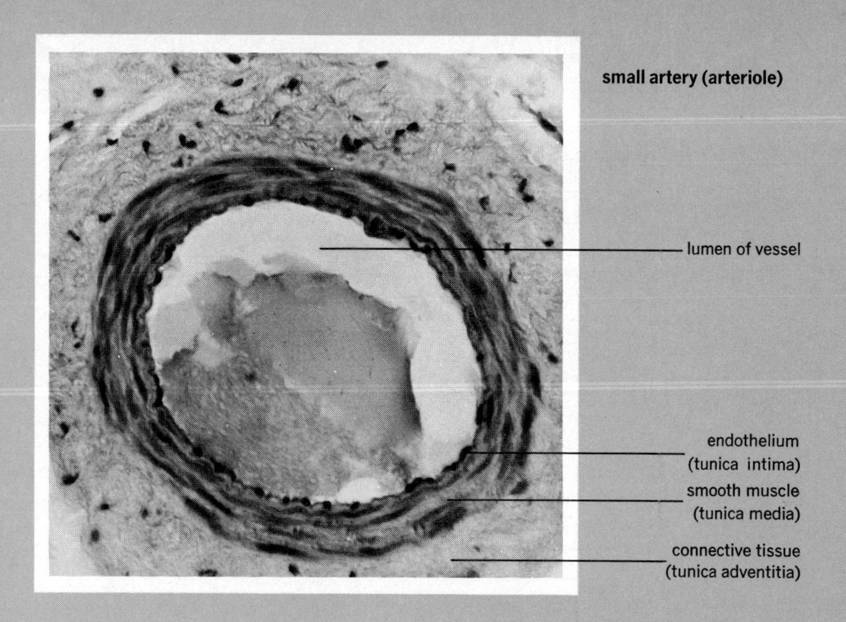

FIGURE 25.3 The histology of arteries.

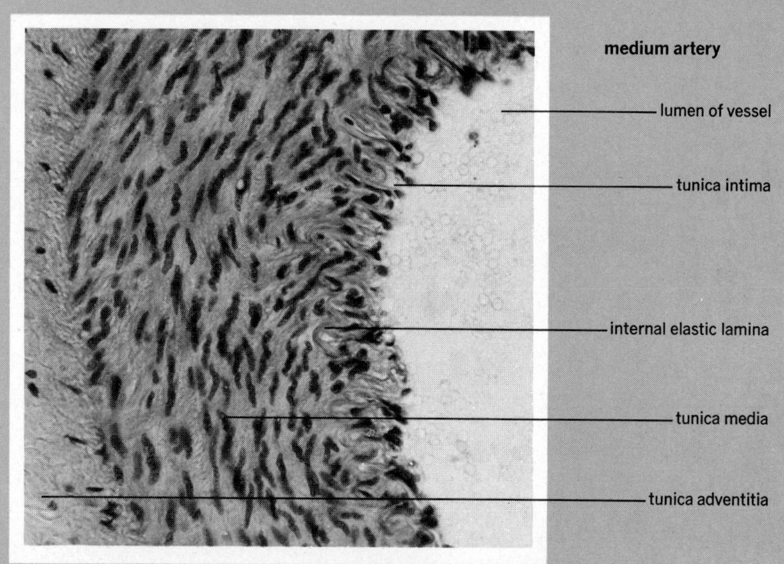

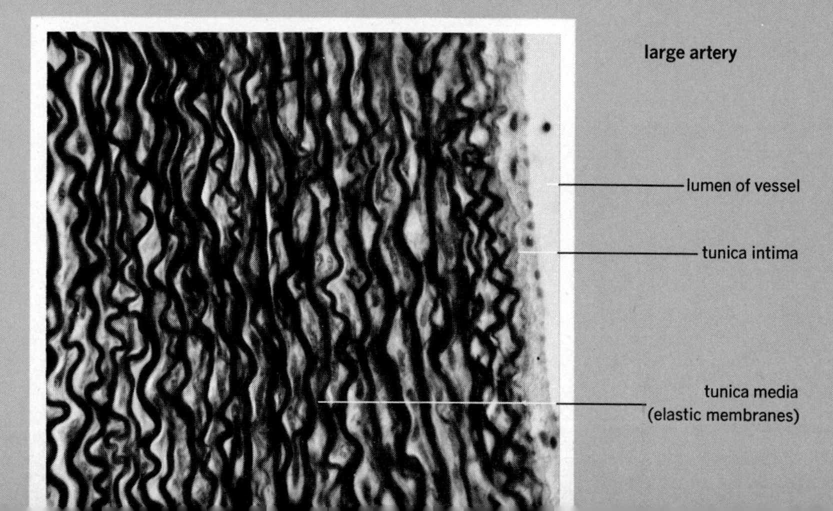

Capillaries

Capillaries (Fig. 25.4), as the vessels where exchange of materials occurs between blood and cell, must have thin walls to permit exchange by filtration and diffusion. They are usually 7–9 μm in diameter and consist of only an endothelial tube of simple squamous epithelium supported by a few loosely arranged connective tissue fibers. Their great surface area, 800 times that of the aorta, results in a slow flow of blood so that exchange has time to occur. In certain body areas (spleen, liver), large vessels similar to capillaries called sinusoids serve the exchange function; such vessels may have a lining composed of phagocytic cells.

FIGURE 25.4 The histology of a capillary as seen under the electron microscope (×40,000). pv = pinocytic vesicles, col = collagen fibrils, ec = endothelial cell, ecn = endothelial cell nucleus, rbc = red blood cell. (Photograph courtesy Norton B. Gilula, The Rockefeller University.)

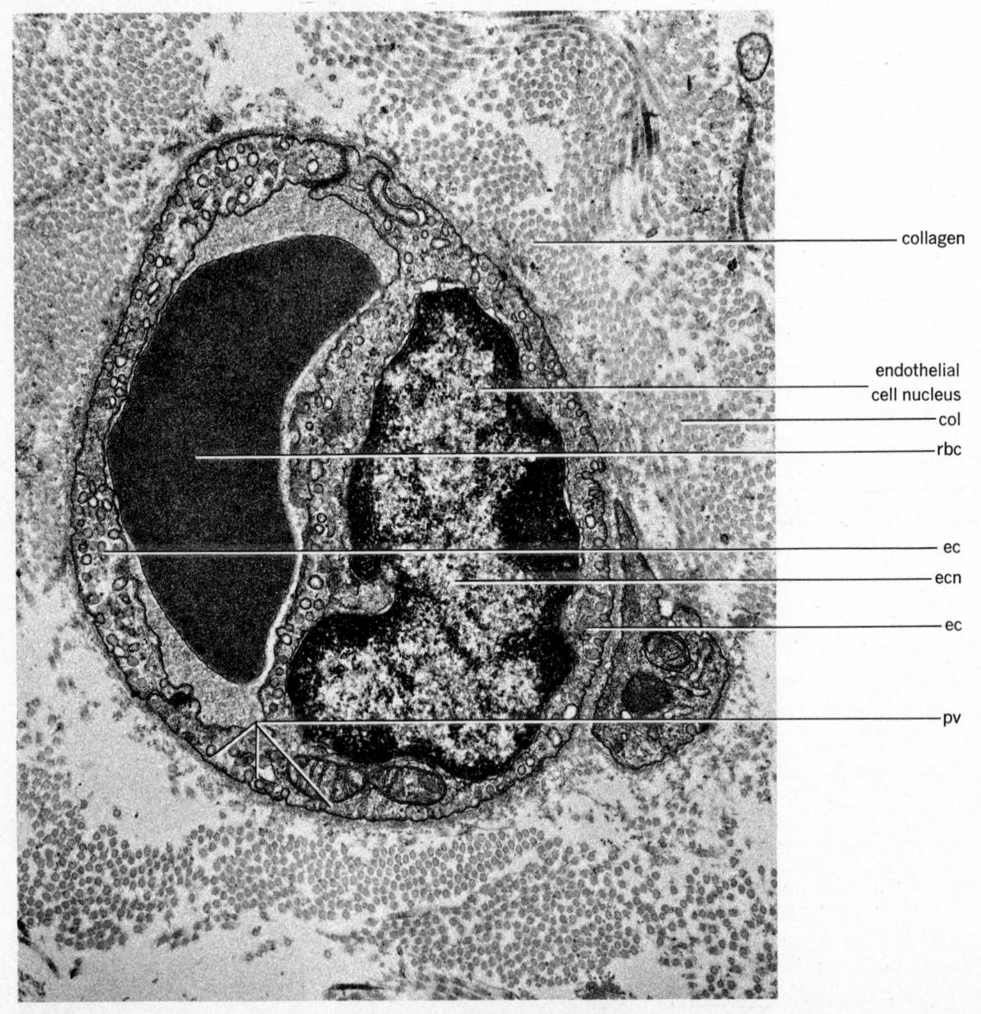

FIGURE 25.5 The histology of veins.

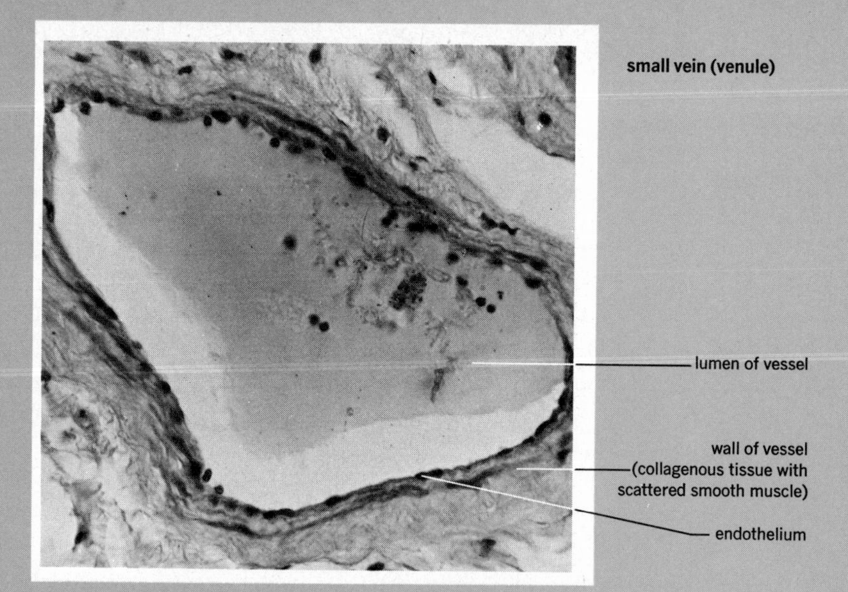

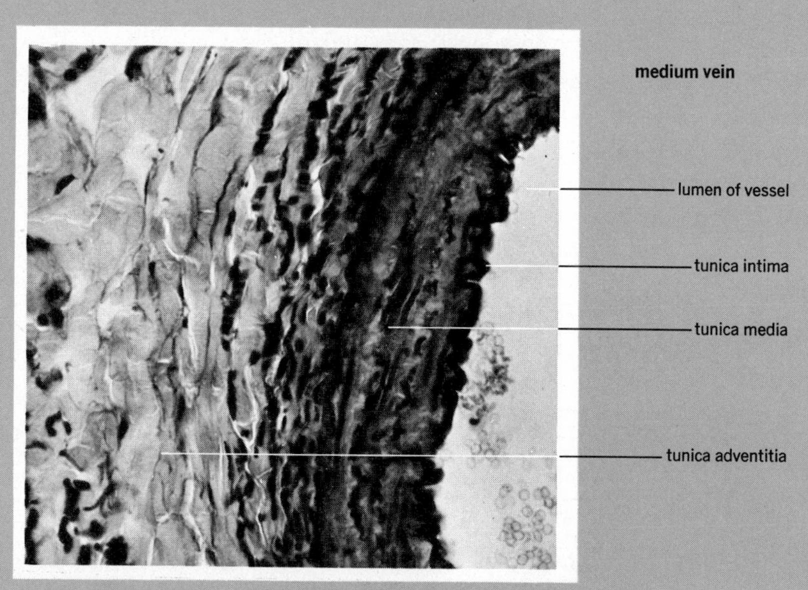

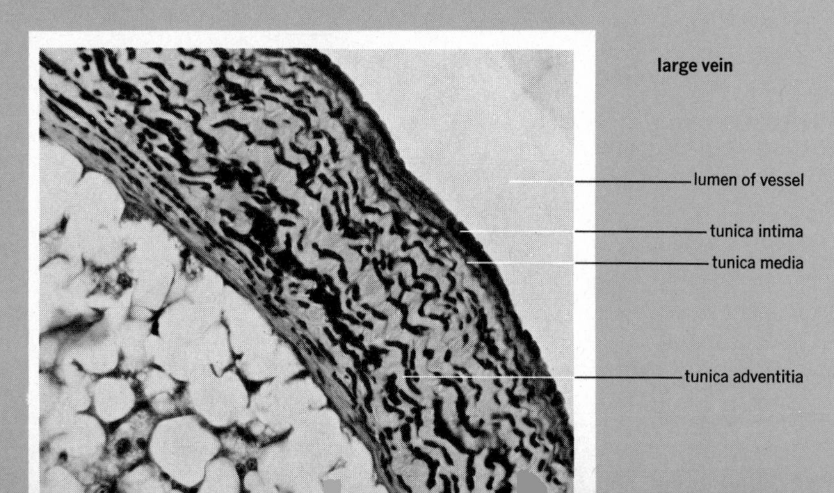

Veins

Large and *medium sized veins* (Fig. 25.5) have three layers of tissue in their walls, as do arteries. In general, total wall thickness is less in a vein than in its corresponding artery, and diameter is greater. In such a vein, the layer having the greatest thickness is the adventitia, which is two to five times wider than the media. Larger veins have a vasa vasorum and contain valves that allow flow of blood only toward the heart. *Venules* are two-layered, with an endothelial tube surrounded by collagenous connective tissue that may have a few scattered smooth muscle cells in it. Table 25.2 summarizes the structure of veins.

CLINICAL CONSIDERATIONS. Superficial veins, having thinner walls and less muscle than arteries and less supporting tissue around them, are more easily enlarged or stretched than deep veins.

VARICOSITIES or varicose veins are superficial veins that have become abnormally elongated or dilated. The human's upright posture, combined with the effect of gravity, tends to cause pooling of blood in the superficial veins of the legs. Valves in these vessels and the massaging action of the skeletal muscles as they contract and relax tend to move the blood toward the heart. Accumulation of blood in these veins, accompanied by lack of muscular activity, may cause enlargement of the veins to a point where the valves become incompetent. Once the cycle of enlargement and valvular incompetence is established, it tends to cause the development of larger and larger varicosities. Support hose and elevation of the feet may aid the condition, or the veins may be removed surgically.

TABLE 25.2 Summary of vein structure

Type of vessel	Average size (mm)	Example(s) or location	Tissue in: Tunica intima	Tunica media	Tunica adventitia	Comments
Large vein	20–30	Vena cava	Endothelium and c.t.	C.t. and scattered smooth muscle cells	Loose c.t.	Few in number, little change in size
Medium vein	2–20	Named veins of appendages and viscera	As above	As above	As above	As above
Small vein (venule)	<2 mm	In tissues and organs	As above	Collagenous c.t. and a few smooth muscle cells	Lacking	Most numerous; can significantly alter venous volume by change in size

The major systemic blood vessels

Arteries

AORTA. The left ventricle gives rise to the aorta, which forms the major vessel of the systemic circulation. The aorta is divided into a superiorly directed portion called the ASCENDING AORTA, a curved portion known as the ARCH, and an inferiorly directed portion known as the DESCENDING AORTA. The latter is subdivided again into a THORACIC (in the chest) and ABDOMINAL (in the abdominopelvic cavity) PORTION. The major branches of the aorta are shown in Fig. 25.6 and summarized in Table 25.3.

THE ARTERIES OF THE UPPER APPENDAGE. The arteries of the upper appendage (Fig. 25.7) are derived from the brachiocephalic and the left subclavian arteries (see Fig. 25.6). A right SUBCLAVIAN ARTERY rises from the brachiocephalic artery and not separately from the aorta. Each subclavian artery gives rise to four main branches: VERTEBRAL, INTERNAL THORACIC *(mammary)*, THYROCERVICAL, and COSTOCERVICAL. As the vessel passes from the thorax to the armpit, the name changes to AXILLARY, and three major branches are given off: LONG THORACIC, VENTRAL THORACIC, and SUBSCAPULAR. Passing onto the upper arm, the name again changes to BRACHIAL, which gives off MUSCULAR and DEEP BRACHIAL branches to the muscles and skin of the upper arm. Just below the elbow, the brachial artery divides to form the RADIAL and ULNAR arteries that follow the respective bones of the forearm. These give off muscular branches to the forearm muscles. At the wrist, the radial artery is a common site where the pulse may be taken. Also at the wrist, radial and ulnar arteries are joined by superficial and deep VOLAR ARTERIAL ARCHES, from which rise the DIGITAL arteries to fingers and thumb. These vessels and their branches are summarized in Table 25.4.

THE ARTERIES TO THE LOWER APPENDAGE. The arteries to the lower appendage (Fig. 25.8) are derived from the COMMON ILIAC arteries that branch to form EXTERNAL and INTERNAL ILIAC arteries. The external iliacs pass across the hip and become the FEMORAL arteries. The two major branches of the femoral artery are the DEEP FEMORAL and the EXTERNAL PUDENDAL. Just above the knee, the femoral becomes the POPLITEAL artery that gives off CUTANEOUS, MUSCULAR, and GENICULATE branches to posterior thigh muscles and knee structures. Below the knee, two vessels are formed from the popliteal artery.

An ANTERIOR TIBIAL artery courses along the anterior-lateral aspect of the tibia and gives off muscular branches.

A POSTERIOR TIBIAL artery follows along the posterior tibial surface and gives muscular, fibular, and cutaneous branches. A branch from the posterior tibial artery known as the PERONEAL artery courses down the lateral side of the leg and gives muscular and cutaneous branches.

The anterior tibial artery becomes the DORSIS PEDIS artery of the upper surface of the foot; the posterior tibial artery becomes the PLANTAR artery of the sole of the foot. The PLANTAR ARCH, or arcuate artery, connects the foot arteries and gives rise to DIGITAL arteries to the toes. The vessels of the lower appendage are summarized in Table 25.5.

Veins

In general, veins (Figs. 25.9 and 25.10) accompany arteries and are named in the same fashion as the arteries. There are two "sets" of veins in the appendages.

Those that accompany the arteries and have similar names are the DEEP VEINS.

Those that lie just beneath the skin are named differently than the arteries and form the SUPERFICIAL VEINS. On the upper appendage, the

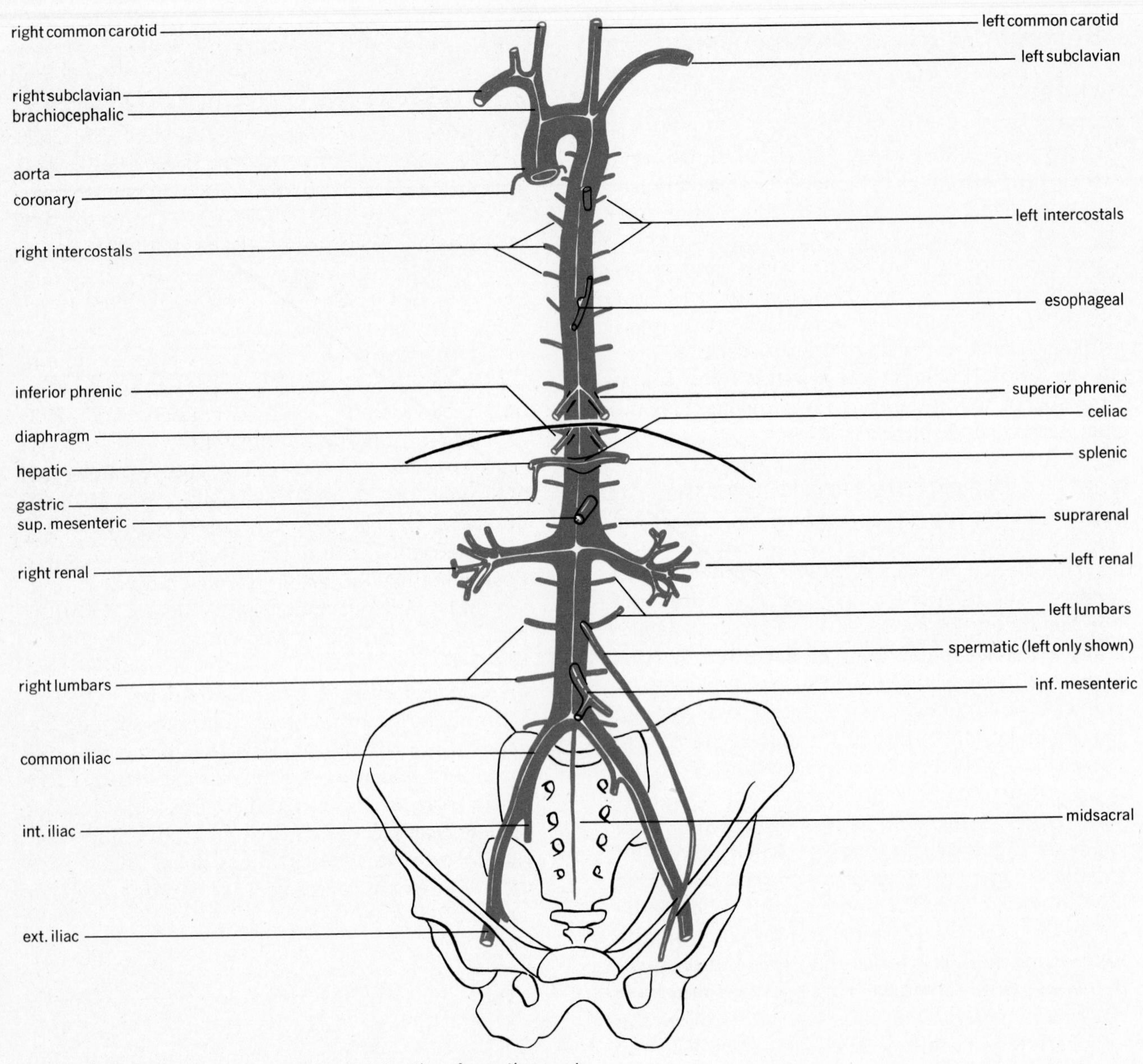

FIGURE 25.6 The major arteries originating from the aorta.

TABLE 25.3 Branches of the aorta and area(s) supplied

Arising from	Branch	Area supplied by branch
Ascending aorta	Coronary	Heart
Arch of aorta	Brachiocephalic; gives rise to: Right common carotid Right subclavian	 Right side head and neck Right arm
	Left common carotid	Left side head and neck
	Left subclavian	Left arm
Descending aorta: Thoracic portion	 Intercostals	 Intercostal muscles, chest muscles, pleurae
	Superior phrenics	Posterior and superior surfaces of the diaphragm
	Bronchials	Bronchi of lungs
	Esophageals	Esophagus
	Inferior phrenics	Inferior surface of diaphragm
Abdominal portion	Celiac; gives rise to: Hepatic Left gastric Splenic	 Liver Stomach and esophagus Spleen, part of pancreas and stomach
	Superior mesenteric	Small intestine, cecum, ascending and part of transverse colon
	Suprarenals	Adrenal glands
	Renals	Kidneys
	Spermatics (male) or ovarians (female)	Testes Ovaries
	Inferior mesenteric	Part of transverse colon, descending and sigmoid colon, most of rectum
	Common iliacs, which give rise to: External iliacs Internal iliacs Midsacral	Terminal branches of aorta Lower limbs Uterus, prostate gland, buttock muscles Coccyx

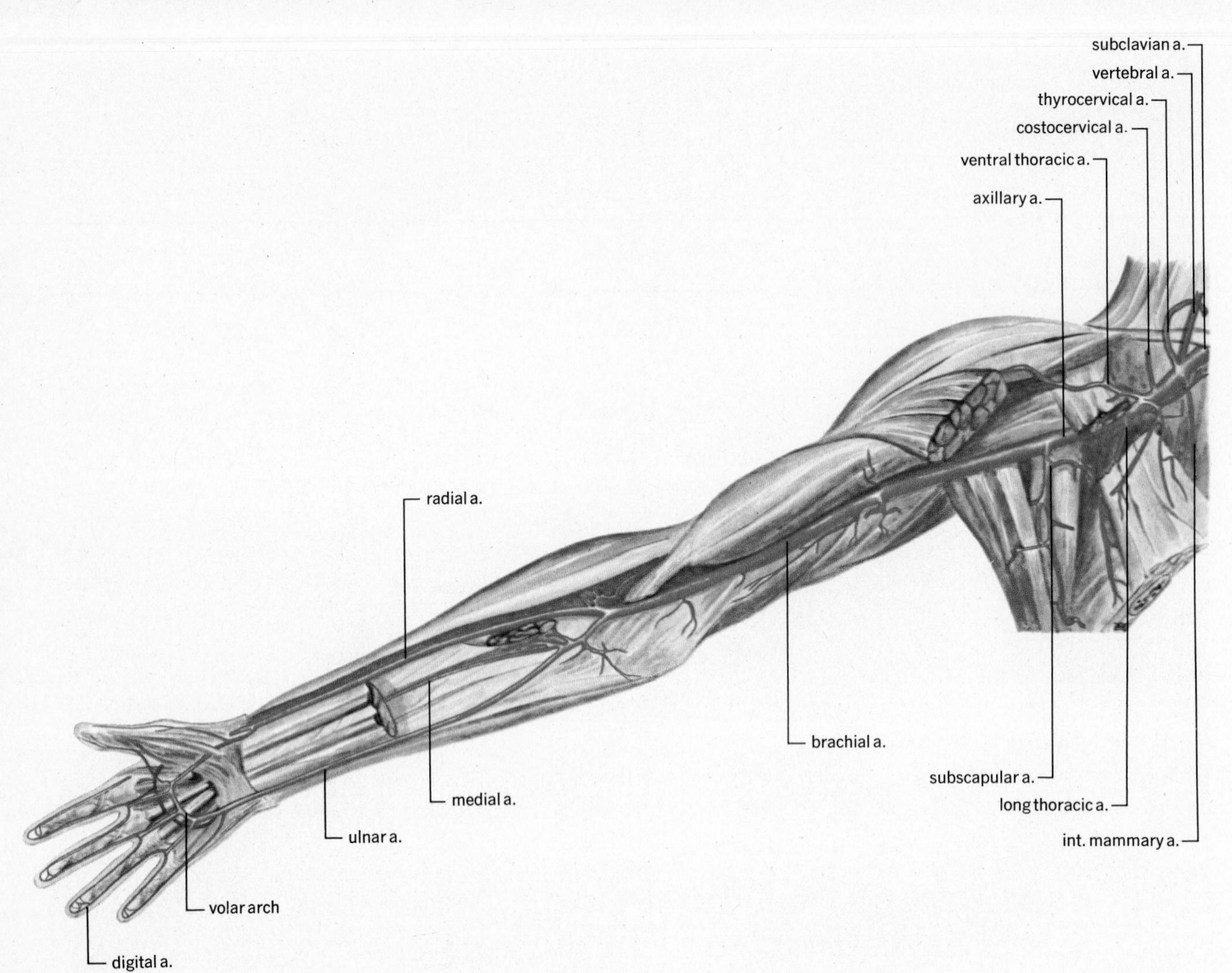

FIGURE 25.7 The major arteries of the upper limb.

TABLE 25.4 Arteries of the upper appendage and area(s) supplied

Artery	Major side branches	Area(s) supplied by branches
Subclavian (beneath clavicle)	Vertebral	Brain and spinal cord
	Internal thoracic (mammary)	Mammary glands, diaphragm, pericardium
	Thyrocervical Costocervical	Muscles, organs, and skin of neck and upper chest
Axillary (armpit)	Long thoracic Ventral thoracic Subscapular	Muscles and skin of shoulder, chest and scapula
Brachial (upper arm)	Muscular	Biceps muscle
	Deep brachial	Triceps muscle. Both also supply skin and other tissues of upper arm
Radial, Ulnar (forearm)	Muscular Muscular	Muscles and skin of the forearm
Palmar (volar) arch	Metacarpals	Muscles and skin of the hand
Digitals	–	Muscles and skin of fingers

TABLE 25.5 Arteries of the lower appendage

Artery	Major side branches	Area(s) supplied by branches
Femoral (thigh)	Deep femoral, pudendals	Muscles and skin of thigh, lower abdomen and external genitalia
Popliteal (knee)	Cutaneous, muscular, genicular	Skin of back of leg, hamstrings, and structures of knee joint
Anterior tibial (anterior side of leg)	Muscular	Anterior crural muscles and skin of leg
Posterior tibial (posterior side of leg)	Muscular, fibular, cutaneous	Posterior crural muscles and skin of leg
Peroneal (lateral side of leg)	Muscular, cutaneous	Lateral crural muscles and skin of leg
Plantar and dorsis pedis Arcuate	Form arcuate arch Metatarsal }	Muscles and skin of foot
Digitals		Muscles and skin of toes

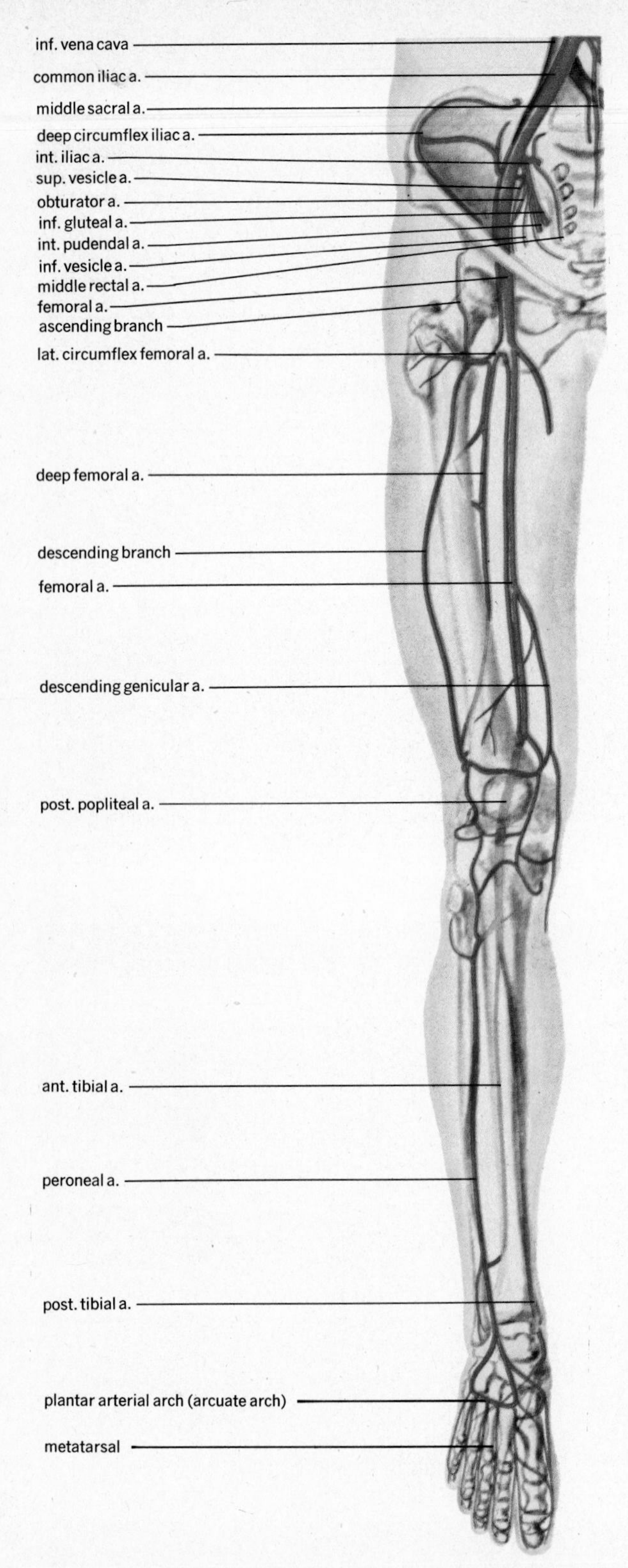

FIGURE 25.8 The major arteries of the lower limb.

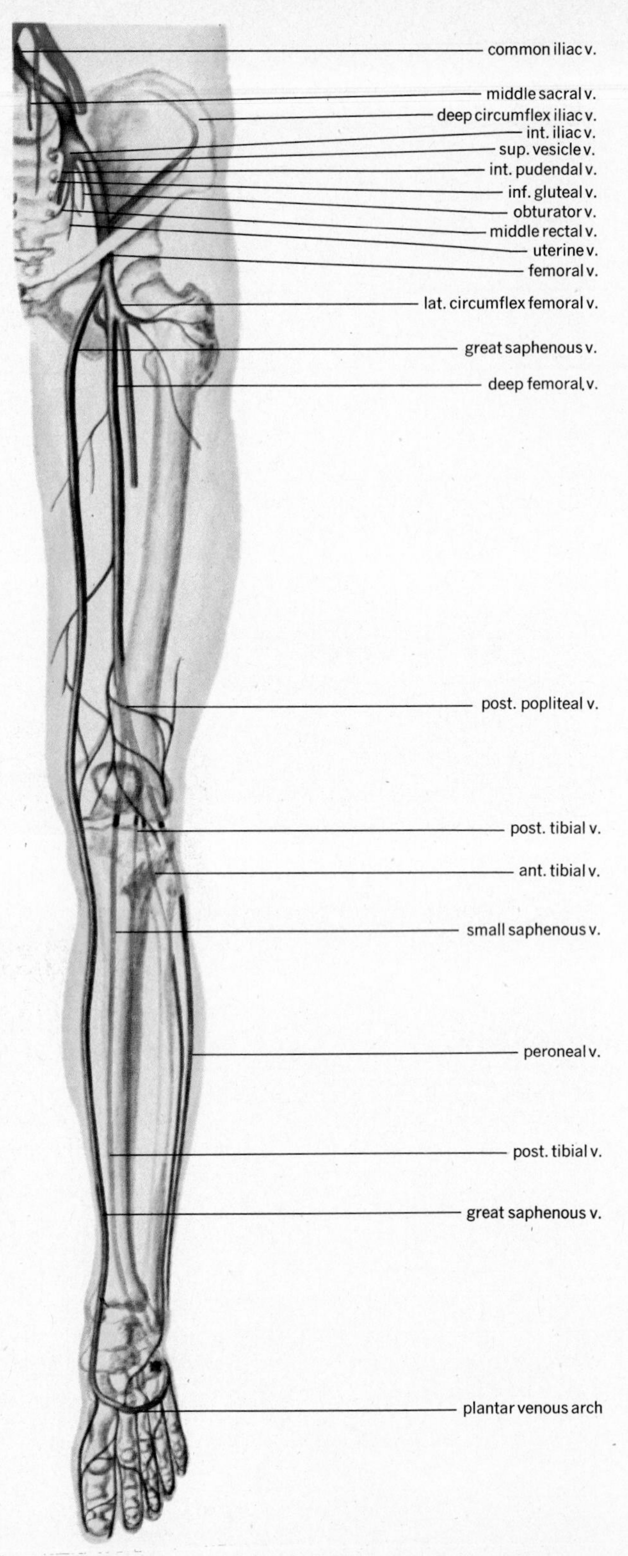

FIGURE 25.9 The major veins of the lower limb.

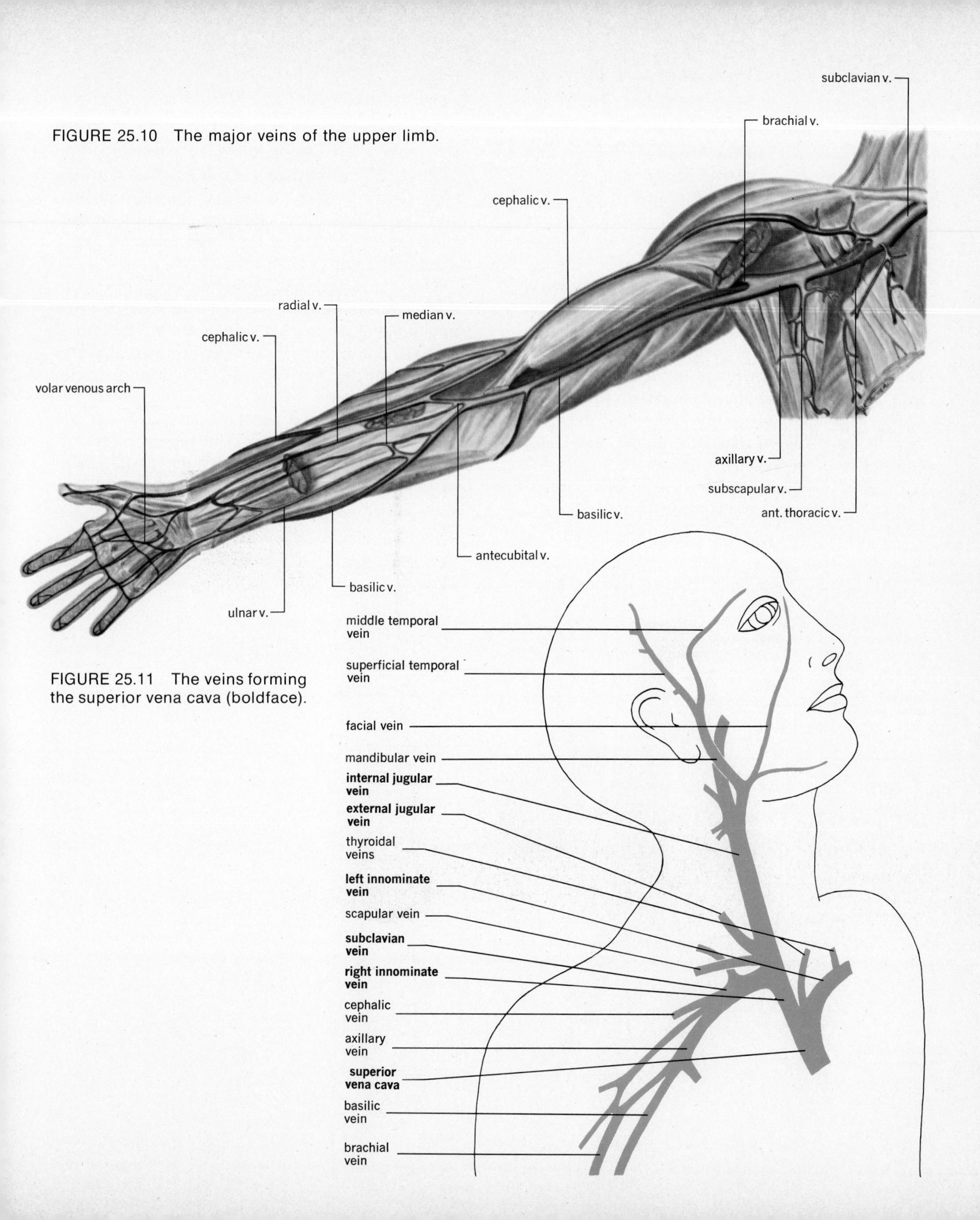

FIGURE 25.10 The major veins of the upper limb.

FIGURE 25.11 The veins forming the superior vena cava (boldface).

CEPHALIC, BASILIC, and ANTECUBITAL veins form the superficial set; on the lower appendage, the SAPHENOUS vein is the superficial one.

The superficial veins of the upper appendage include the *dorsal digital veins* of the fingers that form *metacarpal veins,* seen as the network on the back of the hand. The *cephalic vein* rises from the lateral (radial) side of the metacarpal network, passes proximally along the lateral side of the upper limb, receives vessels from the lateral side of the limb, and empties into the axillary vein. The *basilic vein* originates from the medial (ulnar) side of the metacarpal network, passes proximally along the medial side of the upper limb, receives vessels from the medial side of the limb, and empties into the axillary vein. At the elbow, the two veins are connected by the median cubital vein, a common site of vein puncture.

The AZYGOS VEIN lies along the posterior wall of the thorax; it receives the intercostal veins and veins from the lumbar area. It empties into the superior vena cava just above the right atrium, and can, if necessary, carry the entire blood flow from the lower parts of the body in case of blockage of the inferior vena cava.

The *saphenous* (great saphenous) *vein* begins on the medial margin of the foot and courses proximally along the medial side of the lower limb. It receives tributaries from the entire medial aspect of the thigh, leg, and foot and usually empties into the femoral vein.

The SUPERIOR and INFERIOR VENAE CAVAE are the largest systemic veins. The vessels contributing to their formation are shown in Figs. 25.11 and 25.12 and are summarized in Tables 25.6 and 25.7.

FIGURE 25.12 The veins forming the inferior vena cava (boldface).

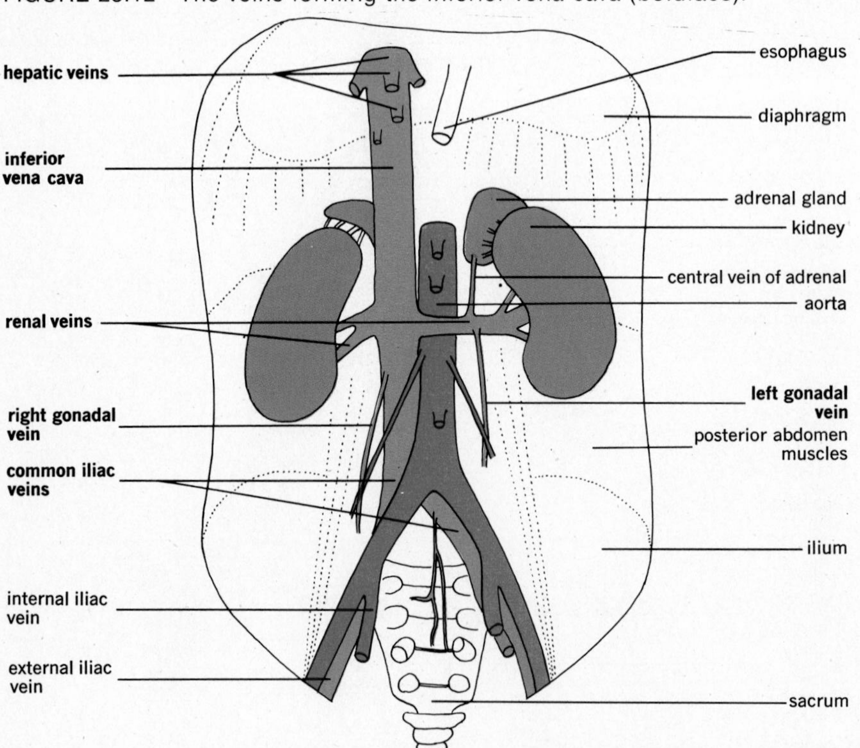

TABLE 25.6 Veins draining into the superior vena cava

Vein	Formed from	Area(s) drained
Internal Jugular	Dural sinuses	Inside of skull and brain
External Jugular	Veins of face	Muscles and skin of scalp and face
Subclavian	Axillary, cephalic[a], basilic[a], and their tributaries, scapular, and thoracic veins	Upper appendage, chest, mammary glands
Innominate (brachiocephalic)	Internal jugular, external jugular, and subclavian	
Azygos	Lumbar and intercostal veins	Posterior aspect of thorax and abdominal cavities

[a] These are the superficial veins of the extremities.

TABLE 25.7 Veins draining into the inferior vena cava

Vein	Formed from	Area(s) drained
Hepatics	Sinusoids of liver	Liver
Renals	Veins of kidney	Kidney
Gonadals	Veins of gonads	Gonads; testes, and ovaries
Common iliac	External iliac	Lower appendage
	Internal iliac	Organs of lower abdomen
	Femoral Saphenous[a] (empties into femoral)	Deep structures of lower appendages. Superficial structures of lower appendages

[a] These are the superficial veins of the extremities.

Circulation in special body regions

Pulmonary circulation (Fig. 25.13)

The PULMONARY TRUNK rises from the right ventricle and branches to form the RIGHT AND LEFT PULMONARY ARTERIES that supply the respective lung with low-oxygen-content blood. The arteries branch to form the arterioles and capillary networks of the lung, in which exchange of gases (O_2 and CO_2) takes place. Four PULMONARY VEINS are formed by the vessels leaving the lungs and

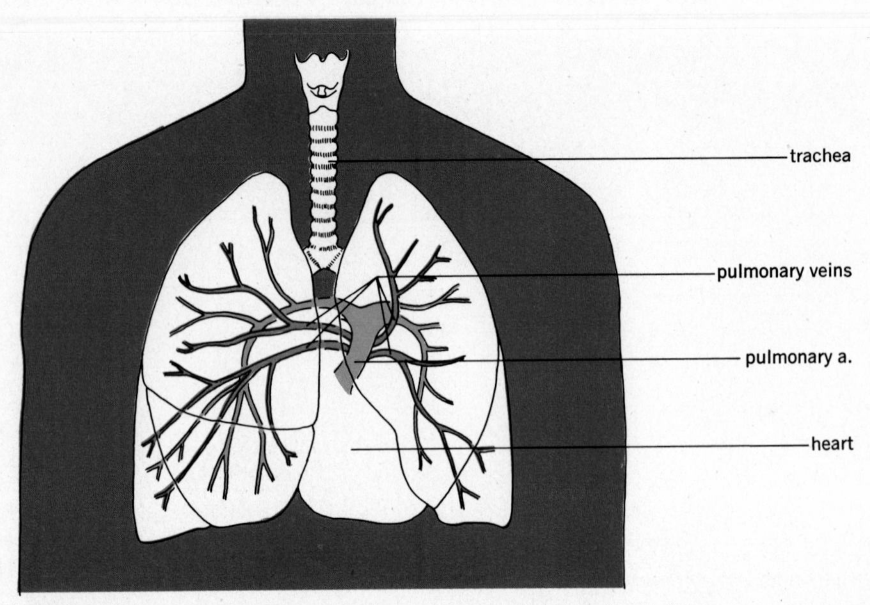

FIGURE 25.13 The major vessels of the pulmonary circulation.

return blood to the left atrium. The system has thinner-walled, larger, and shorter vessels than the systemic circulation and is perfused by pressures about one-fourth those of the systemic circulation (see Fig. 24.20). The pulmonary vessels normally contain about 9 percent of the blood volume, or about 450 ml in the 70 kg adult. It takes about 4 sec to traverse the system. The vessels are sensitive to chemical changes in their environment, but respond only slightly to nervous influences.

Cerebral circulation

The vessels supplying the brain were described in Chapter 14. The cerebral circulation maintains a relative constancy in terms of blood flow, despite rather wide variations in pH, systemic blood pressure, and chemical influences. Volume of blood and cerebrospinal fluid volumes within the skull are inversely related, so that the brain substance is not compressed against the rigid skull. Average blood flow is about 750 ml/min, with the cerebrum receiving about 14 percent of the left ventricular output.

Coronary circulation (Fig. 25.14)

Two coronary arteries rise from the aorta just beyond the aortic valve cups. In most individuals, the right and left coronary arteries supply the respective halves of the heart. About 70 percent of the blood circulated through the heart is returned through the coronary veins and sinus to the right atrium; 30 percent is returned directly to the ventricular cavities through luminal and sinusoidal vessels that gather from capillaries and sinusoids within the myocardium. The coronary circulation receives about 4–5 percent of the left ventricular output.

Pressure in the system is high, as high as aortic pressure (to 135 mm Hg). When the myocardium is contracted, there is no flow through the coronary vessels, since the myocardial capillaries are

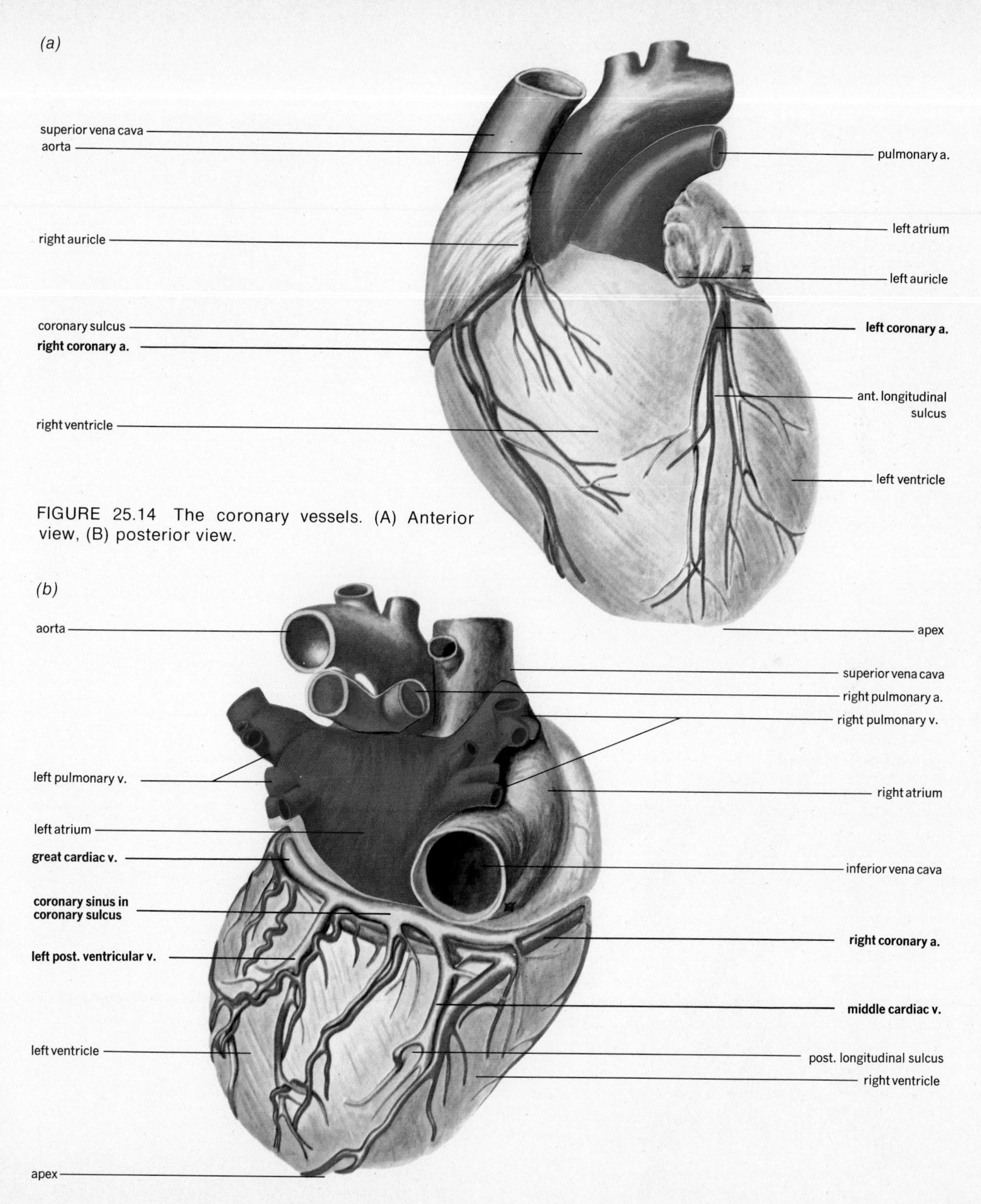

FIGURE 25.14 The coronary vessels. (A) Anterior view, (B) posterior view.

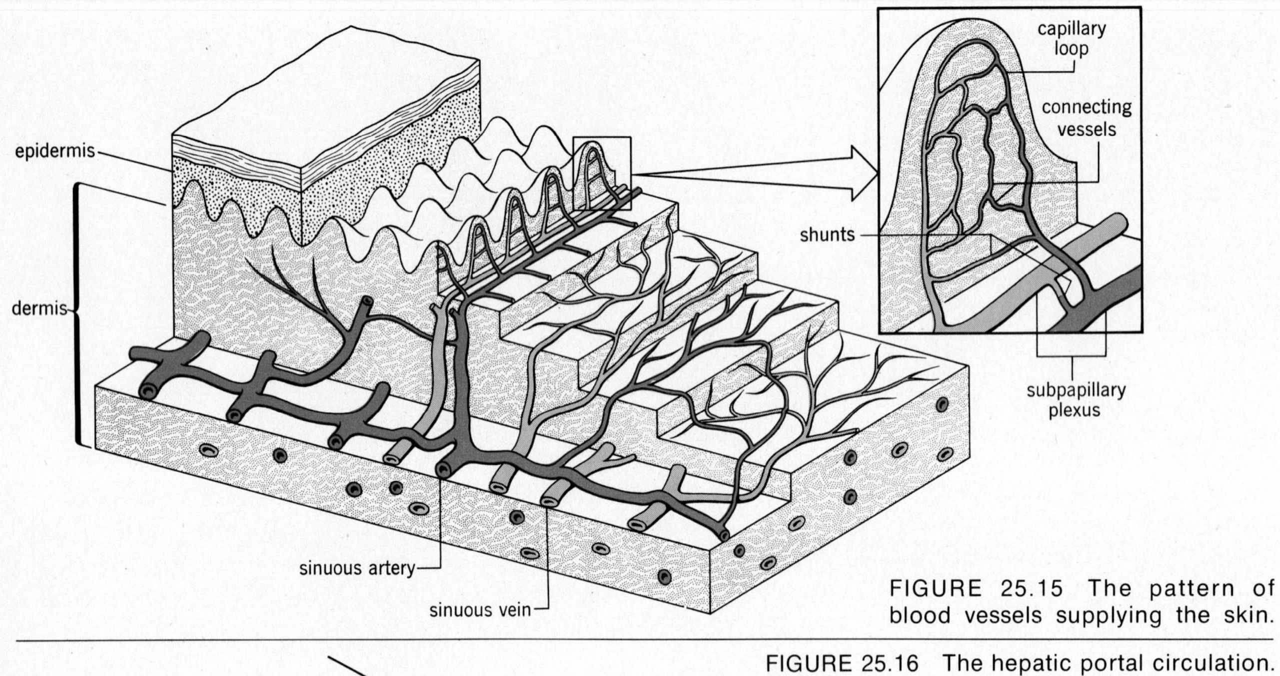

FIGURE 25.15 The pattern of blood vessels supplying the skin.

FIGURE 25.16 The hepatic portal circulation.

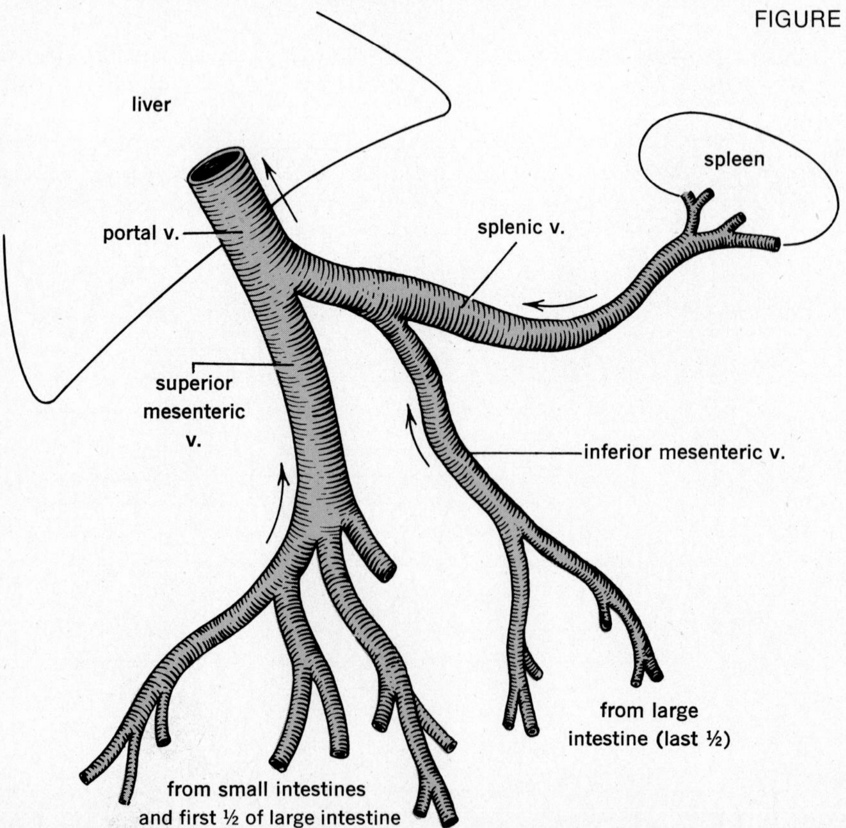

squeezed shut. Relaxation of the myocardium permits flow through the coronary vessels and fills the ventricles.

The heart obtains about 60 percent of its energy through the metabolism of fats, 35 percent from carbohydrates, and 5 percent from amino acids. Fall of blood O_2 tension, rise of blood CO_2 tension, or fall in pH increase coronary flow 200–300 percent by causing coronary vasodilation.

Cutaneous circulation (Fig. 25.15)

The blood vessel arrangement of the skin allows more or less blood to enter the upper dermal layers for purposes of temperature regulation. The ARTERIOVENOUS ANASTOMOSES allow bypassing of the capillary loops and heat conservation, and when open, allow greater heat loss by allowing blood into the capillary loops. Control of flow through the vessels is by sympathetic fibers of the autonomic nervous system.

Hepatic circulation (Fig. 25.16)

The liver receives oxygenated (arterial) blood through the HEPATIC ARTERY (see Fig. 25.6). Venous blood rich in nutrients reaches the liver via the HEPATIC PORTAL VEIN. The vein is part of a *portal system*—that is, a system of veins located between two sets of capillaries (gut and liver). The portal vein is formed by the joining of the gastric, splenic, superior and inferior mesenteric veins, vessels that drain the greater part of the alimentary tract. Combined blood flow is about 1400 ml/min, with 20 percent from the hepatic artery, 80 percent from the portal vein. The liver is drained by the hepatic veins that empty into the inferior vena cava.

Splenic circulation (Fig. 25.17)

The spleen is a lymph organ set in the course of blood vessels. In the embryo, it is a source of blood cells. In the adult, it functions mainly to phagocytose aged red blood cells; it can also "store" about 250 ml of blood that can be injected into the circulation during stress.

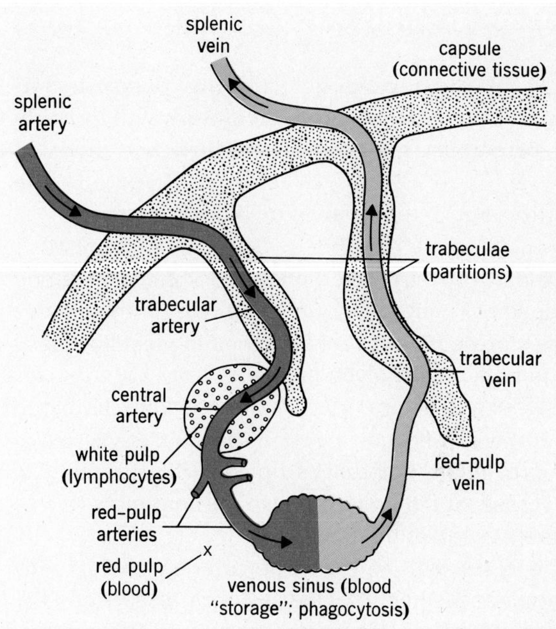

FIGURE 25.17 The scheme of the splenic circulation.

Skeletal muscle circulation

Blood flow through skeletal muscle nourishes the muscle during states of rest and activity. Since the mass of skeletal muscle is so great, the vessels of the muscles play an important role in adjustments or circulatory dynamics. The vessels appear to be provided with sympathetic adrenergic constrictor fibers and some sympathetic cholinergic dilator fibers. The predominant control lies with the constrictors, and the diameter of the vessels is altered by changes in the degree of vasoconstrictor tone. Reflexes that impinge on the medullary vasomotor centers thus bring about changes in caliber that are important to control of blood pressure. Elevated blood pressures cause reflex dilation of skeletal vessels and vice versa.

Pressure and flow in the blood vessels

Study of the factors determining pressure and flow in this portion of the circulation will indicate the principles governing pressure and flow in general. To review, five factors contribute to the creation and maintenance of blood pressure.

CARDIAC OUTPUT determines the volume of blood pumped per minute into the blood vessels and creates a basic pressure and flow designed to meet the needs of the body at any given moment.

PERIPHERAL RESISTANCE, or the difficulty the blood encounters in flowing through the smaller vessels of the circulation, determines not only the pressure in the vessels of the body, but also the amount of flow to different areas of the body.

The ELASTICITY of the large arteries of the body determines the height of the systolic pressure and the degree of fall of diastolic pressure in the system. These vessels expand to contain the ventricular output (stroke volume) at each beat and then recoil as the ventricle relaxes, pushing the blood onward in a pulsatile but continuous flow.

The CIRCULATING VOLUME of blood in the system determines pressure. This factor can be varied by the "storage" of blood in the veins during venous dilation or constriction.

The VISCOSITY of the blood determines pressure in that a "thicker" fluid requires a greater pressure to circulate it.

Blood flow and pressure in the systemic circulation may be said to be determined first by the "basic pressure" and flow created by heart action, which is then modified and/or adjusted by the blood vessels.

Principles governing pressure and flow in tubes

The laws of hydrodynamics that govern the flow of fluids through rigid tubes may be adapted to the flow of blood through blood vessels. The most important laws influencing pressure and flow in the systemic circulation are presented below.

Resistance to flow is directly proportional to the length and inversely proportional to the total cross-sectional area presented by the vessels. The longer the vessel, or the greater the surface presented, the greater the friction encountered by the blood as it flows through the vessel. *Flow is directly proportional to the fourth power of the radius of the tube* and therefore is directly proportional to the cross-sectional area.

Pressure is directly proportional to the cardiac output and inversely proportional to the total cross-sectional area of the vessels. This suggests that pressure should be highest close to the heart and should decrease as the blood traverses the increasing numbers of smaller vessels in the periphery. Marey's law states that arterial blood pressure and heart rate (and therefore cardiac output) are inversely related.

Velocity of flow is directly proportional to pressure and inversely proportional to the total cross-sectional area. Flow should thus be rapid in arteries and veins, slowest in the numerous capillaries.

Pressure, flow, and resistance are related by the equation

$$\text{flow} = \frac{\text{pressure}}{\text{resistance}}$$

which may be solved for any one of the three components if given the other two.

Resistance to blood flow is directly proportional to viscosity of the blood, since the greater the viscosity, the greater the friction encountered during flow. Poiseuille's law takes the viscosity into account as a determinant of blood flow and states

$$\text{flow} = \frac{\text{pressure} \times \text{radius}^4}{\text{length} \times \text{viscosity}}$$

The interrelationships of pressure, surface area, velocity, and flow are shown in Fig. 25.18. Note

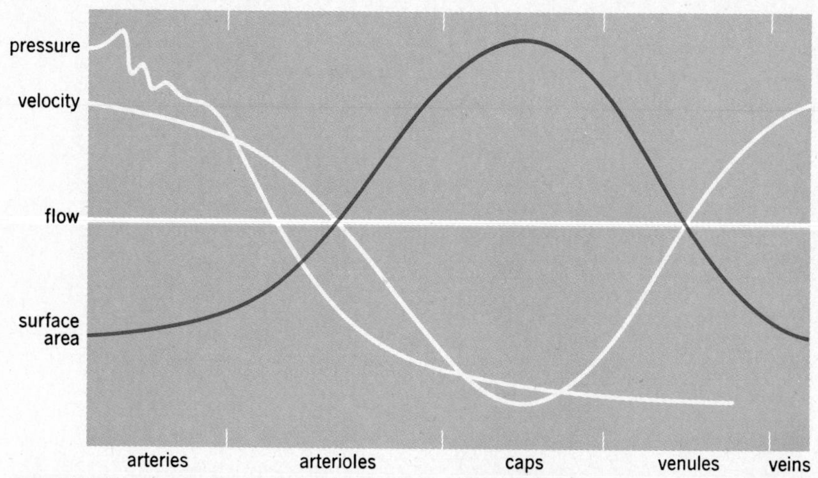

FIGURE 25.18 Interrelationships of surface area, pressure, flow, and velocity in various portions of the vascular system. The factor determining the shapes of the other curves is surface area presented to the blood as it flows through the vessels.

that the surface area appears to be the factor determining the pressure and velocity. In obese persons, each pound of fat is said to add one mile of capillaries to the circulation. Thus, the heart in such an individual must work harder to circulate the blood through the additional resistance imposed by the larger numbers of capillaries. Athletes tend to have more vascular muscles but achieve most of their increased cardiac output by elevations of stroke volume and by "opening up" capillaries normally closed at rest.

Pressure and flow in arteries

Blood is pumped by the left ventricle into the aorta at a SYSTOLIC PRESSURE of approximately 125 mm Hg (range 100–150 mm Hg). Elastic recoil ensures that the DIASTOLIC PRESSURE does not fall much below 75 mm Hg (range 60–90 mm Hg). The PULSE PRESSURE or difference between systolic and diastolic pressure is about 50 mm Hg. The MEAN ARTERIAL BLOOD PRESSURE, represented by (systolic pressure + 2 diastolic pressure)/3, is about 100 mm Hg. Resistance to flow in the aorta is slight because the surface area is small (2.5 cm^2), so velocity is high (100–140 cm/sec) and flow is great (5–6 liters/min). At the terminal end of the aorta in the abdomen, the cross-sectional area is approximately what is was initially, but some 25 cm of vessel length has been traversed so that mean pressure has fallen to about 95 mm Hg. The blood next flows into the increasing number of smaller muscular arteries. Taken together, these vessels present some 60 cm^2 of surface area and cause a drop in mean pressure to about 40 mm Hg at the beginning of the capillary beds. The systemic arteries contain, on the average, about 15 percent of the total blood volume.

Pressure and flow in capillaries

At the arteriolar ends of capillaries, pressure is approximately 40 mm Hg and decreases to about 10 mm Hg at the venous ends of the vessels (Fig. 25.19). A higher "entrance pressure" ensures filtration, a lower "exit pressure" allows osmotic return of water to the capillaries by the plasma protein (oncotic) pressure. Capillaries contain

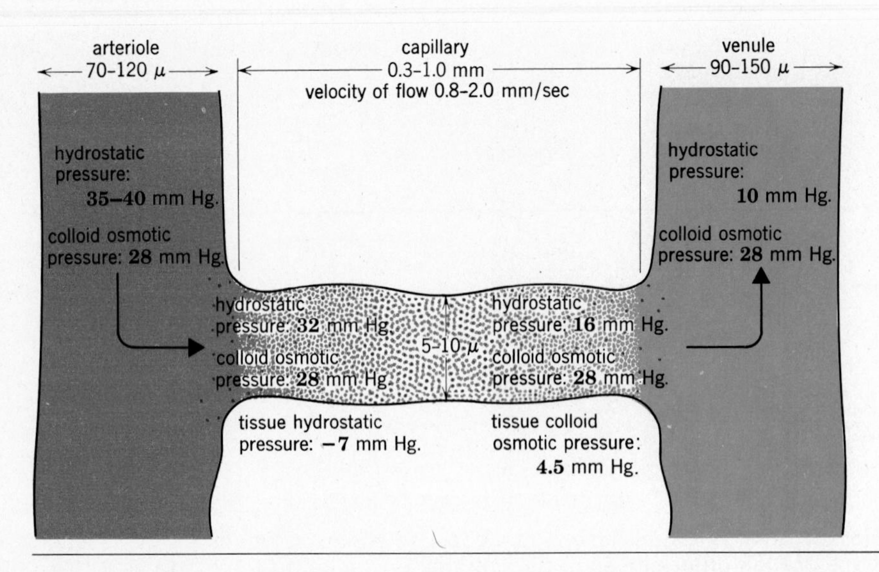

FIGURE 25.19 Capillary dynamics.

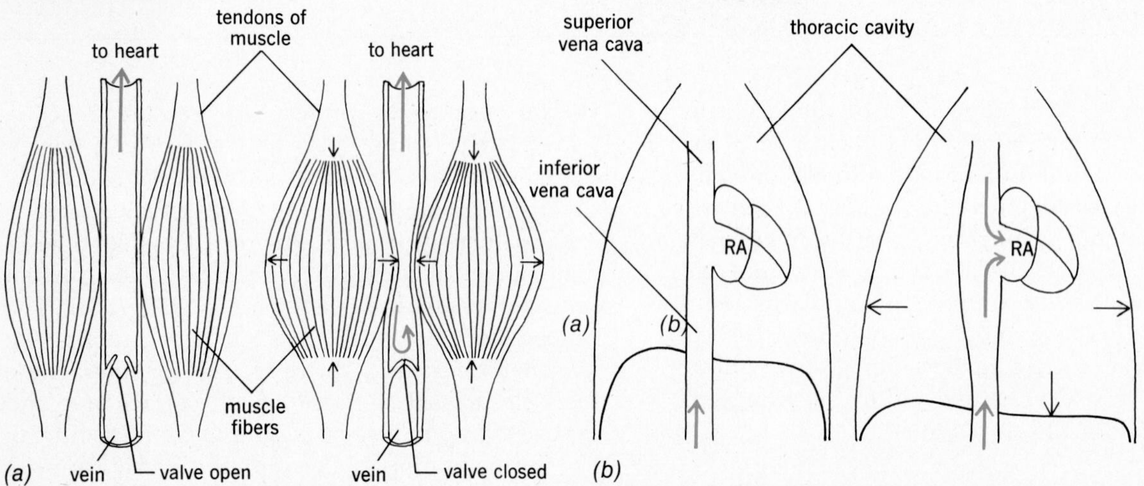

FIGURE 25.20 The skeletal and respiratory pumps. *(a)* The skeletal muscle pump. Relaxed muscle (left) allows blood to flow through open valve. Contracted muscles (right) become shorter and thicker, compressing the vein. Blood attempts to go away from the heart closing the valve. Blood is thus massaged toward the heart. *(b)* The respiratory pump. Inspiration increases thoracic volume, decreases intrathoracic pressure, and expands the thin-walled vena cave. Blood is drawn into the larger vessels and speeds the flow through the venous system, increasing blood flow to the right atrium (RA).

about 5 percent of the blood volume. Velocity here is lowest due to a larger surface area (2500 cm^2).

Not all capillaries in a tissue are necessarily filled with blood when the tissue is at rest. Activity causes many unfilled capillaries to fill passively as their precapillary sphincters (metarterioles) relax, and in this manner exchange of materials is speeded. For example, resting skeletal muscle is estimated to utilize only 1/100 of its available capillaries.

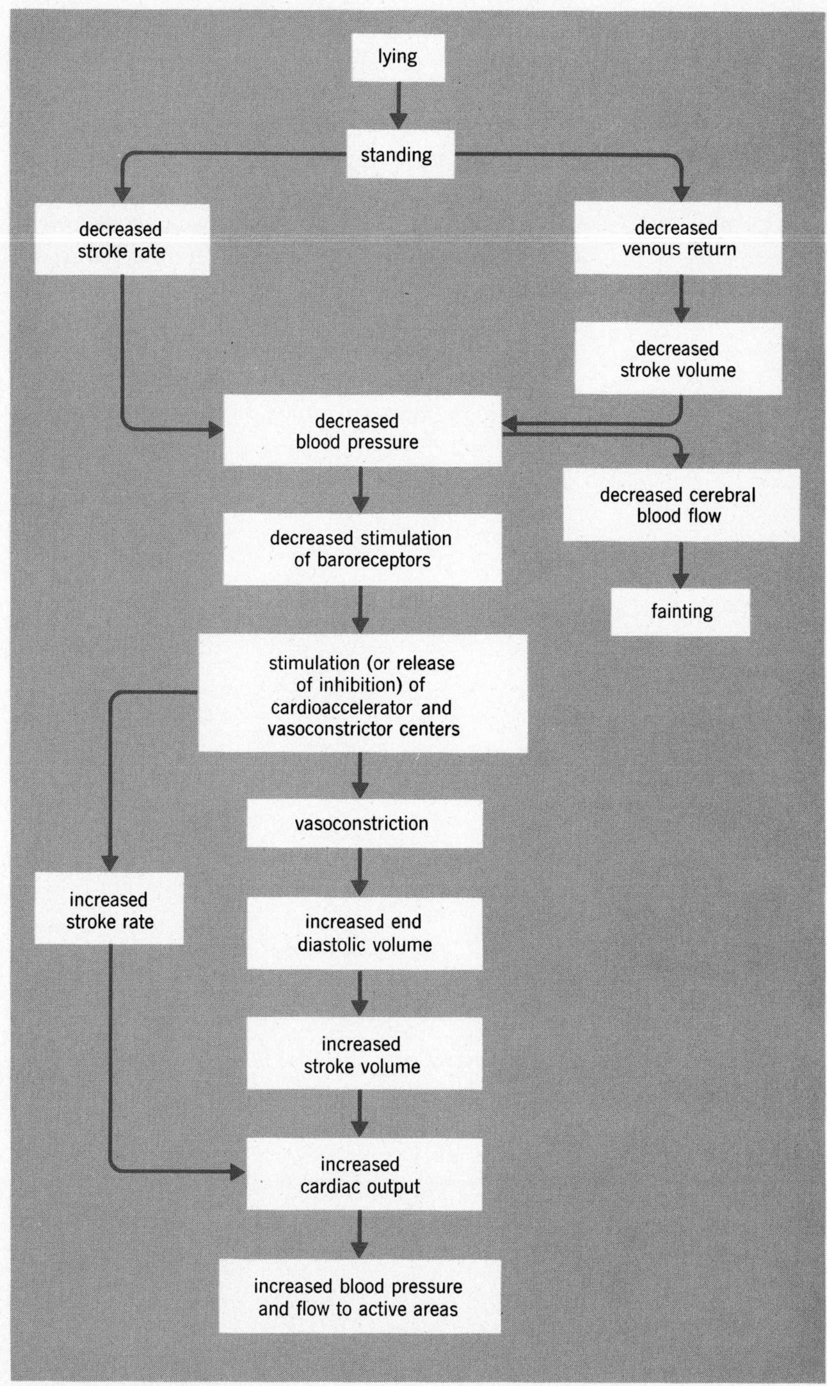

FIGURE 25.21 Vascular and cardiac responses to postural changes.

Pressure and flow in veins

As the veins flow together to form fewer and larger vessels beyond the capillary beds, total surface area decreases to about 325 cm^2. Pressure continues to fall as the length component continues to operate, and since there is no pump in this area to increase the pressure, pressure may fall to zero at the right atrium. However, velocity increases as surface decreases. The veins contain some 60 percent of the blood volume, emphasizing their role as volume and not pressure vessels. It is this percentage that may be altered by change in vein diameter. Obviously, if blood is not to accumulate in some area of the body, the total milliliter/minute flow must be equal in the different portions of the system, in spite of the great differences in volume.

FACTORS GOVERNING FLOW THROUGH VEINS. Blood does flow rapidly through the veins in spite of the low pressures in this portion of the circulation (Fig. 25.20). Among the factors aiding flow in veins and determining the pressure therein are:

The MASSAGING ACTION OF SKELETAL MUSCLES (the "venous pump") as they contract and compress the blood within the veins. Valves in the veins prevent backflow, and the blood is moved toward the heart.

FIGURE 25.22 Vascular and cardiac responses to exercise.

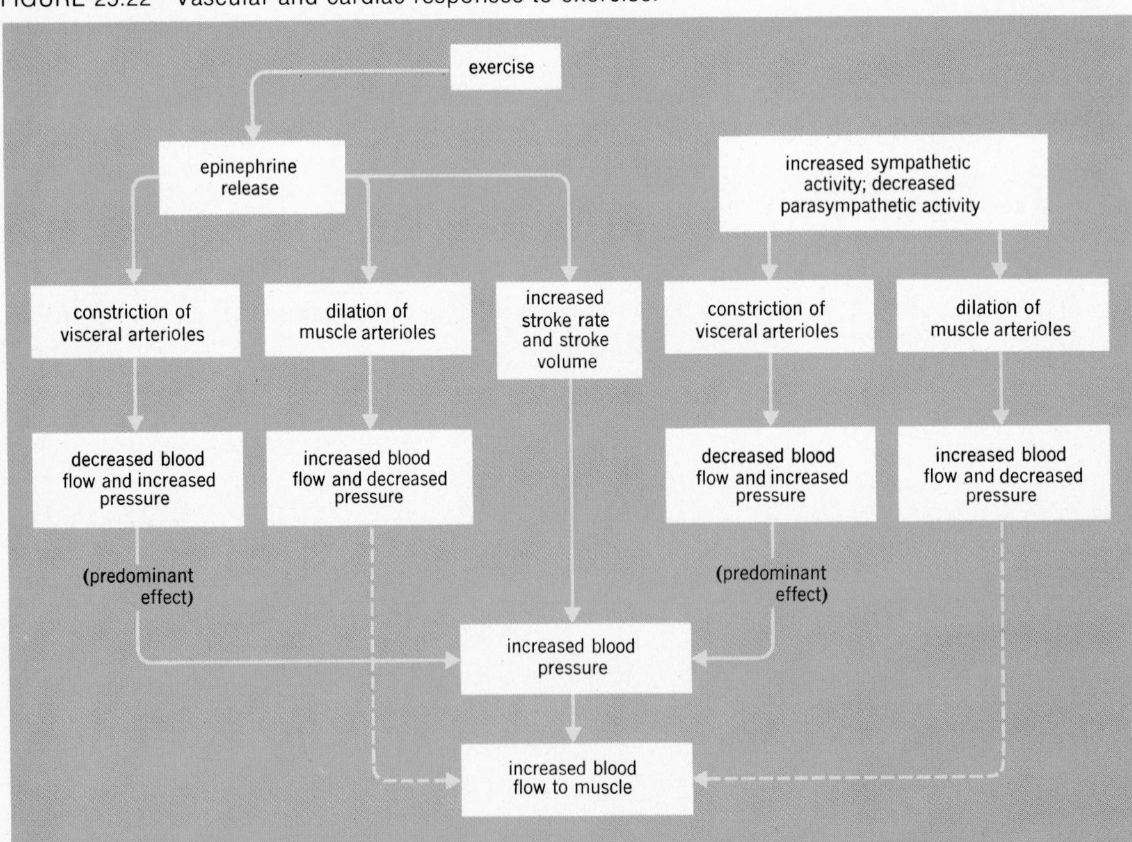

RESPIRATORY MOVEMENTS, particularly inspiration, decrease the pressure in the thorax, expand the venae cavae, and "draw" blood into the thorax.

VENTRICULAR RELAXATION during the cardiac cycle draws blood from the right atrium, which is then replaced by blood from the venae cavae.

The RESIDUAL HEART PRESSURE remaining after blood has traversed the capillary beds (*vis a tergo*) aids movement.

Coordination of heart action and vessel response

Normal systemic circulation responses illustrate the coordinated operation of some of the factors just discussed. Pressure and flow change normally in the body according to changes in four basic factors.

POSTURE. When a person goes from a lying to standing position, the force of gravity tends to cause pooling of blood in veins below the heart. Fall of venous return and blood pressure results. COMPENSATORY NERVOUS MECHANISMS that operate to counteract these changes are:

An increase in CONTRACTION OF SKELETAL MUSCLES compressing the veins.

VENOUS CONSTRICTION, to maintain volume of the venous circulation and to ensure adequate venous return.

A DECREASE IN CARDIAC OUTPUT (about 20 percent) that lowers the amount of blood pumped into the circulation. Rate is 5–10 beats higher in the erect position, but stroke volume is lowered 10–50 percent.

Failure of these compensatory mechanisms to maintain blood flow may lead to POSTURAL HYPOTENSION, failure of the brain to be adequately perfused, and fainting. The sequence of events occurring in response to postural change is shown in Fig. 25.21.

EXERCISE. A REDISTRIBUTION OF BLOOD to active muscles and away from the viscera is the outstanding circulatory adjustment to exercise, in addition to a rise of cardiac output. This redistribution is achieved by vasodilation of the vessels to skeletal muscles and constriction of visceral vessels. A 20-fold increase of the blood supply to active skeletal muscle may be achieved, with cardiac output increasing only 5-fold. These changes are summarized in Fig. 25.22.

TEMPERATURE. Changes in the blood flow through the skin account for the pronounced pressure effects of exposure to a hot environment. Due to cutaneous vasodilation, diastolic pressure may fall 40 mm Hg or more, with rise of cardiac output of only 10–20 percent. This may give rise to feelings of fatigue and drowsiness. Exposure to cold causes cutaneous vasoconstriction and an increase in cardiac output to maintain circulation in the face of an increased peripheral resistance.

CARBON DIOXIDE. Increase in CO_2 levels of the tissues (during exercise, for example) is associated with localized dilation of arterioles and increased flow. The capillaries dilate passively with increased arteriolar flow. Independent of nerves, this effect automatically ensures an increased blood flow to an active tissue.

Many of the changes in systemic circulation described above depend on changes in size of the arteries and veins of the system.

Control of systemic blood vessel size; effects on circulation

Control of diameter of muscular arteries and veins may be said to depend on two categories of stimuli: those provided by nervous factors and those provided by nonnervous factors, including chemicals and the nature of the muscle itself.

Nervous control

Two functional types of nerve fibers may be traced to most arteries and veins that have smooth muscle in their wall. Collectively, they are termed VASOMOTOR NERVES.

VASOCONSTRICTOR FIBERS, derived from the sympathetic division of the autonomic nervous system, tonically stimulate contraction of smooth muscle in the walls of vessels, narrowing them and increasing both peripheral resistance and blood pressure. Their effect is mediated by the production at the nerve endings of norepinephrine (noradrenalin), and the fibers are thus called adrenergic.

VASODILATOR FIBERS, derived from the sympathetic division of the autonomic nervous system, inhibit contraction of vascular smooth muscle and allow an increase of vessel diameter with a consequent decrease of peripheral resistance and blood pressure. The effect of these nerves is mediated by the production of acetylcholine (ACh) at the nerve endings, and these fibers are called cholinergic.

Vasoconstrictor fibers are more numerous and more widely distributed than vasodilator fibers. It should be kept in mind that while vasoconstriction is brought about by stimulation of vasoconstrictor nerves, vasodilation may occur either by active vasodilation through nerves or by inhibition of vasoconstrictor nerves. Thus, it appears that vasodilator fibers are of lesser importance to the body in carrying out normal control of cardiovascular function. Thus, when the sympathetic nervous system discharges impulses, vessels other than in heart and skeletal muscles are constricted, while those to the heart and skeletal muscles may be dilated. Dilation is due to inhibition of tonic vasoconstrictor activity. The blood will take the path of least resistance, and a redistribution of blood to the muscles must then occur.

Vasomotor nerves originate from VASOMOTOR CENTERS located in the medulla of the brainstem. Separate vasoconstrictor and vasodilator areas appear to exist. These centers appear to be influenced by impulses arriving from several body areas.

Pressure-sensitive receptors (baroreceptors) are located in the aortic arch, subclavian and carotid arteries, and lungs. These receptors discharge when pressure in the designated areas rises above some critical value. The baroreceptors send impulses to the medulla, and inhibition of vasoconstrictor activity and heart rate occurs. Large numbers of blood vessels dilate, cardiac output decreases, and blood pressure is reduced.

Baroreceptors located in the right atrium are thought to serve the Bainbridge reflex and cause an acceleration of heart rate by stimulation of the cardioaccelerator centers. Connections between this center and the vasoconstrictor center give a slight vasoconstriction.

The midbrain appears to have a relay station in it serving the sympathetic vasodilator reflex to skeletal muscle.

The hypothalamus contains areas concerned with emotional behavior and temperature regulation. A "heat loss area" is located in the anterior portion of the hypothalamus and causes, among its other effects, a dilation of cutaneous blood vessels, constriction of visceral vessels, and perspiring. Another area, in the posterior portion of the hypothalamus, brings about essentially the opposite reactions to those described above. These areas, then, through effects exerted on blood vessels, determine the degree of heat radiated through the skin, and can affect blood pressure by the change in vessel diameter. The emotional involvement of the hypothalamus is reflected in flushing or blanching of the skin in association with anger, fright, or embarrassment. Little effect is seen on blood pressure since relatively few peripheral vessels are involved.

The cerebral cortex, if stimulated in the motor region, results in increased heart rate and vasoconstriction.

These vasomotor reflexes are summarized in Fig. 25.23.

The changes in vessel size just described occur primarily in the arterial side of the circulation but

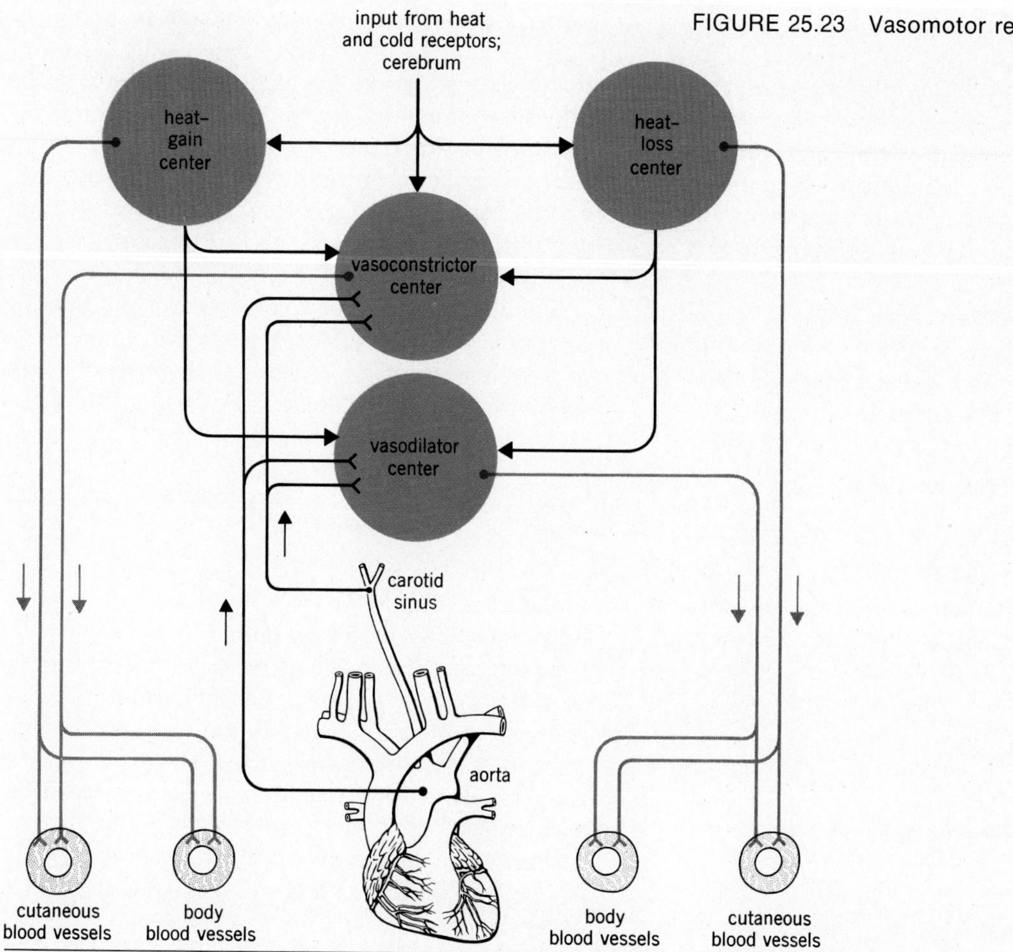

FIGURE 25.23 Vasomotor reflex pathways.

may also affect veins, which usually respond in the same manner as the arteries. Decrease in the diameter of veins results in changes in two parameters (a particular measure of physiological response) of circulatory dynamics:

The volume of the venous return to the right atrium is increased. Constriction of veins increases venous return by raising venous pressure. Constriction also avoids an increase in the 60 percent of blood normally present in the veins and minimizes "pooling."

The pressure in the capillaries is raised. Venous constriction increases capillary pressures and may increase fluid filtration in the capillaries. Edema may result if the pressures in the capillaries exceed the oncotic pressure of the plasma.

Nonnervous control

Nonnervous control of blood vessels is exerted mainly by chemical, thermal, and mechanical factors.

CHEMICAL SUBSTANCES. EPINEPHRINE, produced by the adrenal medulla, and NOREPINEPH-

RINE, produced by the adrenal medulla and sympathetic nerve endings on most effectors, are released in physiologically significant quantities when stress is applied to the body. The body vessels respond differentially to the chemicals. In the skin, skeletal muscle, kidney, and spleen, the effect of both chemicals is vasoconstrictive. In heart, liver, and lungs, a vasodilating effect of both agents is seen.

CARBON DIOXIDE excess causes local vasodilation. It would appear the contraction of smooth muscle is inhibited by CO_2 excess; that is, relaxation of smooth muscle is promoted by excess CO_2.

DIRECT EFFECTS OF HEAT AND COLD. Application of either heat or cold to denervated skin produces a vasodilation. Damage to the vessels or the release of histamine by the stimulus is thought to account for this effect.

The production of BRADYKININ, a polypeptide, results from the action of enzymes released from certain gland cells on tissue fluid proteins. It is a potent vasodilator.

Smooth muscle responds to STRETCHING by contraction. Sudden increases in pressure within a muscular vessel thus cause vasoconstriction. Protection from overdistension results.

Shock

SHOCK is a condition characterized by the development of certain characteristic signs. Among these are:

- Pallor.
- Cold extremities.
- Decreased body temperature.
- Depression of the central nervous system.
- Rapid, weak pulse.
- Low blood pressure.
- Reduced cardiac output.

Regardless of the cause of the symptoms, the most threatening developments are the low cardiac output and low blood pressure. Inadequate cerebral and renal circulation may result. In order to overcome the effects of shock, the body exhibits a number of compensatory mechanisms that attempt to restore blood flow and pressure to within normal limits. This will be considered under the specific types of shock described below.

Shock is usually classified on the basis of the cause.

HEMORRHAGIC SHOCK results from loss of blood or plasma. This is also termed HYPOVOLEMIC SHOCK since circulating blood volume is reduced.

CARDIOGENIC SHOCK, resulting from disturbance in heart action, is characterized by low cardiac output and low blood pressure.

VASCULAR SHOCK, resulting from alterations in blood vessel size from nervous, hormonal, or chemical influences, alters blood pressure.

TRAUMATIC SHOCK, resulting from trauma to muscle and other tissues as in automobile accidents.

More than one type of shock may occur in a given individual at a time, depending on the cause.

In HEMORRHAGIC SHOCK internal or external bleeding through ruptured or damaged vessels may result in rapid or slow losses according to location. Among the responses the body makes are:

Prevention of continued loss by coagulation and vasoconstriction.

Restoration of lost volume by shifts of fluids from interstitial to plasma compartments, by increased intake of fluids, and by increased secretion of ADH and aldosterone.

Decrease in volume of the vascular system is achieved by nerve impulses from the sympathetic nervous system that constrict arteries and veins and by epinephrine secretion, which has the same effect.

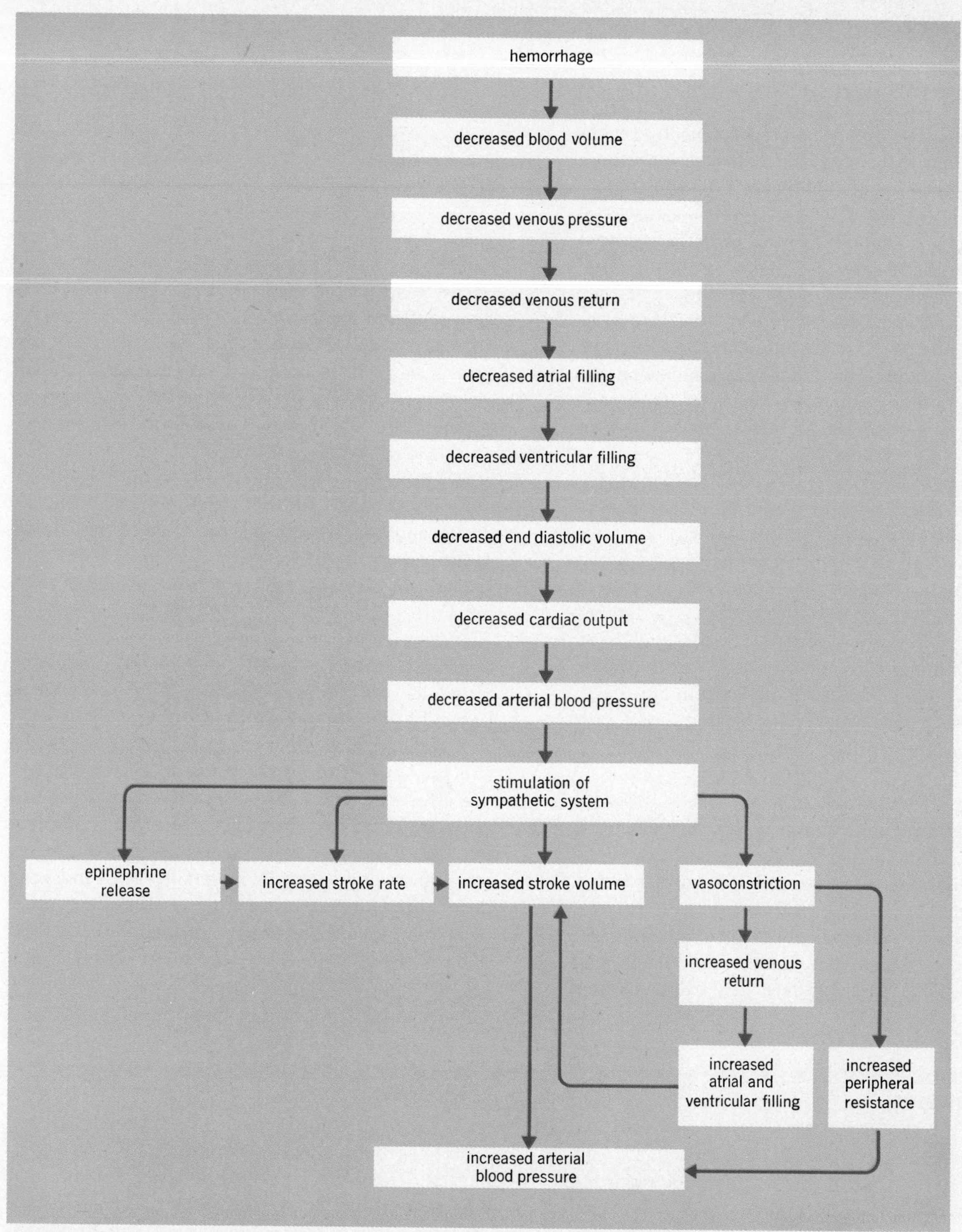

FIGURE 25.24 Some factors involved in production and compensation of hemorrhagic shock.

Not only is vascular volume reduced, but pressure is also elevated unless blood loss is severe. These reflexes are initiated through pressure-sensitive receptors in the body that are not stimulated when pressure falls. The normal effect of impulses from such receptors is to inhibit the activity of the vasoconstrictor center. Thus, if no impulses are present, the vasoconstrictor center is released from inhibition, and muscular vessels are diminished in diameter.

Heart rate is stimulated by epinephrine and sympathetic discharge. Cardiac output is raised.

Shunting of blood from areas such as the skin and kidney to the brain and heart. Failure of the organs from which blood is removed may occur.

The compensatory reactions are so efficient that a loss of about 5 percent of the blood volume is required before changes in circulatory dynamics can be detected. Compensation may be achieved if loss does not exceed 10–15 percent. Above 15 percent, compensation becomes more and more difficult, and death will result with a loss of 25–30 percent. Changes and compensatory mechanisms involved in hemorrhage are depicted in Fig. 25.24.

BURN SHOCK resembles hemorrhagic shock with several important differences.

Plasma rather than whole blood is lost. Leakage into the tissues produces edema and hemoconcentration.

Burned tissue releases toxins and is extremely subject to infection.

Toxins may influence the operation of other organs. For example, toxins may cause vasodilation and increase venous and capillary permeability. Renal blood flow may decrease to the point where filtration ceases and kidney tubules are damaged.

CARDIOGENIC SHOCK results when the action of the heart is disturbed. The heart may fail to fill properly, as when the pericardial sac fills with fluid; it may lose pumping efficiency when there is an infarction (blockage of coronary supply and tissue death); there may be decreased emptying. The common denominators in this disorder are decreased cardiac output and raised venous pressure. Cyanosis is common as circulation slows.

VASCULAR SHOCK is neurogenic if, as a result of sympathetic discharge, vasodilation of skeletal muscle diverts blood from the brain. Pooling of blood in veins occurs, retarding venous return. It may be humoral if the toxins released during tissue damage, histamine, or other chemicals are produced in quantities sufficient to cause widespread vasodilation and consequent fall in blood pressure.

TRAUMATIC SHOCK always involves damage to large masses of skeletal muscle and results in blood loss at the site of injury. The response is basically the same as for hemorrhagic shock.

A final word is perhaps appropriate regarding shock. Compensatory reactions do exist to combat the effects of shock. Most involve sacrificing one body function to aid another. These mechanisms may therefore set up vicious circles that aggravate the shock. For example, diversion of blood from the kidney to the brain may result in retention of toxins that aggravate the condition. Restoration of blood volume, pressure, and heart actions therefore become the primary aims of shock treatment.

Hypertension

Systolic pressures of 140–150 mm Hg and diastolic pressures of 90–100 mm Hg are generally regarded as the upper limits of normal. Sustained elevation of systolic pressures, diastolic pressures, or both above these limits is termed HYPERTENSION or high blood pressure. Sustained elevation of the systolic pressure alone is termed *systolic hypertension;* elevation of diastolic pressure alone is termed *diastolic hypertension.* If a cause of the hypertension can be determined, the hypertension is designated as secondary hypertension; that is, it occurs secondary to some other demonstrable disorder. If no specific cause for the hypertension is evident, it is designated as *essential* or *primary hypertension*.

Among the causes of secondary systolic hypertension are increased cardiac output and rigidity of the walls (arteriosclerosis) of the aorta and main arteries. If systolic pressure rises to 200 mm Hg or above, a danger of rupture of the blood vessels exists, particularly in the brain (cerebrovascular accidents or CVA, strokes).

Diastolic hypertension appears to be a particularly dangerous condition and results in vascular damage that may affect the operation of the organs served by the affected vessels. Vessels serving the kidneys, liver, pancreas, brain, and retina appear to be particularly prone to damage. High pressures in these vessels cause mechanical damage to the vessels that lead to sclerotic (hardening and thickening) changes, fibrosis, and reduction in diameter of the vessels. Blood flow diminishes, and the supplied area undergoes degenerative changes secondary to deprivation of nutrients. A common denominator present in diastolic hypertension, regardless of cause, is generalized peripheral vasoconstriction. Investigation of the causes of the vasoconstriction has led to the development of several hypotheses to explain the vascular spasm.

Goldblatt demonstrated many years ago that partial or complete occlusion of the RENAL arteries, which causes renal ischemia, resulted in the production of an enzyme designated *renin*. The source of renin is believed to be the juxtaglomerular apparatus of the kidney (Fig. 25.25). The stimulus to renin release is diminished oxygen supply and/or pressure to the kidney. Renin acts on a plasma globulin named angiotensinogen produced by the liver and converts it to angiotensin I. A plasma-converting enzyme then changes angiotensin I to angiotensin II, the most powerful vasoconstrictor known. If plasma renin levels are shown to be elevated, hypertension of this type is designated renal hypertension. Angiotensin II also stimulates adrenal cortical production of aldo-

FIGURE 25.25 A representation of the juxtaglomerular apparatus. *(a)* Diagram of kidney blood vessels and nephron showing the location of the juxtaglomerular apparatus. *(b)* The juxtaglomerular apparatus responsible for renin secretion. It consists of two portions: granular cells of the afferent arteriole (juxtaglomerular cells), and the large cells of the distal tubule (macula densa).

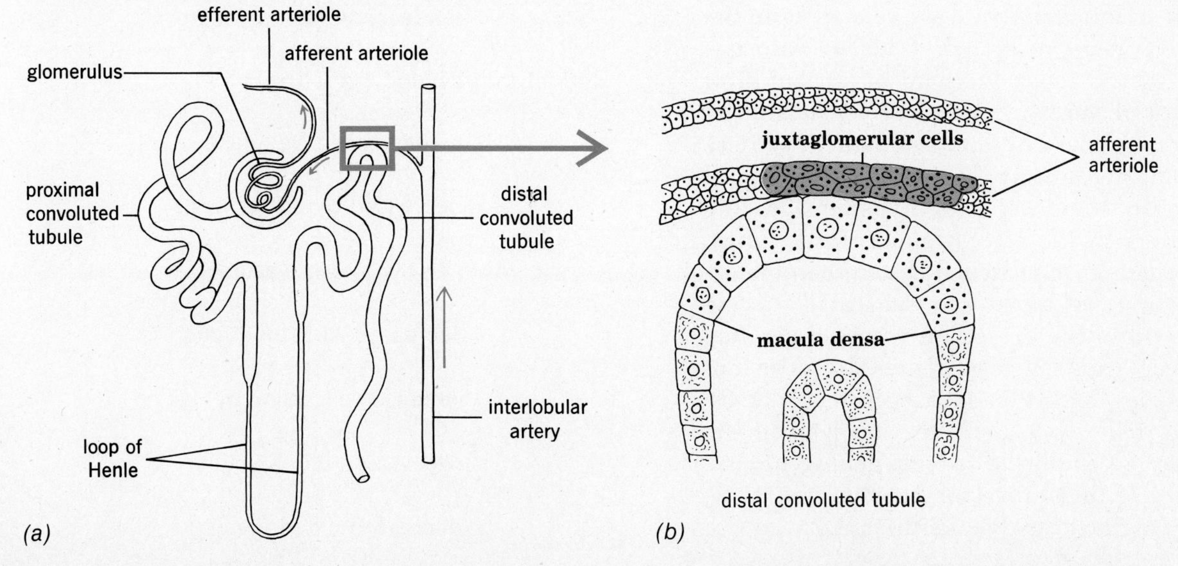

sterone, a hormone that increases kidney reabsorption of sodium. Sodium retention is associated with retention of sufficient water to maintain the blood isotonic. Thus, blood volume increases and blood pressure rises. As time passes, blood renin levels may diminish, without a corresponding fall in blood pressure. The hypertension may thus be said to have become irreversible.

As an explanation of irreversibility, it has been suggested that there has been interference with a renal hormonal mechanism, whose function is to cause arteriolar dilation. Chemical compounds named prostaglandins, which are derivatives of unsaturated fatty acids, have been shown to be produced by the kidney, lungs, brain, seminal vesicles, spleen, adrenal, liver, stomach, small intestine, and other organs. Among the effects of prostaglandins is a potent vasodilating action on peripheral arterioles. The chemicals are theorized to be diminished in production in renal ischemia and thus fail to counteract the vasoconstriction brought about by renin and sympathetic activity. A sudden onset of hypertension, onset before age 20 or after age 50, a history of renal trauma, and lack of a hypertensive history in the immediate family all suggest failure of an antihypertensive mechanism to operate.

Another theory to account for peripheral vasoconstriction is that it is due to NEUROGENIC causes. According to this theory, excessive activity in the sympathetic nervous system leads to vasoconstriction, or the reflex mechanisms that inhibit the activity of the vasoconstrictor center are not operating properly. Excessive sympathetic activity may accompany anxiety, stress, or psychological disorders. In some cases of hypertension, the baroreceptors in the carotid sinus, aorta, lungs, and elsewhere in the vascular system become insensitive or have a reduced sensitivity to the pressure within the vessels in whose walls they are located. Thus, increased pressure does not result in reflex inhibition of vasoconstrictor activity. Drugs that block the effect of sympathetic nerves may be employed to treat neurogenic hypertension. If this proves ineffective, the sympathetic ganglia from the tenth thoracic to the first or second lumbar may be removed. Such a surgical procedure is termed a sympathectomy and deprives blood vessels of the abdomen and lower limbs of vasoconstrictor impulses. They dilate, lowering blood pressure. Disadvantages of this procedure include: the vessels cannot undergo constriction, and blood may tend to pool in the lower parts of the body as position becomes more erect; there is a reduced response to cold environment stress; the organism has a reduced capacity to respond to hemorrhage and hypoglycemia.

EXCESSIVE INTAKE OF SODIUM, as sodium chloride (common table salt), requires retention of sufficient water to render the blood isotonic. In some hypertensive individuals, simple restriction of salt intake will aid treatment since the kidney will excrete the excess water. Diuretics have the effect of promoting both sodium and water excretion. The mechanism of production of hypertension by excessive sodium intake is

increased Na intake
↓
elevated blood Na levels
↓
increased H_2O retention
↓
increased blood volume
↓
increased blood pressure

Reduction of blood pressure is by

decreased Na intake
↓
decreased blood Na levels
↓
hypotonic plasma
↓
decreased ADH production
↓
increased excretion of H_2O
↓
decreased blood volume
↓
decreased blood pressure

Excessive production of adrenal steroids, or the use of steroids in the treatment of inflammation (e.g., rheumatoid arthritis), is associated with sodium and water retention. Again, increased blood volume raises blood pressure.

Prognosis in hypertension does not depend on the height to which the pressure rises but on the appearance of complications. In turn, appearance of complications depends on interactions between factors of diet, genetics, and stress.

Lymphatic vessels

Plan, structure, and function

The lymphatic vessels of the body (Fig. 25.26) carry lymph or tissue fluid from the body tissues and organs to the blood vascular system. Filtered water, protein, and white blood cells are therefore returned or added to the bloodstream, maintaining water and protein homeostasis.

The LYMPH CAPILLARIES are tiny vessels found in all body areas except the brain, spinal cord, and eyes. They have the same basic structure as blood capillaries but are larger and irregular in size. Dense networks of capillaries form lymphatic plexuses in most organs and collect tissue fluid from the interstitial fluid compartment. Lymph capillaries are much more permeable to large solutes than are blood capillaries, and thus filtered proteins, or large molecules such as enzymes that are produced by cells, easily enter the small lymph vessels. Capillaries form larger lymphatics known as COLLECTING VESSELS along which the lymph nodes are placed. The collecting vessels resemble small or medium-sized veins in structure, having three thin tunics or coats named as in blood vessels. Valves are numerous in the collecting vessels and give a "beaded" appearance to the vessel as they control movement only toward the neck.

The LYMPH DUCTS are formed from the collecting vessels; they resemble large thin-walled veins in structure. They, too, have three tunics and valves. The THORACIC DUCT collects lymph from all but the right upper chest, right arm, right side of the neck, and head. It empties into the left subclavian vein at its junction with the left internal jugular vein. Its lower end is called the cisterna chyli, although it does not form an enlarged bulb as in other animal species. The RIGHT LYMPHATIC DUCT collects lymph from the areas listed above and empties into the right innominate (brachiocephalic) vein.

Flow in lymphatics

The contraction and relaxation of skeletal muscles, breathing, and the expansion of lymph vessels by radially arranged elastic fibers around them contribute to the entry and movement of lymph in the vessels. Flow averages about 1.5 ml/min from the cut end of a medium-sized lymph vessel.

Clinical considerations

Since the lymphatics return about one-tenth of the filtered plasma water to the blood vessels, blockage of a major lymph vessel is associated with the development of EDEMA in the area served by the vessel.

Elephantiasis is the result of blockage of lymph vessels by a parasite; it used to be common in Africa. Blockage of lymphatics by cancer cells that have metastasized and been trapped may cause mild edema in the body part served by the plugged vessel.

Lymphatic drainage of the mammary glands (breasts) is of interest in terms of the metastasis of cancer cells from a malignant breast tumor. Lymph plexuses around the secretory lobules of the organ pass to a subcutaneous lymph plexus beneath the areola, toward the axillary lymph nodes, the substernal lymph nodes, and the lower

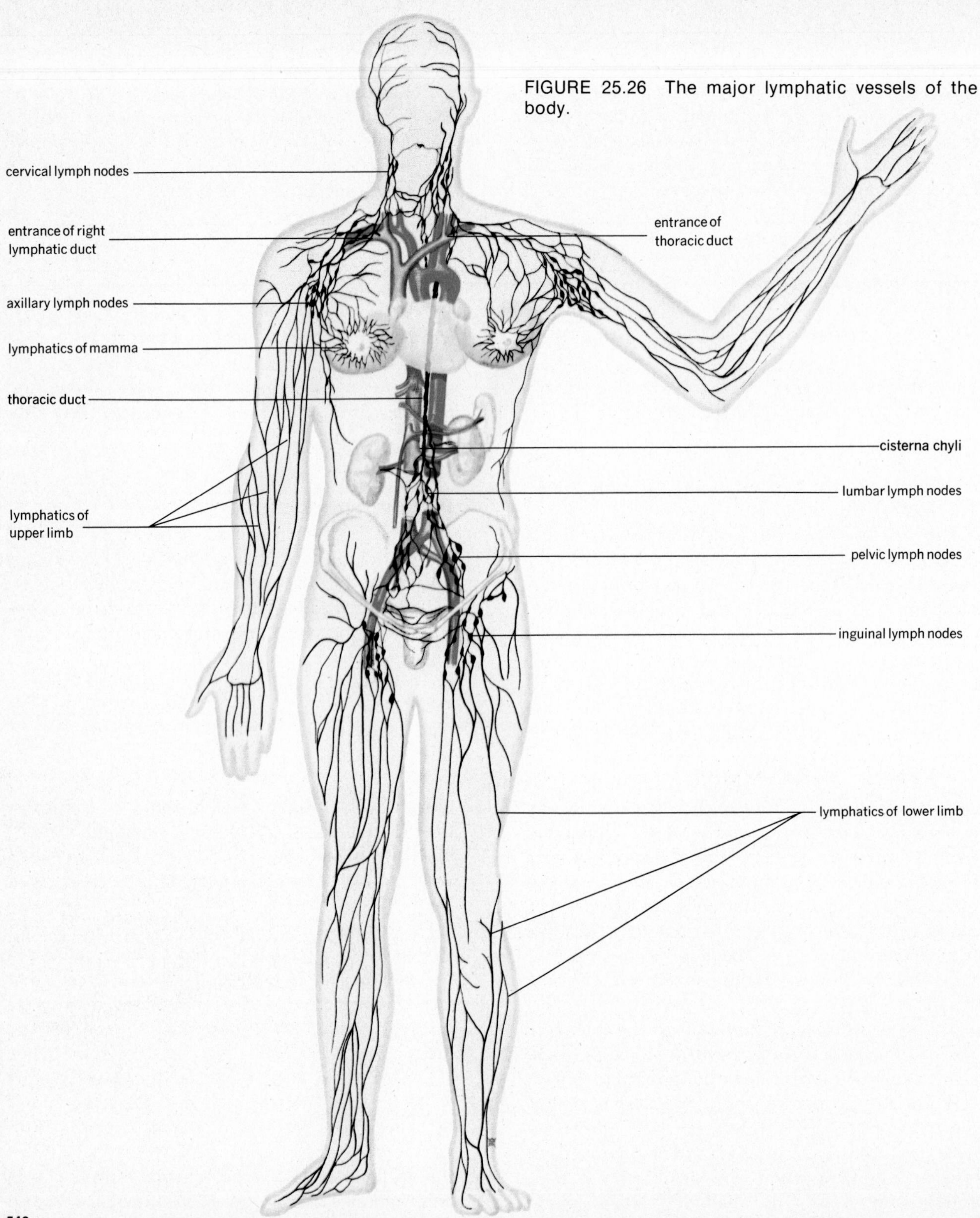

FIGURE 25.26 The major lymphatic vessels of the body.

cervical lymph nodes. This extensive drainage is what makes a radical mastectomy necessary if metastatic cancer cells are present in the axillary nodes. In a radical procedure, the muscles, lymph nodes, and breast tissue on the entire anterior chest wall are removed.

Summary

1. Blood vessels develop as cavities in mesoderm. The cavities grow and join to form networks of vessels. The muscle and connective tissue around the vessels are also of mesodermal origin.
2. There are three major types of vessels in the body.
   a. Arteries carry blood away from the heart.
   b. Capillaries allow for exchange of materials.
   c. Veins carry blood to the heart.
3. Blood vessels have a typical structure.
   a. Arteries have three layers in their wall (intima, media, adventitia); the media is thickest and contains much smooth muscle or elastic tissue or both.
   b. Arteriosclerosis and atherosclerosis are changes occurring in arteries that narrow their lumina and reduce blood flow. High levels of lipid in the blood accelerate atherosclerosis.
   c. Veins usually have three layers in their walls (intima, media, adventitia). The adventitia is the thickest.
   d. Varicosities develop if the veins are stretched and blood collects in them.
   e. Capillaries are very thin-walled tubes through which nutrients and wastes are exchanged. They permeate most tissues, are very small, and give a great surface area for exchange.
4. The major named vessels of the body are presented.
5. The pulmonary circulation is characterized by the low pressures required to perfuse it, by the small decrease in pressure as the blood passes through it, and by the relative independence of the vessels to nervous stimulation.
6. The cerebral circulation is maintained nearly constant in spite of changes in pH, systemic blood pressure, and chemicals. The most potent factor increasing cerebral circulation is increase in CO_2.
7. The coronary circulation to the heart has intermittent flow, since heart contraction compresses the vessels, and relaxation allows flow to occur. The most potent stimuli to increased coronary flow are a fall in blood O_2, a rise in blood CO_2, and decreased pH.
8. The circulation to the skin is controlled nervously and is correlated with blood and environmental temperature. High temperature in either area causes dilation to increase heat loss and vice versa.
9. The circulation to the liver is large in volume and low in pressure. The organ is important in many metabolic reactions, and many nutrients are supplied at a pressure that ensures time for exchange.
10. The spleen may store some blood and is an important organ for removal of worn-out red cells.
11. The skeletal muscles play an important role in adjustments of circulatory dynamics. The large mass of the muscle ensures that dilation is associated with large flow to the muscles and vice versa.
12. Five factors contribute to the creation and maintenance of blood pressure in the vascular system.
    a. Cardiac output determines pressure according to the volume of blood ejected into a system of given capacity.
    b. Peripheral resistance, or the resistance to blood flow through the arterioles, determines pressure and flow in arteries and capillaries.
    c. Elasticity of the large arteries determines the height to which systolic pressure rises and the depth to which diastolic pressure falls. Expansion of the vessel on systole cushions the shock of blood entering the arteries; elastic recoil moves the blood onward during diastole and prevents diastolic pressure fall to low values.
    d. The circulating volume of blood is a determinant of pressure; decreased volume lowers pressure and vice versa.
    e. The viscosity of the blood, if raised, requires a higher pressure to circulate it.
13. Certain laws govern flow and pressure in tubes.
    a. Resistance increases as the tube gets longer or if there are more tubes. Therefore, pressure falls in the area of the arterioles and capillaries.
    b. The cardiac output determines the pressure, and pressure is directly proportional to output; pressure falls as the number of vessels (and surface) increases.
    c. The greater the vessel surface, the slower the rate of blood flow.
    d. The thicker (more viscous) the blood, the more pressure required to pump it.
14. Since the larger arteries are closest to the heart, pressure is highest here, flow is more rapid, and, because of a small area, resistance is least.
15. In arterioles, surface increases, pressure falls, and rate of flow decreases.

16. In capillaries, surface area is greatest, pressure fall is greatest, and resistance is greatest. Thus, flow is slowest. This allows time for exchange of materials across the vessel wall.

17. In veins, pressure continues to fall, flow increases, and resistance falls because of fewer and larger vessels.

18. Blood is caused to flow through veins by four factors.
   a. Muscle contraction massages blood through the veins toward the heart; the direction is controlled by valves.
   b. Inspiration draws blood into the thorax from areas outside the thorax.
   c. Relaxation of the ventricles draws blood into the heart and through the vessels.
   d. Some pressure remains in the blood due to heart action.

19. As the body changes activity or position, heart action and vessel diameters change to ensure adequate blood circulation.
   a. When a person goes from a lying to standing position, heart action decreases, but vasoconstriction occurs. Thus, blood pressure rises or is maintained.
   b. In exercise, heart action increases, vessels to muscles dilate, and those to viscera constrict. Thus, active tissue receives more blood, and blood pressure rises.
   c. Dilation of skin vessels in a hot environment may cause great decrease in blood pressure.
   d. Increase in CO_2 levels of active tissue causes vasodilation locally and ensures active tissue of an increased blood supply.

20. Changes in blood vessel size are controlled by nervous and chemical factors.
   a. Vasoconstrictor and vasodilator centers in the brainstem are reflexly controlled from baroreceptors, located in the aorta, carotid sinus, and other body areas. In general, the heart rate and vessel changes complement each other; that is, if the heart rate decreases, vessels dilate and pressure falls.
   b. Epinephrine causes vasoconstriction and therefore rise in blood pressure; CO_2 causes vasodilation locally and increases local blood flow.

21. Shock is characterized by low cardiac output, low blood pressure, and diversion of blood from vital organs. Recovery depends primarily on restoration of adequate blood pressure to ensure perfusion of vital organs.

22. Hypertension implies elevated systolic and/or diastolic blood pressures.
   a. Elevated systolic pressure is associated with possibility of blood vessel rupture.
   b. Elevated diastolic pressure is associated with vascular damage resulting in narrowed vessels and decrease of blood flow.

23. Lymphatic vessels return tissue fluid and protein to the blood vessels.
   a. Lymph capillaries collect fluid from the tissues. They are very permeable to large molecules.
   b. Collecting vessels are larger and have nodes along their course.
   c. Lymph ducts empty the lymph into the subclavian veins. The two ducts are the thoracic duct and the right lymphatic duct.
   d. Blockage of a major lymphatic is associated with development of edema in the body area served by the vessel.
   e. The breast has an extensive lymph drainage that may favor the spread of cancerous cells.

26 The Respiratory System

Development of the system

The purposes of a respiratory system

Organs of the system
- The conducting division
- The respiratory division
- The lungs
- Blood supply
- Nerves

The physiology of respiration
- Pulmonary and alveolar ventilation
- Exchange of air
- The work of breathing
- Composition of the respired air; gas diffusion

Transport of gases
- Oxygen transport
- Carbon dioxide transport

Acid-base regulation by the lungs

Other functions of the lungs

Control of respiration
- Central influences
- Peripheral influences

Development of the system

The nose and nasal cavities are formed at 3½ to 4 weeks of embryonic life as the face develops (Fig. 26.1). An OLFACTORY PLACODE, or area of thickened ectoderm, appears on the front and lower part of the head. An OLFACTORY PIT develops in the placode and grows posteriorly to fuse with the front part of the primitive gut (foregut). This latter structure forms the PHARYNX. The pit extends above the mouth cavity and is separated from it by a thin membrane. This membrane ruptures and creates a single large NASAL-ORAL CAVITY. At about 7–8 weeks of age, division of the cavity into an upper nasal cavity and a lower oral cavity is begun by plates growing horizontally across the cavity. At about the same time, a vertical plate grows downward from the roof of the nasal cavity. These plates meet and fuse by three months of development. The horizontal plate becomes the HARD PALATE, and the vertical plate the NASAL SEPTUM.

If the horizontal processes fail to meet in the midline, a CLEFT PALATE will result. Cleft palate makes it very difficult for the infant to swallow and create suction to nurse.

The development of the rest of the system (Fig. 26.2) begins as a LARYNGOTRACHEAL GROOVE in the floor of the foregut, at about 26 days. The groove deepens, and the walls come together to form a

FIGURE 26.1 The development of the nasal region from the fourth to twelfth weeks. Numbers 1–5 all occur in the *fourth week.*

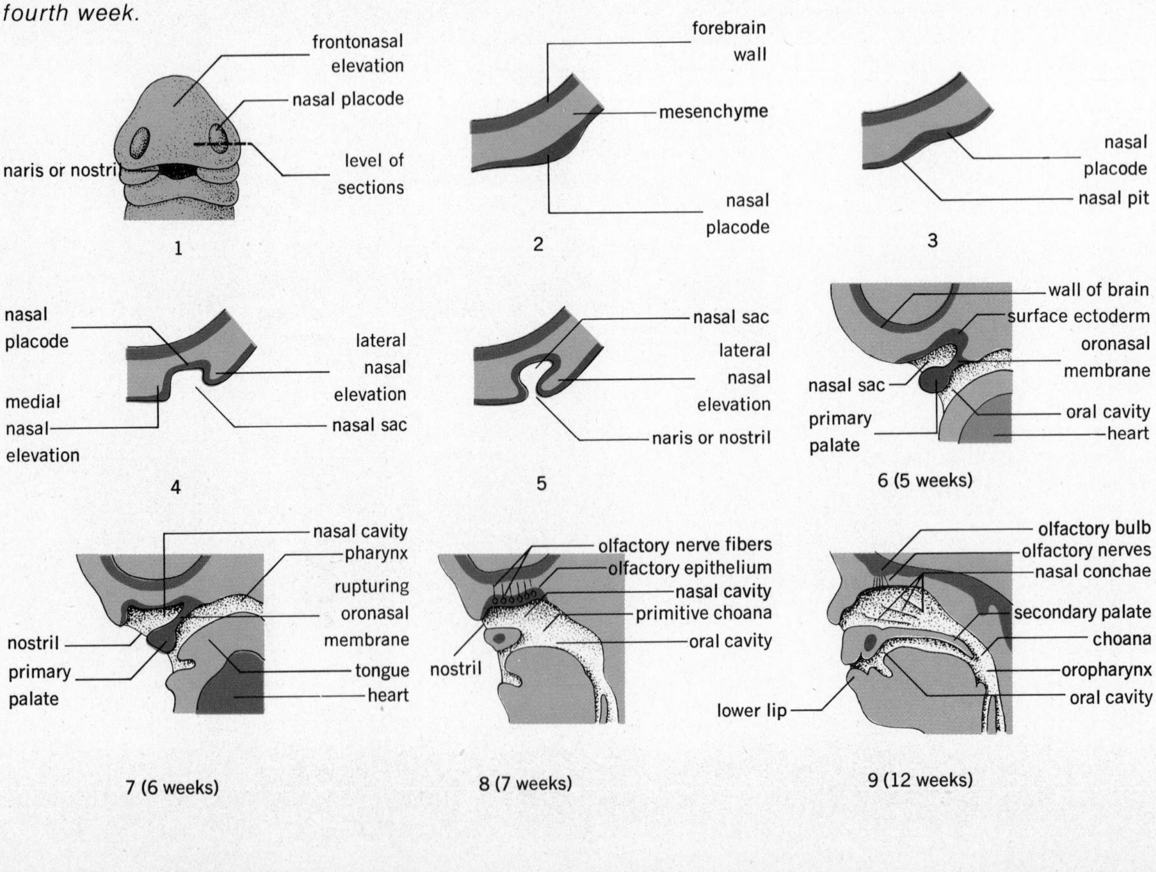

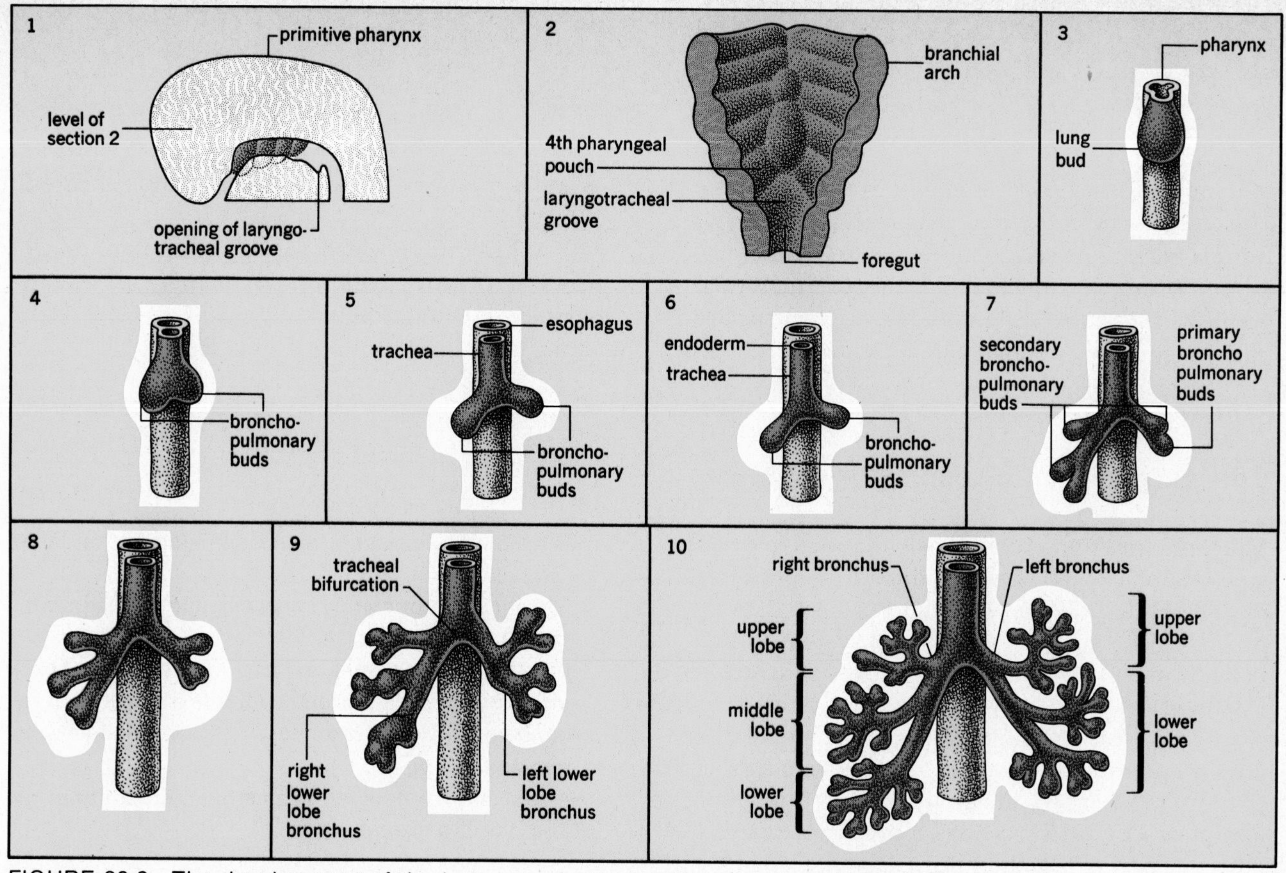

FIGURE 26.2 The development of the lower respiratory system. Numbers 1 and 2, 3½ weeks; 3–6, 4 weeks; 7 and 8, 5 weeks; 9, 6 weeks; 10, 8 weeks.

FIGURE 26.3 The development of the alveoli at 24 and 26 weeks and in the newborn.

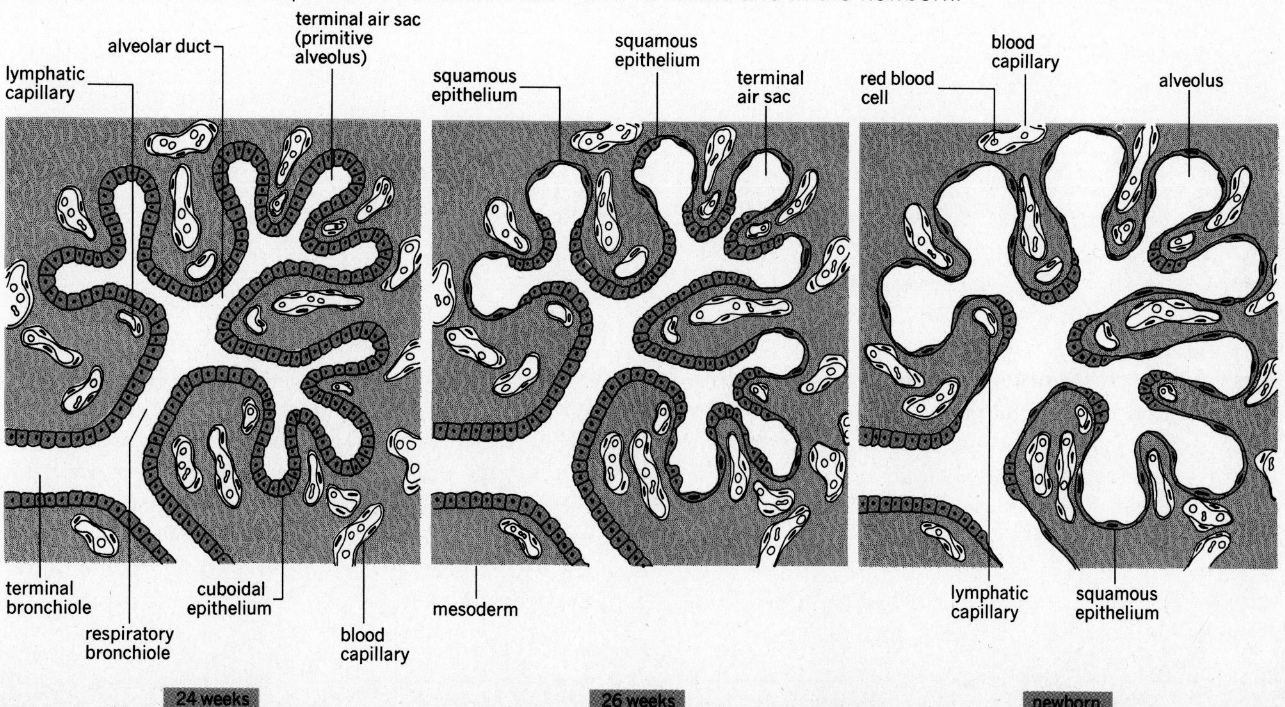

tube. The lower end of the tube forms a LUNG BUD that undergoes growth and repeated branching. The various tubes of the "respiratory tree" and the air sacs are derived from these branchings. In all, some 24 divisions of the original tube occur, to create the vast number and great surface of the respiratory organs. The muscular, cartilagenous, and connective tissue coats begin to form from mesoderm surrounding the tube at about 10 weeks.

Until about 26 weeks of age the lungs do not contain ALVEOLI. After this time, there are usually enough alveoli developed to sustain life if the child is born prematurely (Fig. 26.3).

The purposes of a respiratory system

Metabolic reactions within cells require continual supplies of oxygen. Carbon dioxide is produced as one of the products of that metabolism. It is the primary function of a respiratory system to provide a means of ACQUISITION OF OXYGEN and ELIMINATION OF CARBON DIOXIDE. The system actually is exposed to the external environment and must be provided with a means to PROTECT against changes or toxins in the environment. The elimination of CO_2 also provides a means of REGULATION OF THE pH of the internal environment. A respiratory system thus provides:

A maximum surface area for diffusion of the gases O_2 and CO_2.

A means of constantly renewing the gases in contact with that surface (ventilation).

Means to protect that important surface membrane from harmful environmental factors such as microorganisms, airborne toxic particles, adverse temperatures, and drying.

A first line of defense against sudden shifts in pH of the blood and body fluids.

Organs of the system

The organs of the respiratory system (Fig. 26.4) may be divided into:

A CONDUCTING DIVISION, whose walls are too thick to permit exchange of gases between the air in the tubes and the bloodstream. The nostrils (nares), nasal cavities, pharynx, larynx, trachea, bronchi, and bronchioles are included in this division.

A RESPIRATORY DIVISION, whose walls are thin enough to permit exchange of gases between tube and blood capillaries surrounding them. The respiratory bronchioles, alveolar ducts, atria (the space from which the alveoli of the sacs arise), and alveolar sacs comprise this division.

Essential to the exchange of air within both divisions are the muscles of respiration (diaphragm and intercostal muscles), the ribs, and the sternum.

The conducting division

NASAL CAVITIES. The nasal cavities (Fig. 26.5) open to the exterior through the NARES or nostrils. The VESTIBULE is a dilated area bearing large hairs, lying immediately posterior to the nares. The nasal cavity proper is delimited anteriorly by the nares; posteriorly by the CHOANAE or internal nares opening into the throat; laterally by the ethmoid, maxillae, and inferior conchae; medially by the vomer and ethmoid perpendicular plate; and the floor is formed by the hard palate. The

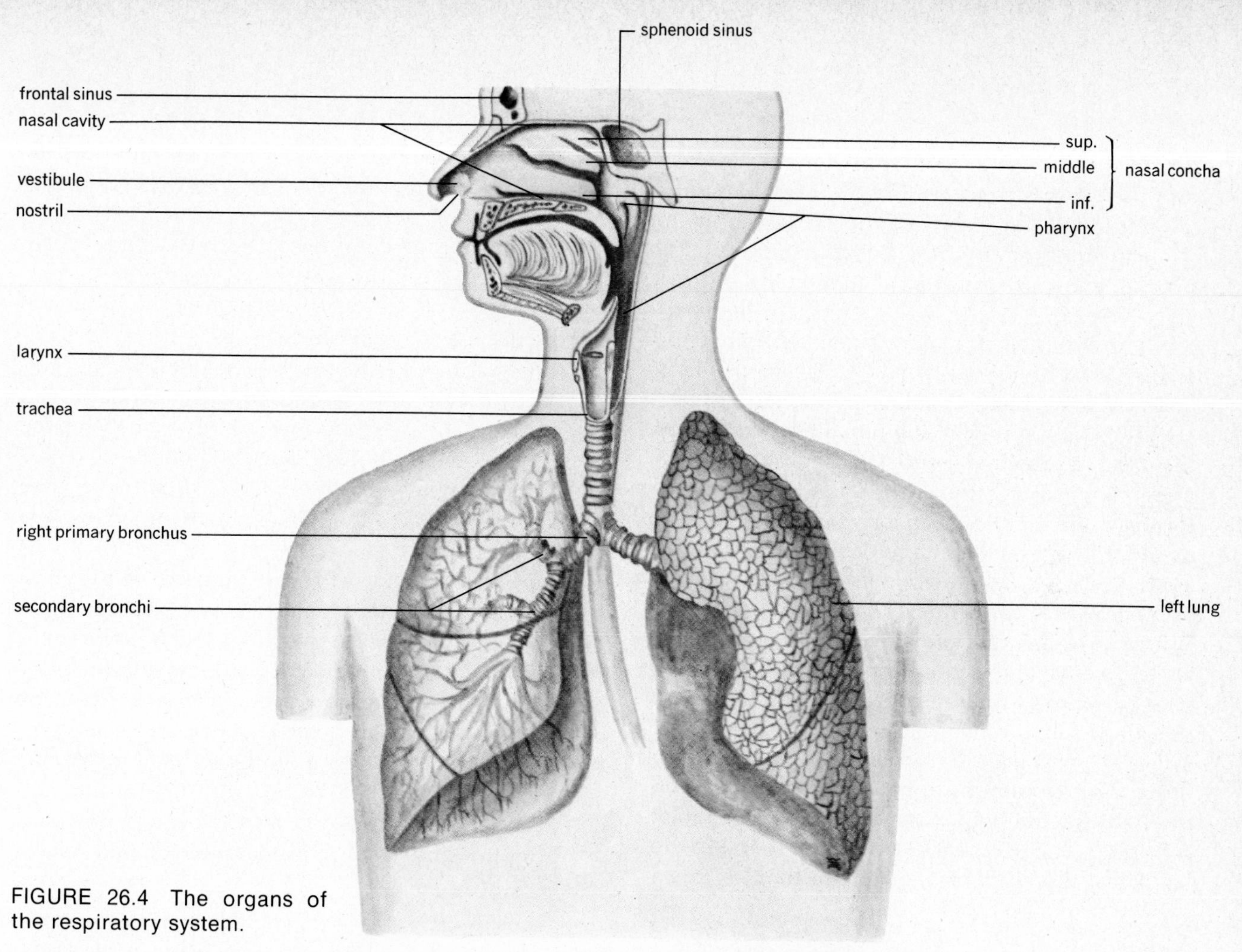

FIGURE 26.4 The organs of the respiratory system.

FIGURE 26.5 The left nasal cavity.

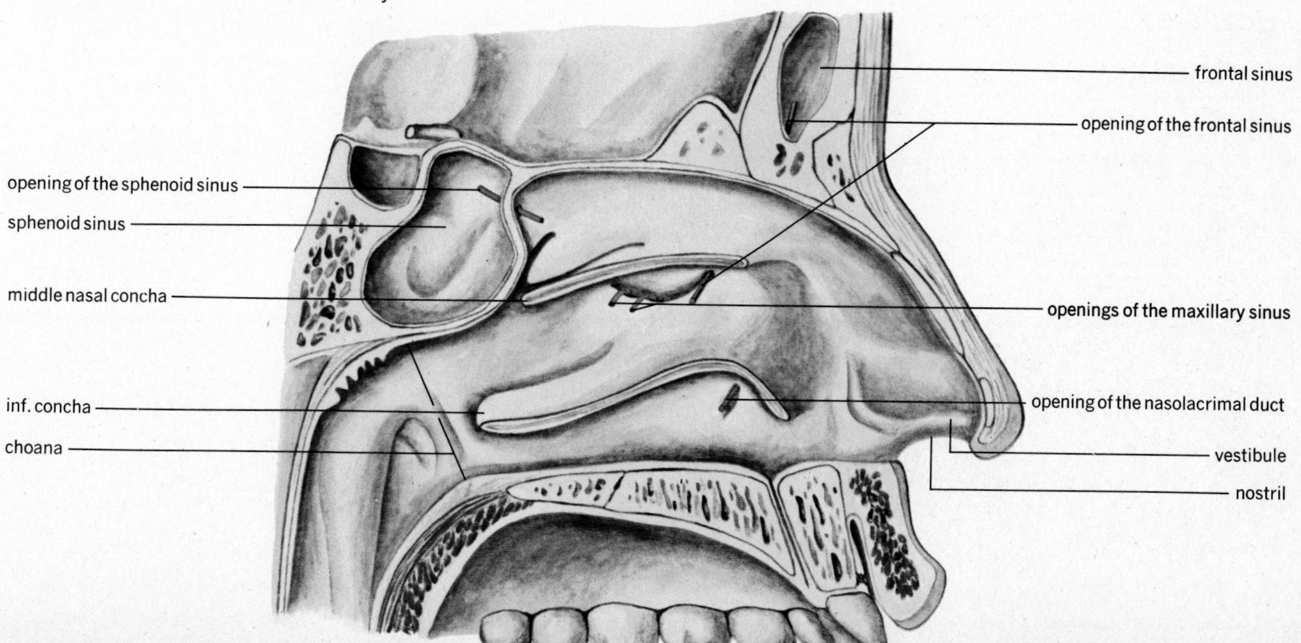

ethmoid conchae and inferior conchae project into the cavity proper and create a turbulent airflow through the cavity, a device essential to cleansing of incoming air. The PARANASAL SINUSES, (frontal, sphenoid, maxillary, ethmoid) open into the cavity (see Fig. 26.5). In the apex of the nasal cavity is the olfactory area wherein are found the receptors for the sense of smell. The vestibule is lined by a stratified squamous epithelium. Except for the olfactory region, the remaining portion of the nasal cavities and sinuses is lined by a mucous membrane closely applied to underlying bone or cartilage. The epithelium of the nasal mucous membrane is pseudostratified ciliated, containing goblet cells. A prominent basement membrane separates the epithelium from the underlying lamina propria. The lamina contains many thin-walled veins and seromucous glands, giving a spongy nature to the membrane. The blood vessels allow radiation of heat to warm incoming air. The viscous mucus traps particles, and the cilia move them toward the throat. The olfactory region has a thicker epithelium devoid of cilia and goblet cells (Fig. 26.6). *Glands of Bowman* are found in the lamina of the olfactory region, and they secrete a watery fluid that moistens the olfactory surface. The mucous membranes rest on the bone or cartilage of the cavity wall.

Clinical considerations. The most common condition affecting the nose is, of course, the "common cold." It is a viral infection of the linings of the nose, and produces RHINITIS (*rhin,* nose, + *itis,* inflammation). Rhinitis is associated with swelling of the linings ("plugged nose"), oversecretion by the glands ("runny nose"), fever, and general tiredness. It is hoped that, within the next few years, vaccines may be made against the viruses causing colds, thus giving protection against the organisms.

SINUSITIS refers to the inflammation of the paranasal sinuses. It may be caused by microorganisms or allergic reactions. Swelling of the membranes that line the openings of the sinuses into the nasal cavities blocks drainage of secretions from the sinuses, and fever, headache, dizziness, and tenderness when pressing on the affected sinus may develop. Sinusitis may be acute or chronic, and treatment is directed toward opening the exits from the sinuses to promote drainage and relieve pressure.

FIGURE 26.6 The olfactory epithelium.

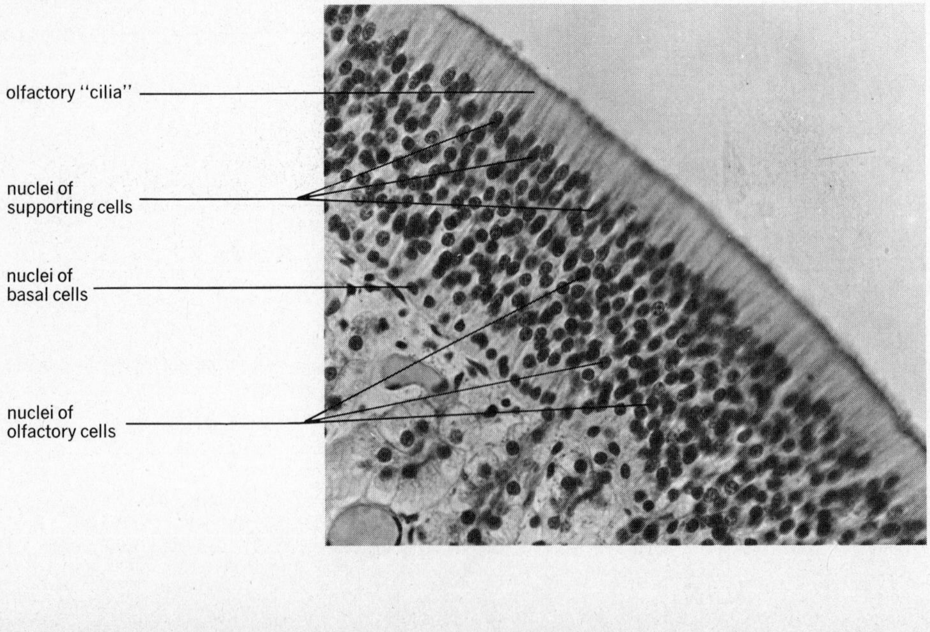

olfactory "cilia"
nuclei of supporting cells
nuclei of basal cells
nuclei of olfactory cells

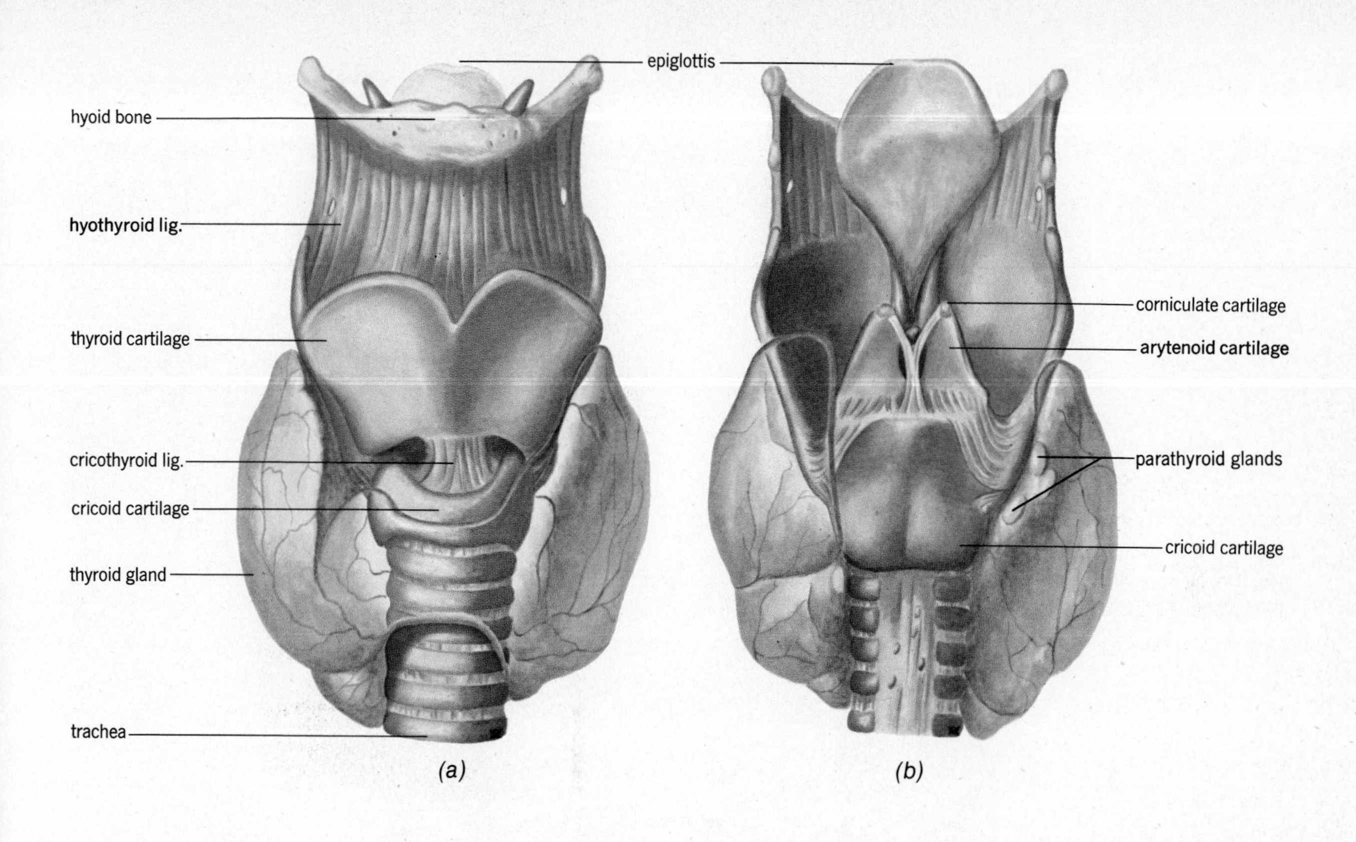

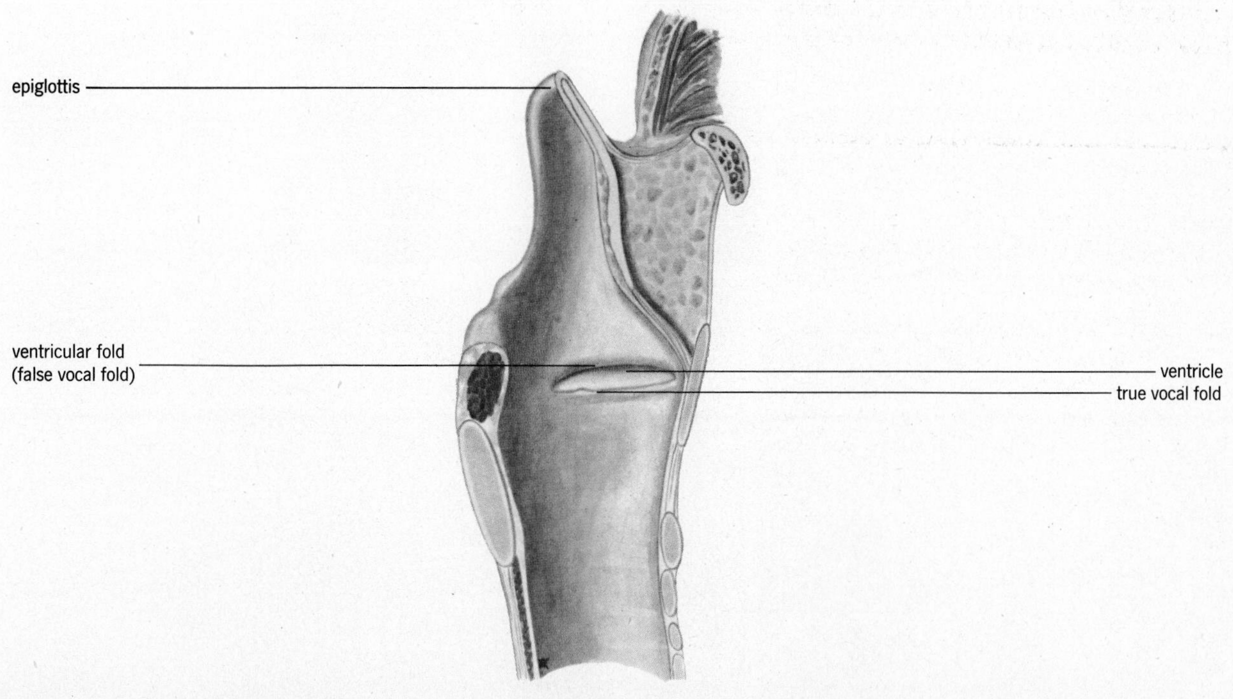

FIGURE 26.7 The larynx. *(a)* Anterior view, *(b)* posterior view, *(c)* the vocal folds.

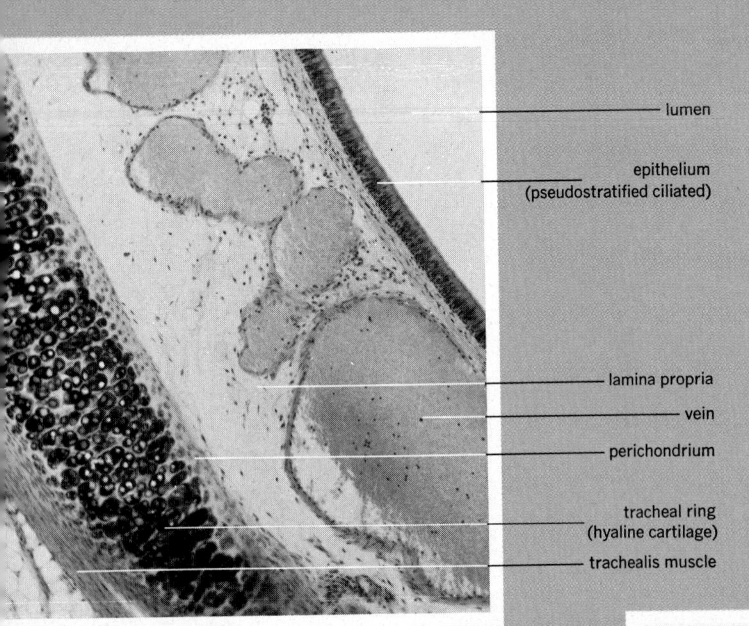

FIGURE 26.8 The histology of the trachea.

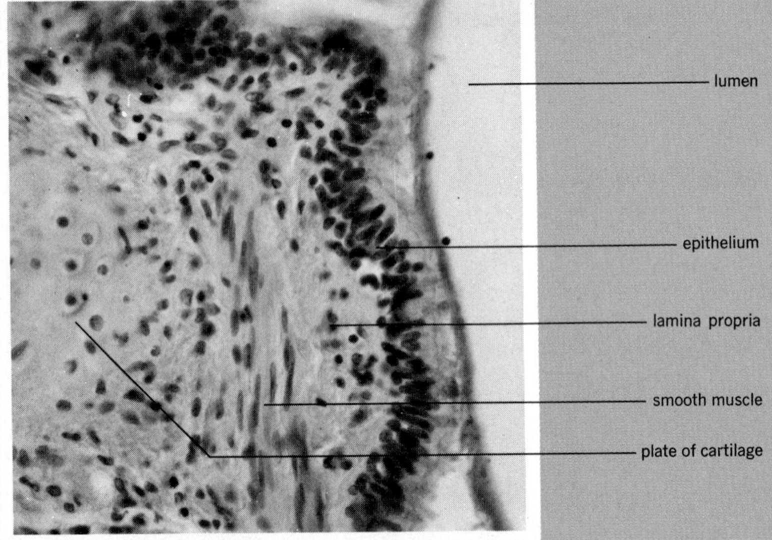

FIGURE 26.9 The histology of a bronchiole.

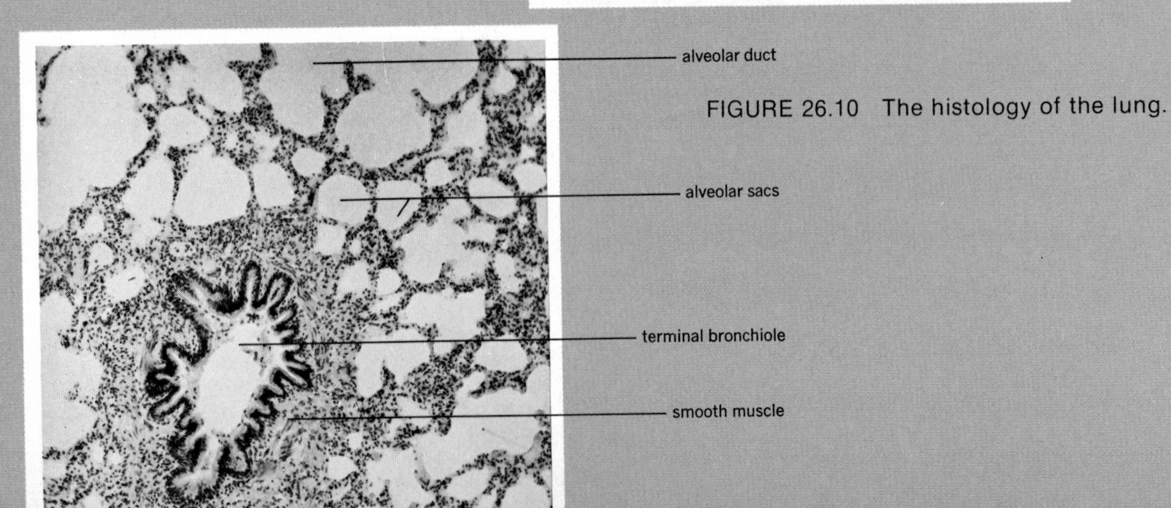

FIGURE 26.10 The histology of the lung.

EPISTAXIS refers to hemorrhage or bleeding from the nose. Trauma, nose picking, fractures of the skull bones, and foreign bodies in the nose are the most common causes of nasal bleeding. Systemic disorders such as leukemia, low platelet levels, and hemophilia are also associated with epistaxis.

PHARYNX. The nasal cavities communicate with the pharynx (Fig. 26.4) through the choanae. The pharynx is an organ common to both respiratory and digestive systems. It extends from behind the nasal cavities to the level of the larynx and is divided into three portions: NASAL PHARYNX, ORAL PHARYNX, and LARYNGEAL PHARYNX.

The nasal pharynx extends from the choanae to the inferior border of the soft palate and is respiratory in function. It receives in its lateral walls the openings of the AUDITORY *(Eustachian)* TUBES from the middle ear. In its posterior wall is found the PHARYNGEAL TONSIL *(adenoid)*. The mucous membrane of the nasal pharynx consists of a pseudostratified ciliated epithelium with goblet cells and a typical lamina. The superior pharyngeal constrictor muscle forms the third layer of its wall.

The oral pharynx extends from the soft palate to the level of the hyoid bone. It is both respiratory and digestive in function. Its mucous membrane consists of a stratified squamous epithelium and a typical lamina. The middle constrictor of the pharynx forms the third layer of this region.

The laryngeal pharynx extends from the level of the hyoid bone to the beginning of the esophagus. It, too, is lined with stratified squamous epithelium and possesses a lamina. The inferior constrictor forms the third layer of the wall.

Clinical considerations. ENLARGEMENT OF THE ADENOIDS, due to inflammation, may interfere with breathing and may convert an individual to "mouth breathing" with great loss of the warming, moistening, and cleansing functions the nasal cavities serve. Surgical removal of the enlarged organ is usually indicated. Since the body has many other lymphoid organs, loss of the adenoids is not severely missed, although a portion of Waldeyer's ring has been removed. (Waldeyer's ring is the circular arrangement of lymphoid tissue formed around the pharynx by the palatine, lingual, and pharyngeal tonsils.)

LARYNX. The larynx is the anterior opening from the inferior portion of the laryngeal pharynx. Three single major cartilages establish the basic shape of the organ (Fig. 26.7). The largest, or THYROID CARTILAGE, consists of two plates or laminae joined in the anterior midline. It forms the "Adam's apple." The CRICOID CARTILAGE lies inferior to the thyroid cartilage and is ring-shaped, being wider posteriorly. Thyroid and cricoid cartilages are composed of hyaline cartilage. The EPIGLOTTIS is a large, leaf-shaped mass of elastic cartilage attached to the superior internal surface of the thyroid cartilage.

Three pairs of accessory cartilages support the vocal folds (see Fig. 26.7). The ARYTENOID CARTILAGES attach to the superior aspect of the cricoid cartilage in the posterior larynx. The CORNICULATE CARTILAGES attach to the apices of the arytenoids, and the CUNEIFORM CARTILAGES are elongated cartilages lying anterior to the arytenoids.

The HYOTHYROID LIGAMENT connects the hyoid bone to the thyroid cartilage; the CRICOTHYROID LIGAMENT connects thyroid to cricoid cartilage. Posteriorly, the larynx is closed by muscle. The vocal folds consist of superiorly placed ventricular folds, or false vocal folds, and inferiorly placed true vocal folds. The true vocal folds or vocal cords have an internal support of collagenous fibers. The epithelium of the larynx is pseudostratified, except over the vocal folds where stratified squamous is found. The lamina contains many seromucous glands. The GLOTTIS is the slit-like opening between the vocal folds. EXTRINSIC MUSCLES of the larynx connect the organ to surrounding structures and aid in elevating and depressing it during swallowing. INTRINSIC MUSCLES connect the various cartilages, control the tension on the vocal cords, and open and close the glottis.

Clinical considerations. LARYNGITIS, or inflammation of the larynx, may result in swelling of the vocal folds and cords to where the air passageway is obstructed, and breathing becomes difficult or impossible. A hole may be made in the trachea *(tracheotomy)* below the larynx to permit air movement to occur. LARYNGECTOMY, or removal of the larynx, is sometimes necessary in cancer of the larynx. Speech is still possible, even though the larynx is removed, by belching air through the esophagus and shaping it into speech (esophageal speech), or by applying a special vibrator to the neck and shaping the sound into speech. LARYNGOSCOPY is a term referring to visual examination of the larynx through an instrument called a LARYNGOSCOPE.

If the superior laryngeal orifice is blocked, as by a large piece of food, the individual may "choke to death" or suffocate. The *Heimlich maneuver* has been developed as a means of dislodging such pieces of food. In this maneuver, a person stands behind and places the arms around the choking individual with the hands locked just below the xiphoid process. By a sudden squeezing movement that drives the hands suddenly inward and upward, a pressure is created on the lungs and stomach that is usually sufficient to "blow" the piece of food out of the lower pharynx, thus permitting resumption of breathing.

TRACHEA AND BRONCHI. The trachea extends about 11 cm into the thorax and terminates by dividing into the right and left bronchi. The TRACHEA is 18–25 mm (¾–1 in.) in diameter. Its wall (Fig. 26.8) consists of a pseudostratified ciliated epithelium containing goblet cells and a lamina with glands. The third layer is composed of C- or Y-shaped RINGS of hyaline cartilage, incomplete posteriorly. The rings support the trachea, keeping it open during breathing. The TRACHEALIS MUSCLE connects the open ends of the tracheal rings. The BRONCHI are similar to the trachea in structure and are about 12 mm (½ in.) in diameter. The right bronchus is about one-half as long as the left; it leaves the trachea at a lesser angle. Further branching of the bronchi results in a treelike arrangement of more numerous and smaller tubes. These tubes are the SECONDARY BRONCHI, BRONCHIOLES, and TERMINAL BRONCHIOLES. Some 13 generations of branchings occur within the lung itself. Histologically, several major trends are seen as the branching occurs.

Rings of cartilage are replaced by plates of cartilage. The plates become smaller and finally disappear as the terminal bronchiole is reached.

As cartilage decreases, the proportion of smooth muscle increases, so that in the terminal bronchiole muscle forms the greatest thickness of the wall.

The epithelium changes from pseudostratified ciliated with goblet cells, to pseudostratified ciliated without goblet cells, to simple columnar ciliated as the terminal bronchiole is reached.

The diameters of the tubes decrease from about 12 mm in the bronchi to 0.7 mm in the terminal bronchioles. Surface area, because of branching, increases from 2.3 cm^2 to about 115 cm^2. Sections of bronchus, bronchiole, and terminal bronchiole are shown in Figs. 26.9 and 26.10.

Clinical considerations. Since the terminal bronchioles have a wall that has no cartilage and mostly smooth muscle, contraction of that muscle may severely obstruct airflow through the tubes. In asthma, spasm of the muscle due to irritants may cause extreme difficulty in filling and emptying the alveoli. Any foreign body or tumor that interferes with airflow through these tubes also leads to labored breathing. EMPHYSEMA causes, as one of its many effects, a collapse and kinking of small and terminal bronchioles, thus interfering with airflow through the tubes.

The respiratory division

The respiratory division (Fig. 26.11) is composed of about nine generations of branchings. The chief trends in this division are a thinning of the epithelium to simple squamous and the appear-

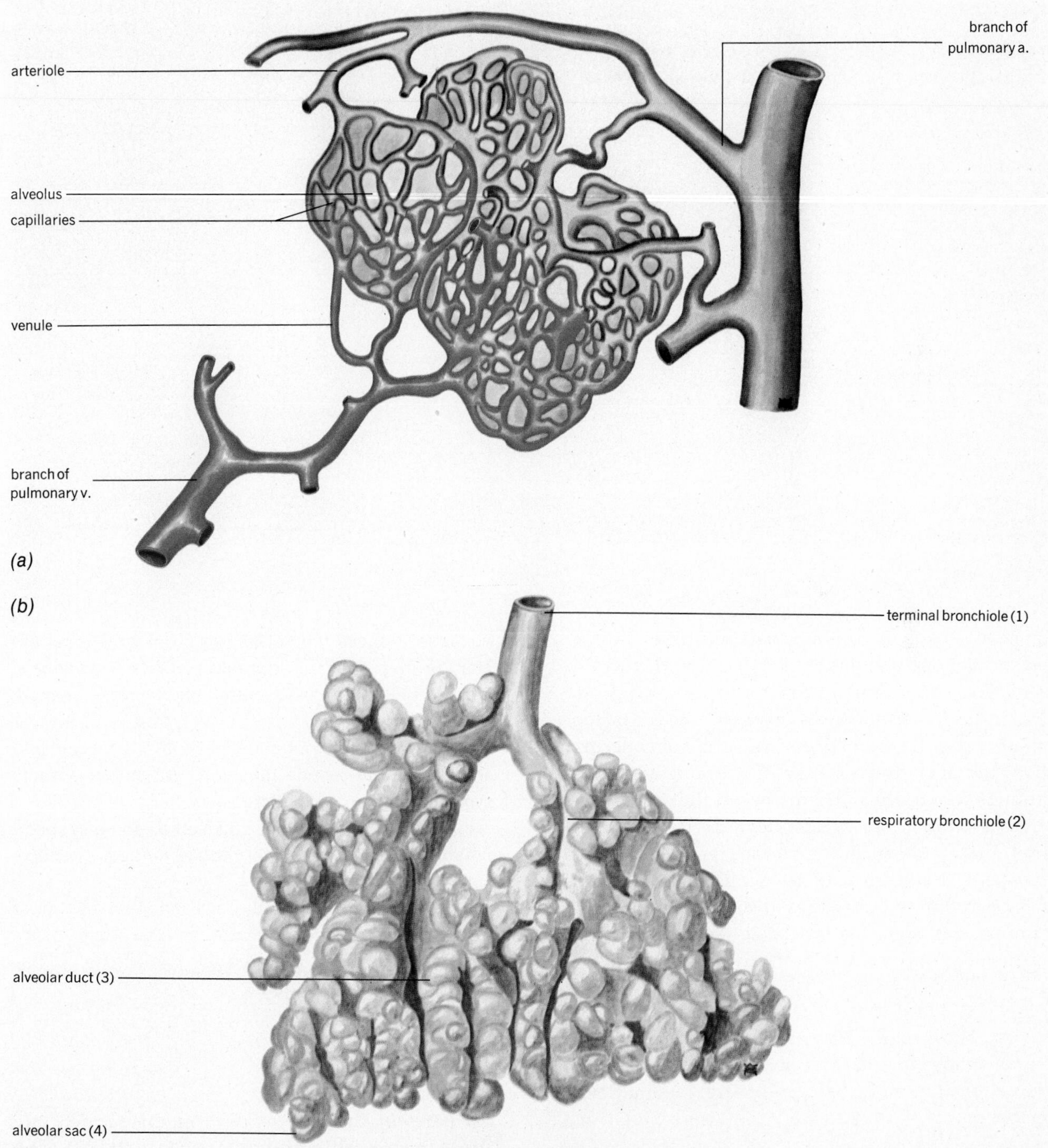

FIGURE 26.11 *(a)* The blood supply of the alveoli. *(b)* The lobules of the lung. 1–4, secondary lobule; 2–4, primary lobule.

TABLE 26.1 Some characteristics of the respiratory system

Organ	Generation of branching	Number of organs	Diameter (mm)	Total cross-sectional area (cm^2)
Trachea	0	1	18–25	2.5
Bronchus	1	2	12	2.3
Lobe bronchi	2	4–5	8	2.1
Small bronchi	5–10	1024	1.3	13.4
Terminal bronchioles	14–15	32,768	0.7	113.0
Respiratory bronchioles	16–18	262,000	0.5	534
Alveolar ducts	19-22	4.2 million	0.4	5880
Alveoli	23–24	300 million	0.2	(50–70m^2)

ance of alveoli in the sides of the tubes. Much elastic connective tissue is found in the walls, replacing the muscle of the conducting division. Increasing vascularity is also apparent in the respiratory division since gas exchange occurs between this division and the branches of the pulmonary arteries. Diameter of the tubes decreases little in the respiratory division, a fact that maintains airflow. Surface area, however, increases tremendously to achieve the surface necessary for efficient gas diffusion. Table 26.1 summarizes some of the basic facts concerning size and surface area of both divisions.

CLINICAL CONSIDERATIONS. The respiratory division, because of its alveoli, is subject to a number of conditions that interfere with gas exchange. Some of the more important are summarized below.

In ATELECTASIS, there is collapse of alveoli, or incomplete expansion in the newborn. Because alveolar collapse reduces the surface through which gas exchange may occur, symptoms of O_2 lack (cyanosis, labored breathing), or CO_2 retention (hyperventilation, acidosis) may occur.

In PNEUMONIA, production of fluid fills alveoli with secretions that reduce diffusion. Basically the same symptoms occur as in atelectasis.

In EMPHYSEMA, the walls between alveoli disappear, creating large cavities that reduce diffusing surface.

One may rightly conclude, then, that the most important conditions affecting the respiratory division are those that reduce gas diffusion between alveoli and blood.

The lungs

All parts of the system beyond the bronchi are contained within the LUNGS (Fig. 26.12). The paired lungs lie within the two lateral PLEURAL CAVITIES of the thorax. A serous membrane, the

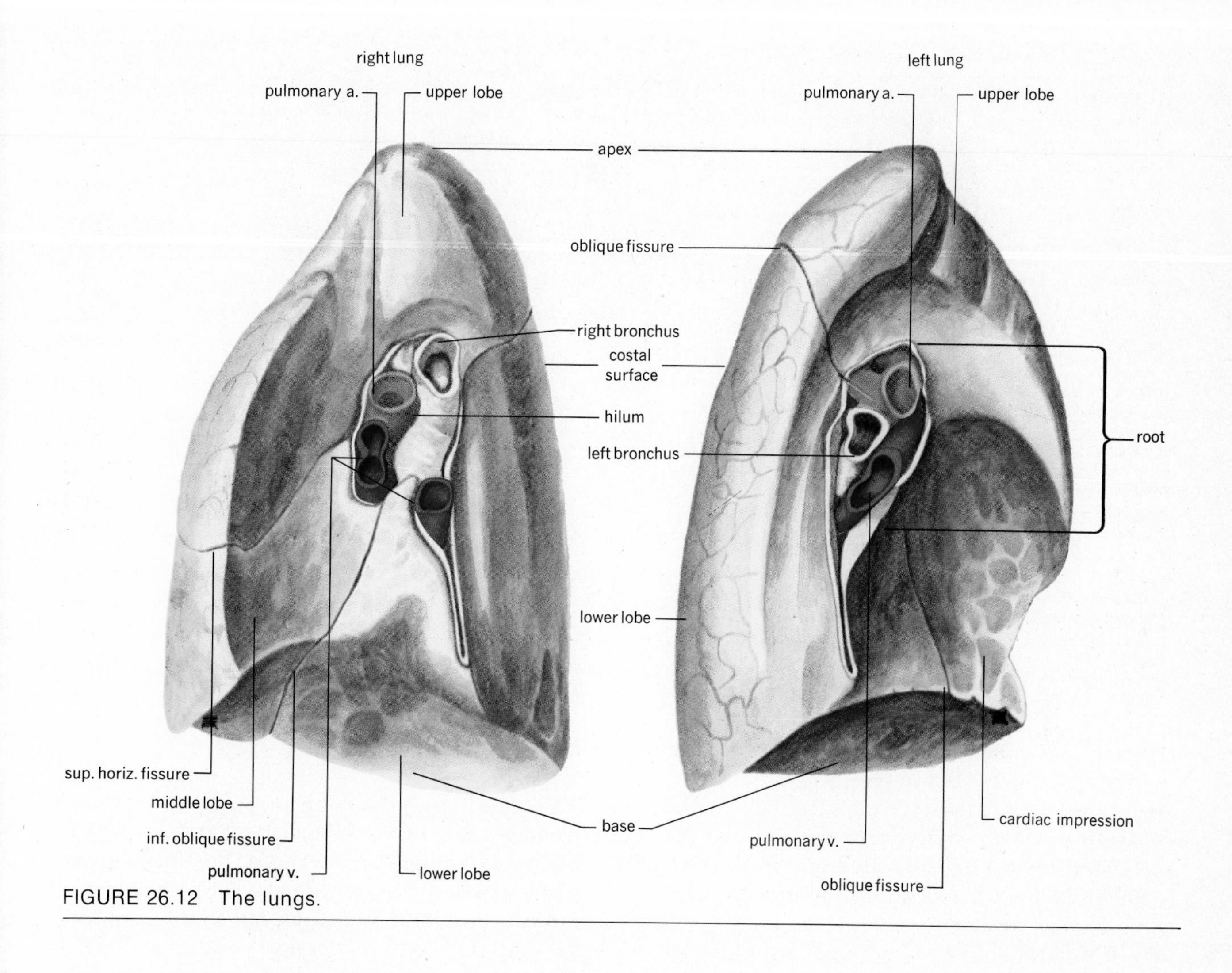

FIGURE 26.12 The lungs.

VISCERAL PLEURA, covers the lung surface, and a similar membrane, the PARIETAL PLEURA, lines the thoracic cavity. These membranes are continuous with one another and are reflected upon each other at the root of the lung, where the bronchi enter the organ. With the lung in place and inflated, the pleural cavity is reduced to a fluid-lined potential space between the two membranes. The fluid allows easy slippage between lung and chest wall as breathing occurs and, by the cohesive effect it creates, aids in keeping the lung expanded. The parietal pleurae are named specifically according to the surface of the thorax they line. Costal pleura lines the inner surface of the ribs and intercostal muscles; diaphragmatic pleura covers the superior aspect of the diaphragm; the cervical pleura rises into the neck. The right lung averages 625 g in weight, the left 562 g. In the male, the lungs form about 1/37 of the body weight, in the female 1/43. The lungs are conical in shape with a narrow superiorly directed APEX and a broad inferiorly directed BASE. The COSTAL SURFACES are rounded to match the curvature of the ribs. The MEDIAL SURFACE is indented to form

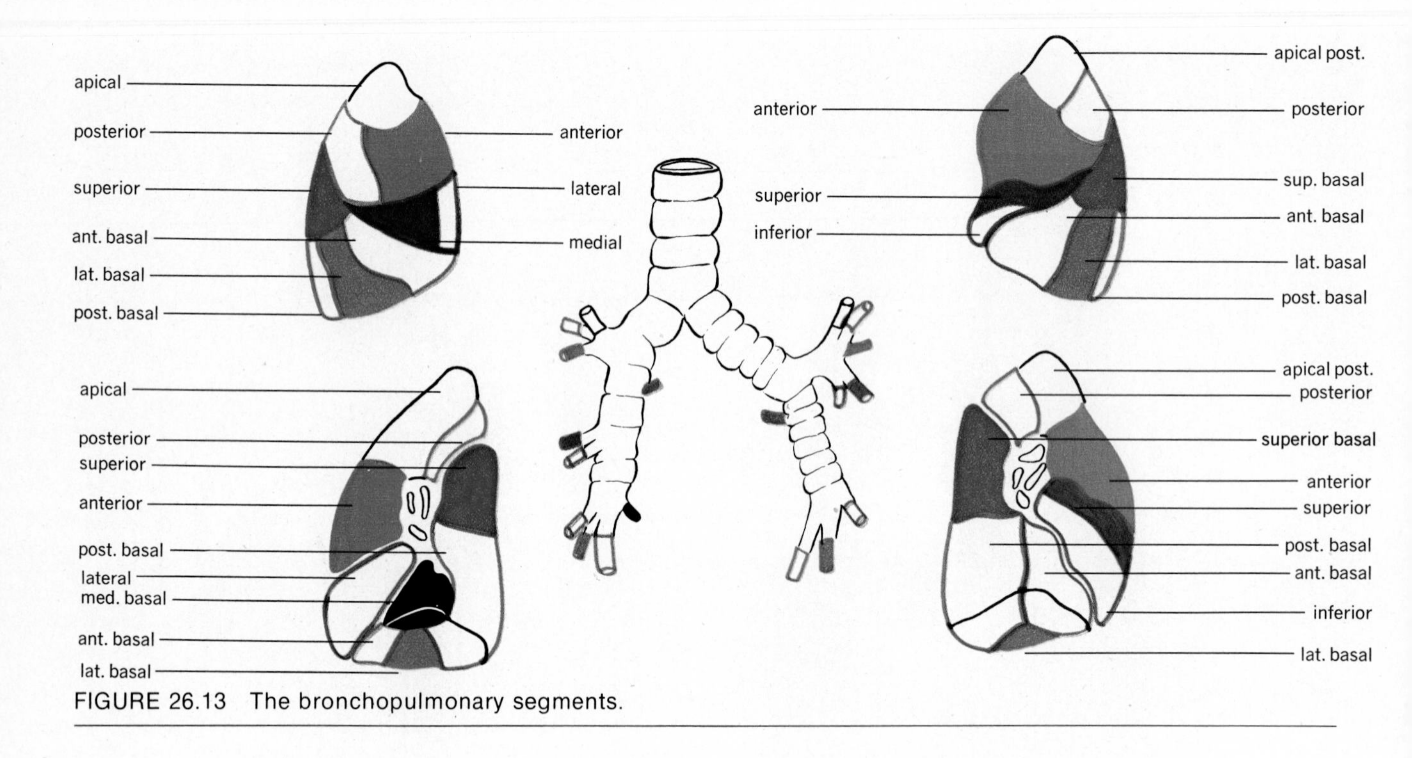

FIGURE 26.13 The bronchopulmonary segments.

a HILUM, the point of entry and exit of bronchi, pulmonary vessels, and nerves. The medial surface of the left lung bears a concavity, the CARDIAC IMPRESSION, which conforms to the shape of the heart. The right lung is divided into three lobes (upper, middle, and lower) by a superior horizontal fissure and an inferior oblique fissure. The left lung is divided into an upper and lower lobe by an oblique fissure only. Further subdivision of the lung substance into BRONCHOPULMONARY SEGMENTS is provided by connective tissue and the branching of the bronchi (Fig. 26.13). The lung may also be subdivided into LOBULES (see Fig. 26.11). A secondary lobule is an anatomical unit created by the fibrous septae that separates a group of terminal bronchioles and all their branches from other groups. A primary lobule is a functional unit consisting of one respiratory bronchiole and its divisions.

Blood supply

The conducting division receives nourishing blood by way of the BRONCHIAL ARTERIES, which rise from the upper aorta and/or the upper intercostal arteries. The bronchial arteries follow the bronchi and bronchioles as far as the terminal bronchioles. The PULMONARY ARTERIES bring blood to the lungs for oxygenation and nourish the tissues of the respiratory division. The pulmonary arteries do not have well-developed muscular arterioles as are found in the systemic circulation. Therefore, much less pressure is required to move the blood through the pulmonary circulation. The capillaries rise directly from the pulmonary arteries in the respiratory zone and form rich networks in the alveolar walls, where oxygenation of the blood occurs. Union of the CAPILLARIES from both the bronchial and pulmonary arteries forms the PULMONARY VEINS which carry the oxygenated blood to the left atrium.

Nerves

Branches of the VAGUS, PHRENIC, and INTERCOSTAL nerves reach the lungs and respiratory muscles. Their distribution and function will be considered in connection with the physiology of the system.

CLINICAL CONSIDERATIONS. Total or partial lung collapse may follow wounds or disease that cause communication of a pleural cavity with the outside *(pneumothorax)* or with the lumina of the system's tubes. Cancers usually require that a part of a lung be removed; if so, a lobe is usually removed because it is more or less a separate unit, and chances of bleeding are reduced. If a whole lung requires removal, the procedure is called a pneumonectomy.

The effects of SMOKING on the respiratory system have been described in detail in the Surgeon General's report on smoking and health. In general, it may be stated that there are changes in the epithelium and loss of cilia that reduce the cleansing and protective functions of the system; there is loss of elastic tissue in the lungs and the respiratory tree that causes collapse of tubes and inability to ventilate the lungs adequately; there is loss of walls between alveoli that results in decrease of diffusing surface. All in all, smoking appears to alter the protective, ventilatory, and diffusing capabilities of the system.

The physiology of respiration

Acquisition of gases by the respiratory system and delivery to cells may be said to occur in stages.

PULMONARY and ALVEOLAR VENTILATION is getting air into *(inspiration)* and out of *(expiration)* the lungs and is a prerequisite to gas exchange. Pulmonary ventilation implies filling the system with air, but not all of it may reach the alveoli. Thus, alveolar ventilation means air that actually reaches the exchange surfaces of the alveoli.

EXTERNAL RESPIRATION is exchange of gases between lung and bloodstream.

INTERNAL RESPIRATION is exchange of gases between bloodstream and cells.

TRANSPORT is the methods by which the gases are carried by the blood.

The utilization of oxygen by cells is termed cellular respiration and is discussed in Chapter 28.

All of these activities require control and may need to be altered according to body activity levels. These activities depend primarily on chemical and nervous factors.

Pulmonary and alveolar ventilation

THE MECHANICS OF BREATHING. INSPIRATION, or intake of air, is brought about by contraction of the diaphragm and external intercostal muscles. Contraction of the diaphragm enlarges the chest cavity in a vertical direction. Contraction of the intercostal muscles causes the ribs to swing up and outward, enlarging the front-to-back and side-to-side dimensions of the chest. Since the chest cavity is a closed cavity with no outside opening, any enlargement of it will cause a pressure drop inside it. The pressure in the chest *(intrathoracic pressure)* is normally slightly less than atmospheric pressure and falls even more on inspiration. Because of a thin film of fluid *(pleural*

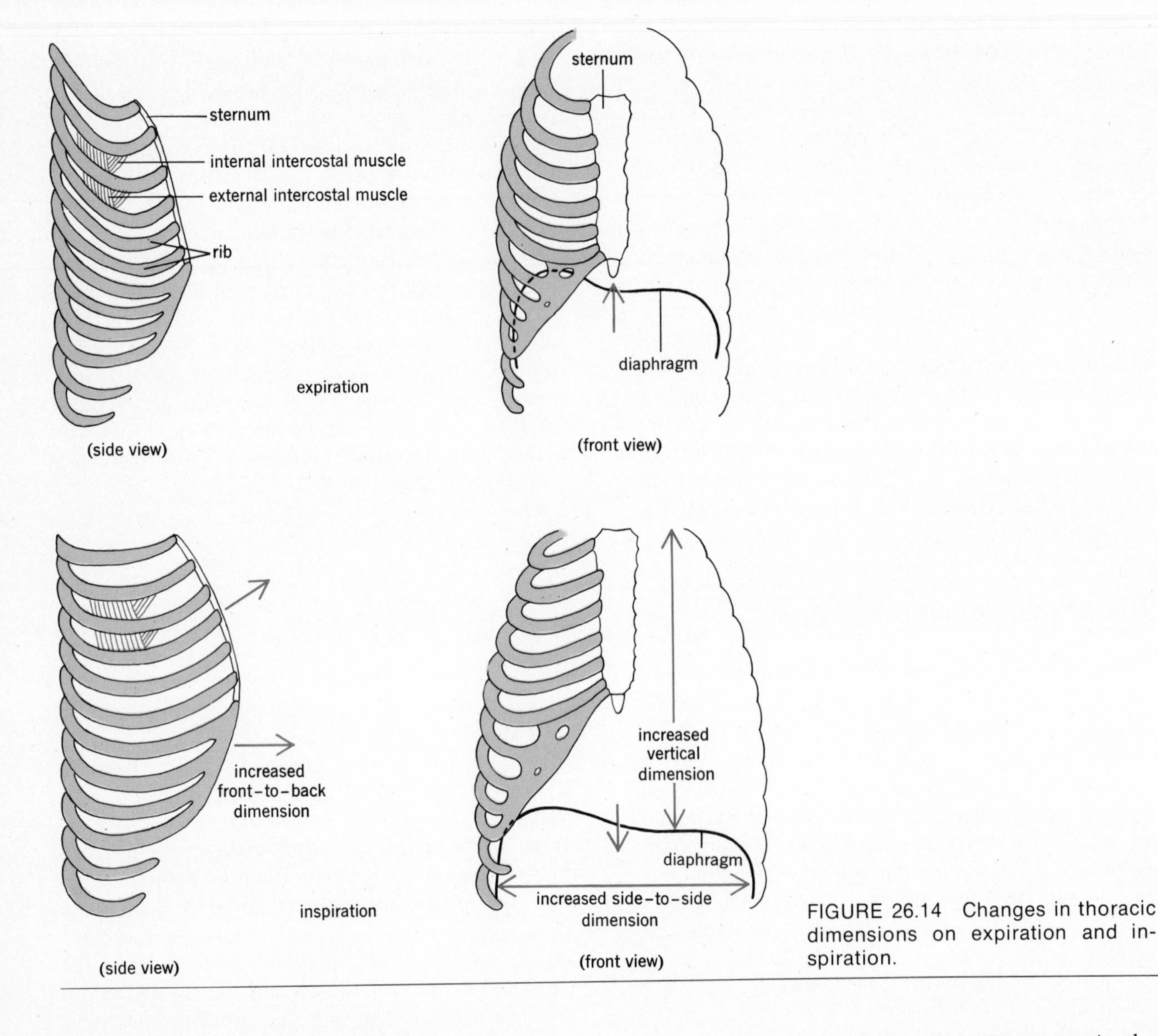

FIGURE 26.14 Changes in thoracic dimensions on expiration and inspiration.

fluid) lying between the visceral and parietal layers of the pleurae, the lung coheres (sticks to) the chest wall, and follows it as the chest enlarges. The pressure within the lungs *(intrapulmonic pressure)* is equal to atmospheric pressure when no breathing is occurring, since the lungs communicate with the atmosphere by way of the bronchial tree. Thus, as the lung expands during inspiration, the pressure falls to less than atmospheric pressure. Air rushes into the lowered pressure area in the lungs. Flow is rapid at first, then slows as the pressures are equalized. Inspiration also stretches the elastic tissue of the lungs as they enlarge.

Normal EXPIRATION is a process not requiring muscular contraction. Relaxation of the diaphragm and external intercostal muscles returns the chest to its original size. Elastic recoil of the lungs creates a higher than atmospheric intrapulmonic pressure that forces air out of the lungs. Again, pressure in the lungs is equalized as the air goes out. Expiration may be made active by contraction of the internal intercostal muscles that depress the ribs and by contraction of abdominal muscles. The latter press the viscera against the diaphragm, speeding its return to normal position.

Figures 26.14 and 26.15 show the changes occurring in chest size and pressures as breathing occurs.

Clinical considerations. Not all the air entering the system reaches the alveoli; some remains in the conducting division. The volume of air in the conducting division is known as DEAD AIR, or air not available for gas exchange with the blood. In certain conditions, such as asthma and emphysema, air cannot enter the alveoli because of blockage of the small bronchioles, and the total surface for diffusion is greatly reduced. In this case, the nonventilated alveoli constitute a physiological DEAD SPACE, and their volume is added to the previous one. In the normal adult individual, dead air in the conducting division amounts to about 150 ml, and the alveoli are all ventilated to some degree. With disease, the nonventilated alveoli and tubes may combine to produce a volume of 1–2 liters.

SURFACE TENSION OF THE ALVEOLI; SURFACTANT. Alveoli behave like bubbles of gas in the fluids of the lung; that is, they have a tendency to collapse, depending on the surface tension of the bubble. Surface tension is caused by the attraction between molecules at the fluid-air surface, and tends to cause a bubble to assume the smallest diameter for its volume. If the molecules at the fluid-air surface are water molecules, the surface forces are quite high, and the tendency is for the surface tension to pull the bubble into a collapsed condition—that is, to assume the smallest size. Additionally, if two bubbles are connected to one another, the smaller bubble, having higher surface tension forces, tends to empty into the larger bubble (Fig. 26.16). In the lung, where alveoli are arranged side by side, the tendency of one alveolus to empty into another cannot be tolerated, since this will lead to the collapse of alveoli. Solving the problem of the tendency of alveoli to collapse depends on the presence on the alveolar surface of a material other than water. A phospholipid, designated SURFACE ACTIVE AGENT or SURFACTANT, has been demonstrated on the alveolar surface.

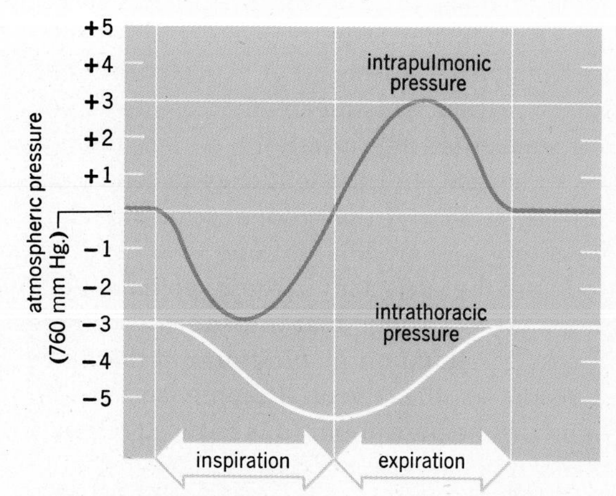

FIGURE 26.15 Fluctuations in intrathoracic and intrapulmonic pressures during breathing.

the pressure exerted by a bubble in a liquid is given by:

$$P = \frac{2T}{r} \quad \begin{cases} P = \text{pressure} \\ T = \text{surface tension} \\ r = \text{radius} \end{cases}$$

the smaller the radius, the greater is the pressure.

if two bubbles are connected as in the following diagram;

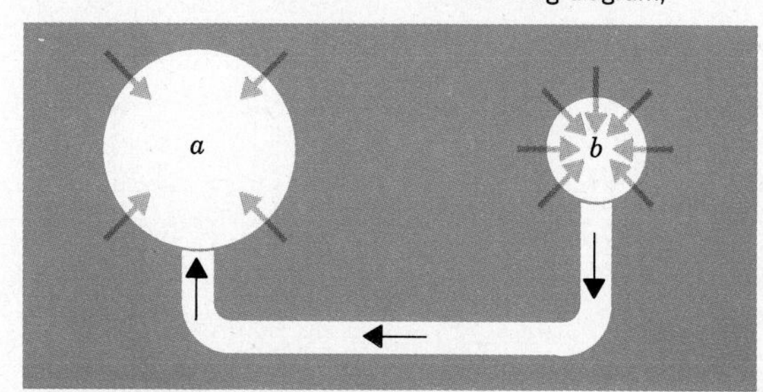

the one having the smaller radius (b) will tend to empty into the larger (a).

FIGURE 26.16 The forces tending to cause alveolar collapse.

It has the remarkable property of lowering alveolar surface tension 7–14 times that expected from an air-water junction. This means that, as the alveoli become smaller (as on expiration), surface tension does not increase, and the tendency to collapse is minimized; similarly, the tendency for one alveolus to empty into an adjacent one is minimized. The action of the surfactant in the alveoli may thus be compared to springs that are stretched as the alveoli are expanded on inspiration but that have less tension as the alveoli become smaller on expiration. If the phospholipid is reduced or missing, alveolar collapse may occur.

Clinical considerations. In infants, insufficient amounts of surfactant may be present, particularly if the infant is born prematurely, and the lungs are not fully developed or expanded. A RESPIRATORY DISTRESS SYNDROME (RDS) known as HYALINE MEMBRANE DISEASE (HMD) develops, and alveoli collapse since surface tension of the alveoli is very high due to lack of surfactant.

Exchange of air

VOLUMES. The amount of air exchanged during normal breathing depends on the size, age, and sex of the breather. Obviously, a small person or child will exchange less air than an adult or large person. In the "standard" 70-kg male, who is breathing normally at rest, about 500 ml of air is moved with each respiration. This volume is called TIDAL AIR (tidal volume). Additional air may be inspired above tidal volume, if forced or during exercise. This volume is called RESERVE INSPIRATORY AIR (reserve inspiratory volume), and amounts to about 3000 ml. The RESERVE EXPIRATORY AIR (reserve expiratory volume) may be forcibly exhaled after a normal expiration. It amounts to about 1100 ml. Even after the most forcible expiration, the lungs do not collapse, and air remains within the alveoli. This air is called RESIDUAL AIR (residual volume) and amounts to about 1200 ml. Even if the lungs are caused to

FIGURE 26.17 Subdivisions of lung air according to minute volume of respiration.

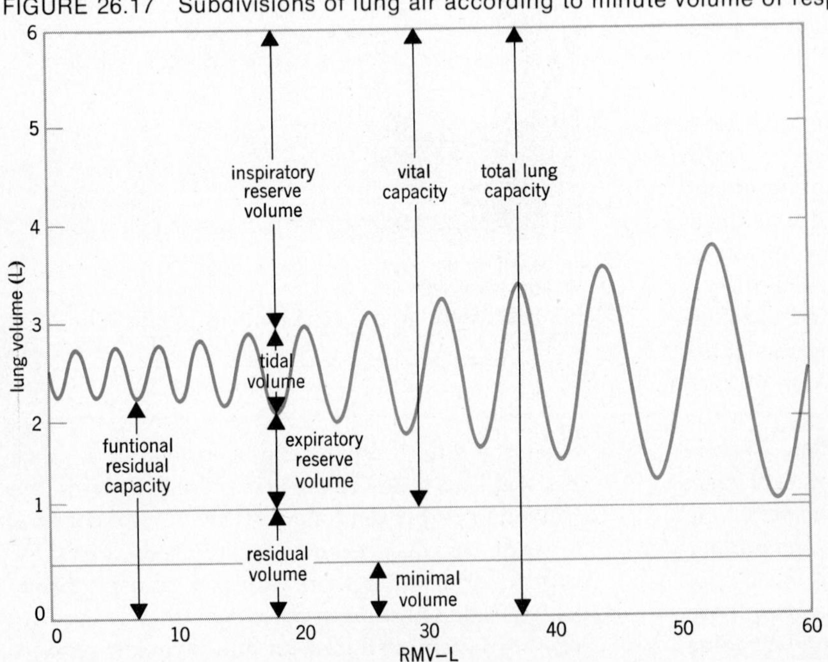

collapse, air remains in small alveoli and in the tissues of the lung. This air is MINIMAL AIR (minimal volume) and amounts to about 600 ml.

CAPACITIES. Adding the volumes in various ways creates capacities that have physiological significance. Among the important capacities are the following.

Vital capacity. The sum of tidal, reserve inspiratory, and reserve expiratory air, the vital capacity represents the maximum amount of air that can be moved in the system.

Functional residual capacity. The sum of reserve expiratory air and residual volume, functional residual capacity assures continual gas exchange with the blood during expiration.

Total lung capacity. The total of all volumes, total lung capacity gives an idea of the lung size.

The relationships of lung volumes and capacities are shown in Fig. 26.17.

The work of breathing

Exchange of air is usually achieved with a minimum amount of effort. EUPNEA is a term referring to normal, quiet breathing. In HYPERVENTILATION, or an increase in rate and depth of breathing, the work involved may increase to where DYSPNEA or labored breathing results. Dyspnea may also imply respiratory distress resulting from inadequate alveolar ventilation. The work of breathing is largely determined by airway resistance, lung compliance, and lung elastance.

AIRWAY RESISTANCE refers to the ease with which air flows through the tubular structures of the respiratory system. Ease of flow depends on the length of the tubes and their diameter. A smaller tube requires more work to cause air flow through it, in much the same way that it takes more "pull" to draw fluid through a small straw than through a larger one. Resistance is also created if alveoli have not emptied properly, and the next inhalation finds air trying to enter alveoli already filled with air. This situation occurs in asthma and emphysema. Thus, any condition that narrows the tubes or causes incomplete emptying of alveoli will require more effort to maintain flow.

LUNG COMPLIANCE refers to the ability of the alveoli and lung tissue to be expanded on inspiration. Normally, the lungs are easily expanded, due to their content of easily stretched elastic tissue. If the elastic tissue is replaced by fibrous nonelastic tissue (as in some infections and in emphysema) the work required to expand the lung will be greatly increased, and labored breathing usually results.

LUNG ELASTANCE refers to the ability of the elastic tissues of the lung to recoil during expiration and force air out of the lungs. If elastic tissue is diminished, the ability of the lungs to recoil is decreased, and increased abdominal muscle activity is required to aid in emptying the lungs.

Figure 26.18 indicates the effects of some of these factors on the work of breathing, with work indicated by oxygen consumption.

MEASUREMENT OF PULMONARY FUNCTION. Several tests may be performed to assess an individual's level of pulmonary function or to discover if a derangement of function is present.

The MAXIMUM VOLUNTARY VENTILATION (MVV) is measured by having the subject breathe as deeply and rapidly as possible into a spirometer for 15 sec and converting the volume of exchange to liters per minute. Normal values should be between 120–200 liters/min. This test gives an indication of the status of the respiratory muscles and the ease of airflow through the system.

The FORCED EXPIRATORY VOLUME, TIMED (FEV_t) is determined by having the subject inhale as deeply as possible, then exhale as rapidly as possible into a spirometer that not only measures volume but the time required to exhale. If the lungs are normal, 68 percent of the vital capacity should be emptied in the first 0.5 sec, 77 percent in 0.75 sec, 84 percent in 1.0 sec, 94 percent in 2 sec, and 97 percent in 3 sec. Times lengthened beyond these values suggest decreased elastance or increased expiratory airway resistance.

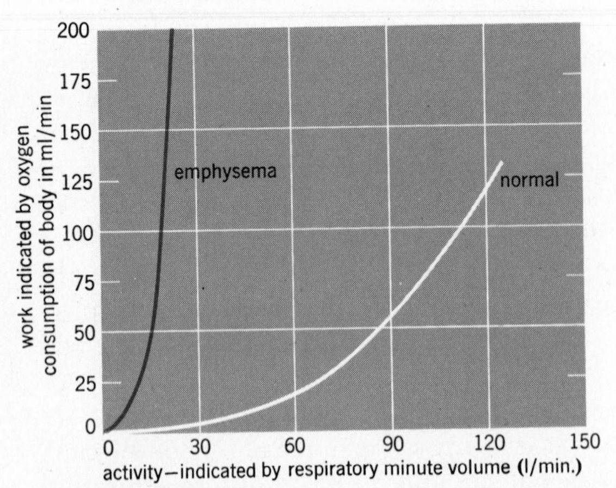

FIGURE 26.18 The oxygen consumption in a normal individual, and in one suffering from emphysema.

TABLE 26.2 Composition of earth's atmosphere

Gas	Percent in dry air
Nitrogen	78.09
Oxygen	20.94
Carbon dioxide	0.03
Rare gases (argon, neon, helium, xenon)	0.94

A MAXIMUM EXPIRATORY FLOW RATE (MEFR) may be measured by a flow meter, with the subject exhaling forcibly through the device. Values of 350–500 liters/min are considered normal. This test is related primarily to airway resistance.

Composition of the respired air; gas diffusion

Humans breathe an atmosphere that normally contains the proportions of gases shown in Table 26.2.

Air with this composition is inhaled and mixed with air already in the lungs and respiratory passageways. In reporting values of gases in the body, the term PARTIAL PRESSURE is conventionally used. Partial pressure is expressed in mm Hg and refers to the portion (or part) of the total pressure of a mixture of gases contributed by one of the gases present in the mixture. It may be calculated by multiplying the percentage of the gas in the mixture times the total pressure. For oxygen and carbon dioxide, the symbols PO_2 and PCO_2 are used to designate partial pressure of oxygen and carbon dioxide, respectively.

Composition of the air in various parts of the respiratory system and in the blood and tissues may be determined. The values in various areas are presented in Table 26.3. Composition of air in the alveoli is controlled by ventilation and diffusion from the bloodstream; levels in the bloodstream are determined by lung values and tissue activity. The table indicates that PO_2 decreases from atmosphere to alveoli to blood to tissues. Thus, diffusion of oxygen will always proceed from "outside" to "inside." Conversely, PCO_2 is highest in the tissues and decreases from tissues to blood to lung to atmosphere. Therefore, CO_2 will always diffuse from "inside" to "outside." Nitrogen concentration is nearly constant in all areas, indicating its noninvolvement in metabolic reactions.

The behavior of gases in the body is explained by assuming that mixtures of gases consist of individual particles moving at velocities determined by temperature. The "gas laws" cited below are related to movement of gases across the alveolar-capillary and capillary-tissue membranes.

At the same temperature and volume, equal numbers of gas molecules will exert the same pressure. (Avogadro's hypothesis). Thus, pressure and therefore diffusion depend only on numbers of molecules, provided that temperature and volume remain essentially constant. In the body, diffusion of a given gas is therefore most rapid when its diffusion gradients are high due to adequate ventilation.

In a mixture of gases, each gas exerts a pressure independent of all other gases (Dalton's law). Total pressure is thus the sum of individual pressures, and

TABLE 26.3 Partial pressures in mm Hg of gases in various areas of the body

Gas	Atmosphere	Trachea; mixture of "in" and "out" air	Alveoli	Arterial blood	Venous blood	Tissues
Nitrogen	596	564	573	573	573	573
Oxygen	158	149	100	95	40	40
Carbon dioxide	0.3	0.3	40	40	46	46
Water vapor	5.7 av.	47	47	47	47	47
Total	760	760	760	755	706	706

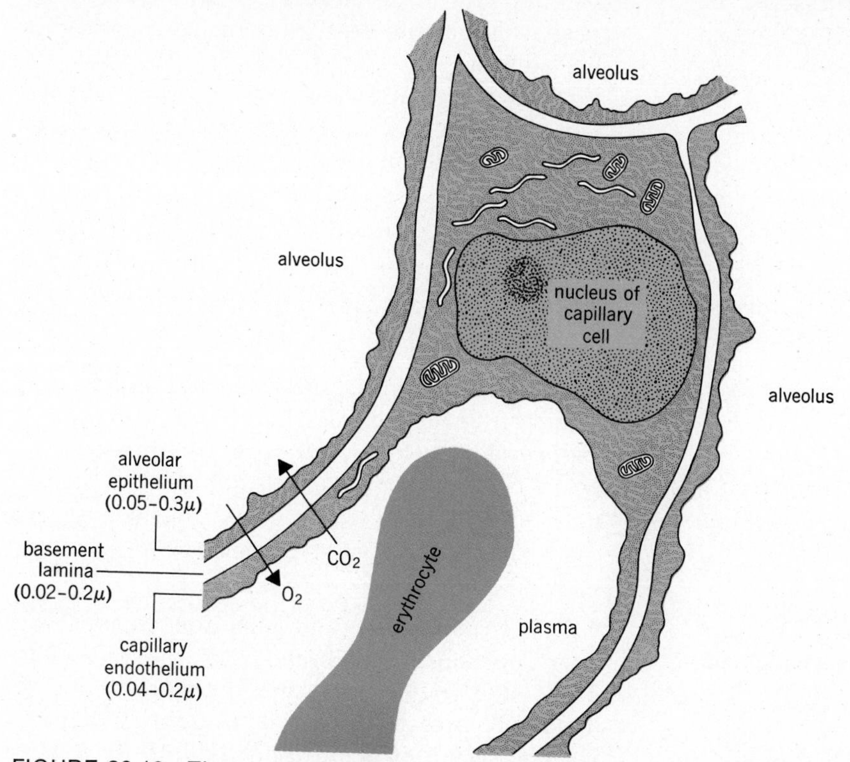

FIGURE 26.19 The alveolar-capillary membrane as viewed in an electron micrograph at 20,000×.

each gas will diffuse according to its own pressure gradient (partial pressure). One thus may consider each gas as behaving as though it alone were present.

When a gas is compressed, its pressure increases proportionately to its volume decrease (Boyle's law). Any decrease of volume must therefore increase gas pressure and speed its diffusion or movement.

If the volume of a gas is kept constant, its pressure is proportional to temperature (Charles' law). Higher temperatures increase kinetic motion of molecules, more collisions per unit time occur, and pressure increases. Thus, rate of diffusion increases to a more-or-less constant value at body temperature.

The solution of a gas in a liquid is directly proportional to the pressure of gas to which the liquid is exposed (Henry's law). Thus, more gas dissolves in a liquid if its partial pressure is greater at the air-liquid interface. The practical application of these gas laws is presented in the section on gas pressure and physiology later in this chapter.

Gas diffusion from blood to lung or blood to tissues occurs through a membrane with a thickness no greater than 1.0 μm. In the case of the alveolar-capillary membranes (Fig. 26.19), the distance is even less. Between blood and tissues, the variable component is the amount of interstitial fluid through which the gas must pass. In any event, little hindrance is exerted on the diffusion of gases by the normal membrane itself. What does determine the rate of diffusion of the gases is the degree of ventilation of the alveoli and thus the steepness of the diffusion gradient.

An additional factor determining overall oxygenation of the blood is the fact that not all alveoli are perfused to the same degree with blood, and not all are ventilated equally. The best index to pulmonary gas exchange efficiency is the ratio of alveolar ventilation to cardiac output. In normal male adults, this ratio is about 0.85, indicating that at rest about 85 percent of maximum diffusion capacity is attained. Exercise causes increases toward 100 percent, but the ratio will probably never reach 100 percent due to both cardiac output and ventilatory increases caused by the exercise.

In any case, diffusion normally proceeds at a rate sufficient to meet body demands for oxygen supply and carbon dioxide removal.

Transport of gases

Oxygen transport

Oxygen is transported primarily through reversible combination of the gas with erythrocyte HEMOGLOBIN. About 95 percent of the oxygen is transported in this manner, the remaining 5 percent IN SOLUTION in the plasma. This relationship may be expressed, in general terms, by the equation

$$\underset{\text{reduced hemoglobin}}{Hgb} + \underset{\text{oxygen}}{O_2} \rightleftharpoons \underset{\text{oxyhemoglobin}}{HgbO_2}$$

FIGURE 26.20 The chemical skeleton of hemoglobin, showing binding sites for gases.

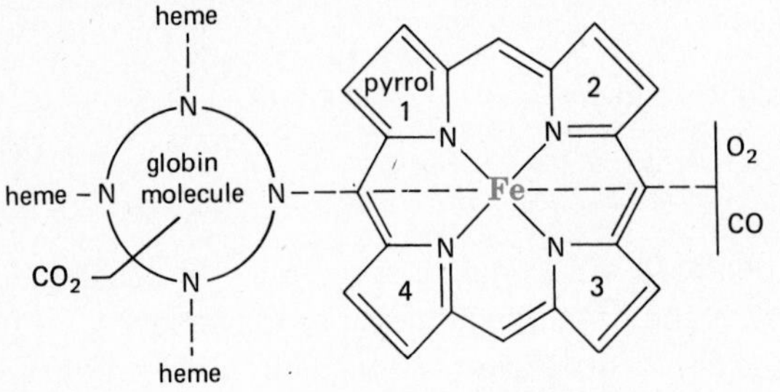

Oxygen combines with the iron atom of the heme portions of the hemoglobin molecule (Fig. 26.20); four heme units and therefore four iron atoms are present in each hemoglobin molecule. Each molecule of hemoglobin is thus capable of carrying eight atoms of oxygen. The four atoms of iron do not become oxygenated or deoxygenated simultaneously but combine with or release oxygen in steps. Each step has its own equilibrium constant (K) that reflects the rate of O_2 uptake. The steps are postulated to occur as

$$\mathrm{Hgb_4 + O_2 \overset{K_1}{\rightleftharpoons} Hgb_4O_2}$$

$$\mathrm{Hgb_4O_2 + O_2 \overset{K_2}{\rightleftharpoons} Hgb_4O_4}$$

$$\mathrm{Hgb_4O_4 + O_2 \overset{K_3}{\rightleftharpoons} Hgb_4O_6}$$

$$\mathrm{Hgb_4O_6 + O_2 \overset{K_4}{\rightleftharpoons} Hgb_4O_8}$$

An OXYGEN-DISSOCIATION CURVE (Fig. 26.21) relates the hemoglobin saturation (average number or percent of hemoglobin molecules combined with oxygen) to PO_2. Several factors influence what the saturation will be.

PO_2. If more O_2 is available for diffusion, one would expect more uptake by the hemoglobin. It is obvious, then, that increased PO_2 will increase saturation. Due to loading of hemoglobin with oxygen in steps, saturation remains virtually constant until alveolar PO_2 falls to about two-thirds normal.

ACIDITY OF ERYTHROCYTE AND PLASMA. The oxygen-dissociation curve is shifted to the right by decrease in pH. This means a decreased saturation at a given PO_2. Thus, at a given PO_2, more O_2 will be driven off of the hemoglobin. The main factor causing a lower pH is an increase in CO_2 production during elevated levels of metabolism and the reaction of CO_2 with H_2O to liberate H ion. Thus, an active tissue automatically ensures itself of an increased oxygen supply.

TEMPERATURE. Increased temperature also shifts the oxygen-dissociation curve to the right and results in decreased saturation at a given PO_2 (Fig. 26.22). Thus, the increased temperature associated with increased activity also drives more O_2 from the hemoglobin.

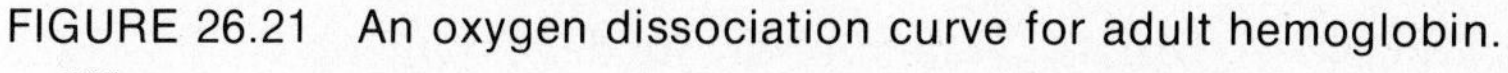

FIGURE 26.21 An oxygen dissociation curve for adult hemoglobin.

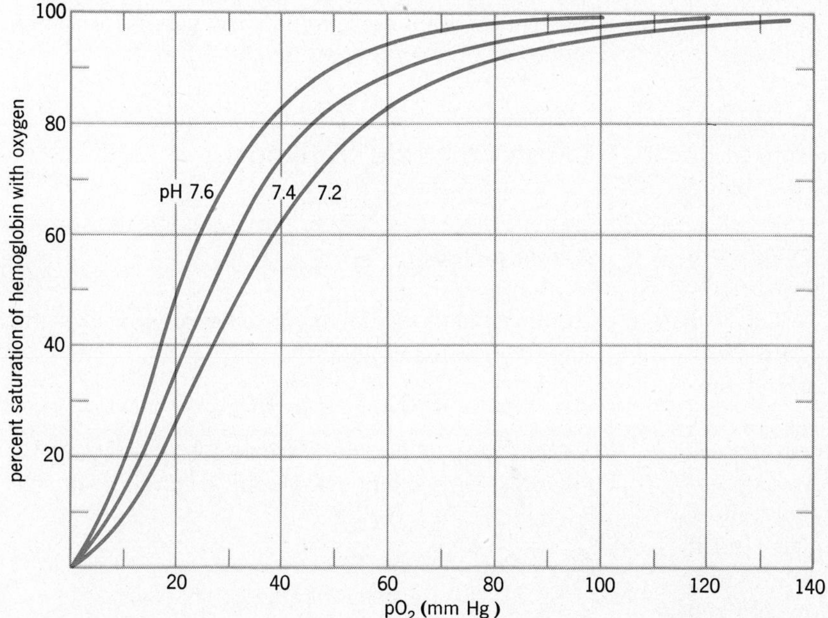

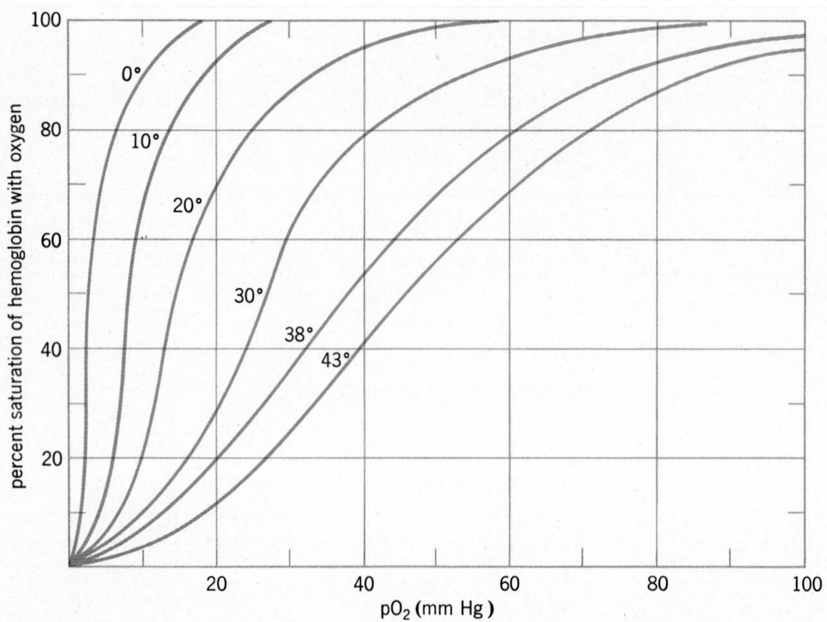

FIGURE 26.22 Oxygen dissociation curves demonstrating the effect of temperature change.

The curve indicates that 100 percent saturation of the hemoglobin does not occur at the PO_2 normally prevailing in the alveoli. This is due to the low solubility of oxygen in body fluids and to the rapid movement of the blood through the lungs. Entry of O_2 into erythrocytes is slowed by the necessity for dissolving of O_2 in the blood plasma before entry into the cell; rapidity of movement of blood means that a few hemoglobin molecules move by the alveoli too quickly to take on a full load of O_2.

The curves presented in Figs. 26.21 and 26.22 are for adult hemoglobin (Hgb A). During embryonic and fetal development, a different type of hemoglobin is present in the erythrocytes, designated fetal hemoglobin (Hgb F). An oxygen-dissociation curve plotted for Hgb F would show it to have the same shape as that for Hgb A, but it would lie to the left of the adult curves. This means that fetal hemoglobin loads more oxygen at lower PO_2. Life *in utero* is life in a low-oxygen environment. Thus, the fetus is ensured of adequate O_2 to support its development in spite of a lower PO_2 delivered to it through its mother's bloodstream and the placenta.

Carbon dioxide transport

Transport of CO_2 occurs by a combination of several methods.

From 64 to 83 percent is carried as BICARBONATE ION according to the equation

$$CO_2 + H_2O \rightleftharpoons H_2CO_3 \rightleftharpoons H^+ + HCO_3^-$$

From 10 to 27 percent COMBINES WITH HEMOGLOBIN to form carbaminohemoglobin according to the equation

$$CO_2 + Hgb \rightleftharpoons \underset{\text{Carbaminohemoglobin}}{HgbCO_2}$$

From 7 to 9 percent is transported in SIMPLE SOLUTION in the plasma.

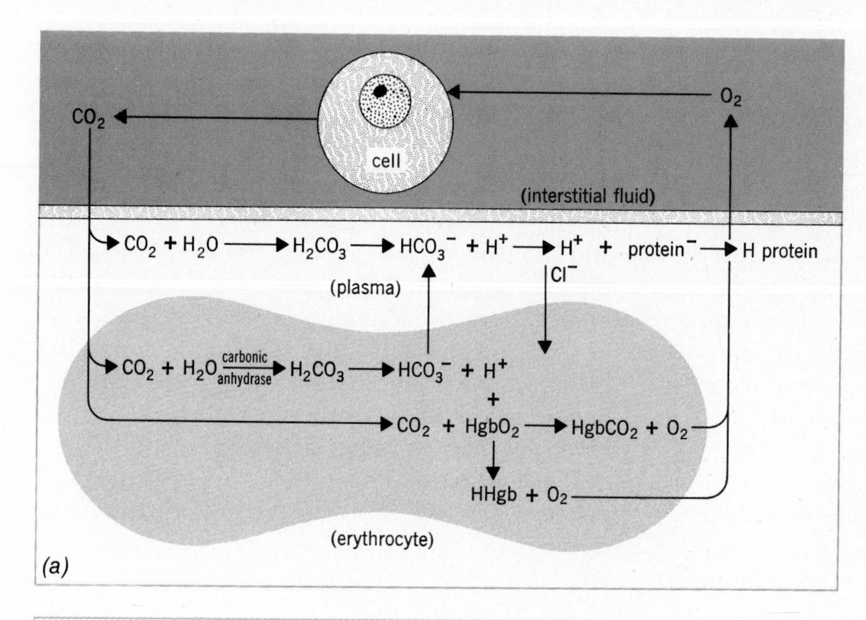

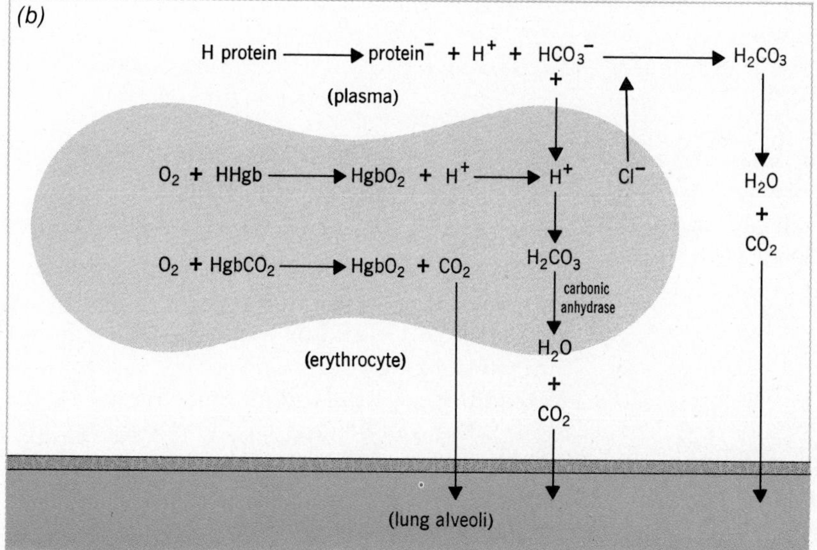

FIGURE 26.23 The reactions occurring at *(a)* blood-tissue interface and *(b)* blood-lung interface.

The reaction of CO_2 and H_2O to form carbonic acid (H_2CO_3), and the release of CO_2 from carbonic acid, is facilitated by the presence in the erythrocytes of an enzyme designated CARBONIC ANHYDRASE. Thus, equilibrium of CO_2 between tissues and erythrocyte and between erythrocyte and lung is rapidly achieved. Also, CO_2 is much more soluble in the body fluids than O_2. These factors result in the highest possible rates of CO_2 exchange.

Figure 26.23 presents a summary of the chemical reactions occurring at the tissues and lungs as gas exchange proceeds, and an explanation of the changes follows.

At the tissues, CO_2 dissolves in plasma water or enters the erythrocytes to react with cellular water. In either case, carbonic acid is formed, which dissociates into hydrogen ion and bicarbonate ion. The hydrogen ion produced in the plasma is buffered by reacting with plasma proteins; the hydrogen ion produced within the erythrocytes does not pass through the cell membrane and is buffered by hemoglobin. Bicarbonate ion accumulates within the erythrocytes and soon begins to diffuse out of the cell into the plasma. A loss of negative charge results, and, to return electrical neutrality, chloride ion moves from plasma into the erythrocytes (chloride shift). Some CO_2 displaces O_2 from the hemoglobin, and the excess H^+ in the erythrocyte causes decrease of pH and also drives O_2 from the hemoglobin.

At the lungs, the action of carbonic anhydrase liberates CO_2 from erythrocyte carbonic acid. To reform carbonic acid requires the hydrogen ion from within the cell and bicarbonate ion from the plasma. As bicarbonate ion moves into the erythrocyte, an excess of negative charges accumulates, and chloride ion moves back into the plasma. The high O_2 levels in this region cause displacement of CO_2 from the hemoglobin, and the formation of oxyhemoglobin results. Hydrogen ion is released from its combination with protein as bicarbonate enters the erythrocyte, and the protein is made available to buffer more H^+.

These processes illustrate the interrelationships between gas transport and the buffering systems of the body.

Acid-base regulation by the lungs

The lung is the most important body organ involved in the regulation of acid-base balance. The production of CO_2 by body cells results in the addition of 13,000–20,000 meq per day of H^+ to the body fluids. The reaction involved in the formation of H^+ from CO_2 is

$$CO_2 + H_2O \rightarrow H_2CO_3 \rightarrow H^+ + HCO_3$$

This reaction is reversed in the lungs by the elimination of CO_2, which shifts the chemical equilibria of the reaction toward the formation of carbonic acid

$$H^+ + HCO_3^- \rightarrow H_2CO_3 \rightarrow H_2O + \underset{\text{(eliminated)}}{CO_2}$$

The rate of elimination of CO_2 depends on the process of ventilation; in turn the arterial PCO_2 is a function of CO_2 production and ventilation according to the equation

$$\text{arterial } PCO_2 = K \frac{CO_2 \text{ production}}{\text{alveolar ventilation}}$$

We thus have a system for self-regulation of arterial PCO_2 and therefore pH. The respiratory centers controlling respiration are stimulated by increase in PCO_2 and/or a decrease of pH, and elimination of CO_2 rises. Conversely, decreases of PCO_2 and/or rise of pH result in decreased stimulation of the centers, and CO_2 elimination is decreased. The entire system is thus seen to be controlled within very narrow limits by a feedback mechanism that monitors the PCO_2 as determined by the balance between CO_2 production and elimination. Adjustments are made rapidly and accurately. The production of states of acidosis and alkalosis were presented in Chapter 21 and should be reviewed at this time.

Other functions of the lungs

The lungs have traditionally been regarded as serving only as areas for gas exchange between the lung and bloodstream. Their functions other than gas exchange have recently been summarized, and include:

SECRETION OF SURFACTANT by alveolar "pneumonocytes." These are large rounded cells, forming part of the lining of the alveoli. They have well-developed Golgi bodies that are instrumental in secretion of surfactant.

PRODUCTION OF KININS, substances that cause vasodilation, KALLIKREINS that release the kinins from an inactivated state (kininogens), and KININASES that destroy the kinins.

INACTIVATION of gastric hormones (e.g., gastrin).

CONVERSION of angiotensin I to angiotensin II in the renal pressor mechanism (see Chapter 29) that elevates blood pressure and stimulates aldosterone release.

METABOLIZING serotonin so that it does not continue to exert vasoconstrictive effects on the body.

SYNTHESIS, STORAGE, and RELEASE of prostaglandins (see Chapter 31).

METABOLISM OF INSULIN.

These metabolic activities appear to be carried out by the endothelial linings of the pulmonary capillaries.

Control of respiration

The basic desire to breathe may be characterized as being due to involuntary spontaneous neuronal activity modified by chemical influences. A basic pattern is established that, within certain limits, may be voluntarily modified to permit talking, singing, hyperventilation, and breath-holding. Reflex control of respiration is afforded by peripheral and central chemoreceptive sites which are triggered by changes in mechanical tension and pH and gas levels in the blood and cerebrospinal fluids. These factors tend to adjust ventilation to levels designed to maintain a balance between gas utilization, production, and elimination.

Central influences

THE RESPIRATORY CENTERS (Fig. 26.24). The neurons responsible for establishing the basic rhythm of breathing are located within the medulla and pons of the brainstem. There appear to be five functional groups of neurons in these areas, each contributing something to the establishment of a normal respiratory rhythm.

Paired INSPIRATORY CENTERS, located in the reticular substance of the medulla, appear to spontaneously initiate inspiration.

Paired EXPIRATORY CENTERS, located dorsally to the inspiratory centers, initiate expiration.

Paired PNEUMOTAXIC CENTERS, located in the anterior pons, contribute reinforcement to the expiratory center to achieve expiration.

Paired NUCLEI OF THE VAGUS NERVES in the medulla provide additional reinforcement for expiration.

An unnamed and unidentified area in the region of the FOURTH VENTRICLE of the brain apparently monitors the pH of the cerebrospinal fluid and stimulates respiration if pH falls.

The primary stimulus for respiration is pro-

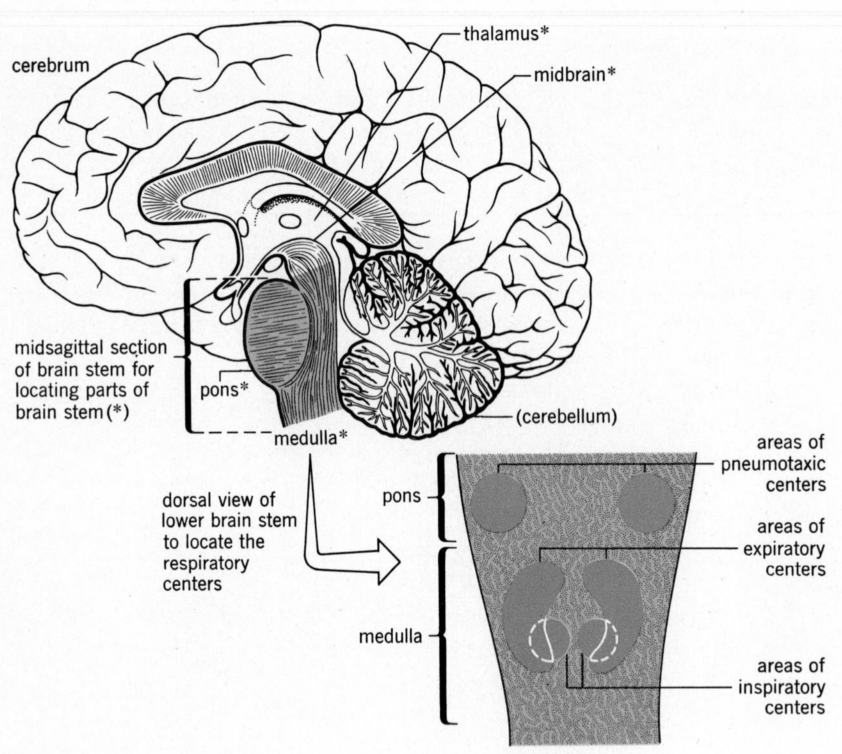

FIGURE 26.24 The location of the respiratory centers.

vided by the presence of CO_2 and H^+ in the cells of the inspiratory center. Discharge of nerve impulses over phrenic and intercostal nerves to diaphragm and intercostal muscles causes inspiration. Inflation and stretching of the lungs causes stimulation of pulmonary receptors, which send impulses over the vagus nerves to inhibit the activity of the inspiratory center (Hering-Breuer reflex). Additional input for expiration is provided by the pneumotaxic and expiratory centers. The result will be a periodic interruption of continuous inspiratory effort, causing expiration. It should be appreciated that the respirations themselves determine arterial PCO_2 and therefore H^+ and so a self-regulating system is created.

Peripheral influences

Additional control of respiration is provided by chemoreceptive cells in the carotid and aortic bodies. These moniter PCO_2 and PO_2 in the circulation to the brain and body generally. Increase of arterial PCO_2 by 0.5 percent, or a fall of pH, reflexly increases rate and depth of breathing, eliminating the excess CO_2. Reflex respiratory pathways are depicted in Fig. 26.25.

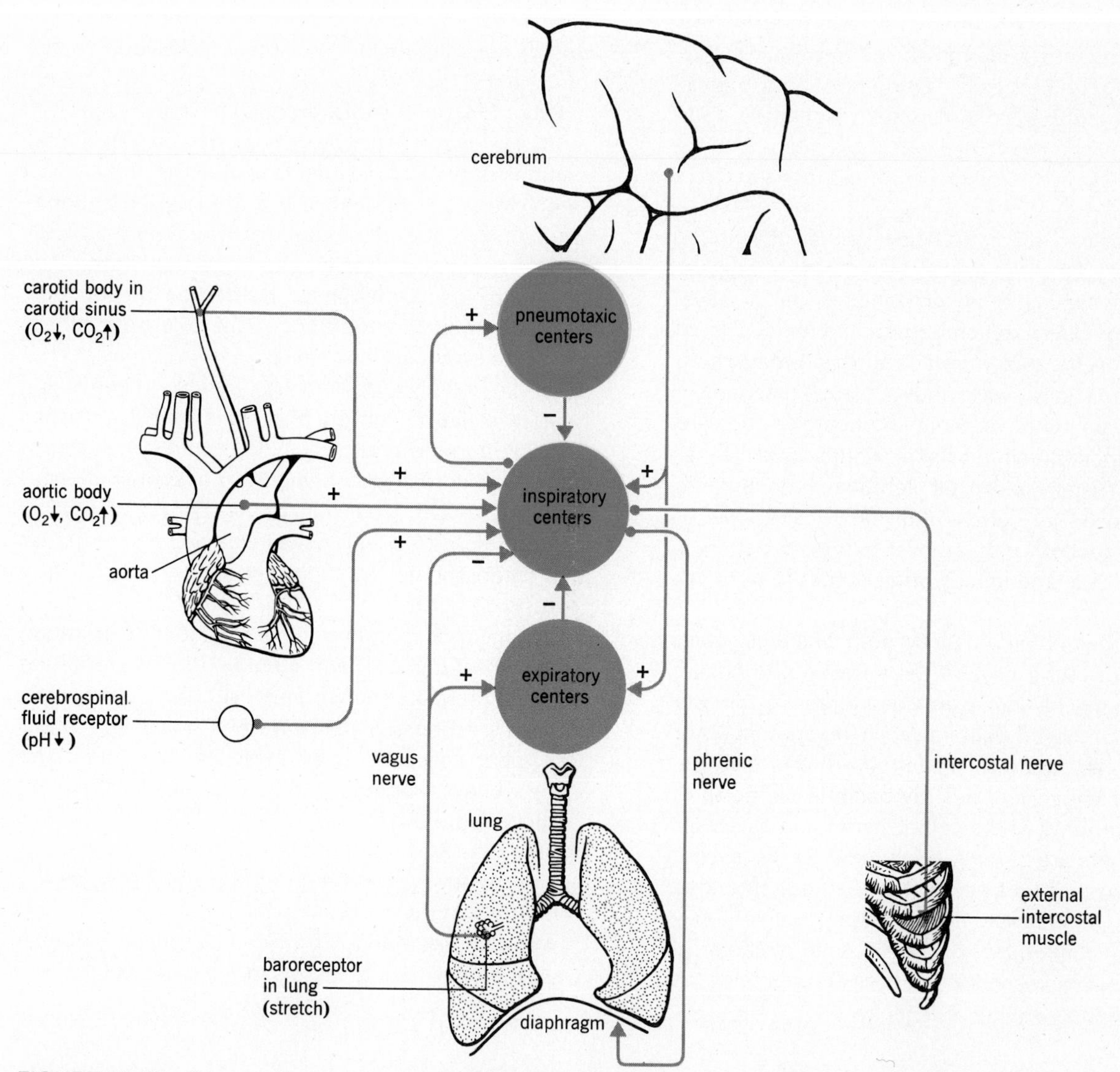

FIGURE 26.25 Some reflex mechanisms affecting breathing (+, stimulation; −, inhibition).

Protective mechanisms of the respiratory system

Since the respiratory system transports gases from the atmosphere to the lungs, a wide variety of potentially dangerous microorganisms, antigens, and particulate material may be inhaled. To aid in understanding the operation of the mechanisms of protection, recall that the greater part of the respiratory system is lined with ciliated epithelium. Also, recall the presence of a mucus coating over the surface epithelium, provided by the externally secreting glands of the lamina propria and by the goblet cells. There are lymphatics, alveolar macrophages, and reflex mechanisms for dislodging large particles from the system.

The specific mechanisms involved are:

The mucociliary escalator. The presence of a mucus layer on the epithelium of the system allows trapping of particles as incoming air contacts the surface. Coordinated ciliary action moves the mucus layer posteriorly in the nasal cavities to the throat where the mass is usually swallowed. A similar mechanism operates from the terminal bronchioles upward through the conducting division. Continuous cleansing of the system above the terminal bronchiole is thus provided. In normal individuals, the rate of mucus production and removal by ciliary action is nicely balanced, and mucus rarely comes to our conscious attention.

Alveolar macrophages. Since both cilia and goblet cells are not found beyond the terminal bronchiole, a cleansing mechanism is provided for the respiratory division. Phagocytic cells (alveolar macrophages or dust cells) are found in the alveoli of this division. These cells engulf and destroy bacteria, particles of dust, foreign antigens, or other harmful substances. Their numbers are partially dependent on the level of contamination of inhaled air, increasing as the load of pollutants increases. Their phagocytic ability may also vary, becoming less as the amounts of chemical contaminants increase. These cells thus provide an extremely important line of defense.

Filtering. The presence of the large hairs around the nostrils tends to restrict the entry of large objects into the system.

Secretion of immune globulins. The presence of immunoglobulin A (IgA) has been detected in the secretions of the laminal glands in all parts of the respiratory tree. The substance is a nonspecific antibody whose production is apparently triggered by a wide variety of antigenic challenges. The amount produced is directly proportional to the degree of antigenic challenge and forms an important defense mechanism against foreign chemicals.

Lymphatics. Numbers of lymphatics are somewhat greater in the respiratory and digestive systems than elsewhere in the body. Recall that both systems contact the external environment. The combination of nodules in the mucous membranes and lymph vessels carrying fluid to lymph nodes aids removal of matter that enters the tissues themselves.

Reflex protective mechanisms. These include sneezing and coughing. Sneezing follows irritation of the nasal mucosa, and coughing follows irritation of the tracheobronchial mucosa. Both responses depend on nervous pathways and result in a sharp inspiration followed by an explosive expiration. The force of the expiration tends to blast the offending particles from the system.

Humidifying. Drying of the diffusing surface is prevented by the humidifying of the incoming air by the secretions of the mucosal glands.

Warming. Warming is provided by radiation of heat from the many blood vessels in the mucosa.

Abnormalities of respiration and gas pressure

Respiratory insufficiency

This condition develops when the respiratory and circulatory systems cannot meet tissue demands for O_2 supply and/or CO_2 removal. Inadequate alveolar ventilation may result from a variety of states. Three common causes of insufficiency are listed below.

NARCOSIS. Narcosis is depression of the central nervous system. Drugs that depress sensitivity of the respiratory centers to CO_2 and H^+ result in CO_2 retention, fall of pH, and decreased PO_2 in arterial blood. Such drugs are termed DEPRESSANTS. Fortunately, such drugs also usually lead to decreased demand for O_2, and thus tissue O_2 levels are not appreciably decreased relative to demands. Respiratory insufficiency due to narcosis therefore has a better prognosis for eventual recovery.

RESPIRATORY DISEASE. EMPHYSEMA, PNEUMONIA, and ATELECTASIS (alveolar collapse) typically reduce the diffusing surface, increase the distance for gas diffusion, prevent normal expansion and/or recoil of the lungs, or combinations of these. The end results will be elevated PCO_2 and depressed PO_2 of the blood. Since tissue demand for O_2 and CO_2 production is usually unchanged or elevated (as with the increased metabolism associated with fever), mild to severe degrees of asphyxia may result. Hypoxia and acidosis at the cellular level may thus prove fatal.

HYPOXIA. Regardless of cause, continued hypoxia (low arterial PO_2) results in the establishment of "vicious circles" that result in increasing degrees of cellular damage. Falls of blood pressure, decreased cardiac output, vasodilation, release of toxins due to cell damage, and other factors combine ultimately to cause irreversible damage to respiratory neurons. Death is the usual result.

Apnea

This term refers to cessation of breathing, and it is usually understood to mean only temporary interruption. Several mechanisms are advanced to explain the development of apnea.

REDUCED STIMULUS TO A NORMAL SET OF RESPIRATORY CENTERS. For example, voluntary hyperventilation reduces blood PCO_2 by "blowing off" alveolar CO_2 and allows greater diffusion from the blood. Until blood PCO_2 returns to normal, the desire to breathe is absent.

INHIBITION OF DISCHARGE OF THE CENTERS, as in prolonging the Hering-Breuer reflex. For example, holding the breath at the end of inspiration prevents deflation of the lungs; apnea will result until blood PCO_2 rises to the point where breathing need overrides the reflex inhibition.

ABNORMAL RESPIRATORY CENTERS require a higher than normal PCO_2 to stimulate them. In periodic breathing (Cheyne-Stokes respiration), periods of hyperpnea alternate with periods of apnea. The hyperpnea reduces alveolar and blood PCO_2, reduced stimulus to respiration is present, and apnea ensues until metabolism restores PCO_2 to higher-than-normal levels.

Dyspnea

This implies LABORED BREATHING and may be associated with a variety of disease entities. It is a symptom rather than a disorder itself. Acidosis, poliomyelitis, emphysema, asthma, pneumonia, atelectasis, cardiac failure, anemia, hemorrhage, fever, and use of stimulants are only a few of the conditions associated with dyspnea.

Polypnea

Polypnea (tachypnea), or ACCELERATION OF RATE OF BREATHING without accompanying increase in depth, occurs during fever (late), pain, and hypoxia. If depth is no greater than dead space volume (150 ml in the normal adult), no alveolar ventilation will occur since all the inspired air remains in tubes that are too thick-walled to allow diffusion. Thus, hypoxia and respiratory acidosis may result.

Resuscitation

Resuscitation of individuals who have suffered respiratory insufficiency or failure requires ventilation of the alveoli to restore arterial PO_2 and assurance of sufficient heart action to deliver the blood to the tissues. Mouth-to-mouth resuscitation (Fig. 26.26) provides a nonmechanical means of lung ventilation. Respirators of a positive-pressure type (Fig. 26.27) force air intermittently into the lungs, and expiration occurs by elastic recoil. Negative-pressure devices (Fig. 26.28)

WHEN BREATHING STOPS SECONDS COUNT
SAVE A LIFE BY ARTIFICIAL RESPIRATION

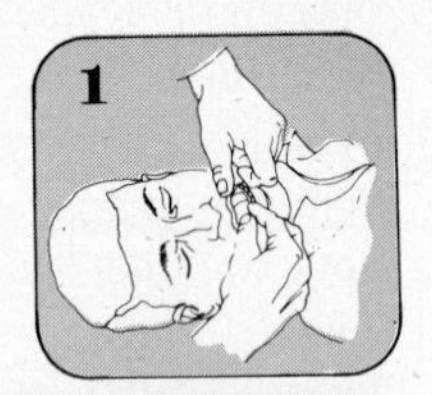

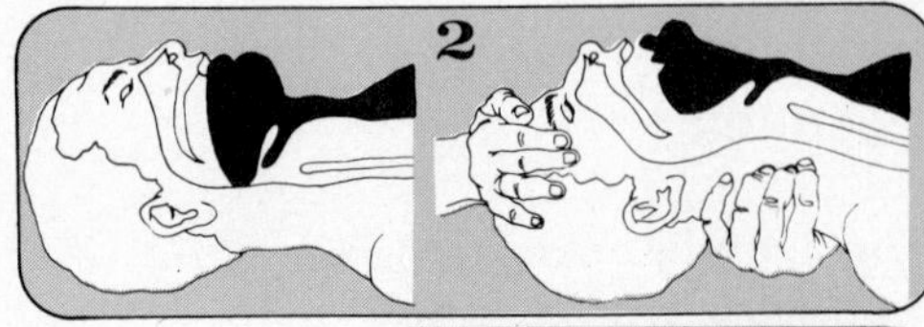

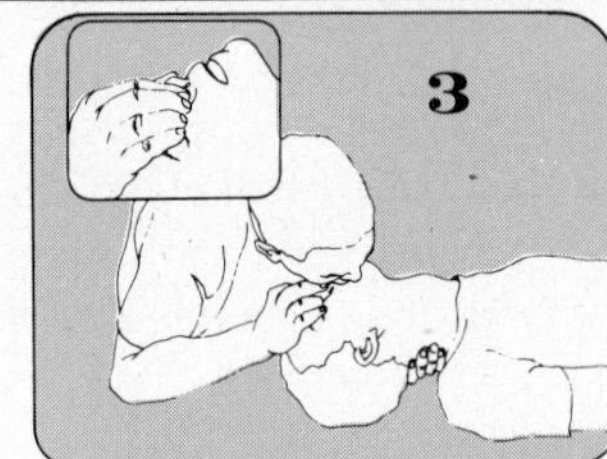

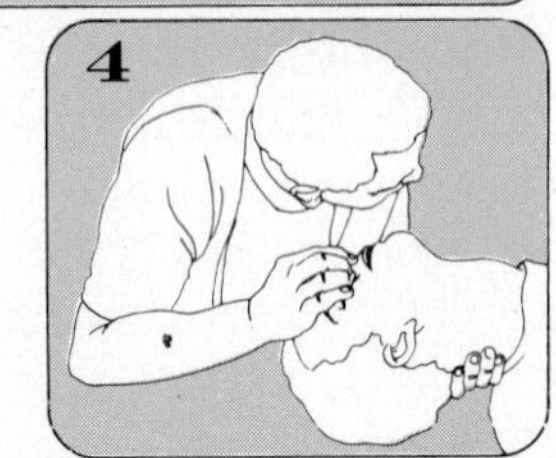

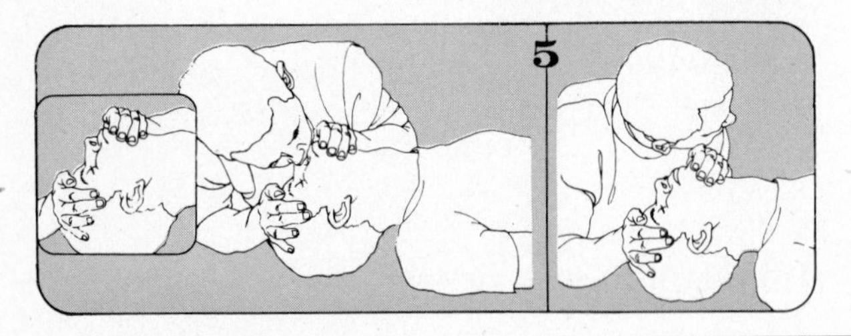

MOUTH-TO-MOUTH METHOD

1. If foreign matter is visible in the mouth, wipe it out quickly with your fingers, wrapped in a cloth, if possible.
2. Tilt the victim's head backward so that his chin is pointing upward. This is accomplished by placing one hand under the victim's neck and lifting, while the other hand is placed on his forehead and pressing. This procedure should provide an open airway by moving the tongue away from the back of the throat.
3. Maintain the backward head-tilt position and, to prevent leakage of air, pinch the victim's nostrils with the fingers of the hand that is pressing on the forehead.

   Open your mouth wide; take a deep breath; and seal your mouth tightly around the victim's mouth with a wide-open circle and blow into his mouth. If the airway is clear, only moderate resistance to the blowing effort is felt.

   If you are not getting air exchange, check to see if there is a foreign body in the back of the mouth obstructing the air passages. Reposition the head and resume the blowing effort.
4. Watch the victim's chest, and when you see it rise, stop inflation, raise your mouth, turn your head to the side, and listen for exhalation. Watch the chest to see that it falls.

   When his exhalation is finished, repeat the blowing cycle. Volume is important. You should start at a high rate and then provide at least one breath every 5 seconds for adults (or 12 per minute).

   When mouth-to-mouth and/or mouth-to-nose resuscitation is administered to small children or infants, the backward head-tilt should not be as extensive as that for adults or large children.

   The mouth and nose of the infant or small child should be sealed by your mouth. Blow into the mouth and/or nose every 3 seconds (or 20 breaths per minute) with less pressure and volume than for adults, the amount determined by the size of the child.

   If vomiting occurs, quickly turn the victim on his side, wipe out the mouth, and then reposition him.

MOUTH-TO-NOSE METHOD

5. For the mouth-to-nose method, maintain the backward head-tilt position by placing the heel of the hand on the forehead. Use the other hand to close the mouth. Blow into the victim's nose. On the exhalation phase, open the victim's mouth to allow air to escape.

---

decompress and compress the environment around the patient and "breathe for him."

Hypoxia

In a preceding section, the term *hypoxia* was used. The word, strictly defined, refers to lowered oxygen tension at the cellular level. Insufficient oxygen at the cellular level may be secondary to diminished levels of oxygen in the blood, known as HYPOXEMIA. Types of hypoxemias and hypoxias are presented below with some causes listed

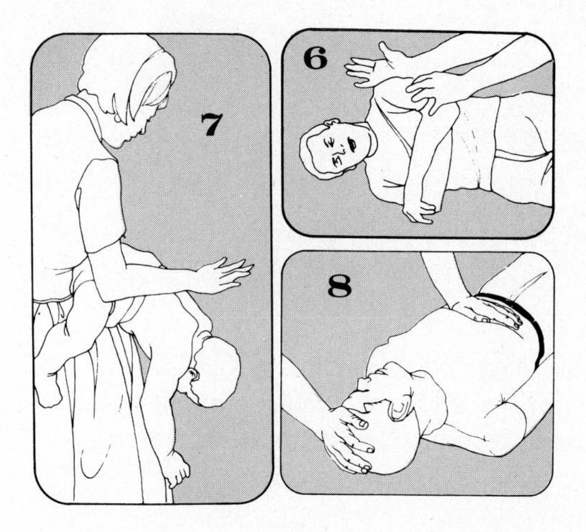

RELATED INFORMATION

6. If a foreign body is prohibiting ventilation, as a last resort, turn the victim on his side and administer sharp blows between the shoulder blades to jar the material free.
7. A child may be suspended momentarily by the ankles or turned upside down over one arm and given two or three sharp pats between the shoulder blades. Clear the mouth again, reposition, and repeat the blowing effort.
8. Air may be blown into the victim's stomach, particularly when the air passage is obstructed or the inflation pressure is excessive. Although inflation of the stomach is not dangerous, it may make lung ventilation more difficult and increase the likelihood of vomiting. When the victim's stomach is bulging, always turn the victim's head to one side and be prepared to clear his mouth before pressing your hand briefly over the stomach. This will force air out of the stomach but may cause vomiting.

When a victim is revived, keep him as quiet as possible until he is breathing regularly. Keep him from becoming chilled and otherwise treat him for shock. Continue artificial respiration until the victim begins to breathe for himself or a physician pronounces him dead or he appears to be dead beyond any doubt.

Because respiratory and other disturbances may develop as an aftermath, a doctor's care is necessary during the recovery period.

THE AMERICAN NATIONAL RED CROSS 

FIGURE 26.26 Mouth-to-mouth resuscitation. (Courtesy American Red Cross.)

(synonomy is presented in the interest of correlation between newer and older terminology; the newer terms are more descriptive of cause and characteristics).

HYPOTONIC HYPOXEMIA *(anoxic anoxia).* Lowered arterial PO_2 is the primary characteristic of this disorder. It is primarily the result of inadequate alveolar ventilation due to obstruction of a respiratory passageway, exposure to altitude with lowered available O_2, or diminished diffusion of O_2 through alveolar walls. It may also result from shunting of blood past the pulmonary capillary beds.

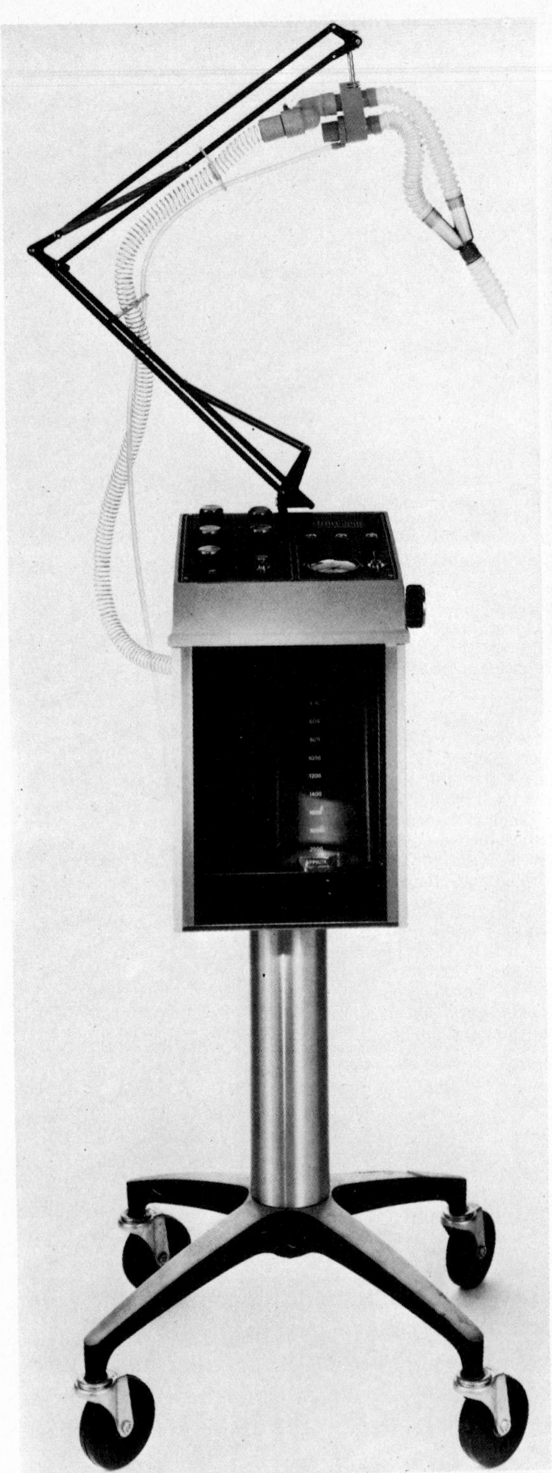

FIGURE 26.27 A positive pressure respirator. (Courtesy Ohio Medical Products, Division of Airco, Inc.)

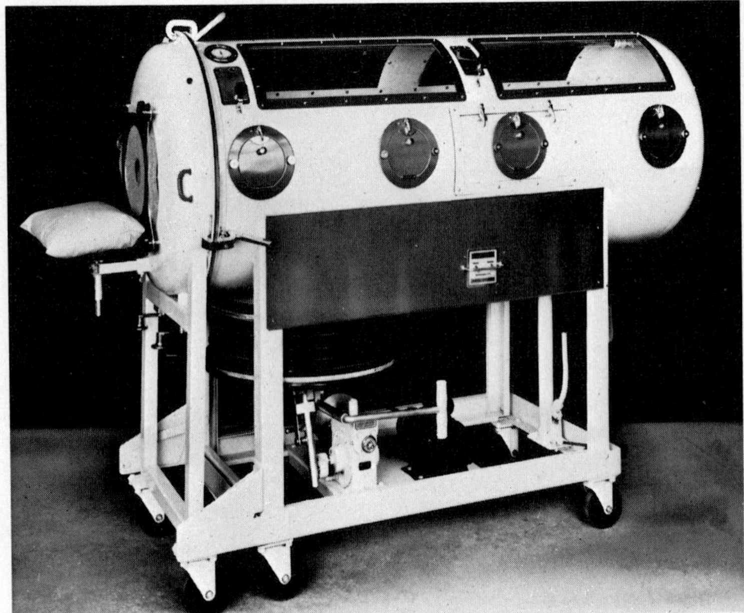

FIGURE 26.28 A negative-pressure respirator. (Courtesy Warren E. Collins, Inc.)

ISOTONIC HYPOXEMIA *(anemic anoxia)*. In the face of normal PO_2 in alveolar air, arterial PO_2 is lowered. A reduction in the hemoglobin or its carrying capacity for O_2 appears to be the cause of this condition. In carbon monoxide poisoning and methemoglobinemia (oxidation of Fe^{++} to Fe^{+++}), total amount of O_2 delivered on hemoglobin to tissues is reduced. Less oxygen combines with ferric (Fe^{+++}) iron. Also, snake venoms or chemicals that destroy erythrocytes may reduce the oxygen-carrying capacity.

HYPOKINETIC HYPOXIA *(stagnant anoxia)*. Basically a circulatory problem, this condition results when blood flow is impeded in some way. Hemoglobin concentrations, alveolar ventilation, and diffusion are all within normal limits. Cardiac failure, embolism or thrombosis, hypotension, and acidosis of metabolic origin may lead to this condition.

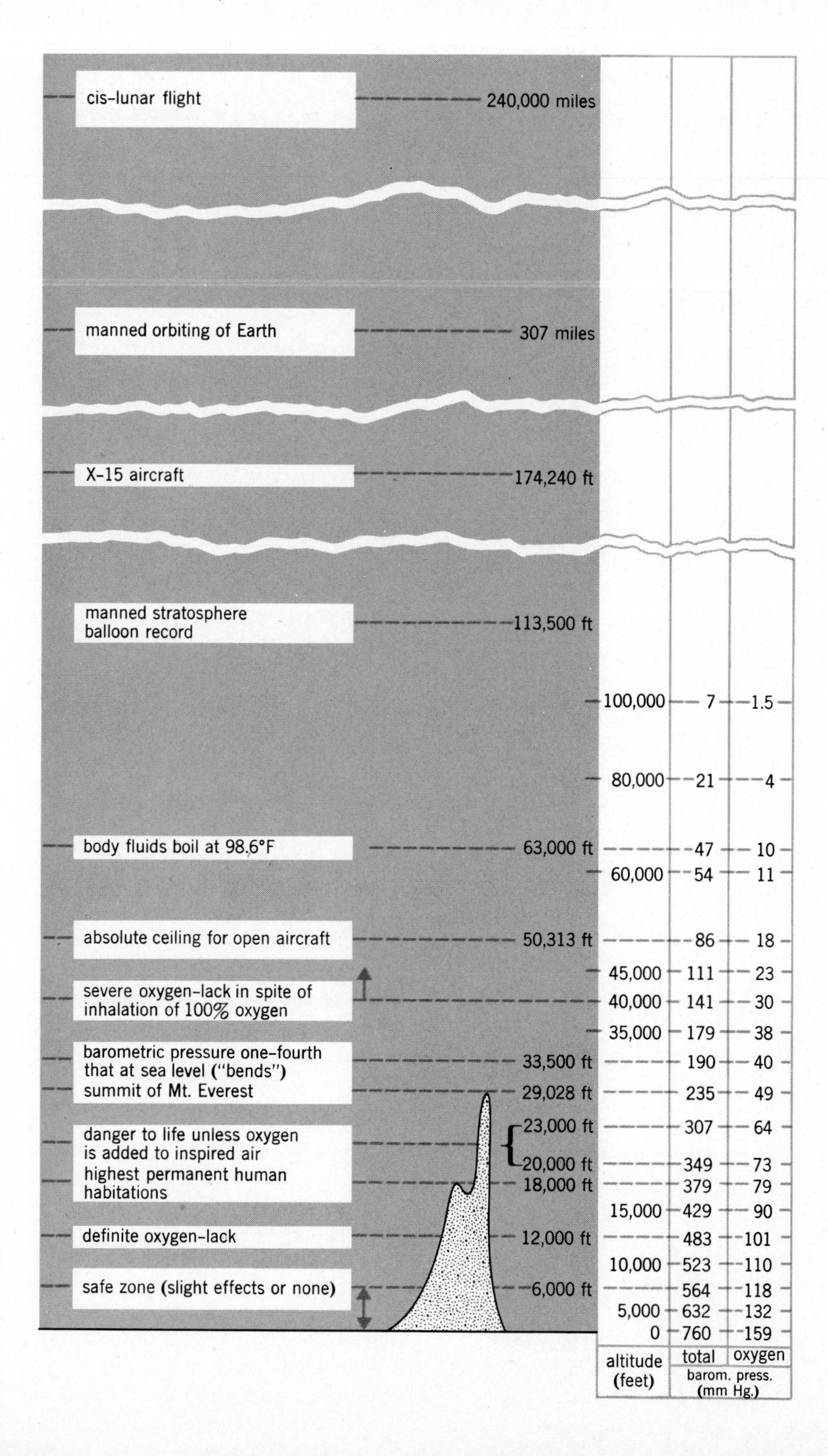

FIGURE 26.29 High altitude. Critical levels of reduced barometric and oxygen pressure are shown in relation to significant stages in the conquest of the upper atmosphere and "near space." Scales at right show total air pressure and pressure of oxygen in the air at increasing altitude. (From Lambertsen, Christian J., "Anoxia, Altitude, and Acclimatization," *Medical Physiology*, V. B. Mountcastle [ed.], Thirteenth Edition, Mosby, St. Louis, 1974.)

OVERUTILIZATION HYPOXIA. In strenuous exercise, convulsions, or where there are narrowed vessels, the demand for O_2 may exceed the supply. Arterial O_2 concentration decreases, and CO_2 rises. O_2 debt results (a temporary shortage of O_2 for oxidation of metabolites).

HISTOTOXIC HYPOXIA. Cellular poisoning, as with cyanide, sulfides, or excessive O_2 levels, causes inability of cells to utilize O_2. Cell death typically results. All other factors (ventilation, diffusion, transport) are normal.

Hypoxemia

Decreased blood PO_2, or hypoxemia, is a common result of exposure to increasing altitude above sea level. The diminishing availability of O_2 in the atmosphere may lead to insufficient oxygenation of arterial blood. Effects on the body are determined primarily by the rate at which the decreased O_2 supply is encountered. In short, if exposure is gradual, the possibility for the body to adjust or acclimatize to the lowered O_2 pressure exists. Figure 26.29 shows some of the relationships between altitude and oxygen pressure. At about 18,000 feet, pure O_2 must be breathed if adequate arterial PO_2 is to be maintained. At about 44,000 feet, pressure must be added to the pure O_2 since total pressure of H_2O and CO_2 in the lungs will equal the environmental pressure and no O_2 will be found in the lungs.

Acclimatization

Adjustment of the body to altitude occurs through changes that ensure continued supply of O_2 to cells in the face of decreased O_2 in the environment. Several changes occur during acclimatization.

INCREASED RATE AND DEPTH OF BREATHING. Relatively greater concentrations of CO_2 stimulate respiration, increasing RMV (respiratory minute volume) and release of CO_2 from blood.

MOVEMENT OF O_2 DISSOCIATION CURVE TO THE LEFT. Increase in pH, caused by ridding the body of more CO_2, causes a left shift; this also means greater affinity for O_2 at lower ambient PO_2.

INCREASE IN RED CELLS. Lowered PO_2 in inspired air and resultant hypoxemia causes release of erythropoietin. Stimulation of red cell production results. More cells per volume of blood increases carrying capacity.

INDIVIDUAL TOLERANCE. Individual tolerance to hypoxia is variable. Probably dependent on circulatory adjustments, some individuals retain their performance better than others in the face of oxygen deprivation.

Adjustments such as those described require time to complete. A resident at high altitude can thus expect to adjust better than one who enters and leaves such environments rapidly (e.g., aviators). Time of acclimatization varies and seems to depend on altitude and individual response. Complete acclimatization to altitudes of 15,000–18,000 feet may take as long as five to nine weeks.

Gas pressure and physiology

The entry of humans into the inner-space (undersea) and outer-space environments has created physiological problems associated with supply of O_2, removal of CO_2, and accumulation of N_2 and CO. In some cases, artificial environments must be created to sustain life, environments that differ in gas composition and pressure from the one to which humans are normally adjusted.

Oxygen

Exposure to oxygen at high partial pressures for periods of time exceeding 12 hours may lead to the development of OXYGEN TOXICITY. The mechanism of the toxicity is believed to be the result of the action of excessive O_2 on intracellular metabolic processes, particularly those reactions concerned with transfer of hydrogen. A number of dehydrogenases that depend for their activity on sulfhydryl (SH) groups may be put permanently into the oxidized state as disulfide (SS) and be rendered incapable of hydrogen transport. Oxygen toxicity is, at least in the earlier stages, reversible. Later, irreversible damage may result. The symptoms of oxygen toxicity are seen in many body areas.

THE LUNGS. IRRITATION of the linings of the respiratory system leads to coughing, bronchopneumonia (inflammation of terminal bronchioles and alveoli), loss of surfactant, pulmonary edema, and pathological changes.

THE CENTRAL NERVOUS SYSTEM. At pressures of 2 atmospheres* or more, toxic effects of O_2 on neural function include the production of nausea, dizziness, tingling of the hands, twitching of muscles of the lips, eyelids, and hands, convulsions, and unconsciousness.

THE EYE. In the adult, VASOCONSTRICTION results that may interfere with the blood supply to the retina. Loss of peripheral vision leads to "gun-barrel vision," where one has a visual field similar to that produced by peering down a gun barrel. Ultimately, damage to retinal cells may result. In infants, RETROLENTAL FIBROPLASIA *(RLF)* has been shown to result from exposure to increased PO_2 (as in an incubator). Extreme constriction of retinal vessels leads to cell death by asphyxia, detachment of the retina, and blindness. It is thus recommended that a limit of 40 percent oxygen be observed when exposure will exceed 12 hours in duration.

Implications of the foregoing discussion as to the use of pure O_2 in diving should be obvious. Starting with 1 atmosphere at the surface and knowing that the pressure of the ocean increases by approximately 1 atmosphere for each 33 feet increase in depth, it becomes clear that toxic effects may be expected to occur beginning at 33 feet, or 2 atmospheres. The rapidity of development of symptoms will depend not only on depth but on time. At 100 feet, for example, a total pressure of 4 atmospheres will exist; a few minutes of exposure to pure O_2 will produce symptoms of toxicity. Solving these problems requires reduction of oxygen pressure. At the present limit (600 feet depth and about 20 atmospheres) of man's diving, an atmosphere containing about 10 percent oxygen is employed.

Nitrogen

Although inert in the metabolic sense, nitrogen exerts important mechanical effects on the body. Under excessive pressure, nitrogen narcosis may result. A EUPHORIA is followed by impairment of higher nervous system functions, unconsciousness, and interference with synaptic function. The mechanism involves many effects; a recent theory suggests that the formation of gas hydrate "microcrystals" stabilizes, or renders incapable of change, the side chains of nerve cell chemicals and results in diminution of neuronal activity. With increased underwater time and depth, another problem is encountered. As atmospheric pressure or exposure time increases, more and more nitrogen is dissolved in the body fluids. If the nitrogen is not allowed to come out of solution slowly as the diver goes toward the surface (decompression), bubbles of nitrogen can form in the tissues, causing DECOMPRESSION SICKNESS ("the

* One atmosphere is equal to 760 mm Hg or approximately 15 $lb/in.^2$ (15 psi).

bends"). Helium, which is often used in mixtures of gases breathed when diving at greater depths, leaves the tissues very rapidly and thus decompression sickness may usually be avoided.

Carbon dioxide

At high partial pressures, CO_2 produces toxic effects on all cells, due to its extremely rapid passage across cell membranes. The most common effect produced is one of DEPRESSION OF CELLULAR ACTIVITY. Vasodilation (with drop in blood pressure), increased heart action, and excitation of the sympathetic nervous system and adrenal medulla are signs of CO_2 toxicity. The mechanism of production of these effects is either by increased amounts of molecular CO_2 or increased $[H^+]$ on cell processes.

Carbon monoxide

A normal product of cellular metabolism, CO is produced at rates that are too low to result in toxic effects. At higher levels, the major effects are produced by the fact that CO binds more strongly to hemoglobin than does oxygen. Oxygen is thus displaced from the hemoglobin; hypoxemia and hypoxia result. Symptoms of CO POISONING include severe frontal headache, fainting, collapse, and the imparting of a cherry-red color to arterial blood as carboxyhemoglobin (HgbCO) is formed.

In space travel, where it is necessary to create artificial environments to sustain astronauts, the problems become those associated with supply of O_2 and removal of toxic gases (CO_2, CO). "Scrubbers," which chemically remove CO_2 and CO from the atmosphere, are easily constructed and are capable of keeping the concentrations of these gases at levels of 0.2–0.3 percent. The United States has employed a 100 percent oxygen atmosphere at 1/3 atmosphere (5 psi) to ensure adequate oxygenation. Upon return to the earth's atmosphere, there are no recompression problems, and the excess oxygen is metabolized. Thus, no embolism problems are encountered. Humidity is also easily controlled by dehumidifying agents that keep the humidity at 36–70 percent. Contamination by trace constituents is the remaining threat as more efficient methods of sealing spacecraft and removal of CO_2 and CO are found.

Summary

1. Development of the respiratory system begins at about 3½ weeks of embryonic life.
   a. The upper part of the system is derived from ectodermal and endodermal portions.
   b. The lower part of the system is derived from the foregut.
   c. Some 24 generations of branchings form the tracheobronchial tree.
   d. Alveoli are not developed in the lung until 26 weeks.
2. A respiratory system provides
   a. A membrane for gas exchange.
   b. Devices to renew gases at the exchange surface.
   c. Devices to protect the body from inhaled materials.
   d. A means of regulating acid-base balance.
3. The organs of the system consist of
   a. A conducting division where no gas exchange occurs.
   b. A respiratory division where gas exchange does occur.
4. The nasal cavities
   a. Warm, moisten, and partially cleanse incoming air.
   b. Contain the openings of the paranasal sinuses.
   c. Contain the olfactory epithelium.
   d. May become inflamed to give rhinitis, and the sinuses may become inflamed to give sinusitis.
5. The pharynx has three parts and is common to respiratory and digestive systems.
   a. The pharynx contains the pharyngeal tonsils (adenoids).
6. The larynx
   a. Is composed of cartilages (main ones: thyroid, cricoid, epiglottis).
   b. Contains the vocal cords.
   c. May become inflamed (laryngitis).
7. The trachea and bronchi
   a. Are the large tubes of the lower respiratory system.

   b. Have a common three-layered structure with cartilage rings.
   c. Conduct, cleanse, and moisten inhaled air.

8. The bronchioles
   a. Are the small tubes of the lower respiratory system.
   b. Have little or no cartilage and much smooth muscle.
   c. Conduct air.
   d. Are the site of muscular spasm that may occlude airflow.

9. The respiratory division is characterized by the presence of alveoli in the walls of the tube; conditions that affect gas diffusion through the walls cause interference with gas exchange.

10. The human has two lungs with a total of five lobes (three right, two left).

11. Ventilation of the lungs is normally provided by muscular contraction (inhalation) and elastic recoil of lung tissue (exhalation) that cause changes in chest volume and pressures.
   a. Air in the conducting division is not available for exchange.
   b. Air in the respiratory division is available for gas exchange.

12. Surfactant is a phospholipid that aids in preventing alveolar collapse.

13. Tidal, reserve inspiratory, and reserve expiratory volumes are measures of the air that can be exchanged during ventilation.

14. Lung capacities are obtained by adding the volumes in various ways.

15. Effort required to ventilate the alveoli depends on airway resistance, lung compliance, and lung elastance. Several of these parameters may be tested for normality.

16. Partial pressure differences in gas concentrations assure "inward" diffusion of oxygen and "outward" diffusion of carbon dioxide.

17. Oxygen is transported primarily on hemoglobin; carbon dioxide is transported as bicarbonate ion, on hemoglobin, and dissolved in the plasma.
   a. An O_2 dissociation curve indicates the relationship of temperature, pH, and PCO_2 to O_2 transport and release.

18. The lungs are capable of regulating acid-base balance by elimination of CO_2.

19. The lungs are active metabolically in secretion, production of kinins and prostaglandins, and metabolism of materials.

20. Respiratory centers and respiratory reflexes control the rate and depth of breathing.

21. Mechanisms that protect the system from airborne toxins include: the mucociliary escalator, alveolar macrophages, filtering devices, secretion of Ig, lymphatics, and reflex mechanisms.

22. A variety of respiratory abnormalities are described.

23. The relationships of high and low O_2, CO_2, CO, and N_2 levels to physiology are discussed.

27 The Digestive System

Cellular activity requires a continual supply of raw materials to sustain that activity. A digestive system provides a means of intake or ingestion of solid and liquid materials, but usually in the form of large complex molecules. By the process of digestion, these large molecules are reduced to their constituent "building blocks" utilizing mechanical and chemical factors. Once reduced to simple units, these units are still not available for cellular use until absorption or passage through the walls of the system has occurred. Undigested or indigestible material, unabsorbed secretions of the tract, bacteria, and excess substances of no use to the body are eliminated from the body in the process of egestion.

To allow orderly action of chemical materials on the foodstuffs, the foods are moved through the system by activity of muscle tissue in the walls; motility is thus provided.

The system is composed of the tubular alimentary tract, generally called the gut, and the accessory glands that develop from the gut.

Development of the system

Formation of the mouth and gut

The mouth is formed as a depression of the surface ectoderm anterior to the first branchial arch of the head at about 22 days of development (Fig. 27.1). The depression is known as the STOMODEUM and is separated from the foregut or primitive pharynx by a two-layered OROPHARYNGEAL MEMBRANE. The stomodeum and the external layer of the membrane are ectodermally lined; the foregut and the internal surface of the membrane are endodermally lined. The membrane ruptures at about 24 days of development and provides a communication of the primitive gut with the amniotic cavity.

During the fourth week of development, a PRIMITIVE GUT is formed as the cavity of the yolk sac is enclosed by folds of the embryo's lateral

FIGURE 27.1 The development of the oral cavity between 3 and 4 weeks.

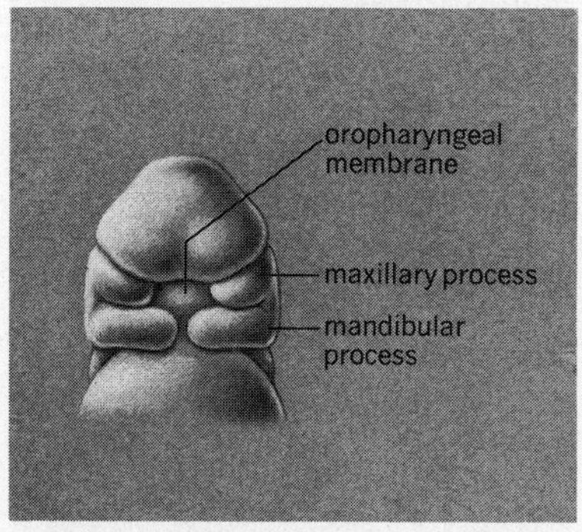

22 ± 1 day

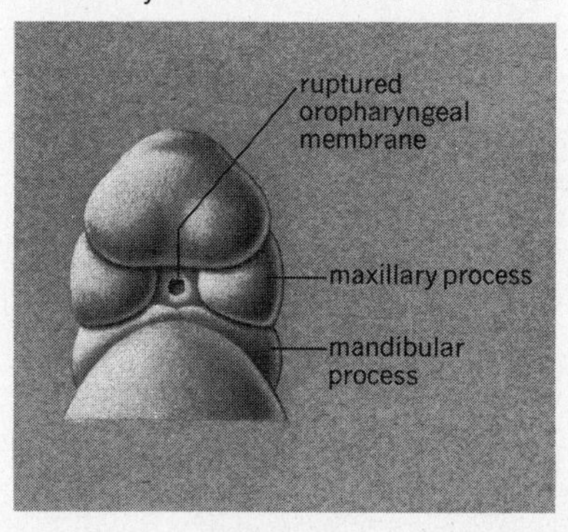

26 ± 1 day

28 ± 1 day

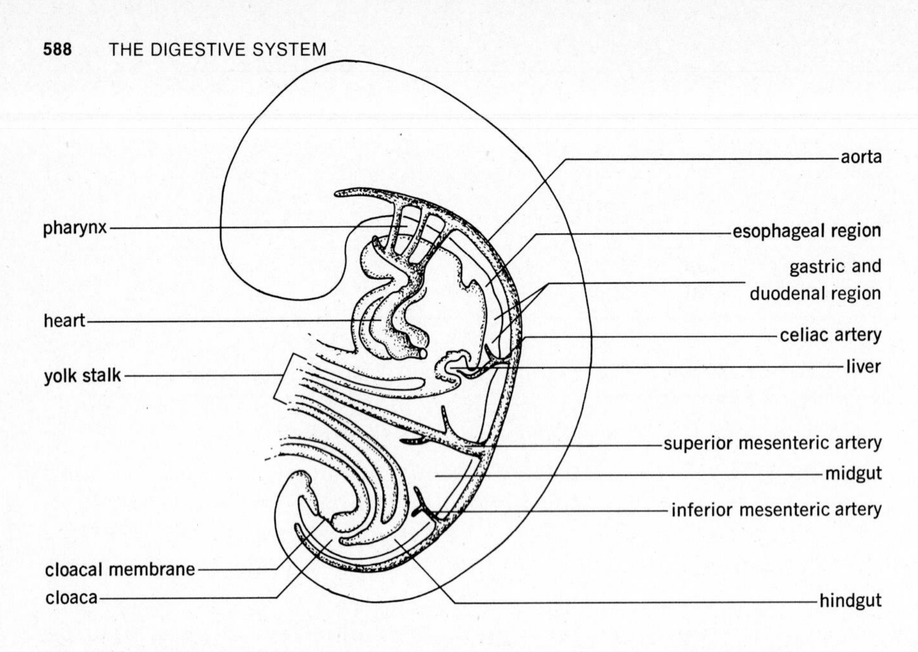

FIGURE 27.2 The development of the digestive system at 3½ weeks.

TABLE 27.1 Derivatives of the gut tube

Part of tube	Derivatives	Comments
Foregut	Esophagus	Recognizable at 4 weeks.
	Stomach	Recognizable at 4 weeks, becomes baglike at about 8–10 weeks, and assumes typical form at about 12 weeks.
	Duodenum to entrance of bile duct	Recognizable at 4 weeks.
	Liver Pancreas	Develop as outgrowths of gut at about 4 weeks; liver lobes form by 6 weeks; pancreas complete by 10 weeks.
	Bile ducts and gall bladder	Connection retained to duodenum forms ducts. Bladder is an outgrowth of duct.
Midgut	Rest of small intestine	Recognizable at 4 weeks; elongates and coils at 5 weeks; villi at 8 weeks.
	Cecum, appendix, one half of large intestine	Separated at 5 weeks; completed by 8 weeks.
Hindgut	Remainder of large intestine, rectum, anal canal	Formed by 4 weeks; completed by 7 weeks.

TABLE 27.2 Some common congenital disorders of the digestive system

Disorder	Frequency (no. per births)	Cause of disorder	Comments
Esophageal atresia	1/2500–3000	No recanalization of esophagus, improper separation from respiratory system. No passage to stomach.	Infant shows excess of saliva and poor nutrition since foods cannot reach stomach.
Pyloric stenosis	1/200 male 1/1000 female	Excessive development of muscle fibers of distal end of stomach	Blockage of stomach exit causes vomiting of feedings, weight loss, and dehydration.
Imperforate anus	1/5000	Failure of anal membrane to rupture. No anus formed.	Surgery necessary.
Megacolon	1/25,000	Failure of nerve cells to innervate a section of the colon.	Part without nerves does not move contents onward, accumulates feces and dilates.

body walls, which are endodermally lined. Three portions of the primitive gut are recognized; they are named, from anterior to posterior, the FOREGUT, MIDGUT, and HINDGUT (Fig. 27.2). Each section is supplied by a major branch of the aorta that will form its blood supply throughout life. The CELIAC ARTERY supplies the foregut, the SUPERIOR MESENTERIC ARTERY supplies the midgut, and the INFERIOR MESENTERIC ARTERY supplies the hindgut. Each portion will differentiate into particular portions of the digestive system, as indicated in Table 27.1.

The ANUS develops as an indentation at the posterior end of the embryo called the PROCTODEUM. It, like the mouth, is separated from the gut by a CLOACAL MEMBRANE that normally ruptures at about seven weeks of development.

The muscular and connective tissue layers of the tract are mesodermally derived and begin formation at about 12 weeks.

Disorders of formation of the gut

As the changes described above are occurring, several other important events take place.

A tracheoesophageal septum separates the esophagus from the trachea (fourth week).

The endoderm of the esophagus proliferates, and the lumen (cavity) of the esophagus is obliterated (seven weeks) to be recanalized at about nine to ten weeks.

The small intestine elongates and "herniates" into the umbilical cord through the as-yet not closed opening from the midgut to the cord (five weeks) and then, as the embryo's body grows and provides more room in the abdominal cavity, the intestine returns to the body (about ten weeks).

The cloaca (the terminal portion of the hindgut) is partitioned into an upper rectum and anal canal and a lower urogenital sinus, thereby effecting a separation of digestive from urogenital system (about seven weeks).

If such processes as these do not occur normally, or development is arrested at some stage, malformations of the tract may occur. Several of the more common congenital disorders of the system are presented in Table 27.2.

The organs of the system

The organs of digestion (Fig. 27.3) include those forming the tube, or ALIMENTARY TRACT, and ACCESSORY STRUCTURES essential to some phase of the digestive process. The tract includes the mouth, pharynx, esophagus, stomach, small and large intestines, rectum, anal canal, and anus. The accessory structures are located outside the limits of the tract proper and empty into it through ducts. They are the salivary glands, liver, and pancreas.

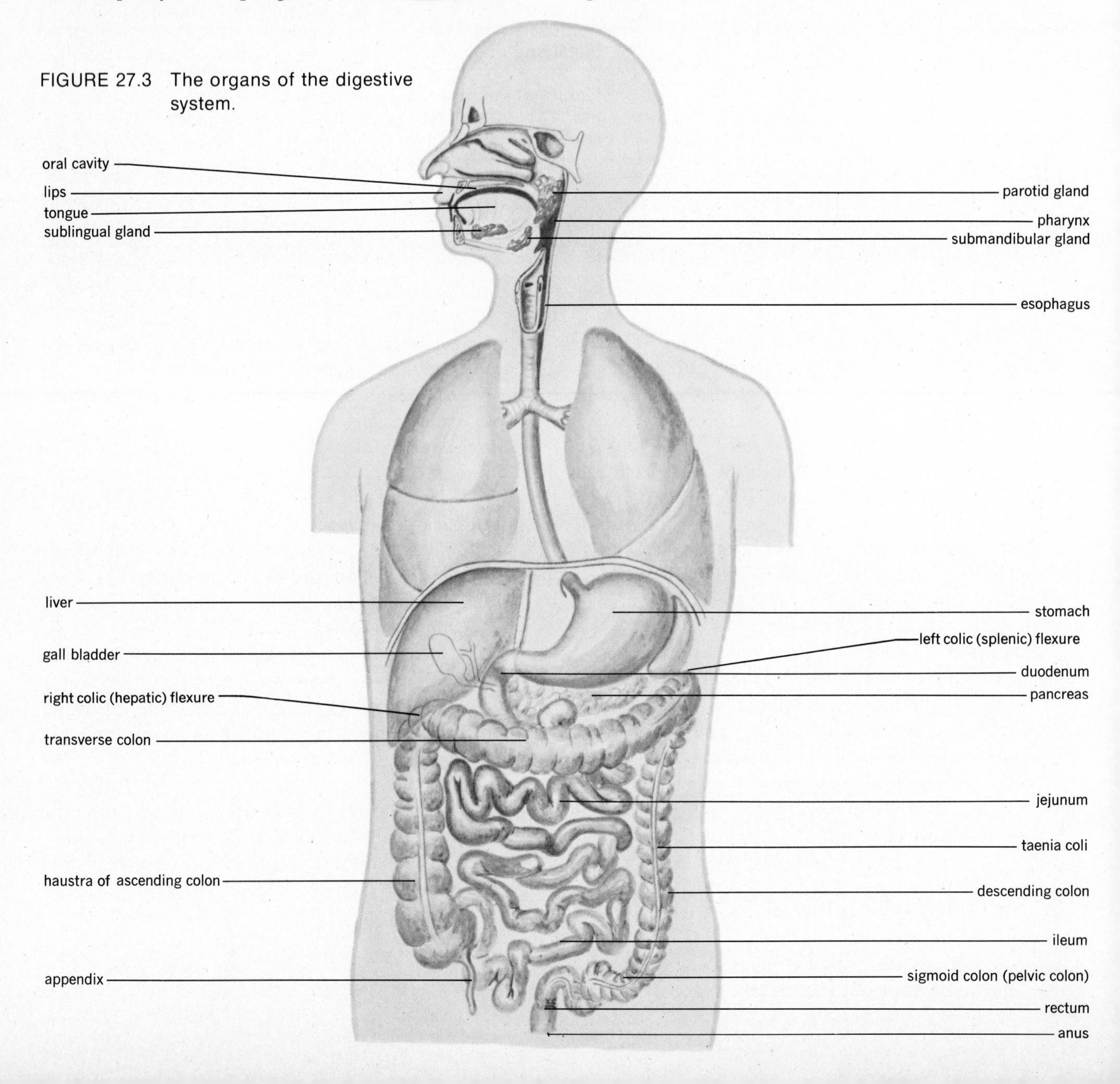

FIGURE 27.3 The organs of the digestive system.

The structure and functions of the organs of the system

The tissue plan of the system

The organs of the alimentary tract, particularly those from the esophagus to the anal canal, have the same basic arrangement of tissue layers in their walls. Each organ, however, has certain special structural features of its own that enable one to identify it histologically as well as grossly. These special features make possible the unique functions performed by each organ. Four basic layers of tissue are described (Fig. 27.4):

MUCOSA *(tunica mucosa).* This layer consists of a lining EPITHELIUM, an underlying layer of connective tissue, the LAMINA PROPRIA, and a thin layer of nonstriated (smooth) muscle, the MUSCULARIS MUCOSAE. The mucosa is glandular except in the front portion of the mouth and in the anal region.

SUBMUCOSA *(tunica submucosa).* This layer consists of vascular connective tissue. In the intestines, the submucosa is thrown into folds, the PLICAE CIRCULARES, which serve to increase the surface area of the gut. The submucosa contains glands in the esophagus and first part of the small intestine. An important nerve plexus, the SUBMUCOSAL PLEXUS *(of Meissner),* is also located within this tissue layer.

MUSCULARIS *(tunica muscularis* or *muscularis externa).* The muscular coat is typically composed of TWO LAYERS of muscle. The inner layer is in the form of a tight spiral *(circular layer),* while the outer coat is a loose spiral *(longitudinal layer).* The muscle is nonstriated (smooth) except in the mouth, pharynx, upper esophagus, and anal canal. Between the two muscle layers is located the second major nerve plexus of the tract, the MYENTERIC PLEXUS *(of Auerbach).*

FIGURE 27.4 A diagram representing the tissue layers of the alimentary tract.

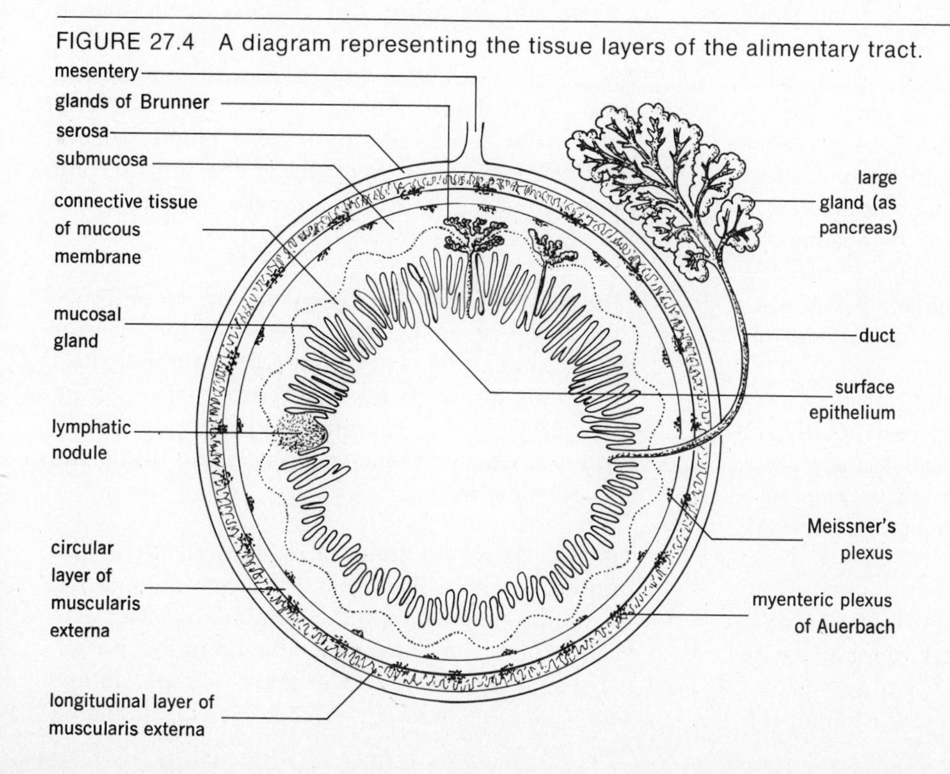

SEROSA or ADVENTITIA. The outermost coat of the tract is the serosa on those organs suspended by mesenteries within a body cavity. Where the organ is not suspended by a mesentery, but is immediately surrounded by other organs or tissue, there is no clear line of demarcation between the organ and the surrounding tissue. In this case, the outer coat is referred to as the adventitia.

The mouth and oral cavity

Food and drink are normally taken in or INGESTED by the mouth. Food intake appears to be controlled by a FEEDING CENTER in the hypothalamus. When the organism is "full" or satiated, ingestion ceases through the activity of a hypothalamic SATIETY CENTER. These centers (Fig. 27.5) work together to balance food intake to activity levels normally, and to maintain a normal body weight for the age of the organism. Control mechanisms for feeding and satiety may involve one or a combination of factors, as listed below.

FALL OF BLOOD GLUCOSE LEVELS may trigger feeding. Absorption of the sugar after digestion of carbohydrates raises blood sugar levels and activates the satiety center. A major weakness of this theory is the development of hunger seen in diabetes mellitus (sugar diabetes) despite very high blood sugar levels.

FEELINGS OF FULLNESS brought on by feeding and stretching of the stomach and abdominal wall may trigger nervous reflexes that inhibit feeding.

RISE OF AMINO ACID LEVELS of the blood after eating may diminish hunger.

The digestion, absorption, and metabolism of foods after eating is associated with increased activity and HEAT PRODUCTION that may stimulate the satiety center and slow feeding. This effect of raising body temperature is called the *specific dynamic action* (SDA) of foods.

Feeding also seems to depend not just on need for food (hunger), but also on desire for a specific food (appetite). Humans, over the long run, appear to eat items that are required for body function if the choice is available, and consumption of a given food at any specific meal should not become a matter for argument.

THE MOUTH AND ORAL CAVITY. The mouth (Fig. 27.6) is bounded anteriorly by the LIPS, laterally by the CHEEKS, and posteriorly by the SOFT PALATE. The VESTIBULE lies just inside of the lips external to the teeth and gums. The ORAL CAVITY forms the major part of the mouth and lies centrally to the gums and teeth.

THE LIPS AND CHEEKS. The lips are two highly mobile, vascular, and sensitive structures surrounding the orifice of the mouth. They are covered externally by thin skin and internally by a soft (uncornified) stratified squamous epithelium. Being thick and stratified, this type of epithelium is the most resistant to mechanical abrasion. The cheeks are similar in structure. Together, these structures are important in moving food between the teeth during chewing, in transferring food to different parts of the oral cavity, and in the articulation of speech. The redness of the mouth linings is due to the translucent epithelium and the highly vascular nature of the underlying connective tissue. Absorption of materials can occur from the mouth into these blood vessels. Medicines are often held under the tongue, and the dissolved products are directly absorbed into the vessels of the mucous membrane.

THE GUMS. The GUMS *(gingivae)* are a continuation of the mucous membrane of the mouth over the alveolar margins of the mandible and maxillae. The gums attach to the teeth, aiding in holding them to their sockets, and are continuous with the PERIOSTEUM *(periodontal membrane)* lining the socket of the tooth.

TEETH. There are four varieties of teeth in the human mouth, reflecting evolutionary adaptation to a diet that consists (or should consist) of a variety of animal and vegetable sources. In each half of the upper and lower jaw there are, anteriorly, two INCISORS. These teeth are chisel-shaped

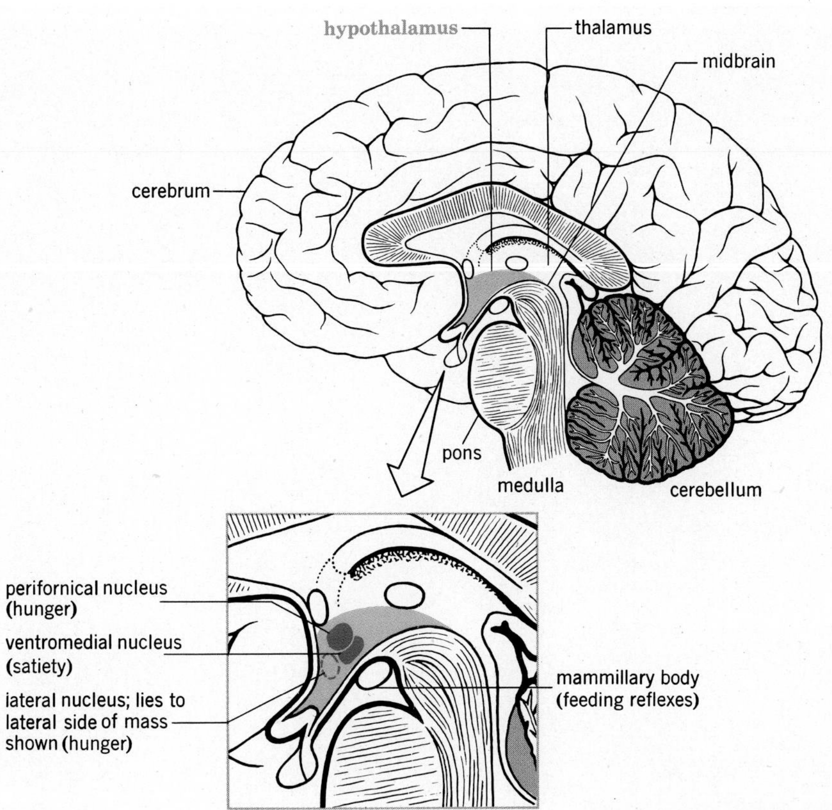

FIGURE 27.5 The control of feeding and satiety by the hypothalamus.

FIGURE 27.6 The mouth.

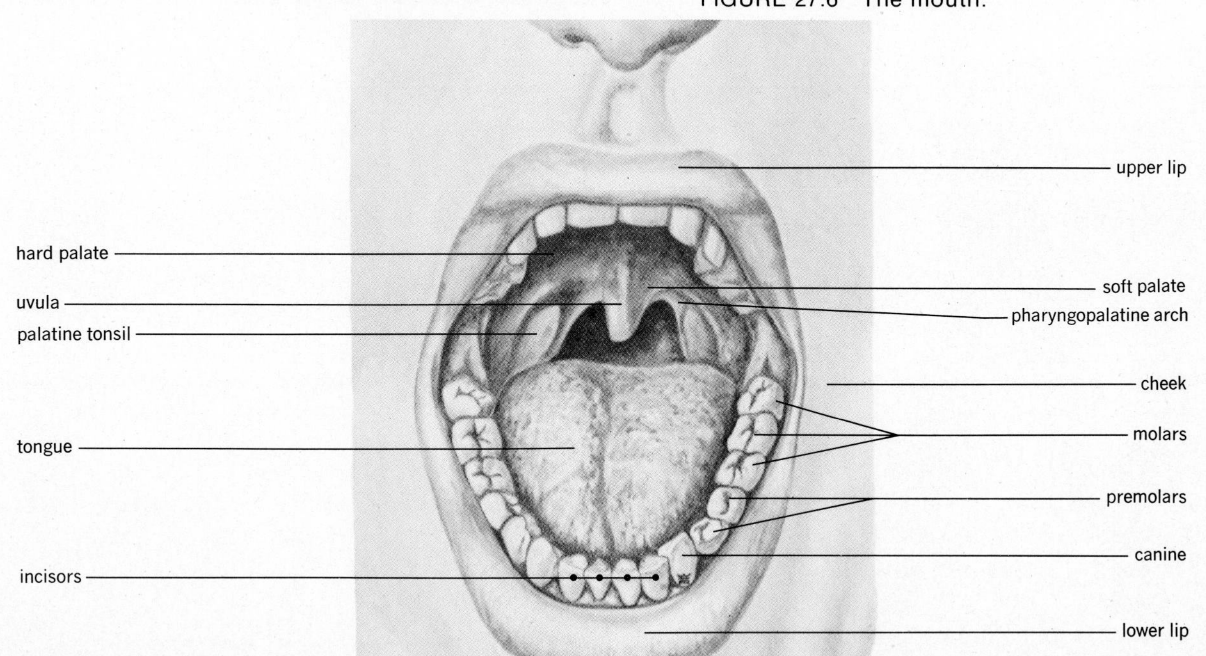

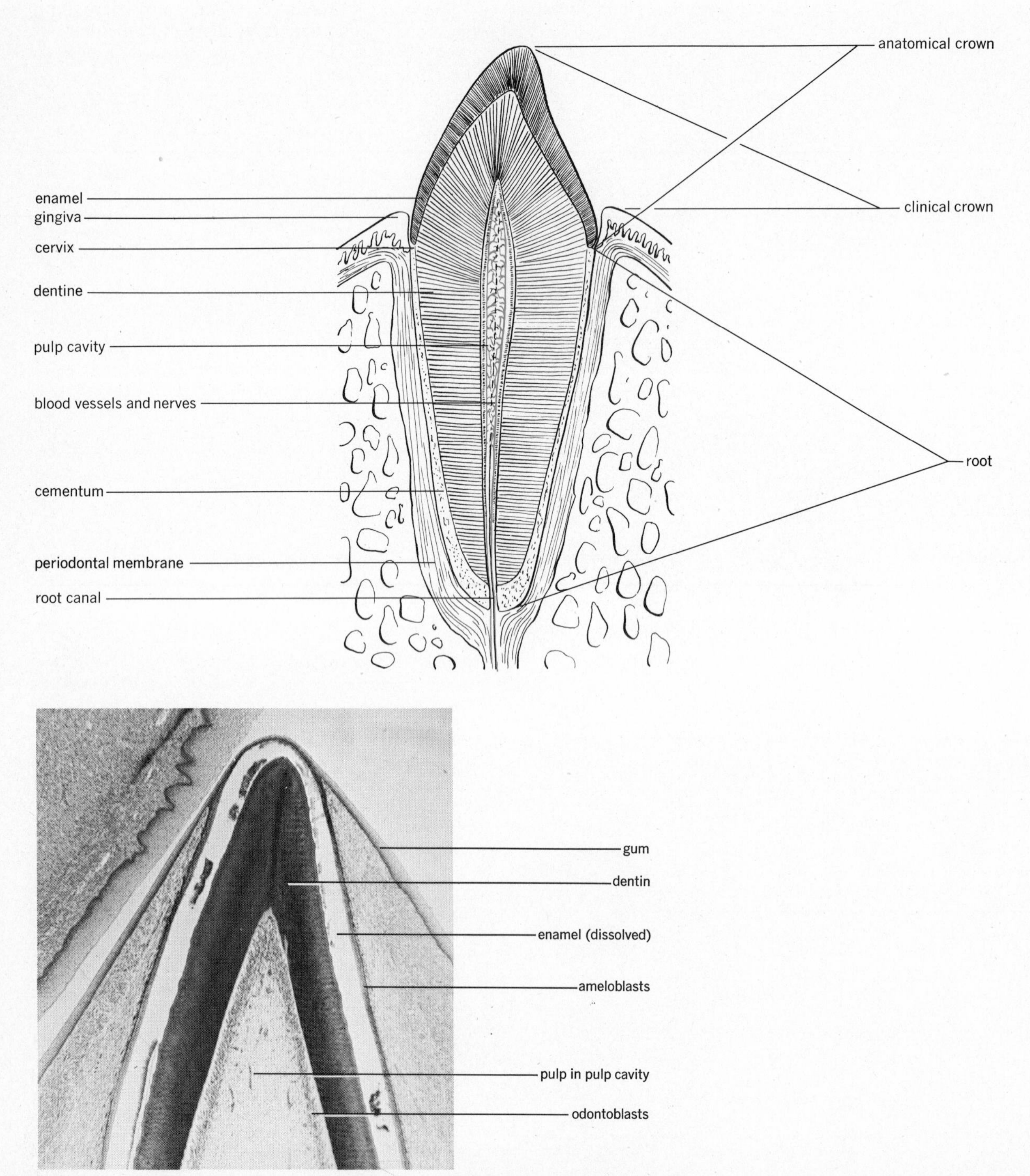

FIGURE 27.7 The structure of a tooth as revealed in longitudinal sections.

and exert a shearing or scissorslike action useful in biting. Next to the incisors is a single canine or eyetooth. Canines are conical, are prolonged in some animal forms into "fangs," and are used for tearing or shredding food, particularly meat. Next are two PREMOLARS *(bicuspids)*, typically having two roots and two grinding surfaces or cusps on their tops. Three MOLARS *(tricuspids)* follow. Premolars and molars are teeth specialized for grinding foods, particularly those rich in cellulose, and are responsible for causing the finest mechanical subdivision of the food. The numbers of teeth indicated lead to the establishment of the dental formula for the adult human of 2 + 1 + 2 + 3, or a total of 8 teeth, in each quarter of the jaw. The first five of the eight teeth *(deciduous teeth)* are replaced by *permanent teeth*. The individual thus

2 deciduous incisors	2 permanent incisors
1 deciduous canine	1 permanent canine
2 deciduous molars	2 permanent premolars
	3 permanent molars
5 × 4 = 20	8 × 4 = 32

total 52

Time of eruption of teeth

Deciduous teeth	Months
Lower central incisors	6–8
Upper central incisors	9–12
Upper lateral incisors	12–14
Lower lateral incisors	14–15
First molars	15–16
Canines	20–24
Second molars	30–32

Permanent teeth	Years
First molars	6
Central incisors	7
Lateral incisors	8
First premolars	9–10
Second premolars	10
Canines	11
Second molars	12
Third molars	17–18

normally acquires a total of 52 teeth during a lifetime.

A section through a typical tooth (Fig. 27.7) shows its parts and the materials of which it is composed. The tooth consists of two major portions, the CROWN and ROOT, connected by a slightly constricted region, the NECK (cervix). The crown may be further subdivided into a CLINICAL CROWN, the visible part, and the ANATOMICAL CROWN, which includes the clinical crown and a part normally covered by the gums. The root carries an APICAL FORAMEN, which opens through a ROOT CANAL into the PULP CAVITY of the tooth. The tooth is composed of the following structural materials:

Enamel. Enamel is produced by cells called AMELOBLASTS and covers the anatomical crown. It consists of 95–97 percent inorganic material (chiefly calcium phosphate) and is the hardest substance in the body. However, it is rather brittle.

Cementum. Cementum is produced by the periodontal membrane, covers the root of the tooth, and is similar to bone in composition.

Dentine. Dentin is produced by cells that line the pulp cavity called ODONTOBLASTS, forms the greater mass of the tooth, and also is bonelike in composition.

Pulp. Pulp fills the tooth cavity and is a vascular connective tissue liberally supplied with nerves and lymphatics. The pulp provides the means of nourishing the tooth during development and in adult life.

THE TONGUE. The tongue (Fig. 27.8) is a muscular organ covered with connective tissue and a stratified, partially cornified, squamous eipthelium. The musculature of the tongue may be divided into EXTRINSIC MUSCLES, which originate outside of the tongue and insert within it, and INTRINSIC MUSCLES, which both originate and insert within the organ. Extrinsic muscles are responsible for the gross movements of the tongue (in and out, side to side). Such movements are important in guiding food between the teeth for chewing and in swallowing. The extrinsic muscles include the hyoglossus, genioglossus, and

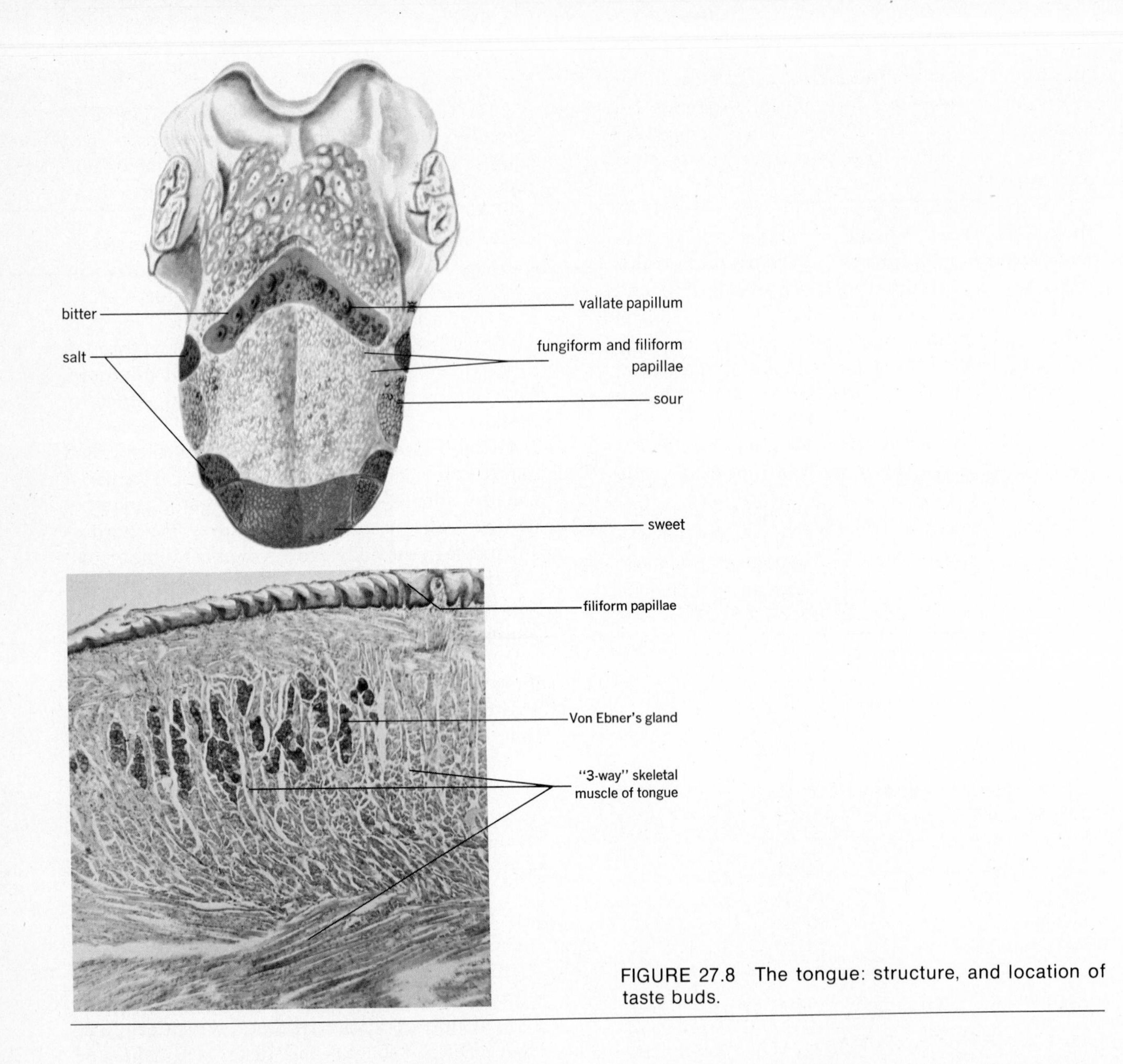

FIGURE 27.8 The tongue: structure, and location of taste buds.

styloglossus. The intrinsic musculature is responsible for the changes in shape of the tongue during speech and swallowing, and includes the transverse, vertical, and longitudinal lingual muscles. It may be recalled that three cranial nerves (VI, IX, XII) supply sensory and motor fibers to the tongue. The tongue is attached to the floor of the oral cavity by the FRENULUM, a membrane lying vertically in the midline of the cavity floor.

The upper surface and sides of the tongue carry a variety of PAPILLAE, which are folds of lamina propria covered with epithelium. FILIFORM PAPILLAE are conical projections distributed evenly over the anterior two-thirds of the tongue. FUNGI-

FORM PAPILLAE are rounded elevations most common on the sides and top of the tongue. These two types of papillae roughen the surface of the tongue, increasing its efficiency during manipulation of food in the mouth. CIRCUMVALLATE (*vallate*) PAPILLAE are embedded in the tongue and are provided with a trench or moat around the papillum proper. Vallate papillae are found on the

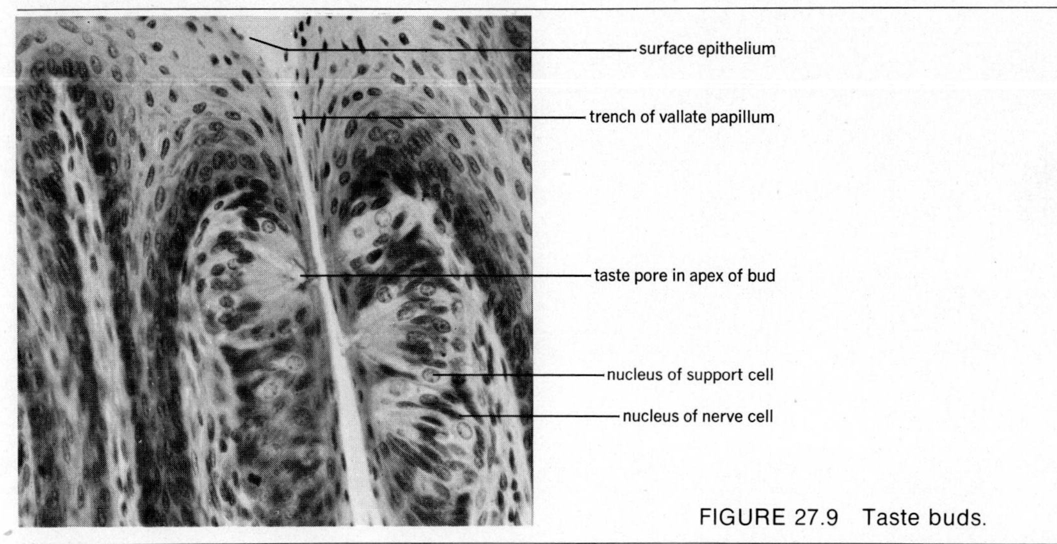

FIGURE 27.9 Taste buds.

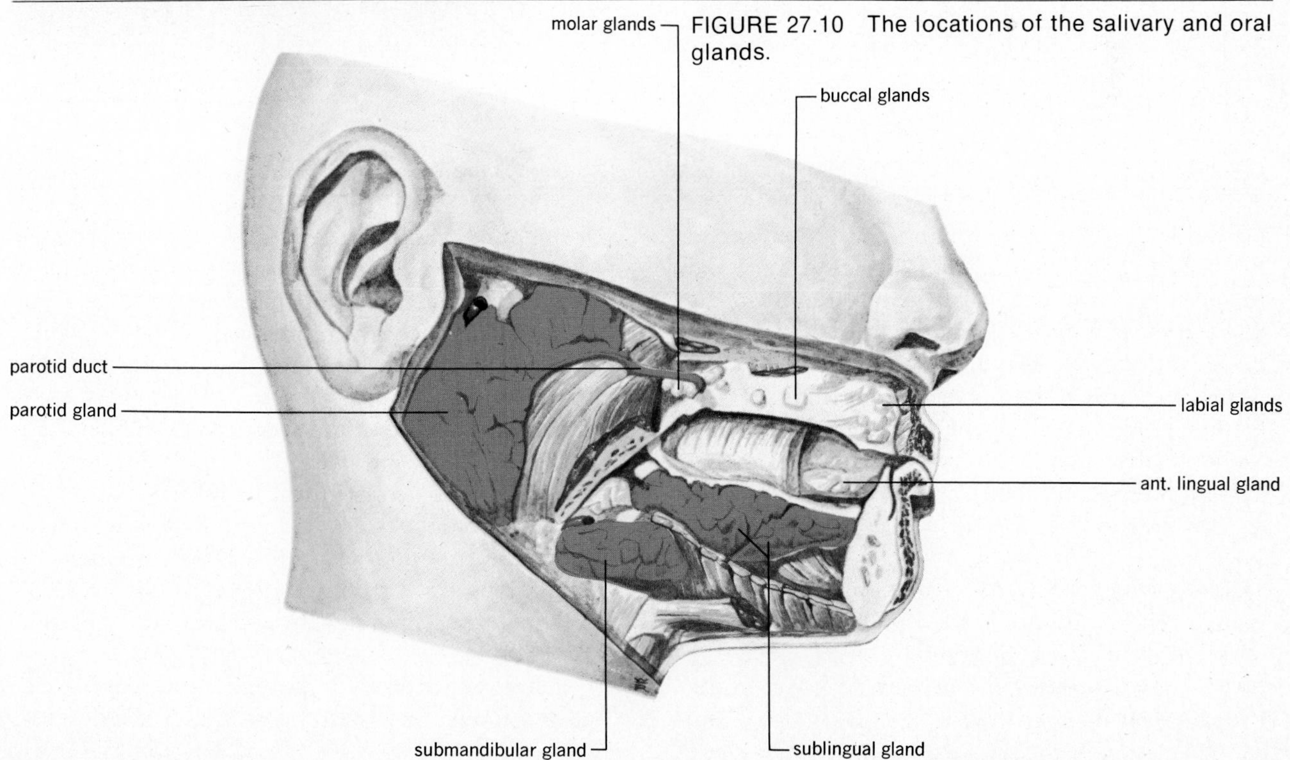

FIGURE 27.10 The locations of the salivary and oral glands.

TABLE 27.3 The salivary glands

Name of gland	Location	Cellular composition	Name of duct	Entry of duct into mouth	Secretion contains
Parotid	Side of mandible in front of ear	All serous	Stensen's	Lateral to upper second molar	Water, salts, enzyme
Submandibular	Beneath the base of the tongue	Mostly serous, some mucous	Wharton's	Papillum lateral to frenulum	Water, salts, enzyme, some mucus
Sublingual	Anterior to submandibular under tongue	Mostly mucous, some serous	Rivinus'	With duct of submandibular	Mostly mucus, a little water, salt, and enzyme

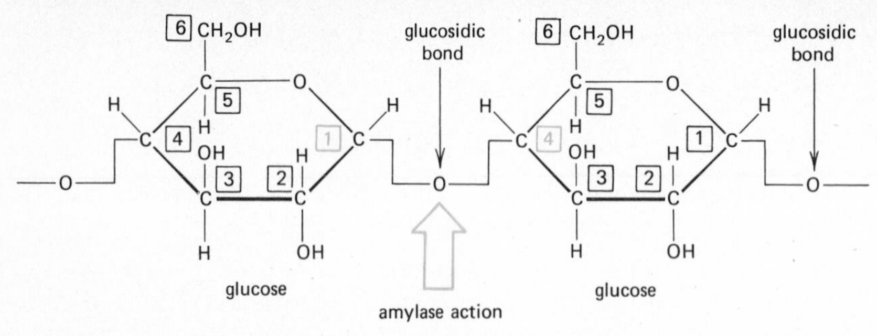

FIGURE 27.11 The glucosidic bonds that link glucose molecules into more complex carbohydrates.

posterior surface of the tongue and are arranged in the form of a V. TASTE BUDS (Fig. 27.9) are found in the fungiform and vallate papillae and on the walls of the soft palate. Emptying into the trenches of the vallate papillae are the ducts of the von Ebner's glands. The secretory portions of the glands lie between the muscle bundles of the tongue. These glands produce a watery fluid that aids in moistening foods and in dissolving them so that they may be tasted.

THE SALIVARY GLANDS. Three pairs of salivary glands (Fig. 27.10) are located outside the mouth and empty their secretions by ducts into the mouth. Two varieties of cells may be found in the glands, SEROUS CELLS and MUCOUS CELLS. The former are small, very granular cells that secrete a watery fluid rich in salivary amylase. The latter are larger, pale-staining cells that produce a viscous mucus lacking in amylase. Table 27.3 summarizes the important facts concerning the salivary glands.

It can be seen that the secretion of each gland is different. The composite fluid of all three sets of salivary glands constitutes the saliva.

Saliva. Saliva is an acidic fluid (pH 6.35–6.85) produced in amounts up to 2500 ml per day. An average figure for daily volume would be 1000–1500 ml. The saliva has the composition shown in Table 27.4.

Saliva functions to MOISTEN and SOFTEN ingested foods, to LUBRICATE foods for swallowing,

TABLE 27.4 Some constituents and characteristics of adult human saliva

Constituent or property	Concentration or value
pH	5.8–7.1
Water	994 g/l
Sodium	6–23 meq/l
Potassium	14–41 meq/l
Bicarbonate	2–13 meq/l
Chloride	15–31.5 meq/l
Phosphorus (inorganic) (organic)	81–217 mg/l 0–133 mg/l
Calcium	2.3–5.5 meq/l
Thiocyanate (–SCN)	24–380 mg/l
Magnesium	0.16–1.06 meq/l
Iodine	0.002–0.202 mg/l
Urea	140–750 mg/l
Ammonia	10–120 mg/l
Amino acids (products of bacterial action on proteins)	Variable–21 have been found
Uric acid	5–29 mg/l
Protein (albumin and globulin)	1.4–6.4 g/l
Lysozyme (a bacteriolytic enzyme)	to 0.15 g/l
Amylase (ptyalin)	0.38 mg/ml
Glucose	100–300 mg/l
Cholesterol	25–500 mg/l
Vitamin C	0.58–3.78 mg/l
Mucin	0.8–6.0 g/l

to CLEANSE the mouth and teeth, to moisten the mucous membranes of the mouth, and as a route for the EXCRETION of many materials.

The digestive function of saliva centers around its content of SALIVARY AMYLASE *(ptyalin)*. This enzyme attacks the chemical bonds (glycosidic 1, 4 bonds, Fig. 27.11) between the simple sugar units in cooked starch (carbohydrates), and, by hydrolysis, splits the starch into smaller units known as dextrins. The amylase also has the power, given sufficient time, to break the dextrins into disaccharides. However, foods do not normally remain in the mouth long enough for the disaccharide stage to be reached by more than 3-5 percent of the dextrins.

$$\text{starch} \xrightarrow[\text{}]{\text{salivary amylase}} \text{dextrins (95–97\%)} \xrightarrow{\text{salivary amylase}} \text{disaccharide (3–5\%)}$$

Control of salivary secretion. A continuous production of saliva is necessary for the moistening and cleansing functions it serves. The greatest volume of saliva is produced when food is present in the mouth. The control mechanism for secretion (Fig. 27.12) is a nervous reflex originating in the TASTE BUDS and the walls of the mouth. The buds act as receptors, initiating nerve impulses when dissolved food enters them. Mechanical stimulation of the walls of the mouth causes nerve impulses to be formed in the touch receptors located there. Afferent (incoming) impulses pass over the FACIAL, GLOSSOPHARYNGEAL, and VAGUS NERVES to SALIVARY CENTERS in the brainstem. Efferent (outgoing) impulses pass from the superior salivary nucleus to the parotid gland, and from the inferior salivary nucleus to the submandibular and sublingual glands. The glands act as effectors and secrete saliva. The nature of the secretion, as well as its quantity, can be altered by

FIGURE 27.12 The control of salivary secretion.

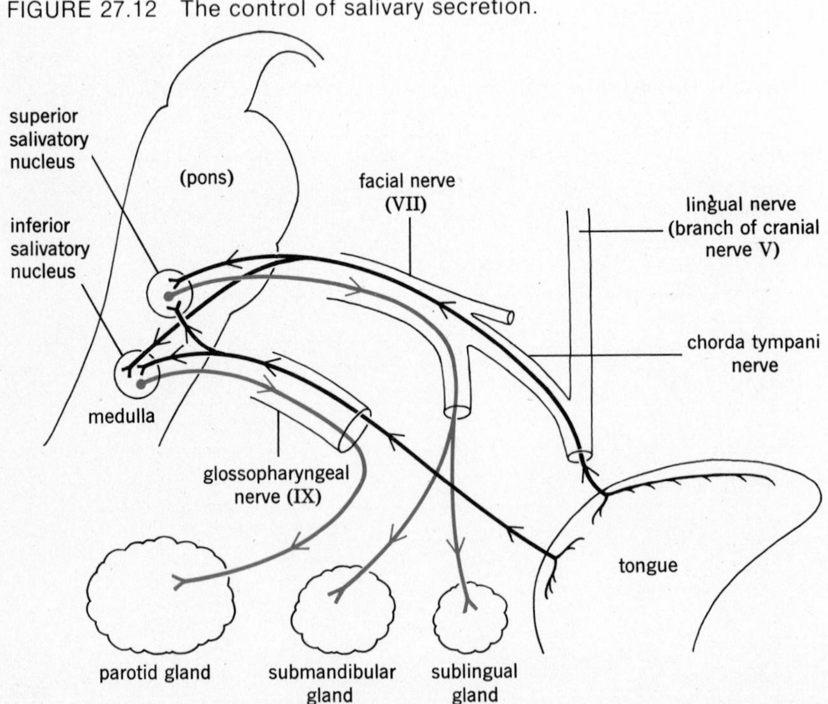

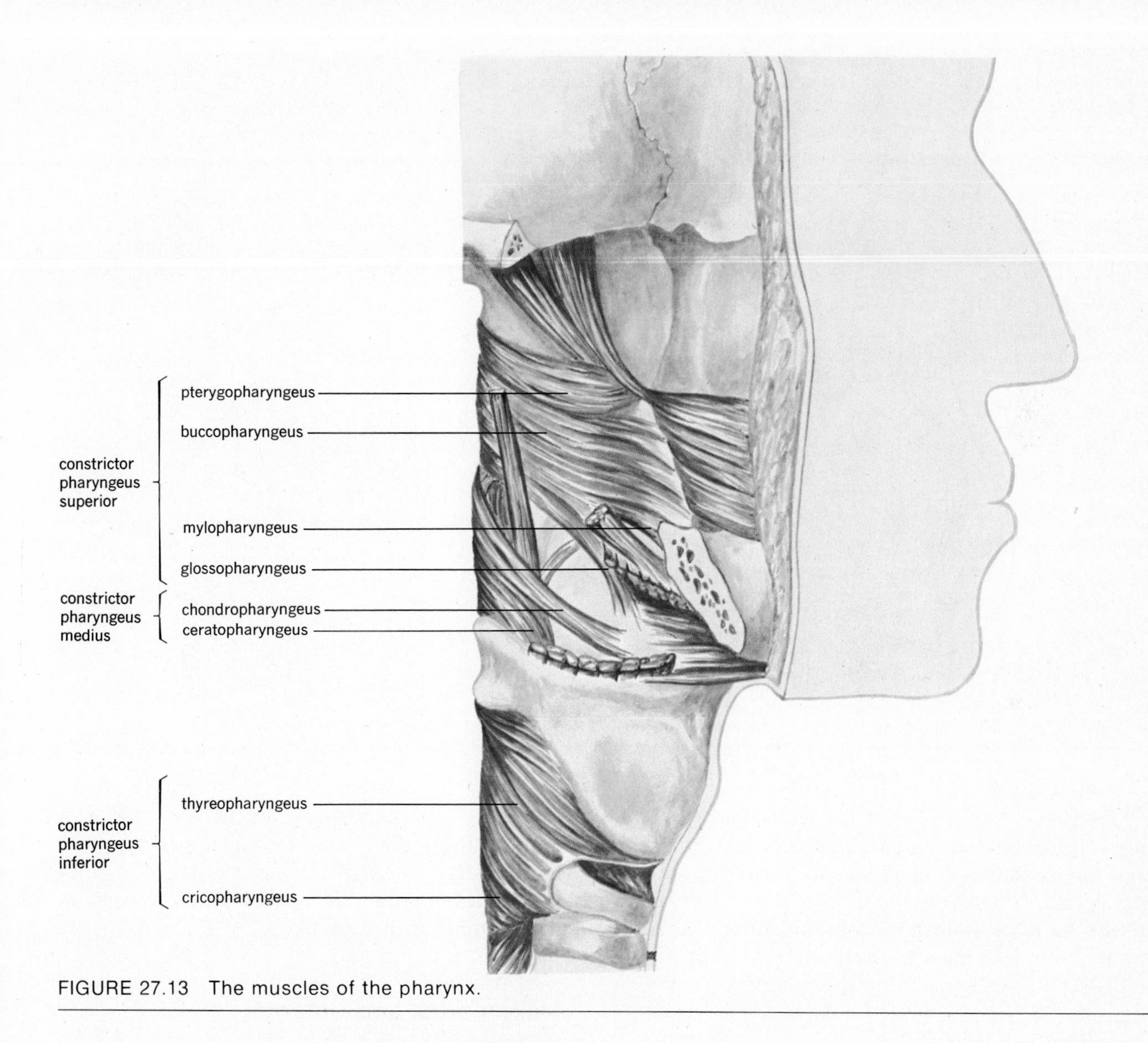

FIGURE 27.13 The muscles of the pharynx.

this mechanism. For instance, soft foods occasion the production of saliva having a smaller volume and lower water content than do dry foods. This mechanism represents the most effective means, but not the only one, by which salivary secretion may be augmented. The sight, sound, and smell of food in preparation, or the mere thought of food, may cause the "mouth to water." Therefore, "higher cerebral influences" exist that may activate the brainstem salivary centers.

The pharynx and esophagus

The pharynx or throat (Fig. 27.13) is a region common to both the respiratory and digestive sys-

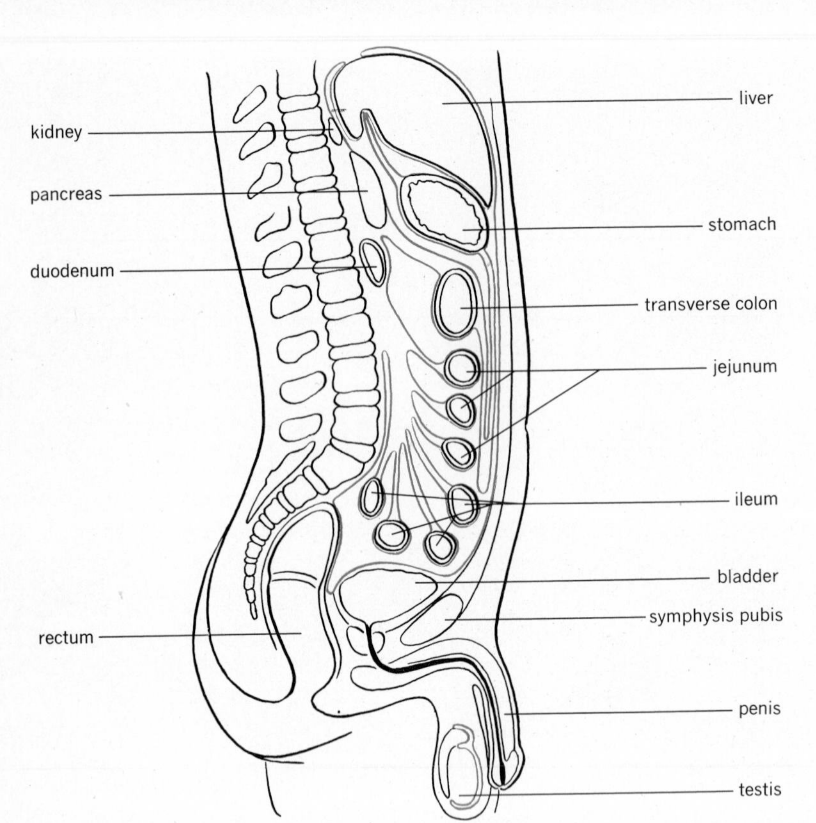

FIGURE 27.14 The arrangement of the mesenteries supporting the abdominal viscera.

tems. It lies behind the nasal cavities, oral cavity, and larynx, and is divided into the corresponding nasal, oral, and laryngeal portions. Three layers of tissue form its wall: a mucous membrane, a fibrous layer, and muscle. The muscle is skeletal in type. The involvement of the pharynx in the digestive process is to act to transfer food from mouth to esophagus.

The esophagus is a muscular tube about 25 cm (10 in.) in length that connects the pharynx to the stomach. The tissue plan follows that stated earlier. The epithelium is stratified squamous. The muscularis is skeletal in the upper portion, mixed skeletal and smooth in the middle portion, and all smooth in the lower portion. The outermost layer is an adventitia except on the lower 1–2 cm, where it is replaced by a serosa. The esophagus conducts food to the stomach and has no digestive function. The remaining organs of the alimentary tract lie below the diaphragm in the abdominopelvic cavity.

Mesenteries and omenta

The abdominopelvic or peritoneal cavity, and some of the organs within it, are lined and covered by serous membranes known generally as the PERITONEUM. The PARIETAL PERITONEUM lines the walls of the cavity proper, while the VISCERAL PERITONEUM covers the organs as part of their serosa. Both membranes consist of a simple squamous mesothelium supported by a submesothelial connective tissue layer.

As the liver, stomach, intestine, and some of the reproductive organs develop, they grow from the tissues lying external to the peritoneum into the cavity, pushing ahead of them the lining membrane (Fig. 27.14). The organs eventually sever their connections with their points of origin and become suspended within the cavity by a double-layered fold of the peritoneum. This double-layered suspending structure is a mesentery. The mesenteries transmit numerous blood vessels to the organs they suspend. Additionally, specific mesenteries are often named according to the organ they suspend; for example, the *mesogastrium* (stomach), *mesocolon* (large intestine), *mesovarium* (ovary).

Between the liver and stomach is a double-layered fold of tissue, the LESSER OMENTUM. Since it does not suspend one organ, it is not a true mesentery. Between the stomach and the first portion of the small intestine is a double-layered fold of tissue, the GREATER OMENTUM. This membrane becomes greatly elongated, folds upon itself, and forms a four-layered curtain overlying the intestines. The greater omentum commonly serves as a

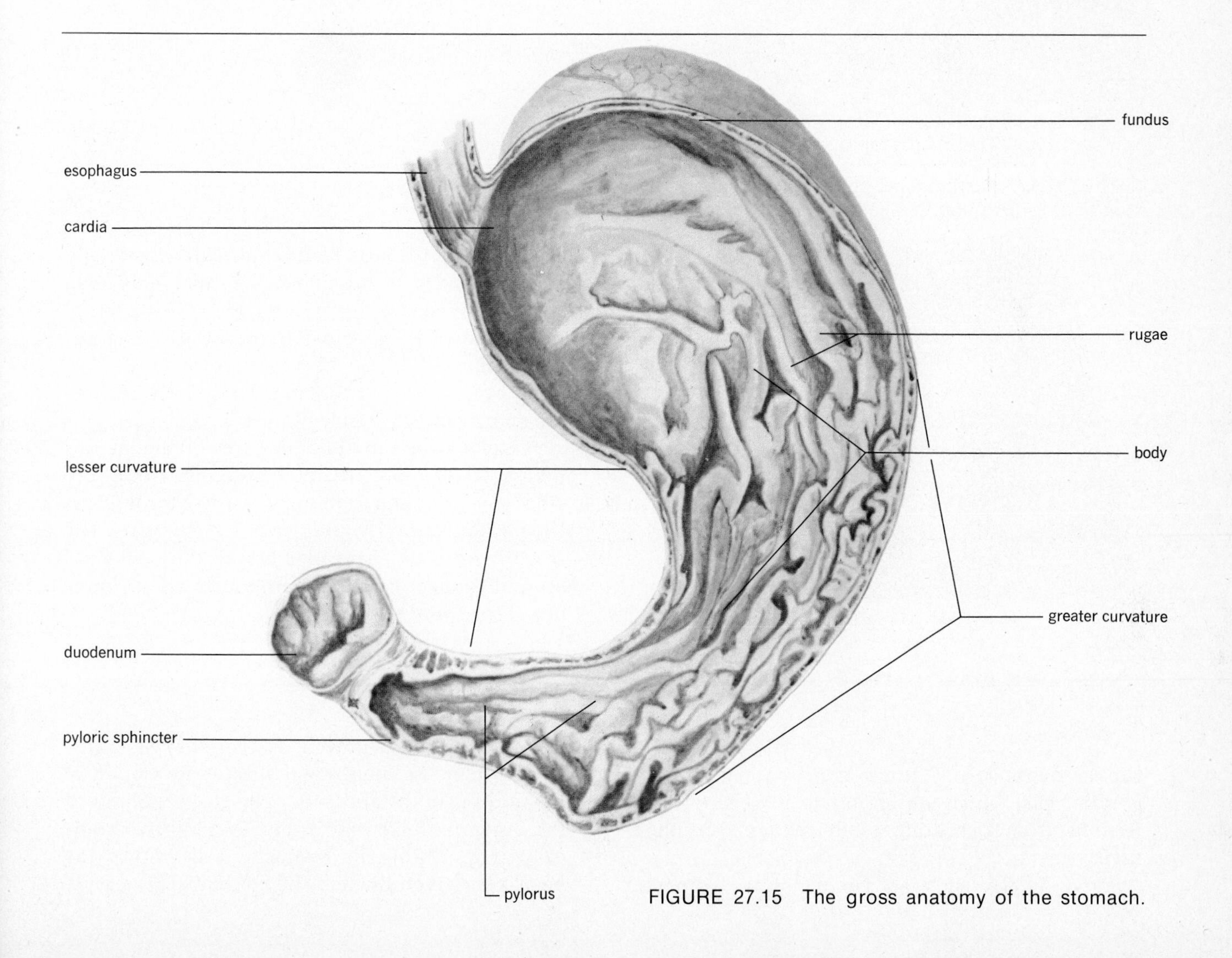

FIGURE 27.15 The gross anatomy of the stomach.

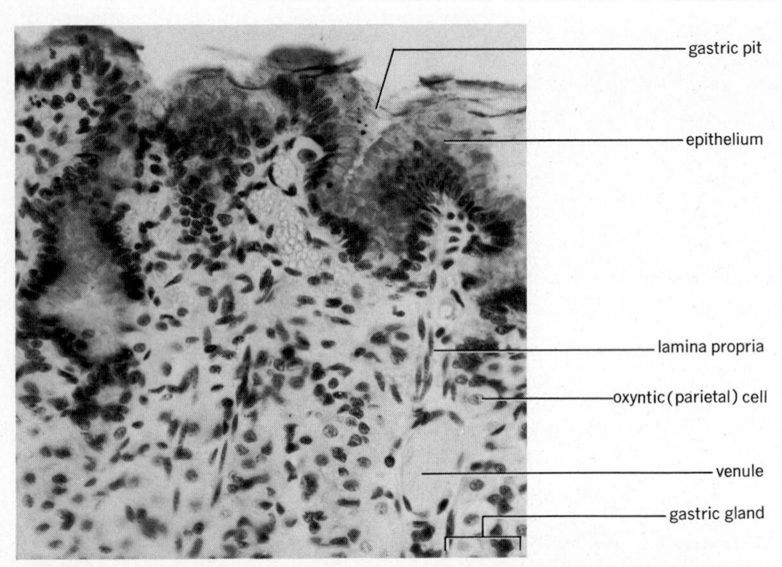

FIGURE 27.16 The microscopic anatomy of the stomach.

FIGURE 27.17 A zymogenic cell of the stomach as it would appear through the electron microscope.

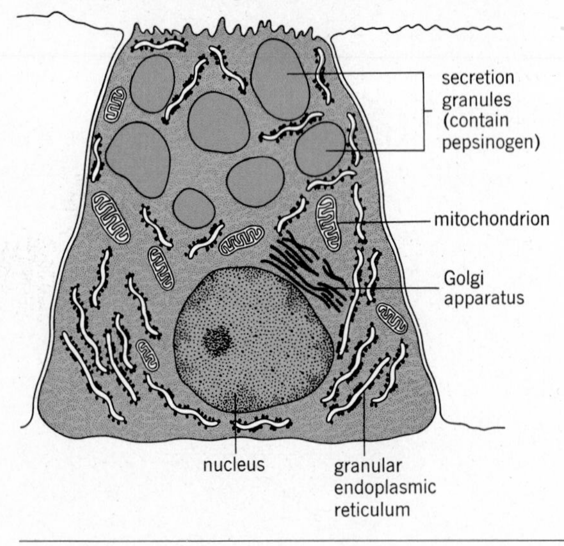

storage depot for fat. It also contains many phagocytic cells, which are important in keeping infections from becoming established within the cavity.

Some organs such as the pancreas, kidney, ureters, and urinary bladder remain essentially against the body wall and are not suspended within the cavity by mesenteries. The parietal peritoneum passes across their surface. Such organs are said to be RETROPERITONEAL—that is, lying behind the peritoneum. These relationships are shown in Fig. 27.14.

PERITONITIS is an inflammation of the peritoneum. It is a potentially dangerous condition, since the inflammation may spread throughout the peritoneal cavity and infect all its organs. The reason for easy spread of infection revolves around the fact that the membranes are continuous throughout the cavity.

The stomach

ANATOMY. The stomach is a J-shaped organ lying under the diaphragm slightly to the left of the midline of the abdomen. The gross anatomy of the organ is illustrated in Fig. 27.15. Microscopically (Fig. 27.16), the stomach wall follows the basic plan given earlier. The epithelium is simple columnar, in which the outer portion of the cells contains mucin (mucus). The mucin provides a continuous covering protecting the interior of the organ from auto- (self-) digestion. Extending into the lamina propria from the *foveolae* or gastric pits are gastric glands. These are estimated to number 35 million and are lined with four cell types:

MUCOUS NECK CELLS. These occur in the upper or neck portion of the gland and are cuboidal cells with flattened or crescent-shaped nuclei. They produce mucus and intrinsic factor.

ZYMOGENIC *(chief)* CELLS (Fig. 27.17). These cells are cuboidal granular units that produce pepsinogen, an inactive form of the enzyme pepsin. The cells also produce rennin and gastric lipase, although in much smaller quantities than pepsin.

OXYNTIC *(parietal)* CELLS (Fig. 27.18). Oxyntic cells produce hydrochloric acid by the reactions shown in Fig. 27.19 and are distinguished by being spherical units with centrally placed nuclei.

ARGENTAFFIN CELLS. Scattered among the other cells, argentaffin cells have basal acidophilic granules and apical nuclei. Their exact function is unknown, but they have been shown to contain serotonin.

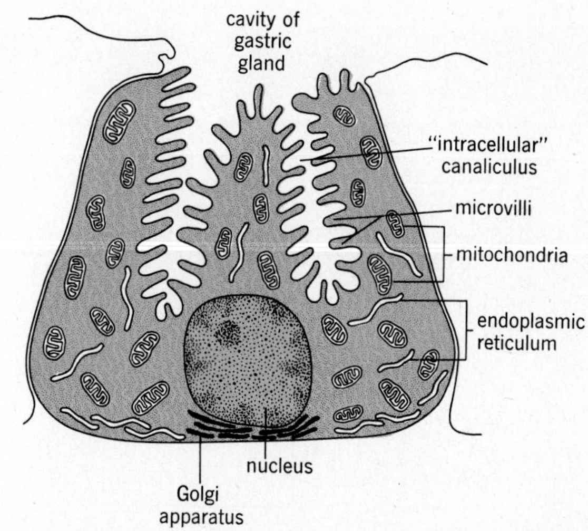

FIGURE 27.18 An oxyntic cell of the stomach as it would appear through the electron microscope.

DIGESTION IN THE STOMACH. The stomach is an organ for STORAGE of the food consumed during a meal. During the time foods are in the stomach, they are subjected to the action of the gastric juice, the constituents of which are shown in Table 27.5. The juice is a water solution of HYDROCHLORIC ACID containing PEPSIN and the enzymes listed below. Pepsin, secreted as pepsinogen, is activated by HCl and breaks proteins into proteoses and peptones (units of 4–12 amino acids).

PEPSIN B^5. Having a molecular weight of 36,000, pepsin B^5 differs structurally from pepsin in the terminal amino acid; it acts more efficiently at pH 3. It is more important in infants, where full HCl production has not been achieved.

GASTRICSIN. Having about one-fourth the activity of pepsin, gastricsin is most active at pH 3.

RENNIN. Coagulating milk protein, and having an optimum pH of 3.7, rennin is more abundant in the stomach of young animals, including humans, where milk is a prominent constituent of the diet.

LIPASE. Small quantities of easily destroyed, fat-digesting lipase are found in the stomach. It exerts little or no digestive action on fats in the adult stomach; digestion of butterfat in the infant stomach occurs by gastric lipase.

AMYLASE. Found in the stomach, but originating from the salivary glands, amylase may continue to digest starch for some 15–30 minutes after food is swallowed. If a single bolus (mass of masticated food) is swallowed, gastric juice quickly penetrates the food and stops amylase action. In the full stomach, gastric juice penetration of the foods is much slower, and 70 percent digestion of starch to disaccharides may occur.

PROTECTION OF THE GASTRIC MUCOSA. Enzymes and acid could lead to autodigestion of the stomach if it were not for the production of mucus by mucous neck and epithelial cells of the stomach wall. Mucus is alkaline in reaction and coats the mucosa, aiding in neutralization of acid at the

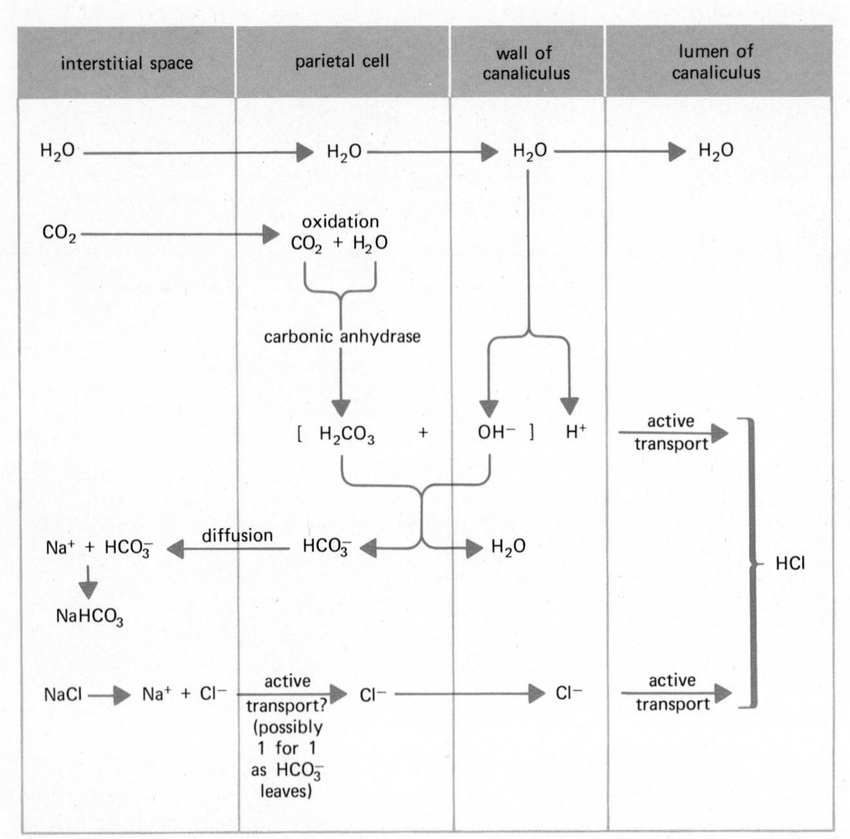

FIGURE 27.19 The steps in the production of HCl by the gastric oxyntic cell.

cell surface. An antipepsin also exists that can neutralize peptic activity.

OTHER FUNCTIONS OF THE STOMACH. Limited ABSORPTION of materials of small molecular weight (water, salts, alcohol) occurs from the stomach. The stomach also produces the INTRINSIC FACTOR that binds vitamin B_{12} (the extrinsic factor) and renders it more absorbable in the intestine. Vitamin B_{12} is essential to the formation and maturation of erythrocytes, and failure to absorb it will result in the development of pernicious anemia.

CONTROL OF GASTRIC SECRETION. Secretion of gastric juice occurs even during fasting and is augmented when food is ingested. Secretion is controlled in three ways or phases:

CEPHALIC *(psychic)* PHASE. Food in the mouth enters taste buds, and afferent impulses pass to the brainstem and from the brainstem to the stomach via the vagus nerve. The stomach responds directly to the vagal stimulation by the secretion of 50–150 ml of gastric juice. Vagal stimulation also releases gastrin (see below).

GASTRIC PHASE. This phase is primarily under hormonal control and occurs when swallowed food contacts the gastric juice produced during the cephalic phase. Proteins in the food are broken down into proteoses and peptones by the action of pepsin. The

TABLE 27.5 Normal constituents and properties of adult gastric juice

Constituent or property	Concentration or value
Fasting volume	50 ml
Secretion rate	74 ml/hr (fasting) 101 ml/hr (after a meal)
Water	994–995 g/l
pH	1.9–2.6
Bicarbonate	none
Chloride	75–150 meq/l
Phosphorus	Av. 70 mg/l
Potassium	6.4–16 meq/l
Sodium	19–69 meq/l
Calcium	2.0–4.8 meq/l
Magnesium	0.3–3.0 meq/l
Total nitrogen (amino acids, ammonia, urea, uric acids, creatinine)	752 mg/l
Proteins	2.8 g/l
Mucin	0.6–15.0 g/l
Carbohydrates (chiefly hexoses)	321 mg/l
Enzymes (pepsin)	28 KU/24 hr[a]

[a] KU—One unit (U) is the amount of any enzyme that will catalyze the transformation of 1 micromole (μM) of substrate per minute at standard conditions. A kilounit (KU) is 1000 times as large, that is, it will catalyze 1 millimole (mM) under the same conditions.

proteoses and peptones act as SECRETAGOGUES (secretory stimulators) and cause cells of the stomach mucosa to produce the gastric hormone called gastrin. Gastrin enters the blood vessels leaving the stomach and is distributed to the entire body including the stomach. When it reaches the stomach, gastrin stimulates the production of 600 ml or more of gastric juice over a period of three to four hours.

INTESTINAL PHASE. Food placed directly into the small intestine in experimental animals has been shown to cause gastric secretion. Secretion has been attributed to the production of a HORMONAL MATERIAL. The nature of the hormone is unknown. It is characterized as being "gastrinlike" and may be secretin. Some investigators contend that no hormone is involved and that the chemical is an absorbed product of digestion acting as a secretagogue.

Inhibition of gastic secretion occurs in several ways. The digesting food mass tends to neutralize the acids of the gastric juice, and the pH rises. Increase of pH results in a decrease of pepsin action, and the gastric phase is essentially diminished. Fats, entering the small intestine from the stomach, cause the production of the hormone gastrone (enterogastrone). This is absorbed into the blood vessels of the intestine and ultimately reaches the stomach, where it inhibits gastric juice secretion and simultaneously decreases gastric motility. The *enterogastric reflex* also slows gastric secretion and motility, thus delaying passage of food to the small intestine. This reflex is triggered when food or irritants enter the duodenum and cause nerve impulses to be carried over sympathetic and vagus nerves to the stomach.

EMPTYING THE STOMACH. The speed of emptying of the stomach contents (chyme) into the small intestine requires three to four hours and is determined by several factors:

THE TONICITY OF THE GASTRIC CONTENTS. Emptying is retarded if contents are more or less concentrated than the plasma. This suggests the presence of osmoreceptors in the stomach wall.

FAT CONTENT OF THE MEAL, if elevated, increases emptying time, presumably by the enterogastrone mechanism that inhibits not only gastric secretion but also motility.

DECREASED VAGAL ACTIVITY reduces motility and secretion, increasing emptying time.

EMOTIONAL STATE. Short-time stress increases emptying time; chronic stress decreases it. Chemicals produced during emotional stress stimulate or inhibit gastric motility and thus influence emptying time. The mechanism of emptying of the stomach depends on peristaltic activity of the antrum, as determined by the factors listed above.

ABSORPTION FROM THE STOMACH. The stomach wall is rather impermeable to the passage of materials. The molecules resulting from digestion are still quite large. Therefore, absorption from the stomach is limited, confined to the passage of water, salts, and small molecules such as alcohol.

Small intestine

ANATOMY. The small intestine extends from the pyloric valve of the stomach some 3.3–4 m (10–12 ft*) to the ileocecal valve of the colon. It is divisible into three parts: the DUODENUM, the first 25–30 cm (10–12 in.); the JEJUNUM, the next 1–1.5 m (3–4 ft); and the ILEUM, the remaining 2–2.5 m (6–7 ft). All parts have a typical and similar structure (Fig. 27.20). The divisions between the parts can be appreciated only microscopically.

Circular folds of the submucosa *(plicae)*, finger-like projections of the mucosa *(villi)*, and mucosal glands *(crypts of Lieberkühn)* are characteristic features of the entire intestine. Plicae and villi provide for increased surface area for absorption and secretion. These two devices are estimated to result in about 10 m^2 of surface area. The surfaces of the epithelial cells lining the intestine are provided with *microvilli*. These are tiny cytoplasmic projections from the free surface of the cell, and they further increase the absorptive surface. The crypts (of Lieberkühn) produce the digestive enzymes that complete food digestion in the small intestine and that secrete mucus in the colon. The three sections of the intestine may be distinguished microscopically by:

The presence of submucosal mucous glands (Brunner's) in the duodenum.

The highest plicae and thinnest wall in the jejunum.

The presence of aggregated lymphoid nodules (Peyer's patches) in the submucosa of the ileum.

DIGESTION IN THE SMALL INTESTINE. The small intestine receives not only the secretions of its

* The small intestine is this long in the living human. Postmortem measurements may be 7 m (20–21 ft).

FIGURE 27.20 The gross anatomy, blood and lymph supply, and microscopic anatomy of the small intestine.

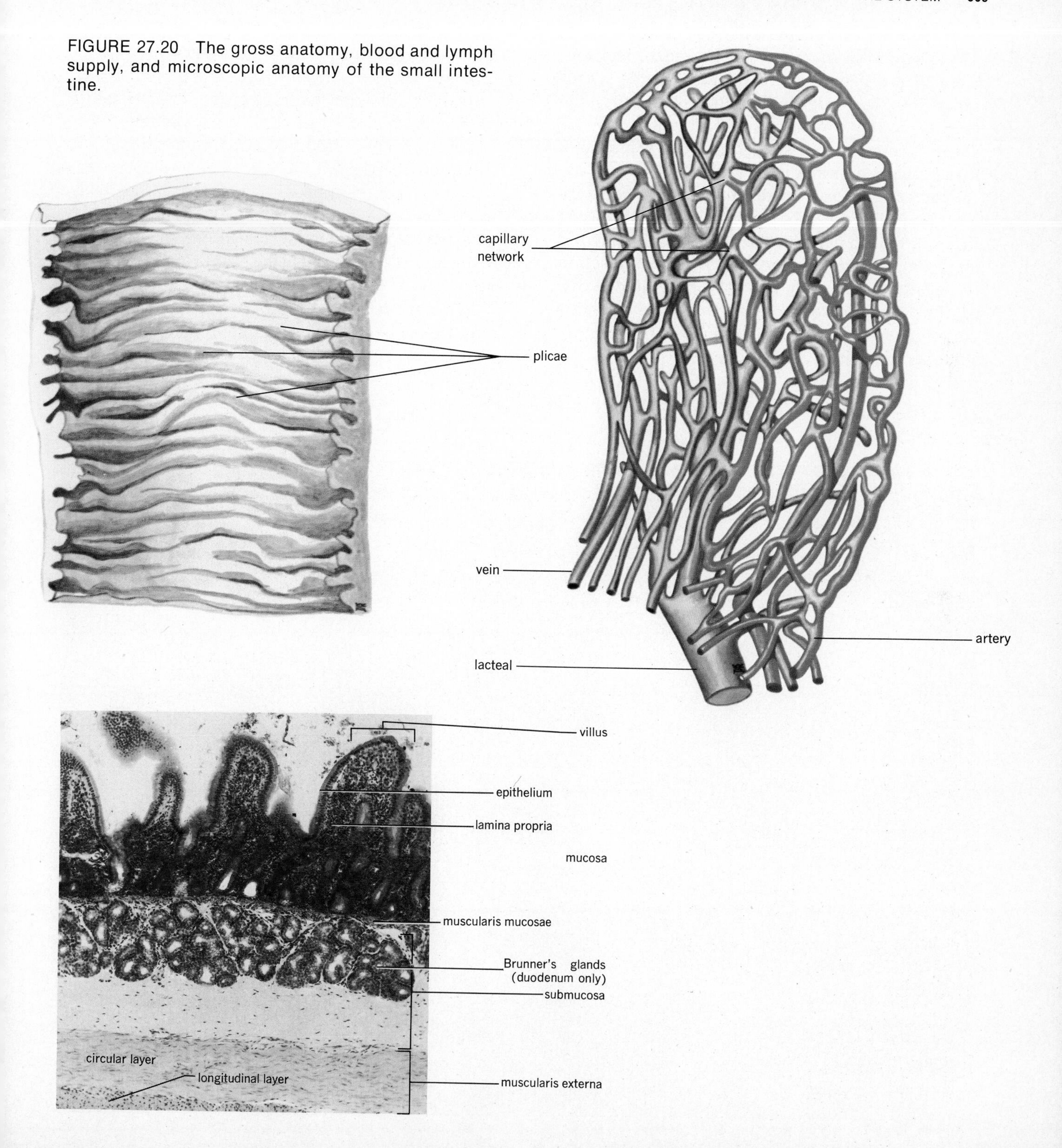

own glands *(succus entericus)* but also the secretions of the pancreas and liver. From the standpoint of logical order in the digestive process, the secretions of the pancreas and liver will be considered here and a discussion of the anatomy of these organs deferred until later.

Pancreatic juice. Pancreatic juice (Table 27.6) is an alkaline fluid (pH 7.1–8.2) that creates the proper environment for the action of all enzymes operating in the intestine, and it stops the action of pepsin. The alkaline reaction of the fluid is due to the bicarbonate content. The juice contains enzymes active on all three foodstuff groups—proteins, carbohydrates, and fats.

Protein and nucleic acid enzymes

TRYPSIN, secreted as trypsinogen, is activated by enterokinase and trypsin itself (autocatalytic).

CHYMOTRYPSIN, secreted as chymotrypsinogen, is activated by trypsin.

CARBOXYPOLYPEPTIDASE, secreted as procarboxypolypepidase, is also activated by trypsin.

DEOXYRIBONUCLEASE, secreted in active form, digests the DNA present in the cells of the foods eaten.

The interrelationships between these enzymes and their effects may be diagrammed as follows:

TABLE 27.6 Some properties and constituents of pancreatic juice

Constituent or property	Concentration or value
pH	7.5–8.8
Secretion rate	6–36 ml/hr (700–2500 ml/day)
Water	987 g/l
Bicarbonate[a]	25–150 meq/l (amount proportional to secretory rate)
Chloride[a]	4–129 meq/l
Potassium	6–9 meq/l
Sodium	139–143 meq/l
Calcium	2.2–4.6 meq/l
Total nitrogen	0.76–0.98 g/l
Glucose	85–180 mg/l
Enzymes (amount/min after stimulation by pancreozymin)	
Amylase	0.29–1.30 mg/min
Carboxypeptidase A	0.36–1.45 mg/min
Chymotrypsin	1.22–7.6 mg/min
Trypsin	0.38–1.42 mg/min
Lipase	0.78–3.50 KU/min

[a] Sum of HCO_3^- and Cl^- concentrations is constant at about 154 meq/kg of water.

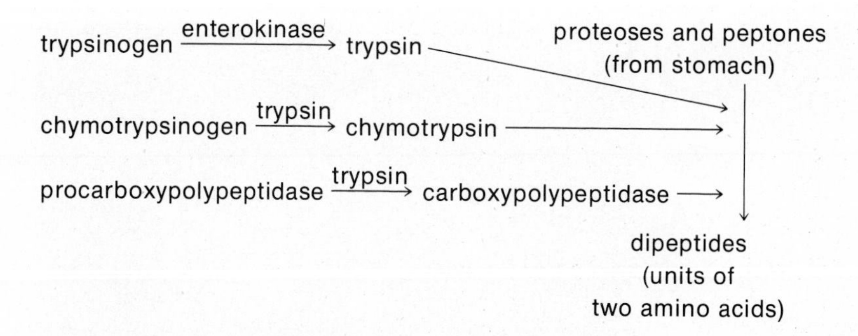

Carbohydrate enzyme. PANCREATIC AMYLASE *(amylopsin)* attacks the dextrins resulting from salivary digestion and converts them to disaccharides.

$$\text{dextrins} \xrightarrow{\text{pancreatic amylase}} \text{disaccharides} \begin{cases} \text{maltose} \\ \text{lactose} \\ \text{sucrose} \end{cases}$$

Lipid enzyme. PANCREATIC LIPASE *(steapsin)* splits triglyceride into its component fatty acids and glycerol in one step.

INTESTINAL JUICE. The intestinal juice produced by the intestinal glands (crypts of Lieberkühn) completes the digestion of the proteins and carbohydrates.

Proteolytic enzymes. EREPSIN, a generic name for a group of enzymes including aminopeptidase and dipeptidase, splits remaining dipeptides into amino acids.

Specific carbohydrases. Specific carbohydrases attack their respective disaccharides and convert them to simple sugars.

$$\text{maltose} \xrightarrow{\text{maltase}} \text{2 glucose molecules}$$

$$\text{sucrose} \xrightarrow{\text{sucrase}} \text{glucose and fructose}$$

$$\text{lactose} \xrightarrow{\text{lactase}} \text{glucose and galactose}$$

CONTROL OF SECRETION OF PANCREAS AND SMALL INTESTINE. Release of pancreatic juice is primarily under the control of hormonal (chemical) mechanisms. The most effective stimulant to pancreatic secretion is the presence in the duodenum of HCl from the stomach. Other materials including water, meat juices, bread, fats, soaps, and alcohol stimulate pancreatic secretion. All these substances cause the intestinal mucosa to produce a material called SECRETIN. This enters the bloodstream and upon reaching the pancreas causes the secretion into the pancreatic duct of a buffered watery solution deficient in enzymes. Fats are the most effective stimulant to the production of another hormone, PANCREOZYMIN (PZ). This is produced by cells of the intestinal mucosa and also enters the blood vessels. Upon reaching the pancreas it stimulates the secretion of enzymes into the pancreatic duct. Stimulation of the vagus nerve also causes enzyme secretion by the pancreas. Thus, two chemicals are necessary for total pancreatic secretion, and enzyme secretion alone is stimulated by nerves.

Secretion of intestinal juice may be elicited by a wide variety of mechanical, nervous and chemical stimuli to the wall of the intestine. A hormone designated as ENTEROCRININ is produced when the intestine is stroked, touched, stretched, or flooded with liquid. The hormone enters the circulation and is distributed to the entire small intestine, where it stimulates succus entericus (intestinal juice) production. A nervous influence is also seen in the production of intestinal juice. Parasympathetic stimulation increases, and sympathetic stimulation decreases intestinal secretion.

THE FUNCTION OF THE LIVER IN THE DIGESTIVE PROCESS. The liver produces the bile, which, although not an enzyme, aids pancreatic lipase in the digestion of fats. Its properties and constituents are shown in Table 27.7. Bile coats the small fat droplets and prevents their coalescence into larger

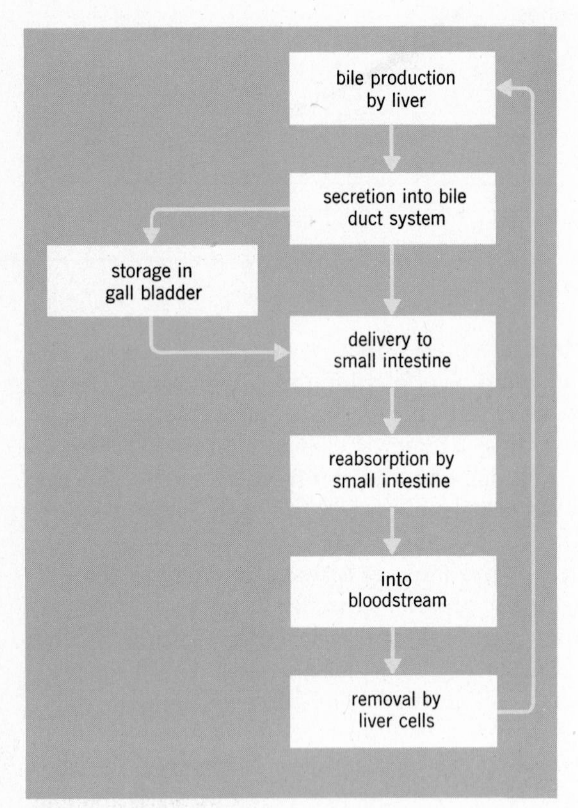

FIGURE 27.21 The circulation of bile in the body.

TABLE 27.7 Some properties and constituents of hepatic bile[a]

Constituent or property	Concentration or value
pH	6.2–8.5
Secretion rate	250–1100 ml/day
Bicarbonate	30 meq/l
Chloride	89–118 meq/l
Potassium	2.6–12 meq/l
Sodium	131–164 meq/l
Calcium	3.3–4.1 meq/l
Iron	0.4–3.1 mg/l
Copper	0.35–2.05 mg/l
Total nitrogen	0.25–1.45 g/l
Urea	236 mg/l
Bilirubin (as glucuronide)	0.12–1.35 g/l
Protein	1.4–2.7 g/l
Cholesterol	0.8–1.8 g/l
Bile acids	6.5–14.0 g/l

[a] In the gall bladder, bile is concentrated by removal of water; thus, constituents may increase in concentration by a factor of 2–10 times.

drops. This enhances the ability of the lipase to digest the drops. This action is termed the EMULSIFYING ACTION of the bile. Fatty acids produced by the action of the lipase are insoluble in the water of the digestive fluid and in this state are difficult to absorb. The bile salts react chemically with the fatty acids and convert them to water-soluble compounds that are much easier to absorb. This action of the bile is termed the HYDROTROPIC ACTION. Although produced continually by the liver, the bile is stored in the gall bladder until required for digestion. Release from the gall bladder is caused by a substance historically designated CHOLECYSTOKININ (CCK). Recently CCK has been shown to be identical to PZ. It causes the contraction of the gall bladder musculature, emptying the stored bile into the small intestine.

Bile undergoes a "circulation," as shown in Fig. 27.21. Table 27.8 summarizes the actions of the various chemicals produced by the organs of digestion.

ABSORPTION FROM THE SMALL INTESTINE. Simple sugars are rapidly absorbed from the intestine at constant rates, even against concentration gradients. This suggests that they are being actively transported. They are not, however,

TABLE 27.8 Chemicals produced by the organs of digestion

Material	Site of production	Stimulus to production	Substrate or organ affected	Effect or end product
Salivary amylase	Salivary glands	Nervous	Cooked starch	Dextrins 95–97%; disaccharides 3–5%
HCl	Stomach, parietal cell	Nervous and chemical	Pepsinogen	Pepsin
Pepsin	Stomach, zymogenic cell (as pepsinogen)	Nervous and chemical	Protein	Proteoses and peptones
Rennin	Stomach, zymogenic cell	Nervous and chemical	Milk protein	Curdles protein
Gastric lipase	Stomach, zymogenic cell	Nervous and chemical	Fats	Fatty acids and glycerol[a]
Enterokinase	Duodenal mucosa	Food entering duodenum	Trypsinogen	Trypsin
Trypsin	Pancreas (as trypsinogen)	Primarily chemical	Proteoses and peptones and inactive pancreatic proteoses	Dipeptides and active proteases
Chymotrypsin	Pancreas (as chymotrypsinogen)	Primarily chemical	Proteoses and peptones	Dipeptides
Carboxypolypeptidase	Pancreas (as procarboxypolypeptidase)	Primarily chemical	Proteoses and peptones	Dipeptides
Pancreatic amylase	Pancreas	Primarily chemical	Dextrins	Disaccharides
Pancreatic lipase	Pancreas	Primarily chemical	Fats	Fatty acids and glycerol[a]
Erepsin	Small intestine	Primarily chemical	Dipeptides	Amino acids[a]
Sucrase	Small intestine	Primarily chemical	Sucrose	Glucose and fructose[a]
Maltase	Small intestine	Primarily chemical	Maltose	2 glucose[a]
Lactase	Small intestine	Primarily chemical	Lactose	Glucose and galactose[a]
Bile	Liver	Chemical	Fats	Emulsification solubilizes fatty acids
Gastrin	Stomach mucosa	Secretagogues	Stomach	Secretion of gastric glands
Unnamed	Duodenal mucosa	Food in duodenum	Stomach	Secretion of gastric glands
Secretin	Duodenal mucosa	HCl in duodenum	Pancreas and liver	Secretion of pancreas stimulate bile production
Pancreozymin	Duodenal mucosa	HCl in duodenum	Pancreas	Secretion of pancreatic enzymes

TABLE 27.8 continued Chemicals produced by the organs of digestion

Material	Site of production	Stimulus to production	Substrate or organ affected	Effect or end product
Enterocrinin	Duodenal mucosa	Stretch or contact in duodenum	Small intestine	Secretion of intestinal glands
Gastrone	Duodenal mucosa	Acid in duodenum	Stomach	Inhibits gastric secretion
Cholecystokinin	Duodenal mucosa	Fats in duodenum	Gall bladder	Expulsion of stored bile
Mucus	All organs	Probably inherent programming in cell	None specific	Lubrication

[a] End products of chemical digestion.

all absorbed at the same rate. Galactose passes most rapidly (77 g/hr), followed by glucose (70 g/hr), fructose (43 g/hr), mannose (19 g/hr), xylose (15 g/hr), and arabinose (9 g/hr). These sugars are actively absorbed into the blood vessels of the villus and are conducted to the liver before passing to the body generally. Disaccharides, probably because there are no carriers to transport them, are absorbed to only a limited extent.

Amino acids are also actively absorbed from the intestinal contents (chyme). To a small extent, dipeptides, proteoses, and other larger units are also absorbed, probably passively. Some undigested protein also apparently passes across the wall, but in extremely small amounts. As with the simple sugars, the amino acids are absorbed into the blood vessels of the villus and are conducted to the liver.

Glycerol, being a water-soluble compound, is easily absorbed passively, chiefly into the lacteals of the villi. Fatty acids as such are water insoluble and are absorbed with difficulty. However, they can and do react with bile acids to form soluble soaps, which are then passively absorbed into the lacteal. Once in the lacteals, glycerol and fatty acid react with one another to form triglycerides again. They are then coated with protein, which prevents the coalescence of the fat droplets formed; as tiny droplets (chylomicrons), they pass through the thoracic duct and empty into the subclavian veins. The fats thus bypass the liver on the first trip. After a meal rich in fats, the blood may actually be turned white from the heavy accumulation of fats. This is called *lipemia.* The chylomicrons are removed from the blood to storage areas or are made ready for metabolism by the clearing factor *(lipoprotein lipase)* of the plasma.

Salts are actively transported through the wall and water follows osmotically. Some pinocytosis may account for a small amount of water absorption. The trace substances also appear to pass actively, especially vitamins.

The large intestine

ANATOMY. The large intestine (Fig. 27.22) is about 1½ m (5 ft) long and is composed of the CECUM, COLON, RECTUM, and ANAL CANAL. It is larger in diameter than the small intestine and is easily recognized by its sacculations or HAUSTRA. The cecum is a large blind pouch below the entrance of the ileum; the VERMIFORM APPENDIX rises from it. The COLON starts with an ASCENDING PORTION passing upward along the right side of the abdominal cavity. Near the liver it bends, forming the RIGHT COLIC *(hepatic)* FLEXURE, and passes across the abdominal cavity as the TRANS-

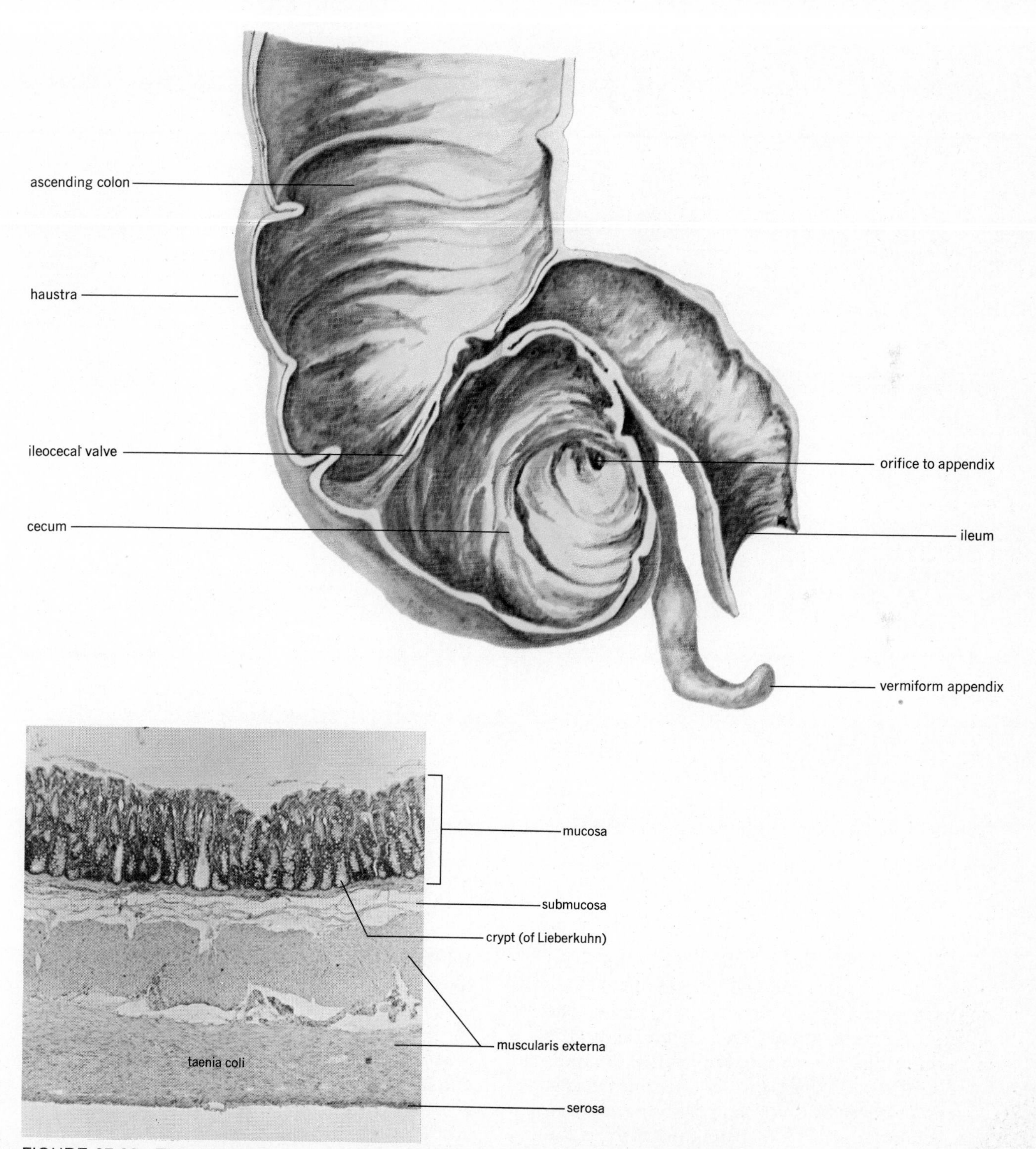

FIGURE 27.22 The gross and microscopic anatomy of the ileocecal region and colon.

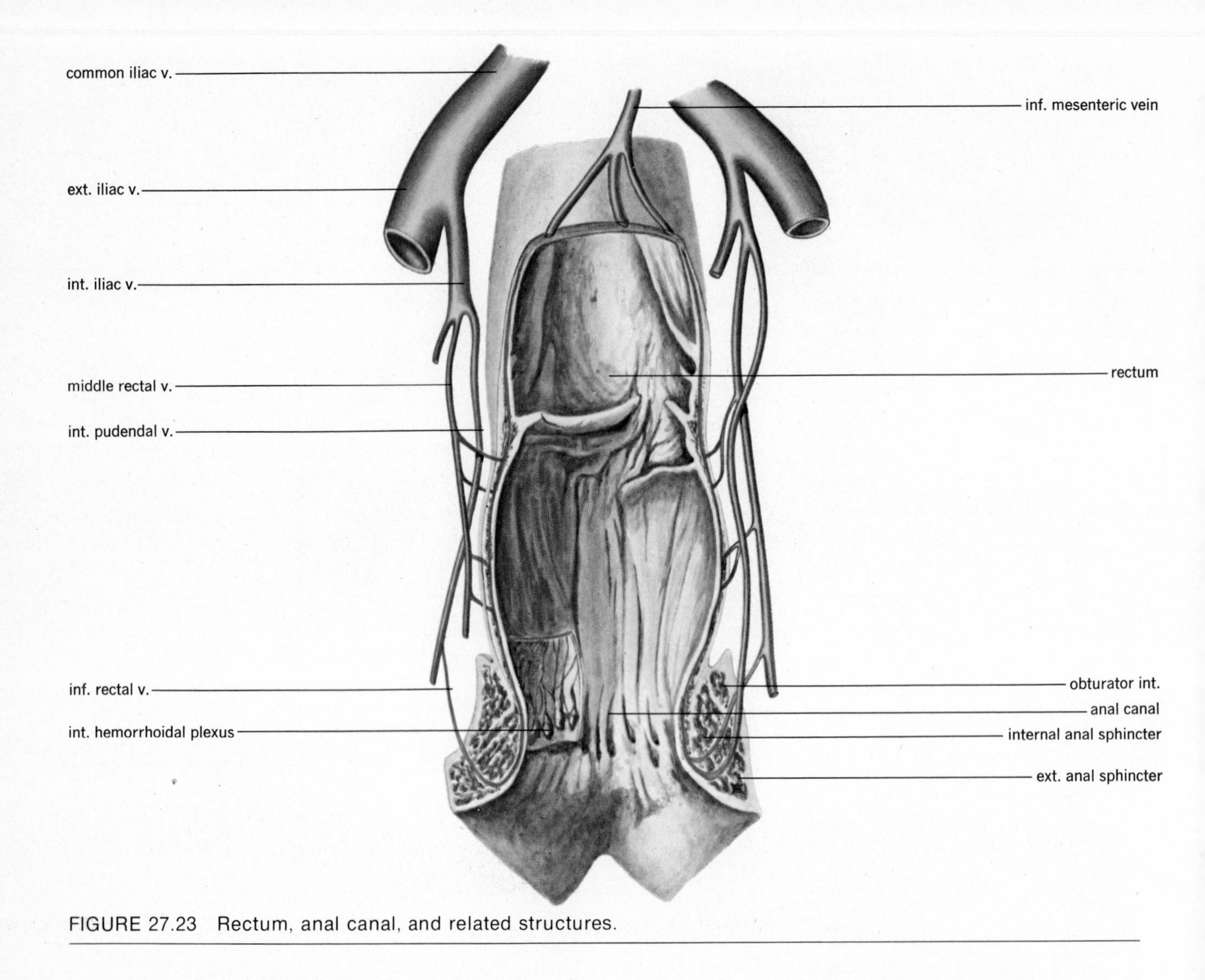

FIGURE 27.23 Rectum, anal canal, and related structures.

VERSE COLON. Near the spleen it again bends, forming the LEFT COLIC *(splenic)* FLEXURE, and passes downward along the left side of the abdominal cavity as the DESCENDING COLON. The SIGMOID COLON begins at the pelvic brim and is an S-shaped bend returning the colon to the midline position. At the level of the third sacral vertebra it becomes continuous with the RECTUM. The rectum (Fig. 27.23) is distinguished from the colon proper by the absence of haustra and by the presence of the RECTAL COLUMNS. The latter are longitudinal folds of tissue in the rectal wall, containing hemorrhoidal arteries and veins. If the veins become enlarged, they may give rise to the condition known as HEMORRHOIDS.

Microscopically, the large intestine shows no villi and a simple columnar epithelium with many goblet cells. The mucus produced by these cells lubricates the drying mass of fecal material as it passes through the colon. The lamina propria

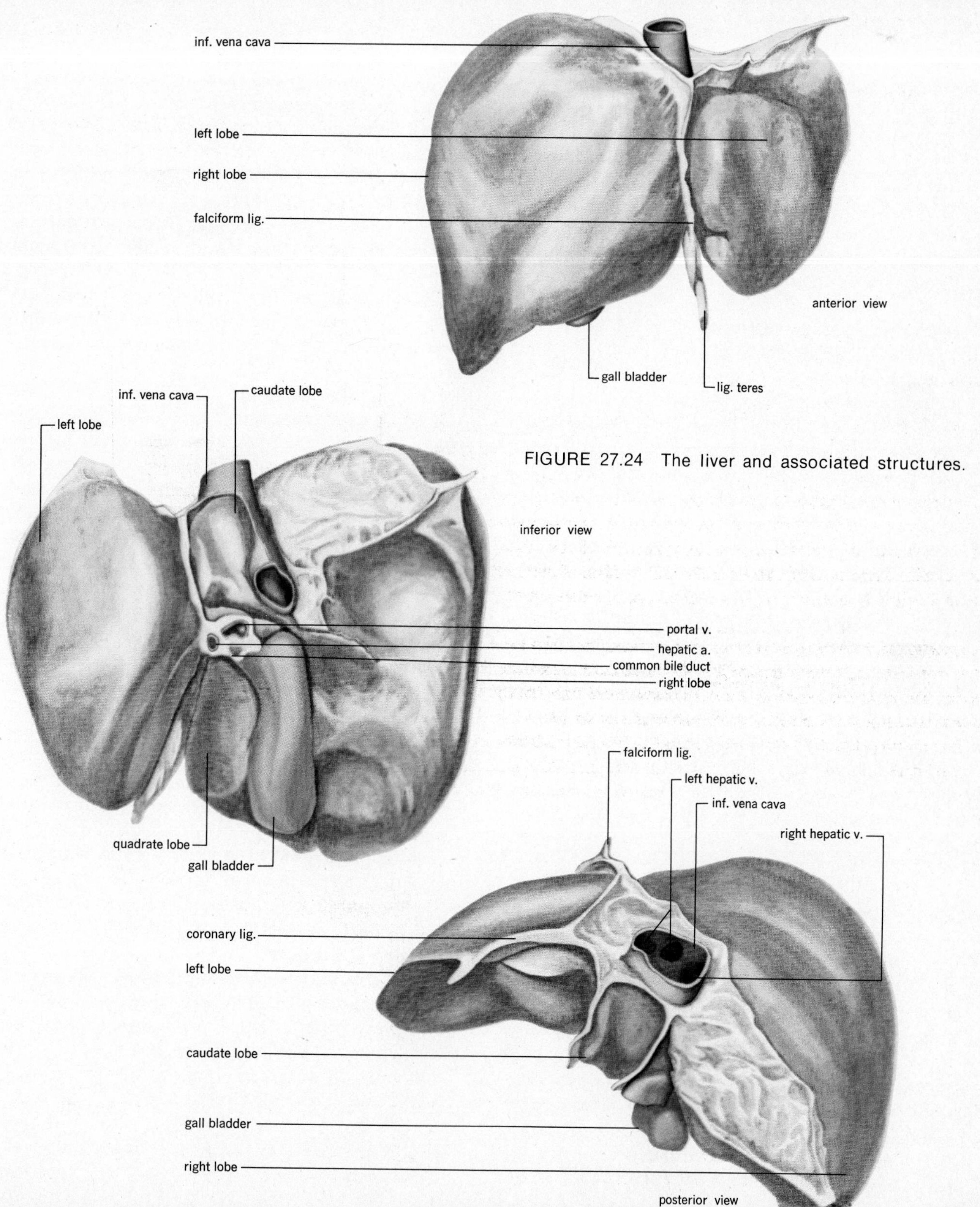

FIGURE 27.24 The liver and associated structures.

contains many long crypts of Lieberkühn. The submucosa is typical, while the muscularis externa has a complete inner circular layer and a discontinuous outer longitudinal layer. The longitudinal layer is in the form of three strips of smooth muscle, the TAENIA COLI. The taenia are shorter than the colon, and to make the colon conform to the taenia length, it must be accordion pleated, hence the haustra.

ABSORPTION IN THE COLON. The chief substance absorbed in the colon is WATER. It has been estimated that in humans, the colon absorbs 300–400 ml of water and about 60 meq of SODIUM per day. The colon also maintains a considerable FLORA OF MICROORGANISMS, and these produce a variety of nutritionally important materials. Vitamins K, B_{12}, thiamin, and riboflavin, and amino acids are synthesized by these organisms and are absorbed by the host. Various gases that contribute to *flatus* are also produced.

FECES FORMATION. Material of no further use to the body is eliminated as feces. Feces consist of approximately 60 percent solid matter, including microorganisms, residues of digestive secretions, and undigested material. The remaining 40 percent is water. Feces are collected in the rectum until sufficient mass has accumulated to cause defecation.

The liver

ANATOMY. The liver (Fig. 27.24) is the largest gland in the body and is located in the right upper quadrant of the abdomen, against the diaphragm. It consists of two main LOBES, a large right one and a smaller left one. Two small lobes, the caudate and quadrate, are associated with the right lobe. Located between the right and left lobes is the FALCIFORM LIGAMENT, which attaches the liver to the anterior abdominal wall and diaphragm. The CORONARY LIGAMENT superiorly and posteriorly, and the right and left TRIANGULAR LIGAMENTS laterally, also fix the liver to the diaphragm. The liver is covered with a FIBROUS CAPSULE, the capsule of Glisson. Associated with the liver is the GALL BLADDER and its ducts. Leaving the liver are two HEPATIC DUCTS, which fuse to form the COMMON HEPATIC DUCT. Joining this duct is the CYSTIC DUCT from the GALL BLADDER. From the junction of the common hepatic and cystic ducts, the COMMON BILE DUCT passes to the small intestine.

The unit of structure of the liver is the LOBULE (Fig. 27.25). These are five- to seven-sided cylindrical units, 1–2½ mm in diameter, that (in the human) are incompletely invested by connective tissue. The center of the lobule is formed by the CENTRAL VEIN, from which radiate interconnecting CORDS OR PLATES OF LIVER CELLS. Between the cords are irregular blood spaces, or SINUSOIDS, lined with endothelial and Kupffer cells. The latter are fixed phagocytes, which are important in the destruction of worn out erythrocytes and in the removal of bacteria reaching the liver from the intestines. Blood reaches the liver through two afferent channels, the hepatic artery and the portal vein. The hepatic artery brings fresh (oxygenated) blood to the organ, while the portal vein, draining the intestines, brings blood rich in amino acids, sugars, vitamins, and other nutrients to the liver for processing and storage. They both empty into the sinusoid. The smaller interlobular branches of these vessels may be seen running together with the INTERLOBULAR BILE DUCT in the loose connective tissue surrounding each lobule. When these three structures are seen together in microscopic sections of the lobule, they are referred to as the portal area. The entire course of these vessels into the liver is known as the PORTAL CANAL.

THE BILE SYSTEM (Figs. 27.25 and 27.26) begins as tiny BILE CAPILLARIES or canaliculi around each and every hepatic cell. These drain into the interlobular duct mentioned above, and ultimately into the small intestine, through the system of extrahepatic ducts already described.

FUNCTIONS OF THE LIVER. The many functions of the liver include:

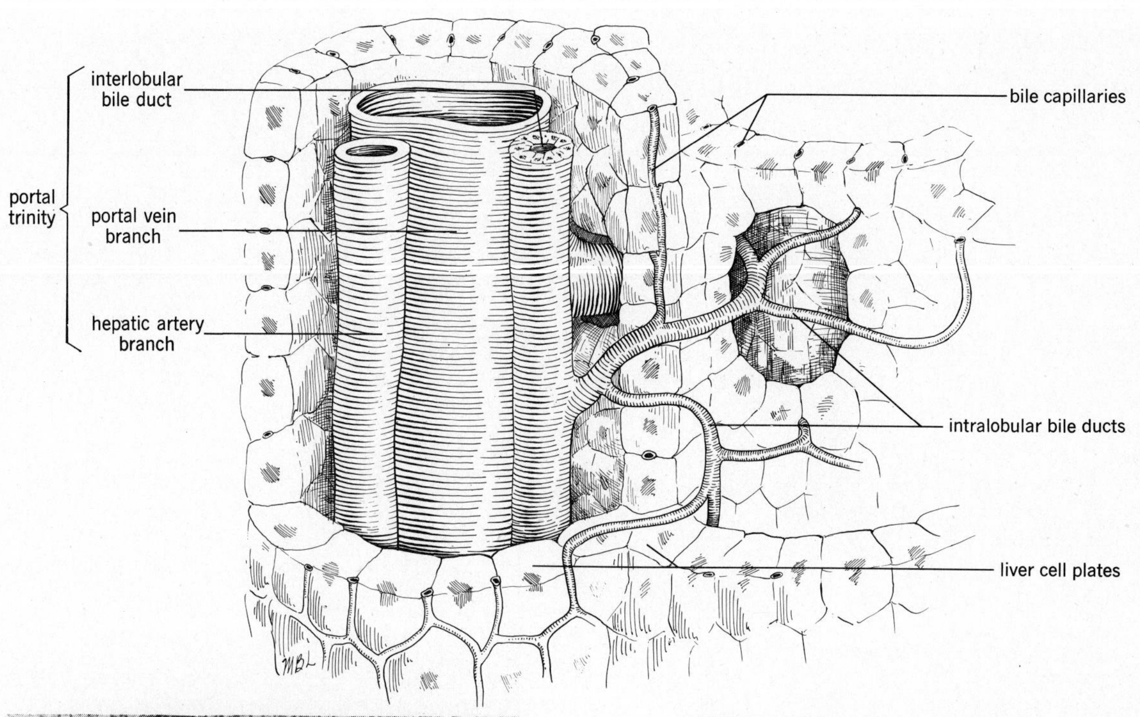

FIGURE 27.25 The microscopic structure of the liver.

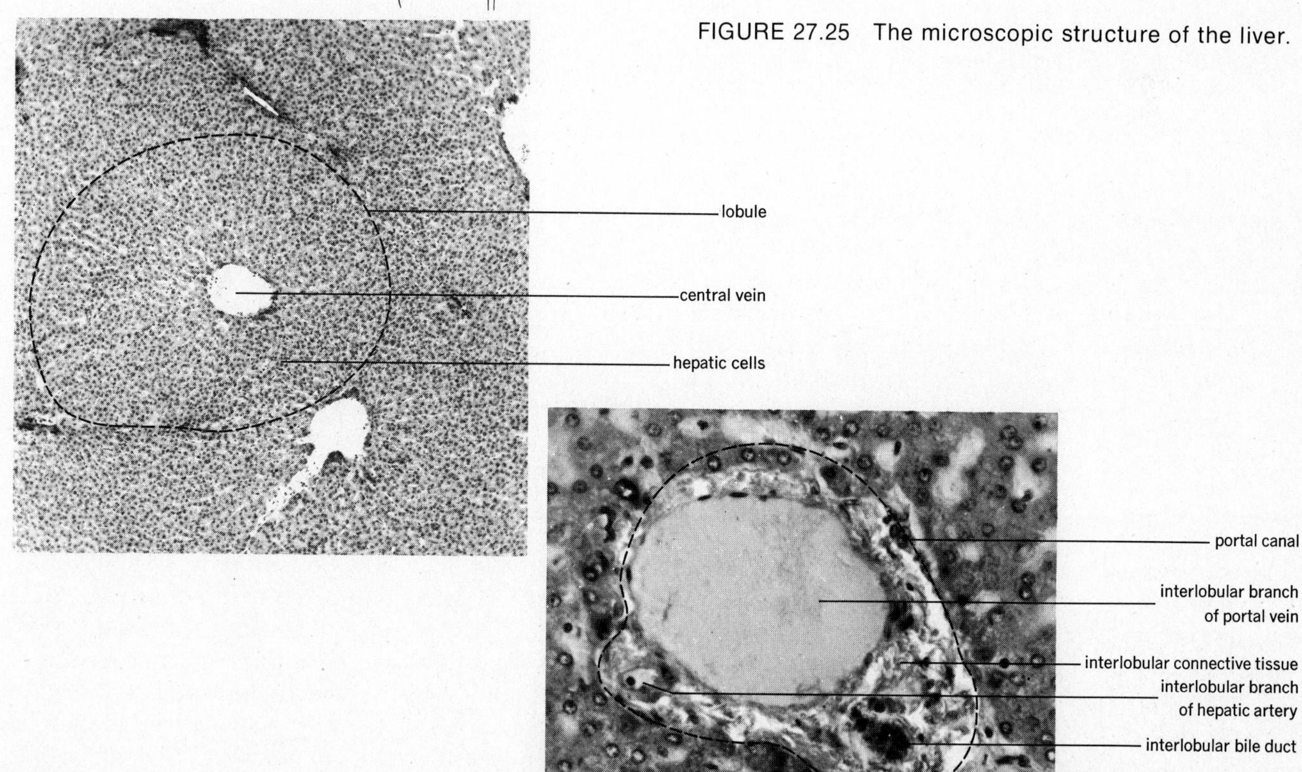

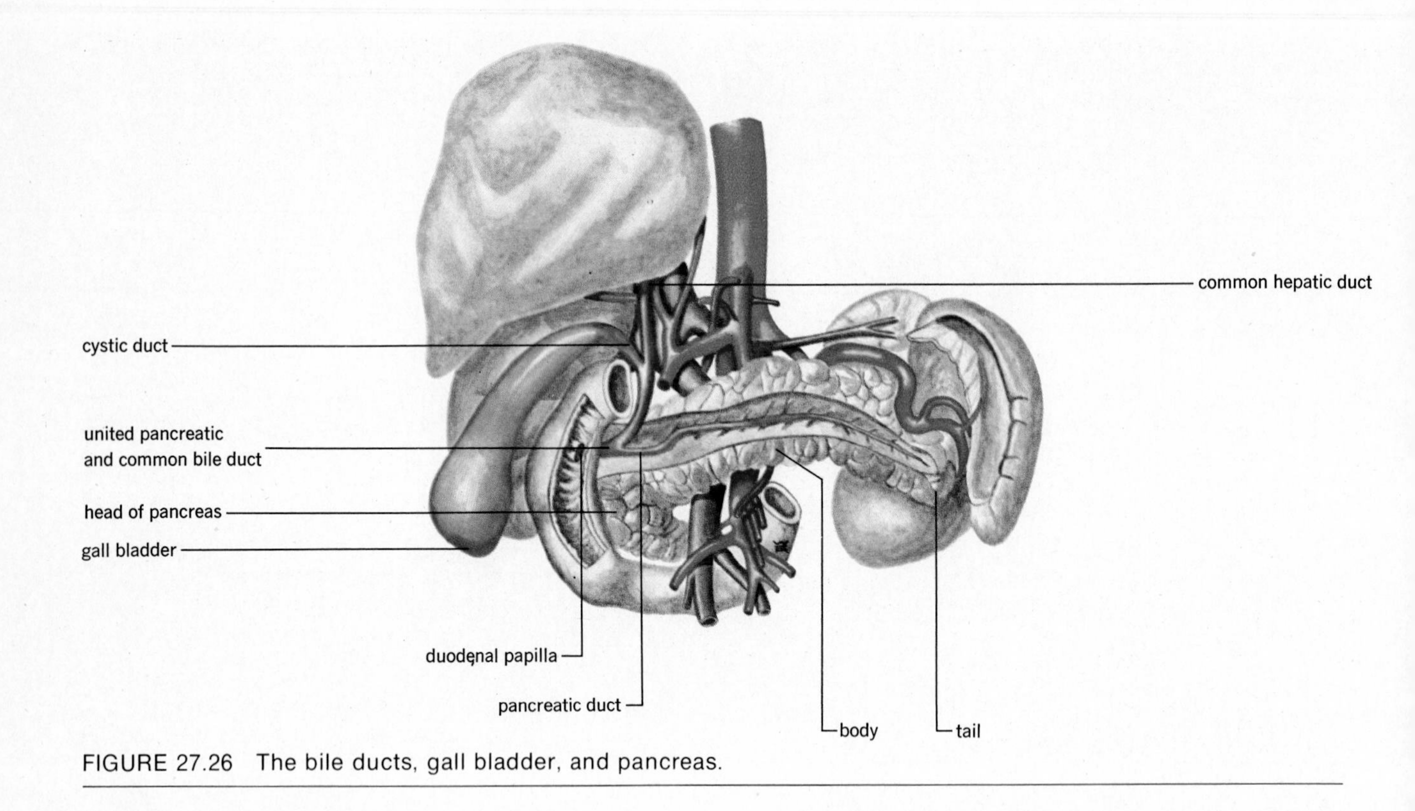

FIGURE 27.26 The bile ducts, gall bladder, and pancreas.

It acts as a BLOOD RESERVOIR and as a means of transferring blood from portal to systemic circulations. A normal volume of about 350 ml of blood is found in the organ. An additional liter may be "stored" in the organ and transferred to the systemic circulation.

FILTERING of the portal blood occurs through the activity of phagocytic reticuloendothelial cells (Kupffer cells) found in the sinusoids. Bacteria, entering the blood through the intestine, are removed by these cells.

PRODUCTION OF BLOOD CELLS is a normal function of the fetal liver and occurs in the adult in certain abnormal states.

DESTRUCTION OF AGED ERYTHROCYTES occurs through the activity of the Kupffer cells.

EXCRETION OF BILE containing bile salts, cholesterol, acids, and various dyes occurs through liver activity.

The liver DETOXIFIES materials that might be harmful to the body by combination with other materials (conjugation, methylation) by oxidation and reduction.

It is involved in METABOLIC REACTIONS including: catabolism of glucose, glycogen, fatty acids, amino acids; synthesis of triglycerides, ATP, glycogen, urea.

It STORES iron, vitamins, and glycogen.

Because of the rather specific functions associated with the liver and the specific enzymes involved, a variety of tests are available to assess normal or pathological conditions of the organ. The basis for the test and the tests themselves are described below.

Bile pigment metabolism produces bilirubin, which is then conjugated with glucuronic acid and excreted into the intestine. Reduction in the intestine produces urobilinogen. Reabsorption of bile by the intestine results in return of bilirubin and/or urobilinogen. Excessive accumulation of bile pigment results in JAUNDICE, which results in yellowish coloration of the tissues. Several causes are described for jaundice.

OBSTRUCTIVE JAUNDICE is caused most commonly by blockage of the common bile duct (Fig. 27.27). It results in excessive passage of bile into the bloodstream.

HEPATIC JAUNDICE is caused by hepatic cell damage due to toxins, viruses, and poisons with release of bile into the circulation.

HEMOLYTIC JAUNDICE results from excessive destruction of erythrocytes, with liberation of excessive amounts of pigment into the bloodstream.

The ICTERUS INDEX test determines the coloration of the plasma in comparison to the color of a solution (1:10,000 dilution) of potassium dichromate. The Ictotest produces a purple color on a test mat in the presence of excessive bilirubin in the urine.

The pancreas

The functions of the pancreas relative to the process of digestion have already been discussed.

ANATOMY. The pancreas (see Fig. 27.26) is a carrot-shaped gland located along the greater curvature of the stomach, between it and the duodenum. It possesses a HEAD, NECK, BODY, and tapering TAIL. It usually empties through a single duct, the PANCREATIC DUCT (of Wirsung), into the small

FIGURE 27.27 The relationship of the bile ducts to the genesis of obstructive jaundice.

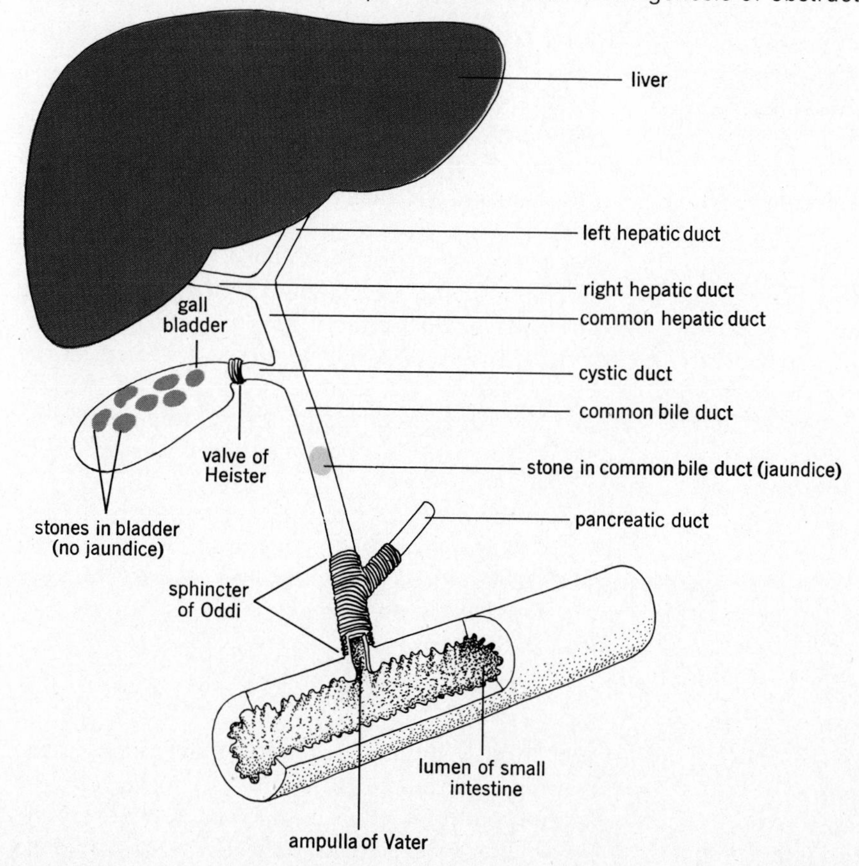

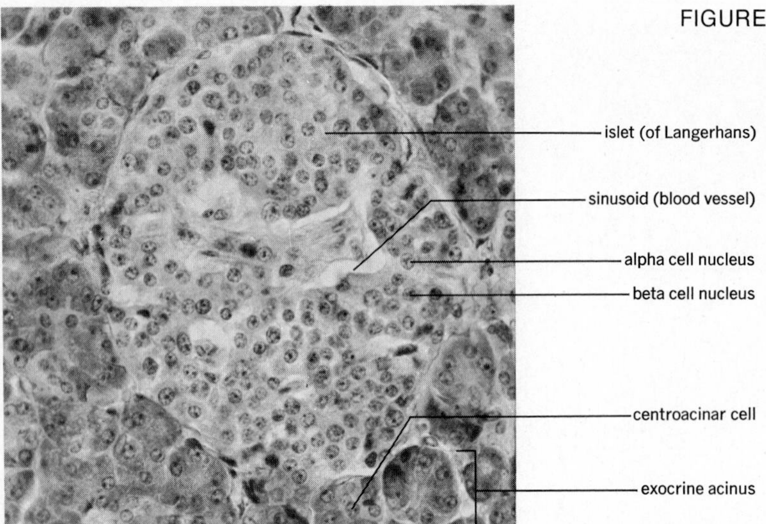

FIGURE 27.28 Microscopic anatomy of the pancreas.

intestine in company with the common bile duct. An accessory duct (of Santorini) may, in some cases, be found emptying separately into the small intestine above the main duct.

The pancreas consists of an EXOCRINE PORTION, producing digestive enzymes, and an ENDOCRINE PORTION, producing hormones. The exocrine portion is supplied with ducts and consists of many small spherical masses of cells *(acini)* attached to the duct system in much the same fashion as grapes hanging in a bunch. Scattered among the acini are islets of endocrine tissue [*islets of Langerhans* (Fig. 27.28)], which are not ducted and which place their products directly into the bloodstream. The islets consist of two main cell types called alpha and beta cells.

FUNCTIONS OF THE PANCREAS OTHER THAN THOSE ASSOCIATED WITH DIGESTION. The ALPHA CELLS produce a hormone known as GLUCAGON. This material tends to cause the conversion of stored glycogen in the liver and muscle into glucose, and thus raises the blood sugar level. The BETA CELLS produce the more familiar hormone INSULIN. Insulin lowers the blood sugar level by causing a greater entry of glucose into the body cells, and by increasing the storage of glucose in the liver as glycogen. A delicate balance exists between the production and action of these two hormones, so that the blood sugar level normally remains within rather narrow limits (80–95 mg/100 ml).

Failure of the islets to produce sufficient insulin leads to a great increase in the blood sugar level (300–1200 mg/100 ml) and the development of DIABETES MELLITUS, or sugar diabetes.

Clinical considerations

Dental caries

Cavity formation in the teeth appears to be one of the chief penalties of civilization and the consumption of soft, sweet foods. While no one cause can be advanced to explain cavity formation, the agreed common denominator is lactic acid production by bacterial action *(Lactobacillus* sp.) on foods in the mouth. Excess acid appears to speed the demineralization process in the teeth, creating areas of weakness that are more susceptible to further change. Many host factors, such as the general nutritional state, the availability of calcium in the diet, dental hygiene practices, and heredity, influence the susceptibility of the teeth to acid damage. These are mainly involved in the formation and maintenance of good tooth materials. Fluoride appears to be of value when administered as part of the diet during the time the teeth are forming. It promotes the formation of more perfect crystalline structures that are more resistant to dislocation and damage. Direct application of the fluorides to tooth surfaces, as in dentrifices, can apparently increase the resistance of the teeth to damage. The main benefit to be derived from any program of oral hygiene is to cleanse the teeth and mouth of adherent food particles, and so remove sites where bacteria and their products can accumulate.

Mumps

Mumps is a disease of viral origin affecting primarily the parotid salivary glands. The gland becomes swollen, painful, infiltrated with lymphocytes, and engorged with blood. The disease is transmitted by droplets of saliva containing the virus. The disease runs its course in about three weeks.

Gastritis

In inflammation of the stomach, the mucosal cell-cementing substance is damaged by irritant foods, alcohol, aspirin, or many other substances. HCl and pepsin may then diffuse into the mucosa causing further damage. Pain and discomfort results, and further action of the stomach contents on the mucosa may create ulcerations in the stomach wall.

Peptic ulcer

The specific causes of local erosion of the stomach and duodenal wall are still a mystery, but there appear to be three common denominators present in most ulcer patients:

The presence of a stressful situation.

Overproduction of gastric juice.

The secretion of excess pepsin and hydrochloric acid in the stomach when food is not present.

Inflammation of the small intestine

DUODENITIS, JEJUNITIS, and ILEITIS refer to inflammation of the corresponding segments of the small intestine. Cause is unknown, but associations between excess gastric secretion (duodenitis), bacterial infection (jejunitis), and allergic reaction (ileitis) have been demonstrated.

Appendicitis

Because it is a blind-end portion of the alimentary tract, the appendix is very likely to accumulate bacteria and food materials. The organ can normally cleanse itself of these collections. When their exit is blocked, inflammation may result, with development of nausea, vomiting, and pain in the right lower quadrant. Surgical removal is usually indicated before rupture and peritonitis result.

Cirrhosis of the liver

Cirrhosis of the liver involves fibroid invasion of the organ with decrease in the number of functioning hepatic cells. Causes include malnutrition, overconsumption of alcohol, and various disease processes (hepatitis) that affect the liver and cause degeneration of cells.

Viral hepatitis

Viral hepatitis is a disease that seems to be reaching near-epidemic proportions. INFECTIOUS HEPATITIS is transmitted mainly by the use of virus-contaminated tools or consumption of tainted food and water. SERUM HEPATITIS is the result of transmission by transfusion of infected blood and by the sharing of contaminated needles by drug addicts. The liver enlarges, and signs of hepatic malfunction (jaundice, abnormal liver function tests) appear.

Pancreatitis

This disease appears to have as its common denominator the release of pancreatic enzymes into the gland, as in obstruction of the duct, or direct injury to the gland. Onset is sudden, and pain is experienced in the upper abdomen.

Typhoid fever

Typhoid fever is due to a bacillus *(Salmonella)* transmitted by contamination from infected feces. The transmitting agent may be a fly going from feces to food, or the unclean hands of an infected food handler. The organism enters through the mouth, multiplies in the tonsils, and invades the bloodstream. From here, it lodges in the lymph nodes, lungs, bone marrow, spleen, and liver. Increase in the size of the nodes and particularly the Peyer's patches of the ileum is seen. With modern antibiotic therapy, the disease is usually controlled.

Diarrhea and constipation

These are usually the result of other abnormal processes in the body and should be considered symptoms rather than disease entities themselves. Diarrhea means the evacuation of a watery, unformed stool. It may be caused by bacterial infection, food sensitization of the tract, poisons (especially metals), vitamin lack (B group), circulatory disturbances, and emotional disturbances. Constipation refers to the passage of hard, dry feces. It should be emphasized that consistency of the mass is the critical criterion, not the frequency of evacuation. Many people pass a stool of normal size and consistency only a few times a week. Constipation appears to be associated with slow passage of feces through the colon, allowing a greater-than-normal water absorption to occur. The condition is generally eased by an increase in the consumption of laxative foods and water. Laxatives themselves more often irritate than aid and should be used sparingly.

The idea that one should have one bowel movement each day ("regularity") represents an ideal situation, rather than a necessary one. Laxatives often disturb the normal flora of the colon, itself a factor in normal colon function, and keep the rectum empty, diminishing the "urge to defecate." It takes about three days for the flora to reestablish itself, and during this time there is a reduced urge to defecate. The habitual laxative-taker has taken another dose within this three-day period, and the colonic flora has a difficult time ever reestablishing itself. Laxatives may also diminish the motility of the gut muscle, resulting in disturbance of normal rhythm.

Motility in the alimentary tract and pharynx

The mouth and pharynx

CHEWING *(mastication)* involves mechanical subdivision of foods by the action of the teeth. The degree of subdivision achieved is largely determined by the length of time the food is chewed. Chewing is begun voluntarily and is continued, largely at the reflex level, as follows. Mechanical stimulation of the walls of the mouth causes a reflex inhibition of the mouth-closing musculature (masseter, temporalis, medial pterygoid) and stimulation of the digastricus. The mandible is depressed, and the mouth opens. The opening of the mouth reduces the inhibition of the closing muscles, and they reflexly contract. The cycle is repeated in rhythmic fashion. Mastication not only reduces the size of the food particles, but also mixes the food with saliva for commencement of digestion and lubrication. SWALLOWING *(deglutition)* is initiated voluntarily by the collection of a formed mass of food (bolus) on the tongue, and the movement of that bolus by the tongue into the pharynx. The mandible is closed, the tongue is elevated to seal the mouth, and the soft palate is elevated to seal the nasal cavities. The tongue then moves backward propelling the food into the pharynx by creating a positive pressure of 4–10 mm Hg. The larynx is elevated, the epiglottis is bent backward over the glottis, and the bolus moves into the lower pharynx. Once into the lower pharynx, further movement of the bolus is through reflex action. Stimulation of the pharyngeal wall results in the initiation of nervous impulses that pass to the medullary swallowing center over the glossopharyngeal (IX) and vagus (X) cranial nerves. The center discharges to the pharyngeal and esophageal musculature over the trigeminal, facial, vagus, accessory, and hypoglossal (V, VII, X, XI, and XII) cranial nerves. These control a complex series of actions that relax the upper esophageal sphincter, contract the lower pharyngeal musculature, and move the bolus into the esophagus. Once in the esophagus and still under this control a ringlike constriction is formed behind the bolus to move it forward.

Esophagus

In the esophagus, movement is completed by a PERISTALTIC WAVE that sweeps from pharynx to stomach in about 9 seconds. The musculature of the esophagus is serially innervated by the vagus nerve. Stretching of the esophagus by the bolus evokes discharges by the vagus, which creates a strong contraction of the circular muscle layer. Stimulation of successive segments of the esophagus causes the wave to traverse the length of the esophagus, stripping the mucosa clean of food particles.

Stomach

The stomach tends to act as a unit in terms of its motility, since the muscle layers are essentially one continuous mass. When food enters the stomach under the impetus of the esophageal peristaltic wave, the stomach musculature undergoes a brief RELAXATION *(receptive relaxation)*, followed by heightened activity. The volume of an empty stomach is about 50 ml, and the initial relaxation allows foods to enter. Further movements of the stomach are of two types: TONIC CONTRACTIONS and PERISTALSIS. Tonic contractions occur at the frequency of 15–20 per minute and are found in all parts of the stomach. They serve to mix and churn the contents of the stomach. Peristaltic waves originate near the upper part of the organ and sweep toward the small intestine at the rate of one or two per minute. Peristalsis moves small amounts of the stomach contents into the small intestine for further digestion. The basic motility of the stomach depends on inherent activity within the submucous and myenteric plexuses and is only modified by the action of extrinsic nerves. Sympathetic stimulation inhibits, and parasympathetic stimulation increases, gastric motility. During fasting, the vagus nerve, in response to a lowered blood sugar level, stimulates the muscle of the stomach to very strong contractions. Such contractions are termed "HUNGER CONTRACTIONS." Feeding is the normal response to these contractions.

Small intestine

Circular and longitudinal muscle layers operate much more independently in this organ than they do in the stomach. In addition, the muscularis mucosae plays an important role in one of the

TABLE 27.9 A summary of motility in the alimentary tract

Area	Type of motility	Frequency	Control mechanism	Result
Mouth	Chewing	Variable	Initiated voluntarily, proceeds reflexly	Subdivision, mixing with saliva
Pharynx	Swallowing	Maximum 20 per min	Initiated voluntarily, reflexly controlled by swallowing center	Clears mouth of food
Esophagus	Peristalsis	Depends on frequency of swallowing	Initiated by swallowing	Transport through esophagus
Stomach	Receptive relaxation	Matches frequency of swallowing	Unknown	Allows filling of stomach
	Tonic contraction	15–20 per min	Inherent by plexuses	Mix and churn
	Peristalsis	1–2 per min	Inherent	Evacuation of stomach
	"Hunger contractions"	3 per min	Low blood sugar level	"Feeding"
Small intestine	Peristalsis	17–18 per min	Inherent	Transfer through intestine
	Segmenting	12–16 per min	Inherent	Mixing
	Pendular	Variable	Inherent	Mixing
	Villus movements shortening and waving	Variable	Villikinin	Facilitates absorption
Colon	Peristalsis	3–12 per min	Inherent	Transport
	Mass movement	3–4 per day	Stretch	Fills pelvic colon
	Tonic	3–12 per min	Inherent	Mixing
	Segmenting	3–12 per min	Inherent	Mixing
	Defecation	Variable 1 per day-3 per week	Reflex triggered by rectal distension	Evacuation of rectum

several types of movement occurring in the intestine.

PERISTALSIS. PERISTALTIC WAVES occur at a frequency of 17–18 per minute. They depend on an intact myenteric (Auerbach's) plexus and occur primarily through activity of the circular muscle layer. The primary stimulus initiating peristalsis is a stretch of the intestine wall. The resulting reflex contraction of the muscularis is designated the MYENTERIC REFLEX. Examination of the fluids leaving the intestine after the myenteric reflex has been elicited shows a high content of a chemical substance serotonin. This suggests that serotonin may play a role in the reflex, but the details are largely unknown.

SEGMENTING CONTRACTIONS. SEGMENTING CONTRACTIONS are ringlike local contractions of the circular muscle layer occurring at the rate of 12–16 per minute. The contractions appear to be due to inherent activity within the plexuses and muscle and serve to mix and churn the intestinal contents.

PENDULAR MOVEMENTS. Occurring primarily within the longitudinal muscle layer, PENDULAR CONTRACTIONS shorten a length of the gut. They do not seem to have a particular frequency. They also serve to mix the intestinal contents.

VILLUS CONTRACTIONS. SHORTENING of the villi, as well as WAVING motions, are observed in the intestine. A material designated villikinin may be isolated from the blood leaving the intestine. Villikinin is produced when the upper part of the intestine is bathed by the digesting food mass. Villus movements stir the intestinal contents and aid the rate of absorption of materials by continually exposing the villus to "fresh" material. The muscularis mucosae is primarily responsible for villus movements.

Colon

Movements of the colon are similar to those seen in the small intestine. TONIC and SEGMENTING movements similar to those of the stomach and small intestine are present. PERISTALSIS moves materials through the colon. Frequency of peristalsis is lower in the colon than elsewhere in the tract, varying from 3–12 per minute. A mass movement, or very strong peristaltic wave, occurs three to four times a day and drives material into the pelvic colon.

DEFECATION. Material moving into the rectum from the pelvic colon distends the rectum and initiates the act of defecation. Distension results in a reflex contraction of the rectal musculature that tends to expel the rectal contents. Expulsion cannot occur until the external sphincter, composed of voluntary striated muscle, is relaxed. Defecation is thus a reflex act which can be voluntarily inhibited. Another reflex can also stimulate defecation. The gastrocolic reflex occurs when the stomach is distended with food and stimulates contraction of the rectal musculature. The uninhibited operation of the reflex is perhaps best seen in an infant, where feeding is almost invariably followed by defecation within 15–20 minutes.

A summary of motility in the alimentary tract is given in Table 27.9.

Summary

1. The digestive system takes in food, digests it to absorbable end products, absorbs those products, and eliminates unusable materials.
2. The system begins its development during the third week of embryonic life with the formation of a fore-, mid-, and hindgut. Each part develops certain portions of the gut and its accessory structures (see Table 27.1).
   a. Abnormal development may cause relatively common congenital disorders of the system, including atresia, stenosis, and imperforate anus.

3. The organs of the system are divided into two groups.
   a. The tubelike alimentary tract includes the mouth, pharynx, esophagus, stomach, small and large intestines, appendix, rectum, and anal canal.
   b. The accessory organs include tongue, teeth, salivary glands, liver, pancreas, and gall bladder.

4. The alimentary tract has a typical layering of tissues in its wall that includes four layers.
   a. The mucosa is internal, glandular, and absorbs or protects.
   b. The submucosa is vascular connective tissue, and it nourishes.
   c. The muscularis externa is muscular and creates the motility in the tract.
   d. An outer serous or fibrous layer surrounds and protects the tube.

5. The mouth receives food.

6. The tongue is a muscular organ surfaced with a variety of papillae for roughening and taste. The organ manipulates food in the mouth and articulates speech.

7. The teeth subdivide food, occur in two sets (deciduous and permanent), are of four types (incisors, canines, premolars, and molars), and have a typical structure.

8. Three pairs of salivary glands secrete saliva.
   a. Saliva moistens and softens foods, cleanses the mouth, and excretes substances.
   b. Salivary amylase (ptyalin) splits starches to dextrins.
   c. Control of salivary secretion is by nerves.

9. The esophagus is a muscular tube conducting food to the stomach.

10. The peritoneum lines the abdominopelvic cavity and covers organs. Mesenteries suspend organs within the cavity; omenta lie between organs and may store fat.

11. The stomach stores food and secretes gastric juice.
   a. Gastric juice contains HCl and pepsin.
   b. Pepsin breaks proteins to proteoses and peptones.
   c. Gastric secretion is controlled in three phases: cephalic, gastric, and intestinal. The first is under the control of the vagus nerve; the last two are hormonally controlled by gastrin and an unnamed hormone.

12. The small intestine completes the digestion of foods and absorbs the end products of digestion.
   a. It has a duodenum, jejunum, and ileum.
   b. Villi and plicae increase surface area for secretion and absorption.
   c. Pancreatic juice, emptied into the small intestine, contains: three protein-digesting enzymes called trypsin, chymotrypsin, and carboxypeptidase that create dipeptides; amylase to create disaccharides; lipase to break down fats to glycerol and fatty acids.
   d. Bile, secreted by the liver and released from the gall bladder, emulsifies fats and renders fatty acids water soluble.
   e. Intestinal juice, secreted by the intestine wall, contains amylases that liberate simple sugars from double sugars and erepsin for liberating amino acids from dipeptides.

13. Control of pancreatic, liver, and intestinal secretion is by hormones.
   a. Secretin and pancreozymin (cholecystokinin) control pancreatic secretion.
   b. Secretin stimulates liver production of bile, and cholecystokinin (pancreozymin) releases bile from the gall bladder.
   c. Enterocrinin stimulates secretion by intestinal glands.

14. The large intestine has haustra and taenia coli, lacks villi, and consists of appendix, cecum, colon, rectum and anal canal. The colon consists of ascending, transverse, descending, and sigmoid portions. It absorbs water, salts, and products of bacterial activity, and forms feces.

15. The liver is the largest gland of the body, has lobes and lobules, and carries on a variety of functions, including synthesis, metabolism, interconversion of substances, formation and destruction of blood cells, and storage of nutrients.

16. The pancreas is carrot-shaped and produces digestive enzymes and true hormones.

17. A variety of disorders of the system are described.

18. Movements (motility) of the tract subdivide foods, mix and churn food with enzymes, and move materials through and out of the tract (see Table 27.9 for types and details).
   a. Control of motility is by nerves, chemicals, and physical factors (osmotic pressure, stretch).

28

Intermediary Metabolism and Nutrition

General characteristics of intermediary metabolism

Simple sugars, amino acids, glycerol, and fatty acids form the major sources of energy for running the body's physiological mechanisms. A variety of METABOLIC CYCLES degrade such substances in a stepwise fashion because cells have no mechanisms by which large single outputs of energy can be captured and directed to particular processes. Energy released may also be used in the synthesis of new substances, and the reactions may also require accessory factors such as hormones, vitamins, and metals.

The various cycles that enable degradation and synthesis of substances within the body cells constitute that area of physiology known as intermediary metabolism.

Determination of the HEAT VALUES of the foodstuffs may be made by igniting specified quantities of the foods in a bomb calorimeter (Fig. 28.1) and measuring the elevation of temperature of the water around the ignition chamber. For carbohydrate, fat, and protein, the heat values are 4.1, 9.4, and 5.6 kcal per gram, respectively. In the body, protein is not completely combusted, and the energy content of the urea formed from its metabolism must be subtracted from the figure given above. This reduces the heat value of protein to 4.1 kcal/g. For ease of calculation, the heat values given above are commonly "rounded off" to 4 KCAL/G OF CARBOHYDRATE, 9 KCAL/G OF FAT, and 4 KCAL/G OF PROTEIN.

FIGURE 28.1 A diagram of a bomb calorimeter.

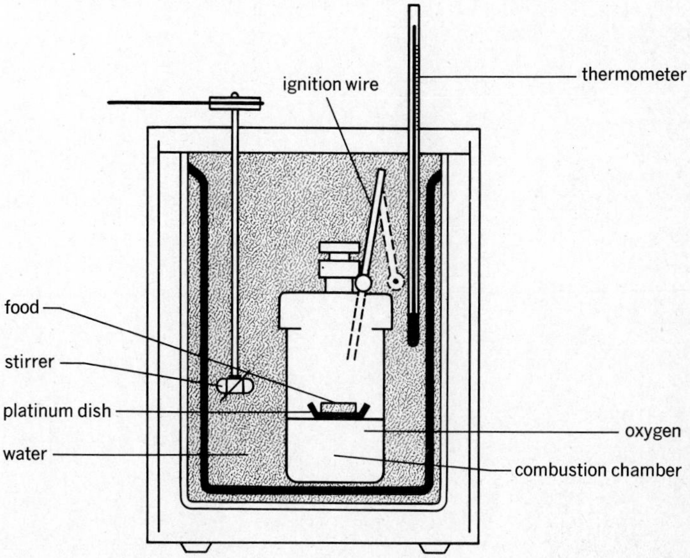

Expression of energy content

The conventional unit used to express energy content of a foodstuff is the CALORIE. There are two designations of calories:

A CALORIE *(small c)* is defined as the amount of heat required to raise the temperature of 1 gram of water 1°C (as from 15° to 16°C).

A CALORIE *(capital C)* or *kilocalorie* (kcal) is 1000 times larger than the small calorie and is the unit generally used in describing the energy changes of intermediary metabolism. It is this Calorie that is commonly used when talking about diets.

The symbol ΔG is employed to express the energy required for or released from a given reaction. If $A \rightarrow B$ and B has a lower energy content than A, ΔG is negative. The reverse situation results in a positive ΔG.

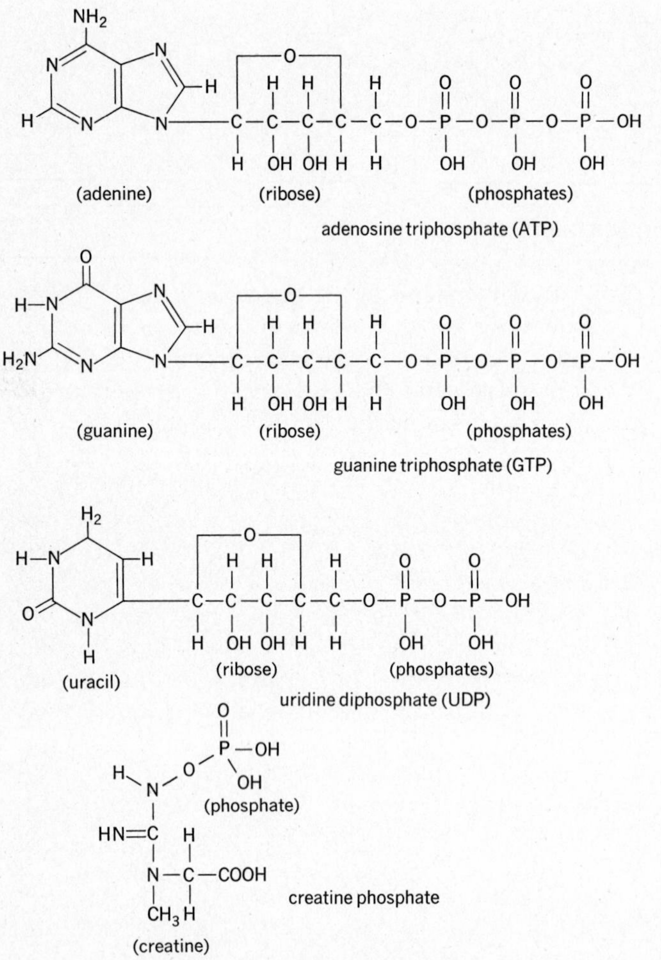

(adenine) (ribose) (phosphates)

adenosine triphosphate (ATP)

(guanine) (ribose) (phosphates)

guanine triphosphate (GTP)

(uracil) (ribose) (phosphates)

uridine diphosphate (UDP)

(phosphate)

(creatine)

creatine phosphate

FIGURE 28.2 The formulas of several high-energy phosphate compounds.

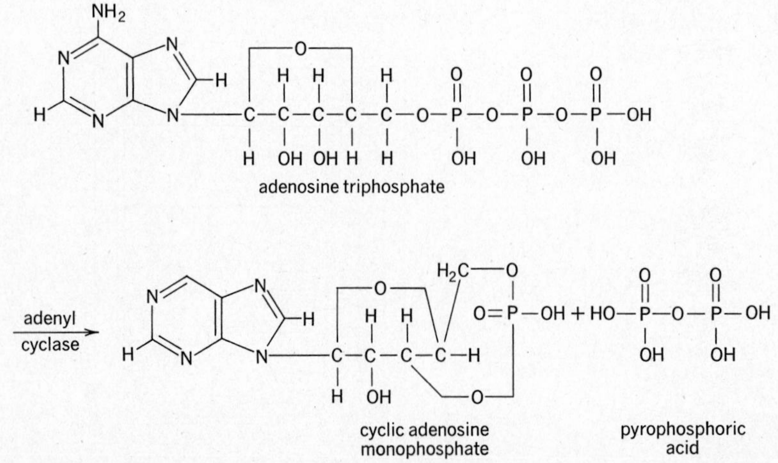

FIGURE 28.3 Cyclic nucleotides.

adenosine triphosphate

adenyl cyclase

cyclic adenosine monophosphate

pyrophosphoric acid

Coupling energy to physiological mechanisms

Although foods are consumed for their energy content, the energy in their chemical bonds is not directly available to such physiological mechanisms as muscle contraction, membrane transport systems, or glandular secretion. The foodstuff energy, released by metabolic cycles, must be transferred to a compound that can pass the energy on to the reaction. A group of large organic molecules known as HIGH-ENERGY PHOSPHATE COMPOUNDS are formed during intermediary metabolism and serve as the immediate sources of energy for running the body machinery. Among these compounds are the nucleotides adenosine triphosphate (ATP), guanine triphosphate (GTP), uridine diphosphate (UDP), and the molecule creatine phosphate (CP) (Fig. 28.2). These compounds have in common one or more phosphate groups that are held to the rest of the molecule by chemical bonds (high-energy phos-

TABLE 28.1 Some effects of cyclic AMP increase

Reaction, tissue, or process affected	Result or effect	Comments
Glycogenolysis (liver) Gluconeogenesis (liver) Ureogenesis (liver) (Urea formation) Ketogenesis (liver) (Ketone formation)	Stimulated Stimulated Stimulated Stimulated	Increase of cAMP caused by glucagon and epinephrine. Insulin decreases cAMP and decreases all these processes
Hepatic enzyme synthesis	Increased synthesis	Transaminase and carboxykinase are increased most
Histone (a type of protein) phosphorylation	Stimulates	Activates histone kinase; insulin stimulates reaction
Enzymes of prostate gland	Activates and elevates levels	Control some phases of gland's CH_2O metabolism
Insulin release	Stimulated	Decreased blood sugar increases cAMP, which stimulates insulin release
Release of anterior hormones	Increased secretion	Thyrotropin releasing factor and T_3 increase cAMP and increase TSH secretion
Thyroxin secretion	Increased secretion	TSH activates cAMP in thyroid; more thyroxin released
Calcitonin secretion	Increased secretion	Calcitonin increases Ca^{++} deposition in bone
Prostaglandin secretion	Increased secretion	Prostaglandins activate many enzymes in body cells
Water movement across membranes	Increased movement	cAMP probably stimulates active solute movement and therefore water

phate bonds or ~) that release more than the usual amount of energy when they are broken. By a process known as PHOSPHORYLATION, these phosphate groups may be transferred to other compounds, and the transfer may result in some visible activity or it may raise the energy level of the phosphorylated molecule, enabling it to begin or continue its progress in a metabolic scheme.

In the last analysis, the function of foods we eat is to be metabolized to supply energy for the synthesis of these energy-rich "middle-men" in the heirarchy of body function. Some 90 percent of foodstuff energy follows this fate.

Compounds essential to intermediary metabolism

In addition to the nucleotides and phosphate compound mentioned above, several other compounds appear necessary in order that intermediary metabolism may proceed.

CYCLIC MONONUCLEOTIDES (Fig. 28.3) are produced by the action of cyclase enzymes on a basic "straight" nucleotide such as ATP or GTP. In the process, a double-phosphate group known as a pyrophosphate is split out, and the original three-phosphate compound (e.g., ATP—T for tri-) is

TABLE 28.1 continued Some effects of cyclic AMP increase

Reaction, tissue, or process affected	Result or effect	Comments
Sodium transport across membranes	Increased transport	Probably stimulates active transport
Gastric secretion	Inhibits secretion	cAMP reduces blood flow; may also increase release of insulin, glucagon and prostaglandins, all of which decrease gastric secretion
Immune response	Activation of immunocompetent cells	
Skeletal muscle	Initiates contraction	cAMP increases permeability of sarcoplasmic reticulum to Ca^{++}; therefore, it initiates contraction
Smooth muscle	Inhibition of contraction	
Platelets	Increased activity	Increase of cAMP stimulates cyclase and increases activity in platelets
Pineal gland	Increased secretion of melatonin	Effect mediated by increased light
Nerve tissue	Increases protein kinase activity	
Parathyroid effects on bone and kidney	Increases kidney reabsorption of Ca^{++}; increases resorption of bone	Stimulates active transport by kidney; activate enzymes of osteoclasts

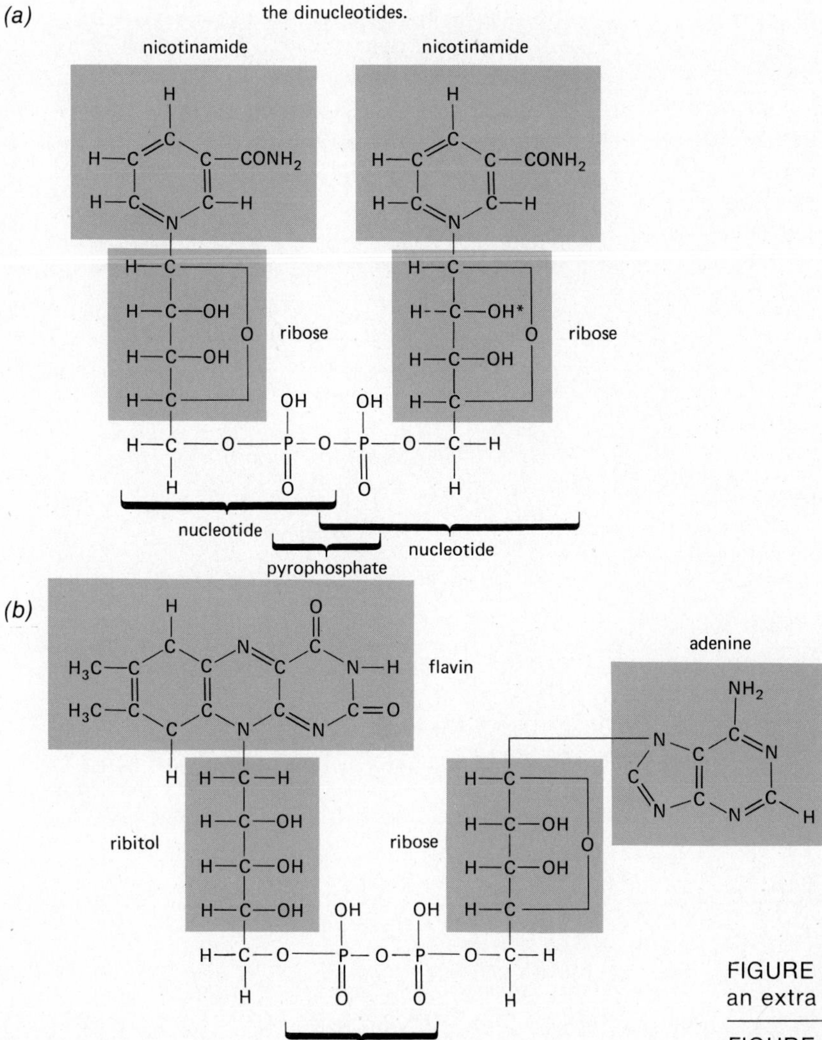

FIGURE 28.4 The dinucleotides *(a)* NAD (NADP has an extra phosphate at the *). *(b)* FAD.

FIGURE 28.5 Coenzyme A. The molecule attaches to other compounds at the -SH group.

reduced to a one-phosphate compound (e.g., AMP—M for mono-). Cyclic mononucleotides have been implicated as activators of enzyme systems that bring about chemical reactions. In turn, it has been shown that the production of cyclic mononucleotides is influenced by hormones, so that a basis for the ability of hormones to alter chemical reactions may be established. Table 28.1 presents some of the processes affected by cyclic nucleotides, with AMP as the example.

DINUCLEOTIDES (Fig. 28.4) are formed by the linking of two nucleotides through their phosphate groups; they are important in oxidation of foodstuff molecules. One form that oxidation takes is through the removal of hydrogen from the molecules of foodstuffs; the hydrogens are transferred to the dinucleotides, which act as HYDROGEN ACCEPTORS, and are ultimately carried to a scheme that forms water and ATP.

COENZYME A (Fig. 28.5) is an energy-rich compound that combines with and activates compounds so that they may enter a metabolic scheme or continue their process through a cycle. It activates acetic acid for entry into the Krebs cycle, and fatty acids for entry into beta oxidation.

Carbohydrate metabolism

Transport into cells and phosphorylation

It may be recalled that the processes of digestion produce a variety of monosaccharides, including glucose, fructose, and galactose. These are absorbed into the bloodstream and distributed to body cells. To be utilized by cells, these sugars must pass through the cell membrane. Passage is by means of active transport systems, and, once into the cell, the sugar is PHOSPHORYLATED in the presence of ATP and specific enzymes, as shown in Fig. 28.6.

The phosphorylation of the sugar is irreversible except in liver and kidney tubule cells. In these areas, a phosphatase enzyme is present that splits the phosphate group from the sugar. This particular reaction frees the sugar to leave the cell and go elsewhere in the body, as shown in Fig. 28.7.

FIGURE 28.6 The phosphorylation of several simple sugars.

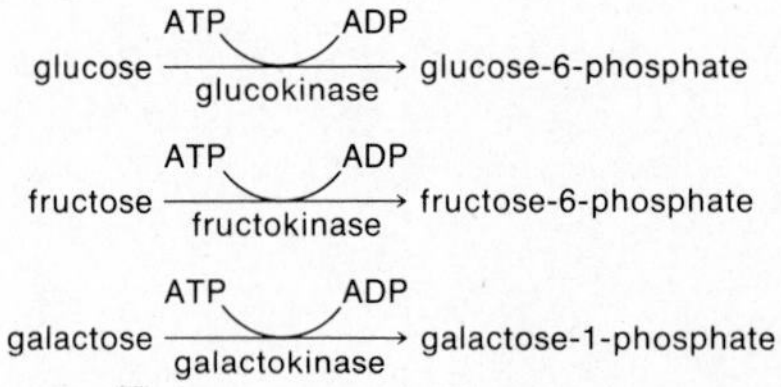

glucose —(ATP → ADP; glucokinase)→ glucose-6-phosphate

fructose —(ATP → ADP; fructokinase)→ fructose-6-phosphate

galactose —(ATP → ADP; galactokinase)→ galactose-1-phosphate

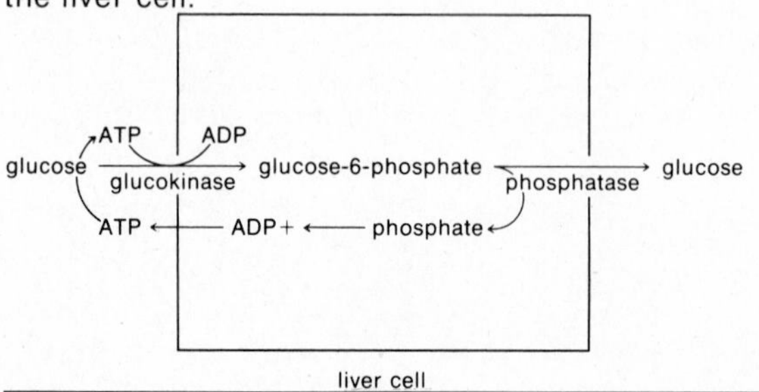

FIGURE 28.7 The dephosphorylation of glucose in the liver cell.

Interconversion of simple sugars

Glucose and fructose form the preferred sources of energy for cellular activity. There are no enzyme systems capable of metabolizing galactose efficiently. Galactose forms about one-third of the monosaccharides liberated by carbohydrate digestion in the alimentary tract. In order to use galactose in the body economy, it must be converted to a form that is metabolizable—that is, glucose. Liver cells possess a system of enzymes capable of achieving this conversion. An energy-rich

FIGURE 28.8 The conversion of galactose into glucose.

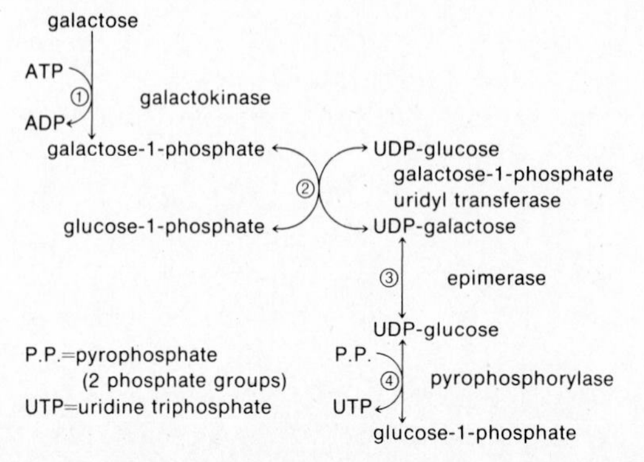

phosphate compound, uridine diphosphoglucose (UDPG) is required in addition to the enzymes. The reactions occurring are shown in Fig. 28.8.

Reaction 1 activates the galactose, using ATP. In reaction 2, galactose replaces glucose on the uridine compound. In reaction 3, conversion of galactose to glucose occurs by rearrangement of atoms in the sugar molecule. In reaction 4, glucose is split from the uridine molecule with formation of another high-energy compound, UTP.

In the inherited metabolic disease GALACTOSEMIA, there is accumulation of galactose in the blood due to an apparent inability to convert it to glucose. Deficiency of the enzyme catalyzing step 2 may be demonstrated.

Fructose may be metabolized by the same series of enzymes metabolizing glucose, and thus does not require conversion.

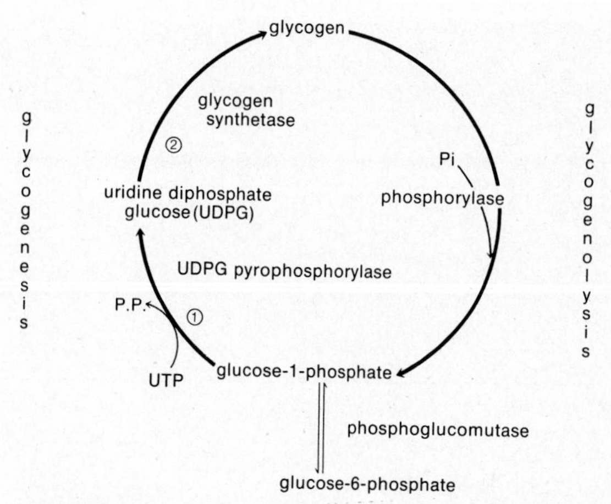

FIGURE 28.9 The reactions of glycogenesis and glycogenolysis.

Formation of glycogen

Glycogen ("animal starch") (Fig. 28.9) represents a polysaccharide used as a storage form of glucose when the latter is present in the blood in greater amounts than the body can utilize. The formation of glycogen from glucose is designated as GLYCOGENESIS. Glucose-1-phosphate* is the starting material for glycogenesis. The latter material may be formed from galactose, or by interconversion from glucose-6-phosphate, according to the reaction:

$$\text{glucose-6-phosphate} \underset{\text{phosphoglucomutase}}{\rightleftharpoons} \text{glucose-1-phosphate}$$

UDPG serves as the source of glucose to be polymerized into glycogen. It is formed, in reaction 1, by combining UTP and glucose-1-phosphate. Reaction 2 is the actual polymerization; it is influenced, in direct fashion, by the hormone insulin.

Recovery of glucose from glycogen

Glucose may be recovered from glycogen by the process known as GLYCOGENOLYSIS (see Fig. 28.9). This reaction is directly influenced by the hormones glucagon and epinephrine. So far, the following facts have been established:

The main use to which carbohydrates and fats are put in the body is to release energy to synthesize ATP, the directly utilizable source of energy for cellular reactions.

Transport of sugars into cells requires ATP in order to activate the sugar.

Sugars other than glucose, specifically galactose, may be converted into glucose.

Glycogen may either be formed from glucose or

* Hexose sugars contain six carbons, which are numbered from the aldehyde group, 1 to 6. A number appearing in the name of a compound designates the particular carbon to which the chemical group following the number is attached. For example:

glucose (ring form, carbons ① to ⑥) or glucose (open chain: ①CHO, H–②C–OH, HO–③C–H, H–④C–OH, H–⑤C–OH, ⑥CH_2OH) + phosphoric acid (HO–$PO(OH)$–OH) → glucose-6-phosphate + H_2O

glucose | phosphoric acid | glucose-6-phosphate

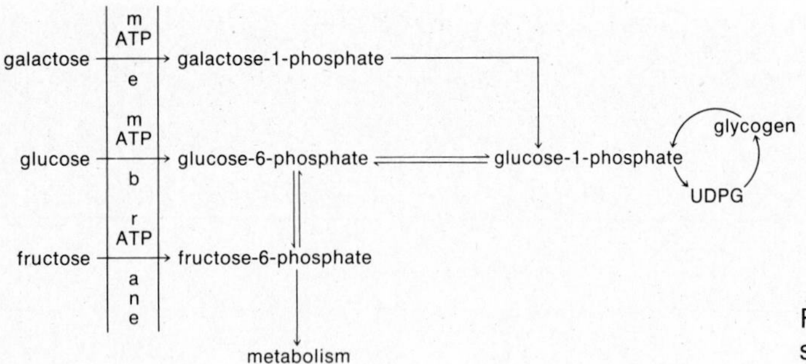

FIGURE 28.10 The interrelationships of simple sugars and glycogen.

may be broken down into glucose. Both conversions are hormone-dependent.

Specific enzymes are required for each step in these reactions.

The interrelationships of the reactions presented so far are shown in Fig. 28.10. Enzymes are not included.

Metabolic disorders associated with glycogen metabolism

The genes in the nucleic acids of the nuclear DNA determine the synthesis of specific enzymes involved in control of metabolic cycles. Thus, mutation or alteration in the genes concerned with production of enzymes necessary for glycogenesis and glycogenolysis may result in missing or defective enzymes. Seven types of GLYCOGENOSES (glycogen storage diseases) are described in Table 28.2. All involve the formation of an abnormal glycogen, or an abnormal amount of glycogen is deposited in muscle and liver cells. Of particular interest are the disorders associated with muscle phosphorylase deficiency, brancher enzyme deficiency, debrancher enzyme deficiency, and synthetase deficiency. These cause, respectively, an inability to raise blood glucose levels to the point where useful muscular work may be done, little storage of glycogen in the liver, an inability to degrade glycogen, and little or no formation of glycogen in the liver and muscles. While occurrence of these disorders is described as "rare," when they occur, there is a great effect on carbohydrate metabolism.

Glucose and fructose metabolism

Conversion of glucose and fructose to pyruvic acid is known as GLYCOLYSIS. The process is anaerobic; that is, it can be carried out in the absence of oxygen. Glycolysis releases about 56 kcal of energy per mol of glucose, and results in a net gain of two ATP molecules to the body. The reactions involved are shown in Fig. 28.11.

Reaction 1 utilizes one ATP molecule and activates the sugar to enable it to continue its metabolism. Reaction 2 rearranges glucose into fructose. It is at this step that absorbed fructose may enter the scheme without being changed into glucose. Reaction 3 uses another ATP molecule, forming a 6-carbon compound with phosphate on either end. Reaction 4 splits the 6-carbon fructose into two different 3-carbon compounds, each with a phosphate attached. The dihydroxy compound is then changed into the glyceraldehyde compound, and from this point onward, two molecules are carried simultaneously. Reaction 5 phosphorylates each compound again, but uses inorganic phosphate (Pi) and not ATP. Reaction 6 releases enough energy to resynthesize two ATP molecules. At this point two ATP molecules have been used, so the

TABLE 28.2 Glycogenoses (glycogen storage diseases)

Type and/or name of disorder	Characteristics	Enzyme abnormality	Heredity
Type I, glucose-6-phosphatase deficiency	Enlargement of liver due to glycogen accumulation—glycogen is normal; cannot convert G-6-P to glucose ∴ accumulates G-I-P	G-6-phosphatase deficiency	Autosomal recessive
Type II, generalized glycogenosis	Cardiac enlargement due to myocardial glycogen deposits	Lack of glucosidase to degrade glycogen	Autosomal recessive
Type III, limit dextrinosis	Cells packed with glycogen —can degrade glycogen at −1,4− links; not at −1,6− links	Deficiency of debrancher enzyme	Autosomal recessive
Type IV, glycogenosis	Cirrhosis of liver; large amounts of abnormal glycogen (fewer branches and longer chains)	Deficiency of brancher enzyme	Unknown
Type V, myophosphorylase deficiency glycogenosis	Increased amounts of normal glycogen in skeletal muscle; cannot perform work, no glucose available in the muscle	Muscle phosphorylase deficiency	Autosomal recessive
Type VI, hepatophosphorylase deficiency glycogenosis	No description	Liver phosphorylase deficiency	Autosomal recessive
Type VII, UDP–glucose-glycogen synthetase deficiency	Hypoglycemia and convulsions; little or no glycogen in liver	Synthetase deficiency	Unknown

amount is balanced. Reactions 7 and 8 again result in enough energy release to resynthesize two more ATP molecules. This represents a net gain. Reaction 9 occurs without enzymes and forms pyruvic acid.

Since the scheme releases 56 kcal and 2 × 7.7 kcal is trapped in the two ATP molecules formed, the efficiency of the operation in terms of energy trapping is $\frac{15.4}{56.}$ kcal × 100, or 27.5 percent.

Lactic acid formation

Further combustion of pyruvic acid to form, eventually, CO_2 and H_2O requires the presence of oxygen. If the O_2 necessary for this combustion is not immediately available, as during strenuous muscular exertion, pyruvic acid is changed to lactic acid. The lactic acid acts as a temporary storage form for the pyruvic acid, until O_2 becomes available, whereupon lactic acid is converted back

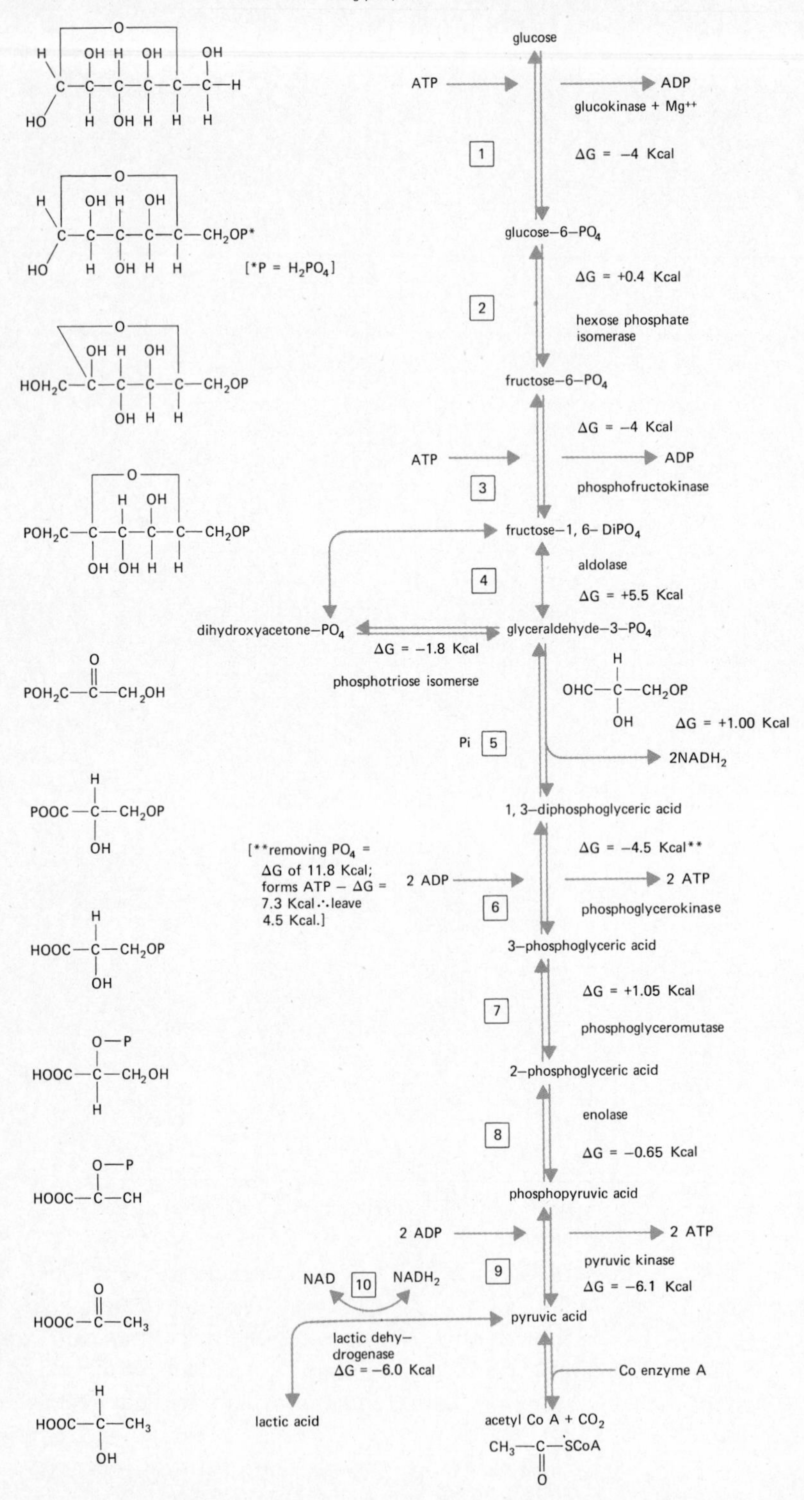

FIGURE 28.11 Glycolysis, the anaerobic breakdown of glucose (Embden-Meyerhof pathway).

FIGURE 28.12 The reaction forming lactic acid from pyruvic acid.

$$\underset{\text{pyruvic acid}}{CH_3-\overset{O}{\overset{\|}{C}}-COOH} \xrightarrow[NADH_2 \;\to\; NAD]{+2H} \underset{\text{lactic acid}}{CH_3-\overset{OH}{\underset{H}{C}}-COOH}$$

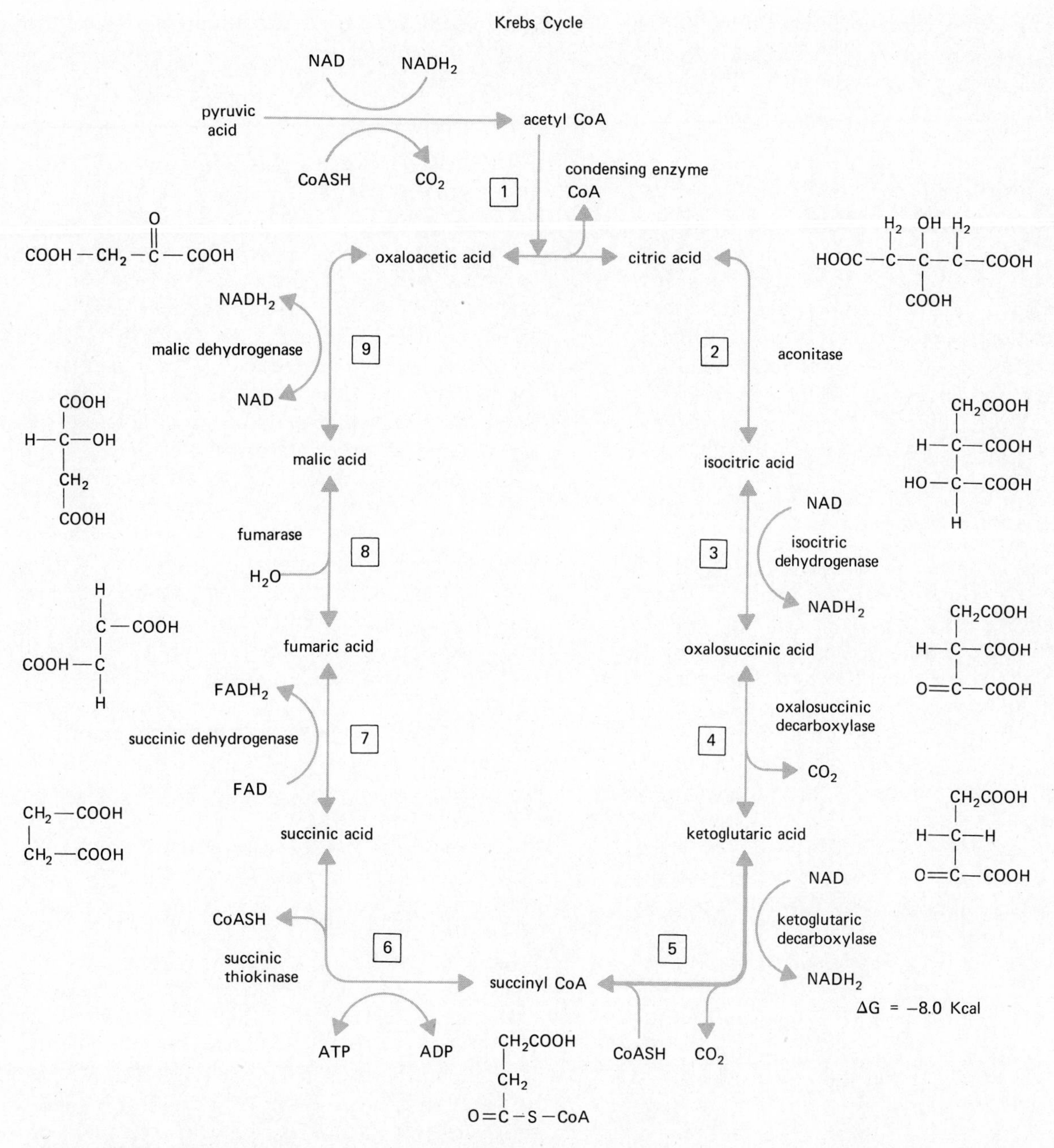

FIGURE 28.13 The Krebs cycle, the aerobic breakdown of pyruvic acid.

to pyruvic acid and combusted. Lactic acid formation occurs as shown in Fig. 28.12. The hydrogens are contributed and picked up by hydrogen acceptors, molecules described previously.

Pyruvic acid degradation; the Krebs cycle (Fig. 28.13)

In order to enter the cycle that further combusts it, pyruvic acid must be changed to acetic acid. This is accomplished by removal of CO_2 *(decarboxylation)* from the pyruvic acid molecule. In the transformation, a pair of H^+ is released.

The acetic acid then reacts with another molecule, coenzyme A (CoA), which activates the acetic acid molecule

acetic acid + CoA → acetyl CoA

Acetyl CoA then enters the Krebs cycle (citric acid cycle, tricarboxylic acid or TCA cycle). The main job that the Krebs cycle carries out is to produce 1 ATP molecule, 2 CO_2, and 8 H^+ per revolution.

Reaction 1 couples the 2-carbon acetyl CoA to the 4-carbon oxaloacetic molecule to form a 6-carbon unit, citric acid. Reaction 2 rearranges the molecule. Reaction 3 removes 2 H. Reaction 4, removing 1 CO_2, reduces the 6-carbon oxalosuccinic acid to the 5-carbon alpha-ketoglutaric acid. Reaction 5 removes an additional CO_2, as well as 2 H, and creates the 4-carbon compound, succinyl CoA. Studies involving isotope-labeled acetyl CoA have proved that the 2 CO_2 molecules released originate from the acetyl CoA. We can therefore state that the acetyl CoA (derived from pyruvic acid) is undergoing degradation. Reaction 6 produces enough energy to synthesize a molecule of ATP from ADP. Reaction 7 produces 2 H. Reaction 8 is a rearrangement of the molecule. Reaction 9, in addition to releasing 2 H, regenerates oxaloacetic acid, which is then ready to accept another acetyl CoA molecule and repeat the cycle. It should be noted that the cycle is not completely reversible, owing to the fact that reactions 4 and 5 can proceed in only one direction. The cycle would be traversed twice in combusting the 2 pyruvics resulting from glycolysis, with a total production of 16 H, 2 ATP, and 4 CO_2 molecules.

Fate of hydrogens released in the various cycles

Examination of the scheme of glycolysis, pyruvic → acetic transformations, and Krebs cycle discloses that a total of 24 hydrogens have been liberated from one molecule of glucose (4 from glycolysis, 4 from pyruvic acid → acetic acid transformations, and 16 by two revolutions of Krebs). The hydrogens are picked up by molecules known as hydrogen acceptors, described earlier.

Hydrogens released by glycolysis and the pyruvic → acetic transformation are picked up by NAD, as are 6 of the 8 released by one revolution of Krebs cycle. FAD accepts hydrogens only from step 7 of the Krebs cycle. Remembering that to combust 2 pyruvic acid molecules requires two revolutions of the Krebs cycle, we can state that 20 of the 24 hydrogens produced are accepted by NAD, and 4 by FAD. The hydrogens are then carried to still another cycle known as OXIDATIVE PHOSPHORYLATION (Fig. 28.14), which converts the hydrogens to H_2O and releases energy for ATP formation.

We may note that each compound on the left side of the scheme accepts hydrogens (or electrons), and each one on the right has hydrogen (or electrons) on it. Also, if hydrogens enter this scheme on NAD, they begin at the top and thus give rise to 3 ATP molecules for each pair of hydrogens. Remembering that 20 hydrogens arrive at the scheme on NAD, we can calculate that 30 ATP molecules (20 H = 10 pair; 3 ATP per pair or 3×10) will be produced. Four hydrogens come to this cycle on FAD. These give rise to 2 ATP per pair, a total of 4 ATP. A total of 34 ATP are thus produced by oxidative phosphorylation from the hydrogens released in other cycles.

Recalling that the starting material was glucose, and remembering that 2 ATP were produced by glycolysis and 2 by the Krebs cycle in combusting

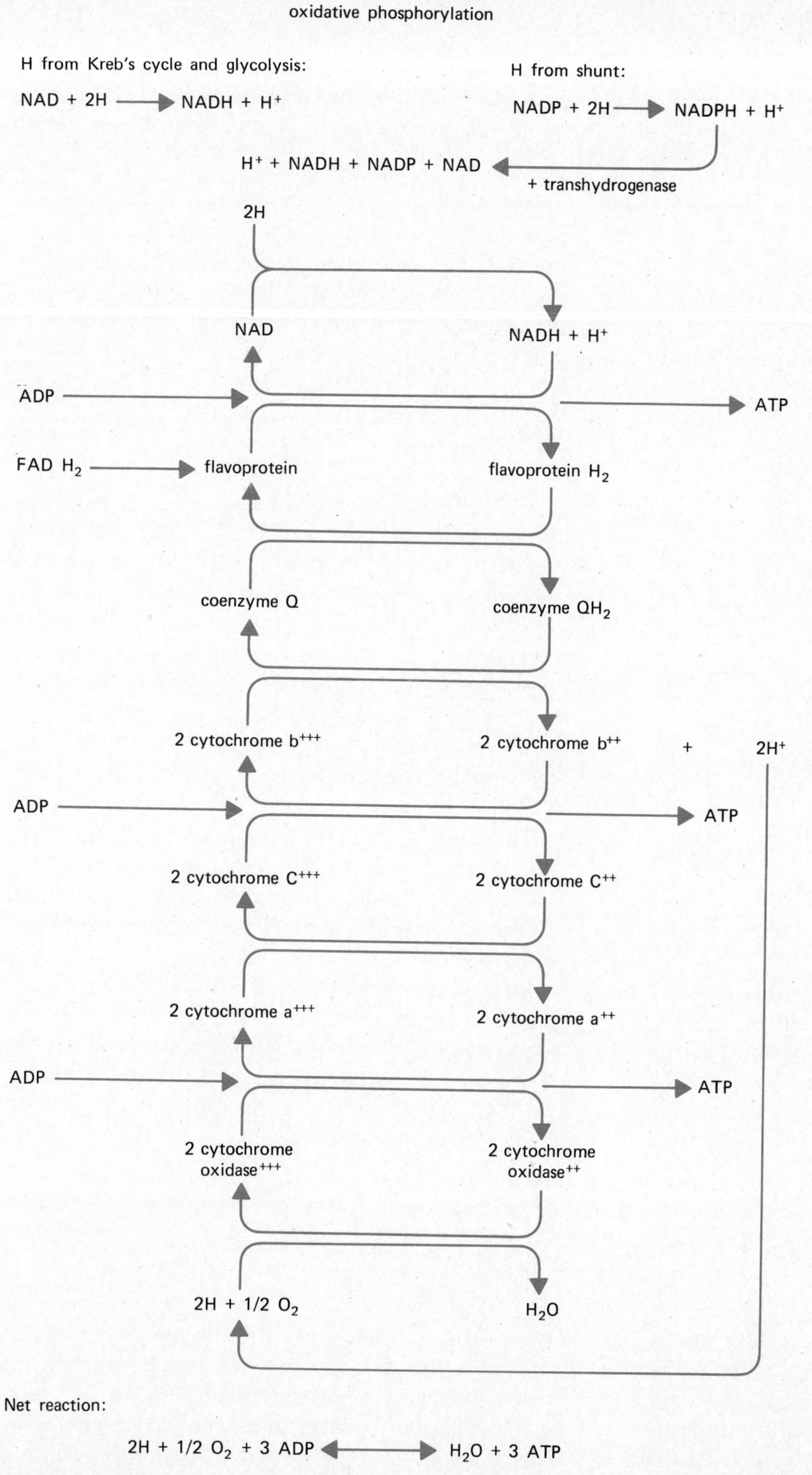

oxidative phosphorylation

H from Kreb's cycle and glycolysis:

NAD + 2H → NADH + H^+

H from shunt:

NADP + 2H → NADPH + H^+

H^+ + NADH + NADP + NAD ← + transhydrogenase

2H

NAD → NADH + H^+

ADP → ATP

FAD H_2 → flavoprotein → flavoprotein H_2

coenzyme Q → coenzyme QH_2

2 cytochrome b^{+++} → 2 cytochrome b^{++} + $2H^+$

ADP → ATP

2 cytochrome C^{+++} → 2 cytochrome C^{++}

2 cytochrome a^{+++} → 2 cytochrome a^{++}

ADP → ATP

2 cytochrome oxidase^{+++} → 2 cytochrome oxidase^{++}

2H + 1/2 O_2 → H_2O

Net reaction:

$$2H + 1/2\ O_2 + 3\ ADP \longleftrightarrow H_2O + 3\ ATP$$

FIGURE 28.14 The fate of hydrogens released by metabolic cycles.

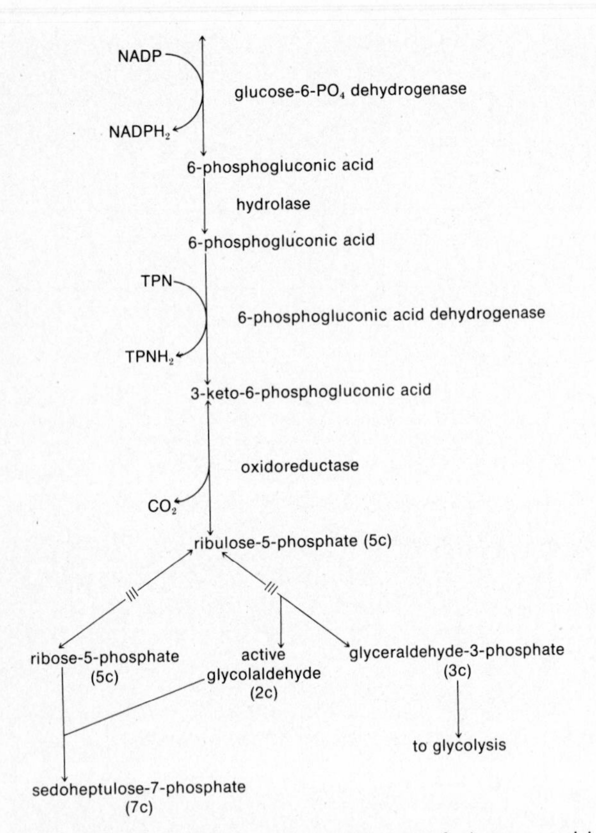

FIGURE 28.15 The direct pathway of glucose oxidation; the hexose monophosphate shunt.

2 acetyl CoA molecules, we can conclude that the complete combustion of a molecule of glucose will create a net total of 38 ATP molecules to the body. These 38 molecules represent a storage of 292.6 kcal of energy (38 × 7.7 kcal per phosphate bond) in the body cells. Because there is inherent in a glucose molecule 686 kcal of energy, the trapping represents an efficiency of 42.6 percent $\left(\frac{292.6}{686}\text{ kcal} \times 100\right)$—a good figure. Recall again the statement that a main aim of combusting foodstuffs is to release energy for ATP synthesis. The combustion of a single glucose molecule creates a significant increase in the body's store of ATP.

Alternative pathways of glucose metabolism

Two additional metabolic pathways exist for the degradation of glucose. The DIRECT OXIDATIVE PATHWAY (hexose monophosphate shunt or HMS) degrades glucose into 5-carbon sugars, glyceraldehyde, a 2-carbon fragment known as active glycolaldehyde, and utilizes NADP as the hydrogen acceptor. The 5-carbon sugars are utilized in nucleic acid synthesis. $NADPH_2$ serves as the hydrogen donor for fatty acid synthesis. The reactions of the direct pathway are shown in Fig. 28.15.

The URONIC ACID PATHWAY creates UDPG for utilization in galactose → glucose conversion and glycogenesis, and creates glucuronic acid for utilization into connective tissue ground substance. This is shown in Fig. 28.16.

Of the two pathways, the shunt must be regarded as more important as a metabolic pathway for glucose degradation. It is an important route of glucose metabolism in the lactating mammary gland. It can handle up to 30 percent of the glucose metabolism in other cells.

Lipid metabolism

Digestion of fats in the intestine results in the production of fatty acids and glycerol as the major end products. Absorption of these end products occurs primarily by physical processes, and the products are carried from the gut by way of the lymphatics (lacteals). As the fatty acids and glycerol pass through the intestinal mucosa, they are resynthesized into triglycerides (3 fatty acid molecules plus 1 glycerol molecule) and enter the lacteal as tiny droplets of fat known as CHYLO-

glucose–6–phosphate

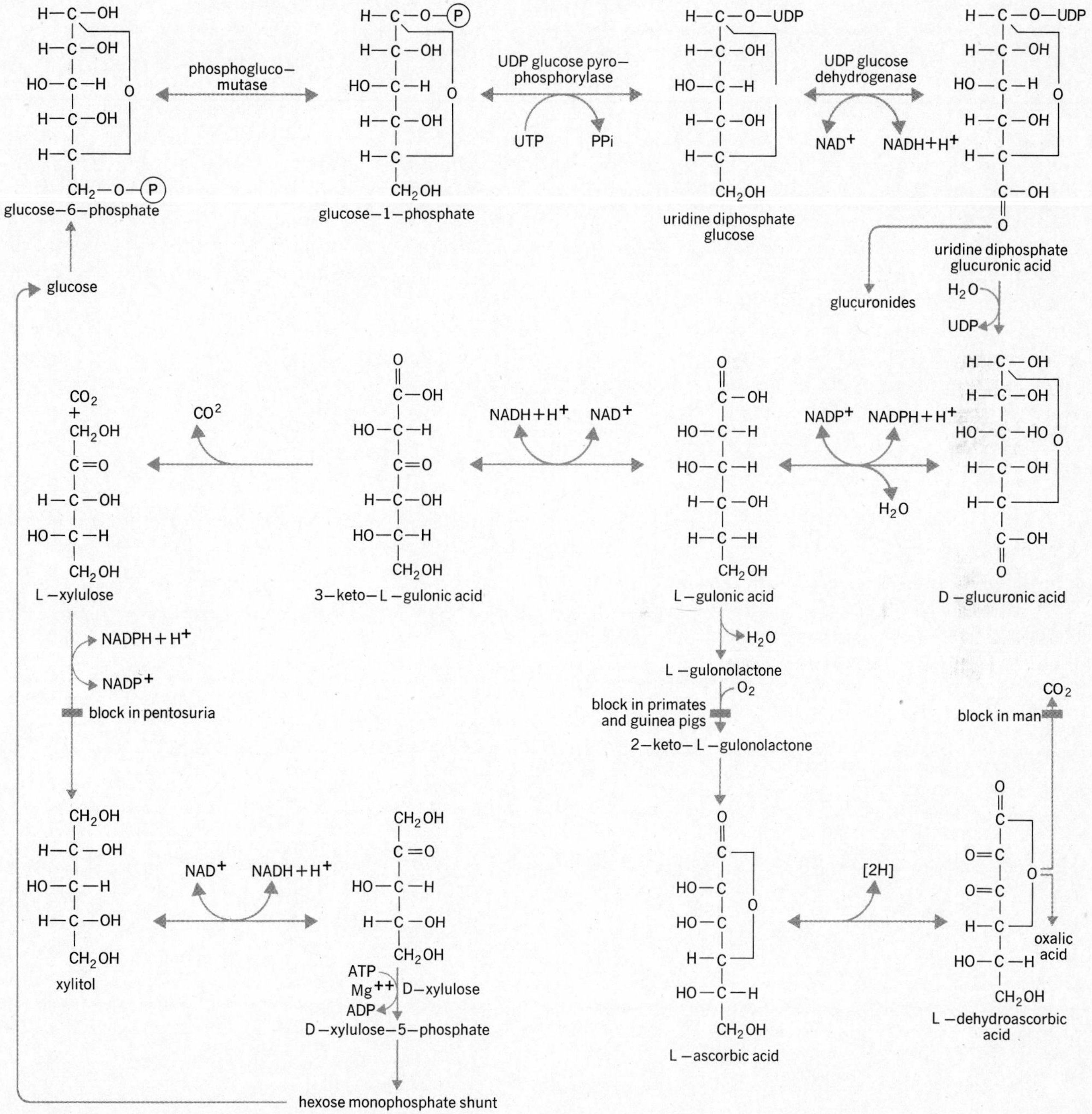

phosphoglucomutase

glucose–1–phosphate

UDP glucose pyrophosphorylase

UTP PPi

uridine diphosphate glucose

UDP glucose dehydrogenase

NAD^+ $NADH + H^+$

uridine diphosphate glucuronic acid

glucuronides

H_2O UDP

D–glucuronic acid

$NADP^+$ $NADPH + H^+$ H_2O

L–gulonic acid

$NADH + H^+$ NAD^+

3–keto–L–gulonic acid

CO^2

CO_2 + L–xylulose

glucose

$NADPH + H^+$ $NADP^+$

block in pentosuria

xylitol

NAD^+ $NADH + H^+$

D–xylulose

ATP Mg^{++} ADP

D–xylulose–5–phosphate

hexose monophosphate shunt

H_2O

L–gulonolactone

O_2

block in primates and guinea pigs

2–keto–L–gulonolactone

L–ascorbic acid

[2H]

L–dehydroascorbic acid

block in man

CO_2

oxalic acid

FIGURE 28.16 The glucuronic acid pathway.

MICRONS. Coalescence of the chylomicrons is prevented by adsorption of lacteal protein upon the fat, creating a lipoprotein. The chylomicrons are then transferred to the venous circulation via the thoracic duct. After consumption of a meal rich in fat, the fat content of the blood may rise to where the blood becomes "milky." A LIPEMIA has thus occurred. Over a period of 4 to 6 hours, the lipemia will disappear. Disappearance is due to absorption of chylomicrons by liver cells and subsequent metabolism, and to the action of a plasma enzyme, LIPOPROTEIN LIPASE *(clearing factor)*, which hydrolyzes the fats again into free fatty acids and glycerol. The acids and glycerol may then enter storage depots in the body. The chief storage areas are in the subcutaneous tissue *(panniculus adiposus)*, in the omentum, and in muscles.

The fate of glycerol

Glycerol, a 3-carbon compound, is easily converted to glyceraldehyde-3-phosphate, a compound found in the scheme of glycolysis, as shown in Fig. 28.17.

The reactions are reversible, and since the compound glyceraldehyde-3-phosphate is common to the metabolism of two different foodstuff groups, it forms a method of linking the two schemes. In short, it is possible to convert lipid sources to carbohydrate and vice versa.

The ultimate fate of the glyceraldehyde will be either conversion to glucose or metabolism via glycolysis.

The fate of fatty acids

Fatty acids undergo synthesis into neutral fats in

FIGURE 28.17 The conversion of glycerol to glyceraldehyde.

$$\text{glycerol} \underset{\text{phosphatase}}{\rightleftharpoons} \text{dihydroxyacetone phosphate} \underset{\text{triose isomerase}}{\rightleftharpoons} \text{glyceraldehyde-3-phosphate}$$

FIGURE 28.18 Beta oxidation, the degradation of a fatty acid.

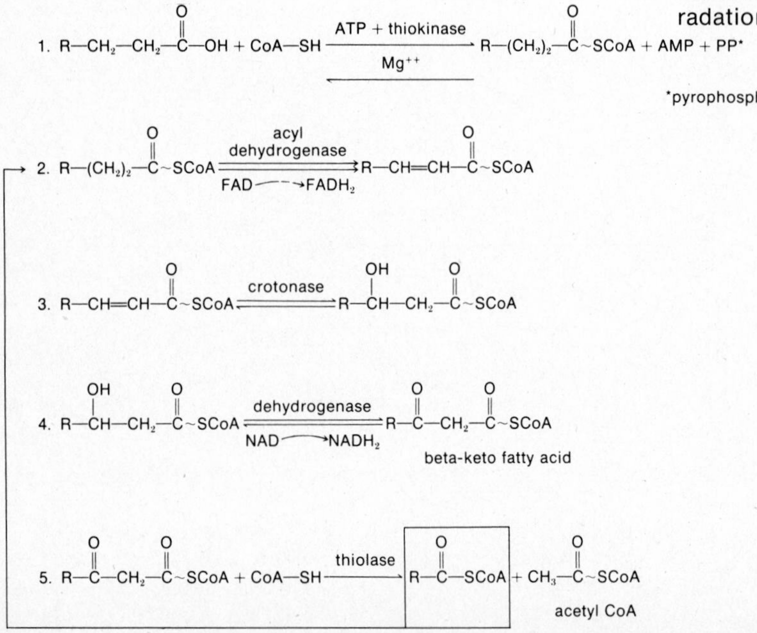

1. $$R{-}CH_2{-}CH_2{-}\overset{O}{\overset{\|}{C}}{-}OH + CoA{-}SH \underset{Mg^{++}}{\xrightarrow{\text{ATP + thiokinase}}} R{-}(CH_2)_2{-}\overset{O}{\overset{\|}{C}}{\sim}SCoA + AMP + PP^{*}$$

2. $$R{-}(CH_2)_2{-}\overset{O}{\overset{\|}{C}}{\sim}SCoA \underset{FAD \rightarrow FADH_2}{\xrightleftharpoons{\text{acyl dehydrogenase}}} R{-}CH{=}CH{-}\overset{O}{\overset{\|}{C}}{\sim}SCoA$$

3. $$R{-}CH{=}CH{-}\overset{O}{\overset{\|}{C}}{\sim}SCoA \xrightleftharpoons{\text{crotonase}} R{-}\underset{}{\overset{OH}{\overset{|}{C}H}}{-}CH_2{-}\overset{O}{\overset{\|}{C}}{\sim}SCoA$$

4. $$R{-}\overset{OH}{\overset{|}{C}H}{-}CH_2{-}\overset{O}{\overset{\|}{C}}{\sim}SCoA \underset{NAD \rightarrow NADH_2}{\xrightleftharpoons{\text{dehydrogenase}}} R{-}\overset{O}{\overset{\|}{C}}{-}CH_2{-}\overset{O}{\overset{\|}{C}}{\sim}SCoA$$

beta-keto fatty acid

5. $$R{-}\overset{O}{\overset{\|}{C}}{-}CH_2{-}\overset{O}{\overset{\|}{C}}{\sim}SCoA + CoA{-}SH \xrightarrow{\text{thiolase}} \boxed{R{-}\overset{O}{\overset{\|}{C}}{-}SCoA} + CH_3{-}\overset{O}{\overset{\|}{C}}{\sim}SCoA$$

acetyl CoA

*pyrophosphate

depot areas, or they undergo degradation via a process known as BETA OXIDATION.* In general terms, the process involves the degradation of a fatty acid by the progressive removal of 2-carbon segments in the form of acetyl CoA. Five steps occur, as shown in Fig. 28.18.

Acetyl CoA is a compound found in the carbohydrate metabolism scheme at the beginning of the Krebs cycle. Again, a compound linking two footstuff metabolism schemes is present. Accordingly, AcCoA formed from fats may enter Krebs for combustion or be synthesized into glucose.

Synthesis of fats (triglycerides)

To synthesize a triglyceride, glycerol and fatty acids are required. Glycerol may be made from glyceraldehyde-3-phosphate. Fatty acids are synthesized by two routes.

BETA REDUCTION is a reversal of the steps in beta oxidation. $NADPH_2$ from the shunt provides the source of hydrogen needed to create the fatty acid (the step when FAD picked up H_2). This pathway is found in the mitochondria.

A CYTOPLASMIC SYSTEM, independent of the mitochondria, carboxylates (adds CO_2) to acetyl CoA to form a 3-carbon unit, malonyl CoA. Malonyl CoA is then added to a fatty acid chain (even number of carbons) and, during the addition, CO_2 is removed. A beta-keto fatty acid results. From here, reversal of steps 4 to 1 in beta oxidation occurs.

Phospholipid and other complex lipid metabolism

The major phospholipids consist of glycerol, two fatty acids, and a phosphate that may be present by itself or in combination with choline or ethanolamine.

If phosphate alone is present, phosphatidic acid is formed.

If phosphate and choline are present, lecithin is formed.

If phosphate and ethanolamine are present, cephalin is formed.

Sphingomyelins are compounds of fatty acids, phosphate choline, and an alcohol sphingosine. Cerebrosides are composed of sphingosine, fatty acid, and galactose. The last two major categories of materials are important constituents of the nervous system. Degradation of these substances is accomplished by appropriate enzyme systems that reduce the materials to their constituent building blocks.

In certain disease states, there may be defective formation or degradation of these materials.

MULTIPLE SCLEROSIS is associated with excessive loss of phospholipids and sphingolipids from myelinated portions of the central nervous system.

In GAUCHER'S DISEASE, NIEMANN-PICK DISEASE, and TAY-SACHS DISEASE, deposition of abnormal phospholipids occurs in the cells of the cerebral cortex, spleen, liver, and lymph nodes. A disorder of synthesis is suspected. All of these conditions are characterized by developmental and mental retardation, changes in neurons, and enlargement of liver and spleen.

Cholesterol metabolism

Some 1.3 grams of cholesterol is added to the body economy each day, with approximately 1 gram of that being synthesized and the remainder entering via the diet. It is eliminated from the body by way of the bile acids and in the feces.

Synthesis of cholesterol occurs from acetyl CoA, followed by ring formation to give the characteristic four-ring formation of the compound.

* A fatty acid is an even-numbered (16 or 18) chain of carbon atoms. At one end is a carboxyl (−COOH) group. Carbons in the chain are named alpha, beta, gamma, delta, etc., carbons, commencing with the first carbon beyond the carboxyl carbon. The changes occurring in beta oxidation take place on the beta carbon.

Dietary cholesterol is absorbed from the intestine into the lacteals and forms chylomicrons. From the blood it is removed chiefly by the liver and is incorporated into bile acids.

Cholesterol has been implicated in the genesis of coronary heart disease and atherosclerosis. Elevated blood cholesterol levels contribute to deposition of the lipid in the walls of blood vessels (atherosclerosis), narrowing their lumina and decreasing blood flow beyond the area of deposition. If narrowing of the vessels of the heart occurs, diminished coronary flow and heart attacks may result. The substitution of unsaturated fats for cholesterol in the diet has been advanced as a possible remedy for high blood cholesterol and as a means to slow deposition of fat in blood vessel walls. These fats are supposed to hasten the metabolism of cholesterol and to slow its synthesis. Such measures may temporarily reduce blood cholesterol levels, but the body quickly resumes its production of cholesterol, essential for synthesis of hormones and bile acids.

Essential fatty acids

Certain fatty acids present in the body contain —C=C— bonds (double bonds) and are known as UNSATURATED FATTY ACIDS. They differ from the saturated fatty acids thus far described in that they contain less hydrogen than the saturated acids and are found in animal and vegetable oils, rather than adipose tissue. Some examples are as follows:

Palmitoleic acid
$CH_3(CH_2)_5CH{=}CH(CH_2)_7COOH$

Oleic acid
$CH_3(CH_2)_7CH{=}CH(CH_2)_7COOH$

Linolenic acid
$CH_3CH_2CH{=}CHCH_2CH{=}CHCH_2CH{=}CH(CH_2)_7COOH$

Linoleic acid
$CH_3(CH_2)_4CH{=}CHCH_2CH{=}CH(CH_2)_7COOH$

Arachidonic acid
$CH_3(CH_2)_4(CH{=}CHCH_2)_4(CH_2)_2COOH$

Of these, palmitoleic and oleic acids may be synthesized in the body and are not essential acids. The others cannot be synthesized, must be taken as such in the diet, and are thus known as essential fatty acids. These acids function in the prevention of fatty liver and aid in the degradation of cholesterol. They also may be used in the treatment of skin lesions in infants on low-fat diets.

ATP production during degradation of fatty acids

Degradation of fatty acids produces a net gain of far more ATP to the organism than does combustion of carbohydrate. The main reason for this is the production of more acetyl CoA molecules and hydrogens than are produced during carbohydrate breakdown. Taking as an example the breakdown of stearic acid (18 carbons), we can calculate ATP production.

Four hydrogens are produced for each CoA formed in beta oxidation (see steps 2 and 4). Two of these are picked up by FAD, two by NAD.

To degrade an 18-carbon acid into 2-carbon units, 8 splits are required. Therefore, 8×4 or 32 carbons will be produced, 16 going to FAD, 16 to NAD.

Nine acetyl CoA molecules will be produced, which, going through the Krebs cycle, will produce 8 H and 1 ATP per molecule. Nine revolutions will take place, producing 72 H (9×8) and 9 ATP (9×1). Of the hydrogens, 18 will be picked up by FAD and 54 by NAD.

A total of 34 H (16 + 18) will be carried on FAD to oxidative phosphorylation. The 34 H on FAD will give 2 ATP per pair, or 34 ATP.

The 70 H (16 + 54) on NAD will give 3 ATP per pair, or 105 ATP. A total of 148 ATP will be produced (9 from Krebs, 34 on FAD, 105 on NAD).

Formation of ketone bodies

The combustion of excessive amounts of fat may

$$CH_3-\overset{O}{\overset{\|}{C}}-SCoA + CH_3-\overset{O}{\overset{\|}{C}}-SCoA \xrightleftharpoons{\text{thiolase}} CH_3-\overset{O}{\overset{\|}{C}}-CH_2-\overset{O}{\overset{\|}{C}}-SCoA + CoA-SH$$

acetoacetyl CoA

$$CH_3-\overset{O}{\overset{\|}{C}}-CH_2-\overset{O}{\overset{\|}{C}}-SCoA \xrightarrow[\text{liver only}]{\text{deacylase in}} CH_3-\overset{O}{\overset{\|}{C}}-CH_2-\overset{O}{\overset{\|}{C}}-OH + CoA-SH$$

acetoacetic acid

$$CH_3-\overset{O}{\overset{\|}{C}}-CH_2-\overset{O}{\overset{\|}{C}}-OH \xrightarrow{-CO_2} CH_3-\overset{O}{\overset{\|}{C}}-CH_3 \quad \text{(acetone)}$$

$$CH_3-\overset{O}{\overset{\|}{C}}-CH_2-\overset{O}{\overset{\|}{C}}-OH \xrightarrow{+2H} CH_3-CHOH-CH_2-COOH$$

(beta-OH butyric acid)

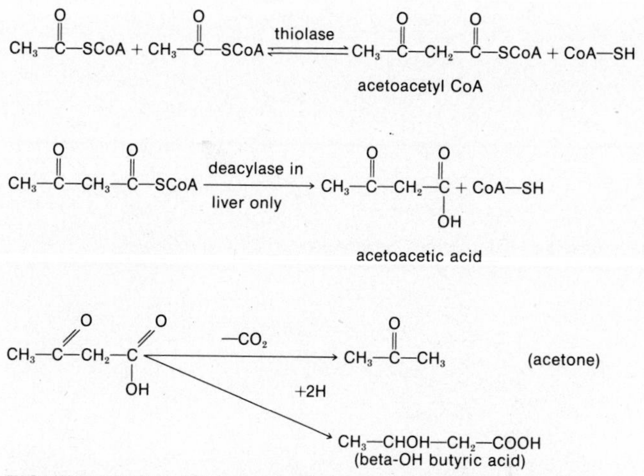

FIGURE 28.19 The formation of the ketone bodies.

cause "oversaturation" of the Krebs cycle. If Krebs cannot combust the AcCoA as fast as it is produced, a compound known as acetoacetic acid is formed, and, from this, acetone and beta-hydroxybutyric acid will be produced, as shown in Fig. 28.19.

In uncontrolled diabetes mellitus (sugar diabetes), KETOSIS is common due to the reliance of the body on the combustion of fats. These acids (acetoacetic and beta-hydroxybutyric) are also threats to the body's buffering capacity. Buffering the organic acids may not leave enough buffer to handle the normal H_2CO_3 production. Hence, ketosis and ACIDOSIS are commonly seen to occur together.

Amino acid and protein metabolism

Reactions of amino acids

Amino acids that are taken into cells may suffer one of two fates: they may be synthesized into proteins, or they may undergo reactions that lead ultimately to their degradation or conversion into other compounds.

Protein synthesis

Synthesis of proteins was discussed in Chapter 5. The five basic steps are summarized below.

By the process of TRANSCRIPTION, nuclear DNA is used to synthesize m-RNA and t-RNA.

m-RNA moves to the ribosomes, and t-RNA moves to the cytoplasm.

t-RNA attaches to activated amino acids and carries them to the m-RNA "on" a ribosome.

The ribosome moves along the m-RNA, attracting t-RNA molecules with attached amino acids.

The amino acids are joined by peptide bonds to form proteins that are released into the cytoplasm, along with free t-RNA molecules.

While many cells synthesize proteins for their own specific use, the liver is the major organ that supplies the proteins circulating in the bloodstream.

Ammonia and the formation of urea

If not synthesized into proteins, amino acids undergo deamination in the liver. Deamination is a process whereby a hydrogen and the amine group of the amino acid are removed and formed into ammonia (NH_3)and a residue called an alpha-keto acid is formed (Fig. 28.20). The alpha-keto acid may be one that is a component of glycolysis or the Krebs cycle, and it may follow, either in a degradative or synthetic course, the fate of the molecules of those cycles. For example, the relationships of several amino acids to components of glycolysis and the Krebs cycle are

$$\left.\begin{array}{l}\text{glycine} \xrightarrow{-NH_3} \text{acetic acid} \\ \text{alanine} \xrightarrow{-NH_3} \text{pyruvic acid}\end{array}\right\} \begin{array}{l}\text{glycolysis and con-} \\ \text{version to Krebs} \\ \text{cycle compound}\end{array}$$

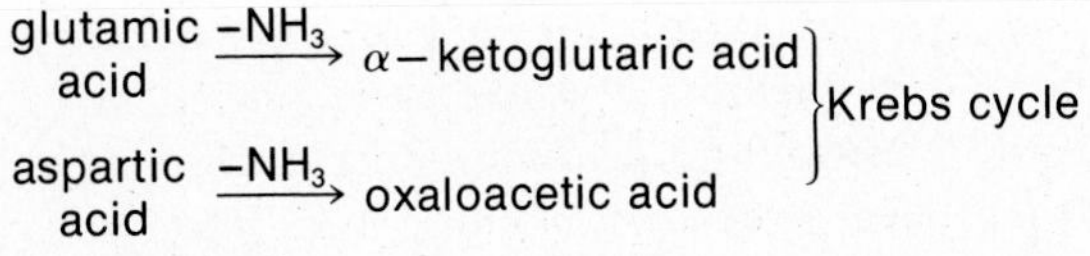

$$\left.\begin{array}{l}\text{glutamic acid} \xrightarrow{-NH_3} \alpha\text{-ketoglutaric acid} \\ \text{aspartic acid} \xrightarrow{-NH_3} \text{oxaloacetic acid}\end{array}\right\}\text{Krebs cycle}$$

$$CH_3-\underset{H}{\overset{NH_2}{C}}-COOH + \tfrac{1}{2}O_2 \xrightarrow[\text{oxidase}]{\text{amino acid}} NH_3 + CH_3-\overset{O}{\overset{\|}{C}}-COOH$$

alanine — ammonia — pyruvic acid (an α keto acid)

FIGURE 28.20 The deamination of an amino acid.

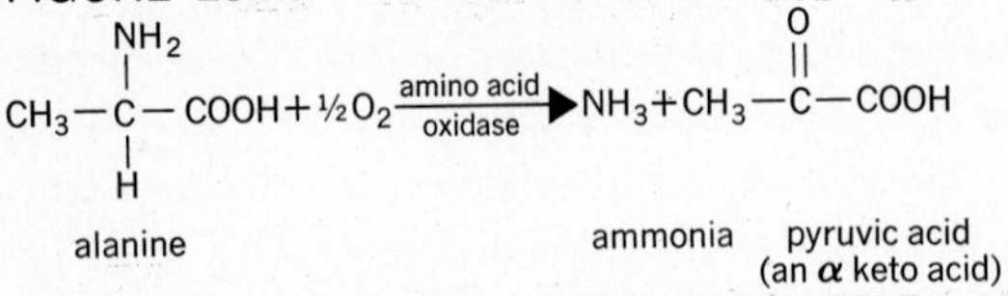

FIGURE 28.21 The formation of urea; the ornithine cycle.

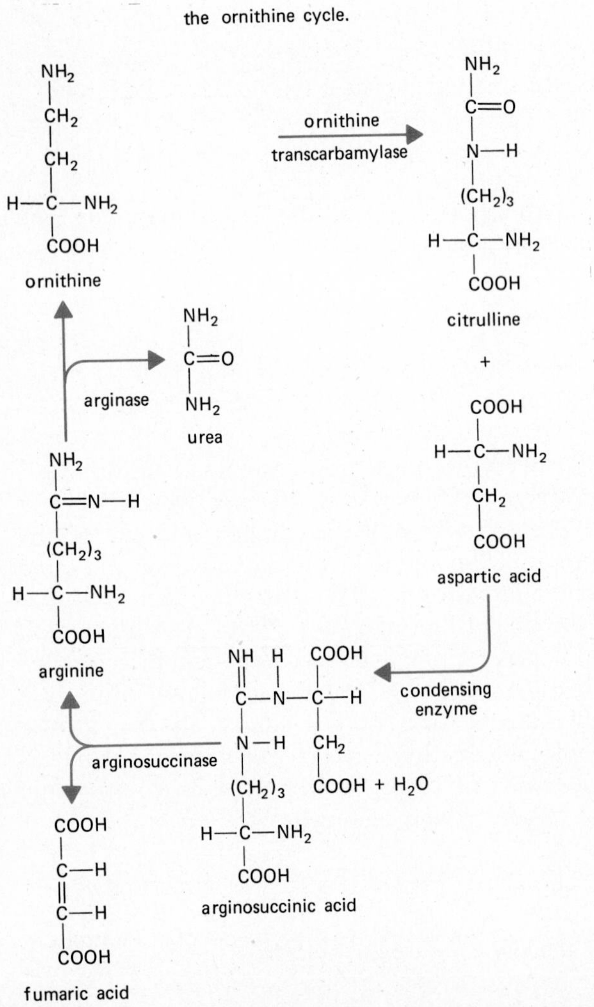

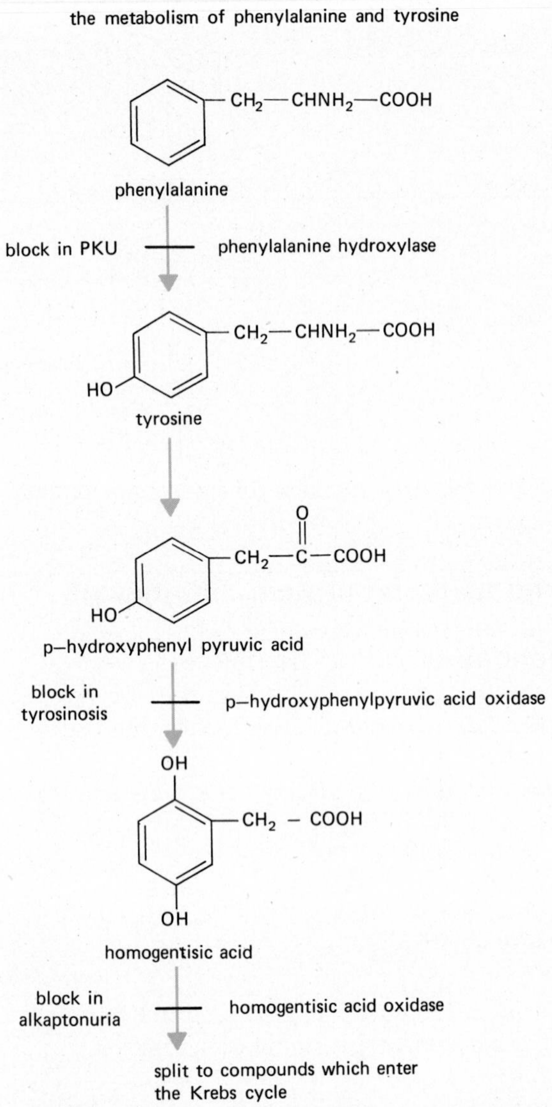

the ornithine cycle.

ornithine ($NH_2-CH_2-CH_2-CH(NH_2)-COOH$) → ornithine transcarbamylase → citrulline ($NH_2-C(=O)-NH-(CH_2)_3-CH(NH_2)-COOH$)

citrulline + aspartic acid ($COOH-CH(NH_2)-CH_2-COOH$) → condensing enzyme → arginosuccinic acid ($HN=C(-NH-CH(COOH)-CH_2-COOH)-NH-(CH_2)_3-CH(NH_2)-COOH$) + H_2O

arginosuccinic acid → arginosuccinase → arginine ($NH_2-C(=N-H)-(CH_2)_3-CH(NH_2)-COOH$) + fumaric acid ($COOH-CH=CH-COOH$)

arginine → arginase → ornithine + urea ($NH_2-C(=O)-NH_2$)

the metabolism of phenylalanine and tyrosine

phenylalanine ($C_6H_5-CH_2-CHNH_2-COOH$)

block in PKU — phenylalanine hydroxylase

tyrosine ($HO-C_6H_4-CH_2-CHNH_2-COOH$)

↓

p-hydroxyphenyl pyruvic acid ($HO-C_6H_4-CH_2-CO-COOH$)

block in tyrosinosis — p-hydroxyphenylpyruvic acid oxidase

homogentisic acid ($(OH)_2C_6H_3-CH_2-COOH$)

block in alkaptonuria — homogentisic acid oxidase

split to compounds which enter the Krebs cycle

FIGURE 28.22 The metabolism of tyrosine and phenylalanine.

The ammonia produced by deamination is an extremely toxic substance that would require a large volume of fluid to dilute it effectively to nontoxic levels for excretion. The problems of toxicity and water conservation are solved by way of the ORNITHINE CYCLE (Fig. 28.21). Two hereditary disorders of the cycle are recognized.

CITRULLINURIA is characterized by elevated blood, urine, and cerebrospinal fluid levels of citrulline with occurrence of mental retardation, alkalosis, and disturbance of blood ion levels. Although the exact site of the defect is not known, a deficiency of condensing enzyme is the result.

ARGINOSUCCINIC ACIDURIA is characterized by accumulation of arginosuccinic acid in the blood and urine. A deficiency of arginosuccinase is the cause.

If certain alpha-keto acids are available, synthesis of the corresponding amino acids can take place by the process of TRANSAMINATION. This process involves the addition of an amine group, derived from an amine donor, to the keto acid. The nonessential amino acids (see Fig. 3.1) are synthesized in the body in this manner.

Since specific enzymes are required for the metabolism of amino acids, the possibility of genetically determined disorders of metabolism exists.

MAPLE SYRUP URINE DISEASE. Leucine, valine, and isoleucine, all branched-chain amino acids, fail to be decarboxylated during their metabolism and appear in the urine, giving it a characteristic maple syrup odor. A decarboxylase enzyme is deficient.

In metabolism of histidine, HISTIDINEMIA results when deficiency of histidinase is present.

In the metabolism of phenylalanine and tyrosine, several enzyme defects are possible (Fig. 28.22).

ALKAPTONURIA results when homogentisic acid oxidase is deficient and homogentisic acid is excreted in the urine. The urine oxidizes and darkens on exposure to air.

PHENYLKETONURIA (PKU) results from deficiency of phenylalanine hydroxylase. Phenylalanine accumulates and may result in irreversible mental damage.

TYROSINOSIS results from deficiency of p-hydroxyphenylpuruvic acid oxidase, and the acid accumulates.

Calorimetry

The calorie and heat value of foodstuffs

To review, it may be useful to state again that energy is most easily measured by the large Calorie (kilocalorie), which is that amount of heat necessary to raise the temperature of 1 kg of water 1°C. To determine the heat content of any given amount of a foodstuff, it may be ignited or combusted in a bomb calorimeter, and the heat liberation is measured directly. Such determinations made for carbohydrate, fat, and protein show heat values of 4.1, 9.4, and 5.6 kcal/g of substance, respectively. Within the body, protein is not completely combusted, the urea being excreted. Since urea has a small energy content that is not available for use, the figure of 5.6 kcal/g must be reduced to 4.1 kcal/g of protein. For ease of calculation, the heat values are commonly rounded off to 4 kcal/g carbohydrates, 9 kcal/g fat, and 4 kcal/g protein.

Direct and indirect calorimetry

Since all of the energy produced by body activity is ultimately dissipated as heat, measurement of heat production by a living organism can give clues as to the amount of activity occurring and the caloric requirements that must be met to keep the animal in a state of caloric equilibrium. In short, caloric intake must balance output. Direct calorimetry involves placing an animal in a chamber similar to that shown in Fig. 28.23, measuring its heat production O_2 consumption, and CO_2 production, and determining the energy content of feces and urine. The apparatus is expensive and cumbersome. Indirect calorimetry is based on the fact that the combustion of foodstuffs is attended by a more or less fixed requirement for oxygen to be used in the combustion, and by the production of a fixed amount of CO_2 by the combustion. Knowing the volume of O_2 required and the vol-

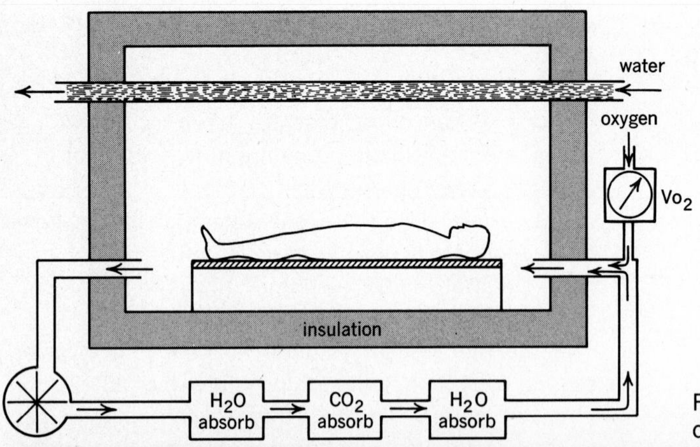

FIGURE 28.23 A diagram illustrating the basics of direct calorimetry.

TABLE 28.3 The percents of carbohydrate and fat combusted at different RQs

Nonprotein RQ	1.0	0.95	0.9	0.85	0.8	0.75	0.71
Percent CH_2O combusted	100	82	65	47	29	11	0
Percent fat combusted	0	18	35	53	71	89	100

ume of CO_2 produced, one may calculate the RESPIRATORY QUOTIENT (RQ). $RQ = CO_2/O_2$.

Carbohydrate (glucose) combustion may be represented as

$$\underset{\text{glucose}}{C_6H_{12}O_6} + 6\ O_2 \rightarrow 6\ CO_2 + 6\ H_2O$$

$$RQ = \frac{6}{6} = 1$$

Fats, having less oxygen in the molecule, require more from the outside in order to liberate CO_2.

$$\underset{\text{tristearin}}{2\ C_{57}H_{110}O_6} + 163\ O_2 \rightarrow 114\ CO_2 + 110\ H_2O$$

$$RQ = 114/163 = 0.70$$

Proteins have a calculated average RO of 0.8. Because of unknown structure, most proteins cannot be specifically represented as to O_2 and CO_2 required and produced.

The particular RO exhibited by the body as a whole is a reflection of the relative amounts of basic foodstuffs that are being combusted. If an RQ at rest was measured as 1.0, one could assume that the foodstuff being combusted was entirely carbohydrate. A similar conclusion could be reached for fats if the resting RQ was 0.7. A normal RQ for the body is about 0.82, indicating that a mixture of carbohydrate and fat is being combusted. Protein is not normally utilized for energy if carbohydrate and fat are present, so that one may assume that a given RO is the result of CH_2O and fat combustion. Tables have been worked out relating the percentage of CH_2O and fat combusted at different ROs. Table 28.3 presents such data.

Knowing what the RO is, we can next calculate a quantity known as the CALORIC EQUIVALENT OF OXYGEN. It represents the number of calories produced by fuel combustion per liter of O_2 at a given RO. This quantity, once determined, will enable calculation of caloric production if only O_2 consumption is known, and thus we can express energy production and requirements in terms of calories. Again, tables have been produced relating oxygen equivalents to calories. Table 28.4 presents such data.

TABLE 28.4 Caloric value per liter of O_2

RQ	0.707	0.75	0.80	0.85	0.90	1.0
kcal	4.686	4.739	4.801	4.862	4.924	5.047

A determination of heat production by the indirect method would therefore require the following information:

The RQ under the particular conditions specified.

The caloric equivalent of O_2 at that RQ.

The liters of O_2 consumed during the time the experiment was proceeding.

As a sample calculation:

$$\text{RQ} = 0.8$$
$$\text{caloric equivalent at } 0.8 = 4.801 \text{ kcal/l}$$
$$O_2 \text{ consumption} = 20 \text{ l/hr}$$

Question: What is subject's caloric output per day?

O_2 consumption of 20 l/hr = 20 × 24 hr or 480 l/day
at RQ of 0.8, 1 l of O_2 = 4.801 kcal/l
480 l/day × 4.801 kcal/l = 2304.48 kcal/day

The basal metabolic rate

If determinations, as have just been illustrated, were to be taken with the subject at complete mental and physical rest, at least 12 hours after the last meal, the resulting number of calories could be said to represent the subject's basal metabolic rate (BMR), or the rate of calories produced to maintain all basic body processes, extra activity eliminated. Actual BMR is determined by several factors.

AGE. There is a progressive decrease in rate associated with aging.

SEX. Females have about 10 percent lower BMR than equivalent-sized males. This represents a smaller percentage of muscle in the female body.

RACE. Orientals have been shown to have a slightly lower BMR than Caucasians.

EMOTIONAL STATE. Anxiety or stress elevates BMR.

CLIMATE. Individuals adapted to cold climates have higher BMRs.

HORMONE LEVELS in bloodstream. Thyroxin appears to be a substance that "sets the thermostat" of oxidative processes in the body. BMR rises and falls in direct proportion to circulating levels of this substance.

SURFACE AREA. Caloric expenditure appears best related to body surface, inasmuch as the skin represents a radiation surface for heat loss. Caloric release is usually expressed in kilocalories per square meter of surface per hour (kcal/m^2/hr).

Nutritional considerations

Maintenance of adequate daily caloric intake as well as providing proper balance of nutrients may be assured by choosing a diet that contains 33–42 g of protein, 250–500 g of carbohydrate, and 66–83 g of fat. Nutritional authorities recommend that selection be made from the categories of milk and milk products, meats, poultry and fish, vegetables and fruits, and breads and cereals. Table 28.5 presents recommendations for daily intake of calories and protein, according to several factors.

TABLE 28.5 Daily recommendations for calories and protein intake according to sex, age, weight, and height[a]

	Age (years)	Weight kg (lb)	Height cm (in.)	Total calories	Protein (g)
No difference by sex	0–1/6	4 (9)	55 (22)	kg × 120	kg × 2.2
	1/6–1/2	7 (15)	63 (25)	kg × 110	kg × 2.0
	1/2–1	9 (20)	72 (28)	kg × 100	kg × 1.8
	1–2	12 (26)	81 (32)	1100	25
	2–3	14 (31)	91 (36)	1250	25
	3–4	16 (35)	100 (39)	1400	30
	4–6	19 (42)	110 (43)	1600	30
	6–8	23 (51)	121 (48)	2000	35
	8–10	28 (62)	131 (52)	2200	40
Males	10–12	35 (77)	140 (55)	2500	45
	12–14	43 (95)	151 (59)	2700	50
	14–18	59 (130)	170 (67)	3000	60
	18–22	67 (147)	175 (69)	2800	60
	22–35	70 (154)	175 (69)	2800	65
	35–55	70 (154)	173 (68)	2600	65
	55–75+	70 (154)	171 (67)	2400	65
Females	10–12	35 (77)	142 (56)	2250	50
	12–14	44 (97)	154 (61)	2300	50
	14–16	52 (114)	157 (62)	2400	55
	16–18	54 (119)	160 (63)	2300	55
	18–22	58 (128)	163 (64)	2000	55
	22–35	58 (128)	163 (64)	2000	55
	35–55	58 (128)	160 (63)	1850	55
	55–75+	58 (128)	157 (62)	1700	55
	Pregnant			+200	65
	Lactating			+1000	75

[a] The table assumes sufficient calories to permit normal growth in subadults and to maintain weight in an adult. Normal activity is assumed.

Intake of specific foods

CARBOHYDRATES. After weaning, cereals (in bread and breakfast foods) and fruits and vegetables provide the major sources of carbohydrate. Sufficient carbohydrate should be provided to assure normal weight gain or maintenance according to the individual's activity levels. Table 28.6 is presented as a guide to desirable adult weight.

PROTEIN. As structural materials, components of enzymes, and contractile tissues, proteins assume primary importance in the body economy. Egg and milk sources provide the most complete proteins in terms of essential and nonessential amino acids, closely followed by meats, poultry, and fish. Plant proteins are deficient in some amino acids or contain too low an amount to sustain optimum nutrition. Determination of the nitrogen balance of the individual (amount of nitrogen intake as compared to excretion) may indicate if protein intake is adequate. (Starvation is associated with greater output than intake; growth, just the reverse.)

TABLE 28.6 Desirable weights in pounds

	Height ft.	in.	Small bones	Medium bones	Heavy bones
Male (in typical indoor clothing; shoes with 1-in. heels)	5	2	112–120	118–129	126–141
	5	3	115–123	121–133	129–144
	5	4	118–126	124–136	132–148
	5	5	121–129	127–139	135–152
	5	6	124–133	130–143	138–156
	5	7	128–137	134–147	142–161
	5	8	132–141	138–152	147–166
	5	9	136–145	142–156	151–170
	5	10	140–150	146–160	155–174
	5	11	144–154	150–165	159–179
	6	0	148–158	154–170	164–184
	6	1	152–162	158–175	168–189
	6	2	156–167	162–180	173–194
	6	3	160–171	167–185	178–199
	6	4	164–175	172–190	182–204
Female (in typical indoor clothing in shoes with 2-in. heels)	4	10	92– 98	96–107	104–119
	4	11	94–101	98–110	106–122
	5	0	96–104	101–113	109–125
	5	1	99–107	104–116	112–128
	5	2	102–110	107–119	115–131
	5	3	105–113	110–122	118–134
	5	4	108–116	113–126	121–138
	5	5	111–119	116–130	125–142
	5	6	114–123	120–135	129–146
	5	7	118–127	124–139	133–150
	5	8	122–131	128–143	137–154
	5	9	126–135	132–147	141–158
	5	10	130–140	136–151	145–163
	5	11	134–144	140–155	149–168
	6	0	138–148	144–159	153–173

FATS. Fats form important energy sources and aid the absorption of fat-soluble vitamins (A, D, E, K). Intake must include the essential fatty acids to ensure proper metabolism of other lipids. Evidence is gathering to suggest that heavy fat intake early in life results in the formation of larger-than-normal numbers of adipose cells that accumulate fat and may lead to obesity. Insufficient intake of fats may retard myelination of nerves and cause retarded neural development.

VITAMINS. Vitamins are substances essential to normal cellular metabolism. None are synthesized in the body, which thus depends on dietary intake. Two groups of vitamins are recognized.

FAT-SOLUBLE VITAMINS (A,D,E,K) are ingested and absorbed with fats in the diet. WATER-SOLUBLE VITAMINS (B complex, C, folic acid, pantothenic acid, biotin) are widely distributed in foods, but are easily destroyed by heat and oxidation. Table 28.7 presents facts related to the vitamins and certain other food factors considered vital to normal cellular function.

TABLE 28.7 Vitamins

Designation letter and name	Major properties	Requirement per day	Major sources	Metabolism	Function	Deficiency symptoms
A–Carotene	Fat soluble yellow crystals, easily oxidized	5000 I.U.	Egg yolk, green or yellow vegetables and fruits	Absorbed from gut; bile aids, in liver	Formation of visual pigments; maintenance of normal epithelial structure	Night blindness, skin lesions
D_3–Calciferol	Fat soluble needlelike crystals, very stable	400 I.u. much made through irradiation of precursors in skin	Fish oils, liver	Absorbed from gut; little storage	Increase Ca absorption from gut; important in bone and tooth formation	Rickets (defective bone formation)
E–Tocopherol	Fat soluble yellow oil, easily oxidized	Not known for humans	Green leafy vegetables	Absorbed from gut; stored in adipose and muscle tissue	Humans–maintain resistance of red cells to hemolysis	Increased RBC fragility
					Animals–maintain normal course of pregnancy	Abortion, muscular wastage
K–Naphthoquinone	Fat soluble yellow oil, stable	Unknown	Synthesis by intestinal flora; liver	Absorbed from gut; little storage; excreted in feces	Enables prothrombin synthesis by liver	Failure of coagulation
B_1–Thiamine	Water soluble white powder, not oxidized	1.5 mg	Brain, liver, kidney, heart; whole grains	Absorbed from gut; stored in liver, brain, kidney, heart; excreted in urine	Formation of cocarboxylase enzyme involved in decarboxylation (Krebs cycle)	Stoppage of CH_2O metabolism at pyruvate, beri-beri, neuritis, heart failure, mental disturbance
B_2–Riboflavin	Water soluble; orange-yellow powder; stable except to light and alkalies	1.5–2.0 mg	Milk, eggs, liver, whole cereals	Absorbed from gut; stored in kidney, liver, heart; excreted in urine	Flavoproteins in oxidative phosphorylation (hydrogen transport)	Photophobia; fissuring of skin
Niacin	Water soluble; colorless needles; very stable	17–20 mg	Whole grains	Absorbed from gut; distributed to all tissues; 40% excreted in urine	Coenzyme in H transport, (NAD, NADP)	Pellagra; skin lesions; digestive disturbances, dementia

INORGANIC SUBSTANCES. Fourteen elements of the periodic table have been shown to be essential for health. Table 28.8 presents information concerning these elements.

Hyperalimentation and obesity

Malnutrition is usually associated with undernutrition. A greater problem in many parts of the

TABLE 28.7 continued Vitamins

Designation letter and name	Major properties	Requirement per day	Major sources	Metabolism	Function	Deficiency symptoms
B_{12}—Cyanocobalamin	Water soluble; red crystals; stable except in acids and alkalies	2–5 mg	Liver, kidney, brain. Bacterial synthesis in gut	Absorbed from gut; stored in liver, kidney, brain; excreted in feces and urine	Nucleoprotein synthesis (RNA); prevents pernicious anemia	Pernicious anemia; malformed erythrocytes
Folic acid (Vitamin B_c, M, pteroyl glutamate)	Slightly soluble in water; yellow crystals; deteriorates easily	500 micrograms or less	Meats	Absorbed from gut; utilized as taken in	Nucleoprotein synthesis; formation of erythrocytes	Failure of erythrocytes to mature; anemia
Pyridoxine (B_6)	Soluble in water; white crystals; stable except to light	1–2 mg	Whole grains	Absorbed from gut; one-half appears in urine	Coenzyme for amino acid metabolism and fatty acid metabolism	Dermatitis; nervous disorders
Pantothenic acid	Water soluble; yellow oil; stable in neutral solutions	8.5–10 mg	?	Absorbed from gut; stored in all tissues; urine	Forms part of coenzyme A (CoA)	Neuromotor disorders, cardiovascular disorders; GI distress
Biotin	Water soluble; colorless needles; stable except to oxidation	150–300 mg	Egg white; synthesis by flora of GI tract	Absorbed from gut; excreted in urine and feces	Concerned with protein synthesis, CO_2 fixation and transamination	Scaly dermatitis; muscle pains, weakness
Choline (maybe not a vitamin)	Soluble in water; colorless liquid; unstable to alkalies	500 mg	?	Absorbed from gut; not stored	Concerned with fat transport; aids in fat oxidation	Fatty liver; inadequate fat absorption
Inositol	Water soluble; white crystals	No recommended allowance	?	Absorbed from gut; metabolized	Aids in fat metabolism; prevents fatty liver	Fatty liver
Para-amino benzoic acid (PABA)	Slightly water soluble; white crystals	No evidence for requirement	?	Absorbed from gut; little storage; excreted in urine	Essential nutrient for bacteria; aids in folic acid synthesis	No symptoms established for humans
Ascorbic acid (Vitamin C)	Water soluble; white crystals; oxidizable	75 mg/day	Citrus	Absorbed from gut; stored; excreted in urine	Vital to collagen and ground substance	Scurvy—failure to form c.t. fibers

world is overnutrition, or the intake of calories in excess of body needs, leading to the development of obesity. Obesity is a more threatening form of mal- (poor) nutrition than deficiency of caloric intake. A definition of obesity is that a person is 20 percent over "ideal weight" (see Table 28.6). In the United States it has been estimated that 15 million people are 20 percent over their ideal

TABLE 28.8 Inorganic substances							
Substance	Requirements per day	High level sources	Where absorbed	Where found in body	Functions	Effects of	
						Excess	Deficiency
Calcium	About 1 gm	Dairy products, eggs, fish, soybeans	Small intestine	Bones, teeth, nerve, bloodstream, muscle	Bone structure, blood clotting, muscle contraction, excitability, synapses	None	Tetany of muscles, loss of bone minerals
Phosphorus	About 55mg/kg	Dairy products, meat, beans, grains	Small intestine	Bones, teeth, nerve, bloodstream, muscle, ATP	Bone structure, intermediary metabolism, buffers, membranes	None	Unknown; related to rickets, loss of bone mineral
Magnesium	Estimated 13 mg	Green vegetables, milk, meat	Small intestine	Bone, enzymes, nerve, muscle	Bone structure, factors with enzymes, regulation of nerve and muscle action	None	Tetany
Sodium	Newborn 0.25 gm Infant 1 gm Child 3 gm Others 6 gm	All foods, table salt	Stomach, small and large intestine	Extracellular fluids	Ionic equilibrium, osmotic gradients, excitability in all cells	Edema hypertension	Dehydration, muscle cramps, renal shutdown
Potassium	1–2 gms	All foods, meats, vegetables, milk	Stomach, small and large intestine	Intracellular fluids	Buffering, muscle and nerve function	Heart block (>10 meq/L)	Changes in ECG, alteration in muscle contraction
Sulfur	0.5–1 gm ?	All protein containing foods	Small intestine, as amino acids primarily	Amino acids, bile acids, hormones, nerve	Structural as amino acids are made into proteins	Unknown	Unknown
Chlorine	2–3 gm	All foods, table salt	Stomach, small and large intestine	Extracellular fluids	Acid-base balance, osmotic equilibria	Edema	Alkalosis, muscle cramps
Iron	Infant 0.4–1 mg/kg Child 0.4 mg/kg Adult 16 mg	Liver, eggs, red meat, beans, nuts, raisins	Small intestine	Respiratory proteins (hemoglobin, myoglobin, cytochromes)	O_2 and electron transport	May be toxic	Anemia (insufficient hemoglobin in red cells
Copper	Infant and Child 0.1 mg/kg Adult 2 mg	Liver, meats	Small intestine	Bone marrow	Necessary for hemoglobin formation	None	Anemia
Cobalt	Unknown	Meats	Small intestine	Liver (Vitamin B_{12})	Essential to hemoglobin formation	None	Pernicious anemia
Iodine	Children 40–100 micrograms Adult 100–200 micrograms	Iodized table salt, fish	Small intestine	Thyroid hormone	Synthesis of thyroid hormone	None	Goiter, cretinism
Manganese	Unknown	Bananas, bran, beans, leafy vegetables, whole grains	Small intestine	Bone marrow, enzymes	Formation of hemoglobin, activation of enzymes	Muscular weakness, nervous system disturbance	Subnormal tissue respiration
Zinc	Unknown	Meat, eggs, legumes, milk, green vegetables	Small intestine	Enzymes, insulin	Part of carbonic anhydrase, insulin, enzymes	Unknown	Unknown
Fluorine	0.7 part/million in water is optimum	Fluoridated water, dentifrices, milk	Small intestine	Bones, teeth	Hardens bones and teeth, suppresses bacterial action in mouth	Mottling of teeth	Tendency to dental caries

weight and are thus obese. What causes obesity? Overeating because of boredom, unhappiness, or other emotional problems accounts for 95 percent of obesity. Genetic causes account for less than 5 percent. The penalties paid for obesity include a higher mortality (death rate) from heart disease, the development of diabetes mellitus, digestive disorders, cerebral hemorrhage, and a higher morbidity (sickness) rate.

Basically, the control of obesity involves motivation to reduce weight and a caloric intake less than that required for activity and general living. A person serious about losing weight must have an appreciation of the caloric costs of activity and adjust calorie intake accordingly. A loss of one pound per week requires about 500 kcal/day reduction in intake, or an increase in activity sufficient to increase caloric consumption by 500 kcal per day. All in all, it seems easier to reduce the caloric intake than to increase the exercise level. For example, dietary omission of an ice cream sundae, or a piece of pie à la mode, eliminates the necessary calories; to burn up an equivalent amount of calories would involve running for an hour, walking (at four miles per hour) for nearly two hours, or washing dishes for over seven hours.

The use of diet pills, usually amphetamines, is effective for about two weeks. Such drugs diminish appetite and elevate the spirits. After two weeks, the body adjusts to the drug, and appetite tends to return to previous levels. To get the effect again, a user may take more drug or take it more often, leading to a dependency on it (see Appendix).

Summary

1. A variety of metabolic cycles are available within body cells to combust foods and synthesize new substances from basic building blocks. These reactions constitute intermediary metabolism.
   a. Such cycles synthesize or degrade products in a steplike fashion to avoid large energy releases or demands.
   b. Energy content of a food is conventionally expressed by the Calorie (kilocalorie) or the amount of heat required to raise the temperature of 1 kg of water 1°C.
2. A variety of large molecules are essential to intermediary metabolism.
   a. Nucleotides, combinations of nitrogenous bases, pentose, and phosphate, serve as energy sources for physiological activities and as sources of cyclic nucleotides to control chemical reactions. ATP is the best known nucleotide; cyclic AMP (cAMP) is the best known cyclic nucleotide.
   b. Dinucleotides act as hydrogen acceptors in various metabolic schemes. There are three: NAD, NADP, and FAD.
   c. Coenzyme A (CoA) is an activator of chemical compounds and ensures intermediary metabolism of most carbohydrates and lipids.
3. Carbohydrate metabolism involves the utilization of at least nine different metabolic pathways or reactions.
   a. Phosphorylation is a necessary step for any monosaccharide as it enters a cell. This assures further metabolism.
   b. Galactose and fructose may be converted to glucose for cell use. This assures that the energy in these sugars is not lost.
   c. Glycogen may be formed (glycogenesis) from glucose by liver and muscle cells.
   d. Glucose may be released from glycogen (glycogenolysis) by enzymatic action. (**3***c* and **3***d* aid in regulating blood sugar levels.)
   e. Glycolysis is a cycle not requiring oxygen, and it reduces glucose to two pyruvic acid molecules, four hydrogens, two ATP, and about 56 kcal of energy. The two new ATP molecules trap about 15 kcal of energy; about 41 kcal is released as heat.
   f. The hexose monophosphate shunt is an alternate path for glucose combustion; it also produces $NADPH_2$ for triglyceride synthesis; it produces five carbon sugars (e.g., ribose) for RNA synthesis.
   g. The Kreb cycle degrades pyruvic acid to CO_2 and produces ATP, hydrogen, and over 600 kcal of energy. It requires oxygen.
   h. Oxidative phosphorylation is a cycle that utilizes hydrogens released from other cycles, and converts these to water and ATP.
4. Gluconeogensis is a scheme involving the synthesis of glucose from (mainly) amino acids.
   a. An amino acid has its amine group removed and then enters glycolysis and is converted to glucose by a reversal of the glycolysis steps.

b. Amino acids are not normally combusted for energy in any quantity. They are "spared" by the presence of glucose and fatty acids.

5. Triglyceride metabolism involves three basic schemes.

a. Glycerol may be converted into compounds in glycolysis and made into glucose. Fats and carbohydrates are interconvertible.

b. Beta oxidation degrades a fatty acid by the progressive removal of two carbon units (acetyl CoA) from the fatty acid.

c. Beta reduction results in fatty acid production by putting acetyl CoA molecules together.

6. Phospholipids are found in cell membranes and in nerve cells, and include lecithins, cephalins, myelin, and cerebrosides.

7. Cholesterol is a lipid that forms the basis for bile acids and steroid hormones.

a. It is synthesized from acetyl CoA.

b. It has been implicated in heart disease and in vascular occlusion.

8. Essential fatty acids are unsaturated fatty acids; they aid in the metabolism of other lipids.

9. Ketone bodies are formed from acetyl CoA when the Krebs cycle is not able to handle all the acetyl CoA produced by metabolism. Their production may result in ketosis, ketonemia, ketonuria, and acidosis.

10. Amino acids, if metabolized and not synthesized into proteins (see Chapter 4), are usually deaminated and enter glycolysis or the Krebs cycle.

a. Amine groups may form ammonia, which is converted to urea by the ornithine cycle.

11. A variety of genetically determined metabolic disorders are associated with the metabolism of carbohydrates, lipids, and amino acids.

12. Calorimetry is the study of energy requirements and energy production.

a. Each basic foodstuff has a per gram calorie release when combusted. This is known as its heat value. Carbohydrates release 4 kcal/g, fats 9 kcal/g, and protein 4 kcal/g.

b. The respiratory quotient (RO) relates the O_2 required to combust a food to the CO_2 in the combustion and is related to the type of foodstuff being combusted. Carbohydrates have an RO of 1.0, fats 0.7, and proteins 0.8.

c. The caloric equivalent of oxygen relates the calories produced when 1 liter of O_2 is used to combust a foodstuff. It varies with RO.

d. The basal metabolic rate (BMR) reflects energy production associated with maintenance of basic body activity, with no exercise involved. It varies according to many factors (age, sex, climate, race, emotional state, hormones, skin surface area, chemicals).

13. Nutrition is concerned with quality and quantity of food intake as related to normal and optimal growth, development, metabolism, and weight.

a. Specific recommendations are presented as to intake of carbohydrates, fats, proteins, vitamins, and minerals.

b. Overnutrition and obesity and their consequences of greater morbidity and mortality are considered.

The Urinary System

Pathways of excretion

Metabolism of nutrients by the body cells results in the production of a great variety of substances. Carbon dioxide, nitrogenous byproducts (such as ammonia, urea, and uric acid), heat, and excess water must be eliminated lest they disrupt the homeostasis of the body. The lungs account for elimination of the greater part of the carbon dioxide, along with small quantities of water and heat. The alimentary tract accounts for the loss of some carbon dioxide, water, heat, some salts, and the unabsorbed secretions of the digestive glands. The skin plays the major role in the elimination of excess body heat, but a minor role in elimination of solid wastes. Also treated as wastes, in that they will ultimately be eliminated from the body, are the substances present in greater amounts than are required for normal cellular function; they have, for the most part, entered the body in the diet. Water, salts, hydrogen ion, sulfates, and phosphates are examples of physiologically important substances often present in excess amounts. The kidney is the most important organ in regulating excretion of solutes resulting from metabolism. Table 29.1 summarizes the excretory organs of the body and the substances with which they deal.

TABLE 29.1 Excretory organs of the body and substances with which they deal

Organ	Substance(s)	Comments
Kidney	Water	Regulates body hydration
	Nitrogen containing wastes of metabolism (NH_3, urea, uric acid)	Originate from protein metabolism. Toxic if retained.
	Inorganic salts (Na^+, Cl^-, K^+, $PO_4^{\equiv}$, $SO_4^=$), H^+, HCO_3^-	Regulates osmotic pressure, pH of ECF; maintains excitability of cells.
	Drugs	
	Detoxified substances	Detoxification occurs in the liver; kidney rids body of end product.
Lungs	Carbon dioxide	Regulation of acid-base balance
	Water	Fixed loss
	Heat	Fixed loss
Skin	Heat	Regulates temperature
	Water	Cooling
Alimentary tract	Digestive wastes	–
	Salts (Ca^{++})	–

Development of the system

Three separate and successive excretory organs develop in the human (Fig. 29.1).

At about 3½ weeks of embryonic development, a ridge of mesoderm called the UROGENITAL RIDGE forms along the posterior aspect of the celom (primitive body cavity). Early in the fourth week of development, a series of tubular structures develops in the ridge. This series of tubules constitutes the nonfunctional PRONEPHROS or "forekidney." Paired pronephric ducts develop that grow posteriorly to connect with the cloaca. The pronephros degenerates quickly, but its duct remains to be utilized by the second kidney.

The MESONEPHROS or "midkidney" develops later in the fourth week in a position posterior to that of the pronephros. The mesonephros develops MESONEPHRIC TUBULES within its substance, and these grow laterally until they contact and fuse with the pronephric duct, now known as the mesonephric duct. The medial ends of the tubules establish a cuplike indentation that becomes associated with mesonephric blood vessels to form, together, mesonephric or RENAL CORPUSCLES that can function to remove wastes from the bloodstream. The mesonephric tubules develop from anterior to posterior, and as the posterior ones form, the anterior ones degenerate so that there are usually no more than 40 tubules in each kidney at a time. The mesonephros degenerates completely by the end of the embryonic period.

The METANEPHROS or "hindkidney" begins to develop at about five weeks, and becomes functional at seven or eight weeks. It forms the functional kidney of the human. The metanephros develops from two separate sources. The RENAL PORTION develops from the metanephric or posterior portion of the urogenital ridge, and its growth and differentiation result in the formation of tubules that are associated with the blood vessels as described for the mesonephros. These tubules form the proximal and distal tubules and the loops of Henle. The number of tubules ulti-

FIGURE 29.1 The early development of the urinary system. *(a)* Lateral view, *(b)* ventral view.

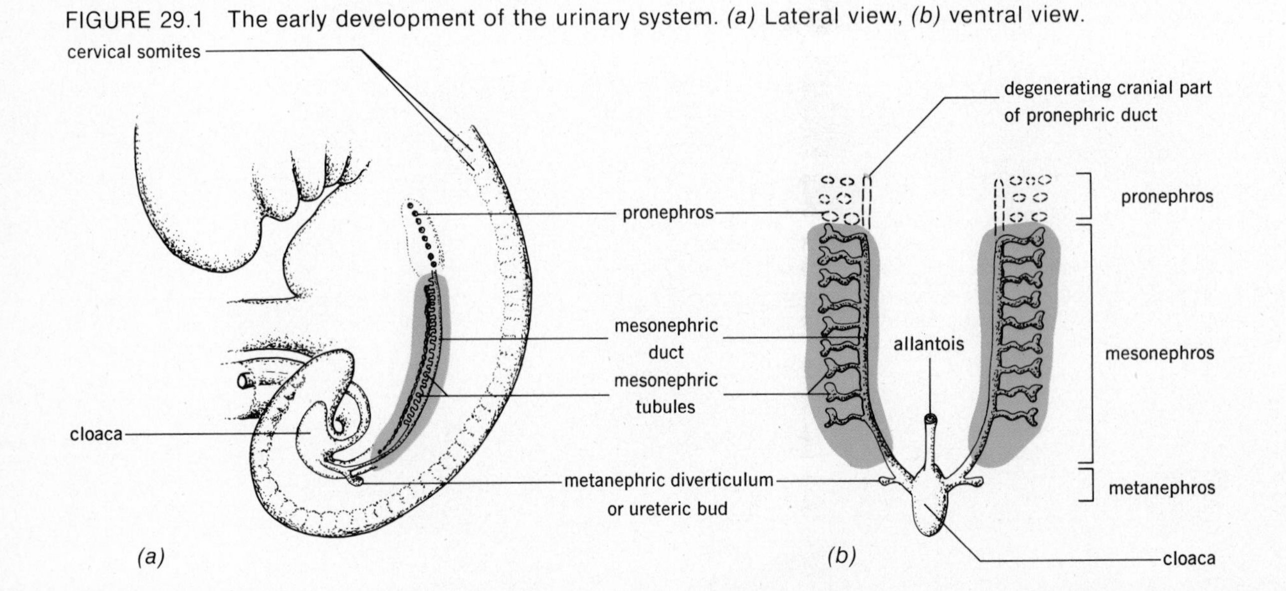

TABLE 29.2 Some congenital anomalies of the kidney

Disorder	Frequency	Cause of disorder	Comments
Agenesis Bilateral	1/6000 autopsies	Failure of both kidneys to develop	Fetus makes it to term, is using placenta to excrete. Dies several days after birth of uremia.
Unilateral	1/1200 autopsies	Failure of one kidney to develop	One kidney suffices to maintain life.
Aberrant blood vessels	1/105 autopsies	Extra blood vessels develop to supply kidney	Most common disorder. Not serious.
Horseshoe kidney	1/1200 individuals	Lower poles of kidney fused across midline	No symptoms.
Bifid (two) pelves	10% of individuals	2 pelves develop on a single kidney	No symptoms.

mately reaches 1–1.5 million in each metanephros. As the metanephric tubules are forming, a URETERIC BUD grows out from the posterior side of the mesonephric duct. It gives rise to the ureter, the renal pelvis, the calyces, and the collecting tubules of the kidney. The mesonephric duct then degenerates except for certain parts that form portions of the ducts of the male reproductive system.

The BLADDER is formed as the cloaca is divided into an upper urogenital sinus and the lower hindgut (rectal) portion. The urethra opens to the exterior from the posterior end of the bladder.

Congenital anomalies

Anomalies of the kidney are relatively common, but may not be severe enough to cause symptoms. As long as a fetus is within the uterus, the bloodstream is being cleared of wastes via the placenta; thus, if the kidneys have failed to develop, no problems will become evident until after birth, when requirements for excretion and regulation are placed on the offspring. Table 29.2 presents several of the more common congenital anomalies of the kidney.

Organs of the system

The urinary system (Fig. 29.2) includes two KIDNEYS, two URETERS, a single URINARY BLADDER, and a single URETHRA to the exterior.

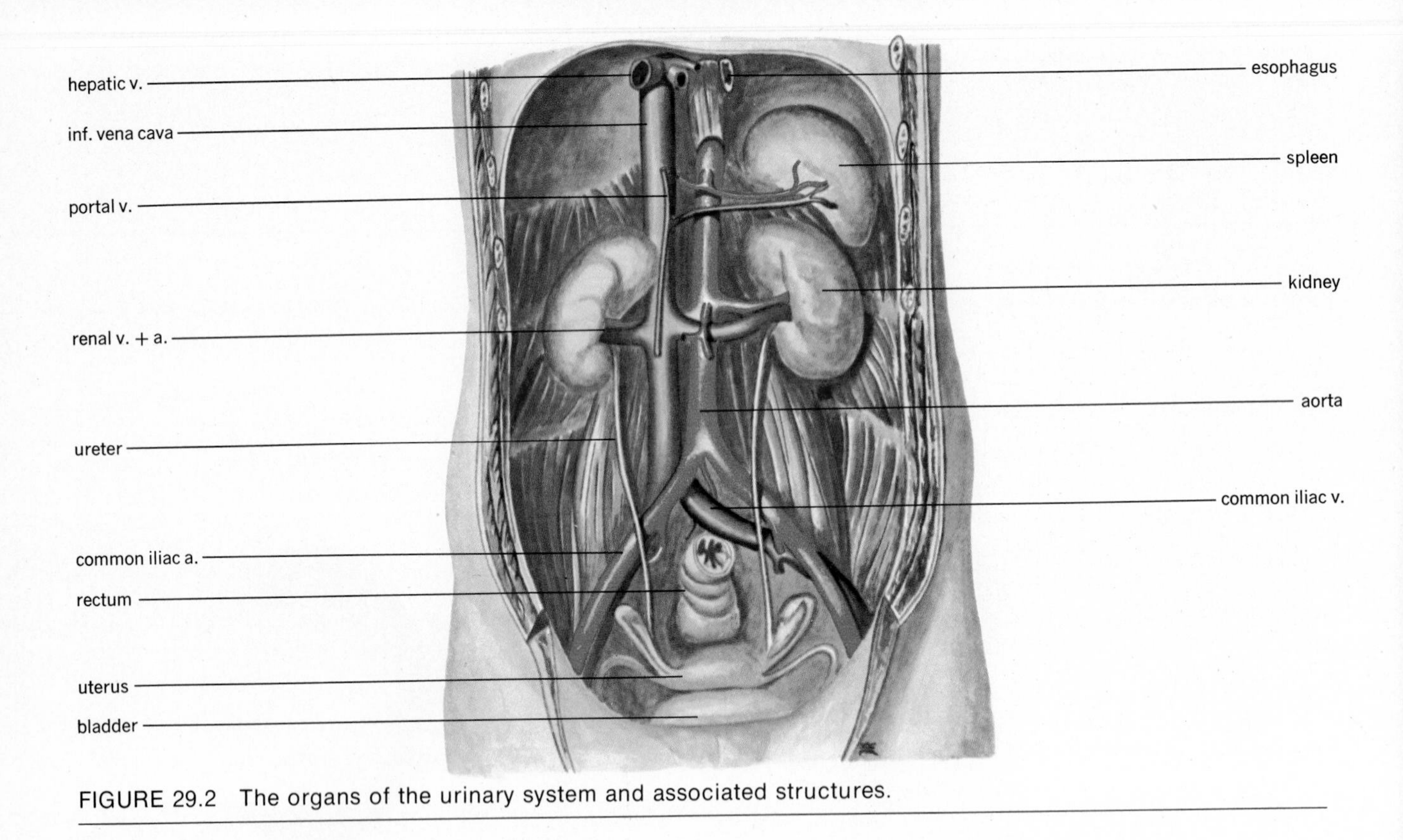

FIGURE 29.2 The organs of the urinary system and associated structures.

General functions of the system

The kidneys are responsible for the REMOVAL OF METABOLIC WASTES and excess materials from the blood by the processes of excretion and urine formation. The kidney also REGULATES THE COMPOSITION AND PHYSICAL PROPERTIES of the blood and hence the whole internal environment. It monitors acid-base balance, osmotic relationships, and the content of organic and inorganic solutes in the blood. The other organs of this system are involved only in the relatively simple tasks of transport, storage, and elimination of urine.

The kidneys

Size, location, and attachments

Living human kidneys are bean-shaped, reddish organs, located to either side of the twelfth thoracic to third lumbar vertebral bodies. Lacking mesenteries, the organs are RETROPERITONEAL; that is, they lie behind the lining of the abdominal cavity. The left kidney is usually placed slightly higher than the right. The kidneys measure about 11 cm long, 5–7.5 cm wide, and 2.5 cm thick.

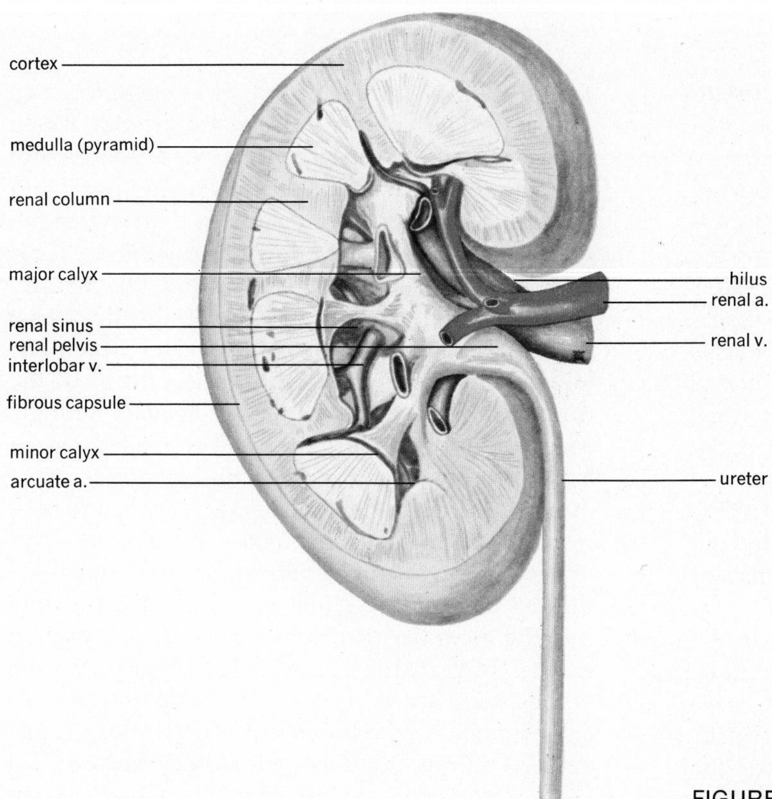

FIGURE 29.3 A frontal section of the kidney.

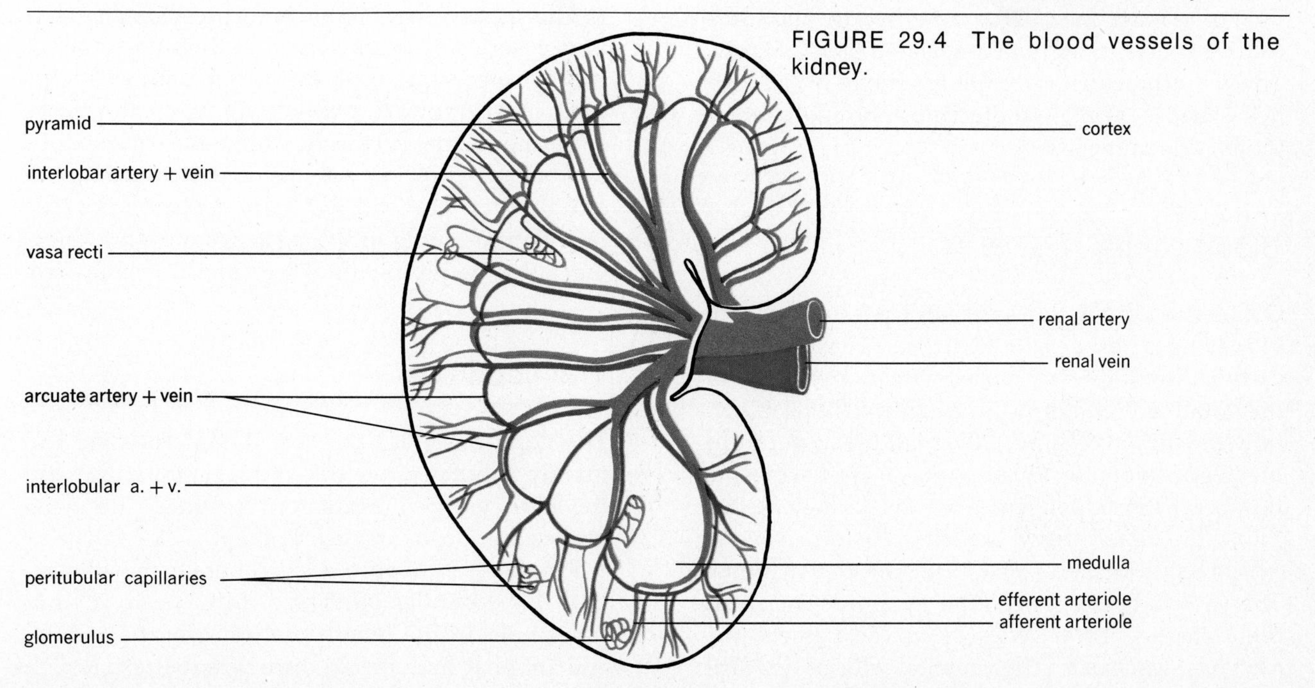

FIGURE 29.4 The blood vessels of the kidney.

Gross anatomy

A medially directed concavity is termed the HILUS; it leads to the RENAL SINUS. The blood vessels, nerves, and ureters enter and exit the organ at this region. Affixing the kidneys in position behind the parietal peritoneum are the ADIPOSE CAPSULE *(perirenal fat)* and the double layers of the SUBSEROUS *(renal)* FASCIA. A FIBROUS CAPSULE forms the external covering of the kidney itself. A frontal section of the organ (Fig. 29.3) reveals the upper expanded end of the ureter (the RENAL PELVIS) and the calyces lying within the renal sinus. The MAJOR and MINOR CALYCES are the primary and secondary subdivisions of the pelvis. An inner MEDULLA is composed of 8 to 18 RENAL *(medullary)* PYRAMIDS, with bases directed toward the periphery of the kidney, and apices or PAPILLAE projecting into the minor calyces.

The CORTEX arches around the bases of the pyramids and projects between the pyramids, where it is designated as the RENAL COLUMNS. Cortical and medullary portions together constitute the PARENCHYMA or cellular portion of the kidney. The parenchyma is formed, in each kidney, of at least one million microscopic nephrons (renal tubules) with their associated blood supply. The nephron carries out the functions of the kidney—that is, formation of urine and regulation of the blood composition.

Blood supply (Fig. 29.4)

Originating from the aorta and passing to each kidney is a large RENAL ARTERY. These arteries carry to the kidneys approximately one-fourth of the total cardiac output, a quantity (renal blood flow or RBF) averaging 1200 ml of blood per minute. Of this volume, 650 ml is plasma (renal plasma flow or RPF). Upon entering the kidney at the hilus, the renal artery branches to form several INTERLOBAR ARTERIES that enter the parenchyma. The vessels pass between the pyramids and, near their bases, form a series of curved vessels, the ARCUATE ARTERIES. The arcuate arteries in turn give rise to a series of radially directed vessels running through the cortical region of the kidney. These vessels are known as INTERLOBULAR ARTERIES. The complex of arcuate arteries rising from any one given interlobar artery does not interconnect with similar vessels rising from other interlobar arteries. Because of this arrangement, the arcuate arteries are known as *end arteries.* This arrangement is one factor ensuring the maintenance of a high level of pressure on the blood as it reaches the outer portions of the cortex. The interlobular arteries give off nutrient branches to the cortical substance and to the capsule of the kidney. Also, and most important, branches called AFFERENT ARTERIOLES go to the ball-like networks of capillaries called the GLOMERULI. From the glomeruli, EFFERENT ARTERIOLES lead out to form either a network of capillaries around the renal tubules or a looped blood vessel that dips deeply into the medullary portion of the kidney and then returns toward the cortex. The efferent arteriole carries a smaller volume of blood than the afferent and is smaller in diameter. If a network of capillaries is formed, it is known as the PERITUBULAR CAPILLARY NETWORK; the vascular loop is known as the VASA RECTI. From here the blood vessels form a series of veins named in the same manner as the incoming arteries. INTERLOBULAR VEINS drain into ARCUATE VEINS lying along the bases of the medullary pyramids. Unlike the corresponding arteries, the arcuate veins do connect with one another. These in turn form INTERLOBAR VEINS that form a single RENAL VEIN from each kidney, the latter emptying into the inferior vena cava.

The nephrons

The nephrons (Fig. 29.5) are the microscopic and mainly tubular structural and functional units of the kidneys. It is estimated that there are 1–1.5 million nephrons in each kidney.

Two types of nephrons are generally recognized. Both have similar parts and differ from one another only in the length of certain of the tubules and in what happens to the efferent arteriole as it

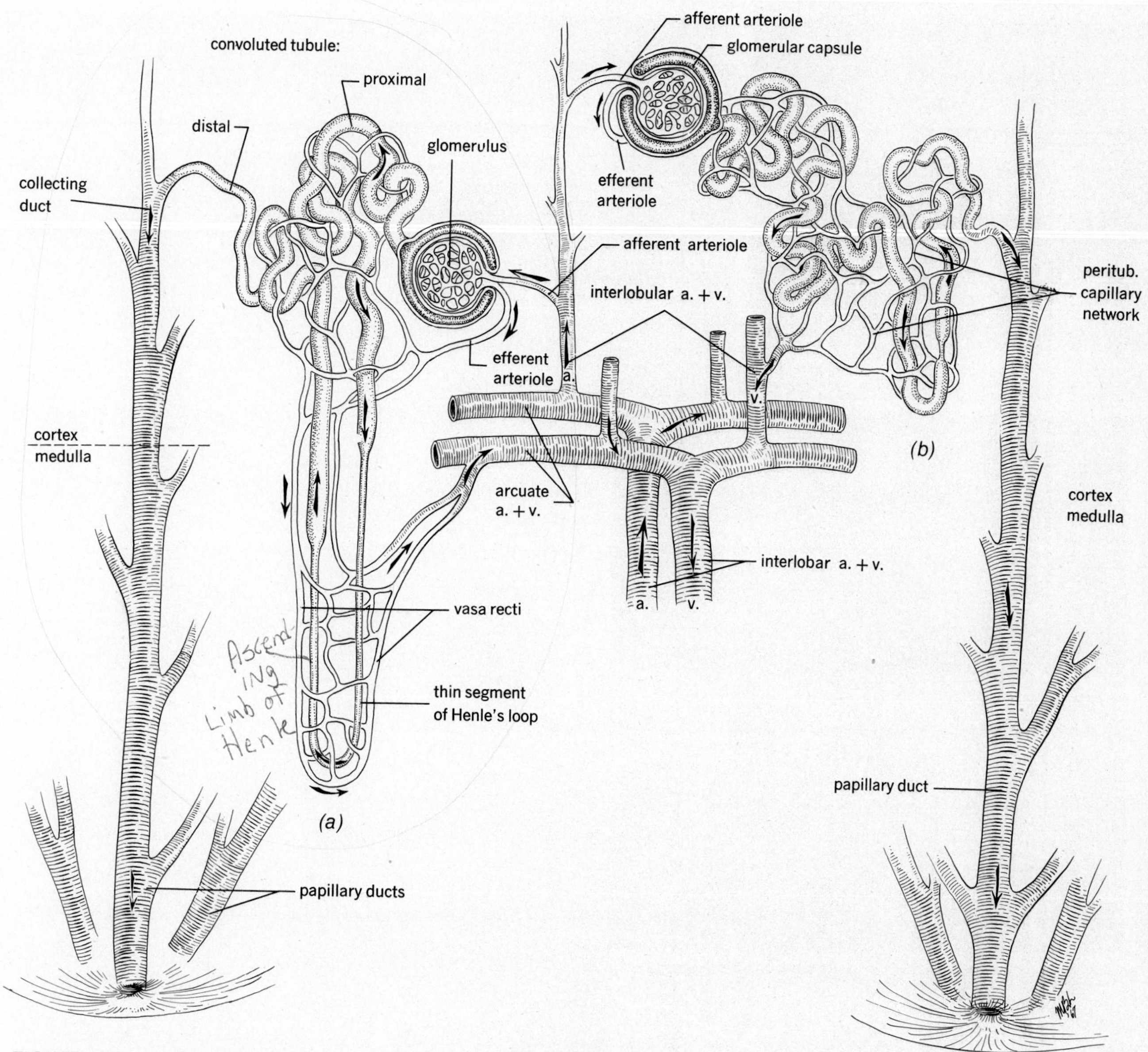

FIGURE 29.5 The nephrons of the kidney and their associated blood vessels. *(a)* Juxtamedullary nephron, *(b)* cortical nephron.

leaves the glomerulus. The two types are known as CORTICAL and JUXTAMEDULLARY NEPHRONS. The cortical nephron usually has its glomerulus in the outer portion of the cortex, and the rest of its parts rarely reach deeply into the medulla. The juxtamedullary nephron, on the other hand, has its glomerulus close to the corticomedullary junction, and has part of its structure located deep within the medulla.

A nephron begins with a double-walled cup

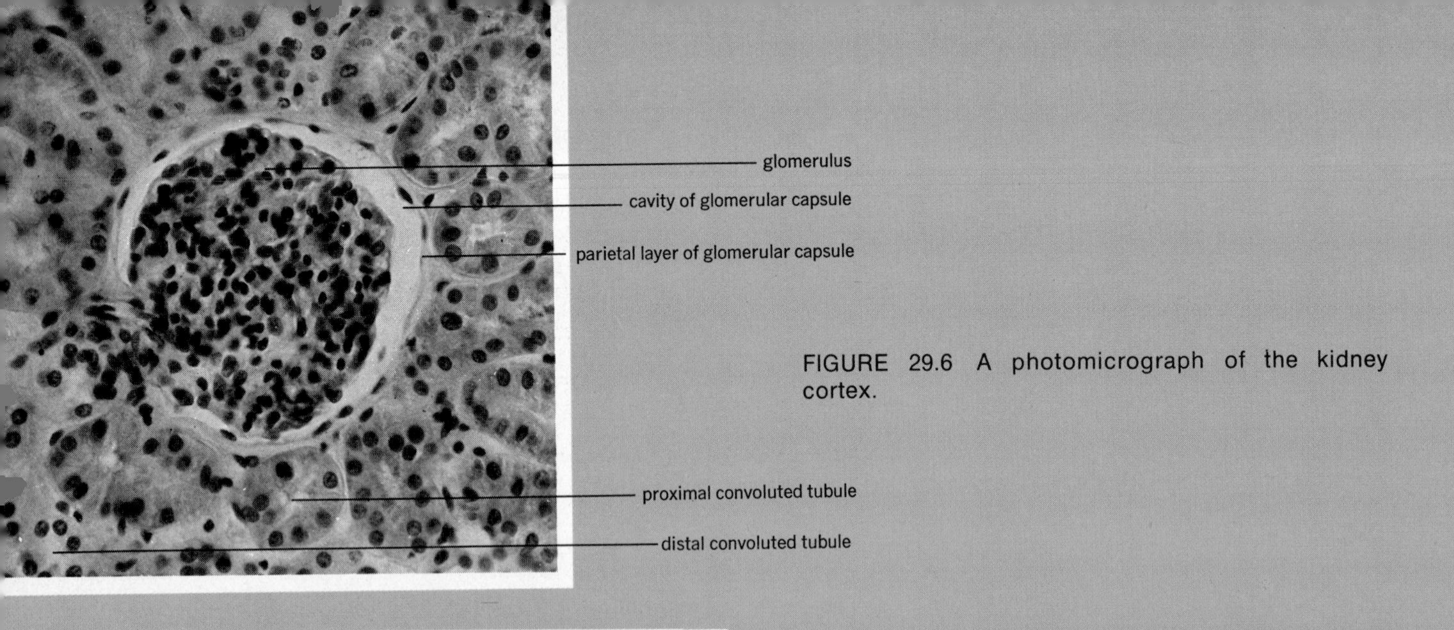

FIGURE 29.6 A photomicrograph of the kidney cortex.

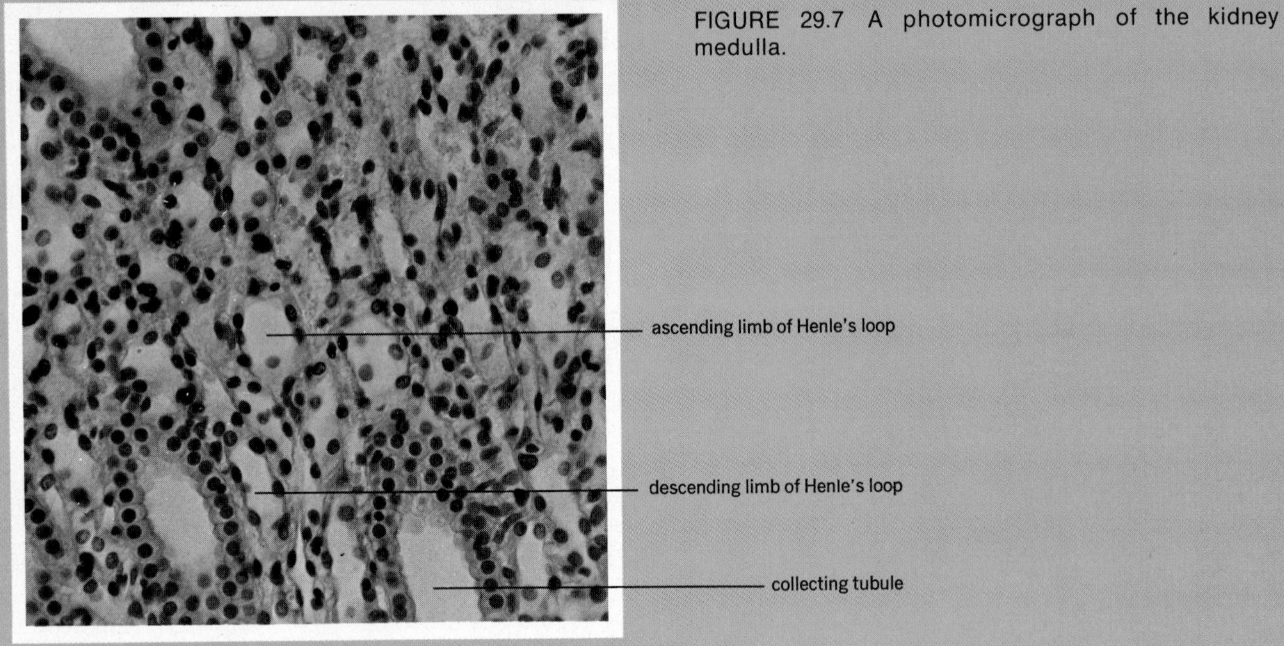

FIGURE 29.7 A photomicrograph of the kidney medulla.

termed the GLOMERULAR *(Bowman's)* CAPSULE. The inner wall of this capsule is known as the VISCERAL LAYER, and it follows intimately the twists and turns of the glomerular capillary network. It is formed of highly modified cells known as PODOCYTES. The outer or PARIETAL LAYER of the capsule is a simple squamous epithelium and lies a short distance from the visceral layer so that an actual space between the two layers is created (Fig. 29.6). The capsule and the contained glomerulus form a unit designated as the RENAL CORPUSCLE. Leading from the capsule is a PROXIMAL CONVOLUTED TUBULE in which the cells are cuboidal, have central nuclei, and have a "brush border" on the lumen side. The brush border consists of minute cytoplasmic extensions of the cell called microvilli. These serve to increase the surface area for reabsorption and/or secretion of materials by this part of the nephron. The proximal tubule then becomes straight as it nears the medullary region. The cells become flat, and the tube narrows and dips toward or into a pyramid as the DESCENDING LOOP OF HENLE. Then the tube bends back upon itself, enlarges, its cells becomes rectangular, and, as the ASCENDING LOOP OF HENLE (Fig. 29.7), it returns to or toward the cortical region. In the cortex, the tube again becomes straight and then convoluted and is known as the DISTAL CONVOLUTED TUBULE. Its cells are cuboidal, have central nuclei, have lighter-staining cytoplasm than the proximal tubule, and carry no brush border. This portion joins a COLLECTING TUBULE. The collecting tubules receive the distal terminations of many nephrons. They open into the calyces of the pelvis through the PAPILLARY DUCTS. In a cortical nephron, the loop of Henle is quite short and typically does not reach into the medulla. It is around the tubules of such a nephron that we find the peritubular capillary network. The loops of Henle of the juxtamedullary nephrons, on the other hand, dip very deeply into the medulla, and the loops of these nephrons are followed by the looped vessels known as the vasa recti.

An anatomical difference such as described obviously suggests a functional difference between the two types of nephrons and, indeed, such a difference is found to exist. In general, it may be said that cortical nephrons are reabsorptive and secretory structures. Juxtamedullary nephrons also carry out these functions; in addition, they are responsible for creating an osmotic gradient within the parenchyma (cellular part of the kidney), which makes it possible to elaborate a hypertonic urine.

The formation of urine and regulation of blood composition

The nephron forms urine, which is HYPERTONIC and usually more ACID than the plasma. The nephron also governs, within narrow limits, the composition of the blood leaving the kidney. In accomplishing these tasks, the kidney is aided by hormones secreted by certain endocrine glands and by chemical substances originating within the organ itself.

The processes employed by the nephron in forming urine are FILTRATION, TUBULAR TRANSPORT, the COUNTERCURRENT MULTIPLIER and EXCHANGER, ACIDIFICATION, and CONCENTRATION of the filtrate.

Glomerular filtration

The blood, with its contained load of wastes, excess materials, and substances the body wishes to conserve, is delivered to the glomerulus under a hydrostatic blood pressure (P_b) of 60–75 mm Hg. The end artery structure of the arcuate arteries is instrumental in assuring maintenance of high pressure as the vessels branch and penetrate into the kidney. The glomerular capillaries and the closely applied visceral layer of the glomerular capsule act as a coarse sieve. Providing the blood pressure is higher than the sum of all opposing pressures (see below), all materials small enough to pass through these membranes will be forced through by the pressure differential in a process of filtration, and the fluid formed is called the FILTRATE. The capillaries behave as though they possess pores of small diameter (100Å), and the cells of the visceral layer (podocytes) do not form a continuous stratum (Fig. 29.8a, b, c). The foot processes of the podocytes as they are applied to the capillaries form a series of "slit pores" that govern the passage of materials into the capsule. In general, it may be stated that substances having a molecular weight up to 10,000 pass easily through these layers of tissue. Above this figure, materials pass with increasing difficulty, and a probable upper limit is reached at a molecular weight of about 200,000. Thus, nearly all materials present in the plasma, except the formed elements, are

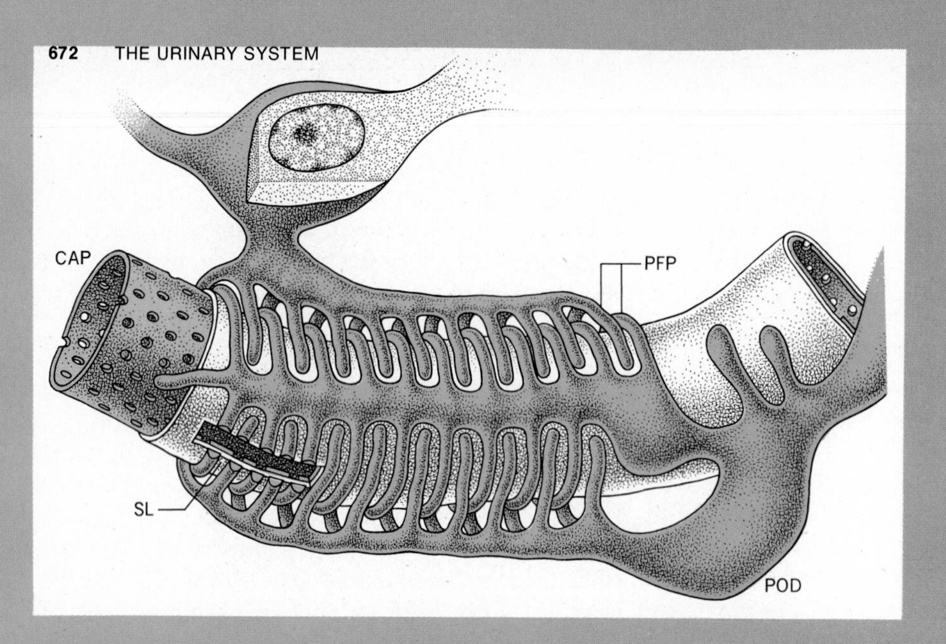

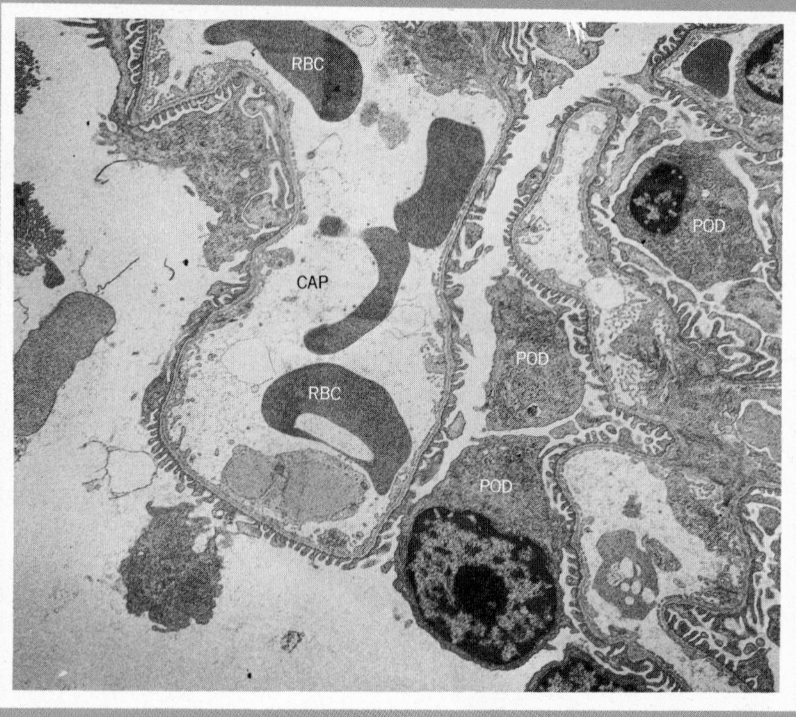

filtered into the cavity of the glomerular capsule. The filtrate contains protein, chiefly albumin, to the extent of 10–20 mg/100 ml. A mathematical calculation would indicate that some 30 g of protein is thus filtered per day. The material (filtrate) formed by this process resembles plasma (see Table 29.5), except that it lacks formed elements and has a lower protein concentration. Since the

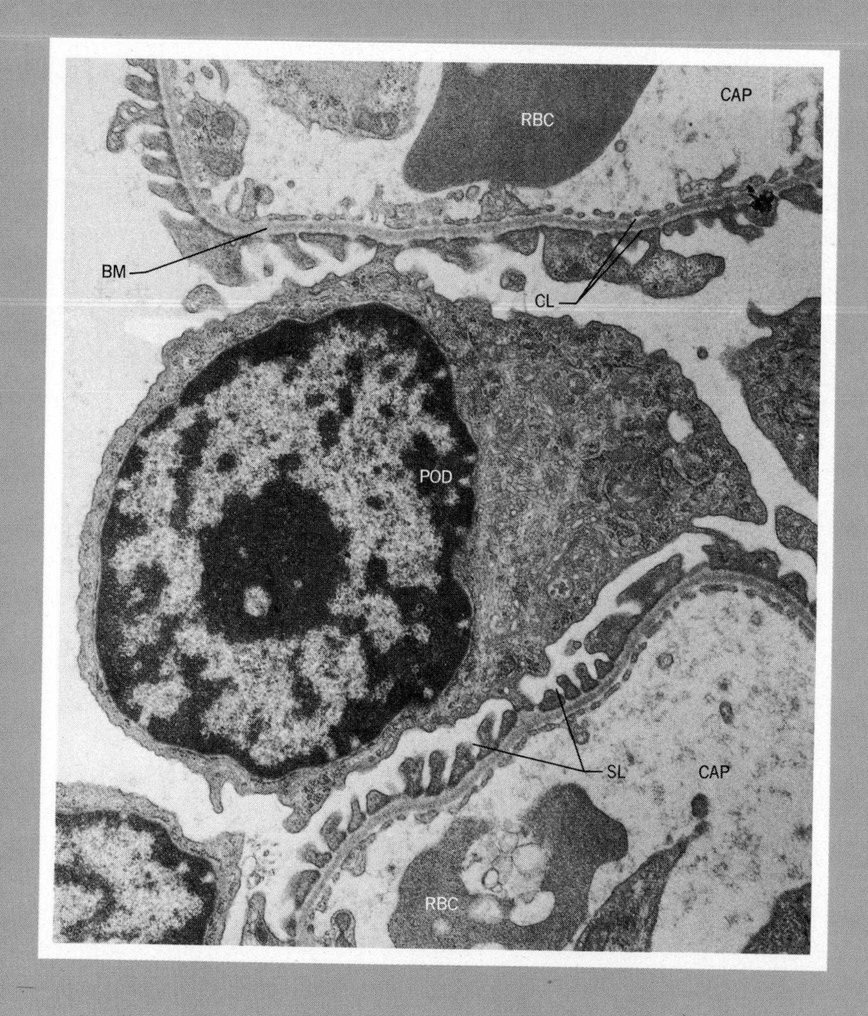

FIGURE 29.8 The ultrastructure of the renal corpuscle. CAP = capillary, POD = podocyte of visceral layer of capsule, RBC = red blood cell, BM = basement membrane, CL = cement layers, PFP = podocyte foot processes, SL = slit pore(s). (Electron micrographs courtesy Norton B. Gilula, The Rockefeller University.)

greater concentration of plasma proteins does not filter through, it exerts an osmotic pull, or back pressure, that tends to cause water movement back into the glomerulus. This osmotic back pressure (P_o) normally amounts to about 30 mm Hg. Also opposing the filtration of material through the glomerulus is the renal interstitial pressure (P_{rip}) or the pressure within the interstitial space around the kidney tubules, and the intratubular pressure (P_{it}). This latter quantity is the resistance the filtrate encounters as it attempts to push into the kidney tubules from the glomerulus. Because the tubules are already full of fluid, a resistance to the further movement of fluid is encountered. P_{rip} and P_{it} amount to about 10 mm Hg each. By adding together the opposing forces and subtracting them from the original blood pressure, a quantity called the effective filtration pressure (P_{eff}) may be calculated. This pressure represents the net pressure, or the force actually causing filtration through the glomerular capsule. Filtration will not proceed normally, nor will the wastes be removed effectively from the blood, unless a normal effective filtration pressure is maintained. The relationships described above may be expressed mathematically by the following formula:

$$P_{eff} = P_b - (P_o + P_{rip} + P_{it})$$

$$P_{eff} = (60\text{–}75) - (30 + 10 + 10)$$

$$P_{eff} = 10\text{–}25 \text{ mm Hg}$$

The rate of glomerular filtration (GFR) can be measured experimentally and averages 120 ml/min for both kidneys. The total plasma flow was 650 ml/min. If only 120 ml/min is filtered, a filtration fraction of 18.5 percent may be calculated (120/650 × 100). That is, only about one-fifth of the fluid arriving in the glomerulus is actually filtered.

Tubular reabsorption and secretion

The filtrate contains not only the waste materials that must ultimately be eliminated from the body, but also quantities of substances the body finds useful, such as water, hormones, vitamins, enzymes, glucose, inorganic salts, proteins, and amino acids. In order that these materials not be lost from the body, a second process must occur at this time—tubular transport, which involves the use of active processes to transport through the cells of the kidney tubule materials that would not otherwise be removed by passive processes. If the direction of tubular transport is from the lumen or cavity of the tubule through the cell and ultimately into the surrounding blood vessels, it is called REABSORPTION. The materials the body finds useful are reclaimed in this fashion. Some materials considered as wastes are not removed completely enough from the blood by filtration and are actively transported from the blood through tubule cells and into the tubule lumen. Movement in this direction is termed SECRETION.

In the proximal convoluted tubule an average reabsorption of 80–90 percent of the physiologically useful solutes is achieved. Among the materials actively transported through this region are glucose, amino acids and protein, phosphate, sulfate, uric acid, vitamin C, beta-hydroxybutyric acid, and calcium, sodium, and potassium ions. As these solutes are moved out by active processes, water follows osmotically, so that it too achieves an 80–90 percent reabsorption in this area. Chloride, a negatively charged ion, follows positive ions (Na^+ and K^+) out of the tubule by electrostatic attraction. Because of the water movement accompanying the solute transport, the NET TONICITY OF THE FLUID AT THIS POINT DOES NOT CHANGE. Its volume has been reduced, but its osmolarity and pH have not been affected.

The reabsorption of sodium ion by the proximal tubule is directly proportional to the concentration of aldosterone secreted by the adrenal cortex. Reabsorption of calcium ion is increased by parathyroid hormone. Glucose usually achieves a 100 percent removal in the proximal tubule because the carrier system transporting glucose has a transport maximum (T_{MG}) of 300–375 mg glucose per minute. The amount of glucose in the filtrate is normally 80–120 mg/100 ml. Therefore, the carrier system has no difficulty in removing all the filtered glucose. In certain diseases where the concentration of glucose in the blood is high, such as diabetes mellitus, the amount of filtered glucose may exceed the ability of the transport system to reabsorb it, and under these conditions glucose will spill over into the urine. Protein is also absorbed completely in the proximal tubule, probably by pinocytosis. As water passes out of the tubule, an increase in the concentration of urea in the tubule occurs. Urea (a nitrogenous waste) does not possess a carrier system and would not normally be reabsorbed. However, its concentration rises to a level at which some will pass out of the tubule by simple diffusion. The reabsorption of urea is an example of *obligatory* ("have to") *reabsorption* over which the nephron has no control. The movement by carriers can be controlled and is referred to as *facultative* (optional) *reabsorption*.

The proximal tubule is also an important area for secretion. A variety of organic acids (phenol red, hippuric acid, creatinine, para-aminohippuric acid, penicillin, Diodrast) and strong organic bases (choline, guanidine, histamine) are secreted into the filtrate by the proximal tubule cells.

THE CONCEPT OF CLEARANCE. The term *clearance* implies the ability of the kidney to eliminate a given substance from the plasma in a specific length of time. It is expressed as ml/min and is a hypothetical volume equivalent to the amount of a substance that would be contained in that volume of blood. For example, if a clearance is 100 ml/min, this implies that the amount of that substance removed per minute would be equivalent to that contained within 100 ml of blood. Additionally, clearance of certain substances may give an indication of how the kidney is treating a substance. For example, the glomerular filtration rate (GFR) may be measured by a substance that meets the following criteria: it is cleared or removed from the blood only by filtration by the kidney; it is not stored or metabolized by any body cell; it is not bound to proteins in the plasma; it is not toxic to the body. Such a substance is the polysaccharide *inulin*. The material is infused intravenously in a large initial dose, followed by a continuous infusion to keep the plasma level constant. A timed urine specimen is then collected and analyzed for inulin. A plasma sample is also collected and its inulin level determined. The GFR is calculated by the equation:

$$C_{in} = \frac{U_{in}V}{P_{in}}$$

Where C_{in} = clearance of inulin (GFR), or rate of removal from the blood (ml/min)
U_{in} = concentration of inulin in urine (mg/ml)
V = rate of urine formation (ml/min)
P_{in} = plasma concentration of inulin (mg/ml)

With typical values substituted in the equation, inulin clearances (and GFR) are about 120 ml/min, or about 180 liters per day. If a material has a clearance equal to that of inulin, it may be concluded that the kidney is treating it in the same manner as inulin—that is, by filtration only. If a substance has a clearance less than that of inulin, it must have been removed (i.e., reabsorbed) from the filtrate because it appears in the urine at a slower rate than did inulin. If a substance has a clearance greater than that of inulin, it is being added to the filtrate (i.e., secreted) to give a greater concentration than could have appeared due to filtration alone. Clearance tests can thus be performed to assess the level of kidney function.

The filtrate next enters the loop of Henle, and here, in the juxtamedullary nephron, is exposed to the countercurrent multiplier and the countercurrent exchanger. These mechanisms create the conditions necessary for achieving a concentrated or hypertonic urine.

The countercurrent multiplier

This physiological mechanism is so named because of the hairpinlike arrangement of Henle's loop and the fact that fluid flows in opposite directions in the two limbs (countercurrent). *It has as its purpose the creation of an increasing concentration of salt as one progresses from the periphery to the tips of the medullary pyramids.* As the mechanism operates, the concentration of solute is increased (multiplied) in the descending limb and renal interstitial fluid (interstitium) by the activity of the ascending limb. In order for the multiplying effect to be achieved, several conditions must be met. First, the ascending portion of the loop of Henle must be capable of actively transporting ions from the filtrate in its lumen into the inter-

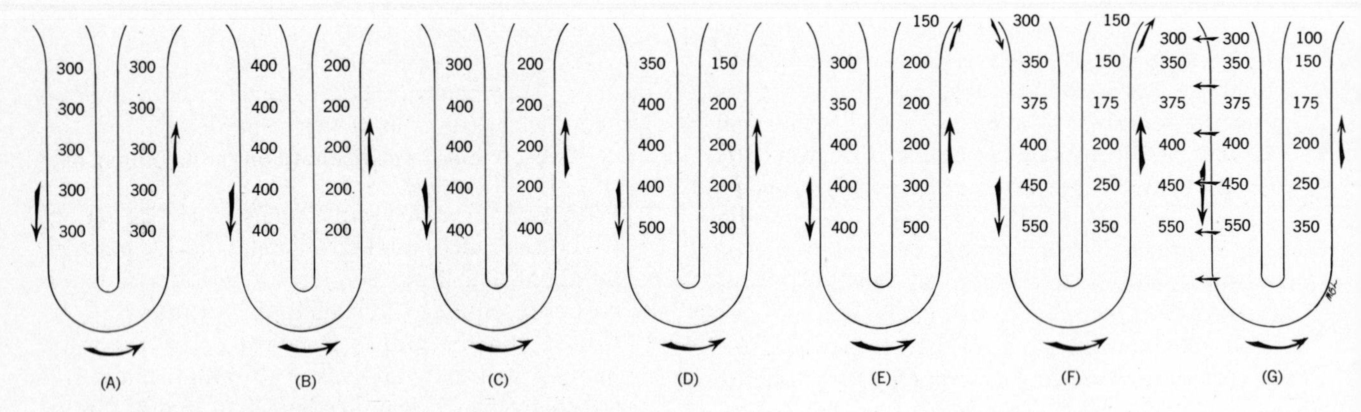

FIGURE 29.9 The operation of the countercurrent multiplier.

stitium and into the descending portion of the loop. A carrier system for sodium ion exists in these cells, and the predominant extracellular ion is sodium. Therefore, one may say that the creation of the osmotic gradients depends on the transport of sodium ion. Second, the ascending portion of the loop must be impermeable to water, so that as sodium is transported out, water does not follow it osmotically. Third, the ascending portion of the loop must be capable of transporting sodium to the extent that it can achieve a 200 milliosmolar (mOs) difference in the concentration of sodium between it and the descending portion of the loop. Assuming an initial input of fluid to the top of the descending portion of the loop of 300 mOs concentration, and viewing what happens next much in the same fashion as a movie film run slowly enough that one can see the individual frames, what happens is described below (Fig. 29.9).

Assuming the entire loop to first be filled with 300 mOs fluid

a. The ascending loop will transport sodium into the descending loop until a 200 mOs difference has been attained.

b. This activity will place a 400 mOs solution in the descending loop opposite a 200 mOs solution in the ascending loop.

If we now advance the whole mechanism one step,

c. A new mass of fluid of 300 mOs concentration will enter the top of the descending loop, and a 200 mOs solution will exit from the top of the ascending loop. In the lower part of the loop, moving a 400 mOs area around the tip and into the ascending loop will place it opposite another 400 mOs region, which descends into the space vacated by the one that moved around the tip of the loop. The 200 mOs difference has been lost at the top and tip of the loop. Sodium ion will again be transported,

d, until a 200 mOs difference has been attained. In the lower part of the descending loop, a concentration of 500 mOs will be attained, with a 300 mOs area opposite it. Advancing the mechanism still another step,

e, will place a 300 mOs solution opposite a 200 mOs solution in the ascending portion of the loop. Again, sodium ion will be transported until the 200 mOs difference is attained.

f,g. As this process progresses, it can be seen that a gradually increasing milliosmolar concentration will be created as the descending loop is traversed, and a gradually decreasing milliosmolar concentration will result in the ascending portion.

Remembering that the descending portion of the loop permits free diffusion of materials, the changes created in the descending loop will result in similar changes in the interstitium. Therefore, an increasing concentration of solute (Na^+) will be found the deeper one penetrates into the medulla.

The maximum figure to which the concentration can rise, assuming that the previously mentioned conditions are met, is 1200–1400 mOs (Fig. 29.10). The material thus delivered to the distal convoluted tubule is hypotonic, containing about one-third the solute originally fed into the loop.

The countercurrent exchanger

If the sodium ion that diffused or was actively transported into the interstitium is to exert any osmotic activity, it must remain within this area of the kidney. The countercurrent exchanger

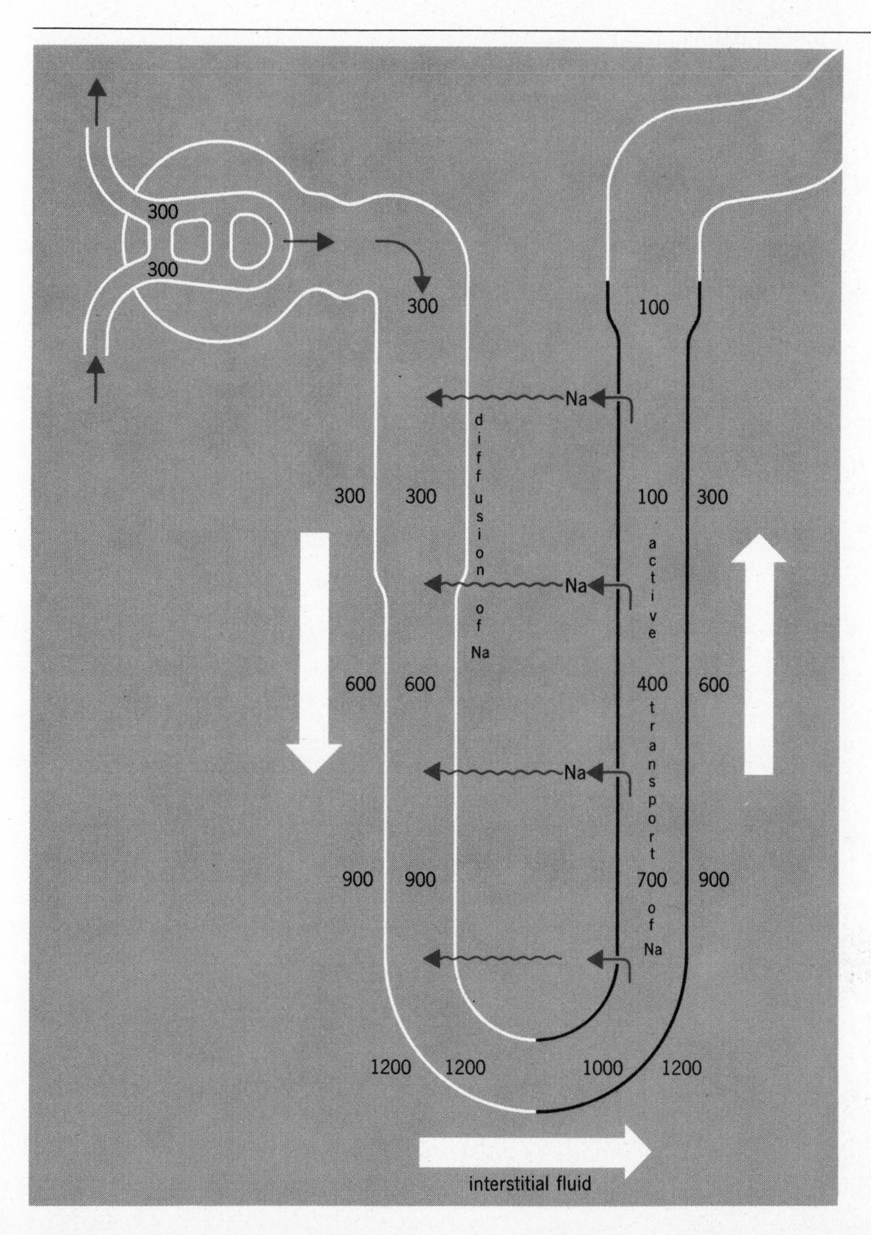

FIGURE 29.10 The end result of the countercurrent multiplier's operation.

operates between the interstitial fluid of the kidney and the vasa recti. Blood in the vasa recta is forced to pass through the medullary region, which has the increasing concentration of solutes created by the multiplier. Into this vascular loop will diffuse sodium ion, and water will diffuse out, so that the changes occurring in the osmolarity of this blood follow the same pattern as the fluid in the loop of Henle. As the vascular loop climbs back out of the medulla, sodium ion diffuses out and water in. *This tends to keep the sodium ion within the medullary region and maintains the concentration gradient.*

Acidification

The distal convoluted tubule receives a hypotonic fluid from which it may reabsorb any remaining physiologically useful solutes. Recall that the proximal tubule achieved only 80–90 percent reabsorption of materials. The distal tubule is also an important area of secretion, primarily of hydrogen ion, ammonia, and potassium ion. The secretion of these three materials is important in the ability of the kidney to participate in the regulation of acid-base balance.

FIGURE 29.11 Methods of increasing H^+ secretion by the kidney. *(a)* Secretion of ammonia, *(b)* secretion of potassium.

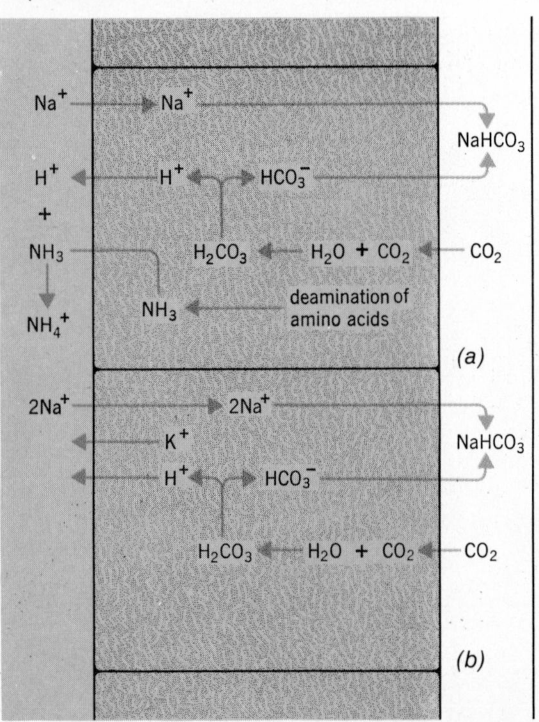

Acid-producing foods predominate over alkali-producing foods under normal dietary conditions. Much free hydrogen ion also results from the production of HCl by the stomach cells and subsequent reabsorption lower in the tract. The body therefore faces the problem of maintaining average pH (7.4) in the face of forces tending primarily to lower the pH. Acids entering the extracellular fluids first react with the chief buffer in the extracellular fluid, sodium bicarbonate. Carbonic acid is formed, which is subsequently carried to the lungs and disposed of as carbon dioxide, as shown in the following:

$$HA + NaHCO_3 \longrightarrow NaA + H_2CO_3 \xrightarrow[\text{in lungs}]{\text{carbonic anhydrase}} H_2O + CO_2 \uparrow$$

A = unspecified anion ($SO_4^=$, $PO_4^=$, Cl^-)

The anion combined with sodium will eventually be excreted in the urine. It would be disadvantageous to the body to lose the anion as the sodium salt since sodium, in the form of sodium bicarbonate, is the primary buffering agent of the extracellular fluid. The kidney rids the body of the anion and reabsorbs the sodium, exchanging it for a hydrogen ion.

Ammonia, derived from the deamination of amino acids and glutamine, diffuses into the tubule lumen and there reacts with a hydrogen ion to form an ammonium ion. The ammonium ion may then react with a variety of anions and thus carry out not only an excess hydrogen ion, but an anion as well. Ammonia production normally occurs at a relatively low level unless the body is presented with large amounts of hydrogen ion. Ammonia production therefore acts as an

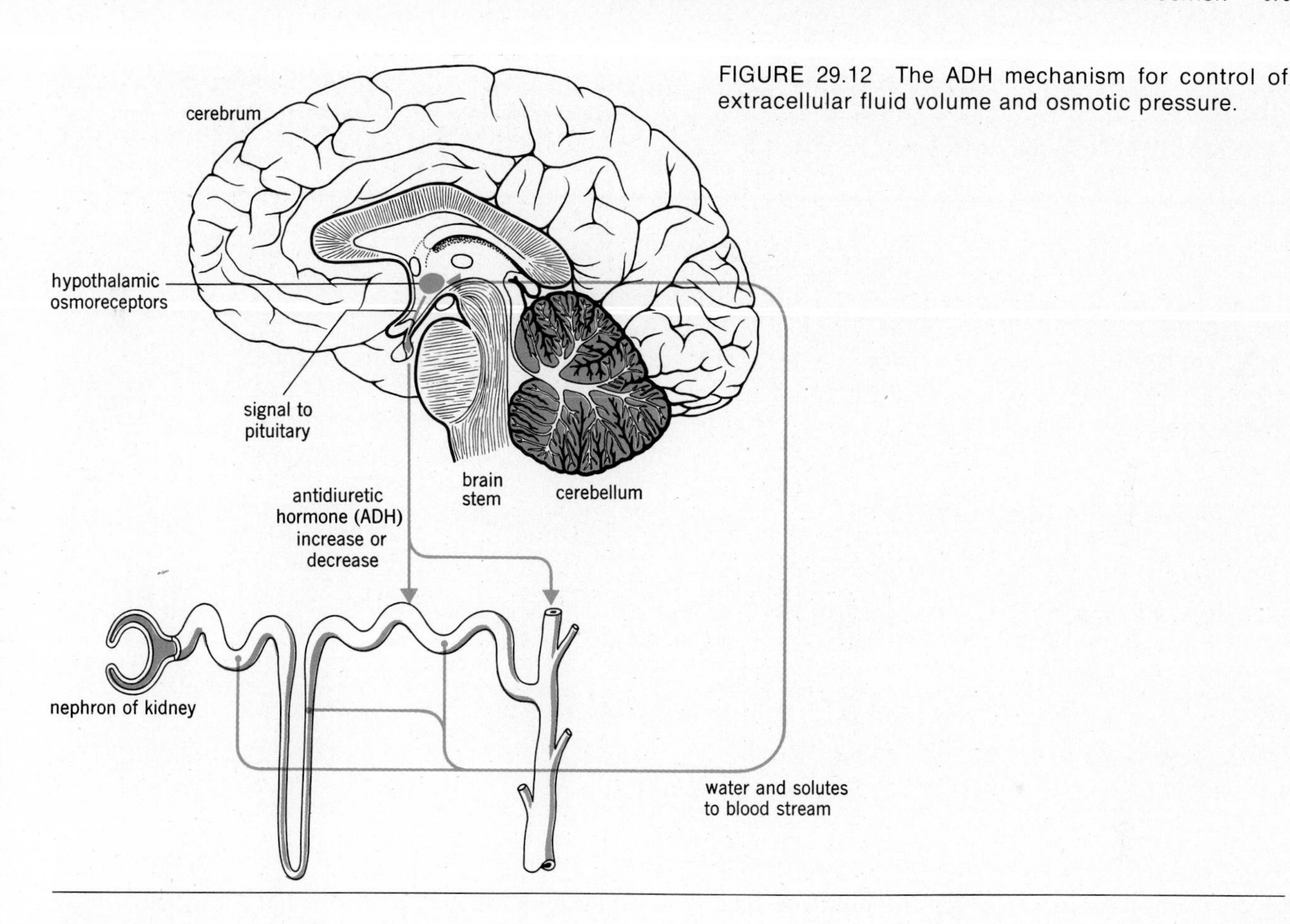

FIGURE 29.12 The ADH mechanism for control of extracellular fluid volume and osmotic pressure.

additional means of disposing of hydrogen ion, in the form of ammonium ion. Figure 29.11 illustrates the operation of some of these processes.

The urine leaving the distal tubule has thus been acidified, but is still hypotonic, since any exchange of materials has been made primarily on a one-for-one basis.

Production of a hypertonic urine

The collecting tubules receive the hypotonic solution from the distal tubule. These tubules run through the medulla toward the tips of the pyramids. Remember that in the medulla the countercurrent multiplier has operated to create an increasing degree of solute concentration toward the pyramid tip. The tubule therefore has around it a solution of much greater solute concentration than the fluid in the tubule. There is thus a tendency for water to leave the collecting tubule by osmosis, but this is not permitted unless the tubule becomes permeable to water.

The tubule becomes permeable to water in the presence of ADH or ANTIDIURETIC HORMONE. This substance is produced in the hypothalamus and is stored in and released from the posterior lobe of the pituitary gland (Fig. 29.12). It exerts a permissive effect on the cells, causing them to allow passage of water into the interstitium. The amount of ADH secreted is determined by the osmotic pressure of the blood reaching the hypothalamus. As the fluid passes through the collecting tubules and into the papillary ducts, it loses more and

TABLE 29.3 Summary of the processes occurring in urine formation

Process	Where occurring	Force responsible	Result
Filtration	Renal corpuscle	Blood pressure, opposed by osmotic, interstitial, and intratubular pressures	Formation of fluid having no formed elements and low protein concentration
Tubular transport Reabsorption	Proximal tubule Distal tubule Loop of Henle	Active transport	Return to bloodstream of physiologically important solutes
Secretion	Proximal tubule Distal tubule	Active transport	Excretion of materials Acidification of urine
Acidification (acid and base regulation)	Distal tubule	Active transport and exchange of alkali for acid	Excretion of excess H^+ Conservation of base (Na^+ and HCO_3^-)
Countercurrent multiplier and exchanger	Loop of Henle and vasa recta	(Multiplier) active transport (Exchanger) diffusion	Creates conditions for hypertonic urine formation
ADH mechanism	Collecting tubule and papillary ducts	Osmosis of water under permissive action of ADH	Formation of hypertonic urine

more water and will become hypertonic. The highest concentration to which the fluid can rise is 1200–1400 mOs, which is the maximum concentration that the multiplier can create. Normally, a urine of about 800 mOs is produced, indicating that the cells of the collecting tubules never become freely permeable to water. Table 29.3 summarizes the processes involved in urine formation.

Factors controlling urine volume

The volume of urine produced by the kidneys depends on the balance between several of the basic physiological processes discussed in the previous sections.

The amount of filtrate formed varies directly with the HYDROSTATIC PRESSURE of the blood. Assuming that osmotic back pressure remains constant, then the higher the blood pressure, the greater the rate of filtration. In shock or cardiac failure the blood pressure may fall below that required for adequate filtration. Uremia may result and require artificial removal of wastes. The principle of the artificial kidney is shown in Fig. 29.13.

The volume of the filtrate may be increased by an INCREASE IN CONCENTRATION OF SOLUTES in the filtrate, or by a decrease in plasma protein content. In diabetes mellitus, for example, the large amount of glucose present in the filtrate draws water osmotically into the tubule, or prevents its loss as sodium is transported out.

The volume of urine formed varies inversely with the amount of ANTIDIURETIC HORMONE (ADH) secreted, and in turn, ADH secretion varies directly with the solute concentration of the blood reaching the hypothalamus. Excess fluid intake poses a threat of dilution of the internal environment and is relieved by decrease in water reabsorption. Diabetes insipidus results from loss or greatly diminished production of ADH.

The volume of fluid excreted by the urinary system varies inversely with WATER LOSS BY OTHER SYSTEMS, especially the digestive system, and water loss through perspiration. The body must maintain a reasonable balance between fluid intake and loss, and if vomiting, diarrhea, or excessive sweating increase fluid loss, urine volume will decrease. This again is effected mainly by variation in ADH secretion.

Alteration in the efficiency of operation of the countercurrent multiplier, particularly in the transport of sodium, will change the urine volume. Certain drugs, such as diuretics, increase urine volume by decreasing the active removal of sodium ion from the filtrate. This allows these ions to remain as osmotically active particles in the filtrate. Diuretics (Table 29.4) also cause the countercurrent multiplier to equilibrate at a lower maximum concentration, so that the amount of water passing through the walls of the collecting tubules will be less.

The kidney and acid-base balance

The kidneys play a vital role in the regulation of acid-base homeostasis by their ability to secrete H^+ in the proximal and distal convoluted tubules. Hydrogen ion accumulates in the body as a result of the production of CO_2 from metabolic processes, and the reaction of CO_2 with H_2O is according to the following equation:

$$CO_2 + H_2O \rightleftharpoons H_2CO_3 \rightleftharpoons H^+ + HCO_3^-$$

The body normally is capable of preventing the accumulation of excessive H^+ in the form of H_2CO_3 by elimination of CO_2 through the lungs. Additional sources of H^+ include such metabolic sources as the production of organic acids during incomplete combustion of foodstuffs, the production of acids from dietary intake of phosphorous (as phosphate), sulfur (as sulfate), and the metabolism of compounds containing these elements. The lungs play only a minor role in the elimination of H^+ from these sources; elimination depends on the kidney.

Hydrogen ion is secreted by the proximal and distal tubule cells in exchange for sodium ion, the latter being returned to the bloodstream. The exchange is made on a one-for-one basis. Secreted H^+ reacts with bicarbonate in the tubular lumen to form carbonic acid; carbonic acid dissociates to form CO_2 and H_2O, with the CO_2 diffusing into the tubule cells according to its concentration gradient where carbonic anhydrase catalyzes the formation of carbonic acid. The dissociation of

TABLE 29.4 Some diuretics and their mechanisms of action

Mechanism of action	Examples	Site of action and result
Inhibition of ADH secretion	Water	Hypothalamus; urine volume ↑
	Ethylalcohol	Hypothalamus; urine volume ↑
Inhibition of Na^+	Caffeine	Glomerulus and tubules; GFR; ↓ tubular Na^+ reabsorption ∴ volume
reabsorption	Mercurial salts	Beyond proximal tubules; volume ↑
	Thiazides (Diuril)	Loop of Henle and distal tubule. Volume ↑
	Furosemides (Lasix)	Loop of Henle. Volume ↑
Decrease H^+ secretion and thus Na^+ reabsorption	Carbonic anhydrase inhibitors (Diamox)	Proximal and distal tubules; since H^+ and Na^+ are exchanged one for one, results in Na^+ remaining in tubules. Volume ↑

this carbonic acid provides more H^+ for secretion, and the bicarbonate that results is transported into the bloodstream. The process is illustrated in Fig. 29.11 and results in maintenance of body bicarbonate and sodium reserves at the same time it rids the body of H^+. When H^+ excretion and $NaHCO_3$ reabsorption alone are insufficient to maintain acid-base balance, the kidney may secrete ammonium ion (NH_4^+) in addition to H^+, and it may secrete potassium ion (K^+) in addition to H^+. Deamination of amino acids in the tubule cells produces ammonia, which combines with an H^+ to form the ammonium ion, according to the equation

$$NH_3 + H^+ \longrightarrow NH_4^+$$

The ammonium ion then diffuses into the tubule lumen. Active potassium secretion results in the passage of two positive ions (H^+ and K^+) into the tubule lumen and results in the reabsorption of 2 Na^+ in exchange. Both mechanisms essentially double the number of $NaHCO_3$ molecules reabsorbed from the tubule lumen, restoring the alkali of the bloodstream to a normal ratio of H^+. These processes are also depicted in Fig. 29.11.

The reactions described above do not occur instantaneously; a day or two is required to restore normal pH in the face of acid-base disturbances. As presented in Chapter 21, there are four major disturbances of acid-base balance. RESPIRATORY ACIDOSIS occurs from diminished elimination of CO_2 through the lungs. Accumulation of H^+ stimulates breathing, which may suffice to eliminate the excess CO_2 producing the disorder. Additionally, increased kidney secretion of H^+ aids in returning the pH to normal. On its own, over a period of about two days, the kidney can fully compensate the condition. RESPIRATORY ALKALOSIS results from an excessive loss of CO_2 that may occur during hyperventilation. Loss of CO_2 lowers the H^+ available from the reaction of CO_2 and H_2O. Loss of CO_2 slows breathing, and kidney secretion of H^+ is lowered. Diminished secretion of H^+ results in less HCO_3^- available for reabsorption, and the pH falls. METABOLIC ACIDOSIS develops from loss of bicarbonate from the body, as in diarrhea, or from excessive H^+ produced by exaggerated or abnormal metabolic processes, as in diabetes mellitus. The kidney increases its secretion of H^+ and elimination of organic acids and reabsorbs increased amounts of bicarbonate ion. METABOLIC ALKALOSIS may result from loss of H^+, as in vomiting, or as a result of excessive intake of alkalies, as in antacid medications. Here again, decreased secretion of H^+ by the kidney tubules and lowered reabsorption of bicarbonate ion aids in returning pH to normal. Therefore, the kidney is the last—but most powerful—line of defense in control of acid-base homeostasis.

The urine

Composition and characteristics

Normal urine is an amber or yellow transparent fluid having a pH between 5 and 7, a characteristic odor, and a specific gravity varying between 1.015 and 1.025.

The COLOR is due to the presence of urobilinogen, a pigment derived from the destruction of hemoglobin by cells of the reticuloendothelial system, as follows:

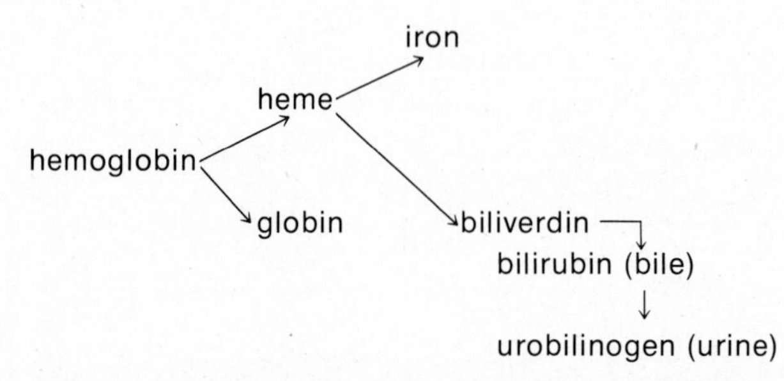

The ACIDIC REACTION is due primarily to secre-

TABLE 29.5 Comparisons of the properties of plasma, filtrate, and urine[a]

Property or substance	Plasma	Filtrate	Urine	Degree of change between filtrate and urine
Osmolarity (mOs)	300	300	600–1200	2–4 fold
pH	7.4 ± 0.02	7.4 ± 0.02	4.6–8.2	300–30,000 fold
Specific gravity	1.008–1.010	1.008–1.010	1.015–1.025	–
Sodium	0.3	0.3	0.35	–
Potassium	0.02	0.02	0.15	7 fold
Chloride	0.37	0.37	0.60	2 fold
Ammonia	0.001	0.001	0.04	40 fold
Urea	0.03	0.03	2.0	60 fold
Sulfate	0.002	0.002	0.18	90 fold
Creatinine	0.001	0.001	0.075	75 fold
Glucose	0.1	0.1	0	–
Protein	7–9 g %	10–20	0	–

[a] Values, unless otherwise indicated are grams %.

TABLE 29.6 Some normal constituents, origins, and amounts excreted per day in normal urine

Constituent	Origin	Amount per day
Water	Diet and metabolism	1200–1500 ml
Urea	Ornithine cycle	30 g
Uric acid (purine)	Catabolism of nucleic acids	0.7 g
Hippuric acid	Liver detoxification of benzoic acid	Trace
Creatinine	Destruction of intracellular creatine phosphate of muscle	1–2 g
Ammonia	Deamination of amino acids	0.45 g
Chloride (as NaCl)	Diet	12.5 g
Phosphate	Diet and metabolism of phosphate containing compounds	3 g
Sulfate	Diet, metabolism of sulfate containing compounds, formation of H_2SO_4 in kidney tubules	2.5 g
Calcium	Diet	200 mg

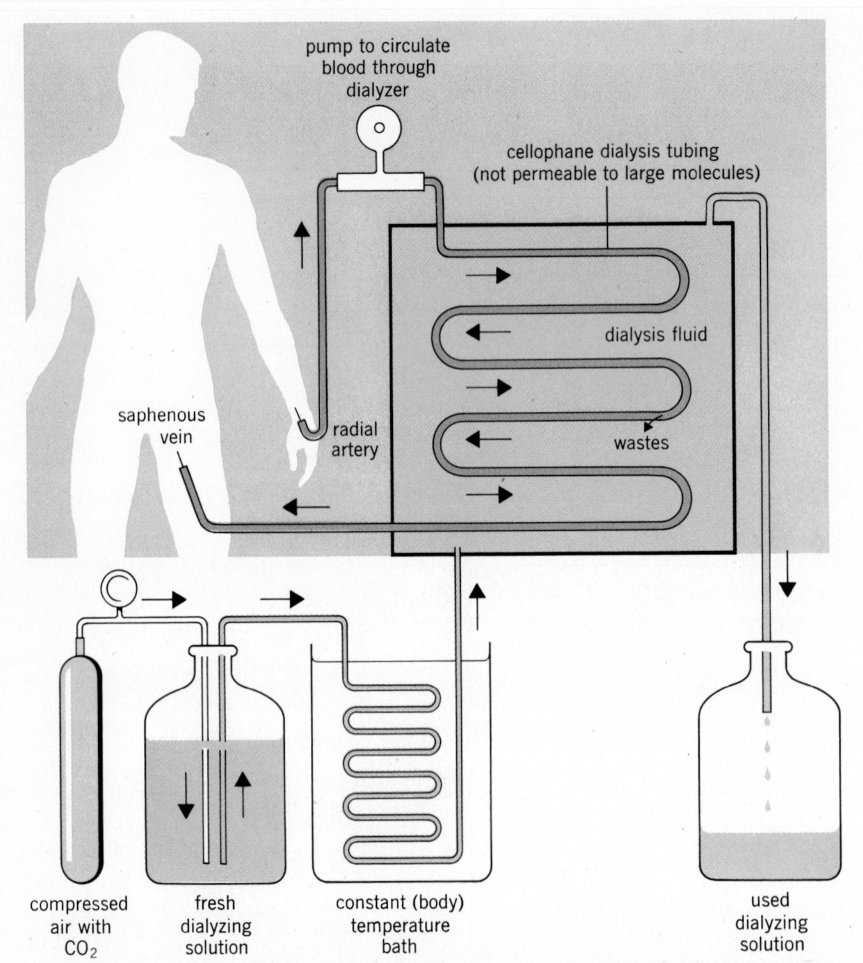

FIGURE 29.13 The principle of operation of the artificial kidney (hemodialysis unit). The dialysis fluid contains electrolytes and essential substances in the same proportions as in the blood.

tion of H^+ into the filtrate by the proximal and distal tubules. The ODOR of freshly voided urine is due chiefly to the presence of organic acids, while the odor of stale urine is ammoniacal due to the decomposition of urea with release of ammonia. The specific gravity depends on the total solute concentration per unit volume of fluid and accounts for the variability of this characteristic.

The composition of normal urine varies little in terms of materials present. Tables 29.5 and 29.6 indicate some comparisons between plasma, filtrate, and urine and some normal constituents of urine, their origins, and amounts excreted per day.

Glucose normally does not appear in the urine except after a meal rich in carbohydrates because it is usually completely reabsorbed.

Protein is filtered only to a limited extent and is also normally completely absorbed.

The ketone bodies, acetone and beta-hydroxybutyric acid, are present in the urine only when large amounts of fatty acids are being catabolized, as in diabetes mellitus. Only when these materials are present in the urine above trace levels can their presence be considered abnormal.

Casts may be found in the urine and are evidence of renal lesions. They are named according to composition, and include epithelial, fatty, pus, blood, or hyaline (clear) casts. The casts maintain the shape of the kidney tubules where they are formed.

CALCULI *(stones)* are precipitated masses of inorganic material. They may grow to large size and bizarre shapes, and may require surgery for their removal.

Signs of disordered renal function

Prolongation of POLYURIA (excessive urine volume), OLIGURIA (scanty urine volume), or ANURIA (cessation of urine production) beyond a few hours is usually indicative of abnormal renal function. DECREASED CLEARANCES of inulin or other test materials and increased blood urea levels (uremia) also indicate malfunction. In damage to the nephrons, PROTEINURIA typically occurs due to increased permeability of glomerular membranes or trauma and destruction of tubule cells. The proteins appearing in the urine are most commonly albumins from the blood. They are the smallest and pass through membranes most easily. Accumulation of urea as uremia results in lethargy, pruritus (itching), muscle twitching, mental confusion, and coma. If not lowered, urea may cause death through toxic effects on the brain and liver. The artificial kidney may be employed in the process of HEMODIALYSIS. EDEMA or excessive fluid retention is also characteristic of certain renal disorders, particularly those secondary to heart failure and decreased filtration of solutes. ACIDOSIS, secondary to low filtration of metabolic acids or secondary to decreased H^+ secretion, may also occur in kidney malfunction.

The artificial kidney (Fig. 29.13) relies on the principle that individual chemical substances in the bloodstream will diffuse through selective membranes if a gradient is created to cause that diffusion. In the artificial kidney are many feet of cellophane dialyzing tubing, surrounded by a fluid having a composition similar to blood except for zero concentrations of wastes (e.g., urea, creatinine, uric acid). The machine receives its blood from the radial artery of the patient. Blood entering the "kidney" is infused with heparin to keep it from coagulating in the machine. Diffusion of wastes into the dialyzing fluid occurs in the machine. Outflowing blood is infused with an antiheparin substance (e.g., protamine) to minimize bleeding in the patient as blood is returned to the saphenous vein of the lower limb. About 500 ml (1 pint) of the patient's blood is in the machine at any given time, a volume loss to which the body adjusts easily. Usually, a patient is "on the machine" for 8–12 hours every three to four days. Actual time spent depends on the severity of damage to the patient's kidneys or if the organs are present. The machine is thus used to "rest" damaged kidneys or to substitute entirely for missing organs.

Clinical considerations

The most common functional disorders affecting the kidney are associated with genetic and infectious causes.

GENETIC CAUSES. At least seven errors of metabolism associated with defects in renal transport systems are recognized. Gene defects, which result in an abnormal enzyme or carrier involved in an active transport system, are postulated as the causes of the defects. Several of the more important disorders are presented in Table 29.7.

WILMS' TUMOR is a malignant embryonic renal tumor of genetic origin that forms the most common malignant abdominal tumor of infancy and childhood. It usually occurs in only one kidney, and removal of the affected organ, followed by radiation and/or chemotherapy, results in 70–90 percent survival rates.

INFECTIOUS CAUSES. PYELONEPHRITIS is a bacte-

TABLE 29.7 Some disorders of renal transport mechanisms

Condition	Heredity	Disturbance of	Controlled by	Comments
Nephrogenic diabetes insipidus	Autosomal dominant	Water excretion; tubule cells fail to respond to ADH	Reduction of dietary solutes and protein intake	Control measures extend 1–2 years; then child can usually control own diet
Renal glucosuria	Autosomal dominant	Glucose reabsorption; failure to reabsorb normal amounts of glucose in the kidney	None required	Glucose appears in urine when blood sugar level is normal
Hartnup disease	Autosomal recessive	Tryptophane (an essential amino acid) reabsorption and metabolism; gross amino aciduria	Administration of nicotinamide	May lead to shortage of nicotinic acid for NAD and NADP synthesis
Cystinuria	Autosomal recessive	Transport of the four dibasic amino acids (cystine, lysine, arginine, ornithine)	Administration of bicarbonate to prevent stone formation in kidney; diuresis	Cystine stones are most common complication; treatment minimizes formation

rial infection of the kidney in which the causative organism is delivered to the kidney via the bloodstream, or there is a retrograde infection from the bladder up the ureters to the kidney. Antibiotic therapy depends on the sensitivity of the causative organism to the drug. GLOMERULONEPHRITIS results from an infectious agent, or its product(s), that damages the glomerular basement membrane. Chemical alterations in its structure render it antigenic, antibodies are produced, and the resultant antigen-antibody reaction results in cracking and splitting of the basement membrane. Loss of plasma albumin is common in this disorder. The causative organism involved is a streptococcus. Infections of the kidney are the most common causes leading to renal failure, which implies an inability of the kidney to perform its normal tasks adequately. Filtration rate falls, urea elimination is decreased, and signs of uremic poisoning ("odor of urine" to the perspiration, generalized itching, low urine output) appear. The NEPHROTIC SYNDROME is the result of glomerular damage and tubular necrosis (death of the tissues). The primary symptom is a massive edema that develops secondarily to plasma protein loss through the damaged nephrons. Treatment is directed toward prevention of secondary complications (e.g., infections) while the damage is repaired by the kidney itself.

Other functions of the kidney

At the point shortly before the afferent arteriole enters the glomerulus, there is an interesting group of modified smooth muscle cells called *the juxtaglomerular cells.* An adjacent region on the distal tubule, called the *macula densa*, consists of modified tubule cells. The two areas together constitute the JUXTAGLOMERULAR APPARATUS (see Fig. 25.25). This particular region of the arteriole has been shown to be sensitive to changes in blood flow. It has been hypothesized that if the kidney becomes ischemic, the apparatus secretes an enzymelike material called *renin*. This particular substance works on a substrate in the plasma called hypertensinogen (angiotensinogen) and converts it into a weakly active material, hypertensin I. Hypertensin I is in turn converted to hypertensin II by a plasma enzyme. Hypertensin II is an active vasoconstrictor, bringing about a narrowing of arterioles over the entire body. This particular effect raises the blood pressure and ensures a continuance of high blood pressure to the kidney, which is a necessary requisite for filtration to occur. Renal hypertension, a chronic elevation of the blood pressure, may occur if the kidney suffers continual ischemia and renin production is not diminished. Hypertensin II also affects the outer zone of the adrenal cortex which secretes aldosterone. The hormone is instrumental in governing reabsorption of sodium by the kidney tubules. Therefore, the kidney governs its own activity to a small degree.

A great degree of AUTOREGULATION is demonstrated by the kidney in controlling blood flow to the glomerulus. The afferent arteriole has been shown to undergo constriction or dilation in response to a wide range (90–220 mm Hg) of renal artery pressure so as to maintain renal blood flow constant. The response occurs in the absence of all nerves to the kidney and in a kidney removed from the body and artificially perfused. It is therefore an inherent mechanism within the kidney itself.

Three theories have been advanced to explain autoregulation.

The THEORY OF CELL SEPARATION suggests that as blood passes through the interlobular arteries into the afferent arterioles, a separation of red cells and plasma occurs that allows greater perfusion of some glomeruli, while others suffer a reduction. Things "balance out" to maintain an overall nearly constant rate of filtration.

The THEORY OF INTRARENAL PRESSURE suggests that increases in renal arterial pressure to the kidney are matched by increases in the pressure of the interstitial fluid of the kidney. This compresses the capillary networks and veins of the kidney to impede flow and maintain pressure.

The MYOGENIC THEORY suggests that all autoregulation is the result of changes in tone of the smooth muscle of the afferent and efferent arterioles. It is known that smooth muscle cells respond to stretching (as with an increased perfusion pressure) by a stronger contraction. This maintains vessel diameter, flow, and pressure. Decrease of pressure causes relaxation that increases diameter and maintains flow and pressure. If the afferent arteriole dilates and the efferent arteriole does not change, flow and pressure to the glomeruli will increase. Dilation of the efferent arteriole or constriction of the afferent arteriole reduces flow and pressure in the glomeruli. Also, pressure changes in the afferent arteriole may influence the juxtaglomerular apparatus to secrete chemicals (e.g., renin) that result in vascular diameter changes. The myogenic theory would appear to have more direct proof to sustain it because autoregulation is eliminated when the kidney is perfused by substances (cyanide, procaine) that paralyze smooth muscle.

The ureters

Gross anatomy and relationships

The ureters extend from the hilus of the kidney to the urinary bladder, a distance of 28–35 cm (about 11–14 in.). They are retroperitoneal in placement and have a decreasing diameter as they course toward the bladder.

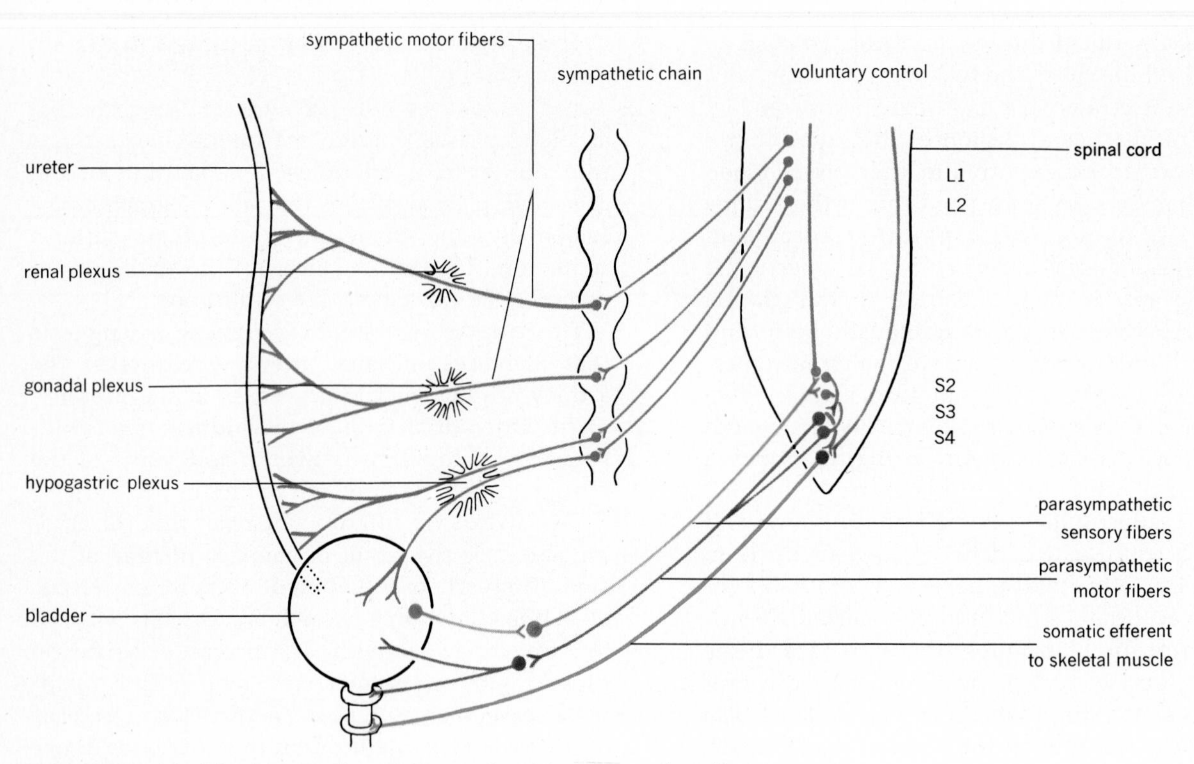

FIGURE 29.14 The innervation of the ureter, bladder, and urethra.

Miscroscopic anatomy

Three coats of tissue form the wall of the ureter. A MUCOSA, composed of transitional epithelium and lamina propria, forms the inner layer. The central MUSCULARIS is composed of inner longitudinal and outer circular layers of smooth muscle throughout most of the length of the ureter. On the lower one-third of the organ, a third layer of muscle (outer longitudinal) is added. The outer FIBROUS COAT *(adventitia)* blends without demarcation into the surrounding subserous fascia.

Innervation

The ureters in their upper part receive sympathetic fibers from the renal plexus; in the middle part, they receive fibers from the ovarian or spermatic plexus; near the bladder, they receive fibers from the hypogastric nerves (Fig. 29.14). These fibers exert primarily a motor effect. They cause the ureters to exhibit rhythmic peristaltic contractions traveling at a speed of 20–25 mm/sec and at a frequency of 1–5/min. The urine therefore enters the bladder, not in a continuous stream, but in separate squirts synchronous with the arrival of the peristaltic wave.

The urinary bladder

Gross anatomy and relationships

The urinary bladder serves as a reservoir for the urine until it is voided. The organ is located posterior to the symphysis pubis and is separated from the symphysis by a PREVESICULAR SPACE. The space, filled with loose connective tissue, allows for expansion of the filling bladder. Internally, three openings may be found in the bladder wall: the two ureters and the urethra. An imaginary line drawn to connect these three openings outlines the TRIGONE.

Microscopic anatomy

While similar to the ureter in structure, the bladder has more cell layers in the transitional epithelial lining, a submucous layer of loose tissue between mucosa and muscularis, and three heavy layers of smooth muscle in the muscularis, disposed longitudinally, circularly, and longitudinally. Around the urethral opening, a dense mass of circularly oriented smooth muscle forms the internal sphincter of the bladder. A serous layer is formed by the peritoneum over the superior surface of the organ.

Innervation

The efferent nerves to the bladder and urethra (Fig. 29.15) are from both sympathetic and parasympathetic divisions. The sympathetic fibers

FIGURE 29.15 The innervation of bladder illustrating the genesis of bladder dysfunction.

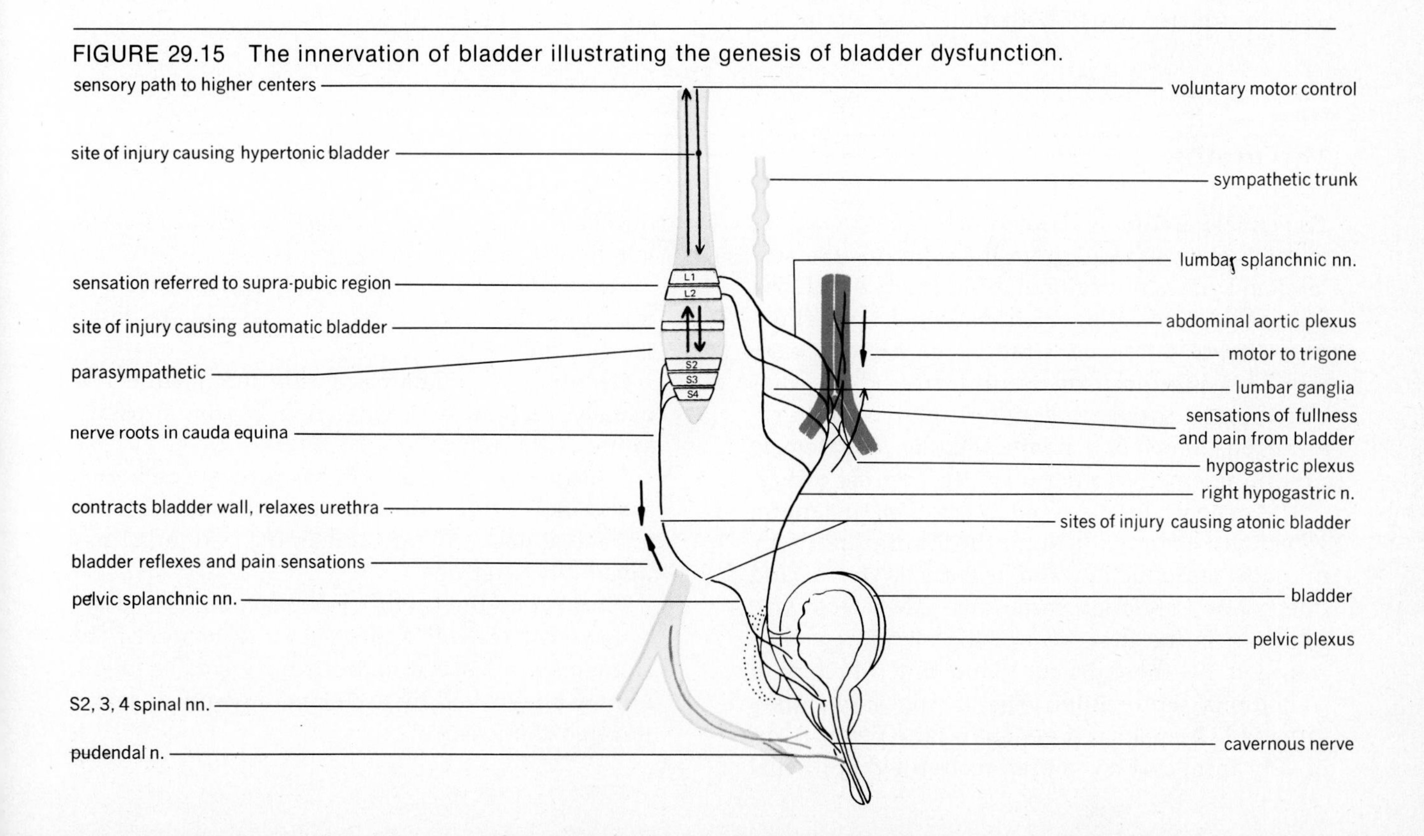

furnish inhibitory fibers to the muscle of the bladder; they furnish motor fibers to the trigone and internal sphincter, and to the muscle of the upper part of the urethra. These fibers rise in the lumbar spinal segments and pass to the bladder via the inferior hypogastric plexus. The parasympathetic nerves supply motor fibers to the detrusor muscle (the muscle of the bladder) and inhibitory fibers to the internal sphincter. The desire to urinate occurs when a volume of 200–300 ml of urine has accumulated in the bladder. Stretch on the muscle of the bladder, brought about by filling, evokes an afferent impulse in the pelvic nerves. This impulse ascends through the spinal cord to a center in the hindbrain. An efferent discharge of motor impulses to the muscle of the bladder is accomplished through descending pathways in the cord and through the pelvic motor nerves. The same efferent or motor fibers bring about a simultaneous relaxation of the internal sphincter so that urine may be emptied from the bladder into the urethra (micturition). Urination may occur by reflex action not involving the hindbrain center. Filling evokes a reflex contraction of the detrusor and relaxation of the sphincter, which is served by lower cord segments only. The reflex is seen in infants, where the voluntary control over sphincters has not yet been achieved. Nerve injuries produce three types of bladder dysfunction.

ATONIC BLADDER. Interruption of sensory supply results in loss of bladder tone, and the organ may become extremely distended with no development of an urge to urinate.

HYPERTONIC BLADDER. Interruption of the voluntary pathways results in excessive tone, and very small distentions create an uncontrolled desire to urinate.

AUTOMATIC BLADDER. Complete section of the cord above the first sacral nerve exit (S1) produces automatic emptying in response to filling, by the above-described cord reflex.

The urethra

The female urethra is about 4 cm (1¾ in.) in length, is completely separate from the reproductive system, and bears no regional differences. A MUCOSA is present, consisting of transitional epithelium near the bladder and stratified squamous elsewhere; a lamina propria underlies it. A MUSCULARIS is present, consisting of circularly arranged fibers of smooth muscle. The female urethra opens just anterior to the vaginal orifice.

The male urethra is about 20 cm (8 in.) in length, serves as a common tube for the terminal portions of both reproduction and urinary systems, and does show regional variations. The PROSTATIC URETHRA is the first 3 cm of the organ; it is surrounded by the prostate gland and lined with transitional epithelium. The MEMBRANOUS URETHRA is 1–2 cm long; it penetrates the pelvic floor, is very thin, and has a pseudostratified epithelial lining. The CAVERNOUS URETHRA is about 15 cm in length and lies within the penis.

Disorders associated with the lower urinary tract include CYSTITIS, URETHRITIS, and OBSTRUCTION of the urethra.

Cystitis, or inflammation of the bladder, is usually secondary to infection of the prostate, kidney, or urethra.

Infection of the male urethra is most commonly associated with gonorrheal infections; in the female, almost any infection of the perineum may invade the urethra.

Blockage of the urethra is most commonly associated with calculi (stones) in the bladder. The stone may be voided in the urine, if small enough, or may be crushed by use of a cystoscope inserted through the urethra.

Summary

1. Lungs, skin, alimentary tract, and kidneys are pathways of excretion for wastes of metabolism and materials of no use to the body.
2. The kidneys develop from the mesoderm at about 3½ weeks.
   a. Three kidneys form in series with the last (metanephros) forming the permanent kidney.
   b. The excretory ducts and bladder develop from the cloaca.
3. Anomalies of the kidney are common because of the nature of the developmental processes (three kidneys, different origin of kidney and ducts).
4. The bean-shaped kidneys
   a. Regulate homeostasis of the ECF and excrete wastes.
   b. Are located in the posterior upper abdominal cavity.
   c. Have a hilus, sinus, cortex, medulla, and contain functional units called nephrons.
   d. Are supplied with blood from the renal arteries.
5. Nephrons
   a. Are the functional units of the kidney.
   b. Contain a capsule, convoluted tubules, loop of Henle, and collecting tubules.
6. Urine is formed by a series of processes.
   a. Glomerular filtration creates a fluid similar to plasma but lacking cells and low in protein.
   b. Tubular transport adds materials to the filtrate (secretion) and removes essential substances (reabsorption).
   c. The countercurrent multiplier and exchanger create and maintain a high salt level in the renal ECF.
   d. Acidification occurs by active transport of H^+ in the convoluted tubules.
   e. A concentrated urine is produced when ADH permits water to leave the nephron osmotically.
7. Urine volume is influenced by several factors.
   a. Increase of blood pressure increases volume.
   b. Increase of plasma solutes increase volume.
   c. ADH increase decreases volume.
   d. Loss by other routes decreases volume.
8. The kidney is involved in acid-base balance and can compensate an imbalance in two or three days.
9. Urine is the yellow, usually transparent, acidic, waste-containing product of nephron activity.
10. The symptoms and signs of renal lesions are presented.
11. The kidneys produce renin, a factor in vasoconstriction, and autoregulate their blood flow through wide ranges.
12. Renal pelvis, ureters, and urinary bladder have a three-layered structure including transitional epithelium, muscularis, and an outer fibrous coat.
13. The urethra carries urine to the exterior and shows differences in the two sexes.
   a. The female urethra is short and separate from the reproductive system.
   b. The male urethra serves both urinary and reproductive systems and shows regional differentiation.

The Reproductive Systems

The term *reproduction* implies to most persons the act of mating between two members of opposite sex of one species, with consequent production of offspring. However, the term should be understood to include those processes that result in replacement of damaged or dead cells, as in the mitotic replacement of epithelial cells. In either case, the uncontrolled production of new units can result in neoplasms, or stress, crowding, and unhealthy competition for living space, with the possibility of creation of famine and pestilence.

The reproductive systems of the male and female provide the bases for production of sex cells (eggs or ova, and sperm), and for the housing and nourishment of offspring until they are sufficiently mature for independent survival. This chapter presents mainly the anatomical considerations associated with the systems; hormonal relationships appearing in the production of offspring are considered in Chapter 33.

Development of the systems

The development of the reproductive systems (Fig. 30.1) is associated with the development of the urinary system. It may be recalled that the mesodermal masses developing in the posterior portion of the celom are known as the urogenital ridges, and that parts of the developing kidneys serve as portions of the reproductive systems, particularly in the male.

During early development, the reproductive systems pass through similar series of stages. The gonad (sex gland; ovary or testis) shows first indication of development at about five weeks. At this time, a GERMINAL EPITHELIUM develops on the medial side of the urogenital ridge, which sends cords of epithelial cells, the PRIMARY SEX CORDS, into the underlying mesenchyme. An outer cortex and inner medulla may be distinguished at this time, and the gonad is said to be in an INDIFFERENT STAGE, capable of developing into either a testis or an ovary. In females XX chromosome constitution assures development of the CORTEX into an ovary, while in males, with an XY chromosome constitution, the MEDULLA differentiates into the testicular tissue, and the cortex undergoes regression.

In the male, at about seven weeks of fetal development, the primary sex cords branch and form the seminiferous tubules and the straight and rete tubules of the duct system. Mesenchymal cells between the cords give rise to the interstitial (Leydig) cells that synthesize the testicular hormone (testosterone). The epididymis, efferent ductules, and vas deferens are formed from the mesonephric tubules and duct that remain from the development of the mesonephros. Accessory glands (prostate, seminal vesicles, bulbourethral) form as outgrowths of the deferens and urethra. Testes descend into the scrotum at about eight months of fetal development.

Ovaries develop more slowly and can be recognized as such at about 10 weeks. In the female, some of the cells of the sex cords undergo mitosis to form some 400,000 primordial follicles before birth. These then undergo meiosis as far as first meiotic prophase (see Chapter 5). There is no postfetal formation of follicles. The primordial

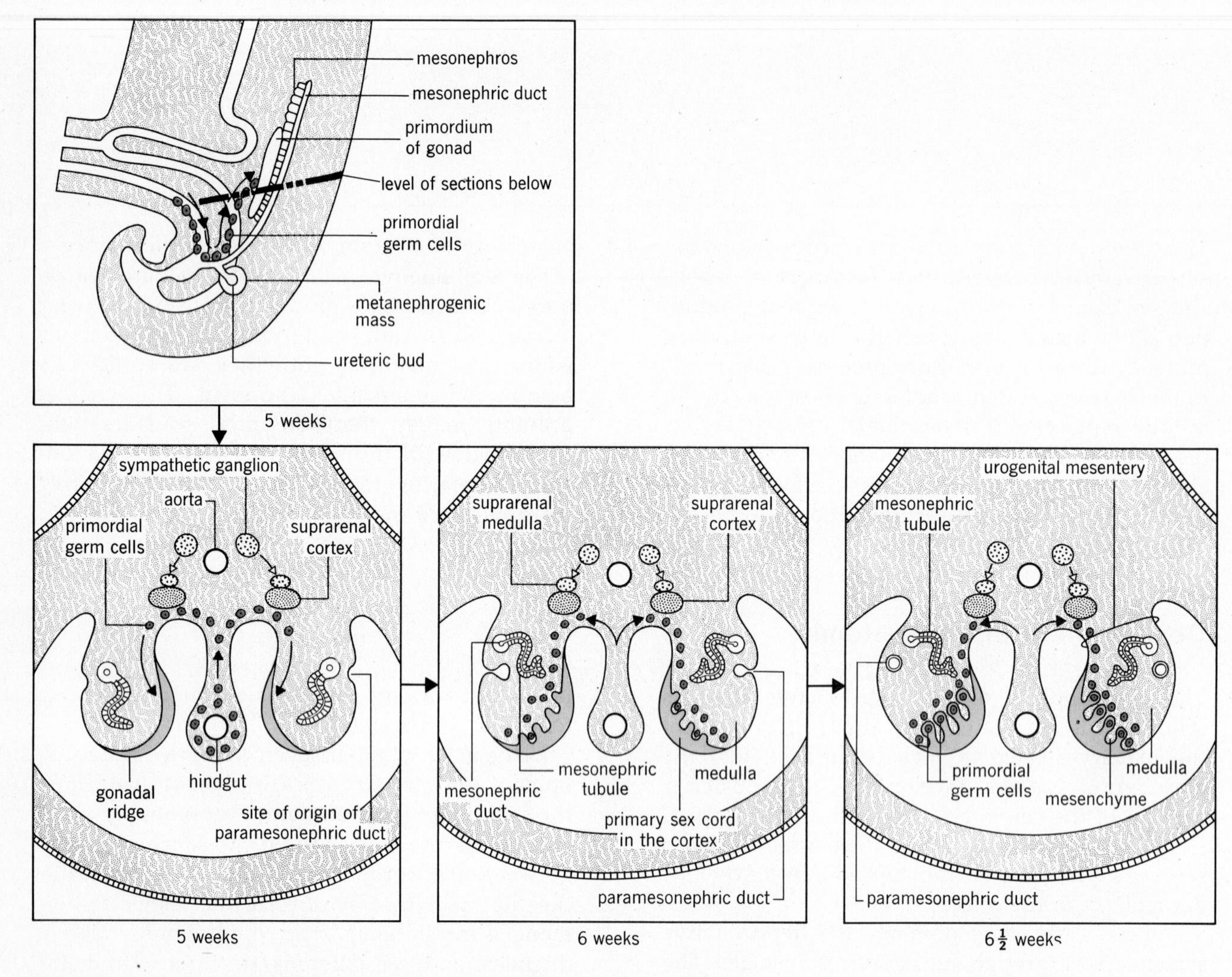

FIGURE 30.1 The early development of the gonads and their ducts.

follicles consist of the oogonium (primitive egg or ovum) and a surrounding layer of follicular cells derived from the cord. The uterine tubes, uterus, and vagina develop from the paramesonephric (Mullerian) ducts, tubes that form from the celomic epithelium on the lateral sides of each mesonephros. Originally paired, these ducts normally fuse in their distal ends to form the single uterus and vagina.

EXTERNAL GENITALIA (Fig. 30.2) also pass through an indifferent stage at about four or five weeks. A genital tubercle presages development of scrotum or labia majora; a medial phallus will form the penis or clitoris. At about eight weeks of fetal development, the two sexes can be differentiated by the characteristic appearance of the external genitalia. Table 30.1 summarizes some of the homologies in the male and female reproductive systems.

Congenital malformations of the genital sys-

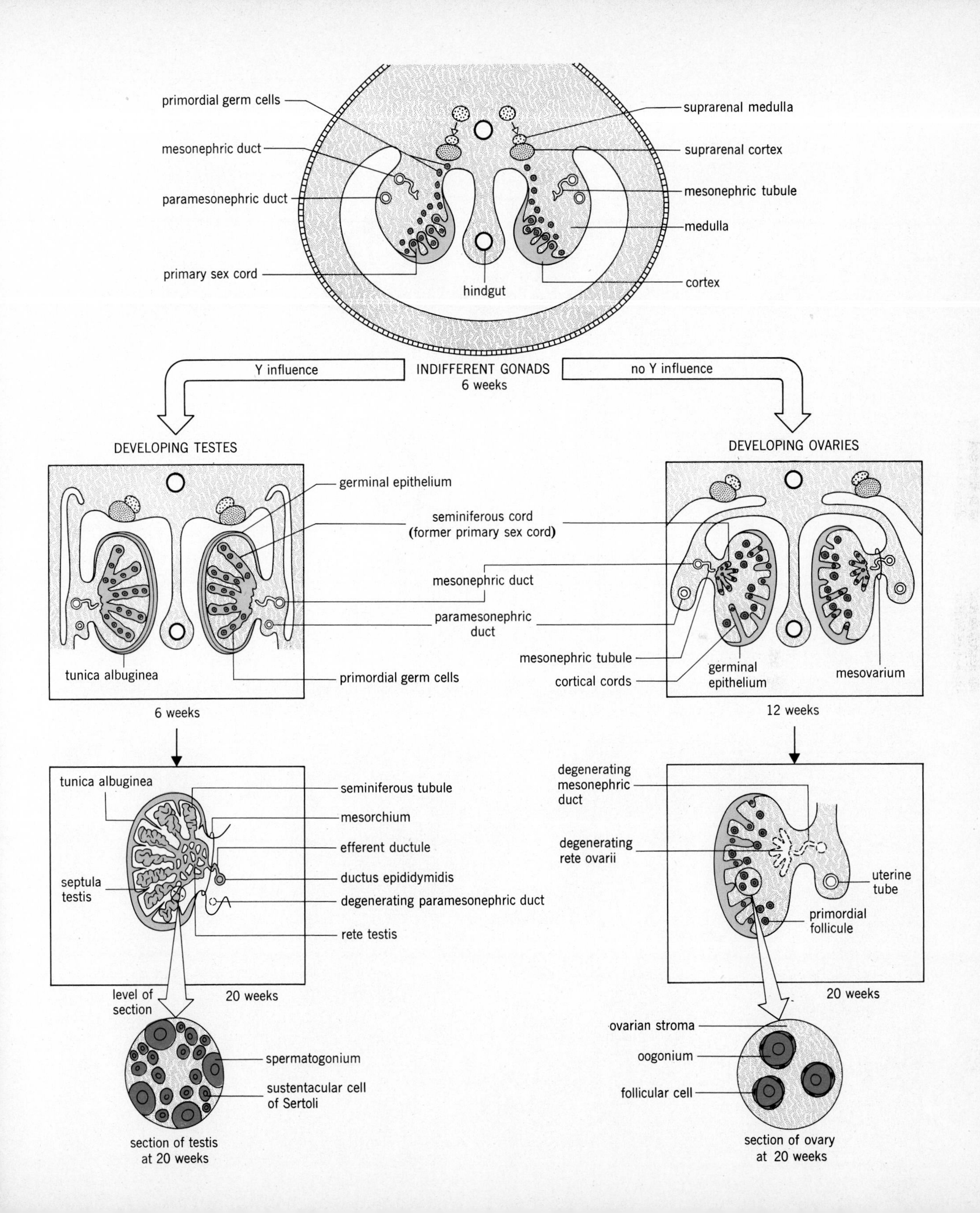

primordial germ cells
mesonephric duct
paramesonephric duct
primary sex cord
suprarenal medulla
suprarenal cortex
mesonephric tubule
medulla
cortex
hindgut
Y influence
INDIFFERENT GONADS
6 weeks
no Y influence
DEVELOPING TESTES
DEVELOPING OVARIES
germinal epithelium
seminiferous cord
(former primary sex cord)
mesonephric duct
paramesonephric
duct
mesonephric tubule
cortical cords
tunica albuginea
primordial germ cells
germinal
epithelium
mesovarium
6 weeks
12 weeks
tunica albuginea
seminiferous tubule
mesorchium
efferent ductule
ductus epididymidis
degenerating paramesonephric duct
rete testis
septula
testis
level of
section
20 weeks
degenerating
mesonephric
duct
degenerating
rete ovarii
uterine
tube
primordial
follicle
20 weeks
spermatogonium
sustentacular cell
of Sertoli
section of testis
at 20 weeks
ovarian stroma
oogonium
follicular cell
section of ovary
at 20 weeks

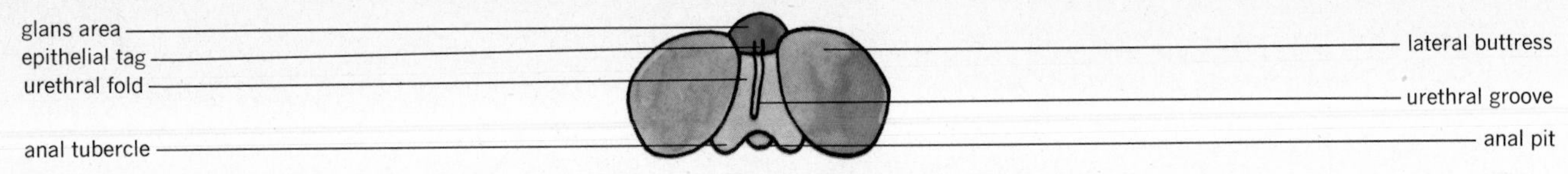

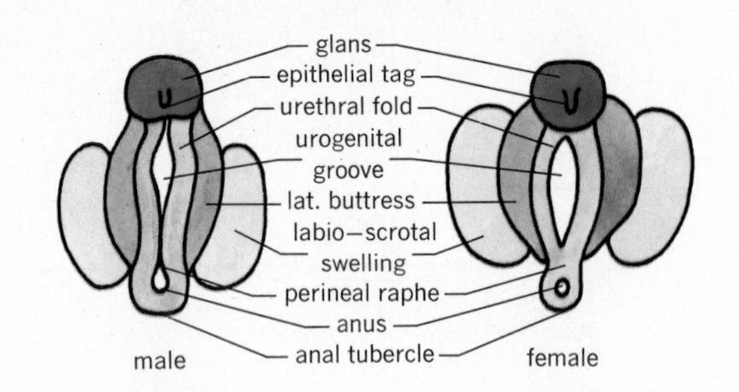

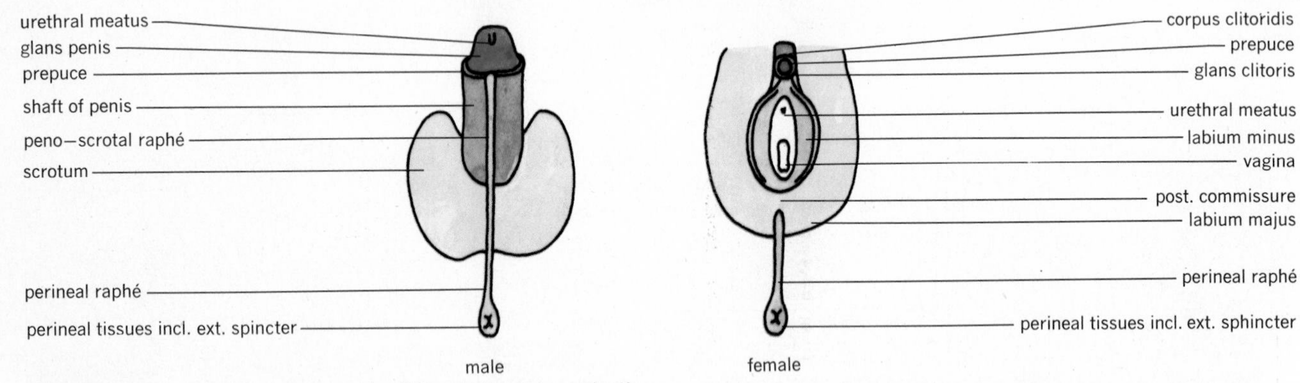

FIGURE 30.2 The development of the external genitalia.

TABLE 30.1 Some homologies of the reproductive system

Indifferent stage	Male	Female
Gonad	Testis	Ovary
Mesonephric tubules (cranial)	Efferent ductules and head of epididymis	Epoöphoron (vestige)
Mesonephric duct	Ductus epididymis	Duct of epoöphoron (vestige)
Mesonephric duct	Ductus deferens	Gartner's duct (vestige)
Mullerian duct, upper part	Appendix testis (vestige)	Uterine tube
Mullerian duct, middle part	–	Uterus
Mullerian duct, lower part	–	Vagina (upper)
Urogenital sinus	Prostatic urethra	Vestibule (middle)
Phallus	Penis	Clitoris
Lips of urogenital groove	Urethral penile surface	Labia minora
Genital swellings (tubercles)	Scrotum	Labia majora

TABLE 30.2 Some malformations of the reproductive system			
Condition	-€ constitution[a]	Frequency	Comments
Gonadal agenesis (Turner's syndrome)	XO	15-50% of females with no menstruation	Ovaries only streaks, infantile genitalia, and associated skin and skeletal defects (see general references); sterility
Hermaphroditism	Normal for sex	Rare	Both ovarian and testicular tissue occur in one person; sex organs reflect the conflict of both male and female hormone production
Kleinfelter's Syndrome	XXY	Unknown	Mental retardation common; testes atrophic
Pseudo-hermaphroditism	Normal for sex	1/61,000 births	Sex chromatin tests normal for sex; genitalia may appear similar to those of opposite sex due to absence of hormone secretion (males look like females) or excess adrenal sex hormone secretion (females develop male type genitalia)
Cryptorchidism (undescended testes)	Normal (XY)	4% of births show one or both testes undescended	By 12 months of age, 80% of undescended testes have descended
Bicornate uterus	Normal (XX)	Unknown	Failure of Mullerian ducts to fuse in uterus formation

[a]€ = chromosome

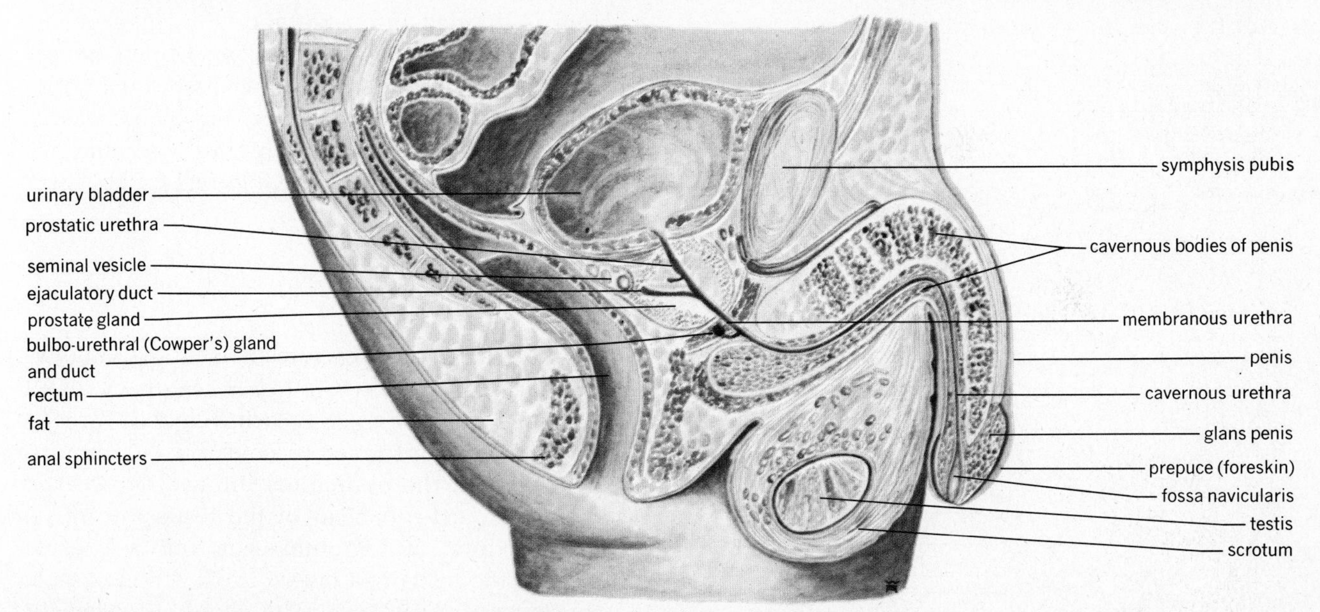

FIGURE 30.3 The male reproductive organs and associated structures.

tems are based primarily on incomplete development of a given system by sex, or simultaneous development of organs of both systems. In either case, both internal and external reproductive organs may be ambiguous, or characteristic of neither system. Table 30.2 presents information related to such malformations.

The male reproductive system

Organs

The organs of the male reproductive system (Fig. 30.3) may be conveniently grouped under three headings.

THE ESSENTIAL ORGANS. The TESTES, because of their production of male sex cells (spermatozoa or sperm), constitute the essential organs of the system. The testes also produce the male sex hormone (testosterone), which is responsible for development of certain of the remaining organs of the system and of the male secondary sex characteristics.

THE EXCRETORY DUCTS. The STRAIGHT TUBULES, RETE TUBULES, EFFERENT DUCTULES, EPIDIDYMIS, DUCTUS *(vas)* DEFERENS, EJACULATORY DUCTS, and URETHRA store sperm or convey sperm to the exterior.

FIGURE 30.4 A diagram showing the internal structure of the testis.

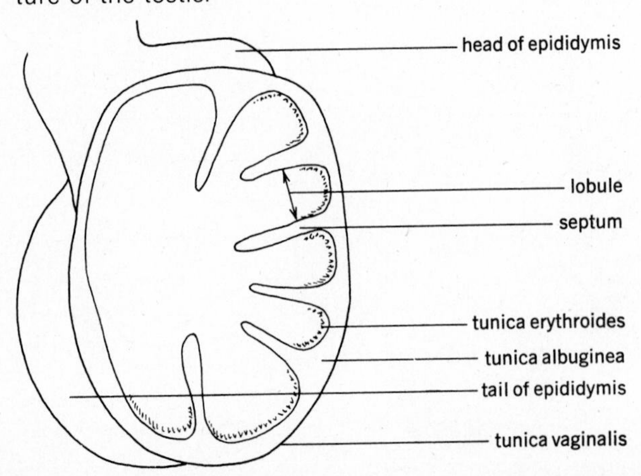

THE ACCESSORY GLANDS. The SEMINAL VESICLES, PROSTATE, and BULBOURETHRAL *(Cowper's)* GLANDS, contribute secretions which, together with the sperm, comprise the semen.

After about eight months of intrauterine life, the testes are located within the SCROTUM (Fig. 30.3). The scrotum is an evagination of skin and superficial fascia from the lower anterior abdominal wall. It is divided into lateral portions by the SUPERFICIAL RAPHE. Within the fascia is found the DARTOS TUNIC, several layers of smooth muscle. Exposure to cold causes contraction of the tunic and the scrotum becomes wrinkled. Warmth reverses the process. Internally, the scrotum is divided into right and left compartments by the dartos and fascia. Each compartment houses a testis, epididymis, and other associated structures, including ducts, nerves, and blood vessels. Each compartment communicates with the abdominal cavity through the inguinal canal.

The testes

BASIC STRUCTURE. Each testis is an ovoid body, measuring 4–5 cm (1¾–2 in.) in length by about 2.5 cm (1 in.) in diameter. Each weighs 5–7 g. The outer covering is formed by the TUNICA VAGINALIS, a portion of the abdominal lining *(mesothelium)* that was pushed ahead of the testis during its descent into the scrotum. Forming a capsule around the testis is a layer of collagenous tissue, the TUNICA ALBUGINEA. The albuginea is thick-

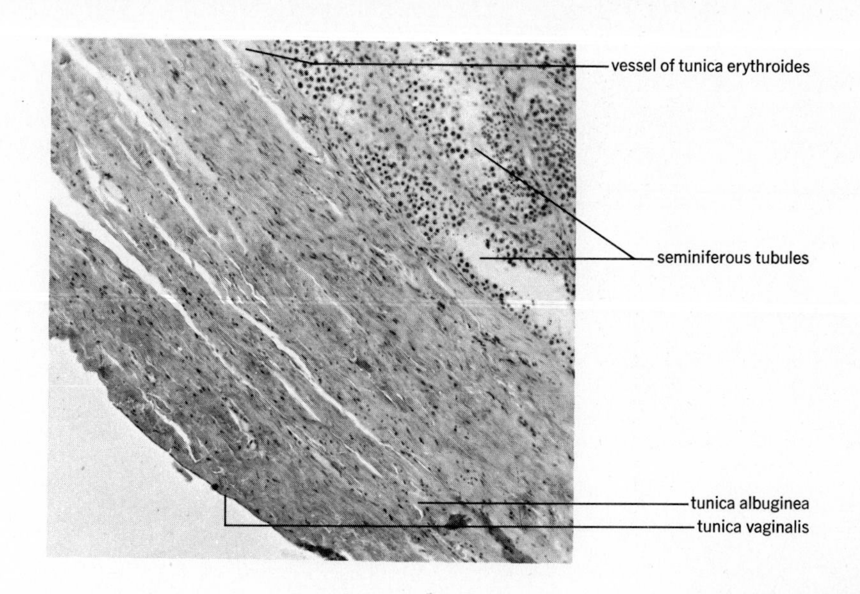

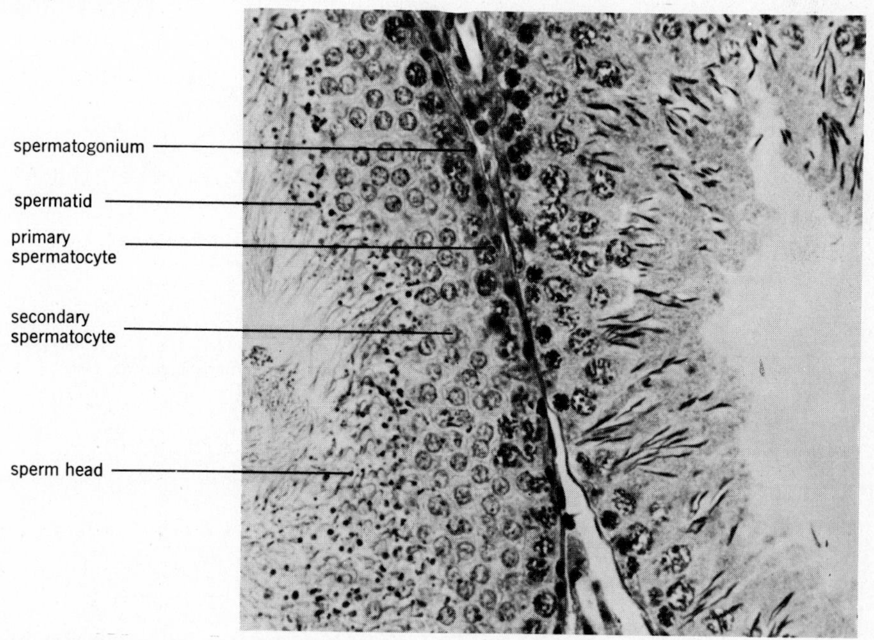

FIGURE 30.5 Photomicrographs of the structure of the testis.

ened on the medial side of the testis to form the MEDIASTINUM TESTIS. A third and vascular coat, the TUNICA ERYTHROIDES, lies deep to the albuginea. From the albuginea rise connective tissue partitions or SEPTAE, which penetrate radially into the testis and divide it into about 250 separate LOBULES. Each lobule contains one to four coiled SEMINIFEROUS TUBULES. The seminiferous tubules connect with straight tubules, which run to the mediastinum and join the rete tubules

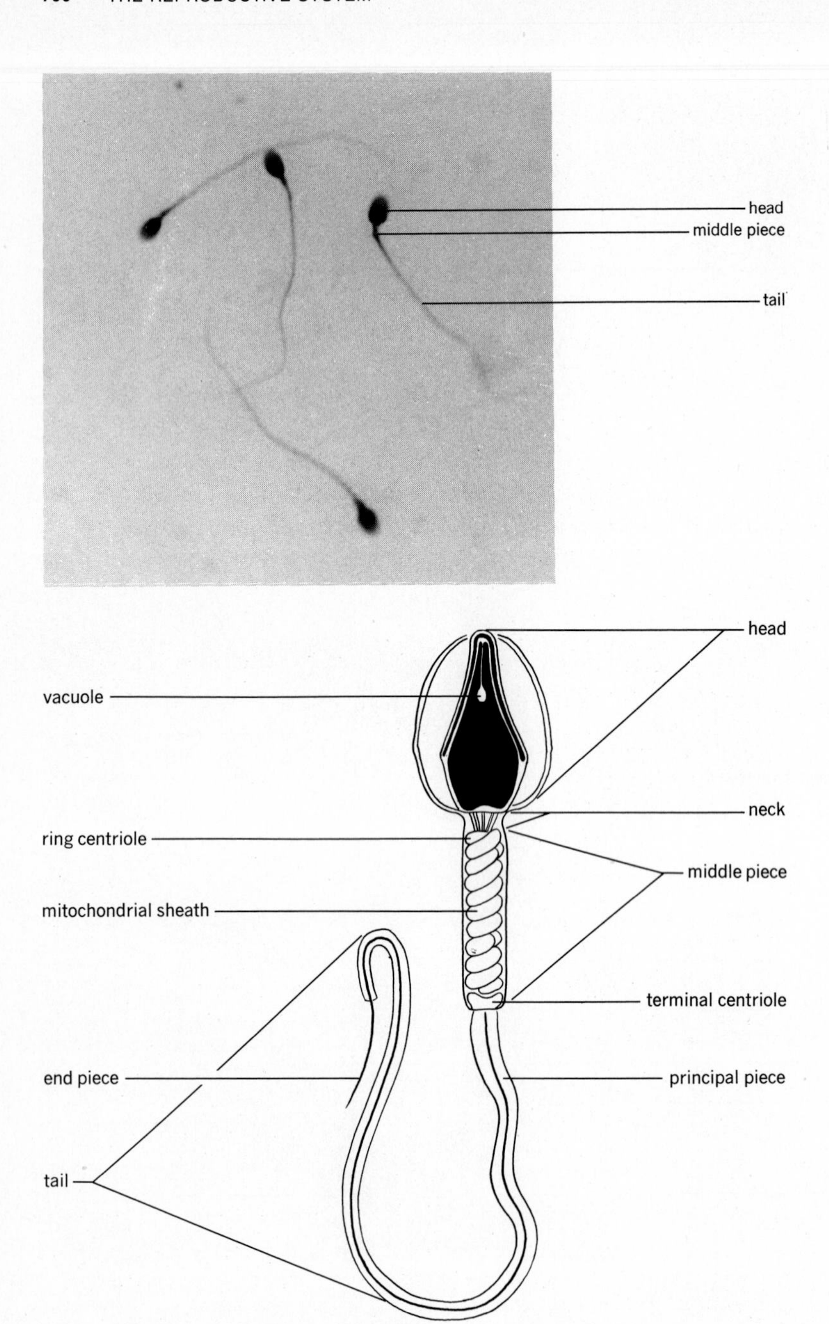

FIGURE 30.6 The appearance and fine structure of sperm.

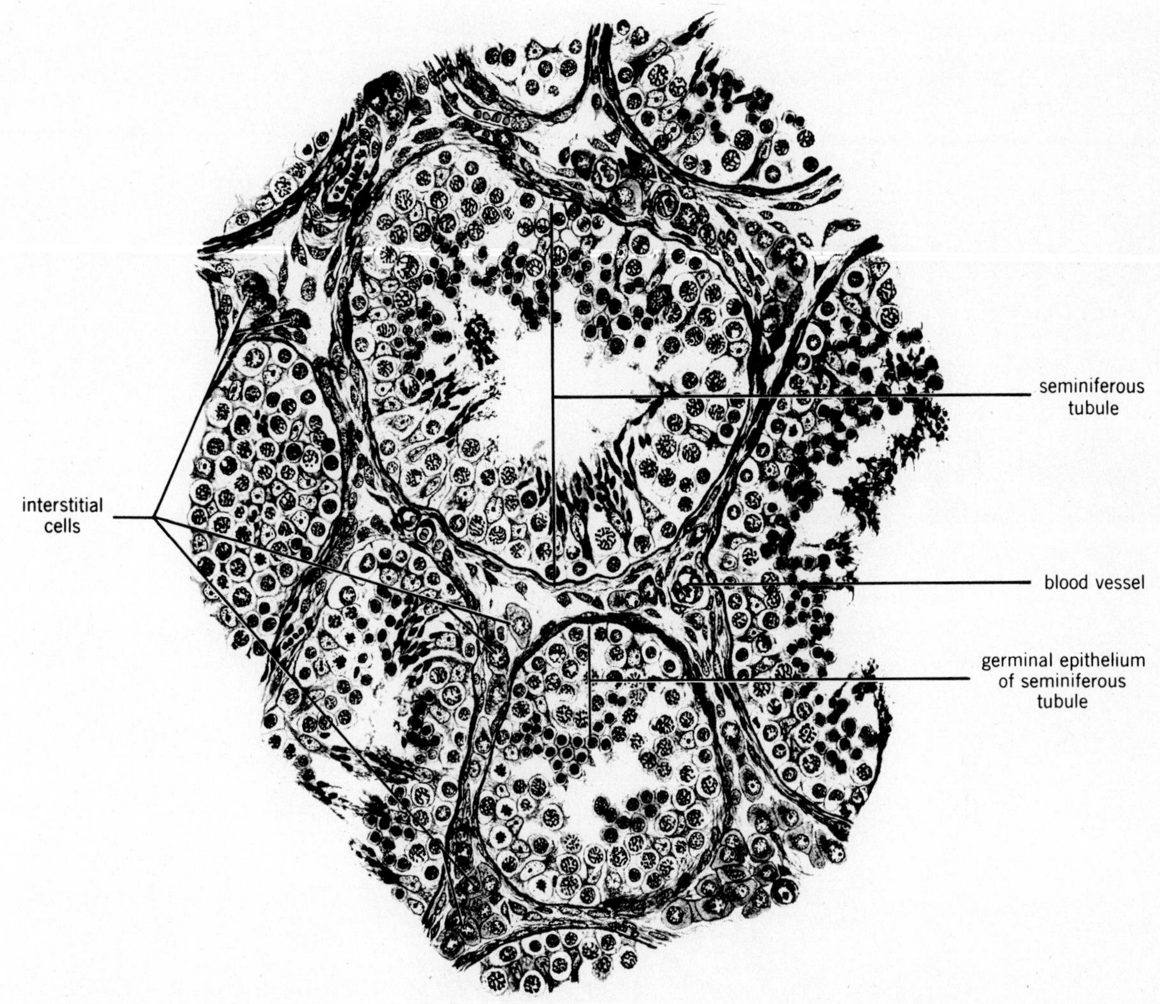

FIGURE 30.7 The location of the interstitial cells of the testis (Courtesy W. B. Saunders Co., from W. Bloom and D. W. Fawcett, *Testbook of Histology*, Ninth Edition, Saunders, Philadelphia, 1968).

located therein. These features are depicted in Fig. 30.4.

MICROSCOPIC ANATOMY. A section of the testis shows it to be composed of many cut seminiferous tubules with intertubular connective tissue, blood vessels, and cells (Fig. 30.5). The seminiferous tubule is lined with the germinal epithelium. The epithelium contains two types of cells.

Sertoli cells. Recognizable by their oval nuclei with prominent nucleoli, the Sertoli cells are actually columnar cells that produce secretions designed to nourish the spermatozoa during their final stages of development.

Spermatogenic cells. Spermatogenic cells, arranged in several rows from the periphery of the tubule inward, represent spermatozoa in various stages of development. It may be noted that sper-

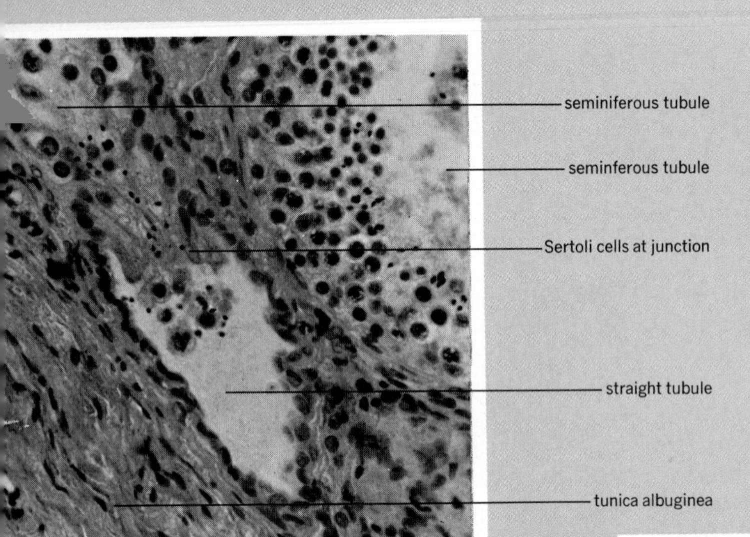

FIGURE 30.8 The junction of the seminiferous and straight tubules.

FIGURE 30.9 The rete tubules.

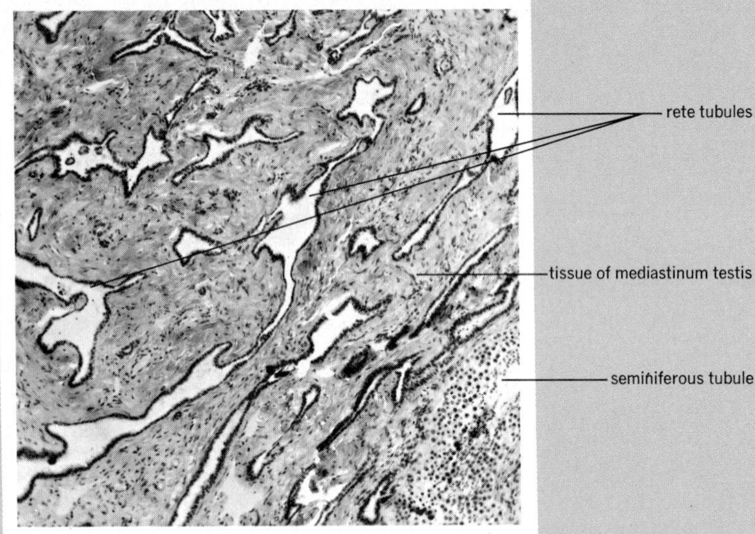

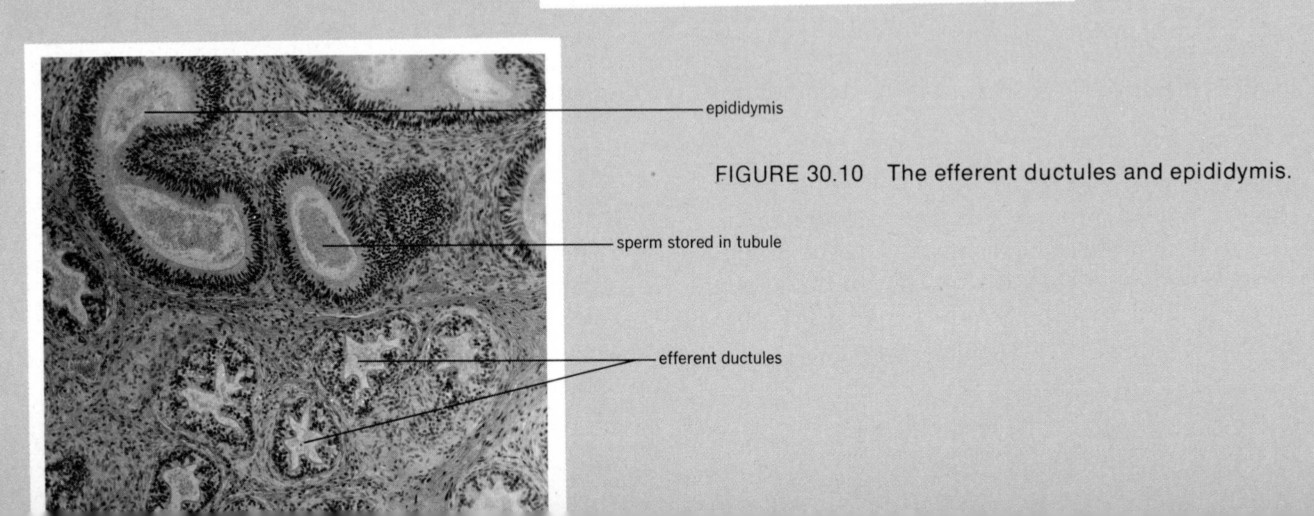

FIGURE 30.10 The efferent ductules and epididymis.

matogenesis occurs by meiosis so that the sperm cells are haploid. The most immature cells are located against the basement membrane and are called SPERMATOGONIA. Next is found a layer or two of large cells with large round nuclei, the PRIMARY SPERMATOCYTES. Smaller cells, having the same general appearance as the primary spermatocytes, are found still closer toward the tubule lumen and are SECONDARY SPERMATOCYTES. The SPERMATIDS are more internal yet and show small dark nuclei. Spermatids metamorphose into SPERMATOZOA (Fig. 30.6). Sections of testes taken before sexual maturity show only cords of epithelial cells. The stages described above appear at puberty.

Between the coils of the seminiferous tubules are groups of INTERSTITIAL CELLS (Fig. 30.7). These cells produce the male sex hormone testosterone, which controls the appearance and development of the male secondary sex characteristics and the growth of some of the accessory organs of the male reproductive system.

The duct system

The seminiferous tubules connect with the STRAIGHT TUBULES. Straight tubules are lined with a low, simple, cuboidal epithelium. The junction between a seminiferous and straight tubule is marked by the presence of only Sertoli cells, which project into the straight tubule (Fig. 30.8). Straight tubules pass to the mediastinum testis where they join the RETE TUBULES (Fig. 30.9). Rete tubules are irregular in diameter and are lined with simple squamous or simple cuboidal epithelium. All remaining portions of the duct system lie outside the limits of the testis.

EFFERENT DUCTULES, numbering 12 to 16, convey spermatozoa to the epididymis for storage. They are lined with alternating groups of tall, ciliated cells and short, nonciliated cells. The lumen of the tubule thus appears folded (Fig. 30.10). A thin coat of smooth muscle appears around the tubule. The EPIDIDYMIS is a single, coiled tube about 7m (20–21 ft) in length, which is held to the testis by connective tissue. Its superior end, or head, receives the efferent ductules; the centrally placed body continues into the elongated, inferiorly placed tail. The lining of the epididymis is a pseudostratified epithelium, consisting of tall columnar elements bearing microvilli, the stereocilia, and basal cells that do not reach the lumen (Fig. 30.10). It is thought that the stereocilia act to convey nutrient materials to the sperm stored in the epididymis. Again, a thin layer of smooth muscle surrounds the epithelium. The DUCTUS *(vas)* DEFERENS (Fig. 30.11) is continuous with the tail of the epididymis. It passes from the scrotum through the INGUINAL CANAL, arches behind the urinary bladder, and joins the ejaculatory duct to empty into the male urethra. The deferens is lined by a folded mucous membrane consisting of pseudostratified epithelium and an underlying lamina propria. A heavy coat of smooth muscle is next, consisting of inner longitudinal, middle circular, and outer longitudinal layers. An adventitia of connective tissue surrounds the muscle. The heavy muscular coat propels the sperm through the deferens with considerable force.

The adventitia of the deferens not only serves as the outer coat of that duct, but also serves to bind the deferens and its accompanying arteries, veins, and nerves into the structure known as the SPERMATIC CORD. A small band of skeletal muscle, the cremaster muscle, is also found in the cord. It elevates the testis during sexual stimulation and during exposure to cold.

Accessory glands (Figs. 30.12 and 30.13)

The SEMINAL VESICLES are paired, tortuous sacs lying on the posterior aspect of the urinary bladder. The vesicles do not, as was once thought, store sperm. Occasionally, a few sperm may be found within the lumina of the vesicles, but they have simply been "eddied" into the vesicles. The ducts of the vesicle and deferens join to form the ejaculatory duct, emptying into the urethra. The PROSTATE GLAND surrounds the upper urethra. It consists of 30–50 separate secretory units, emptying

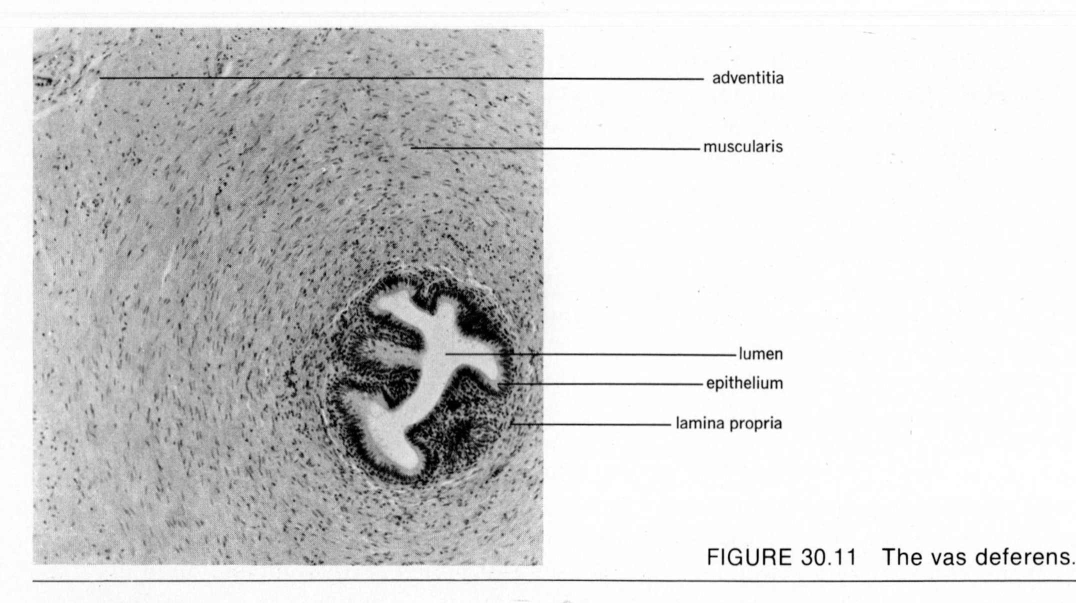

FIGURE 30.11 The vas deferens.

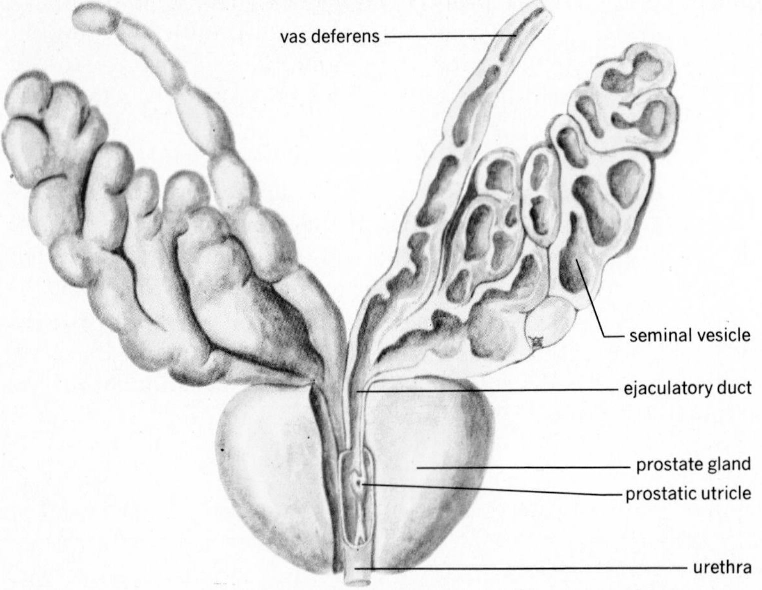

FIGURE 30.12 The prostate, seminal vesicles, and associated structures.

by 16–32 separate ducts into the urethra. The gland has much smooth muscle between the secretory portions, which is an aid to the emptying of the organ. The secretion of the prostate is alkaline, slimy, and has a characteristic odor. It is thought to activate the heretofore immobile sperm.

The BULBOURETHRAL *(Cowper's)* GLANDS empty into the urethra below the prostate. Their secretion is mucoid in nature, and its alkaline reaction aids in neutralization of any acid urine in the urethra.

Semen

The spermatozoa, normally numbering 300-400 million per ejaculate, are combined with the secretions of the seminal vesicles, prostate, and bulbourethral glands to form the SEMEN. The average volume of the semen in an ejaculate is about 3 ml, the pH 7.35-7.50. Fructose and ascorbic acid are the major components of the seminal vesicle contribution, which comprises about 60 percent of the total volume. The prostate contributes about 30 percent of the total semen volume. Prostatic secretion is rich in cholesterol, phospholipids, buffering salts, and prostaglandins. The latter are fatty acids having many effects on body function (see Chapter 31). The bulbourethral glands contribute about 5 percent of semen volume, and the sperm themselves about 5 percent. Semen as a whole gives bulk to suspend the sperm, aids in neutralization of the acidity in the female tract, and activates the sperm. Sperm placed in the female reproductive tract have a maximum life expectancy of three days.

Male urethra and penis

The male urethra is about 20 cm (8 in.) in length and is subdivided into three portions.

PROSTATIC URETHRA *(pars prostatica).* This portion is approximately the first 3 cm below the bladder and is surrounded by the prostate gland. It is lined with transitional epithelium and receives the ejaculatory ducts and the prostatic ducts. A bulge on the posterior wall marks the position of the prostatic utricle, the male homolog of the uterus.

MEMBRANOUS URETHRA *(pars membranacea).* This portion is 1-2 cm in length and passes through the pelvic floor. It is lined with a pseudostratified epithelium and receives the ducts of the bulbourethral glands.

CAVERNOUS URETHRA *(pars cavernosa).* The re-

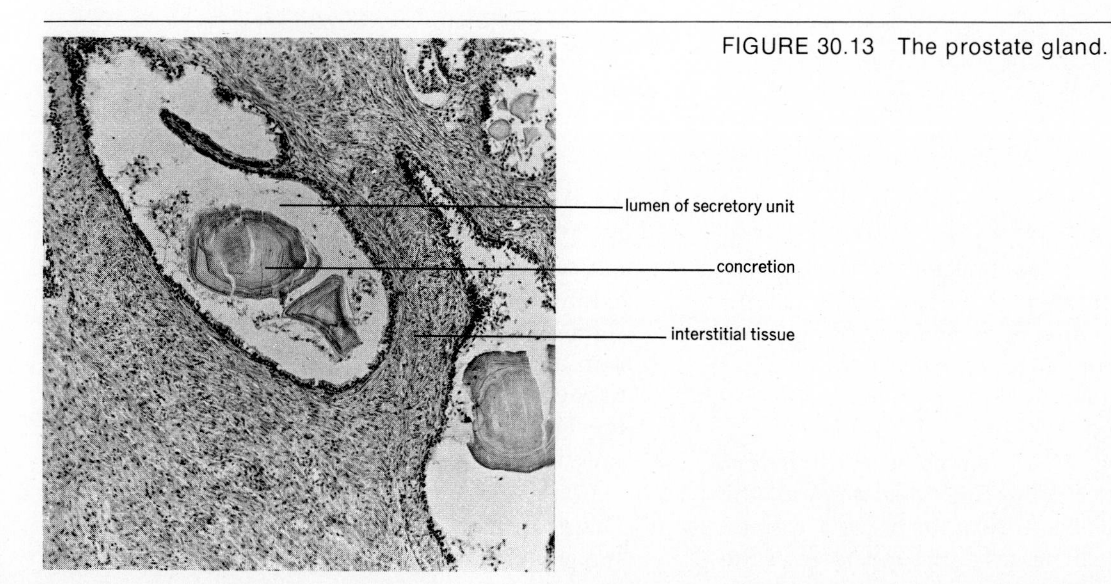

FIGURE 30.13 The prostate gland.

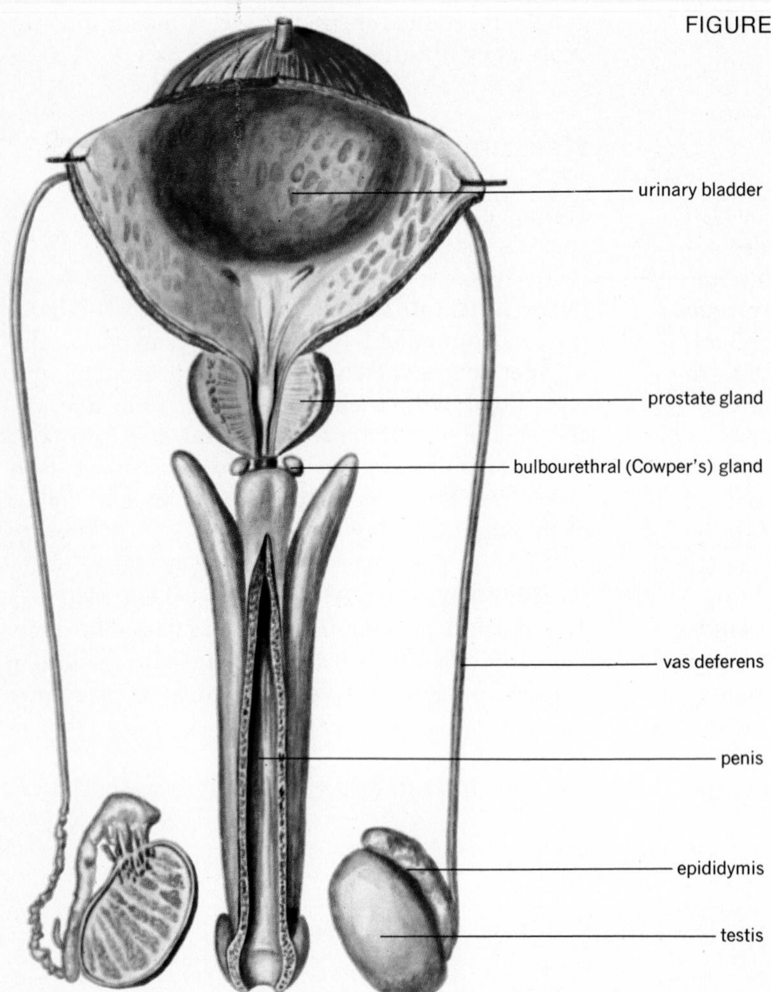

FIGURE 30.14 The penis and associated structures.

maining portion of the urethra lies within the penis and is about 15 cm in length. It is lined with pseudostratified epithelium. Fissurelike evaginations from the lumen form the *lacunae of Morgagni,* into which empty ducts from mucus-secreting *glands of Littré.*

The PENIS (Fig. 30.14) serves as a copulatory organ to introduce sperm into the vagina of the female. The basic structure of the organ is formed by three cylindrical masses of tissue known as the CAVERNOUS BODIES. The bodies are supported by a tunica albuginea and are bound together by connective tissue and covered by thin skin. The skin at the neck of the penis folds back on itself to cover the head or glans of the penis and forms the *prepuce* or *foreskin.* Circumcision allows the prepuce to draw away from the glans.

Two of the cavernous bodies are dorsally placed. These are the CORPORA CAVERNOSA PENIS. The single ventrally placed member is the CORPUS

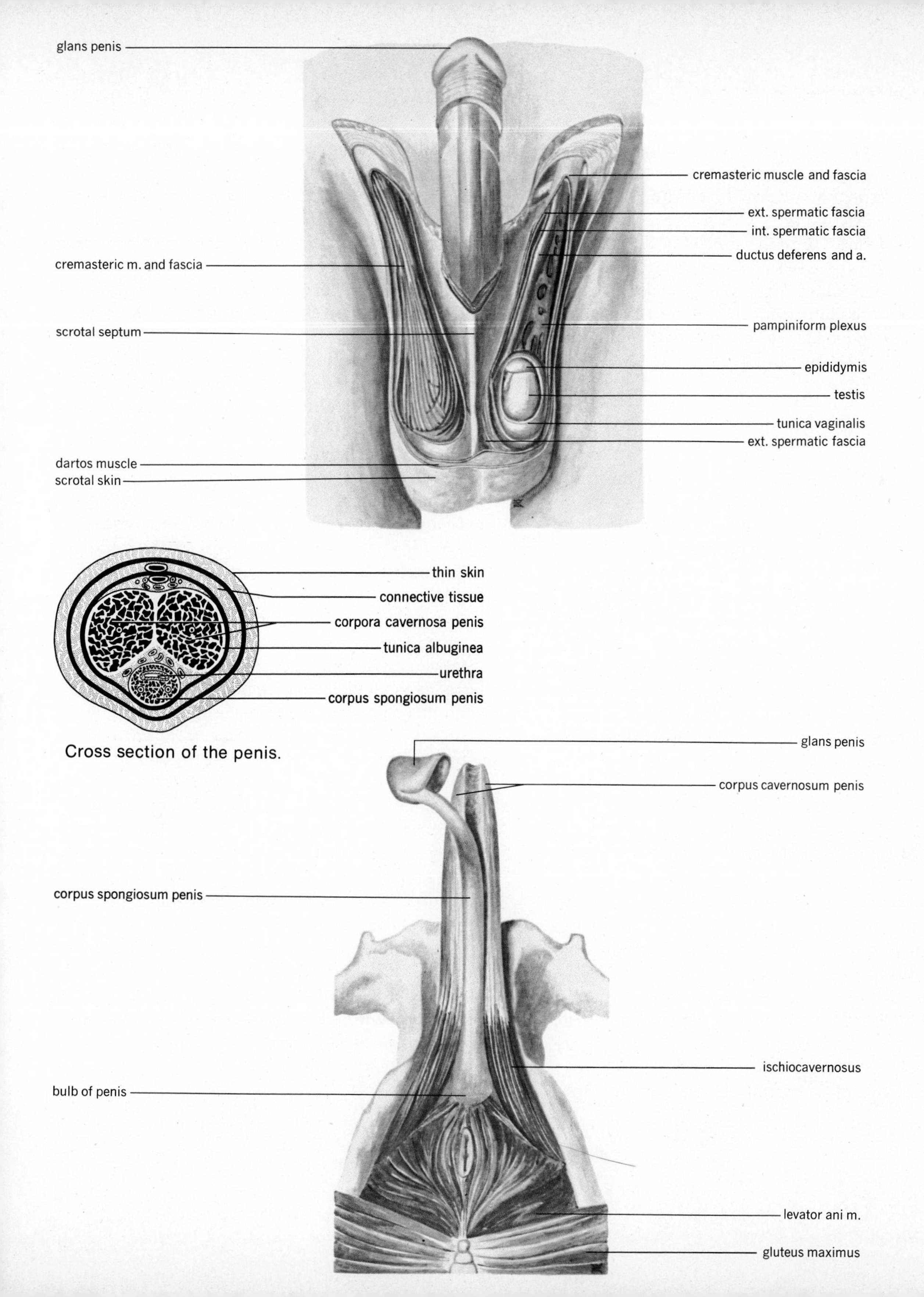
glans penis
cremasteric muscle and fascia
ext. spermatic fascia
int. spermatic fascia
ductus deferens and a.
cremasteric m. and fascia
pampiniform plexus
scrotal septum
epididymis
testis
tunica vaginalis
ext. spermatic fascia
dartos muscle
scrotal skin
thin skin
connective tissue
corpora cavernosa penis
tunica albuginea
urethra
corpus spongiosum penis
glans penis
corpus cavernosum penis
corpus spongiosum penis
ischiocavernosus
bulb of penis
levator ani m.
gluteus maximus

Cross section of the penis.

SPONGIOSUM PENIS (corpus cavernosum urethrae). It transmits the cavernous urethra. The distal end of the spongiosum is expanded into the glans. The blood vessels to the organ are muscular arteries that empty into the sinuses of the corpora. Dilation of these vessels allows a greater inflow of blood into the bodies. They swell, the veins draining the corpora are compressed, and the penis becomes turgid or erect—an effective copulatory organ.

Endocrine relationships

Full discussion of endocrine control of the testis and the effects of testosterone will be given in Chapter 31. However, it may be stated at this time that the FOLLICLE-STIMULATING HORMONE (FSH) of the anterior pituitary is the controlling factor of spermatogenesis. The INTERSTITIAL-CELL-STIMULATING HORMONE (ICSH), also a product of the anterior pituitary, controls production of testosterone. Testosterone in turn assures full development of the external aspects of maleness, such as hair growth, voice change, and muscular development, and assures the complete growth of the ducts and accessory glands of the male reproductive system.

Clinical considerations

MALE INFERTILITY is defined as absence of the ability to induce conception, or inability of the male to produce sperm capable of fertilizing the ovum. Congenital causes, such as maldevelopment of the reproductive organs, may result in infertility, as may diseases (e.g., venereal diseases, mumps) that attack the testes and destroy or reduce their ability to produce sperm.

Among the most common infections affecting the male reproductive system are gonorrhea and syphilis. The microorganisms causing these diseases may be transmitted by sexual intercourse, body-to-body contact, or oral contact with an infected person. Recent evidence indicates that live gonorrhea organisms may be isolated from contaminated toilets, silverware, or doorhandles and doorknobs up to eight hours after deposition of the bacteria.

IMPOTENCE, or an inability to create and sustain an erection, may result in inability of the male to copulate successfully with the female. Impotence appears to be an increasingly common affliction of the male in "civilized" societies.

Sperm counts less than about 50 million per milliliter of semen make it difficult for conception to occur because not enough sperm are provided to:

Resist the loss of numbers due to acids in the vagina.

Survive the 6-in. trip through the uterus and uterine tubes to fertilize the ovum.

Provide sufficient hyaluronidase, an enzyme that liquefies the intercellular cement holding several layers of follicular cells to the released ovum.

Sperm are capable of surviving up to three days within the female reproductive tract, so that pregnancy can occur even though copulation has not occurred for several days. If ovulation occurs and sperm are present, pregnancy may result.

ENLARGEMENT OF THE PROSTATE GLAND is a common occurrence with aging. If the gland enlarges greatly, it may compress the urethra to the point where both reproductive function and urination become difficult or impossible. Surgery is usually indicated to remove the enlarged organ. Cancer of the prostate is the second most common neoplasm in the male and is the sixth leading cause of cancer death in the male.

The female reproductive system

Organs

The organs of the female reproductive system (Fig. 30.15) may be grouped under two headings.

INTERNAL ORGANS

OVARIES. Like testes, the paired ovaries are responsible for production of female sex cells, or ova, and for the production of two major female hormones.

UTERINE TUBES. These paired structures convey ova to the uterus.

UTERUS. The single organ wherein the development of new individuals begins.

VAGINA. The externally opening tubular organ that serves as the receptacle for sperm and as the birth canal.

EXTERNAL ORGANS

CLITORIS. A single organ of sensory nature.

LABIA MAJORA and MINORA. Folds of skin and mucous membrane surrounding the vaginal and urethral openings.

MAMMARY GLANDS. Actually modified apocrine sweat glands, these organs are so closely related functionally to the other reproductive organs that they are usually discussed as parts of the reproductive system.

The ovaries

The ovaries are paired ovoid bodies measuring about 3 cm long, 2 cm wide, and 1 cm deep. They are pelvic in location, lying lateral to the uterus. A mesentery, the mesovarium, suspends the ovaries on the posterior aspect of the uterine broad ligament. Each ovary has a hilus marking the entrance of the blood vessels and nerves supplying the organ. The ovarian ligament attaches the ovary to the uterus, while the suspensory ligament attaches it to the pelvic wall.

Microscopically, the ovary may be seen to be composed of:

The GERMINAL EPITHELIUM, a surface epithelium.

The TUNICA ALBUGINEA, a capsule of collagenous connective tissue.

A STROMA of connective tissue internal to the capsule.

OVARIAN FOLLICLES, representing ova and surrounding tissues in various stages of development.

The germinal epithelium is a layer of simple cuboidal or high squamous cells on the ovarian surface. During fetal life, it may be the source of some ova, but is functionless in this respect after birth. Recall that the sex cords account for the production of the great majority of ova.

The tunica albuginea confers some strength and support to the entire organ. The stroma may be subdivided into an outer, more dense layer, the cortex, and an inner, looser medulla, or zona vasculosa. The latter term reflects the vascular nature of the inner region.

The ovarian follicles

The follicles (Fig. 30.16) are estimated to number about 400,000 in the ovaries prior to birth. They remain in a quiescent condition until puberty, when hormonal changes initiate follicle development and menstruation. A total of about 400 ova will develop and be discharged over the reproductive life of the female. The remainder will undergo a degenerative process known as atresia.

PRIMORDIAL FOLLICLES are the most numerous of the follicles and are also the smallest. These follicles are 40–50 μm in diameter and may be found in the external portions of the cortex. They consist of an ovum surrounded by a single layer of stromal cells. Further development of the follicle results in the formation of a PRIMARY FOLLICLE. Primary follicles are larger, about 80 μm in diam-

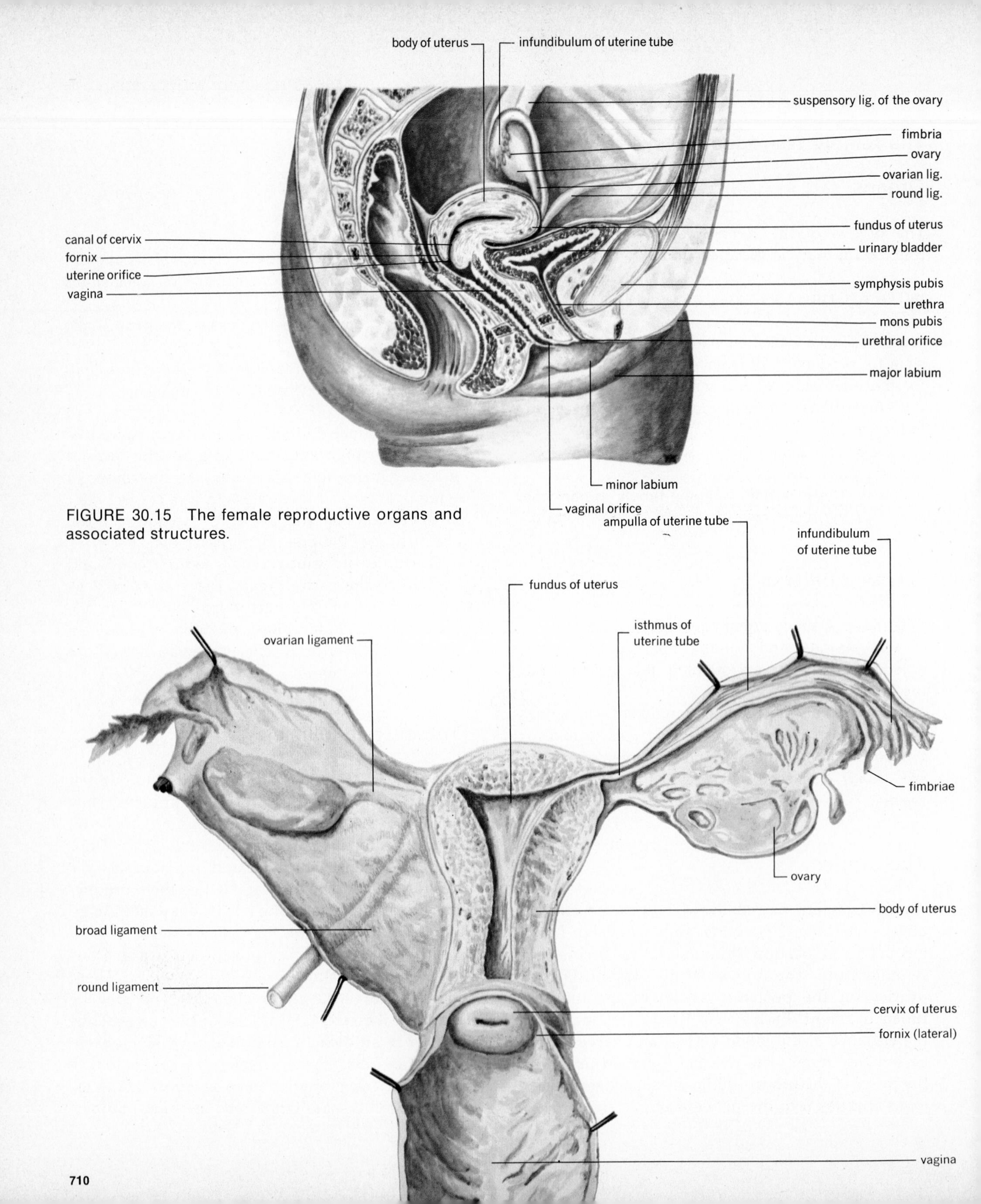

FIGURE 30.15 The female reproductive organs and associated structures.

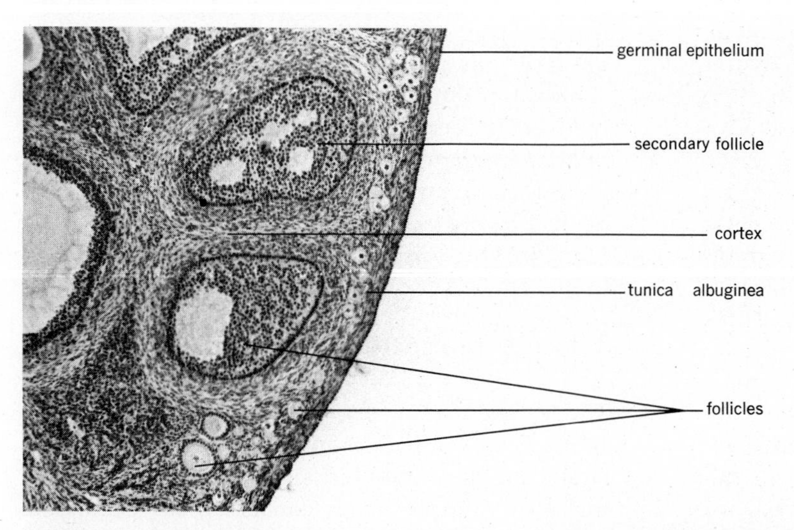

FIGURE 30.16 The microscopic structure of the ovary.

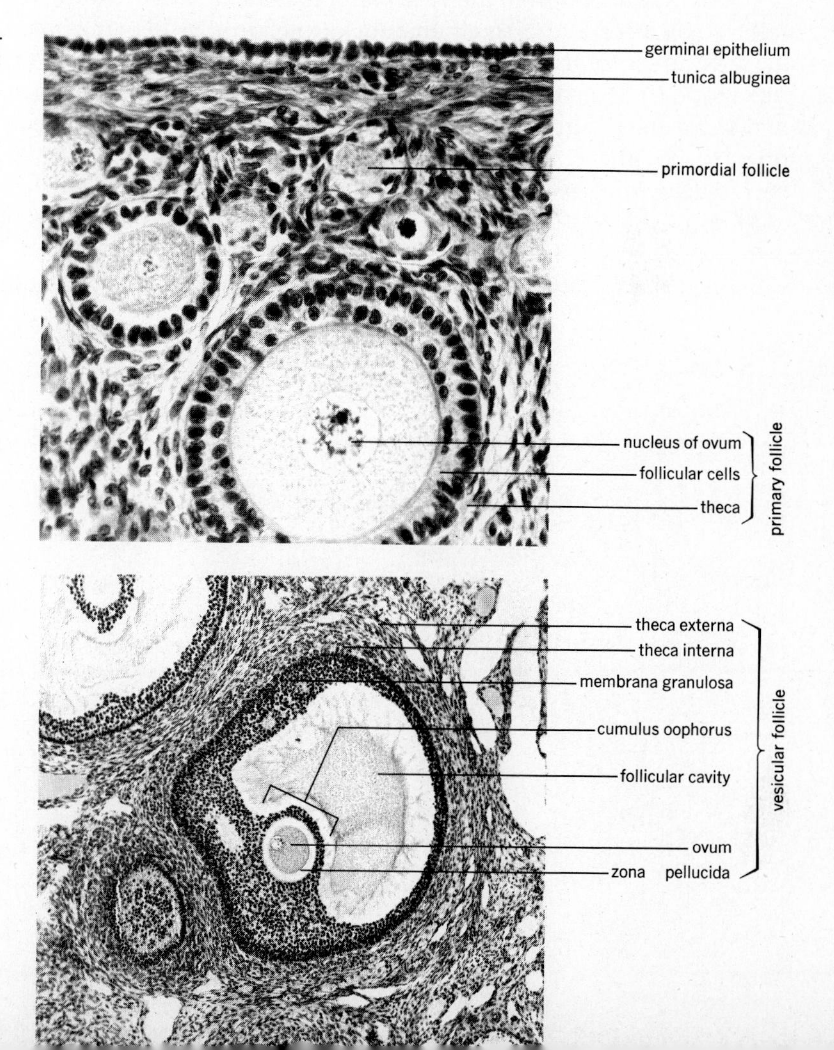

FIGURE 30.17 Primary and vesicular follicles of the ovary.

eter, and have several layers of what may now be designated as follicular cells around the ovum. Next, several small cavities develop within the layers of follicular cells, and the unit is called a SECONDARY FOLLICLE. These cavities enlarge, fuse, and eventually form a single follicular cavity. The ovum is forced to the periphery of the follicular cavity. The entire follicle increases to a maximum size of 10–13 mm and is known as a VESICULAR or GRAAFIAN FOLLICLE. It possesses many parts labeled in Fig. 30.17.

The developing follicle is the source of ESTROGEN, a female sex hormone responsible for the development of female secondary sex characteristics and for the development of several of the organs of the female system. OVULATION, release of the ovum, occurs from the vesicular follicle. The follicle collapses and a small amount of bleeding may occur into the follicle, forming a CORPUS HEMORRHAGICUM. The tissues remaining in the ovary are next caused to form a CORPUS LUTEUM (Fig. 30.18), the source of the second hormone, PROGESTIN *(progesterone)*. The latter hormone maintains and advances the changes initiated by estrogen. The fate of the corpus luteum will be the same regardless of whether or not the ovum is fertilized. With no fertilization, the luteum undergoes degeneration in about three weeks; with fertilization and successful uterine implantation, the luteum remains functional for about 6 months of the pregnancy before degenerating. In either case, the luteum is replaced by collagenous connective tissue (scar tissue) and a CORPUS ALBICANS (Fig. 30.19) is formed.

The uterine tubes

The paired UTERINE *(Fallopian)* TUBES are about 10 cm (4 in.) in length and attach to the superior-lateral aspect of the uterus. They are supported by the mesosalpinx. They consist of an expanded funnel-shaped outer end, the INFUNDIBULUM, bearing many fingerlike FIMBRAE. The AMPULLA lies between the infundibulum and the uterine wall; the ISTHMUS of the tube penetrates the uterine wall to open into the uterine cavity. The uterine tube receives the discharged ovum and conveys it to the uterus. Microscopically (Fig. 30.20) the uterine tube shows three layers in its walls.

An inner MUCOSA consists of a simple columnar

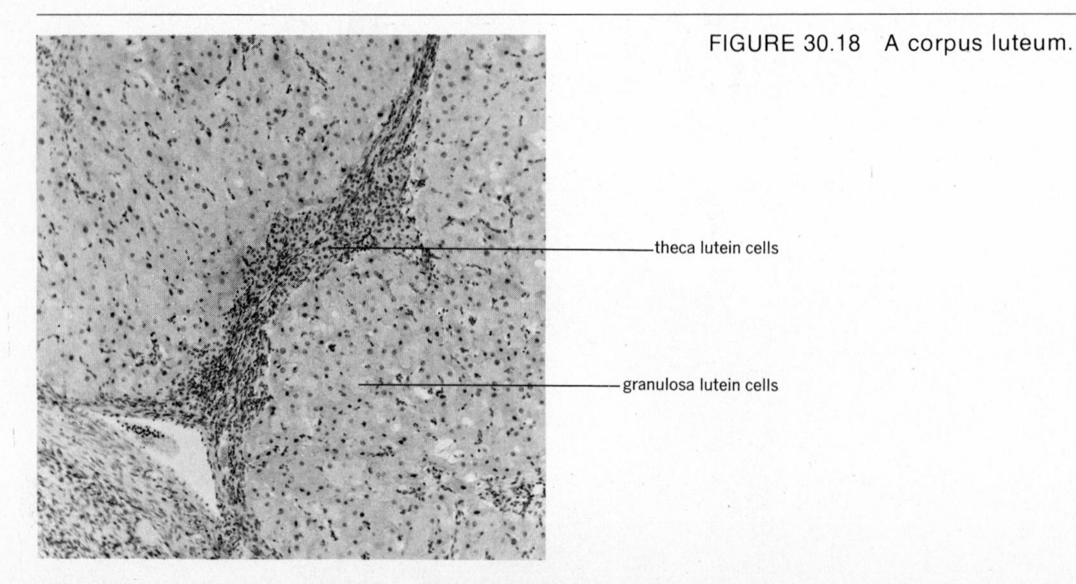

FIGURE 30.18 A corpus luteum.

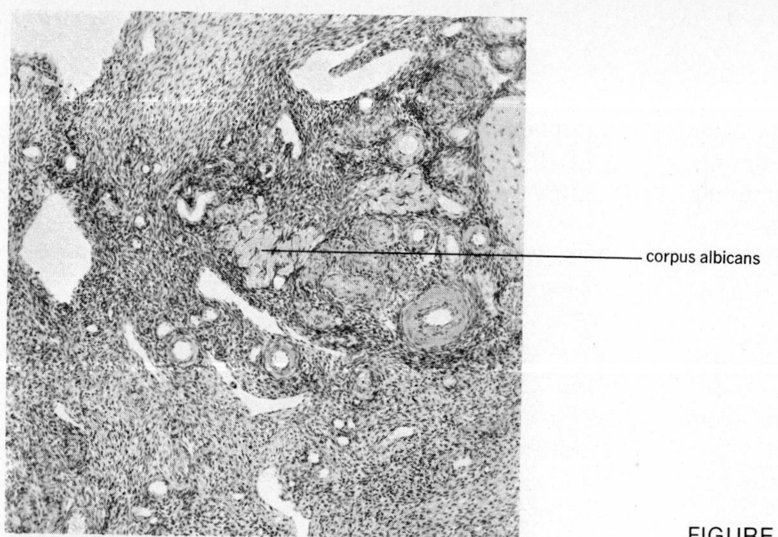

FIGURE 30.19 A corpus albicans.

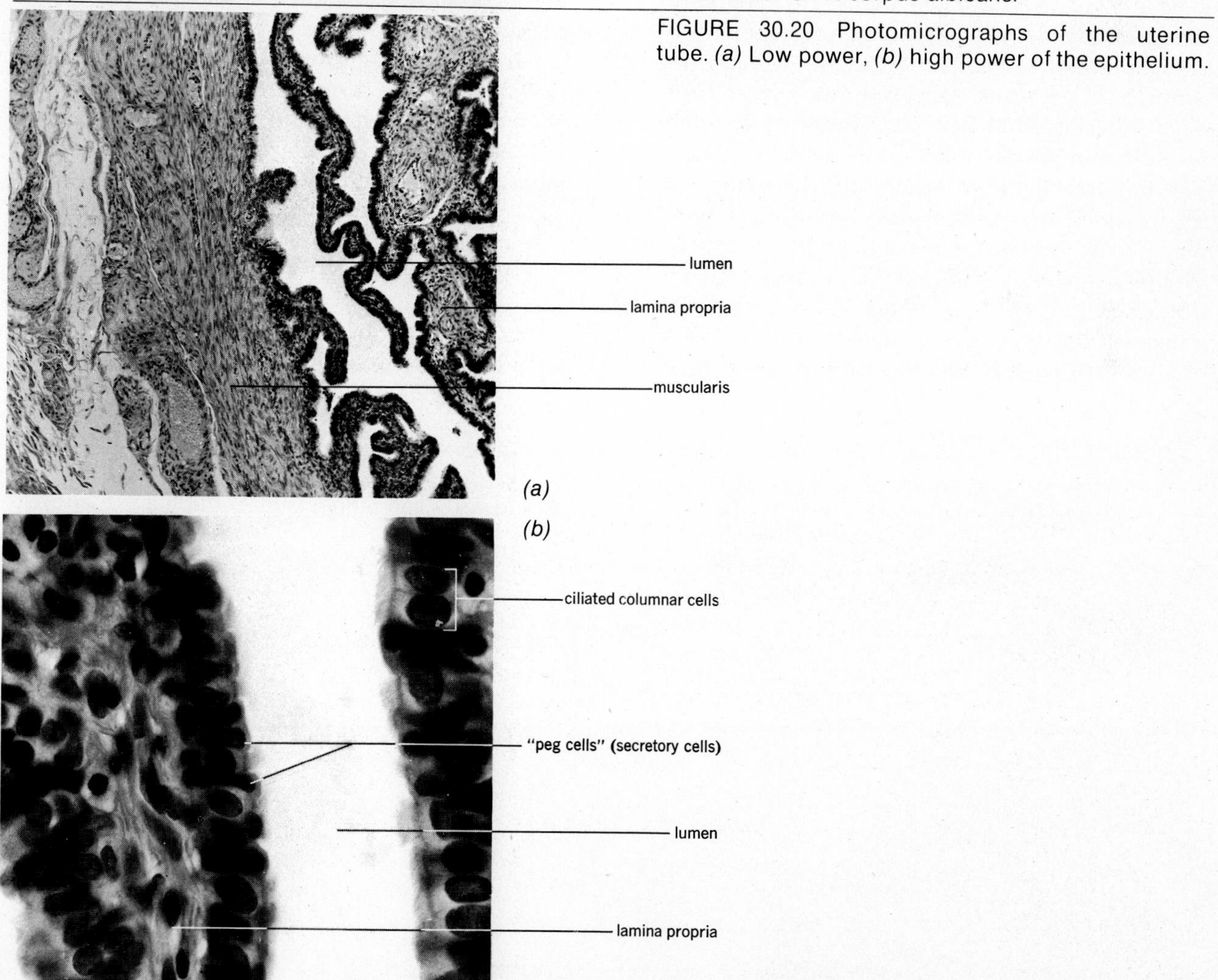

FIGURE 30.20 Photomicrographs of the uterine tube. *(a)* Low power, *(b)* high power of the epithelium.

epithelium having two types of cells: ciliated cells, believed responsible for creating a "current" in the fluids of the pelvic cavity that draws the egg into the tube, and nonciliated secretory cells (peg cells) inserted between the ciliated cells and thought to contribute nutrients to the ovum. A lamina propria underlies the epithelium.

A MUSCULARIS consists of a heavy, inner circular, and a thin, outer longitudinal coat of smooth muscle. The muscle is responsible for movement of the ovum through the tube.

An outer SEROSA has a typical structure.

Eggs ovulated from the ovary are capable of being fertilized and undergoing development for a maximum of 24 hours after release. The journey through the uterine tube requires about three days. If fertilization is to occur, it must therefore take place in the outer portion of the uterine tube. The implications of these facts for the sperm will be discussed later. Occasionally, a fertilized ovum may not be drawn into the tube, but will implant in the pelvic cavity. Or, it may implant in the uterine tube itself. Such implantations are designated ECTOPIC. Pelvic implantations generally fail due to inadequate vascular connections; tubal implantations are usually surgically terminated (if discovered) before the growing embryo ruptures the tube.

The uterus

The single uterus is a pear-shaped organ measuring 7.5 cm in length, 5 cm wide, and 2.5 cm thick (2½ × 2 × 1 in.) in the female who has never borne children. It is supported behind the bladder and anterior to the rectum by eight ligaments. The paired BROAD LIGAMENTS run laterally from the uterus to the side walls of the true pelvis; the paired ROUND LIGAMENTS pass anteriorly from the superior lateral margins of the uterus; four UTEROSACRAL LIGAMENTS pass from the uterus to the posterior pelvic wall.

The uterus itself has a broad upper FUNDUS, a tapering BODY, and a CERVIX, or neck, that projects into the upper portion of the vagina.

Microscopically, the uterus shows three major layers of tissue (Fig. 30.21):

The PERIMETRIUM or outer serous coat covers all parts of the uterus except the cervix, and consists of

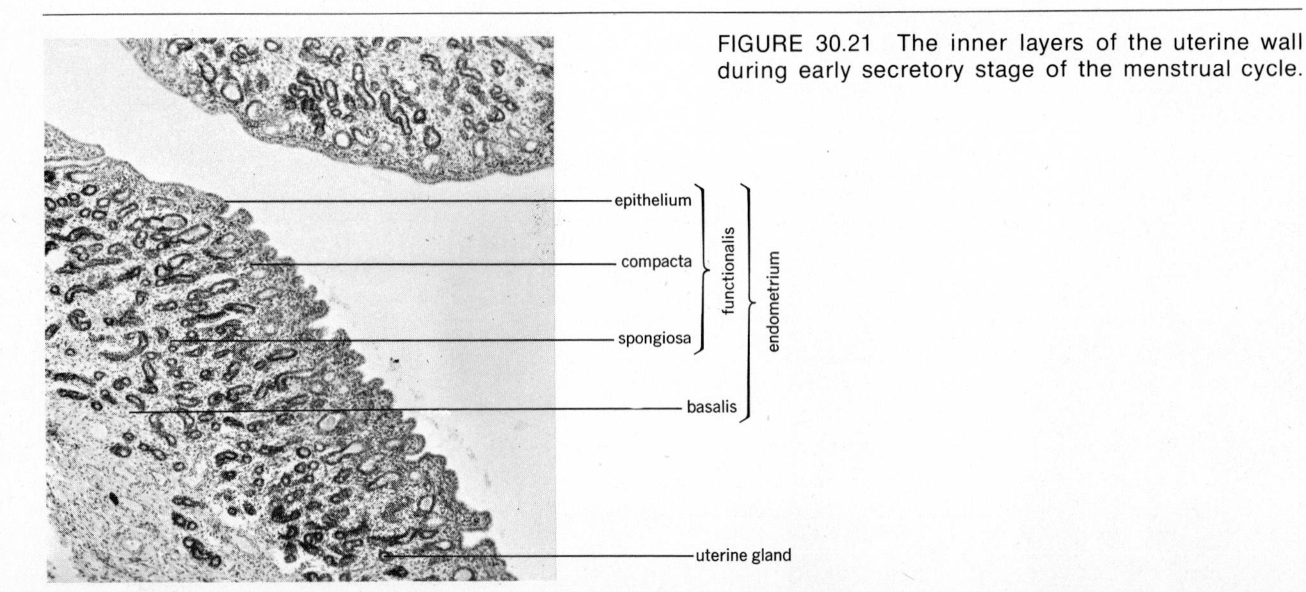

FIGURE 30.21 The inner layers of the uterine wall during early secretory stage of the menstrual cycle.

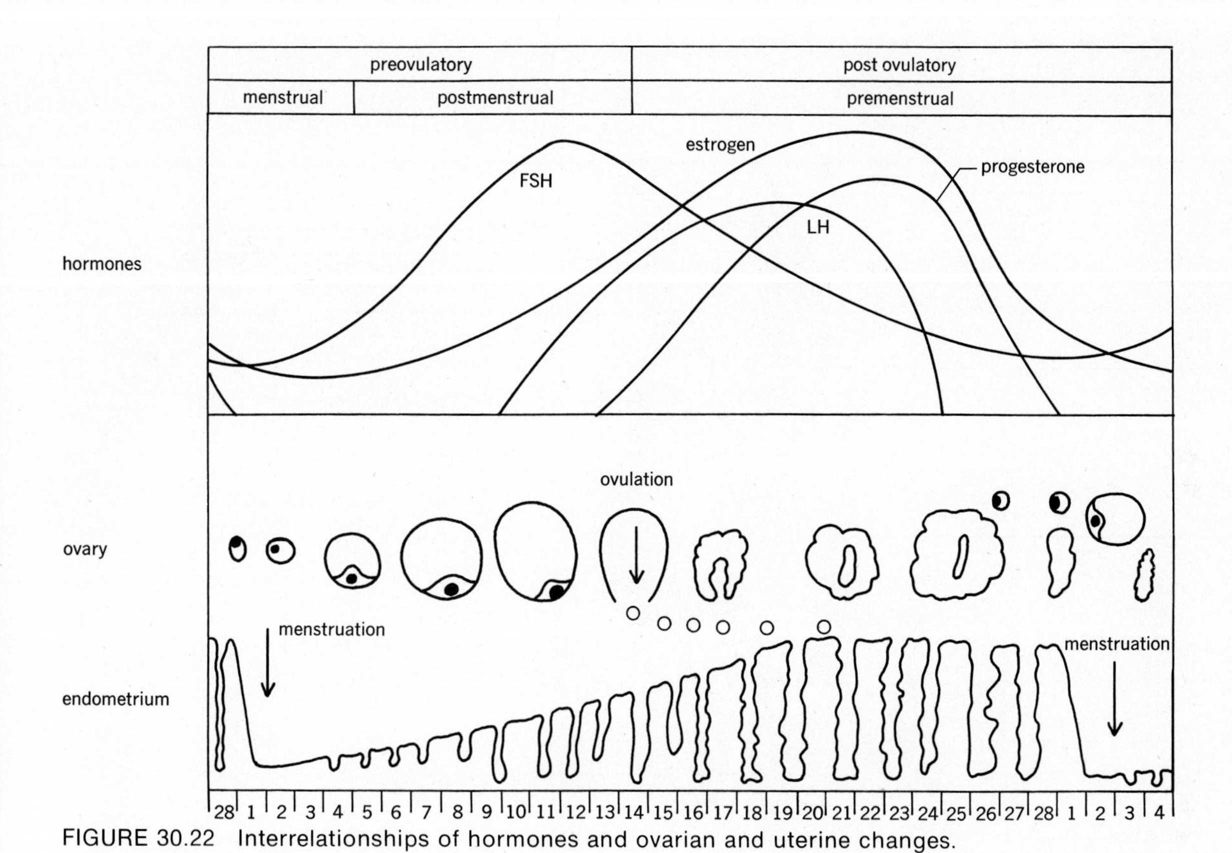

FIGURE 30.22 Interrelationships of hormones and ovarian and uterine changes.

a mesothelial lining supported by underlying connective tissue.

The MYOMETRIUM or muscular coat forms about 75 percent of the thickness of the uterine wall. It consists of four ill-defined layers of smooth muscle.

The ENDOMETRIUM or mucous membrane, shows cyclical changes in appearance, after puberty, with the assumption of menstrual cycles. This layer is composed of two sublayers: a deep lying basalis, which is not shed during menstruation, and a superficial functionalis, which is shed during menstruation.

The menstrual cycle

The MENSTRUAL *(estrus)* CYCLE is hormonally controlled, depending on ovarian production of estrogen and progestin. The hormonal interrelationships involved will be discussed in greater detail in Chapter 31. The cycle is usually subdivided into four stages.

MENSTRUAL STAGE, the first three to five days of the cycle. This stage commences with the first external show of blood and is characterized by involution of the blood vessels of the thickened endometrium, tissue degeneration, and sloughing of the lining.

PROLIFERATIVE *(follicular)* STAGE, the next seven to fifteen days. Under the influence of estrogen from the growing follicle, epithelial repair is begun, and there is multiplication of connective tissue elements and proliferation of glands and blood vessels. A thickness of about 2 mm is attained. Ovulation occurs during the latter portion of this stage.

SECRETORY *(luteal)* STAGE, the next fourteen to fifteen days. After ovulation, corpus luteum develop-

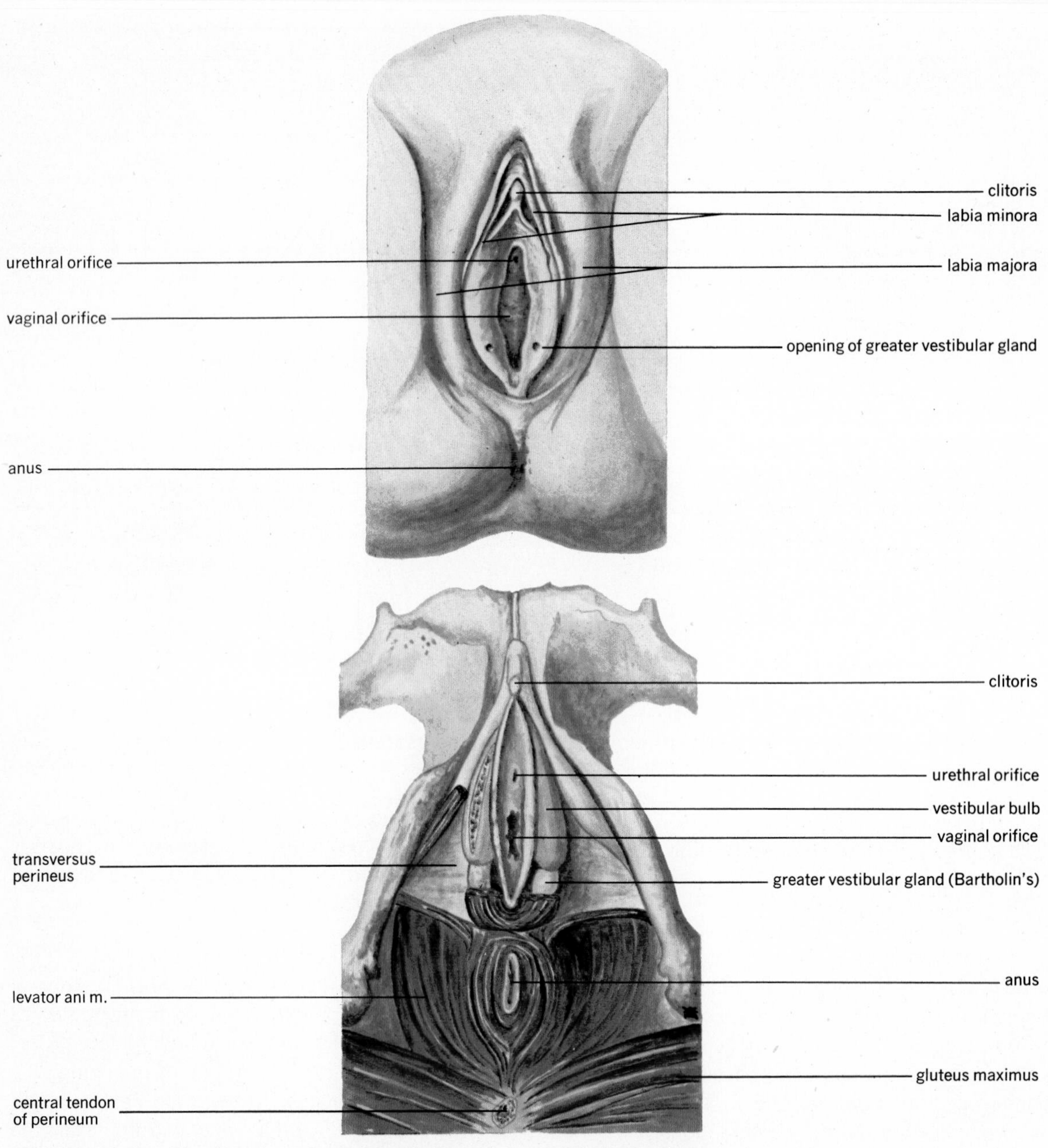

FIGURE 30.23 The female external genitalia and associated structures.

ment and progestin production cause increased glandular and vascular proliferation, and some glandular secretion. A thickness of 4–5 mm is attained preparatory to receiving the ovum if it is fertilized. Implantation of a fertilized ovum "signals" the luteum to maintain itself. If no implantation occurs, the luteum degenerates and precipitates the fourth stage.

The PREMENSTRUAL STAGE. Arteries begin to involute; tissue breakdown is initiated. The stage terminates with external show of blood.

The length of a cycle is variable according to the individual. It may be as short as 24 days or as long as 35 days. If there is one timespan that remains more or less constant, it is the time from ovulation to menstruation—14–15 days. It should not be assumed that all cycles are 28 days in length, or that ovulation always occurs 14 days after menstruation. Some interrelationships of hormones and the ovarian and uterine changes are shown in Fig. 30.22.

The vagina

The vagina is a tubular organ about 8 cm in length. Its lumen is normally not maintained in an open state, but is collapsed. At the upper end, the cervix of the uterus projects about 1 cm into the vagina. A series of moatlike fornices (posterior, anterior, lateral) surround the cervix.

Microscopically, the vagina may be seen to possess three coats of tissue.

An outer ADVENTITIA of connective tissue fixes the vagina to surrounding tissue.

A middle MUSCULARIS of circularly and longitudinally arranged smooth muscle is present.

An inner MUCOUS MEMBRANE consists of a lamina propria and a stratified squamous epithelium. The lamina is rich in glycogen, which decomposes into organic acids. These create an acidic environment in the vagina useful in retarding microorganism growth.

The vagina receives the penis, and sperm are deposited in its upper end. The acid nature (pH 4.5) of the vagina is spermicidal. Recall that semen is alkaline. If sperm are to survive in the vagina, temporary neutralization of acidity is necessary.

Acidity is only the first hazard a sperm encounters if it is to fertilize an ovum in the uterine tube. It must next make a trip of some 6 in. through the uterus and "upstream" against the ciliary current in the uterine tube. Finally, only one sperm may fertilize the egg; the remaining sperm die. These hazards may account, in part, for the necessity of having many millions of sperm in the ejaculate.

The external genitalia

Collectively, the external genital organs (Fig. 30.23) are designated as the VULVA or PUDENDUM. The LABIA MAJORA are two fat-filled, hair-covered (during and after puberty) folds of skin extending from a point about 1 cm anterior to the anus forward to the pubis. They appear to be protective in function.

The LABIA MINORA are smaller folds of skin, lacking hair and fat, that intimately surround the vaginal and urethral orifices.

The CLITORIS is a single midline organ lying anterior to the urethra and surrounded by the anterior portion of the labia minora. It is composed of cavernous tissue similar to that of the penis. The clitoris becomes engorged with blood during sexual excitement and contributes to the achievement of female orgasm.

The mammary glands

The MAMMARY GLANDS (Fig. 30.24) are modified apocrine sweat glands. In infants, children, and men, they are present in rudimentary or undeveloped form. Under hormonal influences, chiefly that of estrogen, they begin their development in the adolescent female. The glands reach their maximum development in the pregnant female. Each gland lies superficial to the pectoralis major muscle and is composed of some 20 lobes of secretory tissue. These lobes empty through three to

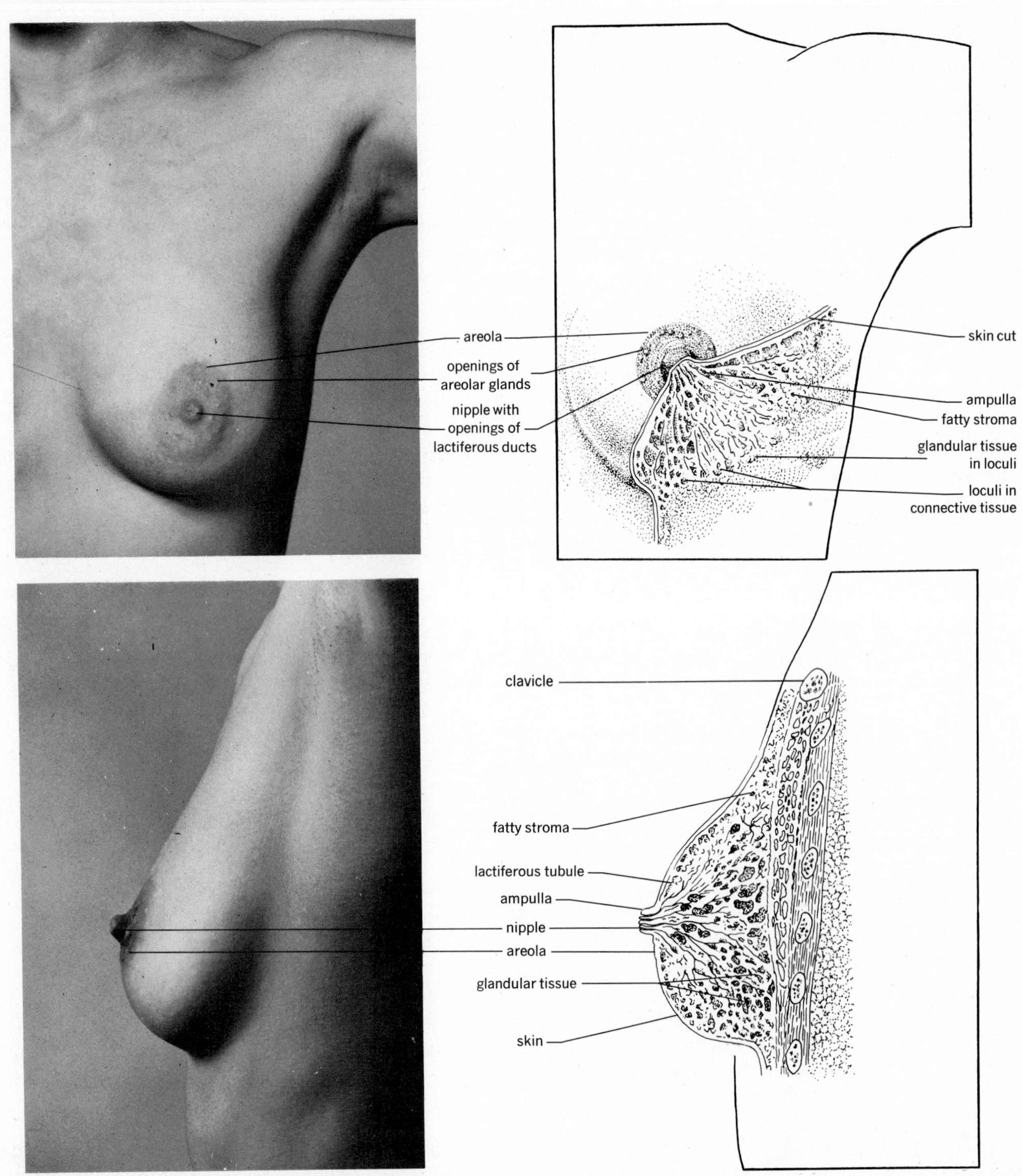

FIGURE 30.24 The mammary gland.

TABLE 30.3 Comparison of male and female reproductive systems

	Male	Female
Essential organs	Testes—paired	Ovaries—paired
Position of essential organs	Scrotal—outside the abdominal cavity; descend to scrotum before birth	Pelvic—inside abdominal cavity
Structure of essential organ	Tunics: Vaginalis—mesothelial outer covering Albuginea—collagenous capsule around organ; thickened medially to form mediastinum Erythroides—vascular coat deep to albuginea Subdivision: Internal division into lobules by septae from albuginea; each lobule contains a seminiferous tubule lined with germinal epithelium Sex cell development: After puberty, spermatogenesis proceeds in 5 stages: 1. Spermatogonia 2. Primary spermatocyte 3. Secondary spermatocyte 4. Spermatid 5. Spermatozoan	Tunics: Tunica albuginea forms capsule; covered by germinal epithelium Internal Structure: Connective tissue stroma forms main mass of organ; may be subdivided into outer cortex, and inner vascular medulla; stroma contains follicles Sex cell development: At birth, 400,000 immature eggs already present in ovary; at puberty, ovum development proceeds in 3 stages: 1. Primary follicle 2. Secondary follicle 3. Vesicular (Graafian) follicle Ovulation occurs next (release of ovum from follicle); tissues remaining in ovary form corpus luteum, then corpus albicans
Ducts	Intratesticular: Straight tubules connect seminiferous tubules to rete tubules Rete tubules—in mediastinum; connect to efferent ductules Extratesticular: Efferent ductules—12 to 16; connect to epididymis Epididymis—20-ft tube for storage Ductus deferens—ejaculatory duct Urethra	Uterine tubes: Portions: infundibulum with fimbrae; ampulla; isthmus Convey egg to uterus Uterus: Single organ; parts include fundus, body, cervix; after puberty, undergoes cyclical changes Vagina: single organ opening to exterior

TABLE 30.3 continued Comparison of male and female reproductive systems

	Male	Female
Accessory organs of the system	Seminal vesicles: Paired structures contributing alkaline secretion to semen Prostate: Surrounds upper urethra; secretion activates sperm Bulbourethral glands: Secretes mucin Penis: Erectile copulatory organ; three cavernous bodies compose it	Bulbourethral glands: Empty into area of vaginal orifice Secrete mucin External organs: Labia majora—large folds of skin around urethral and vaginal openings Labia minora—small folds of skin closer to vagina Clitoris—single sensory and erectile structure to urethra Mammary glands—modified skin glands for nursing young
Cyclical changes after puberty	Not all seminiferous tubules are active at the same time; waves of spermatogenesis sweep over tubules	Ovary ripens (on the average) one egg per month Uterus shows menstrual cycle—phases: 1. Menstrual—shedding of tissue 2. Proliferative—repair and regrowth—ovulation during this stage 3. Secretory—increase in glands and vessels 4. Premenstrual—ischemia and tissue degeneration Vagina shows changes associated with uterine phases
Endocrine functions	Interstitial cells: Produce testosterone	Follicle: Produces estrogen Corpus luteum: Produces progestin
Endocrine control by pituitary	FSH: Control spermatogenesis ICSH: Controls interstitial cells	FSH: Control follicle maturation LH: Controls corpus luteum formation and secretion
Comments	Semen: Total secretions (including sperm) of male tract. Vol: 3 ml—300,000,000 sperms. Alkaline to neutralize acidity in female tract. Urethra common to reproductive and urinary systems; 3 divisions	Female reproductive system separate from urinary

TABLE 30.4 Methods of contraception currently available

Method	Mode of action	Effectiveness if used correctly	Action needed at time of coitus	Requires resupply of materials used	Requires instruction in use	Requires services of physician	Suitable for menstrually irregular women	Side effects
Oral pill 21 day administration	Prevents follicle maturation and ovulation	Highest	None	Yes	Yes–timing	Yes–prescription	Yes	Early–some water retention, breast tenderness. Late–possible embolism, hypertension
Intrauterine device (coil, loop)	Prevents implantation	High	None	No	No	Yes–to insert	Yes	Some do not retain device. Some have menstrual discomfort
Diaphragm with jelly	Prevents sperm from entering uterus, plus jelly spermicidal	High	Previous to coitus	Yes	Yes–must be inserted correctly each time	Yes–for sizing and instruction on use	Yes	None
Condom (worn by male)	Prevents sperm entry into vagina	High	Yes	Yes	Not usually	No	–	Some deadening of sensation in male
Temperature rhythm	Determines ovulation time by noting body temperature at ovulation. T↑	Medium	No	No	Definitely. Must learn to interpret chart correctly	No. Physician should advise	Yes, if skilled in reading graph	None. Requires abstinence during part of cycle
Calendar rhythm	Abstinence during part of cycle	Medium to low	No–no coitus	No	Definitely. Must know when to abstain	No. Physician should advise	No!	None (pregnancy?)
Vaginal foams	Spermicidal	Medium to low	Yes. Requires application before coitus	Yes	No	No	Yes	None usually. May irritate
Withdrawal	Remove penis before ejaculation	Low	Yes. Withdrawal	No	No	No	Yes	Frustration in some
Douche	Wash out sperm	Lowest	Yes. Immediately after	No	No	No	Yes	None

Vasectomy (Tying or cutting the vas deferens) or tubal ligation (tying or cutting the uterine tubes) affords additional means of preventing fertilization. These procedures are generally not reversible, and sterility will result.

five ducts at the apex of the nipple. The pigmented area around the NIPPLE, the AREOLA, lacks mammary ducts, but may have several small papillae representing the openings of several large sebaceous glands. The glands are covered with skin, which is very thin over the areolae and nipples. Hair is absent in these two areas. The mature glands are very vascular organs, with a lymphatic drainage toward the axillary, substernal, and diaphragmatic lymph nodes. The extensive lymph drainage favors widespread metastases in malignancy of the breast.

Table 30.3 compares the male and female reproductive systems.

TABLE 30.5 Other methods of contraception under investigation

Method	Mode of action	Effectiveness if used correctly	Action needed at time of coitus	Requires resupply of materials used	Requires instruction in use	Requires services of physician	Side effects
"Mini-pill"—very low content of progesterone (¼ mg.)	Inhibits follicle development	High	No	Yes	Yes	Yes—prescription	Irregular cycles and bleeding (25%)
"Morning-after pill" (DES, diethylstilbesterol)	Arrests pregnancy probably by preventing implantation. 50 × Normal dose of estrogenic material	By currently available data, high	No. For 1–5 days after coitus	Yes	No	Yes	Breast swelling, nausea, water retention. Use by mother can cause rare type of vaginal cancer in female offspring
Vaginal ring—inserted in vagina; contains progesteroid in it	"Leaks" progesteroid into bloodstream through vagina at constant rate. Thereby inhibits follicle maturation	Studies are "promising"	No	Yes. Perhaps at yearly intervals	Yes	Yes	Spotting, some discomfort
Once-a-month pill	Injected in oil base into muscle. Slow passage of birth control drug into circulation inhibits follicle maturation	Said to be 100 percent	No	Yes. On monthly basis	No	Yes	Similar to oral pill
Depo-Provera (DMPA) 3 month injection	Injected. Inhibits follicle development	By currently available data, high	No	Yes. On 3 month basis	Yes	Yes	Similar to oral pill

Birth control

Tables 30.4 and 30.5 are presented in the interest of acquainting the student with methods of contraception currently available. The physiological principles of the "Pill" are presented in Chapter 31.

Clinical considerations

CARCINOMA (cancer) of the breast is the most common form of cancer in the female, and is the leading cause of death from cancer in the female. Cancer of the uterine cervix is the third most common type of cancer in the female. Breast cancer may be easily discovered by routine self-examination of the breasts, and the Papanicolaou (Pap) test can discover many cervical cancers before they become life-threatening. Chronic irritation, trauma, and viral infections appear to be the most common causes of cervical cancer.

AMENORRHEA refers to failure of a female to menstruate by 18 years of age or cessation of menstruation in a menstruating female; it calls for a complete investigation of the functional status of the hypothalamus, pituitary, ovaries, and uterus.

DYSMENORRHEA refers to painful menstruation. Its cause is not known; it is associated with cramps and pain 24 to 48 hours before menstrual flow.

Venereal diseases

The venereal diseases (VD) are the number-one communicable diseases in the United States today. The continuing increase in incidence of VD may be due to several basic causes.

TABLE 30.6 Some characteristics and effects of gonorrhea and syphilis

	Gonorrhea	Syphilis
Causative organism	Neisseria gonorrhoeae	Treponema pallidum
Incubation period	2–14 days (usually 3 days)	7–90 days (usually 3 weeks)
Method of transmission	Sexual, oral, or physical contact with an infected person. Contaminated object up to 8 hours after organisms deposited. *Infants*—during birth through vagina.	Sexual, oral, or physical contact with an infected person. Blood transfusion. On contaminated objects, dies quickly by drying. *Infants*—may acquire during birth, or through the placenta if mother not treated before third trimester and adequately.
Contact examination *(all sex contacts exposed within the following time periods)*	2 weeks (male) 1 month (female)	*Primary.* 3 months (+ duration of symptoms). *Secondary.* 6 months (+ duration of symptoms). *Early latent.* 1 year. *All syphilis.* "Family" contacts as indicated.

There have been and will continue to be changes in attitudes toward sexual intercourse that separate it from its reproductive function. Increased sexual contact has therefore increased the possibility of spread of VD among the population.

The increasing number of younger people in the population has increased the incidence of sexual contact with greater possibility of spread of VD (the highest incidence is presently in the 15–30 age group).

A large part of the social stigma associated with VD has disappeared, and reporting of cases has increased. It may be hoped that increased reporting of cases has been an incentive for the establishment or increase in funding of programs de-

TABLE 30.6 continued Some characteristics and effects of gonorrhea and syphilis

	Gonorrhea	Syphilis
Clinical characteristics	Discharge; burning, pain, swelling of genitals and glands. **Male.** Purulent urethral discharge, hematuria, chordee. Urethritis, prostatitis, seminal vesiculitis, epididymitis, occasional involvement of testes. If untreated, symptoms disappear in about 6 weeks, and organism persists in the prostate gland (Gc carrier). **Female.** *Child,* vaginitis. Leukorrhea, tubal abscess, urethritis, cervicitis, pelvic inflammation (Peritonitis). Tubal stricture, possible sterility due to closure of tubes. In both male and female, healing is by scar tissue formation. Strictures and closing of tubular structures may result. Arthritis, endocarditis, meningitis may occur.	**Primary.** Chancre present, solitary, nonpainful ulcer on genital or mucous membranes. **Secondary.** Rashes or mucous patches. Macules or papules on hands, feet, oral cavity, genitoanal area, trunk, extremities. **Tertiary:** *Latency.* No symptoms, positive serology. Profound changes are produced in the skin, mucous membranes, skeleton, GI tract, kidney, brain; heart and blood vessels show destructive changes (abscesses, scarring, tissue destruction). Tabes dorsalis in spinal cord destroys dorsal columns. Paresis and psychosis may result. *Relapse.* Recurrence of infectious lesions after disappearance of secondary lesions. *Late.* Cardiovascular, central nervous system, gummata. Obvious systemic damage appears.

signed for case finding and treatment of venereal diseases.

The discontinuance of indiscriminate use of antibiotics for "colds" and other infections that do not warrant their use has resulted in the removal of a "brake" on the spread of VD, in that many previously undiagnosed cases are no longer being treated inadvertently. Use of antibiotics has resulted in the development of antibiotic-resistant strains of the VD organisms.

Better epidemiological studies have resulted in the discovery of reservoirs of infection that may not have been brought to light previously.

The belief of many segments of the public that a "shot" of antibiotic will cure VD has perhaps led to the development of complacency, as well as

TABLE 30.6 continued

	Gonorrhea	Syphilis
Diagnostic procedures	Culture of discharge. Smear. Fluorescent antibody test (FAT). Currettage to get tissue containing cocci. Several cultures may be necessary as not all tests consistently show cocci. History, clinical and contacts. (Serologic test for syphilis).	Darkfield examination – microscopic for spirochete. Serologic (blood) test for antibody (reagin) to organism. [(Wasserman, VDRL – Venereal Disease Research Lab) False positive tests may be reported by smallpox antibodies, hepatitis, mononucleosis, and high fevers.] Spinal fluid test. X-rays of long bones of infants. History, clinical and contacts.
Treatment	Penicillin Broad spectrum antibiotics (e.g. sulfonamides, streptomycin). Organism must be sensitive to the drug of choice.	Penicillin Broad spectrum antibiotics (e.g., Erythromycin, Tetracycline). Organism must be sensitive to drug of choice. Reexamination at 6 months and 1 year to evaluate treatment results.

a false sense of security. All patients must be rechecked at a certain interval following treatment to ensure that the tests for organisms are negative. Also, after an initial set of symptoms, several of the venereal diseases may enter an asymptomatic period that may persist for years, leading to the belief that the disease has disappeared. During this time, damage may continue te be wrought in various body systems, although no overt changes become apparent.

Reinfection is common since no lasting immunity is developed to the organisms. Contacts must be examined and adequately treated if they are infected to prevent further spread. Sexual partners, diagnosed as infected, must be treated at the same time or they will be playing "ping-pong" with the organism. Persons having a gonococcal infection must also be checked for syphilis because both infections may be present at the same time but the amount and type of treatment may differ.

Gonorrhea ("clap") and syphilis ("bad blood") are presently in epidemic proportions throughout the United States and some other countries. These diseases have no respect for race, creed, or color, and one cannot tell an infected person by just looking at him or her.

All personal information regarding cases, contacts, and suspects of VD is legally confidential, which should encourage infected persons to seek early treatment of the disease(s).

Free treatment is available through local health departments. Other clinic facilities, private physicians, or both are available in most communities. Information on a state's law regarding the treatment of minors may be obtained through the health department. Some states now treat minors without the parents' consent, since the child's future welfare is at stake, and as an effort to cut down the VD incidence in sexually active teenagers.

Some characteristics and effects of gonorrhea and syphilis are presented in Table 30.6.

Summary

1. The reproductive systems begin development at about five weeks, pass through an indifferent stage, and assume a form characteristic of the proper sex at eight to ten weeks.
   a. All ova (eggs) the ovary will produce are formed prior to birth.
   b. Sperm are not produced until puberty.

2. The male reproductive system includes the essential organs (testes), a series of ducts, accessory glands, and external genitalia.
   a. Each testis is ovoid and surrounded by three tunics or coats of tissue.
   b. Each lobule (250) contains one to four coiled seminiferous tubules, lined with germinal epithelium in various stages of development of sperm.
   c. Interstitial cells produce male sex hormone.
   d. The ducts transport and store sperm and propel semen during ejaculation.
   e. The accessory glands contribute fluids to suspend and nourish sperm and neutralize acidity.
   f. The penis is a copulatory and excretory organ composed of erectile tissue (cavernous bodies).
   g. Anterior pituitary hormones govern sperm production and hormone secretion.

3. The female reproductive system is composed of the internal organs (ovaries, uterine tubes, uterus, vagina) and the external genitalia.
   a. The ovaries are ovoid and contain a capsule, stroma, and follicles in various stages of development (primordial, primary, secondary, vesicular, Graafian). After ovulation, the follicular tissue forms a corpus luteum and then a corpus albicans.
   b. The uterine tubes conduct ova to the uterus and are the usual sites of fertilization.
   c. The uterus has three layers (endo-, myo-, and perimetrium). The endometrium undergoes cyclical changes during menstrual cycles. The latter has four stages.
   d. The vagina has a three-layered structure, receives the penis during copulation, and acts as the birth canal.
   e. The external genitalia include the two pairs of labia (majora, minora) that act to protect the vaginal and urethral openings, the clitoris, and the mammary glands that provide nourishment for offspring.

4. Tables showing methods for prevention of pregnancy are presented.

5. Tables and discussion of venereal diseases as to symptoms, characteristics, and potential dangers are presented.

31 The Endocrine System

Development of the major endocrines

General principles of endocrinology

- Criteria for determining endocrine status
- Hormones and parahormones
- Chemical nature of hormones
- Secretion, storage, transport, and use of hormones
- Function, effect, and action of hormones
- Control of secretion of hormones

The hypophysis (pituitary)

- The pars distalis
- Hormones of the distalis
- Control of secretion of pars distalis hormones
- The pars intermedia
- The pars tuberalis
- Neurohypophysis
- Disorders of the hypophysis

The thyroid gland

- Hormone synthesis
- Control of the thyroid
- Effects of thyroxin
- Other factors influencing thyroid activity
- Thyrocalcitonin
- Disorders of the thyroid

The parathyroid glands

- The parathyroids and ion metabolism
- Effects of the hormone
- Effects on other organs and cells
- Disorders of the parathyroids

Development of the major endocrines

During the fourth to fifth week of development, the embryo shows a series of cartilagenous arches, the BRANCHIAL ARCHES, in the neck region. Outpocketings of the lining of the foregut, the PHARYNGEAL POUCHES, will develop between arches. In the roof of the foregut, a single median outpocketing, RATHKE'S POUCH, appears. The HYPOPHYSIS (pituitary) is a gland whose anterior portion (anterior lobe) is derived from Rathke's pouch. This portion migrates up and backward to fuse with a downgrowth of the hypothalamus, the latter representing primarily the posterior lobe of the gland.

The THYROID GLAND is derived from a single medial outgrowth of the floor of the foregut at the level of the second pharyngeal pouch. It migrates inferiorly to assume a final position around the larynx and upper trachea.

The PARATHYROIDS develop from the anterior portions of the third and fourth pharyngeal pouches. These migrate inferiorly to assume a final position on the dorsal side of the thyroid.

The THYMUS, not a true endocrine, originates from the posterior portions of the third and fourth pharyngeal pouches and migrates to lie behind the sternum.

The ADRENAL MEDULLA is a derivative of neural crest material. A covering of urogenital ridge mesoderm forming the CORTEX of the gland is applied to the adrenal medulla.

The endocrine tissue of the PANCREAS (islets) arises from the ends of the endodermal outpocketings of the gut. The outpocketings form the ducts, and the islets lose their connection with the ducts.

The endocrine cells of TESTIS and OVARY differentiate from the mesodermal cells as the gonads develop.

The diversity of origins of the endocrines, in terms of both site and germ layer is impressive. The origins are shown in Fig. 31.1.

FIGURE 31.1 Origins of some of the endocrine organs from branchial pouches.

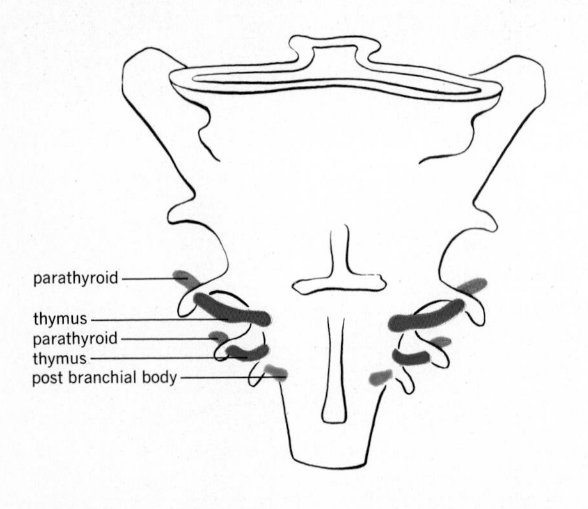

General principles of endocrinology

We have examined the role of the nervous system in stimulus-response and homeostatic control. The endocrine system is primarily concerned with GOVERNING BODY PROCESSES such as growth, differentiation, and metabolism. Endocrine glands utilize the bloodstream to distribute their chemical messengers, the HORMONES, and in this way achieve contact with all body cells. It is becoming increasingly evident that the two "control systems" operate together to control body activity. The nervous system may, for example, respond to environmental changes and influence an endocrine to secrete a product that has the effect of restoring homeostasis.

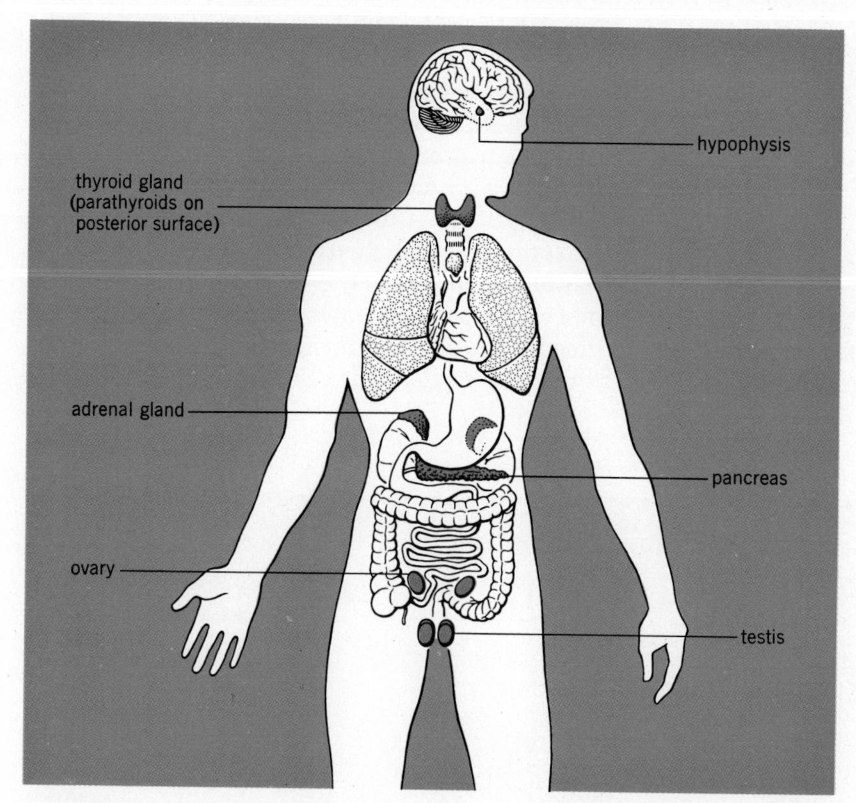

FIGURE 31.2 The locations and names of the major endocrine organs.

Criteria for determining endocrine status

Endocrine glands are designated as being CIRCUMSCRIBED, or specific groups of cells, that PRODUCE SPECIFIC CHEMICAL SUBSTANCES having well-defined effects on body function. Removal of a group of cells suspected of being an endocrine causes clearly defined alterations of body function, and administration of the active chemical of the gland likewise alters body function. Utilizing these criteria, we recognize the endocrine structures shown in Fig. 31.2.

Hormones and parahormones

The product of an endocrine gland (as defined above) may properly be called a HORMONE. There are many chemical substances that have specific effects on the body, but are not traceable to specific cells. Examples include carbon dioxide, gastrin, secretin, enterocrinin, and the prostaglandins. Such substances are commonly called "hormones"; the term PARAHORMONE, however, serves to distinguish such substances from true hormones.

Chemical nature of hormones

Mammalian hormones fall into three classes: PROTEINS, POLYPEPTIDES, or compounds such as GLYCOPROTEINS; AMINES; and STEROIDS. A general correlation exists between the germ layer of origin of the endocrine and the class of hormone produced. Most amine hormones are produced by ectodermally derived endocrines. Steroids are produced by mesodermal derivatives. Polypeptide or protein hormones are produced by endodermal derivatives.

Secretion, storage, transport, and use of hormones

A given endocrine cell synthesizes its hormones from raw materials present in the blood reaching the gland. Some glands (e.g., thyroid) produce more hormone than the body requires at a time and may store excess hormone within the gland. Secretion into the bloodstream occurs as the result of a chemical or nervous stimulus. In the bloodstream, most steroid and amine hormones attach to a specific plasma protein "carrier" and are transported to the body cells in this combined fashion. At the cell, the hormone may attach to the cell surface and enter by pinocytosis, or pass directly through the membrane to influence enzymes, the nucleus, or cellular organelles. As the

TABLE 31.1 General summary of endocrine glands

Item	Comments
Functions of endocrines	To integrate, correlate, and control body processes by chemical means
Criteria for establishing function as endocrine	Cells are morphologically distinct Cells produce specific chemicals not produced elsewhere Chemicals exert specific effects; effect is lost if gland removed, restored if chemical administered
Hormones vs. parahormones	Hormones are produced by cells qualifying as endocrines Parahormones are chemicals having specific effects but not produced by endocrines
Chemical nature of hormones	Steroid – mesodermal origin (e.g., gonads) Polypeptides – endodermal origin (e.g., pancreas) Small MW amines – ectodermal origin (e.g., posterior pituitary, adrenal medulla)
"Life history" of hormones	Secreted into blood, go to cells and there (probably) influence chemical reactions. Are inactivated as used, inactivated product eliminated
Methods of control available	
1. Neurohumor	Nerve cells produce a chemical; it goes to endocrine and controls secretion. Stimulus is chemical
2. Nerves	Nerve fibers pass to endocrine. Stimulus is electrical (nerve impulse)
3. Feedback	Target organ hormone influences secretion of another endocrine which stimulated target organ
4. Nonhormonal organic substances in blood	Glucose acting on pancreatic islets. Rise causes increased secretion (usually)
5. Total osmolarity of blood	Requires nervous system to detect it. Nervous system then signals endocrine.
6. Inorganic substances in blood	Ca^{++} on parathyroid. Effect usually direct, not inverse

cell utilizes the hormone, it generally alters the hormone's structure, partially or completely destroying the physiological activity of the chemical. This process is termed "inactivation." The inactivated product passes from the cell to the bloodstream and is usually excreted. Inactivation of hormones by the organ it affects *(target organ)* results in the requirement that hormones be continually secreted by the endocrine. In short, hormones are not reused, but are used up by the target organ.

Function, effect, and action of hormones

The FUNCTION of a hormone is what we conclude the purpose of a hormone to be, in terms of whole body function. EFFECT refers to measurable alterations in function occasioned by deprivation or administration of a hormone. ACTION refers to the cellular mechanism influenced by the hormone. For example, thyroxin, the product of the thyroid gland, is known to increase glycolysis within the mitochondria (action) and to bring about an increase in BMR and heat production(effect). It is concluded that the hormone controls BMR (function).

Hormone action appears to depend on the presence within the cell of a chemical reaction whose rate can be influenced by the hormone. Some hormones are without action on certain cells, suggesting the absence of an influencable reaction.

Control of secretion of hormones

In general terms, SIX DIFFERENT METHODS exist by which both rate and quantity of hormone secretion may be determined.

NEUROHUMORS. A nerve cell or group of nerve cells produces a chemical (neurohumor) that is sent to the endocrine and causes secretion.

DIRECT NERVOUS CONNECTION. The endocrine receives nerve fibers from the autonomic nervous system and responds to a nerve impulse.

RECIPROCAL OR FEEDBACK CONTROL. One endocrine produces a hormone that affects a second endocrine. The second endocrine produces its hormone, which influences secretion of the first endocrine. If the effect of the second hormone is to inhibit secretion of the first endocrine, the effect is said to be a *negative feedback*. If the effect is to increase secretion it is a *positive feedback*.

BLOOD LEVELS OF ORGANIC SUBSTANCES other than hormones.

BLOOD LEVELS OF SOLUTES in general.

BLOOD LEVELS OF SPECIFIC INORGANIC SUBSTANCES.

Understanding of the discussions of individual endocrines will be aided if the remarks on the previous pages are well understood. Table 31.1 summarizes the preceding sections.

The hypophysis (pituitary)

The hypophysis is a gland of double origin lying in the sella turcica of the sphenoid bone. It measures about 10 × 13 mm and weighs approximately ½ g. The gland is composed of a portion derived from the roof of the oral cavity, the adenohypophysis, and a portion derived from the hypothalamus, the neurohypophysis. Each region has several subdivisions, as shown in Fig. 31.3 and Table 31.2.

Both major divisions of the gland maintain connections with the hypothalamus (see Fig. 31.3).

TABLE 31.2 Divisions of the hypophysis

Major divisions	Subdivisions
Adenohypophysis	Pars distalis (anterior lobe) Pars tuberalis Pars intermedia
Neurohypophysis	Infundibulum (stalk) Pars nervosa (posterior lobe)

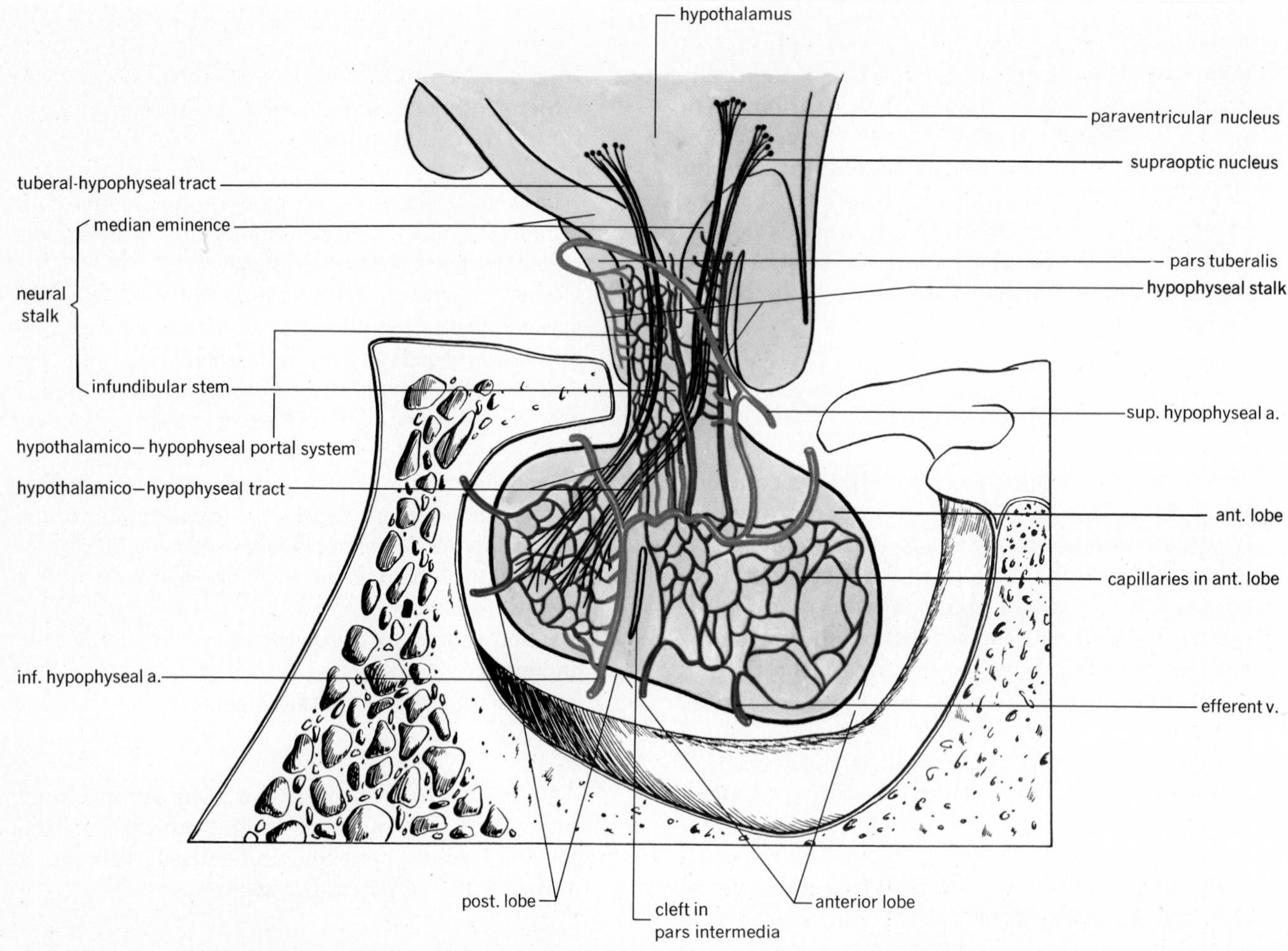

FIGURE 31.3 The hypothalamic-hypophyseal connections.

The ADENOHYPOPHYSIS is connected to the hypothalamus by a system of blood vessels originating in the hypothalamus and terminating as sinusoids in the pars distalis. This system is designated as the HYPOTHALAMICO-HYPOPHYSEAL PORTAL SYSTEM (HHPS) or PITUITARY PORTAL SYSTEM. The NEUROHYPOPHYSIS is connected to the hypothalamus by nerve fibers originating in the hypothalamus and terminating in the pars nervosa. These fibers form the HYPOTHALAMICO-HYPOPHYSEAL TRACT (HHT).

These blood and nervous connections emphasize at least two important facts relative to the activity of the hypophysis: first, these connections enable nervous and endocrine systems to work in an integrated fashion to control body activity; second, endocrine response to nerve input may be achieved rapidly.

The pars distalis

With ordinary stains, three cell types, designated ACIDOPHILS, BASOPHILS, and CHROMOPHOBES, may be recognized in the distalis. These cells are named according to their preference for red, blue, or no stain, respectively. With special stains, particularly PAS (periodic acid Schiff, a glycoprotein stain) and aldehyde thionine (protein polysaccharide stain), seven cell types may be differentiated. The production of at least six hormones has been attributed to the distalis, and a tentative assignment of hormones to a particular cell type has been made. The morphology and function of these cells is shown in Fig. 31.4 and described in Table 31.3.

Hormones of the distalis

GROWTH HORMONE [*somatotropin, somatotrophic hormone (STH), human growth hormone (HGH)*]. Growth hormone from human hypophysis is a single chain of 188 amino acids linked in two places by disulfide (—S—S—) bonds (Fig. 31.5). It has been synthesized in the laboratory. The normal pituitary contains 4–10 mg of hormone, and there is little difference in hormone content with age. Secretion of growth hormone is such as to maintain a blood level of about 3 μg (3/1,000,000 g) per ml.

FIGURE 31.4 The cells of the pars distalis.

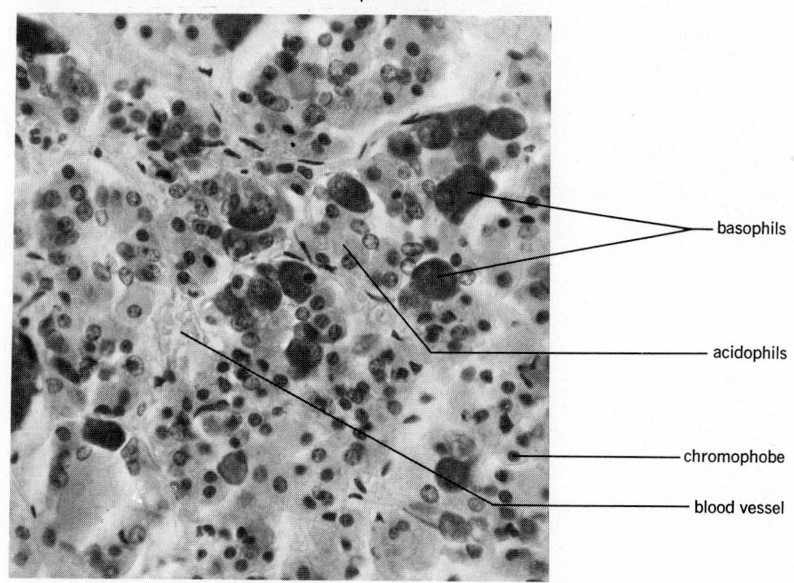

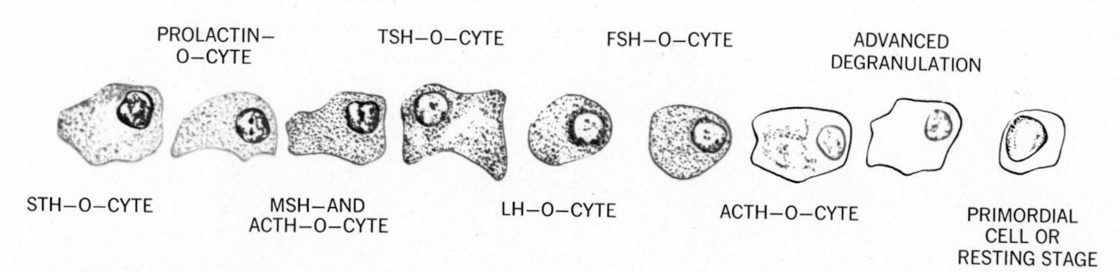

azocarmine—orange—G stain

TABLE 31.3 Distalis cell types

Names of cells under:				
Ordinary stain	Special stain	Description	Staining preference	Hormone produced
Acidophils	Somatotropic cell (α_1-orangophil)	Round or ovoid shape with small round acidophilic cytoplasmic granules	Orange G	Human growth hormone (HGH)
	Lactotropic cell (α_2-carminophil)	Round or ovoid, contains large round or oval acidophilic cytoplasmic granules	Azocarmine	Prolactin
Basophils	Corticotropic cell (β1)	Irregular shape, eccentric nucleus	PAS	Adrenocortico-tropic hormone (ACTH)
	Thyrotropic cell (β2)	Irregular shape, eccentric nucleus	PAS and aldehyde thionine	Thyroid-stimulating hormone (TSH)
	Luteotropic cell (Δ_1)	Round outline, nearly central nucleus	PAS and aldehyde thionine	Luteinizing hormone (LH)
	Follicle stimulating hormone cell (Δ_2)	Round outline, nearly central nucleus	PAS	Follicle-stimulating hormone (FSH)
Chromophobe	Chromophobe	Irregular outline, eccentric nucleus, very pale staining cytoplasm	None, stains little if at all	ACTH?

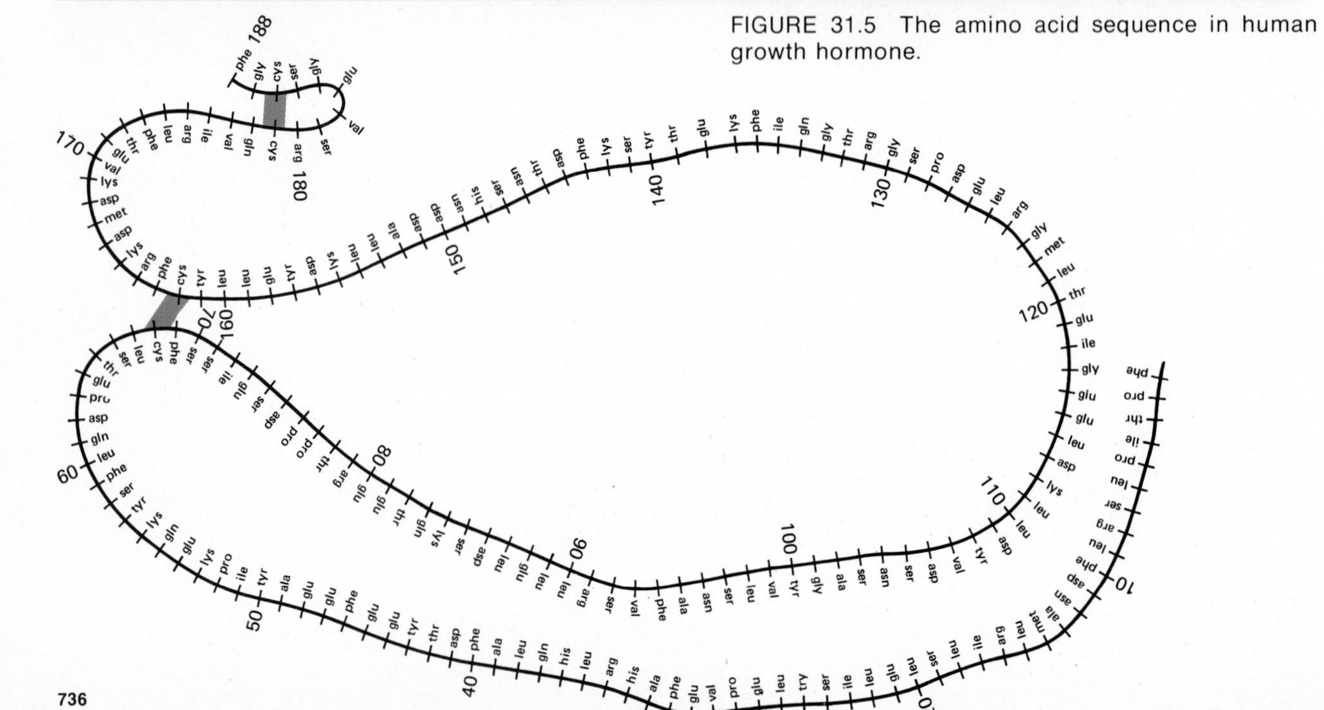

FIGURE 31.5 The amino acid sequence in human growth hormone.

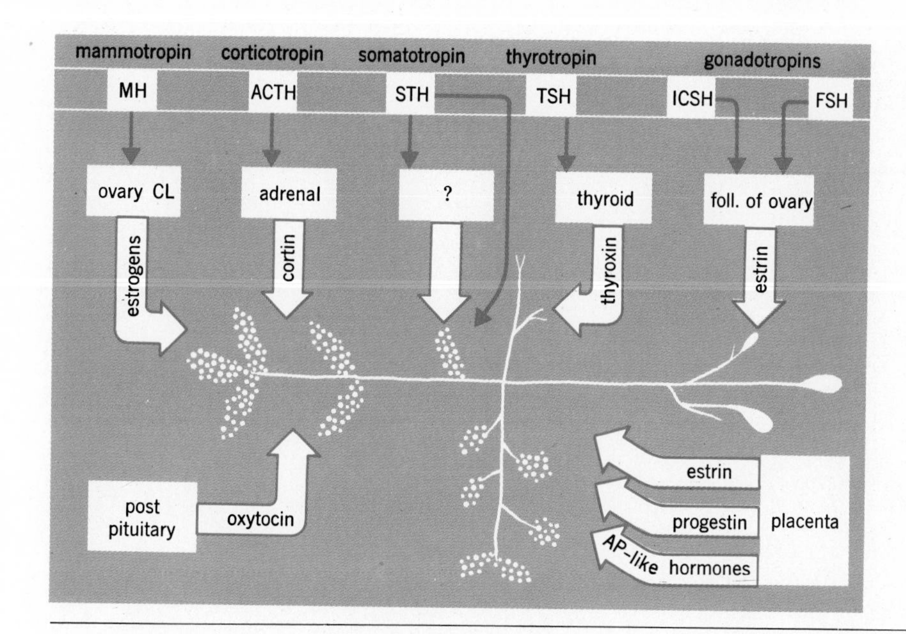

FIGURE 31.6 The hormones involved in lactation.

The hormone exerts its effect primarily on the hard tissues of the body, secondarily on the soft tissues. In these two tissues, growth hormone increases rate of growth and maintains the size of body parts once maturity has been reached. Metabolic effects include: increasing the conversion of carbohydrates to amino acids (transamination); increasing the cellular uptake of amino acids; mobilization of fats from storage areas; increasing fat metabolism, as evidenced by a fall in respiratory quotient (RQ); increasing blood sugar level (diabetogenic effect). There is some evidence to suggest that the fat-mobilizing effect may be separated from the other effects; some investigators claim a separate lipid mobilizing hormone exists.

PROLACTIN *(lactogenic hormone).* Prolactin is a polypeptide hormone of 205 amino acids. It is one of several hormones involved in milk production by the mammary glands. If the gland has been "primed" by sex hormones, thyroid hormone, and adrenal hormones, prolactin will cause milk secretion (Fig. 31.6). In birds, other effects of the hormone include aiding the expression of maternal behavior (nesting, egg hatching).

ADRENOCORTICOTROPIC HORMONE *(ACTH).* ACTH is a single unbranched chain of 39 amino acids. It is synthesized as the body requires it, since the gland has less than ¼ mg in it. ACTH controls the synthetic and secretory activity of the two inner zones of the adrenal cortex. Metabolic effects on cells in general include mobilization of fats, production of hypoglycemia, and increase in muscle glycogen. Acting through the adrenal cortex, ACTH is also involved in body resistance to stress.

THYROID-STIMULATING HORMONE *(thyrotropin, TSH).* TSH is a glycoprotein having a molecular weight of about 25,000. TSH influences all phases of thyroid gland activity: accumulation of material, synthesis, and secretion. Acting through the thyroid, TSH becomes involved in regulation of basal metabolic rate. Effects on other tissues include increasing the breakdown of fat, working with growth hormone, and increasing water content of loose connective tissue.

LUTEINIZING HORMONE *(LH, luteotropin; in the male, it is called ICSH for interstitial-cell-stimulating hormone).* LH is a glycoprotein with a molecular

TABLE 31.4 Hormones of the pars distalis

Hormone	Characteristics	Effects on body
Growth Hormone (GH)	Molecular weight 21,500; unbranched peptide chain	Controls body hard and soft tissue growth
Prolactin	Molecular weight about 23,500	Causes milk secretion in "primed" gland
Adrenocorticotropic hormone (ACTH)	Molecular weight 4567; contains 39 amino acids in single peptide chain	Controls activities of adrenal cortex
Thyroid stimulating Hormone (TSH)	Molecular weight 25,000, composed of 200 to 300 amino acids	Controls all aspects of thyroid activity
Luteinizing hormone (LH) or Interstitial cell stimulating hormone (ICSH)	Molecular weight about 30,000; a glycoprotein	Female: Corpus luteum formation, ovulation, implantation Male: Controls interstitial cell activity
Follicle stimulating hormone (FSH)	Molecular weight about 30,000; a glycoprotein	Female: Controls follicular maturation Male: Controls spermatogenesis

weight of about 30,000. In the female LH occasions changes leading to the formation of the corpus luteum in the ovary and is involved in readying the mammary gland for secretion. It is necessary for ovulation and implantation of the zygote. In the male, ICSH controls secretion of the interstitial cells of the testis.

FOLLICLE-STIMULATING HORMONE *(FSH).* FSH is also a glycoprotein with molecular weight about 30,000. In the female, FSH controls the maturation of primordial follicles to vesicular follicles; in the male, it controls spermatogenesis. Table 31.4 presents a summary of the characteristics and effects of pars distalis hormones.

Control of secretion of pars distalis hormones

The role of the hypothalamus in maintenance of homeostasis was discussed in Chapter 13. It was emphasized that many other areas of the brain have nervous connections with the hypothalamus, and that the bloodstream is also monitored and can influence hypothalamic activity. One of the efferent connections of the hypothalamus is to the pituitary gland, and it has been shown that a system of blood vessels and nerve tracts connects the hypothalamus and the pituitary. Such systems of connections enable emotional disturbances, sensory input and environmental changes to be translated into endocrine response.

Control of secretion and synthesis of hormones by the adenohypophysis of the pituitary is believed to be mediated by chemical factors produced by the hypothalamus in neurosecretory cells and placed in the pituitary portal system for delivery to the adenohypophysis. Such chemicals are most properly termed REGULATING FACTORS, for some may stimulate release of adenohypophyseal hormones (releasing factors), and others

TABLE 31.5 Hypothalamic factors controlling secretion of adenohypophyseal hormones

Hypothalamic factor	Designation/ abbreviation	Amino acid sequence (if known)	Comments
Corticotropin (ACTH) releasing factor (or hormone)	CRF (CRH)	Ac-Ser-Tyr-Cys-Phe-His ($AspNH_2$-$GluNH_2$) Cys (Pro-Val)-Lys-$GlyNH_2$	Chemical instability and difficulty of assay make chemical formula uncertain
Thyrotropin (TSH) releasing factor (or hormone)	TRF (TRH)	Glu-His-$ProNH_2$	It is believed that the factor causes increase in cAMP in pituitary cells, and this increases release of TSH. TSH in turn increases thyroid activity by increasing thyroid cAMP
Luteinizing hormone (LH) releasing factor (or hormone) and Follicle-stimulating hormone (FSH) releasing factor (or hormone)	LH-RF (LH-RH) and FSH-RF (FSH-RH)	Glu-His-Trp-Ser-Tyr-Gly-Leu-Arg-Pro-$GlyNH_2$	One decapeptide appears to have both LH and FSH releasing ability. Effects not yet separated
Growth hormone (GH) releasing factor (or hormone) or Somatotropic hormone (STH) releasing factor (or hormone)	GH-RF (GH-RH) or STH-RF (STH-RH)	Val-His-Leu-Ser-Ala-Glu-Glu-Lys-Glu-Ala	May be species specific
Growth hormone (GH) inhibiting factor (or hormone) or Somatotrophic hormone (STH) release inhibiting factor or Somatostatin	GIF (GH-RIF) or STH-RIF	HAla-Gly-Cys-Lys-Asn-Phe-Phe-Trp-Lys-Thr-Phe-Thr-Ser-CysOH	–
Prolactin-inhibiting factor (or hormone) or Prolactin-releasing inhibiting factor (or hormone)	PIF (PIH) or PRIF (PRIH)	Unknown	Believed to be a small MW substance, similar to oxytocin. Concentration reduced by suckling

TABLE 31.5 continued Hypothalamic factors controlling secretion of adenohypophyseal hormones

Hypothalamic factor	Designation/ abbreviation	Amino acid sequences (if known)	Comments
Prolactin-releasing factor (or hormone)	PRF (PRH)	Unknown	Produced after birth of offspring. TRH also has a prolactin stimulating effect
Melanocyte-stimulating hormone (MSH) inhibiting factor (or hormone)	MIF (MIH)	Pro-Leu-GlyNH$_2$ and/or Pro-His-Phe-Arg-GlyNH$_2$	—
Melanocyte-stimulating hormone (MSH) releasing factor (or hormone)	MRF (MRH)	Cys-Tyr-Ile-Glu-Asn-CysOH	—

may inhibit secretion (inhibiting factors). Some investigators believe these factors qualify as true hormones.

Today, many hypothalamic factors have been isolated or postulated to exist, and in several cases their chemical structure has been determined and the compound synthesized (Table 31.5).

Other factors than the hypothalamus appear to be involved in distalis secretion. For example, growth hormone secretion is increased by a fall in blood sugar level. This effect may be direct on the gland or through the hypothalamus. Negative feedback exists between thyroxin and thyroid-stimulating hormone, cortical steroids and adrenocorticotropic hormone, progestin and luteinizing hormone, and estrogen and follicle-stimulating hormone. Again, the problem is whether the effect is direct on the distalis or through the hypothalamus.

The pars intermedia

In the human hypophysis, the pars intermedia is virtually nonexistent, being reduced to a few cells and spaces (cysts) between the distalis and neural lobe. A hormone consisting of 22 amino acids and differing from those of lower animals is secreted by the human intermedia. It is called beta-MSH, and its physiological significance is uncertain. In lower vertebrates (fishes and amphibians), the intermedia is the source of MELANOCYTE-STIMULATING HORMONE *(MSH)*, which causes expansion of pigment cells (melanophores) in the skin of such animals. A protective effect of "matching" the color of the animal to the environment is the result. In man, adrenocorticotropic hormone has a function similar to MSH. Release of MSH is controlled by hypothalamic regulating factors (see Table 31.5).

The pars tuberalis

No known hormone is produced by the cells of the tuberalis.

Neurohypophysis

The neurohypophysis cannot be considered a separate endocrine gland, for it produces no hormones. The hypothalamus produces the hormones of the neural lobe and passes them over the hypothalamico-hypophyseal nerve tract merely to be stored in and released from the nerve terminals in the nervosa.

The characteristic cell of the neural lobe appears to be the pituicyte (Fig. 31.7).

HORMONES. Two hormones may be isolated from the neural lobe, VASOPRESSIN-ADH *(VADH)* and OXYTOCIN. Both are small molecules containing eight amino acids in a ring (Fig. 31.8).

Vasopressin-ADH exerts a stimulating effect on smooth muscle of arteries, causing vasoconstriction and an elevation of blood pressure. The hormone also exerts an antidiuretic effect (hence ADH for antidiuretic hormone) on the kidney, resulting in increased reabsorption of water from the kidney tubules. Control of VADH secretion is determined by osmosensitive cells within the hypothalamus that continually monitor blood osmotic pressure.

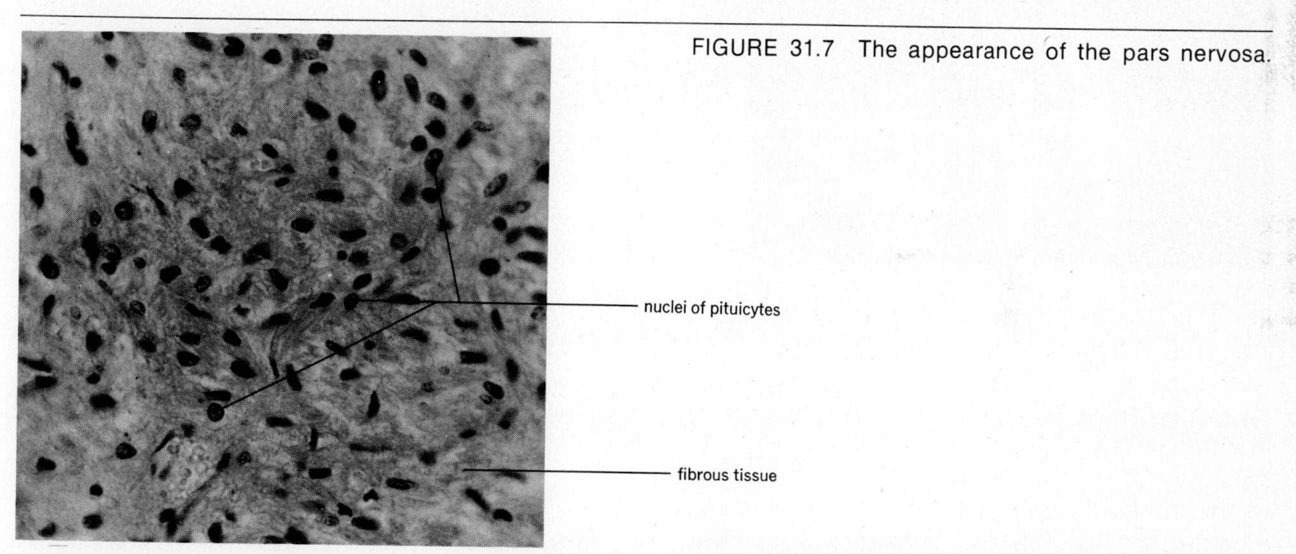

FIGURE 31.7 The appearance of the pars nervosa.

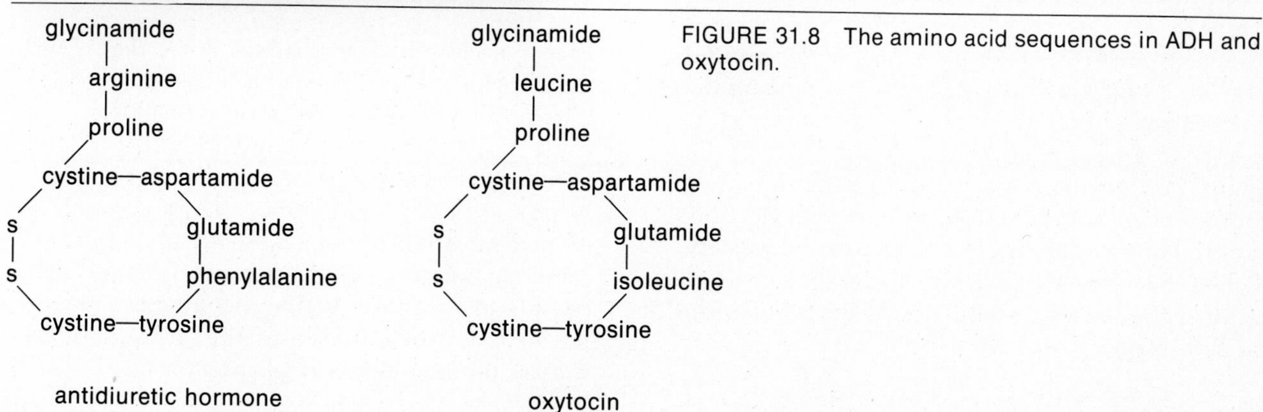

FIGURE 31.8 The amino acid sequences in ADH and oxytocin.

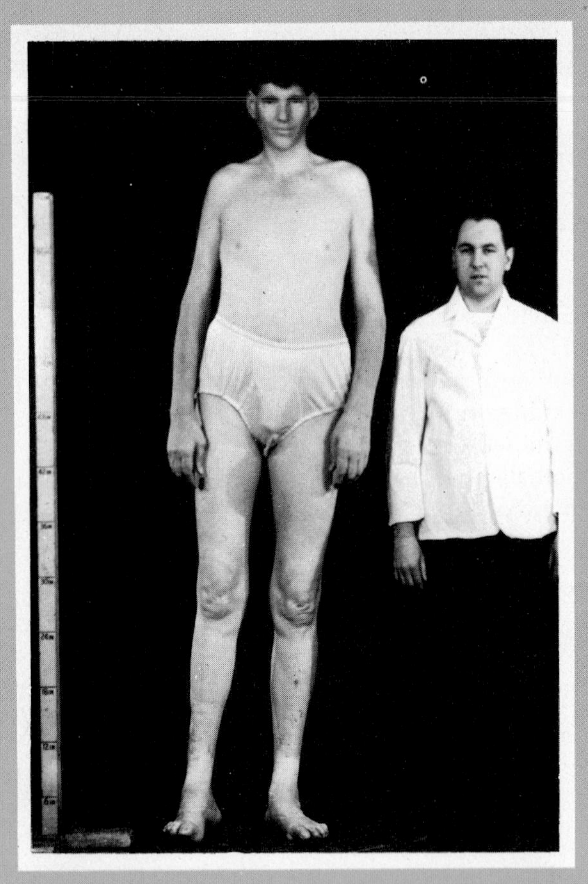

FIGURE 31.9 Giantism in a 42-year-old man. He is 7 feet, 6 inches tall. His companion is normal. The stick is 6 feet tall. (From the teaching collection of the late Dr. Fuller Albright. Courtesy Endocrine Unit & Department of Medicine, Massachusetts General Hospital.)

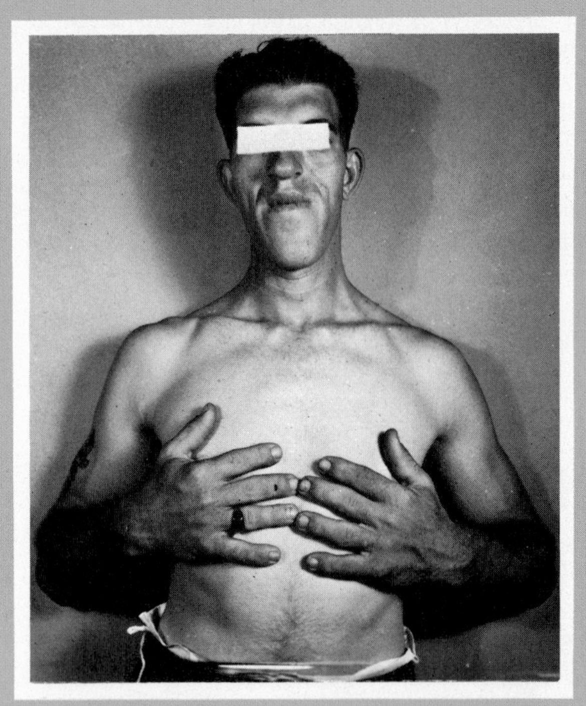

FIGURE 31.10 Acromegaly in an adult. Note the coarseness of the facial features, the enlarged mandible, and the large hands with thick, blunt fingers. (Armed Forces Institute of Pathology.)

Oxytocin exerts its effect primarily on the smooth muscle of the pregnant uterus and the contractile cells *(myoepithelial cells)* around the ducts of the mammary glands. It stimulates the contraction of both types of cells, and is the active principle given in injections to induce or speed labor. Oxytocin is not essential to either function, but apparently aids childbirth and milk ejection. Control of secretion has not been definitely established. Suckling by the infant increases secretion, as does the sex act, and dilation of the cervix prior to childbirth.

Disorders of the hypophysis

Excess secretion of hormones from the adenohypophysis may, in theory, result in overproduction of all hormones, or of only one. The usual cause of oversecretion of any hormone is a tumor involving the cells in which a particular hormone is produced. Chromophobe tumors account for 85 percent of all hypophyseal tumors; alpha cell tumors account for 10–14 percent; beta and delta cell tumors are rare. Visual disturbances may be the first sign of a tumor, as the enlarging gland presses on the neighboring optic tract.

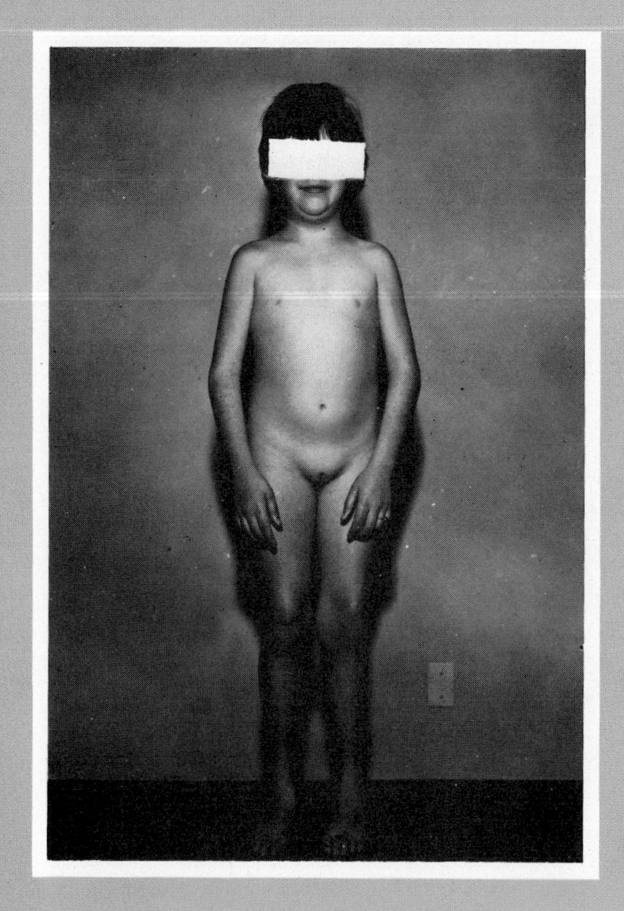

FIGURE 31.11 Hypophyseal infantilism in a 15-year-old female. Note shortness of stature (4 feet, 3-3/4 inches tall) and failure to develop sexually. (Armed Forces Institute of Pathology.)

GIANTISM (Fig. 31.9) is the result of excess growth hormone secretion occurring before skeletal maturity has been achieved. The body continues to grow and may reach heights in excess of 8 feet and weights in excess of 400 pounds. Excess growth hormone production after maturity results in ACROMEGALY (Fig. 31.10). The bones can no longer grow in length, but can increase in width or thickness through the activity of the periosteum. The jaw, hands, and feet are most commonly affected. Treatment of hypophyseal tumors may be carried out by radiation delivered to the gland from the outside; the gland is nearly inaccessible to surgery.

Deficient secretion of growth hormone results in HYPOPHYSEAL INFANTILISM (Fig. 31.11). The body is well proportioned but juvenile in appearance. Sexual characteristics are minimally developed and stature is small. There is apparently no effect on mental capacity. Treatment of this condition requires administration of human growth hormone. Hormones from other species are ineffective.

Excess secretion of vasopressin, secondary to hypothalamic tumor, results in EXCESSIVE WATER RETENTION, dilution of body fluids, and weight gain. Limitation of water intake and administration of diuretics aid treatment. Failure of vaso-

TABLE 31.6 Some disorders of the hypophysis

Condition	Cause	Hormone(s) involved	Secretion is: Excess	Secretion is: Deficient	Characteristics	Comments
Giantism	Tumor before maturity	Growth hormone (GH)	X		Large body size	Proportioned but large
Acromegaly	Tumor after maturity	Growth hormone (GH)	X		Misshapen bones, hands, face, feet	Disproportional growth
Hypophyseal infantilism	Destruction of acidophils by disease, accident, etc.	Growth hormone (GH)		X	Juvenile appearance, sexually and in height	Body properly proportioned
Water retention + *demonstration of high VADH blood levels*	Hypothalamic tumor	Vasopressin-antidiuretic hormone	X		Dilution of body fluids, weight gain	*Must* differentiate from H_2O retention due to other causes
Diabetes insipidus	Hypothalamic damage	Vasopressin-antidiuretic hormone		X	Excessive urine excretion	No threat to life
Sheehan's syndrome	Atrophy of distalis	All		X	Sexual characteristics involute, no menses, anemic	Occurs mostly in the female. Give target organ hormones
Hypophyseal myxedema	Basophil degeneration	Thyroid-stimulating hormone		X	Dry flaky skin, lethargy, anemia	May be controlled by thyroxin

pressin secretion as a result of hypothalamic damage results in DIABETES INSIPIDUS. Excessive dilute urine production (to 56 liters per day in severe cases) is characteristic. Apparently there are no disorders associated with oxytocin.

Table 31.6 presents a summary of hypophyseal disorders. The table includes disorders not described in the text.

The thyroid gland

The thyroid gland (Fig. 31.12) is a TWO-LOBED organ lying on the lateral aspect of the lower larynx and upper trachea. A small ISTHMUS of thyroid tissue connects the two lobes across the midline. The gland weighs about 20 g in the adult and is one of the most vascular of all endocrines. The gland receives 80–120 ml of blood per minute, a fact important in maintaining adequate supplies of building blocks to the organ. The gland is composed of many small spherical units known as

THYROID FOLLICLES. Each follicle is lined with a simple cuboidal epithelium carrying microvilli, and is filled with THYROID COLLOID. The colloid represents a storage form of hormone. Parafollicular cells (C cells), distinguished by a lighter-staining cytoplasm, are interspersed with the follicular epithelial cells and may be found between follicles. These features are shown in Fig. 31.13.

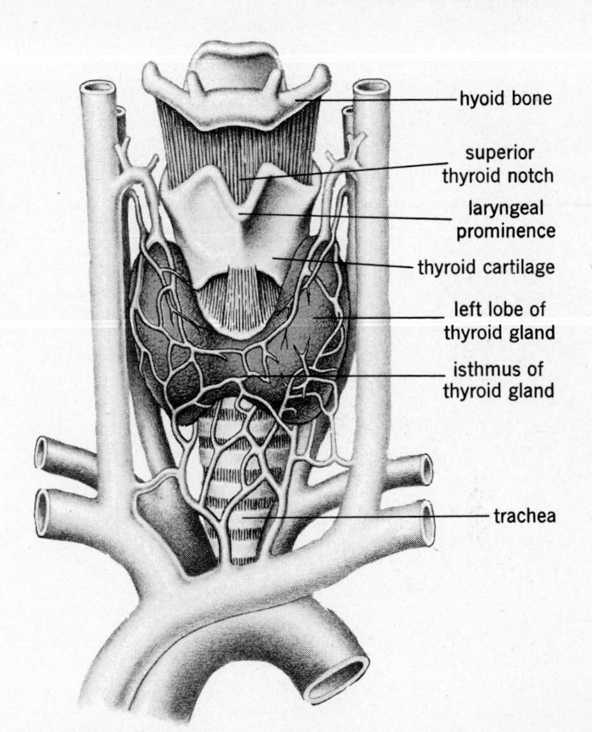

FIGURE 31.12 Gross anatomy and location of the thyroid gland.

Hormone synthesis

The substance generally regarded as the true hormone of the thyroid is THYROXIN. Thyroxin (tetraiodothyronine, T_4) is an iodinated amino acid with the formula shown in Fig. 31.14. Removal of one iodine results in triiodothyronine or T_3, even more active physiologically than T_4. Thyroxin and T_3 are synthesized (see Fig. 31.14) only after the gland has accumulated iodide and the amino acid tyrosine. By iodinating the molecule of tyrosine, then coupling two acids together, thyroxin is formed. Depending on body demands for hormone, the hormone may be released into the bloodstream, or linked to the protein thyroglobulin in the colloid. If linked, the thyroxin must be hydrolyzed from the thyroglobulin before entering the blood. Once into the blood, thyroxin combines with a specific plasma protein designated thyroid-binding globulin (TBG) for transport.

FIGURE 31.13 Microscopic anatomy of the thyroid gland.

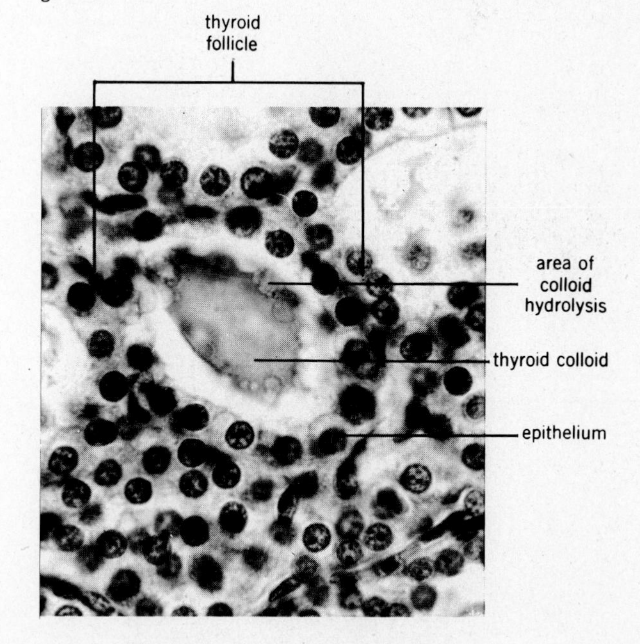

Control of the thyroid

Accumulation of the necessary building blocks, the synthesis, and the release of thyroxin are all controlled by pituitary THYROID-STIMULATING HORMONE (TSH). Thyroxin in turn exerts a negative feedback on TSH production.

Effects of thyroxin

Thyroxin has no particular target organ; those cells

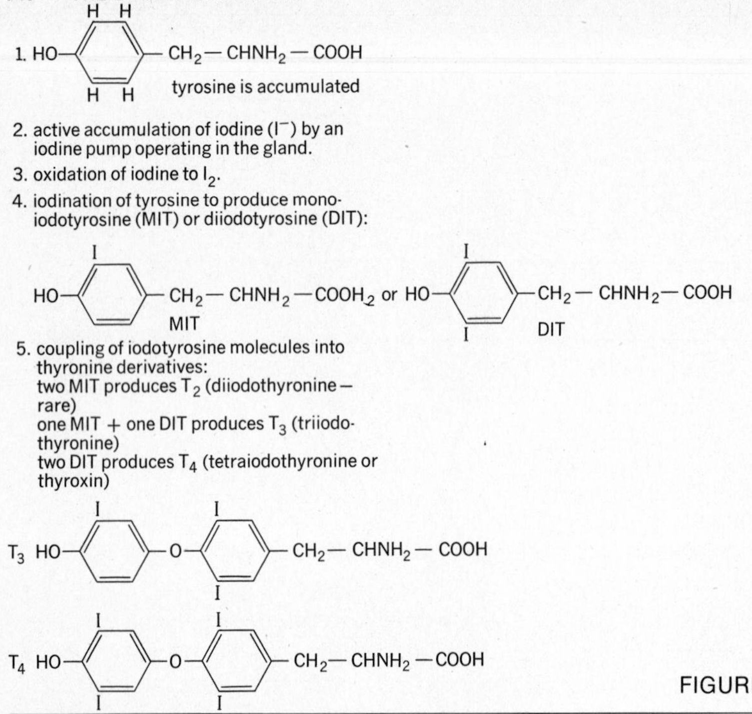

1. HO–(benzene ring)–CH_2—$CHNH_2$—COOH

   tyrosine is accumulated
2. active accumulation of iodine (I^-) by an iodine pump operating in the gland.
3. oxidation of iodine to I_2.
4. iodination of tyrosine to produce mono-iodotyrosine (MIT) or diiodotyrosine (DIT):

   HO–(ring, I)–CH_2—$CHNH_2$—$COOH_2$ or HO–(ring, I, I)–CH_2—$CHNH_2$—COOH

   MIT DIT
5. coupling of iodotyrosine molecules into thyronine derivatives:
   two MIT produces T_2 (diiodothyronine—rare)
   one MIT + one DIT produces T_3 (triiodo-thyronine)
   two DIT produces T_4 (tetraiodothyronine or thyroxin)

T_3 HO–(ring, I)–O–(ring, I, I)–CH_2—$CHNH_2$—COOH

T_4 HO–(ring, I, I)–O–(ring, I, I)–CH_2—$CHNH_2$—COOH

FIGURE 31.14 The steps in the synthesis of T_3 and T_4.

FIGURE 31.15 Hyperthyroidism. Note the marked protrusion of the eyeball (exophthalmos) that may accompany the disease. (R. W. Carlin, Medical Photographer.)

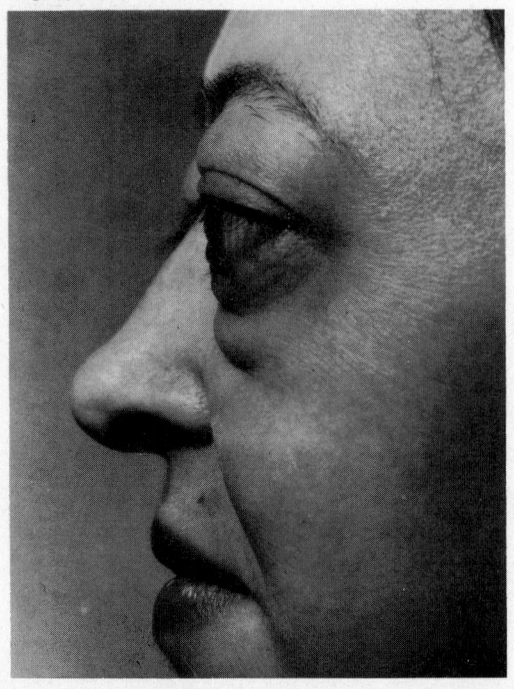

possessing catabolic systems of enzymes respond to the hormone. Its effects may be divided into four general categories.

CALORIGENIC EFFECT. Thyroxin accelerates the catabolic reactions of glycolysis, Krebs cycle, beta oxidation and oxidative phosphorylation. One mg of thyroxin can raise heat production by 1000 Calories and CO_2 production by 400 g.

GROWTH AND DIFFERENTIATION. Working with growth hormone, thyroxin ensures proper development of the brain. Deficiency of thyroxin during development leads to smaller and fewer neurons and defective myelination.

METABOLIC EFFECTS. Thyroxin acts as a diuretic, increasing urine production. It increases protein breakdown, increases uptake of glucose by cells, enhances glycogenolysis, and depresses blood cholesterol levels.

MUSCULAR EFFECTS. Thyroxin in excess interferes with ATP synthesis and may thus speed exhaustion of energy in muscle. Both skeletal and cardiac muscle are affected in this manner.

Other factors influencing thyroid activity

A wide variety of influences may alter the basic activity of the gland. Among these are the following.

ANTITHYROID AGENTS. Generally called GOITROGENS, because they cause goiter or enlargement of the thyroid, such materials inhibit one or more of the steps in synthesis of hormone. Included are the thioureas, phenols, and thiocyanates. The thiocyanates are present in cabbage, turnips, rutabaga, and mustard; hence, these foods are said to be goitrogenic.

GONADAL HORMONE LEVELS. Excessive gonadal hormone appears to decrease thyroxin transport.

PREGNANCY. Pregnancy increases all aspects of thyroid activity due, apparently, to fetal competition for maternal iodide and amino acids.

AGE. Activity of the thyroid decreases with age, although the decrease is small.

STRESS. Stress, particularly that of a cold environment, increases thyroid activity.

LONG-ACTING THYROID STIMULATOR (LATS). LATS is a gamma globulin antibody produced during certain disease states of the thyroid. It originates from plasma cells in autoimmune disorders. It acts on the thyroid in a manner similar to TSH, resulting in stimulation of catabolic reactions.

Thyrocalcitonin

In 1965 a hormone from the thyroid gland was shown to lower plasma concentrations of calcium and phosphate. It is secreted in response to a high blood calcium level. The name calcitonin was applied to the substance, and it was originally thought to be secreted by the parathyroids. Subsequently, the name thyrocalcitonin was applied, reflecting its thyroidal origin. Thyrocalcitonin apparently works with parathyroid hormone to control body calcium and phosphate balance. The effect is mediated primarily through the kidneys (increased Ca^{++} absorption) and possibly on bones (increased mineralization). It is believed to be produced by the parafollicular cells.

Disorders of the thyroid

Overproduction of thyroxin due to thyroid tumor, or overstimulation by TSH, brings about the condition known as HYPERTHYROIDISM *(Grave's disease)* (Fig. 31.15). There are different degrees of a

FIGURE 31.16 Childhood myxedema; note protruding umbilicus and puffy appearance of face.

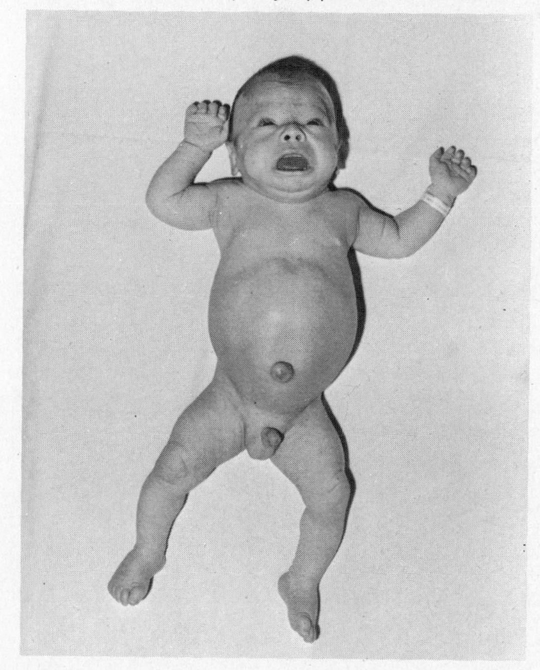

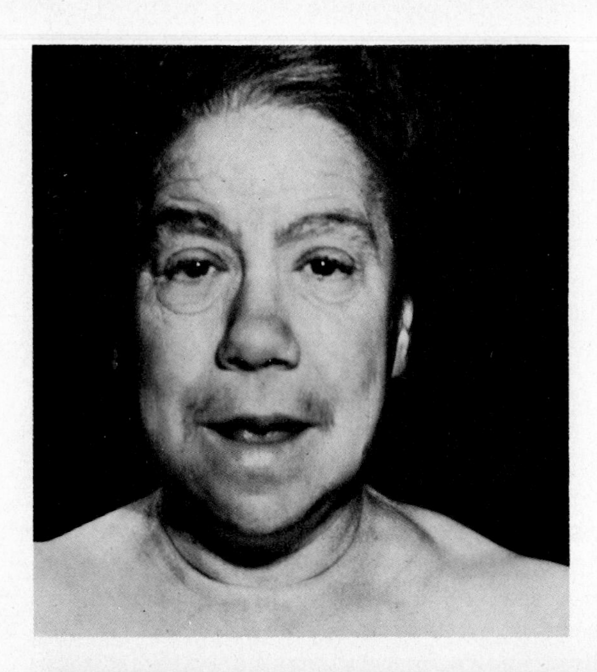

FIGURE 31.17 Myxedema in a 55-year-old female. Her PBI was less than 1 μg/100 ml (normal 4–8), and metabolic rate was 51 percent below normal. Note puffy appearance of face and generalized fatigued appearance. (From the teaching collection of the late Dr. Fuller Albright. Courtesy Endocrine Unit & Department of Medicine, Massachusetts General Hospital.)

TABLE 31.7 Summary of the Thyroid

Item	Comments
Location and parts; weight	Two-lobed, plus connecting isthmus; located on lower larynx and upper trachea; 20 grams in weight
Requirements for hormone synthesis	The amino acid tyrosine; iodide in diet; TSH to control all steps in synthesis
Hormones produced and effects:	
Thyroxin	Main controller of catabolic metabolism
Thyrocalcitonin	Lowers blood Ca^{++} and $PO_4^{\equiv}$ levels
Control of hormone secretion	
Thyroxin	By thyroid stimulating hormone
Thyrocalcitonin	By blood Ca^{++} level
Disorders	
Hypersecretion	Creates hyperthyroidism: Elevated BMR Elevated heart action Exophthalmus goiter
Hyposecretion	Creates: Goiter—enlarged gland Hypothyroidism (cretinism, myxedema) Low BMR Low heart rate, blood pressure, body temperature

basically similar set of symptoms in the disease, so the following list will not attempt to separate the various specific diseases.

Excessive nervousness and excitability.
Elevated heart rate and blood pressure.
Elevated BMR (40 to 100 percent).
Weakness.
Weight loss.
Bulging of the eyes; exophthalmus.

The first five symptoms may be accounted for by considering the catabolic stimulating effect of the hormone; in short, everything is accelerated. The EXOPHTHALAMUS (see Fig. 31.15) is due to increases in mass of tissue behind the eye, pushing it forward. Deficient thyroid activity results in a wider variety of disorders, depending on the cause of the deficiency.

SIMPLE GOITER is generally the result of insufficient dietary iodide. The gland enlarges, "hoping" to trap more of the available iodide, which, of course, creates more demand, a greater enlargement, and so on. Untreated goiters can reach extreme sizes. Dietary deficiency of iodide in a pregnant woman may result in insufficient thyroxin production by the embryo, and cretinism may occur.

CRETINISM results in idiocy (remember the role of thyroxin in nervous system development) and retarded growth. The mental damage cannot be overcome. MYXEDEMA is seen after birth, in both children and adults (Figs. 31.16 and 31.17) Myxedema is associated with dry skin, low BMR, and intolerance to cold.

Treatment of these conditions usually involves the administration of dried thyroid gland, by mouth, in the form of a tablet. Only cretinism will fail to respond to such treatment.

Table 31.7 summarizes facts about the thyroid gland.

The parathyroid glands

There are usually four parathyroid glands in the human located on the posterior aspect of the thyroid (Fig. 31.18). They measure about 5 mm in diameter, and have a combined weight of about 120 mg. Histologically, the glands consist of densely packed masses of principal (chief) cells and oxyphil cells (see Fig. 31.18). The PRINCIPAL CELLS are regarded as the source of parathyroid hormone (PTH), while the OXYPHIL CELLS are thought to be reserve cells, capable of assuming hormone production if need arises. PTH is a polypeptide chain containing 74 to 80 amino acids. Regulation of PTH secretion is determined by the blood calcium level, a fall of blood calcium increasing PTH secretion, and vice versa.

The parathyroids and ion metabolism

Calcium, phosphate, magnesium, and citrate are ions that assume an extremely important role in the body economy. As constituents of the bones and teeth, these ions play structural roles in the body. The phenomenon of muscular contraction apparently depends on calcium ion, as does normal membrane permeability. Calcium is also involved in nervous irritability and in the clotting of the blood. Regulation of the body content of these ions therefore becomes very important in terms of whole body function.

CALCIUM. The average adult body contains about 1100 g of calcium, 99 percent of which is found in bone. Intake occurs by absorption of the ion from the gut, and it is then placed in the extracellular fluids (plasma, interstitial fluid). Removal of calcium from extracellular fluid occurs by filtration in the kidney (99 percent reabsorbed), exchange with tissues (chiefly muscle and bone), and secretion of digestive fluids into the gut and out via the

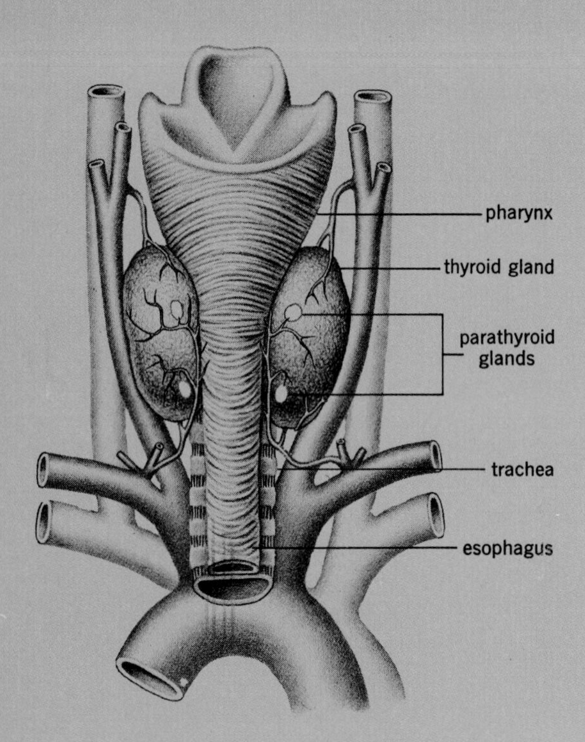

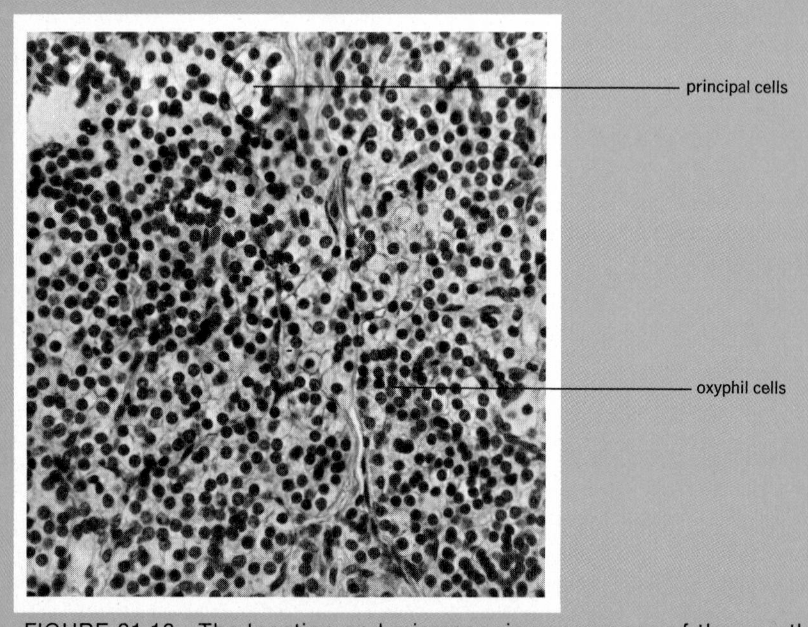

FIGURE 31.18 The location and microscopic appearance of the parathyroid glands.

feces. In the adult, about 1 g of calcium per day will keep all forces in balance.

PHOSPHATE. The body contains about 500 g of phosphate, 85 percent of which is in the skeleton. Other compounds containing phosphate include ATP, ADP, and creatine phosphate, all important energy sources. DNA and RNA also contain phosphate. Phosphate is absorbed through the gut, apparently in inverse proportion to calcium. The absorbed phosphate is also placed into the extracellular fluids, from which it is withdrawn by cells, bony tissue, and the kidney. The kidney route constitutes the major path for excretion of phosphate.

MAGNESIUM. Total body content of magnesium is about 200 meq with 50 percent of this in the skeleton. Absorption occurs in the gut, apparently in inverse proportion to calcium absorption. Magnesium is essential as a cofactor in many enzymatic reactions, particularly those involving ATP synthesis and degradation and ribosomal protein synthesis. Renal excretion is the major pathway for magnesium loss.

CITRATE. Citrate ion is an important intermediate in glycolysis and the Krebs cycle. It forms a major source of energy for ATP synthesis.

This brief discussion of these ions indicates their importance in body function. The regulation of cellular and fluid levels of these substances is under the primary control of the parathyroid glands.

Effects of the hormone

If the hormone controls the metabolism of the previously mentioned ions, it must be involved in at least three phases of this metabolism: absorption, exchange between cells and body fluids, and excretion. Accordingly, it is suggested that the gut, bones, and kidney are the three tissues or organs primarily affected by PTH.

EFFECTS ON THE GUT. PTH INCREASES ABSORPTION of calcium and magnesium from the intestine, provided adequate amounts of vitamin D are present. Phosphate absorption is also increased by the hormone.

EFFECTS ON BONE. PTH acts directly on bone. The effect is thought to be mediated via osteoclasts (bone-destroying cells) and the connective tissue fibers of bone. PTH causes increase in numbers of osteoclasts and occasions proliferation of fibers. The result will be removal of the inorganic phase of bone (with consequent RISE OF BLOOD Ca^{++} and $PO_4^{\equiv}$) and increase in organic content. Action of the hormone on bone may revolve around lysosome dissolution in bone cells and consequent enzymatic destruction of bony tissue.

EFFECTS ON THE KIDNEY. PTH controls REABSORPTION of calcium and magnesium in direct proportion to PTH levels, while increasing phosphate excretion.

The combination of effects on intake, use, and excretion tends to keep plasma levels of these materials within normal limits.

Effects on other organs and cells

The mammary gland responds to PTH by lowering secretion of calcium in milk. In cells such as those of the kidney, liver, or intestine grown in tissue culture, PTH increases decarboxylation, releases H^+, causes swelling of mitochondria, and stimulates the Krebs cycle. A caution: PTH is not the only hormone exerting an effect on Ca^{++} metabolism. It might be predicted that any hormone concerned with growth would lead to, at the very least, increased absorption of Ca^{++} and $PO_4^{\equiv}$. Thus, growth hormone, thyroxin, and certain sex hormones work with PTH on the absorptive aspect. If one is concerned with maintenance, rather than growth, PTH and thyrocalcitonin are the two hormones most concerned with controlling metabolism of these ions.

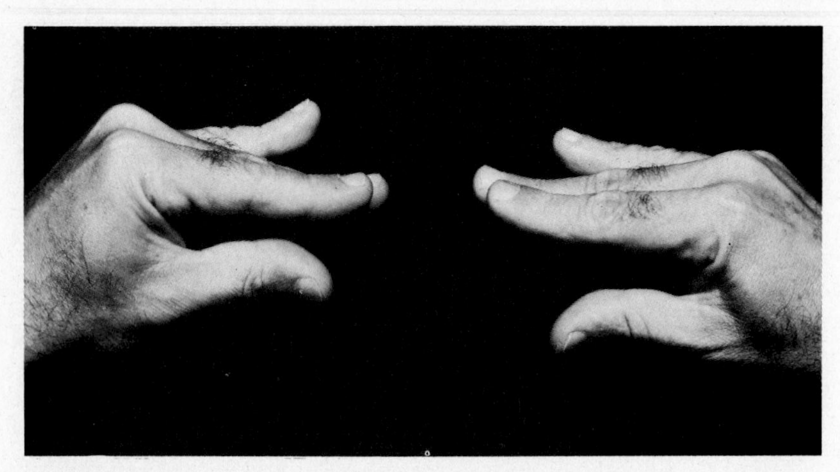

FIGURE 31.19 Carpospasm (Trousseau's sign) in hypoparathyroidism. Occluding the blood flow to the hands, with a blood pressure cuff pumped to about 100 mm Hg, causes muscle spasm of the hands. Position is characteristic for the disease. (Camera, M.D. Studios.)

Disorders of the parathyroids

HYPERPARATHYROIDISM *(von Recklinghausen's disease)* usually results from tumor formation in the glands. Excess PTH results in destruction of bone tissue and subsequent formation of excess fibrous tissue and cysts in the bones *(osteitis fibrosa cystica).* The bones may become softened to the point that they collapse. Blood Ca^{++} levels are very high, resulting in muscular weakness, mental disorder, and cardiac irregularities. The kidneys may form calculi, and vomiting, constipation, and other intestinal disturbances are usually present.

HYPOPARATHYROIDISM *(tetany)* usually results from damage to or removal of parathyroids during thyroid surgery or destruction of parathyroids by disease, infection, or hemorrhage. The term *tetany* reflects the primary symptom of the disease; muscular rigidity and paralysis due to low blood calcium levels. Characteristic positions of the hand *(carpospasm)* are seen (Fig. 31.19). The primary consideration in treatment is to achieve an elevation of the blood calcium level. This may be achieved immediately by intravenous injection of a solution such as 5 percent calcium gluconate. This relieves the symptoms, and further treatment may then be instituted for control of the disease.

The pancreas

The endocrine portion of the pancreas, the ISLETS *(of Langerhans),* develops from the terminal portions of the ducts of the exocrine (digestive portion) and subsequently separates from the ducts. Estimates of the number of islets vary from 500 thousand to 2 million. The tissue becomes functional at about three months *in utero.* Microscopically, the islets show two major cell types, designated alpha and beta. ALPHA CELLS comprise about 25 percent of the islet population, and to them is attributed the production of the hormone GLUCAGON (hyperglycemic glycogenolytic factor or HGF). Glucagon is a single, long-chain polypeptide with a molecular weight of 3485. Its secre-

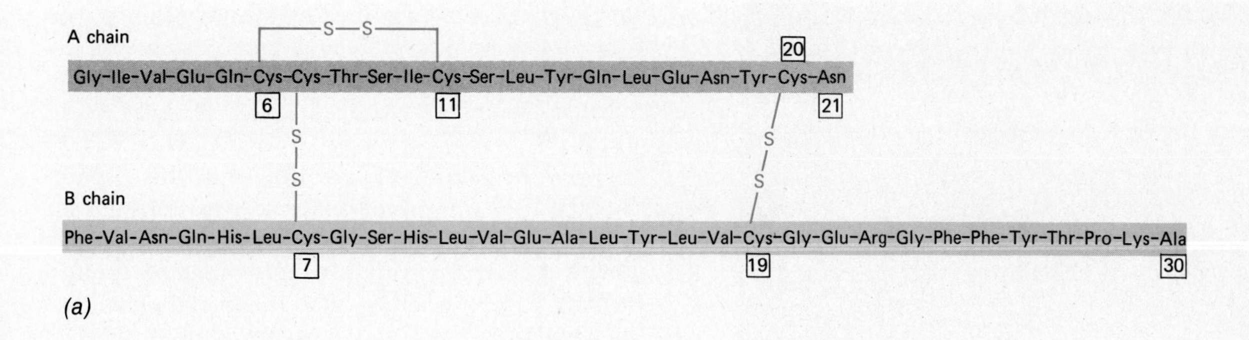

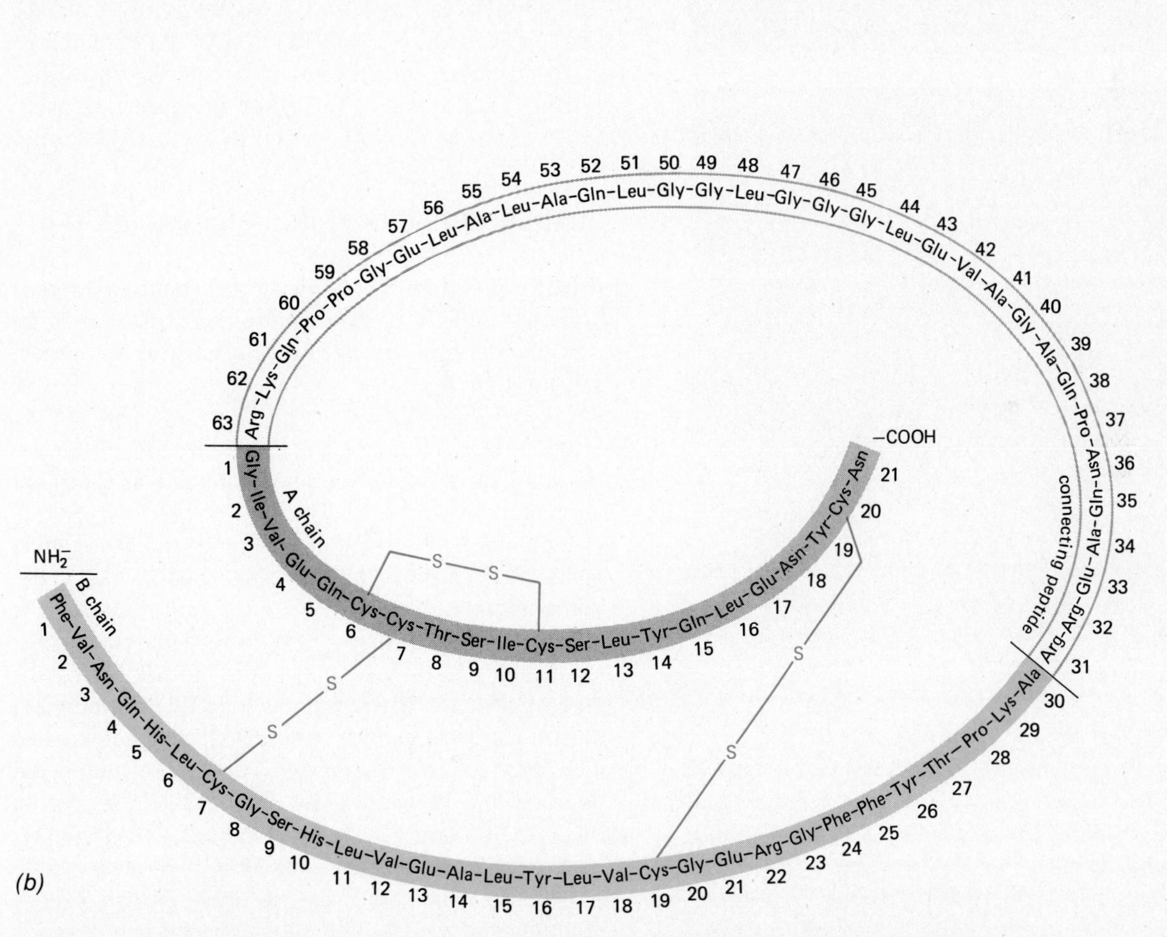

FIGURE 31.20 *(a)* The amino acid sequence in insulin. *(b)* The structure of proinsulin.

tion from the alpha cells is caused by a fall in blood sugar and by a rise of growth hormone secretion. The hormone exerts its action on the enzymes stimulating conversion of glycogen to glucose (glycogenolysis) by stimulating formation of cAMP that increases the activity of phos-

TABLE 31.8 Tissue in which insulin affects glucose uptake

Tissue affected	Tissue not affected
Skeletal, cardiac, and smooth muscle	Brain
Leucocytes	Intestinal epithelium
Lens	Erythrocytes
Hypophysis	Testis
Mammary	Islet cells
Liver	Kidney tubules
Adipose tissue	

TABLE 31.9 Factors affecting insulin secretion

Factors increasing insulin secretion	Factors decreasing insulin secretion
Glucose intake (hyperglycemia)	Epinephrine
Increase in intake of certain amino acids	Certain diuretics (thiazides)
cAMP	Glucose metabolism blocking agents (e.g., 2-deoxyglucose, mannoheptulose)
Vagal stimulation	
Intestinal parahormones	
Sulfonylureas	
Glucagon	

phorylase. Such conversion raises the concentration of blood glucose and may, if secretion continues, create a hyperglycemia (excess blood sugar level).

The BETA CELLS, constituting about 75 percent of the islet cells, produce the more familiar hormone INSULIN. Insulin is a complex protein hormone consisting of four protein chains of two types. It has a molecular weight of about 6000. Insulin is produced as proinsulin and is hydrolyzed to insulin prior to secretion (Fig. 31.20). Insulin is secreted in response to a rise in blood sugar level (as after a meal); it lowers the blood sugar by stimulating the conversion of glucose to glycogen (glycogenesis) and by stimulating uptake of glucose by certain cells (Tables 31.8 and 31.9).

Two conclusions may be drawn at this point: the blood sugar level reflects a balance of the effect of the two hormones, and the effects of both hormones are exerted primarily on the metabolism of carbohydrate. Recalling that the metabolism of carbohydrates, fats, and proteins is interrelated, we may further conclude that a disturbance of carbohydrate metabolism must cause disruption in the metabolism of the other substances as well.

Insulin deficiency; diabetes mellitus

Insulin deficiency results from the inability of the beta cells to produce enough insulin for body requirements. A series of reactions is set in motion by the deficiency.

BLOOD SUGAR LEVEL RISES *(hyperglycemia)*. Glucose polymerization into glycogen is reduced, as is cellular uptake.

More sugar is filtered in the kidney than the tubule cells can reabsorb. GLUCOSE APPEARS IN THE URINE *(glucosuria)*.

The cells of the body, unable to acquire glucose, shift their energy source to FATS. The RQ decreases; excess acetyl-CoA (from beta oxidation) is produced; the Krebs cycle cannot handle it all, and acetoacetic acid production is increased. The ketone bodies are increased in concentration, and KETOSIS may result. Also, the buffering systems of the body may be overtaxed and ACIDOSIS may occur.

Glucose eliminated in the urine carries excess quantities of water with it. URINE VOLUME IS INCREASED *(polyuria)*.

Excess urinary loss of water dehydrates the body, and a great thirst develops, with RISE OF WATER INTAKE *(polydipsia)*.

Glucose loss triggers the desire to eat, and EXCESS QUANTITIES OF FOOD may be consumed *(polyphagia)*.

High levels of fats in the blood may lead to narrowing of blood vessel caliber through ATHEROSCLEROSIS. Occlusions of vessels in the legs may give rise to diabetic gangrene as tissues are deprived of blood, die, and become infected. The events described, along with their causes, constitute the symptoms of diabetes mellitus.

Treatment of diabetes may depend on the severity of the disease. Mild disease may often be controlled by dietary measures, while severe disease may require insulin administration.

Diabetes is one of the oldest-known endocrine disorders, the sweetness of the urine having been described in Egypt in 1500 B.C. Its major symptoms were described at the beginning of this section. Two types are recognized today. Juvenile-onset type occurs usually before 15 years of age, is of rapid onset, is the more severe type, and results in dependence on insulin for control. The beta cells are atrophic (wasted or degenerated). Maturity-onset type occurs usually after age 35–40. It develops slowly, is usually not severe, may be controlled by diet alone in some cases, and usually is associated with vascular complications (atherosclerosis, arteriosclerosis, hemorrhage). Several factors appear to predispose to the development of diabetes.

HEREDITY. Twenty to fifty percent of individuals developing diabetes show a history of the disorder in their family. Some geneticists believe the disorder is inherited as an autosomal recessive trait. Others believe it to be multifactored.

OBESITY. If an individual is predisposed to the development of diabetes, any condition that stresses the beta cells may trigger the appearance of symptoms. Such an individual is often on the border line of deficiency and is "tipped over" by a stressor. Obesity may be one such factor.

AGING. Tolerance to intake of glucose diminishes with age as if the beta cells' capacity to produce insulin diminishes with age. Dietary restriction of carbohydrate with advancing age would thus seem to be a reasonable suggestion. It might also reduce the tendency of older individuals to put on weight.

INFLUENCE OF OTHER HORMONES. Growth hormone is known to increase an existing diabetes, as do adrenal cortical hormone (cortisol) and thyroxin. They cannot, however, produce diabetes in the absence of beta-cell weakness.

INFECTION. Infections in general appear to act as stressors and may cause appearance of diabetic symptoms in a latent diabetic. The presence of diabetes seems to reduce resistance to infection.

The development of diabetes appears to proceed in STAGES. Three to seven stages are described according to the authority involved. The sequence presented below is that according to Danowski.

Stage I. Prediabetes. No symptoms are shown; a predisposition to develop diabetes is present.

Stage II. Stress diabetes. Appearance of diabetic symptoms with stress; disappearance with removal of stressor.

Stage III. Appearance of abnormal glucose tolerance tests (prolonged elevation of blood sugar after ingestion of glucose solution). Fasting blood glucose levels are normal. Normally, blood sugar rises after glucose ingestion and returns to normal (80–120 mg percent) within three to four hours.

Stage IV. Fasting blood glucose is elevated and there is prolongation of the glucose tolerance test.

Stage V. Ketosis appears. Ketone bodies appear in the urine (ketonuria).

Stage VI. Ketonuria is marked, with ketonemia developing.

Stage VII. Acidosis appears, coma may develop, and ketone levels are high in blood and urine.

Children are usually in stages V, VI, or VII when diagnosed. Adults are usually in stages II, III, or IV when diagnosed.

VASCULAR COMPLICATIONS in diabetes occur mainly as a result of deposition of lipids in the intima of the blood vessels. This narrows luminal

TABLE 31.10 Types of insulin available for use in diabetes

Type	pH[d]	Time of maximum effectiveness (hours)	Duration of effect (hours)
Semilente	7.1–7.5	4–6	12–16
Lente	7.1–7.5	8–12	18–24
Ultralente	7.1–7.5	16–18	30–36
Crystalline Zn[a]	3.0	4–6	6–8
NPH[b]	7.1–7.4	8–12	18–24
Protamine Zn[c]	7.1–7.4	14–20	24–36
Globin insulin	3.6	6–10	12–18

[a] Almost pure insulin—quick results, often used in combination with longer acting form.
[b] "*N*eutral *P*rotamine *H*agedorn"—insulin and equivalent amount of protein.
[c] Excess protein; stretches effect.
[d] pH determines the state of the insulin. At low pH, the insulin goes into solution, is absorbed from the injection site rapidly, and is shorter in duration of effect. At higher pH, the insulin tends to remain crystalline and is absorbed more slowly with longer duration of effect.

diameter, reduces blood flow, and starves tissues. Often, the tissues become gangrenous. Examination of the retinal vessels by the ophthalmoscope may discover the presence of diabetic changes in blood vessels (diabetic retinopathy).

Insulin as the treatment for diabetes

Table 31.10 summarizes some types of insulin available for treatment of diabetes.

Insulin must be given by injection inasmuch as it is a protein and would be destroyed by the digestive enzymes if given by mouth. One must also recognize that it is virtually impossible to obtain normal insulin levels by injection, and that only control, rather than cure, can be achieved in the disease.

Oral treatment of diabetes

If any beta cells survive in the islets, the administration by mouth of certain substances may control diabetes. Most belong to the group of chemicals known as SULFONYLUREAS (Fig. 31.21). They act in several ways:

Tend to INHIBIT GLYCOGEN BREAKDOWN; an effect exerted directly on the liver.

INCREASE SECRETION OF INSULIN by any surviving beta cells.

INCREASE CELLULAR UPTAKE of glucose.

Side effects of the sulfonylureas include gastrointestinal disturbances, decrease of white cells and platelets, jaundice, hypoglycemia, and a cumulative effect on lowering blood sugar in the presence of alcohol. A prospective patient for oral medication should be carefully evaluated as to which preparation and what dosage is used.

Hypoglycemia

Low blood sugar levels may result from starvation, failure of sugar absorption, high activity levels, hyperinsulinism due to islet tumors (insulin shock), or overdosing with insulin or oral medication. Regardless of cause, hypoglycemia affects one body area primarily—the brain. Brain cells normally utilize only glucose and are therefore immediately affected by low blood sugar levels. The higher brain centers are affected first, causing:

Disturbances in locomotion.
Mental confusion.
"Dull-wittedness."
Decreased muscle tone.

Effects on lower centers include:

Disturbances in respiration.
Loss of consciousness.
Depression of reflexes.
Depression of body temperature.

If the blood sugar level continues to decline, death will result. Treatment involves raising the blood sugar level. If the individual can recognize the onset of symptoms, consumption of sugar or a candy bar may arrest their development. If unconscious, intravenous injection of glucose solution is necessary.

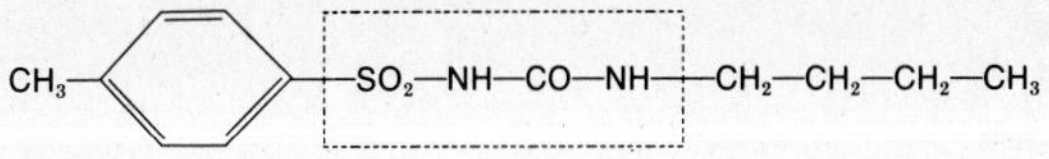

CH_3—C_6H_4—[SO_2—NH—CO—NH]—CH_2—CH_2—CH_2—CH_3

tolbutamide
(1-butyl-3-p-tolylsulfonylurea)

CH_3—C_6H_4—[SO_2—NH—CO—NH]—N< (CH_2—CH_2—CH_2 / CH_2—CH_2—CH_2)

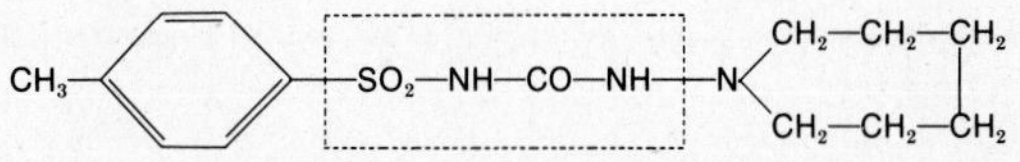

tolazamide
(1-(hexahydro-1-azepinyl)-3-p-tolylsulfonylurea)

CH_3—CO—C_6H_4—[SO_2—NH—CO—NH]—C_6H_{11}

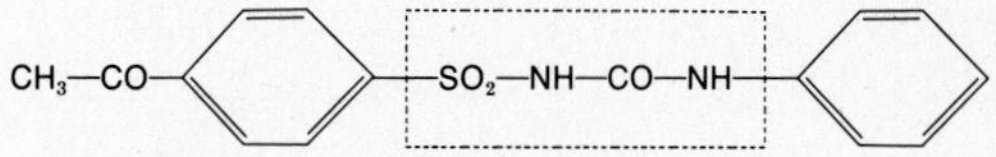

acetohexamide
(N-(p-acetylbenzenesulfonyl)-N[1]-cyclohexylurea)

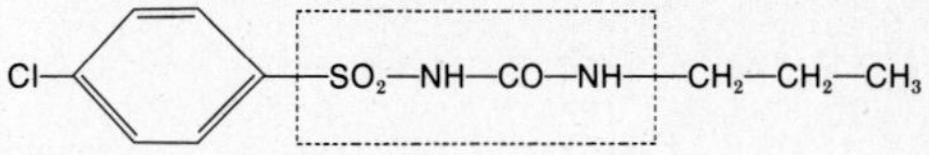

Cl—C_6H_4—[SO_2—NH—CO—NH]—CH_2—CH_2—CH_3

chlorpropamide
(1-propyl-3-p-chlorobenzenesulfonylurea)

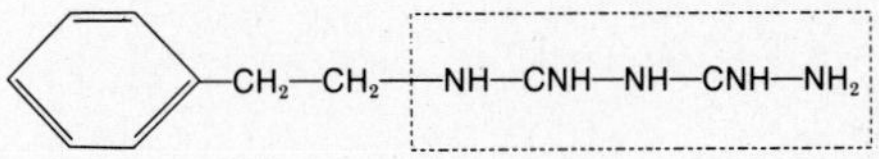

C_6H_5—CH_2—CH_2—[NH—CNH—NH—CNH—NH_2]

phenformin
CN[1]-B-phenethylformamidinyliminourea
or phenethylbiguanide)

FIGURE 31.21 Formulas of compounds orally effective in control of some types of diabetes. Active portion of the molecule is enclosed in the rectangle.

The adrenal glands

The adrenal glands are paired, hat-shaped organs located superior to the kidneys. Each gland consists of an inner MEDULLA of ectodermal origin and an outer CORTEX of mesodermal origin (Fig. 31.22). The medulla is not essential for life; at least one-sixth the total cortical substance is necessary to maintain life.

The medulla

The hormone-producing cells of the medulla are known as the chromaffin cells and are arranged in the form of cords. The cords of cells are separated by large venous sinuses. Upon stimulation, the chromaffin cells secrete into the sinuses, which empty very rapidly into the circulation. Control of medullary secretion is by way of preganglionic neurons of the sympathetic nervous system that terminate without synapse on the medullary cells. Response is therefore very rapid.

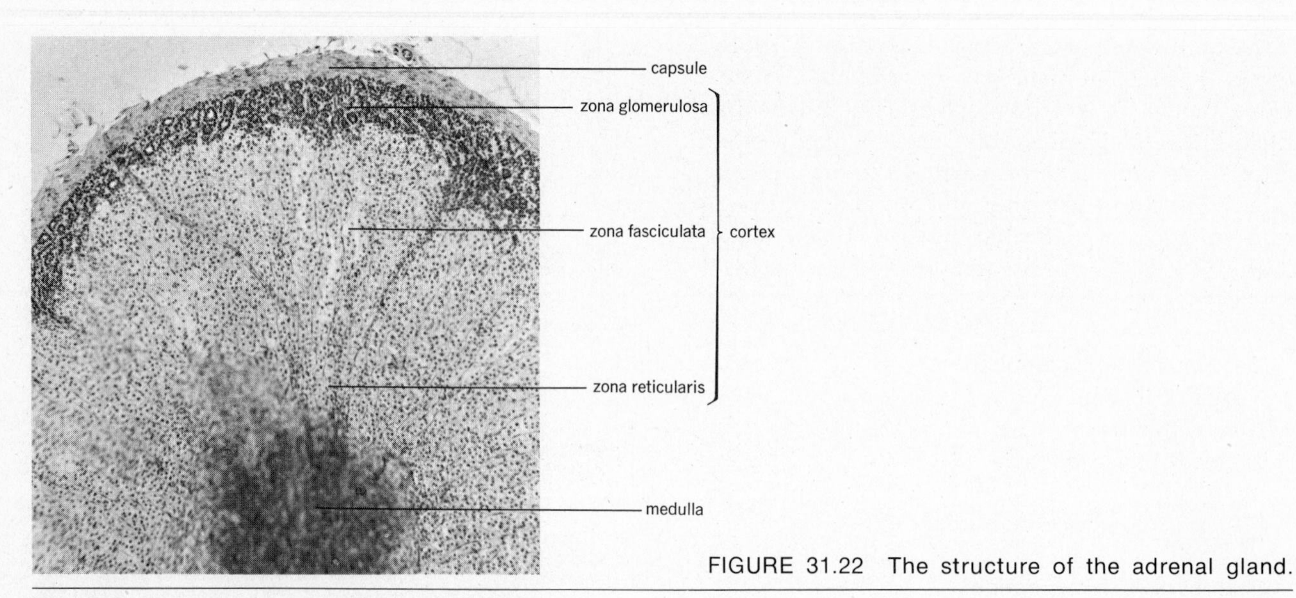

FIGURE 31.22 The structure of the adrenal gland.

FIGURE 31.23 The chemical formulas of epinephrine and norepinephrine.

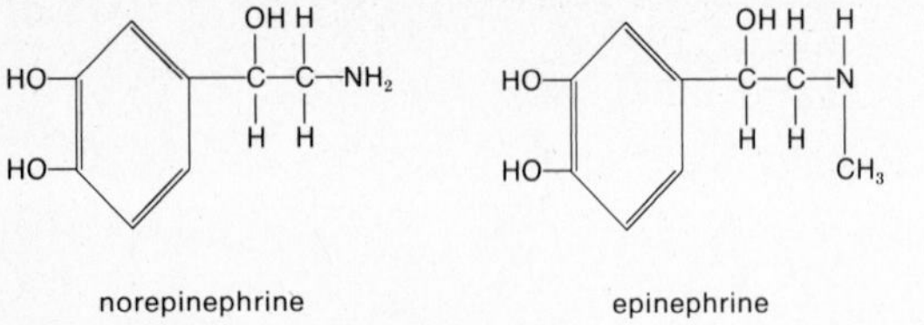

norepinephrine

epinephrine

Hormones of the medulla

Two active substances, NOREPINEPHRINE and EPINEPHRINE *(adrenalin)* (Fig. 31.23), may be isolated from the medulla. Norepinephrine composes about 20 percent of the medullary secretion, epinephrine the remaining 80 percent. Both hormones are "sympathomimetic"; that is, their effects are similar to those obtained by stimulation of the sympathetic nervous system. The effects of the two hormones are presented in Table 31.11.

The list of effects of epinephrine appears to create a set of conditions to enable the body to "run or fight" (the "fight or flight" reaction).

Fortunately, the effects, although great, are short-lived. The hormone is inactivated by the liver in about three minutes.

Disorders of the medulla

No disease of hypofunction exists. As stated before, the medulla is not essential for life. Its functions can be taken over by the sympathetic nervous system. Occasionally, a medullary tumor (phaeochromocytoma) may occur, resulting in overproduction of hormone. This is a very dangerous condition, because of the extremely high blood pressures that may be achieved (to 300 mm Hg systolic pressure). Surgical removal of the tumor is mandatory before circulatory damage occurs.

The cortex

The adrenal cortex is subdivided into three zones, based primarily on cellular arrangement (see Fig. 31.22). The outer ZONA GLOMERULOSA contains cortical cells in ball- or knotlike masses (glomeruli). The central ZONA FASCICULATA contains radially arranged cords (fascicles) of cells. The inner ZONA RETICULARIS has branching cords of cells. Each zone produces a different type of steroid hormone having widely different effects on the body. The

TABLE 31.11 Effects of adrenal medullary hormones

Hormone	General effect	Mechanism of production of effect
Norepinephrine	Raises blood pressure	Causes constriction of muscular arteries
Epinephrine	Raises blood pressure	Increases rate and strength of heartbeat Causes constriction of all muscular arteries except those to heart and skeletal muscles, the latter dilating Causes contraction of spleen
	Stimulates respiration	Increases rate and depth of breathing by direct effect on respiratory centers
	Dilates bronchi and bronchioles	Causes relaxation of smooth muscle of respiratory system
	Slows digestive process	Inhibits muscular contraction of stomach, intestines
	Postpones fatigue in skeletal muscle, and increases efficiency of contraction	Increases muscle glycogenolysis Creates more ATP
	Causes hyperglycemia and glucosuria	Increases liver glycogenolysis; kidney tubule Tm_g is exceeded
	Increases O_2 consumption, CO_2 production	Stimulates general metabolic activity of cells
	Miscellaneous effects:	
	Stimulates sweating	Stimulates eccrine secretion
	Increases lacrymation	Stimulates secretion of lacrimal glands
	Dilates pupil	Causes contraction of dilator pupillae
	Increases salivary secretion	Stimulates secretion (nervous)
	Increases coagulability of blood	Accelerates clotting reaction

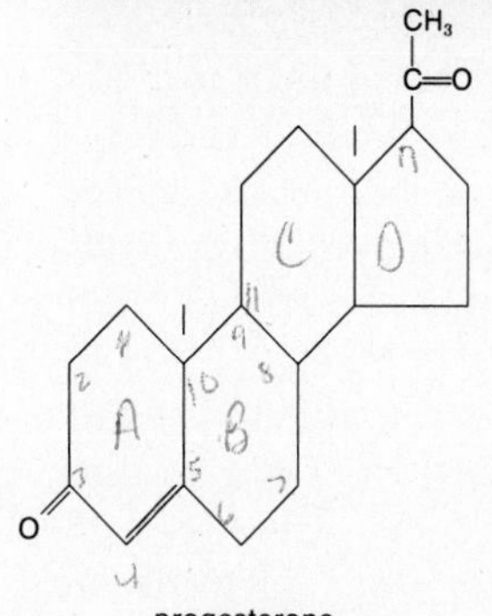

progesterone
(estrogen)

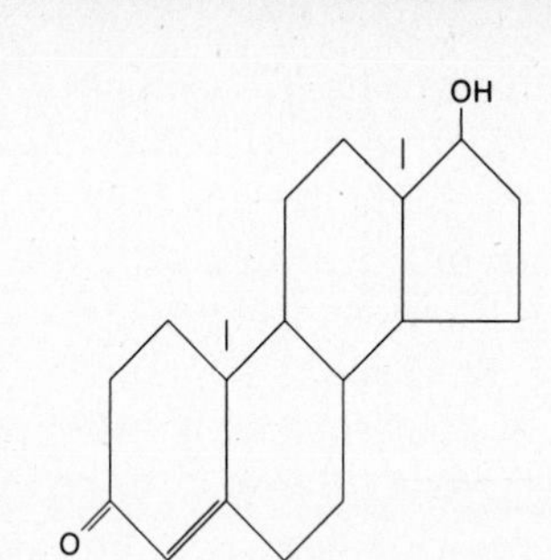

testosterone
(androgen)

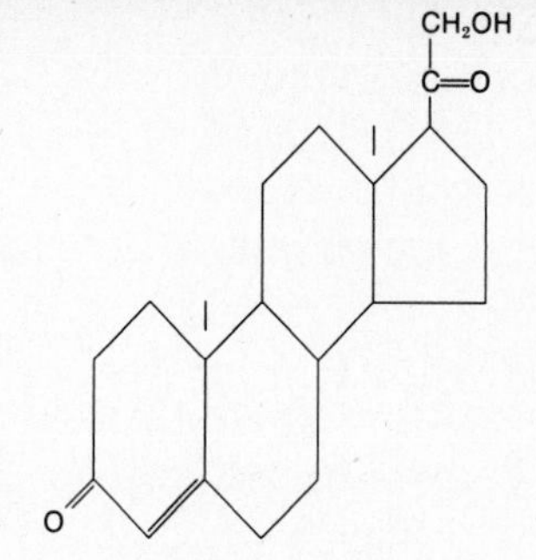

11-desoxycorticosterone
(mineralocorticoid)

OH
NO

estradiol
(estrogen)

CH_2OH
C=O
HO
OH
O

cortisol
(glucocorticoid)

H CH_2OH
O=C C=O
HO
O

aldosterone
(mineralocorticoid)

FIGURE 31.24 The formulas of some important adrenal steroids.

TABLE 31.12 Adrenal cortical hormones

Zone of production	General category of hormones	Specific hormonal examples	Effects of hormone—how produced
Zona glomerulosa	Mineralocorticoids	Aldosterone	Increases tubular transport of sodium, decreases transport of potassium; sodium reabsorption causes chloride, bicarbonate, and H_2O reabsorption by physical attraction
Zona fasciculata	Glucocorticoids	Cortisone Cortisol Corticosterone	Increases gluconeogenesis, stimulates deamination Maintains muscular strength; keeps adequate glucose and ATP available Maintains proper nerve excitability Exerts anti-inflammatory effects, possibly by rendering lysosome more resistant to disruption by disease Increases resistance to stress
Zona reticularis	Cortical sex	Androgens Estrogens	Androgens exert antifeminine effects; accelerate maleness; estrogen, just the reverse

glomerulosa produces MINERALOCORTICOIDS, affecting sodium metabolism; the fasciculata produces GLUCOCORTICOIDS, affecting foodstuff metabolism (chiefly carbohydrates); the reticularis produces SEX HORMONES, chiefly androgens (male).

Table 31.12 presents some of the specific effects of these hormones. Fig. 31.24 shows the formulas of the more important hormones in each group.

Glucocorticoids and stress

Chronic stress of any sort causes increased production of glucocorticoids. This increase in hormone levels may serve as a call to arms to equip the body to resist the stress. As a result of the increased production, amino acids are redistributed in the body, antiinflammatory effects are produced, and all of the repair mechanisms of the body are stimulated. Continued high levels of glucocorticoids are damaging to the body. For example, ulcer formation, high blood pressure, and atrophy of lymphatic tissue will occur. The individual may, in the long run, be rendered more susceptible to disease. In conclusion, increased hormone production is observed during stress, but the exact role this plays in the body's adaptive mechanisms is in doubt.

In 1953, Hans Selye proposed that such stimuli be called *stressors* and advanced the theory that the body responded to any or all stressors by a more-or-less consistent series of reactions he termed the GENERAL ADAPTATION SYNDROME *(G.A.S.)*. Selye divided the G.A.S. into three phases.

The ALARM REACTION, in which the stressor causes initial nervous and circulatory depression, followed by ACTH secretion and development of resistance to the stressor.

The STAGE OF RESISTANCE, in which full resistance to the stressor has been developed as cortisol secretion is elevated, and the body is functioning at a higher level than before.

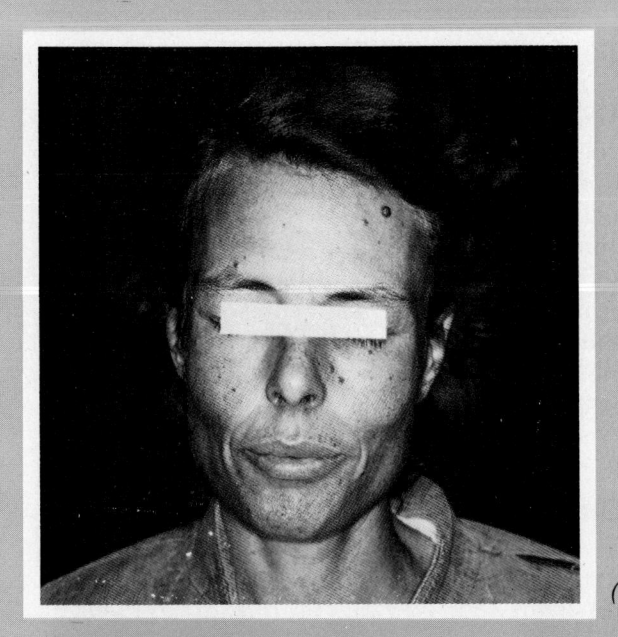

(a)

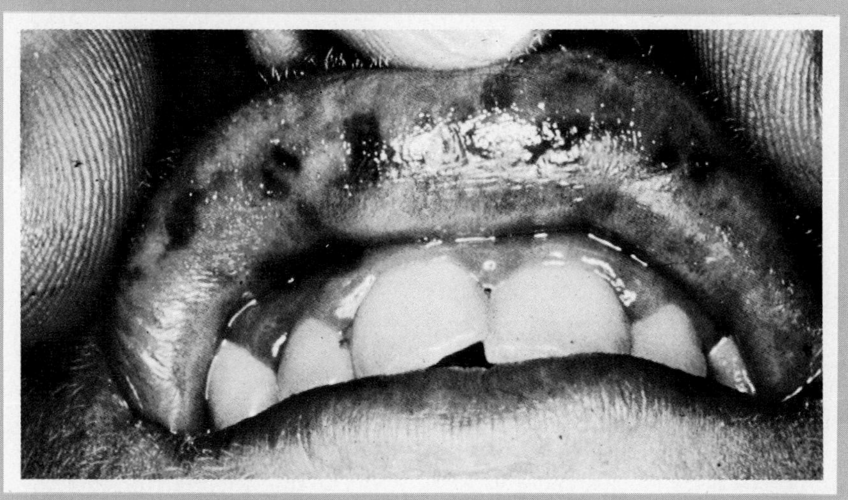

(b)

FIGURE 31.25 Addison's disease. *(a)* Shows generalized fatigued and dehydrated appearance of a patient having the disease. (Armed Forces Institute of Pathology.) *(b)* The lips and gums of an Addison's patient showing characteristic dark (actually brown) pigment deposits. (The Center for Disease Control, Atlanta, Georgia.)

The STAGE OF EXHAUSTION, in which high levels of cortisol in the body begin to wreak havoc on digestive, circulatory, and immune systems. ULCERS, DECREASED RESISTANCE TO INFECTION, HEMOCONCENTRATION, and SHOCK appear as the adaptation can no longer be maintained. While there is no strict proof of Selye's hypothesis, it is an attractive one to explain the "diseases of civilization" present in today's world.

Glucocorticoids inhibit inflammatory responses to injury when given in large doses. The mechanism appears to be one of stabilization of lysosome breakdown, reduction of fibroblast activity, and reduction of swelling. However, the amounts required to produce the antiinflammatory effect are so large that they produce the typical effects of glucocorticoid excess. Additionally, corticoid use may mask the appearance of symptoms that

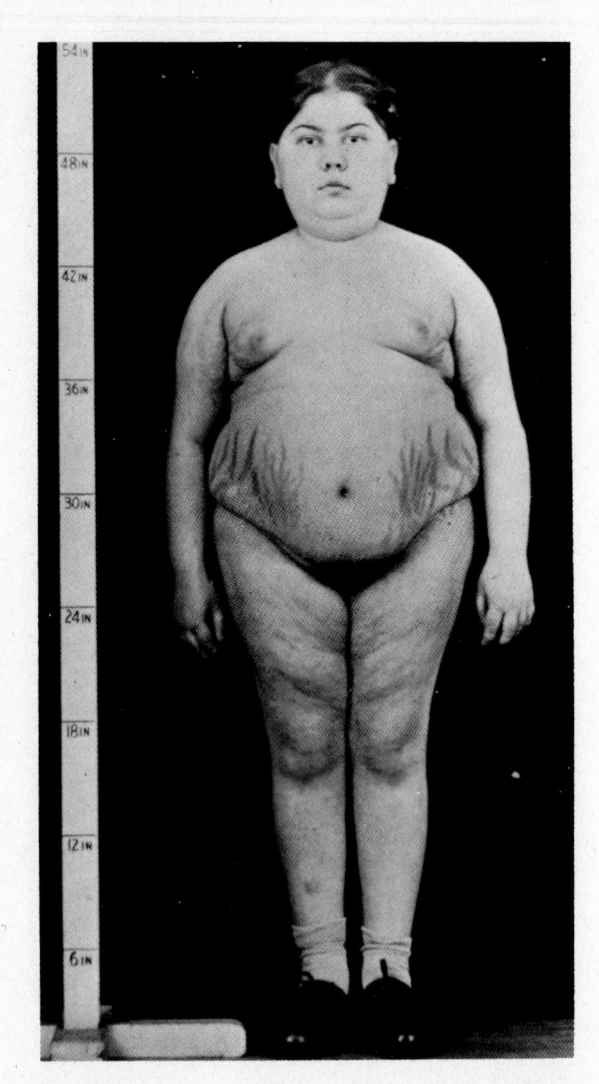

FIGURE 31.26 Cushing's syndrome in a 12-year-old female. Note the moon face, pendulous abdomen, abdominal striae, and concentration of adiposity on the trunk. (From the teaching collection of the late Dr. Fuller Albright. Courtesy Endocrine Unit & Department of Medicine, Massachusetts General Hospital.)

would call for further treatment. The use of glucocorticoids should thus be tempered by the knowledge that they are dangerous.

Control of cortical secretion

Activity of the glomerulosa appears to be controlled by the BLOOD SODIUM LEVEL and by ANGIOTENSIN. Increase in either factor stimulates aldosterone secretion, leading to increased sodium reabsorption, water retention, and chloride and bicarbonate reabsorption. ACTH plays little or no role in glomerulosa activity. The fasciculata and reticularis are under ACTH control. The hormones produced from the latter two zones also exert a negative feedback upon ACTH production.

Disorders of the cortex

HYPOFUNCTION, particularly of the fasciculata, results in ADDISON'S DISEASE (Fig. 31.25). Anemia, muscular weakness, fatigue, elevated blood potassium, and lower blood sodium evidence deficiency of glucocorticoids and mineralocorticoids. A peculiar bronzing of the skin is apparently due to the excess ACTH production occasioned by the removal of the negative feedback system as cortical hormone production decreases. (ACTH has melanocyte-expanding properties in the human.) Administration of cortisol controls the condition.

HYPERFUNCTION creates several disorders, depending on the zone involved. In most cases, hyperfunction is the result of a cortical tumor. Two of these conditions are of interest. CUSHING'S DISEASE (Fig. 31.26) is due to overproduction of glucocorticoids. The individual shows regional adiposity sparing the extremities, stripes on the abdomen, and diabetic tendencies. Surgical removal of the tumor followed by cortisol administration controls the disease. The ADRENOGENITAL SYNDROME results from a fasciculata tumor with resultant overproduction of sex hormones. In general, overproduction of this category of hormones will masculinize a female, or accelerate the sexual development of the male. If the excess production occurs *in utero*, when the child's reproductive organs are developing, bizarre alterations in

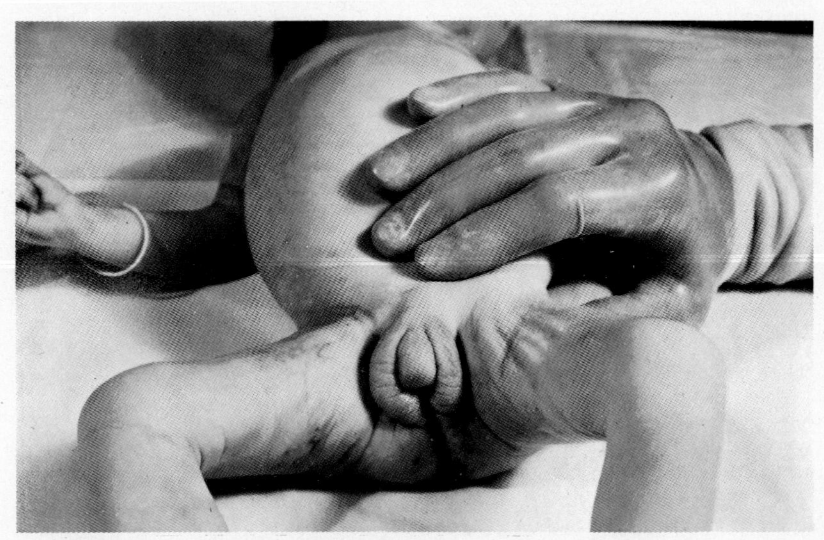

FIGURE 31.27 Adrenogenital syndrome in a 24-day-old female. Note masculinization, with enlarged clitoris (center of genitalia) and scrotumlike development of the labia. (Armed Forces Institute of Pathology.)

appearance of the organs may result. For example, if a genetic female is subjected to excess male hormone, the result will be conflicting orders to the developmental mechanism. Figure 31.27 shows a sufferer from adrenogenital syndrome.

The gonads and sexual differentiation

The basis of sex

The ultimate maleness or femaleness of an individual depends on the simultaneous action of several factors.

The GENETIC CONTRIBUTION, determined by the sex chromosomes received from the parents (XX or XY).

The GONADAL CONTRIBUTION, expressed by the appearance of the gonads and by the development of the external genitalia and secondary sex characteristics.

The HORMONAL CONTRIBUTION, determined by the secretion of estrogens or androgens by the developing and mature gonad and, to a degree, by the influence of hormones on the sexual organs.

The SOCIAL CONTRIBUTION, determined by how the parents view and treat the offspring and by the offspring's view of itself. Often termed the gender role, this component is neither male nor female at birth and is a result of environment alone.

THE GENETIC CONTRIBUTION. Human chromosomes normally consist of two sex chromosomes and 22 pairs of autosomes. Two sex chromosomes of the same type (XX) are characteristic of the female, while two different sex chromosomes (XY) are characteristic of the male. During meiosis, production of ova and sperm normally results in a halving of the diploid chromosome number and the production of cells having 22 plus X or 22 plus Y chromosomes. In some individuals, failure

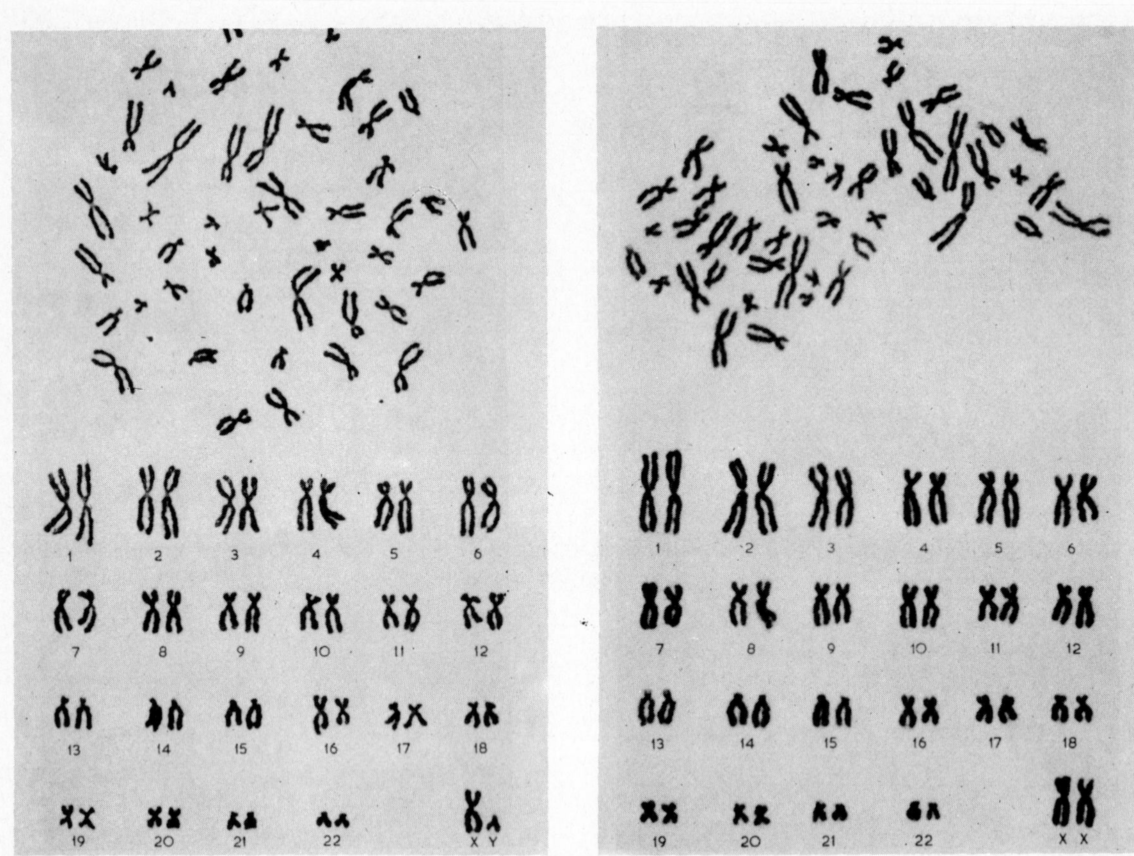

FIGURE 31.28 Normal human chromosomes (top), and karyotype (bottom). Male to the left; female to the right. (Courtesy Little Brown & Company, from L. S. Penrose, *Recent Advances in Human Genetics*, Little Brown, Boston, 1961.)

of chromosomes to separate (nondisjunction) during meiosis may result in a cell receiving both members of a chromosome pair, while another cell receives none. Deletion may result from a chromosome break followed by a disjunction that carries most of the chromosome to one cell, with only a fragment to another. Nondisjunction may occur in either autosomes or sex chromosomes. Congenital disorders may result that are reflected in the body generally or in sexual development. Some of the conditions that may result from such disorders are presented in Table 31.13. A normal human karyotype is shown in Fig. 31.28.

THE HORMONAL CONTRIBUTION. Development of the accessory organs of the female reproductive system depends primarily on genetic influences. Development of the male organs requires an *in utero* production of testosterone by the fetal gonad at about one month of development. Subsequent secretion does not occur until puberty at approximately 9–17 years of age. Thus, the appearance of the male external genitalia depends on embryonic secretion of sex hormone.

THE SOCIAL CONTRIBUTION. Included here are the psychosocial factors associated with the rela-

TABLE 31.13 Annotated list of chromosomal disorders

Disorder and frequency per number of live births	Chromosome (€)		Mechanism of production	Characteristics
	No.	Abnormality		
Autosomal disorders				
Down's syndrome (Mongolism); 1:600	47	3 € in group 21 (trisomy 21)	Nondisjunction	Simian (monkey) type finger and hand prints; low-set ears
Trisomy of group 18; 1:4,000	47	3 € in group 16–18	Nondisjunction	Index finger crossed over middle finger, mental retardation, do not thrive, renal and cardiac abnormalities
Deletion of group 17–18	46	Deletion of long or short arms € 17–18	Chromosome break	Small head, short stature, receding chin, gamma globulin deficiency
Trisomy 13–15 syndrome	47	Extra € in group 13–15	Nondisjunction	Cleft palate and lip, deafness, mental retardation, extra digits, simian lines
Cri-du-chat (cat cry) syndrome	46	Deletion of part of € 5	Deletion	Catlike cry in infancy, do not thrive, mental retardation
Sex chromosome disorders				
Turner's syndrome 1:10,000	45/46	45 O or 45 X	Nondisjunction	Cardiac abnormalities webbed neck, sexual infantilism
Kleinfelter's syndrome; 1:500	47	XXY (most common)	Nondisjunction	Small nonfunctional gonads, sterility
Pseudohermaphroditism	46XY or 45XO	Missing X €	Nondisjunction	Masculinization of genitalia
True hermaphrodite	46	46XX/XY	Ovum fertilized by two sperm	Presence of both functioning ovarian and testicular tissue

tionships between the individual and members of the same or opposite sex, mannerisms of dress, and orientation of sexual impulses. At present, all attempts have failed to show any correlation between chromosome abnormalities and gender role; thus, as stated earlier, gender role seems to depend solely on environment. The gender role is usually firmly established between 18–30 months of age.

PUBERTY. The assumption by the female of secondary sex characteristics and menses (menstruation) usually marks her first "change of life." It is associated with hypophyseal output of FSH and LH, with the maturation of ovarian follicles and production of ovarian hormones. In the male, sperm production and assumption of male secondary sex characteristics mark onset of puberty.

Age of onset is 9–17 years in females and 13–16 years in males. Resultant production of gonadal hormones causes maturation of the sex organs.

Ovarian hormones

The ovary produces three hormones.

ESTROGEN *(estrin)* is a steroid hormone produced by the vesicular follicle. Estrogen causes the proliferative phase of the menstrual cycle, increases mammary growth, stimulates contraction of the uterus, prevents atrophy of the accessory organs, and develops the female secondary characteristics. It also primes the body for progestin.

PROGESTIN *(progesterone)* is produced by the corpus luteum. Progestin is responsible for the secretory phase of the menstrual cycle, further develops the mammary glands, is necessary for placenta formation, and is required for ovulation.

RELAXIN, produced in very small quantities before birth, softens the pelvic ligaments. This softening theoretically allows enlargement of the birth canal. In the human, the effect is negligible.

Control of ovarian secretion

FOLLICLE-STIMULATING HORMONE (FSH) from the anterior pituitary initiates development of follicles, and, as a consequence, estrogen levels rise. Estrogen inhibits FSH production.

FIGURE 31.29 The relationships of hormones to ovarian and uterine activity during the menstrual cycle.

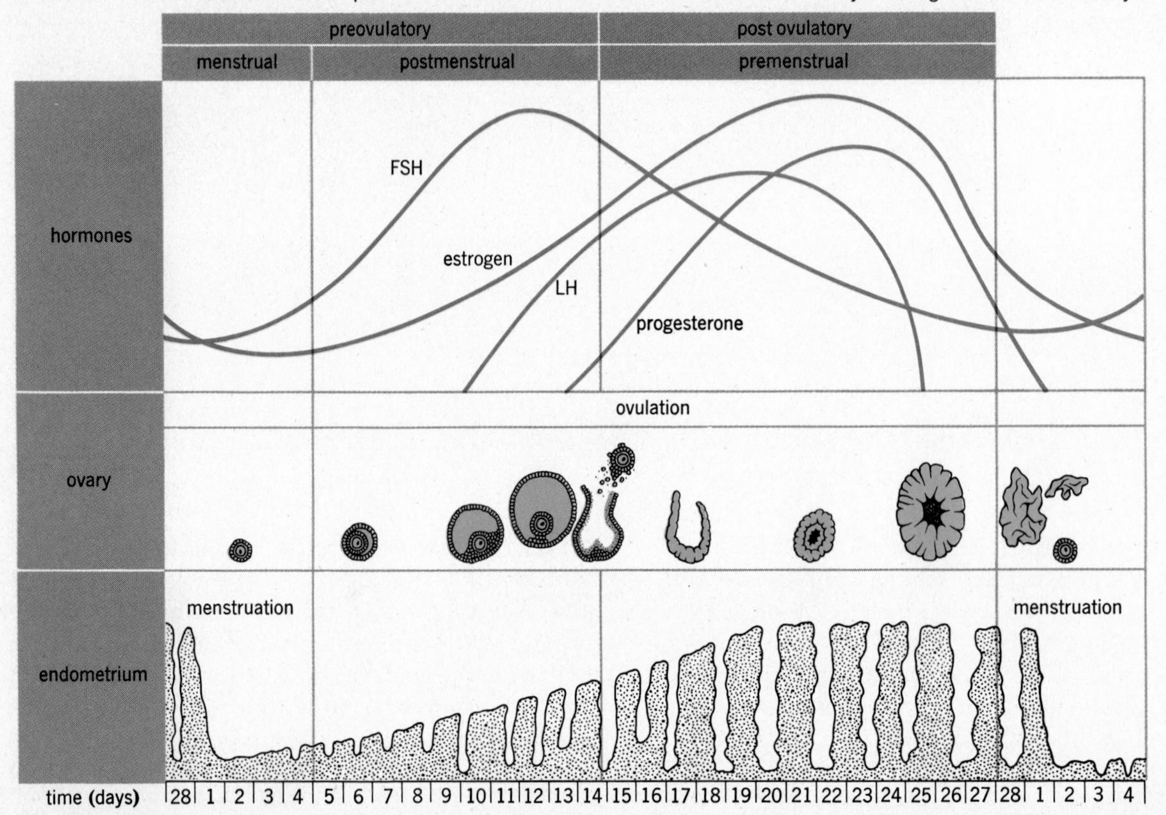

LUTEINIZING HORMONE (LH) from the pituitary causes corpus luteum formation with progestin production. Progestin also inhibits FSH production. The four hormones work together, as shown in Fig. 31.29. We may note the reciprocal relationships of these substances.

Oral contraception

The use of "the Pill" to control conception is based on the above effects of estrogen and progestin in inhibiting FSH production. The first substances utilized were similar to progestin. Although they inhibited FSH, and therefore follicle development, they also created a number of uncomfortable side effects, including mammary swelling, water retention, and a state of "pseudopregnancy." Newer compounds have the effects of estrogen, creating far fewer side effects. Since oral contraceptives act to inhibit follicular growth, consumption must begin after menstrual flow, when follicles are prone to begin their growth, and continue uninterrupted for the full course of pills (normally about 21 days). A pill taken before intercourse inhibits nothing and is ineffective in preventing conception.

Fertility drugs

"Fertility drugs" are estrogens, progesteroids (progesteronelike substances), or gonadotropins. Most cases of failure to conceive result from menstrual cycles that do not release ova *(anovulatory cycles)*. Administration of gonadotropins causes ripening and release of ova, often several at a time, so that multiple births may result. Estrogens and progesteroids are employed if the uterus fails to develop its lining to the point where successful implantation may occur.

The placenta

In the event of a pregnancy, formation of a placenta and hormone secretion by the placenta ensure the continued secretion of progesterone by the corpus luteum and produce female sex hormones that can, if necessary, sustain the pregnancy alone after three months. At least five hormones are known to be secreted by the placenta.

ESTROGEN ensures the growth of the uterine lining, the development of the mammary duct tissue, and the sustenance of secondary characteristics.

PROGESTERONE ensures the vascularity and glandularity of the uterine wall, ensures the development of the secretory tissue of the mammary glands, and quiets the muscular activity of the uterus.

HUMAN CHORIONIC GONADOTROPIN (HCG) is similar to LH and ensures stimulation of the corpus luteum, therefore resulting in continued secretion of progesterone by that structure.

CHORIONIC GROWTH-HORMONE-PROLACTIN (GCP) has lactogenic and growth-stimulating activity.

An unnamed hormone with TSH ACTIVITY is also secreted by the placenta.

The placenta seems to be insurance that the pregnancy will not suffer deficiencies of hormones necessary to continuation of development.

The testis

The INTERSTITIAL CELLS of the testis produce TESTOSTERONE, a steroid. This hormone serves the same function in the male that estrogen serves in the female; that is, it develops and maintains the accessory organs and the secondary sexual characteristics.

Control of testicular secretion

INTERSTITIAL CELL STIMULATING HORMONE (ICSH) from the pituitary governs testosterone secretion. The latter hormone in turn exerts a negative feedback on ICSH production.

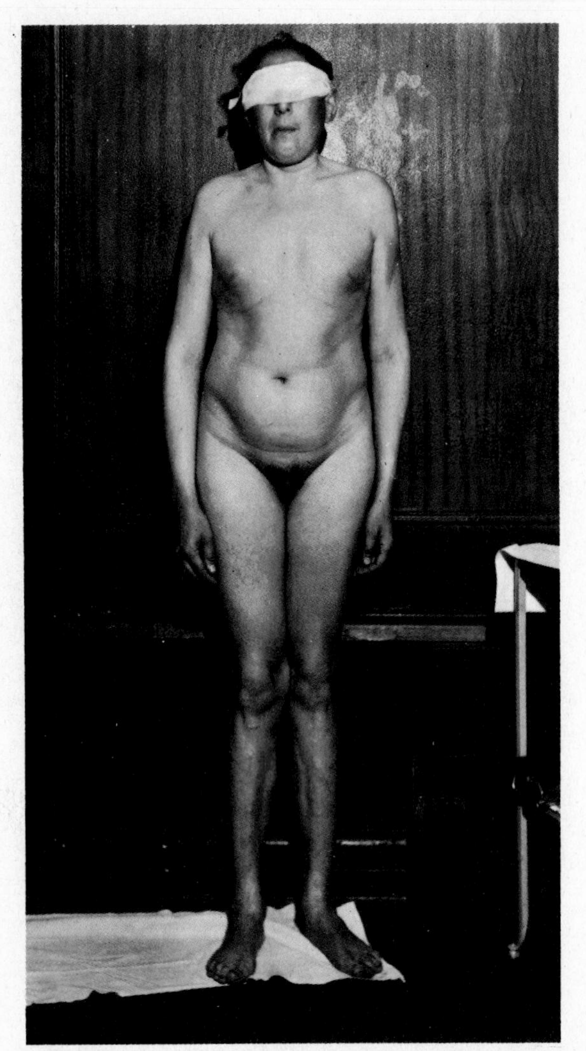

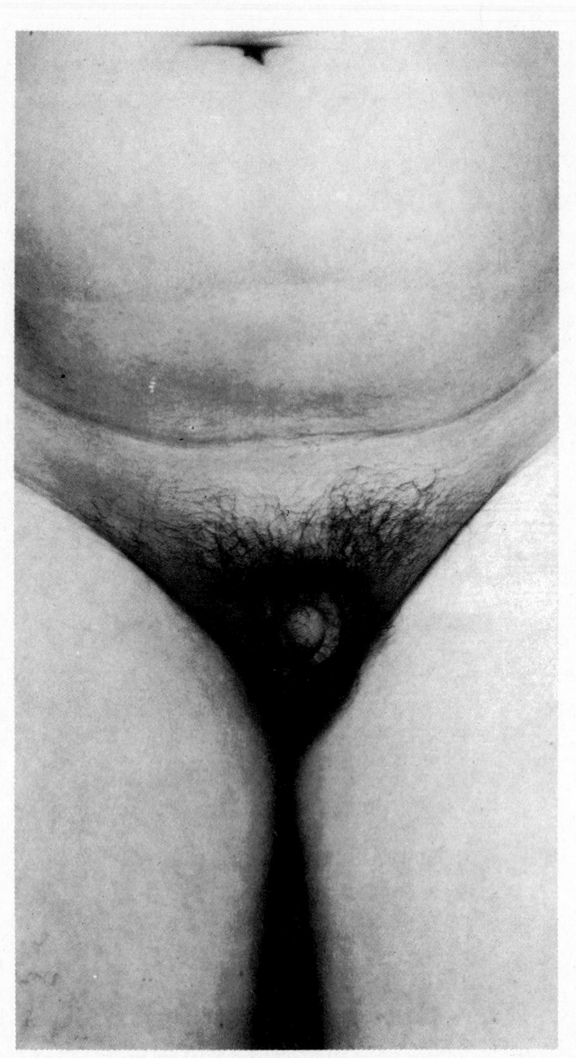

FIGURE 31.30 Eunuchoidism in a 40-year-old male. Note disproportionately long arms and legs for height (6 feet, 1 inch) and failure of the genitalia and secondary characteristics to develop normally. (Lester V. Bergmann & Associates.)

Disorders of the ovaries and testes

Both sexes may suffer from deficient production of hormones in the condition known as EUNUCHOIDISM (Fig. 31.30). As expected, the most obvious symptoms will be atrophy of accessory organs and disappearance of secondary characteristics. Complete loss of hormones produces eunuchism, similar to eunuchoidism, but presenting more severe symptoms. Treatment with the appropriate hormone will control symptoms.

HYPERFUNCTION, due to gonadal tumors, creates PRECOCIOUS PUBERTY. In this condition, accessory organ development and characteristics appear at earlier-than-normal ages. Precocious females may actually become fertile at very tender ages (the

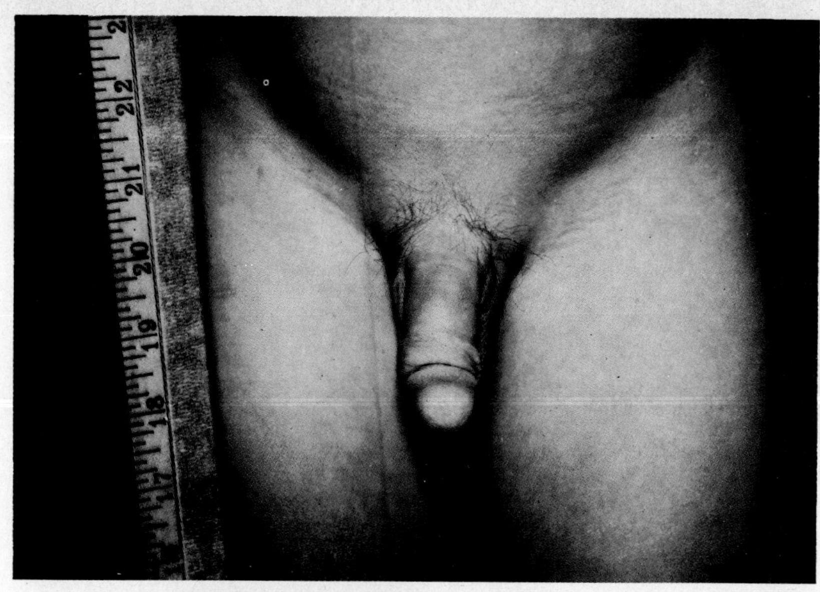

FIGURE 31.31 Precocious puberty in a 3-year-old male. Note development of genitalia equivalent to that of an adolescent, and appearance of pubic hair. (Armed Forces Institute of Pathology.)

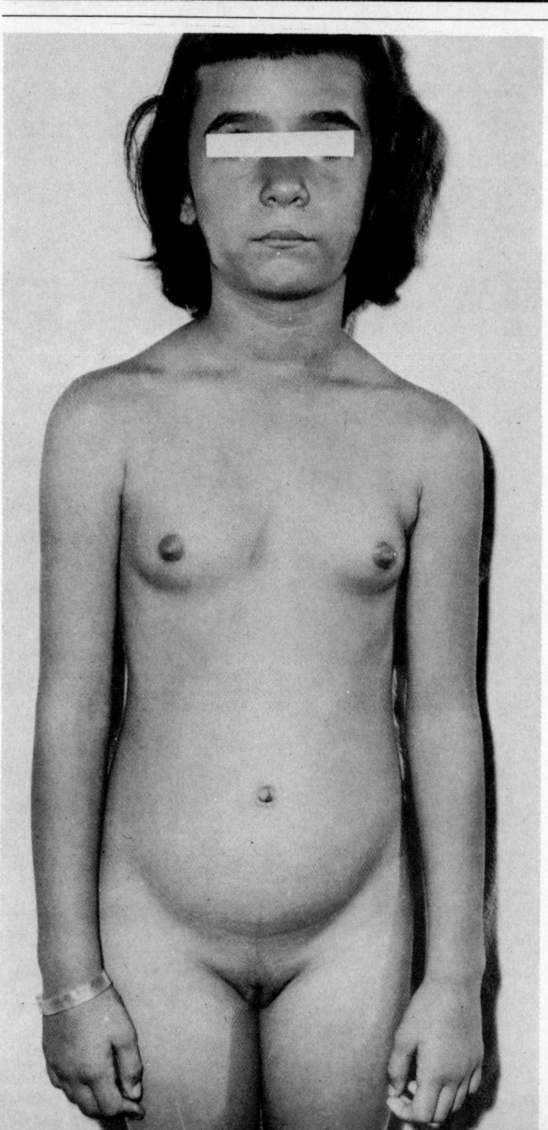

FIGURE 31.32 Precocious puberty in a 7-year-old female. Height was 4 feet, 3-1/2 inches. Note development of pubic hair and mammary glands. (Lester V. Bergmann & Associates.)

earliest on record became pregnant at five years old). Figures 31.31 and 31.32 show male and female precocious puberty patients. Surgical removal of the tumor is required, and the development may be arrested and partially reversed.

Climacterics

As they age, both males and females undergo a cessation or diminution of gonadal hormone production. In the female, this period is often called the MENOPAUSE, and it usually occurs between 45 and 50 years of age. In males, similar slowdown is termed the male CLIMACTERIC, and it occurs gradually, beginning at about 60 years of age.

In both sexes the following symptoms may develop:

PSYCHIC SYMPTOMS (more common in female). Nervousness, irritability, crying spells, some loss of mental acuity.

VASOMOTOR SYMPTOMS. Sweating, headache, hot and cold flashes.

CONSTITUTIONAL SYMPTOMS. Fatigue, muscular weakness, "lack of ambition."

SEXUAL SYMPTOMS. Loss of sex drive; in female, ultimate sterility.

It should be emphasized that the climacterics will occur as aging proceeds. Understanding of the condition and acting accordingly is often the most important factor in getting through the period. Appropriate sex hormones are often used to treat those individuals in whom these symptoms are causing problems in their activities, the general idea being to make the disappearance of the hormones more gradual and less disturbing.

The prostaglandins

The prostaglandins (PGs) are a series of cyclic, oxygenated derivatives of prostanoic acid (Fig. 31.33). They are formed by the action of a prostaglandin synthetase enzyme system located on the microsomal (ER fragments) fraction of the cellular organelles of a wide variety of body cells (e.g., lungs, liver, muscle, reproductive organs). A long-chain polyunsaturated fatty acid is caused to assume a cyclic form under the influence of the synthetase.

The PGs are categorized by the chemical groups found attached to the cyclic portion of the molecule, and by their degree of unsaturation. Utilizing the chemical groups on the ring, four

TABLE 31.14 Tissues where prostaglandins increase accumulation of cyclic 3′5′-AMP

Tissue	Species	Prostaglandin (μM) most active	
Lung	Rat	PGE_1	2.8
Spleen	Rat	PGE_1	2.8
Diaphragm	Rat	PGE_1	2.8
Adipose	Rat	PGE_1	2.8
Leucocytes	Human	PGE_1	–
Platelets[a]	Human	PGE_1	0.28
	Human	PGE_1	0.01
	Human	PGE_1	0.1
	Human	PGE_1	0.1
	Rabbit	PGE_1	0.1
	Rabbit	PGE_1	0.15

[a] Indicates those tissues where prostaglandins increase adenylcyclase activity.

basic types may be distinguished, as shown in Fig. 31.34.

Further description of the molecule is provided by a subscript number that indicates the number of unsaturated carbon-carbon bonds in the molecule. Additionally, the designation α (alpha) or β (beta) is used to describe the direction of projection of the chemical groupings at carbons 8 or 9, 11, and 15. Alpha designates a bond or bonds projecting "below" the average plane of the 5-carbon ring; beta implies bond projection "above" the average plane of the 5-carbon ring. The primary PGs and their precursors are presented in Fig. 31.35.

Prostaglandins are widely distributed in the animal kingdom. They have been found in corals and within the "higher" animal organisms, and have been demonstrated in most animal cells and

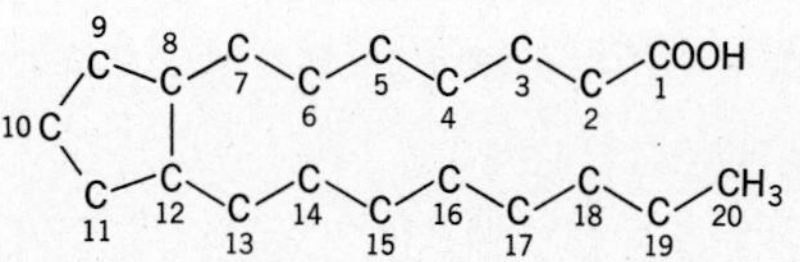

FIGURE 31.33 The chemical structure of prostanoic acid.

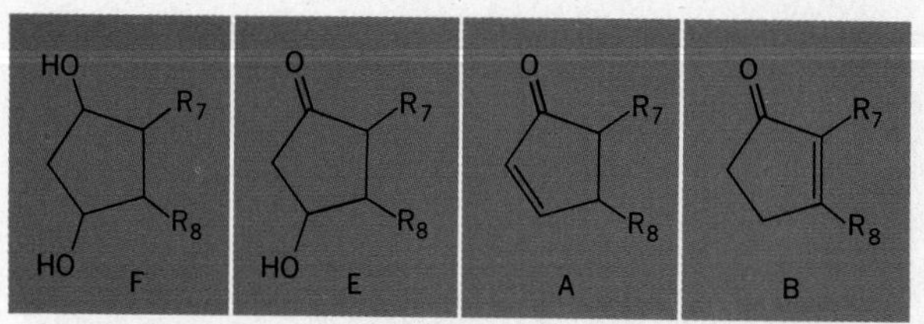

FIGURE 31.34 The four basic ring structures of the prostaglandins. (Courtesy N. Anderson and H. Benson, Annals, N.Y. Acad. Sci., Vol. 180, p. 15, Fig. 1.)

TABLE 31.14 continued

Tissue	Species	Prostaglandin (μM) most active	
Liver	Rat	PGE_1	–
Anterior pituitary[a]	Rat	PGE_1	2.8
Aorta	Rat	PGE_1	2.8
Bone	Rat	PGE_1	–
Gastric mucosa[a]	Guinea pig	PGE_1	–
Kidney	Dog, rat	PGE_1	–
Heart	Guinea pig	PGE_1	0.01
		$PGF_{1\alpha}$	0.01
Corpus luteum[a]	Bovine	PGE_2	28
Thyroid	Dog	PGE_2	–
Erythrocytes[a]	Rat	PGE_2	0.03

TABLE 31.14 (J. E. Shaw, Annals, N.Y. Acad. Sci., Vol. 180, p. 242, Table 1.)

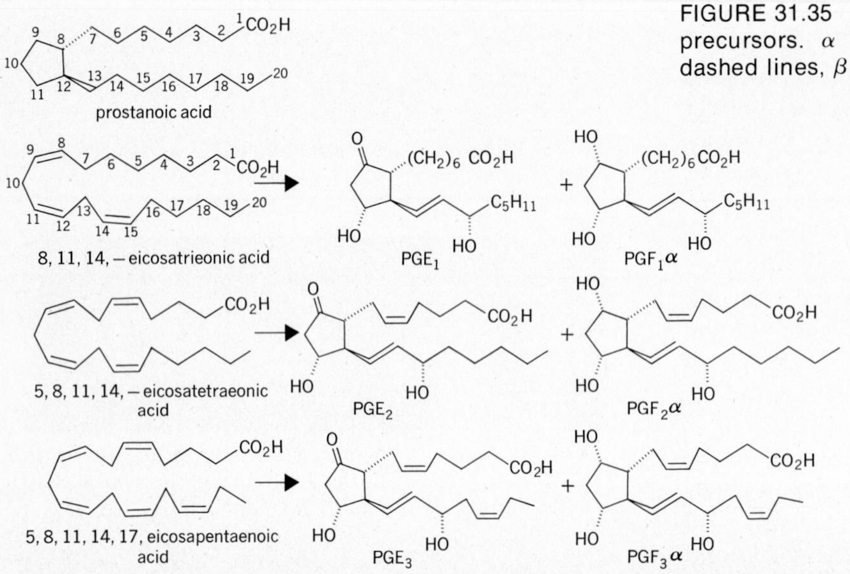

FIGURE 31.35 The primary prostaglandins and their precursors. α bond configuration is indicated by dashed lines, β by solid lines.

TABLE 31.15 Tissues where prostaglandins[a] inhibit hormonally induced responses

Tissue	Hormone	Response
Toad bladder	Vasopressin	Water transport
Rabbit kidney tubules	Vasopressin	Water transport
Rat adipocytes[b]	Epinephrine ACTH TSH Glucagon Growth hormone	Lipolysis
Cerebellar Purkinje cells	Norepinephrine	Inhibition of discharge frequency

[a] PGE_1 most effective at $<0.28\ \mu M$.
[b] Inhibition associated with decreased cyclic AMP accumulation (Butcher & Baird, 1968).

TABLE 31.15 (J. E. Shaw, Annals, N.Y. Acad. Sci., Vol. 180, p. 243, Table 2.)

tissues. They were first demonstrated in human seminal fluid in 1930. Since that time, their distribution has been shown to be nearly ubiquitous.

With the discovery of the presence of cyclic nucleotides and their role in control of enzymatic reactions, PGs have been investigated as controllers of the adenylcyclase or guanylcyclase enzyme systems, and therefore the amount of cyclic nucleotides in cells. Evidence to date indicates that PG effect is indeed exerted via the cyclase enzymes, and may be either stimulatory or inhibitory in effect. The inhibitory effect may additionally be exerted through the mechanism of blocking the receptor site for a hormone that normally increases cyclic nucleotide formation. All this suggests that PGs may act as "intracellular switches" to turn chemical reactions off or on.

Tables 31.14 and 31.15 show some tissues in which PGs increase cAMP concentrations, or where hormone effect is inhibited.

Release of PGs from tissues or cells is caused by a wide variety of stimuli, as shown in Table 31.16, and may be associated with disturbances of any cell membrane by normal or pathological processes. Cells may release PGs to resist change, as when tension is applied to smooth muscle or when inflammation occurs. A primary result appears to be to resist tearing or rupture of the cell membrane.

Effects of PGs are nearly as widespread as the tissues in which they are found. The F and E groups have been most extensively investigated as to effects. The list presented below indicates some, but certainly not all, of the effects of PGs on the body.

SMOOTH MUSCLE. Uterine smooth muscle is

TABLE 31.16 The release of prostaglandins by various stimuli

Species	Tissue	Stimulus
Rabbit	Eye (ant. chamber)	Mechanical
	Spleen	Catecholamines Serotonin
	Epigastric fat pad	Hormones
	Somatosensory cortex	Neural Analeptics, etc. Reticular formation
	Spleen	Catecholamines
	Adrenal	Acetylcholine
Dog	Spleen	Neural Catecholamines Colloids
	Bladder	Distension
	Cerebral ventricles	Serotonin
Rat	Phrenic diaphragm	Neural Biogenic amines
	Epididymal fat pad	Neural Biogenic amines
	Stomach	Neural Stretch Secretagogues
	Skin	Inflammation
	Liver	Glucagon
	Lung	Air embolus Infusion of particles
Guinea pig	Lung (whole, perfused)	Anaphylaxis Particles Histamine Tryptamine Serotonin Massage Air embolus Distension
	Lung (chopped)	Stirring
Human	Thyroid	Medullary carcinoma
	Uterus	Parturition Distension
	Platelets	Thrombin
Frog	Intestine	Distilled water
	Skin	Isoproterenol
	Spinal cord	Neural Analeptics

TABLE 31.16 (Courtesy P. J. Piper and J. Vane, Annals, N.Y. Acad. Sci., Vol. 180, pp. 376–377, Table 2.)

stimulated to increase its contractions by PGE. Vascular smooth muscle is relaxed by PGE and PGA_2. PGF causes contraction of vascular smooth muscle.

SODIUM EXCRETION. PGA, PGA_2, and PGE_2 are all natriuretic and can lower blood pressure by decreasing blood volume via Na and water excretion. The effect appears to be mediated via the kidney.

ION AND WATER EXCHANGE. PGE increases calcium release from binding sites and inhibits the ADH effect on toad bladder and perhaps also on kidney tubules.

SKIN. The highest concentrations of PGs are found in the epidermis, where they are postulated to be involved in the synthesis of the waterproofing substances of the upper epidermal cell layers. If deficient in PG, these cells have been shown to become more permeable to water.

THE OVARY. $PGE_{2\alpha}$ has been shown to be luteolytic—that is, capable of inducing the destruction of the corpus luteum.

GASTRIC SECRETION. PGA_1 and PGE_1 reduce gastric secretion.

INFLAMMATORY RESPONSE. PGE appears to increase the response of leucocytes in reaching an area of inflammation or injury. This substance also appears to increase blood flow and the edematous response that usually follows trauma to a tissue. It also causes increased breakdown of tissue basophils (mast cells).

NEURONS. PGF_2 has been shown to depress synaptic transmission and elevate discharge rates of neurons. It is not a transmitter itself, but influences those processes that synthesize or degrade the normal transmitters. PGE also elevates discharge by neurons, but has no effect on synaptic transmission.

Therapeutic uses of PGs will probably revolve around their effects on smooth muscle and the ovary. The use of PGs to CONTROL HYPERTENSION (through vasodilating effects), nasal congestion and asthma (through vasoconstrictive and smooth muscle relaxing effects respectively), ULCERS (through decreased gastric secretion), and as ABORTIFACIENTS (through luteolysis) has reached the experimental stage in humans. Minimal side effects appear to be produced, and the PGs may, in the future, achieve the fame accorded the antibiotics at their appearance.

The pineal gland

The pineal gland, located on the posterior aspect of the midbrain (Fig. 31.36), has, in recent years, acquired new status as an endocrine structure. Several chemically active substances have been isolated from the pineal, including MELATONIN, norepinephrine, serotonin, and histamine. Of these substances, melatonin is synthesized only in the pineal. Melatonin has the property of causing agglomeration of pigment granules in pigment cells (melanophores) in amphibian skin. In such animals, the secretion of melatonin may be greatly diminished or abolished by enucleation of the eye. Secretion of the substance thus appears to respond to changes in light as detected by the eye and may have protective value to the organism in terms of camouflaging it from predators. Effects of the hormone in man are uncertain. Pineal secretion appears to respond to gonadotropin secretion, and melatonin may inhibit the ovary in its hormone secretion. The role of the pineal in the human body remains to be determined.

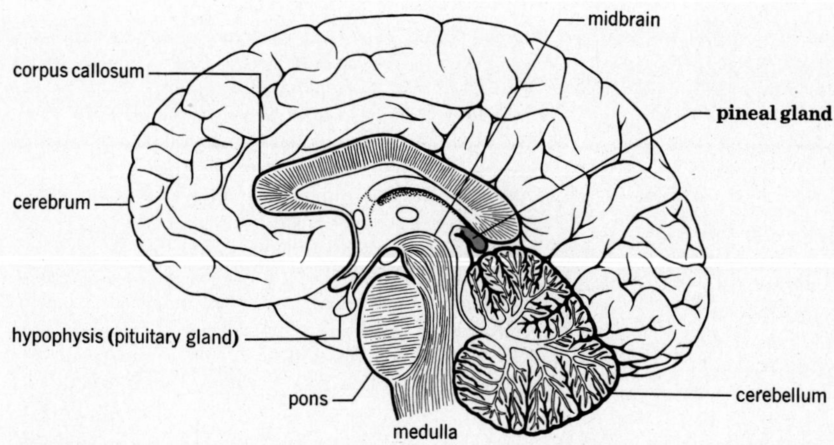

FIGURE 31.36 The location of the pineal gland as shown on a midsagittal section of the brain.

Summary

1. The endocrine glands develop as outgrowths of the pharynx, gut, or from mesoderm or neural crest material.
2. Endocrines provide a means of controlling homeostasis, growth, development, and metabolism by blood-borne hormones.
3. Endocrines are specific groups of cells that secrete specific chemicals.
4. There are three classes of hormones: proteins, amines, and steroids.
5. Hormones are synthesized by endocrines, are produced in amounts as required, are inactivated by their use, and influence chemical reactions in the body.
6. There are six basic control devices for governing secretion of hormones.
7. The pituitary has two main parts: anterior and posterior lobes.
   a. The anterior lobe produces six hormones (GH, prolactin, ACTH, TSH, LH, FSH).
   b. The anterior lobe is controlled by hypothalamic regulating factors and feedback mechanisms.
   c. The posterior lobe makes no hormones, but stores ADH (antidiuretic hormone) and oxytocin (smooth muscle stimulant) produced by the hypothalamus.
   d. The release of posterior lobe hormones is controlled by nerves.
   e. Disorders include giantism, acromegaly, and diabetes insipidus.
8. The thyroid gland lies in the neck and produces thyroxin and calcitonin.
   a. Thyroxin controls BMR and is involved in growth, especially of the brain cells.
   b. Synthesis of thyroxin is controlled by TSH.
   c. Calcitonin influences metabolism of calcium and phosphate, and aids their deposition in bony tissue.
   d. Calcitonin production is directly related to blood calcium level.
   e. Thyroid disorders include hyperthyroidism, goiter, and cretinism.
9. The four parathyroid glands lie on the posterior aspect of the thyroid.
   a. The hormone (PTH) influences metabolism of Ca, PO_4, Mg, and citrate by increasing absorption and removal from bony tissue.
   b. Secretion is inversely related to blood calcium level.
   c. Disorders of the parathyroids include hyperparathyroidism and tetany.
10. The pancreas islets produce two hormones: glucagon and insulin.
    a. Glucagon increases conversion of glycogen to glucose and raises blood sugar levels.
    b. Insulin increases conversion of glucose to glycogen, increases glucose uptake by certain cells, and thus lowers blood sugar levels.
    c. Insufficient insulin production results in diabetes mellitus.

d. Diabetes results in hyperglycemia, glucosuria, and other characteristic symptoms; it develops in stages.
e. Diabetes may have a hereditary basis that results in disease production when the body is stressed by a variety of factors.
f. Hypoglycemia results in abnormal brain function.
g. Secretion of both hormones is controlled by blood sugar levels.

11. The adrenal glands contain two functional areas: medulla and cortex.
a. The medulla produces epinephrine and norepinephrine that accelerate body activity and energy expenditure.
b. The cortex produces: mineralocorticoids (e.g., aldosterone) that deals with Na metabolism, glucocorticoids (e.g., cortisol) that are anabolic hormones, and sex hormones.
c. Glucocorticoids also confer resistance to stress but may damage body organs.
d. Cortical secretion is controlled by blood Na level and angiotensin (mineralocorticoids) and by ACTH (others).
e. Disorders of the cortex include Addison's disease (deficiency), Cushings syndrome (excess), and the adrenogenital syndrome.

12. The ovaries and testes produce hormones that, with genetic and social factors, determine maleness and femaleness.
a. Ovarian hormones include: estrogen, which develops female sex organs and characteristics, initiates puberty, and menses; progesterone, essential for ovulation, endometrial growth, and placenta formation.
b. Ovarian activity is controlled by FSH and LH.
c. The use of female hormones or materials having similar effects is the basis for oral contraception.
d. "Fertility drugs" stimulate ovum development.
e. The placenta produces estrogen, progesterone, LH-like materials, and other hormones.
f. Testosterone, produced by the testes, ensures development of male sex organs and characteristics.
g. Testosterone secretion is controlled by ICSH (LH).
h. Both sexes may suffer deficiency of hormone(s) and loss of characteristics; excess hormone causes precocious puberty or accentuation of characteristics.

13. Climacterics are associated with decrease of gonadal hormone production, the development of nervous and constitutional symptoms, and atrophy of genital tissues.

14. Prostaglandins are derivatives of fatty acids.
a. They are widely distributed in animal cells.
b. They appear to act to "turn off or on" certain chemical reactions in the body, notably cAMP production.

15. The pineal gland produces melatonin and may be involved in ovarian hormone production.

Circadian Rhythms

Rhythms, their extent and significance

Rhythmic or cyclic changes in body function appear to be a fundamental property of human life. One's life is considered by some to be the ultimate cycle. Within this time period are fluctuations that are annual (certain hormones), monthly (menstrual), daily (sleep-waking), hourly (temperature), in minutes (cell division), in seconds (heartbeat), or in fractions of seconds (EEG rhythms).

Humans appear to be "tuned" most often to daily rhythms that repeat in a CIRCADIAN (*circa*, about + *dies*, a day) fashion, with a time period of 23–25 hours. Such rhythms are not usually evident externally, but can be determined by analysis of body fluids and other physiological functions. Strength and weakness, susceptibility to stress, emotional stability, test performance, and many other functions change with time of day; we are indeed different persons at different times of the day. In terms of effects of drugs and medications, the time of day of ingestion may influence body response to the substance.

Human cyclic rhythms were, until relatively recently, the subject of only anecdotal reports. With the advent of jet travel throughout the world, and the subsequent dislocations of sleep and waking cycles, the disruption of normal rhythms or "habits" has resulted in increased interest and research into the subject of human rhythms. Nevertheless, a considerable amount of folklore still permeates many theories as to what is responsible for the origin of many rhythms.

Not only does it appear that a particular cycle must be stable for optimum health, but several cycles may need to be integrated in a specific fashion for normal function. A ratio of 4:1 for heartbeat and breathing frequency appears to be normal. We all know the feelings of distress that occur when the heart races or its rhythm is disturbed. Functions alternate from right to left sides of the body. For example, temperature is slightly higher on the left side of the body during sleep and higher on the right during waking hours; one breathes more through one nostril than the other, and then changes about every three hours.

Alterations in timing of cycles of cell division may signal the presence of malignancy. Other indications of cyclical function in the human are presented in Table 32.1.

Determination of circadian rhythms

Newborn humans do not have circadian rhythms to their body functions. As any new parent knows, the only real predictable quality of an infant is its unpredictability. By 14–20 weeks of age, most newborns have acquired circadian patterns. What is responsible for the assumption of rhythmical oscillations?

Exogenous stimuli as rhythm setters

A wide variety of influences originating outside the body have been correlated with cyclic patterns in both humans and "lower animals." Crab metabolism is inversely related to cosmic radiation; earth's magnetic field appears to determine

TABLE 32.1 Human circadian rhythms in day-adapted subjects

Body cell, tissue, organ or fluid monitored	Parameter	Time of cycle peak[a]
Brain	Total EEG	1 pm
	Delta waves	12 m
	Theta waves	11 am
	Alpha waves	1 pm
	Beta waves	12 m
Epidermis	Mitosis	11:30 pm
Urine	Volume; rate of excretion	8:30 am
	Potassium excretion	1 pm
	Sodium excretion	9 am
	Hydroxycorticosteroid excretion	11 am
	Cortisol excretion	9:30 am
	17 ketosteroid excretion	11 am
	Epinephrine excretion	1 pm
	Norepinephrine excretion	1 pm
	Aldosterone excretion	1 pm
	Magnesium excretion	1 am
	Phosphate excretion	8 pm
	pH	1 pm

turning and movement in snails and worms; humans have had their circadian rhythms shortened by the application of strong electric fields under controlled conditions; water "dowsers," who claim to locate underground water, appear to have extraordinary sensitivity to small changes in the earth's magnetic field. Other examples of exogenous stimuli that may be involved in rhythm setting include moon cycles and the behavior of mental patients (lunatics), sunspot activity and increased excitement in mental wards, and cycles of light and darkness.

Of the various possible exogenous rhythm setters, cycles of light and dark appear to have the best correlation to human cyclic function. This correlation has led to a search for a mechanism by

TABLE 32.1 continued

Body cell, tissue, organ or fluid monitored	Parameter	Time of cycle peak[a]
Blood	Neutrophils	12:30 pm
	Lymphocytes	11:30 pm
	Monocytes	3 am
	Eosinophils	11:30 pm
	Hematocrit	10 am
	Ca^{2+}	9:30 pm
	Na^{1+}	5 pm
	PCO_2	11 am
Plasma or serum	Proteins	2 pm
Whole body	Temperature (oral)	3 pm
	Physical vigor	3:30 pm
	Weight	6 pm
	Heart rate	4 pm
	Systolic blood pressure	7 pm
	Diastolic blood pressure	10 pm
	Respiratory rate	2:30 pm

[a] Times quoted reflect the average of the peak values of the particular function; the range of a function is not indicated, and the low point of a given function may be assumed to occur approximately 12 hours from the time indicated.

which light and dark might influence physiological and psychological behavior. Interest has centered on light-dark cycles because both plant and animal forms appear to react to them, and, as discussed below, certain animals, including humans, appear to have a receptor-effector mechanism that can cause nervous and hormonal reaction to light and darkness.

Light and darkness

Response to light and dark cycles appears to be widespread throughout the entire living world. In nineteenth-century Europe, formal gardens were sometimes planted in the form of a clockface, with different plants blooming, or opening and closing, at different hours. Insects may assume greater activity levels that correlate with light-dark patterns. Birds appear, in part, to govern the timing of their migrations according to length of day.

A possible receptor-effector mechanism that is available, if not functional, in humans for detection of changes in length of light and darkness is the eye-pineal system.

As mentioned in the previous chapter, the pineal secretes melatonin, a chemical that alters skin color. In lower animals, the survival value of a system that enables camouflaging of the body is obvious. In humans, stimuli appear to be delivered to the pineal from offshoots of the visual pathways. One way of testing the hypothesis that light-dark cycles are responsible for human circadian rhythms is to put an individual into a cave for varying periods of time and to determine if physiological parameters alter. It has been demonstrated that sleep cycles, menstrual cycles, and temperature cycles are altered by such procedures.

Implications and prognostications

Recent investigations have indicated that cycles do change with age, and that synchronization of various cycles is also altered. It would seem that a great deal more experimentation designed to elucidate changes and control mechanisms of circadian rhythms should be done. It may then become possible to alter or restore rhythms to normal pattern or, at least, to learn to avoid those stresses that may change a pattern. Thus, longer life and more quality to life may result, as well as elucidation of the role of circadian rhythms in aging.

Summary

1. Many human physiological activities exhibit cyclical alterations that have about a 24-hour periodicity.
   a. Such a cycle is termed a circadian rhythm.
   b. Circadian rhythms occur in brain activity, urine excretion of metabolites, water, salts, hormones, blood cell numbers, and temperatures.

2. Assumption of a circadian rhythm may be the result of hereditary factors, environmental influences, or combinations of these.
   a. External factors such as cosmic radiation, geomagnetism, electromagnetism, electrical fields, sunspots, and light and darkness have been investigated as circadian time-setters.
   b. The eye appears to be important as a receptor for light-dark alterations; such alterations appear to be a primary determinant of circadian rhythms.
   c. Rhythms are not present at birth. They appear at 14–20 weeks of age.
   d. Aging does not cause disappearance of circadian rhythms; it may desynchronize certain rhythms.
   e. Susceptibility to drugs and other substances is altered during the day. Emotional and physical resistance to stress varies according to time of day.

33

Human Growth and Development; An Overview

This text has emphasized the mechanisms that operate to maintain the homeostasis of body systems and the interrelationships between those systems. A general overview of the human organism from before conception to death may provide a broad consideration of those factors that determine growth and development in both the physical and mental spheres, and determination of the time of death.

The production of sex cells and offspring

Ripening or maturation of ova (eggs) and spermatozoa are processes that are initiated at puberty. The female will produce, at a normal rate of one per month, some 400 mature ova during her fecund years. The male produces, in more or less continuous although decreasing numbers, spermatozoa throughout life. Both ova and spermatozoa are haploid cells; that is, they contain one-half the normal chromosome number characteristic of the species. For humans, the haploid chromosome number is 22 autosomes plus either an X or Y sex chromosome.

The restoration of normal chromosome number (44 autosomes plus XX or XY) requires the union of an ovum and a sperm by either natural or artificial insemination.

The role of the male

The organs of the male reproductive system were described in Chapter 30. Of these organs, the penis serves as the organ for the natural introduction *(insemination)* of sperm into the vagina of the female. To become an effective organ for insemination, the penis must become firm or erected.

Erection of the penis occurs as a result of psychic stimuli, or tactile stimuli applied to the genital area or elsewhere on the body's erogenous zones. Stimulation of the male external genitalia is believed to serve as the primary stimulation pathway. From the penis, nerve impulses pass over the pudendal nerve to the sacral portion of the spinal cord. Parasympathetic impulses then pass over the nervi erigentes that supply the arteries of the penis. As a result of these impulses, the smooth muscle of the arteries relaxes, allowing a greater inflow of blood into the cavernous bodies that form the organ. They fill, compressing the veins of the penis to retard outflow of blood. Continued penile stimulation results in secretion of mucus by the glands of Littré and bulbourethral glands. The mucus provides for neutralization of urine that may remain in the urethra and for a small degree of lubrication of the glans penis. Ejaculation is accomplished by rhythmic sympathetic impulses originating in the lumbar spinal cord and reaching the smooth muscle of the ducts of the male system.

SEMEN has a pH of about 7.2, a pH that reflects the reaction of prostatic fluid (pH 6.45) and that of seminal vesicle secretions (pH 7.3). Both secretions suspend the spermatozoa that normally number 250–300 million per ejaculate. The secretions provide nourishing substances for the sperm, activate the sperm to make them independently motile (prostatic secretion), and aid in neutralizing the normally acidic condition (pH

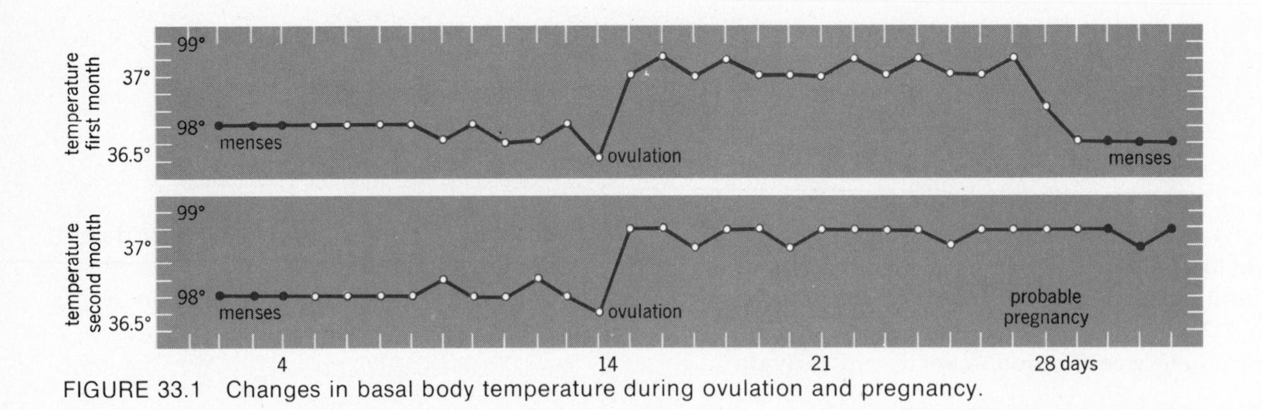

FIGURE 33.1 Changes in basal body temperature during ovulation and pregnancy.

about 3.8) in the upper vagina. Each ejaculate averages about 3.5 ml, 60 percent of which is contributed by the seminal vesicles and about 38 percent by the prostate. The remaining 2 percent consists of sperm.

Ejaculation is accompanied by widespread body sensations, body movement, and elevations of heart rate, arterial pressure, and respiration. These phenomena constitute the male orgasm, after which there is usually a return of the penis to the flaccid state. The role of the male in production of offspring is basically one of insemination or deposition of sperm in the vagina of the female. Passage of the sperm to the distal ends of the uterine tubes must take place before fertilization of the ovum can occur.

The role of the female

The mature ovum is released approximately halfway through a menstrual cycle; the event is called OVULATION and is correlated with a measurable rise in body temperature. If fertilization occurs, the body temperature tends to remain elevated (Fig. 33.1). The female reproductive organs were discussed in Chapter 30, and it is restated that the vagina normally serves as the organ for reception of the penis and sperm. (In artificial insemination, sperm are placed by a pipette into the uterus.)

There appears to be little difference from that of the male in female response to sexual stimulation. The female erogenous zones include the breasts, perineum, and clitoris. As a result of psychic and/or tactile stimulation, the clitoris becomes erect, and the spongy tissues around the vaginal orifice also become engorged with blood. This latter phenomenon is believed to result in a tightening effect that speeds male orgasm. The impulses that originate from stimulation of the female genitalia pass over the pudendal nerve to the sacral cord, and parasympathetic impulses pass to the clitoral blood vessels, dilating them and causing erection, and to the Bartholin's glands that secrete fluids around the vaginal orifice. These secretions supply most of the lubrication for insertion of the penis into the vagina. Thrusting of the male alternately tightens and releases tension on the labia minora that end anteriorly around the clitoris. This action is believed to provide a massaging effect on the clitoris that aids in attainment of female orgasm. Also, during intercourse, the uterus shifts backward slightly, possibly because of an increased blood volume, and the vagina enlarges in the upper end to form a "pool" for the collection of semen. After female orgasm—attended by feelings of release of tension and body movement, elevations of heart rate, blood pressure, and respiration—the uterine orifice may be seen to open, and there are rhythmic contractions of the uterine muscle that may aid in sperm transport through the organ. The role of the female thus becomes one of reception of sperm, the site of FERTILIZATION, and the PROVISION OF NUTRIENTS to assure development of the offspring to a state where it can cope with the external environment on its own.

Factors affecting embryonic and fetal development

The general events of fertilization and early development were described in Chapter 5. The development of the fetus depends on four factors or areas as determinants of adequate sources of nutrients and routes of elimination of metabolic wastes:

The state of maternal nutrition.

The structural and functional status of the placenta.

The genetic makeup of the offspring.

The presence of physical, chemical, or mechanical insults that may be delivered to the mother and/or fetus during the course of the pregnancy.

Maternal nutrition

In general terms, the embryo or fetus will sustain or maintain itself at the expense of the mother. Thus, the mother must possess adequate reserves of nutrients to ensure that embryonic or fetal demands are met, while sustaining herself at the same time. Chronic maternal malnutrition, particularly of protein, iodine, carbohydrate (calories), and other substances, must be reflected in quantitative and qualitative effects on the development of the offspring, both mentally and physically. Research in the field of maternal nutrition

FIGURE 33.2 The scheme of placental circulation.

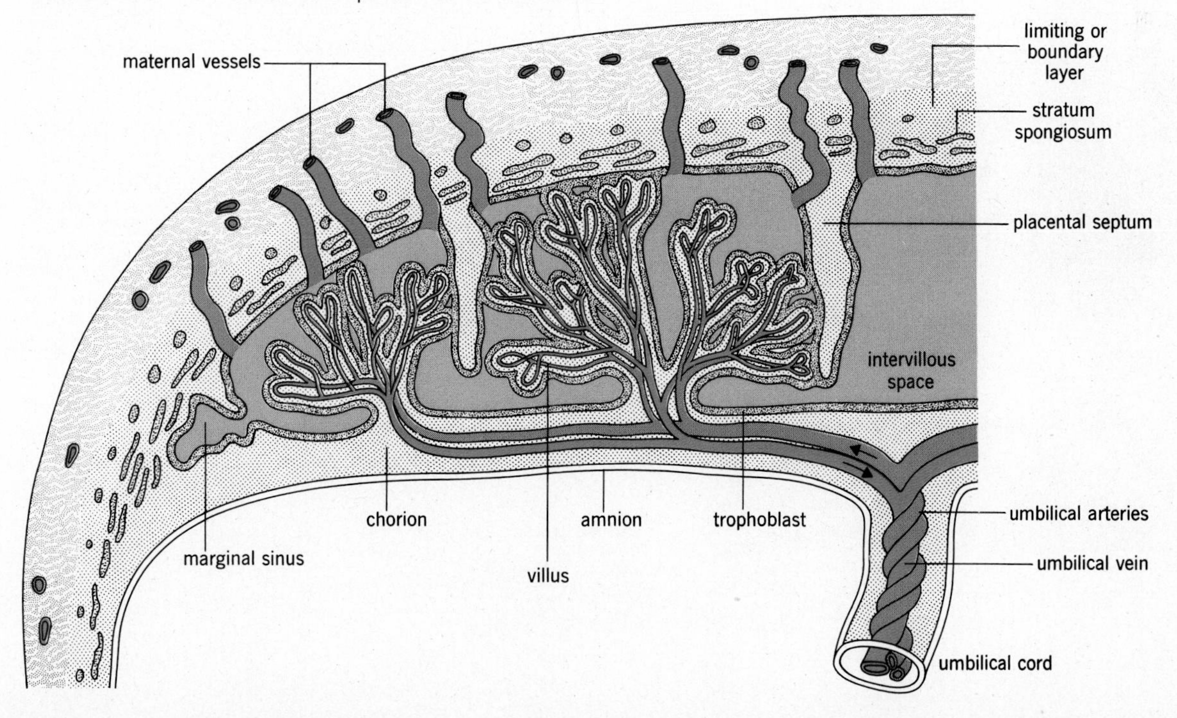

and its effects on offspring growth and development indicates that maternal deprivation must be rather severe (at least for the basic foodstuffs) before obvious effects are seen in the offspring; this area of research should be expanded to assure the prevention of fetal abnormalities related to nutritional status.

The status of the placenta

The placenta (Fig. 33.2) is composed of a fetal component, the chorion, and a maternal component consisting of the endometrial portion of the uterine wall. It contains a number of sections, each called a COTYLEDON; each cotyledon develops countless numbers of fingerlike extensions called the placental VILLI (Fig. 33.3). The villi are of two types: ANCHORING VILLI are immediately surrounded by endometrial tissue and aid in holding the placenta to the uterine wall; FLOATING VILLI lie in blood sinuses containing maternal blood and serve as the route of exchange of materials between offspring and mother.

In a fetus that is born below weight and with little adipose tissue, the placenta is often small and thin for its time of development, may be fibrotic in texture, contains a smaller-than-normal umbilical cord, and may have fewer than the normal 12–15 cotyledons. Obviously, large defects in the placenta influence the ability of the offspring to acquire those materials necessary for growth and to eliminate wastes of metabolism.

The genetic makeup of the offspring

The effects of chromosome additions or deletions on development were discussed in Chapter 31 (see Table 31.13). If chromosome number is normal, hereditary influences may operate to determine rates of growth and development, time of matur-

FIGURE 33.3 Several free villi of the placenta.

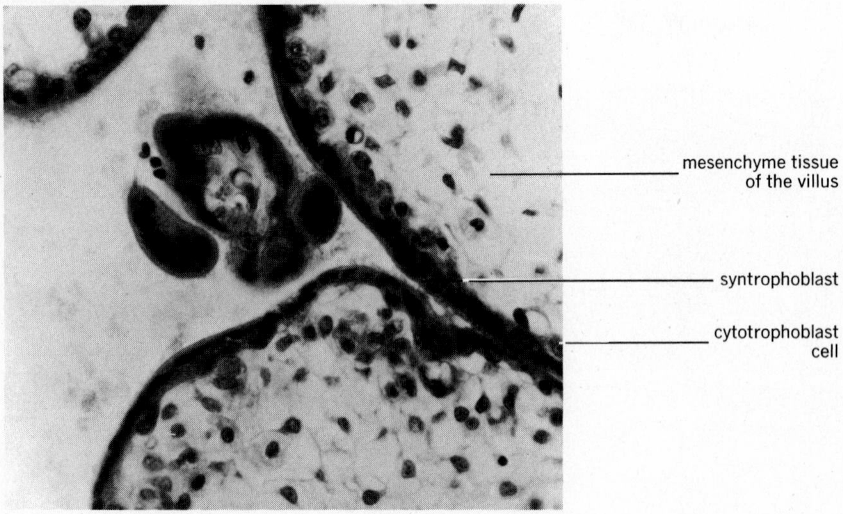

mesenchyme tissue of the villus

syntrophoblast

cytotrophoblast cell

ity, and other aspects of development. In short, some fetuses are "programmed" to develop at different rates than others, and, in most cases, all arrive at the proper end point at the proper time.

Insults and trauma

Trauma may be delivered to the offspring from the outside through the mother's body surfaces, or via the bloodstream. PHYSICAL AGENTS, such as radiation, may cause gene mutations or cell death; abnormal development may be the result. CHEMICAL AGENTS, including viruses and bacteria or their products, may interfere with crucial developmental processes (e.g., rubella contracted during the first trimester of pregnancy can cause heart, visual, and hearing defects). DRUGS such as LSD may cause chromosome breakage, and antiepileptic drugs may increase the incidence of cleft lip and palate. MECHANICAL COMPRESSION of the umbilical cord may lead to asphyxia and death or may reduce the amount of oxygen essential to normal development with the development of abnormalities. The elective surgical termination of a pregnancy in an abortion causes immediate cessation of development in an embryo, or may result in death of the fetus if it is not developed to a point where it can sustain life on its own (is viable).

Birth

At some point, usually about 280 days from the last menstrual period (LMP) or some 267 days after conception, a new individual is born. The series of events preceding birth constitutes LABOR, and the appearance of the baby itself is termed DELIVERY or BIRTH.

Causes of labor and birth

Ten to fourteen days before birth is imminent, the fetal head settles into the pelvis in the process called LIGHTENING. False labor pains (irregular contractions of the uterus) may occur 3–4 weeks before birth. REGULAR CONTRACTIONS of the uterus, discharge of the cervical mucus plug, and RUPTURE OF THE AMNION (bag of waters) with discharge of the amniotic fluid surrounding the fetus signal the approach of birth.

Among the factors suggested to be responsible for onset of labor are: stretching of the uterus with release of prostaglandins and subsequent stimulation of uterine contraction; increased sensitivity of the uterus to oxytocin; decrease of progesterone levels in the bloodstream. Although the reason is not definitely known, labor is usually initiated at a time when the fetus is sufficiently mature in body systems to sustain itself apart from the mother.

Duration of labor is usually longest in full-term first pregnancies (primiparous females)—about 14 hours—and decreases to 7–8 hours in second or more births (multiparous females).

Labor proceeds in three stages (Fig. 33.4). The first stage is EFFACEMENT and DILATION, in which the cervical canal is shortened and thinned (effacement) and the cervix opens to a maximum of about 10 cm (dilation). The second stage is BIRTH of the infant, due to uterine contractions 50–70 seconds in length and occurring every 2–3 minutes. Stage three involves separation and delivery of the placenta *(afterbirth)*. The uterus then normally contracts strongly, closing the uterine vessels and preventing hemorrhage.

Dangers of birth

Prolonged labor, or DYSTOCIA, may result in separation of the placenta before birth, with asphyxia of the fetus. CORD COMPRESSION has the same effect. ABNORMAL PRESENTATION (breech) of the

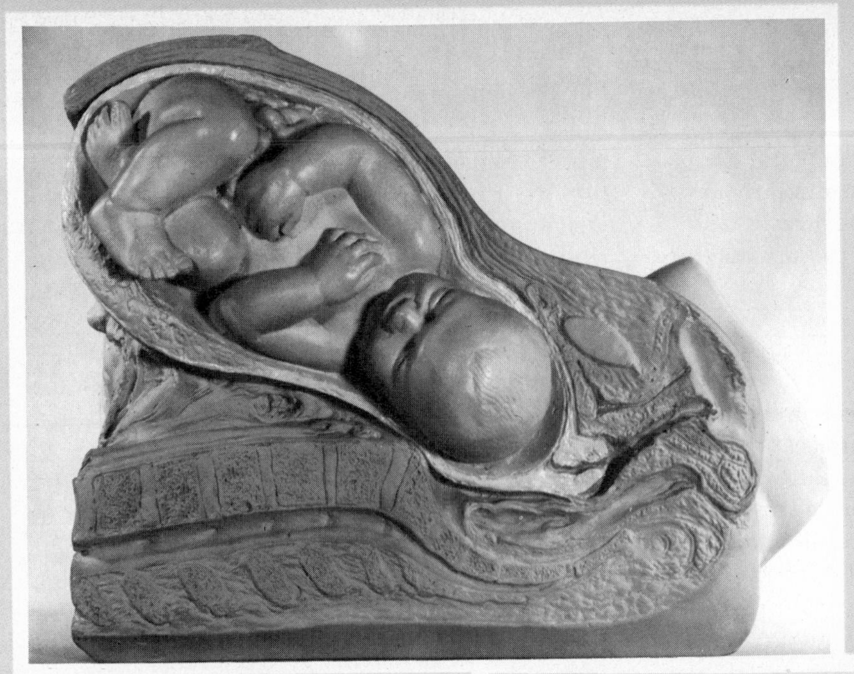

(b)

(a)

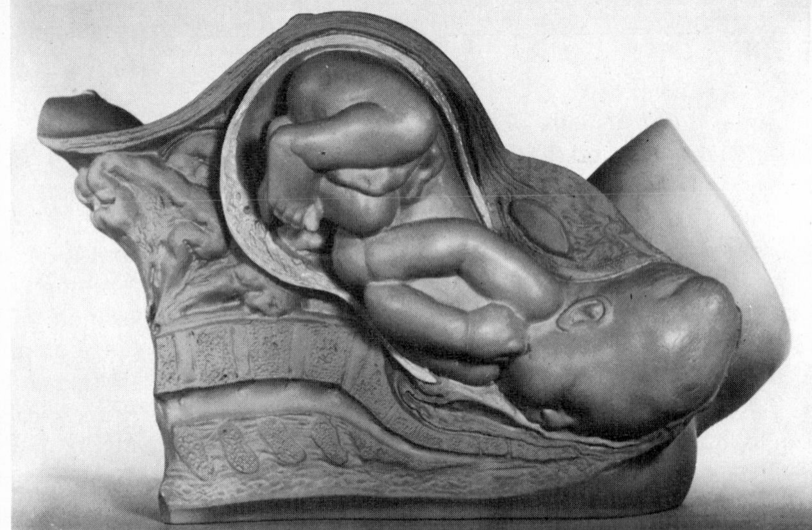

(c)

FIGURE 33.4 The three stages of labor. *(a)* First stage of labor. The rhythmic contractions of the uterus aid the progressive effacement and dilation of the cervix as well as the descent of the infant. *(b)* Second stage of labor. ''Caput,'' or top of infant's head, begins to appear through the vulvar opening. *(c)* Third stage of labor. Separation of placenta. (Dickinson-Belskie models, Cleveland Health Museum.)

fetus may result in difficulty of passage of a body part through the birth canal. The normal "order" of birth is head, shoulders, trunk, and lower limbs. Attempts to extract the fetus in a breech position may result in trauma, as by the use of forceps, or by excessive traction on a body part.

ASPHYXIA may be signaled by a high heart rate (above 160 beats/min) or by slowing of rate (below 120 beats/min). Increased movement followed by depression of movement or the presence of colon contents (meconium) in the amniotic fluid all signal distress, with sympathetic nervous system discharge caused by the asphyxia.

Assessment of status of the newborn

If birth proceeds normally, the newborn individual must be evaluated as to physiological status immediately after birth. The APGAR RATING SYSTEM looks at five basic physiological parameters and assigns points for each of three levels of function. The parameters and scoring are presented in Table 33.1. Good correlation between high Apgar scores and neonatal (newborn) survival have been demonstrated. While the rating system is very general, it indicates general vigor and status of the organism as determined at one and five minutes after birth, and is an assessment for earliest possible medical intervention.

The neonate is usually born slightly acidotic as a result of accumulation of CO_2 in the bloodstream. With the assumption of breathing, the condition usually corrects itself. The ability of the infant to adjust body temperature precisely is not mature at birth, and fluctuations of temperature are more extreme. Kidney function is immature, with low filtration rates and lack of concentrating ability present. The neonate thus loses relatively more water per unit of solute than an older child, and water and electrolyte balances must be monitored more closely. Fever indicates excessive metabolism as compared to heat loss or excessive loss of fluids. Immature immune response may be a source of inadequate reaction to bacterial or viral agents. Breast feeding is believed to provide the neonate with some maternal antibodies that may help protect the infant. The neonate therefore should be guarded against contact with infected persons.

The NEONATAL PERIOD ends after four weeks of life. The next period is INFANCY, extending from four weeks to two years of age. CHILDHOOD is the period between infancy and puberty. ADOLES-

TABLE 33.1 Apgar scoring of newborn infant

Sign	Score		
	0	1	2
A Appearance (color)	Blue or pale	Body pink; extremities blue	All pink
P Pulse (heart rate)	None	Below 100	Over 100
G Grimace (reflex irritability in response to sole of foot stimulation)	No response	Some movement; grimace	Active with crying
A Activity (muscle tone)	Flaccid	Some flexion of extremities	Active motion
R Respiration (respiratory effort)	None	Slow and irregular; no crying	Good strong cry

TABLE 33.2 Physiological changes during maturation

Parameter	Stage of development				
	Neonate	Infant	Child	Adolescent	Adult
Hemoglobin (grams %)	17–20	10.5–12.5	12–14	14–16	14–16
Heart rate, range (beats/min)	120–140	80–140	70–115	80–90	70–75
Blood pressure, average (systolic/diastolic, mm Hg)	75/40	80/50	85/60	112/70	120/80
Breathing rate (breaths/min)	30–50	20–30	16–20	14–16	14–16
Urine specific gravity	1.002–1.008	→	1.015–1.025	→	→
Daily urine volume (ml)	100–300	400–500	600–1000	800–1400	1000–1500
Motor development (landmarks)	Raise head	Sit; crawl; walk; run; talk	Hop; write; read	Fully developed gross and fine motor skills	→

CENCE extends from puberty to maturity. MATURITY extends from adolescence to SENESCENCE OR OLD AGE. Progressing through these states, the individual achieves physiological function equivalent to that of the adult, grows to adult stature, and becomes an independent instead of a dependent creature in the psychological and social realms. Table 33.2 shows the changes in physiological advancement.

Growth and development

Patterns of growth

GROWTH is commonly defined as an increase in size and/or weight that occurs when synthetic or anabolic processes occur at a faster rate than catabolic ones. The definition stresses the concept that new protoplasmic substance is added to preexisting material to achieve growth. Mere increase in weight may reflect only water accumulation or addition of adipose tissue, and should not be used as a criterion of growth. Growth may occur by increase in size of cells, by increase in number of cells that grow between divisions, or by a combination of both processes. The latter option is the one employed to achieve the great increase in body size that occurs from conception to maturity.

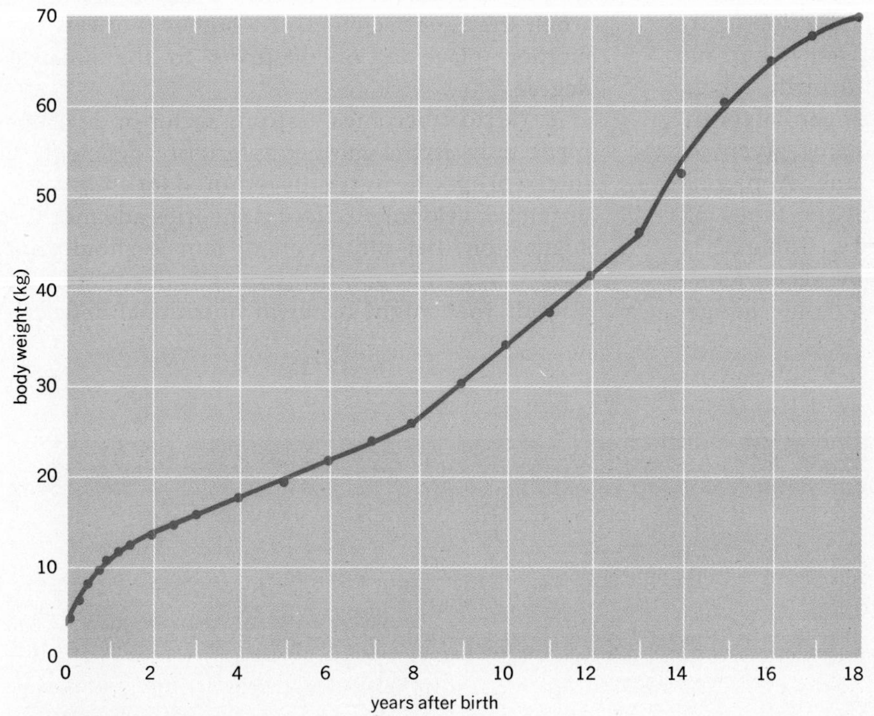

FIGURE 33.5 The human growth curve after birth. (From A. K. Laird, "Evolution of the Human Growth Curve," *Growth*, Vol. 31, 1967, p. 345.)

It must further be emphasized that not all body organs or systems grow at the same rate, particularly after birth; some parts are growing while others are regressing. The progression, either up or down, is not a linear process, but is marked by plateaus, peaks, and valleys. The curve most representative of overall body growth is sigmoid in shape (Fig. 33.5).

Regulation of growth; prenatal development

A wide variety of factors affect growth and development. HEREDITY appears to set a basic rate and direction of development that may be altered by environmental factors such as NUTRITION, ENDOCRINE STATUS, DRUG INTAKE, PARENTAL AGE, SMOKING, DISEASE, and whether an individual is a member of a multiple birth. Environmental factors have a greater or lesser effect on the developing organism, depending on its maturity, the process involved, and the severity of the effect. The "critical-period concept" suggests that those times when changes are occurring rapidly represent points at which environmental factors may cause the greatest effect. The first trimester (three months) of pregnancy constitutes such a time, and drug effects, viral infections, and other insults wreak the greatest havoc on developmental processes at that time. The concept of critical periods may aid medical personnel in deciding if intervention to correct or modify an abnormal pattern of growth is warranted.

GENETIC FACTORS appear to be most important in the prenatal (before birth) period as the influence that directs the differentiation and organization of the offspring. Genetics also sets the general tone in terms of rate and direction of development. If there is a genetic defect that is not lethal, its presence may not be detected until birth, when the child will be said to present a congenital (born with) defect. TERATOGENS is the name given to substances that cause malformations. A partial list of teratogens and examples of the types of defects they produce are presented in Table 33.3. Exactly what effect is produced depends on what system was undergoing the most rapid change when the agent acted, and whether it affects the mother, placenta, or offspring to the greatest degree.

NUTRITION becomes a strong factor in development only if the mother is grossly deficient in intake of specific nutrients required for embryonic or fetal development. Recent theories advance the suggestion that the pregnant female should not limit her weight gain to some specific number of pounds that might result in nutritional deficien-

TABLE 33.3 Some teratogens and their effects

Agent	Types of effects on embryo/fetus
Aminopterin (folic acid antagonist) used in: attempted abortion or treatment of leukemia	Widespread anatomical malformations
Antibiotics (tetracyclines, actinomycin, streptomycin, others)	Deafness, growth retardation, hemolysis
Cortisone	Cleft palate
Dicumarol	Fetal bleeding
Estrogens (DES, diethylstilbesterol)	Masculinization of female fetus; occasionally causes vaginal cancer later in life
Herpes virus	Nervous system defects
Hypoxia	Patent ductus arteriosus
Progestational agents (e.g., nortestosterone)	Masculinization of female fetus (labial fusion, clitoral enlargement)
Quinine	Congenital deafness
Radiation	Skeletal deformities, microcephaly
Rubella virus (german measles)	Blindness, deafness, congenital heart disease, inflammation of many body organs
Salicylates	Hemorrhage, anatomical malformations
Syphilis (untreated)	Blindness, deafness, congenital heart disease, others
Thiouracil derivatives (Iodine, Iodines)	Goiter
Tranquilizers (e.g., thalidomide)	Anatomic malformations, especially of the appendages

TABLE 33.4 Some nutrients essential for normal development of embryo or fetus, and effects of their deficiency

Substance	Deficiency effects
Vitamin D	Abnormal calcification of bones
Vitamin A	Visual disturbances
Vitamin E	Increased red cell fragility
Riboflavin	Defective H^+ transport
Fatty acids	Skin lesions, fatty liver
Iodine	Faulty nervous system development
Protein	General development retarded, small brain

TABLE 33.5 Some drugs influencing growth and development

Analgesics (pain relievers) and antipyretics (fever reducers) (e.g., salicylates)	Hemorrhage, fetal death due to large doses
Antibiotics	
Penicillin	Possible growth retardation
Streptomycin	Deafness
Anticancer agents	Anatomical changes
Anticoagulants (e.g., dicumarol)	Hemorrhage
Antihistamines (e.g., meclizine)	Anatomical changes
Antihypertensives (e.g., reserpine, thiazides)	Impaired neonatal adjustment
Barbiturates	CNS depression or excitation, neurological changes

cies for the fetus. Gain should not be so great to cause difficulty for the mother because of her size or the size of the child, and nutrition should be adequate at all times. Several important substances essential for embryonic and fetal development, and effects of their deficiency, are presented in Table 33.4.

ENDOCRINE STATUS may be reflected by changes that become evident at birth or later in life. ACTH given during the first trimester increases the incidence of cleft palate; adrenal sex hormone hypersecretion may cause adrenogenital syndrome; hyposecretion of thyroid hormone by the mother before the fetus produces its own (at about five months) retards brain growth.

DRUGS are usually thought of as materials that are abused. Any substance other than nutrients placed in the body for a therapeutic purpose may be considered a drug. Drugs act to produce a variety of effects on development, as shown in Table 33.5.

Any drug that must be taken in large doses during pregnancy should be checked for possible effects on the offspring.

Drug effects are generally more severe on the offspring *in utero* than on the mother because of the limited ability of the immature liver and kidneys to detoxify and excrete drugs.

AGE OF THE MOTHER exerts an influence on growth and development in that younger mothers tend to have smaller babies. A higher incidence of malformations is seen in very young and older mothers, and recent evidence suggests that mental development is slower in later children of larger families due to "dilution" of the learning environment by more persons.

Maternal SMOKING increases the incidence of premature births and results in the birth of smaller babies. The effect is believed to be due to the nicotine passed from mother to fetus.

As indicated earlier, INFECTIOUS AGENTS such as viruses may compromise placental function (e.g., mumps) to where life cannot be sustained, or affect the offspring directly (e.g., rubella). Congenital syphilis causes a variety of changes, many of which becomes evident only later in life.

Characteristics of the newborn *(neonate)*

At birth the full-term fetus averages 3405 g (7½ lb) in weight and about 50 cm (20 in.) in length. Five to ten percent of the body weight is lost in the first 24–48 hours after birth, representing a loss of water. Newborns usually appear chubby due to deposition of adipose tissue during the last months of pregnancy. Temperature regulation is imperfect and is handled by alterations in metabolic rate and not by sweating and shivering. The kidney lacks the ability to secrete a concentrated urine, and large volumes of dilute urine are produced; thus, dehydration poses a greater threat to the newborn than to an older child. The immune system is immature, and it makes sense to guard the neonate against contact with infected persons. Behavior at birth is almost entirely controlled by lower cerebral centers and the spinal cord. Many stimuli are received, but their localization is poor, and motor response is generally nondirected. Body functions are not circadian, and the newborn appears to be governed by sleep, hunger, and discomfort.

The neonatal period ends after four weeks of life, and infancy begins. Infancy extends until two years of age, followed by childhood extending from infancy to puberty.

Growth in infancy and childhood

Three primary measures are employed to assess the "normality" of growth in this period of life: gain in weight, gain in height, and increase in head circumference. It must be emphasized that all children are individuals and are different in their rates of growth; thus, there is a wide range of "normality," and *average* or *mean* is the proper term to employ instead of *normal.*

Weight gain averages 200 g (7 oz) per week for the first three months, 140 g (5 oz) per week for months 4–6, 85 g (3 oz) per week for months 7–9, and 70 g (2½ oz) per week for months 10–12. At one year of age, the infant averages 20 lb, and at years 3, 5, and 7, weight averages 30, 40, and 50 lb, respectively. Food intake thus slows after the first half-year of life and should be of no concern to parents if advancement is steady. If the early rate of gain were maintained, an individual would weigh some 322 lb at age 14. Height tends to increase about 3 in. between years 2–5 and 2½ in. between years 5–10. A puberal growth spurt adds 3½–4 in. to the height.

Head circumference indicates growth of the brain and may reveal rapid enlargement due to the presence of hydrocephalus. By the sixth month, the brain has reached 50 percent of the adult size, 60 percent by one year, and 75 percent by two years of age.

Figs. 33.6–33.9 present curves depicting changes in the three parameters discussed.

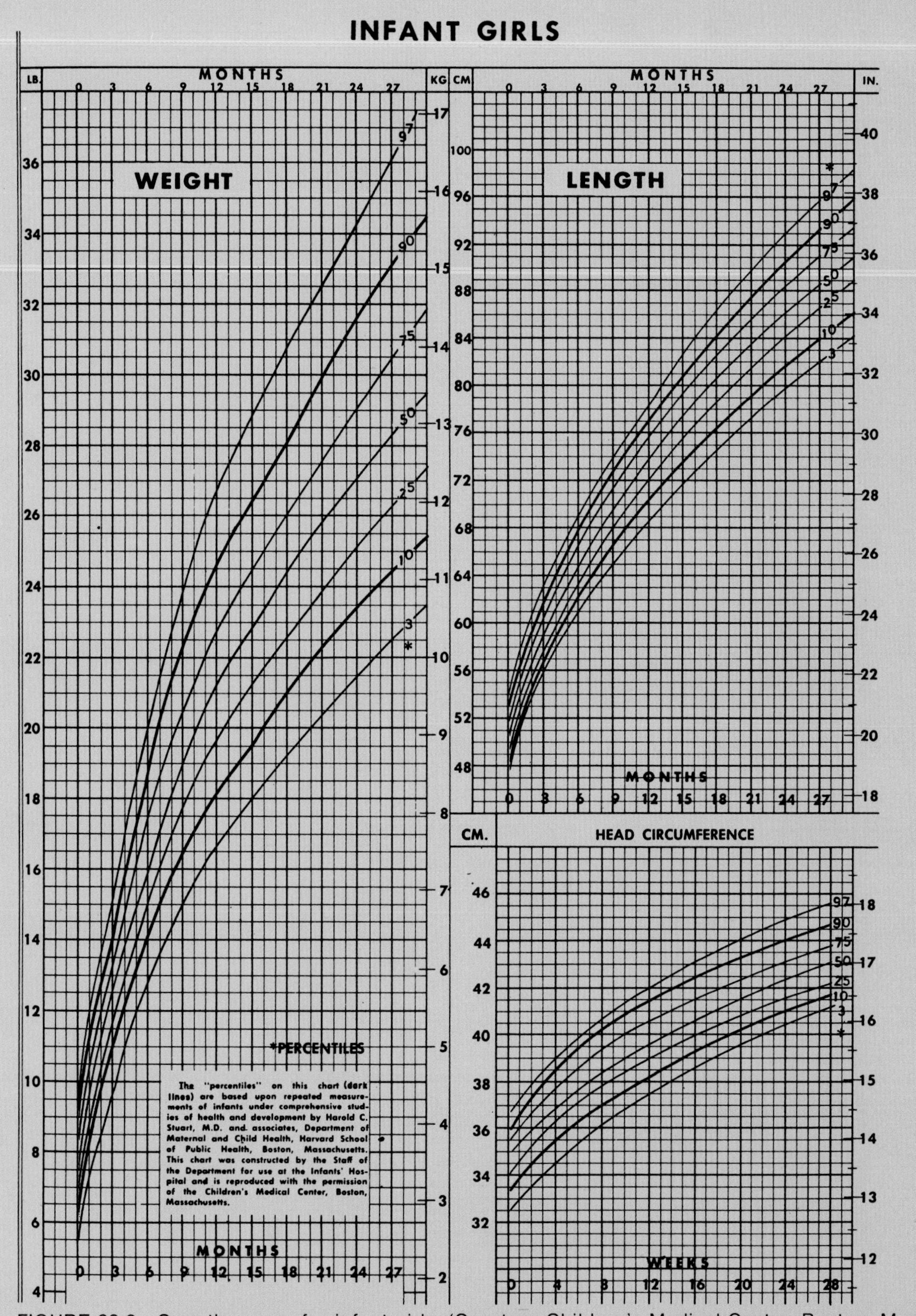

FIGURE 33.6 Growth curves for infant girls. (Courtesy Children's Medical Center, Boston, Mass.)

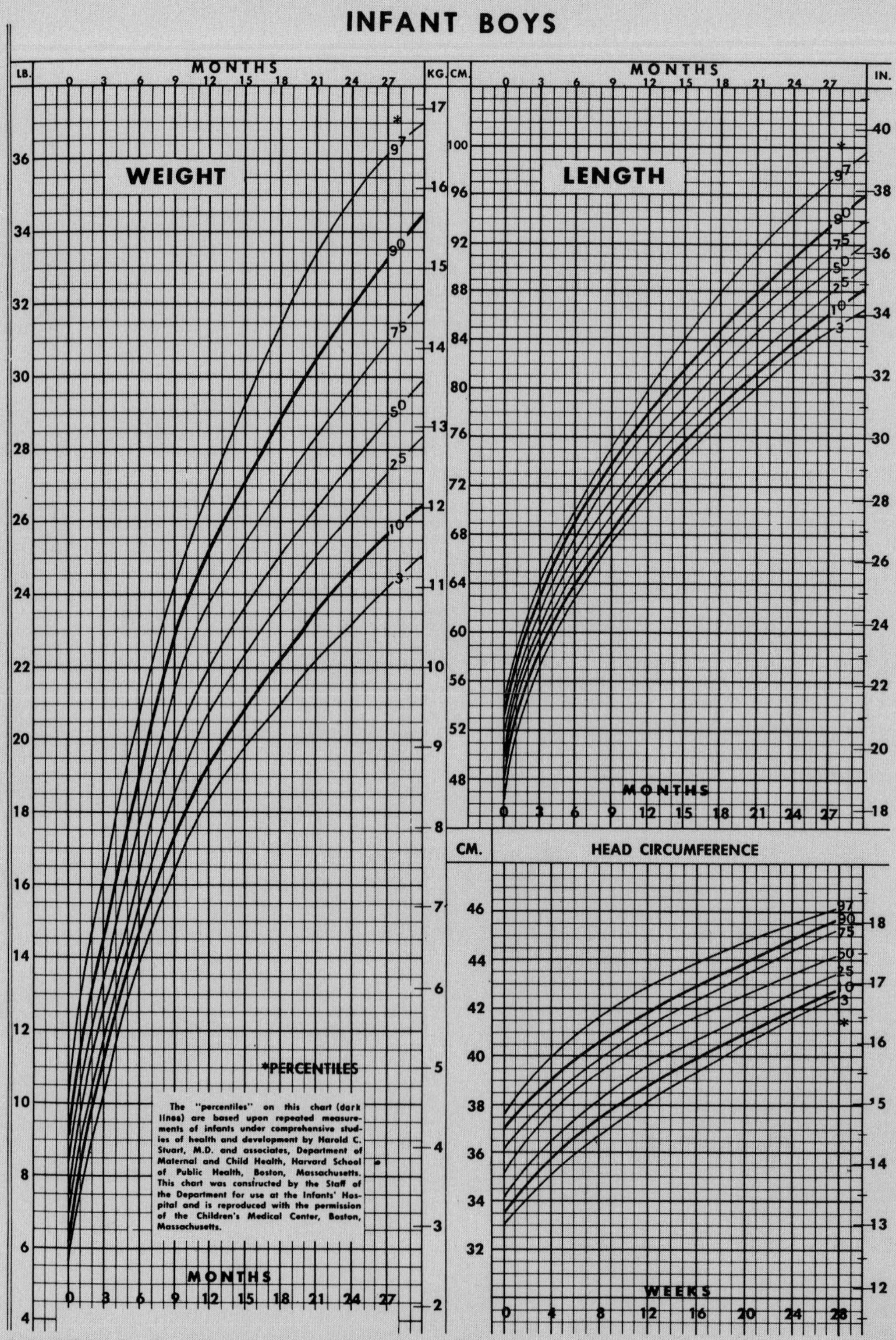

INFANT BOYS

WEIGHT

LB. MONTHS KG.

0 3 6 9 12 15 18 21 24 27

*PERCENTILES

The "percentiles" on this chart (dark lines) are based upon repeated measurements of infants under comprehensive studies of health and development by Harold C. Stuart, M.D. and associates, Department of Maternal and Child Health, Harvard School of Public Health, Boston, Massachusetts. This chart was constructed by the Staff of the Department for use at the Infants' Hospital and is reproduced with the permission of the Children's Medical Center, Boston, Massachusetts.

LENGTH

CM. MONTHS IN.

0 3 6 9 12 15 18 21 24 27

HEAD CIRCUMFERENCE

CM. WEEKS

0 4 8 12 16 20 24 28

FIGURE 33.7 Growth curves for infant boys. (Courtesy Children's Medical Center, Boston, Mass.)

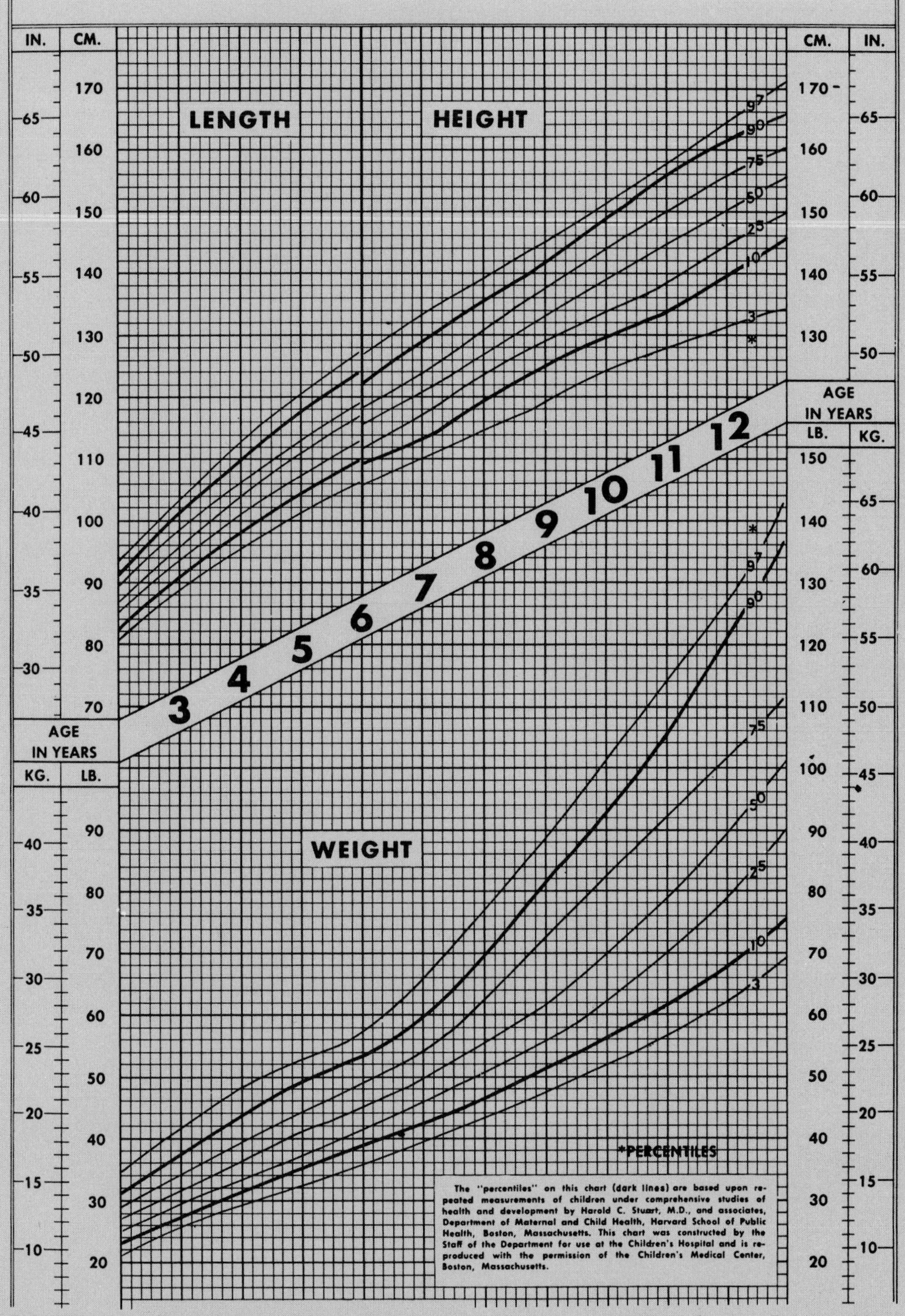

FIGURE 33.8 Growth curves for girls. (Courtesy Children's Medical Center, Boston, Mass.)

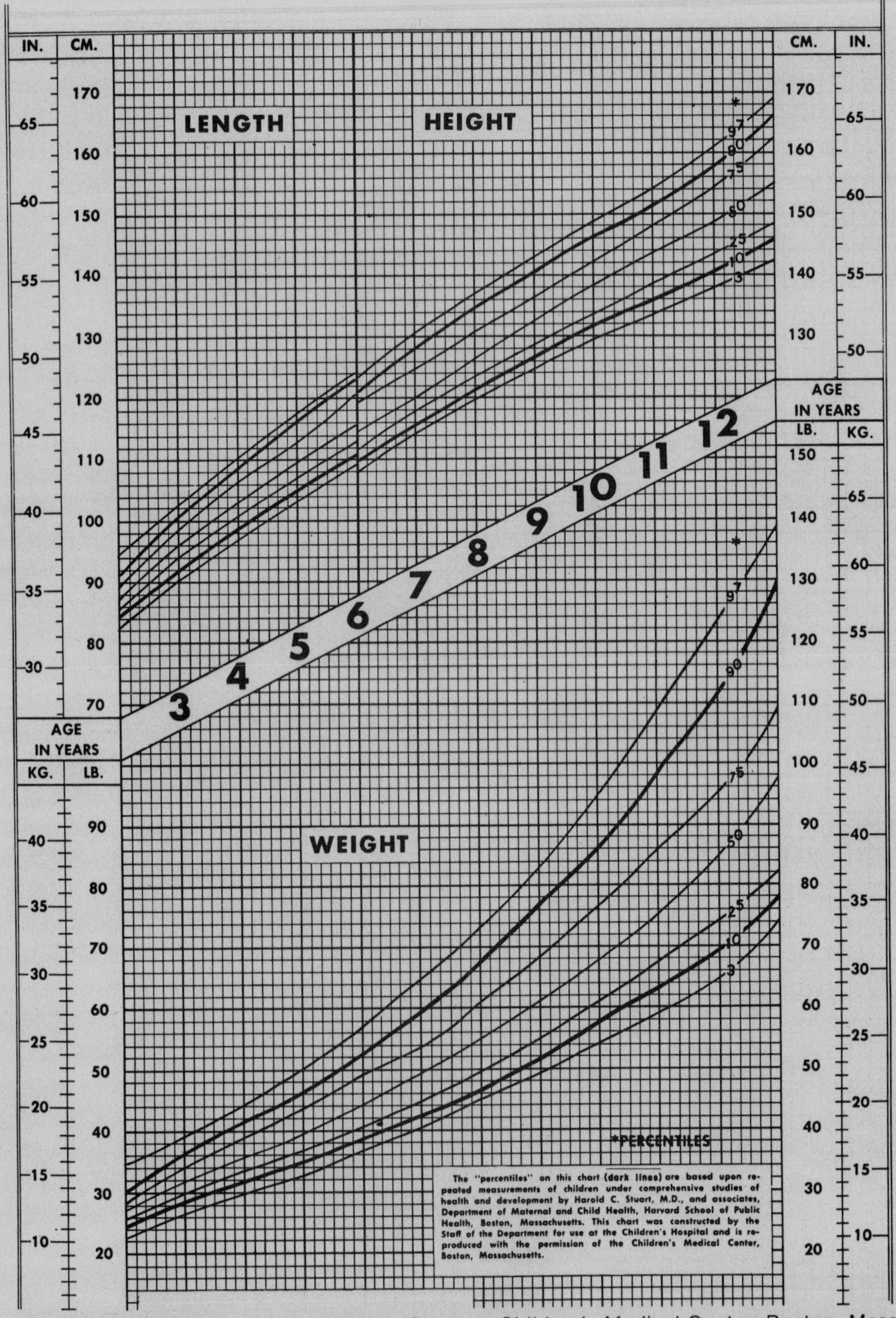

FIGURE 33.9 Growth curves for boys. (Courtesy Children's Medical Center, Boston, Mass.)

In any one individual, increase in these parameters is subject to change according to genetic, nutritional, and disease factors as in the prenatal period.

Individual organ or system growth diverges as the person grows older. General, neural, lymphoid, and genital types of development are shown in Fig. 33.10. Thus, enlargement of tonsils or lack of growth of genitals during certain time periods should not become causes for concern.

Mental and motor development also follows a more-or-less typical pattern, as shown in Fig. 33.11. This figure presents a means of screening a child as to its standing relative to the average; it should not be used as a rigid guide to development. Again, all children are different, they may be behind in one area and ahead in another, development is continuous from conception to maturity, and order of development tends to remain the same in most children, but rate varies.

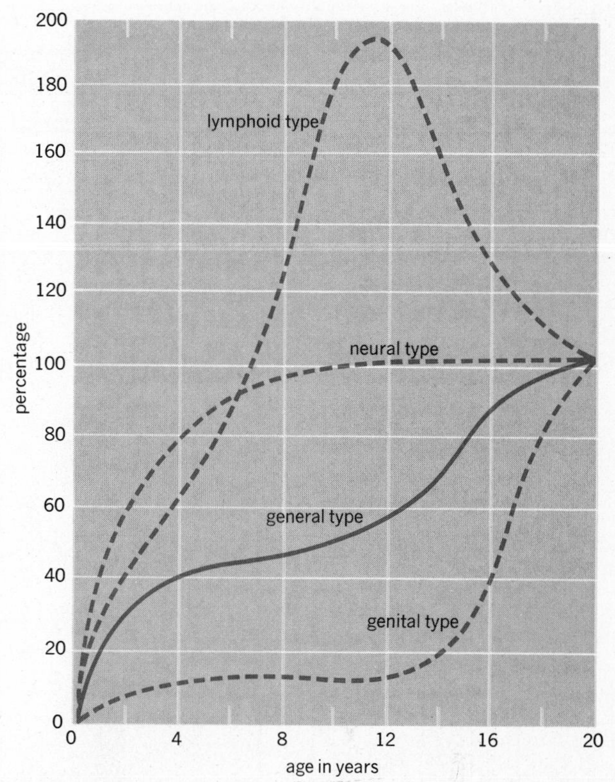

FIGURE 33.10 The major types of growth occurring in the body systems and organs. (From R. E. Scammon, "The Measurement of the Body in Childhood," *The Measurement of Man*, U. of Minn. Press. Used with permission.)

Adolescence

Adolescence is the period extending from puberty to maturity; it is primarily characterized by development of sexual organs, secondary sexual characteristics, and skeletal and muscular growth.

A growth spurt (Fig. 33.12) normally occurs between 10½–13 years of age in females and between 12½–15 years in males. Height growth usually stops at 17 years in females and 18 years in males. The peak of the growth spurt in females is usually followed, within two years, by menarche (menstruation). In females, the usual order of change is:

Rapid increase in height and weight.
Breast changes (pigmented areolae, enlargement of nipples, increase in mass of breast tissue).
Increase in pelvic girth.
Growth of pubic hair, becoming curly about a year after first appearance.
Function of the axillary apocrine sweat glands.
Growth of axillary hair.
Menstruation.
Attainment of mature stature.

In males, the usual order of changes is:

Increase in height and weight.
Enlargement of penis and testicles.
Development of pubic hair, becoming curly about one year after first appearance.
Growth of axillary hair.
Hair development on upper lip, groin, thighs, abdomen (on other parts of face about two years after pubic hair).
Voice changes due to larynx growth.
Nocturnal emissions.
Attainment of mature stature.

Several of these changes are summarized in Fig. 33.13.

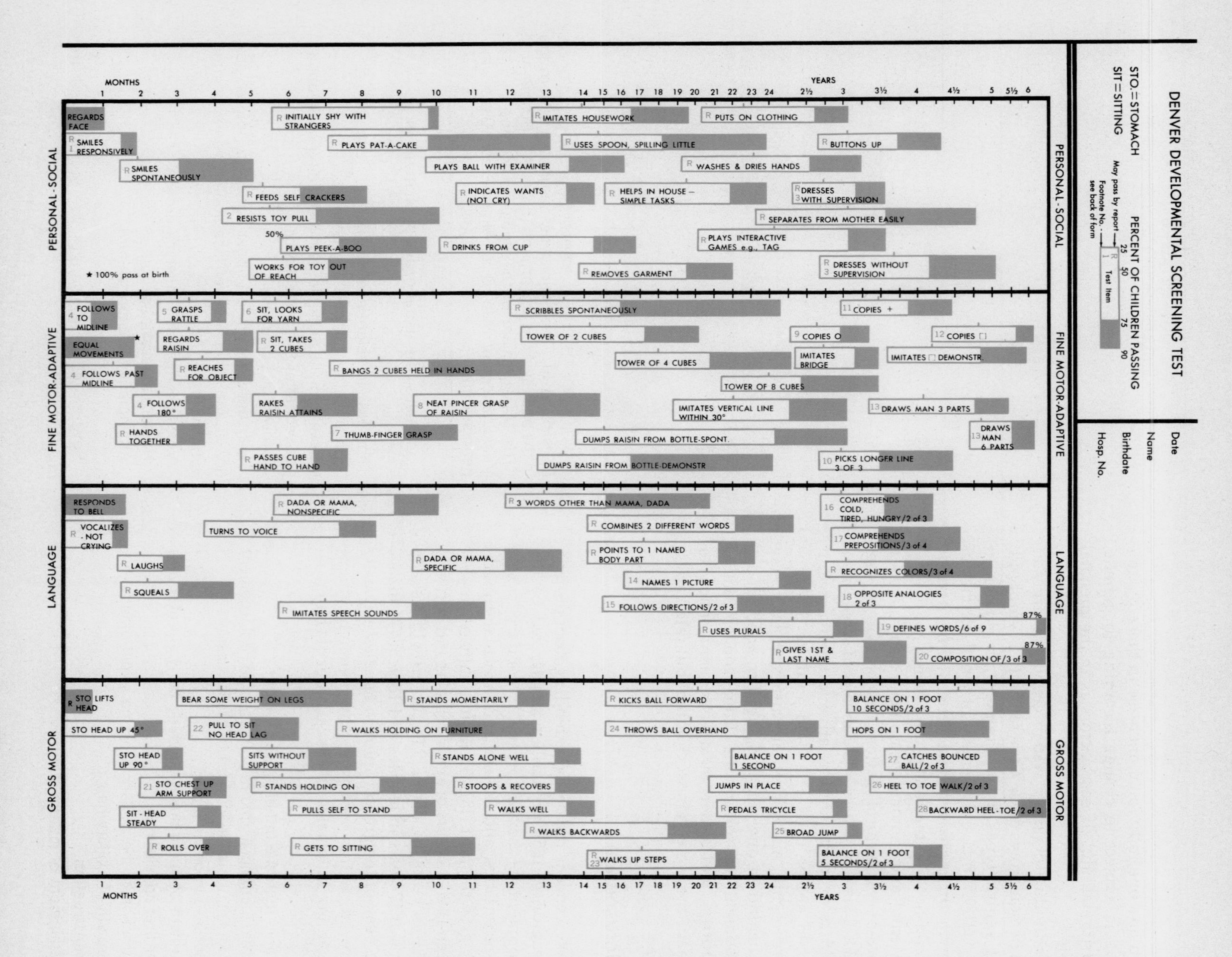

DENVER DEVELOPMENTAL SCREENING TEST
STO. = STOMACH
SIT = SITTING
PERCENT OF CHILDREN PASSING
25 50 75 90
May pass by report
Test Item
Footnote No. - see back of form
Date
Name
Birthdate
Hosp. No.
MONTHS
YEARS
1 2 3 4 5 6 7 8 9 10 11 12 13 14 15 16 17 18 19 20 21 22 23 24 2½ 3 3½ 4 4½ 5 5½ 6
PERSONAL - SOCIAL
REGARDS FACE
SMILES RESPONSIVELY
SMILES SPONTANEOUSLY
FEEDS SELF CRACKERS
RESISTS TOY PULL
50%
PLAYS PEEK-A-BOO
WORKS FOR TOY OUT OF REACH
★ 100% pass at birth
INITIALLY SHY WITH STRANGERS
PLAYS PAT-A-CAKE
PLAYS BALL WITH EXAMINER
INDICATES WANTS (NOT CRY)
DRINKS FROM CUP
IMITATES HOUSEWORK
USES SPOON, SPILLING LITTLE
HELPS IN HOUSE – SIMPLE TASKS
REMOVES GARMENT
PUTS ON CLOTHING
WASHES & DRIES HANDS
PLAYS INTERACTIVE GAMES e.g., TAG
BUTTONS UP
DRESSES WITH SUPERVISION
SEPARATES FROM MOTHER EASILY
DRESSES WITHOUT SUPERVISION
FINE MOTOR-ADAPTIVE
FOLLOWS TO MIDLINE
EQUAL MOVEMENTS
FOLLOWS PAST MIDLINE
FOLLOWS 180°
HANDS TOGETHER
GRASPS RATTLE
REGARDS RAISIN
REACHES FOR OBJECT
SIT, LOOKS FOR YARN
SIT, TAKES 2 CUBES
RAKES RAISIN ATTAINS
PASSES CUBE HAND TO HAND
BANGS 2 CUBES HELD IN HANDS
THUMB-FINGER GRASP
NEAT PINCER GRASP OF RAISIN
SCRIBBLES SPONTANEOUSLY
TOWER OF 2 CUBES
TOWER OF 4 CUBES
DUMPS RAISIN FROM BOTTLE-SPONT.
DUMPS RAISIN FROM BOTTLE-DEMONSTR
IMITATES VERTICAL LINE WITHIN 30°
TOWER OF 8 CUBES
COPIES O
IMITATES BRIDGE
PICKS LONGER LINE 3 OF 3
COPIES +
DRAWS MAN 3 PARTS
IMITATES □ DEMONSTR.
COPIES □
DRAWS MAN 6 PARTS
LANGUAGE
RESPONDS TO BELL
VOCALIZES - NOT CRYING
LAUGHS
SQUEALS
TURNS TO VOICE
DADA OR MAMA, NONSPECIFIC
IMITATES SPEECH SOUNDS
DADA OR MAMA, SPECIFIC
3 WORDS OTHER THAN MAMA, DADA
COMBINES 2 DIFFERENT WORDS
POINTS TO 1 NAMED BODY PART
NAMES 1 PICTURE
FOLLOWS DIRECTIONS/2 of 3
USES PLURALS
GIVES 1ST & LAST NAME
COMPREHENDS COLD, TIRED, HUNGRY/2 of 3
COMPREHENDS PREPOSITIONS/3 of 4
RECOGNIZES COLORS/3 of 4
OPPOSITE ANALOGIES 2 of 3
87%
DEFINES WORDS/6 of 9
87%
COMPOSITION OF/3 of 3
GROSS MOTOR
STO LIFTS HEAD
STO HEAD UP 45°
STO HEAD UP 90°
STO CHEST UP ARM SUPPORT
SIT - HEAD STEADY
ROLLS OVER
BEAR SOME WEIGHT ON LEGS
PULL TO SIT NO HEAD LAG
SITS WITHOUT SUPPORT
STANDS HOLDING ON
PULLS SELF TO STAND
GETS TO SITTING
STANDS MOMENTARILY
WALKS HOLDING ON FURNITURE
STANDS ALONE WELL
STOOPS & RECOVERS
WALKS WELL
WALKS BACKWARDS
WALKS UP STEPS
KICKS BALL FORWARD
THROWS BALL OVERHAND
BALANCE ON 1 FOOT 1 SECOND
JUMPS IN PLACE
PEDALS TRICYCLE
BROAD JUMP
BALANCE ON 1 FOOT 5 SECONDS/2 of 3
BALANCE ON 1 FOOT 10 SECONDS/2 of 3
HOPS ON 1 FOOT
CATCHES BOUNCED BALL/2 of 3
HEEL TO TOE WALK/2 of 3
BACKWARD HEEL-TOE/2 of 3

FIGURE 33.11 The Denver Developmental Screening Test for motor and mental development in infants and children. (Courtesy W. K. Frankenburg and J. B. Dodds, Univ. of Colorado Medical Center.)

DIRECTIONS

DATE

NAME

BIRTHDATE

HOSP. NO.

1. Try to get child to smile by smiling, talking or waving to him. Do not touch him.
2. When child is playing with toy, pull it away from him. Pass if he resists.
3. Child does not have to be able to tie shoes or button in the back.
4. Move yarn slowly in an arc from one side to the other, about 6" above child's face. Pass if eyes follow 90° to midline. (Past midline; 180°)
5. Pass if child grasps rattle when it is touched to the backs or tips of fingers.
6. Pass if child continues to look where yarn disappeared or tries to see where it went. Yarn should be dropped quickly from sight from tester's hand without arm movement.
7. Pass if child picks up raisin with any part of thumb and a finger.
8. Pass if child picks up raisin with the ends of thumb and index finger using an over hand approach.

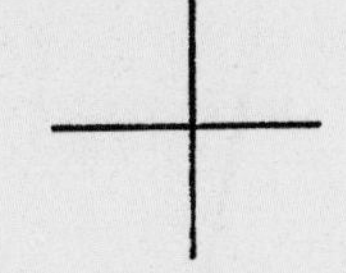

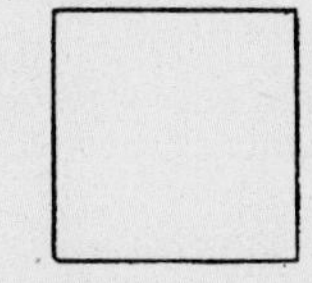

9. Pass any enclosed form. Fail continuous round motions.
10. Which line is longer? (Not bigger.) Turn paper upside down and repeat. (3/3 or 5/6)
11. Pass any crossing lines.
12. Have child copy first. If failed, demonstrate

When giving items 9, 11 and 12, do not name the forms. Do not demonstrate 9 and 11.

13. When scoring, each pair (2 arms, 2 legs, etc.) counts as one part.
14. Point to picture and have child name it. (No credit is given for sounds only.)

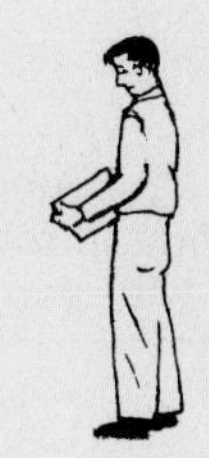

15. Tell child to: Give block to Mommie; put block on table; put block on floor. Pass 2 of 3. (Do not help child by pointing, moving head or eyes.)
16. Ask child: What do you do when you are cold? ..hungry? ..tired? Pass 2 of 3.
17. Tell child to: Put block <u>on</u> table; <u>under</u> table; <u>in front</u> of chair, <u>behind</u> chair. Pass 3 of 4. (Do not help child by pointing, moving head or eyes.)
18. Ask child: If fire is hot, ice is ?; Mother is a woman, Dad is a ?; a horse is big, a mouse is ?. Pass 2 of 3.
19. Ask child: What is a ball? ..lake? ..desk? ..house? ..banana? ..curtain? ..ceiling? ..hedge? ..pavement? Pass if defined in terms of use, shape, what it is made of or general category (such as banana is fruit, not just yellow). Pass 6 of 9.
20. Ask child: What is a spoon made of? ..a shoe made of? ..a door made of? (No other objects may be substituted.) Pass 3 of 3.
21. When placed on stomach, child lifts chest off table with support of forearms and/or hands.
22. When child is on back, grasp his hands and pull him to sitting. Pass if head does not hang back.
23. Child may use wall or rail only, not person. May not crawl.
24. Child must throw ball overhand 3 feet to within arm's reach of tester.
25. Child must perform standing broad jump over width of test sheet. (8-1/2 inches)
26. Tell child to walk forward, heel within 1 inch of toe. Tester may demonstrate. Child must walk 4 consecutive steps, 2 out of 3 trials.
27. Bounce ball to child who should stand 3 feet away from tester. Child must catch ball with hands, not arms, 2 out of 3 trials.
28. Tell child to walk backward, toe within 1 inch of heel. Tester may demonstrate. Child must walk 4 consecutive steps, 2 out of 3 trials.

<u>DATE AND BEHAVIORAL OBSERVATIONS</u> (how child feels at time of test, relation to tester, attention span, verbal behavior, self-confidence, etc,):

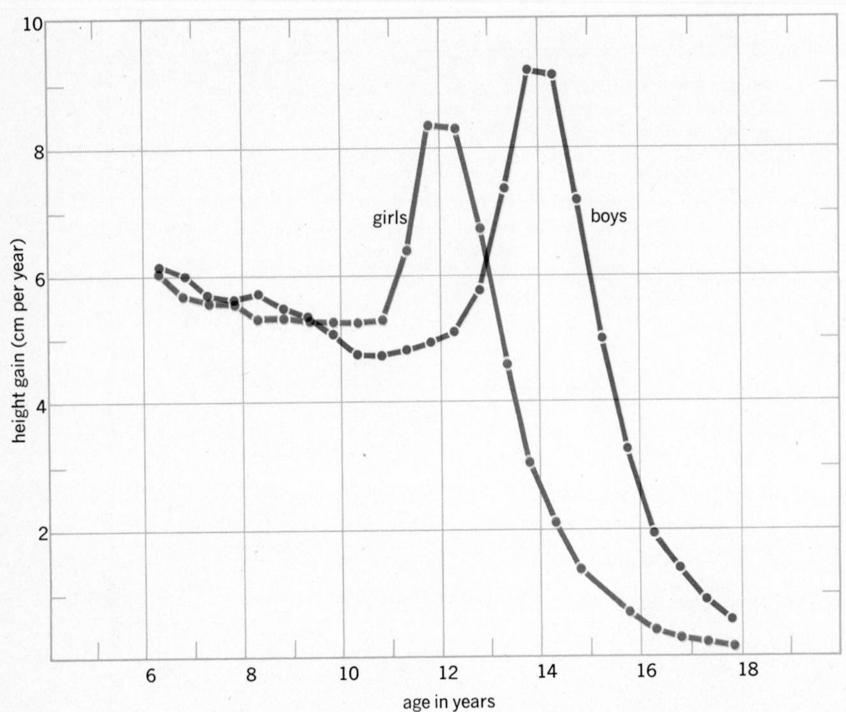

FIGURE 33.12 Curves depicting the adolescent growth spurt in boys and girls. Ages represent averages and do not indicate range. (From J. M. Tanner, *Growth and Adolescence*, Second Edition, Blackwell Scientific Pubs., Oxford, 1962. Used with permission.)

Other physiological changes occurring during adolescence and extending into maturity include:

Progressive decrease of metabolic rate.

Leveling off of cardiovascular function, with attainment of adult levels of heart rate and blood pressure.

Blood cell formation reaches adult levels.

Breathing rates reach adult levels with increase of vital capacity and higher alveolar CO_2 levels.

Myelination of nervous structures continues into the 30s and possibly beyond (Fig. 33.14).

When most of these structural and functional changes have achieved adult values, the individual is considered to be "mature"; this is usually about 25–30 years of age.

Maturity

Maturity is difficult to define, for it includes physical, mental, emotional, and chemical factors. Other than ability to reproduce, long considered a definition of (sexual) maturity, any definition should include mention of attainment of optimum integration between physiological processes and of a steady state in body functions.

Aging and senescence

Aging is a term that should be used to denote the changes, some beginning in infancy, that occur as one grows older. *Senescence* is usually understood

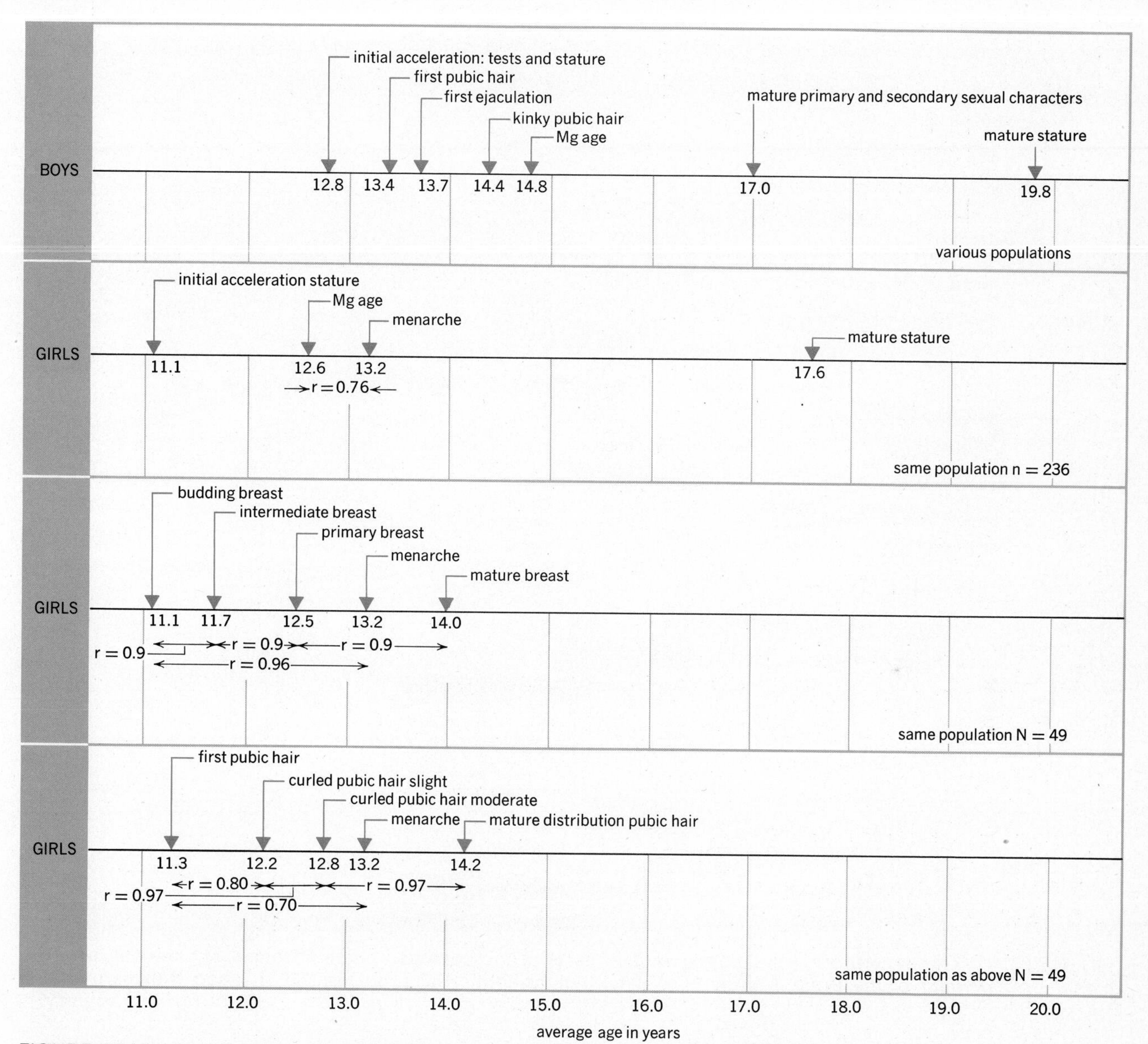

FIGURE 33.13 Average sequence of events occurring in development of secondary sex characteristics in boys and girls. (From E. Hurlock, *Developmental Psychology*, McGraw-Hill, New York, 1968.)

to mean the "state of old age encountered in the last years of life." There is no reason to associate either term with a particular age or mental state, as seems commonly to be done. There is as yet no scientific validation that death is a *natural* event that must be accepted with stoic resignation. It

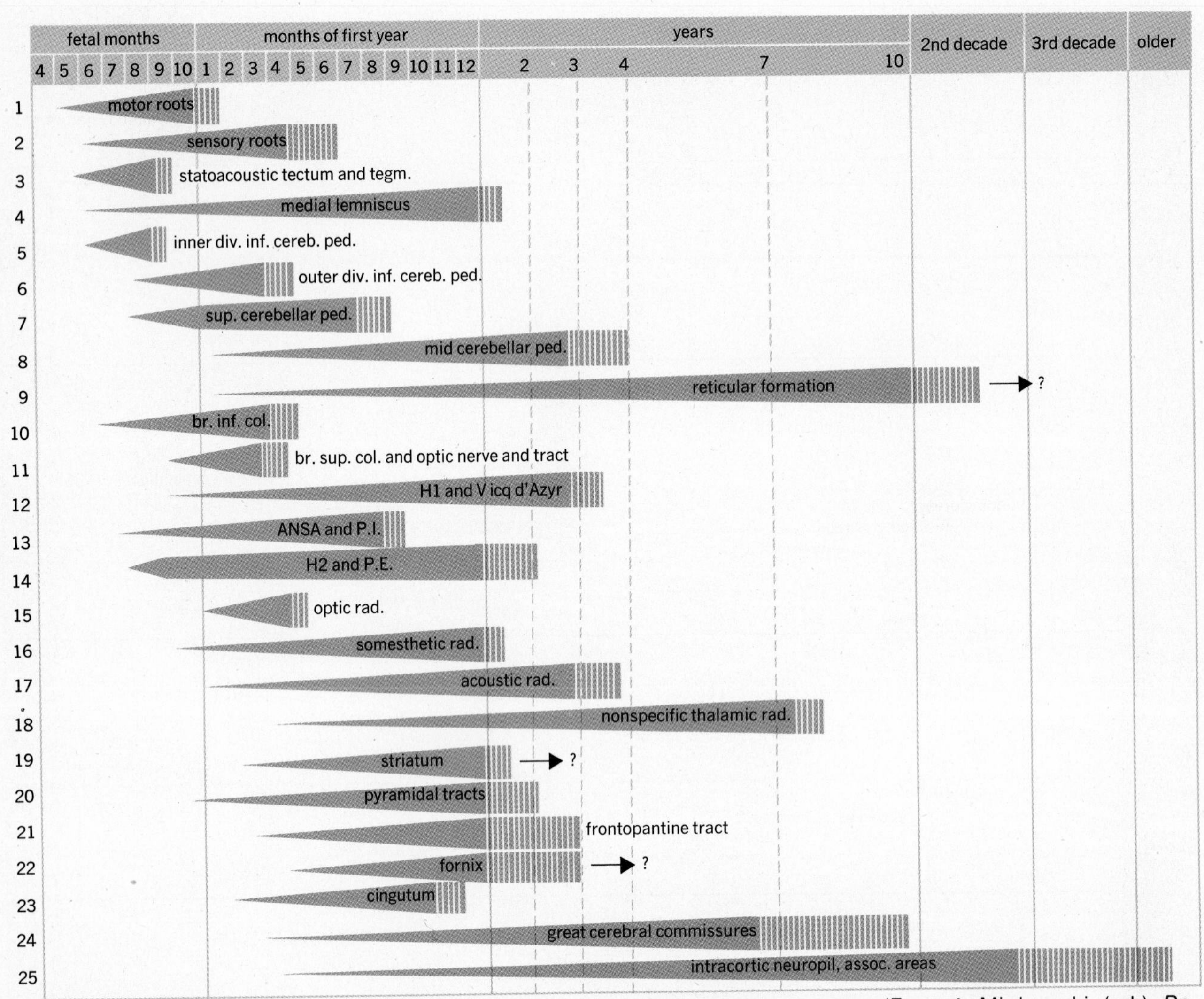

FIGURE 33.14 The pattern of myelination in various parts of the nervous system. (From A. Minkowski (ed.), *Regional Development of the Brain in Early Life,* Blackwell Scientific Pubs., Oxford, 1967. Used with permission.)

appears that death may result from a compromising of functions that *may be* largely preventable. Adding years to life must also be accompanied by attempts to add life to years—continuing the use of the talents of productive individuals.

Aging and death

That body structure and function change with age is an obvious fact. Some changes begin in infancy and may ultimately compromise integration and function to where life can no longer be sustained. Systems age, as they develop, at different rates; in the last analysis, it is perhaps an inability to adapt that kills us.

Theories of aging

There are many ideas as to what causes aging. EXTERNAL THEORIES postulate that outside factors such as radiation, pollutants, nutrition, and disease are responsible for aging. INTERNAL THEORIES suggest programmed genetic changes, failure of the immune system to combat chemicals within the body, or accumulation of toxic materials and "using up" of some "youth potion." In animals, certain procedures have been shown to lengthen life; whether or not they are applicable to humans remains debatable. Among these are:

Restriction of caloric intake lengthens life.

Cooling of poikilothermic (cold-blooded) animals lengthens life, perhaps by slowing metabolic rate.

Increased intake of vitamins C and E, by their ability to pick up free chemical radicals in the body, lengthens life slightly.

All theories of aging are highly speculative; there is little evidence to enable their application to the human organism.

Changes during aging

Physiological processes reach a peak during maturity (about 30 years of age) and thereafter undergo gradual decline. At some point, one or more functions decline below the point necessary to sustain life and death ensues. Heredity, environment, diet, and "speed of living" all influence the rate of decline and time of death.

Three general types of changes contribute to the aging process: SECULAR changes are the result of natural wear and tear; SENESCENT changes are the aging of tissues and organs, especially those with low mitotic rate; PATHOLOGIC complications are those resulting from disease processes developing in the aging organs.

SECULAR CHANGES. As one grows older, tissues appear to demand more metabolic support than the body can supply. In short, secular changes appear to be the result of an IMBALANCE BETWEEN VASCULAR SUPPLY AND TISSUE DEMAND. Thus, the amount of active tissue declines in proportion to its vascularity, and body weight tends to be reduced. Little attempt appears to be made by the body to restore the balance. The changes are usually seen first in endocrine-dependent structures such as breast, prostate, internal reproductive organs, and thyroid.

SENESCENT CHANGES. All body systems share in these changes. Some of the more noteworthy changes in systems are represented below.

Aging of the INTEGUMENT becomes apparent when the epidermis thins and becomes more translucent and dry, and the dermis becomes dehydrated and suffers loss of elastic fibers; the skin thus tends to "sag" on the body. There is usually loss or thinning of hair as a result of lowered sex-hormone levels and lowered synthesis of proteins. Sweat gland secretion diminishes, making the older individual more susceptible to the effects of high environmental temperature. Skin lesions (cancer, ruptured blood vessels, for example) are more common in the aged. Diabetic vasculopathy is also more common in the aged.

Aging of the EYE is reflected by a high incidence in the aged of cataracts, occlusion of retinal vessels with subsequent retinal changes, glaucoma, and changes in the transparency of the cornea. About one person in six in the over–65 age group exhibits some ocular pathology.

Aging of the EAR is evidenced by sclerotic alterations in the eardrum and ossicles that may result in loss of hearing. Vascular blockage may result in sudden hearing loss or loss of ability to maintain body equilibrium.

NEUROLOGICAL disorders are most commonly associated with changes in vascular supply, either gradually (Parkinson's disease) or suddenly ("stroke"). Peripheral loss of sensation acuity, neuritis, and neuralgia appear to be more common in the aged.

CIRCULATORY disorders may include hardening of arteries (arteriosclerosis) and deposition of lipids in the vessel walls (atherosclerosis). If deposition occurs in coronary vessels, the individual becomes a candidate for a "heart attack." Heart action lessens, with fall in cardiac output. Shortness of breath may develop as a result of low cardiac output and low pulmonary perfusion. Hypertension is common.

At the CELLULAR LEVEL, there is diminution of DNA and RNA synthesis, with failure to replace worn-out body proteins.

The ALIMENTARY TRACT undergoes changes in function to a greater degree than changes in structure. Over one-half (56 percent) of aged persons show functional disorders such as heartburn, belching, nausea, diarrhea, constipation, and flatus. Many of these complaints have an emotional basis, based on fear of death and disease and loss of contact with offspring. An organic basis may be demonstrated in some cerebral arteriosclerosis patients. Malignancy of the lower tract is more common in elderly persons (11 percent).

RENAL DISORDERS include lowered filtration rates, reduced reabsorption, and renal hypertension. Pyelonephritis (infection) is the most common renal disease. Calculi (stones) in the kidney and ureter are also more common in the aged.

GONADAL FUNCTION decreases. The ovary atrophies, producing menopause in females; testicular function declines more slowly. Internal organs atrophy and often prolapse ("drop down"), as with the uterus entering the upper vagina, for example.

Loss of the inorganic component of the SKELETON produces osteoporosis and greater liability to fracture. The MUSCLES become smaller and weaker and are prone to cramps and effects of altered electrolyte balance (hypokalemia, low blood potassium).

ARTICULATIONS are affected by arthritis. Osteoarthritis is a noninflammatory condition in which joints undergo degenerative changes in cartilage and synovial membranes. Rheumatoid arthritis is typically inflammatory in nature. Gouty arthritis is a metabolic disorder in which uric acid crystals accumulate in the joints.

ENDOCRINES may increase activity (as in acromegaly) or decrease activity (diabetes, hypothyroidism). Gonadotropins appear to show the widest secretion ranges.

The CENTRAL NERVOUS SYSTEM shows the greatest change in the brain where neuron loss (estimated at 30 cells/min from 30 years onward) may cause increased central reaction times, loss of memory, and personality alterations.

PATHOLOGIC CHANGES. The DISEASES OF OLD AGE are too numerous to consider in this text. Suffice it to say that the aged person is more susceptible to the diseases we all are heir to, and that the presence of a disease process tends to accelerate the aging changes described earlier.

Death

While old age itself does not cause death, organs and systems age at different rates. If the function of a vital organ drops below the critical level for maintenance of life, life ceases. The operation of bodily systems is interrelated so that if one vital organ fails, all fail. "The chain is only as strong as its weakest link." Loss of recovery ability is obvious, so that alterations that would be rapidly returned to normal by a younger person become life-threatening to the aged.

Definitions of death are extremely varied, and a strict definition is almost impossible. Among the definitions considered are the following.

Total, irreversible cessation of cerebral function (as evidenced by a "flat" electroencephalogram) for at least 24 hours; no spontaneous heartbeat or respiration; permanent cessation of all vital functions. At one time, stoppage of kidney function, cardiac standstill, or stoppage of respiration may have been considered signs of death. The use of an artificial kidney, respirators, and heart-lung machines has caused a reevaluation of the definition of death, as has the problem of securing organs from a donor for transplant. At the present time, the spontaneous brain-activity criterion appears to be the most valid criterion of death and one to which specific legal definition has been advanced in some states.

While old age and eventual death is in every person's future, efforts must be made by all concerned not only to add years to life, but also life to years. Age alone should not determine when one ceases or is forced to cease as a productive member of society. It is becoming increasingly clear that care of the body must begin with conception, and that care should cease only with the death of a given individual. Prevention rather than treatment of disease and disorders should become the final goal for each of us.

Summary

1. Production of ova begins at puberty and terminates at menopause; production of sperm begins at puberty and continues throughout most of life.
2. The male provides a means of insemination of the female; the female provides for reception of sperm, a site for fertilization, and provision for protection and nutrition of the developing individual.
3. Several factors determine the course of embryonic and fetal development.
   a. Maternal nutrition provides the substances essential for development.
   b. The status of the placenta determines exchange of materials between embryo and mother.
   c. Genetic makeup of the offspring sets rate of development.
   d. Trauma and other insults may cause deviant development.
4. Birth occurs after a variable period of labor.
   a. Labor occurs in three stages: effacement and dilation of the cervix, delivery of the infant, and delivery of the placenta.
   b. Causes of labor may involve hormones and prostaglandins.
   c. Hazards of birth include asphyxia and abnormal presentation of the fetus.
   d. Assessment of infant status at birth is made by the Apgar system.
5. Growth and development
   a. Occur by increase in size and number of cells.
   b. Do not occur in all organs or systems at the same rate.
   c. Are influenced, after birth, by heredity, nutrition, endocrine status, drugs, maternal age, smoking, and infectious agents.
6. A neonate is physiologically immature in terms of kidney function, temperature regulation, and immune response.
7. Assessment of growth in infancy and childhood is based on gain in height, weight, and head circumference.
8. Adolescence is characterized primarily by sexual development, menarche in the female, and maturation of organ function.
9. Maturity is the time when optimum integration of physiological function occurs.
10. Aging and senescence imply changes that occur from infancy onward to old age.
    a. Theories of causes of aging include action of external agents on the body and internal changes that affect vital reactions.
    b. During aging, there are changes due to wear and tear, aging of organs, and pathological processes.
11. Death occurs when integration of function is lost, and when a vital organ fails.

Appendix

Calculation of osmotic pressure

osmotic pressure in atmosphere $= P_{id} = 0.082055 \times T \times \dfrac{M}{V_m} \times m_2 \times v$

where

0.082055 = the gas constant R in liter atmosphere

T = absolute temperature in kelvin (K) = 273.15 + °C

$\dfrac{M}{V_m}$ = ratio of molar mass to molar volume for water (usually = 1)

m_2 = molality of the solute (number of moles of undissociated solute per 1000 g water)

v = number of particles into which the solute dissociates

Nernst equation

$$E = \frac{RT}{FZ} \log e \frac{[\text{ion}]_o}{[\text{ion}]_i}$$

where

E = potential in millivolts

R = the gas constant (8.317 joules/degree mole)

T = temperature in absolute degrees (273 + degrees centigrade)

F = the Faraday (96,520 coulomb/mole)

Z = the ion valence

$[\text{ion}]_o$ = ion concentration outside the cell

$[\text{ion}]_i$ = ion concentration inside the cell

Note: joule = volt coulomb; thus, answer will be in volts (joule/coulomb).

Terms and symptoms of drug abuse

This chart indicates the most common symptoms of drug abuse. However, all of the signs are not always evident, nor are they the only ones that may occur. The reaction produced by any drug will usually depend on the person, the mood, the environment, the dosage of the drug, and how the drug interacts with other drugs the abuser has taken, or contaminants within the drug. (Reprinted from "Drugs of Abuse," U.S. Government Printing Office, 1972. Courtesy of the Drug Enforcement Administration.)

	SLANG TERMS	DROWSINESS	EXCITATION & HYPERACTIVITY	IRRITABILITY & RESTLESSNESS	BELLIGERENCE	ANXIETY	EUPHORIA	DEPRESSION	HALLUCINATIONS	PANIC	IRRATIONAL BEHAVIOR	CONFUSION	TALKATIVENESS	RAMBLING SPEECH	SLURRED SPEECH	LAUGHTER
MORPHINE	M, dreamer, white stuff, hard stuff, morpho, unkie, Miss Emma, monkey, cube, morf, tab, emsel, hocus, morphie, melter	•		●		● •	•	●	•	●		●			•	
HEROIN	Snow, stuff, H, junk, big Harry, caballo, Doojee, boy, horse, white stuff, Harry, hairy, joy powder, salt, dope, Duige, hard stuff, schmeek, shit, skag, thing,	•		●		•	•	●		●		●			•	
CODEINE	Schoolboy	•		●		● •	•	●		●		●			•	
HYDROMORPHONE	Dilaudid, Lords	•		●		● •	•	●		●		●			•	
MEPERIDINE	Demerol, Isonipecaine, Dolantol, Pethidine	•		●		● •	•	●	•	●		●			•	
METHADONE	Dolophine, Dollies, dolls, amidone	•		●		● •	•	●		●		●			•	
EXEMPT PREPARATIONS	P.G., P.O., blue velvet (Paregoric with anthistamine), red water, bitter, licorice	•		●		●	•	●		●		●				
COCAINE	The leaf, snow, C, cecil, coke, dyna–mite, flake, speedball (when mixed with Heroin), girl, happy dust, joy powder, white girl, gold dust, Corine, Bernies, Burese, gin, Bernice, Star dust, Carrie, Cholly, heaven dust, paradise		•	•		•	•		•				•			
MARIHUANA	Smoke, straw, Texas tea, jive, pod, mutah, splim, Acapulco Gold, Bhang, boo, bush, butter flower, Ganja, weed, grass, pot, muggles, tea, has, hemp, griffo, Indian hay, loco week, hay, herb, J, mu, giggles–smoke, love weed, Mary Warner, Mohasky, Mary Jane, joint sticks, reefers, sativa, roach,	•	•	•		•	•	•	•	•			•			•
AMPHETAMINES	Pep pills, bennies, wake–ups, eye–openers, lid poppers, co–pilots, truck drivers, peaches, roses, hearts, cart–wheels, whites, coast to coast, LA turnabouts, browns, footballs, greenies, bombido, oranges, dexies, jolly–beans, A's, jellie babies, sweets, beans, uppers		•	•		•	•		•	•			•			
METHAMPHETAMINE	Speed, meth, splash, crystal, bombita, Methedrine, Doe		•	•		•	•		•	•			•			
OTHER STIMULANTS	Pep pills, uppers		•	•			•		•				•			
BARBITURATES	Yellows, yellow jackets, nimby, nimbles, reds, pinks, red birds, red devils, seggy, seccy, pink ladies, blues, blue birds, blue devils, blue heavens, red and blues, double trouble, tooies, Christmas trees, phennies, barbs	•		● •	•	●	•	•	●		•	•			•	•
OTHER DEPRESSANTS	Candy, goofballs, sleeping pills, peanuts	•		● •	•	●	•	•	●		•	•			•	
LYSERGIC ACID DIETHYLAMIDE (LSD)	Acid, cubes, pearly gates, heavenly blue, royal blue, wedding bells, sugar, Big D, Blue Acid, the Chief, the Hawk, instant Zen, 25, Zen, sugar lump		•			•	•	•	•	•	•			•		
STP	Serenity, tranquility, peace, DOM, syndicate acid						•	•	•		•			•		
PHENCYCLIDINE (PCP)	PCP, peace pill, synthetic marihuana	•				•				•		•				•
PEYOTE	Mescal button, mescal beans, hikori, hikuli, huatari, seni, wokowi, cactus, the button, tops, a moon, half moon, P, the bad seed, Big Chief, Mesc.		•	•		•			•					•		
PSILOCYBIN	Sacred mushrooms, mushrooms		•	•		•	•	•	•		•			•		
DIETHYLTRYPTAMINE (DMT)	DMT, 45–minute psychosis, business–man's special		•			•	•	•	•		•			•		

• SYMPTOMS OF ABUSE

TREMOR	STAGGERING	IMPAIRMENT OF COORDINATION	DIZZINESS	HYPERACTIVE REFLEXES	DEPRESSED REFLEXES	INCREASED SWEATING	CONSTRICTED PUPILS	DILATED PUPILS	UNUSUALLY BRIGHT SHINY EYES	INFLAMED EYES	RUNNY EYES AND NOSE	LOSS OF APPETITE	INCREASED APPETITE	INSOMNIA	DISTORTION OF SPACE OR TIME	NAUSEA AND VOMITING	ABDOMINAL CRAMPS	DIARRHEA	CONSTIPATION	PHYSICAL DEPENDENCE	PSYCHOLOGICAL DEPENDENCE	TOLERANCE	CONVULSIONS	UNCONSCIOUSNESS	HEPATITIS	PSYCHOSIS	DEATH FROM WITHDRAWAL	DEATH FROM OVERDOSE	POSSIBLE CHROMOSOME DAMAGE	ORALLY	INJECTION	SNIFFED	SMOKED
●		●			●	●	●	●			●	●		●		●	●	●	●	●	●	●	●	●	●			●			●		
●		●			●	●	●	●			●	●		●		●	●	●	●	●	●	●	●	●	●			●			●	●	
●		●			●	●	●	●			●	●		●		●	●	●	●	●	●	●			●			●		●	●		
●		●			●	●	●	●			●	●		●		●	●	●	●	●	●	●		●	●			●		●	●		
●		●			●	●	●	●			●	●		●		●	●	●	●	●	●	●		●	●			●		●	●		
●		●			●	●	●	●			●	●		●		●	●	●	●	●	●	●		●	●			●		●	●		
●		●			●	●	●	●			●	●		●		●	●	●	●	●	●	●			●					●	●		
●				●				●				●		●							●		●		●			●			●	●	
		●								●			●		●						●									●			●
●			●	●		●		●	●			●		●							●	●			●	●		●		●	●		
●			●	●		●		●	●			●									●	●			●	●		●		●	●		
●				●				●													●	●			●			●		●	●		
●	●	●			●	●	●							●		●	●			●	●	●	●	●	●		●	●		●	●		
●	●	●			●									●		●	●			●	●	●	●		●			●		●	●		
●						●		●							●						●	●				●			●	●			
●								●							●						●				●					●	●		
		●	●																		●			●	●					●	●		
								●								●					●	●			●					●	●		
						●		●							●						●	●								●			
						●		●							●						●	●			●						●		●

● SYMPTOMS OF WITHDRAWAL ● DANGER OF ABUSE ● HOW TAKEN

Some normal physiological values (adult)

Blood

Plasma proteins	total 7–9% (6.5 g %)
Albumins	3.75 g % } A/G ratio 1.6 : 1
Globulins	2.3 g % } A/G ratio 1.6 : 1
Ig G	800–1500 mg %
Ig A	140–260 mg %
Ig M	70–130 mg %
Ig D	300 mg %
Ig E	100 mg %
Hemoglobin	
Female	12.5–15 g %
Male	13.0–16 g %
Red cells	
Female	4.5–5 million per mm^3
Male	5.0–6 million per mm^3
Leucocytes	
Total	5000–9000 per mm^3
Neutrophils (segs)	60–70%
Lymphocytes	20–25%
Monocytes	3– 8%
Eosinophils	2– 4%
Basophils	0.1% or less
Platelets	250,000–300,000 per mm^3
Clotting time	
Capillary tube method	2–4 min
Bleeding time	1–3 min
Prothrombin time	12–15 sec
Chemicals	
Serum Na	136–142 meq/L
Serum Cl	100–110 meq/L
Serum K	4–5.4 meq/L
Serum PO_4	1.7–2.6 meq/L
Serum Ca	4.8–5.2 meq/L
Glucose	65–220 mg %
Serum cholesterol	150–250 mg %
Urea	8–25 mg %
PBI	5–8 μg %

17 Ketosteroids	
Female	9–22 mg per day
Male	6–15 mg per day
Hematocrit (packed cell volume—PCV)	
Female	40–45
Male	45–50

Kidney and urine

Glomerular filtration rate	120 ml per min
Renal blood flow	1200 ml per min
Renal plasma flow	650 ml per min
Filtration fraction	18.5%
Urine volume	1500 ml per day
Specific gravity of urine	1.015–1.025
pH of urine	5–7

Chemicals in urine (g/L)	
Urea	23
Hippuric acid	0.6
Uric acid	0.6
Creatine	1.5
NaCl	9
KCl	2.5
H_2SO_4	1.8
H_3PO_4	1.8
NH_3	0.6

Albumin, sugar, ketones, bilirubin, red cells, pus—negative

Gases

Arterial PO_2	75–100 mm Hg
Arterial PCO_2	40 mm Hg
Venous PO_2	60– 65 mm Hg
Venous PCO_2	46 mm Hg

Cerebrospinal fluid

Pressure	70–180 mm H_2O
Specific gravity	1.006–1.008
Protein	15–45 mg %
Glucose	50–75 mg %
Leucocytes	0–6 cells/mm^3
Red cells	none

Respiratory

Breathing rate	16–18 per min
Tidal volume	500 ml
Inspiratory reserve volume	3000 ml
Expiratory reserve volume	1100 ml
Residual volume	1200 ml
Minimal volume	600 ml
Alveolar surface area	50–70 m^2

Heart and vessels

Heart weight	300 g
Resting rate	70 beats per min
Resting stroke volume	60–65 ml
Cardiac cycle	0.8 sec
Pulmonary systolic pressure	30 mm Hg
Aortic systolic pressure	130 mm Hg
Pulmonary diastolic pressure	18 mm Hg
Aortic diastolic pressure	80 mm Hg
O_2 consumption at rest	9 ml per 100 g per min
Systemic venous pressure	14 mm Hg or less
Systemic capillary pressure	20–25 mm Hg

Digestive

Saliva	
Volume	1500 ml per day
pH	5–7
Gastric juice	
Volume	2000 ml per day
pH	0.9–1.5
Intestinal juice	
Volume	3000 ml per day
pH	6.5–7.6
Pancreatic juice	
Volume	2000 ml per day
pH	7.5–8.8
Bile	
Volume	1000 ml per day
pH	6.2–8.5

Miscellaneous

Water, 65% of body weight	
Extracellular fluid	25% of body weight
Plasma	5% of body weight
Intracellular fluid	40% of body weight
Average body water pH	7.4 ± 0.02
Adult skin surface area	1.75 m^2
Chromosome number	46
Gestation period (human)	280 days

Glossary

A glossary, while it is a great help to the reader, is not a substitute for a good dictionary. Only two diacritics (marks over vowels to indicate pronunciation) are used here, the *macron* (¯), showing the long sound of the vowels, and the *breve* (˘), showing the short sounds, as:

macron	**breve**
a in rāte	**a** in căt
e in ēqual	**e** in ĕver
i in īsle	**i** in ĭt
o in ōver	**o** in nŏt
oo in mo͞on	**oy** in Floyd
u in ūnit	**u** in cŭt

The following abbreviations are used:

abbr.	abbreviation
appl.	applied to
A.S.	Anglo-Saxon
Colloq.	colloquial
dim.	diminutive
e.g.	for example
esp.	especially
Fr.	French
G.	Greek
Ger.	German
i.e.	that is
L.	Latin
M.E.	Middle English
opp.	as opposed to; opposite
pert.	pertaining to
plu.	plural
sing.	singular
syn.	synonym

a-, an- [G. alpha]. Prefix: without, away from, not.

ab- [L.]. Prefix: from, away from, negative, absent.

abducens (ăbdo͞o'sĕnz) [L. drawing away from]. The sixth cranial nerve; innervates the lateral rectus muscle of the eyeball.

abduction (ăbdŭc'shŭn) [L. *abducere,* to lead away]. Movement away from the median plane of the body or part.

absorption (ăbsŏrp'shŭn) [L. *absorbere,* to suck in]. Passage of material into or through living cells.

acetabulum (ăsētăb'ūlŭm) [L. *acetabulum,* vinegar cup]. The cotyloid cavity or socket in the pelvic girdle for the head of the femur.

acetylcholine (ă'sĕtĭlkō'lēn). An ester of choline occurring in various organs and tissues of the body. It is thought to play an important role in the transmission of nerve impulses at synapses and myoneural junctions.

Achilles tendon (ăkĭl'ēz tĕn'dŏn). Achilles, hero of the *Iliad,* whose vulnerable spot was his heel because it was the place where his mother held him when she immersed him in the river Styx. The tendon of the gastrocnemius and soleus muscles of the leg.

achondroplasia (ăkŏn'drōplā'sēă) [G. *a-,* not; *chondros,* cartilage; *plasis,* a molding]. Defect in the formation of cartilage at the epiphyses of long bones, producing a form of dwarfism; sometimes seen in rickets. Due to mutation or chromosome abnormalities.

acidosis (ăs'ĭdō'sĭs) [L. *acidum,* acid; G. *ōsis,* condition]. A disturbance in the acid-base balance of the body in which there is an accumulation of acids (as in diabetic acidosis or renal disease) or an excessive loss of bicarbonate (as in renal disease).

acinus (ăs'ĭnŭs) (plu. *acini*) [L. grape]. A saclike group of secretory cells surrounding a cavity.

acne (ăk'nē) [G. *acme,* point]. Chronic inflammatory disease of the sebaceous glands and hair follicles of the skin.

acoustic (ăko͞os'tĭk) [G. *akoustikōs,* hearing]. Pertaining to sound or the sense of hearing. The cochlear portion of the eighth cranial nerve.

acromegaly (ăk'rōmĕg'ălē) [G. *akron,* extremity; *megas,* big]. A chronic disease of middle-aged persons characterized by elongation and enlargement of bones of the extremities and certain head bones, esp. frontal bone and jaws. Accompanied by enlargement of nose and lips and thickening of soft tissues of the face. Due to oversecretion of the somatotrophic or growth hormone from the anterior pituitary — after full growth of individual.

acromion (ăkrō'mēōn) [G. *akros,* summit; *ōmos,* shoulder]. Lateral prolongation of scapular spine.

actin (ăk'tĭn) [G. *aktis,* ray]. The protein of thin myofibrillae, as in the isotropic or I disk. Responsible for the shortening of the muscle when it contracts. Forms a combination with protein myosin when muscle contracts.

action potential (pōtĕn'shăl) [L. *actio,* from *agere,* to do; *potentia,* power]. The measurable electric changes associated with conduction of a nerve impulse or contraction of a muscle.

active transport [L. *trans,* across; *porta,* to carry]. Using carriers, energy (ATP), and enzymes of cell to cause a substance to cross a membrane.

adaptation [L. *adaptare,* to adjust]. The ability of an organism to adjust to environmental change.

Adam's apple. The laryngeal prominence formed by the two laminae of the thyroid cartilage.

Addison's disease (Thomas Addison, Eng. physician, 1793–1860). Disease resulting from deficiency in the secretion of adrenocortical hormones.

adeno- (ăd'ĕnō-) [G. *adēn,* gland]. Prefix: denotes gland or glandular.

adenohypophysis (ăd'ĕnōhīpŏf'ĭsĭs) [G. *adēn,* gland; *hypo,* under; *phyein,* to grow]. The anterior lobe or glandular portion of the hypophysis cerebri (pituitary).

adenoids (ăd'ĕnoydz) [G. *adēn,* gland; *eidos,* shape]. Nasopharyngeal tonsils; lymphoid tissue.

adenosine triphosphate (ATP) (ădĕn'ăsēn). A compound of adenosine (a nucleotide containing adenine and ribose) containing three phosphoric acid groups. An enzyme found in all cells but particularly in muscle cells.

adipose (ăd'ĭpōs) [L. *adiposus,* fatty]. Fat. Of or pertaining to fat; fatty; a connective tissue whose cells contain fat.

adrenal (ăd-rē'năl) [L. *ad,* to; *renes,* kidney]. A gland of internal secretion located superior to the kidney (suprarenal). Secretes hormones from its cortex that are essential to life; a hormone, adrenalin, from its medulla, not essential to life.

adrenocorticotrophic (adrē'nōkŏr'tĭkōtrŏf'ĭk) [L. *ad,* to; *renes,* kidneys; *cortex,* bark; G. *trophe,* nourishment]. *Appl.* hormone secreted by anterior lobe of pituitary gland that controls activity of adrenal cortex; ACTH.

adrenergic (ădrĕnĕr'jĭk) [L. *ad,* to + *renes,* kidney; G. *ergon,* work]. Term used to describe nerve fibers that release norepinepherine (noradrenalin or sympathin) at their endings.

adsorption (ădsŏrp'shŭn) [L. *ad,* to; *sorbere,* to suck in]. The adhesion of molecules to solid bodies; formation of unimolecular surface layer.

adventitia (ăd'vĕntĭsh'ēă) [L. *adventicius,* coming from abroad]. The outermost covering of a structure or organ, such as the tunica adventitia, or outer coat of an artery.

afferent (ăf'ĕrĕnt) [L. *ad,* to; *ferre,* to bring]. Conveying to. Used for nerves, blood vessels, and lymphatic vessels that lead to some central structure; afferent nerves to spinal cord or to brain.

agglomeration (ăglŏmĕrā'shŭn) [L. *ad,* to; *glomus,* ball]. Clustered, as a head of flowers. Adhering mass of protozoa, as in agglomeration of trypanosomes.

agglutination (ăglo͞o'tĭnā'shŭn) [L. *agglutinans,* gluing]. Clumping of microorganisms when a specific immune serum is added to a bacterial culture. Clumping of blood corpuscles when incompatible bloods are mixed. Adhesion of surfaces of a wound.

agglutinin (ăgloo͞'tĭnĭn) [L. *agglutinans,* gluing]. An antibody in the red blood plasma capable of causing clumping of specific antigens on red blood cells.

agglutinogen (ăgloo͞tĭn'ōjĕn) [L. *agglutinans,* gluing; G. *gennan,* to produce]. A substance that stimulates the development of a specific agglutinin, thereby acting as an antigen. A specific antigen used in agglutination tests.

agnosia (ăgnō'zhă) [G. *agnōsia,* ignorance]. Partial or complete loss of ability to recognize familiar objects, especially by seeing, hearing, or touching.

agonist (ăg'ōnĭst) [G. *agōnistēs,* champion]. A prime mover or muscle directly responsible for change in position of a part.

albumin (ălbū'mĭn) [L. *albus,* white; *albumen,* coagulated egg white]. One of a group of simple proteins in both vegetable and animal cells or fluids.

aldosterone (ăldŏs'tĕrōn, ăl'dōstĕr'ōn). The most biologically active mineralocorticoid hormone secreted by the adrenal cortex; regulates sodium, potassium, and chloride metabolism.

alexia (ă'lĕksēă) [L. *a,* from; *legere,* to read]. Aphasia characterized by loss of ability to read.

-algia (ăl'jēă) [G. *algeis,* sense of pain]. Suffix denoting pain.

alkalosis (ălkălō'sĭs) [Arabic *al-aulīy,* ashes of salt wort; G. *ōsis,* condition of]. A condition in which the alkalinity of the body tends to increase beyond normal due to excess of alkalies or withdrawal of acid or chlorides from the blood.

allantois (ălăn'tōis) [G. *allantos,* sausage; *eidos,* resemblance]. A kind of elongated bladder, between the chorion and amnion of the fetus, that grows out from the caudal extremity of the embryo and communicates with the bladder of the urachus. In primates including man the allantois has no function, but its blood vessels develop, becoming the umbilical vein and paired umbilical arteries.

allele (ălēl') [G. *allelon,* reciprocally]. One of two or more genes occurring at the same position on a specific pair of chromosomes and controlling the expression of a given characteristic.

allograft (ăl'ōgrăft) [G. *allos,* other; L. *graphium,* grafting knife]. Graft or transplant between genetically dissimilar members of the same species.

alveolar (ălvē'ălĕr) [L. *alveolus,* small pit]. Pertaining to an alveolus (a tooth socket or small depression). The smallest subdivision of the air system of a lung — the alveoli.

ameboid (ămē'boyd) [G. *amoibē,* change; *eidos,* shape]. Resembling an ameba in shape, in properties, or in locomotion; motion by use of pseudopods.

amenorrhea (ămĕn'ōrē'ă) [G. *a-,* without; *mēn,* month; *rein,* to flow]. Absence or suppression of menstruation.

amino acid A compound containing amino (NH_2) and carboxyl (COOH) groups; constituents of proteins; synthesized in autotrophic organisms, produced from proteins by hydrolysis or obtained from food in heterotrophic organisms.

amnion (ăm'nēōn) [G. amnion, little lamb]. A fetal membrane enclosing amniotic fluid and embryo. By eight weeks it fuses with chorion to form the bag of waters or caul.

amphiarthrosis (ăm'fēărthrō'sĭs) [G. *amphi,* both; *arthron,* joint]. A slightly movable articulation, as a symphysis or a syndesmosis.

ampulla (ămpŭl'lă) [L. *ampulla,* little jar]. Saclike dilatation of a canal or duct, as the semicircular canals or ductus deferens.

amylase (ăm'ĭlās) [L. *amylum,* starch]. An enzyme that converts starch into dextrin or dextrin into maltose; amylolytic enzyme.

anabolism (ănăb'ōlĭzm) [G. *anabolē,* a building

up]. Synthetic or constructive chemical reactions.

anaerobic (ănĕrōb'ĭc) [G. *an-*, without; *aer*, air; *bios*, life]. Not requiring oxygen for life or function.

anastamosis (ănăs'tămō'sĭs) [G. opening]. An opening of one vessel into another, or the joining of parts to form a passage between any two spaces or organs.

anatomy (ănăt'ămē) [G. *ana*, up; *tomē*, cutting]. The science treating the structure of plants and of animals, as determined by dissection.

androgen (ăn'drōjĕn) [G. *anēr*, *andros*, man; *gen*, to produce]. A substance stimulating or producing male characteristics, as testosterone, the male sex hormone.

anemia (ănē'mēă) [G. *an*, not; *haima*, blood]. A condition in which there is a reduced number of erythrocytes or erythrocytes with a reduced amount of hemoglobin.

aneurysm (ăn'ūrĭzm) [G. *aneurysma*, a widening]. Abnormal dilation of a blood vessel due to weakness of the vessel wall or to a congenital defect.

angina (ănjī'nă) [L. *angere*, to choke]. Any disease characterized by suffocation or choking. Angina pectoris occurs when the heart muscle is deprived of blood flow and includes severe pain referred to the chest and left arm.

angiotensin (ăn'jēōtĕn'sĭn) [G. *angeion*, vessel; L. *tensio*, tension]. A pressor substance formed in the body by interaction of renin and a serum globulin fraction. It increases arterial muscle tone. Formerly called angiotonin or hypertensin.

angstrom (ăng'strĕm) [A. J. Ångström, Swedish physicist, 1814–1874]. One ten-millionth part of a millimeter, symbol Å.

anion (ăn'ĭōn, ăn'ĭōn) [G. *ana*, up; *ienae*, to go]. A negatively charged particle or ion that moves up toward the anode or positive pole.

anisocytosis (ănī'sōsītō'sĭs) [G. *anisos*, unequal; *kytos*, cell; *ōsis*, condition]. Inequality in size of cells, esp. erythrocytes. An abnormal condition.

annulus (an'yĕlĕs) [L.]. A fibrous ring surrounding an opening.

anoxemia (ănŏksē'mēă) [G. *an*, without; *oxia*, oxygen; *haima*, blood]. Deficiency of oxygen in the arterial blood.

antagonist (ăntăg'ōnĭst) [G. *antagōnistēs*, adversary]. A muscle or other entity acting in opposition to the action produced by a prime mover or other agonist.

antibody (ăn'tĭbŏd'ē) [G. *anti*, against; A.S. *bodig*, body]. Any protein substance (usually a globulin) that reacts against a specific antigen or inactivates or destroys toxins.

antidiuretic (ADH) (ăn'tĭdīūrĕt'ĭk) [G. *anti*, against; *dia*, intensive; *ouresis*, urination]. Lessening urine secretion. A drug having such an action.

antigen (ăn'tĭjĕn) [G. *anti*, against; *gennan*, to produce]. A substance that induces the formation of antibodies.

antrum (ăn'trŭm) [G. *antrom*, cavity]. A nearly closed cavity or chamber; e.g., the antrum of the stomach; the maxillary sinus (antrum of Highmore).

aorta (āor'tă) [G. *aortē*, the great artery]. The great trunk artery that carries pure blood to the body through arteries and their branches. Arises from left ventricle of heart.

aortic valve (āŏr'tĭk vălv) [G. *aortē*, the great artery; L. *valva*, a fold]. The semilunar valve preventing regurgitation at the exit from the heart into the aorta, composed of three segments.

aphasia (ăfā'zhă) [G. *a-*, not; *phasis*, speaking]. Loss of verbal comprehension, or inability to express oneself properly through speech.

apnea (ăp'nēă) [G. *a-*, not; *pnoē*, breath]. Temporary cessation of breathing.

apocrine (ăp'ŏkrĭn) [G. *apo*, from; *krinein*, to separate]. *Appl.* glands secreting only part of cell contents; secretions gather at outer end of the gland cell and are pinched off; *appl.* mammary gland.

aponeurosis (ăp'ŏno͞orō'sĭs) [G. *apo*, from; *neuron*,

sinew]. The flattened tendon for insertion of, or membrane investing, certain muscles.

appendix (ăpĕn'dĭks) [L. *ad,* to; *pendere,* to hang]. An outgrowth, especially the vermiform appendix. A collection of material at the end of a book.

apposition (ăp'ōzĭ'shŭn) [L. *ad,* to; *ponere,* to place]. Development by accretion. Addition of parts. Fitting together, as the edges of two surfaces.

apraxia (ăprăx'ēă) [G. *a,* not; *prassein,* to do]. The inability to use or to understand the uses of objects or to make purposeful movements owing to lesions in the cortex of the brain.

arachnoid (ărăk'noyd) [G. *arachne,* spider; *eidos,* form]. The intermediate meninx of brain and spinal cord that is thin, weblike, and transparent.

arcuate (ăr'kūāt) [L. *arcuatus,* curved]. Curved or shaped like a bow. *Appl.* arcuate (iliopectineal) line on the os coxae.

areola (ărē'ōlă) [L. a small place]. A small cavity in a tissue; around the nipple, the ringlike pigmented area; the part of the iris enclosing the pupil.

argentaffin (ărjĕnt'ăfĭn) [L. *argentum,* silver; *affinis,* associated with]. Taking a silver stain.

arteriole (ărtē'rēōl) [L. *arteriola,* small artery]. An artery under 0.5 mm in diameter. Its distal end leads into a capillary or capillary network.

arteriosclerosis (ărtēr'ēōsklĕrō'sĭs) [L. *arterio,* artery; G. *sklērōsis,* hardening]. A general term referring to many conditions where the walls of the blood vessels thicken, harden, or lose elasticity.

arthritis (ărthrī'tĭs) [G. *arthron,* joint; *ītis,* inflammation]. Inflammation of a joint, usually accompanied by pain and frequently by changes in structure.

arthrodial (ărthrō'dēăl) [G. *arthrōdēs,* well-jointed]. *Appl.* joints admitting of only gliding movements.

arthrosis (arthrō'sĭs) [G. *arthron,* joint; *osis,* increased]. A degenerative condition of a joint.

-ase. Suffix used in forming the name of an enzyme; added to the part of or name of substance on which it acts; i.e., lactase.

arytenoid (ăr'ĭtē'noyd) [G. *arytoina,* pitcher; *eidos,* form]. A pair of small cartilages of the larynx articulating with cricoid cartilage.

asphyxia (ăsfĭk'sēă) [G. *a,* not; *asphyxis,* pulse]. Decreased amount of oxygen and an increased amount of carbon dioxide in the body as a result of some interference with respiration.

asthenia (ăsthē'nēa) [G. *a,* not; *sthenos,* strength]. Loss or lack of strength; weakness in muscles, cerebellar disease.

asthma (ăz'mă) [G. *asthma,* panting]. Paroxysmal dyspnea accompanied by wheezing and coughing caused by spasms of the bronchioles or due to swelling of their mucous membrane.

astigmatism (ăstĭg'mătĭzm) [G. *a,* not; *stigma,* point; *ismos,* condition of]. Condition in which refractive surfaces of the eyeball are different in one or more planes; usually due to unequal curvature of the cornea.

astrocyte (ăs'trōsīt) [G. *astēr,* star; *kytos,* hollow]. A common neuroglia cell; astroglia; macroglia; Deiters' cell; a neuroglial cell with branching protoplasmic processes in grey matter; a fibrillar or spider cell in white matter.

astrology (ăstrŏl'ōjē) [G. *astron* < *astēr,* star; *legō,* speak]. Anciently, the science of the stars; practical astronomy. The investigation of the aspects, etc., of the planets and their imagined influence on the destinies of men; star-divination.

ataxia (ătăks'ēă) [G. lack of order]. Muscular incoordination when voluntary muscular movements are attempted.

atelectasis (ăt'ēlĕk'tăsĭs) [G. *atelēs,* imperfect; *ectasis,* expansion]. Condition in which lungs of a fetus remain unexpanded at birth. A collapsed or airless condition of the lung.

athero- (ăth'ērō) [G. *athērē,* porridge]. Prefix referring to lipid (fatty) substances.

atherosclerosis (ăth'ērō'sklĕrō'sĭs) [G. *athērē,* porridge; *sklērōsis,* hardness]. A form of arteriosclerosis in which there are localized accumulations of lipid-containing material (atheromas) within or beneath the intimal surfaces of blood vessels.

atomic weight. The weight of an atom of a chemical element as compared with that of an atom of oxygen or hydrogen. The atomic weight of oxygen is taken as 16. On this basis the atomic weight of hydrogen is 1.008.

atresia (ătrē'zhă) [G. *a*, not; *tresia*, a perforation]. Congenital absence or pathological closure of a normal opening; degeneration of ovarian follicles.

ATP. Abbr. for adenosine triphosphate.

atrioventricular (AV) node (āt'rēōvĕntrĭk'ūlăr nōd) [L. *atrium*, corridor; *ventriculus*, a little belly; *nodus*, knot]. A mass of modified cardiac muscle located in lower part of interatrial septum from which the atrioventricular bundle (bundles of His) arises. Passes impulse to interventricular bundle (bundle of His).

atrium (ā'trēŭm) [L. *atrium*, chamber]. A superior cavity of the heart that acts as the receiving chamber; a part of the tympanic cavity of the ear; a space opening into air sacs of the lungs.

atrophy (ă'trĕfē) [G. *a*, not; *trophē*, nourishment]. Wasting or decrease in size of a part due to lack of nutrition, loss of nerve supply, or failure to use the part.

auditory tube (awd'ĭtōrē tūb) [L. *audire*, to hear; *tuba*, a tube]. Eustachian tube (ū stā'kēăn, -shĕn). (Bartolommeo Eustachio, Italian anatomist, 1524–1574). The auditory tube (from the middle ear to the nasopharynx).

auricularis (ŏrĭkūlă'rĭs) [L. *auricula*, small ear]. Superior, anterior, posterior, extrinsic muscles of the external ear.

auscultation (awskultā'shŭn) [L. *auscultāre*, listen to]. Listening for sounds in the body cavities, especially the chest and abdomen, to detect and judge normal or abnormal conditions.

autocatalytic (awt'ŏkătălĭt'ĭk) [G. *autos*, self; *kata*, down; *lysis*, loosing]. Hastening the dissolution or reaction of a cell or substance due to influence of a product or secretion of its own.

autograft [G. *autos*, self; L. *graphium*, grafting knife]. A graft or transplant from one part of a body to another part of the same body.

autoimmune [G. *autos*, self; L. *immunis*, safe]. A condition in which antibodies are produced against a person's own tissue.

autonomic (awt'ōnŏm'ĭk) [G. *autos*, self; *nomos*, law]. Autonomous; self-governing, spontaneous; *appl.* the involuntary nervous system as a whole, comprising parasympathetic and sympathetic systems.

autosome (aw'tōsōm) [G. *autos*, self; *soma*, body]. Any of the paired chromosomes other than the sex (X and Y) chromosomes.

axilla (ăksĭl'ă, ăk'sĭlă) [L. *axilla*, armpit]. The armpit.

axillary (ăk'sĭlărĭ, ăksĭl'ărĭ) [L. *axilla*, armpit]. *Pert.* armpit.

axon (ăk'sŏn) [G. *axon*, axle]. Nerve cell process, limited to one per cell, involved in conducting away from the cell body.

azygos (ăz'ĭgŏs) [G. *a*, without; *zygon*, yoke]. An unpaired muscle, artery, vein, process.

baroreceptors (băr'ōrēsĕp'tŏrz) [G. *baros*, weight; L. *receptor*, a receiver]. Receptor organs sensitive to pressure or stretching.

basal ganglia (nuclei) *(plu.)* (bā'săl găng'lēă) [G. *basis*, base; *ganglion*, knot]. Mass of gray matter beneath third ventricle, consisting of the caudate, lentiform, and amygdaloid nuclei and the claustrum.

basal metabolism [G. *basis*, base; *metabole*, change]. The amount of energy needed for maintenance of life when the subject is at physical, emotional, and digestive rest.

basilar membrane (băs'ĭlăr mĕm'brān) [L. *basilaris*, pert. to a base; *membrana*, membrane]. Membrane extending from the tympanic lip of the osseous spiral lamina to the crest of the spiral ligament in the cochlea of the ear.

basophil(e) (bā'sōfĭl, -fīl [G. *basis*, base; *philein*, to love]. In histology, cells or parts of cells that are readily stained with basic dyes such as methylene blue; a type of white blood cell (leucocyte) characterized by possession of coarse granules, which stain intensely with basic dyes; a type of cell found in the anterior lobe of the hypophysis.

benign (bĕnīn') [L. *benignus*, mild]. Not malignant.

beta oxidation (ŏk'sĭdā'shŭn) [G. β, second letter of the alphabet; *oxys*, sour, pungent]. A metabolic scheme that catabolizes fatty acids by removal of two carbon units.

biceps (bī'sĕps) [L. *bis,* two; *caput,* head]. A two-headed muscle; i.e., biceps brachii.

bicuspid valve (bīkŭs'pĭd vălv) [L. *bis,* twice; *cuspis,* point; *valva,* a fold]. A valve having two cusps or leaflets, between the left atrium and the left ventricle of the heart.

bifid (bī'fĭd) [L. *bis,* two; *findere,* to cleave]. Cleft or split into two parts.

bilirubin (bĭlĭroo'bĭn) [L. *bilus,* bile; *ruber,* red]. The orange or yellow pigment in bile derived from breakdown of heme ($C_{33}H_{36}O_6N_4$).

biliverdin (bĭlĭvĕr'dĭn) [L. *bilus,* bile; G. *virdis,* green]. The green pigment of bile formed by oxidation of bilirubin ($C_{33}H_{34}O_6N_4$).

bio- [G. *bios,* life]. Prefix: life.

-blast [G. *blastos,* germ or bud]. Suffix: immature or primitive.

blastocele (blăs'tōsēl) [G. *blastos,* germ or bud; *koilos,* hollow]. An embryonic stage of development; the cavity of the blastula.

blastocyst (blăs'tōsĭst) [G. *blastos,* bud; *kystis,* bladder]. The germinal or blastodermic vesicle; a stage in embryonic development following the morula.

blastomere (blăs'tōmēr) [G. *blastos,* bud; *meros,* part]. One of the cells formed during primary divisions of an egg; cleavage cell. The cells composing the morula.

BMR Abbreviation for *basal metabolic rate.*

bolus (bō'lŭs) [L. *bolus,* from G. *bolos,* lump]. A rounded mass; lump of chewed food.

boutons terminaux (bootōn' tĕrmĭnō') *(plu.)* [Fr. terminal buttons]. Bulblike expansions at the tip of axons that come into synaptic contact with the cell bodies or dendrites of other neurons.

Bowman's capsule (Sir William Bowman, English physician, 1816–1892) [L. *capsula,* small box]. The capsule containing the glomeruli of the kidney; the cuplike depressions on the expanded ends of renal tubules or nephrons that surround the capillary tufts (glomeruli).

brachialis (brăkēā'lĭs) [L. *brachialis, pert.* arm]. A flexor muscle of the forearm: brachialis anticus.

brachiocephalic (brăk'ēōkĕfăl'ĭk, -sĕf-) [G. *brachium,* arm; *kephalē,* head]. *Pert.* arm and head; *appl.* artery, veins.

brachium (brā'kēŭm, bră-) [G. brachiōn, arm]. The upper arm from shoulder to elbow.

brady- [G. *bradys,* slow]. Prefix: slow, as in bradycardia or slow heart.

bradykinin (brăd'ĭkī'nĭn) [G. *bradys,* slow; *kinēsis,* motion]. A plasma kinin.

branchial arches (brăng'kēăl) [L. *branchia,* gills]. Five pairs of arched structures that form the lateral and ventral walls of the pharynx of the embryo.

Broca's area (brō'kă) [Pierre Paul Broca, Fr. surgeon, 1824–1880]. On left side of brain, controlling movements of tongue, lips, vocal cords, and motor speech area.

bronchiectasis (brŏnkēĕk'tăsĭs) [G. *bronchos,* windpipe; *ektasis,* dilatation]. Condition characterized by dilatation of the bronchus(i) with secretion of large amounts of foul substance.

bronchiole (brŏng'kēōl) [L. *bronchilus,* air passage]. One of the smaller subdivisions of the bronchi.

bronchus (brŏng'kŭs) [G. *bronchos,* windpipe]. One of the two large branches of the trachea.

buccal (bŭk'ăl) [L. *bucca,* cheek, mouth]. *Pert.* cheek or mouth.

buffer (bŭf'ĕr) [Fr. *buffe,* blow]. A substance, esp. a salt of the blood, tending to preserve original hydrogen-ion concentration of its solution when an acid or base is added. A substance tending to offset reaction of an agent administered in conjunction with it.

bulbourethral (bŭl'bōūrē'thrăl) (Cowper's) glands [William Cowper, English anatomist, 1666–1709; L. *bulbus,* bulb; G. *ourethra,* urethra]. Glands lying in the pelvis floor to either side of the membranous urethra of the male; open into cavernous urethra.

bulbus (bŭl'bŭs) [L. *bulbus,* bulb]. A bulb. The knoblike part found in connection with various nerves; a dilatation of base of aorta.

bundle of His [M.E. *bundel,* bind; W. His, G.

anatomist, 1863–1934]. A bundle of fibers of the impulse-conducting system that initiates and controls the contraction of heart muscle. It extends from the AV node and becomes continuous with the Purkinje fibers of the ventricles.

bunion (bŭn'yŭn). Inflammation and thickening of the bursa of the joint of the great toe, usually associated with marked enlargement of the joint and displacement of the toe, laterally.

bursa (bĕr'să) [G. *bursa,* a leather sack]. A fluid-filled sac or cavity found in connecting tissue usually in the vicinity of joints where friction may occur. A blind sac or cavity.

bursitis (bĕrsītĭs) [L. *bursalis,* pert. to a bursa; *itis,* inflammation]. Inflammation of a bursa, esp. those located between bony prominences and muscle or tendon, as the shoulder, knee, etc.

buttocks (bŭt'ŭks) [AS. *buttuc,* end]. The external prominences posterior to the hips, formed by the gluteal muscles and underlying structures. The rump or seat.

calcaneus (kălkā'nēŭs) [L. calcaneus, heel bone]. The heel bone, or os calcis.

-calcemia (kălsē'mēă) [L. *calcarius,* pert. to calcium or lime; G. *haima,* blood]. The calcium level of the blood.

calcification (kăl'sĭfĭkā'shŭn) [L. *calx,* lime; *ferre,* to make]. Deposit of lime salts in the tissues, and commonly in bone.

calcitonin (kălsĭtōn'ĭn) [G. *çalci, pert.* calcium; *tonos,* tone]. A hormone secreted by the thyroid gland that aids in regulating calcium-phosphorus metabolism.

calculus (kăl'kūlŭs) [L. pebble]. A "stone" formed within the body usually composed of mineral salts.

callous (kăl'ŭs) [L. *callus,* hard]. Hard, like a callus. A circumscribed area of the horny layer of the skin that has thickened; osseous material formed between ends of a fractured bone.

calorie, small (kăl'ōrē) [L. *calor,* heat]. A unit of heat. The amount of heat required to raise the temperature of one gram of water 1°C. Abbr. cal.

Calorie, large [L.]. A large calorie or kilocalorie; 1000 times as large as a calorie; the amount of heat required to raise one kilogram of water 1°C. Distinguished from a small calorie by a capital C. Abbr. Cal or Kcal.

canaliculi (kănălĭk'ūlī) *(plu.)* [L. *canaliculus,* small channel]. Microscopic canals through which processes of the bone cells pass.

cancellous (kănsĕl'ŭs) [L. *cancellus,* a grating]. Having a reticular or latticework structure, as the spongy tissue of bone.

cancer (kăn'sĕr) [L. a crab; ulcer]. A malignant tumor or neoplasm (new growth); a carcinoma or sarcoma.

capillaries (kăp'ĭlĕrēz, kăpĭl'ĕrēz) *(plu.)* [L. *capillus,* hairlike]. Minute, thin-walled vessels that form networks in various parts of the body; e.g., blood, lymph, or biliary capillaries. Connect arterioles to venules to carry blood. Allow passage of oxygen and nutrients into blood to be carried to tissues; allow wastes to pass from tissues into blood, etc.

capitate (kăp'ĭtāt) [L. *caput,* head]. Enlarged or swollen at tip; gathered into a mass at apex; *appl.* a bone, os capitatum.

caput [L]. The head, upper part of an organ.

carbaminohemoglobin (kărb'ămē'nōhē'mōglō'bĭn). The compound formed by the combination of carbon dioxide with the amine groups of the globin molecule of hemoglobin.

carbohydrate (kăr'bōhī'drāt) [L. *carbo,* carbon; G. *hydōr,* water]. A compound of carbon, hydrogen, and oxygen, aldehydes or ketones constituting sugars, or a condensation product thereof.

carbonic anhydrase. An enzyme catalyzing the reaction of CO_2 and water to form carbonic acid (H_2CO_3) and vice versa; present primarily in red blood cells.

carcinogen (kărsĭ'nĭjĕn) [G. *karkinos,* cancer; *genere,* to produce]. Substance known to cause cancer or neoplasms.

caries (kă'rēz) [L. rottenness]. Tooth or bone decay.

carotene (kăr'ōtēn) [G. *karoton,* carrot]. A yellow

crystalline pigment present in various plant and animal tissues, stored and converted to vitamin A in liver.

carotid (kărŏt'ĭd) [G. karos, heavy sleep]. *Pert.* chief arteries in the neck; *appl.* arch, ganglion, nerve, etc.

carotid sinus (kărŏt'ĭd sī'nŭs) [G. *karos,* deep sleep; L. *sinus,* a curve, hollow]. A dilated area at the bifurcation of the common carotid artery that is richly supplied with sensory nerve endings of the sinus branch of the vagus nerve.

carrier [Fr. *carier,* to bear]. A large molecule transporting substances across cell membranes in active transport.

cartilage (kăr'tĭlĭj) [L. *cartilagō,* gristle]. A type of dense connective tissue consisting of cells embedded in a ground substance or matrix. Usually bluish-white, firm, and elastic; cells placed in groups in spaces called lacunae. It has no blood or nerve supply of its own.

catabolism (kătăb'olĭzm) [G. *katabolē,* a casting down]. The destructive phase of metabolism; breaking down of complex chemical compounds into simpler ones, usually with a release of energy.

catalyst (kăt'ălĭst) [G. *kata,* down; *lysis,* loosing]. A substance that accelerates or retards a chemical reaction without itself being altered in the process.

cataract (kăt'ărăkt) [G. *katarraktēs,* a rushing down]. Opacity or loss of transparency of the lens of the eye, its capsule, or both.

cation (kăt'īŏn) [G. *kation,* descending]. A positively charged ion or radical.

cauda equina (kaw'dāēkwīn'ă) [L. *cauda,* tail; *equus,* horse]. The terminal portion of the roots of the spinal nerves and spinal cord below the first lumbar nerve; resembles a horse's tail.

caudate nucleus (kaw'dāt nū'klēŭs) [L. *caudatus,* having a tail; *nucleus,* little kernel]. A comma-shaped mass of gray matter forming part of the corpus striatum. Constitutes part of the basal ganglia.

cavernous bodies (kăv'ĕrnŭs) *(plu.)* [L. *cavernosus,* chambered, hollow; *corpus,* body]. Structures of the penis and clitoris containing blood spaces; involved in erection of the organs.

cecum (sē'kŭm) [L. *caecum*]. The blind pouch that forms the first portion of the large intestine.

-cele [G. *hernia,* tumor]. Suffix: a swelling.

cell (sĕl) [L. *cella,* compartment]. A small cavity or hollow; a loculus; a unit mass of protoplasm, usually containing a nucleus or nuclear material; originally, the cell wall.

celom (sē'lŏm) (also coelom) [G. *koiloma,* a hollow, fr. *koilos,* hollow]. A cavity formed within the mesoderm and generally lined by mesothelium. Develops into peritoneal, pleural, and pericardial cavities.

centriole (sĕn'trēōl) [L. *centrum,* center]. A cell organelle that is involved in cell division; forms a spindle.

centromere (sĕn'trōmēr). The structure at the junction of the two arms (chromatids) of a chromosome.

centrosome (sĕn'trōsōm) [G. *kentron,* center; sōma, body]. A cell organ, the center of dynamic activity in mitosis, consisting of one or two centrioles and an attraction sphere.

cephalic (kĕfăl'ĭk, sĕf-) [G. *kephalē,* head]. *Pert.* head; in head region.

cephalin (kĕf'ălĭn, sĕf-) [G. *kephalē,* head]. A phospholipid present in nerve fibers and egg-yolk.

cerebellum (sĕr'ĕbĕl'ŭm) [L. *cerebrum,* brain]. The fourth division of the brain, arising from differentiation of the anterior part of the third primary vesicle. Involved in coordination of skeletal muscles.

cerebral aqueduct (sĕr'ĕbrăl, sĕrē'brăl, ăk'wĕdŭkt) [L. *cerebrum,* brain; *aqua,* water; *ductus,* duct]. Canal in midbrain connecting third and fourth ventricles.

cerebral peduncle (sĕr'ĕbrăl, sĕrē'brăl, pĕdŭn'kl) [L. *cerebrum,* brain; *pedunculus,* a little foot]. A pair of white bundles from the upper part of the pons to the cerebrum. It constitutes the ventral (anterior) portion of the midbrain.

cerebrum (sĕrē'brŭm, sĕr'ĕbrŭm) [L. *cerebrum,* brain]. The forebrain, arising from differentiation of first primary vesicle. Contains the motor and sensory areas, and association areas concerned with the higher mental faculties.

ceruminous glands (sĕro͞o'mĕnĕs) [L. *cera*, wax]. Wax glands of the external auditory meatus.

cervical (sĕr'vĭkăl) [L. *cervix*, neck]. The neck region or pertaining to the neck of an organ.

cervix (sĕr'vĭks) [L. *cervix*, neck]. The neck or a part of an organ resembling a neck. Cervix uteri: neck of the uterus.

chelation (kēlā'shŭn) [G. *chēlē*, claw]. Combining of metallic ions with certain heterocyclic-ring structures so that the ion is held by chemical bonds from each of the participating rings. Triggered by osteoclasts.

chemotaxis (kĕm'ōtăk'sĭs) [G. *chēmeia*, transmutation; *taxis*, arrangement]. The reaction of cells or freely motile organism to chemical stimuli; also chemiotaxis.

chemoreceptor. A sense organ or sensory nerve ending that is stimulated by a chemical substance or change; i.e., taste buds.

chemotactic (kĕm'otăk'tĭk) [G. *chēmeia*, transmutation; *taxis*, arrangement]. Responding to a chemical stimulus be being attracted or repelled.

chiasma (kiăz'mă) chiasmata (kiăz'mătă) *plu.* [G. *chiasma*, cross]. A decussation of fibers, as optic chiasma; in paired chromatids, an exchange of partners in meiosis.

chloride shift (klō'rīd) [G. *chlōros*, green; A.S. *sciftan*, to divide]. The movement of chloride ions into red blood cells as bicarbonate ions are formed and leave the cell during the reaction of CO_2 and water in the cell.

chlorocruorin (klō'rōkroo'ōrĭn) [G. *chlōros*, grass green; L. *cruor*, blood]. A green respiratory pigment; a green-sensitive pigment found in some cones of the retina.

choana (kō'ăña) (choanae, *plu.*). [G. *choanē*, funnel]. A funnel-shaped opening; the internal or posterior naris.

cholecystokinin (kō'lēsĭs'tōkī'nīn) [G. *cholē*, bile; *kystis*, cyst; *kinein*, to move]. A hormonelike chemical secreted by the duodenal mucosa that stimulates contraction of the gall bladder.

cholesterol (kōlēs'tĕrōl) [G. *cholē*, bile; *stereos*, solid]. Cholesterin, a white fatty monohydric alcohol found in protoplasm, nerve tissue, bile, yolk, and other animal substances.

cholinergic (kō'lĕnĕr'jĭk). Term applied to nerve endings that liberate acetylcholine.

cholinesterase (kō'līneš'tĕrās'). An enzyme that hydrolyzes acetylcholine into choline and acetic acid.

chondroblast (kŏn'drōblăst) [G. *chondros*, cartilage; *blastos*, germ]. A primitive cell that forms cartilage.

chondrocyte (kŏn'drōsīt) [G. *chondros*, cartilage; *kytos*, hollow]. A cartilage cell.

chordae tendineae (kŏr'dī tĕn'dĭnī) *(plu.)* [G. *chordē*, cord; L. *tendo*, tendon]. Small tendinous cords that connect the free edges of the atrioventricular valves to the papillary muscles.

Chordata (kŏr'dātă) [G. *chordē*, cord; having a notochord]. The phylum in which vertebrate animals are placed.

chordée (kōrdā') [Fr. corded]. Downward painful curvature of the penis on erection as in gonorrhea.

chorea (korē'ă) [G. *choreia*, dance]. Rapid, irregular, and wormlike movement of body parts; a nervous affliction.

chorion (kō'rēŏn) [G. *chorion*, skin]. An embryonic membrane external to and enclosing the amnion. Has villi that connect with the endometrium to form the placenta.

choroid (kŏr'oyd) [G. *chorion*, skin]. The middle layer of the eyeball between the retina and the sclera.

choroid plexus (kō'royd plĕk'sŭs) [G. *chorion*, skin; L. *plexus*, interwoven]. Vascular structures in the roofs of the four brain ventricles that produce cerebrospinal fluid.

chromatid (krō'mătĭd) [G. *chrōma*, color; *tid*, time]. One of two bodies resulting from the longitudinal separation of duplicated chromosomes; each goes to a different pole of the dividing cell.

chromatin (krō'mătĭn) [G. *chrōma,* color]. A substance in the nucleus that contains nucleic acid proteids and stains with basic dyes. Consists mainly of DNA and is considered to be the physical basis of heredity; forms chromosomes during mitosis and meiosis.

chromomere (krō'mōmēr) [G. *chrōma,* color; *meros,* part]. One of the chromatin granules of which a chromosome is formed; corresponds to an id or a gene.

chromophobe (krō'mōfōb) [G. *chrōma,* color; *phobos,* fear]. Any cell or tissue that stains either poorly or not at all. A type of cell found in pars distalis of the pituitary gland.

chromosome (krō'mōsōm) [G. *chrōma,* color; *sōma,* body]. One of the deeply staining bodies, the number of which is constant for the cells of a species, into which the chromatin resolves itself during karyokinesis and meiosis.

chronaxie (krō'năksē) [G. *chronos,* time; *axia,* value]. A number expressing the sensitivity of a nerve fiber to stimulation. Specifically, the time that a current of two times the threshold value must last to cause depolarization.

chyle (kīl) [G. *chylos,* juice]. The contents of intestinal lymph vessels. Consists mainly of absorbed products of fat digestion.

chylomicrons (kī'lōmī'krŏnz) *(plu.)* [G. *chylos,* juice; *mikros,* small]. Tiny droplets of fat in lacteals and blood.

chyme (kīm) [G. *chymos,* juice]. The mixture of partially digested foods and digestive juices found in the stomach and small intestine.

chymotrypsin (kī'mōtrip'sĭn) [G. *chymos,* juice; *tripsai,* to rub down; *pepsis,* digestion]. An enzyme that, in the small intestine, splits the various protein products of the action of pepsin and trypsin.

chymotrypsinogen (kī'mōtrĭpsĭn'ŏjĕn) [G. *chymos,* juice; *tripsai,* to rub down; *pepsis,* digestion; *-genēs,* producing]. A pancreatic enzyme activated by trypsin and converted into chymotrypsin.

cilia (sĭl'ēă) *(plu.)* [L. *ciliaris, pert.* eyelash]. Hairlike vibratile outgrowths of ectoderm, or processes, of many cells, usually microscopic. In anatomy, eyelash.

ciliary body (sĭl'ēăry) [L. *ciliaris, pert.* eyelash]. Thickened part of vascular tunic of eye between base of iris and anterior part of choroid.

circadian (sĭrkā'dēēn) [L. *circa,* about; *dies,* day]. Describes events that repeat in a length of time approximating a 24-hour cycle.

circumduction (sĭr'kŭmdŭk'shŭn) [L. *circum,* around; *ducere,* to lead]. The action or swing of a limb, such as the arm, in such a manner that it describes a cone-shaped figure, the apex of the cone being formed by the joint at the proximal end, while the complete circle is formed by the free distal end of the limb. Circular movement of the eye.

cirrhosis (sĭrō'sĭs) [G. *kirros,* yellow; *ōsis,* infection]. A chronic liver disease characterized by degenerative changes in the liver structure.

cisterna (sĭstĕr'nă) [L. *cisterna,* cistern]. Closed space containing fluid, as any of the subarachnoid spaces.

cisterna chyli (sĭs'tĕrnă kīl'ē) [L. *cisterna,* cistern; G. *chylos,* juice]. A dilated sac into which empty the intestinal, two lumbar, and two descending lymphatic trunks; the origin of the thoracic duct.

cleido (klī'dō) [G. *kleis,* clavicle]. Prefix: *pert.* clavicle (collarbone).

climacteric (klīmăk'tĕrĭk, klīmăktĕr'ĭk) [G. *klimaktēr,* a rung of a ladder]. That period marking the cessation of a woman's reproductive period (female climacteric or menopause); a corresponding period of lessening of sexual activity in the male (male climacteric).

clitoris (klī'tĕrĭs) [G. *kleitoris,* clitoris]. An erectile organ of the female homologous to the penis in the male.

cloaca (klōā'kă) [L. *cloaca,* sewer]. The common chamber into which intestinal, genital, and urinary canals open; in vertebrates except most mammals.

clone (klōn) [G. *klōn,* a cutting used for propagation]. In tissue culture, a group of cells descended from a single cell.

Co-A. A dinucleotide that activates many substances in metabolic cycles.

coccygeal (kŏksĭj'ēăl) [G. *kokkyx,* coccyx]. *Pert.* coccyx, tailbone, last four fused spinal bones.

cochlea (kŏk'lēă) [G. *kochliās,* a spiral]. Anterior part of labyrinth of the inner ear; spirally coiled like a snail shell. Contains the organ of Corti, the receptor for hearing.

codon. The sequence of three nitrogenous bases of messenger RNA that specifies a given amino acid and its position in a protein.

coenzyme (kōĕn'zīm) [L. *co,* together; G. *en,* with; *zymē,* leaven]. Enzyme activators. A diffusible, heat-stable substance of low molecular weight that, when combined with an inactive protein called apoenzyme, forms an active compound or a complete enzyme called holoenzyme. Examples are adenylic acid, riboflavin, and coenzymes I and II.

cofactor. A substance essential to or necessary for the operation of another material such as an enzyme.

coitus (kō'ĭtŭs) [L. a uniting]. Sexual intercourse between man and woman. Also, coition, copulation.

collagen (kŏl'ăjĕn) [G. *kolla,* glue; *genos,* descent]. A sclero-protein, occurring as chief constituent of white connective tissue fibers and organic part of bone.

colloid (kŏl'oyd) [G. *kolla,* glue; *eidos,* form]. A gelatinous substance that does not readily diffuse through an animal or vegetable membrane. Particles are large and remain suspended. Particle sizes range 1–100 millimicrons.

coma [G. *kōma,* a deep sleep]. An abnormal deep stupor from which the person cannot be aroused by external stimuli.

compound (kŏm'pownd) [L. *cum,* together; *ponere,* to place]. Made up of several elements and having properties different from its parts.

concha (kŏng'kă) [G. *kongche,* shell]. A structure resembling a shell, as the nasal conchae (turbinates) or the hollow of the external ear.

condyle (kŏn'kīl) [G. *kondylos,* knuckle]. A process on a bone for purposes of articulation; a rounded structure adapted to fit into a socket.

condyloid (kŏn'dĭloyd) [G. *kondylos,* knuckle; *eidos,* form]. Shaped like, or situated near, a condyle.

cone (kōn) [G. *kōnos,* cone]. Retinal flask-shaped figure in layer of rods and cones; a receptor cell of the retina concerned with color vision.

congenital (kŏnjĕn'ĭtăl) [L. *congenitus,* born together]. Present at birth.

conjunctiva (kŏnjŭnktī'vă) [L. *con,* with; *jungere,* to join]. Mucous membrane lining of the eyelid that is reflected onto the eyeball.

contraception (kŏn'trăsĕp'shŭn) [L. *contra,* against; *conceptiō,* a conceiving]. The prevention of conception, impregnation, or implantation.

conus arteriosus (kō'nŭs ărtērēō'sŭs) [G. *kōnos,* a cone; *artēria,* artery]. Right cardiac ventricle's upper rounded anterior angle, where pulmonary artery arises.

copulation (kŏpūlā'shŭn) [L. *copulātiō*]. Sexual intercourse.

coracoid (kŏr'ăkoyd) [G. *korax,* crow; *eidos,* form]. *Appl.* bone or part of the pectoral girdle between scapula and sternum; *appl.* ligament stretching over the suprascapular notch.

corium (ko'rēŭm) [G. *chorion,* skin]. The dermis or true skin that lies immediately under the epidermis; contains nerve endings, capillaries, and lymphatics, and is composed of connective tissue.

cornea (kŏr'nēă) [L. *corneus,* horny]. The transparent anterior portion of the fibrous coat of the eye; continuous with the sclera.

corniculate (kŏrnĭk'ūlāt) [L. *corniculum,* little horn]. Having small horns. A cartilage of the larynx.

cornified (kŏr'nĭfīd) [L. *cornu,* horn; *facere,* to make]. Formed into an outer horny layer of epidermis.

coronal (kŏr'ŏnăl) [L. *corona,* crown]. A plane vertical to the median plane, dividing the body into anterior and posterior parts. (Same as frontal.)

coronary sinus (kŏr'ŏnărē sī'nŭs) [L. *corona,* crown; *sinus,* curve or hollow]. The vessel cavity or passage that receives the cardiac veins from the heart. It opens into the right atrium.

corpora quadrigemina (kŏr'pŏră kwŏd'rĭjĕm'ĭnă) *(plu.)* [L. *corpus,* body; *quadri,* four; *geminus,* twin]. The superior portion of the midbrain consisting of two pairs of rounded bodies, the superior and inferior colliculi.

corpus albicans (kŏr'pŭs ăl'bĭkăns) [L. *corpus,* body; *albicans,* white or whitish]. A mass of fibrous tissue that replaces the regressing corpus luteum following rupture of the graafian follicle.

corpus callosum (kŏr'pŭs kălō'sŭm) [L. *corpus,* body; *callosus,* hard]. Broad sheet of white matter uniting the two cerebral hemispheres below the longitudinal fissure.

corpus luteum (kŏr'pŭs lo͞o'tēŭm) [L. *corpus,* body; *luteus,* yellow]. A small yellow body that develops within a ruptured ovarian follicle. It is an endocrine structure secreting progesterone.

corrugator (kŏr'ĕgātŏr) [L. *corrugare,* to wrinkle]. Wrinkled or wrinkling; *appl.* muscles.

cortex (kŏr'tĕks) [L. *cortex,* bark]. Outer or more superficial part of an organ, as the cortex of the adrenal gland or of the cerebrum.

corticoid (kŏr'tĭkoyd). One of many adrenal cortical steroid hormones. Also, corticosteroid.

cortisone (kŏr'tĭsōn). A hormone isolated from the cortex of the adrenal gland and also prepared synthetically. It is closely related to cortisol. It is important for its regulatory action in metabolism of fats, carbohydrates, sodium, potassium, and proteins. Exerts antiinflammatory effects; increases resistance to stress.

costal (kŏs'tăl) [L. *costa,* rib]. Pertaining to ribs or riblike structures.

creatine (krē'ătēn) [G. *kreas,* flesh]. A crystalline substance found in organs and body fluids; combines with phosphate and serves as a source of high-energy phosphate released in anaerobic phase of muscle contraction.

creatinine (krēăt'ĭnēn) [G. *kreas,* flesh]. A katabolic product in muscle and other tissues, excreted in urine.

cremaster (krēmăs'tĕr) [L. *cremaster,* to suspend]. One of the fascialike muscles suspending and enveloping the testicles and spermatic cord.

cretin (krē'tĭn) [Fr. One afflicted with congenital myxedema; a mentally retarded dwarf]. Characterized by lack of growth and mental development; rarely if ever exceeds the mental age of 10.

cribriform (krĭb'rĭfŏrm) [L. *cribrum,* sieve; *forma,* shape]. Sievelike. *Appl.* part of the ethmoid bone.

cricoid (krī'koyd) [G. *krikos,* ring; *eidos,* form]. Shaped like a signet ring; a cartilage of the larynx.

crista (krĭs'tă) [L. *crista,* crest]. A crest or ridge for muscle attachment.

crista galli (krĭs'tă găl'ē) [L. *crista,* crest; *gallus,* chicken, cock]. A process on the superior surface of the ethmoid bone.

crura (kro͞o'ră) (*sing.* crus) [L. legs]. The legs; a pair of long bands or masses resembling legs.

crypt (krĭpt) [G. *kryptos,* hidden]. A simple glandular tube or cavity; pit of stoma; depression in uterine mucous membrane; crypt of Lieberkühn.

cryptorchidism (krĭptŏr'kĭdĭzm) [G. *kryptos,* hidden; *orchis,* testis; *ismos,* condition of]. Failure of testicles to descend into scrotum.

crystalloid (krĭs'tăloyd) [G. *krystallos,* clear ice; *eidos,* form]. A substance that is capable of forming crystals, and that in solution diffuses rapidly through membranes; less than 1 millimicron in diameter; smaller than colloid.

cuneiform (kūnē'ĭfŏrm) [L. *cuneus,* wedge; *forma,* shape]. Wedge-shaped. A cartilage of the larynx.

curettage (kūrĕt'ĭj, kūrĕtahzh') [Fr. *curette,* a cleanser instrument]. Scraping of a cavity.

Cushing's disease (Syndrome) (Harvey Cushing, Amer. surgeon, 1869–1939). A syndrome resulting from hypersecretion of the adrenal cortex in which there is excessive production of glucocorticoids.

cuticle (kū'tĭkl) [L. *cutis,* skin]. An outer skin or pellicle; the epidermis; cuticula.

cyanosis (sīănō'sĭs) [G. *kyanos,* dark blue; *ōsis,* state of]. A condition where skin and mucous membranes have a blue color due to excessive reduction of hemoglobin in the bloodstream.

cybernetics (sībĕrn'ĕt'ĭks) [G. *kybernētikos,* skilled in governing]. Science of communication and control, as by nervous system and brain; *cf.* kybernetics. The study of self-monitoring and regulating mechanisms for maintenance of near-constant values of a function.

cystitis (sĭstī'tĭs) [G. *kystis,* bladder, sac; *ītis,* inflammation]. Inflammation of the bladder usually occurring secondarily to infections of associated organs (kidney, prostate, urethra).

cystoscope (sĭst'ĕskōp) [G. *kystis,* bladder, sac; *skopein,* to examine or view]. An instrument for examining or viewing the interior of the bladder.

-cyte (sīt) [G. *kytos,* cell]. Suffix denoting cell.

cyto- [G]. Prefix indicating the cell.

cytokinesis (sīt'ĕnē'sĭs) [G. *kytos,* cell; *kinēsis,* movement]. Division of cytoplasm in latter stages of mitosis.

cytology (sītŏl'ŏjē) [G. *kytos,* cell; *logos,* discourse]. The science dealing with structure, functions, and life history of cells.

cytoplasm (sī'tĕplăzm) [G. *kytos,* cell; *plasma,* mold]. Substance of cell body exclusive of nuclus.

damping. Steady decrease in amplitude of successive vibrations, as in the cochlea.

dartos tunic (dar'tŏs to͞o'nĭk) [G. *dartos,* skinned; L. *tunica,* a sheath]. The muscular, contractile tissue beneath the skin of the scrotum.

de-. Prefix: down, from.

deamination (dēăm'ĭnā'shŭn) [L. *de,* down; G. *ammōniakon,* resinous gum]. Removal of an amine group (—NH_2) from an amino acid to form ammonia by enzymatic action.

decussate (dēkŭs'āt) [L. *decussāre,* to cross, as an X]. To undergo crossing; crossed.

defecation (dĕf'ĭkā'shŭn) [L. *defaecatio,* voiding of excrement]. The explusion of feces from the large bowel.

deglutition (dēglo͞otĭsh'ŭn) [L. *de,* down; *glutire,* to swallow]. The process of swallowing.

dehydrocholesterol (dēhī'drŏkōlēs'tĕrŏl). A sterol found in the skin and other tissues that forms vitamin D after activation by radiation.

deltoid (dĕl'toyd) [G. Δ delta; *eidos,* form]. More or less triangular in shape. A triangular muscle of the shoulder and upper arm.

demifacet (dĕmĭfās'ĕt) [L. *dimidius,* half; *facies,* face]. Part of parapophysis facet when divided between centra of two adjacent vertebrae.

dendrite (dĕn'drīt) [G. *dendron,* tree]. Nerve cell process that normally conducts impulses toward cell body.

denervate (dēnĕr'vāt) [L. *dē,* from; G. *neuron,* nerve]. To block, remove, or cut the nerve supply to a structure.

de novo (dīnō'vō) [L]. Once more; anew; again.

dens (dĕnz) [L. *dens,* tooth]. Tooth or toothlike process; odontoid process of axis or epistropheus.

dentin (dĕn'tĭn) [L. *dens,* tooth]. A hard, elastic substance constituting the greater part of the tooth.

deoxyribonucleic acid (DNA). (dēŏk'sĕrī'bōno͞oklā'ĭk) Stable nucleic acid component of kinetoplasts, chromosomes, bacterial cells, and phages; consists structurally of two spirals linked transversely and constitutes a pattern or template for replication.

depolarization (dēpō'lărīzā'shŭn) [L. *dē,* from; *polus,* pole]. Loss of polarity or the polarized state.

-derm (dŭrm) [L. *derma,* skin or covering]. A word used either as a prefix or suffix referring to the skin or a covering.

dermatitis (dĕr'mătī'tĭs) [L. *derma,* skin; *ītis,* inflammation]. Inflammation of skin evidenced by itching redness and various skin lesions.

dermis (dĕr'mĭs) [G. *derma,* skin]. The skin; cutis vera or true skin. Corium or derma.

desquamate (dĕs'kwĕmāte) [L. *de,* from; *squamāre,* to scale off]. Shedding of the epidermal surface cells.

detoxify (dētŏks'ĭfī) [L. *de,* from; G. *toxikon,* poison; L. *facere,* to make]. To remove the toxic or poisonous quality of a substance.

dextrose (dĕks'trōs) [L. *dexter,* right; *ose,* sugar]. Another name for glucose, a simple sugar (monosaccharide). Also, grape sugar.

di- [G]. Prefix: twice or double.

diabetes insipidus (dī'ăbē'tēz ĭnsĭp'ĭdŭs) [G. *dia-*

betes, passing through; L. *in,* not; *sapidus,* tasty]. Polyuria and polydipsia caused by inadequate secretion of vasopressin, the antidiuretic hormone, by the neurohypophysis (main portion of the posterior lobe of the pituitary gland).

diabetes mellitus (dī'ăbē'tēz mĕllī'tŭs) [G. *diabētēs,* passing through; *meli,* honey]. A disorder of carbohydrate metabolism, characterized by hyperglycemia and glycosuria and resulting from inadequate production or utilization of insulin.

dialysis (dīăl'ĭsĭs) [G. *dia,* asunder; *lysis,* loosening]. Separation of dissolved crystalloids and colloids through semipermeable membrane, crystalloids passing more readily; permeation.

diapedesis (dī'ăpēdē'sĭs) [G. *diapēdesis,* leaping through]. Emigration of white blood corpuscles through undamaged walls of capillaries into surrounding tissues.

diaphragm (dī'ăfrăm) [G. *diaphragma,* midriff]. A partition, partly muscular, partly tendinous, separating cavity of chest from abdominal cavity in mammals; a most important organ of breathing.

diaphysis (dīăf'ĭsĭs) [G. *diaphysis,* a growing through]. The shaft or middle part of a long cylindrical bone.

diarrhea (dīărē'ă) [G. *dia,* through; *rein,* to flow]. Passage of frequent watery stools: a symptom of gastrointestinal disturbance.

diarthrosis (dī'ărthrō'sĭs) [G. *dis,* twice; *arthron,* joint]. An articulation allowing considerable movement.

diastasis (dīăs'tăsĭs) [G. a separation]. In the cardiac cycle, the time when there is little change in length of muscle fibers and filling of the chamber is very slow. It is followed by contraction of the muscle of the chamber.

diastole (dīăs'tōlē) [G. *diastellein,* to expand]. The relaxation phase of the cardiac cycle, when fibers are elongating and the filling of the chamber is rapid.

dicumarol (dīkōō'mărōl). Proprietary name for bishydroxycoumarin (USP), an anticoagulant that decreases activity of prothrombin in the blood plasma and hence increases prothrombin time.

diencephalon (dī'ĕnsĕf'ălŏn) [G. *dia,* between; *engkephalos,* brain]. Hindpart of forebrain; lies between telencephalon and mesencephalon; includes thalamus and hypothalamus.

diffusion (dĭfū'zhŭn) [L. *dis,* apart; *fundo,* pour]. The act or process of scattering, dissemination, dispersion, circulation. The spontaneous intermingling of the molecules of two fluids, as of gases; distinguished from mixture by mechanical force or by the action of gravity.

digastric (dīgăs'trĭk) [G. *di,* two; *gastēr,* belly]. Two-bellied, *appl.* muscles fleshy at ends, tendinous in middle; biventral; *appl.* one of the suprahyoid muscles; *appl.* branch of the facial nerve; *appl.* a lobule of cerebellum; *appl.* fossa of mandible and of temporal bone.

digitalis (dĭjĭtăl'ĭs) [L. *digitus,* finger]. The dried leaves of the purple foxglove plant; used in powdered form as a heart stimulant.

diopter (dīŏp'tĕr) [G. *dioptron,* something that can be seen through]. The refractive power of a lens with a focal distance of one meter.

diploë (dĭp'lōē) [G. *diploē,* double]. The cancellous tissue between outer and inner tables of certain skull bones.

diploid (dĭp'loyd). Cell having twice the number of chromosomes present in the egg or sperm of a given species.

disaccharide (dīsăk'ărĭd) [G. *dis,* twice; L. *saccharum,* sugar]. A sugar composed of two simple sugars, e.g., lactose, maltose, sucrose.

dissect (dĭsĕct') [L. *dissectus,* fr. *dis,* apart; *seco,* cut]. To cut apart or divide, as an animal body or a plant, in order to examine the structure; anatomize. To analyze and discuss critically.

distal (dĭs'tăl) [L. *distāre,* to be distant]. Farthest from the center, from a medial line, or from the trunk.

diuresis (dīūrē'sĭs) [G. *dia,* through; *ourein,* to urinate]. Secretion and passage of abnormally large amounts of urine.

diuretic (dīūrĕt'ĭk). A substance causing diuresis.

diverticulum (dīvĕrtĭk'ūlŭm) [L. *diverticulare,* to turn aside]. An outpocketing, sac, or pouch in wall of a canal or other hollow organ.

DNA. Deoxyribonucleic acid, a fraction of the nucleoprotein of chromosomes that is the material of the gene.

dogma (dawg'mă) [G. *dogma(t-),* opinion, fr. *dokeo,* think]. A doctrine or system of doctrine concerning religious truth as maintained by the Christian church or any portion of it; hence, a statement of religious faith or duty formulated by a body possessing or claiming authority to decree or decide. Doctrine asserted and adopted on authority, as distinguished from that which is the result of one's own reasoning or experience; a dictum.

dominance (dŏm'ĭnĕns) [L. *dominare,* to rule]. In inheritance of characteristics, the effects of one allele overshadow the effects of the other, and the character determined by the dominant (stronger) gene prevails.

ductus arteriosus (dŭk'tŭs ărtērēo'sŭs) [L. *ducere,* lead; G. *artēria,* artery]. A channel of communication between main pulmonary artery of the fetus and aorta.

ductus deferens (dŭk'tŭs dĕf'ĕrĕns) [L. *ducere,* to lead; *deferens,* to carry away]. The excretory duct of the testis leading from the testis to the ejaculatory duct.

ductus venosus (dŭk'tŭs vĕnō'sŭs) [L. *ducere,* to lead; *vena,* vein]. Smaller, shorter, and posterior of two branches into which the umbilical vein divides after entering the abdomen; empties into the inferior vena cava.

duodenum (dōōēdē'nŭm, dōōah'dĕnŭm) [L. *duodeni,* twelve each]. The short upper portion of the small intestine.

dura mater (dū'ră mā'tĕr) [L. *dura,* hard; *mater,* mother]. The outermost and toughest meninx of the spinal cord and brain.

dysmenorrhea (dĭsmĕnĕrē'ă) [L. *dys,* bad, difficult, or painful; *men,* month; *rein,* to flow]. Painful or difficult menstruation.

dyspnea (dĭspnē'a) [L. *dys,* bad, difficult; *pnoe,* breathing]. Labored or difficult breathing.

dystrophy (dĭs'trĕfē) [L. *dys,* bad, difficult; G. *trephein,* to nourish]. Degeneration of an organ resulting from poor nutrition, abnormal development, infection, or unknown causes.

eardrum (ēr drŭm) [A.S. *eāre + drum*]. The tympanum (tympanic membrane) that forms the lateral wall of the middle ear cavity. It is composed of collagenous tissue, covered laterally by thin skin and medially by the mucous membrane of the middle ear. It is set into vibration by sound waves striking it.

ectoderm (ĕk'tĕdĕrm) [G. *ektos,* outside; *derma,* skin]. The outer germ layer of a multicellular animal from which develop the nervous system, special senses, and certain endocrine glands.

-ectomy (ĕk'tĕmē) [G. *ektomē,* a cutting out]. Suffix *pert.* surgical removal of an organ or gland.

ectopic (ĕktŏp'ĭk) [G. *ek,* out; *topia,* place]. In abnormal place or position.

eczema (ĕk'sĕmă) [G. *ekzein,* to boil out]. A chronic or acute inflammation of the skin characterized by the development of skin lesions.

edema (ĕdē'mă) [G. *oidēma,* swelling]. A condition in which the body tissues contain an excessive amount of tissue fluid. Dropsy.

EEG. Abbr. for electroencephalogram.

effector (ĕfĕk'tŏr) [L. *efficere,* to carry out]. A muscle or gland that responds to impulses carried to it by nerves.

efferent (ĕf'ĕrĕnt) [L. *ex,* out; *ferre,* to carry]. Conveying from a center or specific point of reference. Used for nerves, blood, and lymphatic vessels.

egestion (ējĕs'chŭn) [L. *ex,* out; *gerere,* to carry]. Elimination at the inferior end of the digestive tube.

ejaculation (ējăkūlā'shŭn) [L. *ex,* out; *jacere,* to throw]. Ejection of semen from the male urethra during sexual excitement, or discharge of secretions from vaginal orifice.

ejaculatory duct (ējăk'ūlătōrē dŭkt) [L. *ex,* out; *jacere,* to throw]. A continuation of the ductus deferens from the point of entrance of the seminal vesicles to the the prostatic urethra.

elastance (ēlăst'ănce) [G. *elasticos,* elastic]. The power of elastic recoil after being stretched. Also, elasticity.

elastin (ēlăs'tĭn) [G. *elaunein,* to draw out]. The scleroprotein of which elastic fibers are composed.

electrocardiogram (ECG) EKG) (ēlĕk'trōkăr'deogrăm) [G. *ēlektron,* amber; *kardia,* heart; *gramma,* writing]. A record of the electrical activity of the heart; gives important information concerning the spread of excitation to the different chambers of the heart and is of value in the diagnosis of cases of abnormal cardiac rhythm and myocardial damage.

electrolyte (ēlĕk'trōlīt) [G. *ēlektron,* amber; *lytos,* soluble]. A substance that dissociates into electrically charged ions; a solution capable of conducting electricity because of the presence of ions.

electron (ēlĕk'trŏn) [G. *ēlektron,* amber]. An electrically charged particle that is a component of the atom and of matter; a corpuscle; a particle of negative electricity.

element [L. *elementum,* a rudiment]. A substance that cannot be separated into its components by conventional chemical change.

elephantiasis (ĕl'ĕfăntī'ăsĭs) [G. *elephas,* elephant]. A chronic condition characterized by pronounced hypertrophy of the skin and subcutaneous tissues resulting from obstruction of the lymphatic vessels. The lower extremities and the scrotum are parts most frequently involved.

eliminate (ēlĭm'ĭnāte) [L. *e,* out; *limen,* threshold]. To rid the body of wastes by emptying of a hollow organ; to expel.

embolus (ĕm'bōlŭs) [G. *embolos,* plug]. A mass of undissolved matter present in a blood or lymphatic vessel brought there by the blood or lymph current.

embryo (ĕm'breō) [G. *embryon,* embryo]. A young organism in early stages of development. The first two months of embryological development in man.

embryonic disk (ĕm'brēŏn'ĭk dĭsk) [G. *embryon,* embryo; *diskos,* disk]. An oval, two-layered disk of cells in the blastocyst of a mammal from which the embryo proper develops. Forms about two weeks after fertilization.

emeiocytosis (ēmēōsītō'sĭs) [G. *emein,* to vomit; *cyte,* cell; *-ōsis,* state of]. Elimination of materials from a cellular vacuole by fusion of the vacuole with the cell membrane and release of contents to outside of the cell.

emesis (ĕm'ĕsĭs) [G. *emein,* to vomit]. Vomiting.

-emia (ē'mēă) [G. *haima,* blood]. Suffix: blood.

emmetropic (ĕmĕtrōp'ĭk) [G. *emmetros,* in due measure; *opsis,* sight]. Normal in vision.

emphysema (ĕm'fĭsē'mă) [G. *emphysan,* to inflate]. Distention of tissues by gas or air in the interstices. A condition in which the alveoli of the lungs become distended or ruptured, usually the result of an interference with expiration, or loss of elasticity of the lung.

emulsify (ēmŭl'sĭfī) [L. *emulsio,* emulsion; *facere,* to make]. To form into an emulsion (a mixture of two liquids not mutually soluble).

en- [G. *en, in*]. Prefix: in.

encephalo- (ĕnsĕf'ălō) [G. *enkephalos,* brain]. Prefix *pert.* brain or cerebrum.

endo [G. *endon,* within]. Prefix: within; inner.

endocardium (ĕn'dōkăr'dēŭm) [G. *endon,* within; *kardia,* heart]. The inner layer of the heart wall consisting of endothelium and a subendothelial connective tissue of collagenous fibers.

endochondral (ĕn'dōkŏn'drăl) [G. *endon,* within; *chondros,* cartilage]. Beginning or forming inside the cartilage, *appl.* ossification.

endocrine (ĕn'dōkrĭn) [G. *endon,* within; *krinein,* to separate]. A ductless gland. Internally secreting into the bloodstream.

endoderm (ĕn'dōdĕrm) [G. *endon,* within; *derma,* skin]. The innermost of three germ layers of the embryo.

endolymph (ĕn'dōlĭmf) [G. *endon,* within; L. *lympha,* water]. The fluid found inside the membranous labyrinth.

endometrium (ĕndōmē'trēŭm) [G. *endon,* within; *mētra,* uterus]. The mucous membrane lining the inner surface of the uterus.

endomysium (ĕn'dōmĭz'ēŭm) [G. *endon,* within; *mys,* muscle]. The connective tissue binding muscle fibers.

endoskeleton (ĕn'dōskĕl'ĕtŏn) [G. *endon,* within; *skeletos,* dried up]. Internal skeleton, *opp.* exoskeleton.

endosteum (ĕndŏs'tēŭm) [G. *endon,* within; *osteon,* bone]. The internal periosteum lining the cavities of bones.

endothelium (ĕn'dōthē'lēŭm) [G. *endon,* within; *thēlē,* nipple]. A simple squamous epithelium lining the heart, blood, and lymphatic vessels.

entero- [G. *enteron,* intestine]. Prefix *pert.* intestines.

enteroceptive (ĕntĕrōsĕp'tĭv) [G. *enteron,* intestine; L. *capere,* to take]. Originating within body viscera.

enterocrinin (ĕn'tĕrŏk'rĭnĭn) [G. *enteron,* intestine; *krinein,* to separate]. Hormone from animal intestines that aids digestion by stimulating the secretion of intestinal juice by the intestinal glands.

enterogastric (ĕn'tĕrōgăs'trĭk) [G. *enteron,* intestine; *gastēr,* belly]. *Appl.* intestines and stomach.

enterokinase (ĕn'tĕrōkī'nās) [G. *enteron,* intestine; *kinein,* to move]. Incomplete enzyme of intestinal juice which converts trypsinogen into trypsin.

enzyme (ĕn'zīm) [G. *en,* in; *zyme,* leaven]. An organic catalyst produced by living organisms and acting to cause alteration in rates of chemical reactions without being destroyed in the reaction.

eosinophil (ē'ēsĭn'ĕfĭl) [G. *eos,* dawn (rose-colored); *philein,* to love]. A cell or cellular structure that stains readily with acid stain, eosin; specifically an eosinophil leukocyte.

epi- [G.]. Prefix: upon, at, outside of, in addition to.

epicardium (ĕp'ĭkăr'dēŭm) [G. *epi,* upon; *kardia,* heart]. The thin transparent outer layer of the heart wall, also called visceral pericardium.

epicondyle (ĕp'ĭkŏn'dīl) [G. *epi,* upon; *kondylos,* knuckle]. A medial and a lateral protuberance at the distal end of the humerus and femur.

epicranius (ĕp'ĭkrā'nēŭs) [G. *epi,* upon; *kranion,* skull]. The scalp muscle, consisting of occipitalis and frontalis, connected by galea aponeurotica; occipitofrontalis.

epidermis (ĕp'ĭdĕr'mĭs) [G. *epi,* upon; *derma,* skin]. Cuticle, or outer layer of skin.

epididymis (ĕpĭdĭd'mĭs) [G. *epi,* upon; *didymos,* testes]. A long convoluted tubule resting on the testis and conveying sperm to the ductus deferens.

epiglottis (ĕp'ĭglŏt'ĭs) [G. *epi,* upon; *glotta,* tongue]. A leaf-shaped elastic cartilage, between root of tongue and entrance to larynx. Aids in closing the larynx when swallowing.

epileptic (ĕpĭlĕp'tĭk) [G. *epilēptikos,* pert. to a seizure]. *Pert.* disturbance of consciousness occurring during epilepsy; an individual suffering from the disorder of epilepsy.

epimysium (ĕp'ĭmĭz'ēŭm) [G. *epi,* upon; *mys,* muscle]. The sheath of areolar tissue investing the entire muscle.

epinephrine (ĕp'ĭnĕf'rĭn) [G. *epi,* upon; *nephros,* kidney]. This substance and norepinephrine are the two active hormones produced by the adrenal medulla.

epiphyseal cartilage (ĕp'ĭfĭz'ēăl) [G. *epi,* upon; *phyein,* to grow]. A cartilage separating a parent bone from a secondary bone-forming (ossification) center in the developing infant and child. As growth proceeds it becomes a part of the larger (or parent) bone. Enables a bone to grow in length.

epiphysis (ĕpĭf'ĭsĭs) [G. *epiphysis,* a growing upon]. One of the two ends of a long bone. A center for ossification at each extremity of long bones.

epispadias (ĕp'ĭspā'dēăs) [G. *epi,* upon; *spadōn,* a rent]. Congenital opening of urethra on dorsum of penis; in the female, opening by separation of the labia minora and a fissure of the clitoris.

epistaxis (ĕp'ĭstăk'sĭs) [G. *epistaxein,* to bleed from nose]. Hemorrhage from the nose.

epithelium (ĕp'ĭthē'lēŭm) [G. *epi,* upon; *thēlē,* nipple]. Any cellular tissue covering a free surface or lining a tube or cavity.

eponychium (ĕp'ōnĭk'eŭm) [G. *epi,* upon; *onych,* nail]. The fold of stratum corneum overlapping the lunula of nail.

erythroblastosis fetalis (ĕrĭth'rōblăstō'sĭs fētăl'ĭs) [G. *erythros,* red; *blastos,* germ; *ōsis,* condition; L. *fetus,* fetus; G. *ismos,* condition]. A hemolytic disease of the newborn characterized by anemia, jaundice, and enlargement of the liver and spleen, generalized edema; due to the development in an Rh negative mother of antibodies against an Rh positive fetus.

erythrocruorin (ĕrĭth'rōkrōō'ōrĭn) [G. *erythros,* red; L. *cruor,* blood]. Red iron-containing respiratory pigment; found in some cones of the retina and is red-sensitive.

erythrocyte (ĕrĭth'rōsīt) [G. *erythros,* red; *kytos,* hollow]. A red blood corpuscle.

erythropoiesis (ĕrĭth'rōpoyē'sĭs) [G. *erythros,* red; *poiesis,* making]. The formation of erythrocytes or red blood cells.

erythroprotein. A substance produced by the kidney; stimulates the production of red blood cells.

eschar (ĕs'kăr) [G. *eschara,* scab]. A slough (dead matter or tissue); debris developing from a burn.

esthesia (ĕsthē'zēă) [G. *aisthēsis,* sensation or feeling]. Perception, sensation, or feeling.

estrogen (ĕs'trōjĕn) [G. *oistros,* mad desire; *gennan,* to produce]. A female sex hormone; stimulates development of female primary and secondary sex characteristics; the follicular hormone.

estrus, oestrus (ĕs'trŭs) [G. *oistros,* mad desire]. The recurrent period of sexual activity in mammals other than primates, called heat, characterized by congestion of and secretion by the uterine mucosa, proliferation of vaginal epithelium, swelling of the vulva, ovulation, and acceptance of the male by the female.

ethics (ĕth'ĭks) [G. *ēthikos,* fr. *ēthos,* character]. The science of human duty; moral science. The basic principles of right action.

ethology (ēthŏl'ăjē) [G. *ēthos,* custom; *logos,* discourse]. Bionomics; study of habits in relation to habitat; study of behavior.

eu- (ū) [G. *eu,* well]. Prefix: normal, well, good.

eunuchoidism (ūnŭk'oydĭzm) [G. *eunē,* bed; *eidos,* form; *echein,* to guard; *ismos,* condition]. Deficient production of male hormone, androgen, by the testes; causes atrophy of accessory organs of reproduction and of secondary sex characteristics. Similar action in female.

euphoria (ūfō'rēă) [G. *eu,* well; *pherein,* to bear]. An exaggerated feeling of well-being.

eustachian tube (ūstā'kēăn, -shĕn) [Bartolommeo Eustachio, It. anatomist, 1524–1574]. The auditory tube from the middle ear to the pharynx 3–4 cm long and lined with mucous membrane.

evolution (ĕv'ēlōō'shŭn) [L. *evolvere,* to unroll]. The gradual development or organisms from preexisting organisms since the dawn of life.

ex- [L.] Prefix: out, away from, completely.

excretion (ĕkskrē'shŭn) [L. *ex,* out; *cernere,* to sift]. The passage of waste products and other material from the internal to the external environment through living membranes.

exocrine (ĕks'ĭkrĭn) [G. *exō,* outside, *krinein,* to separate]. The external secretion of a gland. Opposed to endocrine. Term applied to glands whose secretion reaches an epithelial surface either directly or through a duct.

exogenous (ĕksŏj'ĕnŭs) [G. *exō,* out, away from; *gennan,* to produce]. Originating outside of the cell or organism.

exoskeleton (ĕk'sōskĕl'ĕtŏn) [G. *exō,* without; *skeletos,* hard]. A hard supporting structure secreted by ectoderm or by skin.

expiration (ĕk'spīrā'shŭn) [L. *ex,* out; *spirare,* to breathe]. Expulsion of air from the lungs. Death.

exteroceptive (ĕk'stĕrōsĕp'tĭv) [L. *exterus,* outside; *receptus,* having received]. *Pert.* end organs receiving impressions from without.

extracellular (ĕkstrăsĕl'ūlăr) [L. *extra,* outside; *cellula,* little room]. Occurring outside the cell; diffused out of the cell.

extrinsic (ĕkstrĭn'sĭk) [L. *extrinsecus,* on outside]. Acting from the outside, not wholly within the part; *appl.* muscles, etc.

exudate (ĕks'ūdāt) [L. *ex,* out; *sudāre,* to sweat]. Accumulation of fluid in a body cavity or on a body surface. Pus, serum, or the passing of same.

facet (făs'ĕt) [F. *facette,* small face]. A smooth, flat, or rounded surface for articulation.

facilitation (făsĭl'ĭtā'shŭn) [L. *facilis,* easy]. An increased ease of passage of a nerve impulse across a synapse.

FAD. *Abbr.* for flavine adenine dinucleotide, one of several hydrogen acceptors.

fascia (făsh'ēă, făsh'ēă) [L. *fascia,* band]. An ensheathing band or layer of connective tissue, covering, supporting, and separating muscles. Superficial fascia is the hypodermis; deep fascia is around muscles.

fasciculus (făsĭk'ūlŭs) [L. *fasciculus,* small bundle]. A fascicle; a group, bundle, or tract of nerve or muscle fibers, as of medulla spinalis.

fasciculation (făsĭck'ūlā'shŭn). A localized contraction of one or a few motor units in a skeletal muscle.

feces (fē'sēz) [L. *faeces,* refuse]. Stools; excreta; dejecta; excrement. Body waste such as food residue, bacteria, epithelium, and mucus, discharged from the bowels by way of the anus.

feedback (reciprocal or feedback control). One endocrine produces a hormone that affects a second endocrine. The second endocrine produces its hormone, which influences secretion of the first endocrine. If the effect of the second hormone is to inhibit secretion of the first endocrine, the effect is said to be a negative feedback. If the effect is to increase secretion, it is a positive feedback.

femoral (fĕm'ĕrăl) [L. *femur,* the thigh bone]. Pertaining to the thigh bone or femur.

fertilization (fĕr'tĭlĭzā'shŭn) [L. *fertilis,* fertile]. The union of male and female pronuclei.

fetus (fē'tŭs) [L. *foetus,* offspring]. Product of conception after second month of gestation, when it has assumed human form.

fiber (fī'bĕr) [L. *fibra,* thread]. A large threadlike or ribbonlike structure; a muscle cell; usually composed of smaller units.

fibril (fī'brĭl) [L. *fibrilla,* a little fiber]. A very small threadlike structure, often the component of a muscle or nerve cell, or a connective tissue fiber.

fibrillate (fī'brĭlāt, fī'brĭlāt) [L. *fibrilla,* a little fiber]. Spontaneous uncoordinated quivering of a muscle fiber, as in cardiac muscle.

fibrin (fī'brĭn) [L. *fibra,* band]. An insoluble protein found in blood after coagulation, readily digested in gastric juice.

fibrinogen (fībrĭn'ĕjĕn) [L. *fibra,* band; G. *-genes,* producing]. A soluble protein of blood that, by activity of thrombin, yields fibrin and produces coagulation.

fibroblast (fī'brĕblăst) [L. *fibra,* band; G. *blastos,* bud]. A primordial connective tissue cell.

fibrosis (fībrō'sĭs) Abnormal deposition or increase in fibrous tissue in a body part, organ, or tissue.

filiform (fĭ'lĭfŏrm) [L. *filum,* thread; *forma,* shape]. Threadlike or hairlike in shape.

filter [L. *fīltrāre,* to strain through]. To strain or separate on the basis of size; a device to strain liquids.

filtrate (fĭl'trāt). The name given to the fluid that has passed through a filter.

filtration (fĭltrā'shŭn) [L. *fīltrāre,* to strain through]. Passing a fluid under pressure through a filter paper, or a selective membrane; a filtrate is formed.

fimbria (fĭm'brēă) [L. *fimbria,* fringe]. Any fringelike structure; as on the infundibulum of the uterine tube.

fistula (fĭs'chōōlă) [L. *fistula,* pipe]. Pathological or artificial pipelike opening; trachea or water-conducting vessel.

fixators (fĭksā'tŏrz) [L. *fixatio,* to hold]. Muscles that maintain the position of the body; that fix one part to support the movement of another.

flagella (flăjĕl'ă) *(plu.)* [L. *flagellum,* whip]. The lashlike processes of many Protista and of cells, as in choanocytes and certain male gametes.

flatus (flā'tŭs) [L. *flatulentia,* a blowing]. The gas in the digestive tract.

flora (flō'ră) [L. *flos,* flower]. Plant life; the microorganisms of the bowels, adapted to life in that organ.

follicle (fŏl'ĭcŭl) [L. *folliculus,* a little bag]. A small hollow structure containing cells or secretion; an ovarian or hair follicle.

fontanel (fŏn'tănĕl') [F. *fontanelle,* little fountain]. A gap or space between bones in the cranium, closed only by membrane.

foramen (fŏrā'mĕn) [L. *foramen,* opening]. Any small perforation; the opening through coats of ovule; aperture through a shell, bone, or membranous structure.

fornix (fŏr'nĭks) [L. *fornix,* vault, arch]. A moatlike area around the cervix where it protrudes into the superior end of the vagina.

fossa (fŏs'ă) [L. *fossa,* ditch]. A pit or trenchlike depression.

fossa ovalis (fŏs'ă ōvăl'ĭs) [L. *fossa,* a furrow or shallow depression; *ovalis,* egg-shaped]. An opening in the thigh through which the large saphenous vein passes. An oval depression in the interatrial septum marking the position of the foramen ovale of the fetus.

fovea (fō'vēă) [L. *fovea,* depression]. A small pit, fossa, or depression.

frenulum (frĕn'ūlŭm) [L. *frenum,* bridle]. A fold of membrane, as of tongue, clitoris, etc.

FSH. *Abbr.* for follicle-stimulating hormone, secreted by the anterior lobe of the hypophysis.

fulcrum (fŭl'krŭm) [L. *fulcire,* to prop or support]. The point around which a lever turns or pivots.

fundus (fŭn'dŭs) [L. *fundus,* sling or base]. The larger, usually blind end of an organ or gland.

fungiform (fŭn'jĭfŏrm) [L. *fungus,* mushroom; *forma,* shape]. Fungoid or shaped like a fungus; *appl.* tongue papillae.

fusiform (fū'sĭfŏrm) [L. *fūsus,* spindle; *forma,* shape]. Tapering at both ends; spindle-shaped.

galactose (gălăk'tōs) [G. *gala,* milk; *-ose,* sugar]. Milk sugar, $C_6H_{12}O_6$, a monosaccharide or simple hexose sugar. Galactose is an isomer of glucose and is formed, along with glucose, in the hydrolysis of lactose.

galea aponeurotica (găl'ēă ăp'ōnōōrŏt'ĭkă) [L. *galea,* helmet; G. *apo,* from; *neuron,* sinew]. A helmet-shaped structure; epicranial aponeurosis of the scalp muscle or occipitofrontalis.

gamete (găm'ēt) [G. *gamete,* a wife or spouse]. A male or female reproductive cell; that is, a sperm or egg (ovum).

gametogenesis (gămē'tōjĕn'ĕsĭs) [G. *gamete,* wife or spouse; *genesis,* birth or origin]. The formation of gametes; meiosis, spermatogenesis; oogenesis.

gamma (găm'ŭ) [G. γ, letter g of alphabet]. The third item of a series; a microgram (one-millionth gram).

gamma globulin (gă'mŭ glŏb'ūlĭn). A plasma protein carrying most of the blood-borne antibodies.

ganglion (găng'glēŏn) [G. *ganglion,* knot]. A mass of nerve cell bodies giving origin to nerve fibers; a nerve center outside of the central nervous system. A cystic tumor developing on a tendon.

gangrene (găn'grēn) [G. *gangraina,* an eating sore]. A necrosis, or death, of tissue, usually due to deficient or absent blood supply.

gastric [G. *gastēr,* stomach]. *Pert.* stomach.

gastrocnemius (găs'trŏknē'mēŭs) [G. *gastēr,* stomach; *knēmē,* tibia]. Large muscle of calf of leg.

gemellus (jĕmĕl'ŭs) [L. *gemellus,* twin]. Either of two muscles, superior and inferior, from ischium to greater trochanter and to trochanteric fossa, respectively.

gene (jēn) [G. *genos,* descent]. A unit hereditary factor in the chromosome; DNA.

genetics (jĕnĕt'ĭks) [G. *genesis,* descent, origin]. That part of biology dealing with heredity and variation.

genotype (jĕn'ōtīp) [G. *genos,* descent; *typos,* type or kind]. Basic hereditary combination of genes of an organism. A type species. Group marked by same hereditary characteristics.

germ layer. One of the three basic tissue layers (ectoderm, endoderm, and mesoderm) formed in the embryo that give rise to all body tissues and organs.

gerontology (jĕrŏntŏl'ōjē) [G. *geron,* old man; *logos,* study of]. The study of the phenomena of old age.

gestation (jĕstā'shŭn) [L. *gestāre,* to bear]. The period of intrauterine development of a new organism; the time of pregnancy.

giantism (jī'ăntĭsm) [G. *gigas,* giant]. Abnormal development of the body or its parts.

gill slits (gĭl) [M.E. *gille,* gill]. A series of perforations leading from pharynx to exterior, persistent in lower vertebrates, embryonic in higher.

gingivae (jĭnjī'vē) *(plu.)* [L. *gingivae,* gums]. The gums.

ginglymus (gĭng'glĭmŭs) [G. *gingglymos,* hinge-joint]. An articulation constructed to allow of motion in one plane only.

glabella (glăbĕl'ă) [L. *glaber,* bald]. The space on forehead between superciliary ridges.

glands of Bowman (Sir W. Bowman, English histologist and surgeon, 1892). Glands found in the lamina of the olfactory region; secrete a watery fluid. Also Bowman's capsule of kidney trouble.

glenoid (glē'noyd [G. *glēnē,* socket; *eidos,* form]. Like a socket; *appl.* cavity into which head of humerus fits, the mandibular fossa, and various ligaments.

glia (glī'ă) [G. *glia,* glue]. A cell of the neuroglia; nonnervous tissues of the nervous system; supporting and nutritive cells of various types.

globin (glō'bĭn) [L. *globus,* globe]. The basic protein constituent of hemoglobin.

globulin (glŏb'ūlĭn) [L. *globulus,* globule]. One of a group of simple proteins insoluble in pure water but soluble in neutral solutions of salts of strong acids with strong bases.

globus pallidus (glō'bŭs păl'ĭdŭs) [L. *globus,* a globe or sphere; *pallidus,* pale]. Pale section within the lenticular nucleus of the brain.

glomerulonephritis (glŏmĕr'ūlōnĭfrī'tĭs) [L. *glomerulus,* little ball; G. *nephros,* kidney; *-ītis,* inflammation]. A form of nephritis in which the lesions involve primarily the glomeruli.

glomerulus (glŏmĕr'ūlŭs) [L. *glomerulus,* little ball]. A small structure in the malpighian body of the kidney made up of capillary blood vessels in a cluster. Plexuses of capillaries. Twisted secretory parts of sweat glands.

glottis (glŏt'ĭs) [G. *glottis,* back of tongue]. The slitlike opening into the larynx between the vocal cords.

glucagon (glo͞ok'ăgŏn) [G. *glykys,* sweet; *agōn,* assembly]. A pancreatic hormone formed in alpha cells of islets of Langerhans; stimulates glycogenolysis in the liver, causing increase in blood-sugar hyperglycemic-glycogenolytic factor, HGF.

glucocorticoid (glo͞o'kōkŏrt'ĭkoyd) [G. *gleukos,* sweet; L. *cortex;* cortex; G. *eidos,* form]. A general classification of adrenal cortical hormones that are primarily active in protecting against stress and in affecting protein and carbohydrate metabolism.

gluteal (glo͞o'tēăl) [G. *gloutos,* buttock]. Pertaining to the buttocks.

-glyc- (glĭk) [G. *glykys,* sweet]. *Pert.* glucose.

glycine (glī'sēn) [G. *glykys,* sweet]. Amino acetic acid or glycocoll, constituent of various proteins, particularly of collagen, elastin, and fibrin; plays part in the formation of creatine and other compounds.

glycogen (glī'kōjĕn) [G. *glykys,* sweet; *gennan,* to produce]. A polysaccharide commonly called animal starch, a whitish powder which can be prepared from mammalian liver and muscle and other animal tissues.

glycogenesis (glī'kōjň'sĭs) [G. *glykys,* sweet; *genesis,* formation]. The formation of glycogen, as occurs in humans after the eating of a carbohydrate meal.

glycogenolysis (glī'kōjĕnŏl'ĭsĭs) [G. *glykys,* sweet; *gennan,* to produce; *lysis,* dissolution]. Conversion of glycogen into glucose in body tissues.

glycolysis (glīkŏl'ĭsĭs) [G. *glykys,* sweet; *lysis,* dissolution]. Conversion of glucose and fructose to pyruvic acid.

glycoproteins (glī'kprō'tēnz) *(plu.)* [G. *glykys,* sweet; *prōteion,* first]. Compounds of protein with a carbohydrate, including mucins and mucoids; mucoproteins.

glycosuria (glī'kōsū'rēă) [G. *glykys,* sweet; *ouron,* urine]. The presence of sugar (glucose) in the urine.

goblet cell. A type of secretory cell found in the epithelium of the intestinal and respiratory tracts; a unicellular gland that secretes mucus.

goiter (goy'tĕr) [L. *guttur,* throat]. An enlargement of the thyroid gland.

Golgi apparatus or complex (gōl'jē) (C. Golgi, Italian histologist, 1844–1926). Cell constituents, localized or diffuse, often consisting of separate elements, the Golgi bodies. Concerned with cellular synthesis.

gomphosis (gŏmfō'sĭs) [G. *gomphos,* bolt]. Articulation by insertion of a conical process into a socket, as of roots of teeth into alveoli.

gonad (gōn'ăd) [G. *gonē,* birth]. A sexual gland, male or female. Produces eggs or sperms. Name used for the embryonic sex gland before differentiating into testis or ovary.

gout (gowt) [L. *gutta,* drop]. Paroxysmal metabolic disease marked by acute arthritis and inflammation of the joints. Joints affected may be at any location but gout usually begins in the knee or foot. Caused by excess uric acid.

-gram. Combining form *pert.* record drawn by a machine (e.g., cardiogram, telegram).

Graves' disease (Robert J. Graves, Irish physician, 1797–1853). Exophthalmic goiter.

gut [A.S.]. The primitive or embryonic digestive tube (fore-, mid-, hindgut); colloq. for intestine.

gyrus (jī'rŭs) [L. *gyrus,* circle]. A cerebral convolution; a ridge between two grooves.

[H+]. Symbol for hydrogen ion concentration.

hallucination (hăloōsĭnā'shŭn) [L. *alūcinārī,* to wander in mind]. A false perception having no basis in reality; may be auditory, visual, olfactory, etc.

hallux (hăl'ŭks) [L. *hallux,* great toe]. First digit of foot.

hamate (hā'māt) [L. *hamatus,* hooked]. Hooked or hook-shaped at the tip. A bone of the wrist.

haploid (hăp'loyd) [G. *haplous,* single, simple; *eidos,* form]. Having one half the normal number of chromosomes characteristic of the species; sperm and ova are haploid cells.

hapten (hăp'tĕn) [G. *haptein,* to seize]. The portion of an antigen containing the grouping on which the specificity depends.

haptoglobin (hăp'tōglōbĭn) [G. *haptein,* to seize]. A protein in the plasma that combines specifically with hemoglobin released from red blood cells.

haustra (haws'tra) [L. *haurire,* to draw, drink]. The sacculated pouches of the colon.

Haversian system (Clopton Havers, English physician and anatomist, 1650–1702). A central Haversian canal around which are concentric circles of bony matrix, the concentric lamellae, and also rings of osteocytes.

helix (hē'lĭks) (*plu.* helices) [G. *helix,* a coil]. A coil or spiral.

hematin (hĕm'ătĭn) [G. *haima,* blood]. A pigment formed by decomposition of hemoglobin, containing iron and having the property of carrying oxygen; protohem; $C_{34}H_{33}O_5N_4Fe$.

hematocrit (hēmăt'ōkrĭt) [G. *haima,* blood; *krinein,* to separate]. Centrifuge for separating solids from plasma in the blood. The volume of erythrocytes packed by centrifugation in a given volume of blood.

heme (hēm) [G. *haima,* blood]. A blood substance, oxidizing to hematin. Iron-containing red pigment.

hemiazygos (hĕm'ēăz'ĭgŏs) [G. *hemi,* half; *a,* without; *zygon,* yoke]. A tributary of the azygos vein draining dorsal thoracic wall.

hemiplegia (hĕmĭplē'jēă) [G. *hemi,* half; *plēgē,* a stroke]. Paralysis of one half of the body in a right-left direction.

hemocytoblast (hē'mĕsī'tōblăst) [G. *haima,* blood; *kytos,* hollow; *blastos,* bud]. Primitive stem cell from which all blood cells are derived.

hemoglobin (hē'mōglō'bĭn) [G. *haima,* blood; L. *globus,* sphere]. The red respiratory pigment of blood of vertebrates and a few invertebrates, differing according to species, consisting of hematin united to globin.

hemolysis (hēmŏl'ĭsĭs) [G. *haima,* blood; *lysis,* dissolution]. Destruction of red blood cells with release of hemoglobin into the plasma.

hemophilia (hĕm'ōfēl'ēă, hē'mō-) [G. *haima,* blood; *philein,* to love]. Hereditary blood disease characterized by greatly prolonged coagulation time.

hemopoiesis (hē'mōpoyēs'ĭs) [G. *haima,* blood; *poiēsis,* making]. The formation and development of blood cells.

hemorrhage (hĕm'ĕrĭj) [G. *haima,* blood; *rhēgnynai,* to burst forth]. Abnormal loss of blood from blood vessels, either internally or externally.

hemostasis (hēmŏs'tăsĭs) [G. *haima,* blood; *stasis,* stopping]. Arrest of bleeding or of circulation. Stagnation of blood.

Henle's loop (Freidrich G. J. Henle, Ger. anatomist, 1809–1885). A U-shaped portion of a renal tubule lying between the proximal and distal convoluted portions. Consists of a thin descending limb and a thicker ascending limb.

hepar (hē'păr) [G. *hēpar,* liver]. The liver.

heparin (hĕp'ărĭn). A mucoitin polysulfuric acid isolated from the liver, lung, and other tissues. It is produced by the mast cells of the liver and by basophil leukocytes. It inhibits coagulation by preventing conversion of prothrombin to thrombin by forming an antithrombin, and by preventing liberation of thromboplastin from blood platelets.

hepatic (hĕpăt'ĭk) *Pert.* the liver.

hernia (her'nēă) [G. *ernos,* a young shoot]. Protrusion or projection of an abdominal organ through the wall that normally contains it.

heterogeneous (hĕt'ĕrĕjē'nēŭs) [G. *heteros,* other; *genos,* a kind]. Consisting of dissimilar parts.

heterograft (hĕt'ĕrĕgrăft) [G. *heteros,* other; *graphium,* grafting knife]. Grafting of tissues or organs between different species.

heterophil (hĕt'ĕrĕfĭl, -fīl) [G. *heteros,* other; *philein,* to love]. *Pert.* antibody reacting with other than the specific antigen. *Pert.* tissue or microorganism that takes a stain other than the ordinary one.

hexose (hĕk'sōs) [G. *hex,* six; *ose,* sugar]. A six-carbon sugar (e.g., glucose, fructose).

hilum, hilus (hī'lŭm, -ŭs) [L. *hilum,* trifle]. Small notch, opening, or depression, usually where blood vessels enter and/or leave. Hilus of kidney.

hippocampus (hĭp'ōkăm'pŭs) [G. *hippos,* horse; *kampē,* bend]. Part of rhinencephalon forming an eminence extending throughout length of floor of inferior cornu of lateral ventricle; hippocampus major.

histamine (hĭs'tămēn) [G. *histos,* tissue; *ammōniakon,* resinous gum]. Product of the basic amino acid and food constituent histidine, in ergot and animal tissues; stimulates autonomic nervous system, gastric juice secretion, and capillary dilatation.

histio, histo [G. *histos,* tissue]. Prefix: tissue.

histiocyte (hĭs'tēōsīt) [G. *histos,* tissue; *kytos,* hollow]. A primitive blood cell giving rise to a monocyte; a conocyte of reticular origin, or a clasmatocyte derived from endothelium, a reticulo-endothelial cell; fixed macrophage in loose connective tissue; adventitial cell; rhagiocrine cell.

histology (hĭstŏl'ōjē) [G. *histos,* tissue; *logos,* discourse]. The science treating the detailed structure of animal or plant tissues; microscopic morphology.

histotoxic (hĭs'tōtŏks'ĭk) [G. *histos,* tissue; *toxikon,* poison]. Pertaining to a poisonous condition within the cells.

holocrine (hŏl'ĕkrĭn) [G. *holos,* whole; *krinein,* to separate]. *Appl.* glands in which secretory cells disintegrate and form part of secretion, as sebaceous glands.

homeodynamic (hōm'ēōdīnăm'ĭk) [G. *homoio,* alike; *dynamis,* power]. Refers to the constant adjustments necessary to maintain near-constant values in the body.

homeokinesis (hōm'ēōkĭnē'sĭs) [G. *homoios,* alike; *kinein,* to move]. Synonym for homeodynamics.

homeostasis (hōm'ēōstă'sĭs) [G. *homoios,* alike; *stasis,* standing]. The balance of nature; maintenance of equilibrium between organism and environment; the constancy of the internal environment of the body, as in birds and mammals.

homogenous (hōmŏj'ĕnŭs) [G. *homo,* likeness; *genos,* kind]. Like or uniform in structure or composition.

homograft (hō'mōgrăft) [G. *homo,* likeness; L. *graphium,* grafting knife]. Use of tissues or organs of the same species for grafting purposes. Isograft.

homologous (hōmŏl'ōgŭs) [G. *homo,* likeness; *logos,* relation]. Similar in origin and structure.

hormone (hŏrmōn') [G. *hormaein,* to excite]. A secretion of endocrine (ductless) glands that is carried in the circulation and that is an agent of control and coordination.

humor (hū'mŭr) [L. *humor,* fluid, moisture]. Any fluid or semifluid substance in the body.

humanist (hū'mănĭst) [L. *humanus,* belonging to a man]. One versed in the study of humanities; a medieval classical scholar. One who is versed in human nature. A system of thinking in which the interests and development of humans are made dominant.

hyaline (hī'ălēn) [G. *hyalos,* glass]. Crystalline, glassy, translucent. The most abundant cartilage in the human body.

hyalomere (hīăl'ōmēr) [G. *hyalos,* glass; *meros,* part]. The clear, homogeneous part of a blood platelet.

hyaluronidase (hī'ălōōrōn'īdās). An enzyme that increases tissue permeability by diminishing viscosity of hyaluronic acid; Duran-Reynals spreading factor. Liquefies the intercellular element.

hydration (hīdrā'shŭn) [G. *hydōr,* water]. The act or process of causing a chemical compound to become a hydrate (a compound formed by the union of molecules of water with other molecules or atoms).

hydrolysis (hīdrŏl'ĭsĭs) [G. *hydōr,* water; *lysis,* dissolution]. Any reaction in which water is one of the reactants, more specifically the combination of water with a salt to produce an acid and a base, one of which is more dissociated than the other. The reverse of neutralization. A chemical decomposition in which a substance is split into simpler compounds by the addition of and the taking up of the elements of water.

hydrolytic (hī'drōlĭt'ĭk) [G. *hydōr,* water; *lysis,* dissolution]. Pertaining to or causing the reaction between a chemical compound and the hydrogen and hydroxyl ions of water (hydrolysis).

hydrotropic (hī'drōtrŏp'ĭk) [G. *hydōr,* water; *tropē,* turn]. *Appl.* curvature of a plant organ toward a greater degree of moisture; *appl.* substances that make insoluble substances water-soluble.

hydrostatic pressure [G. *hydōr,* water; *statikos,* standing]. Pressure exerted by liquids.

hyperemia (hīpĕrē'mēă) [G. *hyper,* above; *haima,* blood]. An extra amount of blood in the area.

hyperesthesia (hīpĕrĕsthē'zēă) [G. *hyper,* above; *aisthēsis,* sensation]. Excessive sensitivity to sensory stimulation.

hyperglycemia (hī'pĕrglīsē'mēă) [G. *hyper,* above; *glykys,* sweet; *haima,* blood]. Increase of blood sugar, as in diabetes.

hyperhidrosis (hī'pĕrhīdrō'sĭs) [G. *hyper,* above; *hidrōs,* sweat; *-ōsis,* condition]. Excessive sweating.

hypermetropia (hī'pĕrmtrō'pēă) [G. *hyper,* above; *metron,* measure; *ōps,* eye]. Farsightedness; focal point falls behind retina.

hyperphagia (hīpĕrfā'jēă) [G. *hyper,* above; *phagein,* to eat]. Excessive intake of food.

hyperplasia (hī'pĕrplā'zhēă) [G. *hyper,* above; *plassein,* to mold]. Overgrowth; excessive or hyperplastic development due to increase in number of cells.

hypertensinogen (hī'pĕrtĕnsĭn'ōgĕn) [G. *hyper,* above; L. *tensiō,* tension; G. *genna,* to produce]. A globulin present in blood plasma that, when acted upon by the enzyme renin, forms angiotensin. Elaborated in the liver.

hypertension (hī'pĕrtĕn'shŭn) [G. *hyper,* above; L. *tensiō,* tension]. Tension or tonus above normal. A condition in which patient has a higher blood pressure than that judged to be normal.

hypertonic (hī'pĕrtŏn'ĭk) [G. *hyper,* above; *tonos,* intensity]. Having a higher osmotic pressure than another fluid such as the blood.

hypertrophy (hīpĕr'trōfē) [G. *hyper,* above; *trophē,* nourishment]. Excessive growth due to increase in size of cells.

hypnotic (hĭpnŏ'tĭk) [G. *hypnos,* sleep]. An agent producing sleep or depression of the senses.

hypoglycemia (hī'pōglīsē'mēă) [G. *hypo,* under; *glykys,* sweet; *haima,* blood]. Deficiency of sugar in the blood. A condition in which the glucose in the blood is abnormally low.

hypokalemia (hī'pōkălē'mēă) [G. *hypo,* under; L. *kali,* potash, potassium; G. *haima,* blood]. Low blood potassium.

hypokinetic (hī'pōkĭnĕt'ĭk) [G. *hypo,* under; *kinesis,* motion]. *Pert.* decreased motor reaction to stimulus.

hyponychium (hī'pōnĭk'ēŭm) [G. *hypo,* under, *onyx,* nail]. A thickened layer of stratum corneum at the distal end of digit under the free edge of nail.

hypospadias (hī'pŏspā'dēăs) [G. *hypo,* under; *span,* to draw]. Congenital opening of the male urethra upon the undersurface of the penis; also a urethral opening into vagina.

hypotension (hī'pōtĕn'shŭn) [G.]. Low blood pressure.

hypothalamus (hī'pōthăl'ămŭs) [G. *hypo,* under; *thalamos,* chamber]. Region below thalamus; structures forming greater part of floor of third ventricle.

hypothetical (hī'pōthĕt'ĭcăl) [G. *hypo,* under, less than; *tithēmi,* place]. Having the nature of or based on hypothesis; assumed conditionally or tentatively as a basis for argument or investigation.

hypotonic (hī'pōtŏn'ĭk) [G. *hypo,* under; *tonos,* tension]. Having a lower osmotic pressure than that of another fluid, as of serum.

hypoxemia (hī'pŏksē'mēă) [G. *hypo,* under; *oxys,* acid; *haima,* blood]. Lowered blood oxygen.

hypoxia (hīpŏks'ēă) [G. *hypo,* under; *oxys,* acid]. Anoxia; lack of an adequate amount of oxygen in inspired air such as occurs at high altitudes; reduced oxygen content or tension.

ICSH (interstitial cell stimulating hormone). *Syn:* Luteinizing hormone. A hormone produced by the anterior lobe of the hypophysis that induces ovulation and the formation of the corpus luteum. Also stimulates development of interstitial cells of the testes.

icteric (ĭktĕr'ĭk) [G. *ikteros,* jaundice]. *Pert.* jaundice; excessive accumulation of bile pigments in the body tissues.

idiopathic (ĭdēōpăth'ĭk) [G. *idio,* own; *pathos,* disease]. Any condition arising without a clear-cut cause; of spontaneous origin.

ileum (ĭl'ēŭm) [L. *ileum,* groin]. The terminal two-thirds of the small intestine; about 12 feet.

iliac (ĭl'ēăk) [L. *ilia,* flanks]. Pertaining to the ilium, one of the bones of each half of the pelvis.

immune (ĭmūn') [L. *immunis,* safe]. Protected from getting a given disease.

immunoglobulins (ĭm'mūnōglŏb'ūlĭnz) [L. *immunis,* safe; *globulus,* globule]. The system of closely related though not identical proteins capable of acting as antibodies.

impermeable (ĭmpĕrm'ēăbl) [L. *in,* not; *permeāre,* to pass through]. Not allowing passage.

implantation (ĭm'plăntā'shŭn) [L. *in,* into; *planta,* plant]. The act of inserting or grafting; embedding or nidation of fertilized ovum in lining of uterus.

impulse (ĭm'pŭls) [L. *impellere,* to drive out]. A physiochemical or electrical change transmitted along nerve fibers or membranes.

incisura (ĭnsī'zhŭră) [L. *incidere,* to cut into]. Notch, depression, or indentation, as in bone, stomach, liver, etc.

incus (ĭn'kŭs) [L. *incus,* anvil]. In the middle ear, the middle of the three ossicles in the tympanum, the anvil.

infundibulum (ĭn'fŭndĭb'ūlŭm) [L. *infundibulum,* funnel]. A funnel-shaped passage or structure. Tube connecting the frontal sinus with the middle nasal meatus. Stalk of the pituitary gland. Any renal pelvis division. Cavity formed by fallopian fimbriae.

ingest (ĭnjĕst') [L. *ingestus,* taken in]. To convey food material into the alimentary canal or food cavity.

ingestion (ĭnjĕs'chŏn) [L. *ingestus,* taken in]. The taking in of food at the mouth.

inguinal canal (ĭn'gwĭnăl) [L. *inguinalis, pert.* groin]. The canal carrying the spermatic cord in the male and the round ligament in the female.

innervate (ĭnŭr'vāt [L. *in,* in; *nervus,* nerve]. To supply with nerves; to stimulate to action.

innominate (ĭnŏm'ĭnāt) [L. *in,* not; *nomen,* name]. Nameless; *appl.* various arteries and veins.

inorganic [L. *in,* not; G. *organon,* an organ]. A chemical compound not containing both hydrogen and carbon.

insertion (ĭnsĕr'shŭn) [L. *in,* into; *serere,* to join]. The manner or place of attachment of a muscle to the bone that it moves. A putting into.

in situ [L]. In position.

insulin (ĭn'sĕlĭn) [L. *insula,* island]. The antidiabetic endocrine product of pancreas, formed in B-cells of islets of Langerhans; regulator of sugar metabolism.

integument (ĭntĕg'ūmĕnt) [L. *integumentum,* covering]. A covering, investing, or coating structure or layer. *Appl.* skin.

inteferon. A large protein molecule formed by some viral-infected cells; may inactivate the virus.

internuncial (ĭntĕrnŭn'sēăl) [L. *inter,* between; *nuncius,* messenger]. A connector between two other items, as neurons.

interoceptive (ĭn'tĕrōcĕp'tĭv) [L. *inter,* between; *capere,* to take]. In nerve physiology, concerned with sensations arising within the body itself, as distinguished from those (e.g., sight) arising outside the body.

interosseous (ĭn'terŏs'ēŭs) [L. *inter,* between; *os,* bone]. Occurring between bones; *appl.* arteries, ligaments, membranes, muscles, nerves.

interphase. The resting state of a cell, when it is dividing.

interstitial (ĭn'tĕrstĭsh'ăl) [L. *interstitium,* thing standing between]. Placed or lying between. *Pert.* interstices or spaces within an organ or tissue.

interstitial cell. (Leydig's cells; Franz von Leydig, Ger. anatomist, 1821–1908.) Cells located in groups between the seminiferous tubules; produce the internal secretion (testosterone) of the testes.

intervertebral (ĭn'tĕrvĕr'tĕbrăl) [l. *inter,* between; *vertebra,* vertebra]. Occurring between vertebrae; *appl.* disks, fibrocartilages, foramina, veins.

intima (ĭn'tĭmă) [L. innermost]. The inner coat of a blood vessel.

intracellular (ĭn'trăsĕl'ūlăr) [L. *intra,* within; *cellula,* small room]. Within the cell.

intramembranous (ĭn'trămĕm'brănŭs) [L. *intra,* within; *membrana,* film]. Within a membrane; *appl.* bone development.

intravenous (ĭntrăvē'nŭs) [L. *intra,* within; *vena,* vein]. Within or into a vein; i.e., an intravenous injection.

intrinsic (ĭntrĭn'sĭk) [L. *intrinsecus,* inwards]. Inward; inherent; located or originating within a structure, an intrinsic muscle.

in utero [L]. Within the uterus.

in vivo [L]. Within the body.

involution (ĭnvōlū'shŭn) [L. *in,* into; *volvere,* to roll]. Change in a backward or diminishing direction.

ion (ī'ŏn) [G. *iōn,* prp. of *ienai,* going]. An atom or a group of atoms bearing an electric charge; one of the electrified particles formed, according to the theory of electrolytic dissociation, when electrolytes are dissolved in water and certain other solvents.

ionize (ī'ŏnīz) [G. *iōn,* prp. of *ienae,* going]. To convert, totally or in part, into ions; divide into ions.

ipsilateral (ip'sĭlăt'ĕrăl) [L. *ipse,* same; *latus,* side]. On the same side; opposite of contralateral (*contra,* opposite).

iris (ī'rĭs) [L. *iris,* rainbow]. A thin, circular, muscular diaphragm of the eye; contains the eye color.

irritable (ĭr'ĭtăbl) [L. *irrito,* to excite]. Responding easily to the action of stimuli.

ischemia (ĭskē'mēă) [G. *ischein,* to hold back; *haima,* blood]. Temporary decrease of blood supply to a body part.

islets of Langerhans (Lŏng'ĕrhănz) [Paul Langerhans, Ger. pathologist, 1847–1888]. Clusters of cells in the pancreas. The cells are of three types; alpha, beta, and delta cells. The beta cells are found in greatest abundance and produce insulin.

iso [G. *isos,* equal]. Prefix: equal to.

isoagglutinin (ī'sōăglōōt'ĭnĭn) [G. *isos,* equal; L. *agglutinare,* to glue to]. Fertilizin or agglutinin of eggs that reacts on sperm of same species.

isograft (ī'sĕgrăft) [G. *isos,* equal; *graphium,* grafting knife]. A graft or transplant from another animal of the same species. Also, homograft.

isometric (ī'sōmĕ'trĭk) [G. *isos,* equal; *metron,* measure]. Contraction of a muscle in which shortening or lengthening is prevented. Tension is developed but no mechanical work performed, all energy being liberated as heat.

isotonic (īsētŏn'ĭk) [G. *isos,* equal; *tonos,* strain]. Of equal tension; a muscle contraction in which the fibers shorten but the tension remains the same; a solution having the same osmotic pressure as blood plasma.

isthmus (ĭs'mŭs) [G. *isthmos,* isthmus]. A narrow passage connecting two cavities. A narrow structure connecting two larger parts. A construction between two larger parts of an organ, or anatomical structure.

-itis (ī'tĭs) [G]. Suffix: inflammation of.

jaundice (jŏn'dĭs) [L. *galbinus,* greenish yellow]. A condition characterized by yellowness of skin and whiteness of eyes, mucous membranes, and body fluids, due to deposition of bile pigment resulting from excess bilirubin in the blood.

jejunum (jĭjōōn'ŭm) [L. *jejunus,* empty]. The central portion of the small intestine.

juxtaglomerular apparatus (jŭks'tăglōmĕr'ūlăr ap'ără'tus) [L. *juxta,* near; *glomus,* ball; *apparāre,* to prepare]. A structure consisting of myoepithelioid cells forming a cuff surrounding the arteriole leading to a glomerulus of the kidney.

karyoplasm (kar'ēōplăsm) [G. *karyon,* nucleus; *plasma,* a thing formed]. Nuclear protoplasm.

karyotype (kăr'ēōtīp) [G. *karyon,* nuclues; *typos,* form]. A grouping or arrangement of the 46 human chromosomes based on the size of the individual chromosomes.

keratin (kĕr'ătĭn) [G. *keras,* horn]. A scleroprotein forming the basis of epidermal structures such as horns, nails, hairs.

keratolytic (kĕrătōĭt'ĭk) [G. *keras,* horn; *lyein,* to dissolve]. Relative to or causing the loosening of the horny layer of the skin.

keto (kē'tō). Prefix: denotes the presence in an organic compound of the carbonyl or keto group.

ketone (kētōn) [L. *acēt,* vinegar]. Any substance (e.g., acetone) having the carbonyl group in its molecule.

ketosis (kētō'sĭs) [*ketone;* G. *-ōsis,* disease]. Accumulation of ketone bodies in blood or body.

kilocalorie (kĭl'ĕkăl'ĕrē) [G. *chilioi,* one thousand; L. *calor,* heat]. The amount of heat necessary to raise the temperature of one kilogram of water one degree centigrade; the equivalent of 1000 calories (small calories), or one large Calorie.

kinesthesia (kĭn'ĭsthē'zhă) [G. *kinēsis,* motion; *aisthēsis,* sensation]. The sensation concerned with appreciation of movement and body position.

-kinin. Suffix denoting causing of motion or action.

Krause's corpuscle (krow'sĕs) (Wilhelm Krause, German anatomist, 1833–1910). An encapsulated sensory receptor found widely distributed in connective tissue underlying the skin and mucous membranes. It is the end organ for cold sensations.

Kupffer's cells (kŭp'fĕrs) (Karl N. Kupffer, Ger. anatomist, 1829–1902). Fixed phagocytic cells found in the sinusoids of the liver.

kyphosis (kīfō'sĭs) [G. *kyphos,* humpback]. Abnormal posterior convexity of the vertebral column — hunchback.

L-tubule. Longitudinally arranged tubules of the sarcoplasmic reticulum in muscle cells.

labia majora (lā'bēă mă'jōră [L. *labium,* lip; *magnus,* great]. Two large folds of skin constituting the outer lips of the vulva; homologous to scrotum of male.

labia minora (lā'bēă mĭ'nōră) [L. *labium,* lip; *minor,* less, smaller]. Two small folds lying between the labia majora.

labium (lā'bēŭm) [L. *labium,* lip]. A lip or liplike structure.

labrum (lā'brŭm) [L. *labrum,* lip]. Fibrocartilaginous ring surrounding an articular socket, as around acetabulum; circumferential fibrocartilage.

lacrimal (lăk'rĭmăl) [L. *lacrima,* tear]. *Pert.* tears.

lactase (lăk'tās) [L. *lac,* milk; *ase,* enzyme]. An intestinal sugar-splitting enzyme converting lactose into dextrose and galactose; found in intestinal juice.

lactation (lăktā'shŭn) [L. *lactatio,* a suckling]. The function of secreting milk.

lacteal (lăk'tēăl) [L. *lac,* milk]. Lymphatic vessel of small intestine. Seen in the villi.

lactose (lăk'tōs) [L. *lac,* milk; *ose,* sugar]. A milk sugar; a double sugar or disaccharide.

lacuna (lăkū′nă) [L. *lacuna,* cavity]. A space in cartilage or bony matrix, in which the cells lie.

lambdoidal (lăm′doydăl) [G. *lambda,* L; *edios,* form]. v-shaped; *appl.* the cranial suture joining occipital and parietal bones.

lamella (lămĕl′ă) [L. *lamella,* small plate]. Any thin platelike or scalelike structure; e.g., rings of bony tissue around a Haversian canal.

lamina (lăm′ĭnă) [L. *lamina,* plate]. A thin, flat plate or layer.

lamina propria (lăm′ĭnă prō′prēă) [L. *lamina,* plate; *proprius,* one's own, special, proper]. A thin layer of fibrous connective tissue lying immediately beneath the surface epithelium of mucous membranes.

lanugo (lănōō′gō) [L. *lanugo,* wool]. The first hair to appear in the fetus.

larynx (lăr′ĭnks) [G. *larynx,* larynx]. The organ of voice, the enlarged upper end of trachea; musculocartilaginous structure lined with mucous membrane.

latent [L. *latēre,* to be hidden]. Not active; quiet.

lateral (lăt′ĕrăl) [L. *latus, later-,* side; *ad,* toward]. *Pert.* side.

lemniscus (lĕmnĭs′kŭs) [G. *lemniskos,* a filet]. A bundle of sensory nerve fibers in the brainstem.

lens (lĕnz) [L. *lens,* lentil]. A transparent part of the eye that focuses rays of light on retina; crystalline lens.

lesion (lē′zhŭn) [L. *laesio,* a wound]. A wound, injury, or infected area.

leuco-, leuko- (lōō′kō) [G. *leukos,* white]. Prefix: white.

leukemia (lōōkē′mēă) [G. *leukos,* white, growing white; *haima,* blood]. A disease of unknown cause characterized by rapid and abnormal proliferation of leukocytes in the blood-forming organs (bone marrow, spleen, lymph nodes) and the presence of immature leukocytes in peripheral circulation. May be acute or chronic but inevitably fatal.

leukocytes (lōō′kōsīts) [G. *leukos,* white; *kytos,* hollow]. Colorless blood corpuscles.

leukocytosis (lōō′kōsītō′sĭs) [G. *leukos,* white; *kytos,* cell; *-ōsis,* condition of increase]. Increase in number of leukocytes (above 10,000 per mm^3) in the blood, generally caused by the presence of infection.

leukopenia (lōō′kōpē′nēă) [G. *leukos,* white; *penia,* lack]. Abnormal decrease of white blood corpuscles usually below 5000 per mm^3

levarterenol bitartrate. A sympathomimetic. *Syn:* noradrenalin; norepinephrine.

lever (lĕv′ĕr, lē′vĕr) [M. E. *lever,* to raise]. A rigid, elongated structure used to change direction, force, or movement.

ligamentum arteriosum (lĭg′ămĕn′tŭm ărtē′rēō′-sŭm) [L. *ligamentum,* bandage; G. *arteria,* artery]. A fibrous cord, from pulmonary artery to arch of aorta, the remains of the ductus arteriosus of the fetus.

ligamentum flavum (lĭg′ămĕn′tŭm flā′vŭm) [L. *ligamentum,* bandage; *flavum,* yellow]. One of a series of ligaments connecting lamina of adjacent vertebrae.

ligamentum nuchae (lĭg′ămĕn′tŭm nū′kī) [L. *ligamentum,* bandage; *nucha,* nape of neck]. The upward continuation of the supraspinous ligament, extending from seventh cervical vertebra to occipital bone.

limbic system (lĭm′bĭk sĭs′tĕm) [L. *limbus,* border; G. *systēma,* composite whole]. Specifically, a group of structures of the diencephalon and cerebrum that influence emotional stress.

limbose suture (lĭm′bōs sū′chōōr) [L. *limbus,* border; *sutura,* seam]. Overlapping suture.

liminal (lĭm′ĭnăl) [L. *līmen,* threshold]. Threshold; just perceptible.

lingula (lĭng′gūlă) [L. *lingula,* little tongue]. A small tonguelike process of bone or other tissue, as of cerebellum or sphenoid.

lipase (lī′pās, lī′pās) [G. *lipos,* fat; *-ase,* enzyme]. A lipolytic or fat-splitting enzyme found in the blood, pancreatic secretion, and tissues.

lipemia (lĭpē′mēă) [G. *lipos,* fat; *haima,* blood]. Milky appearance of blood due to fat content.

lipid (lĭp′ĭd) [G. *lipos,* fat]. Any one of a group of fats or fatlike substances, characterized by their insolubility in water. Includes true fats (esters of fatty acids and glycerol), lipoids (phospholipids, cerebrosides, waxes), sterols (cholesterol, ergosterol), and hydrocarbons (squalene, carotene).

-lith [G. *lithos,* stone]. Stone, or presence of stones; lithiasis.

locus (lō'kŭs) [L. *locus,* place]. A place; in genetics the location or position of a gene in a chromosome; a place in the brain or heart where abnormal electrical charges may originate.

lordosis (lŏrdō'sĭs) [G. *lordos,* bent so as to be convex in front]. Abnormal posterior concavity of vertebral column — hollow or swing back.

lumen (lū'mĕn, lo͞o-) [L. *lumen,* light]. The cavity of a tubular part or organ.

lunate (lo͞o'nāt) [L. *luna,* moon]. Somewhat crescent-shaped, semilunar. *Appl.* bone of the carpus (wrist).

lunula (lo͞o'nūlă) [L. *lunula,* small moon]. The whitish area at the proximal end of the nail.

luteinizing hormone (lo͞o'tēinīz'ĭng) [L. *luteus,* orange-yellow]. A pituitary hormone that stimulates corpus luteum cell formation and interstitial cells of testis; LH, prolan B, ICSH. Aids ovulation and implantation.

lymph (lĭmf) [L. *lympha,* water]. An alkaline, colorless fluid contained in lymphatic vessels.

lymphocyte (lĭm'fōsīt) [L. *lympha,* clear water; G. *kytos,* cell]. Lymph cell or white blood corpuscle without cytoplasmic granules. They normally number from 20–25 percent of total white cells. May increase to 90 percent in lymphatic leukemia.

-lysis (lī'sĭs) [G. *lysis,* dissolution]. Destruction of blood cells, etc., by a lysin, as when a rabbit's red corpuscles are dissolved by dog's serum.

lysosome (lī'sēsōm) [G. *lysis,* loosing; *sōma,* body]. A particle in cytoplasm, smaller than mitochondria, consisting of a membrane enclosing several enzymes. Capable of breaking almost any large molecule into smaller units.

lysozyme (lī'sēzīm) [G. *lysis,* loosing; *zymē,* leaven]. A globulin found in mammalian tissue secretions, white of egg, and some microorganisms, and having mucolytic and bacteriolytic properties.

macro- (măk'rō). Prefix: large or long.

macrocosm (măk'rĕkŏzm) [G. *makros,* long, large; *kosmos,* world]. The great world; the universe, opposed to microcosm. The whole of any sphere or department of nature or knowledge to which man is related.

macroglia (măk'rĕglī'ă) [G. *makros,* large; *glia,* glue]. Supporting cells of neuroglia that are ectodermal in origin.

macrophage (măk'rĕfāj) [G. *makros,* large; *phagein,* to eat]. A large phagocytic cell, fixed or wandering; a large mononuclear leucocyte; a histiocyte, clasmatocyte, pericyte, etc.

macula (măk'ūlă) [L. *macula,* spot]. One of the sensitive areas in the walls of the saccule and utricle.

mal- [L. *malum,* an evil]. Prefix: ill, bad, or poor.

malignancy (mălĭg'nănsē) [L. *malignus,* of bad kind]. A severe form of something tending to grow worse; e.g., cancer.

malleolus (mălē'ōlŭs) [L. *malleolus,* little hammer]. The protuberance on both sides of the ankle joint.

malleus (măl'ēŭs) [L. *malleus,* hammer]. The largest of the three auditory ossicles in the middle ear, attached to the eardrum, and articulating with the incus.

maltase (mawl'tās) [A.S. *mealt,* grain]. A pancreatic enzyme that acts on maltose converting it by hydrolysis to glucose.

maltose (mal'tōs) [A.S. *mealt,* grain]. Malt sugar, a double sugar (disaccharide) found in malt and seeds. $C_{12}H_{22}O_{11}$.

mantle layer. A layer of embryonic medulla spinalis representing the future gray columns.

manubrium (măno͞o'brēŭm) [L. *manubrium,* handle]. Presternum or anterior part of sternum; handlelike part of malleus of ear.

marrow (măr'ō) [A.S. *mearg,* pith]. Connective tissue filling up cylindrical cavities in bodies of long bones and spaces of cancellous tissue, differing in composition in different bones.

masseter (măsē'tĕr) [G. *masētēr,* one that chews]. Muscle that raises lower jaw and assists in chewing.

mast cells. Spheroid or ovoid cells of very granular protoplasm, numerous in connective tissue where fat is being laid down.

mastication (măs'tĭkā'shŭn) [L. *masticāre*, to chew]. Process of chewing food with teeth until reduced to small pieces or pulp.

mastoid (măs'toyd) [G. *mastos*, breast; *eidos*, form]. Nipple-shaped; *appl*. process of temporal bone, cells, foramen, fossa, notch.

materialism (mătē'rēălĭzm) [L. *materia*, matter]. The doctrine that the facts of experience are all to be explained by reference to the reality, activities, and laws of physical or material substance. Undue regard for material interests.

matrix (măt'rĭks) [L. *mater*, mother]. Ground substance of connective tissue; intercellular material; i.e., cartilage; formative portion of tooth or nail.

mean (mēn) [L. *medius*, in the middle]. In statistics, a number derived from a series of other numbers by a prescribed method of computation. See median.

meatus (mēā'tŭs) [L. *meatus*, passage]. A passage or channel, as acoustic, nasal, etc.

median, -ad (mē'dēăn, -ăd) [L. *medium*, middle; *ad*, toward]. Middle, central; toward the center.

mediastinum (mē'dēăstī'nŭm) [L. *mediastinus*, middle]. Cleft between right and left pleura in and near median sagittal thoracic plane.

mediastinum testis (mē'dēăstī'nŭm tĕs'tĭs) [L. *mediastinus*, middle; *testis*, testicle]. Incomplete vertical septum of testis.

medulla (mĕdŭl'ă) [L. *medulla*, marrow, pith]. Central part of an organ or tissue, as the medulla of the adrenal gland.

medullated (mĕd'ūlātĕd) [L. *medulla*, marrow]. Having a myelin sheath.

mega- [G]. Prefix: large; one million.

megakaryocyte (mĕg'ăkăr'ēōsīt) [G. *megas*, large; *karyon*, nut; *kytos*, hollow]. An amoeboid giant cell of bone marrow, with one large annular lobulated nucleus, containing a number of nucleoli.

meiosis (mīō'sĭs) [G. *meion*, to make smaller]. Process of reduction division of germ-cell chromosomes from diploid to haploid number at maturation. Occurs in formation of sex cells, sperm, and egg.

Meissner's corpuscles (Georg Meissner, Ger. histologist, 1829–1905). Tactile corpuscles, associated with sense of pain, in skin of digits, lips, nipples, and certain other areas.

melanin (mĕl'ănĭn) [G. *melas*, black]. Black or dark brown pigment.

melanocytes (mĕl'ănōsītz) [G. *melas*, black; *kytos*, hollow]. Pigmented cells located in the basal cells of the spinosum; produce melanin.

melanoid (mĕl'ănoyd) [G. *melano*, black; *eidos*, form]. Concerning or resembling melanin.

melanophore (mĕl'ănōfŏr) [G. *melano*, black; *phoros*, bearing]. Cell carrying dark pigment(s).

membrane (mĕm'brān) [L. *membrana*, membrane]. A thin, soft layer of tissue lining a tube or cavity, or covering or separating one part from another; i.e., mucous or serous membranes, etc.

membranous labyrinth (mĕmbrăn'ŭs lăb'ĕrĭnth) [L. *membrana*, membrane; G. *labyrinthos*, a labyrinth]. Structure in osseous labyrinth consisting of utricle and saccule of vestibule, three semicircular ducts, and the cochlear duct.

meninges (mĕnĭn'jēz) (*sing*. meninx) [G. *mēninx*, membrane]. The three membranes enclosing brain and spinal cord, from without inward: dura mater, arachnoid, and pia mater.

meniscus (mĕnĭs'kŭs) (*plu*. menisci) [G. *mēniskos*, a crescent]. Interarticular fibrocartilage found in joints (knee) exposed to violent concussion; semilunar cartilage; intervertebral disc; a tactile disc, being terminal expansion of axis cylinder in tactile corpuscles.

menopause (mĕn'ōpawz) [G. *mēn*, month; *pausis*, cessation]. That period marking the permanent cessation of menstrual activity.

menstruation (mĕnstrōōā'shŭn) [L. *menstrualis*, pert. to menstruation]. The periodic discharge of a bloody fluid from the uterus occurring at more or less regular intervals during the life of a woman from puberty to menopause.

mental (mĕn'tăl) [L. *mentum*, chin]. *Pert*. or in region of chin; *appl*. foramen, nerve, spines, tubercle, muscle.

merocrine (mĕr'ōkrĭn) [G. *meros*, part; *krinein*, to

separate]. *Appl.* glands in which secreting cells are able to function repeatedly, no cell destruction; e.g., sudoriferous and lactiferous glands.

mesencephalon (mĕs'ĕnsĕf'ălŏn) [G. *mesos,* middle; *en,* in; *kephalē,* head]. The midbrain.

mesenchyme (mĕsĕng'kīm) [G. *mesos,* middle; *engchein,* to pour in]. A primitive, diffuse, embryonic tissue derived largely from the mesoderm.

mesentery (mĕs'ĕntĕr'ē) [G. *mesos,* middle; *enteron,* intestine]. A double-layered peritoneal fold connecting the intestine to the posterior abdominal wall.

mesoderm (mĕz'ōdĕrm, mĕz-) [G. *mesos,* middle; *derma,* skin]. The mesoblast or embryonic layer lying between ectoderm and endoderm. Gives rise to connective tissue, muscle, blood, and the cellular parts of many organs.

mesonephric duct (mĕzĕnĕf'rĭk dŭkt) [G. *mesos,* middle; *nephros,* kidney; L. *ducere,* to lead]. Embryonic duct that gives rise in the male to reproductive ducts (ductus epididymidis, ductus deferens, seminal vesicle, and ejaculatory duct). In the female, it gives rise to Gartner's duct of the epoophoron, a rudimentary structure.

mesonephros (mĕz'ĕnĕf'rŏs) [G. *mesos,* middle; *nephros,* kidney]. A type of kidney that develops in all vertebrate embryos of classes above the Cyclostomes. It is the permanent kidney of fishes and amphibians but is replaced by the metanephros in reptiles, birds and mammals.

mesosalpinx (mĕz'ōsăl'pĭnks) [G. *mesos,* middle; *salpinx,* tube]. The free margin of the upper division of the broad ligament, within which lies the oviduct.

mesothelium (mĕzĕthē'lēŭm) [G. *mesos,* middle; *thēlē,* nipple]. An epithelial tissue lining celomic cavities, covering the surfaces of mesenteries and omenta, and forming the outermost layer of many of the viscera.

meta- [G]. Prefix: after or beyond, later or more developed.

metaarteriole (mĕt'ăärtē'rēol) [G. *meta,* after, beyond, over; L. *arteriola,* small artery]. A transition vessel between arteries and capillaries.

metabolism (mĕtăb'ōlĭzm) [G. *metabolē,* change; *ismos,* state of]. Any chemical change, constructive and/or destructive, occurring in living organisms.

metanephros (mĕtănĕf'rŏs) [G. *meta,* after, beyond; *nephros,* kidney]. The permanent kidney of amniotes (reptiles, birds and mammals). A portion of it develops from caudal portion of intermediate cell mass or nephrotome; the remaining portion is derived from a bud of the mesonephric duct.

metaphase (mĕt'ăfāz) [G. *meta,* after, beyond; *phasis,* a shining out]. A stage in mitosis in which chromosomes line up on the equator of a dividing cell.

metastasis (mĕtăs'tăsĭs) [G. *meta,* after, beyond; *stasis,* a standing]. Movement from one part of the body to another; e.g., cancer cells.

metopic (mĕtŏp'ĭk) [G. *metōpon,* forehead]. *Pert.* forehead; *appl.* frontal suture.

micelle (mĭsĕl') [L. a little crumb]. A small complete unit, usually of colloids.

micro- [G]. Prefix: small; one millionth.

microcosm (mī'krōkŏsm) [G. *mikros,* small; *kosmos,* world]. A little world; the world or universe on a small scale; hence, in the theory of Paracelsus, humans combining in themselves all the elements of the great world; *oppo.* macrocosm.

microglia (mīkrŏg'lēă) [G. *mikros,* small; *glia,* glue]. Small phagocytic cells of mesodermal origin in gray and white matter of central nervous sytem.

micron (mī'krŏn) [G. *mikros,* small]. Metric unit of length; equals one-thousandth of a millimeter. Symbol: μ.

microsome (mī'krōsōm) [G. *mikros,* small; *sōma,* body]. Granule of protoplasm as opposed to ground-substance; a minute particle or vesicle in cytoplasm, containing a number of enzymes and partaking in the protein synthesis of the cell.

microvillus (mīkrōvĭl'ŭs) [G. *mikros,* small; *villus,* tuft or hair]. Very small fingerlike extensions of a cell surface.

micturition (mĭktūrĭ'shŭn) [L. *micturīre,* to urinate]. Voiding of urine; urination.

milliequivalent (mĭl'ēēkwĭv'ălĕnt) [L. *mille,* thousandth; *aequis,* equal; *valere,* to be worth]. One-thousandth of an equivalent weight. (An equivalent weight is the quantity, by weight, of a substance that will combine with one gram of H, or 8 grams of oxygen.) *Abbr.* meq.

millimeter (mĭl'ĭmētĕr) [L. *mille,* thousandth; *metron,* measure]. One-thousandth of a meter.

millimicron (mĭlĭmī'krŏn). One-thousandth of a micron. *Abbr.* $m\mu$.

milliosmol. One-thousandth of an osmol. (An osmol is the molecular weight of a substance in grams divided by the number of particles each molecule releases in solution.) *Abbr.* mos.

mimetic (mĭmĕt'ik) [G. *mimētikos,* to imitate]. Imitating or causing the same effects as something else.

miotic (mīō'tĭk) [G. *meiōn,* less]. An agent causing pupillary contraction.

mitochondria (mī'tōkŏn'drēă) *plu.)* [G. *mitos,* thread; *chondros,* grain]. Granular, rod-shaped, or filamentous self-replicating organellae in cytoplasm, consisting of an outer and inner membrane containing phosphates and numerous enzymes, varying in different tissues and functioning in cell respiration and nutrition; chondriosomes and numerous other synonyms. "Powerhouse" of cell.

mitosis (mītō'sĭs) [G. *mitos,* thread]. Indirect or karyokinetic nuclear division, with chromosome formation, spindle formation, with or without centrosome activity. Maintains constant number and quality of chromosomes from one cell generation to another.

mitral valve (mī'trăl vălv) [L. *mitra,* headband; *valva,* a fold]. Bicuspid valve of the left atrioventricular orifice of the heart.

modality (mōdăl'ĭ'tē) [L. *modus,* mode]. The nature or type of stimulus; a property of a stimulus distinguishing it from all other stimuli.

molar solution. A solution in which one gram molecular weight of a substance is present in each liter of the solution.

molecular weight. The sum of the weights of the constituent atoms of a molecule; the weight of a molecule of any gas or vapor as compared with some standard gas, such as oxygen.

molecule (mŏl'ĕcūl) [L. *molecule,* little mass]. The smallest part of a substance that can exist separately and still retain its composition and properties.

monamine oxidase. An enzyme that destroys norepinephrine at synapses.

monitor (mŏn'ĭtĕr) [L. *monere,* to warn]. To watch, check or keep track of; one or something that watches.

mono, mon- [G]. Prefix designating one, single.

monosaccharide (mŏn'ōsăk'ărīd) [G. *monos,* single; L. *saccharum,* sugar]. Simple sugar; e.g., glucose, fructose, galactose; one that cannot be further reduced by hydrolysis.

morphologist (mŏrfŏl'ōjĭst) [G. *morphe,* form; *logos,* discourse]. One who studies or is versed in the science of form and structure of plants and animals, as distinct from consideration of functions.

morphology (mŏrfŏl'ōjē) [G. *morphe,* form; *logos,* discourse]. The science of form and structure of plants and animals, as distinct from consideration of functions.

morula (mŏr'o͞olă) [L. *morum,* mulberry]. A solid cellular globular mass, the first result of fertilized ovum segmentation.

motor (mō'tŏr) [L. *motus,* moving]. Refers to movement or those structures (e.g., nerves and muscles) that cause movement.

motor end plate. The specialized ending of a motor nerve on a skeletal muscle fiber.

motor unit. One motor nerve fiber (axon) and the skeletal muscle fibers it supplies.

mucosa (mūkō'să) [L. *mucus,* mucus]. A mucous membrane.

mucous (mū'kŭs) [L. *mucus,* mucus]. A mucus-secreting structure.

mucous membrane. A thin film, skin, or layer of tissue lining all hollow organs and cavities that open on the skin surface of the body. Secretes mucus.

mucus (mū'kŭs) [L. *mucus,* mucus]. The slimy, glairy substance secreted by goblet cells of a mucous membrane or by mucous cells of a gland.

Müllerian (Müller's) ducts (Johannes B. Müller, Ger. physician, 1801–1858). Embryonic tubes from which the oviducts, uterus, and vagina develop in the female; in the male they become atrophied.

multi- [L]. Prefix: many.

multifidus (mŭl'tĭfĭd'ŭs) [L. *multus,* many; *findere,* to cleave]. The musculotendinous fasciculi lateral to spinous processes from sacrum to axis vertebra.

musculi pectinate (mŭs'kūlī pĕk'tĭnā'tē) (*plu.*) [L. *musculus,* muscle; *pecten,* comb]. Muscle on inner surface of right atrium giving it a ridged appearance.

mutation (mūtā'shŭn) [L. *mutatio,* to change]. A change or transformation of a gene; the evidence of a gene alteration.

mydriatic (mĭdrēăt'ĭk) [G. *midriasis,* dilation]. An agent causing pupillary dilation.

myelencephalon (mī'ĕlĕnsf'ălŏn) [G. *myelos,* marrow; *en,* in; *kephalē,* head]. The lower part of hindbrain.

myelin (mī'ĕlĭn) [G. *myelos,* marrow]. A white fatty material forming medullary sheath of nerve fibers. Speeds impulse conduction.

myelo- [G]. Prefix *pert.* spinal cord or to bone marrow.

myeloid (mī'ĕloyd) [G. *myelos,* marrow; *eidos,* form]. Formed in bone marrow. Like marrow in appearance or structure; *appl.* cells, as megakaryocytes, monocytes, and parenchymal cells; resembling myelin.

myenteric (mīĕntĕr'ĭk) [G. *myo,* muscle; *enteron,* gut]. *Appl.* nerve plexus controlling movement of food towards anus, Auerbach's plexus; *appl.* reflex.

myocardium (mī'ōkăr'dēŭm) [G. *myo,* muscle; *kardia,* heart]. The thick muscular layer of the heart wall.

myoepithelial (mī'ōĕp'ĭthē'lēăl) [G. *myo,* muscle; *epi,* upon; *thēlē,* nipple]. Pertaining to contractile epithelial cells.

myofibril (mīōfī'brĭl) [G. *myo,* muscle; L. *fibrilla,* a small fiber]. A tiny fibril found in muscular tissue, running parallel to the cellular long axis, from one cell to another.

myofilaments (mī'ōfĭl'ămĕnts) (*plu.*) [G. *myo,* muscle; L. *filum,* thread]. Thin, threadlike components of a myofibrilla.

myoglobin (mīōglō'bĭn) [G. *myo,* muscle; L. *globus,* globe]. A respiratory pigment of muscle that binds oxygen.

myometrium (mīōmē'trēŭm) [G. *myo,* muscle; *mētra,* uterus]. Muscular wall of the uterus forming the main mass of the uterus.

myoneural junction (mī'ōnoo͞răl jŭnk'shŭn) [G. *myo,* muscle; *neuron,* nerve; L. *junctiō,* a joining]. Ending of a nerve in a muscle.

myopia (mīō'pēă) [G. *myein,* to shut; *ōps,* eye]. Nearsightedness; image comes to focus ahead of retina. Also, hypometropia.

myosin (mī'ōsĭn) [G. *myo,* muscle]. A muscle globulin acting as an enzyme to aid in initiating muscle contraction.

myotactic (mīōtăk'tĭk) [G. *myo,* muscle; L. *tactus,* touch]. Refers to muscle or kinesthetic sense (kinesthesia).

mystic (mĭs'tĭk) [G. *mystikos,* fr. *myō,* close the lips or eyes]. One who professes direct divine illumination; one who relies chiefly on meditation in acquiring truth; a believer in mysticism; one who has been initiated into a mystery.

myxedema (mĭksĕdē'mă) [G. *myxa,* mucus; *oidēma,* swelling]. Condition resulting from hypofunction of the thyroid gland.

nasal (nā'zl) [L. *nasus,* nose]. *Pert.* nose.

NAD. *Abbr.* nicotinamide adenine dinucleotide, one of several hydrogen acceptors. (NADP. NAD phosphate, another hydrogen acceptor.)

narcotic (năr'kŏt'ĭk) [G. *narkōtikos,* benumbing]. A drug that depresses the central nervous system, relieving pain and producing sleep.

nares (nā'rēz) [L. *nares,* nostrils]. The openings into the nasal cavities.

natremia (nătrē'mēă) [L. *natrium,* sodium; G. *haima,* blood]. Blood sodium.
navicular (năvĭk'ūlar) [L. *navicula,* boat]. Shaped like a boat. Scaphoid bones in the carpus and in the tarsus.
necrosis (nĕkrō'sĭs) [G. *nekrōsis,* a killing]. Death of areas of tissue or bone surrounded by healthy parts; death in mass as distinguished from necrobiosis, a gradual degeneration.
negative pressure. Pressure less than atmospheric pressure; i.e., < 760 mm Hg.
neoplasm (nē'ōplăsm) [G. *neo,* new, recent; *plasma,* a thing formed]. A new and abnormal formation of tissue as a cancer or tumor.
nephron (nĕf'rŏn) [G. *nephros,* kidney]. The structural and functional unit of the kidney consisting of a renal (malpighian) corpuscle (a glomerulus enclosed within Bowman's capsule) and its attached tubule consisting of the proximal convoluted portion, loop of Henle, and distal convoluted portion. Forms urine and regulates blood composition.
nephrotic syndrome (nĕfrŏt'ĭk sĭn'drōm) [G. *nephros,* kidney; *syndrome,* a running together]. Term applied to renal disease of whatever cause, characterized by massive edema, proteinuria, and usually elevation of serum cholesterol and lipids.
nerve (nĕrv) [L. *nervus,* sinew]. A bundle of nerve fibers outside the central nervous system.
nerve center (nĕrv sĕn'tĕr) [L. *nervus,* sinew; *centrum,* center]. One of many cerebrospinal or ganglionic systems originating or controlling vital functions.
neural crest (noo͞'răl krĕst) [G. *neuron,* sinew; L. *crista,* crest]. A band of cells extending longitudinally from which cells forming cranial, spinal, and autonomic ganglia arise.
neural fold (noo͞'răl fōld) [G. *neuron,* sinew]. One of two longitudinal elevations of the neural plate of an embryo which unite to form the neural tube.
neural groove (noo͞'răl grūv) [G. *neuron,* sinew; Middle Dutch, *groeve,* ditch]. A longitudinal groove on dorsal surface of the embryo lying between the neural folds.
neurilemma (noo͞'rĭlĕm'ă) [G. *neuron,* sinew; *lemm,* husk, sheath]. Flattened cells found on fibers of peripheral and autonomic system (also sheath of Schwann) that aid in myelin formation and nerve fiber regeneration.
neuroglia (noo͞rŏg'lēă) [G. *neuron,* sinew; *glia,* glue]. Supporting and protecting tissue of the central nervous system consisting, in part, of macroglial and microglial cells and their processes.
neurohumor (noo͞'rōhū'mŏr) [G. *neuron,* sinew; L. *humor,* moisture]. "Hormone" produced by nervous tissues that activates or inhibits other nervous tissue or its effectors; neurohormone.
neurohypophysis (noo͞'rōhīpŏf'ĭsĭs) [G. *neuron,* sinew; *hypo,* under; *physis,* growth]. Posterior portion or pars nervosa of the pituitary gland.
neuromuscular junction (noo͞'rōmŭs'kūlăr jŭnk'shŭn) [G. *neuron,* sinew, nerve; L. *musculus,* a muscle; *junctiō,* a joining]. The place of union or coming together of nerves and muscles.
neuron (noo͞'rŏn) [G. *neuron,* sinew]. The nerve cell with its outgrowths; structural unit of the nervous system; is excitable and conductile.
neurosis (noo͞rō'sĭs) [G. *neuron,* sinew; *-ōsis,* disease]. A disorder of the mind in which contact with the real world is not maintained.
neutralize (noo͞'trălīz) [L. *neuter,* not; *uter,* either]. To counteract, make inert, or destroy the properties of something.
neutrophil (e) (noo͞'trōfĭl, -fīl) [L. *neuter,* neither; G. *philein,* to love]. A leucocyte that stains easily with neutral dyes.
nigra (nī'gră) [L. *nigra,* black]. Black or blackness.
Nissl substance (Nissl bodies or granules). [Franz Nissl, neurologist in Heidelberg, 1860–1919]. Angular protein particles found in cytoplasm of nerve cell; related to cell metabolism.
node (nōd) [L. *nodus,* knot]. A constricted region; a knob, protuberance or swelling; a small rounded organ. Aggregation of specialized cardiac cells.
nodes of Ranvier (Louis Antoine Ranvier, Fr. pathologist, 1835–1922). Constrictions at intervals of the medullary sheath of nerve fiber.

nomenclature (nō'mĕnklā'chĕr) [L. *nomen,* name; *calare,* to call]. System of naming plants, animals, organs, etc.; binomial nomenclature.

norepinephrine (nŏrĕp'ĭnĕf'rĭn) A hormone produced by the adrenal medulla, similar in chemical and pharmacologic properties to epinephrine, but it is chiefly a vasoconstrictor and has little effect on cardiac output.

nostril (nŏs'trĭl) [A.S. *nosu,* nose; *thyrl,* a hole]. External opening of the nasal cavity.

notochord (nō'tōkŏrd) [G. *nōton,* back; *chordē,* cord]. The dorsal supporting axis of lowest vertebrates, transitory in others; chorda dorsalis.

nuchal (nōō'kăl) [L. *nucha,* the back of the neck]. *Pert.* neck or nucha.

nuclear (nōō'klēăr) [L. *nucleus,* kernel]. *Pert.* nucleus (cell) or a central point.

nucleic acids (nōōklā'ĭk). Large molecules formed by nucleotides. They form the basis of heredity and protein synthesis. DNA; RNA.

nucleolus (nōōklē'ĕlŭs) [L. *nucleolus (dim. of nucleus)* little kernel]. A dense, rounded mass in a cell nucleus, consisting of protein and ribonucleic acid granules, and functioning in RNA and protein synthesis controlled by a special region or nucleolar organizer in the chromosome; a plasmosome or a karyosome.

nucleotide (nōō'klēētĭd) [L. *nucleus,* kernel]. A unit formed from a nitrogenous base, a five-carbon sugar and a phosphoric acid radical. It is the unit of structure of DNA and RNA.

nucleus (nōō'klēŭs) [L. *nucleus,* kernel]. Complex spheroid mass essential to life and controlling center of most cells; mass of grey matter in central nervous system; a nidulus.

nucleus pulposus (nōō'klēŭs pŭlpō'sŭs) [L. *nucelus,* kernel]. The soft core of an intervertebral disc; probably a remnant of notochord.

nystagmus (nĭstăg'mŭs) [G. *nystazein,* to nod]. Involuntary cyclical movements of the eyeball.

obese (ōbēs') [L. *obesus,* fat]. Excessively fat; overweight.

obligatory (ŏblĭg'ătōrē) [L. *obligatorius,* to bind to]. The idea of no choice; bound to or fixed in function.

obturator (ŏb'tĕrā'tŏr) [L. *obturare,* to close]. *Pert.* any structure in neighborhood of obturator foramen.

occlude (ŏklōōd') [L. *occludere,* to shut up]. To block or obstruct something such as a blood vessel.

-oid [G. *eidos,* form]. Suffix: like, resembling.

olecranon (ōlĕk'rănŏn) [G. *olekranon,* point of elbow]. A large process at the upper end of the ulna.

olfactory (ŏlfăk'tōrē) [L. *olere,* to smell; *facere,* to make]. *Pert.* sense of smell; *appl.* stimuli, structures, reactions.

olfactory pit. One of two horseshoe-shaped depressions on the ventrolateral surface of the head of the embryo bounded by lateral and median nasal processes. It gives rise to nostrils and a portion of the nasal fossa. *Syn.* nasal pit.

oligo- (ŏl'ĭgō) [G. *oligos,* little]. Prefix: small, few, scanty, or little.

oligodendrocyte (ŏl'ĭgōdĕn'drōsīt) [G. *oligos,* little; *dendron,* tree; *cyte,* cell]. Neuroglial cells having few and delicate processes.

oliguria (ŏlĭgū'rēă) [G. *oligos,* little; *ouron,* urine]. Formation of small amounts of urine.

-ology [G. *logos,* study]. Suffix: study of; science of; knowledge of.

omentum (ōmĕn'tŭm) [L. *omentum,* a covering]. A folded double layer of peritoneum attached to the stomach and connecting it with certain of the abdominal viscera. Stores fat.

omohyoid (ō'mōhī'oyd) [G. *ōmos,* shoulder; *hyoeidēs,* T-shaped]. *Pert.* shoulder and hyoid; *appl.* a muscle.

oncotic (ŏngkŏt'ĭk) [G. *onkos,* tumor]. Concerning, caused, or marked by swelling. *Pert.* colloid osmotic pressure created by the plasma proteins.

ontogeny (ŏntŏj'ĕnē) [G. *on,* being; *genesis,* descent]. The history of development and growth of an individual.

operon (ŏp'ĕrŏn) [L. *operatiō,* a working]. The combination of an operator gene and a structural gene.

operon concept. The current theory of how genes operate to control body functions.

ophthalmic (ŏfthăl'mĭk) [G. *opthalamos,* eye]. *Pert.* eye.

-opsin (ŏp'sĭn) [G. *opson,* food]. Suffix: *pert.* protein component of the visual pigments.

oral (ō'răl) [L. *os, or-,* mouth]. *Pert.* or belonging to mouth; on side of which mouth lies.

orbicularis (ŏr'bĭkūlā'rĭs) [L. *orbiculus,* little circle]. Muscle surrounding an orifice; a sphincter muscle.

orchido- [G. *orchis,* testicle]. Combining form meaning testicle.

organ (ŏr'găn) [G. *organon,* implement]. Any part or structure of an organism adapted for a special function or functions; composed of two or more tissues.

organelle (ŏrgănĕl') [G. *organelle,* instrument, a small organ]. A part of a cell that has a special function; an organoid.

organ of Corti (kŏr'tē) [G. *organon,* organ]. (Alfonso Corti, It. anatomist, 1822–1888.) An elongated spiral structure running the entire length of cochlea in the floor of the cochlear duct and resting on the basilar membrane. It is the end organ of hearing, containing hair cells, supporting cells, and neuroepithelial receptors stimulated by sound waves.

organology (ŏr'gănŏl'ŏjē) [G. *organon,* implement; *logos,* discourse]. The study of organs in plants and animals and their development.

origin (ŏr'ĭjĭn) [L. *origo,* beginning]. The source of anything; a starting point. The beginning of a nerve. The more fixed attachment of a muscle.

os (ŏs), ossa (ŏs'ă) *plu.* [L. *os,* bone]. A bone; bones; mouth; opening.

-ose (ōs). Suffix: *pert.* sugar.

-osis [G]. Combining form, *pert.* smell, osmosis.

osmo- [G]. Combining form, *pert.* smell, osmosis.

osmol (ŏz'mŏl). The molecular weight of a substance, in grams, divided by the number of particles it releases in solution.

osmolarity (ŏzmōlăr'ĭtē). The number of osmols per liter of solution.

osmoreceptor (ŏzmōrēcĕp'tŏr). A receptor sensitive to osmotic pressure of a fluid.

osmosis (ŏzmō'sĭs) [G. *ōsmos,* impulse; *osis,* intensive]. Diffusion of a solvent through a selective membrane.

osmotic pressure. The force developing when two solutions of different concentrations are separated by a membrane permeable to the solvent.

osseous labyrinth (ŏs'ēŭs lăb'ĕrĭnth) [L. *osseus,* bony; G. *labyrinthos,* a labyrinth]. Consists of vestibule, three semicircular canals, and cochlea. Channeled out of petrous portion of temporal bone.

ossicle (ŏs'ĭkl) [L. *os,* bone]. Any small bone; specifically, the small bones of the middle ear; malleus, incus, and stapes.

ossification (ŏs'ĭfĭkā'shŭn) [L. *os,* bone; *facere,* to make]. Formation of bone substance. Conversion of other tissues into bone.

osteo- [G]. Prefix: bone(s).

osteoblast (ŏs'tēōblăst) [G. *osteon,* bone; *blastos,* bud]. A bone-forming cell.

osteoclast (ŏs'tēōklăst) [G. *osteon,* bone; *klan,* to break]. A cell that absorbs or breaks up bony tissue or cartilage matrix.

osteocyte (ŏs'tēōsīt) [G. *osteon,* bone; *kytos,* hollow]. A bone cell, developed from osteoblast.

osteogenic (ŏs'tēōjĕn'ĭk) [G. *osteon,* bone; *genesis,* descent]. *Pert.* or causing formation of bone.

osteoid (ŏs'tēoyd) [G. *osteon,* bone; *eidos,* form]. Bonelike. A mixture of collagenous fibrils and a semisolid ground substance in which calcium is deposited to form bonelike tissue.

osteomalacia (ŏs'tēōmălā'shă) [G. *osteon,* bone; *malakia,* softening]. Soft bones, due to the failure of calcium deposition. Vitamin D deficiency is often a factor.

osteomyelitis (ŏs'tēōmīĕlī'tĭs) [G. *osteon,* bone; *myelos,* marrow; *-itis,* inflammation]. Inflammation of bone, esp. the marrow, caused by a pathogenic organism.

osteon (ŏs'tēŏn) [G. *osteon,* bone]. A Haversian system. A unit of structure of compact bone.

osteoporosis (ŏs'tēōpŏrō'sĭs) [G. *osteon,* bone; *poros,* a passage; *-osis,* condition]. Abnormal porosity of bone due to osteoblastic failure or deficiency of other necessary building blocks of bone.

oto- [G]. Combining form; the ear.

otolith (ō'tōlĭth) [G. *ous,* ear; *lithos,* stone]. Calcareous particle or platelike structure found in auditory organ of many animals.

-otomy (ŏt'ōmē) [G. *tomē,* incision]. Suffix: *pert.* opening or repair of an organ without its removal.

otosclerosis (ō'tōsklĕrō'sĭs) [G. *ous,* ear; *sklerosis,* hardening]. A condition characterized by chronic progressive deafness, esp. for low tones. Caused by formation of spongy bone, esp. around the oval window with resulting ankylosis of stapes.

ovulate (ŏv'ūlāt) [L. *ovulum,* little egg]. Release of an egg from the ovary.

ovum (ō'vŭm) (*plu.* ova) [L. *ovum,* egg]. A female germ cell; mature egg cell.

oxidation (ŏk'sĭdā'shŭn) [G. *oxys,* sour]. The combining of something with oxygen; loss of electrons.

oxidative phosphorylation. A metabolic scheme that transfers H ions and electrons to produce ATP and water.

oxygen debt. A temporary shortage of oxygen necessary to combust products (lactic acid, pyruric acid) of glucose catabolism.

oxyhemoglobin (ŏksĭhē'mōglō'bĭn). The combination of oxygen and hemoglobin.

oxytocin (ŏk'sĭtō'sĭn) [G. *oxys,* swift; *tokos,* childbirth]. A pituitary hormone that stimulates the uterus to contract, thus inducing parturition. It also acts on the mammary gland to stimulate the release of milk.

pacemaker (pās'mākr) [L. *passus,* a step; A.S. *macian,* to make]. The sinoartrial node, so named because cardiac rhythm commences here, taking place near the spot where the large veins empty into the right atrium. An object that influences the rate of occurrence of an event. Cardiac pacemaker (artificial or electric), an electrical device that can substitute for a defective natural pacemaker and control the beating of the heart by a series of rhythmic electrical discharges.

Pacinian corpuscles (păsĭn'ēăn kŏr'pŭsls) [L. *corpus, -culum,* little body]. (Filippo Pacini, It. anatomist, 1812–1883.) The largest of the nerve end organs of the subcutaneous layer of the skin; nerve fibril covered by a series of concentric lamellae.

palate (păl'ăt) [L. *palatum,* palate]. The roof of the mouth, composed of hard and soft parts.

palmar (păl'măr) [L. *palma,* palm of hand]. *Pert.* palm of hand; *appl.* aponeurosis, nerve, muscle, reflex.

palpate (păl'pāt) [L. *palpāre,* to touch]. To examine by feeling or touching.

pancreozymin (păn'krēōzī'mĭn) [G. *pan,* all; *kreas,* flesh; *zyme,* leaven]. A hormone, extracted from the duodenal mucosa, that stimulates the secretion of pancreatic juice, esp. increasing its enzymatic concentration.

panniculus adiposus (pănĭk'ūlŭs ădĭpō'sŭs) [L. *panniculus,* a small piece of cloth; *adiposus,* fatty]. The subcutaneous layer of fat, esp. where fat is abundant; the superficial fascia, which is heavily laden with fat cells.

papilla (păpĭl'ă) [L. *papilla,* nipple]. A small elevation or nipplelike protuberance.

papillary duct (păp'ĭlărē dŭkt) [L. *papilla,* nipple; *ducere,* to lead]. A short duct that opens on the tip of the renal papilla. Formed by union of straight collecting tubules.

papillary muscles (păp'ĭlarē mŭsls) [L. *papilla,* nipple; *musculus,* muscle]. Muscles on the inner surface of ventricles of the heart to which chordae tendinae are attached. They hold cuspid valves in closed position during ventricular contraction.

para- [G]. Combining form: near, past, beyond, the opposites, beside, abnormal, or irregular.

parahormone (păr'ahŏr'mōn) [G. *para,* beyond; *hormaein,* to set in motion]. A substance conveyed through the circulatory system that exerts a stimulating effect like hormones, yet does not originate in endocrine tissue. *Appl.* carbon dioxide, gastrin, secretin, etc.

paralysis (părăl'ĭsĭs) [G. *paralyein,* to disable at one side]. Temporary suspension or permanent loss of function, esp. loss of sensation or voluntary motion.

paranasal (păr'ănā'zăl) [G. *para,* near; L. *nasalis, pert.* nose]. Situated near or alongside the nasal cavities.

parasagittal (părăsăj'ĭtăl) [L. *para,* beside; *saggita,* arrow]. Any plane parallel to the median plane.

parasympathetic (păr'ăsĭmpăthĕt'ĭk) [G. *para,* beside; *sympathēs,* of like feelings]. Enteral; *appl.* craniosacral portion of the autonomic nervous system; connects to some sacral and some cranial nerves; controls normal body functions.

parathyroid hormone (părăthī'royd hŏr'mōn) PTH. [G. *para,* beside; *thyreos,* shield; *eidos,* form; *horman,* to urge on]. Hormone secreted by the parathyroid glands; regulates calcium and phosphorus metabolism.

parenteral (părĕn'tĕrăl) [G. *para,* opposite; *enteron,* intestine]. Situated or occurring outside of the intestines.

paresthesia (păr'ĕsthē'zēă) [G. *para,* abnormal; *aisthēsis,* sensation]. Abnormal sensation without demonstrable cause.

parietal (părī'ĕtăl) [L. *paries,* wall]. *Pert.* or forming part of wall of a structure; *appl.* cells of stomach, membrane, layer, lobe, bone.

parietal pericardium (părī'ĕtăl pĕr'ĭkăr'dēŭm) [L. *paries,* wall; G. *peri,* around; *kardia,* heart]. The outer serous and fibrous layer of the pericardium.

parotid (părŏt'ĭd) [L. *para,* beside; A.S. *eare,* ear]. Located near the ear, esp. the parotid salivary gland.

paroxysm (păr'ŏksĭzm) [G. *para,* beside; *oxynein,* to sharpen]. A sudden periodic attack or recurrence of a disease.

pars distalis (părz dĭstăl'ĭs) [L. *pars,* a part; *distāre,* to be distant]. That part of the hypophysis forming the major portion of the anterior lobe.

pars intermedia (părz ĭn'tĕrmē'dēă) [L. *pars,* a part; *inter,* between; *medius,* middle]. The intermediate lobe of the hypophysis cerebri (pituitary).

pars tuberalis (părz tū'bĕrăl'ĭs) [L. *pars,* a part; *tuber,* a swelling]. The portion of the anterior lobe of the hypophysis cerebri that invests the infundibular stalk.

parthenogenesis (păr'thĕnōjĕn'ĕsĭs) [G. *parthenos,* virgin; *genesis,* descent]. Reproduction without fertilization by a male element.

patella (pătĕl'ă) [L. *patella,* small pan]. The kneecap.

patent (păt'ĕnt, pā'tĕnt) [L. *pateus,* to be open]. Wide open; not closed.

Paul-Bunnel antibody. (John R. Paul and W. W. Bunnell, Am. physicians.) A test for heterophil antibody used in diagnosis of infectious mononucleosis.

pectineal (pĕktĭn'ēăl) [L. *pecten,* comb]. *Appl.* ridge line on femur and attached muscle.

pedicle (pĕd'ĭkl) [L. *pediculus,* small foot]. A backward-projecting vertebral process; a part of the vertebral arch.

peduncle (pēdŭng'kŭl) [L. *pedunculus,* a little foot]. A band of nerve fibers connecting parts of the brain.

penis (pē'nĭs) [L. *penis*]. The male copulatory organ.

pepsin (pĕp'sĭn) [G. *pepsis,* digestion]. The chief enzyme of gastric juice; converts proteins into proteoses and peptones.

pepsinogen (pĕpsĭn'ōjĕn) [G. *pepsis,* digestion; *gennan,* to produce]. The zymogen or antecedent of pepsin existing in the form of granules in the chief cells of the gastric glands.

peptide (pĕp'tīd) [G. *peptein,* to digest]. A compound containing two or more amino acids formed by cleavage of proteins.

peptone (pĕp'tōn) [G. *pepton,* digesting]. A secondary protein formed by the action of proteolytic enzymes, acids, or alkalies on certain proteins.

perforate (pŭr'fŏrāt) [L. *perforāre,* to pierce through]. To puncture, make a hole through.

perfuse (pŭrfūz') [L. *perfundere,* to pour through]. To pass a fluid through something.

peri- [G]. Prefix: around or about.

pericardial cavity (pĕrĭkăr'dēăl căv'ĭtē) [G. *peri,* around; *kardia,* heart; L. *cavitas,* hollow]. The potential space between the epicardium (visceral pericardium) and the parietal pericardium.

pericardial sac (pĕr'ĭkăr'dēăl săk) [G. *peri,* around; *kardia,* heart; L. *saccus,* sack]. The pericardium, a membrane enveloping the heart.

perichondrium (pĕr'ĭkŏn'drēŭm) [G. *peri,* around; *chondros,* cartilage]. A fibrous membrane that covers cartilage.

pericyte (pĕr'ĭsīt) [G. *peri,* around; *kytos,* hollow]. A macrophage in adventitia of small blood vessels; a pericapillary cell; Rouget cell.

peridontal (pĕr'ĭdŏn'tăl) [G. *peri,* around; L. *dens,* tooth]. Surrounding a tooth or part of one.

perikaryon (pĕr'ĭkăr'ēŏn) [G. *peri,* around; *karyon,* nucleus]. A nerve cell body, as distinct from its axon and dendrites.

perilymph (pĕr'ĭlĭmf) [G. *peri,* around; L. *lympha,* water]. A fluid separating membranous from osseous labyrinth of ear.

perimetrium (pĕrĭmē'trēŭm) [G. *peri,* around; *mētra,* uterus]. Serous coat of the uterus.

perimysium (pĕr'ĭmĭz'ēŭm) [G. *peri,* around; *mys,* muscle]. Connective tissue binding numbers of fibers into bundles of muscles and continuing into tendons; alternatively, *appl.* only to fasciculi envelopes.

perineum (pĕr'ĭnē'ŭm) [G. *perinaion,* part between anus and scrotum]. The region of the outlet of the pelvis.

periosteum (pĕr'ēŏs'tēŭm) [G. *peri,* around; *osteon,* bone]. The fibrous membrane investing the surface of bones.

peripheral (pĕrĭf'ĕrăl) [G. *peri,* around; *pherō,* bear]. Of or pertaining to the outer surface of a periphery. Distant from the center; distal; external.

periphery (pĕrĭf'ĕrē) [G. *peri,* around; *pherein,* to bear]. Outer part or surface of a body; part away from the center.

peristalsis (pĕr'ĭstăl'sĭs) [G. *peri,* round; *stellein,* to place]. Progressing waves of contraction along a muscular tube by action of circular muscles; moves material through tube.

peritoneum (pĕr'ĭtōnē'ŭm) [G. *periteinein,* to stretch around]. A serous membrane partly applied to inner side of abdominal walls, partly reflected over contained viscera.

peritonitis (pĕr'ĭtōnī'tĭs) [G. *peritonaion; -itis,* inflammation]. Inflammation of the peritoneum, the membranous coat lining the abdominal cavity and investing the viscera.

permeability (permēăbĭl'ĭtē) [L. *per,* through; *meo,* pass; *habilis,* able]. The quality or condition of being permeable; allowing passage, especially of fluids.

peroneal (pĕr'ōnē'ăl) [G. *peronē,* pin, fibula]. *Appl.* peroneal artery, nerve.

petrous (pĕt'rŭs) [G. *petros,* stone]. Very hard or stony; *appl.* pyramidal portion of temporal bone between spenoid and occipital; *appl.* ganglion on its lower border; petrosal.

Peyer's patch (pī'ĕrz) (Johann Conrad Peyer, Swiss anatomist, 1653–1712). An aggregation of solitary nodules or groups of lymph nodules found chiefly in the ileum near its junction with the colon.

pH. A symbol used to express the acidity or alkalinity of a solution. Strictly, the logarithm of the reciprocal of H ion concentration.

-phag- [G. *phagein,* to eat]. Combining form: An eater, or *pert.* engulfing or ingestion.

phagocyte (făg'ōsīt) [G. *phagein,* to eat; *kytos,* hollow]. A colorless blood corpuscle or other cell that ingests foreign particles.

phagocytosis (făg'ōsītō'sĭs) [G. *phagein,* to eat; *kytos,* hollow]. The ingestion and destruction of microparasites and cells by phagocytes.

phalanges (fălăn'jēz) *(plu.)* [G. *phalangx,* line of battle]. The bones of the digits.

phallus (făl'ŭs) [G. *phallos,* penis]. The embryonic structure that becomes the penis or clitoris.

pharyngeal pouch (fărĭn'jēăl) [G. *pharyngx,* gullet]. An evagination of the lateral pharyngeal walls.

-phil (fĭl). Combining form: love for; having an affinity for.

phlegmatic (flĕgmăt'ĭk) [G. *phlegma,* inflammation, phlegm]. Sluggish, indifferent. From one of the four natural humors (the cold and moist) in ancient physiology.

phospholipid (fŏs'fōlĭp'ĭd) [G. *phos,* light; *pherein,* to carry; *lipos,* fat]. A lipoid substance containing phosphorus, fatty acids, and nitrogenous base, as lecithin.

photopic (fōtŏp'ĭk) [G. *phōtos,* light; *opsis,* vision]. Pertaining to bright light.

photopic vision. Vision in bright light involving the formation of images and discrimination of color; the cones are photopic.

phrenic (frĕn'ĭk) [G. *phrēn,* diaphragm, mind]. *Pert.* or in region of diaphragm; *appl.* artery, ganglion, nerve, plexus, vein. *Pert.* mind.

phylogeny (fīlŏj'ĕnē) [G. *phylon,* race; *genesis,* descent]. History of development of species or race.

phylum (fī'lŭm) [G. *phylon,* race or tribe]. A group of animals or plants constructed on a similar general plan; a primary division in classification.

physiology (fĭzēŏl'ōjē) [G. *physis,* nature; *logos,* discourse]. That part of biology dealing with functions and activities of organisms.

pia (pī'ă) [L. tender]. Tender or delicate; the inner meninx carrying blood vessels to cord and brain (pia mater).

pia mater (pī'ă mā'tĕr) [L. *pia mater,* kind mother]. The innermost meninx; a delicate membrane closely investing brain and spinal cord.

pineal body (pĭn'ēăl) [L. *pinea,* pine cone]. A median outgrowth from the roof of the diencephalon.

pinna (pĭn'ă) [L. *pinne,* feather]. Auricle or outer ear.

pinocytosis (pĭn'ōsītō'sĭs) [G. *piein,* to drink; *kytos,* hollow]. The ingestion of droplets by cells, by a "sinking in" of the cell membranes.

pisiform (pī'sĭfôrm) [L. *pisum,* pea; *forma,* shape]. Pea-shaped; *appl.* carpal bone, os pisiforme.

pituicyte (pĭtū'ĭsīt) [L. *pituita,* phlegm; G. *kytos,* cell]. A branched, modified neuroglia cell characteristic of the pars nervosa of the posterior lobe of the pituitary gland; also present in the infundibular stalk.

placenta (plăsĕn'tă) [L. *placenta,* flat cake]. A double structure derived in part from maternal uterine tissue and in part from the embryo through which the fetus gets nourishment and excretes its wastes.

plantar (plăn'tăr) [L. *planta,* sole of foot]. *Pert.* sole of foot; *appl.* arteries, ligaments, muscles, nerves, veins, etc.

-plasia (plā'zēă) [G. *plasis,* a molding]. Combining form; *pert.* change or development. Also plastic.

-plasm (plăz'm) [G. *plasma,* a thing formed]. Combining form; *pert.* fluid substance of a cell.

plasma (plăz'mă). The liquid part of the blood.

plasma cell. One capable of producing antibodies in response to antigenic challenge. Also plasmocyte.

plasma membrane. The cell membrane.

plasmin (plăz'mĭn) [G. *plasma,* a thing formed]. Fibrinolytic enzyme derived from its precursor plasminogen.

plasmocyte (plăz'mōsīt) [G. *plasma,* a thing formed; *kytos,* hollow]. Cells found in bone marrow, connective tissue, and sometimes in blood plasma. Considered by some to be abnormal leukocytes.

platelet (plāt'lĕt) [F. *plat,* G. *platys,* flat]. A minute place; a blood platelet.

platysma (plătĭz'mă) [G. *platysma,* flat piece]. Broad sheet of muscle beneath superficial fascia of neck.

pleura (ploo͞r'ă) (*sing.* pleuron) [G. *pleura,* side]. A serous membrane lining thoracic cavity and investing lung.

plexus (plĕk'sŭs) [L. *plexus,* interwoven]. A network of interlacing vessels, nerves, or fibers.

plica (plī'kă) (*plu.* plicae) [L. *plica,* a fold]. Fold of skin, membrane, or lamella.

-pnea (nē'ă). Suffix: *pert.* breathing. Also, pneo.

pneumo- (noo͞'mō) [G]. Combining form: *pert.* lungs or air.

PO_2. Symbol for partial pressure of oxygen.

-pod, -poda, -podo- [G]. Combining forms referring to foot or feet.

poikilocytosis (poy'kĭlōsītō'sĭs) [G. *poikilos,* varied; *kytos,* cell; *-ōsis,* intensive]. Variation in shape of red corpuscles; a condition characterized by poikilocytes in the blood.

pollex (pŏl'ĕks) [L. *pollex,* thumb]. The thumb, or first digit of the normal five in the anterior limb.

poly- [G]. Prefix: many, much, or great.

polycythemia (pŏl'ēsīthē'mēă) [G. *polys,* many; *kytos,* cell; *haima,* blood]. Any condition marked by an abnormal increase in the number of circulating red blood cells.

polydipsia (pŏl'ēdĭp'sēă) [G. *polys,* many; *dipsa,* thirst]. Excessive thirst.

polymer (pŏl'ĭmĕr) [G. *polys,* many; *meros,* a part]. A substance formed by combining two or more molecules of the same substance.

polymerize (pōlĭmĕr'īz) [G. *polys,* many; *meros,* part]. To change into another compound having same elements in same proportions, but a higher molecular weight.

polyphagia (pŏl'ēfā'jēă) [G. *polys,* many; *phagein,* to eat]. Eating abnormally large amounts of food at a meal.

polypnea (pŏl'ĭpnē'ă) [G. *polys,* many; *pnoia,* breath]. Very rapid breathing; panting.

polysaccharide (pŏl'ēsăk'ărīd) [G. *polys,* many; L. *saccharum,* sugar]. A polymer of sugars, having large molecules; e.g., vegetable gums, starches, cellulose, hemicellulose, etc.

polyuria (pŏl'ēū'rēă) [G. *polys,* many; *ouron,* urine]. Excessive secretion and discharge of urine.

pons (pŏnz) [L. *pons,* bridge]. A process of tissue connecting two or more parts. Pons varolli (Costanzo Varolio, It. anatomist, 1544–1575). A rounded eminence on the ventral surface of the brainstem.

popliteal (pŏplĭtē'ăl, pŏplĭt'ēăl) [L. *poples, poplit-,* ham]. Concerning the posterior surface of the knee.

porphyrin (pŏr'fĭrĭn) [G. *porphura,* purple]. One of a group forming the basis of animal and plant respiratory pigments, obtained from hemoglobin and chlorophyll.

portal (pŏr'tăl) [L. *porta,* gate]. An entryway. Concerning a porta or entrance to an organ, esp. that through which the blood is carried to liver.

postganglionic fiber (pōstgăng'glēŏn'ĭk fībr) [L. *post,* after; G. *gangglion,* tumor; L. *fibra,* band]. Autonomic nerve fiber issuing from ganglia; the second neuron in an autonomic pathway.

potential (pōtĕn'shăl) [L. *potentia,* power]. An "electrical pressure." Implies a measurable electric current flow or state between two areas of different electric strength.

pre- [L]. Prefix: in front of, before.

precipitate (prēsĭp'ĭtāt) [L. *praecipitāre,* to cast down]. Something usually insoluble that forms in a solution; to cause precipitation or a casting down of an insoluble mass.

precocious (prēkō'shŭs) [L. *praecoquere,* to mature before]. Matured or developed earlier than normal.

precursor (prēkŭr'sŭr) [L. *praecurrere,* to run ahead]. Anything preceding something else or giving rise to.

preganglionic fiber (prēgăng'glēŏn'ĭk fībr) [L. *prae,* before; G. *gangglion,* tumor; L. *fibra,* band]. Medullated fiber from the spinal cord, ending in synapses around sympathetic ganglion cells; also similar neurons from the sacral spinal cord and brainstem to the parasympathetic ganglia.

primordial (prīmŏr'dēăl) [L. *primordium,* the beginning]. Existing first or in undeveloped form.

pro- [L.; G.]. Prefix: for, from, in favor of.

proaccelerin (prō'ăksĕl'ĕrĭn). The fifth factor (factor V) in blood coagulation. Present in normal plasma but deteriorates rapidly at room temperature. Its function in blood coagulation is unclear.

procreation (prō'krēā'shŭn) [L. *pro,* before; *creo,* create]. Reproduction; generation.

proctodaeum (prŏk'tōdē'ŭm) [G. *prōktos,* anus; *hodos,* way]. The latter part of the embryonic alimentary canal, formed by anal invagination.

progestin (prōjĕs'tĭn) A corpus luteum hormone that prepares the endometrium for the fertilized ovum.

prognosis (prŏg'nō'sĭs) [G. foreknowledge]. Prediction of the course and outcome of a disease or abnormal process.

prolactin (prōlăk'tĭn) [L. *pro,* for; *lac,* milk]. The lactogenic prepituitary hormone; causes milk secretion in "primed" gland.

pronation (prōnā'shŭn) [L. *pronare,* to bend forward]. Medial rotation of the forearm which brings the palm of the hand downward.

pronephros (prōnĕf'rŏs) [G. *pro,* before; *nephros,* kidney]. The earliest and simplest type of excretory organ of vertebrates, functional in simpler forms (cyclostomes) and serving as a provisional kidney in some fishes and amphibians. In reptiles, birds, and mammals, it appears in the embryo as a temporary, functionless structure.

prophase (prō'fāz) [G. *pro,* before; *phasis,* an appearance]. A stage in mitosis characterized by nuclear disorganization and formation of visible chromosomes.

proprioceptive (prō'prēōsĕp'tĭv) [L. *proprius*, one's own; *capio*, to take]. *Pert.* awareness of posture, movement, and changes in equilibrium, and the knowledge of position, weight, and resistance of objects in relation to the body.

prosencephalon (prŏs'ĕnkĕf'ălŏn, -sef-) [G. *pros*, before; *engkephalos*, brain]. The forebrain, comprising telencephalon and diencephalon; the first primary brain vesicle.

prostate (prŏs'tāt) [L. *pro*, before; *stare*, to stand]. A muscular and glandular organ found ventral to the rectum and inferior to the urinary bladder in males.

protein (prō'tēĭn) [G. *prōtos*, first]. Albuminous substance; a nitrogenous compound of cell protoplasm; a complex substance characteristic of living matter and consisting of aggregates of amino acids and generally containing sulphur.

proteolytic (prō'tēōlĭt'ĭk) [G. *prōtos*, first; *lysis*, dissolution]. Hastening the hydrolysis of proteins.

proteose (prō'tēōs) [G. *prōtos*, first]. One of the class of intermediate products of proteolysis between protein and peptone.

proton (prō'tŏn) [G. *prōtos*, first]. An electrically charged particle that is a component of the atom and of matter and carries a positive charge of electricity. Its mass varies; it is much smaller than the atom. It is complementary to the electron.

proximal (prŏk'sĭmăl) [L. *proximus*, next; *ad*, toward]. Nearest the point of attachment, center of the body, or point of reference.

psoas (sō'ăs, psō'ăs) [G. *psoa*, loins]. Name of two loin muscles, major and minor, formerly magnus and parvus.

pseudo- (soo'dō) [G. *pseudēs*, false]. Prefix: false.

psyche (sī'kē) [G. *psyche*, mind]. The human soul; the mind; the intelligence.

psychosis (sī'kō'sĭs) [G. *psyche*, mind]. A disorder of the mind in which contact with reality is lost.

pterygoid (tĕr'ĭgoyd) [G. *pteryx*, wing; *eidos*, form]. A cranial bone. Winglike; *appl.* winglike processes of sphenoid, canal, fissure, fossa, plexus, muscles.

ptyalin (tī'ălĭn) [G. *ptyalon*, saliva]. A salivary amylolytic enzyme converting starch into maltose and dextrin.

puberty (pū'bĕrtē) [L. *pubertas*, puberty]. Period in life at which one becomes functionally capable of reproduction.

pudendum (pūdĕn'dŭm) [L. *pudere*, to be ashamed]. External female genitalia.

pulmo- [L]. Combining form: lung.

pulse (pŭls) [L. *pulsāre*, to beat]. A throbbing caused in an artery by the shock wave resulting from ventricular contraction.

pupil (pū'pĭl) [L. *pupilla*, pupil]. The opening in the center of the iris of the eye.

purine (pū'rēn) [L. *purum*, pure; *uricus*, uric acid]. Parent of a group of hetercyclic nitrogen compounds including purine itself, $C_5H_4N_4$; caffeine, adenine, guanine, xanthine, and uric acid.

Purkinje fibers (pŭrkĭn'jē fībĕrz) (Johannes E. von Purkinje, Hungarian physiologist, 1787–1869). Atypical muscle fibers lying beneath endocardium constituting the impulse-conducting system of the heart ventricles.

putamen (pūtā'mĕn) [L. *putamen*, shell]. The darker, outer layer of the lenticular nucleus.

pyelo- [G]. Combining form: the pelvis.

pyelonephritis (pī'ĕlōnĕfrī'tĭs) [G. *pyelos*, pelvis; *nephros*, kidney; *-ītis*, inflammation]. Inflammation of kidney substance and pelvis.

pyramidal cell (pĭr'ămĭdl sĕl) [G. *pyramis*, a pyramid; L. *cella*, a chamber]. Pyramid-shaped cell of cerebral cortex.

pyrogen (pī'rō'jĕn) [G. *pyr*, fire; *gennan*, to produce]. An agent causing a rise in body temperature.

radial (rā'dēăl) [L. *radius*, spoke]. To pass outward from a specific center, like the spokes of a wheel.

radiation (rādēā'shŭn) [L. *radiāre*, to emit rays]. Sending out heat or other forms of energy as electromagnetic waves; emission of rays from a center; to treat a disease by using a radioactive material.

radical (răd'ĭkăl) [L. *radix,* root]. A group of atoms that does not exist in the free state but as a unit in a compound, as OH, NH_4, C_6H_5, etc.

ramus (rā'mŭs) *(plu.* rami) [L. *ramus,* branch]. Any branchlike structure; mandible, or its proximal part, of vertebrates; branch of a spinal nerve.

raphe (rā'fē) [G. *raphe,* seam]. A ridge or seamlike structure.

Rathke's pouch (Martin H. Rathke, Ger. anatomist, 1793–1860). A diverticulum of ectoderm from the roof of the stomodaeum.

rationalist (răsh'ŏnălĭst) [L. *rationalis,* reason]. One who believes in the formation of opinions by relying on reason alone, independently of authority or of revelation.

re- (rē) [L]. Prefix: again; back.

receptor (rēsĕp'tŏr) [L. *recipere,* to receive]. A sense organ that receives stimuli from the environment; may be complex or simple.

recessive (rē'sĕs'ĭv) [L. *recessus,* to go back]. In genetics, a characteristic that does not usually express itself because of suppression by a dominant gene (allele).

reciprocal (rēsĭp'rŏkăl) [L. *reciprocus,* turning back and forth]. The opposite; interchangeable in nature.

Recklinghausen's disease or syndrome (rĕk'lĭnghŏw'zĕn) (Friedrich D. von Recklinghausen, Ger. pathologist, 1833–1910). Pigmentation of the skin; multiple small fibrous tumors on same with tenderness along nerves, pain in joints, sluggishness, multiple neurofibromatosis.

rectum (rĕk'tŭm) [L. *rectus,* straight]. The continuation of the digestive tract from the pelvic colon to anal canal.

red nucleus (rĕd nōō'klēŭs) [A.S. *read,* a primary color of the spectrum; L. *nucleus,* little kernel]. Large, oval pigmented mass in upper portion of midbrain and extending upward into subthalamus. Receives fibers from cerebral cortex and cerebellum; efferent fibers give rise to rubrospinal tracts.

reduction (rēdŭk'shŭn) [L. *reductio,* a leading back]. Uptake of H by a compound, gain of electrons; restoring a broken bone to normal relationships.

referred (rēfĕrd') [L. *re,* back; *ferre,* to bear]. Sent to another place, as referred pain.

reflex (rē'flĕks) [L. *reflectere,* to turn back]. An involuntary response to stimulus; a stereotyped response.

refract (rēfrăkt') [L. *refractus,* to break or bend]. To bend or deflect a light ray.

refractory period (rēfrăk'tŏrē) [L. *refractiō,* break back]. Period of relaxation of a muscle during which excitability is depressed. If stimulated it will respond, but a stronger stimulus is required and response is less.

regeneration (rējĕnĕrā'shŭn) [L. *re,* again, back; *generāre,* to beget]. Regrowth or repair of tissues or restoring a body part by regrowth.

regurgitate (rēgĕr'jĭtāt) [L. *re,* again, back; *gurgitāre,* to flood]. To return to a place just passed through; backflow.

relaxin (rēlăk'sĭn). A polypeptide hormone secreted in the corpus luteum during pregnancy. Obtained commercially from ovaries of pregnant sows. It has been used as a uterine relaxant and to soften the cervix. Has no estrogenic or progestational effect.

religion (rēlĭj'ŭn) [L. *religio,* connected with *relego,* read over, or with *religo,* bind]. A belief binding the spiritual nature of man to a supernatural being, as involving a feeling of dependence and responsibility, together with the feelings and practices naturally flowing from such a belief. Any system of faith and worship, as the Christian religion.

renal (rē'năl) [L. *renalis,* kidney]. *Pert.* kidney.

rennin (rĕn'ĭn). A coagulating enzyme found in the stomach of ruminants which curdles milk.

replicate (rĕp'lĭcāt) [L. *replicatio,* a folding back]. To duplicate.

repression (rē'prĕshŭn) [L. *repressus,* to check]. To put down or prevent from expressing something; in genetics, the way a gene shuts off its effects.

resorption (rēsŏrp'shŭn) [L. *resorbēre,* to suck in]. The act of removal by absorption, as resorption of an exudate or pus. Removal of hard parts of a tooth as a result of lysis and phagocytic action.

respiration (rĕspĭrā'shŭn) [L. *respirare,* to breathe]. Cellular metabolism; the act of exchange of gases between the body and environment.

respiratory quotient. The ratio between the volume of carbon dioxide produced and the volume of oxygen used.

resuscitation (rēsŭsĭtā'shŭn) [L. *resuscitātus,* to revive]. Bringing back to consciousness.

rete (rē'tē) [L. *rete,* net]. A net or network; a plexus.

rete testis (rē'tē tĕs'tĭs) [L. *rete,* net; *testis,* testicle]. Network of tubes formed by the tubuli recti.

reticular (rētĭk'ūlăr) [L. *reticula,* net]. Interlacing; in the form of a network.

reticular formation (rētĭk'ūlăr fŏrmā'shŭn) [L. *reticula,* net; *formatio,* a structure with definite arrangement and shape]. Groups of cells and fibers arranged in a diffuse network throughout the brainstem. These both fill the spaces and connect the tracts that ascend and descend through the area. Important in controlling or influencing alertness, waking, sleeping, and various reflexes.

retina (rĕt'ĭnă) [L. *rete,* net]. The inner, nervous membrane of the eye; receives images, contains visual receptors (rods and cones).

retro- [L]. Prefix: behind or backward.

retroperitoneal (rĕ'trōpĕrĭtōnē'ăl) [L. *retro,* backward; G. *peritonaion,* peritoneum]. Located behind the peritoneum.

rheobase (rē'ōbās) [G. *rheos,* current; *basis,* step]. In unipolar testing with the galvanic current using negative as the active pole, the minimal voltage required to produce a stimulated response. The threshold of excitation.

rhin-, rhino- [G]. Combining form: nose.

rhinitis (rīnī'tĭs) [G. *rhis, rhin-,* nose; *-ītis,* inflammation]. Inflammation of the nasal mucous membranes.

rhodopsin (rōdŏp'sĭn) [G. *rhodo,* rose; *opsis,* vision]. Visual purple, a pigment in outer segment of retinal rods; light decomposes it and creates an impulse.

rhombencephalon (rŏmb'ĕnsĕf'ălŏn) [G. *rhombos,* wheel; *engkephalos,* brain]. Hindbrain.

ribo- (rī'bō). Combining form: *pert.* ribose, a five-carbon sugar (pentose).

ribonucleic acid (rī'bōnōōclā'ĭc) (RNA). A nucleic acid containing adenine, guanine, cytosine, and uracil in nucleolus, mitochondria, and ribosomes, and taking part in cytoplasmic protein synthesis.

ribosome (rī'bōsōm) [G. *sōma,* body]. A spherical granule or microsomal particle containing ribonucleic acid, or nuclear membrane and membranes of endoplasmic reticulum; takes part in protein synthesis.

rickets (rĭk'ĕts). A form of osteomalacia in children. Results from deficient deposition of lime salts in developing cartilage and newly formed bone, causing abnormalities in shape and structure of bones. Caused primarily by vitamin D deficiency.

risorius (rīsō'rēŭs) [L. *risorius,* laughing]. Muscular fibrous band rising over masseter muscle and inserted into tissues at the corner of the mouth.

RNA. Ribonucleic acid, a fraction of the nucleoprotein of chromosomes; RNA determines the specificity of the proteins synthesized in a cell or organism.

rod (rŏd) [A.S. *rodd,* club]. Slender, straight bar. One of the slender, long sensory bodies in retina responding to faint light. Bacterium shaped like a rod.

Ruffini's corpuscles (rōōfē'nē) (Angelo Ruffini, Italian anatomist, 1864–1929). Encapsulated sensory nerve endings found in subcutaneous tissue, thought to mediate sense of warmth.

saccharide (săk'ărīd) [G. *sakcharon,* sugar]. A sugar; a carbohydrate containing two or more simple sugar units.

saccular (săk'ūlăr) [L. *sacculus,* small bag]. Shaped like a small sac.

sacculus (săk'ūlŭs) [L. *sacculus,* a little bag]. A saccule or little sac; specifically the lower and smaller of the two chambers of the vestibular portions of the membranous labyrinth.

sagittal (săjĭt'ăl) [L. *sagitta,* arrow]. The median plane or any plane parallel to it that divides the body into right and left parts. A suture in the skull; a venous sinus in the head.

salicylate (săl'ĭsĭl'āt, sălĭs'ĭlāt). Any salt of salicylic acid.

saliva (sălī'vă) [L. *saliva,* spittle]. A fluid containing ptyalin, secreted by the salivary and other glands of mouth.

saltatory conduction (săl'tătō'rē cŏndŭk'shŭn) [L. *saltātiō,* leaping; *conducere,* to lead]. Skipping from node to node; said of movement of the potential along myelinated neurons.

sanguine (săn'gwĭn) [L. *sanguis,* blood]. Of buoyant disposition; hopeful; confident; originally, having qualities or a temperament supposed to be due to active blood. One of the four natural humors in ancient physiology.

saphenous (săfē'nŭs) [G. *saphēnēs,* manifest]. *Pert.* or associated with a saphenous vein or nerve in the leg.

sapiens (sā'pēēns) [L. *sapiens,* wise]. The species name of man.

sarcolemma (săr'kōlĕm'ă) [G. *sarx, sarkos,* flesh; *lemma,* a rind]. A delicate membrane surrounding each striated muscle fiber.

sarcoma (sărkō'mă) [G. *sarx,* flesh- *-ōma,* tumor]. A neoplasm of cancer rising from muscle, bone, or connective tissue.

sarcomere (săr'kōmēr) [G. *sarkos,* flesh; *meros,* part]. The portion of a myofibril lying between two Z lines.

sarcoplasm (săr'kōplăzm) [G. *sarkos,* flesh; *plasma,* a thing formed]. Semifluid interfibrillary substance of muscle cells. The cytoplasm of muscle cells.

sarcoplasmic reticulum (săr'kōplăzmĭk rĕtĭk'ūlŭm) [G. *sarkos,* flesh; *plasma,* a thing formed; L. *reticulum,* small net]. A network of ultramicroscopic filaments in the longitudinal interstitial substance between fibrils of muscular tissue.

satellite cells (săt'ĭlīt) [L. *satelles,* attendant]. Flat epitheliumlike cells forming the inner portion of a double-layered capsule that covers a neuron. Neuroglial cells enclosing the cell bodies of neurons in spinal ganglia.

saturated (săt'ūrātĕd) [L. *saturāre,* saturate]. Holding all it can. A saturated fat has all the hydrogen it can hold on its chemical bonds.

scalene (skā'lēn) [G. *skalēnos,* uneven]. Having unequal sides and angles, said of a triangle. Designating a scalenus muscle.

scalpel (skăl'pĕl) [L. *scalpo,* carve]. A small knife used in dissections and in surgery.

scaphoid (skăf'oyd) [G. *skaphē,* boat; *edios,* form]. Shaped like a boat; *appl.* carpal and tarsal bones; *appl.* fossa above pterygoid fossa. Os naviculare.

scheme (skēm) [L. *schema,* a plan]. An orderly series of events or changes, as a metabolic scheme.

schindylesis (skĭn'dĭlē'sĭs) [G. *schindylein,* to cleave]. Articulation in which a thin plate of bone fits into a cleft or fissure, as that between vomer and palatines.

Schwann's sheath (shvŏn) (Theodor Schwann, Ger. anatomist, 1810–1882). The neurilemma of a nerve fiber.

sciatic (sīăt'ĭk) [G. *ischion,* hip-joint]. *Pert.* hip region; *appl.* artery, nerve, veins, etc.

sclera (sklĕ'ră) [G. *sklēros,* hard]. The tough, opaque, fibrous tunic of the eyeball; sclerotic coat, sclerotica.

sclerosis (sklĕrō'sĭs) [G. *sklerosis,* a hardening]. Hardening or toughening of a tissue or organ, usually by increase in fibrous tissue.

scoliosis (skōlēo'sis) [G. *skoliosis,* crookedness]. An abnormal lateral curvature of the vertebral column.

scotopic (skōtōp'ĭk) [G. *scotos,* darkness; *ōps,* eye]. Pertaining to the adjustment of vision for darkness; the rods are scotopic receptors.

scrotum (skrō'tŭm) [L. *scrotum,* a bag]. A medial pouch of loose skin containing testes in mammals.

sebaceous (sēbā'shŭs) [L. *sebaceous,* fatty]. *Pert.* sebum, a fatty secretion of the sebaceous (oil) glands.

sebum (sē'bŭm) [L. *sebum,* tallow]. The secretion of sebaceous glands, consisting of fat and isocholesterin.

secretin (sēkrē'tĭn) [L. *secernere,* to separate]. A parahormone formed in the mucous membrane of the duodenum through the influence of acid contents from the stomach whose function is to stimulate the secretion of pancreatic juice and bile.

secretion (sēkrē'shŭn) [L. *secretio,* separation]. Substance or fluid that is separated and elaborated by cells or glands; process of such separation.

seizure (sē'zhĕr) [M.E. *seizen,* to take possession of]. A sudden attack of a disease or pain.

selective (sēlĕk'tĭv) [L. *sēlectus,* having chosen]. Exhibiting choice, as a selective membrane that passes some materials but not others.

semen (sē'mĕn) [L. *semen,* seed]. A thick, opalescent, viscid secretion discharged from the urethra of the male at the climax of sexual excitement (orgasm). Contains the spermatozoa.

semi- [L.]. Prefix: half.

semilunar (sĕm'ĭlo͞o'năr) [L. *semi,* half; *luna,* moon]. Half-moon shaped; *appl.* fibrocartilages of knee, ganglia, fasica, lobules of cerebellum, valves; *appl.* notch, greater sigmoid cavity between olecranon and coronoid process of ulna.

semilunar valve (sĕm'ĭlo͞o'năr vălv) [L. *semi,* half; *luna,* moon; *valva,* a fold]. Valve between the heart and aorta and valve between the heart and pulmonary artery.

seminal vesicle (sĕm'ĭnăl vĕs'ĭkl) [L. *semen,* seed; *vesicula,* bladder]. A convoluted, saccular outgrowth of the ductus deferens, behind bladder; produces fluid for sperm.

seminiferous tubules (sēmĭnĭf'ĕrŭs to͞o'būlz) *(plu.)* [L. *semen,* seed; *ferre,* to carry; *tubulus,* a tubule]. The structures in which the spermatozoa and seminal fluids are produced.

semipermeable (sĕmĭpĕr'mēăbl) [L. *semi,* half; *per,* through; *meo,* pass]. Partly allowing passage, especially of liquids; "selective."

senescence (sĕnĕs'ĕns) [L. *senescere,* to grow old]. The process of growing old. The period of old age.

sensory (sĕn'sō-rē) [L. *sensorius*]. *Pert.* sensation or the afferent nerve fibers from the periphery to the central nervous system.

serological (sĕrĕlŏj'ĭkăl) [L. *serum,* whey; G. *logos,* study]. *Pert.* study of serum.

serosa (sĕrō'să) [L. *serum,* whey]. Any serous membrane or tunica serosa; visceral peritoneum; false amnion or outer layer of amniotic fold.

serotonin (sĕrĕtō'nĭn) [L. *serum,* whey; G. *tonos,* tightening]. A vasoconstrictor compound in blood platelets, also in brain cells, which causes contraction of smooth muscle.

serous membrane (sē'rŭs) [L. *serum,* serum]. A thin membrane of connective tissue, lining a closed cavity of the body and reflected over the viscera, as mesentery. Mesothelium on free surface.

Sertoli's cells (sĕrtō'lē) (Enrico Sertoli, It. histologist, 1842–1910). Supporting, elongated cells of seminiferous tubules that nourish spermatids.

serum (sē'rŭm) [L. *serum,* whey]. Any serous fluid, esp. the fluid that moistens the surfaces of serous membranes. The watery portion of the blood after coagulation; a fluid found when clotted blood is left standing long enough for the clot to shrink.

sesamoid (sĕs'ămoyd) [G. *sesamon,* sesame; *eidos,* form]. *Appl.* bone developed within a tendon and near a joint, as patella, radial or ulnar sesamoid, fabella. A sesamoid bone.

sex (sĕks) [L. *sexus*]. The quality that distinguishes between male and female.

sex chromosomes. Chromosomes determining sex (XX female, XY male).

sex, nuclear. The genetic sex as determined by the presence or absence of sex chromatin (Barr bodies) in the nuclei of body cells.

shock (shŏk) [M.E. *schokke*]. A state resulting from circulatory collapse (low blood pressure and weak heart action).

shunt (shŭnt) [M.E. *shunten,* to avoid]. A shortcut; passage between arteries and veins that bypasses capillaries; a scheme for metabolism of a foodstuff that eliminates certain steps found in a scheme that metabolizes the same substance.

sickle cell. An erythrocyte that is crescent-shaped because it contains an abnormal hemoglobin.

sinoatrial (SA) node (sĭn′oa′trēăl nōd). A mass of modified cardiac muscle tissue at the junction of superior vena cava with right cardiac atrium, regarded as starting point of the heartbeat. *Syn.* pacemaker.

sinus (sī′nŭs) [L. *sinus,* curve, gulf, or hollow]. A cavity, depression, recess, or dilation; a groove or indentation. *Appl.* cavity in a bone; a large channel for venous blood flow.

sinusoids (sī′nŭsoydz) *(plu.)* [L. *sinus,* a hollow, a curve; G. *eidos,* like]. Resembling sinuses. Minute blood vessels found in such organs as the liver, spleen, adrenal glands, and bone marrow. They are slightly larger than capillaries and have a lining of reticuloendothelium.

solubility (sŏlūbĭl′ĭtē) [L. *solubilis,* to dissolve]. Capable of being dissolved or going into solution.

solute (sŏl′ūt) [L. *solutus,* dissolved]. The dissolved, suspended, or solid component of a solution.

solution (sŏlōō′shŭn) [L. *solvo,* loosen]. A clear, transparent, homogeneous liquid formed by dissolving a substance, whether solid, liquid, or gaseous, in a solvent, as salt dissolved in water.

solvent (sŏl′vĕnt) [L. *solvens,* to dissolve]. A dissolving medium.

soma (sō′mă) [G. *sōma,* body]. *Pert.* body, a cell body, or the body as a whole.

somatic (sōmăt′ĭk) [G. *sōma,* body]. *Pert.* body exclusive of the reproductive cells; *pert.* skeletal muscles and/or skin.

somatotrophin (sō′mătōtrōf′ĭn) [G. *sōma,* body; *trephein,* to increase]. Growth hormone of somatotrophic hormone, STH.

somesthesia (sōm′ĕsthē′zhēă) [G. *sōma,* body; *aisthēsia,* sensation]. The awareness of body sensations.

somite (sō′mīt) [G. *sōma,* body]. An embryonic blocklike segment of mesoderm formed along the neural tube.

spasm (spăzm) [G. *spasmos,* convulsion]. A sudden, involuntary, often painful contraction of a muscle or muscles.

spastic (spăs′tĭk) [G. *spastikos,* convulsive]. Contracted or in a state of continuous contraction.

specific gravity (spēsĭf′ĭk grăv′ĭtē). Weight of a substance compared to that of an equal volume of pure water, with water assigned a value of one.

sperm (spŭrm) [G. *sperma,* seed]. The male sex cells produced in testes. Also spermatozoon; spermatozoa.

spermatid (spĕr′mătĭd) [G. *sperma,* seed]. A cell produced by division of the secondary spermatocyte to become a spermatozoon.

spermatocyte (spĕrmăt′ōsīt) [G. *spermatos,* seed; *kytos,* cell]. A cell originating from a spermatogonium, which forms by division the spermatids giving rise to spermatozoa.

spermatogenesis (spĕrmătōjĕn′ĕsĭs) [G. *spermatos,* seed; *genesis,* produce]. The formation of mature functional spermatozoa.

spermatogonia (spĕr′mătōgō′nēă) *(plu.)* [G. *sperma,* seed; *gonos,* offspring]. Sex cells derived from cords of epithelial cells of the testes.

spermatozoa (spĕr′mătōzō′ă) *(plu.)* [G. *sperma,* seed; *zoon,* animal]. Male reproductive cells, consisting usually of head, middle piece, and locomotory flagellum.

sphincter (sfĭng′kter) [G. *sphinggein,* to bind tightly]. A muscle that contracts and closes an orifice.

spirometer (spīrŏm′ĕtĕr) [L. *spirāre,* to breathe; G. *metron,* measure]. An apparatus that measures air volumes in lungs.

spleen (splēn) [G. *splēn,* spleen]. A vascular organ in which lymphocytes are produced and red blood corpuscles destroyed, in vertebrates.

squama (skwā′mă) [L. *squama,* scale]. A squame or scale; a part arranged like a scale; vertical part of frontal bone; part of occipital bone above and behind foramen magnum; anterior and upper part of temporal bone.

squamous (skwā′mŭs) [L. *squama,* scale]. Consisting of scales; *appl.* simple epithelium of flat nucleated cells, scaly or pavement epithelium.

stapes (stā'pēz) [L. *stapes,* stirrup]. Ossicle in the middle ear that articulates with the incus and contacts the fenestra ovalis of the inner ear.

stasis (stā'sĭs) [G. *stasis,* halt]. Stoppage of fluid flow.

stellate (stĕl'āt) [L. *stella,* star]. Star-shaped; asteroid; radiating; *appl.* hair, cells of Kupffer, ganglion of sympathetic system, ligament of rib, veins beneath fibrous tunic of kidney, etc.

stenosis (stĕnō'sĭs) [G. *stenosis,* a narrowing]. Narrowing of an opening or passage.

stereocilia (stĕr'eoĭl'ēă) *(plu.)* [G. *stereos,* rigid; L. *cilium,* eyelash]. Nonmotile secretory projections on epithelium of duct of epididymis.

steroid (stĕr'oyd). A lipid substance having as its chemical skeleton the phenanthrene nucleus.

stimulus (stĭm'ūlŭs) [L. *stimulus,* goad]. An agent that causes a reaction or change in an organism or in any of its parts.

stomodeum (stōm'ēdē'ŭm) [G. *stoma,* mouth; *hodaios,* on the way]. Anterior invaginated portion of embryonic gut. Ectodermal in origin.

strabismus (străbĭz'mŭs). When the eyes do not both focus at the same point. Crossed eyes or "squint."

stressor. An agent tending to upset homeostasis and to produce strain or stress on the organism.

stridor (strī'dŏr) [L. a harsh sound]. A harsh sound occurring during breathing; like the sound of wind "whining."

stroke (strōk) [A.S. *strāk,* a going]. The total set of symptoms resulting from a cerebral vascular disorder. Also, apoplexy.

stroma (strō'mă) [G. *strōma,* bedding]. Transparent filmy framework of red blood corpuscles; protoplasmic body of a plastid; connective tissue binding and supporting an organ; in ovary, a soft, vascular, reticular framework in meshes of which ovarian follicles are imbedded.

structural gene. A gene that directs the synthesis of specific protein.

styloid process (stī'loyd) [G. *stylos,* pillar; *eidos,* form]. A pointed process of a bone. *Appl.* process of temporal bone.

stylus (stī'lŭs) [L. *stylus,* pricker]. A style; stylet; simple pointed spicule; molar cusp; pointed process.

sub- [L.]. combining form: under, beneath, in small quantity, less than normal.

subclavian (sŭbklā'vēăn) [L. *sub,* under; *clavis,* key]. Below clavicle; *appl.* artery, vein, nerve.

subconscious (sŭbkŏn'shŭs) [L. *sub,* under; *conscius,* aware]. Operating at a level of which one is not aware.

subcutaneous (sŭb'kūtā'nēŭs) [L. *sub,* under; *cutis,* skin]. Under the cutis or skin; *appl.* parasites living just under the skin; *appl.* inguinal or external abdominal ring.

subliminal (sŭblĭm'ĭnăl) [L. *sub,* under; *limen,* threshold]. Less than that required to get a response; not perceptible.

submucosa (sŭb'mūkō'să) [L. *sub,* under; *mucosus,* mucous]. The layer of aerolar connective tissue under a mucous membrane.

substantia nigra (sŭbstăn'shēă nī'gră) [L. *substantia,* substance; *nigra,* black]. Mass of gray matter between the dorsal and pedal parts of the crus cerebri.

substrate (sŭb'strāt) [L. *substratum,* a strewing under]. The substance on which an enzyme acts.

succus entericus (sŭk'ŭs ĕntĕr'ĭcŭs) [L. *succus,* juice; G. *enteron,* gut]. Digestive juice of small intestine in vertebrates.

sucrase (sōō'krās) [Fr. *sucre,* sugar]. An enzyme in the intestinal juice that splits cane sugar into glucose and fructose, the two being absorbed into the portal circulation.

sudoriferous (sōō'dŏrĭf'ĕrŭs) [L. *sudor,* sweat; *ferre,* to carry]. Conveying, producing, or secreting sweat; *appl.* glands and their ducts.

sulcus (sŭl'kŭs) [L. *sulcus,* furrow]. A groove; *appl.* cerebral grooves; those of heart, tongue, cornea, bones, etc.

super-, supra- [L.]. Combining form: above, beyond, superior.

superciliary (sōōpĕrsĭl'ēărē) [L. *super,* over; *cilia,* eyelids]. *Pert.* eyebrows; above orbit.

supination (sōōpīnā'shŭn) [L. *supinus,* bent backward]. Lateral rotation of the forearm that brings the palm of the hand upward.

supraorbital (sōō'prăōr'bĭtăl) [L. *supra,* above; *orbis,* eye-socket]. Above orbital cavities; *appl.* process, artery, foramen, nerve, vein, etc.

surface tension (alveolus). The phenomenon whereby liquid droplets tend to assume the smallest area for their volume. The surface acts as though it had a "skin" on it as a result of cohesion among the surface molecules.

surfactant (sŭrfăc'tănt) [Fr. *sur,* above, over; *faciēs,* face]. A lipid-protein agent that lowers surface tension. Reduces surface tension in the lung alveoli and thus reduces chances of alveolar collapse. Also, surface active agent.

suspension (sŭspĕn'shŭn) [L. *suspensio,* a hanging[. A state in which solute molecules are mixed but not dissolved in a solvent and do not settle out.

suture (sōō'chĕr) [L. *sutura,* seam]. Line of junction of two bones immovably connected, as in the skull.

sym-, syn- [G.]. Combining form: with, along, together with, beside.

sympathetic (sĭmpăthĕt'ĭk) [G. *syn,* with; *pathos,* feeling]. *Appl.* system of nerves supplying viscera and blood vessels and intimately connected with some spinal nerves; part of autonomic nervous system that controls response to stressful conditions.

sympathomimetic (sĭm'păthōmĭmĕt'ĭk) [G. *sympathētikos,* sympathy; *mimētikos,* imitating]. Producing effects resembling those resulting from stimulation of the sympathetic nervous system, such as effects following the injection of epinephrine. Sympathicomimetic.

symphysis (sĭm'fĭsĭs) [G. *symphysis,* a growing together]. The coalescence of parts; the line of junction of two pieces of bone separate in early life, as pubic symphysis; slightly movable articulation with bony surfaces connected by fibrocartilage.

synapse (sĭn'ăps) [G. *synapsis,* union]. The area of functional continuity between neurons.

synaptic vesicle (sĭnăp'tĭk vĕs'ĭkl). Small granular structures found in "end feet" at ends of axons containing chemical substances for transmission across synapses.

synchrondrosis (sĭn'kŏndrō'sĭs) [G. *syn,* with; *chondros,* cartilage]. A synarthrosis in which the connecting medium is cartilage.

syncytium (sĭnsĭt'ēŭm) [G. *syn,* with; *kytos,* hollow]. Condition in which no membrane separates the cells.

syndrome (sĭn'drōm) [G. *syndromē,* a running together]. A complex group of symptoms.

syndesmosis (sĭn'dĕsmō'sĭs) [G. *syndesmos,* ligament]. Articulations with fibrous tissue between the bones.

syneresis (sĭnĕr'ĕsĭs) [G. *synairesis,* drawing together]. Contraction of a gel resulting in its separation from the liquid, as a shrinkage of fibrin when blood clots.

synergist (sĭn'ĕrjĭst) [G. *syn,* together; *ergon,* work]. A muscle or organ functioning in cooperation with another muscle to "firm" an action.

synovial (sĭnō'vēăl) [G. *syn,* with; L. *ovum,* egg]. Secreting a lubricating fluid; synovial membrane, inner stratum of articular capsule, connective tissue secreting synovia for joints.

synovial joints. Joints characterized by one or more synovial cavities.

synovial membrane. Inner stratum of articular capsule, connective tissue secreting a lubricating fluid for joints.

synthesis (sĭn'thĕsĭs) [G. *syn,* together; *tithemi,* place]. The putting of different things together; combination of separate substances, elements, or subordinate parts into a new form.

syphilis (sĭf'ĭlĭs) [Origin uncertain; G. *syn,* with; *philos,* love; or from *Syphilus,* shepherd who had the disease]. An infectious, chronic venereal disease transmitted by physical contact with an infected person.

system (sĭs'tĕm) [G. *systēma,* fr. *syn,* together; *his-*

temi, stand]. In biology, an assemblage of organ structures composed of similar elements and combined for the same general functions, as the nervous system. The entire body, taken as a functional whole.

systemic (sĭstĕm'ĭk) *Pert.* whole or greater part of the body.

systole (sĭs'tōlē) [G. *systolē,* contraction]. Contraction of the muscle of a heart chamber.

tachy- [G. *tachys,* swift]. Combining form: rapid, fast.

tachycardia (tăk'ĭkăr'dēă) [G. *tachys,* fast, speedy; *kardia,* heart]. Increased heartbeat.

tachypnea (tăk'ĭpnē'ă) [G. *tachys,* fast, speedy; *pnoē,* to breathe]. Increased rate of breathing.

taenia coli (tē'nēă kō'lī) [L. *taenia,* tape; G. *kōlon,* colon]. One of three bands of the large intestines into which muscular fibers are collected.

talus (tā'lŭs) [L. *talus,* ankle]. The ankle bone articulating with the tibia, fibula, calcaneus, and navicular bone.

tamponade (tămpŏnăd') [Fr. *tampon,* plug]. To plug up. Cardiac tamponade: a condition resulting from accumulation of excess fluid in the pericardial sac.

taxis (tăk'sĭs) [G. arrangement]. Response to an environmental change; a turning toward (positive) or away from (negative) the change.

telencephalon (tĕl'ĕnsĕf'ălŏn) [G. *tele,* far, *engkephalos,* brain]. Anterior terminal segment of brain.

teleological (tĕl'eolŏj'ĭcăl) [G. *teleos,* complete; *logos,* discourse]. Pertaining to the doctrine of adaptation to a divine purpose, and that evolution is purposive.

telodendria (tĕl'odĕn'drēă) *(plu.)* [G. *telos,* end; *dendrion, dim.* of *dendron,* tree]. The terminal arborization of an axon; end-brush; teleneurite.

telophase (tĕl'ōfāz) [G. *telos,* end; *phasis,* a phase]. The final stage of mitosis characterized by cytoplasmic division and reformation of nuclei.

tension (tĕn'shŭn) [L. *tensio,* a stretching]. The state of being stretched; a concentration of gas in a fluid; the force developed by a muscular contraction as measured by a gauge.

teres (tē'rēz) [L. *teres,* round]. Round and smooth; cylindrical. A cylindrical muscle.

testosterone (tĕstŏs'tĕrōn) [L. *testis,* testicle; G. *stear,* suet]. Testicular hormone.

tetanus (tĕt'ănŭs) [G. *tetanos,* tetanus]. A sustained contraction of a muscle; a disease caused by a bacterium characterized by a sustained contraction of jaw muscles (lockjaw).

tetany (tĕt'ăhnē) [G. *tetanos,* stretched]. A condition marked by muscular spasms, sharp flexion of wrist and ankle joints, cramps, and convulsions; due to abnormal calcium metabolism. Parathyroid hormone deficiency. Also hypoparathyroidism.

thalamus (thăl'ămŭs) [G. *thalamos,* inner chamber]. The largest subdivision of the diencephalon on either side, consisting chiefly of an ovoid, gray nuclear mass in the lateral wall of the third ventricle.

Thebesian valve (thēbē'zēăn vălv) (Adam Christian Thebesius, Ger. physician, 1686–1732). An endocardial fold at the entrance of the coronary sinus into the right atrium.

therapy (thĕr'ăpē) [G. *therapein,* treatment]. The treatment of a condition or disease.

thermal (thĕr'măl) [G. *thermē,* heat]. *Pert.* heat.

thermostat (thĕr'mōstăt) [G. *thermē,* heat; *histēmi,* stand]. A device for the automatic regulation of temperature by utilizing the expansion and contraction caused by changes of temperature in certain metals.

thorax (thō'răks) [G. *thōrax,* chest]. In higher vertebrates, that part of the body between neck and abdomen containing the heart, lungs, etc. The chest.

threshold *(thr*ĕsh'ōld) [A.S. *therscold*]. Just perceptible; the lowest strength stimulus that results in a detectable response or reaction. Also, liminal.

thrombocyte (thrŏm'bōsīt) [G. *thrombos,* clot; *kytos,* hollow]. Blood platelet.

thromboplastin (thrŏm'bōplăs'tĭn) [G. *thrombos,* clot; *plassein,* to form]. The third blood coagulation factor (factor III). A substance found in both blood and tissues; accelerates the clotting of blood.

thrombosis (thrŏmbō'sĭs). The formation of a clot, or the result of blocking a vessel.

thrombus (thrŏm'bŭs) [G.]. A blood clot obstructing a vessel.

thymus (thī'mŭs) [G. *thymos,* thymus]. An endocrine gland in the lower anterior part of the neck, or surrounding the heart, regressing after maximum development at puberty.

thyrotropin (thīrō'trŏpĭn) [G. *thyra,* door; *tropē,* a change]. The thyroid-stimulating prepituitary hormone, increasing the formation of thyroxine from thyroglobin.

thyroxine (thī'rŏksēn) [G. *thyra,* door; *oxys,* sharp]. A compound isolated from the thyroid gland; controls catabolic metabolism.

tissue (tĭs'ū, tĭsh'o͞o) [F. *tissu,* woven]. The fundamental structure of which animal and plant organs are composed; an organization of like cells and intercellular material.

-tome (tōm) [G.]. Combining form: a cutting or a cutting instrument.

tone (tōn) [L. *tonus,* a stretching]. A state of slight constant tension or contraction exhibited by musclar tissue.

tonic (tŏn'ĭk) [G. *tonikos, pert.* tone]. Having tone; continual.

tonicity (tōnĭs'ĭtē) [G. *tonos,* tone]. The property of having tone; a reference to the osmotic pressure of a solution.

tonsil (tŏn'sĭl) [L. *tonsilla,* tonsil]. One of aggregations of lymphoid tissue in pharynx or near tongue base.

tonus (tō'nŭs) [L. *tonus,* tone]. That partial, steady contraction of muscle which determines tonicity or firmness.

tortuous (tŏr'cho͞oŭs) [L. *tortuosis,* twisted]. Turned and twisted; full of windings.

toxic (tŏk'sĭk) [G. *toxikon,* poison]. Poisonous.

toxin (tŏk'sĭn) [G. *toxikon,* poison]. A poisonous substance derived from plant or animal sources; e.g., bacterial toxin: toxin produced by bacteria.

toxoid (tŏk'soyd) [G. *toxikon,* poison; *eidos,* form]. A toxin treated so as to lower or decrease its toxicity, but that will still cause antibody production.

trabeculae (trăbĕk'ūlē) *(plu.)* [L. *trabecula,* little beam]. Septa of connective tissue or muscle extending from a capsule or wall into the enclosed substance or cavity of an organ, as in lymph nodes, trabeculae carneae of heart, etc.

trabeculae carneae (trăbĕk'ūlē kär'nei) *(plu.)* [L. *trabecula,* a little beam; *carneus,* fleshy]. Thick muscular tissue bands of the inner walls of the ventricles of the heart.

trace (trās) [L. *tractus,* a drawing]. A very small amount.

trace element. Metals or organic substances essential in minute amounts for normal body function.

trachea (trā'kēă) [G. *tracheia,* rough]. A duct composed of membranes and incomplete cartilaginous rings conducting air from pharynx to bronchi.

tracheotomy (trā'kēŏt'ōmē) [G. *tracheia,* rough; *tome,* incision]. To open the trachea to allow breathing when upper respiration passages are blocked.

tract (trăkt) [L. *tractus,* a tract]. A course or pathway. A group or bundle of nerve fibers within the spinal cord or brain that constitutes an anatomical and functional unit. A group of organs or parts forming a continuous pathway.

trans- (trăns) [L.]. Prefix: across, through, over, beyond.

transamination (trăns'ămĭnā'shŭn). Transfer of an amino ($—NH_2$) group from one compound to another without formation of ammonia.

transcription (trănskrĭp'shŭn). The process in which DNA gives rise to RNA.

transduce (trănsdo͞os') [L. *trans,* across; *ducere,* to lead]. To change one form of energy to another.

transferrin (trănsfĕr'ĭn) [L. *trans,* across; *ferrum,* iron; F. *in,* more]. A beta globulin in blood plasma that is capable of combining with ferric ions and of transporting iron to various parts of the body.

transient (trăn'zhĕnt) [L. *trans,* across; *ire,* to go]. Temporary, not permanent.

translation (trănslā'shŭn) [L. *translatio,* to translate]. The process by which the code in messenger RNA is utilized to synthesize a specific protein.

trapezius (trăpē'zēŭs) [G. *trapezion,* small table]. A broad, flat, triangular muscle of neck and shoulders.

trapezoid (trăp'ĕzoyd) [G. *trapezion,* small table; *eidos,* form]. Trapezium-shaped; *appl.* ligament, ridge. Lesser multangular bone.

trauma (traw'mă) [G. *trauma,* wound]. An injury or wound.

treppe (trĕp'ĕh) [Ger. *treppe,* staircase]. Increase in height of contractions when a muscle is stimulated rapidly at regular intervals.

tri- [G.]. Combining form: three.

tricuspid valve (trīkŭs'pĭd vălv) [G. *treis,* three; *cuspis,* a point; L. *valva,* a fold]. Valve between the right cardiac atrium and right ventricle.

triglyceride (trīglĭs'ĕrīd). Combination of glycerol with the three fatty acids — stearic, oleic, palmitic. Most animal and vegetable fats are triglycerides.

trochanter (trōkăn'tĕr) [G. *trochantēr,* runner]. *Appl.* processes of prominences at upper end of thigh-bone — greater (major), lesser (minor), and third (tertius).

trochlea (trŏk'lēă) [G. *trochilia,* pully]. A pulleylike structure through which a tendon passes, as of humerus, femur, orbit.

trochoid (trō'koyd) [G. *trochos,* wheel; *eidos,* form]. Wheel-shaped; capable of rotating motion, as a pivot joint.

trophic (trō'fĭk) [G. *trophē,* nourishment]. *Pert.* hormones that control activity of other endocrine structures (e.g., ACTH, TSH, etc.).

-tropic (trō'pĭk) [G. *tropos,* turning]. Combining form: turning, changing, responding to stimulus.

trophoblast (trŏf'ōblăst) [G. *trophē,* nourishment; *blastos,* bud]. The outer layer of cells of epiblast, or of morula.

tropocollagen (trŏpōkŏl'ăjĕn) [G. *tropos,* mode; *kolla,* glue; *gennaein,* to produce]. A long molecule secreted in a fibrocyte; outside the cell, unites with others to form a collagen.

trypsin (trĭp'sĭn) [G. *tryein,* to rub down; *pepsis,* digestion]. Proteolytic enzyme of pancreatic juice.

trypsinogen (trĭpsĭn'ōjĕn) [G. *tryein,* to rub down; *pepsis,* digestion; *-genēs,* producing]. Substance secreted by cells of pancreas converted into trypsin by enterokinase of succus entericus.

T-tubule. Transversely arranged microscopic tubules that convey ECH to the interior of muscle cells.

tunica albuginea (too'nĭkă ălbūjĭn'ēa) [L. *tunica,* a sheath; *albus,* white]. The white, fibrous coat of the eye, testicle, ovary, or spleen.

tunica erythroides. A vascular coat around the testis.

tunica propria (too'nĭkă prō'prēă) [L. *tunica,* a sheath]. Deep portion of the corium containing blood vessels, nerves, glands, and hair follicles.

tunica vaginalis (too'nĭkă văj'ĭnăl'ĭs) [L. *tunica,* a sheath; *vagina,* sheath]. Serous membrane surrounding the front and sides of the testicle.

turbinate (tŭr'bĭnāt) [L. *turbo,* a whirl]. Nasal concha.

twitch (twĭch) [M.E. *twicchen*]. A single muscular contraction in response to a single stimulus.

tyrosinase (tīrō'sĭnās) [G. *tyros,* cheese]. An enzyme that acts on tyrosine to produce melanin.

ulcer (ŭl'sĕr) [L. *ulcus, ulcer-,* ulcer]. An open sore or lesion of the skin or mucous membrane of the body, with loss of substance, sometimes accompanied by the formation of pus.

ultramicroscopic (ŭl'trămī'krōskŏp'ĭk) [L. *ultra,* beyond; G. mikos, small; G. *skopos,* watcher]. So small as to be invisible through an ordinary microscope but visible through an ultramicroscope with powerful side illumination.

umbilical cord (ŭm'bĭl'ĭkăl) [L. *umbilicus,* navel]. The cord formed from the yolk sac and the body stalk; connects the embryo with the placenta and carries the umbilical arteries and veins.

umbilicus (ŭm'bĭl'ĭkŭs) [L. *umbilicus,* navel]. The navel; central abdominal depression at place of attachment of umbilical cord.

un- [A.S.]. Prefix: not, back, reversal.

unconscious (ŭnkŏn'shŭs) [A.S. *un,* not; L. *conscius,* conscious]. Insensible, not aware of surroundings.

uni- [L.]. Combining form: one.

unit (ū'nĭt) [L. *unus,* one]. One of anything; a single distinct object or part.

urea (ū'rēă, ūrē'ă) [G. *ouron,* urine]. Carbamide, a crystalline excretory substance, chief organic constituent of urine. Formed in liver, excreted by kidney.

uremia (ūrē'mēă) [G. *ouron,* urine; *haima,* blood]. A toxic condition associated with elevated blood urea levels; result of poor kidney function.

urethra (ūrē'thră) [G. *ourethra,* urine]. Duct from the urinary bladder to body surface; carries urine.

-uria (ūr'ēă). Suffix: *pert.* urine.

uric acid (ū'rĭk) [G. *ouron,* urine]. An end product of purine (a nitrogenous base) metabolism found in urine. $C_5H_4N_4O_3$.

urine (ū'rĭn) [L. *urina,* urine]. The fluid formed by the kidney and eliminated by the urethra.

uro- [G.]. Combining form: *pert.* urine.

urobilinogen (ū'rōbĭlĭn'ōjĕn) [G. *ouron,* urine; L. *bilis,* bile; G. *genēs,* produced]. A colorless compound derived from bilirubin, oxidized to urobilin, and excreted in urine.

urogenital ridge (ū'rōjĕn'ĭtăl) [G. *ouron,* urine; L. *genitalia,* genitals]. Ridge on dorsal wall of celom; gives rise to genital and mesonephric ridges.

urogenital sinus. Duct into which, in the embryo, the wolffian ducts and bladder empty and which opens into the cloaca. The common receptacle of genital and urinary ducts.

uterine (Fallopian) tube (ū'tĕrĭn, -īn, Fălō'pēăn tūb) (Gabriele Falloppio, It. anatomist, 1523–1562). [L. *uterus,* womb; *tuba,* a tube]. Small tubes attached to either side of the uterus and leading from the region of the ovary.

utriculus (ūtrĭk'ūlŭs) [L. *utriculus,* a little bag]. One of two sacs of the membranous labyrinth in the bony vestibule of the inner ear. A utricle.

vaccinate (văk'sĭnāt) [L. *vacca,* cow]. To inoculate with a vaccine to produce immunity against disease.

vaccine (văksēn') [L. *vacca,* cow]. Killed or attenuated bacteria or viruses prepared in a suspension for inoculation.

vacuole (văk'uol) [L. *vacuus,* empty]. One of spaces in cell protoplasm containing air, partially digested food, or other matter.

vagina (văjī'nă) [L. *vagina,* sheath]. Canal leading from uterus to vestibule.

vagus (vā'gŭs) [L. wandering]. The tenth cranial nerve.

vallate (văl'āt) [L. *vallatus,* surrounded by a rampart]. With a rim surrounding a depression; *appl.* papillae with taste buds on back part of tongue.

valve (vălv) [L. *valva,* a fold]. A structure for temporarily closing an opening so as to achieve one-way flow of fluid.

varicose (văr'ĭkōs) [L. *varix,* a twisted vein]. *Pert.* distended, knotted veins.

vasa recta (vă'să rĕc'tă) [L. *vas,* vessel; *rectus,* straight]. Tubules that become straight prior to entering the mediastinum testis. Straight collecting tubules of the kidney.

vasa vasorum (vā'să văsō'rŭm) [L. *vas,* vessel]. Tiny blood vessels distributed to the walls of larger veins and arteries.

vascular (văs'kūlăr) [L. *vasculum,* a small vessel]. *Pert.* or having blood vessels.

vasoconstrictor (văs'ōkŏnstrĭk'tŏr) [L. *vas,* vessel; *constrictor,* a binder]. Causing constriction of blood vessels. That which constricts or narrows the caliber of blood vessels, as a drug or nerve.

vasodilator (văs'ōdīlā'tŏr) [L. *vas,* vessel; *dilatāre,* to widen]. Causing relaxation of the blood vessels. A nerve or drug that dilates the blood vessels.

vasomotor (văs'ōmō'tŏr [L. *vas,* vessel; *motor,* a mover]. *Pert.* nerves controlling activity of smooth muscle in blood vessel walls. May cause vasodilation (enlargement) or vasoconstriction (narrowing) of the vessel.

vasopressin (văs'ōprĕs'ĭn) [L. *vas,* vessel; *press,* stem of *premere,* to press]. A hormone formed in supraoptic and paraventricular nuclei of the hypothalamus and transported to the posterior lobe of the hypophysis through the hypothalamo-hypophyseal tract. It has an antidiuretic and a pressor effect, elevating blood pressure.

vein (vān) [L. *vena*]. A vessel carrying blood toward the heart or away from the tissues.

vena cava (vē'nă cā'vă) [L. *vena,* vein; *cavum,* hollow]. The principal vein draining into the right atrium of the heart.

venereal (vĕnē'rēăl) [L. *venerus,* from *Venus,* goddess of love]. *Pert.* or resulting from sexual intercourse.

ventilation (vĕntĭlā'shŭn) [L. *ventilāre,* to air]. To move air, especially into and out of lungs. Alveolar ventilation: supplying air to the alveoli; pulmonary ventilation: supplying air to the conducting division of the respiratory system.

ventral (vĕn'trăl) [L. *venter,* the belly]. *Pert.* belly or anterior side of body.

ventricle (vĕn'trĭkl) [L. *ventriculus, dim.* of *venter,* belly]. A cavity or chamber, as in heart or brain; *appl.* fusiform fossa of larynx. *Appl.* dispensing chamber of the heart.

venule (vĕn'ūl) [L. *venula,* a little vein]. A small vein; gathers blood from capillary beds.

vermiform (vĕr'mĭfōrm) [L. *vermis,* worm; *forma,* shape]. Shaped like a worm; *appl.* certain structures, especially appendix.

vermis (vĕr'mĭs) [L. *vermis,* worm]. The middle lobe of the cerebellum.

Vesalius (vēsā'lēŭs), Andreas (1514–1564). Belgian physician; founder of modern anatomy: *Fabrica.*

vesicle (vĕs'ĭkl) [L. *vesicula,* a little bladder]. A small, fluid-filled sac.

vesicular (Graafian) follicle (Regnier de Graff, Dutch physician and anatomist, 1641–1673). (vēsĭk'ūlăr fŏl'ĭkl) [L. *vesicula,* a little bladder; *folliculus,* little bag]. Follicle containing a cavity; a mature ovarian follicle.

villikinin (vĭlĭkī'nĭn) [L. *villus,* shaggy hair; G. *kinein,* to move]. A factor in yeast and duodenal mucosa that stimulates contractility of intestinal villi.

villus (vĭl'ŭs) [L. *villus,* shaggy hair]. Minute fingerlike projections of the mucosa into the lumen of the small intestine.

virus (vī'rŭs) [L. *virus,* poison]. An ultramicroscopic infectious organism that requires living tissue to survive or reproduce.

vis a tergo (vĭs ā tĕr'gō) [L. *vis,* force; *tergum,* the back]. The residual pressure on the blood after it has passed through the capillaries.

visceral. *Pert.* viscera or an outer lining of an organ.

visceral pericardium (vĭs'ĕrăl pĕr'ĭkăr'dēŭm) [L. *viscus, viscer-,* body organ; G. *peri,* around; *kardia,* heart]. The serous inner layer of the pericardium; also called the epicardium.

viscous (vĭs'kŭs) [L. *viscosus,* sticky]. Sticky or of thick consistency.

visual (vĭzh'o͞oăl) [L. *visio,* a seeing]. *Pert.* sight or vision.

vital (vī'tăl) [L. *vitalis, pert.* life]. Essential to or contributing to life.

vitalist (vī'tălĭst) [L. *vita,* life]. One who believes that phenomena exhibited in living organisms are due to a special force distinct from physical and chemical forces.

vitamin (vī'tămĭn) [L. *vita,* life, *amine*]. Any of a group of organic substances, other than proteins, carbohydrates, fats, minerals, and organic salts, that are essential for normal metabolism, growth, and development of the body.

vitreous (vĭt'rēŭs) [L. *vitreus,* glassy]. Glassy; the vitreous body or humor of the eyeball found between lens and retina.

vivisection (vĭv'ĭsĕk'shŭn) [L. *vivus,* alive; *seco,* cut]. The dissection of a living animal; experimentation on living animals by means of ligatures, drugs, etc.

Volkmann's canals (A. W. Volkmann, Ger. physiologist, 1830–1889). Simple canals piercing circumferential or periosteal lamellae of bone, for blood vessels, and joining Haversian canal system.

voluntary (vŏl'ŭntārē) [L. *voluntas,* will]. *Pert.* being under willful control.

vomer (vō'mĕr) [L. *vomer,* plowshare]. The plowshaped bone forming the lower and posterior portion of the nasal septum.

vulva (vŭl'vă) [L. *vulva,* vulva]. The external female genitalia.

Wallerian degeneration (A. V. Waller, Eng. physician-physiologist, 1816–1870). Describes the destructive changes occurring in an axon that has been severed from the cell body.

waste (wāst) [L. *vastāre,* to devastate]. Useless products of body activity; to shrink or become smaller in size and/or strength.

Wharton's jelly (Thomas Wharton, Eng. anatomist, 1614–1673). The gelatinous core of the umbilical cord.

white matter (wīt măt'tĕr). That part of the central nervous system composed primarily of myelinated nerve fibers.

Willis, circle of (Thomas Willis, Eng. physician, 1621–1675). An arterial circle on the base of the brain composed of anterior cerebral, anterior communicating, interior carotid, posterior communicating, and posterior cerebral arteries.

Wormian (wŭr'mēăn) (*O. Worm* or *Wormius,* Danish anatomist, 1588–1654). Sutural bones; irregular isolated bones occurring in the course of sutures, especially in lambdoidal suture and posterior fontanelle; ossa suturarum, Wormian bones.

xeno- (zē'nō) [G. *xeno,* foreign]. Combining form: strange, foreign.

xenograft (zē'nōgrăft) [G. *xeno,* foreign; L. *graphium,* grafting knife]. A graft of tissues or organs between different species.

xiphoid (zĭf'oyd) [G. *xiphos,* sword; *eidos,* shape]. Sword-shaped. *Appl.* xiphoid process — last segment of sternum.

xiphoid process. The lowest portion of the sternum, a sword-shaped cartilaginous process supported by bone.

XX. Complement of sex chromosomes giving rise to a female.

XY. Complement of sex chromosomes giving rise to a male.

yolk sac (yōk săk). A membranous sac surrounding the yolk in an embryo; in the human, a transitory structure incorporated into the gut.

Z-line. An intermediate line in the striation of skeletal muscle; lies within the I-line and serves as the limits of a sarcomere.

zygomatic (zĭg'ōmăt'ĭk) [G. *zygoma,* yoke]. Malar; *pert.* zygoma; *appl.* arch, bone, fossa, processes, muscle, nerve.

zygote (zī'gōt) [G. *zygote,* yolked]. Cell formed by union of two gametes; fertilized egg.

Index

Italicized entries refer to pages on which illustrations appear.
t refers to pages on which tables appear.

$17 50

4.50

16|72.00

64

80

80

assisstant

assistant